Pathophysiology

Contributors

Charles Antzelevitch
Frederic C. Bartter
Sylvia S. Bottomley
Michael J. Brody
Turner E. Bynum
William A. Cain
Reuben M. Cherniack
James Christensen
Gilbert H. Daniels
Leonard J. Deftos
Virginia H. Donaldson
Francis G. Dunn
Stanley Fahn
Arnold J. Felsenfeld
Charles W. Francis
Edward D. Frohlich
John E. Gerich
Julie Glowacki
Antonio M. Gotto
Clarence A. Guenter
John B. Harley
Donald D. Heistad
Neville E. Hoffman
Tah-Hsiung Hsu
Eugene D. Jacobson
Oliver W. Jones
Simon Karpatkin
Christian E. Kaufman
David C. Kem
Jack G. Kleinman
Jacob Lemann, Jr.
Mortimer B. Lipsett
Francisco Llach
Eric W. Lothman

Farahe Maloof
Victor J. Marder
Franz H. Messerli
Gordon K. Moe
Erwin B. Montgomery, Jr.
Béla Nagy
Robert L. Ney
Samuel R. Oleinick
Solomon Papper
Carlos A. Pellegrini
Audrey S. Penn
Amadeo J. Pesce
Janice M. Pfeffer
Marc A. Pfeffer
Victor E. Pollak
Oscar D. Ratnoff
Richard N. Re
Morris Reichlin
Eugene D. Robin
William A. Robinson
Sami I. Said
Frederick J. Samaha
Lawrence M. Simon
Konrad H. Soergel
Dilip L. Solanki
Travis E. Solomon
Carlos A. Vaamonde
Jerry B. Vannatta
Lawrence W. Way
James H. Wells
Robert Whang
William A. Whitelaw
Ellison W. Wittels

Pathophysiology

Altered Regulatory Mechanisms in Disease

Third Edition / With 66 Contributors

Edited by

Edward D. Frohlich, M.D.

Vice President, Education and Research
Alton Ochsner Medical Foundation
New Orleans, Louisiana

J. B. Lippincott Company

Philadelphia

London   Mexico City
New York   St. Louis
São Paulo   Sydney

Sponsoring Editor: Richard Winters
Manuscript Editor: Don Shenkle
Indexer: Gene Heller
Art Director: Maria S. Karkucinski
Designer: Arlene Puttermann
Production Supervisor: Tina Rebane
Production Coordinator: Susan Caldwell
Compositor: Bi-Comp, Incorporated
Printer/Binder: Murray Printing Company

6   5   4   3   2   1

Library of Congress Cataloging in Publication Data
Main entry under title:

Pathophysiology : altered regulatory mechanisms in
   disease.

   Bibliography: p.
   Includes index.
   1. Physiology, Pathological. I. Frohlich, Edward D.
[DNLM: 1. Pathology.   QZ 4 P299]
RB113.P36   1984   616.07   83-821
ISBN 0-397-52103-0

The authors and publisher have exerted every effort to
ensure that drug selection and dosage set forth in this
text are in accord with current recommendations and
practice at the time of publication. However, in view of
ongoing research, changes in government regulations,
and the constant flow of information relating to drug
therapy and drug reactions, the reader is urged to
check the package insert for each drug for any change
in indications and dosage and for added warnings
and precautions. This is particularly important when
the recommended agent is a new or infrequently employed drug.

D e d i c a t i o n

The clinical investigator should not be considered the young physician of the "shining ivory tower" who is unconcerned with reality and practicality. To the contrary, he is any physician, deeply concerned with mechanisms of disease, the vastly complicated interrelationships of function in the total human being, and a refocusing of knowledge to the specific management of his patient. This concern for the "whys" and "hows" of disease is as old as Medicine itself, and indeed, this is its life blood and vitality.

It is for these reasons that we dedicate this volume to continued basic and clinical knowledge—to our esteemed clinical investigators of the Past and to those of the Future.

Contributors

Charles Antzelevitch, Ph.D.
Research Scientist, Masonic Medical Research Laboratory; Adjunct Associate Professor of Pharmacology, State University of New York, Syracuse, Utica, New York

Frederic C. Bartter, BA, M.D.
Professor of Medicine, University of Texas Health Science Center at San Antonio; Associate Chief of Staff for Research, Audie L. Murphy Memorial Veterans Hospital, San Antonio, Texas

Sylvia S. Bottomley, M.D.
Professor of Medicine and Adjunct Associate, Professor of Pathology, University of Oklahoma College of Medicine; Staff Physician, Veterans Administration Medical Center, Oklahoma City, Oklahoma

Michael J. Brody, Ph.D.
Professor of Pharmacology and Associate Director of Cardiovascular Center, College of Medicine, University of Iowa, Iowa City, Iowa

Turner E. Bynum, M.D.
Associate Professor of Medicine, Harvard Medical School; Director, Clinical Gastroenterology, Brigham and Women's Hospital, Boston, Massachusetts

William A. Cain, Ph.D.
Associate Professor of Microbiology and Immunology, Department of Medicine, University of Oklahoma College of Medicine, Oklahoma City, Oklahoma

Reuben M. Cherniack, M.D.
Cetalie and Marcel Weiss Chairman, Department of Medicine, National Jewish Hospital and Research Center, Nathal Asthma Center; Professor and Vice Chairman, Department of Medicine, University of Colorado, Denver, Colorado

James Christensen, M.D.
Professor and Director, Division of Gastroenterology–Hepatology, Department of Internal Medicine, University of Iowa Hospitals and Clinics, Iowa City, Iowa

Gilbert H. Daniels, M.D.
Director, Thyroid Clinic, and Associate Physician, Massachusetts General Hospital; Assistant Professor of Medicine, Harvard Medical School, Boston, Massachusetts

Leonard J. Deftos, M.D.
Professor of Medicine and Chief of Endocrinology, University of California and San Diego Veterans Administration Medical Center, La Jolla, California

Virginia H. Donaldson, M.D.
Professor of Pediatrics and Medicine, University of Cincinnati College of Medicine; Children's Hospital Research Foundation, Cincinnati, Ohio

Francis G. Dunn, M.B.Ch.B., M.R.C.P. (U.K.), F.A.C.C.
Staff Member, Department of Internal Medicine, Division of Hypertensive Diseases, Ochsner Clinic; Assistant Professor of Medicine, Louisiana State University School of Medicine, New Orleans, Louisiana

Stanley Fahn, M.D.
H. Houston Merritt Professor of Neurology, Columbia University College of Physicians and Surgeons, New York, New York

Arnold J. Felsenfeld, M.D.
Assistant Professor of Medicine, University of Oklahoma Health Sciences Center and the Veterans Administration Medical Center, Oklahoma City, Oklahoma

Charles W. Francis, M.D.
Assistant Professor of Medicine, Hematology Unit, University of Rochester School of Medicine and Dentistry, Rochester, New York

John E. Gerich, M.D.
Professor of Medicine and Physiology, Director, Endocrine Research Unit, Mayo Medical School and Mayo Clinic, Rochester, Minnesota

Julie Glowacki, Ph.D.
Assistant Professor of Surgery, Harvard Medical School; Children's Hospital Medical Center, Boston, Massachusetts

Antonio M. Gotto, D.Phil., M.D.
Rob and Vivian Smith Professor and Chairman, Department of Internal Medicine, Baylor College of Medicine; Chief of Internal Medicine Service, J.S. Abercrombie Chair of Atherosclerosis and Lipoprotein Research, Methodist Hospital, Houston, Texas

Clarence A. Guenter, M.D.
Professor and Head, Department of Medicine, University of Calgary Foothills Hospital, Calgary, Alberta, Canada

John B. Harley, M.D., Ph.D.
Department of Medicine, University of Oklahoma Health Sciences Center and Oklahoma Medical Research Foundation, Oklahoma City, Oklahoma

Donald D. Heistad, M.D.
Professor of Internal Medicine, University of Iowa, College of Medicine, Iowa City, Iowa

Neville E. Hoffman, M.D., Ph.D., F.R.A.C.P.
Gastroenterologist, St. John of God Medical Centre, Wembley, Western Australia 6014

Tah-Hsiung Hsu, M.D.
Associate Professor of Medicine, Johns Hopkins University and Hospital, Baltimore, Maryland

Eugene D. Jacobson, M.D.
Vice Dean for Academic Affairs, University of Cincinnati College of Medicine, Cincinnati, Ohio

Oliver W. Jones, M.D.
Professor of Medicine and Pediatrics, Director, Division of Medical Genetics, Department of Medicine, University of California, San Diego School of Medicine, La Jolla, California

Simon Karpatkin, M.D.
Professor of Medicine, New York University Medical Center, New York, New York

Christian E. Kaufman, M.D.
Associate Professor of Medicine, Department of Medicine, The University of Oklahoma College of Medicine, Oklahoma City, Oklahoma

David C. Kem, M.D.
Professor of Medicine, Department of Medicine, The University of Oklahoma College of Medicine; Chief, Section of Endocrinology, Metabolism, and Hypertension, Oklahoma City Veterans Administration Medical Center, Oklahoma City, Oklahoma

Jack G. Kleinman, M.D.
Associate Professor of Medicine, Medical College of Wisconsin, Assistant Chief, Renal Disease Section, Veterans Administration Medical Center, Wood, Wisconsin

Jacob Lemann, Jr., M.D.
Professor of Medicine, Chief, Nephrology Section, Medical College of Wisconsin, Milwaukee, Wisconsin

Mortimer B. Lipsett, M.D.
Director, National Institute of Child Health and Human Development; Clinical Professor of Medicine, United States University of Health Sciences, Bethesda, Maryland

Francisco Llach, M.D.
Professor of Medicine and Chief, Nephrology Section, University of Oklahoma Health Sciences Center and Veterans Administration Medical Center, Oklahoma City, Oklahoma

Eric W. Lothman, M.D., Ph.D.
Assistant Professor of Neurology, Washington University School of Medicine, St. Louis, Missouri

Farahe Maloof, M.D.
Professor of Medicine, Massachusetts General Hospital and Harvard Medical School, Boston, Massachusetts

Victor J. Marder, M.D.
Professor of Medicine, Co-Chief, Hematology Unit, University of Rochester School of Medicine and Dentistry, Rochester, New York

Franz H. Messerli, M.D.
Director, Hemodynamic Laboratory, Ochsner Medical Institutions; Associate Professor of Medicine, Tulane University, New Orleans, Louisiana

Gordon K. Moe, M.D., Ph.D.
Director of Research, Masonic Medical Research Laboratory; Professor Emeritus of Physiology, State University of New York, Syracuse, Utica, New York

Erwin B. Montgomery, Jr., M.D.
Assistant Professor of Neurology, Washington University School of Medicine, St. Louis, Missouri

Béla Nagy, Ph.D.
Associate Professor of Neurology and Pharmacology and Cell Biophysics, University of Cincinnati College of Medicine, Cincinnati, Ohio

Robert L. Ney, M.D.
Professor of Medicine, Director, Division of Endocrinology and Metabolism, Johns Hopkins University School of Medicine, Baltimore, Maryland

Samuel R. Oleinick, M.D., Ph.D.
Professor of Medicine, University of Oklahoma College of Medicine; Veterans Administration Medical Center, Oklahoma City, Oklahoma

Solomon Papper, M.D.
Distinguished Professor and Head, Department of Medicine, University of Oklahoma College of Medicine; Staff Physician, Veterans Administration Medical Center, Oklahoma City, Oklahoma

Carlos A. Pellegrini, M.D.
Assistant Professor of Surgery, University of California at San Francisco, San Francisco, California

Audrey S. Penn, M.D.
Professor of Neurology, Columbia University College of Physicians and Surgeons, New York, New York

Amadeo J. Pesce, Ph.D.
Professor of Experimental Medicine and Professor of Pathology and Laboratory Medicine, University of Cincinnati College of Medicine, Cincinnati, Ohio

Janice M. Pfeffer, Ph.D.
Assistant Professor of Medicine, Harvard Medical School; Brigham and Women's Hospital, Boston, Massachusetts

Marc A. Pfeffer, M.D., Ph.D.
Assistant Professor of Medicine, Harvard Medical School, Brigham and Women's Hospital, Boston, Massachusetts

Victor E. Pollak, M.D.
Division of Nephrology, Department of Internal Medicine, University of Cincinnati Medical Center, Cincinnati, Ohio

Oscar D. Ratnoff, M.D.
Professor of Medicine, Case Western Reserve University School of Medicine; Career Investigator, American Heart Association, Cleveland, Ohio

Richard N. Re, M.D.
Head, Division of Hypertensive Diseases, Ochsner Clinic; Associate Professor of Medicine, Louisiana State University School of Medicine; Associate Clinical Professor of Medicine, Tulane University School of Medicine, New Orleans, Louisiana

Morris Reichlin, M.D.
Chief, Combined Immunology Section, Department of Medicine, University of Oklahoma Health Sciences Center, Oklahoma City, Oklahoma

Eugene D. Robin, M.D.
Professor of Medicine and Physiology, Department of Medicine, Stanford University School of Medicine, Stanford, California

William A. Robinson, M.D., Ph.D.
Professor of Medicine and Head, Medical Oncology University of Colorado Health Sciences Center, Denver, Colorado

Sami I. Said, M.D.
Professor of Medicine and Chief, Pulmonary Disease and Critical Care Section, Department of Medicine, University of Oklahoma Health Sciences Center; Veterans Administration Medical Center, Oklahoma City, Oklahoma

Frederick J. Samaha, M.D.
Professor and Chairman, Department of Neurology, University of Cincinnati Medical Center, Cincinnati, Ohio

Lawrence M. Simon, M.D.
Assistant Professor of Medicine, Stanford University School of Medicine, Stanford, California

Konrad H. Soergel, M.D.
Professor of Medicine and Chief, Section of Gastroenterology, Medical College of Wisconsin, Milwaukee, Wisconsin

Dilip L. Solanki, M.D.
Associate Professor of Medicine, University of Oklahoma College of Medicine; Hematologist, Oklahoma Memorial Hospital and Veterans Administration Medical Center, Oklahoma City, Oklahoma

Travis E. Solomon, M.D., Ph.D.
Associate Chief of Staff for Research, Harry S. Truman Veterans Administration Hospital; Associate Professor of Physiology, Medicine and Surgery, University of Missouri Medical School, Columbia, Missouri

Carlos A. Vaamonde, M.D.
Professor of Medicine, University of Miami School of Medicine; Chief, Nephrology Section, Miami Veterans Administration Medical Center, Miami, Florida

Jerry B. Vannatta, M.D.
Assistant Professor of Medicine, Department of Medicine, University of Oklahoma College of Medicine, Oklahoma City, Oklahoma

Lawrence W. Way, M.D.
Professor of Surgery, University of California at San Francisco, San Francisco, California

James H. Wells, M.D.
Associate Professor, Department of Internal Medicine, Allergy Service, Immunology Section, Department of Medicine, The University of Oklahoma College of Medicine, Oklahoma City, Oklahoma

Robert Whang, M.D.
Professor and Vice Head, Department of Medicine, University of Oklahoma College of Medicine; Chief, Medical Service, Veterans Administration Medical Center, Oklahoma City, Oklahoma

William A. Whitelaw, M.D., Ph.D., F.R.C.P. (C)
Associate Professor of Medicine, University of Calgary, Calgary, Alberta, Canada

Ellison Wittels, M.D.
Assistant Professor, Department of Medicine, Baylor College of Medicine and Methodist Hospital, Houston, Texas

Foreword

It hardly seems possible that a third edition is needed after only 7 years. But when I reflect on what has happened in the sciences during that period, I can see why. First, there is much talk in some medical circles about downgrading the teaching of biomedical science in medical school. It is true that too often medical students are subjected to much irrelevant material. It is important to sort out the intellectual wheat of modern medicine from the chaff. The authors of *Pathophysiology* have succeeded in doing just that!

Second, bioscience is progressing at an overwhelming rate. Just how fast knowledge is accumulating and advancing is illustrated by perusing the *Bibliography of Hypertension*, a bimonthly publication of the American Heart Association. A rough calculation shows that about 1,300 papers appeared in November and December 1982, and perhaps 8,000 during the whole year. It is a frightening prospect for the younger generation of physicians. Unfortunately, the problem is ignored by older physicians.

Third, I would urge physicians as they age not to forget the scaffolding on which their knowledge and experience must be built. With it in place, a continuity is created in which history becomes a living part of the present. It is the main advantage in the process of becoming older. You no longer have to "pull rank" but can exhibit maturity.

Many of us are in danger of being overwhelmed by burgeoning specialties we cannot hope to understand. A large and still growing cadre of peripheral bioscientists threatens to take over leadership in medicine with the help of the news media. Many of us wonder why we worked so hard for an M.D. degree. The answer is that we and not the others have been given a fundamental broad education in medicine upon which an important future body of knowledge can be built. Too many of us do not use the opportunity this creates. We become "pill pushers," not physicians.

It is the purpose of *Pathophysiology* to see to it that the intellectual menopause does not set in prematurely. In the old days we developed "cerebral arteriosclerosis," "senility," and laziness. Now we have Alzheimer's disease and laziness. But you will have neither if you work at keeping your circulation and your brain in a useful and usable condition. If you try thinking, this book will be most helpful. Good luck.

Foreword to the First Edition

Countless biomedical scientists—physiologists, biochemists, pharmacologists and physicians—have attempted to breathe enthusiasm into teaching clinicians basic science. With what degree of success no one knows, except a few clinicians. I say "a few" because only a few know how to measure success.

In my view, when each patient's problem conjures in the physician's mind logical basic mechanisms which may describe how the patient "got that way" and how he may get out of it, success has been achieved. Medicine is then no longer a vocation, an application of cookbook directions, an uninspired routine of diagnosing and treating instead of a true profession in which the questions why and how may be answered. The extraordinary experiments that both nature and man perform otherwise go unnoticed, the chance to add to medical knowledge is lost, and the first two years of medical school are a waste.

Basic and applied sciences are exciting, albeit demanding. Clinical medicine is equally rewarding but each is incomplete without the other.

Much of the failure of the two sides of this coin to appreciate the other lies in the attitudes of clinicians and scientists. The former tend to disdain what they think to be the needless complexities of science. They imagine themselves much too busy curing people to be bothered with science. And the scientists imagine the clinician too dull and unappreciative to be taught. It seems so much more fun to overwhelm him with erudition, and not bother with the difficult problem of simplicity, clarity and screening of the relevant from the irrelevant.

Over the years there has been much discussion of how to order biomedical knowledge and to assign sensible priorities. Few have agreed on any single system.

I have long opposed the plan whereby learning is compartmentalized. Rather, I much prefer to think and teach according to functional systems of the body (Are You Listening and Reading? Modern Medicine, August 28, 1967, pp. 22–24). They provide a framework on which my mind can hang as few, or many details as I care to remember. Each system can be thought of as a vignette which encompasses and unifies all the classified subdivisions of medicine ranging from anatomy to bedside care. By providing structure and continuity, priorities are much more easily ordered.

Medicine today is much like organic chemistry was a century ago. Until chemical structure was introduced into the body of knowledge, the growing number of compounds being discovered was creating a jungle, the many components of which no one could remember. But the ring systems, valence, and stereochemistry brought order, aiding in associative memory and logical progression of thought. Facts are not remembered out of context but fall into a system with meaning. The same may well happen in medicine. Knowledge will be remembered through use of basic principles and building from them a logical structure which is relevant to the functions of the body. It was the wholeness of biomedical knowledge that suggested, for example, the use of a mosaic to describe the varied mechanisms of hypertension.

Fortunately times are changing and, with them, attitudes. I see much greater willingness, even eagerness of younger physicians to close the gap between chaotic empiricism and logical synthesis to make themselves more complete physicians.

I hope this book will help. It is written by some very capable young authors and its editor, Dr. Frohlich, has had experience helping to launch the series called "Physiology for Physicians" published on behalf of the American Physiological Society by the New England Journal of Medicine.

I suspect one of the greater lacks today is not the availability of good pedagogic material but adequate time for the physician to read selectively and learn. He must be careful to protect himself from being overwhelmed by the sheer volume of printed and spoken words and my only advice is learn what you learn well enough so that the patient's problems conjure a logical image of causal mechanisms in your mind. Science should be an integral part of diagnosis and treatment. The patient and you will be the better for it.

IRVINE H. PAGE, M.D.

Preface

When I arranged with the publishers for the first edition of *Pathophysiology*, it was with mixed zeal and ambition. I truly believed that a textbook detailing how the major physiological mechanisms could be deranged by disease was necessary. However, there was the natural possibility of rejection by the reader—the student or practicing physician. For their interest, confidence, and support, the contributors and I are deeply appreciative. In the ensuing years, a second edition, detailing newer thinking and consideration of additional mechanisms, seemed indicated. And now a third edition is necessary, as over these intervening years knowledge has expanded, disciplines have become more sophisticated, and clinical application continues. Still, the initial concept motivating our text remains.

A large number of our major diseases remain without known cause. Whether they be related to cardiology (arteriosclerosis, hypertension, or myocardiopathies), endocrinology (diabetes mellitus, obesity, thyrotoxicosis), respirology (emphysema, chronic bronchitis, bronchiectasis), gastroenterology (peptic ulcer, colitis, cholecystitis), or the broad area of oncology, we are usually unable to ascribe a specific etiology. Nevertheless, with modern diagnostic tools and therapy the physician is usually quite effective in managing patients who suffer from these problems. Often life has been remarkably prolonged and the patient's well-being considerably enhanced.

For the most part, enlightened treatment involves specific correction of those normal regulatory mechanisms altered by the disease process. For example, in hypertension arterial pressure is reduced to normotensive levels; in arteriosclerosis and myocardiopathies ventricular failure is reversed, the myocardium becomes compensated, and normal rhythm is restored; in diabetes mellitus the carbohydrate and lipid metabolism is controlled; and in thyrotoxicosis abnormal metabolism is rendered normal.

Often this mechanistic therapeutic approach used by the physician is so "natural" that it verges on empiricism and might even seem to be reflex or intuitive. However, the rationale for such therapeutic maneuvers has been well thought out and was at one time or another painstakingly documented in the great variety of journals of clinical investigation. Most important to students and practicing physicians is the continuing need to understand the mechanisms underlying the disease and therapeutic programs; and this rationale should be applied generally in daily medical practice. Such thinking leads to good medicine, improved health care, and fewer complications from treatment. The purpose of this book, then, is to present a mechanistic view of pathology which, it is hoped, will provide the student and physician with a means of conceptualizing disease.

As stated a dozen years ago, this book is not intended to be another textbook of medicine, or pathology, physiology, or biochemistry. There are already many well-written textbooks that satisfy the need for an all-encompassing *tour-de-force* of disease or physiology. Nor is this book intended to be used to provide an explanation for each described illness, sign, or symptom; such texts are also already available.

This book therefore is intended to create a way of thinking about disease. It is divided into eight sections concerned with the physician's classical orientation by systems or disciplines. Each chapter has been edited by a well-respected and knowledgeable section editor who has a broad clinical and academic background. Rather than subdivide each section into consideration of the respective component organs and their diseases, well-defined mechanisms of normal function have been selected. In each of the chapters the authors describe the role of the pertinent mechanisms necessary to achieve physiological homeostasis and how, when disease ensues, the regulation of these mechanisms goes awry. Each mechanism is described by an author with broad clinical competence in his respective area who also has been acknowledged for his fine ability as a teacher and writer. Just as the book is not intended to be all-encompassing with respect to disease, it is not all-inclusive with respect to mechanisms. Other mechanisms or areas might well have been covered and may be discussed in future editions. The mechanisms included were chosen to provide the reader with the broadest approach to clinical thinking without unduly entering those areas that still remain speculative or controversial; there was a deliberate attempt to keep the book within a reasonable length. However, the individual chapters dealing with mechanisms are not intended to cover all diseases concerned with the mechanisms under discussion. Rather, the authors were requested to present the underlying physiological concept and those pertinent diseases that best exemplify the

disarray of that mechanism in order to provide a way of thinking that will enable the reader to conceptualize other clinical problems not covered by the subject material.

In planning for this third edition I was most cognizant of those areas not covered in the first two editions and was continually reminded of my initial promise to myself and the publishers; if at all possible, future editions should not suffer from obesity. The present edition has been increased in magnitude by three chapters, but it has eight chapters new in theme and eighteen with new authors having different approaches. Chapters have been totally rewritten concerning the role of the heart as a pump and the newest concepts of the modern pandemic of civilized man, arthrerosclerosis; the mechanisms underlying the neuromuscular diseases; hepatic and biliary mechanisms; and the control of electrolyte balance. Other chapters have been extensively updated by new authors who have provided a different emphasis to the underlying disease mechanism. This is as it should be, since different authorities offer new perspectives to current problems.

Some of the chapters may present overlapping discussions and, at first thought, may seem to duplicate considerations. This was constructed intentionally to provide the reader with an alternative approach to the complexities of the integrative role of the entire body. By discussing these areas separately, an alternative manner of conceptualization is provided that may be more comfortable to some readers. Hopefully, the reader will profit from another way of thinking.

To compensate for the additional pages that come with adding new chapters or separating one former chapter into two new ones (e.g., gastrointestinal absorption of water and lipid substances), we have deleted the two appendices that have followed each section. When deemed necessary by the contributors, normal laboratory values and "classical" references were included in their chapters.

It is hoped that our approach to physiology, pathology, and internal medicine will be used on many levels in medical education. Early in medical school training, courses in pathophysiology or correlative medicine continue to appear in the curriculum as a bridge from the two preclinical years to the two clinical years. This volume is intended to be a companion text for such instructional programs. As the student continues further with medical training, either in school or at the postgraduate level, it is hoped that this text will continue to provide the base upon which broader clinical experiences will be developed. Ultimately, we hope that this approach to disease will provide the learning reinforcement to the practicing physician in his continuing experience and thinking about clinical medicine.

I have been deeply heartened by the reception of the first two editions. Intrinsic to your appreciation and understanding of the fundamental physiological concepts is *relevance.* Basic physiological processes, to be assimilated best by medical students and the practicing physician, must be related to their daily clinical associations. Understanding of the patient and his disease reinforces comprehension of the physiological processes altered by the clinical disorder. And with the personal satisfaction derived by this insight, the learning process is reinforced; thereby, the clinical practice of medicine advances and clinical excellence is maintained.

EDWARD D. FROHLICH, M.D.

Acknowledgments

No scientific publication is the work of one person. It certainly is not so in this multiauthored textbook nor is it even in a single-authored book. Both are dependent on many people. In this regard, the editor expresses the deep appreciation of the contributors to associates who have permitted them to collate their experiences, thoughts, and work into the written word. We are deeply grateful for their support and advice, for the time to reflect and to write, and for their inspiration to work as a team. We also acknowledge the contributions of our scientific predecessors of years past that provided the fundamental knowledge to permit the work of the present and of the future. I express my personal appreciation to my former and present associates not only for these reasons but also for permitting me to take time to organize and collate this volume and for providing me the ambience, moral support, and assistance so that I could continue to pursue this pleasant and satisfying endeavor.

I particularly want to acknowledge the tireless and devoted assistance of my capable secretary and associate, Mrs. Elizabeth Murray. Her advice and editorial suggestions were invaluable. In addition, the comments of students, fellows, colleagues, and "reader-friends" of the first two editions served as excellent "sounding boards" and resources for this third edition. I also thank Messrs. Stuart Freeman, Richard Winters, and Don Shenkle of J. B. Lippincott Company for their fine cooperation in making the publication of this book possible.

The vast accumulation of clinical knowledge and the rapid surge of medical progress during the past generation resulted in no small way from the material encouragement of fundamental and applied research by the United States Public Health Service's National Institutes of Health and by the many volunteer health organizations. Without these agencies the contributors to this volume would not have received much of the support for their work that has permitted them to share their experience and knowledge about pathophysiology. However, rather than citing each separately with each chapter, we collectively express our deep appreciation with the fervent hope that this close cooperation will continue for the benefit of all mankind.

Finally, some personal words of appreciation are allowed by claiming whatever editorial prerogatives are afforded me. I must express my deep appreciation and love to my parents for their encouragement and for instilling in me the fundamental attitudes and principles for a medical and personal life that I hold dear. With time, loved ones pass on; and so it was with my father whose values, love of life, and respect for education and new knowledge remain instilled in his children and grandchildren. Words cannot express my deep love for my wife, Sherry, and my children, Margie,

Bruce, and Lara. As they have grown into young adulthood and have matured over the years of these editions of *Pathophysiology*, the respect for education and scholarly achievement is maintained to be passed on to succeeding generations. I can never repay them for their patience, understanding, loyal support, and encouragement. Many hours and days have been taken from family in this work, but we accept this volume as a product of love, too. I am sure all of us share these thoughts, each in his own way, for our respective professional commitments.

Contents

Introduction   xxvii

Section I
Circulatory Mechanisms   1
EDWARD D. FROHLICH, M.D., SECTION EDITOR

1 **Mechanisms of Dynamic Cardiac Performance**   5
MARC A. PFEFFER, M.D., PH.D. / JANICE M. PFEFFER, PH.D.

2 **Tissue Perfusion**   23
MICHAEL J. BRODY, PH.D. / DONALD D. HEISTAD, M.D.

3 **Mechanisms Controlling Arterial Pressure**   45
EDWARD D. FROHLICH, M.D. / FRANZ H. MESSERLI, M.D. /
RICHARD N. RE, M.D. / FRANCIS G. DUNN, M.D.

4 **Mechanisms of Cardiac Dysrhythmias**   83
GORDON K. MOE, M.D., PH.D. / CHARLES ANTZELEVITCH, Ph.D.

5 **Atherogenic Mechanisms**   107
ELLISON W. WITTELS, M.D. / ANTONIO M. GOTTO, JR., M.D.

Section II
Respiratory Mechanisms   119
CLARENCE A. GUENTER, M.D., SECTION EDITOR

6 **Control of Respiration**   125
WILLIAM A. WHITELAW, M.D.

7 **Ventilation, Perfusion, and Gas Exchange**   143
REUBEN M. CHERNIACK, M.D.

8 **Oxygen Transport and Cellular Respiration**   163
EUGENE D. ROBIN, M.D. / LAWRENCE M. SIMON, M.D.

9 **Metabolic and Endocrine Functions of the Lung**   183
SAMI I. SAID, M.D.

Section III
Renal Mechanisms   203
SOLOMON PAPPER, M.D., SECTION EDITOR

10 **Maintenance of Body Protein Homeostasis**   211
VICTOR E. POLLAK, M.D. / AMADEO J. PESCE, PH.D.

xxiii

11 **Maintenance of Body Fluid and Sodium Volume**   **231**
FREDERIC C. BARTTER, M.D.

12 **Maintenance of Body Fluid Potassium, Calcium,**
Magnesium, and Phosphorus   **249**
CHRISTIAN E. KAUFMAN, M.D. / ARNOLD J. FELSENFELD, M.D. /
JERRY B. VANNATTA, M.D. / ROBERT WHANG, M.D. /
FRANCISCO LLACH, M.D.

13 **Maintenance of Body Fluid Tonicity**   **271**
CARLOS A. VAAMONDE, M.D.

14 **Maintenance of Acid-Base Homeostasis**   **299**
JACK G. KLEINMAN, M.D. / JACOB LEMANN, JR., M.D.

Section IV
Endocrine-Metabolism Mechanisms   **319**
DAVID C. KEM, M.D., SECTION EDITOR

15 **Regulatory Mechanisms of the Hypothalamic-Pituitary**
and of the Pituitary-Adrenal Axes   **323**
TAH-HSIUNG HSU, M.D. / ROBERT L. NEY, M.D.

16 **Regulatory Mechanisms of the Thyroid**   **359**
GILBERT H. DANIELS, M.D. / FARAHE MALOOF, M.D.

17 **Metabolism and Energy Mechanisms**   **387**
JOHN E. GERICH, M.D.

18 **Genetic Mechanisms of Metabolism**   **403**
OLIVER W. JONES, M.D.

19 **Endocrine Mechanisms of Reproduction**   **423**
MORTIMER B. LIPSETT, M.D.

20 **Mechanisms of Bone Disease**   **445**
LEONARD J. DEFTOS, M.D. / JULIE GLOWACKI, PH.D.

Section V
Gastrointestinal Mechanisms   **469**
TURNER E. BYNUM, M.D. / EUGENE D. JACOBSON, M.D., SECTION EDITORS

21 **Motility**   **475**
JAMES CHRISTENSEN, M.D.

22 **Gastric Secretion**   **497**
TRAVIS E. SOLOMON, M.D., PH.D.

23 **Absorption of Water and Water-Soluble Solutes**   **523**
KONRAD H. SOERGEL, M.D.

24 **Absorption of Lipid Solutes**   **547**
NEVILLE E. HOFFMAN, M.D., PH.D.

25 **Hepatic Mechanisms**   **561**
TURNER E. BYNUM, M.D.

26 **Biliary Tract Mechanisms**   **581**
CARLOS A. PELLEGRINI, M.D. / LAWRENCE W. WAY, M.D.

Section VI
Hematological Mechanisms 619
OSCAR D. RATNOFF, M.D., SECTION EDITOR

27 **Erythropoiesis** 623
SYLVIA S. BOTTOMLEY, M.D. / DILIP L. SOLANKI, M.D.

28 **Leukopoiesis** 655
WILLIAM A. ROBINSON, M.D., PH.D.

29 **Thrombopoiesis** 667
SIMON KARPATKIN, M.D.

30 **Coagulation and Fibrinolysis: Mechanisms of the Fluid Interphase** 681
Part 1 Hemostasis—OSCAR D. RATNOFF, M.D.
Part 2 Complement—VIRGINIA H. DONALDSON, M.D.
Part 3 Fibrinolysis—CHARLES W. FRANCIS, M.D. /
VICTOR J. MARDER, M.D.

Section VII
Neuromuscular Mechanisms 737
FREDERICK J. SAMAHA, M.D., SECTION EDITOR

31 **Control of Motor Activity by the Cerebrum and Cerebellum** 741
ERIC W. LOTHMAN, M.D., PH.D. / ERWIN B. MONTGOMERY, JR., M.D.

32 **Extrapyramidal System** 771
STANLEY FAHN, M.D.

33 **Neuromuscular Junction** 789
AUDREY S. PENN, M.D.

34 **Physiology of Normal and Diseased Muscle** 805
BÉLA NAGY, PH.D. / FREDERICK J. SAMAHA, M.D.

Section VIII
Immunological Mechanisms 831
MORRIS REICHLIN, M.D., SECTION EDITOR

35 **Adaptive Immunity** 839
MORRIS REICHLIN, M.D. / JOHN B. HARLEY, M.D., Ph.D.

36 **Allergic Mechanisms** 861
JAMES H. WELLS, M.D. / WILLIAM A. CAIN, Ph.D.

37 **Autoimmune Diseases and Mechanisms of Immunological Injury** 881
MORRIS REICHLIN, M.D.

38 **Mechanisms of Tumor Immunology** 895
SAMUEL R. OLEINICK, M.D., PH.D.

Index 917

Introduction

Arthur C. Guyton, M.D.

Department of Physiology and Biophysics
University of Mississippi School of Medicine
Jackson, Mississippi

During all my professional life I have been devoted to the proposition that the practice of good medicine is a science. Therefore, I feel rewarded personally with each significant new contribution to the synthesis of medical science. When I first read the list of authors who had agreed to contribute to this book and then further learned the topics to be covered, I awaited with anticipation until I should actually read the manuscripts. As these began to assemble it soon became clear that I would not be disappointed, because the topics and the detailed subject matter were chosen well to fill voids between the individual disciplines of physiology and clinical medicine. They were chosen to teach, to illustrate the good sense that prevails when a clinician treats a patient with full knowledge of what is happening in the diseased body, knowledge that allows a multiplicity of approaches rather than the practice of stereotyped routines. For these reasons, even I, though engaged most of my life in interfacing basic physiology with clinical medicine, find a host of new ideas and principles that will carry over into my future teaching.

Shortly, I wish to speak of the new horizons in clinical physiology, but before doing so I think it worthwhile to tell the readers something about the person who has spent so much time and intelligence organizing, directing, and producing this book, Dr. Edward Frohlich. For those who do not already know, this is not Dr. Frohlich's first major contribution to medicine nor to basic physiology. It was evident more than a decade ago that Dr. Frohlich, who is even now still young, would be a major force in teaching basic physiological concepts in the practice of clinical medicine both to students and to professionals in the fields of physiology and clinical medicine.

Dr. Frohlich is an accomplished clinician. He is also thoroughly versed in basic animal physiological experimentation; he has achieved the art of clinical research measurement; and he has an introspective mind that delves into the deeper aspects of basic physiological concepts while searching constantly for ways to make this knowledge fruitful to the patient. It is this rare combination of talents that is necessary for the successful undertaking of a book of this type.

Now, let us look at the borderland that lies between basic medical science and the practice of clinical medicine. For those readers who are clinicians, how often have you had to treat a patient with a disease that still falls within one of our scientific voids, and you have said to yourself, if only I could understand what I am doing? For those of you who are students and are studying the basis of medicine and by now know the molecular configurations of DNA, how does this fit into your life goal of understanding and treating disease? And for those of you who are working on the forefronts of detailed basic physiological concepts, how often have you been driven to your experiments because of your desire to explain and to contribute in your way to the final solution of human problems? The answers to all of these questions can best be given in the form of examples.

One of the most exciting fields for the fertile physiological and clinical mind during the past fifty years has been that of hypertension. In the nineteenth century and in the early part of the twentieth century there was a strong general feeling that hypertension was in some way related to fluids and electrolytes of the body, though most of the reasons for these beliefs were circumstantial. Then, in the early 1930s, most of the momentum shifted to humoral mechanisms as the cause of hypertension, mainly because of the very successful animal preparation made by Goldblatt to cause well-controlled hypertension in animals and because of the subsequent isolation, characterization, and physiological study of the renin-angiotensin system. Later, with still more knowledge, the role of aldosterone in hypertension and the welding of the renin-angiotensin system with aldosterone as well as with salt and water mechanisms came to the forefront. Through all of these periods there also ran an undercurrent of belief among many physicians and experimentalists that hypertension, especially essential hypertension, has a neurogenic basis. Finally, in recent years the systems physiologists, the biomedical engineers and a host of clinicians have come to believe strongly that hypertension can be a manifestation of any

one of many different abnormalities affecting one or more specific parts of an overall and complex arterial pressure regulatory system. Furthermore, our knowledge of these regulatory systems, even of the mathematics of most of their aspects, is beginning to emerge, so much so, indeed, that many types of hypertension can be simulated almost in detail by mathematical models on computers. Much of the story of the science of hypertension will be told in future chapters of this book, and references from these chapters will lead one into one of the most rewarding of stories of the scientific approach to medical problems, a story that interweaves the deepest of our basic physiological concepts with one of the most important of clinical problems.

Another field that captures the imagination, not only of the medical scientist but even of the public, is the pathophysiology of immunological mechanisms, especially their relation to autoimmunity, transplantation, and even survival of the human race in the face of rising population density with its attendant exponential growth of epidemiologic potential. It was hardly ten years ago that we wrote in our textbooks that blood lymphocytes might be nothing more than cast-off products with probably little function once they had been "excreted" by the lymph nodes into the blood, merely waiting their arrival in some tissue where they would be cleansed from the blood by phagocytosis. Yet, at the same time we spoke in terms of phenomena such as tissue immunity or other vague types of immunity that could not be explained on the basis of circulating protein antibodies in the plasma. Then, suddenly, the vista of the lymphatic system opened wide to entwine together all these segments of half-knowledge, displaying the beauty of our body's defense against disease and at the same time beginning to explain the abnormalities that lead to autoimmunity, allergic phenomena, and so forth. And simultaneously came the saga of man's attempts to obviate this defense system so that he can practice his cherished dream of reassembling human beings from organs no longer needed by others. The story of our defense system is still unfolding; a major segment of this scientific forefront will be told in the appropriate section of this book.

I could continue indefinitely to detail the excitement, the fun, and the intellectual reward that characterize what is happening today in the area between basic medical science and clinical medicine. I could tell how basic mathematicians have helped to solve problems in respiratory patients such as asthmatics, emphysematous patients, and patients with capillary-alveolar block. I could tell how the mechanisms for regulation of our body fluids, their electrolyte compositions, their acid-base balance, their pressures, and their volumes, are all beginning to fall into place. Indeed, modern practice of renology and daily control of body fluids in sick patients are both highly dependent on this new and exacting knowledge.

The basic chemistry and genetic control of both normal and abnormal hormones secreted in the body are emerging as a new science and are becoming basic to the practice of endocrinology. And presiding over almost all systems of the body—over the cardiovascular system, the gastrointes-

tinal system, the kidneys, and the lungs— is always the nervous system. Its primary role is to control our rapid bodily functions such as our quick motor acts, motility of the gut, degree of constriction of the blood vessels, and much of secretion by the endocrine glands. But in the background are literally thousands of other control systems that act more slowly, sometimes subservient to the nervous system, sometimes not; sometimes opposing the nervous system, sometimes supporting it; sometimes weak in comparison with the nervous controls, sometimes very potent. These other controls include even the genes themselves, each of which is a basic block in a much larger system of control.

It is the body's control systems that makes it possible for the human being to live, indeed, that demand that he live, that drive him to live, so much so that he has to perform a positive act to make himself die. And it is abnormalities in these control systems or in their target organs that are the bases of disease. The modern doctor is inheriting a new and formidable task to understand these systems so that he might practice a better type of medicine, but the compensations are commensurately great, for a doctor's pleasure in his work is directly related to the extent that he understands what he is doing.

Pathophysiology

Section I

Circulatory Mechanisms

Introduction

Much information in regard to the cardiovascular system has been amassed the past four decades, greatly augmenting knowledge gained over the preceding 350 years. Nevertheless, the most common cardiovascular diseases remain, for the most part, of unknown cause. True, much knowledge has been gained concerning pressor mechanisms, and with the introduction of antihypertensive drugs great inroads have been made into reversing the morbidity and mortality from the hypertensive diseases; but, in the final analysis, the bulk of patients with hypertension have chronic illnesses of unknown cause. Arteriosclerosis, although less well understood, is unfolding as a multifactorial problem involving hemodynamics, biochemistry, metabolism, genetics, and immunology. Likewise, with development of more refined biochemical methods, diagnostic techniques, and pharmacological tools, we have come to think of the various possibilities underlying the ischemic heart diseases and the manifold possible types of myocardiopathies. We are even studying means to preserve diseased myocardium.

Through association with engineers and biophysically oriented physiologists, as well as investigators with a strong interest in biochemistry, we have come to think in terms of total integration of the cardiovascular system. We no longer consider the heart as the central organ "calling the plays," as it were, for the entire systemic circulation and its associated organs. We are at least sophisticated enough clinically to appreciate such concepts as: local regulation of flow; local tissue demands regulating the output of blood from the heart; the important interdependency of the autonomic nervous, endocrine, and renal systems with the cardiovascular system; and the role of the various endocrine activities of the cardiovascular system as it pertains to the fluxes of vasoactive agents liberated from the various organs. Only in recent years have we learned of the production of a substance in the atrium that can produce natriuresis. This concept is most exciting since it provides a logical mechanism whereby the heart can sense an overfilling of the circulation and respond to this by an appropriate physiological means.

The lessons learned from the various diseases comprising what we so glibly diagnose as "hypertension" are beginning to pay off. We now realize that what we have lumped collectively as essential hypertension is a host of unhomogeneous problems manifested by varying degrees of dysregulation of the many controlling pressor mechanisms. From this physiological sorting-out process will come descriptions of other discrete diseases. Hence, by detecting operable pressor mechanisms in a particular patient we are now better able to understand his particular problem and apply more intelligently specific, life-prolonging therapy.

3

After considerable thought we have outlined five particular mechanisms, as they remain uniquely under the purview of the physician–clinical physiologist concerned with the cardiovascular system. The first of the following five chapters presents a discussion of the contractile mechanisms of the heart and their pertinence to myocardial performance. This discussion is concerned not only with concepts of muscle mechanics, relationships of pressure, flow, and volume, and dilatation and hypertrophy but also— in biochemical terms, wherever possible—as they are related to the diseases of cardiac "inflow," cardiac "outflow," and intrinsic "pump failure." A natural follow-through discussion is concerned with the local aspects of tissue perfusion, the mechanisms of local regulation of flow, and the relationships of systemic vascular, endocrine, metabolic, and biochemical alterations to the concepts of tissue nutrition, and how they pertain to disease. Not only should we be concerned with the problems of mechanical obstructions to the flow of blood, we are now concerned with the role of circulating chemical and vasoactive agents, and the interrelationships of the cardiovascular with the nervous and endocrine systems and systemic metabolism in various pathological states. The following chapter concerns mechanisms controlling arterial pressure that serve to maintain arterial pressure at normotensive levels. It describes how, when these delicately interrelated mechanisms are disrupted and go awry, various diseases develop, manifested either by increases or by decreases in arterial pressure. No cardiovascular discussion would be complete without elucidation of the peculiar and intrinsic property of cardiac muscle to contract rhythmically. Rather than a discussion of the myriad of specific arrhythmias, a concept of rhythmicity and the nature of the various electrophysiological mechanisms whereby rhythm disorders may develop—impulse propagation, block, escape, reentry, ectopic pacemakers—are presented, so application to specific rhythm disturbances can be drawn therefrom. An entirely new approach to the problem of atherosclerosis is presented that, as suggested earlier, involves consideration of the interplay of a number of atherogenic mechanisms. Thus, in two broad cardiovascular disease areas we can now see evidence supporting the Mosaic theory of Irvine H. Page.

Classically, clinical cardiology and cardiovascular physiology have been presented in terms of physical diagnostic techniques and various radiographic and electronic diagnostic tools. Following such an introduction, the specific congenital, rheumatic, hypertensive, endocardial, myocardial, and pericardial diseases are discussed in terms of the specific pathophysiological characteristics of each disease. Presentations of this type are abundantly accessible. It is therefore genuinely hoped that with a more complete understanding of disease mechanisms, an enlightened approach to the physiological problems in clinical practice will further stimulate the student and physician to practice with the goal of overcoming our present ignorance of disease causes.

EDWARD D. FROHLICH, M.D.

1 Mechanisms of Dynamic Cardiac Performance

Marc A. Pfeffer, M.D., Ph.D.
Janice M. Pfeffer, Ph.D.

Cardiac performance responds both to the short- and long-term needs of the body. Short-term adjustments in cardiac performance are regulated by alterations in the rate, force, and extent of contraction of the heart. These beat-to-beat adjustments in cardiac performance reflect a close coupling between the heart and the systemic and pulmonary circulations. Less well understood are the adjustments in cardiac performance that are suited to the long-term alterations in metabolic demand. Indeed, it is the capacity of the heart for growth, atrophy, and remodeling that makes this biological pump beyond compare with the finest inanimate machine.

The dramatic changes in cardiac morphology and function that occur in the early neonatal growth period are striking examples of the capacity of the heart to adapt to long-term alterations in functional demands. As a result of the patency of the ductus arteriosus and the high resistance to blood flow in the collapsed lungs of the fetus, the right and left ventricles function in parallel, reflecting this evenly shared work load in weights that are equal. At birth the abrupt gaseous distention of the lungs produces a marked reduction in pulmonary vascular resistance and the simul-taneous, sudden loss of the placenta raises systemic vascular resistance. Within minutes the ductus arteriosus begins to close, and for the first time the two ventricles begin to function in series and against unequal loads. The postnatal pattern of growth of the heart, in which there is a preponderant development of the left ventricle, reflects the intrinsic capacity of this biological pump to remodel itself in response to long-term functional demands. This adaptive ability occurs in an organ which loses the capacity to generate new myocytes early in neonatal life. Therefore, new contractile units must be added to existing cells. The increased fiber size of the left ventricular myocytes in relation to those of the right ventricle reflects the relative hypertrophy of the left ventricle in response to the augmented load that commences at birth. Ironically, it is this unique capacity of the ventricle to respond to functional demands with growth and remodeling that provides important information as to the pathophysiological loading conditions. Indeed, in many instances abnormal alterations in ventricular mass and geometry provide the key clues as to the etiology of the underlying cardiovascular disease. This chapter provides an understanding of the cardiac contractile process and a framework for the approach to disorders of cardiac performance.

Morphology and mechanism of contraction

Structure

The contractile units of the myocardium are contained in cells termed myocytes which branch and interconnect across intercalated discs to form the network of muscle fibers of the myocardium (Fig. 1–1A). This interdigitating network of individual cell functions as a syncytium to produce the coordinated contraction and relaxation of the muscular elements of the heart. The myocytes are composed of myofibrils which traverse the length of the cell, and the parallel arrangement of the myofibrils imparts the characteristic striated appearance to cardiac muscle (Fig. 1–1B). The myofibril consists of repeating units of contractile elements called sarcomeres, the fundamental contractile units of striated muscle (Fig. 1–1C). A sarcomere is defined as the distance from Z line to Z line and ranges in length from 1.6 μ to 2.4 μ. The sarcomere is composed of myosin and actin myofilaments. The thin actin myofilament attaches to the Z line and extends toward the center of the sarcomere. The relatively thick myosin myofilament is located in the middle of the sarcomere and is surrounded by six actin filaments at the areas of overlapping of thick and thin filaments (Fig. 1–1D). The movement of one myofilament relative to another via crossbridges produces the shortening and lengthening of Z-band distances that generates muscle contraction and relaxation.

In addition to the contractile proteins, actin and myosin, the sarcomere contains regulator proteins, such as troponin and tropomyosin, that are located along the actin myofilament at potential regions for combination with myosin. In the absence of these regulator proteins the actin and myosin would react spontaneously as a result of the ATPase activity of the myosin molecule and the intracellular levels of ATP. The association of the troponin-tropomyosin complex with actin is thought to alter the ATPase activity of the myosin molecule so as to prevent interaction between actin and myosin. The inhibition of the troponin-tropomyosin regulator proteins by calcium confers calcium sensitivity on the actin-myosin reaction and provides a mechanism whereby contraction may be regulated.

Excitation-contraction coupling

As the membrane potential of the myocyte decreases from resting to threshold level, the membrane conductance for Na^+ dramatically increases, producing the sudden depolarization of the cell membrane. During the plateau phase of the action potential of cardiac muscle there is a relatively slow inward movement of calcium (Fig. 1–2). The small amount of extracellular Ca^{++} that enters the cell during the action potential causes the release of sufficient intracellular Ca^{++} from the sarcoplasmic reticulum to saturate the troponin-tropomyosin complex and thereby remove the inhibition of myosin ATPase. In the presence of high intracellular levels of ATP and creatine phosphate the actin and myosin then interact. This interaction is generally conceived as the head of the myosin molecule attaching to the actin and producing a relative movement of the actin filament past the myosin (Fig. 1–3). This movement produces the shortening, or force, of the contraction.

Relaxation is produced by the active (ATP-requiring) reuptake of intracellular Ca^{++} by the sarcoplasmic reticulum. A reduction of the Ca^{++} level at the myofilaments leads to the dissociation of Ca^{++} from the troponin-tropomyosin system. This reactivation of the troponin-tropomyosin system inhibits the actin-myosin interaction and produces relaxation.

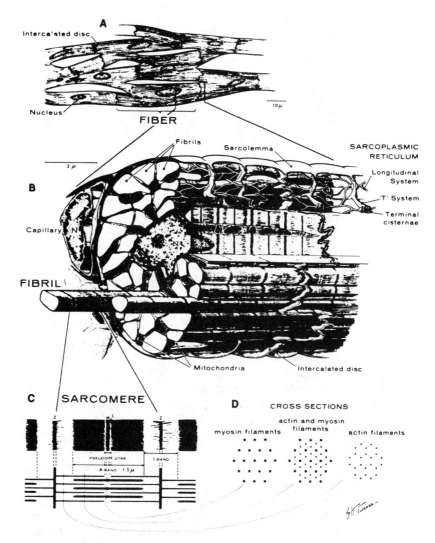

FIG. 1—1. Cellular and subcellular organization of the myocardium (*A*) Branching arrangement of myocytes. (*B*) Subcellular arrangement of myocyte with a representation of relationships of myofibrils, mitochondria and sarcoplasmic reticulum. (*C*) Longitudinal representation of a sarcomere, the fundamental contractile unit of striated muscle. (*D*) Cross-sectional arrangement of sarcomere with sections through the center M zone, which contains only myosin filaments, the A band, where myosin and actin overlap (each thick myosin filament is surrounded by six thin actin filaments), and the I band, which is nearest the Z line and contains only actin filaments. (Braunwald, E., Ross, J. Jr., and Sonnenblick, E. H.: Mechanisms of Contraction of the Normal and Failing heart. New York, Little, Brown & Co., 1976.)

Cardiotonic drugs

Because of the pivotal role of Ca^{++} in the excitation-contraction coupling process of cardiac, skeletal, and vascular smooth muscle, it is of little surprise that the recent clinical availability of drugs that interfere with the entry of Ca^{++} into the muscle cell has generated such intense interest. These so-called Ca^{++} channel blockers, or Ca^{++} entry blockers, are a heterogeneous class of

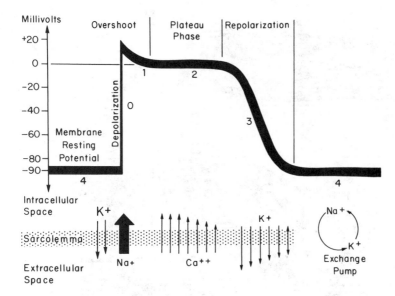

FIG. 1–2. Schematic representation of the major cation fluxes occurring during the various phases of the action potential of cardiac muscle.

compounds that share the property of inhibiting the entry of Ca^{++} into the cell through the slow channels during the plateau phase of the action potential. These compounds vary greatly in their potency, mechanism of action, and, most importantly, tissue specificity. Nifedipine is believed to reduce the number of channels by which Ca^{++} enters the cell. In contrast, verapamil alters the kinetics of Ca^{++} entry into the channel. By inhibiting the entry of Ca^{++} into the myocyte, both nifedipine and verapamil produce a depression of contractile function as assessed in preparations of isolated cardiac muscle. These agents also reduce vascular tone and have direct negative chronotropic and dromotropic (conduction) effects. Many of the direct effects of the calcium channel blockers observed in isolated tissue preparations are offset by reflexive adjustments when these compounds are systemically administered and therefore are usually not clinically apparent. Nifedipine is an effective arteriolar dilator and, as a result of a reflexive increase in sympathetic activity, its net clinical effect may be an increase in heart rate and contractile state despite its direct negative chronotropic and inotropic properties. Nifedipine also has been shown to

be extremely effective in preventing coronary arterial spasm in patients with Prinzmetal's angina. The decrease in arterial blood pressure and in coronary arterial resistance by nifedipine also makes it attractive for use in patients with classical exertional angina pectoris. Although the direct effects of verapamil are qualitatively similar to those of nifedipine, arteriolar dilatation is not as prominent. In addition, the direct myocardial effects of a depression in contractility and a slowing of atrioventricular conduction are more apparent. Diltiazem, another Ca^{++} channel blocker, shares the features common to both verapamil and nifedipine. Although diltiazem has not been used as extensively as the other two Ca^{++} channel blockers in the clinical setting, it appears to have both cardiac and vascular smooth muscle effects that are intermediate to those of verapamil and nifedipine. These differences in drug actions underscore the importance of integrating the knowledge of their direct pharmacologic properties with that of their physiologic responses to understand more fully their clinical effects.

Familiarity with the role of cytosolic Ca^{++} in the excitation-contraction coupling process of cardiac muscle also pro-

LOW Ca⁺⁺

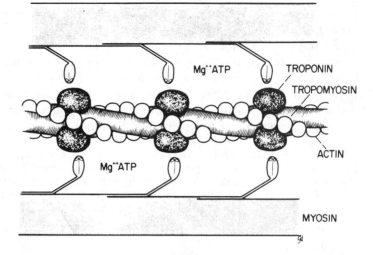

HIGH Ca⁺⁺

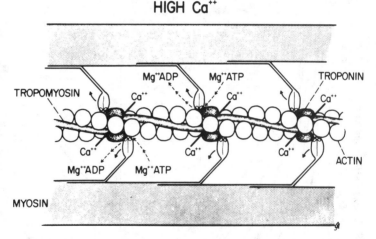

FIG. 1–3. Interaction of myosin head with actin. In low-Ca⁺⁺ environment (*top*) troponin and tropomyosin prevent myosin and actin interaction. In higher Ca⁺⁺ state (*bottom*) there is an alteration of troponin-tropomyosin complex (schematically shown as size change), which no longer inhibits the movement of myosin with respect to actin. (Alpert, N. R., and Hamrell, B. B.: Cardiac hypertrophy: A compensatory and anticompensatory response to stress. *In* Vassalle, M.: Cardiac Physiology for the Clinician. New York, Academic Press, 1976.)

vides a framework for understanding the mechanism whereby the cardiac glycosides may augment myocardial contractility. Digitalis has long been known as an inhibitor of the sodium-potassium exchange pump of the cell membrane. This energy-requiring transport system maintains the low intracellular sodium concentration. Digitalis-induced inhibition of monovalent-cation active transport results in increased intracellular sodium content and, consequently, enhanced Ca⁺⁺ influx. Indeed, recent studies of spontaneously beating cells from chick ventricles have shown a direct relationship between cardiac glycoside-induced augmentation of contraction and increased intracellular sodium and Ca⁺⁺ content. Increased cytosolic Ca⁺⁺ facilitates the disinhibition of the troponin-tropomyosin complex, the interaction of the actin and myosin myofilaments, and the more forceful contraction of cardiac muscle.

Mechanical performance

Isolated muscle

Studies of the contractile properties of isolated cardiac muscle have generated the concepts and terminology that are the foundation for an understanding of the performance of the intact heart. Papillary muscles from small mammals are generally used because the parallel arrangement of fibers from this region of cardiac muscle avoids the confounding effect of fiber orientation. Moreover, their small cross-sectional area greatly diminishes core tissue ischemia. Simply described, these studies are performed by placing a papillary muscle in an oxygenated bath of physiologic salt solution and then electrically stimulating it to produce contractions. A fulcrum-lever system permits control of the loading conditions both before and during each contraction. Prior to each contraction the muscle is stretched to a resting length by a weight called the preload. The *preload*, then, is the passive force which imparts a given stretch, or length, to the muscle prior to the contraction.

FIG. 1—4. A series of force-velocity curves constructed against different preloads (initial muscle length). An increase in preload resulted in an augmentation of the actively developed force with no change in the initial velocity of shortening. (Braunwald, E., Ross, J. Jr., and Sonnenblick, E. H.: Mechanisms of Contraction of the Normal and Failing Heart. Boston, Little, Brown & Co. 1976.)

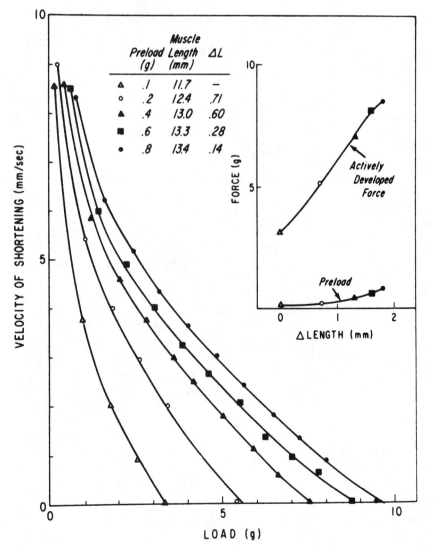

The *length-tension* relationship is a fundamental property of striated muscle that can readily be demonstrated in the isolated papillary muscle preparation by altering the preload and hence the initial muscle stretch, or length. As initial muscle length is increased, the actively developed force is augmented until an optimal length (Lmax) is achieved (Fig. 1–4). Isometric contractions from an initial length of Lmax develop more force than do contractions initiated from either greater or lesser degrees of initial fiber distention.

Another fundamental property of striated muscle, the inverse *force-velocity* relationship, is obtained from a series of isotonic contractions at constant preloads. As the *afterload* (i.e., the force that the muscle must generate before shortening can occur) is increased, the velocity and extent of shortening decline (Fig. 1–4). When the afterload is too great for the muscle to shorten, the contraction is isometric and a force is developed without muscle shortening. For a given preload (initial muscle length) a particular force-velocity relationship is obtained. Increasing preload until the optimal initial muscle length is achieved results in a shift in the force-velocity relationship and an augmentation in the maximal *force* developed. However, the maximal *velocity* of shortening at low afterloads is not altered by changes in the initial muscle length.

Superimposed on the fundamental length-tension and force-velocity relationships of striated muscle, cardiac muscle has the relatively unique capability of altering its *contractile state*. In the presence of an increased $[Ca^{++}]$ or exposure to positive inotropic agents, such as catecholamines or digitalis, the force-velocity relationship of cardiac muscle is shifted upward and to the right (Fig. 1–5). This increase in both the velocity of shortening and the peak isometric force at comparable loads cannot be attributed, by definition, to changes in the initial muscle length. This property of cardiac muscle to alter its contractile, or inotropic, state has been likened to a change in the *rate of reaction* at each crossbridge site, whereas the underlying mechanism of

FIG. 1–5. Alteration of the force-velocity relationship by a positive inotropic agent (norepinephrine). The solid circles represent contractions against the same preload in the presence of norepinephrine. A positive inotropic agent induces an increase in the velocity as well as maximum force of contraction. (Braunwald, E., Ross, J. Jr., and Sonnenblick, E. H.: Mechanisms of Contraction of the Normal and Failing Heart. New York, Little, Brown & Co., 1976.)

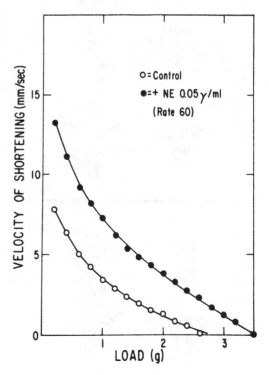

the length-tension relationship is attributed to an alteration in the *number of reactive sites* available for the contraction (Fig. 1–6). Since the heart functions as a syncytium, with all fibers stimulated during each contraction, cardiac performance cannot be regulated by adjustments in the number of stimulated fibers, such as occurs in skel-etal muscle. However, the unique property of cardiac muscle to alter its contractile state provides an important dimension to the regulation of cardiac performance.

Intact heart

The intertwining of cardiac muscle and connective tissue to form the ventricular

FIG. 1–6. Analogy to characterize alterations in preload and contractile state. Preload alterations are represented as a change in the number of horses (reactive sites) with no change in the speed characteristics of the horses. Therefore, a greater force can be generated (C_2) but the velocity against a minimal load in unaltered (A_1 and A_2).

In contrast, a change in contractile state is represented as an alteration in the type but not the number of horses. With a positive inotropic intervention, the speedier, stronger horse has both a greater initial velocity (D_2 vs. D_1) and increased maximal force (E_2 vs. E_1). (Braunwald, E., Ross, J. Jr., and Sonnenblick, E. H.: Mechanisms of Contraction of the Normal and Failing Heart. New York, Little, Brown & Co., 1967.)

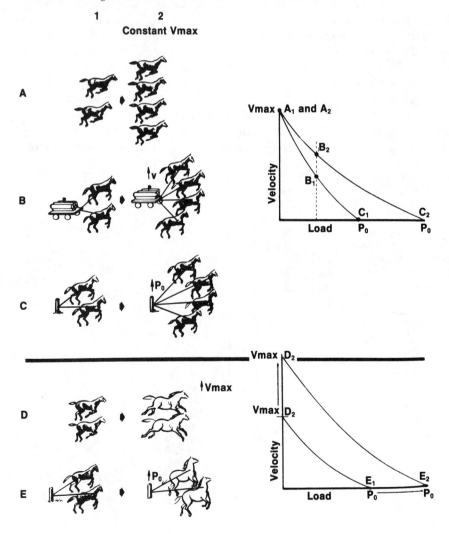

cavity greatly increases the complexity of determining the mechanical properties of the intact heart. Nonuniform fiber orientation, chamber geometry, and nonisotonic contractions are important components of the intact heart that need not be considered in isolated papillary muscle preparations. Nevertheless, the contractile behavior of the intact heart is also characterized by the assessment of pre- and afterload conditions as well as the contractile state (Fig. 1–7).

The determination of loading conditions in the intact heart requires the calculation of *wall stress,* or force per unit area. Several components of wall stress (i.e., circumferential, meridional, and radial stresses) exist in the wall of a three dimensional structure such as the left ventricle, and complex formulae based on idealized ventricular geometries of either a sphere, ellipsoid, or cylinder have been developed to determine wall stress. For practical purposes, however, it is important to understand the relative contributions of the components used to determine wall stress so that directional changes in loading conditions can be assessed readily. The *Laplace law* for quantitation of wall stress was first applied to physiological systems by Woods in 1892

and provides the basis for the determination of ventricular loading conditions. The Laplace relationship for a sphere, as it is applied to the heart, states that wall tension (T, or stress) is a function of intracavitary pressure (P) times the quotient of chamber radius (r) and two times the wall thickness (h), or, $T = P \times r/2h$.

End-diastolic wall stress is analogous to *preload* in the isolated muscle preparation, as both characterize the force that distends fibers and determine the sarcomere length just prior to each contraction. The Laplace relationship shows clearly that preload depends upon ventricular volume and mass as well as end-diastolic pressure. The *Frank-Starling* relationship, whereby an increase in preload produces an augmented stroke volume or stroke work, is the expression of the length-tension relationship in the intact heart. This intrinsic property of cardiac muscle plays a vital role in adjusting the beat-to-beat output of the left and right ventricles. A diseased ventricle tends to operate chronically at a higher preload (an increased end-diastolic pressure and volume) in order to maintain stroke volume. Although such chronic adjustments in preload may help sustain forward output at rest, the reverse capacity of the ven-

FIG. 1–7. Components of ventricular performance of the intact left ventricle. Ventricular mass and geometry play a central role in determinations of both loading conditions (afterload as well as preload) and the stroke volume. The dashed line from *afterload* to *myocardial fiber shortening* represents the inverse relationship between these two variables. The open circles from *afterload* and *preload* to *LV mass and geometry* indicate that long-term alterations in loading conditions can lead to specific remodeling of the ventricular chamber.

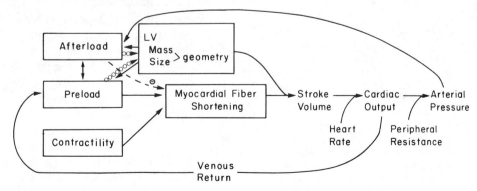

tricle to increase filling further with exercise or stress is therefore diminished, since preload is already increased at rest.

Afterload, the stress in the ventricular wall during ejection, is also related to ventricular geometry and mass as well as arterial pressure. For example, the dilated ventricular chamber of a patient with a congestive cardiomyopathy operates against a greater afterload than the heart of a patient with the same arterial pressure but normal ventricular chamber dimensions. Conversely, the thick-walled, small-volume left ventricle of a patient with aortic stenosis may preserve systolic wall stress at normal levels despite markedly elevated intraventricular pressures. Thus, afterload is *not* synonymous with arterial pressure.

As opposed to the constancy of afterload in the isotonically contracting papillary muscle, the afterload of the intact ventricle

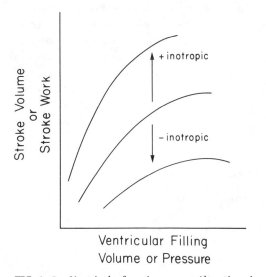

FIG. 1–8. Ventricular function curves. Alterations in preload produce changes in stroke volume or work as described by the classic Frank-Starling relationship. However, an alteration in the contractile state shifts the stroke volume-filling pressure relationship. Therefore, although the Frank-Starling relationship describes a fundamental property of the ventricle, superimposed alterations in contractile state produce a "family of ventricular function curves." Upward or downward movements at a given filling volume represent positive or negative changes, respectively, in ventricular contractility.

varies during ejection. Ventricular volume decreases during ejection, and this reduction of the radius term of the Laplace relation produces a decline in afterload. Alterations in pressure and wall thickness during ejection also contribute to the nonuniformity of afterload in the intact ventricle. The pattern of contraction of the intact ventricle against a varying load is termed *auxotonic*.

As in the isolated muscle, an increase or decrease in afterload in the intact heart results in an inverse change in the extent and velocity of muscle shortening, which is expressed as a change in stroke volume and velocity of fiber shortening. In contrast to isolated muscle studies, it is difficult experimentally to vary afterload in the intact heart without simultaneously altering preload; however, under carefully controlled conditions the inverse relationship between afterload and stroke volume can be demonstrated. Recently, the impaired ventricle with a depressed contractile state has been shown to be more sensitive to an increase in afterload than has the normal ventricle. This sensitive inverse relationship can be taken advantage of clinically by using afterload-reducing therapy to increase the stroke volume of the failing ventricle.

The unique ability of cardiac muscle to alter its *contractile state* (i.e., the rate and intensity of the fiber-shortening reaction from a given pre- and afterload state) is also readily apparent in the performance of the intact ventricle. The relationship of left ventricular filling pressure (or volume) to stroke volume (or work) has been termed a ventricular function, or Starling, curve. As stated earlier, this relationship is analogous to the length-tension relationship observed in isolated muscle whereby an increase in resting length exposes more contractile sites, eventuating in a more forceful contraction. Each ventricle can manifest a series, or family, of ventricular function curves (Fig. 1–8). This capacity to augment or impair ventricular perfor-

TABLE 1–1
Clinically Useful Indices of Ventricular Contractility

Index	Comment
Ventricular function curve	Classical relationship of ventricular output (stroke volume or work) to preload (end-diastolic volume or pressure) (Fig. 1–8). Changes in ventricular output not produced by alterations in pre- or afterload indicate augmentation or depression of contractile state. Although measurements require an invasive procedure, the construction of ventricular function curves has been extremely useful in the management of critically ill patients.
Ejection fraction (EF) EF = stroke volume/end-diastolic volume	EF may be determined by a variety of invasive (ventriculography) and noninvasive (echocardiographic, radioactive tracer) studies. EF is sensitive to changes in loading conditions (see text); however, the ease with which EF can be obtained noninvasively makes it an attractive tool for initial and serial assessments of global ventricular function. Angiographic, radionuclide, and two-dimensional echocardiographic ventriculography all provide additional information concerning regional wall motion.
Fractional shortening (FS) FS = end-systolic dimension/end-diastolic dimension	Similar to EF, except based on ventricular dimensions, not volumes. Therefore, fewer manipulations of primary measurements. Assumes uniform wall motion.
Mean velocity of circumferential fiber shortening (Vcf) Vcf = fractional shortening/ejection time	As with EF and FS, Vcf can be obtained from invasive or noninvasive procedures. Theoretically, Vcf is more independent of changes in loading conditions.
Maximum rate of change in ventricular pressure (dP/dt)	Invasive measurement, requiring high fidelity recordings of ventricular pressure. Although dP/dt is very sensitive to changes in contractile state, it is also sensitive to loading conditions. More sophisticated indices such as dP/dt/common diastolic pressure and dP/dt/P have been developed which maintain sensitivity to contractility and attempt to normalize for differences in loading conditions.
Systolic time intervals (STI) Pre-ejection period (PEP) and ejection time	PEP can provide a useful noninvasive index of the duration of isovolumic systole and is inversely related to dP/dt. STIs can provide important supplemental information to the echocardiographic assessment of LV function. STIs may be of value in following the chronic course of patients with LV dysfunction.
End-systolic pressure-volume relations	The slope of the end-systolic pressure-volume relationship appears to be an expression of ventricular contractile state. The greater the slope, the more complete the systolic emptying for any given end-systolic pressure. This quantification of contractility has been shown to be relatively independent of loading conditions. Indeed, alterations in loading conditions are required to generate the slope. Although most promising, the clinical application is currently cumbersome. Recent attempts to derive the end-systolic pressure-volume relation from noninvasive measurements are of great potential value.
Ejection fraction vs. afterload	EF as mentioned is sensitive to and varies inversely with afterload. However, combining EF with afterload provides an excellent means of assessing whether or not the EF is appropriate for the level of afterload. The clinical difficulty arises in assessing afterload, which varies during systole and is a function of ventricular pressure, wall thickness, and chamber dimensions (see text).

mance, which cannot be explained by alterations of either pre- or afterload, characterizes a change in contractility. Agents which produce alterations in ventricular performance that are independent of changes in pre- and afterload are termed *inotropic agents*. Positive inotropic agents such as norepinephrine, isoproterenol, dopamine, digitalis, and amrinone produce a more vigorous contraction, and thus eject a greater stroke volume under the same pre- and afterload conditions (Fig. 1–8). On the other hand, an acute reduction in contractility can be produced by negative inotropic agents such as the β-adrenergic receptor inhibitors, most antiarrhythmics, anesthetic compounds, as well as insufficient oxygen delivery to the myocardium.

Although the determinants of ventricular performance reflect intrinsic properties of muscle, the complex interrelationship of preload and afterload, as well as of myocardial contractility in the intact circulation, greatly limits the ability to identify which of these specific aspects of cardiac performance may be predominantly altered. As a prerequisite to the assessment of the contractile state, the loading conditions under which the intact ventricle is functioning must be ascertained. Isolated indices of pump function, such as cardiac output or stroke volume, may provide little insight into the underlying properties of the myocardium. The low stroke output observed in profound hemorrhage is clearly the result of a reduction in venous return and filling pressure from the contracted circulatory volume despite a frequent enhancement of ventricular contractility that may be initiated through adrenergically mediated reflexive adjustments. Ventricular ejection fraction (stroke volume/end-diastolic volume), a frequently used index of contractile state, is also sensitive to loading conditions. A normal ejection fraction value in a relatively unloaded ventricle, as in mitral regurgitation, may in fact suggest an impairment of contractility. Conversely, a reduced ejection fraction may be associated with a normal contractile state if the afterload is high, as observed, for example, in some patients with aortic valvular stenosis.

There are many indices of the contractile state of the ventricle, all of which have conceptual and practical limitations (Table 1–1). However, each provides an index for the assessment of the contractile behavior of the heart, which is relatively independent of loading conditions. For the most part, these indices are extremely valuable in assessing directional changes in the contractile state of the ventricle. However, none provides a discrete quantitative term whereby the contractile state of ventricles of varying mass and geometry can be directly compared.

Circulatory failure

Despite the present limitations in quantifying the contractile state and loading conditions of the ventricle in the clinical setting, the principles derived from experimental studies using papillary muscles and isolated and intact heart preparations provide an important framework for an enlightened approach to circulatory disorders. Albeit limited, the clinical assessment of preload, afterload, and contractile status permits important distinctions between diverse cardiac and noncardiac causes of circulatory dysfunction. This assessment should be the initial step taken in directing therapeutic measures.

Circulatory failure is an extremely broad term for a clinical syndrome in which the delivery of nutrients and removal of waste products from metabolizing tissue is inadequate and which can be produced by both cardiac and noncardiac events (Fig. 1–9). Regardless of its etiology, the compensatory responses to hypoperfusion produce a characteristic pattern of circulatory insufficiency. Reflex sympathetic stimulation of the heart is increased, resulting in tachycardia and a selective redistribution of blood flow in which there is an increase

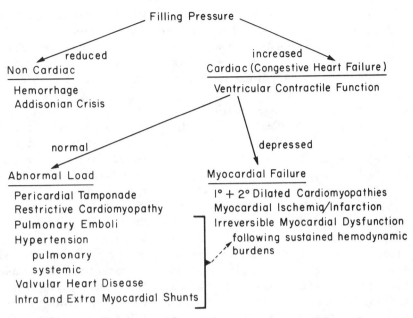

FIG. 1–9. Cardiac and noncardiac etiologies of circulatory insufficiency. Assessments of the loading conditions and contractile state are key discriminators in establishing an etiology. The clinical syndrome of congestive heart failure can be the result of abnormal loading conditions with a normal contractile state as well as a primary depression of ventricular contractility. Although chronic abnormal loading conditions, such as hypertension and valvular heart disease, can lead to true depression of ventricular contractility, the syndrome of congestive heart failure may also be produced by the abnormal load prior to the development of myocardial failure.

in resistance to less vital vascular beds (*e.g.,* skin and gastrointestinal regions). Other compensatory adrenal and renal responses act to retain sodium and water and are manifested as oliguria with an extremely low urinary sodium content. Despite these compensatory mechanisms, cognitive functions can be impaired when the hypoperfusion is severe.

An estimate of preload by the clinical determination of intravascular volume, though at times difficult, provides an important initial distinction between the cardiac and noncardiac etiologies of hypoperfusion. Hypovolemia with cardiogenic shock may produce a pattern of circulatory failure characterized by a reduced systemic arterial pressure, tachycardia, cool extremities, oliguria, and cognitive dysfunction. Increased central venous pressure (CVP), neck vein distention, peripheral edema, and evidence of pulmonary vascu-

lar redistribution or interstitial edema by chest roentgenography strongly suggest a cardiac cause for this low-output state. The absence of the latter features in the presence of dehydration, a reduced CVP, and the associated features of volume contraction suggest volume depletion as the mechanism for the circulatory failure. All too often, however, considerable difficulty and uncertainty are involved in the clinical assessment of volume status. Quantitation of right and left heart filling pressures by bedside right-heart catheterization with advancement of the catheter tip to the pulmonary wedge position has been extremely valuable for directing therapy in these critically ill patients.

Congestive heart failure

Congestive heart failure should be viewed as a clinical syndrome characterized by an elevation of left and/or right ventricular filling pressures and a suboptimal cardiac output. Although an elevated preload in the setting of circulatory insufficiency points to a likely cardiac disorder, it does not necessarily imply that cardiac muscle function is inherently impaired (i.e., *myocardial failure*). Indeed, cardiac disorders that produce circulatory dysfunction should be subclassified on the basis of the loading and contractile status of the myocardium (Fig. 1–9). For example, pericardial tamponade can produce circulatory failure with the clinical picture of hypoperfusion and elevated filling pressures despite a normal or even augmented myocardial contractile state. As in most instances, the assessment of preload, afterload, and contractile status permits a fuller understanding of the etiology of the circulatory disorder and points to an appropriate mode of therapy. In the case of pericardial tamponade, an echocardiographic evaluation would reveal a vigorously contracting ventricle with a small cavitary volume. The reduced cardiac output thus could not be due to either a depressed contractile status or a markedly augmented afterload. In this instance, even though filling pressures are elevated, an inadequate preload as a consequence of a reduced (i.e., restricted) ventricular diameter accounts for the reduced stroke volume. Therapy is therefore directed toward increasing ventricular filling volume most readily by pericardiocentesis or, in the case of a thickened and diseased pericardium, its surgical removal.

Marked elevations in afterload can also produce a pattern of circulatory insufficiency which is not necessarily associated with an impairment in myocardial contractility. In severe systemic arterial hypertension, the elevation in arterial pressure and total peripheral resistance may be so acute as to impose an inordinate load on a left ventricle that has not yet adapted structurally to an elevated afterload by concentric hypertrophic growth. Without this form of adaptive hypertrophy (a relative increase in wall thickness) the often dramatic eleva-

tions in arterial pressure may produce a marked increase in afterload in accord with the Laplace relation. As discussed previously, both the isolated papillary muscle and intact ventricle exhibit an inverse relationship between afterload and fiber shortening. Indeed, in each preparation isometric (or isovolumic) contractions can be demonstrated at an afterload against which the muscle cannot shorten. Therefore, the afterload imposed during an acute episode of severe hypertension may be so great that ventricular ejection is markedly limited. In this circumstance, when the load exceeds the capacity of even the normal heart to respond, a circulatory pattern of congestive heart failure can develop. Once again, it is critical to recognize that herein the syndrome of congestive heart failure is produced by an excessive afterload and not an intrinsic contractile abnormality. The appropriate and most effective therapeutic measures for this form of circulatory insufficiency are directed toward decreasing the afterload (i.e., reducing arterial pressure and total peripheral resistance), thus permitting the heart to perform within its normal limits.

This state has been frequently confused with an even lesser increase in arterial pressure and total peripheral resistance associated with an expanded extracellular (and plasma) volume as a result of severe renal functional impairment (e.g., hypertension associated with the oliguric or anuric phase of acute glomerulonephritis). The circulation is *congested* in that arterial, central venous, and pulmonary wedge pressures are elevated and cardiac output may be reduced, yet the myocardium is not failing. Appropriate therapy in this instance is immediate contraction of extracellular fluid volume through hemodialysis.

Myocardial failure occurs when the circulatory pattern of congestive heart failure is produced by a true impairment of ventricular contractility. The reduced ventricular ejection observed in myocardial failure cannot be attributed to either a greatly reduced preload or a markedly increased afterload. Although afterload is generally increased in the dilated, chronically failing ventricle even at normotensive systemic perfusion pressures as a consequence of an altered ventricular geometry, the magnitude of this abnormal load is usually not sufficient to account for the reduced ventricular ejection.

The approach to patients with myocardial failure should be multifaceted with diagnostic, therapeutic, and preventive measures vigorously pursued. Although the causes of true myocardial failure are diverse and in many instances poorly understood, the establishment of a diagnosis (even if only of the involved altered physiological indices) can lead to a more specific and effective therapy. As in patients with fever from a variety of infectious and noninfectious sources, the clinical features of patients with severely impaired ventricular performance of different causes are often indistinct. However, it must be determined by clinical history, physical examination, and laboratory evaluations whether the impairment in cardiac performance is the result of a prior sustained hemodynamic burden (hypertension, valvular heart disease), loss of contractile tissue (myocardial infarction), or cardiomyopathic process (primary or secondary) (Fig. 1–10). Moreover, the therapeutic approach to the patient with severe ventricular dysfunction as a consequence of a large myocardial infarction and ventricular aneurysm will differ substantially from that of the patient in whom the ventricular dysfunction is associated with hemochromatosis.

Therapeutic measures for myocardial failure are nonspecific, but apply to a vast number of patients. While therapy is directed toward improving contractile performance with positive inotropic agents, the loading conditions of the impaired ventricle must be optimized. Diuretic therapy and restriction of sodium intake, long known to decrease symptoms of central venous and pulmonary congestion, also reduce ventricular afterload as a consequence of the reduction in ventricular volume. In more refractory cases of myocardial failure, vasodilating agents may be added to the inotropic and diuretic therapy in order directly to reduce ventricular afterload. Each form of therapy has inherent risks for which particular care must be maintained. With diuretic therapy, for example, too vigorous a volume depletion might occur, resulting in an excessive re-

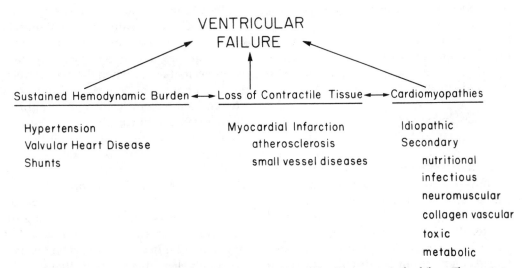

FIG. 1–10. Broad mechanistic representation of factors leading to ventricular failure. These categories are not mutually exclusive. The between-group interaction may play an important role in the manifestation of ventricular failure.

duction in ventricular filling. Thus, the Frank-Starling adaptation may be removed from the impaired ventricle, resulting in a reduction in cardiac output. In addition, the loss of potassium in a state of secondary hyperaldosteronism with hypokalemia may predispose the patient to ventricular dysrhythmias and less efficient cardiac function. All too often maintaining the balance between symptoms of congestion and those of systemic hypoperfusion is an extremely delicate task. Such fine tuning of therapy requires a nonregimented approach with strict attention to details. Despite the risks of each form of therapy, the judicious use of positive inotropic agents, diuretics, and vasodilators can be effective in reducing the symptoms of congestive heart failure as well as improving functional capacity.

In addition to the use of pharmacological measures to improve cardiac muscle function and to optimize loading conditions, a careful search for treatable conditions (anemia, infections, arrhythmias) that exacerbate cardiac failure should be pursued. Prevention of further insults to the damaged myocardium should be stressed as another important facet in the management of patients with myocardial failure. Maintaining adequate nutrition, preventing thromboembolic events and dysrhythmias, and avoiding alcohol, cigarettes, and drugs with even minor adverse effects on the myocardium add an often overlooked dimension to the therapy of patients with myocardial failure.

Ventricular hypertrophy

The association between myocardial failure and ventricular hypertrophy is so common that ventricular hypertrophy is frequently viewed as a marker of ventricular failure rather than as an anatomic marker of a pathologic hemodynamic burden that can result in muscle dysfunction. Myocardial failure is most often preceded by a sustained increase in ventricular load. In conditions of systemic or pulmonary hypertension and valvular or congenital

heart disease, the ventricular chamber upon which the abnormal load is imposed hypertrophies and thereby becomes capable of sustaining the hemodynamic burden without evidence of dysfunction. Indeed, these "remodeled" ventricles often develop intraventricular pressures or eject stroke volumes that exceed the capacity of the normal ventricle. The ability to remodel, i.e., to alter ventricular mass and geometry, in response to prolonged demands provides the ventricle with an important, potentially long-term adaptive mechanism. In conditions of chronic excessive afterload, such as systemic hypertension or aortic stenosis, left ventricular hypertrophy occurs in a concentric pattern, i.e., the ventricular mass increases out of proportion to chamber volume. From the Laplace relationship it is apparent that this remodeling effectively permits the thick-walled left ventricle to sustain the elevated pressure without a marked increase in afterload. In conditions of volume overload, such as congenital or acquired arteriovenous shunts or valvular insufficiency, ventricular hypertrophy occurs in an eccentric pattern, i.e., the muscle mass increases in proportion to or less than chamber volume. This type of remodeling is highly effective for producing large stroke volumes, since a greater volume output is produced from a larger cavity for the same degree of muscle shortening. However, as with any adaptation, there are limitations and important trade-offs that occur when the normal relationship between structure and function is altered. The eccentrically hypertrophied, volume-overloaded ventricle that develops in aortic or mitral regurgitation, although a highly efficient pump, is extremely sensitive to increases in afterload. Conversely, the concentrically hypertrophied, pressure-overloaded ventricle that appears in aortic stenosis is capable of developing supranormal intraventricular pressures but may have a limited flow-generating capacity. Nonetheless, this ability to alter ventricular mass and geometry in response to long-term demands is an important adaptation that is superimposed on the short-term beat-to-beat alterations in rate, force, and degree of ventricular contraction.

Thus, in many diseases ventricular hypertrophy provides an effective early, but frequently long-term, compensation by which an abnormal workload can be sustained without clinical symptoms of cardiac dysfunction. However, it is most important to recognize the presence of the imposed abnormal workload, since the sustained hyperfunctional state, without correction, may eventually progress to true myocardial failure. Indeed, epidemiological studies indicate that systemic hypertension is one of the most common causes of congestive heart failure. To date, however, the precise mechanisms or factors that are responsible for this transition from compensated hypertrophy to ventricular dysfunction remain poorly understood. Antihypertensive therapy in systemic hypertension and the proper timing of valve replacement in valvular heart disease are the most important preventive measures in the therapy for myocardial failure.

The *cardiomyopathies* represent a diffuse group of diseases of the myocardium that may be classified on the basis of distinctive morphological and functional abnormalities into three major groups: dilated (formerly congestive), hypertrophic (formerly idiopathic hypertrophic subaortic stenosis), and restrictive. Patients with dilated cardiomyopathies present with symptoms of cardiac failure and cardiomegaly, yet the diffuse depression of their contractile state cannot be attributed to a current or prior hemodynamic burden known to induce myocardial dysfunction. However, in addition to hypertensive, valvular, and congenital heart disease, significant coronary artery and pericardial diseases must be excluded before the diagnosis of a primary disease of heart muscle can be established. Idiopathic dilated cardiomyopathy is more common than the secondary forms in which the same morphological and functional pattern as the dilated cardiomyopathy exists. These secondary forms are associated with coexisting metabolic, infectious, connective tissue, or neuromuscular diseases. Even in the group that is not associated with systemic diseases, excessive alcohol intake, antecedent viral infections, and, less commonly, recurrent rheumatic fever, puerperium, and a family history of cardiomyopathy can be incriminated in the development or potentiation of dilated cardiomyopathy.

The loss of normally functioning myocardial tissue is a relatively common cause of ventricular failure in North America and Europe. The etiology and precise mechanism for the ventricular dysfunction associated with the loss of myocardium are more straightforward than those of the heart failure observed in the cardiomyopathies or following a sustained hemodynamic burden. Although myocytes have the capacity to hypertrophy, the ability of heart muscle to regenerate new cells is lost in the early neonatal period, and thus replacement of myocytes with fibrous connective tissue in the healing phase of a *myocardial infarction* represents an irreversible reduction in the number of myocytes. In many instances the remaining viable myocardium is sufficient to maintain adequate circulatory function. Not infrequently, however, the loss of contractile tissue may be so extensive that the burden placed upon the remaining myocytes is insurmountable and cardiac failure ensues. In addition to the loss of myocardium from infarction, other factors such as the status of the nonoccluded coronary arteries, the integrity of the valvular structures and conducting system, and the loading conditions of the ventricles must be considered in assessing overall cardiac performance. Nonetheless, a spectrum of dysfunction exists following myocardial infarction that is related in large part to the extent of tissue necrosis.

The evaluation of ventricular loading conditions and contractile performance provides only the basic framework from which to treat the circulatory failure. It is common in clinical settings for the resulting systemic tissue hypoperfusion to result from a combination of contributory factors. Too vigorous a diuresis in aortic stenosis with or without contractile dysfunction may produce a state of hypoperfusion that would not be present without the combined cardiac and noncardiac factors (Fig. 1–9). Clusters of disorders can manifest severe congestive heart failure in moderate, otherwise well-compensated heart disease. Dysrhythmias, anemia, and the increased metabolic demands of fever are well known to exacerbate congestive heart failure. Abnormal thyroid function can also contribute to the development of congestive heart failure in individuals with previously compensated heart disease. Similarly, the coexistence of coronary artery disease and valvular heart disease appears to have a synergistic effect in producing congestive heart failure from lesions that independently would be well tolerated.

Cardiac performance is ultimately characterized by the loading conditions and contractile state of the heart, components of myocardial function which can be only roughly approximated in the clinical setting. Nevertheless, these indices of cardiac performance permit the important distinction as to whether congestive heart failure is produced mainly by abnormal loading conditions or by true myocardial failure. Thus, a more precise physiological definition of the cardiovascular disorder must be sought in order to provide more specific and effective therapy.

Annotated references

Biedert, S., Barry, W. H., and Smith, T. W.: Inotropic effects and changes in sodium and calcium contents associated with inhibition of monovalent cation active transport by ouabain in cultured myocardial cells. J. Gen. Physiol. 74:479, 1979. (Provides evidence that the inotropic action of cardiac glycosides is accompanied by a net uptake of sodium and calcium.)

Braunwald, E. (ed.): Heart Disease. A Textbook of Cardiovascular Medicine. Philadelphia, W. B. Saunders, 1980. (A comprehensive textbook of clinical cardiology which provides a solid background of cardiovascular physiology and pharmacology.)

Braunwald, E., Ross, J., Jr., and Sonnenblick, E. H.: Mechanisms of Contraction of the Normal and Failing Heart. Boston, Little Brown & Company, 1976. (This monograph provides the fundamental information concerning cardiac performance from the subcellular level to the actions of an integrated circulatory system.)

Chapman, C. B., and Mitchell, J. H.: Starling on the Heart. London, Dawsons of Pall Mall, 1965. (Delightful collection of some of Starling's work, including the Linacre Lecture on the Law of the Heart.)

Grossman, W.: Cardiac hypertrophy: Useful adaptation or pathologic process? Am. J. Med. 69:576, 1980. (Discusses types of hypertrophy with emphasis on contractile performance in both experimental and clinical conditions of chronic hemodynamic overloads.)

Guyton, A. C., Jones, C. E., and Coleman, T. G.: Circulatory Physiology: Cardiac Output and its Regulation. Philadelphia, W. B. Saunders, 1973. (Provides basic concepts and integrated approach to cardiac and peripheral factors which regulate cardiac output.)

Johnson, R. A., and Palacios, I.: Dilated cardiomyopathies of the adult. N. Engl. J. Med. 307:1051, 1982. (Well-presented and referenced review of causes and clinical features of dilated cardiomyopathies.)

McAllister, R. G., Jr.: Clinical pharmacology of slow channel blocking agents. Prog. Cardiovasc. Dis. 25:83, 1982. (Succinct summary of clinical pharmacology of currently available Ca^{++} entry blockers.)

Nayler, W. G., and Grinwald, P.: Calcium entry blockers and myocardial function. Fed. Proc. 40:2855, 1981. (Provides a clear description of the current view of the mechanisms of action of Ca^{++} entry blockers on the myocardium.)

Sagawa, K.: The ventricular pressure-volume diagram revisited. Circ. Res. 43:677, 1978. (Presents the rationale as well as data for the use of the slope of the end-systolic pressure-volume relationship as a measure of the contractile state of the intact ventricle.)

Sarnoff, S. J.: Myocardial contractility as described by ventricular function curves. Physiol. Rev. 35:107, 1955. (Clear description of the use of the "family of ventricular function curves" to assess contractile state.)

Weber, K. T., Janicki, J. S., Hunter, W. C., Sanjeev, S., Pearlman, E. S., and Fishman, A. P.: The contractile behavior of the heart and its functional coupling to the circulation. Prog. Cardiovasc. Dis. 24:375, 1982. (Up-to-date review of the mechanical properties of the intact ventricle with clear presentation of coupling of heart to arterial and venous systems.)

2

Tissue Perfusion

Michael J. Brody, Ph.D.
Donald D. Heistad, M.D.

Cellular functions depend upon a system for delivery of substrate and for removal of the waste products of that activity. In the human body, this delivery and removal function is provided by the cardiovascular system. One can judge the adequacy with which the cardiovascular system performs its functions by the magnitude of tissue perfusion. If tissues are perfused with a sufficient amount of normally constituted blood to meet metabolic requirements and to provide for disposal of waste materials, the purpose of the cardiovascular system is satisfied.

Regulation of tissue perfusion

Tissue perfusion is regulated by two major factors: arterial blood pressure, which provides the energy for driving blood through the vascular channels; and the resistance to blood flow provided by the physical characteristics of the blood vessels and of the blood itself.

Blood pressure

Regulation of arterial blood pressure is considered in detail in other chapters. Briefly, arterial pressure is the result of cardiac output and the resistance to flow provided by the vessels. In the cardiovascular system pressure is developed by the active contraction of the heart. This contraction produces the cardiac output which is distributed throughout the system in a manner dependent upon the resistance to flow provided by individual segments of the vascular tree. Two elementary considerations should be emphasized: first, without adequate myocardial contractile activity, perfusion is severely restricted. Second, although blood pressure is determined by both cardiac output and the resistance to blood flow, in vascular beds with low potential for producing active changes in vascular resistance (e.g., when neurogenic influences to the vessel are low and the vessels themselves respond poorly to vasomotor influences) blood pressure becomes the major determinant of perfusion. As discussed below, in certain circumstances adequate tissue perfusion may be related directly to the magnitude of arterial pressure.

Vascular resistance

Emphasis will be placed in this chapter upon those factors that play prominent roles in determining the adequacy of tissue perfusion by alterations in the resistance to blood flow. Vascular resistance may be broadly defined as the resistance to blood flow provided by the geometry of the vascular tree and by the physical properties of the fluid circulating through the vascular compartment. According to the Poiseuille relationship, the length and the radius of a tube are the geometric components providing resistance to flow through the tube. Although this relationship does not hold strictly for the distensible vascular compartment, it can be concluded that radius (which is an exponential function) is the major determinant of resistance to flow. Radius is contributed to and altered by passive elastic components of the vessel wall, muscular elements of the vessel wall that contract actively, and structural alterations in the vessel wall that provide obstruction to flow.

The resistance to blood flow provided by the total cardiovascular system can be estimated by dividing the difference between arterial pressure and right atrial pressure by the cardiac output. This value, referred to as total peripheral resistance, is a useful index of the average or overall radius of the vascular compartment. Total peripheral resistance does not, however, provide any estimate of the distribution of blood flow between organs or of the adequacy of perfusion of any or all organs. In addition, total peripheral resistance does not provide any direct measure of the contribution of active or passive tension of vascular smooth muscle to the level of radius in the vessel wall. For example, the administration of a ganglionic blocking agent, by interfering with neurogenic vasoconstrictor influence on the vessel wall, may be expected to reduce active tension in the vessel wall and thus reduce vascular resistance. However, total peripheral resistance is unchanged following administration of the ganglionic blocker because cardiac output and blood pressure are reduced in approximately the same proportion. The average caliber of the vascular tree remains unchanged because of the reduction in arterial (i.e., distending) pressure which allows the vessels to adjust passively to unchanged caliber. This example is used to illustrate the point that active tension may change when there is no change in computed total peripheral resistance, if at the same time arterial pressure is altered.

Methods for estimating perfusion

Vascular resistance of individual organs can be determined with degrees of difficulty that vary, depending upon the organ under consideration. These measurements of vascular resistance depend upon accurate determination of blood flow to the organ. Plethysmographic techniques can be used to measure flow to the forearm, which is primarily muscle, and to a digit, which is primarily skin. Determination of blood flow to the forearm after epinephrine iontophoresis gives a more accurate measurement of muscle blood flow, since the iontophoresis technique virtually eliminates blood flow to skin by causing intense cutaneous vasocontriction. Depot injections of radioactive xenon or sodium iodide into muscle or skin allows measurement of "capillary" blood flow at the site of injection by external counting techniques and estimation of disappearance rate of the isotope.

Renal blood flow can be estimated with the clearance of p-amino hippurate (PAH). Accurate measurement of blood flow to the kidney, however, requires that the extraction of PAH be determined. Virtually all interventions that alter renal blood flow also change intrarenal distribution of that flow, so that clearance measurements per se can be misleading. Dye dilution has been used clinically to measure renal blood flow, but the variability and error associated with this measurement may be large. Splanchnic blood may be estimated in man from the clearance of bromsulphalein or indocyanine green by the liver. The green dye is almost completely extracted by the liver in a single passage, and the extraction coefficient is stable. The Doppler velocity probe has proven to be useful in assessment of the severity of extracranial occlusive disease of the carotid arteries.

Distribution of radioactive microspheres to various organs following their injection into the left atrium provides a measurement of flow to all organs and allows the estimation of intraorgan distribution of flow at the time of injection. Six to eight estimates of flow are currently possible in each animal, as spheres labeled with elements of different isotopic radioactivity emission energies are used. Measurements of coronary flow are possible in man using a thermodilution technique whereby the thermal indicator (cold saline) is infused into the coronary sinus and sensed by a thermistor on the same catheter 2 cm. downstream. The dilution of the indicator is a function of coronary sinus flow. Using ^{201}thalium the regional distribution of myocardial perfusion can be determined with external counting techniques. This method is useful in studying patients with ischemic heart disease. Recently, a Doppler velocity probe has been used to measure coronary blood flow in patients undergoing cardiac surgery. The method has been used to examine coronary vascular reserve and thereby to examine the functional significance of vascular lesions which are demonstrated by angiography. Virtually all methods presently available for the determination of blood flow and vascular resistance in individual organs give limited information concerning adequacy of tissue perfusion. It is essential also to provide estimates of how effectively the available flow reaches the portion of the circulation where exchange between blood and tissues may occur by estimating intraorgan distribution of flow.

From a functional point of view, circulation within organs can be divided into two compartments: the first of these a series of channels in which diffusion between blood and tissues may occur, and the second a series of channels in which there is rela-tively little surface area available for diffusion to occur. Thus, any given level of blood flow to an organ can be divided into its *nutritional* and *nonnutritional* components. Adequate tissue perfusion depends upon the existence of nutritional blood flow and is not necessarily related to the total blood flow delivered to an organ. It is convenient to think of nonnutritional flow as passing through an arteriovenous shunt, although there may be little anatomical evidence for the existence of such shunts.

Techniques have been developed for studying shifts in the distribution of blood flow within organs. Changes in the extraction of oxygen provide a convenient index for determining whether the amount of blood flow reaching sites where diffusion may occur is altered. More sophisticated techniques involve determination of the clearance of a freely diffusible isotope such as rubidium. Coronary, skeletal muscle, and skin circulation can all be shown to possess significant potential for functional shunting of blood away from nutritional vascular channels.

Coronary atherosclerosis is an example of a condition in which nutritional circulation may be inadequate despite the existence of coronary blood flow in the normal range. After myocardial infarction, although blood flow to the infarcted region is decreased, blood flow to the normal myocardium may be increased, so that total coronary blood flow is normal. Angina pectoris usually is precipitated by stimuli such as exercise that increase myocardial oxygen requirements. Angina occurs when increases in blood flow are not sufficient to meet the increased metabolic needs of the heart. During angina, increases in total coronary blood flow may be within the normal range, but inadequate increases in flow to a limited area of the subendocardium may produce focal ischemia.

Nitroglycerin, which is the most effective antianginal agent, can relieve anginal pain effectively without altering total coronary blood flow. Its efficacy may be related to a reduction in myocardial oxygen demands and to redistribution of blood flow towards subendocardium.

The kidney is an organ in which shifts of blood flow can also cause significant loss of function. Blood supply to functional nephrons is concentrated in the outer portion of the renal cortex. A variety of studies have suggested that interventions that alter total renal blood flow can produce significant redistribution of flow within the kidney. For example, renal vasoconstriction during adrenergic discharge tends to shunt blood flow away from outer cortex, whereas renal vasodilators such as furosemide and ethacrynic acid tend to shunt blood flow toward outer cortical nephrons. In experimental congestive heart failure, the percentage of renal blood flow supplying outer cortical nephrons is reduced. This observation suggests that shifts in intrarenal blood flow distribution may well be important in the retention of sodium and the formation of edema associated with heart failure. In several forms of renal disease the fraction of blood flow to the outer cortex may be reduced.

Changes in vascular muscle contraction

The contractile state of vascular muscle is perhaps the major physiological determinant of tissue perfusion. Three factors regulate the contractile activity of vascular muscle. For purposes of this discussion these are referred to as *humoral, neurogenic,* and *intrinsic* factors.

Humoral factors regulating muscle contraction

A large number of substances with vasoactive properties are carried in the blood and exert important effects on vascular muscle tension and, thus, on tissue perfusion. In any portion of the vascular tree the level of active tension in smooth muscle effected by humoral factors is the algebraic sum of stimulant and relaxant influences exerted upon the muscle.

ADRENERGIC. Catecholamines found in the blood arise from several sources. The major source of epinephrine is the adrenal medulla; another source is extraneous chromaffin tissue. Norepinephrine is derived primarily from its secretion from adrenergic nerve terminals. Catecholamines may produce either vasodilation or vasoconstriction, depending on the affected organ. Thus, epinephrine is a powerful human skeletal muscle vasodilator. Epinephrine circulating in the blood is subject to uptake by sympathetic nerve terminals through the same active process by which the norepinephrine is released into the synaptic cleft. It has been demonstrated recently that epinephrine that is taken up into the nerve terminal from the bloodstream can subserve a neurotransmitter function identical to norepinephrine in those vascular beds where the amines are vasoconstrictors. However, in skeletal muscle, where epinephrine is a potent vasodilator, selective neurogenic vasodilator stimuli may cause responses that are mediated by epinephrine through activation of β_2-adrenergic receptors.

Epinephrine and norepinephrine have the potential to constrict coronary blood vessels; however, because they increase myocardial contractility and heart rate and thereby increase myocardial metabolism, their net effect is coronary vasodilation. Prominent vasoconstrictor effects of the catecholamines are observed in the cutaneous, renal, and mesenteric circulation, whereas the cerebral vascular bed is much less reactive to adrenergic stimulation.

Ordinarily, the levels of catecholamines circulating in the blood probably exert minimal effects on tissue perfusion. However, under certain conditions increased blood levels of catecholamines may produce important effects on distribution of blood flow and on tissue perfusion. Prominent examples are pheochromocytoma, hypotensive shock, anxiety state, exercise, hypertension, and cardiac failure.

The development of selective agonists and antagonists of adrenergic receptors has improved the potential for therapeutic efficacy. Alpha$_1$-adrenergic receptors are localized in vascular smooth muscle and promote vasoconstriction. Alpha$_2$-receptors, localized primarily on the presynaptic nerve terminal membrane, inhibit the release of the sympathetic transmitter. Activation of alpha$_2$-receptors in the central nervous system decreases central sympathetic outflow. Alpha$_1$-antagonists such as prazosin and alpha$_2$-receptor agonists such as clonidine have proven to be effective antihypertensive agents. Selective beta-adrenergic agonists and antagonists have also been developed. Beta$_1$-adrenergic receptors mediate cardiac inotropic and chronotropic actions of catecholamines, whereas beta$_2$-receptors are found in vascular and bronchial smooth muscle. Selective beta$_1$-antagonists, such as metoprolol and atenolol, as well as nonselective beta-receptor antagonists, are useful in the treatment of angina pectoris. The former selective agents exhibit fewer bronchoconstrictor or peripheral vasospastic side effects than observed with nonselective beta-adrenergic receptor blocking agents.

The catecholamine dopamine has been the subject of considerable recent interest. Although it probably plays little role as a circulating humoral agent, it has considerable potential for therapeutic application to the treatment of hypotensive states. The agent is a cardiac stimulant and, unlike other catecholamines, is a renal vasodilator through activation of selective dopaminergic receptors in the renal vascular bed. Several dopamine receptor agonists have been developed with selective renal vasodilator activity. Their efficacy as antihypertensive agents is under investigation.

PHEOCHROMOCYTOMA. Arterial pressure may be increased in this disorder by vascular effects or cardiac effects or a combination of the two. Vasoconstriction alone may raise arterial pressure, and this increased pressure may be reduced dramatically by the administration of an alpha-adrenergic receptor blocking agent such as prazosin. Intense stimulation of the heart by high levels of circulating catecholamines may also lead to hypertension. In such a situation the administration of a beta-adrenergic receptor blocking agent would lower arterial pressure.

HYPOTENSIVE SHOCK. Cardiovascular reflexes help compensate for reduced arterial pressure by increasing sympathetic discharge in this condition. If these reflexes are at all intact, the secretion of catecholamines from adrenergic nerves and from the adrenal medulla is enhanced and the circulating blood levels of these substances will be increased. Poor tissue perfusion in the shock state may be the result primarily of two contributing factors. Reduced arterial pressure alone will limit the amount of flow available for adequate perfusion of the tissues. If in vital organs this is combined with vasoconstriction promoted by the high circulating levels of catecholamines, perfusion is further compromised. In hypotensive shock the level of arterial pressure alone is not an informative indicator of the adequacy of tissue perfusion, since intense vasoconstriction may be supporting arterial pressure at the expense of adequate perfusion. Treatment of hypotensive shock should always be directed at improving tissue perfusion and raising arterial pressure. Ordinarily, this requires that the output of the heart be improved in addition to whatever measures are taken to relieve vasoconstriction produced by exaggerated sympathetic activity. Alpha-adrenergic receptor blocking agents such as phentolamine and phenoxybenzamine have been used in conjunction with plasma expanders to improve tissue perfusion by relieving vasoconstriction. Since these agents do not have receptor selectivity, they also block alpha$_2$-adrenergic receptors, and actually increase circulating levels of catecholamines and thus secondarily increase cardiac performance. Although catecholamines probably play a role in producing vasoconstriction in hypotensive shock, they may be used with considerable efficacy in the treatment of shock, if special attention is paid to their cardiostimulant properties.

ANXIETY STATES. These may be associated with increased blood levels of catecholamines. Several cardiovascular signs observed in anxiety are probably attributable to the catecholamines—for example, cold clammy palms, tachycardia, dilated pupils, and hypertension.

EXERCISE. In *exercise* an extraordinary demand for increased perfusion is made by skeletal muscles and heart. The distribution of blood flow in exercise is appropriate to the demands for flow required by skeletal muscle. Flow to muscle is increased dramatically while flow to other organs such as the kidneys, skin, and splanchnic circulations is reduced or remains unchanged. In any event, the *fraction* of cardiac output delivered to an organ such as the kidney is reduced considerably. Although the blood levels of catecholamines are increased during exercise, it is not really known whether these substances contribute to the redistribution of blood flow to skeletal muscle.

HYPERTENSION. Except for its demonstrated role in pheochromocytoma, an adrenergic humoral mechanism has not been shown conclusively to play a significant role in the maintenance of hypertension. Using improved and more precise methods for the detection of catecholamines in blood, it has been shown by some, but not all, investigators that the level of circulating catecholamines in the blood of essential hypertensives is slightly higher than in normotensives. This finding may reflect change in the turnover rate or metabolism of catecholamines as well as in the amount released at the nerve endings.

It has been suggested that the enzyme responsible for catecholamine biosynthesis, dopamine β-hydroxylase (DBH), may be a marker for sympathetic activity. Measurements of DBH activity in hypertension have not been consistent, although there are several reports of increased levels which might be attributed to increase in sympathetic discharge. A subgroup of patients with essential hypertension was reported in which the subjects exhibited signs of sympathetic hyperactivity including increased heart rate, increased urinary excretion of dopamine, and catecholamine metabolites.

CARDIAC FAILURE. Tissue perfusion is impaired in congestive heart failure. A compensatory response to cardiac failure is an increase in adrenergic drive manifested by elevated blood and urine levels of catecholamines and their metabolites. Adrenergic drive to the decompensated myocardium is derived from a humoral rather than a neurogenic source, since the heart in severe failure is depleted of catecholamines.

The increased peripheral vascular resistance in congestive heart failure may be related to at least three factors: (1) increased sympathetic tone to resistance as well as capacitance vessels, (2) increased levels of circulating catecholamines, and (3) structural changes in the vessel wall which can be demonstrated in the presence of adrenergic blockade and which might be related to sodium and water retention in the vessel wall. Other circulating vasoconstrictors including angiotensin II and vasopressin are also elevated. The relative extent to which these vasoconstrictors reduce tissue perfusion in a given subject is not predictable and improvement in cardiac performance may be achieved with vasodilator therapies that include alpha-adrenergic receptor blockade and inhibition of the renin-angiotensin system.

Renin-angiotensin system

The pathophysiological role of the renin-angiotensin system in hypertension is considered in detail elsewhere in this book. With specific reference to tissue perfusion it seems likely that this system plays an important role only in malignant stages of hypertensive disease. Here the high blood levels of renin and angiotensin II undoubtedly contribute significantly to the high vascular resistance. It is also likely that the vasoconstrictor properties of angiotensin are prominent in renal vascular hypertension and in eclampsia.

The availability of competitive antagonists of angiotensin II, such as saralasin, and inhibitors of angiotensin-converting enzyme such as captopril and enalapril, has allowed for a more specific classification of patients in whom arterial pressure is sustained by the vasoconstrictor properties of angiotensin II; those subjects with hypertension and high plasma renin activity respond with significant pressure lowering to these inhibitors of angiotensin formation or action. These might be the same subjects who would be expected to show a significant hypotensive response to beta-adrenergic receptor blockers, since these agents inhibit the release of renin by the kidney. The effectiveness of the competitive blocker does not necessarily mean that increased vascular resistance in such subjects derives solely from the vasoconstrictor properties of angiotensin. Indirect effects such as facilitation of norepinephrine release from sympathetic nerve terminals and central facilitation of sympathetic discharge may just as well be involved as vasoconstrictor mechanisms.

It can be anticipated that new therapeutic modalities based upon interference with the renin-angiotensin system will be developed. Although only in experimental stages, several peptide inhibitors of renin have been shown to reduce arterial pressure in experimental high-renin states. These agents, which block the enzymatic conversion of angiotensinogen to angiotensin I, are of great theoretical interest because they offer the opportunity of interfering with the earliest stage of angiotensin biosynthesis. A separate and distinct renin-angiotensin system has been identified in central nervous system tissue. Both renin and angiotensin have been localized in central neurons using immunocytochemical techniques. The significance of this system in physiological and pathophysiological states remains to be identified.

Prostaglandins

Prostaglandins are endogenous, acidic, lipid-soluble substances with the potential to alter tissue perfusion through a variety of mechanisms. The complex biosynthesis of prostaglandins originates from the conversion of arachidonic acid, a 20-carbon, tetra-unsaturated long-chain fatty acid that is a component of plasma membrane phospholipids. The first reaction that leads to the production of a series of unstable intermediates is catalyzed by the enzyme system referred to as prostaglandin synthetase or cyclo-oxygenase. Virtually complete inhibition of prostaglandin synthesis can be achieved with a variety of compounds, including aspirin, nonsteroidal antiinflammatory agents such as indomethacin, and structural analogues of arachidonic acid. There is now ample evidence that the anti-inflammatory effect of these agents depends in large part upon the inhibition of prostaglandin synthesis.

There is an enormous diversity of biological actions found in the various products of arachidonic acid metabolism. The major agents with the potential to influence tissue perfusion are PGE, PFG, thromboxane (TXA_2), prostacyclin (PGI_2) and leukotriene. For some time it was believed that the major vasodilator effects of prostaglandins resided in PGE_2; however, it is now appreciated that the unstable product PGI_2 has the greatest physiological significance. Vasoconstrictor effects can be observed with $PGF_2\alpha$ and TXA_2. TXA_2 is a profoundly potent stimulant of platelet aggregation, and there is great interest in the possibility that thrombus formation that may precede acute cerebral or myocardial ischemia is mediated by increased thromboxane formation or increased platelet sensitivity to the agent. A number of thromboxane synthetase inhibitors have been developed that appear to have both antithrombotic and antihypertensive efficacy. The ability of a thromboxane synthetase inhibitor to lower arterial pressure is postulated to result from shunting of the arachidonic acid metabolic cascade from thromboxane synthesis to synthesis of prostacyclin.

Since prostaglandins are degraded very rapidly in blood to inactive metabolites, it is rather unlikely that they play any major physiological or pathophysiological role as circulating substances. The bulk of evidence suggests that substances are formed locally in response to a variety of stimuli and that they produce their major effects on the cardiovascular system at or near their site of production. The agents are capable of modulating and/or mediating the vascular actions of a number of vasoactive substances. For example, the vasoconstrictor effects of angiotensin and norepinephrine are modulated by the formation of vasodilator prostaglandins. Thus, the administration of a cyclo-oxygenase inhibitor will facilitate the effects of a variety of vasoconstrictor agents. In the kidney the vasodilator action of bradykinin appears to be mediated in large part by the formation of vasodilator prostaglandins. PGE_2 and PGI_2 have prominent inhibitory effects on the sympathetic nerve terminal, reducing the amount of sympathetic neurotransmitter release per nerve impulse.

Although not directly related to the prostaglandins chemically, several antihypertensive lipids have been isolated from the renal medulla. These are termed antihypertensive polar renal medullary lipid (APRL) and antihypertensive neutral renal medullary lipid (ANRL). Since arterial pressure in experimental states can be lowered by transplantation of renal medullary tissue, or more specifically cultures of renal interstitial cells that contain these lipids, these substances may well be involved in the maintenance of arterial pressure. It has been demonstrated that ANRL has unique cardiovascular effects. Unlike most pure vasodilators that reflexly increase sympathetic activity in response to the lowering of arterial pressure, ANRL simultaneously lowers sympathetic nervous system discharge and arterial pressure.

VASOPRESSIN (ANTIDIURETIC HORMONE). Vasopressin is an endogenous octapeptide liberated from the posterior pituitary gland, the prominent physiological role of which is in control of water balance. However, in addition to its effects on permeability of renal tubules, vasopressin possesses vasoconstrictor properties which may come into play if the blood levels of the material are elevated sufficiently. Conditions in which these levels are attained are general anesthesia, cardiopulmonary bypass, and hypotension. Lower levels of vasopressin, closer to the physiologic range, may alter distribution of blood flow to various organs without significantly affecting arterial pressure and may do so either through direct vasoconstrictor effects or indirectly by altering responsiveness of vascular smooth muscle to other vasoconstrictors such as norepinephrine.

Release of vasopressin is inhibited by stretch of left atrial receptors that are sensitive to changes in circulating blood volume and by activation of the arterial mechanoreceptors.

The vasoconstrictor effects of vasopressin are diminished through an unusual interaction with the baroreceptor reflex. The potentiation of the pressor effect of vasopressin by baroreceptor deafferentation is significantly greater than that for any other endogenous pressor substance. It appears that vasopressin, through actions in the brain stem, greatly facilitates baroreflex compensation for a rise in arterial pressure. Thus, the potential for vasopressin to promote vasoconstriction and hypertension is offset by this unusual interaction with the baroreceptor reflex. It can be predicted that in any situation in which the circulating levels of vasopressin are elevated and in which the baroreceptor reflex is obtunded such as in severe hypotensive states, the vasoconstrictor effects of vasopressin will become much more prominent. The physiological role of vasopressin is not restricted to its humoral effects upon release from the posterior pituitary. Vasopressinergic neurons have been identified to project from parvocellular portions of the paraventricular nucleus to both brain stem and spinal cord levels. These projections appear to play a physiological role in modulating sympathetic nervous system discharge.

When reflex inhibition of vasopressin is reduced or lost, for example when the left atrium collapses during cardiopulmonary bypass, high circulating levels of vasopressin ensue. The coronary vessels are quite sensitive to the vasoconstrictor action of vasopressin. In animals, high levels of vasopressin reduce coronary blood flow and, secondarily, cardiac output. A depressant effect of vasopressin on the heart remains to be demonstrated in clinical situations in which the levels of the polypeptide are high.

The availability of sensitive and selective radioimmunoassays for vasopressin has accelerated studies on the role of the peptide in pathophysiological states in which relatively modest changes in blood levels may occur. Increased circulating levels of vasopressin have been found in virtually all experimental forms of hypertension and in many human subjects with essential hypertension. The increments in vasopressin are probably insufficient to account for the elevation in arterial pressure, yet the possibility exists that the change represents a marker of an abnormality that contributes to high arterial pressure. Although not yet available for clinical use, a number of competitive antagonists of vasopressin have

been developed. It can be anticipated that their application to study of the role of vasopressin in hypertension and other cardiovascular disorders will be substantial.

KININS. A variety of potent vasodilator polypeptides are formed by the actions of certain proteolytic enzymes on plasma protein precursors. The substance bradykinin serves as the prototype for this class of endogenous substances. The kallikreins, enzymes found in glandular sources such as pancreas, as well as in plasma, produce active peptides from precursor plasma kininogens. Kallidin, a decapeptide, and bradykinin, a nonapeptide, have similar biological properties. Kallidin can be degraded in plasma by an aminopeptidase to form bradykinin. The peptides are metabolized by kininases, an enzyme system with functional identity to the angiotensin-converting enzyme. The antihypertensive agent captopril is therefore also a kininase inhibitor that raises circulating blood levels of kinins and potentiates their biological effects. It also has been shown that strong neural activation of salivary secretion, with attendant elaboration of kinin into plasma, can lower arterial pressure in the presence of captopril.

The major physiological role of kinins has been proposed to be in the local regulation of blood flow and function of such organs as the salivary glands and pancreas. The increased blood flow needed for metabolic activity of such glands is thought to derive at least in part from the vasodilator action of kinins formed within these glands by the secretion of proteolytic enzymes. On the pathophysiological side, kinins are believed to play a part in hyperemia associated with inflammation and as vasodilators in hypotension produced by anaphylactic reactions. Their most prominent pathophysiological role appears to be in the carcinoid syndrome, where the vasodilator peptide bradykinin may be responsible for the characteristic cutaneous flushing. In these patients, administration of epinephrine causes release of the enzyme kallikrein from metastatic lesions in the liver. Kallikrein liberates kallidin, which is converted to vasoactive bradykinin. Serotonin and histamine may also be involved in the hypertensive flushing attacks of carcinoid syndrome.

Antagonists of the kallikrein-kinin system have been developed. Aprotinin is a large polypeptide that inhibits kallikrein and other enzymes involved in blood coagulation and fibrinolysis. It has been used for the treatment of hyperfibrinolysis and pancreatitis. Competitive antagonists of bradykinin have also been developed; however, at present their potency and specificity is limited.

Neurogenic factors regulating smooth muscle contraction

The sympathetic nervous system is responsible for rapid circulatory adjustments in response to a variety of stressful physiological and pathological conditions such as diving, exercise, temperature changes, heart failure, myocardial infarction, hemorrhagic shock, hypoxia, and emotional stress. A sympathetic discharge causes an adjustment in the peripheral circulation through the regulation of vascular resistance in each organ which, in turn, governs the fraction of cardiac output supplied to that organ, or, in other words, the perfusion of that organ. There are several components to the sympathetic nervous system. The adrenergic component, which has norepinephrine as a neurotransmitter, is the major pathway, and it moderates the level of vasoconstriction in each vascular bed. Other sympathetic components mediate vasodilatation through the release of either acetylcholine or histamine.

ADRENERGIC COMPONENT. The resistance of each vascular bed is determined by the density of sympathetic adrenergic innervation, the responsiveness of resistance vessels to the released neurotransmitter norepinephrine, and the frequency of efferent sympathetic discharge to that bed. Each of these factors varies significantly in different organs. For example, stimulation of the sympathetic nerve supply to the kidney, to the forelimb, and to the heart of dogs causes vasoconstriction in the renal vessels and in vessels of the forelimb (greater in the former than in the latter) and, in contrast, vasodilatation of the coronary vessels. Thus, if cardiac output is constant, a sympathetic discharge would cause a reduction in blood flow to the limbs and kidneys and an increase in flow to the coronary vessels. It appears also that the responsiveness of cerebral vessels to norepinephrine is negligible, except during acute increases in arterial pressure. Thus, when a fall in arterial pressure is prevented by a generalized sympathetic discharge, it would favor the distribution of blood flow to the cerebral and the coronary circulations. Since the heart and brain depend on aerobic metabolism for their function, the peripheral circulatory adjustment is essential to maintain their perfusion in clinical situations in which cardiac output is limited or in which the conservation of oxygen is necessary.

Activation of sympathetic efferent vasoconstrictor pathways is not uniform. Activation of an afferent input, for example through stimulation of chemoreceptors, may result in a selective activation of efferent sympathetic fibers such that vasoconstriction is noted in some beds (i.e., muscle) and vasodilatation in others (i.e., coronary circulation). Such a differential response is certain to modify tissue perfusion in each organ even if cardiac output remained constant.

Afferent impulses originating from a variety of sensors influence the sympathetic efferent discharge to the cardiovascular system. Afferents originating in the arterial baroreceptors, as well as in the cardiopulmonary receptors, are predominantly inhibitory, whereas afferent impulses originating in the chemoreceptors and in mechanoreceptors of skeletal muscle appear to be primarily excitatory. In addition to the afferent input from these sensors, the vasomotor centers in the medulla are modulated by input from other parts of the central nervous system, including the hypothalamus, and the cerebral cortex. Interaction between the afferent inputs plays an important role in determining the final sympathetic efferent output; for example, activation of arterial baroreceptors through a rise in arterial pressure inhibits the respiratory and vasoconstrictor effects of chemoreceptor activation.

Special visceral afferent systems participate in circulatory homeostasis. The kidney contains sensory receptors for changes in renal perfusion pressure, oxygen tension, and tubular sodium gradients. The afferent nerves from the kidney project by way of multisynaptic pathways through the spinal cord and brain stem to the preoptic region of the anterior hypothalamus. Renovascular and mineralocorticoid hypertension appear to depend in part on the participation of renal sensory mechanisms. The increases in overall sympathetic nervous system activity associated with these hypertensive states are attenuated by renal denervation. Although no pathophysiological role has yet been established for them, hepatic osmoreceptors help control thirst and vasopressin secretion. These receptors are ideally located for their role in helping maintain fluid and electrolyte balance since ingested sodium reaches the circulation only after passing to the liver via the portal vein.

Inhibition of neurogenic vasoconstriction such as is observed with stimulation of arterial mechanoreceptors (baroreceptors) during a rise in pressure or stimulation of myocardial receptor during an acute increase in cardiac size tends to reflexly decrease arterial pressure. Depending upon the degree of inhibition of vasoconstrictor tone in different circulations, such reflexes would cause a redistribution of flow to various organs. In the following paragraphs the role of neurogenic factors in producing maldistribution of blood flow and defective perfusion of certain organs in clinical situations are considered.

ORTHOSTATIC HYPOTENSION AND DECREASED CEREBRAL PERFUSION. Upright tilt decreases cardiac filling pressure and cardiac output. Afferent impulses originating in low- and high-pressure baroreceptors trigger the compensatory cardiovascular adjustments consisting of tachycardia and vasoconstriction predominantly in skin, muscles, splanchnic beds, and kidneys. This reflex maintains arterial blood pressure and cerebral blood flow.

In the presence of an intact adrenergic system, orthostatic hypotension may result from severe hypovolemia. Thus, under this circumstance compensatory reflexes are stimulated but are insufficient to oppose the marked reduction in cardiac output. Reflex tachycardia, very low central venous pressure, and increased blood catecholamines in a patient with orthostatic hypotension all suggest that the hypotension is caused by hypovolemia and should be treated with volume replacement.

In the absence of reflex tachycardia and peripheral vasoconstriction during upright tilt, the percentage of cardiac output supplying each organ remains unchanged, but the fall in cardiac output will be reflected uniformly in all organs. The resulting reduction in cerebral blood flow may cause dizziness or syncope. Such a defective adrenergic reflex may be seen, for example, after prolonged recumbency or in diabetic patients with peripheral neuropathy. The presence of anhydrosis, normal or high venous pressure, and the absence of arterial pressure overshoot immediately after termination of the Valsalva maneuver all suggest that the cause of the orthostatic hypotension is autonomic neuropathy rather than hypovolemia.

A defect in adrenergic transmission may also result from a metabolic fault in synthesis of norepinephrine rather than structural damage to autonomic nerves. Such a defect may be found in patients with familial dysautonomia, and may be demonstrated by reduced levels of urinary VMA and manifestations of sympathetic paralysis. In some patients, the orthostatic hypotension may result from a failure of the integration of the baroreceptor reflex in the central nervous system such as may be seen in the Shy-Drager anomaly. This syndrome is characterized by upper motor neuron paralysis, cerebellar damage, impotence, urinary incontinence, constipation, incoordination, and severe orthostatic hypotension.

In some patients vasodilation has been observed during orthostatic hypotension. This dilatation is analogous to that seen in vasodepressor syncope, in which severe bradycardia and a marked increase in forearm blood flow precede the fainting. Thus, not only may the defective adjustment in orthostatic hypotension result from an absence of vasoconstrictor activity in nonvital organs, but it may be aggravated by an active vasodilatation in vessels of skeletal muscle, which tends to reduce still further cerebral perfusion by diverting an already reduced cardiac output away from the cerebral circulation. The mediator of this vasodilatation is not known, but it could very well be acetylcholine. This dilatation may also represent activation of vascular beta-adrenergic receptors.

AORTIC STENOSIS. The exertional syncope of aortic stenosis may be related to activation of stretch receptors in the left ventricle which relay inhibitory afferent impulses to the vasomotor center, preventing the reflex vasoconstriction known to occur with exercise. Failure of vasoconstriction in the nonexercising extremity of man during exercise has been demonstrated in patients with aortic stenosis. This abnormal response was reversed following replacement of the aortic valve with a prosthesis.

MYOCARDIAL INFARCTION. Shock following acute myocardial infarction appears to involve two types of peripheral vascular response. In the face of reduced arterial pressure from depression of myocardial performance, vascular resistance increases, presumably due to activation of the baroreceptor reflex. In some subjects vascular resistance fails to increase in the face of hypotension. This is probably the result of activation of ventricular receptors which inhibit the baroreceptor reflex and thus normal compensatory reflex vasoconstriction. There are more ventricular receptors in the inferior wall of the heart than in the anterior myocardium; experimental and clinical evidence indicates that ischemia or infarction of the inferior wall produces greater activation of ventricular receptors, with more hypotension and bradycardia.

HEMORRHAGE. Numerous factors, in addition to reduced blood volume, contribute to the reduced tissue perfusion associated with hemorrhage. In the absence of replacement of blood volume, compensatory reflex constriction maintains arterial pressure; but in certain vascular beds intense vasoconstriction is a compensation at the expense of adequate tissue perfusion. Skeletal muscle, kidney, skin, and splanchnic vascular beds exhibit this major increase in resistance to flow and reduction of perfusion. The reflex vasoconstriction is not mediated solely through the arterial baroreceptors. Reduction in stretch of receptors in the cardiopulmonary area also plays a major role in controlling incremental sympathetic discharge, especially when arterial pressure is below the range where the arterial baroreceptors function efficiently. Humoral factors also contribute significantly to vasoconstriction associated with hemorrhage. Release of renin, generation of angiotensin II, and release of catecholamines and vasopressin are three such humoral factors that come into play.

ISCHEMIC VASOSPASTIC DISEASE IN THE EXTREMITIES (RAYNAUD'S PHENOMENON). It has been proposed that excessive adrenergic stimulation directed selectively to the vessels of the upper limb is responsible for the intermittent vasospasm and associated pallor and cutaneous ulcerations seen in the hands of young patients with Raynaud's phenomenon. More recent evidence, however, favors a different mechanism such as a defect in basal heat production limiting the ability of such patients to dilate their cutaneous vessels. The unusual sensitivity to cold that these patients exhibit supports this contention.

Reserpine administered into the brachial artery of patients suffering from Raynaud's phenomenon causes a sustained vasodilatation without interruption of sympathetic transmission. These results have been particularly satisfactory in young patients with clear-cut vasospastic manifestations. The relief of pain and the rapid healing of ulcers have been the most obvious manifestations of clinical improvement. Recent reports suggest that the slow-channel calcium entry blockers also are useful in treatment of Raynaud's phenomenon.

CHOLINERGIC COMPONENT. Cholinergic innervation in skeletal muscle and skin has been identified in both man and animals, but it does not appear to be involved to any great extent in baroreceptor reflexes. Activation of this system is seen in responses to emotional stimuli, under stressful situations, during vasovagal syncope, and possibly in anticipation of exercise. Activation of this cholinergic pathway increases blood flow primarily in skeletal muscle and, in the absence of increased cardiac output (in vasovagal syncope in contrast to exercise), compromises cerebral perfusion. There is evidence that a vagal cholinergic vasodilator system supplies the coronary vessels. This system may be involved in mediating coronary vasodilatation during stimulation of chemoreceptors, since there does not appear to be any sympathetic cholinergic innervation of the coronary vessels.

OTHER VASODILATOR SYSTEMS. A histaminergic pathway supplying the limbs of experimental animals is activated by stimulation of carotid baroreceptors. This system may have pathophysiological significance, since the vasodilator response to its activation by stimulation of carotid baroreceptors is reduced in animals with renal hypertension. One might postulate that a defective vasodilator mechanism may contribute to the increased peripheral vascular resistance found in hypertension. The release of histamine appears to be from a nonneuronal, nonmast cell pool associated with the blood vessels under the reciprocal control of adrenergic nerves.

Another sympathetic vasodilator pathway has been identified by direct nerve stimulation in cutaneous vessels of experimental animals. The pathway is noncholinergic and nonhistaminergic and appears to be activated during stimulation of chemoreceptors. Conceivably, a defect in this cutaneous vasodilator system may contribute to the exaggerated vasoconstrictor responsiveness in patients with ischemic vascular disease of the limbs.

Using immunocytochemistry, a number of putative *peptidergic* transmitters have been identified in sympathetic innervation to vascular smooth muscle. These transmitter candidates include substance P and the vasoactive intestinal peptide (VIP). These agents are potent vasodilators, but their neurogenic vasodilator role has not yet been established.

Purinergic and *dopaminergic* neurogenic vasodilator systems also have been postulated. The purinergic system is proposed to use ATP as the transmitter. There is much better histochemical, biochemical, and functional evidence for purinergic modulation of gastrointestinal smooth muscle tone than for the proposed vasodilator mechanism. Dopaminergic innervation to renal vascular smooth muscle is well established. Activation of this special catecholaminergic system leads to renal vasodilatation, presumably by interaction of neurally released dopamine with the same receptors excited by exogenous dopamine and its analogues.

Intrinsic tissue factors regulating smooth muscle contraction

AUTOREGULATION. Blood vessels possess the intrinsic ability to regulate the level of active tension of their smooth muscle. The ability to regulate tissue perfusion independently of neurogenic and humoral influences has been called autoregulation of blood flow. Different vascular beds vary with respect to their ability to regulate blood flow over a wide range of perfusion pressures. For example, skeletal muscle exhibits relatively poor autoregulatory capacity as compared to the kidney, which can autoregulate blood flow almost perfectly over a range of pressures between 80 and 200 mm. Hg. By this ability, the kidney can, by increasing its vascular resistance, keep blood flow and glomerular filtration rate constant when arterial pressure is raised.

The mechanism for this intrinsic response of vascular smooth muscle remains obscure. Some earlier explanations were based on physical factors such as changes in viscosity brought about by plasma skimming, or changes in tissue pressure brought about by increased filtration of fluid at elevated hydrostatic pressures. It is generally agreed, however, that autoregulation primarily involves active responses of vessels. The rate of oxygen delivery to arterioles and precapillary sphincters could regulate blood flow by altering contraction of these small vessels. Accumulation of metabolites, when flow is reduced, may cause vasodilatation and allow flow to be restored. Finally, it has been suggested that the activity of smooth muscle pacemaker cells in vessel walls may be increased by stretch, causing greater smooth muscle contraction. Current evidence suggests that, although vascular muscle in several organs responds to changes in wall tension ("myogenic" autoregulation), the predominant autoregulatory mechanism is metabolic.

Autoregulation of renal blood flow has been implicated in the pathogenesis of hypertension. Several authors have suggested that the first phase of hypertension involves increased cardiac output without any change in peripheral vascular resistance. The hypothesis suggests that renal vessels especially increase their resistance in an effort to keep blood flow constant in the face of increased arterial pressure. Local renal vasoconstriction might lead to the elaboration of renin and to sodium and water retention, and further accelerate the development of hypertension. Although this hypothesis is intriguing, there is at present little experimental evidence to support the concept that autoregulatory responses are involved in either initiating or sustaining high arterial pressure.

METABOLIC FACTORS. Such factors may be released in response to absolute or relative tissue ischemia and cause vasodilatation to restore optimal blood flow, adequate tissue perfusion, and oxygenation. The mediators of this metabolic vasodilatation are not known, but it is likely that the combination of changes in oxygen and carbon dioxide tension, hydrogen ion and other cation concentrations, changes in osmolarity, and the amount of adenosine compounds released, and accumulated Krebs intermediate metabolites in the immediate environment of the blood vessels contribute to adjustments in vascular tone. The net effect of these metabolic factors is a "tight coupling" between metabolism and blood flow. The cerebral and coronary vessels are particularly sensitive to changes in their metabolic environment. Excess carbon dioxide is a potent cerebral vasodilator; hypoxia causes significant coronary and cerebral vasodilatation.

Deficiency of tissue oxygenation may result either from the failure of supply of oxygen to tissues when their oxygen demand is normal or from an increased oxygen consumption in the hypermetabolic state, creating a situation in which there is relative oxygen deficiency.

FAILURE OF DELIVERY OF OXYGEN TO THE TISSUES. Many causes account for this failure. A right-to-left shunt, which may be large enough to cause significant depression of arterial oxygen saturation to levels of 70 to 75 per cent, produces cyanosis, polycythemia and digital clubbing.

HIGH ALTITUDE EXPOSURE. Hypoxia resulting from high altitude exposure could result in a net peripheral vascular response which is the result of at least four effects. One is the direct vasodilator effect of hypoxia seen mostly in skeletal muscle vessels but which would also be expected to occur in coronary and cerebral vessels; the second is a sympathetic discharge causing peripheral vasoconstriction predominantly in skin and, possibly, in mesenteric and renal vessels; third, a direct inhibitory effect of hypoxia on adrenergic responses is possible whereby the vasoconstrictor action of sympathoadrenal stimulation, possibly at a peripheral site, is inhibited or blocked; and fourth, hyperventilation modulates vasoconstrictor responses. The degree to which these various effects influence the different vascular beds is presently unknown but is of critical importance if we are to understand the mechanisms whereby peripheral blood flow is distributed to various organs in hypoxic states.

THIAMINE DEFICIENCY. Along with several other diseases in which coenzymes necessary for oxidative decarboxylation of pyruvic acid are lacking thiamine deficiency is associated with failure of tissue oxygen delivery. Deficiency of this and other respiratory enzymes resulting from severe chronic vitamin deficiency may trigger the release of vasodilator metabolites and produce clinical manifestations of a high cardiac output state as seen in beriberi. It should be remembered that excessive and too rapid replacement of thiamine reverses the vasodilation and constricts the peripheral circulation, and may precipitate cardiac decompensation.

LAENNEC'S CIRRHOSIS. The hyperkinesis seen in Laennec's cirrhosis may be caused in part by a deficiency in respiratory enzymes analogous to that found in beriberi. There may also be small portal and pulmonary arteriovenous communications which may contribute to the low peripheral resistance and high cardiac output.

ANEMIA. All forms of anemia result in reduced ability of the blood to deliver adequate amounts of oxygen to the tissues. Because of the large oxygen reserve in the venous effluent of most organs, tissue oxygenation may be partially maintained (despite the presence of significant anemia) through the process of more complete extraction of oxygen. However, venous oxygen content is low in the coronary circulation, and in the presence of severe anemia an increased myocardial oxygen requirement must be met by an increased coronary blood flow, achieved by either dilatation or increased perfusion pressure, or both. However, anemia must be rather severe, with a hemoglobin level of less than 7 g./100 ml. of blood, before significant cardiovascular changes are evident.

An abnormal hemoglobin also may reduce oxygen-carrying capacity of the blood significantly. Such situations arise in methemoglobinemia, carboxyhemoglobinemia and sulfhemoglobinemia. Tissue hypoxia, as estimated by the plasma lactic acid-pyruvic acid ratio, does not occur in resting patients who have levels of hemoglobin above 6 g./100 ml. In such patients, compensatory mechanisms must take place to deliver more oxygen to tissues. These may include vasodilatation, increased cardiac output, and a rise in 2,3-diphosphoglycerate which facilitates oxygen dissociation from hemoglobin.

INCREASED OXYGEN CONSUMPTION. This may be the result of increased metabolism, which may or may not be associated with the formation of high-energy phosphate bonds in adenosine triphosphate (ATP). During exercise, pregnancy, and anabolic processes following acute illness, a hyperkinetic state provides for delivery of greater than normal amounts of oxygen to tissues, which is then utilized for the formation of high energy phosphate bonds. This represents an efficient oxidative phosphorylation. However, in certain states hypermetabolism may occur without the formation of ATP. This may take place in non-shivering thermogenesis, pheochromocytoma, adrenergic calorigenesis, diabetic ketosis, idiopathic hyperkinetic heart syndrome, and possibly hyperthyroidism. The increased oxygen utilization may represent an activation of metabolic pathways that are not ATP dependent. This respiration without phosphorylation appears to be wasteful but may be essential for normal cellular processes.

In both idiopathic hyperkinetic heart syndrome and hyperthyroidism there is a high cardiac output and stroke volume, a bounding pulse, and vasodilatation. Idiopathic hyperkinesis may represent a state of neurogenic vasodilatation accompanied by excessive neurogenic cardiac stimulation. It may also reflect increased sensitivity of beta-adrenergic receptors to adrenergic stimulation or an overactivity of such receptors. On the other hand, hyperkinetic states associated with anemia or hyperthyroidism do not appear to be influenced by beta-adrenergic receptor blocking drugs and, therefore, may not be dependent on activation of sympathoadrenal pathways involving such receptors. In fact, one might use the responsiveness to beta-adrenergic blockade as an index of the contribution of the sympathoadrenal system to the hyperkinesis.

There is little information concerning the relative degrees of vasodilatation of various vascular beds in the hyperkinetic states beyond the fact that cutaneous blood flow must be increased for the dissipation of heat. One might expect that the organs consuming the largest amounts of oxygen in these states would be the ones having the greatest degree of vasodilatation. Since cardiac output seems to parallel oxygen consumption in the hyperkinetic states, it may be assumed that the peripheral vasodilatation does not represent autonomic or physiological shunting, and the increase in cardiac output occurs because of increased tissue demand for blood flow.

Structural factors altering resistance to blood flow

Arterial insufficiency of vital organs is the major cause of death in the United States. The largest percentage of such insufficiency results from atherosclerotic lesions in the intima, particularly of the coronary vessels, leading to ischemic heart disease. Other diseases (autoimmune, inflammatory) associated with structural changes of the vessel wall are of lesser importance because of their lower incidence. Diabetes mellitus also may cause a specific degenerative change in the capillaries which results in decreased tissue perfusion and ischemic changes.

Hypertension

Arteriolar hypertrophy is a well-documented feature of prolonged hypertension. This morphological alteration provides the structural contribution to high vascular resistance associated with hypertension. High arterial pressure is usually associated with hyperresponsiveness of vascular smooth muscle. The altered wall-to-lumen ratio seen in the hypertrophied vessel wall can by itself increase the vascular response to a vasoconstrictor through a geometric mechanism alone. There is, however, evidence that a true increase in sensitivity to vasoconstrictors contributes to this hyperresponsiveness. This effect is distinguishable from the structural contribution and may be a function of electromechanical properties of vascular smooth muscle.

The mechanism by which chronic high arterial pressure induces vascular smooth muscle hypertrophy involves the sympathetic nervous system. A trophic effect of the sympathetic innervation is necessary for the full expression of vascular muscle growth in the face of hypertension. When sympathetic denervation is present, vascular hypertrophy is attenuated. This adaptive response to a chronic pressure load may have beneficial effects. In the cerebral circulation, arterial hypertrophy protects the microcirculation. In an experimental model, stroke-prone rats developed hemorrhagic strokes on the sympathetically denervated side of the cerebrum.

Metabolic studies of arterial wall demonstrate that the movements of nutrients, substrates, and catabolic products are directed through the luminal as well as the adventitial sides of the vessel wall. Vasa vasorum in the adventitia apparently nourish the outer layers of arterial and venous media. In arteries with a thick wall, such as the thoracic aorta of dogs and man, vasa vasorum penetrate into the media to nourish the vessel.

The integrity of the endothelium is important for transfer of nutrients to the intimal layer. Inflammatory intimal lesions lead to fibrosis and hyalinoid changes deeper in the vessel wall. These lesions are independent of cholesterol and lipids and should be considered distinct entities separate from atherosclerotic lesions or from Mönckeberg's sclerosis, which is essentially a focal calcification of the media. The initial damage to the endothelial layer may also promote thrombus formation. Organization of such thrombi and formation of new capillaries within them might favor intramural hemorrhage and promote further inflammatory reaction in the wall.

Arteriosclerosis and atherosclerosis

Atherosclerosis is basically an involvement of the intimal surface of the arteries with fatty plaques. The first lesions appear as fatty streaks. As the lesions progress and enlarge, they become fibrous with a necrotic core of fatty debris. The aorta and the left coronary arteries are most extensively involved, and the aortic involvement is frequently intense at its terminal bifurcation into the iliac arteries. Involvement of

the coronary vessels is sometimes out of proportion to atherosclerosis elsewhere, and the epicardial segments of the coronary vessels are more heavily involved than the transmural segments. Narrowing of the coronary vessels occurs predominantly in the more proximal coronary arteries rather than in the intramural section of the coronary vessel. Lesions generally have to decrease the diameter of the coronary vessel by 80 to 90 per cent before a significant reduction in flow occurs. It is important, however, to appreciate that less narrowing may be critical during reactive hyperemia or exercise when coronary flow increases severalfold. Thus, a lesion which is not limiting at rest may produce myocardial ischemia during exercise. In the cerebral vessels distribution of atherosclerosis is patchy and extracranial vessels are usually more severely involved than intracranial vessels.

An important concept that has emerged recently is that vascular spasm is important in the pathogenesis of ischemia in many patients with angina pectoris. This concept has important therapeutic implications. Nitroglycerin is useful in treatment of patients with fixed atherosclerotic lesions, who usually have exercise-induced angina, and in patients with spasm, who often have angina at rest. Beta-receptor antagonists such as propranolol are useful in the treatment of patients with exercise-induced angina, but they are not useful in patients with spasm. Calcium blockers are useful in treatment of spasm-induced angina.

It is difficult to recognize clinically the development of atherosclerosis, except possibly by angiographic visualization of the narrowed lumen. Otherwise, one has to depend upon the clinical manifestation of ischemia in the various organs. The manifestations are often catastrophic: cardiac arrest, following myocardial infarction; paralysis, from occlusion of a cerebral vessel; gangrene, following occlusion of peripheral limb vessels; hypertension, after occlusion of renal vessels; and intestinal gangrene, from occlusion of the mesenteric arteries. Some of the less alarming, but nonetheless serious, manifestations of advanced vascular disease are symptoms of intermittent claudication in the legs, Raynaud's phenomenon in the hands, transient cerebral ischemic attacks, and angina pectoris.

RISK FACTORS. The concepts concerning the prevalence of atherosclerosis and its association with biological and social variables have evolved primarily from epidemiological studies in which a high correlation was found between ischemic heart disease and certain risk factors. In descending order the major factors appear to be hyperlipidemia, cigarette smoking, hypertension, obesity, and diabetes mellitus.

HYPERLIPIDEMIA. Plasma lipids of clinical importance are cholesterol, triglycerides and free fatty acids. Free fatty acids circulate with albumin; cholesterol and triglycerides are bound to lipoproteins.

TYPES OF HYPERLIPOPROTEINEMIA. Hyperlipoproteinemia may be conveniently classified into five types according to the scheme proposed by Fredrickson, Levy and Lee. Type I is a very rare genetic recessive disease associated with deficiency of lipoprotein lipase and appears in early childhood, manifested primarily by marked elevation in chylomicrons. After an overnight fast, the creamy layer on top of a clear plasma will indicate the presence of these fat globules. Clinical manifestations are primarily those of lipemia retinalis, eruptive xanthomas, hepatosplenomegaly, and abdominal pain. Type V hyperlipoproteinemia is also associated with chylomicronemia and similar clinical manifestations, but it usually appears in early adulthood. Both conditions are associated with pancreatitis and Type V with insulin-dependent diabetes mellitus.

Type II hyperlipoproteinemia is common and is associated with elevated serum cholesterol. The appearance of the plasma is clear and the clinical manifestations include xanthelasma, tendon xanthomas, and juvenile corneal arcus. Type III is relatively uncommon. It is associated with elevation of both cholesterol and triglycerides, giving a cloudy plasma; its clinical characteristics include xanthomas on palmar creases and tendon xanthomas, and it is usually seen in adulthood. Type IV disease is common and is associated with elevation of triglycerides and normal cholesterol; the plasma is turbid. Types II, III, and IV are associated with accelerated vascular disease. Types II and IV are associated primarily with coronary vascular disease, Type III with a high incidence of both coronary and peripheral vascular atherosclerotic lesions.

Treatment of hyperlipoproteinemias is based on considerations of lipoprotein particle size and on absolute and relative levels of cholesterol and triglycerides. In general, if there is a massive elevation of triglycerides (>1000 mg./dl.), a marked reduction in dietary fats is recommended to reduce the possibility of pancreatitis. In other hyperlipidemias, reduction in dietary saturated fats and correction of obesity are recommended. Drug treatment is often used, but drugs are clearly indicated only in marked, refractory hypercholesterolemia. When the level of high-density lipoproteins is low, increases in HDL by moderate exercise or by correction of obesity may be beneficial.

DIABETES MELLITUS. Diabetes mellitus predisposes to atherosclerotic changes in large arteries and, in addition, produces characteristic microvascular changes. The capillary basement membrane is thickened in diabetes, but capillary permeability to large and small molecules is increased. There also appears to be a reduction in the number of microvessels and atrophy of arteriolar smooth muscle, so that vasodilator responses are impaired. Thus, in diabetics, the combination of atherosclerotic disease of large arteries and diabetic microvascular disease often leads to profound impairment of tissue perfusion.

Vascular anomalies

CONGENITAL VASCULAR ANOMALIES. Congenital anomalies such as are seen in lungs of patients with hereditary hemorrhagic telangiectasia, have been reported as causes of a hyperkinetic circulatory state. The hyperkinetic circulation in such individuals may be related to the deficiency in tissue oxygenation associated with severe anemia rather than to the vascular dysplasia. Anemia in such patients is caused by frequent attacks of bleeding.

ACQUIRED VASCULAR ANOMALIES. PAGET'S DISEASE. Acquired arteriovenous communication may be seen in Paget's disease of bone, but it is not known whether the increased vascularity is the primary defect in the involved bone. In patients with the hyperkinetic syndrome associated with Paget's disease bony involvement is extensive (>30 per cent of bones). The increase in blood flow to pagetic limbs may be in large part related to increased cutaneous flow over the affected bone.

PREGNANCY. *In pregnancy* high levels of blood flow through the uterine arteries may produce 30 to 40 per cent increases in cardiac output. The increased blood volume observed in the third trimester may also contribute to the high output. The ease with which the heart meets these temporary increases in demand, often despite severe valvular deformities, testifies to the tremendous reserve of the heart muscle.

Another group of acquired vascular abnormalities, which may have significant hemodynamic effects by reducing renal blood flow, are the fibrosing lesions of the renal artery that result in renal hypertension. This problem is discussed in Chapter 3.

Collagen diseases

These diseases involve the medium-sized and small arteries and arterioles in various parts of the body with inflammatory fibrinoid changes. The vascular involvement may be the result of an autoimmune disease which also affects other tissues.

POLYARTERITIS NODOSA. This is a collagen disease that manifests itself primarily in younger men, with fever, abdominal pain, hypertension, polyarteritis, and eosinophilia. Several organs may be involved, but the most frequently damaged is the kidney, resulting in severe hypertension.

SYSTEMIC LUPUS ERYTHEMATOSUS. This is a collagen disease of unknown cause which may involve all organs; its legions are most often seen in the walls of small arteries and arterioles. Here, also, the presence of focal or diffuse glomerulonephritis may be the terminal event in this unrelenting disease.

SCLERODERMA. In scleroderma the involvement of the skin or subcutaneous tissue may be the most visible manifestation; however, the more serious complication involves pulmonary fibrosis and pulmonary vascular obstruction, resulting in emphysema, atelectasis, bronchiectasis, and pulmonary hypertension with cor pulmonale. Scleroderma also involves the gastrointestinal tract and particularly the lower esophagus; in addition, obliterative vascular lesions in the gastrointestinal tract may result in ulceration, perforation, infarction and hemorrhage. In this, as in the other collagen diseases, involvement of the muscular, skeletal, and cardiovascular systems is common.

Rheological and mechanical factors altering resistance to blood flow

Altered viscosity

The physical characteristics of the blood can, under certain circumstances, be important determinants of resistance to blood flow. According to the Poiseuille relationship, vascular resistance is directly related to blood viscosity. Viscosity varies relatively little in the range of hematocrits between 0 and 40 per cent, but it increases rather steeply when hematocrits increase above the normal level. If all other factors remain unchanged, an increase in hematocrit from approximately 40 to 70 per cent will, by doubling the relative viscosity, double the resistance to blood flow. Noncellular constituents may have predictably similar effects on viscosity. For example, chylomicrons present in large quantities in hyperlipidemia or macroglobulins in Waldenstrom's macroglobulinemia could increase blood viscosity substantially. The high hematocrits found in polycythemia undoubtedly contribute to increased resistance to blood flow and, thus, to the increase in blood pressure seen in this condition. There is very little effect on tissue perfusion in anemic states with low hematocrits, because the relationship between hematocrit and viscosity is altered very little at low hematocrit levels.

Red cell wall rigidity
Normally, the red cell is easily deformed and undergoes considerable change in shape as it passes through the capillary. If the wall should change in respect to this physical characteristic so that it resisted deformation, the ease with which cells could pass the capillary would be reduced; thus tissue perfusion could be altered by this physical change in the nature of the cell wall alone. Some recent evidence suggests that the red cell wall becomes more rigid in hypoxia. Further investigation is necessary, however, to determine whether red cell wall rigidity is an important regulator of tissue perfusion.

Clotting, thrombosis, and platelet aggregation
Physical obstruction of vascular channels can be produced by clotting or thrombosis. Obstruction of major vessels, such as pulmonary, coronary, or cerebral arteries, may lead to catastrophic consequences if the obstruction severely compromises perfusion of the organ.

The role of altered blood coagulation in shock has received attention. At the irreversible stage of shock there may be damage to the endothelial lining of small vessels and capillaries, with subsequent fibrin deposition, accumulation of microthrombi, and intravascular coagulation. This, in addition to diffuse and extensive vasospasm, may severely restrict tissue perfusion and cause cellular death. The major example of a tissue lesion that follows disseminated intravascular coagulation is renal cortical necrosis associated with gram-negative septicemia. Other organs may also be involved in this generalized reaction, which mimics the Shwartzman reaction.

Annotated references

Abboud, F. M.: The sympathetic system in hypertension: State-of-the-art review. Hypertension 4(Suppl. II):208–225, 1982. (Hypertension depends to a great extent on augmented neurogenic vasoconstrictor mechanisms. Humoral factors, local vascular abnormalities, and increased central sympathetic discharge all interact in hypertensive states. The therapeutic efficacy of antisympathetic antihypertensive agents is largely attributable to the importance of neurogenic mechanism in hypertension.)

Bayliss, L. E.: The rheology of blood. In Hamilton, W. F. (ed.): Handbook of Physiology. Sec. 2, Circulation, Vol. 1, p. 137. Washington, D.C., American Physiological Society, 1962. (Physical characteristics of blood are often ignored as causes of altered tissue perfusion. Altered viscosity is perhaps the most important of these.)

Berecek, K. H., and Brody, M. J.: Evidence for a neurotransmitter role for epinephrine derived from the adrenal medulla. Am. J. Physiol. 242:H593–601, 1982. (Epinephrine may serve a neurotransmitter function in the sympathetic nervous system. The source for the amine is the adrenal gland, which provides circulating levels that can be taken up and stored in sympathetic nerve endings.)

Berne, R. M., Foley, D. H., Watkinson, W. P., Miller, W. L., Winn, H. R., and Rubio, R. The role of adenosine as a mediator of metabolic vasodilation in the heart and brain: A brief overview and update. In Baer, H. P., and Drummond, G. I. (eds.): Physiological and regulatory functions of adenosine and adenine nucleosides, p. 117. New York, Raven Press, 1979. (Local release of adenosine appears to participate in the vasodilation that helps sustain blood flow in brain and heart in the face of acute or chronic partial obstruction of arterial inflow.)

Bohlen, H. G.: Pathological expression in the microcirculation: Hypertension and diabetes. In The Physiologist, Vol. 25, No. 4, p. 391. Washington, D.C., American Physiological Society, 1982. (In contrast to the vascular hypertrophy of hypertension, which protects the microcirculation, the microcirculatory changes of diabetes are characterized by degeneration.)

Brody, M. J., and Johnson, A. K.: Role of the anteroventral third ventricle region in fluid and electrolyte balance, arterial pressure regulation, and hypertension. *In* Martini, L., and Ganong, W. F. (eds.): Frontiers in Neuroendocrinology, Vol. 6, p. 249. New York, Raven Press, 1980. (The renin-angiotensin system exerts its effects on the cardiovascular system through direct vasoconstrictor and indirect neural actions. Hypothalamic receptor areas mediating the central effects of angiotensin are also involved in the pathogenesis of renal hypertension. Related regions of the brain participate in non-renin dependent hypertensive states.)

Carretero, O. A., and Scicli, A. G.: Possible role of kinins in circulatory homeostasis. State of the art review. Hypertension 3(Suppl. I):14–12, 1981. (The kallikrein-kinin system participates in local blood flow regulation, fluid-electrolyte homeostasis, blood coagulation, and inflammation. Altered excretion of components of the system has been found in some hypertensive subjects.)

Cohn, J. N.: Vasodilatory therapy for heart failure: the influence of impedance on left ventricular performance. Circulation 48:5–8, 1973. (Peripheral organ blood flow is reduced in congestive heart failure and can be increased not only by inotropic interventions, but by vasodilator agents as well. More recently, improvement in subjects with failure has been achieved by vasodilation induced by interruption of the renin-angiotensin system, i.e., with the angiotensin converting-enzyme inhibitor captopril.)

Cowley, Jr., A. W.: Vasopressin and cardiovascular regulation. *In* Guyton, A. C., and Hall, J. E. (eds.): Cardiovascular Physiology IV. International Review of Physiology, Vol. 26, Chap. 6, p. 189. Baltimore, University Park Press, 1982. (Vasopressin is not only an antidiuretic hormone but a vasoconstrictor as well. Blood levels of vasopressin are elevated in many forms of hypertension but, other than extreme conditions such as hemorrhage, deep anesthesia, shock, etc., in which blood levels are very high, the physiological and pathophysiological roles as a vasoconstrictor remain to be established.)

Folkow, B: Physiological aspects of primary hypertension. *In* Physiological Reviews, Vol. 62, No. 2, p. 347. Washington, D.C., American Physiological Society, 1982. (A conceptual framework for the pathogenesis of hypertension is provided. Genetic and environmental factors are proposed to interact with the cardiovascular system to produce structural changes in blood vessels that perpetrate the hypertensive state.)

Heistad, D. D., Marcus, M. L., Larsen, G. E., and Armstrong, M. L.: Role of vasa vasorum in nourishment of the aortic wall. Am. J. Physiol. 9(5):H781–787, 1981. (Ligation of vasa vasorum produces medial necrosis of the aorta in dogs. It is likely that interference with blood flow through vasa vasorum may be important in the pathogenesis of medial necrosis and dissecting aneurysm in chronic hypertension.)

Langer, S. Z.: Presynaptic receptors and the regulation of transmitter release in the peripheral and central nervous system: Physiological and pharmacological significance. *In* Usdin, E., Kopin, I. J., and Barchas, J. (eds.): Catecholamines: Basic and Clinical Frontiers, Vol. 1, p. 387. New York, Pergamon Press. 1979. (Selective adrenergic receptor subtypes help regulate tissue perfusion by modulating the release of neurotransmitter from the sympathetic nerve terminal. Specific pharmacological agonists and antagonists of these receptors are used therapeutically.)

Marcus, M., Wright, C., Doty, D., Eastham, C., Laughlin, D., Krumm, P., Fastenow, C., and Brody, M.: Measurements of coronary velocity and reactive hyperemia in the coronary circulation of humans. Circ. Res. 49:877–891, 1981. (A method is described to measuring phasic coronary velocity in humans. Quantitative measurements of coronary reactive hyperemia are provided.)

Mark, A. L., Kioschos, J. M., Abboud, F. M., Heistad, D. D., and Schmid, P. G.: Abnormal vascular responses to exercise in patients with aortic stenosis. J. Clin. Invest. 52:1138–1146, 1973. (Exercise produces constriction in vascular beds of nonexercising organs. Excessive activation of ventricular receptors in patients with aortic stenosis produces reflex vasodilatation and probably is the primary mechanism of syncope in these patients.)

Muirhead, E. E.: Antihypertensive functions of the kidney. Arthur C. Corcoran Memorial Lecture. Hypertension 2:444–464, 1980. (The kidney is the source of antihypertensive lipids that may be significant in protecting against the development of high arterial pressure.)

Needleman, P., and Isakson, P. C.: Intrinsic prostaglandin biosynthesis in blood vessels. *In* Bohr, D. F., Somlyo, A. P., Sparks, Jr., H. V. (eds.): Handbook of Physiology. Sec. 2, The Cardiovascular System. Vol. II, p. 613. Washington, D.C., American Physiological Society, 1980. (Formation by the arachidonic acid cascade of prostacyclin and thromboxane is enhanced under a large variety of physiological and pathophysiological states that influence blood flow regulation.)

Peach, M.: Renin-angiotensin system: Biochemistry and mechanism of action. Physiol. Rev. 57:313, 1977. (The renin-angiotensin system is a major physiological humoral system for blood pressure and blood flow control. The peptide angiotensin is a very potent vasoconstrictor that is formed in the bloodstream. There is now great interest in its possible formation within the blood vessel wall and central nervous system).

Ross, R., and Kariya, B.: Morphogenesis of vascular smooth muscle in atherosclerosis and cell culture. *In* Bohr, D. F., Somlyo, A. P., and Sparks, Jr., H. V. (ed.): Handbook of Physiology. Sec. 2, The Cardiovascular System, Vol. II, p. 69. Washington, D.C., American Physiological Society, 1980. (Abnormalities in smooth muscle proliferation may be a significant factor in the development of atherosclerotic lesions that ultimately impair tissue perfusion.)

Rowell, L. B.: Active Neurogenic Vasodilatation in Man. *In* Vanhoutte, P. M., and Leusen, I. (eds.): Vasodilatation, Chap. 1, p. 1. New York, Raven Press, 1981. (A variety of neurogenic vasodilator systems are involved in blood flow regulation in man. We have increasing evidence that many special neurotransmitters are released in different organs.)

Shepherd, A. P., and Riedel, G. L.: Optimal hematocrit for oxygenation of canine intestine. Circ. Res. *51*:233–240, 1982. (The optimal hemotocrit for tissue oxygenation is similar to the normal range of hematocrit. In the mesenteric circulation, marked increases or decreases in hematocrit reduce oxygen delivery.)

Thomas, J. E., Schirger, A., Fealer, R. D., and Sheps, S. G.: Orthostatic Hypotension. Mayo Clin. Proc. *56*:117–125, 1981. (Orthostatic hypotension may occur as the result of autonomic dysfunction or despite a normal autonomic nervous system.)

Walker, J. L., Thames, M. D., Abboud, F. M., Mark, A. L., and Klopfenstein, H. S.: Preferential distribution of inhibitory cardiac receptors in left ventricle of the dog. Am. J. Physiol. *4*(2):H188–192, 1978. (Ventricular receptors are located preferentially in the inferior wall of the heart. This observation is thought to explain the observation that reflex hypotension and bradycardia occurs more commonly in inferior myocardial infarction than in anterior myocardial infarction.)

Winternitz, S. R., and Oparil, S.: Importance of the renal nerves in the pathogenesis of experimental hypertension. Hypertension *4*(Suppl. III):108–115, 1982. (A special renal sensory system with afferent neurogenic projections to the central nervous system appears to be capable of activating the sympathetic nervous system in hypertension.)

3   Mechanisms Controlling Arterial Pressure

Edward D. Frohlich, M.D.
Franz H. Messerli, M.D.
Richard N. Re, M.D.
Francis G. Dunn, M.D.

Stated in the simplest of terms, arterial pressure reflects an interplay between blood flow and resistance to that flow. The heart, of course, supplies the energy for the circulation of blood. Each contraction ejects a small volume (about 75 ml.) into the arterial system and by this action forces blood into capillaries. Each contraction produces a pressure pulse which results because the stroke volume is ejected into a reservoir, the aorta, which has limited capacity and distensibility, is elastic, and is already partly filled (i.e., it contains the diastolic arterial volume). The height of the arterial pulse, the systolic pressure, reflects not only the volume ejected and the rate of ejection but also the elasticity of the aorta. In contrast, the level to which pressure falls between contractions, the diastolic pressure, reflects the resistance to ventricular outflow.

Arterial pressure is controlled within the normal range by a variety of mechanisms that affect blood flow and vascular resistance. These mechanisms are also important in the pathophysiological alterations of arterial pressure control: hypotension and hypertension (see list below). Thus, for the most part, abnormalities of pressure result from abnormalities of control mechanisms. As such, then, these alterations may be considered diseases of regulation.

Pressure Control Mechanisms

Hemodynamic
Neural
Catecholamine
Renopressor
Renal
Volume
Sodium
Electrolytes (e.g., K^+, Ca^{++}, Mg^{++})
Hormonal
Depressor (e.g., prostaglandins, kinins, histamine)

Normotension

Circulatory dynamics

BASIC CONSIDERATIONS. Alone among the disorders of the cardiovascular system, hypertension and hypotension are defined by quantitative alterations of a biophysical measurement. But arterial pressure is normally maintained within a relatively narrow range, and in order to understand pressure and its abnormalities, it is helpful to consider some laws of hydrodynamics and their translation into clinical terms.

The flow rate of any fluid along a tube is related to a pressure gradient along that tube and to the resistance it meets. Resistance (R) cannot be measured directly, and therefore is calculated as the ratio of the pressure gradient (P) to the rate of flow (F):

$$R = P/F \ (1).$$

If the flow of a fluid within a cylindrical vessel is laminar, its rate can be predicted, depending on its viscosity (v), from the dimension of the tube and the associated pressure gradient, according to the Poiseuille formula:

$$F = P \left(\frac{r^4}{l.v} \right) k \ (2)$$

where r is radius, l, the length of the tube, and k a constant ($\pi/8$) factor arising from calculus integrations. From operations (1) and (2), it is obvious that

$$R = \frac{v.l}{r^4} \times 8\pi.$$

Within the usual physiological limits of blood viscosity and assuming an unchanging vascular length in the same individual, variations in resistance usually result from active, passive, or structural changes of vessel diameter. Since the radius is expressed as its fourth power in the equation, flow and pressure may be markedly affected by relatively small changes in the caliber of the resistance vessels, the arterioles. Moreover, since vascular capacitance is similarly expressed, small changes in the venular capacity will have profound effects on the distribution of intravascular volume.

Translation of these mathematical symbols into clinical terms yields the basic equation expressing the relationship of arterial pressure, cardiac output, and "total peripheral resistance" (TPR):

$$MAP = CO \times TPR,$$

where CO (cardiac output) is the equivalent of F; MAP (mean arterial pressure) the equivalent of P, is the integrated arterial pressure over one cardiac cycle. This integration can be obtained either by damping pulsatile pressure changes, by electronically integrating the pressure curve, or by calculating it as diastolic pressure $+\frac{1}{3}$ pulse pressure. The marked difference between mean arterial pressure and central venous pressure, as well as the relatively small fluctuations of the latter, allow us to disregard it in calculations of TPR.

The total peripheral resistance is the composite of the vascular resistance for the entire systemic circulation. Resistance to flow obeys the same laws as series and parallel arrangements of electrical resistances. It is clear, therefore, that a change in TPR does not necessarily indicate that similar quantitative or even similar directional changes are occurring in all individual vascular territories. A word of caution is needed against unqualified translation of TPR into an index of peripheral arteriolar vasoconstriction, as this simplistic approach fails to recognize the important role that large and small arteriovenous shunts, precapillary sphincters, passive arterial variations, structural vascular changes, as well as collateral vessels, may sometimes play in that respect.

Despite both of these restrictions and the well-recognized limitations in the application of Poiseuille's equation to the intact organism, these calculations are useful in assessing the relative roles played by cardiac output and vascular (e.g., primarily arteriolar) resistance in changes of arterial pressure. Cardiac output is universally expressed as cardiac index in relation to body surface area (l./min./m.^{2}). In this way it is possible to compare individuals having varying body weights and heights. And, in a similar fashion, total peripheral resistance can be calculated from either cardiac output or index in arbitrary units (either as mm.Hg/ml./min., mm.Hg/l./min./m.^{2}, or simply arbitrary PRU as PRU or PRU/m.^{2}) or in fundamental units of force

$$\frac{(\text{mm.Hg} \times 0.1 \times 13.6 \times 980 \text{ dynes. sec. cm.}^{-5})}{\text{ml./sec.}}$$

where 13.6 is specific gravity of mercury, 980 represents g, and 0.1 is the factor to transform mm. into cm.

AORTA AND LARGE VESSELS. As the left ventricle ejects blood intermittently into a partially filled arterial tree, arterial pressure varies between a maximum at the peak of ejection in systole and a minimum at the end of diastole. Whereas mean arterial pressure depends on the relation between flow and resistance defined above, actual systolic and diastolic levels and their difference (pulse pressure) are influenced by additional factors which may not necessarily affect mean pressure. Wiggers has repeatedly emphasized the importance of large arteries in that respect. The systolic pressure rise in the aorta depends in part on its distensibility (stiffness), because the less its wall stretches to receive the ejected blood, the higher the pressure will be. Obviously, other variables such as speed of ejection, magnitude of arterial blood runoff during systole, presence of localized constrictions (coarctation) or ectasias (aneurysms) will also have important effects. Conversely, the rate and level of pressure decline during diastole will depend not only on blood flow (i.e., the "runoff") out of the arteries but also on extent of aortic elastic recoil on the diminishing blood content. This role of the aorta has been variously described as "aortic compression chamber" or the "windkessel effect."

Variations in heart rate, stroke volume, and total peripheral resistance will thus affect not only mean pressure but also arterial pulse pressure. The latter changes can be deduced from the following equation defining aortic distensibility (K): K = dp/dv V, dp being the increment in pressure per increment in volume (dv), and V the aortic end-diastolic volume. Experimentally, with aortic distensibility constant, increasing peripheral resistance will be associated with declining pulse pressure.

SYSTEMIC HEMODYNAMICS. The relationship MAP = CO × TPR states the basic determinants of arterial pressure and provides a framework for understanding the influences of various factors that control pressure within the normal range and are of importance, as well, in hypertension and hypotension. It is obvious from this equation that if pressure remains unchanged, it must be regulated by these two variables which must be reciprocally related—as one falls the other must rise, and vice versa.

In the final analysis, there are three ways to change *cardiac output* (Fig. 3–1). One is by an increase in heart rate. A second means is through an increase in venous return to the right side of the heart with a resultant increase in pulmonary blood volume. The third is an increase in myocardial contractility; and, in some circumstances, this results in an increased ejection fraction.

FIG. 3–1. Effect of central blood volume (cardiopulmonary volume) on cardiac output (*B*) and the 95 per cent confidence limits of the relationship. *A* indicates that if myocardial contractility is disproportionately increased by cardiac nerve stimulation, output will increase even if central blood volume is unchanged. This is because of increased ejection fraction. (Clinical Hypertension. ©1973, Searle & Co. Reproduced with permission.)

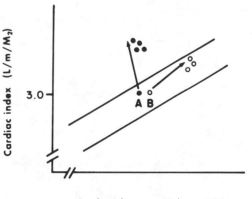

VENOUS RETURN. This is influenced by two factors—total blood volume and the tone of capacitance vessels, particularly those below heart level. Normally, these two are inversely related, because a given change in blood volume (e.g., rapid transfusion or bloodletting of 6 per cent of total blood volume) has little effect because of compensatory change in capacitance vessels. Even without such stimuli of volume expansion or contraction, the capacity of these reservoir veins is important because cardiac output is a direct function of the ratio of cardiopulmonary blood volume to total blood volume (Fig. 3–2), indicating that venous tone plays an ongoing role in determining cardiac output. Thus, as long as the heart is not failing, the venous return to the cardiopulmonary area will be directly related to the cardiac output; and this is another way to express the Frank-Starling relationship.

MYOCARDIAL CONTRACTILITY. This is also modified in two ways. One is the degree of ventricular filling (according to Starling's law) and the other is a change in the contractility of the myocardium. Thus, if the heart is working as a competent pump, it must respond to changes in filling pressure and volume. When diastolic volume increases, the rate of ejection must also increase; this results in a direct relationship between stroke volume (SV) and mean rate of left ventricular ejection (MRLVE) (Fig. 3–3). In addition, myocardial contractility may be modified by a variety of endogenous mechanisms including autonomic control, availability of calcium ions, and circulating humoral factors. Hence, with respect to this same relationship, the effect of adrenergic stimulation would be an inappropriately higher MRLVE for any given SV (Fig. 3–3) (i.e., myocardial contractility is out of proportion to the degree of stretch.) Contractility of course influences the ejection fraction; with diminished contractile strength less of the diastolic volume is ejected and with increased contractility a greater percentage.

VASCULAR RESISTANCE. This is controlled by neural, local, and humoral factors (see Chap. 2). All subdivisions of the vascular bed are richly supplied by adrenergic nerves which provide for vasoconstriction mediated through postsynaptic α-receptor stimulation. Thus vasodilatation, in contrast to vasoconstriction, seems to be primarily a local phenomenon participated in by regional hypoxia, lactic acid, adenosine compounds, histamine, and bradykinin. Reactive hyperemia that follows release of arterial occlusion seems well explained by the local accumulation of vasodilator substances, while active hyperemia that accompanies dynamic exercise cannot be completely explained on that basis. Hypoxia causes vasodilatation as does also the reduced blood pH that accompanies respiratory and metabolic acidosis. Prostaglandins may well play an important role in the local control of vascular resistance; and, in the renal vascular bed, current evidence suggests that the angiotensin

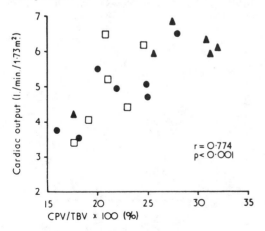

FIG. 3–2. The relationship between cardiac output and the ratio of cardiopulmonary blood volume to total blood volume (CPV/TBV) in normotensive subjects (squares) and renovascular hypertensive (*triangles*) and essential hypertensive (*circles*) patients. (Ulrych, M., et al.: Brit. Heart J. *31*:573, 1969. Reproduced with permission.)

II may be a factor in regulating total renal blood flow as well as distribution of flow to the various zones of the kidney.

Neural control of the circulation

NEURAL REFLEXES. The autonomic nervous system plays a major role in control of the circulation. It does this by influencing cardiac rate and contractility and arteriolar resistance, and by distensibility of the venous compartment. These changes will permit adjustment of circulatory hemodynamics in response to a variety of stimuli so that tissue perfusion may proceed optimally.

These neural circulatory controls are carried out through an afferent, sensing system, central vasomotor centers, and an efferent, effector system. The afferent nerve endings are pressure sensors (baroceptors or mechanoreceptors) because they respond to deformation of the vessel wall. Low-pressure sensors are found in the atria, the lungs, and large veins, whereas the high-pressure mechanoreceptors are found in the left ventricle arterial system above the cardiac level. Chief among the latter are those in the carotid sinus and the aortic arch. Fibers from the carotid sinus travel cephalad in the glossopharyngeal nerve, those from the aortic arch in the vagus. Probably of lesser importance for control of arterial pressure are those receptors in the atria, left ventricle and lungs, the fibers from which are vagal in location. Impulses traveling centrally from these receptors influence the medullary vasomotor centers to increase or decrease control of the heart and peripheral circulation. Impulses emanating from these centers travel along the two components of the efferent system—sympathetic and parasympathetic. Parasympathetic efferent fibers go primarily to the S-A and A-V nodes and both atria, but the sympathetic efferent fibers are distributed to these areas as well as to both ventricles and the peripheral vasculature. There are two types of sympathetic receptors, alpha and beta. The

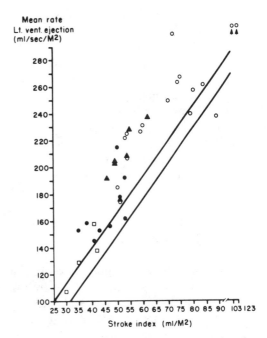

FIG. 3–3. Idiopathic hyperdynamic state: the relationship between stroke index and mean rate of left ventricular ejection: *straight lines* delineate the 95 per cent confidence limits of normal subjects; *open circles* represent normotensive patients with high cardiac index; *triangles* represent normotensive patients with normal cardiac index. Hypertensive patients are represented by either *filled circles* (those with high cardiac output) or *squares* (those with normal cardiac index). The figure illustrates the criteria used to define a hyperdynamic circulation irrespective of the level of cardiac output. (Dustan, H. P., and Tarazi, R. C.: Rev. Med. Therapeutique [Paris] No. 2, 8 Jan. 1973.)

former primarily influence the peripheral circulation and determine the degree of arteriolar and venous tone; the latter affect primarily cardiac rate and contractility.

Simply stated, under normal circumstances, this integrated system acts to raise arterial pressure when it is decreased and to decrease it when it rises. Perhaps the best way to visualize this control is to consider what happens to the circulation with a change from the supine to the upright posture, either by standing up or being tilted upright on a tilt table. The purpose of these circulatory adjustments is to counteract the effect of gravity on the distribution of blood within the vascular system. Thus, when an individual is upright there is, sequentially, a pooling of blood in the highly distensible venous system below the cardiac level, a decrease in venous return and central volume (volume of blood in the heart and lungs), and as a result a fall in cardiac output and systolic pressure. If this sequence were to continue unchecked, fainting would occur because of inadequate cerebral blood flow. However, an intact autonomic nervous system prevents this. The fall in systolic pressure is immediately sensed by the carotid mechanoreceptors, and the result is an almost instantaneous increase in sympathetic vasomotor outflow affecting both venous and arterial systems. It limits venous distention, thereby reducing the peripheral and dependent venous pooling and stabilizing venous return, cardiac output and systolic pressure. It also constricts arterioles and thus raises total peripheral resistance and diastolic pressure. Pulse pressure is narrowed a little, and as a result the mean arterial pressure, in normal persons, is changed only by ±10 mm. Hg. As part of this sympathetic stimulation, heart rate increases slightly. When a tilt table is used, it is possible to study readjustments that occur with return to the supine position. As the table is tilted back there is a transient rise in arterial pressure, the tilt-back overshoot. This occurs when the pooled blood is returned to the central circulation, suddenly increasing cardiac output into the now constricted arterial system. This is normalized quickly, and within seconds arterial pressure is returned to the pre-tilt levels. That the tilt-back overshoot reflects increased sympathetic vasomotor tone is established by its disappearance following sympathectomy or use of drugs that suppress sympathetic vasomotor activity.

Another good example of neural circulatory control is the response to a sudden increase in intrathoracic pressure against a closed glottis, as with the Valsalva maneuver. This sharply diminishes venous return, decreases cardiac output and systolic pressure which, in turn, results in an increased vasomotor outflow. Suddenly, straining is stopped, blood rushes into the thorax, and the augmented cardiac output is thrust into an arterial system whose outflow resistance has been increased, and there is a brief overshoot of arterial pressure. As the pressure rises, the opposite reflex readjustment is seen in a slowing of heart rate and decline of pressure to prestrain levels.

Thus, the sympathetic nervous system works constantly to maintain proper pressure-flow relationships in the various vascular beds, and this knowledge immediately raises the question concerning its inability to prevent hypertension. The answer is not known, but considering the complexities of the system and the demands it must meet, it is not surprising that abnormalities can be found in hypertensive patients. One explanation is that there is an upward resetting of the responsiveness of the carotid sinus-aortic arch mechanoreceptors so that they sense an abnormally high pressure as being abnormally low resulting in an inappropriate sympathetic vasomotor outflow and vasoconstriction even in the face of high pressure. Another is that some deficiency of sympathetic control exists such that increases in intravascular volume or cardiac output are poorly modulated and vasoconstriction results.

ADRENERGIC, NEUROEFFECTOR MECHANISMS—CATECHOLAMINES. Since it is the sympathetic component of the autonomic nervous system that plays the major role in neural control of the circulation, this discussion will be restricted to adrenergic-catecholamine neuroeffector mechanisms. Like the renal pressor system, these are made up of enzyme systems in series (Table 3–1). (For additional Figures and discussion of dopamine metabolism, refer to Figs. 32–2 through 32–4.)

The amino acid tyrosine is taken up by brain, peripheral adrenergic neurons, and

TABLE 3–1
Catecholamine Synthesis and Metabolism

Catecholamine Synthesis

Substrate	Site of Reaction in Sympathetic Nerves	Enzyme	Product
Tyrosine	Mitochondria	Tyrosine hydroxylase	Dihydroxyphenylalanine (DOPA)
DOPA	Cytoplasm	Aromatic L-amino acid decarboxylase	Dopamine
Dopamine	Granulated vesicles	Dopamine β-oxidase	Norepinephrine
Norepinephrine	Cytoplasm	Phenylethanolamine N-methyl transferase*	Epinephrine

*Present in adrenal medulla and nerve endings of heart and uterus.

Fate of Norepinephrine Released From Nerve Endings

Initial Step	Site of Inactivation	Inactivating Enzyme	Compound Formed	Further Metabolism	Urinary Compound
Reuptake	Intraneuronal	Monoamine oxidase	Dihydroxymandelic acid	(COMT in liver and kidneys)	VMA
Release into circulation	None				Free norepinephrine
Release into circulation	Liver, kidney	Catechol-O-methyl transferase (COMT)	Normetanephrine	None	Normetanephrine

chromaffin cells (most importantly those of the adrenal medulla). In the mitochondria of these cells the tyrosine is hydroxylated by tyrosine hydroxylase to L-dopa. This migrates into the cytoplasm, where it is decarboxylated to dopamine. Dopamine, in turn, moves into a specialized, cellular subcompartment, a vesicle, where it is acted upon by dopamine β-oxidase and becomes norepinephrine. The adrenal medulla is the only organ having the capacity to methylate norepinephrine (NE) to form epinephrine (E), since it contains the enzyme, phenylethanolamine-N-methyl transferase.

Our concern here is primarily with NE, because it is the major catecholamine released from adrenergic nerve endings both upon stimulation and as a continuous "leaking" process. It is the neuroeffector agent exerting its physiological effects by attaching to receptor sites that are in close proximity to the nerve terminals.

Typical of Nature's profligacy, more NE is released than is needed, so there are a number of ways it is handled. First, some escapes unchanged into circulating blood and is rendered inactive by an enzyme, catechol-O-methyl transferase (COMT) which is present in large quantities in liver and kidney. Through the action of COMT, normetanephrine is formed and is excreted in the urine as such. Probably, however, the bulk of physiological inactivation occurs through reuptake of NE by nerve terminals where it is again stored in granulated vesicles.

Another process of chemical inactivation, deamination, occurs within the nerves themselves through action of the enzyme, monoamine oxidase (MAO). The compounds thus produced are physiologically inactive and on release into the circulation are methylated in tissues containing COMT to form vanillyl-mandelic acid (VMA) and methoxyhydroxyphenylglycol (MHPG).

Catecholamines and their metabolites are excreted in the urine. The amounts of free NE and E are very small, because most of that which escapes reuptake and is released into the circulation is transformed into normetanephrine and metanephrine, while that metabolized by nerve endings appears mostly as VMA.

The released NE from postganglionic nerve endings stimulates either pre- or postsynaptic adrenergic receptors. Postsynaptic (α_1) receptor stimulation on vascular smooth muscle cells results in vasoconstriction, and of the few on myocardial cells result in a shortened refractory period; stimulation of presynaptic (α_2) receptors acts as a negative feedback to shut off NE release from the nerve ending. In contrast, postsynaptic β-receptor stimulation results in vasodilation or increased chronotropic, inotropic, and metabolic activity of the myocardium. Early evidence suggests that presynaptic β-stimulaton may stimulate NE release, but more information is necessary to establish this hypothesis.

Alpha- and beta-receptors are also located in the brain and the effects of their (central) stimulation are opposite to the peripheral effects. Thus, central α-receptor stimulation reduces arterial pressure (and these receptors are of the postsynaptic α_2 type); central β-receptor stimulation will elevate arterial pressure.

Extracellular fluid volumes

DISTRIBUTION. Total body water accounts for about 60 per cent of body weight and is contained in two compartments—intracellular and extracellular. Extracellular fluid (ECF) has two components, the plasma and interstitial fluid volumes, which are in dynamic equilibrium at the capillary level. The plasma volume (PV) and red cell mass comprise the intravascular volume. Considering these basic features, it should not be surprising to find that the ECF and the vascular system are intimately related because the intravascular volume distends the vascular bed and the interstitial fluid (IF) surrounds it. Although intravascular volume is not usually considered a hemodynamic variable, in hypertension it often is, and in hemorrhagic shock it is at the origin of all the abnormalities.

The extracellular fluid volume is about

20 per cent of body weight; roughly speaking, interstitial fluid comprises about 80 per cent of this and plasma about 20 per cent. Actually, the PV/IF ratio is normally maintained around 0.233. Under normal circumstances, the precise volumes of these fluid compartments depend upon the size of the individual; also, the percentage of body weight they represent is inversely related to the amount of body fat. Thus, men have larger volumes than women and fat people relatively less than lean people. This latter feature creates difficulties in expressing data so that individuals and groups may be compared regardless of how obese or lean they may be. Relating these volumes to lean body mass provides the best expression, but since that measurement is difficult, other ways of expression must be used. The most usual one is in relationship to body weight, even with its recognized limitations. A better way is the use of body height as reference, because it is not influenced by the degree of obesity and because it remains fairly constant over many years. Body surface area has not been used extensively as a reference index but it has the advantage of providing some consideration of the amount of water contained in fatty tissue which use of height does not.

INTRAVASCULAR VOLUME AND THE VASCULAR SYSTEM. Broadly speaking, the vascular system has three components, each having its own characteristics. The arterial segment has limited distensibility and is maintained at high pressure and low volume; it is thought to contain about 20 per cent of the total blood volume. Although the capillary bed is of considerable length, it contains only 5 per cent of the total blood volume. Capillary pressure is determined by the amount of constriction of precapillary arterioles and postcapillary venules. The venous side of the circulation is a low-pressure, highly distensible compartment which contains about 75 per cent of the intravascular volume.

Both the arterial and venous compartments are importantly affected by sympathetic vasomotor outflow but, characteristically, these effects are different. Since the volume of the arterial side is relatively small compared to the venous, its capacity is insignificantly changed by neural influences, although in this vascular segment resistance to blood flow and perfusion pressure can be greatly modified. In contrast, sympathetic vasomotor activity plays a large role in determining the capacity of the venous side, while at the same time affecting pressure but little.

The amount of blood that the heart pumps, in the absence of cardiac failure, depends on the amount returned to it. This, in turn, is a function of the total intravascular volume and the central blood volume. The latter, the volume of blood in the heart and lungs, is determined in large measure by the capacity of the systemic (peripheral) venous compartment. Thus it is possible, on the one hand, to have a large blood volume, venous pooling, low central blood volume and low cardiac output while, on the other hand, a small blood volume, diminished venous capacity, a disproportionately high central blood volume, and a normal or even slightly increased cardiac output.

CONTROL OF EXTRACELLULAR FLUID VOLUMES. Although highly compartmentalized, plasma and interstitial fluid are one system with their locus of contiguity at the capillary level. Their electrolyte compositions are practically identical; the relatively minor concentration differences reflecting the protein content of plasma. Since these two fluids are in dynamic equilibrium, any change in electrolyte composition is quickly transmitted across the capillary endothelium. The volume relationships of plasma and interstitial fluid are controlled by capillary (hydrostatic) filtration pressure, the oncotic pressure of plasma proteins, tissue pressure, and tissue fluid oncotic pressure. The precapillary arterioles determine the amount of arteriolar pressure that is transmitted to the capillary bed. This outward force plus the tissue oncotic pressure minus the tissue pressure and the plasma protein oncotic pressure is the capillary filtration pressure. On the venular side of the capillary bed reabsorption of interstitial fluid is determined by the facilitating effect of oncotic pressure and by venular hydrostatic pressure, which reflects the relationship between the capacity of the systemic venous compartment and the volume of blood it contains. Broadly speaking, interstitial fluid is controlled by the volume and composition of the plasma and this, in turn, by external water and electrolyte exchanges which are primarily a renal function. A subsidiary system, the lymphatics, exists to remove the small amounts of protein filtered and any excess of filtered fluid over that reabsorbed.

Although fluid losses occur through skin, lungs, and bowel, these are normally of lesser importance in homeostasis compared to the control exerted by the kidneys, because of the large volumes of ECF that are constantly being processed. In brief, water and electrolyte balances are achieved through the amount of glomerular filtrate formed and the amounts reabsorbed. According to Pitts, 67 to 87 per cent of glomerular filtrate is reabsorbed in the proximal tubule. With a normal GFR of 150 l./day, as much as 50 and as little as 20 l. leave the proximal tubule to be processed at more distal sites. If the 24-hour urine volume is 1.5 l., this represents only 1 per cent excretion of glomerular filtrate. With these amounts in mind, it is easy to comprehend how changes in GFR, and in proximal or distal tubular reabsorption can greatly modify the amount of plasma water excreted and thus modify the volume and composition of the ECF.

The fractional amount of fluid reabsorbed from the proximal tubule is determined by the volume filtered (GFR), peritubular capillary pressure, and ECF volume. Thus, if GFR per nephron is reduced, if peritubular capillary pressure is reduced, or if the ECF is contracted, a greater pro-

portion of the filtered sodium and water is reabsorbed. Contrarily, if these functions are increased, less tubular fluid is reabsorbed and more escapes into distal segments of the nephron for processing there. Distally, salt and water processing are under the influences of aldosterone and antidiuretic hormone. It seems likely also that locally released angiotensin II and prostaglandins also play a role.

The renal pressor system

DESCRIPTION AND MEASUREMENT. The importance of this system in regard to pressor homeostasis is that its end-product, angiotensin II, is not only a potent pressor substance but it also influences all the other mechanisms that affect arterial pressure. The system itself is really two enzymes in series (Fig. 3–4). In the first reaction, a proteolytic enzyme of renal origin called *renin* reacts with a circulating globulin (renin substrate or angiotensinogen) to release the decapeptide angiotensin I. This serves as substrate for converting enzyme which, by cleaving two terminal amino acids, produces the octapeptide angiotensin II. This is rendered inactive by plasma and tissue angiotensinases. Although renin comes from the kidney, it is active in circulating blood, is present in arterial walls, and has been shown to be synthesized by arterial smooth muscle cells in tissue culture. Additionally, there is good evidence for the presence of renin in brain and uterus. Although converting enzyme is present in plasma and tissues, the major conversion of angiotensin I to II occurs in the lung.

Angiotensin I has no effect on arterial pressure, but mere nanogram amounts of angiotensin II are sufficient to produce substantial elevations. Information currently available suggests that the plasma concentration of angiotensin II is usually less than 100 pg./ml. At these low concentrations angiotensin II exerts its effects on several target organs: it (1) constricts arteriolar smooth muscle; (2) stimulates aldos-

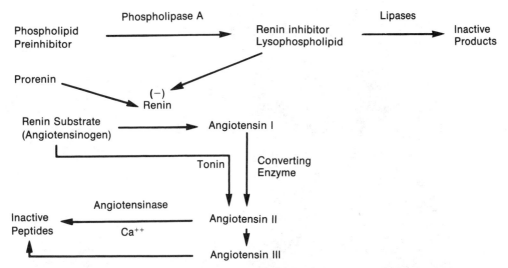

FIG. 3–4. A schematic diagram of the renopressor system.

terone production in the adrenal cortex; (3) releases catecholamines from the adrenal medulla; (4) acts on certain brain centers to increase adrenergically mediated cardiovascular outflow or on other centers to stimulate thirst; and (5) may directly increase myocardial contractility. Each of these actions serves to elevate arterial pressure.

Current information concerning the renal pressor system comes from estimation of plasma renin activity (PRA). Under proper conditions of pH and temperature, the endogenous plasma renin is allowed to act on endogenous substrate. By inhibiting plasma angiotensinases enough angiotensin is formed to allow its quantification by radioimmunoassay. Renin substrate can also be measured by incubating plasma with enough added human renin to completely exhaust the protein and then assaying the angiotensin formed.

It should be emphasized that the methods widely used provide an estimate of the activity of renin but do not measure it directly.

Recently, purification of human renin has been achieved and a direct radioimmunoassay developed. This technique promises to provide greater insight into the

mechanism of action of the renin-angiotensin system. It must be recalled that the currently employed direct radioimmunoassays detect both active and inactive (precursor?) renins, whereas the enzymatic methods (PRA) detect the active form of the enzyme.

It has been appreciated recently that renin exists in the kidney and in plasma in several forms, each having a different molecular weight. Additionally, although the precursor-product relationship, if any, between these forms is unknown, it is clear that PRA can be increased by a variety of *in vitro* manipulations (e.g., acid dialysis, freezing, treatment with trypsin). These laboratory observations suggest the activation of previously inactive forms of renin present in plasma and possible clinically relevant situations still to be described.

SOURCE OF RENIN—THE JUXTAGLOMERULAR APPARATUS. Renin comes from the kidney, hence the term *renal pressor system*. It is formed and stored in the juxtaglomerular apparatus (JGA), or complex, at the vascular pole of the glomerulus (Fig. 3–5). At this point the macula densa portion of the distal tubule is in close proximity to the afferent and efferent arterioles. The juxtaglo-

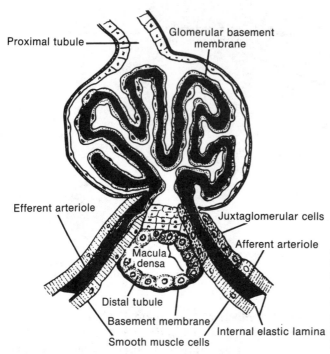

Proximal tubule

Glomerular basement membrane

Efferent arteriole

Juxtaglomerular cells

Macula densa

Afferent arteriole

Distal tubule

Basement membrane

Smooth muscle cells

Internal elastic lamina

FIG. 3–5. A schematic representation of the juxtaglomerular apparatus, showing the relationships of juxtaglomerular cells of the afferent arteriole to macula densa cells of the distal tubule. These granulated juxtaglomerular and macula densa cells form the juxtaglomerular apparatus.

merular complex is composed of granular cells in the afferent arterioles, the macula densa and the polkissen, a group of cells in the triangle formed by the afferent and efferent arteriole and the distal tubule. The complex is richly innervated by sympathetic nerve fibers.

The granular cells of the afferent arterioles seem to be the primary location of renin. Experimentally, there is a close correlation between the granularity of these cells and renal renin content. The role of the macula densa has not been completely defined, but there is evidence to suggest that the amount of sodium reabsorbed at that site is an important determinant of renin release.

Considering the fact that the granular cells are in the afferent arterioles and that the macula densa is in close anatomical proximity to them, it should not be surprising to find that both arterial pressure and sodium excretion are among the factors known to affect renin production and/or release.

RENIN RELEASE. There are a number of stimuli that strongly influence renin release and, therefore, circulating renin. These stimuli include height of arterial pressure, magnitude of intravascular volume, sympathetic vasomotor outflow, the amount of sodium thought to be present at the macula densa, and plasma potassium concentration.

Information concerning renin release in man has come from the clinical application of animal studies. Thus it has been found that rapid arterial pressure reduction with intravenous administration of sodium nitroprusside or diazoxide increases PRA. Rapid reduction of intravascular volume by controlled bleeding also increases PRA. A slower reduction is produced by the negative sodium balance that accompanies a low sodium diet or diuretic drug treatment, and this elevates PRA as well. The converse is also true, because plasma volume expansion turns off the stimulus for renin release and reduces PRA. So predictable are these responses that dietary sodium restriction and intravenous sodium chloride infusion are used as standard tests of the renal pressor system in man.

The sympathetic nervous system also plays a role in renin release. Normal cardiovascular adjustments during upright posture depend on augmented sympathetic vasomotor outflow, and one of the effects of this is a rise in PRA. In addition, both α- and β-adrenergic blocking drugs have been shown to diminish PRA. As well as manipulations of sodium intake, upright posture is frequently used as a clinical test of renin release.

Plasma sodium concentration in some way influences the amount of renin produced or released into the circulation. This has been clearly shown in animal experiments using renal perfusion techniques with which it is possible to vary sodium concentration of the perfusate without varying that of the whole body. Hyponatremia increases renin release and hypernatremia diminishes it. Whether this reflects a direct effect on the afferent arteriolar component of the JGA or on the amount of sodium available for macula densa reabsorption is not yet known. At any rate, there is little information to suggest that plasma sodium concentration has any ongoing control over PRA under normal conditions.

The amount of sodium present in tubular fluid at the level of the macula densa is difficult to measure. Thus, the macula densa theory of renin control cannot be tested in man. Certainly dietary sodium restriction results in diminished sodium excretion and increased PRA, while sodium chloride infusions have the opposite effects. However, the former is associated with decreased intravascular volume, which increases sympathetic vasomotor outflow. This in itself could be responsible for augmenting renin release. It has been suggested that the reason that diuretic drugs increase PRA is not only because they decrease plasma volume but also because they paralyze the sodium reabsorptive capacity of the macula densa. Unfortunately there is no way to study this possibility in man.

Potassium, as well as sodium, may play some role in renin release because it has been shown that potassium loading can diminish PRA and depletion can elevate it independently of change in aldosterone secretion. How important potassium concentration is in the normokalemic individual is yet to be determined.

Hormonal influences

ALDOSTERONE. The chief electrolyte active steroid concerned with normal sodium and potassium homeostasis is aldosterone. Its synthesis and release are primarily controlled by angiotensin, and currently available evidence suggests that the heptapeptide des-asp-angiotensin II, or so-called angiotensin III, may play a role in aldosterone secretion as well. Recently, an enzyme termed *tonin* has been identified in the adrenal gland, and this enzyme has the capacity to generate angiotensin II directly from substrate. However, its role in the control of aldosterone biosynthesis is at present unclear. Additionally, the potassium ion per se can stimulate aldosterone secretion, and adrenocorticotropin (ACTH) has at least a transient effect.

In general, however, factors that stimulate or suppress aldosterone production are, therefore, those that stimulate or suppress renin release. Since aldosterone acts in the distal nephron to increase sodium reabsorption it is understandable that sodium loading is normally associated with suppression of renin release and decreased aldosterone production, so that the excess sodium can be excreted. To protect against ECF depletion during sodium deprivation renin and aldosterone increase and urinary sodium excretion becomes negligible. So closely regulated are these functions normally that there is a significant positive correlation between PRA and either aldosterone production or excretion and a negative correlation of aldosterone with urinary sodium excretion. In fact assessing whether aldosterone production or excretion is normal requires that the steroid value be judged in relationship to the urinary sodium excretion.

Aldosterone is the major steroid influencing potassium homeostasis. It rises as plasma potassium concentration rises, causing enhanced urinary potassium excretion, and falls as serum potassium decreases. Actually, there is substantial experimental evidence to indicate that the concentration of potassium in adrenal zona glomerulosa cells is one of the major factors influencing aldosterone release.

Thus, the renal pressor system and potassium are the stimuli that primarily control aldosterone normally and although ACTH can transiently increase aldosterone production its role is not major.

OTHER STEROIDS. Several other steroidal hormones produced by the adrenal cortex have been implicated in the pathogenesis of hypertension. Notable alterations are discussed elsewhere (Chap. 15) and involve the excessive stimulation of the adrenal cortex by the pituitary (ACTH) in Cushing's disease, the excessive production of corticosteroids by the adrenal cortex (in Cushing's syndrome), or by the abnormal accumulation of mineralocorticoids through inherited recessive genetic defects in adrenal steroidal biosynthesis. Other syndromes of excessive production of specific adrenal steroidal hormones have been described recently involving 18-OH-desoxycorticosterone (18-OH-DOC) and 16-β-hydroxydehydroepiandrosterone (16-βOH-DHEA). Because each of these steroidal hormones seems to raise arterial pressure and operate pathophysiologically through the same pressor mechanisms and manifests itself clinically by greater or lesser degrees of hypokalemic alkalosis, hyporeninemia, expanded intravascular volume, and impaired carbohydrate tolerance, discussion in this chapter is confined primarily to aldosterone.

OTHER HORMONES. Excessive production of thyroid hormone has been associated with primarily systolic hypertension; and hypothyroidism has been associated with diastolic hypertension. Thyrotoxicosis and myxedema are discussed elsewhere in this text (Chap. 16). Noteworthy, however, is the interaction of the thyroid hormone and the sympathetic nervous system, the pharmacological suppression of which is associated with the amelioration of all cardiovascular manifestations of excessive production of thyroid hormone. The precise mechanisms by which thyroid hormone and adrenergic neural functions interact still remain to be completely defined. But recent evidence indicates a direct effect of thyroid hormones on the number of β-adrenergic receptor sites as one contributing factor. In addition, changes in the plasma concentration of catecholamines themselves have been reported in hypothyroidism.

Excessive production of estrogenic hormones also has been associated with elevated arterial pressure; and this altered state is best exemplified by the hypertension associated with oral contraceptives. Recent studies concerned with the mechanisms underlying this form of hypertension relate to the increased production of the protein substrate of renin and the expansion of intravascular volume, although the precise pathophysiological mechanisms of this disorder still remain incompletely understood. Finally, excessive production of growth hormone and of parathyroid hormone have also been associated with an increased incidence of hypertension. While the mechanism for the hypertension associated with growth hormone has not been elucidated, it has been shown that the hypercalcemia of hyperparathyroidism is associated with increased arterial pressure; and with correction of the abnormal calcium metabolism, arterial pressure usually normalizes.

No discussion of hormonal influences on control of arterial pressure would be complete without mention of the possible role of brain renin. There is evidence to indicate that renin is present in the brain and that angiotensin II receptors also exist there. It may well be that a brain renin-angiotensin system plays an important role in arterial pressure control by modulating adrenergic outflow or by influencing the secretion of natriuretic factors (see below). Additionally, abnormalities of vasopressin (ADH) secretion have been detected in some forms of hypertension, suggesting a possible permissive role of this hormone in the pathogenesis of hypertension. It must also be recalled that prostaglandins, usually acting locally (but possibly also systemically), can profoundly affect not only arterial pressure but also regional tissue perfusion and renal function. Thromboxane is a prostaglandin derivative that is synthesized in platelets and is an extremely potent naturally occurring vasopressor substance. In the kidney, as elsewhere, angiotensin II stimulates the secretion of vasodilating prostaglandins that tend to moderate its vasoconstricting effects. Thus, the prostaglandins and possibly other lipid-soluble factors may play an important role in the regulation of arterial pressure.

Hypertension

Resting hemodynamics

Arterial hypertension results from a disproportion between cardiac output and total peripheral resistance. Because early studies of hypertensive patients showed a normal cardiac output in the absence of left ventricular failure, it was generally assumed that the hemodynamic hallmark of hypertensive diseases lay entirely in the peripheral resistance. However, from recent studies of various specific types of hypertension, a more complex hemodynamic picture has emerged (Table 3–2).

ESTABLISHED ESSENTIAL HYPERTENSION. As the earlier studies indicated, patients with well-established essential hypertension (hypertension of unknown cause) are characterized by a normal cardiac output and elevated peripheral resistance. However, with progressing age, and as hypertension becomes more severe, total peripheral re-

TABLE 3–2
Hemodynamic Characteristics of Different Types and Severity of Hypertension

Type	Heart Rate	Cardiac Output	Left Ventricular Contractility	Total Peripheral Resistance
Borderline ("labile")	↑	↑ or N	↑	Inappropriately N or ↑
Essential				
No cardiac involvement	↑	N	N or ↓	↑
Left atrial enlargement	↑	N	↓	↑ ↑
Left ventricular hypertrophy	↑	↓	↓ ↓	↑ ↑ ↑
Cardiac failure	↑	↓ ↓	↓ ↓ ↓	↑ ↑ ↑ ↑
Renovascular	↑	N or ↑	N or ↓	N or ↑
Renal parenchymal disease				
Normal renal function	↑	N	N	↑
Renal failure (and anemia)	↑	↑	N or ↓	↑
Primary aldosteronism	↑	N or ↑	N or ↑	↑
Aortic coarctation	↑	↑	N	↑

(N = normal; ↑ = increased; ↓ = decreased)

sistance rises and there is a progressive reduction of resting cardiac output. Thus, in hypertensive patients with definite left ventricular hypertrophy even before cardiac decompensation occurs, cardiac output and rate of left ventricular ejection decrease as total peripheral resistance rises. Patients without cardiac involvement have a normal output and ventricular contractility despite an obvious increase in cardiac work due to hypertension.

Studies of regional circulations have shown that the elevated resistance in established hypertension seems to be uniformly distributed in practically all vascular territories, except the kidney, where it may be more intense, and in the skeletal muscles, where it is slightly less marked. This pattern has been likened by Brod to a constant preparedness for exercise or response to unspecified stress (fight or flight).

BORDERLINE OR LABILE HYPERTENSION. Repeated reports have established that a large proportion of patients with either borderline (or labile) and mild essential hypertension may have increased cardiac output and normal, near normal, or subnormal values of total peripheral resistance, at least in the younger patients. Borderline (or labile) hypertension is defined in these reports as a state in which arterial pressure is normal at times but above 90 mm. Hg on several other occasions. With age the elevation of cardiac output decreases, and so does the magnitude of intravascular (plasma) volume (see below) and that amount of circulating blood that is distributed to the cardiopulmonary area from the periphery. In those patients with higher cardiac output, if autonomic support (adrenergic and parasympathetic) is inhibited pharmacologically and the patients are compared with normotensive subjects, the slightly reduced, normal, or even slightly elevated total peripheral resistance is then found to be "inappropriately normal," or actually increased. Thus, already at this stage in the development of hypertension

there is evidence of arteriolar constriction. Moreover, the normal or even reduced intravascular volume associated with an increased cardiopulmonary volume strongly suggests peripheral venoconstriction that serves to redistribute the circulating blood to the cardiopulmonary area, thereby augmenting venous return and cardiac output. Indeed, this effect may be augmented further by increased myocardial contractility. That proportion of the cardiac output which is distributed to the organs remains normal and unchanged.

ESSENTIAL HYPERTENSION. As hypertension and the vascular disease progress in severity, arterial pressure steadily rises in direct proportion to the increase in total peripheral resistance. Therefore, arteriolar constriction increases as hypertension becomes more severe and, since intravascular (plasma) volume decreases as pressure and vascular resistance rise (see below), it is reasonable to assume that postcapillary (venular) constriction also occurs. In response to this increasing afterload imposed upon the left ventricle, there is a progressive adaptation. Initially, this is reflected physiologically as a hyperfunctioning left ventricle that performs more work and demonstrates increasing contractility. The ventricular muscle also adapts to the increasing pressure load by anatomically restructuring itself through concentric left ventricular hypertrophy. The precise mechanisms by which the physical afterload stimulus is transformed into a biochemical response are, at present, poorly understood. Nevertheless, the hypertrophy process enables the heart to function normally for a prolonged time ("stable hypertrophy") until eventually the pressure overload results in cardiac failure unless treatment intervenes.

The progressive stages of left ventricular hypertrophy can be ascertained clinically. The earliest evidence by electrocardiographic criteria may be shown by the findings of left atrial abnormality. This finding does not mean that left atrial disease precedes ventricular hypertrophy; it does mean, however, that the early hypertrophying left ventricle is less compliant, reflecting a diastolic filling road on the left atrium, thereby inducing its enlargement. These findings are confirmed echocardiographically by the presence of left atrial enlargement and left ventricular hypertrophy. But even at this early stage of left ventricular hypertrophy, there is evidence of impaired ventricular contractility. This becomes still more obvious as actual left ventricular hypertrophy becomes evident by both conventional electrocardiographic and chest x-ray criteria. And this evidence of increasing left ventricular hypertrophy is confirmed by echocardiographic measurements. At this stage, when left ventricular hypertrophy is clinically evident, resting cardiac output is reduced and blood flow distribution similarly falls.

RENAL AND RENOPRIVAL HYPERTENSION. One of the important results of recent clinical studies has been the differentiation of renal hypertension into at least two types; one is related (at least initially) to activation of the renal pressor mechanism, the other, to loss of renal parenchyma and possibly of an antipressor effect too. The first is exemplified by renal arterial stenosis and the second by the anephric state (renoprival). In a schematic form (Fig. 3–6) that admits of many exceptions, the former is characterized in man by elevated plasma renin activity, especially in severe hypertension, contracted plasma volume, and indices of increased neurogenic activity. Renoprival hypertension, on the other hand, is marked by a direct relationship between blood volume and arterial pressure, very low or absent plasma renin activity, and deficient or impaired neural circulatory reflexes. Before nephrectomy patients with parenchymal disease present varying mixtures of these two extremes with either the "renal" or "renoprival" element predominating. Each of these forms of hypertension is expressed at

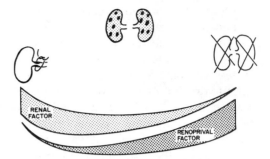

FIG. 3–6. A schematic representation of the relative participations of renopressor (renal) and renoprival factors in renal hypertensions. The renopressor factor seems to be the main influence in hypertension associated with renal arterial disease or experimental renal arterial narrowing. With renal disease both factors may participate, whereas in the absence of kidneys or, often, in terminal renal failure the renoprival, or volume dependent, factor predominates.

either end of the clinical spectrum; and in between are the many instances of renal hypertension that are encountered clinically. Thus, for example, with renal parenchymal damage resulting from an ischemic renal arterial lesion, it is not inconceivable to encounter some expansion of intravascular volume that would tend to lower circulating PRA into the normal range. And, conversely, there are some patients with severe renal parenchymal disease and renal failure with elevated PRA who demonstrate a normal arterial pressure following bilateral nephrectomy despite vigorous hemodialysis plasma ultrafiltration. (The reader is referred to more detailed discussions of these specific mechanisms, below.)

RENOVASCULAR HYPERTENSION. Much knowledge has accumulated concerning renovascular hypertension, since it can be produced experimentally in a variety of ways in different animals. Essential to proper evaluation of experimental studies is an awareness of the differences that may result from variations in experimental techniques, the timing of observations, and the variety of animal species used. Although early studies in man showed that both cardiac output and total peripheral resistance were elevated, continued observations indicated that output was occasionally normal or even decreased. Further, even though arterial pressure was positively correlated with cardiac output, the increased output was not solely responsible for the hypertension; after successful surgical repair of the arterial lesion or nephrectomy, reduction in total peripheral was more often observed than reduction in output. This common participation of varying degrees of increased output and resistance in the maintainance of renovascular hypertension in man corresponds to recent experimental studies in rats and dogs. Following clipping of renal arteries or cellophane wrapping of the kidney, the early rise of arterial pressure is primarily due to increased cardiac output; later, total peripheral resistance rises while output returns toward normal.

Two main hypotheses have been advanced to account for the increased cardiac output noted at the onset of experimental hypertension. Transient expansion of extracellular fluid and plasma volume was noted in rats and though there was no correlation between the increased blood pressure and the volume expansion, the implication was that hypertension was initiated by fluid retention. However, no change in blood volume has been measured during the initial stages of cellophane perinephritic hypertension in dogs when cardiac output was rising; and, indeed, both in man and dogs with chronic renovascular hypertension plasma volume is reduced. The combination of lower intravascular volume and increased cardiac output suggests increased tone of capacitance vessels causing redistribution of blood from the periphery to the central cardiopulmonary circulation. In support of this suggestion are a highly positive correlation between central blood volume and cardiac output in both animals and man and the repeated findings in dogs and rats with clipped renal arteries of increased mean circulatory pressure in the face of normal or reduced plasma volume. This redistribution of blood resulting from a restricted capacity of venous reservoir may be an indirect effect of angiotensin on venous reservoirs through an effect on the sympathetic nervous system. From various studies of the influence of angiotensin II on catecholamine reuptake and storage as well as central vasomotor mechanisms, there is increasing evidence supporting the concept that angiotensin II modulates neural function by affecting neurohumoral transmission centrally, in the brain stem, as well as at the peripheral nerve endings.

RENAL PARENCHYMAL DISEASE. As indicated above, hypertension in patients with renal failure from parenchymal disease is renin dependent, volume dependent, or a mixture of the two. Regardless of the predominant pressor mechanism the total hemodynamic characteristic is a raised peripheral resistance. Hypertension does not invariably accompany end-stage renal parenchymal disease, although a significant degree of anemia does. Therefore, all patients with uremia demonstrate an increased cardiac output, and those with normal arterial pressure have a much lower total peripheral resistance than patients with hypertension. In patients with renin-dependent hypertension associated with parenchymal disease, it is assumed that the basic mechanism is the vasoconstriction of angiotensin II, probably augmented further by fluid volume overload. In support of this is the finding that after bilateral nephrectomy arterial pressure falls because of a decreased participation of the renopressor system on arteriolar resistance. However, cardiac output does not change, as the anemia is not affected by the operation. The volume-dependent component, likewise, is not influenced by nephrectomy. Volume-dependent hypertension is characterized by increased response of pressure to the volume excess.

Its cause is unknown but it may depend in part on the deficient neurocirculatory reflexes that occur in uremia.

Uremia and anemia are not essential features of the volume-dependent hypertension that is often found in patients with end-stage kidney disease. A study of nonazotemic hypertensive patients with pyelonephritis and normal hemoglobin showed normal cardiac output, elevated TPR and positive correlation of arterial pressure with intravascular volume. Naturally, these patients did not have any gross evidence of peripheral neuropathy, and there were no findings to suggest severe impairment of circulatory reflexes.

PRIMARY ALDOSTERONISM. In primary aldosteronism, as in all types of hypertension, arterial pressure ranges from modestly elevated to very high levels. It tends to be associated with a hyperdynamic circulation in that cardiac output and mean rate of left ventricular ejection are often elevated. However, cardiac output may play a lesser role in the hypertension, because arterial pressure and cardiac output are inversely related. This is a salt- and water-dependent hypertension, because it remits with any treatment (low sodium diets and/or diuretics) that produces sufficient negative salt and water balance. Further, in the untreated state, the elevated arterial pressure is often associated with expanded plasma volume, most particularly in those patients with mild to moderate hypertension. This hypervolemia influences cardiac output because output and plasma volume are directly correlated. Presumably the degree of hypertension represents the individual patient's vasoconstrictor response to the fluid volume overload and, perhaps, the cardiac output responds in a compensatory fashion to the increased resistance (i.e., the higher the resistance, the lower the output). With the higher arterial pressures, then, there would be a tendency to maintain intravascular volume at a lower level,

and this would account for the finding of a negative correlation between TPR and intravascular volume.

COARCTATION OF THE AORTA. This disorder produces hypertension that is unique among the various types, not only because of etiology but also because it is unassociated with progressive arteriolar disease. It results from a reduced size of the aortic compression chamber. Cardiac output is usually increased, and the ejection of a large stroke volume into an aorta of diminished capacity and relatively limited diastolic arterial runoff accounts for the systolic hypertension that is so characteristic of coarctation. In the absence of congestive heart failure there is no evidence for a vasoconstrictor element; leg blood flow has been found to be normal or decreased, arm blood flow, normal or increased, and cerebral blood flow, increased. The diminished aortic capacity explains why exercise-induced increases of stroke volume and heart rate can raise systolic pressure to very high levels in patients with coarctation. Still a matter that remains unresolved is the contribution of the renopressor system to the hypertension of aortic coarctation.

Hemodynamic responses to stress and exercise

Studies at rest under basal conditions often do not reveal the full scope of hemodynamic abnormalities in hypertension. Various stressful stimuli such as application of cold, production of ischemic pain, disturbing demands during interviews, exercise, and pharmacological agents have therefore been used to detect hypothetical hyperresponsiveness of hypertensive patients. By and large, few consistent differences between the normotensive individual and the hypertensive patient have been demonstrated with regard to magnitude of pressure rise with stress. However, compared with normotensive controls, the duration of the rise is longer and more importantly perhaps, its hemodynamic pattern is probably different in hypertensive patients.

Thus, in most normal subjects, increased arterial pressure during a stressful interview is due to increased cardiac output, whereas most hypertensive patients respond by an increased TPR. In the absence of cardiac failure, response to dynamic exercise in hypertensive patients is much the same as that of normal subjects. Cardiac output increases proportionately to oxygen uptake and TPR falls. This fall, however, is never to normal levels, so that blood pressure remains high and indeed the extent of reduction in TPR may not be commensurate with the rise in output, especially in young subjects. These observations as well as the greater tendency to respond to stress by raising vascular resistance, both point to an early and persistent alteration in peripheral reactions in hypertension.

Neural mechanisms

CATECHOLAMINES. With the exception of pheochromocytoma, measurements of catecholamines and their metabolites have not yet been found to be of great value in estimating neurocirculatory control abnormalities in hypertension. This is so in spite of the compelling clinical evidence for both hyperactivity and hypoactivity of the nervous system in the various types of hypertension. Reasons for this discrepancy are clear. Recent studies have shown that plasma norepinephrine levels can provide a fallible guide to adrenergic participation, since they depend not only on the release of this neurohumoral substance from the postganglionic nerve ending but also on its metabolic clearance rate. Conceivably, the norepinephrine spillover rate would provide a more reliable index to the overall sympathetic activity than does the circulating plasma level of norepinephrine. Indeed, norepinephrine spillover rates were found (in these studies) to be elevated in 20 per cent of patients with essential hyper-

tension and to be reduced in patients with autonomic insufficiency. In support of these data, the rate of spillover was directly related to the firing rates of the adrenergic nerves, although the precise source of the increased catecholamines has not yet been identified. Therefore, it is possible that measurement of regional norepinephrine spillover rates may reflect more accurately the adrenergic participation from various organs.

Nevertheless, from various reports of measurements of urinary and plasma catecholamines (norepinephrine, more specifically), it seems that about 20 to 25 per cent of patients with essential hypertension have mildly elevated levels. This incidence may be somewhat higher in young, nonobese, white patients with borderline or early established hypertension. The levels are not as high as in patients with pheochromocytoma and may be suppressed by pharmacological agents that directly stimulate central α-receptors, in contrast to failure to suppress in patients with catecholamine-producing tumors.

Neural reflexes

Some evidence of abnormalities of the sympathetic nervous system in hypertension patients has been provided by studying cardiovascular reflex responses.

As indicated previously, the sympathetic nervous system plays a major role in hemodynamic adjustments during upright posture and arterial pressure varies normally ±10 mm. Hg (mean arterial pressure) during 5 minutes of a 50-degree head-up tilt. Most hypertensive patients respond in a normal fashion to this stimulus, but there are some in whom orthostatic hypotension develops, while in others pressure rises abnormally (orthostatic hypertension). In one study of these responses, the patients with orthostatic hypertension had a much greater increase in total peripheral resistance than the others, suggesting that the exaggerated rise in arterial pressure resulted from either increased sympathetic vasomotor outflow or an exaggerated response to normal outflow.

Further evidence for increased neural participation in these patients with orthostatic hypertension was a greater overshoot of diastolic pressure following the Valsalva maneuver. Although this is a qualitative evaluation, it indicates that neural reflex responses are not identical in all hypertensive patients. Further in the study just described, patients in the three groups had different clinical characteristics. Thus, patients with orthostatic hypertension had the lowest arterial pressures, the least severe vascular disease, and the best responses to antihypertensive drugs, most of which suppressed sympathetic adrenergic activity. In contrast, those with orthostatic hypotension, had highest diastolic pressure, most severe vascular disease, and poorest responses to treatment. The patients with normal pressure responses to tilt had intermediate characteristics. At least in these patients, it would seem that exaggerated neural circulatory control was a factor in those with the mildest hypertension, while in those with the severest vascular disease other factors were operating.

As far as centrally mediated neural stimuli are concerned, it has been found that patients with hypertension respond to mental arithmetic and other stressful experiences with exaggerated rises in arterial pressure because of increases in TPR. Exaggerated responses similar to these were found in young patients with borderline hypertension as well as in the normotensive offspring of parents with essential hypertension. Again, however, it should be pointed out that these experiments do not differentiate between increased sympathetic outflow from central vasomotor centers and increased vascular responsiveness to normal outflow.

PHYSIOLOGICAL RELATIONSHIPS OF SYMPATHETIC VASOMOTOR ACTIVITY. Considering the extent of neural control of metabolic functions and the vascular system, it is not surprising to find quantitative relationships in hypertensive patients between estimates of sympathetic vasomotor activity and other systems that influence arterial pressure.

Since pressure within the vascular system can be looked upon as being partly determined by the capacity of the system and the volume it contains and since the former can, in large measure, be controlled by sympathetic activity, it might be expected that intravascular volume would be inversely related to sympathetic activity. The finding of a negative correlation between daily dietary sodium intake and plasma catecholamine levels in patients with essential hypertension lends further credence to this concept. Moreover, using the fall in arterial pressure produced by the ganglion blocking agent, trimethaphan, as an index of vasomotor tone, it has been found in hypertensive patients that, generally speaking, the smaller the plasma volume the greater the depressor effect of trimethaphan. At the present time, there is no way of knowing whether intravascular volume is reduced because increased sympathetic tone has diminished vascular capacity or whether increased sympathetic vasomotor outflow is a compensatory response to a reduced plasma volume.

Renin release is influenced by (among other factors) neural stimulation, and presumably this is one explanation for increased PRA during upright posture. Because of individual variations in height and natural exercise vigor, the usual procedure of 4 hours of standing and walking is certainly not a standard stimulus, and the degree of sympathetic stimulation so produced is impossible to quantify. Moreover, myocardial contractility is also partly determined by neurogenic activity, and therefore certain hemodynamic functions may be used for evaluating neural and renal pressor relationships. In patients with essential, renal parenchymal disease and renovascular hypertension, left ventricular ejection rate and cardiac index were found to be directly related to PRA, suggesting a neural influence on circulating renin. There is also a large body of evidence from animal experiments to indicate that angiotensin II increases sympathetic vasomotor activity by affecting central vasomotor outflow directly and catecholamine handling by peripheral adrenergic nerves. These factors could be interpreted as providing evidence for a neural effect of angiotensin rather than for a common sympathetic stimulus affecting both myocardial contractility and renin release.

Volume mechanisms

PLASMA VOLUME AND EXTRACELLULAR FLUID. It has been known since 1948 that plasma volume is decreased in some hypertensive patients. At first it was not possible to separate patients into various diagnostic groups so the significance of oligemia was not appreciated. More recently, however, plasma volume characteristics have been described for essential and renovascular hypertension and for pheochromocytoma, primary aldosteronism and renal parenchymal disease. In some groups plasma volume is decreased; in others it is increased; in yet others it is inappropriately normal (Table 3–3).

In essential hypertensive men and women plasma volume is lower than normal and in the men, the higher the arterial pressure, the lower the volume. Similar decreases have been found in patients with stenosing lesions of the renal artery and with pheochromocytoma. Primary aldosteronism is often accompanied by a modest expansion of plasma volume, as is acute glomerulonephritis. In each of these instances, since red cell mass is usually normal, plasma volume is a faithful reflection of the total intravascular volume; if it is diminished so is total blood volume and vice versa. In hypertension associated with renal parenchymal disease, and most obvious in chronic renal failure, fluid retention is a common feature and although plasma volume is often much increased, this has occurred, in part, as compensation for the reduction of red cell mass. Even at that, total blood volume is often increased.

The relationship of plasma to interstitial

TABLE 3–3
Plasma Volume (PV) and Red Cell Mass (RCM) in Normal People and in Hypertensive Patients

	PV (ml./cm. of height) Mean (±S.E.M.)*	RCM (ml./cm. of height) Mean (±S.E.M.)*
Men		
Normal	18.4 (0.33)	12.0 (0.22)
Essential hypertension		
Diastolic BP < 105 mm. Hg	18.3 (0.54)	13.0(0.52)
Diastolic BP > 105 mm. Hg	16.0 (0.37)	11.9(0.44)
Renal hypertension		
Arterial stenosis	17.6 (0.82)	12.2(0.46)
Parenchymal disease	18.2 (0.56)	13.0 (0.90)
Primary aldosteronism†	19.8	11.8
Women		
Normal	15.3 (0.25)	8.5 (0.16)
Essential hypertension	14.1 (0.19)	8.3 (0.34)
Renal hypertension		
Arterial stenosis	14.3 (0.39)	7.9 (0.33)
Parenchymal disease	14.7 (0.29)	8.2 (0.64)
Primary aldosteronism†	15.3	9.1

* S.E.M., Standard error of mean.
† Numbers too small for calculation of S.E.M.

fluid volumes in these various hypertensions has recently been explored, and this helps in understanding the mechanisms of plasma volume changes. In one study of essential hypertensive men, extracellular fluid was found to be normal, but since plasma volume was decreased, the ratio of plasma to interstitial fluid volume was also decreased from a normal value of 0.223 to 0.194. This information suggests that in this particular group of hypertensive patients, factors controlling the distribution of fluids in the ECF compartment were abnormal, whereas those controlling total volume were not. One interpretation would be that capillary hydrostatic pressure is slightly elevated, thereby causing a small portion of plasma water normally kept in the intravascular compartment to be translocated to the interstitial space. In support of this interpretation is the common experience that plasma volume expansion is a frequent accompaniment of antihypertensive drug treatment (diuretics and β-blockers excepted) which can occur without weight gain and suggests a transfer of interstitial fluid to the intravascular compartment.

It must be recalled that intravascular volume is not only determined by renal function, the renin-angiotensin-aldosterone system, and other steroidal hormones. There is growing evidence supporting the existence of a *natriuretic factor(s)* (i.e., the so-called third factor) produced either in the brain or in other sites such as the left atrium of the heart. These factors presumably act directly on the proximal tubule of the kidney to produce natriuresis. As originally proposed by Dahl, inherited defects in renal sodium excretion may lead to excessive production of natriuretic factor and attendant increases in vascular resistance by virtue of the effect of this factor on the Na-K-ATPase enzymatic system in the vascular smooth muscle cell. Although this formulation must still be considered hypothetical, it derives considerable support from recent studies demonstrating alterations in cation transport by erythrocytes and leukocytes of patients and experimental animals with genetic forms of hypertension.

PHYSIOLOGICAL RELATIONSHIPS OF PLASMA VOLUME. To consider plasma volume alone provides no insight into its role in hypertension. But to consider it in relationship to other physiological variables such as arterial pressure, cardiac output, circulating renin activity, and neurogenic vasomotor tone allows a glimpse of its contributions to circulatory dynamics. Before proceeding to a discussion of these relationships, two points should be reemphasized. One is that plasma volume is really a reference to total intravascular volume, since the red cell mass is usually normal in hypertension. The other is that since the bulk of intravascular volume is contained in the venous limb, distribution of blood between the systemic and central venous compartments is an important consideration.

DIASTOLIC ARTERIAL PRESSURE. This and TPR are inversely related to plasma volume in men with essential and renovascular hypertension and in patients with pheochromocytoma (Fig. 3–7). Although plasma vol-

FIG. 3–7. Relationship between plasma volume and supine diastolic pressure calculated as weekly average from four daily measurements of 47 essential hypertensive men (r = −0.468, P < 0.001). (Tarazi, R. C., et al.: Arch. Intern. Med. *125*:836, 1970)

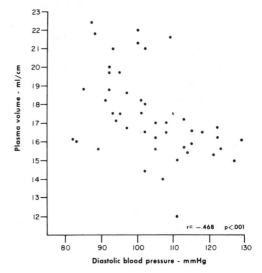

ume is also reduced in women with essential and renovascular hypertension, it does not seem to be correlated with diastolic pressure (perhaps because of variability due to the menstrual cycle). In contrast to this inverse relationship, pressure and volume are directly correlated in patients with chronic renal parenchymal disease. In acute glomerulonephritis, although hypervolemia is characteristic, it has not been found to be directly related to arterial pressure.

CARDIAC OUTPUT. In essential and renovascular hypertension, cardiac output can be normal and sometimes slightly elevated, even in the face of considerable reduction of plasma volume. Since the central blood volume is normal and total blood volume is reduced, a greater proportion must be redistributed to the central circulation from the periphery. Although cardiac output bears no direct relationship to plasma or total blood volume, it is directly correlated with the ratio of central blood volume to total intravascular volume. Since the capacity of the systemic venous compartment seems limited in these hypertensive states it is tempting to ascribe this to increased sympathetic activity, however the evidence is strictly inferential. In contrast, in acute glomerulonephritis cardiac output is directly associated with total blood volume, suggesting that in this congested circulatory state the distribution of blood between the systemic and central venous compartments is normal. This suggestion is compatible with the postulate that this type of hypertension results from failure of peripheral vasodilatation in the presence of a hyperdynamic circulation.

NEURAL MECHANISMS. Since the capacity of the vascular bed is determined by the amount of vasoconstriction in both arterial and systemic venous compartments, and since this can be considered a neurogenic function, it would be anticipated that intravascular volume and sympathetic vaso-

motor tone would be inversely related. Thus, the smaller the blood volume, the greater would be neurogenic activity. This association is difficult to test clinically because, although the measurement of intravascular volume is relatively easy, sympathetic control of the vasculature can only be inferred. It is anticipated that with refinements in techniques for measuring catecholamine levels, clearance, and spillover rates, and using improved clinical pharmacological techniques, more precise means for understanding neural mechanisms will evolve. These techniques will be augmented by more precise classification of patients that takes into consideration the type and severity of hypertension, age, race, sex, medication history, and complications of disease.

PLASMA RENIN ACTIVITY. Also related to intravascular volume is plasma renin activity. Thus, plasma volume and plasma renin activity have been found to be inversely related in normal men, men with essential hypertension, and patients (men and women) with renovascular hypertension (Fig. 3–8). Although renin release is known to be stimulated by hemorrhage, the mechanism whereby intravascular volume and PRA are related in normal and hypertensive individuals has yet to be elucidated. Thus, renin release may be stimulated by adrenergic factors induced by contracted intravascular volume, by less stretch of the juxtaglomerular apparatus in the kidney, by other factors, or a combination of any or all.

CONTROL OF ECF IN HYPERTENSION-EXAGGERATED NATRIURESIS. Hypertensive patients excrete an intravenously administered salt load faster than normotensive individuals. This phenomenon is called exaggerated natriuresis and, as long as GFR is reasonably well maintained, it is directly correlated with the level of pressure. Since normotensives achieve the same rates of urinary sodium excretion as hypertensives

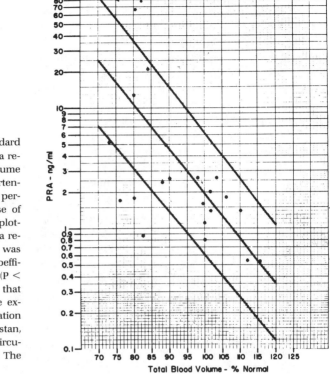

FIG. 3—8. Regression line, (±) 2 standard deviations for the association of plasma renin activity (PRA) with intravascular volume in 21 patients with renal arterial hypertension. Total blood volume, expressed as percentile deviation from normal because of naturally occurring sex differences, is plotted against the logarithm of the plasma renin value because the relationship was found to be exponential. Correlation coefficient for the entire group was −0.584 (P < 0.02). However, when the four values that fell below 2 standard deviations were excluded from the calculation, the correlation coefficient was −0.820 (P < 0.001). (Dustan, H. P., Tarazi, R. C., and Frohlich, E. D.: Circulation 41:560, 1970. By permission of The American Heart Association, Inc.)

after prolonged saline infusion, exaggerated natriuresis can be viewed as a mechanism for protecting the hypertensive from ECF expansion. It would be tempting to relate this to the contracted plasma volume so often found in hypertensives, except that exaggerated natriuresis can also be demonstrated in patients with primary aldosteronism who often have expanded plasma volume.

This facilitated salt excretion cannot be explained by increases in GFR, and since it occurs in the presence of increased aldosterone production, it cannot be related to changes in that hormone. Although a natriuretic hormone, third factor (see above), has been postulated, the available evidence suggests that exaggerated natriuresis has a renal hemodynamic mechanism which may be in part the renal expression of a systemic hemodynamic response to volume expansion. Thus, it has been shown that hypertensive patients have an elevated

renal vein wedge pressure which rises yet higher with saline administration; this is taken to indicate a higher than normal pressure in peritubular capillaries, which would serve to limit sodium reabsorption. In addition, such patients respond to volume expansion with a rise in cardiac output which may reflect hypertensive patients' tendency to maintain a greater proportion of intravascular volume in the central circulation because of restricted capacity of the systemic venous compartment. Relative or absolute deficiency of renin (angiotensin) seems to be another factor because "low-renin" essential hypertensive patients have been found to excrete a saline load faster than patients with high renin.

Renal pressor mechanisms

When it became possible to measure plasma renin activity, it was expected that this would allow definition of a renal component in any hypertension. Generally speaking, at the present time, plasma renin activity has not provided a precise measurement of any mechanism but only indications of one. This is not surprising on two counts. In the first place, knowledge of angiotensin II levels must be available before final conclusions are drawn concerning a renal pressor factor. Also, angiotensin II has a number of physiological effects, in addition to its direct vasoconstrictor action, that could modify arterial pressure. These include both central and peripheral augmentation of sympathetic nervous activity, a role in catecholamine release from the adrenal medulla, and a major regulatory function in aldosterone production.

That there is more than the renal pressor system to any hypertension—even the most fulminant renal type—is shown by

FIG. 3–9. Regression lines and formulas for the relationship of plasma renin activity (PRA) to diastolic arterial pressure in patients with renal arterial stenosis. Line *A* indicates the relationship for all 21 patients of the group; line *B* for 19 patients, excluding the two with the highest values for renin activity. (Dustan, H. P., Tarazi, R. C., and Frohlich, E. D.: Circulation, *41*:563, 1970. By permission of The American Heart Association, Inc.)

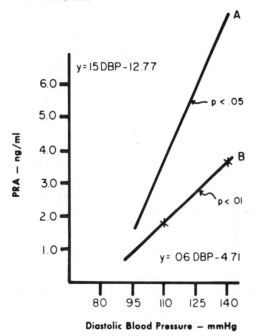

the marked increases in PRA that accompany such normotensive states as hepatic cirrhosis and the nephrotic syndrome.

There are, however, hypertensive states in which PRA is usually elevated, usually normal or usually decreased. In any of these it is not now possible to determine its contribution to the hypertension, although the availability of angiotensin blocking drugs and antirenin antibodies may soon provide a general test.

Peripheral PRA may be elevated in patients with renal arterial stenosis and seems to bear a direct relationship to the height of diastolic arterial pressure (Fig. 3–9). This is important to remember: since these patients in a hospital setting often have mild labile hypertension, the finding of normal PRA does not mean that the hypertension is nonrenal. Because peripheral PRA is inconsistently elevated in renovascular hypertension, measurements in renal venous blood have been advocated. This is based on the likelihood that in unilateral renal arterial stenosis the affected kidney will produce more renin than the unaffected kidney will produce, and that bilateral stenoses, being unequally severe, will have a similar effect. Additionally, comparison of renal venous renin levels with the level found in the inferior vena cava (below the level of the renal veins) will permit detection of the bilateral excessive secretion of renin that characterizes bilateral renal arterial disease. This approach is helpful in diagnosis, but the scope of its value has not yet been completely established. True, in unilateral lesions disparity in renin production usually indicates that the kidney responsible for hypertension will be cured or markedly improved by operative treatment.

At the other end of the spectrum of renal arterial stenosis is the patient with malignant hypertension, very high PRA and secondary aldosteronism as indicated by hypokalemia and normal or slightly depressed serum sodium concentration. These are some of the characteristics of malignant hypertension of any type (except, perhaps, that accompanying adrenal hyperplasia) but they do not eliminate the possibility of renal arterial stenosis or renal parenchymal disease as a cause of hypertension.

In terminally uremic patients maintained by chronic dialysis, measurement of PRA can indicate a renal pressor component. In these patients two types of hypertension exist. One is volume-dependent and can be controlled by hemodialysis, since it occurs only with salt and water excesses; in these patients PRA is not elevated. The other type seems to depend on continued activation of the renal pressor system. These patients have elevated levels of PRA, and their hypertension cannot be controlled by dialysis. This pressor component disappears after bilateral nephrectomy; if hypertension persists, it is volume-dependent and hemodialysis will then provide excellent control of arterial pressure (see Fig. 3–6).

While there is no indication of renal participation in most patients' essential hypertension, there is a group of patients in whom PRA has diagnostic significance in that it is abnormally low. This group is composed of the few patients with primary aldosteronism and a larger number whose hypertension has been termed *low renin essential hypertension*. In both types of hypertensives, the usual stimuli used to increase circulating renin are ineffective. Thus, low-sodium diets, upright posture, and rapid arterial pressure reduction fail to elevate PRA. In both primary aldosteronism and low-renin hypertension of undetermined cause there is no suggestion that the renal pressor system is a factor in the hypertension. In low-renin essential hypertension the indices of aldosterone secretion are normal and hypokalemia is not seen. This contrasts with that which is found in patients with primary hyperaldosteronism. At present, however, it appears that low-renin essential hypertension results, at least in part, from increased ad-

renal sensitivity to angiotensin II. In any event, both groups of patients with low-renin hypertension (i.e., primary hyperaldosteronism and low-renin essential hypertension) usually respond well to diuretic therapy (and/or dietary sodium restriction). It should be emphasized again that those with primary hyperaldosteronism have hypokalemia, and agents that actively conserve potassium are indicated in order to prevent the unwanted effects of hypokalemia. In regard to primary aldosteronism, it seems likely that depressed PRA reflects the modest positive salt and water balance that results from aldosterone excess.

The wide use of oral contraceptive agents has brought attention to another type of hypertension that may depend on the renal pressor system. The estrogen component of these medications increases renin substrate, and although renin itself is not necessarily increased, more angiotensin is formed. Until there are studies of compounds that block the enzymes of the renal pressor system, there is no way of knowing the exact nature of the hypertension that occurs in a few of the women so treated. However, when treatment is discontinued, the components of the renal pressor system return to normal levels and the hypertension disappears.

Hypotension

For the purpose of this discussion, the subject of hypotension seems best presented in its acute and chronic forms. This is in contrast to hypertension, which is primarily a chronic disorder. True, acute hypertension occurs but, with the exception of pheochromocytoma and sudden increase in intracranial pressure (as with intracranial hemorrhage), it is always an intensification of a preexisting disorder. Although there are clinical examples of chronic hypotensions which are certainly disabling to the few patients so afflicted, it is acute hypotension, or shock, that is numerically more significant.

Acute hypotension and shock

Hypotension and *shock* are not synonymous terms. Although shock is traditionally considered in connection with disturbances of arterial pressure, to define it solely in terms of hypotension is not accurate. The basic abnormality in shock is inadequate tissue perfusion, and the seriousness of the clinical condition is determined by the degree of blood flow reduction and the particular needs of the tissues. These considerations, however, do not detract from the importance of close monitoring of arterial pressure levels, since adequate perfusion depends, in part, on an adequate head of pressure to propel blood to the tissues.

In contrast with hypertension, shock is more easily described in terms of its cause than of its underlying mechanisms (see list below). Although the precipitating cause of most clinical hypotension or shock is obvious, we still do not know precisely all the mechanisms that are operative and the factors responsible for eventual irreversibility and death. Recognizing and correcting the cause is essential but defining the pathological mechanisms involved is equally valuable in planning adequate supportive therapy and correcting factors that might interfere with its effectiveness.

The mechanisms underlying shock are multiple. Not all forms have the same hemodynamic patterns and even within the same type of shock patterns differ. Furthermore, the hemodynamic pattern can vary at different stages of shock in the same patient. A multifactorial approach is as relevant to the study of shock as it is to hypertension. Factors operative in shock can be subdivided under similar headings: neurogenic, volume, humoral, and hemodynamic mechanisms. Even when the initial cause can be identified exactly, the mechanisms responsible for shock might be multiple from the very first, as for instance, the slowed circulation and hypotension in cardiogenic shock may result both from depressor coronary reflexes as well as from impaired myocardial contractility. More usually, mechanisms involved in shock grow in complexity with the passage of time as one disturbance leads to secondary effects (e.g., hemorrhage can lead to impaired tissue perfusion with resulting acidosis and accumulation of various metabolites leading to vasodilation and further hypotension). Cardiac function can be impaired in patients with shock from initially noncardiac causes. Conversely, hypovolemia can appear in the course of shock after myocardial infarction and sometimes plays a major role in its persistence. Compensatory mechanisms are also set in action in an effort to compensate for the original disturbance such as intense sympathetic stimulation associated with hypovolemia.

Classification of shock, therefore, is still, for the most part, descriptive, and this has the advantage of focusing attention on the primary cause so that it can be dealt with as rapidly as possible. Because the various factors controlling arterial pressure are as important in hypotension as in hypertension, a discussion of pathophysiology of

---

Pathophysiological Classifications of Clinical Forms of Hypotension or Shock According to Mechanisms and Causes

Causes	Mechanisms
Adrenal insufficiency	Humoral, volume, electrolytic
Anaphyllactic	Histamine, hypersensitivity, hypovolemia
Autonomic insufficiency	Neural
Burns (extensive)	Hypovolemia, toxins?
Cardiac tamponade	Hemodynamic
Faint	Neural
Hemorrhage	Hypovolemia
Myocardial infarction	Hemodynamic; neural
Myxedema	Humoral; +?
Pheochromocytoma	Humoral (catecholamine); volume
Pooling of blood	Hypovolemia
Pulmonary embolus	Hemodynamic; neural
Sepsis	Toxins; +?
Spinal anesthesia	Neural; pharmacological

shock based on the mechanisms that are disturbed seems the most logical way to summarize the available information.

Mechanisms in shock

NEUROGENIC. Relatively pure examples of disturbed neural control of the circulation in shock are the forms complicating spinal cord injury, high levels of spinal anesthesia or excessive doses of adrenergic blocking agents. This type of shock is associated with reduced peripheral arteriolar tone and reduced cardiac output. The latter is produced in part by venous pooling with diminished venous return to the heart and possibly in part by blockade of some cardiac sympathetic fibers in high spinal anesthesia. However, neurogenic disturbances may be involved in other types of shock as well, such as the loss of peripheral arteriolar tone secondary to pain or to initiation of coronary depressor reflexes in myocardial infarction.

As discussed earlier (see Neural Control of the Circulation), the autonomic nervous system controls both arteriolar resistance and distensibility of the venous limb of the circulation. Disturbances in neural control may therefore affect either area with diverse hemodynamic results. Therefore, diminished sympathetic tone can lead to arteriolar dilation and a fall in total peripheral resistance, or at least impaired compensatory responses to a fall in cardiac output. Conversely, excessive sympathetic response to a falling pressure can lead to a high vascular resistance with marked arteriolar constriction; this might help maintain an acceptable level of arterial pressure but at the cost of reduced tissue perfusion, tissue anoxia, increased capillary permeability, and circulatory stasis. Blood volume is usually reduced. On the other hand, the effect of impaired neural control may be more marked on the venous limb rather than on the arteriolar side of the circulation, resulting in loss of venous tone, increased capacitance of veins and hence impairment of venous return, and reduced cardiac output. Blood volume may be normal or even increased in absolute terms; however, it is inappropriate for the increased venous capacity.

These different effects of altered vascular control have been described separately for convenience. One or the other pattern may be prominent in individual instances, but obviously various and changing combinations are more usual than pure forms. It must be noted that arteriolar dilation or loss of venous tone can result from factors other than depressed sympathetic activity, such as sepsis, toxins, or vasodilating metabolites. Nevertheless, the resulting hemodynamic effects are similar.

VOLUME. Hypovolemia can result from hemorrhage, or plasma loss (as in burns, in diffuse peritonitis, or in certain types of anaphylactic reactions), as well as from acute dehydration in patients with severe vomiting, or diarrhea, intestinal obstruction, diabetic acidosis, and Addison's disease. Marked venodilation leading to peripheral translocation of blood volume may produce the same effects as severe hypovolemia, even though blood volume may actually be increased. Hypovolemia may also develop in response to vasopressor agents such as norepinephrine used in the treatment of shock and may then be responsible for diminished response to the medication.

Circulatory failure in hypovolemic shock is due primarily to inadequate filling of the heart, and often the circulation can be restored to normal by increasing the intravascular volume. This is clearly demonstrated in patients with "pure" types of hypovolemic shock, as in the early stages of hemorrhage. The normal response to acute hypovolemia is increased sympathetic tone; hypertension and shock will be intensified by any factor interfering with that response. Thus, in patients with burns or extensive trauma, the circulatory failure caused by the loss of volume may be accentuated by the effect of the products of infection or tissue necrosis on cell metabolism. Acidosis also promotes arteriolar vasodilation and impairs vascular response to sympathetic agents. It has been suggested that impaired myocardial contractility secondary to reduced coronary blood flow in hemorrhagic shock can accentuate further the fall of cardiac output and might be one of the main factors leading to irreversibility.

HEMODYNAMIC AND CARDIAC. The hemodynamic patterns associated with hypotension are as varied as the patterns described for hypertension. In some instances cardiac output is normal, or even slightly elevated, and total peripheral resistance is markedly reduced. More frequently, the low arterial pressure is due to a markedly reduced cardiac output, and total peripheral resistance is normal or even increased. These findings vary with the type of shock, intensity of compensating reactions, incidence of complications, and drug therapy. Further, the hemodynamic pattern is not stable, even in one patient.

Recent studies have demonstrated that shock could be described best as a group of syndromes with varied hemodynamic patterns that unfold with time. Thus, in hemorrhagic shock the initial pattern, low cardiac output and raised TPR, is followed by a progressive fall of resistance in later stages as the vasoconstriction is overcome by vasodilator mechanisms secondary to tissue anoxia. In patients with septic shock, the early hemodynamic disturbance consists of increased cardiac output with a low TPR, possibly secondary to hyperthermia and metabolic demands. Cardiac output may remain normal or elevated until the end unless limited by blood volume deficit or myocardial failure. Simple hypotension after myocardial infarction can be associated with a normal or a low cardiac output, but when clinical signs of shock are evident, output is then most frequently depressed.

The heart plays a central role in the evolution of shock, either as a cause of some types or because of its secondary involvement by intense sympathetic stimulation or by the effects of hypotension and poor tissue perfusion. It may initiate shock through different mechanisms. Rapid tachyarrhythmia, myocardial compression by acute cardiac tamponade, or pump failure secondary to myocardial infarction can lead to marked reduction of cardiac output. On the other hand, a coronary-initiated depressor reflex may lead to hypotension through reduction of arteriolar resistance. Conversely, the heart might suffer in shock which was not initially cardiac, leading to marked changes in hemodynamic and clinical patterns of the disease. When mean aortic pressure falls to hypotensive or shock levels, the coronary vascular bed becomes maximally dilated, and flow is then pressure dependent. Alterations in coronary flow by hypotension can be critical in patients with atherosclerotic vessels, as coronary blood flow through stenotic areas is particularly sensitive to changes in perfusion pressure. Thus, progressive cell injury and tissue necrosis can develop both in cardiogenic shock with extension of the infarction and in noncardiogenic shock; in both conditions, secondary focal necrosis may be extensive.

HUMORAL. Many of the theories advanced at one time or another to explain shock have included suggestions of a "toxic" agent acting to reduce arteriolar tone and injure cellular metabolism. The toxin may be exogenous, as in sepsis, but many endogenous toxins have been reported in late stages of shock from all causes, and they are thought to be responsible for its refractoriness to treatment. It has been suggested that endotoxin can enter the circulation through the intestinal wall rendered permeable by prolonged hypotension; in patients with burns, the skin might be the source of the endotoxin. A myocardial depressant factor (MDF) has been described in blood of animals and man in shock, which is said to be released into circulation as the result of severe pancreatic ischemia and seems to have a direct cardiac depressant effect. In other studies, however, its presence has not been proved.

There is no doubt that a great many vasodilator agents are released with prolonged shock, which, in turn, may enhance further deterioration of the cardiovascular system (see list below). Whether these vasodilating agents are essential for the development of that stage of shock associated with irreversibility, as some have argued, or only a complicating aspect at some unspecified stage has not been clearly defined.

Tissue Factors Implicated in Shock

Acidosis—metabolic or respiratory	Polypeptides: angiotensin, vasoactive intestinal polypeptide
Adrenal steroids	
Altered purine metabolism	Polysaccharides— endogenously released
Catecholamines	Properdin—antibody depletion
Endorphins and encephalins	Prostaglandins
Histamine	Proteolytic enzymes— lysosomes
Iron release (ferritin)	Reticuloendothelial depression
Kinins	Serotonin
Na/K imbalance	

The endogenous release of certain naturally occurring vasoactive biochemical agents may have profound effects on vascular resistance, causing a fall in arterial pressure. Among these agents is bradykinin, a powerful vasodilator. Histamine, another potent vasodilator, affects arterioles primarily, and may act upon capillaries directly or indirectly. The renin-angiotensin system is activated in any circumstance under which arterial pressure falls; this is associated with stimulation of adrenergic mechanisms. The renin-angiotensin system serves to restore arterial pressure through the vasoconstriction by angiotensin II and the sodium- and water-retaining effects of secondarily released aldosterone. Stimulation of the sympathoadrenal system serves to restore arterial pressure through the vasoconstriction produced by released catecholamines and its positive myocardial stimulation. Epinephrine, in low concentrations, is a dilator of skeletal muscle, hepatic, and coronary vessels.

Recent studies in *Escherichia coli* septicemia with endotoxin shock indicate that the specific opiate antagonist naloxone rapidly reversed the hypotension, primarily through its positive inotropic effect on the heart. Since naloxone has minimal cardiovascular effects in the absence of shock, it seems possible that endogenously released β-endorphins could contribute to the myocardial depression. It is also known that β-endorphins and ACTH are released from the pituitary in response to stress. Vasopressin may also play a role in shock, since its release is altered by changes in arterial pressure associated by stimulation of osmolality of blood or of low-pressure mechanoreceptors. As indicated, angiotensin and catecholamines may also be elevated in shock. Prostacyclin (a vasodilator) and thromboxane A, a vasoconstrictor and stimulant of platelet aggregation, may both play important roles in shock.

Humoral factors that, under normal conditions help regulate blood flow distribution locally, may aggravate the vascular disturbances in shock and contribute to its resistance to therapeutic agents. Hypercarbia and local hypoxia cause peripheral vasodilation. It appears that the cardiovascular effects of metabolic or respiratory acidosis arise primarily from the decreased pH and the increased hydrogen ion concentration of blood during hypercarbia, rather than from the alterations in the partial pressure of carbon dioxide per se. Oxygen lack and excess carbon dioxide may also affect the capillaries directly, and prominent capillary dilation may lead to marked depressor responses and possibly loss of plasma volume to the interstitium.

Cellular aspects of shock

Profound cellular and biochemical alterations occur in shock, and these have contributed greatly to our understanding of this complex condition (see list on p. 75, above). Four principal alterations should be considered.

1. Changes in oxygen hemoglobin affinity through hypoxia, acidosis, and accumulation of 2,3-diphosphoglyceric acid (2,3-DPG). Thus, a physiologically induced compensatory increase in oxygen delivery to the tissues occurs. Caution should be used in correcting the acidosis, since this will reduce oxygen delivery to the tissues, potentially dangerous in tissues such as the myocardium, in which oxygen extraction is already at a high level.
2. Intracellular mitochondrial damage occurs predominantly in hemorrhagic and endotoxic shock. Hypoxia per se often does not cause damage unless it is prolonged, since the mitochondria are able to continue to synthesize ATP even in the presence of significant hypoxia.
3. Lysosomal membrane damage permits release of lysosomal enzymes, the serum level of which correlates with the degree of shock. Furthermore, plasma MDF closely parallels lysosomal enzymes in the plasma, and it is thought that this factor is derived from these enzymes. There is also some evidence that endotoxins may cause the formation of a lysosomal-releasing factor through activation of the alternate as well as the classic complement pathway.
4. Activation of the complement system which occurs in certain types of shock can certainly cause cellular damage

through impairment of membrane function. Thus, the mechanism of cellular damage in shock is complex and may also involve kinins, histamine, and facilitation of phagocytoses through opsonin-like activity.

SEQUENTIAL PATTERNS IN VARIOUS CAUSES OF SHOCK. To describe the different types of shock by a simple disturbance such as "hypovolemia" or "normal output-low resistance" or "low output-high resistance" is a simplistic attempt to confine in a static classification a process that is eminently dynamic. The patterns describing a shock state in its initiation might be completely altered in the terminal phases. Factors at work include not only the complications of persistent shock, such as the vasodilator effect of accumulating metabolites secondary to tissue anoxia, but also compensatory mechanisms which can be identified during the progress of the disease. These include increased sympathetic activity with cardiovascular stimulation, increased heart rate and rise in peripheral resistance, increased cardiac output from hyperthermia, and increased central venous pressure behind a failing heart. Various hemodynamic signs of circulatory breakdown may occur in the terminal stages of the several etiological categories and include: a paradoxical fall in peripheral resistance despite a low cardiac output; development of cardiac failure because of poor coronary perfusion; redistribution of blood volume with peripheral pooling or increased capillary filtration; and extravascular loss of fluid. Opening of arteriovenous shunts in the systemic as well as in the pulmonary circulation can lead to poor tissue perfusion and poor oxygenation, despite a high cardiac output and increased velocity of the circulation.

Precise identification of the mechanisms at work depends on careful determination of physiologic indices including both hemodynamic measurements and indices of adequacy of tissue perfusion as blood lactate, arteriovenous oxygen difference, and arterial pH. Thus can therapy be adjusted to the disturbance present in each circumstance rather than being a shotgun reflex approach to isolated signs.

Chronic hypotension

The normal values for arterial pressure cover a wide range. The brain can normally be adequately perfused even at systolic pressure of 60 mm. Hg. Symptoms of fatigue and weakness in patients with systolic pressure even as low as 80 mm. Hg should not be attributed to hypotension unless associated with signs of adrenal insufficiency, impaired venous return resulting in reduced cardiac output, and autonomic deficiency.

Nevertheless, chronic hypotension may be defined as a persistent reduction in arterial pressure, making it inadequate to meet the tissue perfusion demands. It is not surprising, in view of the myriad of factors that operate to control arterial pressure, that there are many underlying mechanisms that could account for this disorder. In this condition, hypotension may be present in both the supine and upright postures, but orthostasis (upright posture) invariably exacerbates both the symptoms (fatigue, light-headedness, and sometimes syncope) and the fall in blood pressure. Two broad categories can be considered as underlying mechanisms for chronic hypotension: hemodynamic and neural.

With respect to hemodynamic factors, the fundamental abnormalities in this group of disorders is a reduction in effective circulating blood volume that leads to a fall in cardiac output. That intact reflexive responses are present can be demonstrated by the appropriate increase in heart rate to the upright posture. In addition, total peripheral resistance should also increase; but in patients with chronic hypotension this is inadequate to maintain the arterial pressure in the presence of reduced systemic flow. Hypovolemia from

any cause produces chronic hypotension. In addition, obstruction to venous return to the heart will produce a similar and sometimes more dramatic picture, occasionally even with a sudden loss of consciousness.

The second mechanism that frequently participates in acute and chronic hypotension involves the nervous system. A vasovagal episode occurs when there is an active dilation of the resistance vessels associated with bradycardia as a result of the cholinergic discharge. In addition to the slow heart rate, there is diaphoresis, nausea, salivation, and increased peristalsis. This acute faint reaction is a common clinical problem. However, a more sinister group of disorders are those attributable to loss of adrenergic function that impairs adaptive cardiovascular responses.

The diseases that involve impaired participation of these mechanisms are discussed below.

ADRENAL INSUFFICIENCY. Chronic hypotension is common in patients with Addison's disease. Although the exact reason for this is not known, hypovolemia certainly plays a role, and hyponatremia cannot be disregarded as having no influence. It has been suggested that the hypotension is related to reduced cardiac performance, and if this is the case it would seem to be the mirror image of the hyperkinetic circulation often found in primary aldosteronism. This is possible because it has been shown that aldosterone has a direct stimulatory effect on myocardial contractility, at least in the perfused rat heart. There is no direct clinical evidence that steroid hormones have a direct effect on arteriolar responsiveness, and sympathetic reflexes appear to be intact. Unfortunately, what evidence is available does not provide an overall view of the circulatory abnormalities that accompany this disorder.

LOW CARDIAC OUTPUT. This certainly, though rarely, causes chronic hypotension, but it deserves emphasis here because it need not present as a life-threatening event such as shock; it is often posturally induced. As indicated, anything that interferes with venous return or impairs ventricular filling can diminish cardiac output so much that hypotension results. This occurs even in the face of marked reflexive increases in sympathetic vasomotor outflow.

Examples of this kind of hypotension are infrequent: women in late pregnancy may faint when lying supine because of inferior vena caval obstruction by the gravid uterus, and patients with atrial myxoma or pedunculated intraatrial thrombus may faint on sitting up or standing. Initially these episodes must be differentiated from vasodepressor syncope of prolonged quiet standing and idiopathic orthostatic hypotension. In the former there is initially a normal response to standing with a modest reduction in cardiac output, a slight increase in heart rate, and enough peripheral reflex vasoconstriction to maintain venous return and arterial pressure. Later, however, there is a sudden fall in TPR, a marked bradycardia, and fainting occurs despite little or no decrease in systemic flow. In contrast, in idiopathic orthostatic hypotension due to autonomic dysfunction there is failure of reflex compensatory response and neither heart rate nor periperal vasoconstriction increases normally.

Other causes of orthostatic hypotension are venodilatation produced by certain drugs, venous stasis associated with extensive varicosities or angiomatous malformations, and, in muscle atrophy, loss of the venous compression in the lower extremities provided by a normal musculature.

IDIOPATHIC ORTHOSTATIC HYPOTENSION—AUTONOMIC DEFICIENCY. Idiopathic orthostatic hypotension is the commonest of the chronic hypotensions. It is caused by a failure of the barostatic mechanism at some point along the reflex arc. Because of this, hemodynamic adjustments to upright posture are not made. Both cardiac output and arterial pressure decrease, often so much that fainting occurs. Function of the baroceptor reflex arc can be interfered with by certain systemic diseases such as syphilis, diabetes melitus, or amyloidosis; by localized central lesions, such as brain tumors or syringomyelia; by normal aging; and by recumbency. The effect of bed rest is of particular importance because it can aggravate orthostatic hypotension from other causes. This explains why orthostatic hypotension produced by antihypertensive sympatholytic drugs is almost always worse in the morning on arising. By the same token, repeated upright tilting or use of a head-up bed may reduce orthostatic hypotension caused by autonomic deficiency.

Diagnosis of idiopathic orthostatic hypotension requires an understanding of the various neurocirculatory reflexes (Table 3–4) and localization of the lesion involves a stepwise reasoned approach. The first step consists of establishing the integrity of the baroceptor reflex arc. This can be accomplished by determining heart rate responses to variations in arterial pressure and, whenever possible, responses of TPR to head-up tilt. If the reflex arc is impaired there will be little or no increase in heart rate with decreases in pressure (i.e., orthostatic hypotension) and impaired cardiac slowing with increases in pressure (i.e., bolus intravenous injections of phenylephrine). Also, there is failure of TPR to rise as the arterial pressure falls on head-up tilt.

Interference with baroceptor reflexes can be due to lesions at any point along the reflex arc or, more rarely, to loss of responsiveness of the effector organ (arteriolar smooth muscle). The commonest are lesions in the efferent sympathetic neurones. These are characterized by loss of pressure responses to all reflex pressor maneuvers; the Valsalva maneuver, cold water immersion, or stressful mental arithmetic will not raise blood pressure as they normally should, since the final common neural

TABLE 3–4
Functional Localization in Autonomic Insufficiency

Test	Response	Note
Integrity of entire baroreflex arc		
Valsalva maneuver ⎱→	*Arterial pressure	→Efferent sympathetic fibers
Head-up tilt ⎰	*Heart rate	→Efferent cardiac sympathetic and parasympathetic
Efferent sympathetic pathway (not dependent on baroceptor reflex)		
Mental arithmetic	↑ BP	Depends on patient's cooperation
Cold pressor test	↑ BP	Depends on pain perception
Deep inspiration	↓ Finger blood flow	Spinal reflex from chest wall
Raise body temperature	Generalized sweating	Hypothalamic (temp. center) stimulation of sympathetic system
Efferent parasympathetic (vagal) fibers		
Atropine injection	↑ HR	Dose-dependent test of vagal influence on the heart
End-organ responsiveness		
Phenylephrine i.v. (α-agonist)	↑ BP	Arteriolar smooth muscle responsiveness
Isoproterenol i.v. (β-agonist)	↑ HR	Responsiveness of β-cardiac receptors

* Both responses depend on entire reflex arc. If heart rate increases (normal response) while arterial pressure falls during head-up tilt (abnormal response) the dissociation helps to localize lesion in efferent sympathetic pathways.

pathway is not functioning. However, arteriolar responsiveness to norepinephrine infusion will be intact or indeed exaggerated (denervation hypersensitivity) in contrast with the impaired responsiveness found in purely arteriolar lesions. Similarly, loss of reflex sweating (warming contralateral limb) despite the presence of responsive sweat glands, is another sign of sympathetic dysfunction. In many but not all cases of efferent limb dysfunction, the cardiac nerves (sympathetic and parasympathetic) are involved; if the heart is functionally denervated its rate will be fixed, unresponsive to atropine injections, carotid sinus pressure, or increase in arterial pressure by phenylephrine. If the disease is not far advanced or in hypotension due to localized lesions (e.g., lumbar sympathectomy) demonstration of cardiac slowing in response to phenylephrine injections or to carotid sinus massage will indicate that afferent nerves, center, and some cardiac nerves are intact.

Pathophysiological localization of central lesions is more difficult; reflex pressor responses are also interfered with. Some help may be derived from the presence of other neural signs (rigidity, parkinsonism, nystagmus, alterations of deep reflexes, etc.) but specially from demonstration of intact peripheral sympathetic innervation. This is most readily achieved by showing increased toe or finger blood flow with local anesthesia of the corresponding nerves. Lesion of afferent baroceptor nerves may be suspected from the association of postural hypotension and unchanged heart rate during head-up tilt, with normal responses of efferent nerves to maneuvers not involving the baroceptor reflex. Blood pressure will rise normally in response to cold and mental arithmetic; peripheral blood flow will increase following nerve blockade, and an atropine injection will produce tachycardia. The loss of baroceptor sensitivity can be demonstrated directly by the contrast between response to atropine and the lack of cardiac slowing with increase in blood pressure.

Annotated references

Books

Genest, J., Koiw, E., and Kuchel, O. (eds.): Hypertension: Physiopathology and Treatment, 208 pp. New

York, McGraw-Hill, 1977. (An encyclopedic multi-author volume on all fundamental and clinical aspects of hypertension.)

Guyton, A. C.: Circulatory Physiology. III. Arterial Pressure and Hypertension. Philadelphia, W. B. Saunders, 1980. (A systems approach to an understanding of the control of arterial pressure.)

Kaplan, N. M.: Clinical Hypertension, ed. 3, 454 pp. Baltimore, Williams & Wilkins, 1982. (A current textbook written for all clinically oriented physicians on the multifactorial aspects of the hypertensive diseases.)

Laragh, J. H., Bühler, F. R., and Seldin, D. W. (eds.): Frontiers in Hypertension Research, 628 pp. New York, Springer-Verlag, 1981. (A current text offering the present state of knowledge on current investigative areas relating to hypertension, with a strong emphasis on the role of the renal pressor system.)

Papers

Abboud, F. M., Heistad, D. D., Mark, A. L., and Schmid, P. G.: Reflex control of peripheral circulations. Prog. Cardiovasc. Dis. 18:371–403, 1976. (This is an excellent review with ample references on cardiovascular control mechanisms and their relationship to shock and postural hypotension.)

Abboud, F. M.: The sympathetic nervous system in hypertension: State-of-the-art review. Hypertension 4(2):208–225, 1982. (The most current review of this subject as of the publication of this text.)

Blaustein, M. P.: Sodium ions, calcium ions, blood pressure regulation, and hypertension: A reassessment and a hypothesis. Am. J. Physiol. 232:C165–C173, 1977. (An integrative hypothesis of the role of the postulated humoral substance and how it relates with the active cellular sodium-pump and calcium-related contractile machinery in vascular smooth muscle contraction.)

Dahl, L. K., Knudsen, K. D., and Iwai, J.: Humoral transmission of hypertension: Evidence from parabiosis. Circ. Res. 24(Suppl. I):21–33, 1969. (Concept for the humoral transmission influencing intracellular transport of sodium.)

Davis, J. O., and Freeman, R. H.: Mechanisms regulating renin release. Physiol. Rev. 56:1–56, 1976. (Although not recent, still an important summary of the factors that regulate renin release.)

De Wardener, H. E., and MacGregor, G. A.: Dahl's hypothesis that a saluretic substance may be responsible for a sustained rise in arterial pressure: Its possible role in essential hypertension. Kidney Int. 18:1–9, 1980. (A statement integrating Dahl's hypothesis with the postulated natriuretic third factor.)

Dustan, H. P., and Page, I. H.: Some factors in renal and renoprival hypertension. J. Lab. Clin. Med. 64:948–959, 1964. (Important clinical studies demonstrating renal and renoprival mechanisms in hypertension.)

Esler, M., Jackman, G., Bobik, A., Leonard, P., Kelleher, D., Skews, H., Jennings, G., and Korner, P.: Norepinephrine kinetics in essential hypertension: Defective neuronal uptake of norepinephrine in some patients. Hypertension 3:149–156, 1981. (An attractive concept for adrenergic participation in essential hypertension.)

Folkow, B.: Physiological aspects of primary hypertension. Physiol. Rev. 62:347–504, 1982. (Current state-of-the-art update on the physiological mechanisms underlying abnormal control of arterial pressure.)

Freis, E. D.: Hemodynamics of hypertension. Physiol. Rev. 40:27–54, 1960. (The review of the state of knowledge of hemodynamic mechanisms in hypertension that was the forerunner of current work in this area.)

Frohlich, E. D.: Hemodynamics of hypertension. In Genest, J., Koiw, E., and Kuchel, O., (eds.): Hypertension: Physiopathology and Treatment, pp. 15–49. New York, McGraw-Hill, 1977. (A review of the hemodynamic factors in the experimental and clinical forms of hypertension.)

Frohlich, E. D.: Hemodynamic factors in the pathogenesis and maintenance of hypertension. Fed. Proc. 41:2400–2408, 1982. (A recent review that integrates the pathogenetic and adaptive hemodynamic factors into an overall concept of the development of hypertension.)

Frohlich, E. D., and Messerli, F. H.: Sodium and hypertension. In Papper, S. (ed.): Sodium. Cations of Biologic Significance, Vol. 2, pp. 144–174. Boca Raton, CRC Press, 1982. (A current review of the role of the sodium ion in hypertension.)

Frohlich, E. D.: The heart in hypertension. In Genest, J., Kuchel, O., Hamet, P., and Cantin, M.: Hypertension: Physiopathology and Treatment, 2nd ed. New York, McGraw-Hill Book Company (in press). (A current review of the role of the heart in hypertension—in its development, as an adaptive organ, and in response to therapy.)

Frölich, J. C., Hill, J. R., McGiff, J. C., Needleman, P., and Nies, A. S.: Prostaglandins. Subgroup Report of the Hypertension Task Force. DHEW Publication No. (NIH) 79-162, pp. 1–98. Washington, D.C., U. S. Government Printing Office, 1979. (A state-of-the-art review of current knowledge of the prostaglandins in hypertension.)

Guyton, A. C., Granger, H. J., and Coleman, T. G.: Autoregulation of the total systemic circulation and its relation to control of cardiac output and arterial pressure. Circ. Res. 28(Suppl. I):93–97, 1971. (A classical presentation of the systems analysis postulating the role of the kidney in the pathogenesis of hypertension through the phenomenon of total body autoregulation.)

Kotchen, T. A., and Guthrie, G. P., Jr.: Renin-angiotensin-aldosterone and hypertension. Endocr. Rev., 1(1):78–99, Winter 1980. (A recent, comprehensive, and unbiased review of a subject of great interest and much controversy.)

Laragh, J. H., Ulick, S., Januszewicz, V., Deming, Q. B., Kelly, W. G., and Lieberman, S.: Aldosterone secretion and primary and malignant hypertension. J. Clin. Invest. *39*:1091–1106, 1960. (An important study that demonstrates the role of the renopressor system and aldosterone in malignant hypertension.)

Muirhead, E. E.: Antihypertensive functions of the kidney. Hypertension *2*:444–464, 1980. (A review of the author's long series of studies that have demonstrated the depressor role of the kidney.)

Nosjletti, A., McGiff, J. C., and Colina-Chourio, J.: Interrelations of the renal kallikrein-kinin system and renal prostaglandins in the conscious rat. Circ. Res. *43*:799–807, 1978. (These are important considerations for the conceptual understanding of the interaction of these two modulating systems.)

Philipp, T., Distler, A., and Cordes, I.: Sympathetic nervous system and blood pressure control in essential hypertension. Lancet *2*:959–963, 1978. (Another well-done clinical study supporting the role of adrenergic mechanisms in maintaining arterial pressure in essential hypertension.)

Tarazi, R. C.: Hemodynamic role of extracellular fluid in hypertension. Circ. Res. *38*(Suppl. II):78–83, 1976. (An excellent discussion of volume mechanisms in hypertension.)

Thomas, J. E., Schirger, A., Fealey, R. D., and Sheps, S. D.: Orthostatic hypotension. Mayo Clin. Proc. *56*:117–125, 1981. (An excellent review of orthostatic hypotension.)

Tosteson, D. C., Adragna, N., Bize, I., Solomon, H., and Canessa, M.: Membranes, ions, and hypertension. Clin. Sci. *61*:5s–10s, 1981. (A review of the evidence demonstrating a red blood cell membrane defect in sodium transport in hypertension.)

Vertes, V., Cangiano, J. L., Berman, L. B., and Gould, A.: Hypertension in end-stage renal disease. N. Engl. J. Med. *280*:978–981, 1969. (Describes two types of hypertension accompanying terminal renal parenchymal disease: one "volume-dependent" and controlled by proper dialysis, the other "renin-dependent" and controlled only by bilateral nephrectomy.)

Wagner, H. N., Jr.: Orthostatic hypotension. Bull. Johns Hopkins Hosp. *105*:322–359, 1959. (Excellent and still timely review of the problem.)

Weil, J. V., and Chidsey, C. A.: Plasma volume expansion resulting from interference with adrenergic function in normal man. Circulation *37*:54–61, 1968. (One of the first papers showing that an antihypertensive drug can increase intravascular volume.)

Wurtman, R. J.: Catecholamines. Boston, Little, Brown & Co., 1966. (An excellent discussion of biosynthesis, storage, and metabolism of catecholamines although a number of years old.)

4   Mechanisms of Cardiac Dysrhythmias

Gordon K. Moe, M.D., Ph.D.
Charles Antzelevitch, Ph.D.

All disorders of cardiac rhythm are the result of abnormal impulse generation, abnormal impulse conduction, or both. Much of what is known about arrhythmias (more properly, dysrhythmias, because some abnormalities are quite precisely rhythmic) has been derived from analysis of ectopic rhythms in the clinic, but models of most of these disorders can be reproduced in the laboratory, many of them in healthy tissue. The relationship between pathology and physiology, between clinic and laboratory, is thus closer than in many other disease states. The recent surge of interest in cardiac dysrhythmias, largely the result of the development of intensive care units, has created an area in which the physiologist, the pharmacologist, and the clinician have come together in forums and conferences of great mutual benefit. The development of cellular cardiac physiology, made possible by the invention of the Ling-Gerard microelectrode, has greatly influenced this marriage; clinical cardiologists talk about membrane potentials, ion fluxes, and dV/dt; while physiologists have discovered the Wolff-Parkinson-White (WPW) syndrome, parasystole, and paroxysmal atrial tachycardial (PAT) with block.

This chapter is an attempt to demonstrate how laboratory models may contribute to the understanding of mechanisms of clinically important disorders of cardiac rhythm. It is not intended to be a clinical treatise on the subject. (For clinical interpretations of dysrhythmias, the reader is referred to the texts by Pick and Langendorf, Harrison, and Schamroth.)

Impulse generation

THE NORMAL SITE. Spontaneous rhythmicity, or automaticity, is a property of embryonic heart cells that develops before cardiac innervation is established. Normally restricted to the sinus node in the adult heart, pacemaker activity can also appear in certain specialized cells of the atria, at the upper and lower margins of the A-V node, and in cells of the His-Purkinje system. Except in grossly abnormal ionic environments, automaticity is said not to occur in *myocardial* cells of atria or ventricles. Ectopic impulses that are not coupled at a fixed interval to a preceding discharge are probably the result of true pacemaker activity in the *specialized conduction tissue* of either the atria or the ventricles. Many instances of premature atrial or ventricular contractions can be ascribed to enhanced automaticity in such sites.

Pacemaker activity is characterized by slow diastolic (Phase 4) depolarization (Fig. 4–1B). Immediately following the Phase 3

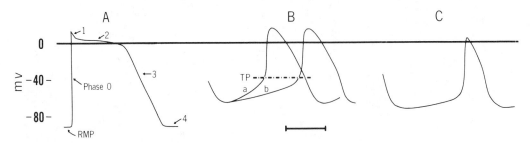

FIG. 4–1. (A) Schematic transmembrane action potential of a ventricular muscle cell, illustrating the various phases as numbered by Hoffman and Cranefield. RMP, the resting membrane potential, is primarily due to the electrochemical potential established by the difference in intracellular versus extracellular potassium concentrations. It depends, of course, on the permeability of the cell membrane to potassium ions. No potential difference could be recorded across a completely impermeable membrane. The horizontal line represents zero potential difference, recorded when the inside and the outside of the cell are at the same voltage. During electrical diastole (Phase 4), the inside of the cell is approximately 90 mV negative with respect to the outside. When this value approaches 60 mV, the cell "fires," an event accompanied by an explosive increase in the permeability of the membrane to sodium ions. This "depolarization" (which is really a reversal of potential) is called Phase 0. The initial brief abortive repolarization is Phase 1, the "plateau" is designated as Phase 2, and the more or less rapid phase of repolarization which restores the resting membrane potential is Phase 3. Except in pacemaker cells, the resting membrane potential (Phase 4) remains constant until the cell is again activated. The duration and configuration of action potentials vary in the several cardiac tissues. The refractory period, in general, lasts from Phase 0 until repolarization is about two thirds complete. In cells of the A-V node, refractoriness lasts beyond full repolarization.

(B) Schematic representation of S-A nodal action potentials. TP, threshold potential; i.e., that membrane voltage at which the cell fires. Note that Phase 4 is not stable, that Phase 0 (depolarization) is much slower than in the muscle cells, and that the membrane potential immediately following Phase 3 is less negative. The action potentials a and b indicate the change in cycle length which would result from a change in the slope of "Phase 4" depolarization. (After Hoffman, B. E., and Cranefield, P. F.: Electrophysiology of the Heart. New York, McGraw-Hill, 1960.)

(C) Action potential recorded from a spontaneously active Purkinje fiber excised from a dog's heart. Note similarity in shape to S-A nodal potentials. In the absence of pacemaker activity, this fiber would exhibit action potentials much more like the schematic one in A.

repolarization of an automatic discharge, the intracellular potential begins to diminish—i.e., it becomes less negative with respect to the external medium. This diastolic depolarization is the result of a change in the balance between inward and outward currents flowing during diastole. When the membrane potential reaches a threshold value (TP, Fig. 4–1B), the cell discharges. It is undoubtedly important that agencies that tend to induce partial depolarization also commonly favor the development of slow oscillations of the membrane potential, which may reach threshold and induce active pacemaker responses. Increased serum K^+ concentration, however, which also reduces the RMP, tends to suppress ectopic pacemakers, probably because increased extracellular K^+ increases the permeability of the membrane to potassium.

The frequency with which a pacemaker fires depends on the slope of Phase 4 depolarization and on the threshold membrane voltage. Modulation of the frequency of the S-A node through cholinergic or adrenergic influences is characterized in the former case by a decrease and in the latter an increase in the slope of Phase 4 depolarization, without a significant change in the threshold at which the spontaneous discharge is triggered. Although both the slope and threshold may be modified by a "direct" action of drugs, many of the cardioactive and antiarrhythmic drugs act, at least in part, through facilitation or inhibition of the effects of the autonomic neurohumors. Digitalis may enhance vagal effects on the S-A node and also antagonize adrenergic effects; the latter effect of digitalis is commonly referred to as its direct or nonvagal influence. Quinidine, procainamide, or lidocaine may increase heart rate by antagonizing cholinergic effects on Phase 4 depolarization. It should be emphasized that a drug that interferes with cholinergic effects cannot cause an acceleration of the heart unless "vagotonic" effects are present, nor can an antiadrenergic agent (e.g., propranolol) cause a decrease of heart rate in the absence of adrenergic influences.

ECTOPIC SITES. An abnormal site of pacemaker activity can be exposed by failure of the normal rhythmicity of the S-A node, by failure of penetration of the normal impulse to a subordinate pacemaker ("entrance block"), or by enhanced activity of an ectopic focus. All three of these mechanisms can be produced in experimental models; no major impairment of normal cardiac physiology is required.

Failure of normal dominance can occur if S-A nodal rhythmicity is severely depressed, as by an episode of enhanced vagotonia. The resultant rhythm, commonly referred to as a vagal escape, may be due to impulse generation in the junction between atrium and A-V node, in the nodal-His junction, or in the His bundle itself. For want of direct evidence on the precise site of origin, such escaped rhythms are referred to as junctional rhythms. There is, however, no compelling evidence that the A-V node in man may not assume a pacemaker role.

The simplest example of ectopic pacemaker function occurring as a result of entrance block is the junctional or His bundle rhythm which occurs in the presence of A-V block. Clearly, without the emergence of a subordinate pacemaker, complete A-V block would be invariably fatal.

Escape rhythms in a "normal" heart are slow—in the range of 30 to 50 beats per minute. However, junctional or idioventricular rhythms may emerge at much faster rates when subordinate pacemaker activity is enhanced. If the ectopic discharge frequency exceeds that of the S-A node, an ectopic tachycardia of atrial, junctional, or ventricular origin may result. Agencies commonly responsible for such rhythms are digitalis, which can enhance subordinate pacemaker activity while depressing normal impulse formation in the S-A node, and adrenergic discharge, which, at least in some conditions, may locally enhance pacemaker activity.

When a subordinate pacemaker is "protected" by entrance block from discharges propagated from the sinus node, it may only occasionally produce an extrasystole recognizable in the electrocardiogram (ECG). A parasystolic rhythm results.

It is of considerable interest that when a tissue (for example, a Purkinje fiber) begins to discharge spontaneously, the contour of its action potential changes remarkably. A pacemaker site in the Purkinje system yields action potentials that resemble those recorded in the normal sinus node (Fig. 4–1C). In spite of this similarity, the behavior of ectopic pacemakers is, at least quantitatively, different from that of the S-A node.

Impulse propagation

NORMAL MECHANISMS. To say that the heart is a syncytium is not to say that the heart is homogeneous; cells specialized for impulse generation connect with cells specialized for conduction, which deliver excitation to cells specialized for contraction. The specialized conduction pathways of the ventricle—the His-Purkinje system—have long been recognized. But only recently has it been functionally demonstrated that specialized conduction pathways, conducting more rapidly than "ordinary" atrial myocardium, connect the S-A and A-V nodes, and the right and left atria. Because these special pathways, anatomically characterized by Thomas James, are more resistant than atrial muscle to the depolarizing effects of hyperkalemia and of digitalis, it is possible for the normal impulse of sinus origin to be transmitted to the A-V node and the ventricles without discernible P waves (sinoventricular rhythm).

Conduction in the atria has an apparent velocity of about 80 cm./sec. On arrival at the A-V node, abrupt deceleration occurs. In isolated preparations of the rabbit A-V node, conduction appears to be continuous, but at a very slow velocity; approximately 50 msec. are required for transmission across a distance of only about 2 mm. (i.e., an average velocity of 4 cm./sec.).

The speed of conduction in excitable tissue is a function of the stimulating efficacy of the action potential and the excitability of the tissue. The stimulating efficacy of the action potential is, in turn, a function of its amplitude (i.e., the total voltage swing from resting potential to peak voltage) and to a lesser extent its rate of rise (dV/dt).

Weidmann showed that the rate of rise is a function of the membrane potential from which the action potential develops. In tissue in which excitation takes place from a high resting potential ("higher" meaning more negative inside) the rate of rise is rapid (approaching 1,000 V/sec. in Purkinje fibers). In a tissue in which the membrane potential, the rate of rise of the action potential, and the excitability are low, the conduction velocity is correspondingly slow. Such is the situation in the A-V node. Conduction velocity through the A-V node, like pacemaker frequency in the S-A node, is enhanced by adrenergic influences and depressed, even to the point of complete block, by cholinergic influences. Electrocardiographically, these effects are expressed respectively as an abbreviation and as a prolongation of the P-R interval. The functional refractory period (RP) of the A-V node is abbreviated by adrenergic and prolonged by cholinergic influences. Here

again, cardioactive drugs may influence A-V conduction, either directly or by altering the effects of the autonomic neurohumors. Digitalis enhances vagal effects and antagonizes adrenergic effects, and, in high enough doses, incomplete or complete A-V block may result. Quinidine, procainamide, atropine, or lidocaine may facilitate A-V transmission and alleviate incomplete block by opposing cholinergic action. It is for this reason that these agents may lead to an undesirable increase in the ventricular rate in cases of supraventricular tachyarrhythmias. Propranolol, by antagonizing adrenergic effects on the A-V node, may enhance the degree of A-V block in atrial tachycardias—an action that can be used therapeutically to reduce the ventricular rate in atrial flutter or fibrillation.

Slowing of propagation within the A-V node may be the result of cell-to-cell depression of conduction velocity, but it may also represent actual but temporary arrest at critical junctions. A slowly rising action potential proximal to a junction may lead, in distal elements, to a "local response" which only slowly reaches the firing threshold. With sufficient depression of conductivity, complete A-V block may occur. Although the terminology may not be quite accurate, depression of this magnitude is commonly referred to as *decremental* conduction.

Upon reaching the His bundle, the normal cardiac impulse accelerates to velocities which may exceed 2 m./sec., only to decelerate again when the ventricular myocardium is reached. Of the total conduction time represented by the P-R interval (say 160 msec.), approximately 40 msec. are accountable to intra-atrial conduction time, about 80 msec. to transnodal conduction, and about 40 msec. to conduction within the His-Purkinje system of the ventricle. In the human heart, estimation of these subdivisions of the P-R interval can be made from His bundle electrograms recorded by intracardiac electrode probes, as first demonstrated by Giraud and co-workers; a considerable improvement in the interpretation of complex arrhythmias has resulted from this approach.

SUBNORMAL AND SUPERNORMAL CONDUCTION. Normal impulse propagation takes place, once the sinus node has discharged, in tissue that has fully recovered from its refractory period. Conduction velocity, like excitability, is a function of the degree of recovery of the tissue. However, the situation is complicated. Tissue that has not yet fully repolarized remains closer to its threshold potential (i.e., its excitability is "supernormal"), but its rate of depolarization is limited. Whether conduction is depressed or supernormal during this stage depends on which of these determinants of conduction velocity is predominant. Purkinje fibers normally go through a phase of supernormal excitability before repolarization is complete; supernormal conduction may accompany this phase. During Phase 4 depolarization similar considerations may apply. As the transmembrane potential diminishes, membrane excitability and resistance increase. Accordingly, conduction velocity may be enhanced as a result of incipient pacemaker activity. When depolarization in Phase 4 becomes excessive, the reverse may occur; the decreased amplitude and rate of rise of the action potential may now become the predominant factors. This possibility has, in fact, been emphasized as a possible explanation of entrance block in parasystolic foci by Singer and co-workers.

Supernormal conduction has been demonstrated only in the specialized conduction system of atria and ventricles; examples of so-called supernormal A-V conduction have been explained by other mechanisms by Moe, Childers, and Merideth.

One of the several possible mechanisms of supernormal A-V conduction is illustrated in Figure 4–2. The ECG was recorded from a patient with A-V block, with an atrial rate of about 83 and an idioventricular rate of 30 beats per minute. Beats of sinus origin were propagated to the ventricles if they occurred about 0.6 sec. after an

FIG. 4–2. An example of so-called supernormal A-V conduction. In the electrocardiogram the third and 11th sinus beats (*P*) are propagated to the ventricles during a period of nearly complete A-V block. The A_3V_3 response represents the "supernormal" event. Other ventricular complexes are of junctional origin; the first QRS complex may be a "fusion" beat.

In the diagram, the horizontal broken line indicates a level of vagal activity above which A-V conduction is impossible. The solid curve represents the reflex vagal activity induced, with a postulated lag of 0.56 sec., by each ventricular beat. Note that A_2, A_6, and A_9 would have been conducted in the absence of idioventricular beats. (EKG from Katz, L. N., and Pick, A.: Clinical Electrocardiography: The Arrhythmias. Philadelphia, Lea & Febiger, 1956. Graph from Moe, G. K., Childers, R. W., and Merideth, J.: An appraisal of "supernormal" A-V conduction. Circulation *38*:5, 1968)

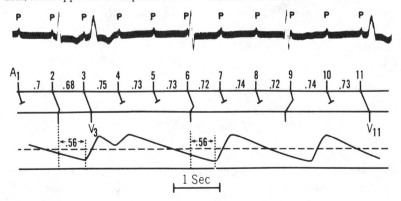

idioventricular QRS complex, or after 1.6 sec. or longer, but not during the intervening period of 1 sec. The earlier propagated responses were ascribed to supernormal A-V conduction. The diagram interprets the intermittent conduction in terms of the phasic vagal discharge resulting from stimulation of the arterial baroreceptors, assuming that the lag between a QRS complex and the effects of the subsequent vagal surge is about 0.6 sec. According to this schema, cholinergic effects on the A-V node would be minimal at 0.5–0.6 sec. after a ventricular discharge. Propagation from atrium to ventricle would be particularly likely to succeed at that moment, while later atrial responses would be blocked.

In general, a response initiated very early in the recovery phase (Phase 3) of an action potential is propagated at subnormal speed. Impulses that encounter tissue in its relatively refractory period will be delayed in transit, and may, if they encounter tissue sufficiently refractory, be altogether suppressed. The phenomena of propagation in relatively refractory tissue have been extensively studied, and numerous models of the generation and maintenance of cardiac arrhythmias have been developed through such studies. Refractoriness can, for example, be used to demonstrate the phenomenon of *unidirectional block*, which may be shown by the following example. If a region, A, has a refractory period intrinsically shorter than an adjacent region, B, then propagation of an early impulse from A may be blocked at the junction; the earliest impulse that could be generated in B would, however, successfully cross the junction in the opposite direction.

Block and reentry

Many clinical examples of ectopic impulse generation are undoubtedly the result of "true" pacemaker activity, but the possibility of reentry or circus movement excitation has been postulated for many years

and is easily demonstrable in the laboratory. Reentry demands the following conditions: (1) unidirectional block at some junctional site; (2) slow propagation over a parallel route in the cardiac syncytium; therefore (3) delayed excitation of the tissue beyond the blocked junction; and (4) reexcitation of the tissue proximal to the block. The time for conduction over the parallel route and for excitation of the distal elements must exceed the refractory period (RP) of the tissue proximal to the block.

The primary event, the *sine qua non* of reentry, is the existence of block. An impulse that sweeps in an orderly fashion through the entire cardiac syncytium cannot, by definition, reenter. Furthermore, the long RP of cardiac cells operates to prevent reentry. Nevertheless, even normal cardiac tissue can be tricked into a situation that permits reentry. We may proceed to describe some experimentally induced reentrant rhythms which almost certainly have their clinical counterparts.

Reciprocal rhythm

Reentry has been proposed as the mechanism of closely coupled premature beats, or of beats with fixed coupling, for many years. Since their first description it has seemed probable that ectopic beats which are attached at a fixed interval to a previous event must somehow be dependent upon that event; one way in which such dependence could be established is through the mechanism of reentry. Within the atria or ventricles, such a reentrant circuit is difficult to demonstrate, but many clinical and experimental studies have established reentry as the mechanism for reciprocal beats, whether isolated or repetitive.

The earliest observations of reciprocal beats, reported by White in 1915 and 1921, suggested that an impulse originating in the A-V node (a *junctional beat*, in modern parlance) could return to the atria over one portion of the node, cause retrograde excitation of the atria ("negative" or "inverted" P wave), and return via an alternate pathway through the node to the ventricles. This combination of events, namely, two QRS complexes of supraventricular configuration bracketing an inverted P wave, has been frequently described. As in many such situations, clinical observation and analysis preceded laboratory investigation. Scherf and Shookhoff demonstrated a similar event experimentally in 1926, and their observations were later confirmed and extended by Moe and co-workers in 1956, and by Rosenblueth in 1958. It was not until recently when, through the use of intracardiac stimulation and recording, confirmation of the animal experiments was achieved by Schuilenberg and Durrer in human subjects that the mechanism of reciprocal beats was accepted by cardiologists in general.

Basically, reciprocal beats ("echoes" to the physiologists) are an illustration of the fundamental conditions for reentry that have long been accepted by clinician and physiologist alike. These conditions are illustrated in Figure 4–3. An impulse initiated in the ventricle (or His bundle) enters

FIG. 4–3. (*Top*) Schematic representation of longitudinal dissociation in the A-V mode. In the sketch at the left it is assumed that the upper portion of the node is functionally dissociated. At the time of arrival of a "junctional" beat, the pathway at the left (α) can support retrograde transmission, but that on the right (β) is blocked. The impulse returns to the atrium, enters the β pathway from above, and returns to the site of block. If the time taken for the passage through α, atrium, and β exceeds the refractory period of the lower node and His bundle (final common pathway, FCP) a reciprocal beat, or "echo," will discharge the ventricles.

In the diagram at the right, the time course of this event is illustrated. A premature beat initiated in the His bundle (H_2) traverses the FCP, leaving it refractory for the time indicated by the stippled area. If propagation back to the atrium over α and down to the FCP over β occurred at full speed (*dotted lines*), no echo would appear. Slow transmission over one or both of the upper nodal pathways (*solid lines*) permits the reciprocal response to clear the R-P of the lower node, leading to an echo (H^*). (Moe, G. K., and Mendez, C.: The physiologic basis of reciprocal rhythm. Prog. Cardiovasc. Dis. 8:461, 1966.)

The diagram illustrates unidirectional block exposed by a premature beat which arrives in the node before complete recovery from a prior response (i.e., while the node is relatively refractory). It is evident that any agency which causes unequal depression of two pathways in the node could also lead to reciprocal beats.

(*Bottom*) An example of reciprocation in a case of junctional rhythm. The first three QRS complexes are followed by retrograde P waves, conducted at increasing R-P intervals (retrograde Wenckebach phenomenon). When the R-P interval reaches 0.28 sec., a reciprocal beat occurs. (After Katz, L. N., and Pick, A.: Clinical Electrocardiography. I. Arrhythmias. Philadelphia, Lea & Febiger, 1956.)

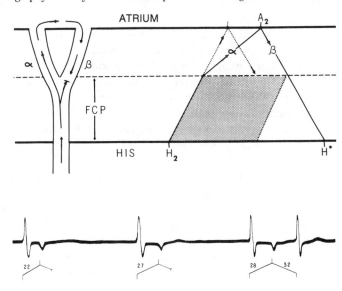

the A-V node at a time when one of the two available pathways in the upper node is still refractory; the impulse is blocked at the refractory junction (Condition 1). Because the alternate or parallel pathway is still partially refractory, retrograde passage through the node is slow (Condition 2). Activation of the atria results in delayed antegrade activation of the tissue above the site of block (Condition 3). Because the total transit time exceeds the RP of the tissue below the site of block, reexcitation of that tissue (final common pathway, or FCP in the figure) occurs, yielding a ventricular echo (Condition 4). This schema does not differ in any important detail from the verbal description in White's case reports; he emphasized that a long R-P interval was necessary to permit recovery from refractoriness of what we have chosen to call the final common pathway. The only difference, and it is not an important one, is that his reciprocal beats occurred "spontaneously" (i.e., unidirectional block or longitudinal dissociation in the node was present as the result of pathology or of medication). In the experimental situation, longitudinal dissociation and unidirectional block occurred as a result of relative refractoriness. An early premature response enters the node during a stage of nonuniform recovery; some elements can conduct while others fail to fire.

If dissociation can occur in the upper node in response to premature beats initiated during the relatively refractory period, it can also occur when nodal conductivity is depressed by other agencies; e.g., digitalis, hypoxia, or ischemic damage. The difference between the clinical situations and the experimental conditions is only a technical one. The important point is that cardiac tissue is not physically homogeneous; depression of function will not be uniformly distributed.

If dissociation can occur in the retrograde direction it should also be demonstrable in the normal direction. Fewer examples of atrial than of ventricular echoes have been described in the clinical literature, but atrial reciprocation has been clearly demonstrated both in the animal and, recently, in the clinical laboratory. It is, in fact, easier to induce atrial echoes than ventricular echoes by intracardiac stimulation in human subjects.

RECIPROCAL TACHYCARDIA. If the events in Figure 4–3 can occur once, they can, given the proper time relations between conduction time and RP, occur repetitively. This would result in a paroxysmal supraventricular tachycardia—a mechanism proposed long ago by Barker and others and described experimentally by Moe, Cohen, and Vick. Bigger and Goldreyer recently reported quite compelling evidence that a number of cases of "junctional" tachycardia are indeed the result of repetitive reciprocation.

Figure 4–4 was assembled from their report. Part A shows the initiation of an episode of tachycardia by a spontaneous atrial premature depolarization. In Part B, a similar paroxysm was induced by an electrically stimulated atrial response; in Part C, a single atrial stimulus terminates the event.

In the cases described by Bigger and Goldreyer, paroxysms of tachycardia were initiated by a strategically timed atrial premature systole, spontaneous or induced, just as in the laboratory models. Similar episodes could also occur as the result of an anatomical anomaly. Suppose that an aberrant communication exists between atria and ventricles. Such a connection need not conduct with equal facility in both A-V and V-A directions. If it regularly conducts in the A-V direction, the WPW configuration of the QRS complex will be manifest. If it fails in the A-V sense, no abnormality will be apparent; retrograde conduction from ventricle to atrium would occur, under "normal" circumstances, at a time when the atria are still refractory. Suppose, now, that A-V conduction is delayed as in Wenckebach periods. If the P-R interval is sufficiently prolonged, retrograde excitation of

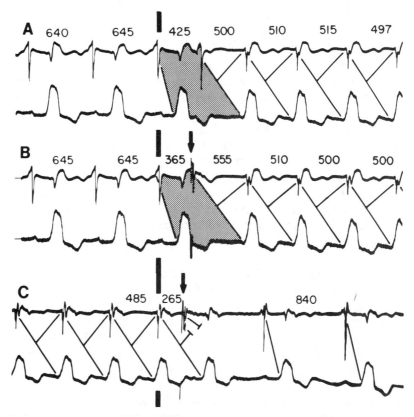

FIG. 4–4. Records obtained from a clinical case of paroxysmal supraventricular tachycardia. This is a composite figure constructed from the series described by Bigger and Goldreyer. The three records, *A, B,* and *C,* were obtained from the same patient and are representative of results recorded in five additional subjects. In each record the upper tracing is an atrial electrogram recorded from an intracardiac electrode probe; the lower record is a body surface ECG. (*A*) A premature atrial systole initiates an episode of reciprocal tachycardia at a rate of about 120. (Numbers indicate cycle lengths in milliseconds.) (*B*) A similar episode is induced by a stimulated atrial premature response (*arrow*). (*C*) An earlier premature response (evoked at the arrow) terminates an episode of tachycardia. (Bigger, J. T., and Goldreyer, B. N.: The mechanism of supraventricular tachycardia. Circulation 42:673, 1970, by permission of The American Heart Association.)

the atria may now be possible (i.e., the atria will have recovered from refractoriness before activation of the aberrant communication is achieved). A retrograde P wave will result, and reentry of the ventricles can now occur over the A-V nodal pathway. This circuit, too, can sustain itself, and much evidence suggests that the patient with intermittent preexcitation is prone to episodes of paroxysmal tachycardia. Thus, in the original report establishing the WPW syndrome as a clinical entity, reference was

made to the common occurrence of supraventricular tachycardias.

The response of a reciprocal rhythm to intervention depends in large measure upon the basic mechanism. If the circuit is completed through the A-V node itself (i.e., down one longitudinally dissociated pathway, up the other) one should expect that vagal stimulation (reflexly induced by carotid sinus stimulation, digitalis, or cholinergic drugs) or any other agency that depresses A-V nodal conductivity should terminate the attack. If one limb of the circuit is through an aberrant A-V communication, vagal intervention would succeed only if complete block of A-V conductivity were achieved. Whatever the communication between atria and ventricles, the introduction of one or more stimuli to the atrium or ventricle (e.g., through a transvenous electrode) should interrupt the circuit and restore normal rhythm. This maneuver, shown to be predictable by Moe and co-workers in 1963, has also been effective clinically (Fig. 4–4). The mode of termination, in fact, supports the diagnosis of the mechanism.

SINOATRIAL RECIPROCATION. Barker and others suggested that some instances of paroxysmal supraventricular tachycardia were due to a self-sustained circuit involving atrial muscle and S-A node. The sinus node, like the A-V node, propagates impulses slowly, and its conductivity can be depressed by vagal stimulation. It is, however, difficult to demonstrate conclusively that reciprocation can occur between atrium and sinus node. Experiments by Han and others strongly suggest this mechanism; and there is evidence that it may also be induced in man. If it does occur in man, then the susceptibility to vagal stimulation can be readily explained.

Atrial flutter: reentry or focus?

We have chosen reciprocal rhythm as the prototype of reentry because it is easy to produce experimentally and because it has been amply confirmed clinically. It is equally easy to produce a circus movement flutter but not so easy to demonstrate that this is indeed a clinical entity.

An atrial tachycardia in the frequency range of flutter (5 or 6 impulses/sec.) could certainly result if an intra-atrial ectopic pacemaker elected to discharge at such a rate. That tachycardias in that frequency range can be produced by local application of alkaloids such as aconitine provides no evidence that ectopic pacemakers are in-

deed the cause of the dysrhythmia; nor does the existence of "continuous" atrial activity in the ECG constitute proof that the mechanism is a circus movement. A more realistic approach is to recognize that both mechanisms may exist and to attempt to define, in each individual case, which of the two (and there can hardly be more than two) applies.

PRODUCTION OF EXPERIMENTAL FLUTTER. Flutter that is clearly the result of a circus movement about an obstacle can be produced in the dog heart, provided that the obstacle is large enough. (The reader is referred to Rosenblueth and Garcia Ramos' classical demonstration of experimental circus movement flutter in the dog heart.) As in the experimental model of reciprocal rhythm, the round trip conduction time must exceed the RP of the tissue in the circuit. This in itself poses no major problem. Thus, if the conduction velocity in atrial muscle is 50 cm./sec. (assuming relative refractoriness), and if the RP of the atrium at the flutter frequency is 150 msec., then the obstacle need be no more than 7.5 cm. in circumference; the resulting frequency will be about 6.6 impulses/sec. The experimental problem is: how can one-way conduction be established around a natural or artificial obstacle of such dimensions?

Practically, this problem is "finessed" by applying stimuli at a frequency grossly exceeding the ability of the atria to follow— namely, in the range of 20 to 50 impulses/ sec. At such stimulus frequencies, atrial fibrillation occurs. Upon termination of the stimuli, chance dictates whether or not flutter will ensue; in the experimental situation, if an obstacle of sufficient size exists naturally, or has been created by crushing atrial tissue, the probability is in the range of 50 per cent for any given period of stimulation.

INITIATION OF ONE-WAY CONDUCTION. For a flutter to be established as a circus movement, the primary condition is unidirectional block. Suppose, as in the original experiments of Mines, we have a slender ring of excitable tissue. It is not difficult to imagine that a single early premature stimulus applied to such a ring would find the tissue on the left refractory, and the tissue on the right excitable. One-way conduction would ensue. If, on the other hand, we cut a hole in the center of an infinite sheet of conducting tissue, it is extremely unlikely that one-way conduction could ever be established. An impulse, initially blocked at the left by refractory tissue, would proceed to the right. Soon after the origin of the impulse, recovery would occur at the initially blocked site. As the impulse progressed it would invade the blocked site and proceed to engage the tissue around the obstacle in both directions; no circuit would be possible. Suppose now that the sheet of excitable tissue is of finite dimensions, and that a relatively narrow isthmus separates one obstacle from another (e.g., the inferior vena cava and the tricuspid orifice). A *single* premature beat might fail to encounter unidirectional block at the isthmus, but the chance is good that the chaotic excitation characteristic of a brief period of atrial fibrillation may result in temporary occlusion of the isthmus. Once unidirectional block is established, and provided that the obstacle is of sufficient size, flutter is the inevitable result (Fig. 4–5). If the isthmus is narrow enough, and if the RP of the tissue within it is long enough, one-way conduction could be established by a single premature beat.

FIG. 4–5. Diagram illustrating how a circus movement could be established around an obstacle. The tissue is assumed to be a closed surface in the form of a square envelope. For visual display of both front and rear surfaces, the envelope has been slit along the edges XX', XY, and YY', and unfolded; the dotted line X'Y' represents one intact seam. Accordingly, the segments XX', YY', and XY on the left are to be regarded as contiguous with the corresponding segments on the right. It is assumed that a premature stimulus is applied at *S* at a time when a band of tissue between the obstacles (indicated by the black bar) has not yet recovered from a prior excitation; this establishes the condition for one-way conduction, which proceeds to envelop both obstacles. At the time indicated by isochron 9, the wave front has nearly completed the circuit around the smaller obstacle, *A*, and by this time the initially refractory band would have long since recovered. As indicated by the "rear view" of the envelope, the remotest parts of the tissue would not yet have been invaded when reentry of the isthmus occurs.

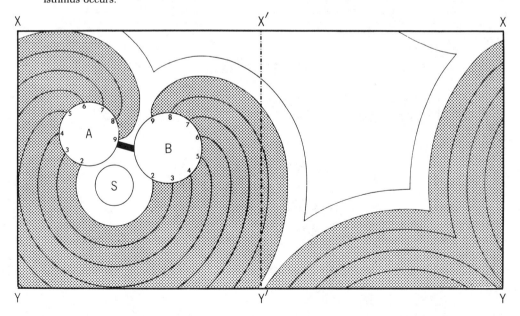

INFLUENCE OF PREFERENTIAL CONDUCTION PATHWAYS. Pastelin et al. recently proposed that the loops formed by the three internodal pathways participate in the experimental flutter induced by the techniques of Rosenblueth and Garcia Ramos, and may even be essential for its maintenance. They demonstrated persistence of flutter in electrograms recorded from these internodal bands at a time when "ordinary" atrial muscle had been rendered inexcitable by a brief infusion of K^+. Also, they were unable to induce flutter after two of three tracts had been severed.

Allessie et al. were able to initiate flutter in small pieces of left atrium excised from rabbit hearts. In their experiments no anatomical obstacles were demonstrable within the flutter circuit; in effect, the refractory period itself provided the barrier, as in the diagram in Figure 4–6. This kind of small circuit is probably more character-istic of unstable local events during fibrillation than of a stable circus movement flutter.

DETERMINANTS OF FLUTTER FREQUENCY. The frequency of a circus movement flutter is determined by the mean conduction velocity of the circulating wave front and the circumference of the obstacle. The conduction velocity, in turn, depends, as in any conducting tissue, upon the stimulating efficacy of the action potential (amplitude and rate of rise) and upon the excitability of the tissue. If the circumference of the obstacle is small, the advancing wave front will encroach upon the relatively refractory wake of the preceding wave, and the conduction velocity will diminish. If the obstacle is large, conduction will proceed at full speed in fully recovered tissue (i.e., about 80 cm./sec.). To support a flutter at a frequency of 300 per minute, full-speed prop-

FIG. 4–6. Schematic representation of a fixed circuit without an anatomic obstacle, generated in a mathematical model of atrial fibrillation. Each of the hexagons represents a "unit" of tissue having six neighbors. The upper numbers in the units indicate the time of firing expressed in time steps (one time step = 5 msec.). Conduction time is assumed to be 15 msec. per unit in relatively refractory tissue. Note that the first unit that would permit reentry is unit 3; it would recover from refractoriness at time 22 and would be reexcited by its neighbor at time 27. The lower numbers in the inner 8 units, which the circuit comprises, represent recovery times, also in time steps, assuming a refractory period (R) of 19 time steps (i.e., 95 msec.). In this model, the refractory period provides the "obstacle." (Stacy, R. W., and Waxman, B. (eds.): Computers in Biomedical Research, Vol. 2 Chap. 9. New York, Academic Press, 1965. Reproduced with permission of the publisher.)

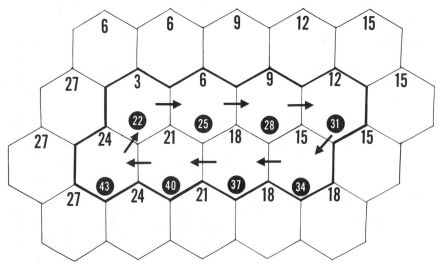

agation would require an obstacle with a perimeter of 16 cm.—a rather large hole for even a dilated human atrium. It is likely, therefore, that the impulse travels at less than full speed through relatively refractory tissue surrounding a smaller orifice.

INFLUENCE OF INCREASED REFRACTORY PERIOD. How do alterations of the RP influence a circus movement flutter? An increase in the duration of the RP must reduce the velocity (and, therefore, the frequency) or terminate the rhythm. The larger the obstacle, the more stable the rhythm, and the less vulnerable it will be to prolongation of the refractory period. In a hypothetical case in which the obstacle is so large that a considerable length of fully recovered tissue lies between the relatively refractory tail of one impulse and the advancing wave front of the next, an increase in the duration of the RP will have no operational effect; the impulse will continue to travel in fully excitable tissue. If, however, the impulse is traveling in relatively refractory tissue (i.e., if the pathway is a "tight fit"), any prolongation of the RP must reduce the circuit speed. Consider now what must happen if we permit conduction velocity to be continuously graded from 80 cm./sec. to zero. This is, of course, unrealistic; but if it were true, an infinite prolongation of the RP would never arrest the arrhythmia; it would merely reduce the frequency to a value approaching zero. There is, however, a lower limit to the conduction velocity that cardiac tissue can sustain repetitively without failure of transmission. The lower limit is, unfortunately, not known; however, a figure of 50 per cent of normal is probably not unrealistic. Accordingly, if we now consider some reasonable parameters for a circus movement, we can estimate whether or not a moderate prolongation of the RP will break the circuit. If the flutter rotates in a tight fit loop, any specific prolongation of RP must slow the wave front, leading to an increase in the cycle length and, in turn, to a further increase of the RP; the larger the initial obstacle, the greater the prolongation of the RP necessary to terminate the arrhythmia. It is possible that some cases of flutter are resistant to quinidine and similar antiarrhythmic agents, not because they are not due to a circus movement, but because the obstacle is large.

Thus far we have not considered specific effects on conduction velocity. Clearly, any agent that prolongs the RP but also depresses conduction velocity will be less effective than an agent that does not depress conduction. This has been admirably demonstrated by Mendez and others in a comparative study of various antihistaminic compounds that have antiarrhythmic properties. One of these, clemizole, prolongs the atrial refractory period without specifically depressing conduction; the "wavelength" of an experimental circus movement is therefore prolonged, and the circuit is promptly interrupted. In a number of clinical cases of flutter, the drug was also effective. As the authors suggest, the agent may provide a tool for distinguishing a circus movement flutter from an ectopic focus.

INFLUENCE OF RP ABBREVIATION. The response of a circus movement flutter to prolongation of the RP is predictable, given logical assumptions, but what should be expected of an agency that abbreviates the atrial RP? A reduction of RP should increase the flutter frequency, provided that the wave front was coursing through relatively refractory tissue. Increased vagal discharge, reflexly or pharmacologically induced, will diminish the *mean* RP of the atria, but it will not do so uniformly. Alessi and co-workers demonstrated that some areas are profoundly affected, whereas neighboring areas exhibit little or no effect. If the circus movement is initially a "tight fit," it will invade some areas of atrial tissue that are just barely able to respond. If the flutter frequency is increased even moderately by increased vagal discharge, these areas will respond irregularly or will be forced to respond 2:1; fractionation of the wave front will occur, and conversion of the flutter to fibrillation will result. This is the textbook picture of the response of flutter to digitalis administration. What happens when digitalis is withdrawn? The situation is closely related to the experimental preparation in which a circus movement flutter can be induced by a brief period of fibrillation. When the digitalis-induced enhancement of vagal tone wanes, the atria may no longer be able to support fibrillation (i.e., the "stimulator" is turned off); chance then dictates, as in the experimental animal, whether flutter resumes, or sinus rhythm supervenes.

It is clear from the above remarks that fibrillation may be converted to flutter and flutter to fibrillation, merely by manipulation of the atrial refractory period. It is not surprising that spontaneous drift from one to the other might occur (flutter-fibrillation—or, more colorfully, "flitter"*) with spontaneous variations of vagal tone. Does it follow, then, that these dysrhythmias are manifestations of the same disorder?

There have been conflicting views of this question for half a century. One view holds that flutter and fibrillation are both due to repetitive discharge from an ectopic pacemaker. When the ectopic focus fires at a frequency that permits regular activation of all areas of the atria, the picture is flutter. If the frequency is so fast that some areas fail to respond to every impulse, fractionation and irregularity of successive wave fronts occur, and the picture will be the chaotic electrical pattern of fibrillation. This mechanism, which can be duplicated in the laboratory by local application of aconitine, is certainly possible. How such an ectopic focus is generated, and how it can be stable enough to sustain an arrhythmia for years, are unanswered questions. The alternate view, that flutter and fibrillation are both examples of circus movement activity, differing only with respect to frequency, is probably not tenable. Flutter can lead to fibrillation, and fibrillation can convert to flutter, but the mechanisms by which the dysrhythmias are sustained are probably different.

Production of fibrillation

Let us first consider in some detail how fibrillation can be produced. Suppose rapidly repetitive stimuli are applied to a site in a sheet of tissue which is perfectly uniform with respect to refractory period duration, excitability, and conduction velocity. Tissue under the stimulating electrodes will respond with a new action potential as soon as excitability has recovered from the RP of a prior response. The new wave front will spread in a smoothly concentric pattern over the whole sheet; the temporal interval between successive responses will be fixed by the minimal RP (determined by the limiting frequency), and the spatial interval (the wavelength) will be determined by the limiting conduction velocity and the frequency. The important point is that an electrical record of such activity would be absolutely regular: fibrillation would not be possible, and activity would cease as soon as the stimulator was turned off.

BEHAVIOR OF A NONUNIFORM MATRIX. Consider now what must happen if the tissue, like any biological system, is *not* absolutely uniform from point to point and from moment to moment. We need consider only variations in the duration of the refractory period, recognizing that nonhomogeneity in architecture, excitability, conduction velocity, ionic environment, and nerve supply must add to the complexity of the conducting matrix. A stimulus applied at a time when this nonhomogeneous matrix is fully excitable (late in electrical diastole) will be propagated in a roughly concentric fashion from the stimulated site. However, the first of a series of premature stimuli, applied at the earliest possible moment after the basic response, will encounter an irregularly ex-

*With apologies to Dr. J. A. Abildskov

citable field. Some immediately surrounding areas will still be refractory and incapable of responding. Other elements, still partially refractory, will conduct the new wave front slowly, and others, more completely recovered, will conduct more rapidly. The premature wave front, in other words, will not be even roughly concentric; it will be irregularly serrated to conform with the retreating edge of refractoriness of the prior response. As the wave front progresses, the initial delay imposed by refractory tissue will permit time for recovery of more distant areas; the wave front will accelerate. Furthermore, areas initially refractory will have recovered, and may be invaded from their distal margins. The possibility of reentry thus exists. To the initial variability (the intrinsic variation in refractory period duration in the nonhomogeneous matrix) will be added an additional variation: those elements that responded promptly to the first premature stimulus will, because of the brief preceding cycle, be subjected to a still further abbreviation of the RP; those that, being refractory, were excited only after a delay will not undergo an equivalent shortening. The initial temporal dispersion will thus be amplified: the system will be, so to speak, divergent. Turbulence will develop near the stimulated site after one or more repetitions of this event, and will spread to include the whole matrix. These concepts have been extensively studied by Moe and co-workers in a computer model; the behavior of the model conforms with the known behavior of fibrillation in dog atria in all respects in which suitable tests have been applied.

SELF-SUSTAINING TURBULENCE. Once turbulence is established in the model, it sustains itself indefinitely. It is not maintained by an ectopic focus (no ectopic impulse generator was incorporated in the program); it is not maintained by a circus movement revolving about an obstacle (no obstacles were built into the matrix). The

arrhythmia sustains itself because multiple small wave fronts course irregularly through the "tissue." These wavelets accelerate in units that have more completely recovered from the refractory state, decelerate in relatively refractory tissue, block at the margins of completely refractory areas, and divide and reunite around refractory islets. The wave fronts will change continuously in number, breadth, velocity, and direction. The process will be sustained indefinitely unless, through intervention or by chance, all wandering wavelets coalesce.

The conditions that determine whether fibrillation will persist or not are: the total mass (or area) and the geometry of the tissue; the mean duration of the refractory period and its range of variation; and the conduction velocity. A large area can support more individual wavelets than a small one; chance coalescence is thus less likely. An unobstructed area will permit greater turbulence than one with many perforations. It can be shown in the model that numerous obstacles separated by less than one wavelength will, if the obstacles are large enough, permit rhythmic activity (flutter) but not fibrillation. Prolongation of the RP will force an increase of the wavelength, thereby reducing the number of wavelets; and if variation of RP duration is eliminated, the initially turbulent activity becomes organized and rhythmic, or may cease altogether.

COMPARISON OF MODEL WITH CLINICAL FIBRILLATION. A table of numbers in a mathematical model is not identical with the dilated atria of mitral stenosis, but the similarities are nevertheless striking. The atria in cardiac insufficiency are dilated (i.e., the surface area is increased); this is, as in a model, an important factor in the maintenance of the arrhythmia. Prolongation of the RP, by quinidine in man, or by manipulation of an equation in the model, results in termination of the disorder. Abbreviation of the RP, especially if irregularly distributed, increases the likelihood of a self-sustained arrhythmia—in man (digitalis), as in the model. Fibrillation, in the presence of suitable obstacles, may convert to flutter, and vice versa—in man and in the model. None of these observations proves the mechanism of fibrillation in man; they do provide an explanation that cannot be excluded.

Drugs that prolong the RP of atrial tissue may be expected to terminate atrial fibrillation, but they need not do so invariably. If the atrial surface area is very large, as in severe mitral stenosis and atrial dilatation of long standing, the chance of conversion is minimized. Larger doses of quinidine, which specifically depress conduction velocity, will not further enhance the probability. Drugs that depress Phase 4 depolarization without prolonging the RP (e.g., propranolol, diphenylhydantoin, and lidocaine) should not be expected to terminate either atrial flutter or atrial fibrillation, unless these are the result of ectopic pacemaker activity. The case for reentrant activity in both of these arrhythmias is supported by the lack of efficacy, in general, of these drugs in supraventricular arrhythmias.

Ventricular premature contractions

Premature beats of the ventricle, unless closely coupled, are commonly regarded as benign manifestations of pacemaker activity. Benignity is not, however, synonymous with simplicity. If these events are in truth the result of pacemaker activity, it is necessary to explain why they escape discharge and resetting by the normal cardiac impulse. If, on the other hand, they result from reentry, it is necessary to consider how reentry may occur within the ventricle.

PARASYSTOLE. When ventricular premature beats interrupt the normal rhythm at various points in the cardiac cycle (i.e., with varying coupling intervals), and when

the interectopic intervals are approximate multiples of a common denominator, presumed to be the cycle length of an ectopic pacemaker, it is assumed that an idioventricular pacemaker fires in a pattern totally independent of the normal rhythm. A necessary condition for the existence of a parasystolic pacemaker is "entrance block." The ectopic generator must be "protected" by an area of unidirectional block that prevents active invasion and discharge but permits escape of the abnormal response.

Total independence of an ectopic pacemaker from the electrical events in the surrounding tissue is unlikely. Although active propagation of a normal impulse into the pacemaker site may be blocked, the fact that a functional exit pathway exists indicates that the pacemaker must undergo subthreshold depolarizations electrotonically induced by the action potentials in the surrounding ventricle.

The nature and amplitude of such electrotonic influences have been demonstrated in a model consisting of a Purkinje fiber pacemaker mounted in a three-chamber bath in which the central chamber, perfused with isotonic sucrose, serves as the area of unidirectional block. Action potentials (or depolarizing current pulses), initiated on one side of the gap, induce partial depolarizations in the pacemaker tissue. When these subthreshold events occur early in the pacemaker cycle, the next spontaneous discharge is delayed; when they fall beyond the midpoint of the cycle, the subsequent discharge is accelerated.

As a result of the electrotonic modulation of the pacemaker cycle, the pacemaker will be entrained; i.e., it will be forced to fire in patterns that depend upon the rhythm of the dominant pacemaker and on the degree of block in the region protecting the ectopic focus. The rules governing the behavior of this interaction have been studied in a computer model, confirmed in biological models, and applied with some success to the analysis of clinical tracings (Fig. 4–7). One important conclusion from such stud-

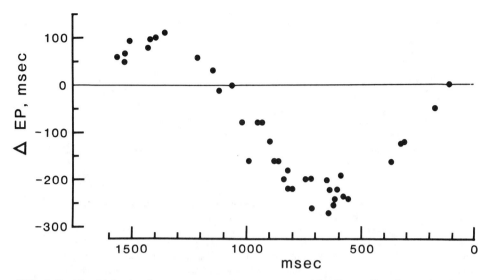

FIG. 4–7. The biphasic phase-response curve reconstructed by "inverse" analysis of a clinical tracing published by Lightfoot (1978) as an example of ventricular parasystole. Abscissae: the temporal position of a normal (supraventricular) QRS complex in the ectopic cycle, expressed in milliseconds preceding an ectopic discharge. Ordinates: delay (above the horizontal line) and abbreviation of the intrinsic ectopic cycle length as a function of the position of the intervening event. Estimated intrinsic cycle length of the ectopic pacemaker, 2160 msec. The crossover point, from delay to acceleration, occurred at approximately 50 per cent of the ectopic cycle. Those ectopic cycles within which a single discharge of sinus origin occurred fell between about 1050 and 600 msec. on the abscissal scale. Points earlier and later represent the influence of two or more beats of sinus origin within manifest ectopic events. During the tracing from which this analysis was constructed, the S-A nodal cycle length averaged about 940 msec.

ies is that ectopic rhythms often ascribed to reentry may in fact be the expression of an ectopic pacemaker subjected to a major influence by events in the surrounding tissue. In other words, while parasystole, as defined by the rigid criteria established decades ago, is a relatively uncommon phenomenon, ectopic pacemaker activity, *modulated* by impulses of supraventricular origin, is much more common.

REFLECTION: A FORM OF REENTRY. When the degree of entrance block is rather low, and whether or not the tissue in the ectopic site is a functionally active pacemaker, the depolarization electrotonically transmitted across the area of block may reach threshold in the distal site after a delay so long that recovery of excitability has been reached in the proximal tissue. In that event, a "reflected" response will occur as a closely coupled premature response (Fig. 4–8). The reflection is a form of reentry, but with no circus movement reentrant loop; there is, instead, a to-and-fro passage of the same impulse across the same area of depressed conductivity. In the models of reflection, active and continuous propagation across the blocked area does not occur. An impulse is arrested at one barrier, after which delayed activation occurs beyond the gap; the delayed distal response reverses the process and, after a further delay, the reexcitation of the proximal tissue occurs. In this situation, just as in the generation of circus movement activity described above, the primary condition is block. Neither a circus loop reexcitation nor a reflected reentry can occur without a region of impaired conductivity.

CIRCUS MOVEMENT REENTRY. Reflection, as described above, is easy to demonstrate in isolated strands of cardiac tissue. In fact, a

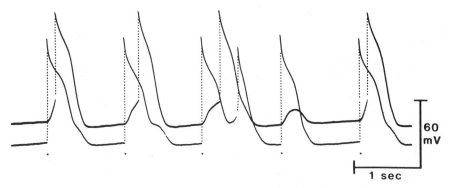

FIG. 4–8. An example of reflected reentry across an area of impaired conductivity. Transmembrane action potentials recorded on either side of a sucrose gap preparation of a strand of canine Purkinje fibers. The proximal end of the fiber (*lower trace*) was stimulated at a cycle length of 1200 msec. Excitation of the distal segment (*upper trace*) occurs after increasing delays, in a Wenckebach sequence, until the third response, delayed by 250 msec., results in reexcitation of the proximal tissue: a reflected reentry. Following the reflection, proximal-to-distal excitation fails, whereupon the sequence is repeated. This sequence illustrates a number of important physiological phenomena: (1) the frequency-dependence of electrotonic propagation across the narrow zone of block (at a slower driving frequency the distal responses would not undergo progressive delays); (2) conduction is impaired long after full repolarization of cells; (3) electrotonic interactions occur on both sides of the gap; prolongation of Phase 3 of the action potential in the proximal cell is apparent in the first and second responses; and (4) while Phase 4 depolarization at a very slow rate was present in both proximal and distal segments, significant pacemaker activity appears not to be essential for reflection.

model of parasystole can be converted to a model of reflection by simply shunting the high impedance of a sucrose gap through an external resistor. It is somewhat more difficult to create a circus movement loop in which reentry can be demonstrated, but it has been accomplished in small loops of Purkinje tissue and must be considered as a possible mechanism of ventricular ectopic activity.

We have already indicated that four conditions must be met before reentry can occur. The prime condition is unidirectional block. The other conditions can be readily accounted for in reciprocal rhythm, but it is difficult to imagine that a sufficiently long loop could account for even a closely coupled premature beat in the ventricles. Sasyniuk and Mendez have removed a major stumbling block. In their experiments, unidirectional block in an isolated Purkinje-muscle preparation was achieved by introducing premature stimuli during the relatively refractory period. Early prema-

ture responses initiated in the Purkinje fibers were blocked at some, but not all, Purkinje-muscle junctions. The important feature of this event is that the action potential of tissue just proximal to a site of block is remarkably abbreviated. This abbreviation, which is the result of an electrotonic current flow between the last elements to fire (terminal Purkinje fibers) and the first elements that fail (underlying muscle), is accompanied by an equivalent abbreviation of the RP. Slow propagation over parallel junctions, and slow propagation through the muscular syncytium, can then result in reexcitation (reentry of the terminal Purkinje fiber). With suitable time relations, reentry can occur as a closely coupled premature beat. Because of the extraordinarily brief action potential in the blocked fiber, the reentrant circuit need be no more than a few millimeters in length. Reentry has also been demonstrated in small loops of isolated Purkinje tissue in which conduction has been severely depressed (Wit et al.). Here, too, unidirectional block is a prerequisite. Activity of this nature could well occur within an ischemic area.

It is possible for a "silent" reentry to occur (silent in the sense that it does not appear in the body surface ECG); repetition of the circuit can then lead to a complete reentry which may or may not be closely coupled. It follows that the length of the coupling interval does not necessarily differentiate between a reentrant and a pacemaker mechanism. It is also important to recognize that, whether reentrant or pacemaker, unidirectional block provides the essential background.

The use of premature beats to expose "spontaneous" reentry may appear to be artifical. It is not. *Any* agency that results in unidirectional block facilitates reentry. Local or general increase in extracellular K^+ can result in localized block; the localized ischemia of small arterial occlusion can also be responsible. In any isolated clinical situation it may be impossible to define the mechanism of a premature beat, but the laboratory demonstration of reentry can hardly be a phenomenon unique to the experimental preparation, never duplicated in nature.

ECTOPIC ACTIVITY INDUCED BY DIGITALIS. Digitalis intoxication may cause single or multiple ectopic beats or sustained tachycardias, more commonly in the ventricles than in supraventricular structures. These have been attributed to enhancement of pacemaker activity (i.e., accelerated Phase 4 depolarization in the specialized conducting tissues of the heart). Microelectrode recordings obtained from isolated strands of Purkinje tissue by Ferrier and co-workers demonstrate that "true" pacemaker activity is depressed by concentrations of cardiac glycosides that do, nevertheless, induce automatic activity. The last of a series of electrically paced responses in such preparations is followed by a transient depolarization which may or may not reach the

firing threshold for the fiber. If an automatic discharge results, the "ectopic" beat may in turn be followed by another depolarization. Single responses or repetitive trains may then occur. Circumstantial evidence suggests that the depolarizations are not due to a time-dependent decrease of K^+ permeability, but rather to a phasic influx of calcium ions. The phenomenon is of interest because it may represent a general mechanism of impulse generation under other abnormal circumstances as well. It is of particular interest because a train of automatic beats, initiated by a single stimulus, may also be "turned off" by a single stimulus (Fig. 4–9).

It had long been assumed that if a dysrhythmia can be terminated by a single low-intensity shock, it must be reentrant. Unfortunately, this is no longer a definitive diagnostic criterion.

Electrical conversion of cardiac arrhythmias

When high-voltage stimuli are applied to the chest wall to terminate ventricular tachycardia or fibrillation, or to arrest atrial fibrillation or flutter, it is probable that the current density through the heart is high enough to depolarize every excitable cell (including, of course, intrathoracic and intracardiac nerves). All cells will therefore be thrown into the refractory state simultaneously, and any wave front or fronts in existence at that instant will be extinguished.

The astonishingly high success rate of this technique does not provide proof that the mechanism of the susceptible arrhythmias is reentry, but it does suggest that many cases of atrial flutter, atrial fibrillation, and ventricular tachycardia and fibrillation are in fact the result of self-maintaining circuits—single and regular in flutter, multiple and irregularly wandering in fibrillation.

More suggestive than the results of massive electrical stimuli are the effects of brief,

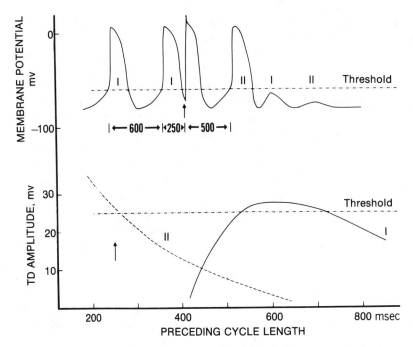

FIG. 4—9. Termination of an ectopic rhythm by a single premature stimulus. This is a simulated series of events based on observations reported by Ferrier et al. (Circ. Res.). When a series of driven responses of a Purkinje fiber is followed by a pause, two or more transient depolarizations (TDs) appear. These are coupled to the last driven response. The first (I) follows at an interval approximately equal to the last preceding cycle length. The second (II) is coupled at approximately 2× the preceding cycle. The amplitude of the depolarizations is also a function of the driven cycle. TD-I reaches a maximum when the preceding cycle is about 600 msec.; the maximum may exceed the firing threshold, in which case a self-sustained ectopic rhythm will develop. This is the situation represented by the simulated action potentials in the upper line. TD-II, which occurs at twice the cycle length, reaches a maximum when the cycle is less than about 250 msec., and may also cross the threshold.

In the sequence shown, a self-sustained "ectopic rhythm" occurs at a spontaneous cycle length of 600 msec.; each spontaneous beat is followed by a suprathreshold "TD-I." When a premature stimulus is applied at an interval of 250 msec., the expected TD-I is abolished, but TD-II (coupled at 500 msec.—i.e., twice the preceding cycle) generates an additional spontaneous response. But when the cycle length is 500 msec., neither TD-I, nor TD-II reaches the threshold. Accordingly, the ectopic pacemaker is "turned off." (Moe, G. K.: Rev. Physiol. Biochem. Pharmacol. 72:55, 1975)

locally applied stimuli at little more than threshold intensity, as in the cases of paroxysmal supraventricular tachycardia described by Bigger and Goldreyer and referred to above.

Some clinical cases of flutter can be "captured" (as they invariably can in the laboratory) by stimulating the atria at a frequency slightly higher than the spontaneous rate. Failure of conversion does not exclude a circus movement mechanism; success strongly supports it.

Drug therapy

In the preceding sections we have described how dysrhythmias may arise as a consequence of abnormalities of conduction and automaticity. Correction of these abnormalities is the aim of drug therapy; the electrophysiological actions of antiarrhythmic agents in isolated tissues have been studied in an attempt to define mechanisms of action and to guide the choice of therapeutic agents. Decades ago, when the circus movement theory was applied to explain almost all disorders of rhythm except conduction block, and when quinidine was the only available effective agent, prolongation of the refractory period was believed to be the one important attribute of an antiarrhythmic drug, but this has been shown not to be universally true.

Table 4–1 lists the effects of therapeutic concentrations of currently employed antiarrhythmic agents on the principal electrophysiological properties of isolated cardiac tissue. In the classification adopted in the table, drugs that decrease membrane responsiveness and prolong the refractory period are designated Class I. They include quinidine, procainamide, and disopyramide. Drugs that have little or no effect on membrane responsiveness and shorten the refractory period are designated as Class II. Propranolol and its congeners, verapamil and other "slow" channel blocking agents, and bretylium fall in neither class. It is generally true that Class I drugs are more effective against atrial arrhythmias (flutter, fibrillation) than Class II drugs, but in the treatment of ventricular arrhythmias the choice of therapeutic agent remains empirical. One drug after the other may be tried until one or a combination of two or more is found to be effective. It is somewhat distressing that after nearly 70 years of research no firmer basis exists for the choice of antiarrhythmic drug therapy.

Membrane responsiveness

Membrane responsiveness is a term used to describe the relationship between the maximal rate of rise of the action potential upstroke (dV/dt_{max}) and the membrane potential at the time of activation. It characterizes the response to stimuli encroaching upon the repolarization phase of the action potential, or falling during Phase 4 depolarization. To the extent that dV/dt_{max} influences impulse conduction, drug-induced changes in membrane responsiveness may alter conduction velocity. In normal tissues, the decrease in membrane responsiveness following exposure to therapeutic levels of drugs such as quinidine or procainamide is attended by only a moderate deceleration of conduction velocity, but in depressed (i.e., partially depolarized) tissues similar concentrations cause a marked depression or block of conduction.

TABLE 4–1
*Effects of Antiarrhythmic Drugs on Electrophysiological Properties of Purkinje Fibers**

	Class I Quinidine Procainamide Disopyramide	Class II Lidocaine Phenytoin	Propranolol	Verapamil	Bretylium
Ectopic automaticity	↓	↓	↓	↓	↑†, →
Membrane responsiveness	↓	↓ , →‡	↓	→	→
Refractory period	↑	↓	↓	→	▲
Conduction velocity (normal tissue)	↓	→	↓	→	→
Conduction velocity (depressed tissue)	▼	▼	▼	▼	↑†, →

*Direction of arrow indicates direction of change; → = no change. Boldface arrows indicate a greater degree of change.
†The result of catecholamine release from sympathetic nerve endings on initial exposure to the drug.
‡Depends on extracellular potassium concentration.

This disparity is even greater with other agents (lidocaine, phenytoin) which do not depress conduction in normal tissue, but severely depress it in tissue in which conduction is already compromised. Depression of conductivity, which plays a major part in the genesis of both reentrant and parasystolic arrhythmias, may, when further depressed, terminate the abnormal rhythms. Circus movement reentry may be aborted as a consequence of the conversion of a unidirectional block to a bidirectional block. Reflected reentry may cease as a result of the development of block in either antegrade or retrograde direction across the common pathway, and a parasystolic pacemaker may be left to fire in total isolation because of the imposition of exit conduction block.

The degree of conduction impairment that permits the manifest expression of dysrhythmias is often critically poised; i.e., a narrow time zone may exist within which a successful reentry can occur, bounded on both sides by zones in which either better or worse conduction prevents the abnormal events. It follows that agents that improve conduction may cause previously silent areas to become overt. It is perhaps not surprising that the treatment of abnormal rhythms is still essentially empirical.

Refractoriness

Refractoriness is still of major significance in the efficacy of antiarrhythmic drugs. Agents that either lengthen or shorten the refractory period may be effective in arresting ectopic activity. Under normal conditions, changes in refractoriness parallel changes in the action potential duration. In pathological states, however, refractoriness (measured as the inverse of excitability) may outlast full repolarization of the action potential by hundreds of milliseconds. The imposition of a narrow zone of block (inexcitability) between two active segments of tissue can cause an apparent increase in the refractory periods of the segments on either side of the block. In pathological states, therefore, prolongation of the functional refractory period may represent a delayed recovery of the conductivity of the system.

Conductivity is not readily dissociable from refractoriness. For example, a prolongation of the relatively refractory period imposes an extension of the time during which conduction is depressed. (For discussion of this interaction and its bearing on circus movement see Moe, G. K.: Rev. Physiol. Biochem. Pharmacol. 72:55, 1975.) In much the same way that depression of conductivity may either suppress or initiate ectopic activity, it is so also with prolongation of refractoriness. For example, prolongation of the A-V nodal refractory period may facilitate the initiation of A-V nodal reciprocation in the presence of an aberrant A-V communication; but prolongation of the refractory period in the Kent bundle may abort such an episode. In general, however, reentrant rhythms, now as in 1914, are suppressed by agents that prolong the refractory period, whether or not the action potential duration is prolonged.

Ectopic automaticity

With the exception of bretylium, all of the agents listed in Table 4–1 are capable of suppressing automaticity of fibers within the His-Purkinje system. Unlike the others, verapamil (a slow channel blocking agent) suppresses pacemaker activity occurring at low levels of membrane potential (e.g., "triggered" activity).

Even when a tachycardia is a reentrant rhythm (reflection or circus movement), the precipitating event may be a premature response arising from an ectopic focus. Accordingly, agents that suppress pacemaker activity may be useful prophylactically, whether or not they can be used to suppress a sustained attack of tachycardia.

Conclusions

Models of many disorders of rhythm can be constructed in "normal" cardiac tissue.

In some cases these are undoubtedly the result of ectopic pacemaker activity.

The characteristics of some experimental arrhythmias strongly point to reentry; the results of drug therapy, discrete intracardiac stimulation, and cardioversion support this interpretation in a variety of clinical situations. The initiation of a tachysystolic rhythm may be due to a single premature impulse; the maintenance of the rhythm need not be due to continuing discharge of the ectopic focus. Accordingly, drugs or other agencies that prevent recurrent attacks may have a mechanism of action different from that of those agencies that terminate the attacks.

Annotated references

General

Cranefield, P.: The Conduction of the Cardiac Impulse. Mt. Kisco, N.Y., Futura, 1975. (A treatment of the physiological basis of arrhythmias with emphasis on the so-called "slow" response.)

Harrison, D. C. (ed.): Cardiac Arrhythmias: A Decade of Progress. Boston, G. K. Hall, 1981. (A collection of essays, basic and clinical.)

Hauswirth, O., and Singh, B. N.: Ionic mechanisms in heart muscle in relation to genesis and the pharmacological control of cardiac arrhythmias. Pharmacol. Rev. *30:*5, 1979. (An excellent and comprehensive review, listing 318 references.)

Pick, A., and Langendorf, R.: Interpretation of Complex Arrhythmias. Philadelphia, Lea & Febiger, 1979. (An atlas of complex disorders, with diagnostic interpretation by the foremost authorities. This book supplements the excellent volume by Katz and Pick published by Lea and Febiger in 1956.)

Schamroth, L.: Disorders of Cardiac Rhythm, ed. 2. Oxford, Blackwell Scientific Publications, 1980.

Zipes, D. P., Bailey, J. C., and Elharrar, V. (eds.): The Slow Inward Current and Cardiac Arrhythmias. The Hague, Martinus Nijhoff Publishers, 1980. (A collection of essays designed to bring the "slow" response up to date.)

Atrial flutter and fibrillation

Alessi, R., Nusynowitz, M., Abildskov, J. A., and Moe, G. K.: Nonuniform distribution of vagal effects on the atrial refractory period. Am. J. Physiol. *194:*406, 1958.

Allessie, M. A., Bonke, F. I. M., and Schopman, F. J. G.: Circus movement in rabbit atrial muscle as a mechanism of tachycardia. Circ. Res. *33:*54, 1973.

Haft, J. I., Kosowsky, B. D., Lau, S. H., Stein, E., and Damato, A. N.: Termination of atrial flutter by rapid electrical pacing of the atrium. Am. J. Cardiol. *20:*239, 1967.

James, T. N.: Am. Heart J., *66:*498, 1963. (Describes the anatomy of the internodal tracts. Special physiological characteristics of these fibers are detailed by Vassalle and Hoffman: Circ. Res. *17:*285, 1965, and by Wagner et al.: Circ. Res. *18:*502, 1966.)

Mendez, R., Kabela, E., Pastelin, G., Martinez Lopez, M., and Sanchez Perez, S.: Antiarrhythmic actions of Clemizole as pharmacological evidence for a circus movement in atrial flutter. Arch. Exp. Pathol. Pharmacol. *262:*325, 1969.

Moe, G. K., Rheinboldt, W. C., and Abildskov, J. A.: A computer model of atrial fibrillation. Am. Heart J. *67:*200, 1964.

Pastelin, G., Mendez, R., and Moe, G. K. Participation of atrial specialized conduction pathways in atrial flutter. Circ. Res. *42:*386, 1978.

Rosenblueth, A., and Garcia Ramos, J.: Studies on flutter and fibrillation. Am. Heart J. *33:*677, 1947. (The classical demonstration of experimental circus movement flutter in the dog heart. A mathematical treatment of impulse transmission in a sheet of excitable tissue with obstacles was also published by Rosenblueth and Norbert Wiener. It is on this treatment that Fig. 4–5 is based.)

Waldo, A. L., MacLean, W. A. H., Karp, R. B., Kouchoukos, N. T., and James, T. N.: Continuous rapid atrial pacing to control recurrent or sustained supraventricular tachycardias following open heart surgery. Circulation *54:*245, 1976.

Automaticity and conduction

Ferrier, G. R.: Digitalis arrhythmias: Role of oscillatory afterpotentials. Prog. Cardiovasc. Dis. *19:*459, 1977. (A selected review of abnormal automaticity induced by digitalis.)

Hoffman, B. F., and Singer, D. H.: Effects of digitalis on electrical activity of cardiac fibers. Prog. Cardiovasc. Dis. *7:*226, 1964. (This review includes an excellent summary of action potential characteristics, including the behavior of normal and ectopic pacemakers.)

Moe, G. K., Childers, R. W., and Merideth, J.: An appraisal of "supernormal" A-V conduction. Circulation *38:*5, 1968.

Sasyniuk, B., and Mendez, C.: A mechanism for reentry in ventricular tissue. Circ. Res. *28:*3, 1971.

Singer, D., Lazzara, R., and Hoffman, B. F.: Interrelationship between automaticity and conduction in Purkinje fibers. Circ. Res. *21:*537, 1967. (It has been suggested that conduction velocity is depressed in the vicinity of pacemaker sites and that the "entrance block" which "protects" a parasystolic focus may be due to local Phase 4 depolarization.)

Weidmann, S.: Effects of calcium ions in local anaesthetics on electrical properties of Purkinje fibres. J. Physiol. *129*:568, 1955. (Discovered the important relationship between membrane potential and rate of rise of the action potential. His studies, together with related observations, are discussed in the somewhat out-of-date but still excellent book by Hoffman and Cranefield: Electrophysiology of the Heart, New York, McGraw-Hill, 1960.)

Wit, A. L., Hoffman, B. F., and Cranefield, P. F.: Slow conduction and reentry in the ventricular conducting system. Circ. Res. *30*:1, 1972.

Parasystole

Since the original description by Kaufmann and Rothberger in 1920, this interesting phenomenon has been described in many clinical studies and case reports. Recent studies in experimental models have contributed to the understanding of the behavioral patterns to be expected of "protected" pacemakers:

Antzelevitch, C., Jalife, J., and Moe, G. K.: Characteristics of reflection as a mechanism of reentrant arrhythmias and its relationship to parasystole. Circulation *61*:182, 1980.

El-Sherif, N., Lazzara, R., Hope, R. R., and Scherlag, B. J.: Re-entrant ventricular arrhythmias in the late myocardial infarction period. Manifest and concealed extrasystolic groupings. Circulation *56*:225, 1977. (One of a series of studies conducted in experimental infarction in dogs.)

Jalife, J., Antzelevitch, C., and Moe, G. K.: The case for modulated parasystole. Pace, 1982 *5*:911, 1982. (A brief view of recent clinical and experimental studies.)

Jalife, J., and Moe, G. K.: A biologic model of parasystole. Am. J. Cardiol. *43*:761, 1979.

Lightfoot, P. R.: Parasystole simulating ventricular bigeminy with Wenckebach-type coupling prolongation. J. Electrocardiol. *14*:385, 1978. (Fig. 4–7 was constructed from the published records in this case.)

Moe, G. K., Jalife, J., Mueller, W. J., and Moe, B. H.: A mathematical model of parasystole and its application to clinical arrhythmias. Circulation *56*:968, 1977.

Reciprocal rhythm

Bigger, J. T., and Goldreyer, B. N.: The mechanism of supraventricular tachycardia. Circulation *42*:673, 1970. (One of the earliest clinical confirmations of A-V nodal reciprocation as a mechanism of paroxysmal supraventricular tachycardia.)

Moe, G. K., and Mendez, C.: The physiologic basis of reciprocal rhythm. Prog. Cardiovasc. Dis. *8*:461, 1966.

White, P. D.: Atrioventricular rhythm following auricular flutter. Arch. Intern. Med. *16*:571, 1915; Bigeminal atrioventricular rhythm. Arch. Intern. Med. *28*:213, 1921. (Two interesting reports on reciprocal rhythm.)

5   Atherogenic Mechanisms

Ellison W. Wittels, M.D.
Antonio M. Gotto, Jr., M.D., D.Phil.

Arteriosclerosis is a general term referring to a variety of changes that affect the arterial wall. It is characterized pathologically by a loss of elasticity and by thickening of the wall. Atherosclerosis is a specific type of arteriosclerosis that affects man. It comes from the Greek "athera," meaning gruel, and "sclerosis," which means hardening. The name is derived from the gruel-like lipid material that is found in the center of the atherosclerotic lesion. The atherosclerotic process was described as early as 1580 B.C. and is now the most common disease affecting Western society. By the time an American man reaches the age of 65, his chances of developing clinical evidence of cardiovascular disease is about 37%. For a woman, there is an 18% chance of developing cardiovascular disease by the age of 65.

Clinically, presentation of the atherosclerotic lesion depends upon its anatomic location and rate of progression. Thus, if there is a gradual narrowing of the lumen that impedes the supply of blood, collateral circulation may develop, depending on the arterial bed affected. Acute occlusion of the diseased artery may result from thrombosis superimposed on a preexisting atherosclerotic plaque. The result may be a myocardial infarction or cerebrovascular accident. In other instances, the atherosclerotic process, in conjunction with hypertension, may lead to weakening of the vessel wall, causing aneurysm formation within the arterial wall. The aneurysm may dissect or rupture. Transient ischemic attacks may result from embolization to a cerebral vessel or from an ulcerated carotid atherosclerotic plaque. Thus, the clinical manifestations of atherosclerosis may include coronary artery disease (CAD), angina pectoris, myocardial infarction, congestive heart failure, transient ischemic attacks, cerebrovascular accidents, peripheral vascular disease, claudication, aneurysm formation, and arterial dissection or rupture. An atherosclerotic lesion of the medium-sized or large muscular arteries may, by its growth, narrow the lumen and decrease the blood supply. The lesion may ulcerate and be the site of hemorrhage and thrombosis.

While the clinical manifestations of atherosclerotic lesions are well documented, the precise details of formation of the atherosclerotic lesion have been difficult to pinpoint. In a monumental review of atherosclerotic lesions, DeBakey emphasized variability in the location and progression of these lesions not only from person to person but also within each individual.

Much recent study has focused on the etiological role of risk factors in atherosclerosis, in the cell biology of the arterial wall, and in the interaction of platelets and lipoproteins from the blood.

Normal artery histology and pathology

The medium-sized and large arteries, where atherosclerotic lesions develop, consist of three layers: intima, media, and adventitia. The intima is bound by a single endothelial cell layer on the luminal side of the vessel wall and is separated from the rest of the arterial wall by the internal elastic membrane. A small number of smooth muscle cells is located between the endothelial cell layer and the internal elastic membrane connective tissue. With increasing age, the number of smooth muscle cells within the intima increases, as does the thickness, until those of the intima and the media are approximately equal. The intimal layer is more prominent in the coronary arteries than in the aorta or other arteries.

The intact endothelial cell wall forms a barrier to the passage of substances from the lumen of the vessel. Two pathways have been identified whereby macromolecules may pass from the lumen to the vessel wall. One is the intracellular junction between endothelial cells. These junctions are 100 to 200 Å in size. A second is by penetration of the endothelial cell through endocytotic vesicles. Molecules of approximately 600 to 1000 Å in diameter are transported through the vesicles.

The medial layer of the arterial wall is bound by the internal elastic membrane on the luminal side and by the adventitia on the other. The media consists of smooth muscle cells surrounded by collagen, elastic fibers, and proteoglycans. It is separated from the adventitia by the external elastic lamina. The adventitia is composed principally of fibroblasts with some smooth muscle cells, collagen, and proteoglycans.

Nourishment of the arterial wall comes from two sources. The intima and approximately the inner one-half of the media are nourished by direct perfusion of metabolites from the lumen. The outer part of the media and the adventitia are nourished by the vaso vasorum, a complex of arterioles and venules. Between these two areas is a vascular area referred to as the *watershed*.

Atherosclerotic process

Atherosclerosis initially involves the intima of the blood vessel, with secondary changes in the media. Several different types of lesions have been recognized: fatty streak, fibrous plaque, complicated lesion, and myointimal hyperplasia.

Fatty streak

The term *fatty streak* describes a pale yellow, flat or slightly raised intimal lesion, in which the predominant abnormality is the accumulation of cholesterol and cholesteryl ester, most of which is intracellular. Foam cells, containing intracellular lipids with vacuolated cytoplasm, are found within the fatty streaks. These cells are derived from either smooth muscle cells or macrophages. The juvenile fatty streak appears independent of lipid levels within a few months after birth, initially in the proximal portion of the aortic arch and later in the remainder of the thoracic aorta and in the abdominal aorta. The extent of aortic intimal involvement increases from roughly 10% to approximately 50% between ages 10 and 25. Fatty streaks accumulate in the coronary arteries 10 to 15 years later than they do in the aorta, and they appear in the cerebral vessels in the third and fourth decades of life. Black children have been described as having more extensive fatty streaks in the aortic intima than do children of other ethnic groups. Fatty streaks are also more extensive in females than in males, which tends to cast doubt as to their significance in the evolution of clinically significant atherosclerosis. A great deal of individual variability exists in the extent of aortic intima with fatty streaks. Juvenile fatty streaks contain principally intracellular lipid within smooth muscle cells and macrophages but little extracellular lipid.

There is significant debate as to whether the juvenile fatty streak is involved in the subsequent development of the atherosclerotic lesion. Aortic atherosclerotic lesions develop in areas where no preceding juvenile fatty streak is found. A better correlation between the fatty streak and the subsequent atherosclerotic lesion exists in the coronary arteries. A "progressive" fatty streak has been described that consists of lesions resembling the juvenile fatty streak. However, this fatty streak frequently contains extracellular lipids as well as necrotic, lipid-filled smooth muscle cells and macrophages. In addition, the progressive fatty streak contains more collagen and elastic fibers than the juvenile fatty streak. Some researchers think that the progressive fatty streak represents a transition from the juvenile fatty streak to the fibrous plaque, a lesion highly characteristic of atherosclerosis.

Fibrous plaques

Fibrous plaques are whitish-to-gray elevated lesions that protrude into the vessel lumen. They have a thick capsule of collagen and elastic tissue along with smooth muscle cells that contain lipid, mostly in the form of cholesterol and cholesteryl ester. Proteoglycans or glycosaminoglycans are also present. The plaques appear 15 to 20 years later than fatty streaks and correlate with the clinical symptoms of atherosclerosis.

Complicated lesion

The complicated atherosclerotic lesion develops from the fibrous plaque. Its hallmark is calcification. It may be derived from the fibrous plaque by cell necrosis and disruption of the plaque, leaving an ulcerated lesion. Microthrombi may form on the plaque, resulting in eventual occlusion of the lumen. The occlusive process leads to ischemia and, ultimately, to infarction.

Fibroelastic lesion

Another type of atherosclerotic lesion is the fibroelastic lesion or internal cushion. This lesion consists of proliferated smooth muscle cells, in a subendothelial area, and of connective tissue. The lesion, which contains very little lipid, has been found in the aortic, iliac, and femoral arteries, especially at sites of branching. The extent of intimal thickening increases with aging. Such lesions have been noted in the epicardial coronary arteries in a 34-week-old fetus as well as in a newborn, and more in males than in females of all ages. Arterial regions of fibromusculoelastic lesions seem more susceptible to plaque formation if other risk factors are present.

Etiology of atherosclerosis

Several risk factors characterize some but not all individuals who develop atherosclerosis. Many theories have been proposed to explain atherosclerosis based on the associated risk factors. It is apparent that no single risk factor can explain the findings on atherogenesis. Therefore, with advances in the field of cell biology and the development of methods for cell culture, a great deal of emphasis has been placed on functional abnormalities of the different components of the blood and the vessel wall that may play a role in the formation of the atherosclerotic plaque. These studies will now be described, beginning with the cellular components of the arterial wall.

Endothelial cell

As described above, plasma constituents gain entry into the wall of the blood vessel by pinocytosis, or through intercellular junctions. The current view of atherosclerosis is that endothelial injury is a very important factor in the initiation of the process. Injury to the endothelium could alter the endothelial barrier and allow for more rapid passage of substances from the blood into the arterial wall. Injury to the endothelial lining could involve any of the following mechanisms:

Mechanical and hemodynamic (e.g., hypertension)
Microbiological (e.g., viruses, bacteria, and bacterial products such as endotoxin)
Metabolic (e.g., hyperlipidemia, thromboxanes, diabetes mellitus, homocysteinemia)
Smoking
Immune injury

Homocysteinuria has been associated with atherosclerosis. A balloon-tip catheter has been used in experimental animals to induce injury to the endothelial wall. Normal endothelium undergoes replacement, and interference with the normal process of replacement or metabolic injury altering the permeability of the endothelial cell seems more likely than actual physical damage or disruption.

Platelets and atherosclerosis

The intact endothelium is nonthrombogenic, and platelets do not adhere to it. Platelet adherence may occur with injury to the endothelial lining, since the subendothelial layer of microfibrils of connective tissue has a high affinity for platelets. Platelets may cause damage to the endothelial cells by release of agents that increase their permeability through the cell wall. A factor that stimulates the proliferation of smooth muscle cells and their migration from the media to the intima has been identified; it is released from the α-granules of platelets and has a molecular weight of approximately 28,000. A peptide that stimulates the formation of connective tissue is also released from the platelets, having a molecular weight of 10,000 to 20,000. Lysosomal enzymes are released during platelet aggregation and may increase the permeability of the endothelium. The platelet-derived growth factor has been extensively studied by Ross and his associates; it and other growth factors may provide a crucial insight into the cellular basis of atherosclerosis.

Following experimental denudation of the endothelium with a balloon catheter, platelet adhesion to the damaged area occurs within 10 minutes of injury. Three to five days thereafter, smooth muscle cells migrate through the internal elastic lamina and into the intima. Maximal thickening of the intima occurs at about 3 months, usually consisting of 5 to 15 layers of new smooth muscle cells surrounded by the connective tissue matrix. The endothelial cover regenerates, and by 6 months the intimal thickness regresses by one or two cell layers.

Smooth muscle cell

The importance of the platelet in the process of smooth muscle proliferation and aggregation has been documented experimentally. Rabbits made thrombocytopenic before endothelial injury do not develop intimal thickening. In swine with a form of von Willenbrand's disease, the platelets do not adhere to collagen and do not develop intimal proliferation. Dipyridamole, which interferes with platelet aggregation, also inhibits intimal smooth muscle proliferation. The smooth muscle cells in culture are able to contribute at least three constituents of the extracellular matrix, namely collagen (cell types I and III), elastin fibers, and glycosaminoglycans. The glycosaminoglycans have a marked propensity to bind low-density lipoproteins. Collagen in atherosclerotic plaque differs in structure from the normal media, about 65% having a structure called type I and 35% type III in the atherosclerotic plaque. In contrast, 70% of the collagen in the normal media is of the type III structure and only 30% is type I.

Benditt and Benditt have proposed the *monoclonal theory of atherogenesis*. According to this theory, the smooth cells within a plaque originate from a single clone of cells. A transformation or mutation induced by a virus or toxic substance presumably would induce the development of the abnormal clone of smooth muscle cells. Evidence to support the theory has been obtained from examination of the cells in atheromatous plaques of females. Cells of females contain one of two types of an isozyme form of glucose-6-phosphate dehydrogenase. Benditt and Benditt have observed that about 80% of the plaques from females contain only one form of the isozyme, which suggests that it was derived from a single type of cell. Experiments with dietary-induced atherosclerosis in swine have not supported a monoclonal origin in this species.

Thromboxane A₂ and prostacyclin

Arachidonic acid is a fatty acid released from the phospholipids of the cell membrane by the action of a phospholipase. The phospholipid phosphatidyl inositol is particularly rich in arachidonic acid in the 2 position. When exposed to platelets, arachidonic acid causes aggregation. Arachidonic acid is converted by a series of chemical reactions to a highly unstable group of substances called endoperoxides (Fig. 5–1). A key enzyme in this pathway is the cyclooxygenase, which is irreversibly inhibited by as little as 50 mg. of aspirin, taken orally. In the platelet, the major product of the arachidonic acid pathway is thromboxane A₂, which causes platelet aggregation and contraction of smooth muscle cells within blood vessels. The major product of the pathway within the endothelial cells is prostacyclin, or PGI₂, which inhibits platelet aggregation and causes relaxation of blood vessels.

The rationale behind low-dose aspirin therapy is to inhibit preferentially the platelet cyclooxygenase. The platelet contains no nucleus, is unable to synthesize protein, and thus cannot replace the irreversably inhibited cyclooxygenase. The endothelial cell contains a nucleus and can synthesize protein and replace the cyclooxygenase. If a deep injury is induced experimentally to the arterial wall, a defective repair can result in a calcified "scar,"

FIG. 5–1. Prostaglandins, thromboxanes and thrombosis. Arachidonic acid release in the blood vessel and platelet, its inhibition, and its metabolism by pharmacological substances. (Copyright © Baylor College of Medicine. Printed with permission.)

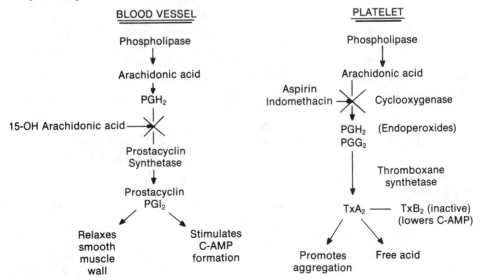

similar to the Mönckenberg sclerosis found in humans. Such an injury exceeds the regenerative capacity of the smooth muscle tissue. A superficial injury induced over a large area may result in regions that lack regrowth of endothelial tissue. These areas may develop lesions that have the characteristics of atherosclerotic lesions. In human beings, the injury to the endothelial cell may be physiological, rather than physical, trauma. The normal replacement of endothelial cells is believed to be a physiological response to wear and tear.

Thrombogenic hypothesis

If rabbits are made thrombocytopenic prior to removal of the endothelium by experimental injury, smooth muscle proliferation does not occur. Animals made thrombocytopenic 24 hours after endothelial injury do develop atherosclerotic lesions. The thrombogenic hypothesis of atherosclerosis has existed for over 100 years, having been proposed by von Rokitansky in 1841. According to the thrombogenic hypothesis, a mural thrombus originally composed of a large number of platelets becomes surrounded by fibrin. Within a short time, white cells and smooth muscle cells infiltrate into the thrombus. Eventually, the mural thrombus is invaded by lipid-rich material and is then organized into the vessel wall. Born has postulated the *erythrocyte-hemodynamic hypothesis*, in which he proposed that the initiating event of the thrombus in an atherosclerotic lesion is a fissure that develops in an atherosclerotic plaque. Rupture of the fissure leads to increased hemodynamic stress that results in red cell rupture and adenosine diphosphate (ADP) release. The ADP activates platelets and contributes to their subsequent aggregation as a mural thrombus. Born's hypothesis is predicated on the fact that the platelet thrombus occurs at sites where blood flow is nonlaminar. The thrombus is the result of hemorrhage into a plaque or a tortuous or stenotic region. Atherosclerotic plaques occur with increased frequency at sites of bifurcation or around orifices. Born has also postulated that the hemodynamic stress associated with hemorrhage is insufficient to activate platelets but could activate red cell ADP, which then would activate the platelets, leading to aggregation and thrombus formation.

The thrombogenic theory, as we have presented it, proposes an involvement of thrombosis in the pathogenesis of an atherosclerotic plaque. Thrombus as a terminal clinical event, superimposed on a pre-existing plaque, is now a well-established phenomenon. Hemorrhage into a plaque is believed to be a precipitating event for occlusive thrombus formation. The recently introduced thrombolysis technique of treating acute myocardial infarction is based on the direct observation that thrombosis occurs during infarction and may be reversed by activation of fibrinolysis through injection of an enzyme such as streptokinase. The efficacy of such treatment is still under investigation. Kadish has postulated that fibrin may cause disorganization of the endothelium upon contact and serve as a nidus for further thrombus formation. Inadequate fibrinolysis would thus contribute to the atherosclerotic process.

To summarize, platelets may influence atherosclerosis in several ways: by increasing permeability of the vascular wall, by secreting a factor that induces smooth muscle proliferation in the arterial wall, and by inducing mural thrombi at the site of atherosclerotic lesions.

Plasma lipids and atherosclerosis

In 1862, Virchow postulated that lipids in the atherosclerotic plaques are derived from the plasma by imbibition or insudation. In 1913, Nichicoff and his colleagues performed a historic experiment in which atherosclerotic lesions were produced in rabbits by a cholesterol-rich diet in which the cholesterol was dissolved in vegetable oil. Many experimental models have been adapted using a high-fat, high-cholesterol diet to induce atherosclerotic lesions.

Lipoproteins

Because lipids are insoluble in an aqueous medium, they are transported in plasma as macromolecular complexes called lipoproteins. There are five classes of lipoproteins: chylomicrons, very low density lipoproteins (VLDL), intermediate-density lipoproteins (IDL), low-density lipoproteins (LDL), and high-density lipoproteins (HDL). The lipoproteins are separated and classified on the basis of their electrophoretic mobility and rate of flotation in the ultracentrifuge. The chylomicrons are made in the intestine and carry the dietary cholesterol and triglyceride. VLDL are formed primarily in the liver and transport the cholesterol and triglyceride made in that organ. LDL are derived mainly from the catabolism of VLDL, and IDL are believed to be intermediates in this pathway. Some LDL are produced by the liver in homozygotes for familial hypercholesterolemia. LDL carry about one-half to two-thirds of the plasma cholesterol, and higher levels of LDL-cholesterol predispose to premature atherosclerosis and coronary heart disease; they are subdivided into two subclasses: HDL$_2$ and the less lipidated and denser HDL$_3$. The HDL$_2$ fraction is thought to correlate more closely with protection against coronary heart disease than HDL$_3$.

Hypercholesterolemia

The mechanism whereby hypercholesterolemia leads to atherosclerosis appears to involve much more than simple imbibition or insudation of the cholesterol molecules through the wall of the blood vessel, although this has been noted to occur. Studies of extracts from artery wall indicate that apolipoprotein B (apoB), the major protein component of LDL, is found in the wall of the blood vessels. In fact, some studies have shown that concentrations of plasma apoB correlate more strongly with coronary heart disease than do concentrations of plasma cholesterol or LDL-cholesterol. Hypercholesterolemia and hyperlipidemia have been found to be toxic to the endothelial cell in some experimental animals. Thus, elevated cholesterol and lipid levels may in some circumstances induce the endothelial injury to initiate the atherosclerotic process.

LDL acts as a mitogen in stimulating the growth and proliferation of smooth muscle cells in tissue culture or *in vitro*. This has been shown in smooth muscle cells from many species including rabbits, monkeys, and humans. Although LDL is less potent *in vitro* than is the platelet-derived growth factor, its importance should not be discounted, since atherosclerosis develops over a period of years. There are important differences between the atherosclerotic model induced by mechanical injury and that induced by hypercholesterolemia. It is important to note, for example, that reendothelialization of the lesions induced by hypercholesterolemia does not lead to regression of the lesion, as it does in the case of mechanical injury. With hypercholesterolemia, the endothelial cells seem to regenerate from cells on the lesions. Hypercholesterolemia appears to interfere with healing of the endothelium after mechanical injury.

The accumulation of lipids, and particularly of cholesterol and cholesteryl ester, within the atheroma is highly characteristic of these lesions. Most of the cholesterol found in the plaque is derived from the plasma lipoproteins, particularly LDL. Thus, elevated cholesterol has been shown to be capable of inducing damage to the endothelial cell, of stimulating proliferation of smooth muscle cells, and of contributing to accumulation of cholesterol and cholesteryl lipid within the developing atheroma. Experimental atherosclerosis may be induced in animals by dietary changes to raise the plasma cholesterol and LDL. Regression of induced atherosclerotic changes in animals has been observed after removing the dietary stimulus or profoundly lowering cholesterol and LDL through diet and/or drugs.

Cellular lipid uptake

Goldstein and Brown and their associates have provided a great deal of information concerning the cellular transport of cholesterol and LDL, especially of the receptor-mediated uptake of LDL. High-affinity receptors for LDL are present on the surfaces of smooth muscle cells, endothelial cells, fibroblasts, and many other cells. Internalization of LDL cholesterol occurs through a system of endocytotic vesicles or coated pits. Pastan and Willingham have shown that the coated pit buds off the cell membrane and forms a structure called a *receptosome*. Within approximately 30 minutes, the ingested particle is found within a lysosome where hydrolysis of LDL protein and of cholesteryl ester occurs. Cholesteryl esters are hydrolyzed by a lysosomal cholesteryl ester hydrolase. Free cholesterol can cross the lysosomal membrane, and the cell may deal with it in a number of ways. The cholesterol may be incorporated into membranes; in fact, cholesterol is necessary to support cellular synthesis of membranes. Once the cell's requirement for cholesterol is met, endogenous synthesis of cholesterol is inhibited. This is achieved by suppressing the rate-limiting enzyme of cholesterol biosynthesis, 3-hydroxy-3-methylglutaryl coenzyme A reductase (HMG CoA reductase). Cellular cholesterol can be stored to a limited extent as cholesteryl ester. The receptor-mediated uptake of LDL activates acyl:cholesterol acyltransferase (ACAT), an enzyme that catalyzes the esterification of cholesterol by a long-chain fatty acid.

Cellular mechanisms for regulation of cholesterol content are summarized in Fig. 5–2. These protective mechanisms appear to break down in the atherosclerotic artery. In severely diseased arteries, a 10-fold or more increase of unesterified cholesterol and a 50-fold increase in cholesteryl ester content can be found. With injury to the endothelial layer and hypercholesterolemia exceeding 400 mg./dl., the rate of transfer of LDL has increased approximately 100 times in experimental animals across the endothelial wall. It may be speculated that the excess esterification of cholesterol by the enzyme ACAT or a decrease in the activity of the lysosomal hydrolase could contribute to cholesteryl ester accumulation. Björkerud has suggested that the cytoplasmic formation of cholesteryl esters as droplets may lead to the formation of "foam cells." Goldstein and Brown hypothesize that certain "scavenger cells," which are represented as macrophages or monocytes and are present in the arterial wall, can take up the cholesterol independent of LDL receptors. The high-affinity LDL receptor on endothelial cells, smooth muscle cells, and fibroblasts binds and internalizes normal or native LDL and hypertriglyceridemic VLDL. The macrophage receptor binds poorly to normal LDL but has a high affinity for acetylation. This finding has given rise to the speculation that LDL may be altered in some way as it circulates in the blood or after it is trapped within the arterial wall. Glycosaminoglycans in the arterial wall readily bind LDL. The altered LDL would then be concentrated within the macrophage and contributed to the arterial accumulation of cholesterol and cholesteryl ester.

The normal LDL receptor is deficient or defective in the genetic disorder familial hypercholesterolemia. This results in a decreased catabolism of LDL and increased circulating cholesterol levels, which overwhelm the arterial wall. Premature atherosclerosis and coronary artery disease occur; the average age of the first myocardial infarction in affected males (heterozygotes) is about 40 years.

These studies of the intracellular metabolism of cholesterol offer new insight into

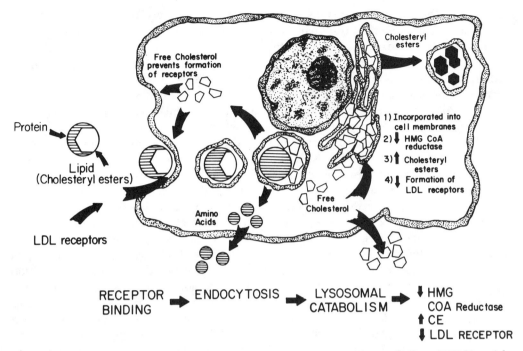

FIG. 5—2. LDL metabolism by nonhepatic tissues. This pathway for metabolism of LDL by peripheral tissues is based on studies with fibroblasts grown in tissue culture. Theoretically, HDL could interfere with LDL uptake by the receptor or could promote the loss of cholesterol from the cell. No *in vivo* evidence exists to support these postulated mechanisms; however, in experiments with isolated cells, HDL can exert both of these actions. (Gotto, A. M., Jr., and Jackson, R. L.: Plasma lipoproteins and atherosclerosis. *In* Gotto, A. M., Jr., and Paoletti, R. (eds.): Atherosclerosis Reviews, Vol. 3. New York, Raven Press, 1978.)

the cell biology of the atherosclerotic plaque. The cholesteryl ester that is found in the fatty streak differs from ester found in plasma; it is intracellular and predominantly cholesteryl oleate. In the advanced atherosclerotic plaque, the predominant cholesteryl ester is extracellular in the lipid core and is cholesteryl linoleate, the major cholesteryl ester in LDL. Possibly, the accumulation of cholesteryl esters is toxic to the cell, ultimately causing cell death with release of the lipid and cellular debris into the matrix of the atheroma.

Cellular lipid metabolism

The arterial cell has no mechanism for metabolizing cholesterol. The gonads and adrenal cortex convert a small amount of cholesterol to steroid hormones. The only

mechanism by which the body can rid itself of excess cholesterol is through the liver, by the biliary excretion of cholesterol or bile acids, which are formed from cholesterol. The cholesterol and bile acids are then excreted into the bile and are eventually eliminated in the feces. Glomset has postulated that HDL serves as a reverse transport pathway for shuttling cholesterol from peripheral tissues back to the liver. It is apparent that the cell must possess some mechanism for eliminating excess cholesterol. In *in vitro* experiments, HDL has been shown to promote cholesterol removal from cells.

In summary, investigations into the etiology of atherosclerosis have focused on building a hypothesis based on injury to the endothelium, the role of the smooth muscle cell, the platelet and substances produced therefrom; on thrombosis; on the plasma lipoproteins; and more recently on the macrophage. Pathophysiological alterations of the various components in the atherosclerotic lesion have been studied, including lipids, collagen, elastin, and glycosaminoglycans. There may be no single unified explanation for atherogenesis. Regardless of the etiology, the atherosclerotic plaque is properly viewed as a proliferative lesion. The smooth muscle cell can be considered proliferative as part of a reparative process. In the absence of significant hypercholesterolemia and elevations of LDL, injury to the endothelial cell may allow for a reparative process to take place, while repeated injury or persistent hypercholesterolemia may lead to further damage and increasing atherosclerotic lesion. This may well represent the "response to injury" hypothesis that has been put forth. On the other hand, the monoclonal hypothesis of Benditt raises the issue of mutation, perhaps resulting from toxins, chemicals, or cholesterol, which induces changes in the smooth muscle cells leading to a "benign" tumor growth that would represent the atherosclerotic plaque.

Since many risk factors contribute to atherosclerosis (see next section), this disease may be viewed as multifactorial. While hypercholesterolemia and high concentrations of LDL are not the only risk factors for coronary artery disease, there appears to be a critical level necessary for atherosclerosis to occur. In many non-Western countries, the plasma concentration of cholesterol is less than 160 mg./dl. and LDL levels are very low. Atherosclerosis and coronary heart disease are very rare under these conditions, even when other risk factors are present. For example, in Japan and China, hypertension and cerebrovascular disease are major causes of death, possibly related to high salt consumption. Cigarette smoking is also common. Levels of plasma cholesterol and LDL tend to be low, and atherosclerosis is much less common than in Western society, despite the presence of the other risk factors.

Risk factors and atherosclerosis

Of the people who develop atherosclerosis, most exhibit identifiable characteristics called *risk factors*. Risk factors are considered to be more frequent among the population who develop atherosclerosis than among the general population. The risk factor hypothesis is as follows: If a person has a risk factor, he or she is more likely to develop clinical manifestations of atherosclerosis and is likely to do so earlier than a person with no risk factors. By definition, this does not imply a cause-and-effect relationship between a risk factor and development of atherosclerosis. There has yet to be scientific documentation that the correction or elimination of a risk factor would reduce the chances of atherosclerosis. The risk factors that are more frequently present among people with coronary artery disease (CAD) are listed below.

Risk Factors for Coronary Artery Disease

Primary	Secondary
Elevated serum cholesterol	Elevated serum triglycerides
Elevated blood pressure	Heredity
Cigarette smoking	Diabetes mellitus
	Obesity
	High saturated fat diet
	Lack of physical activity
	Low vital capacity
	Abnormal electrocardiogram
	Sex (male)
	Age (middle age)
	Blood Type A, AB, or B
	Softness of drinking water
	Personality type A

The Framingham study, begun in 1948, has done much to link the incidence of cardiovascular disease and the risk factors associated with its development and is, in fact, the major study that developed the concept of risk factors. The selected population included over 65,000 persons aged 30 to 59, who were followed for over 20 years. With data obtained from the Framingham study, it is possible to predict with reasonable accuracy the risk of developing CAD based on the presence or absence of a number of risk factors in each individual.

Elevated serum cholesterol has been established as a major risk factor for the development of CAD. There is a statistically significant increased incidence of CAD as the level of serum cholesterol increases. It has long been recognized that there is earlier onset and increased risk of CAD among homozygous and heterozygous type IIA hyperlipoproteinemic patients.

Hypertension is another major risk factor. It has been postulated that mechanical factors play a major role in the disruption of the endothelial barrier. Elevated intra-arterial pressure by itself has been postulated as a cause for atherosclerosis. Atherosclerosis has been found to occur only in those parts of the vascular system perfused at high pressure. A third factor that has been postulated to increase the incidence of atherosclerosis in patients with hypertension is increased endothelial permeability caused by high levels of circulating catecholamines and other vasoactive substances. Interendothelial junctions have been found to open with angiotensin, seratonin, and bradykinin infusion. Tyramine has been shown to break free endothelial plasma membranes. Epinephrine and dopamine have been linked to intercellular and subendothelial extracellular edema, respectively. The mechanical theory of atherogenesis does not incriminate hypertension *per se*.

A significant relationship between atherosclerosis and cigarette smoking has also been established. The level of CO among cigarette smokers is higher than levels in the blood of nonsmokers. Tissue hypoxia with accumulation of lipid in the cell wall has been thought to be the cause of accelerated atherosclerosis among smokers. Cigarette smoking has been thought to increase the filtration of lipids into the arterial wall. The development of hypoxia due to the high level of CO has also been proposed. Hypoxia would have a direct effect upon endothelial permeability. In addition, it may have an effect upon the watershed area, with increased trapping of lipoproteins with relative hypoxia.

Much work remains to be done to establish a direct cause-and-effect relationship between the epidemiologically determined risk factors and atherosclerosis. How each of these risk factors may work in the pathogenesis of atherosclerosis has not yet been determined. This is especially true for the secondary, or minor, risk factors.

Annotated references

Benditt, E. P., and Gown, A. M.: Atheroma: The artery wall and the environment. Int. Rev. Exp. Pathol. *21*:55, 1980. (This chapter explores the epidemiology and causes of vascular disease.)

Björkerud, S. U.: Mechanisms of atherosclerosis. *In* Ioachim, H. L. (ed.): Pathobiology Annual, vol. 9. New York, Raven Press, 1979. (A broad review of theories concerning atherogenesis, covering the past several decades.)

Born, G. V. R.: Fluid-mechanical and biochemical interactions in haemostasis. Br. Med. Bull. *33*:193, 1977. (A look at some questions regarding the effects of the fluid dynamics in and around a vascular leak on the circulating platelets responsible for sealing it and on the chemical agents responsible for making the platelets adhesive.)

Born, G. V. R.: Do platelets contribute to atherogenesis? *In* Gotto, A. M., Jr., Smith, L. C., and Allen, B. (eds.): Atherosclerosis V. New York, Raven Press, 1979. (A concise review of the history of thought concerning the role of platelets in atherogenesis.)

Brown, M. S., and Goldstein, J. L.: Receptor-mediated endocytosis: Insights from the lipoprotein receptor system. Proc. Natl. Acad. Sci. USA, *76*:3330, 1979. (Discussion of the cellular functions of the cholesterol derived from internalized LDL, impact of genetics on the binding and internalization of LDL, and receptor systems in general.)

Brown, M. S., Kovanen, P. T., and Goldstein, J. L.: Regulation of plasma cholesterol by lipoprotein receptors. Science *212*:628, 1981. (A general review of studies of the lipoprotein transport system, coupled with a look at the role of lipoprotein receptors in cholesterol regulation and possible pharmacological manipulations of the receptors.)

Gotto, A. M., Jr., and Paoletti, R. (eds.): Atherosclerosis

Reviews, vols. 1–4. New York, Raven Press, 1971–1979. (An in-depth review of the many facets of atherosclerosis, written for the practicing physician.)

Gotto, A. M., Jr., Smith, L. C., and Allen, B. (eds).: Atherosclerosis V. New York, Springer-Verlag, 1980. (Proceedings of the Fifth International Symposium on Atherosclerosis, containing over 160 papers on a very wide variety of topics related to atherosclerosis.)

Gross, H. L. (ed.): Atherosclerosis. Kalamazoo, Mich., Upjohn, 1977. (Review of research on and clinical manifestations of atherosclerosis.)

Harker, L. A., Ross, R., and Glomset, J.: Role of the platelet in atherogenesis. Ann. N.Y. Acad. Sci. 275:321, 1976. (Discussion of endothelial cell injury and proliferative smooth muscle cell response.)

Kadish, J. L.: Fibrin and atherogenesis: A hypothesis. Atherosclerosis 33:409, 1979. (Presents the view that atherogenesis represents an abnormal reaction to intimal injury.)

Kottke, B. A., and Subbiah, M. T. R.: Pathogenesis of atherosclerosis. Mayo Clin. Proc. 53:35, 1978. (A look at experimental work in the etiology of atherosclerosis.)

McCullagh, K. G.: Revised concepts of atherogenesis. Clev. Clin. Quart. 43:247, 1976. (An attempt to sum up the theories of atherosclerosis and to form a hypothesis. Focus on hyperlipidemia.)

McGill, H. C., Jr.: Atherosclerosis: Problems in pathogenesis. In Gotto, A. M., Jr., and Paoletti, R. (eds.): Atherosclerosis Reviews. New York, Raven Press, 1977. (Review of the history of thought concerning pathogenesis, written for the practicing physician.)

Moncada, S., Needleman, P., Bunting, S., and Vane, J. R.: Prostaglandin endoperoxide and thromboxane generating systems and their selective inhibition. Prostaglandins 12:323, 1976. (Two enzyme systems and their selective inhibition by indomethacin and benzydamine are described.)

Ross, R., and Glomset, J. A.: The pathogenesis of atherosclerosis. N. Engl. J. Med. 295:269, 1976. (A review of hypotheses concerning the basis for intimal smooth muscle proliferation.)

Schwartz, S. M., Gajdusek, C. M., and Selden, S. C., III: Vascular wall growth control: The role of the endothelium. Arteriosclerosis 1:107, 1981. (Reviews general principles of cell growth, current knowledge of the control of endothelial growth, and the role of endothelial injury and repair in the pathogenesis of atherosclerosis.)

Steinberg, D.: Research related to underlying mechanisms in atherosclerosis. Circulation 60:1559, 1979. (Presents a brief resume of advances in atherosclerosis research over the past 30 years.)

Thomas, W. A., Reiner, J. M., Florentin, R. A., Scott, R. F., and Lee, K. T.: Transformation of naturally occurring quiescent intimal smooth muscle cell masses into atherosclerotic lesions in swine fed a hyperlipidemic diet. In Gotto, A. M., Jr., Smith, L. C., and Allen, B. (eds.): Atherosclerosis V. New York, Springer-Verlag, 1979. (A review of studies of atherosclerotic lesions in preadolescent swine, with emphasis on the hypothesis that all intimal cell masses are activated by atherogenic diet.)

Wissler, R. W.: The principles of the pathogenesis of atherosclerosis. In Braunwald, E. (ed.): Heart Disease: A Text of Cardiovascular Medicine. Philadelphia, W. B. Saunders, 1980.

Section II

Respiratory Mechanisms

Introduction

Human respiration may be considered to include all the processes whereby oxygen is removed from the environment and delivered to the metabolizing cells and carbon dioxide from the cells is delivered to the environment. Few body functions are as urgently life sustaining and at the same time as clearly understood, readily evaluated, and reasonably manipulated therapeutically. Abnormal ventilation due to malfunction of the respiratory control centers, neuromuscular disease, or excess work of breathing may be easily documented and therapeutically modified. Problems of gaseous exchange in the lungs are easily assessed, and appropriate oxygenation or ventilation may be established so exquisitely that acceptable levels of arterial gases can be achieved in all but the most severe disease. The laboratory can quantify the blood hemoglobin and its effectiveness as an oxygen carrier. Measurement of total blood flow or cardiac output permits calculation of the total oxygen delivered to the tissues. When we reach the level of tissue blood flow or the even more elusive analysis of gas exchange between systemic capillary and intracellular organelles, the analytical techniques become insensitive, thus preventing clear documentation of the state of cellular respiration.

The successful development of life-sustaining environments for travel through oxygen-free outer space and for the extreme hyperbaric conditions of deep sea exploration (up to 45 atmospheres of pressure), testify to the pragmatic nature of our scientific understanding of the respiratory pathway.

We continue to learn more about the limits of tolerance for the respiratory system. In spite of numerous previous ascents, it was not until 1978 that the first climber reached the peak of Mount Everest without oxygen. At least some individuals had sufficient reserves to climb to an altitude of 29,000 feet (barometric pressure less than 250 mm. Hg) and return. The severe hypoxemia at this altitude results in muscle weakness, easy fatigability, altered central nervous system function, and altered ventilatory control. The benefits of muscle conditioning, altitude acclimitization, and physiological adaptation are now well documented and are beyond the scope of this text. Impaired function of the respiratory system may result in disability or symptoms, but the threshold or tolerable limits of impaired function are not well defined. An intriguing example is seen in recent studies which documented frequent apneic spells in normal individuals during sleep. Men have many more such spells per night than women. When frequent, these spells produce central nervous system and cardiovascular problems, but it is not yet clear how many or how prolonged these episodes may be without producing secondary dysfunction.

Clearly then, a major deficit in our understanding of the pathophysiology of respiration lies in the problem of defining normalcy or disease. If a normal range of vital capacity includes 20 per cent above and below a predicted mean, a given individual may have initially scored at the upper limit of normal; and, subsequently, when he may have deteriorated by one-third to the lower limit of normal, he would still be classified as being normal. Therefore, when the statistically defined ranges of normal are established for a population, they have limited application to the individual. Such limitations cause the investigator to base his understanding of altered function in disease on analysis of the unequivocally or severely abnormal patient. Thus, patients with severe chronic bronchitis or emphysema typically have extreme alterations in blood gas exchange as manifested by wasted ventilation (increased dead space), high alveolar-arterial oxygen gradients, low arterial oxygen tensions, and frequently carbon dioxide retention. They uniformly have decreased maximum expiratory flow rates. While this knowledge is important, the clinician is more frequently called upon to interpret symptoms of early stages of the disease when the definable physiological abnormalities are minimal. Such patients may be symptomatic only during stress (e.g., exercise or associated illness) or they may be more sensitive to altered function and therefore aware of altered function at milder levels of abnormality than generally pertains. Evaluation and treatment of altered function in such mildly abnormal individuals presents a cardinal clinical challenge.

Physicians through the ages have diagnosed and treated dysfunction even though they had little understanding of the cause of a disease. It comes as no surprise, then, that the diseases most commonly associated with abnormal function as outlined in this section do not have well-defined etiologies. Although manipulation of these abnormal functions may improve the patient's status and relieve his symptoms in a way gratifying to both patient and physician or may even be life saving, the underlying disease process is frequently unaltered. For example, when bronchodilators temporarily relax bronchial smooth muscles effectively, both physician and asthmatic patient are relieved, but the underlying disease remains, only to become manifest at another time.

When studying this section, the student should consider several major aspects of pathophysiology involving the lung which are not unique to the lung and therefore covered better elsewhere. These include immune-mediated diseases of the lung and the currently much discussed area of tissue (lung) edema.

As in other organ systems, the activity of the immunological system may be turned against the lung tissues, resulting in tissue damage and abnormal function. The pathophysiology of bronchial asthma, Goodpasture's syndrome, hypersensitivity pneumonitis, and chronic granulomatous infection such as sarcoidosis all represent host responses to immune processes. These four conditions have at least some explanation through the four classical immune mechanisms described by Gel and Coombs. The mechanisms of immune-mediated diseases are discussed in greater detail in Chapter 37.

Lung edema with or without proliferation of cellular elements and connective tissue is a common result of a wide range of lung injuries. This is popularly referred to by a group of synonyms including noncardiogenic pulmonary edema, low-pressure pulmonary edema, permeability edema, capillary leak syndrome, or the adult respiratory distress syndrome. The mechanism whereby fluid leaves the capillary bed and accumulates in the lung tissue follows the same principles as the development of edema in other tissues and other organs. When these factors are not due to hydrostatic and osmotic forces, but are predominantly due to changes in capillary permeability, a mechanism of capillary injury must be considered. Several reproducible experimental models have been widely studied. One of the most popular is the intravascular injection of oleic acid, which reproducibly causes diffuse edema throughout the lungs. The resulting fluid may accumulate in the peribronchial, perivascular or alveolar interstitial space, or in the alveolar air space. The effects on mechanical properties of the lung and gas exchange have been well established. The mechanism whereby the capillary leak developed remains uncertain. The contributing roles of prostaglandins, bradykinin, leukocyte lysosomes, oxygen radicals in leukocytes, and other potentially injurious factors have not been sorted out.

The following chapters highlight those aspects of physiology and abnormal function which are fundamental to understanding the respiratory system in health and common diseases. They merely touch upon large areas of fundamental knowledge regarding the environment, stress such as exercise, infections, and the effects of systemic disease on lung function, intriguing themes developed in other tomes.

CLARENCE A. GUENTER

6

Control of Respiration

William A. Whitelaw

The passive lungs cannot work without the neuromuscular pump which expands and contracts them. The physician dealing with respiratory problems must have an understanding of how this pump functions and how it can be altered in disease either as an adaptive response or because of disease of the pump itself.

Normal function

The traditional physiology of respiratory control concentrates narrowly on the principle of homeostasis of P_{CO_2} and P_{O_2} by means of "negative feedback loops." Arterial blood is sampled by chemoreceptors, which in turn stimulate the respiratory center to increase or decrease alveolar ventilation, restoring P_{CO_2} or P_{O_2} to a predetermined level. It is convenient to begin by explaining how such a simple system operates. Breathing is in fact regulated much better than would be predicted by this theory, and following sections show how the chemical control theory may be made more sophisticated, and how accessory, nonchemical control loops further help to stabilize ventilation. Finally, the elements of the pump—the respiratory centres, muscles, and chest wall—will be described in more detail.

The chemical control system

The main purpose of the respiratory control system is to maintain body acid-base balance and oxygen supply. A major element of the system is thus a set of receptors which measure levels of P_{O_2} and P_{CO_2} and hydrogen ion and stimulate the respiratory centers in a way that tends to restore these values to normal if they are altered by an extraneous process. Of these, the P_{CO_2} receptors have the most potent effect on breathing. A brief description of how a "negative feedback system" works to stabilize CO_2 illustrates some important principles.

SIMPLE FEEDBACK CONTROL. Figure 6–1 is an engineer's schematic diagram of the respiratory chemostat, so called because its principle of operation is analogous to that of a simple household thermostat.

The lungs must remove carbon dioxide from the bloodstream at the same rate as it is produced by metabolism. If carbon dioxide is made faster than it is removed, alveolar P_{CO_2} will rise and acidosis will ensue; the reverse process will lead to alkalosis. The amount of CO_2 exhaled per minute is given by the concentration of CO_2 in alveoli multiplied by the volume of alveolar air exhaled per minute.

$$\dot{V}_{CO_2} = \dot{V}_A \cdot F_{ACO_2} = \dot{V}_A \cdot \frac{P_{ACO_2}}{P_B - 47} \quad (1)$$

where $\dot{V}_A$ is alveolar ventilation per minute, F_{ACO_2} is fraction of alveolar CO_2, P_{ACO_2} is alveolar P_{CO_2}, $P_B - 47$ is barometric pressure

125

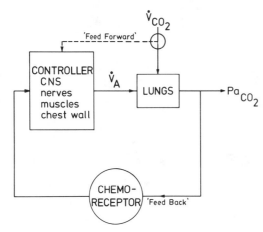

FIG. 6–1. Conceptual model of a body "chemostat," or automatic control system, to stabilize arterial P_{CO_2}. In a "feedback" system the controller sets the level of alveolar ventilation ($\dot{V}_A$) according to information it receives from the chemoreceptor sampling arterial blood. If metabolic rate ($\dot{V}_{CO_2}$) increases, P_{CO_2} rises, and chemoreceptors tell the controller to increase $\dot{V}_A$. In a "feed forward" system the controller is informed directly about changes in $\dot{V}_{CO_2}$ and increases $\dot{V}_A$ accordingly, so there is no need for P_{CO_2} to rise.

minus water vapour pressure, $\dot{V}_{CO_2}$ is carbon dioxide production per minute.

As long as metabolic rate is constant, alveolar ventilation and P_{ACO_2} are reciprocally related by the equation:

$$P_{ACO_2} = \frac{(P_B - 47)\dot{V}_{CO_2}}{\dot{V}_A} \qquad (2)$$

To maintain P_{CO_2} constant when CO_2 production increases, the controlling system must somehow recognize a change in $\dot{V}_{CO_2}$ and adjust $\dot{V}_A$ to match it, keeping the value of the fraction $\dfrac{\dot{V}_{CO_2}}{\dot{V}_A}$ constant. Since arterial blood is equilibrated with alveolar gas, arterial P_{CO_2} is approximately equal to alveolar P_{CO_2}. If the rate of CO_2 production increases (as in exercise for example) but alveolar ventilation remains the same, venous blood will deliver carbon dioxide into alveoli faster than it is being removed, and P_{ACO_2} will rise. If an arterial P_{CO_2} detector is arranged so that when P_{CO_2} rises, it will stimulate the respiratory centers to in-

crease $\dot{V}_A$, then $\dot{V}_A$ will go up and P_{ACO_2} will drop back toward normal. The effectiveness of this kind of controller depends on how big a change in $\dot{V}_A$ follows a given change in P_{ACO_2}. This can be seen quantitatively in Figure 6–2. Figure 6–2a shows equation 2 in graphic form for two different values of $\dot{V}_{CO_2}$. At rest, P_{ACO_2} is related to $\dot{V}_A$ by a hyperbola. During exercise, $\dot{V}_{CO_2}$ is higher and there is a new hyperbola, where to maintain the same P_{CO_2} as at rest $\dot{V}_A$ must be much higher. Figure 6–2b shows how the controller reacts to an increase in P_{CO_2}. This sort of graph is obtained in human subjects by having them breathe gas mixtures that include carbon dioxide in various concentrations and measuring ventilation and alveolar P_{CO_2}. As the carbon dioxide level is artificially raised, respiratory centers are stimulated and $\dot{V}_A$ goes up. The slope of the line is said to describe the sensitivity of the respiratory system to CO_2. Figure 6–2c shows how such a system reacts when $\dot{V}_{CO_2}$ rises. A resting subject starts at point A with normal $\dot{V}_A$ and P_{ACO_2}. When exercise begins he must move to some point on the exercise curve. If ventilation did not increase at all he would go to point B, with a P_{CO_2} of 78. Because the CO_2 receptor stimulates breathing exactly as described in Figure 6–2b, however, he moves up the CO_2 response line to point C. $\dot{V}_A$ is increased and the new value of P_{ACO_2} is higher than normal but not by a great deal. How much the P_{CO_2} rises depends on the sensitivity of the respiratory system to P_{CO_2}. If the response is very steep the subject will move up to point D with even higher ventilation and a P_{CO_2} more nearly normal. On the other hand, a patient who is relatively insensitive to P_{CO_2} will move up to point E, allowing P_{CO_2} to rise rather than increase ventilation. The sensitivity of this respiratory controller has been assessed in many normal subjects and patients. There is a wide range in normal, roughly between the extremes shown in the examples of Figure 6–2. The sensitivity increases when hypoxemia or metabolic acidosis is present but decreases in sleep, in anesthesia, and when sedative or narcotic drugs are given.

Principles that apply to this kind of negative feedback controller are as follows:

1. By its nature it can defend the respiratory system only against stresses that actually change P_{CO_2}.
2. It does not regulate P_{CO_2} perfectly. The only stimulus to increase $\dot{V}_A$ is the abnormal CO_2, so that ventilation does not increase unless P_{CO_2} is at least a little bit elevated.
3. For the same reasons it never over-corrects P_{CO_2}.
4. It can be assessed by measuring the ventilatory response to CO_2.
5. An important point to notice is that the removal of carbon dioxide depends on *alveolar* ventilation *not minute* or *total* ventilation. As the respiratory center increases its output and steps up the pumping action of the lung, the total flow of air is important, but so also is the pattern of breathing because of the effect of dead space. Alveolar ventilation equals minute ventilation minus dead space ventilation as defined in the following equation.

$$\dot{V}_A = (V_T - V_D) \cdot f = \dot{V}_E - V_D \cdot$$
$$= \dot{V}_E(1 - V_D/V_T) \quad (3)$$

where $\dot{V}_A$ is alveolar ventilation per minute, V_T is tidal volume, V_D is dead space per breath, f is respiratory rate (frequency), $\dot{V}_E$ is total ventilation per minute (minute ventilation).

Thus, an important detail of pump activity is the ratio of dead space to tidal volume. When the chemoreceptor works correctly and the respiratory center throttle is turned on appropriately, giving the correct minute ventilation, the removal of carbon dioxide from blood can still be impaired if (1) dead space is abnormally high, or (2) the pattern of breathing is changed to one of rapid, small breaths.

There are additional chemical feedback mechanisms, one that increases ventilation

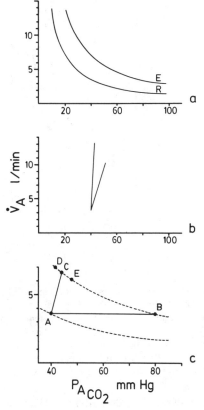

FIG. 6–2. Quantitative aspects of feedback control. See text for a full explanation. *A*. Relation between alveolar ventilation and P_{CO_2} at rest ($\dot{V}_{CO_2}$ = 200 ml./min) and light exercise ($\dot{V}_{CO_2}$ = 400 ml./min). *B*. Two carbon dioxide response curves. In each case a subject began at resting P_{CO_2} of 40 mm Hg., and normal resting ventilation. When CO_2 is artificially raised, ventilation may increase rapidly in a sensitive subject (*steeper line*) or slowly in an insensitive one (*less steep line*). *C*. Magnified graph of relations between ventilation and P_{CO_2}.

in response to hypoxemia, another responding to changes in pH. The way in which they work is very similar to the CO_2 mechanism. Since pH and P_{O_2} can vary independently of P_{CO_2}, it is useful for each to be able independently to stimulate alveolar ventilation to protect the organism from hypoxemia and to stabilize pH in the case of metabolic acidosis. The response of normal subjects to hypoxemia is shown in Figure 6–3. As P_{O_2} drops, ventilation rises, but the change is small before P_{O_2} reaches

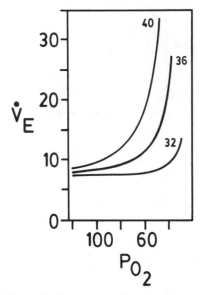

FIG. 6–3. Ventilatory response to hypoxia of a normal subject. This graph describes what happens to total ventilation ($\dot{V}_E$) when a subject is made hypoxic by breathing a gas mixture low in oxygen. Carbon dioxide is added to the inspired gas to keep arterial P_{CO_2} at some fixed level. The three curves are done at different P_{CO_2} levels, as indicated in millimeters of mercury.

about 50 mm. Hg. Sensitivity to P_{O_2} depends on P_{CO_2}. When P_{CO_2} is high, a much smaller drop in P_{O_2} will stimulate $\dot{V}_A$. On the other hand, when P_{CO_2} is below normal, there is almost no response, even to severe hypoxemia.

In some clinical situations the three chemical feedback mechanisms act in the same direction. For example, in asphyxia the drop in pH, the rise in P_{CO_2}, and the drop in P_{O_2} all work together synergistically to stimulate respiration, the low P_{O_2} increasing sensitivity to P_{CO_2} and vice versa. In other situations, the different mechanisms may antagonize each other and tend to cancel out, for example in pneumonia, in which hypoxemia stimulates breathing but the increased ventilation lowers P_{CO_2}, raises pH, and reduces sensitivity to a low P_{O_2}, thus vitiating the original effect of hypoxemia.

CHEMORECEPTORS. Two sets of receptors are recognized. The "peripheral receptors"

in the major central arteries include the carotid and aortic bodies and measure P_{O_2}, P_{CO_2} and pH in arterial blood. The "central receptors" are in the head and behave as if they were mainly sensitive to pH. Although the degree of respiratory stimulation corresponds to pH of the cerebrospinal fluid (CSF) in some circumstances, it does not always do so. It is postulated that the receptors may be sensitive to pH of brain interstitial or even intracellular fluid. Carbon dioxide equilibrates rapidly between blood and cerebral fluids, changing pH and stimulating both carotid bodies and central receptors, but the central effect is the more important. Hypoxemia acts almost entirely through the carotid bodies. For pH the situation is more complex, since the arterial and the central chemoreceptors are more or less equally sensitive to it. Hydrogen ions do not easily cross in and out of cerebral fluids and it may happen that the pH sampled by central receptors is radically different from the arterial pH sampled by the carotid bodies. These two pH receptors may then act in opposite directions creating a confusing clinical picture. Examples of this are discussed below (see Metabolic Acidosis).

THE CAROTID BODIES. The principal oxygen sensors, the carotid bodies are found very close to the carotid sinus at the bifurcation of the common carotid artery. They are small, highly vascular structures which have been isolated in animal experiments and thoroughly studied. They receive arterial blood from the carotid artery and send afferent neural information to the respiratory centers through the glossopharyngeal nerve. The ratio of blood flow to oxygen uptake in the carotid body is very high so that the venous blood draining from them has a P_{O_2} nearly the same as arterial blood. The receptors are clearly sensitive to P_{O_2}, not to oxygen concentration or saturation. Being quite insensitive to loss of oxygen-carrying capacity, they fail to stimulate breathing in anemia or carbon monoxide poisoning, except when these are very severe. When carbon dioxide tension is high the carotid body sensitivity to oxygen is enhanced and vice versa.

Experiments in which carotid bodies are perfused with blood at different oxygen tensions while recordings are made from afferent nerve fibers show that these tiny organs have a steady low-level discharge when P_{O_2} is in the normal range and increase their firing rate progressively as P_{O_2} falls. A graph of carotid body discharge against P_{O_2} very closely resembles the graph of ventilation against P_{O_2} found in normal people (Fig. 6–3). In such preparations, discharge from the carotid body can be altered by stimulating its sympathetic nerve supply or by applying autonomic drugs like atropine and phenoxybenzamine, but such manipulations are not known to have any clinical significance. Under the microscope, the carotid body is seen to consist of nests of cells that contain neurosecretory granules and are connected to each other by synaptic junctions. Although it is small, self-contained, and simple in structure, it has not so far yielded up the secret of how it operates.

The carotid body increases its discharge when arterial pressure falls, which accounts for part of the hyperpnea of shock. It is stimulated by cyanide, giving rise to marked hyperpnea in cases of poisoning. It is also stimulated by nicotine and mediates some of the effects (tachycardia, vasoconstriction) of this drug. Dopamine strongly inhibits its discharge. Increased temperature causes an increase in carotid body discharge, but the hyperpnea of fever is likely mediated by central as well as peripheral mechanisms.

Aortic bodies, similar to carotid bodies in structure and function, are scattered over the surface of the aorta but are not known to be of importance in man.

CENTRAL CHEMORECEPTORS. Most of the response to change in P_{CO_2} is mediated by receptors within the cranium. Ventilation does not seem to correspond directly to the activity of a hypothetical receptor in the intracranial arteries, but when CSF is examined, minute ventilation is found to correlate very well with CSF pH no matter whether this pH is altered through changes in P_{CO_2} of arterial blood or changes in bicarbonate concentration of CSF. From these observations it is concluded that central chemoreceptors are exposed to brain interstitial fluid or CSF, and are exquisitely sensitive to pH. Hyperventilation can be produced by perfusing the ventricles with acid solutions or by diseases which make CSF acid. Since hydrogen ion concentration seems to be the immediate stimulus, changes in P_{CO_2} affect ventilation by shifting hydrogen ion-bicarbonate equilibrium governed by the Henderson-Hasselbalch equation:

$$pH = pKa + \log \frac{(HCO_3^-)}{\alpha \, P_{CO_2}} \qquad (4)$$

If CSF bicarbonate is large, as in chronic respiratory acidosis or metabolic alkalosis, a given change in P_{CO_2} will produce a change in CSF pH which is smaller than normal. The central receptors will therefore receive a smaller stimulus than normal and the ventilatory response to arterial P_{CO_2} will be attenuated. This mechanism accounts for part of the loss in CO_2 sensitivity found in chronic CO_2 retention. The receptors appear to be close to the surface of the medulla oblongata on its ventrolateral aspect. Some observations indicate that these, or possibly a separate set of intracranial receptors are also responsive to arterial pH or P_{CO_2}. Thus, ventilation can be affected to some degree by arterial gases, even when CSF values are held constant. Much remains to be learned about the location, mechanism of action, and physiological role of the central receptors. In some circumstances (at altitude, for example), ventilation goes in a direction opposite to that predicted by measuring CSF pH. This implies that the powerful central chemoreceptor drive to breathe can be overridden by some other control system, and suggests a wary approach to any analysis of ventilatory response to chemical stimuli. However, some otherwise perplexing clinical situations can be understood by considering changes in CSF.

While the chemical control system has been very thoroughly studied for several decades using sophisticated techniques, much remains to be learned about it. As described here, chemical feedback does not satisfactorily explain the fact that P_{CO_2} remains completely within normal limits in the face of the huge changes in metabolic rate that occur in exercise. (Fig. 6–2c shows that to treble or quadruple ventilation by negative feedback requires that P_{CO_2} rise above normal even in subjects with the steepest response lines.) Nor does it explain how patients with extreme muscle weakness or severe airways obstruction contrive to keep their CO_2 normal, making near maximal respiratory efforts without any more than normal chemical stimulus.

Three of the many refinements of the chemical feedback theory that attempt to explain the increased ventilation of exercise are listed below:

1. The carotid bodies are sensitive to the changes in Pa_{CO_2} and Pa_{O_2} that occur with each breath, as well as to the mean value. The amplitude of these fluctuations increases in exercise and could constitute an added stimulus.
2. The sensitivity of the controller to CO_2 seems to be temporarily enhanced in exercise, the ventilatory response curve in Figure 6–2 becoming more vertical.
3. A "feed-forward" system has been postulated in which the body measures $\dot{V}_{CO_2}$ directly, and adjusts ventilation to suit, without waiting for Pa_{CO_2} to change. (One way the body could do this would be to measure mixed venous P_{CO_2} and multiply this by cardiac output, but no evidence can be found for the existence of CO_2 receptors in the venous circulation.) A feed-forward system would very

nicely explain the wide variety of circumstances (exercise, hyperthyroidism, muscle weakness) in which $\dot{V}_{CO_2}$ and $\dot{V}_A$ are closely coupled.

None of these factors has yet been shown to explain convincingly the ventilation of exercise, but the remarkable fact that P_{CO_2} does stay within a very narrow range in almost all circumstances argues forcibly that a chemical regulatory system must be involved, at least as a final reference.

Nonchemical stimuli

It would not be reasonable to confine the regulation of such a crucial life support system as breathing to a single mechanism consisting of a few chemosensitive cells in the medulla and the tiny carotid bodies. In fact, the respiratory system is provided with an interlocking set of accessory control devices designed to stabilize or adjust ventilation in the face of many kinds of interference that can be anticipated in the course of everyday life or common illnesses. Some of these mechanisms are probably unknown and others poorly understood, but they are effective enough to maintain stable ventilation for minutes, hours, or days while ignoring signals from chemoreceptors. Moreover, they can anticipate and prevent changes in ventilation due to changes in mechanical load on the respiratory system and increase ventilation before planned exercise begins. Examples of the influence of nonchemical control systems follow:

1. Conscious subjects continue to breathe normally for a long time when the most potent chemical stimulus, P_{CO_2}, is removed by hyperventilation.
2. While talking, normal people maintain nearly constant normal ventilation in spite of a major increase in P_{CO_2} induced experimentally by raising P_{CO_2} of inspired gas.
3. In rapid eye movement (REM) sleep, ventilation remains close to normal al-

though ventilatory response to P_{CO_2} is severely attenuated.

4. Treated patients with such diseases as pneumonia and pulmonary edema hyperventilate in spite of normal oxygen, low P_{CO_2}, and high pH.

Numerous reflex mechanisms have been identified in animals and probably have a counterpart in man. They come from various receptors and seem well designed to stabilize ventilation in various conditions. In the lung are three kinds of receptors with axons in the vagus nerve, which are discussed in the following paragraphs.

Stretch receptors located in the smooth muscle of trachea and mainstem bronchi fire when lungs expand, sending a signal that is closely related to lung volume. In animals, stretch receptors are responsible for the Hering-Breuer reflex; their role in normal breathing is to switch off inspiration at a certain level, causing expiration to begin, and thus accelerating respiratory rate. Steady inflation of the lungs, as in positive end-expiratory pressure, elicits reflex tonic activity of expiratory muscles through this mechanism. It has been hard to demonstrate that these receptors influence breathing in man, although recordings from vagus nerves exposed at operation show that they are active. Conceivably they may be important in patients on mechanical ventilators who often synchronize their own breathing efforts with those of the machine. This could come about through stretch receptor activity. As the lungs are passively inflated, the receptors send an inhibitory discharge to the inspiratory neurons. The inhibition is maximal at end-inspiration and switches off the patient's own inspiratory effort just as the respirator enters its expiratory phase. As the lungs deflate, inhibition is removed, and the patient begins to inspire again at about the same time as the machine delivers its next impulse of positive pressure.

Irritant and cough receptors are unmyelinated fibers located just beneath the epithelium of larynx, trachea, and bronchi. Upper and lower airway receptors are morphologically identical but have different effects. Stimulation of tracheal and laryngeal *cough receptors* by dust or toxic chemicals leads to cough and bronchoconstriction. The same stimuli applied to *irritant* receptors within the lung lead to tachypnea and increased ventilation. In experimental animals, irritant receptors have been shown to fire in response to atelectasis, bronchoconstriction, pneumothorax, and microembolism. Clearly, they may be invoked to explain hyperventilation in many clinical situations.

J receptors found in the periphery of the lung are stimulated by accumulation of fluid in the interstitial space and cause rapid, shallow breathing. They may be involved in the increased ventilation of pulmonary edema.

Receptors in the muscles and joints are well disposed to identify movements and stresses in the respiratory system and help make adjustments to unexpected loads on muscles. Thus spindles and tendon organs in intercostal muscles and diaphragm probably regulate length and tension in these muscles, making sure that appropriate breathing efforts are made even when extramechanical loads, changes in postural stress, or unexpected chest deformations occur. Joint receptors of the rib cage may inform the respiratory centers about changes in lung volume. In exercise, joint and muscle receptors in the limbs may transmit information about the amount of work being done allowing the respiratory system to anticipate needs for ventilation.

The effect of nonchemical mechanisms can be seen in the *loading response*. If a normal subject or patient is obliged to breathe through a small tube with high resistance simulating added airways obstruction, he immediately increases his inspiratory efforts, managing to maintain nearly normal tidal volume and ventilation even if resistance is raised to two or three times normal. While some of this response could be attributed to chemoreceptors (the load leads to reduction in ventilation, which in turn leads to increased carbon dioxide and increased chemically stimulated effort to breathe) most of it occurs even when changes in blood gases are prevented by manipulating inspired gas tensions. The loading response varies in importance from one subject to the next and is abolished by sleep or anesthesia. The receptors that mediate it are unknown, but probably lie in the chest wall.

The respiratory centers

Respiratory centers are all the parts of the central nervous system that mediate between stimuli and respiratory responses. Many parts of the brain, including the cortex, hypothalamus, pons, medulla, and cerebellum, are able to influence breathing, but the way in which the various parts cooperate to generate normal breathing patterns is poorly understood. It is everyday experience that breathing occurs automatically in waking and sleep, that all the breathing muscles can be activated or put at rest by voluntary will, that emotions affect the amount and pattern of breathing, and that the regular to-and-fro of quiet breathing is frequently interrupted as the respiratory muscles turn to the performance of specialized acts including speech, singing, swallowing, sighing, coughing, hiccuping, and defecation. Moreover, some respiratory muscles have a postural function. Each of these functions is probably served by a different neural circuit in the brain. Each uses slightly different combinations of respiratory muscles (speech and singing are controlled largely with intercostals, for example) and each of the circuits may be differently affected by sleep or drugs. The respiratory centers that have drawn the focus of most research are the circuits which are most active during general anesthesia or after decerebration of experimental animals. They generate a steady rhythmic breathing pattern, respond to changes in blood P_{CO_2} and P_{O_2}, and are quite likely the ones that govern breathing during sleep and quiet breathing in wakefulness.

Even in the anesthetized experimental animal (usually a cat) from which most of our information is drawn, there seems to be a hierarchy of respiratory rhythm generators, just as there is a hierarchy of rhythm generators in the heart (the ventricles, A-V node, and S-A node). A primitive rhythm generator in the spinal cord may be capable of carrying on if all higher centers fail. The medulla can generate a rather irregular rhythm by itself if all higher areas are destroyed. When the pons, which includes the *pneumotaxic center*, is intact, rhythm is faster and more finely controlled. In awake animals, cortical centers become important and may either influence lower centers or bypass them to control respiratory muscles directly. Most clinical observations can be explained satisfactorily by a simple scheme in which there are two major control circuits. One, called the metabolic system, is located in the pons and medulla. It seems to control breathing in sleep and anesthesia and to be mainly responsive to changes in gas tensions of blood or CSF. It has a descending pathway that lies in the anterolateral cord close to the reticulospinal tract. The other, called the behavioural system, includes circuits in the cerebral cortex, governs complex integrative breathing acts such as speech, may control breathing during much of the day, can supersede the metabolic system in many circumstances, and has its own separate descending pathway close to the corticospinal tract. Damage to one or another of these systems can occur in isolation (Fig. 6–4A).

The medullary and pontine centers have been extensively studied in animals, and several recent reviews provide good summaries of the physiological data. Since little of this information has direct clinical relevance, only a few salient features will be

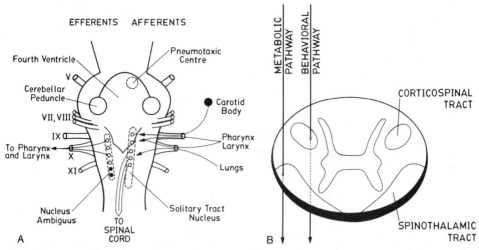

FIG. 6–4. *A.* Schematic diagram of the anatomy of the brainstem respiratory centers. Open circles in the solitary tract nucleus and nucleus ambiguus represent inspiratory neurons; solid circles are expiratory neurons. Connections between neurons are not shown, but some are described briefly in the text. *B.* Schematic diagram of the separate descending pathways for metabolic and behavioral systems.

described here. Although most of this information is derived from studies in cats, there is no reason to doubt that analogous circuits exist in man.

By successively cutting of pieces of the central nervous system until breathing is finally stopped, older experimenters located the irreducible minimum of brain required for breathing in the medulla. Extensive exploration of this region with modern electrophysiological techniques has shown two major pockets of neurons that fire in synchrony with breathing (Fig. 6–4*B*). One of these lies in or alongside the nucleus of the solitary tract, beneath the floor of the fourth ventricle. This nucleus is the receiving station for visceral sensory information, including taste, touch and other stimuli from the tongue, pharynx, larynx, and lungs. The other lies near the nucleus ambiguus, which is the visceral motor nucleus of the tenth nerve. The neurons beside the nucleus of the solitary tract, called the dorsal respiratory group (DRG), fire in inspiration and send axons directly to alpha motoneurons of the phrenic nerve. Some of them also receive afferent information from vagal pulmonary stretch receptors. When lungs expand, these receptors fire and the discharge, feeding directly to DRG cells, inhibits them. This simple circuit provides the neural mechanism for the Hering-Breuer reflex found in animals, where inflation of the lungs can interrupt inspiration. While this reflex controls respiratory rate and tidal volume in some mammals, it has not been shown to be very important in normal people. Inspiratory neurons in the DRG also receive inhibitory impulses from receptors in the upper airways. The physiological role of these is not clear, but they may be responsible for the diving reflex, swallowing reflexes, or the exaggerated tendency of newborn infants to slow their breathing when stimulated in the nose or face. In each of these cases, upper airway stimulation leads to apnea. The respiratory neurons of the nucleus ambiguus (called the ventral respiratory group, or VRG) include some motoneurons that supply pharyngeal and laryngeal muscles which contract in inspiration as well as others whose connections are not known. Some of the neurons in this area fire in expiration.

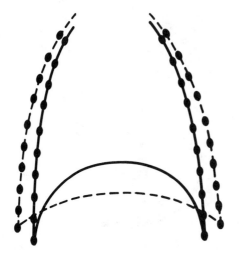

FIG. 6–5. Action of the diaphragm. The diaphragm is shown as a curved line across the bottom of the rib cage. At end-expiration, indicated by solid lines, rib cage diameter is small and the dome of the diaphragm is sharply curved. As the diaphragm contracts (dotted lines), the dome descends and becomes flatter, while the lower border of the rib cage is simultaneously pushed upward and outward.

The *pneumotaxic center* has been localized to the nucleus parabrachialis medialis in the upper pons. Its connections are not clear, but stimulation of the area can shut off inspiration and start expiration just as stimulation of vagal stretch receptors or upper airway receptors can.

In the lower pons are neurons which exert a steady tonic discharge, generally inspiratory. They may play a role by setting the end-expiratory level. That is, if there is a constant inspiratory tone, inspiratory muscles will still be active when all the phasic variation is gone at end-expiration and functional residual capacity will be above the resting level. When higher centers are destroyed, the balance of tonic activity may be altered and an overriding tonic inspiratory activity arising from this area may cause the form of breathing called apneusis. However, the idea that cells of this "apneustic center" play an important part in generating basic respiratory rhythm has been abandoned for the present.

The respiratory muscles

ACTION OF THE MUSCLES. The diaphragm is the principle muscle of respiration. It is attached by its edges to the lower borders of the rib cage and arches up under the lungs as shown in Figure 6–5. When its fibers contract, they generate tension in the whole curved sheet of muscle plus central tendon. This tension results in a pressure gradient across the diaphragm according to LaPlace's law:

$$(P_{abd} - P_{p1}) = \frac{2T}{R} \qquad (5)$$

where T is tension and R is radius of curvature.

Contraction thus generates negative pressure in the pleural space and causes the lungs to expand. The tension is also applied to the attachments of the diaphragm. The result of the tension depends on which part is fixed and which is movable. If the rib cage were solidly fixed in one position, contraction would draw down the dome of the diaphragm. On the other hand, if the diaphragm dome were fixed in position, tension would pull up the rib cage. Whether the diaphragm dome is fixed or can move depends on pressure in the abdomen, which in turn depends on the activity of abdominal muscles. In descending, the diaphragm must displace the contents of the abdomen, and this displacement can only be outward, by distending the anterior abdominal wall. If tense abdominal muscles prevent such motion, diaphragm contraction must raise the rib cage. It is clear that the muscles of the abdominal wall can act as synergists of the diaphragm. They remain relaxed in quiet breathing, but there is some positive pressure in the relaxed abdomen and the rib case is lifted up to some degree while the dome of the diaphragm descends. In some situations the diaphragm seems to be able to expand the rib cage without any support from abdominal pressure, but this action is not yet fully explained. In exercise, abdominal muscles contract in inspiration to support the diaphragm as ventilation increases.

GENERAL PRINCIPLES OF RESPIRATORY MUSCLE FUNCTION. The respiratory muscles are similar in their behaviour to other skeletal muscles. The distance a respiratory muscle moves or the volume it displaces when it contracts depends not only on the neural stimulus applied to it but on many other factors. These include the mechanical load against which it is working (in the case of the respiratory muscles the resistance and compliance of the lungs and chest wall); the resting length of the muscle (through the force-length relationship, similar to the "preload" in cardiac mechanics); the speed of contraction (through the force-velocity relationship); the mechanical advantage of the muscle attachments on the chest wall; the cooperation of other muscles; and muscle fatigue and training.

Both force-length and changes in mechanical advantage make inspiratory muscles less effective at high lung volume than at FRC. This is easiest to visualize in the case of the diaphragm, which becomes very short as it descends but also changes shape from a dome to a flat membrane stretched across the bottom of the rib cage. In that position its contraction will tend to draw the ribs inward, reducing lung volume rather than expanding it. Similarly, the expiratory muscles are most effective at high lung volume.

The importance of muscle synergy is shown by patients with quadriplegia due to high spinal cord lesions. Without contraction of the intercostal muscles during inspiration, the rib cage loses its rigidity so that contraction of the diaphragm, instead of lifting the rib cage as one unit and expanding it, tends to draw in the lower part of the rib cage and decrease thoracic volume. Breathing is thus bound to be less efficient.

MUSCLE FATIGUE. The idea that muscles can become fatigued is familiar from everyday experience, and clinicians have often wondered whether patients who appear exhausted in the final stages of respiratory failure due to emphysema or fibrosis may actually have fatigue of respiratory muscles. It has been difficult to define, detect or measure fatigue in any satisfactory way but recent interest in this subject promises to bring a better understanding. In general, fatigue may be said to be present when after repeated hard contractions a muscle can no longer be made to generate its normal force. When this happens the failure might be in the metabolic machinery of muscle contraction, in excitation contraction coupling, at the neuromuscular junction, in the spinal cord, in the motor cortex, or elsewhere in the central nervous system. It has been shown that it is possible to fatigue the inspiratory muscles of normal subjects by having them breathe repeatedly through a very high resistance until they are no longer able to generate a prescribed inspiratory pressure. Fatigue appears sooner if the effort is greater or if hypoxemia is present.

Just as they can be fatigued, respiratory muscles can be trained. Specific exercises such as breathing against an inspiratory resistor lead to improved strength and endurance of inspiratory muscles. Exercises can be directed specifically at strength or endurance. It might be hoped that specific training would prepare patients with serious lung disease to cope better with episodes of acute respiratory embarrassment, and this principle is being applied to patients with quadriplegia and children with cystic fibrosis. General exercise incidentally trains respiratory muscles and this may be an unsuspected but important benefit of exercise programs. It has been proposed that fatigue of respiratory muscles may give rise to dyspnea. If this is true, it may help to explain how physical conditioning can improve exercise tolerance in

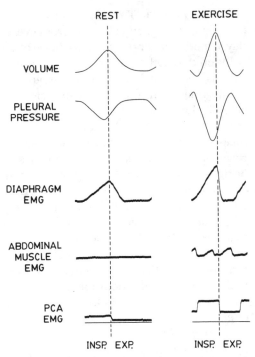

FIG. 6—6. Normal breathing. Volume is shown with inspiration up, pleural pressure with positive up. Pleural pressure and volume are slightly out of phase because of the flow resistance of the respiratory system. Instantaneous electrical activity of the diaphragm, external oblique muscle of the abdomen, and posterior cricoarytenoid (PCA) muscle of the larynx are shown.

patients with lung disease when there is no detectable change in lung function.

The pattern of normal breathing

Breathing is a complex, coordinated act involving simultaneous and sequential activation of many muscles. It is orchestrated by a central pattern generator in the central nervous system and can be adapted to suit many circumstances. Polygraphic tracings of two typical respiratory cycles of a normal seated subject are shown in Figure 6–6, one at rest and one at moderate exercise. At the top, a volume tracing shows inspiration and expiration. In quiet breathing the cycle begins at the end of a relaxed expiratory plateau with all muscles inactive. The diaphragm begins to be stimulated and its activity rises in more or less

straight ramp to the end of inspiration. Intercostal muscles turn on at about the same time to stabilize the rib cage against diaphragm tension and to expand the ribs outward, while the posterior cricoarytenoid muscle contracts to open the glottis. (Recordings of many other upper airway muscles would show similar activity as they contract to stiffen the walls of the oropharynx and keep it patent. Even the alae nasae have phasic activity in quiet breathing). During the inspiratory phase, the diaphragm descends and the belly expands, but the rib cage also expands. At the end of inspiration, diaphragm activity does not stop abruptly but declines gradually so that the diaphragm lets volume out slowly. Activity in posterior cricoarytenoids decreases, allowing the larynx to close partially and slowing expiration by increasing the resistance to flow. Exhalation is thus actively braked. The volume is monitored by stretch receptors and joint receptors while tension and length of respiratory muscles are measured continuously by tendon organs and muscle spindles, and this information is sent back to the central nervous system. Additional phasic motor output is sent to numerous muscles of the trunk to make minor postural adjustments as the mass of abdominal contents moves downward and upward.

This pattern can be modified in many situations. The whole pattern may be preempted by a cough, sigh, or speech pattern. If the subject lies down, subtle adjustments of individual muscle tone will compensate for the new geometry, and upper airway muscles will become more active to counteract the increased tendency of the tongue to fall back in the throat and the pharynx to collapse. The second polygraphic tracing shows what happens when ventilation is increased by the stimulus of exercise or increased carbon dioxide. Tidal volume is increased, as is respiratory frequency, shown by a decrease in the duration of the single breath. Diaphragm activity rises more steeply and reaches a higher

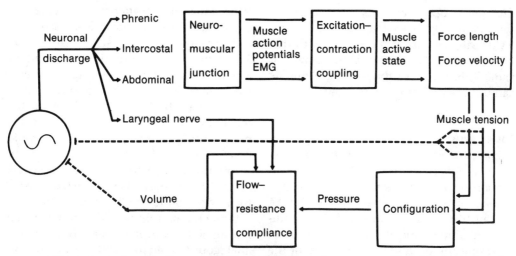

FIG. 6–7. A schematic diagram of the "output" of the respiratory centers. Neural discharge from the medullary or higher centers is integrated in the spinal cord and sent out along peripheral nerves, then goes through a series of transformations (indicated by boxes) before emerging as flow or volume. Along the way, output can be measured as electromyogram voltage, as intrathoracic pressure, or as flow or volume. At several points the output modifies itself through reflex feedback loops (indicated by dashed lines). Muscle spindles and tendon organs are stimulated by muscle length and tension, and modify activity at both the spinal and central levels. Lung volume directly affects lung airway resistance, and can affect neural output through stretch receptors. Ventilation changes blood gas levels and feeds back through the chemoreceptor reflexes.

peak. It then falls much more sharply at the beginning of expiration, while the laryngeal abductors remain more active, keeping the larynx wide open and allowing faster expiratory flow. Abdominal muscles, which were silent during quiet breathing, are now active in expiration. They may accelerate expiratory flow, but the more important effect of their activity at the end of expiration is to force the thorax below resting FRC and thus increase tidal volume. Abdominal muscle activity is also seen in inspiration when abdominal muscle contraction assists inspiration by increasing abdominal pressure, thus supporting the diaphragm and allowing it to life up the rib cage.

Output of the respiratory centers

When considering patients who appear to have some disorder of the ventilatory pump, it is convenient to keep in mind a simple scheme of the processes that take part in transforming the rhythmic discharge of respiratory neurons into movement of air (Fig. 6–7). Neural discharge is transmitted down the spinal cord, integrated at the anterior horn cell, and leaves the spinal cord and brain stem in numerous peripheral nerves. These can be grouped into the nerves to laryngeal and pharyngeal muscles, phrenic nerves, and segmental nerves to intercostal and abdominal muscles. The impulses then cross neuromuscular junctions, depolarize muscle cells, and set off muscle contraction. Tension generated by muscle metabolic machinery depends on fiber length and velocity. It pulls on muscle attachments, generating an amount of pressure that depends on the geometry of the thorax. The pressure in turn causes expansion of the respiratory system, but the extent and speed of the volume change depends on the resistance and compliance of both lungs and chest wall. Resistance of the upper airway can be changed by the muscles located there. At various stages, neural feedback loops can modify the output so that net output depends on peripheral sensory activity as well as respiratory center activity. Disease can interfere with this process at any stage along the way, but careful dissection of the various elements can usually pinpoint the problem.

Dyspnea

Dyspnea or *shortness of breath* is a common complaint, useful in directing the attention of the physician to disease of the respiratory or cardiovascular system. A careful history of dyspnea can be invaluable in pinpointing the diagnosis and in assessing the duration and severity of the disease. In general, dyspnea can be defined as a distressing sensation associated with breathing, but it means different things to different patients and to different clinicians. Shortness of breath and labored breathing often seem necessary for homeostasis, but there are many patients in whom the symptom causes much distress without any apparent benefit. Useful information about the sources and character of dyspnea comes from conversation with patients, from examination of patients with clear-cut neurological lesions and dyspnea, and from reports of physiologists who have experienced dyspnea, usually during breath holding or experiments in which mechanical loads have been added to the respiratory system.

It is customary to separate dyspnea from other readily defined feelings of discomfort in the chest. These include angina pectoris (which may in practice be difficult to separate from dyspnea on exertion), pleuritic or chest wall diseases which cause pain on inspiration, awareness of normal breathing or of normally increased ventilation as in exercise, and neurotic overconcern with patterns of breathing. Clinicians and physiologists who have experienced dyspnea have tried to classify the distressing sensations associated with breathing. Although there is no general agreement, it seems likely that dyspnea may include several

quite different sensations: (1) a vague feeling of tightness in the chest associated with irritation of airways, (2) a discomfort felt in the rib cage during breath-holding experiments and breathing against heavy mechanical loads, (3) anxiety provoked by hypoxemia, (4) dizziness and headaches due to high P_{CO_2}, and (5) a conscious fear of not breathing enough, found, for example, in chronic respirator patients when the respirator is turned off while they are awake. Physiologists have directed most of their attention to the first two of these.

While the neurophysiology of dyspnea is not agreed upon, it is clear that it is not analogous to pain in that there are not specific dyspnea receptors projecting to the central nervous system. Stimuli arising from many different sources appear to give rise to similar sensations. In what part of the central nervous system these are collected and how they reach the conscious level is not known.

Some suggestions as to the source of dyspnea are listed and discussed below.

1. *Awareness of increase in normal breathing movements.* When exposed to increased inspired carbon dioxide levels, subjects are unable to detect the increased excursion of their respiratory muscles until minute ventilation is twice its resting value. Subjects rebreathing CO_2 do not find breathing uncomfortable until they reach about 50 litres per minutes. During exercise, normal subjects reach very high ventilation before becoming aware of their breathing. Patients with Cheyne-Stokes breathing are often not aware of their unusual breathing pattern. The ability to detect increased breathing movements thus seems to be limited, and awareness of them is not at all distressing until very large volumes are reached.

2. *Awareness of abnormal blood chemistry.* Hypoxemia and hypercapnia cause symptoms of faintness, confusion, and anxiety but do not by themselves usually

cause shortness of breath. Inadequate oxygenation of the brain causes confusion and euphoria, often before the drive to ventilation becomes intense enough to cause distress. Unconsciousness and death may therefore occur without warning. This poor correlation between hypoxemia and dyspnea can result in death in aviators, underwater swimmers, and divers. Ignorance of this principle on the part of physicians and nurses can lead to the death of patients with hypoxemia if lack of dyspnea is interpreted as proof of adequate oxygenation. In the presence of hypoxemia (at altitude, for example), exertional dyspnea occurs at lower work loads than when oxygen is normal, but this may be due to awareness of increased breathing effort rather than directly to the low P_{O_2}.

3. *Awareness of decrease in breathing movements.* The ability of normal subjects to notice when their ventilation is reduced has not been measured. Most patients who underventilate without having obstructive or restrictive lung disease appear to be relatively free of distress. When a normal subject was paralyzed with curare, he was not made uncomfortable by the lack of breathing movements. Paralyzed patients seem to suffer little dyspnea while underventilating, although they are subject to unpleasant sensations due to hypoxia and hypercapnia. After a period of overventilation on a respirator such patients may become distressed when they are removed from the respirator and minute volume is lowered and P_{CO_2} raised to normal. Such patients are often unable to tolerate normal levels of CO_2 after becoming accustomed to levels that are too low, and the sensation in these cases seems to arise from the chemical stimulus.

4. *Awareness of stimuli arising in lung receptors.* Hyperpnea and dyspnea are often seen in patients whose airways or lung parenchyma is diseased and whose vagal receptors, particularly irritant or J receptors, could be assumed to be overactive. It has been argued on clinical grounds that the special complaint of tightness in the chest may be specifically caused by lung receptors. Vagal blockade has abolished this symptom in acute asthma and has reduced dyspnea in patients with restrictive lung disease, pulmonary artery obstruction, and pulmonary lymphatic carcinomatosis. Vagal blockade also lengthens the time that normal subjects are able to hold their breath. In this situation, however, the blockade results in a decrease in the involuntary muscle contractions that occur during breath holding and it may be that the sensation of dyspnea comes from these movements of the chest wall rather than directly through the vagus.

5. *Awareness of stimuli arising from the chest wall.* Muscle spindle receptors which seem ideally placed to transmit information about movement, position, and tension in muscles are apparently not responsible for providing any conscious sensation. Joint receptors, on the other hand, probably inform the conscious subject about the movement and position of his respiratory muscles. Patients with high spinal cord sections are often unaware of their breathing movements even though the vagus nerve remains intact. The role of tendon organs in conscious sensation is not known. A case has been described in which severe dyspnea was relieved by local anesthesia of segmental nerves supplying an area of the lower rib cage previously affected by disease.

6. *Awareness of added mechanical loads.* A number of experiments show that normal subjects are capable of detecting very small changes in respiratory system impedance when these are produced by adding external mechanical loads to simulate increased airway resistance or decreased compliance. Receptors for this are probably located in the chest

wall, although pharyngeal pressure sensations may contribute. Such experiments prove that information about mechanical loads can reach the conscious level and make it reasonable that mechanical changes could produce dyspnea.

Many examples of dyspnea can be explained by postulating that the distress comes from awareness of unexpected or undue strain upon the respiratory muscles or the chest wall. As the muscles contract against an unusually high load (or against a thorax in an unusual shape) they generate high tensions within themselves as well as pulling on the rib cage and altering its shape. These strains might be detected by tendon organs or joint receptors. Stimuli such as high P_{CO_2}, low P_{O_2}, and drive from pulmonary receptors can act through this mechanism. Instead of directly stimulating the sensory cortex, they may act by stimulating respiratory centers, causing stronger contraction of respiratory muscles and large strains in the chest wall which secondarily give rise to dyspnea. Such a theory can explain why patients with damaged respiratory centers or weak muscles usually do not complain very much of dyspnea. Similarly, it would predict that subjects who try harder to breathe against obstruction would have more dyspnea, the dyspnea being secondary to effort rather than the other way around.

Voluntary breath holding is a convenient way to produce dyspnea in normal subjects and illustrates some of the principles above. In breath holding, or apnea of any cause, alveolar and arterial gas tensions change in a predictable way, but not as quickly as most people imagine. The rate of change of oxygen and carbon dioxide depend on the rate of production of carbon dioxide and consumption of oxygen on the starting volume of gas available as stores, and on the interaction of P_{O_2} and P_{CO_2} through the Haldane and Bohr effects. As it

turns out, the storage space for carbon dioxide is large, and P_{CO_2} rises only about 5 mm. Hg per minute during breath holding after oxygen breathing and about 15 mm. Hg per minute after air breathing. The body oxygen stores are smaller, being confined to the oxygen in alveolar gas and hemoglobin, so that oxygen falls at a rate of about 1 mm. Hg per second in a breath-hold following air breathing.

Breath holding becomes uncomfortable long before gas tensions reach a level that causes distress, and the discomfort seems to be related to involuntary contractions of the respiratory muscles. Soon after the beginning of a breath-hold the diaphragm begins to make contractions, after which breathing is prevented only by closing off the upper airway. The contractions become stronger and stronger, generate large negative pressures in the thorax and distort the chest wall, probably causing dyspnea by stimulating chest wall receptors. If the contractions are prevented by administering curare, apnea can go on for several minutes without significant discomfort. In breath holding, the abnormal blood gas tensions thus seem to cause dyspnea indirectly, by causing respiratory muscle contractions, which in turn act on some sensory mechanism in the chest wall.

References

General reviews of regulation of respiration

Berger, A. J., Mitchell, R. A., and Severinghaus, J. W.: Regulation of respiration (1, 2, and 3). N. Engl. J. Med. 297:92, 138, 194, 1977.

Hornbein, T. F. (ed.): Regulation of Breathing, parts 1 and 2. In Lenfant, C. (ed.): Lung Biology in Health and Disease, Vol. 17. New York, Marcel Dekker, 1981.

Neural respiratory centers

Cohen, M. I.: Neurogenesis of respiratory rhythm in the mammal. Physiol. Rev. 59:1105, 1979.

Mitchell, R. A.: Neural regulation of respiration. In Williams, M. H. (ed.): Disturbance of respiratory control,

Clinics in Chest Medicine, Vol. 1, No. 1. Philadelphia, W. B. Saunders, 1980.

Newsom, Davis, J.: Control of the muscles of breathing. *In* Widdicombe, J. G. (ed.) Respiratory Physiology. London, Butterworth & Co., 1974.

Wyman, R. J.: Neural regulation of the breathing rhythm. *In* Knobil, E. (ed.): Annual Review of Physiology, p. 417. Palo Alto, Annual Reviews, 1977.

Peripheral and central chemoreceptors

Purves, J. J. (ed): Chemoreceptor mechanisms. Cambridge, Cambridge University Press, 1975.

Loeschke, H. H.: Central nervous chemoreceptors. *In* Widdicombe, J. G. (ed.): Respiratory Physiology, Vol. 2. MTP International Review of Science. London, Butterworth & Co., 1974.

Pappenheimer, J. P.: The ionic composition of cerebral extra-cellular fluid and its relation to control of breathing. Harvey Lectures, Series 61, p. 71. New York, Academic Press, 1967.

Respiratory reflexes

Widdicombe, J. G., and Fillenz, M.: Receptors of the lungs and airways. *In* Neil, E. (ed.): Enteroreceptors. Handbook of Sensory Physiology, Vol. 3. Berlin, Springer-Verlag, 1972.

Respiratory muscles

Derenne, J.-P. H., Macklem, P. T., and Rousson, C. H.: The respiratory muscles: Mechanics, control and Pathophysiology. Parts I, II, and III. Am. Rev. Respir. Dis. *118:*119, 373, 581, 1978.

Campbell, E. J. M., Agostoni, E., and Newsom Davis, J.: The respiratory muscles. London, Lloyd-Luke, 1970.

Macklem, P. T.: Respiratory muscles: The vital pump. Chest *78:*753, 1980.

7

Ventilation, Perfusion, and Gas Exchange

Reuben M. Cherniack, M.D.

The major functions of the lung are to provide oxygen to the blood, perfusing it so that it may be carried to the tissues, and to remove carbon dioxide which has been produced in the tissues. This exchange of gases is accomplished by the convective movement of air into the lungs by the breathing movements (ventilation) and the convective movement of blood through the lungs by the pumping action of the heart (perfusion).

Ventilatory function

Lung volumes

The breathing movements take place within the framework of the total lung capacity (TLC), which can be divided into several components or subdivisions. The absolute values of these subdivisions depend upon the age, sex, and size of the person, although the proportion of the TLC that each occupies is remarkably similar. Normally, the resting level (i.e., the end-expiratory position) or functional residual capacity (FRC) is about 40 per cent of the TLC. The lungs cannot be entirely emptied voluntarily, and the gas that remains at the end of a maximum expiration is the residual volume (RV). An increased RV, particularly if it was greater than 30 per cent of the

TLC, was for many years believed to indicate emphysema. However, an increase in RV indicates only that the lungs are hyperinflated, and this can occur in any condition associated with obstruction of the airways, such as bronchial asthma. In addition, RV increases with age and may be as much as 50 per cent of the TLC in an elderly healthy person.

The most widely used method for calculating lung volume involves having the subject pant while in a body plethysmograph and determining the relationship between the mouth pressure (identical to alveolar pressure) and that in the plethysmograph while the subject makes inspiratory and expiratory efforts against a closed shutter. Under these circumstances decompression and compression of gas in the lungs during the panting maneuver against the closed shutter is reflected by changes in the opposite direction of the pressure in the box. Using Boyle's law, one can then calculate lung volume or thoracic gas volume (TGV) at the point of obstruction.

Mechanical properties

In order to carry out the breathing movements, the respiratory muscles must overcome the elastic and flow-related resistances of the lungs and chest wall. The elastic resistance is related to the amount of distention of the respiratory system or the tidal volume; the flow resistance is related to the rate of change in tidal volume, or airflow. Definitive information about these mechanical resistances can be gained by simultaneous measurement of the volume changes, the rate of airflow, and the variations in pressure that are involved.

ELASTIC RESISTANCE. The elastic properties, or the compliance, of the respiratory system are determined by measurements of the amount of distention induced by a change in pressure under static conditions, when there is no air flowing. The compliance of the lungs can be determined by relating the change in volume to the change in transpulmonary pressure from end-expiration to end-inspiration. Similarly, the compliance of the chest wall can be determined by relating the change in lung volume to the difference between the intrathoracic pressure and the pressure exterior to the chest wall.

Examples of the elastic behavior, or the compliance, of three different types of lung are shown in Fig. 7–1. In the normal lung (A), a change in transpulmonary pressure of 5 cm. H_2O results in an inspiration of 1 liter of air, so that its compliance is 0.200 l./cm. H_2O. When the lung loses its elasticity (B), as in emphysema, the same change in transpulmonary pressure leads to an inspiration of 2 liters of air so that the compliance is 0.400 l./cm. H_2O. If the lungs are stiff (C), as in pulmonary fibrosis or edematous congestion, this change in transpulmonary pressure results in an inspiration of only 0.5 liter, so that the compliance is only 0.100 l./cm. H_2O.

While the calculation of lung compliance is a useful determination, the relationship between the transpulmonary pressure and lung volume is not linear. Therefore, no single value of lung compliance describes the elastic properties in an individual, and any determined value will depend on the lung volume at which it is calculated. This is illustrated in Figure 7–2, in which the pressure-volume relationships of the lung in a healthy individual and in three representative patients with pulmonary disease are shown. It can be seen that the pressure-volume curve is shifted downward and to the right (i.e., lung volume is reduced and transpulmonary pressure is greater at any volume) in the patient with pulmonary fibrosis, and over the tidal volume range the slope of the curve is decreased (i.e., lung compliance is low). In both the patient with emphysema and the one with asthma, the pressure-volume curve is shifted upward and to the left (i.e., lung volume is increased and transpulmonary pressure is less at any lung volume). However, the slope of the curve is different in the two patients; in emphysema, the slope of the curve over the tidal volume range is increased (lung compliance is high), while in asthma (or in bronchitis) the slope is the same as that of a healthy individual.

SURFACE PROPERTIES. Because one cannot empty the lungs completely, it is im-

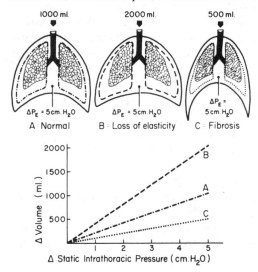

FIG. 7–1. The distention produced by a change in intrathoracic pressure of 5 cm. H_2O in a normal lung (A), a lung which has lost elasticity (B), and a lung which has become fibrosed (C). (Cherniack, R. M., Cherniack, L., and Naimark, A.: Respiration in Health and Disease, ed. 2. Philadelphia, W. B. Saunders, 1972.)

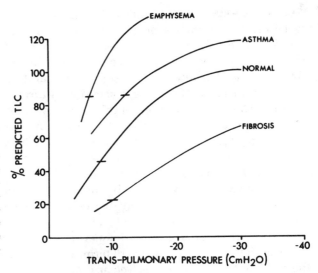

FIG. 7–2. The pressure-volume characteristics in a normal individual and three patients with respiratory disease. In pulmonary fibrosis, the curve is shifted downwards and to the right and its slope is altered. In asthma and emphysema, the curve is shifted upwards and to the left; the slope of the curve is normal in asthma and increased in emphysema.

possible to describe the pressure volume behavior of the lungs completely *in vivo*. However, much can be learned about the elasticity of lungs from study of excised lungs. The pressure-volume characteristics of the excised lung when it is filled with air are different from those of one filled with saline (Fig. 7–3); transpulmonary pressures are much greater at any given volume when the lung is filled with air. This is because the surface tension at the air-liquid interface during air filling is much greater than the negligible amount at the liquid interface during saline filling. More than one half of the measured elastic recoil of the lungs is due to the surface-active forces in the lung. Despite these surface forces, the lungs normally do not collapse completely, even after a maximal expiration. A surface-active material (surfactant) that lines the surface of the terminal lung units (alveoli, alveolar ducts, respiratory bronchioles) allows the lung to remain inflated at low pressures and also allows alveoli of different size to coexist at the same pressure. A deficiency of this material may render the lung unstable and promote collapse (atelectasis). Impaired surface activity has been demonstrated in lungs of children dying from hyaline membrane disease, in the adult respiratory distress syndrome

FIG. 7–3. Static pressure-volume relationships when the lungs are inflated with air and with saline. (Cherniack, R. M., Cherniack, L., and Naimark, A.: Respiration in Health and Disease, ed. 2. Philadelphia, W. B. Saunders, 1972.)

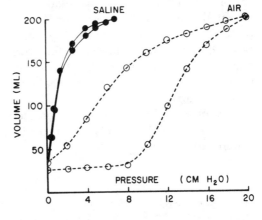

(ARDS), and in patients exposed to high concentrations of oxygen for lengthy periods, and is seen after open heart surgery and in a wide variety of experimental conditions. These conditions may be associated with altered alveolar cell function and a deficiency in available surfactant, which results in alveolar instability and a tendency to develop focal areas of atelectasis, which, in turn, aggravates any gas exchange abnormality.

FLOW RESISTANCE. In addition to inertia, which must be overcome during acceleration and deceleration, airway resistance and tissue viscance must be overcome during breathing. Because the resistance to movement of air in the airways accounts for about 85 per cent or more of the resistance, it is the airway resistance or the total flow resistance that is usually assessed in patients.

To calculate airway resistance, the airway pressure and flow must be determined. Since the relationship between plethysmograph pressure and airway pressure (derived during the estimation of lung volume), is known, one can determine the relationship between the alveolar pressure and related airflow during unobstructed breathing, and the flow resistance of the tracheobronchial tree (the pressure differential between the airway opening and the alveoli per l./sec. of airflow) can be calculated. The use of the body plethysmograph to measure the flow resistance is particularly useful, because like the elastic properties of the lung, the size of the bronchi and thus the airflow resistance vary with lung volume. To correct for the effect of lung volume, conductance (G), which is the reciprocal of resistance, is expressed per unit lung volume and is called specific conductance (SGaw).

FIG. 7–4. The effect of a local bronchial obstruction on the pressure-volume relationship of the lungs when 1000 ml. of air is inhaled. (Cherniack, R. M., Cherniack, L., and Naimark, A.: Respiration in Health and Disease, ed. 2. Philadelphia, W. B. Saunders, 1972.)

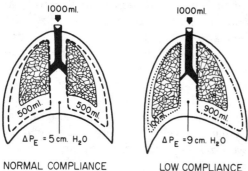

NORMAL COMPLIANCE   LOW COMPLIANCE

In the normal subject, the total flow resistance of the lungs is approximately 1.8 cm. H_2O/l./sec. of airflow, the greater part of this resistance being due to laminar flow, and only about one-tenth of it is due to turbulence. In patients suffering from airway obstruction, the total flow resistance is increased, predominantly because of turbulence, and may be greater than 5 cm. H_2O/l./sec. of airflow. When the ventilation is increased above normal, such as during exertion, this turbulent resistance becomes exceedingly high.

INTERACTION BETWEEN FLOW RESISTANCE AND ELASTIC RESISTANCE. A substantial proportion of the total airway resistance is present in the major airways; the small peripheral airways, which provide a large cross-sectional area for airflow, contribute less than 20 per cent to the total flow resistance. As a result, even though the peripheral airways are considerably diseased and ventilation of the distal air spaces is markedly impaired, the total airway resistance may be virtually unaltered. Fortunately, such defects may be recognized by techniques which evaluate the distribution of ventilation, ventilation-perfusion ratios, and gas exchange in the lungs.

The distribution of inspired gas to individual peripheral lung units depends on the uniformity of the distribution of impedances to inflation of the units. When two or more parallel units of lung are subjected to the same inflation or deflation pressure, each fills or empties at a rate determined by its time constant (i.e., the product of its flow resistance and compliance). If the time constants of the peripheral lung units are equal, they will fill or empty uniformly; conversely, if the time constants are unequal, filling or emptying will be nonuniform. When there is no localized disease, a tidal volume is equally distributed to different areas of lung (Fig. 7–4, *left*), and in normal individuals this appears to be what happens. This indicates that in healthy individuals, the time constants are relatively uniform throughout the lung. However, when there is a localized airway obstruction, the air tends to move into the areas of the lung that offer the least resistance (Fig. 7–4, *right*). Under these circumstances, for the same total volume of air to be inspired, a greater intrathoracic pressure must develop, and as a result the calculated compliance will be lower. When the time constants are unequally distributed in the lung, the inspired air is distributed unequally. This disturbance of distribution increases concomitantly with a fall in lung compliance when the rate of breathing is increased. The same finding would be seen if the flow resistance of peripheral lung units were equal everywhere but their compliance were dissimilar.

Alterations in lung compliance related to increasing respiratory rate have been termed frequency-dependent compliance; and in the absence of an abnormality of standard tests of ventilatory function or the overall pressure-volume characteristics of the lung, this finding is generally considered to represent the presence of "small airway disease." Clearly, since a similar finding would be present if the compliance of the peripheral lung units was nonhomogenous, it would be preferable to speak of peripheral lung unit dysfunction.

Another test which may reflect early airway disease is the single-breath nitrogen test, in which the "closing volume" or "closing capacity" (the volume at which airways presumably close in the lungs) and the slope of Phase III (the change in expired N_2 concentration during the slow expired vital capacity) are measured. It has been suggested that closing volume or the slope of Phase III is abnormal (increased) when all other simple tests of ventilatory function are normal in many cigarette smokers and in asthmatics in remission. It would appear that the volume at which airways close is related to either narrowing of the lumen of the peripheral airways, as in bronchitis, or a loss of radial traction on the airways, i.e., a loss of lung elastic recoil,

as in emphysema. On the other hand, the slope of Phase III is another way to examine the distribution of time constants in the lung. Although the single-breath nitrogen test has been well validated and its variability well studied, its usefulness in recognizing early disease and its prognostic implications are still not established sufficiently to warrant its general use at this time.

Work of breathing

The amount of mechanical work necessary to overcome the resistances offered by the lung and the chest wall during breathing is performed by the respiratory muscles. During quiet breathing, almost all of the muscular work is carried out during inspiration, for the elastic recoil of the lungs is sufficient to overcome the flow resistance of both the air and the tissues during expiration. If the expiratory resistance is high, however, expiratory muscular work may be required. In normal subjects, the total mechanical work performed on the lungs has been estimated to be approximately 0.3 to 0.7 k./m./min. In a patient suffering from airflow limitation (such as bronchial asthma or emphysema), the mechanical work necessary to overcome the flow resistance is increased considerably; in pulmonary fibrosis, much more work must be performed in order to overcome the high elastic resistance of these "stiff lungs."

OXYGEN COST OF BREATHING. In order to perform the mechanical work, the respiratory muscles require oxygen. Figure 7–5 illustrates the change in oxygen consumption associated with increasing ventilation in healthy individuals, as well as in patients suffering from emphysema, congestive heart failure, and obesity. In the healthy subject, the oxygen cost of breathing at rest varies from 0.3 to 1.0 ml./l. of ventilation (about 2 per cent of the total oxygen consumption), and the minute ventilation can be increased considerably with little alteration in the total oxygen consumption. After a certain level of ventilation is reached, the

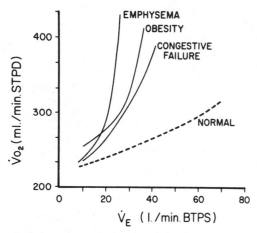

FIG. 7-5. The changes in oxygen consumption ($\dot{V}O_2$) associated with increasing ventilation in a normal subject and patients with congestive heart failure, obesity and emphysema. (Cherniack, R. M., Cherniack, L., and Naimark, A.: Respiration in Health and Disease, ed. 2. Philadelphia, W. B. Saunders, 1972.)

oxygen consumption rises disproportionately with further increases in ventilation. If the mechanical resistances are increased, the oxygen consumption rises considerably even with small increases in ventilation, and it may amount to as much as 50 per cent of the total oxygen consumption. In addition, the disproportionate increase in oxygen consumption with increasing ventilation will occur at a much lower level of ventilation. The high oxygen requirements of the respiratory apparatus means that the proportion of the oxygen consumption available for the other muscles of the body is reduced; this is extremely important clinically, particularly during exertion.

It has been suggested that respiratory muscle fatigue may play a role in the development of respiratory failure. Since skeletal muscle will become fatigued when the rate of energy consumption (oxygen cost) is greater than the rate of energy being supplied to it, this suggestion has validity. A low mechanical efficiency, a reduction in energy stores or a reduced supply to the muscles (i.e., a low O_2 content or cardiac output), or an increase in external mechanical work performed because of flow limitation or stiff lungs may predispose to the development of respiratory muscle fatigue. Indeed, it has been shown that respiratory muscle fatigue will develop in healthy individuals at transdiaphragmatic pressures greater than 40 per cent of the maximum achievable when they breathe through added resistance. The fatigue will come on sooner if the individual breathes a hypoxic gas mixture or at an increased FRC. That fatigue of the respiratory muscles may be an important clinical event is evidenced by the fact that the strength and endurance of the respiratory muscles can be increased by training patients with chronic airflow limitation as in cystic fibrosis and in quadraplegic individuals.

Another important aspect of the oxygen cost of breathing, particularly in respiratory disease, is that, when ventilation is increased, the tendency to lower the alveolar carbon dioxide tension may be offset by the increased metabolic production of carbon dioxide by the respiratory muscles. Even though the ventilation could be increased still further, it would serve no useful purpose, because the greater ventilation would merely increase the tendency toward carbon dioxide retention. This has been estimated to occur in normal individuals at a ventilation of about 140 l./min.; but in patients with an increased oxygen cost of breathing, the level of ventilation that is maximally effective in lowering the carbon dioxide tension may be as low as 15 to 20 l./min.

In addition, there is apparently a relationship between the amount of mechanical work of breathing and the rate at which an individual breathes. For any given alveolar ventilation, both normal subjects and patients with respiratory disease appear to breathe at a rate and depth at which the work of breathing is minimal. When the elastic resistance is increased, as in pulmonary fibrosis or kyphoscoliosis, the respirations tend to become rapid and shallow, probably because of the greater work required to overcome the elastic resistance with even small increases in tidal volume. In contrast, when the flow resistance is increased, the respirations tend to become slower and deeper—because a faster respiratory rate leads to an increase in the resistance to airflow.

Exchange of gases in the lungs

The air that enters the alveoli takes part in gas exchange with mixed venous blood which is arriving in the pulmonary capillaries, so that the gas tensions of the pulmonary capillary blood and the alveolar air come almost into equilibrium. This transfer of gases at the alveolar level is entirely due to diffusion and thus is determined by the partial pressures of the gases on both sides of the alveolocapillary membrane. Normally, the partial pressures of oxygen and of carbon dioxide of the mixed venous blood coming to the pulmonary capillaries are 40 and 46 torr, respectively, while those of the blood leaving the alveoli and entering the pulmonary veins are 100 and 40 torr, respectively. If the amount of air taking part in gas exchange is reduced, or if there is failure of the gas tensions in the alveolar and pulmonary capillary blood to come into equilibrium, the pulmonary venous and arterial blood gas tensions will be abnormal.

Failure of the gas tensions in the alveoli and the pulmonary capillary blood to equilibrate indicates an impairment of diffusion. As a result, the oxygen tension gradient between the alveoli and the artery $(P(A-a)O_2)$ will be increased (>10 torr). The $P(A-a)O_2$ will also be greater than normal when ventilation-perfusion ratios are uneven in different areas of the lungs, or there is true venous admixture.

Distribution of ventilation

Even in young healthy persons, and particularly in older persons, the inspired air is not distributed uniformly throughout the lungs (Fig. 7–6). Because of the effect of

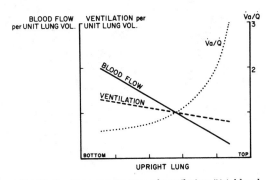

FIG. 7–6. The distribution of ventilation (VA), blood flow (Q̇) and ventilation-perfusion relationships in the upright lung. (Cherniack, R. M., Cherniack, L., and Naimark, A.: Respiration in Health and Disease, ed. 2. Philadelphia, W. B. Saunders, 1972.)

gravity on transpulmonary pressure, the air spaces at the top of the lung are expanded more than those at the bottom. When breathing occurs at the normal resting level, less of the inspired air will go to the air spaces at the top of the lung than to those at the bottom of the lung. In contrast, if one were to breathe near the residual volume, the air spaces at the top of the lung would be ventilated more than those at the bottom of the lung.

In addition to the gravity effect, as we have seen earlier, the distribution of the inspired gas is also affected by regional alterations in the mechanical resistances offered by the lung, the airways, and the extrapulmonary structures. This is illustrated in Figure 7–7, in which the situation depicted in Figure 4–4 is shown again, but it is the effect of the localized obstruction on the nitrogen concentration in the expired air following a single inspiration of pure oxygen which is presented. In the normal situation, the inspired oxygen enters all areas of lungs equally and the air leaves these areas synchronously and equally during expiration (Fig. 7–7A). As a result there is a definite expiratory plateau of the nitrogen concentration curve. On the other hand, in the presence of localized airway disease, the inspired oxygen enters the areas of lung that offer the least resistance before it enters the others, and during

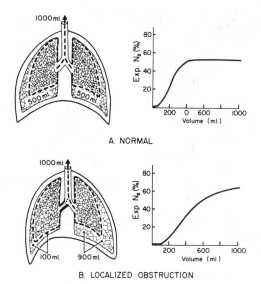

FIG. 7–7. The effect of a localized airway obstruction on the distribution of air. (Cherniack, R. M., Cherniack, L., and Naimark, A.: Respiration in Health and Disease. ed. 2. Philadelphia, W. B. Saunders, 1972.)

expiration, it moves out of the unobstructed areas before it leaves the obstructed areas. This asynchronous delivery of air from the two lungs during expiration results in a rising alveolar nitrogen concentration (Fig. 7–7B). As discussed earlier, provided the function of the central larger airways and the overall elastic characteristics of the lung are normal, the change in the slope of the nitrogen concentration plateau (which has been called Phase III of the curve) during a slow expired vital capacity maneuver following the inhalation of 100 per cent oxygen, or an increase in the volume at which there is a terminal increase in nitrogen concentration (closing volume or closing capacity), likely indicates peripheral lung unit disturbances.

Distribution of perfusion

Like the inspired gas, the distribution of blood flow in the lungs is affected by gravity (Fig. 7–6). The pressures in the arteries and the veins near the top of the lung are lower than those at the bottom, so that blood flow is less at the apex than at the base of the upright lung. The apical-basal blood flow differences are largely abolished in the supine position, since the two parts of the lung are nearly at the same hydrostatic level, but there will still be some differences between the upper and lowermost portions of the lung. Similarly, if one lies on one side, the dependent lung will be better perfused, in comparison to the contralateral lung. During exercise, as a result of a rise in pulmonary artery pressure, the total blood flow increases at the apex more than at the base, so that there may be a slightly more uniform distribution of blood.

Matching of ventilation and perfusion

Even though the total alveolar ventilation and the total pulmonary blood flow may be normal, hypoxemia develops if the distribution of the ventilation in relation to the perfusion of lung units is not uniform throughout the lungs. Figure 7–6 demonstrates that even in the normal upright lung, the ventilation-perfusion ratios are high at the apex (top) and low at the base (bottom) of the lung. Clearly, then, the gas concentrations in the alveoli must also differ from region to region, even in healthy individuals, but these regional differences are not large enough to interfere materially with gas exchange, so that the mixed arterial blood normally has nearly the same composition as the mixed alveolar gas (i.e., the $P(A\text{-}a)O_2$ is normal).

In patients with cardiorespiratory disease there are frequently gross variations in ventilation-perfusion relationships throughout the lung, so that the $P(A\text{-}a)O_2$ is increased. Perfusion of inadequately ventilated alveoli or, in the extreme case, perfusion of nonventilated alveoli (i.e., low ventilation-perfusion areas) will mean that poorly aerated venous blood leaves these pulmonary capillaries and mixes with fully "arterialized" blood coming from other well-ventilated alveoli, so that hypoxemia and hypercapnia are found in the arterial blood. This is called venous-admixture-like perfusion. Hypercapnia will not be present if there is sufficient hyperventilation of the remaining well-perfused alveoli. However, because of the shape of the oxyhemoglobin dissociation curve, hyperventilation does not significantly hyperoxygenate the blood leaving the poorly ventilated regions, and the arterial hypoxemia is not corrected to any significant degree.

Where the ventilation of alveoli is maintained but the blood perfusion is limited, or, in the extreme case, when there is no perfusion (i.e., high ventilation-perfusion areas) the gas leaving such alveoli tends to have the same composition as the air in the tracheobronchial tree and thus contributes to the physiologic dead space. The blood that perfuses these well-ventilated alveoli becomes fully oxygenated and probably excessively depleted of carbon dioxide, so that the arterial gas tensions are usually normal, although the CO_2 tension may be low, as long as normally perfused alveoli are well ventilated (i.e., there is no associated venous-admixture-like perfusion).

By use of techniques employing the inhalation and intravenous injection of radioactive tracer materials such as ^{133}Xe, much has been gained in the understanding of factors that normally influence the distribution of ventilation and perfusion of the lungs (e.g., posture, as well as the impact of respiratory disease). However, it is important to understand that the perfusion calculated using these techniques represents only that blood flow which is perfusing ventilated alveoli. If there is perfusion of nonventilated regions of lung, or of lung units whose conducting airways are obstructed (i.e., venous-admixture-like perfusion) this will not be recognized by the gaseous isotope techniques.

True venous admixture

Mixed venous blood that does not come into contact with alveoli but mixes with blood in the pulmonary or the systemic circulation (true venous admixture) may also cause an elevated $P(A-a)O_2$ and a subnormal arterial oxygen tension. Even in healthy individuals, approximately 2 to 4 per cent of the total pulmonary blood flow is shunted. An increased amount of true venous admixture may be intracardiac (as in some congenital heart diseases) or from the pulmonary artery to the pulmonary vein (as in pulmonary arteriovenous fistulae). A picture resembling that of true venous admixture can also be encountered if areas of lung are nonventilated but still perfused (such as atelectasis or pulmonary edema). Although considerable hypoxemia may be present in true venous admixture, carbon dioxide retention is frequently not encountered, because the lungs are usually healthy and capable of compensatory hyperventilation so that hypocapnia is often present.

Diffusion

Altered diffusion only rarely accounts for hypoxemia. A diffusion defect may develop in disease when the alveolocapillary membrane is markedly thickened (as in diffuse fibrosis), the capillary bed is considerably reduced (as in thromboembolic obstruction) or if a large volume of lung is destroyed by disease or removed by surgery. Even then, the diffusion defect probably only contributes to a high $P(A-a)O_2$ in such patients when they breathe a low concentration of oxygen during heavy exercise. In general, the finding of a low diffusing capacity is the result of marked mismatching of ventilation and perfusion. Similarly, impaired diffusion probably never significantly limits exchange of carbon dioxide in the lungs, since it diffuses 20 times faster than oxygen does.

Alveolar hypoventilation

In a healthy person only about 70 to 80 per cent of each tidal volume reaches the alveoli, supplying oxygen to, and removing carbon dioxide from, the pulmonary capillary blood. The amount of ventilation taking part in gas exchange is called the alveolar ventilation, the remainder being wasted (dead space component). In many patients suffering from pulmonary disease, a greater amount of the inspired air is wasted, so that the proportion of the inspired air that takes part in gas exchange is reduced. Normally, an increase in the tidal volume could compensate for the effect of the increased dead space, but the patient with chronic respiratory disease is frequently unable to increase his ventilation sufficiently to provide an adequate alveolar ventilation because of the mechanical disturbances in the lungs.

Whenever the amount of air taking part in gas exchange is inadequate to cope with the metabolic production of carbon dioxide, the arterial carbon dioxide tension will be elevated. Under these circumstances (i.e., alveolar hypoventilation) the hypercapnia is always associated with hypoxemia (unless the individual is inhaling an oxygen-enriched gas mixture). On the other hand, in pure alveolar hypoventilation the gas tensions in the alveoli and the pulmonary capillaries come into equilibrium so that the $P(A-a)O_2$ is normal (<10 torr) in uncomplicated cases. Hypoxemia and hypercapnia due to alveolar hypoventilation are encountered in conditions in which the total ventilation is decreased (e.g., barbiturate poisoning or muscular paralysis), the respiratory pattern is rapid and shallow (e.g., obesity or kyphoscoliosis), or the dead-space ventilation is increased without a concomitant increase in minute ventilation (e.g., emphysema).

Even if the total ventilation appears to be normal, hypoxemia and hypercapnia will be present whenever the CO_2 production is high for that particular alveolar ventilation. If Figure 7–5 is considered to be representative of the relationship between alveolar ventilation and carbon dioxide production rather than minute ventilation and oxygen consumption, it will be seen that any alveolar ventilation will be associated with a greater than normal CO_2 production in patients suffering from cardiopulmonary insufficiency. Clearly then, the high oxygen cost of breathing found in respiratory disease has important implications in the development of respiratory insufficiency, and any exertion or respiratory insult that requires an increase in ventilation may be associated with aggravation of hypoxemia and hypercapnia.

Respiratory insufficiency

Respiratory insufficiency may develop in a multitude of conditions, and may be manifest by severe hypoxemia alone or in association with carbon dioxide retention.

Hypoxemic respiratory failure

Nowadays the major form of acute respiratory insufficiency encountered is severe hypoxemic failure, which develops in a wide variety of medical and surgical conditions which do not involve the lungs initially. Severe hypoxemic respiratory failure may complicate multiple trauma, prolonged shock or hypotension, fluid overload, sepsis, aspiration pneumonia, viral pneumonia, fat embolism, and burns, or it may occur following cardiopulmonary bypass. In these situations, progressive hypoxemia, a fall in lung compliance, and bilateral roentgenographic pulmonary infiltrations, a condition which has been termed the adult respiratory distress syndrome (ARDS) may develop after 6 to 24 hours. In this syndrome, widespread damage to the lung parenchyma is thought to increase the alveolocapillary membrane permeability and lead to the development of pulmonary edema. The hypoxemia that is encountered in this syndrome persists despite the administration of 100 per cent oxygen (i.e., it resembles true venous admixture) and is due to the continued perfusion of alveoli that are either collapsed or flooded with edema fluid.

Hypercapnic respiratory failure

Respiratory insufficiency resulting in hypoxemia and hypercapnia (i.e., alveolar hypoventilation) develops in patients suffering from bronchopulmonary disease (particularly bronchitis and emphysema), chest wall or thoracic cage disease (i.e., obesity), or following central nervous system depression. The findings in the arterial blood in acute and chronic alveolar hypoventilation are presented in Table 7–1. In acute alveolar hypoventilation, there has not been sufficient time to compensate for the CO_2 retention so that the serum bicarbonate and total carbon dioxide are little elevated, and therefore the arterial pH can be markedly acidemic. In chronic alveolar hypoventilation there has been compensation for the elevated partial pressure of carbon dioxide through elimination of chloride and retention of base and bicarbonate, so that the bicarbonate and carbon dioxide content of the blood are elevated, the serum chloride is low, and the arterial pH is restored towards the normal range.

In patients with chronic bronchopulmonary disease, respiratory insufficiency frequently manifests itself initially by a compensatory increase in ventilation and/or perfusion, so that, while the $P(A-a)O_2$ is increased, near-normal blood gases are maintained at the expense of increased respiratory or cardiac work (patients with obstructive pulmonary disease who do this are sometimes labeled *pink puffers*). At a later stage, or even in the early stages in some patients, particularly if the resistance to airflow is markedly increased, this compensation may be inadequate, and alveolar hypoventilation, with severe hypoxemia and carbon dioxide retention, may develop (patients with obstructive pulmonary disease in whom this develops with associated right-sided heart failure have been labeled *blue bloaters*). The tempo of progression from the first to the second stage is highly variable, and indeed, as indicated, the order in which they appear may vary.

There is little correlation between the degree of gas exchange abnormality and the severity of the patient's symptoms, and many patients who maintain relatively normal blood gas tensions, although at the cost of a marked increase in respiratory

TABLE 7–1
Arterial Blood Gas Findings in Alveolar Hypoventilation

Arterial Blood	Acute	Chronic
Po_2*	↓	↓
P_{CO_2}	↑	↑
HCO_3^-	↔	↑
CO_2 content	↔	↑
pH	↓	↔

* Unless patient is receiving oxygen
↑, Increased; ↓, Decreased; ↔, Unchanged

work, may be much more disabled than others who have chronic hypoxemia and hypercapnia. Nevertheless, progressive decompensation is almost invariably associated with an increasing severity of symptoms and is attended by secondary effects of hypoxia and acidemia, and in some instances hypercapnia, which further compromise respiratory and cardiac function. Severe hypoxemia (and acidosis) leads to vasoconstriction of the pulmonary vasculature, thereby increasing the pulmonary vascular resistance. The high pulmonary vascular resistance leads to pulmonary hypertension, and if this persists, to right ventricular hypertrophy and eventually right ventricular failure—one of the most serious manifestations of chronic respiratory insufficiency.

In patients with chronic carbon dioxide retention, hypoxia may replace carbon dioxide as the major stimulus to respiration. The administration of high oxygen concentrations to such persons may remove the hypoxic stimulus and exaggerate the respiratory failure because ventilation falls. Nevertheless, oxygen should not be withheld from such patients; rather, efforts to reduce the work of breathing should be even more vigorously introduced.

Assessment of pulmonary function

There are a variety of tests that may be used to detect disturbed pulmonary function, and the judicious use of such tests is essential to the evaluation of the patient with respiratory complaints. The tests vary widely in degree of sophistication and complexity, but in general they fall into two groups: those relating to the ventilatory function of the lungs and chest wall and those relating to gas exchange.

Ventilatory function

In practice, one usually assesses ventilatory function by determination of lung volume, which is predominantly a reflection of the elastic resistance or distensibility of the lungs and thorax, and the rate of airflow during a forced expiratory vital capacity, which is predominantly a reflection of the airflow resistance.

LUNG VOLUME. The total lung capacity (TLC) and its subdivisions, which vary according to age, body size, and sex, provide a useful index of the distensibility of the respiratory system. A change in TLC and its compartments is indicative either of an alteration in elastic properties of the lungs or the chest wall, or of respiratory muscle dysfunction. The slow vital capacity (VC) of itself is related to the compliance of the

FIG. 7–8. Total lung capacity and its subdivisions normally, in the obstructive pattern and in the restrictive pattern. V.C. = vital capacity, FRC = functional residual capacity, R.V. = residual volume. (Cherniack, R. M., Cherniack, L., and Naimark, A.: Respiration in Health and Disease. ed. 2. Philadelphia, W. B. Saunders, 1972.)

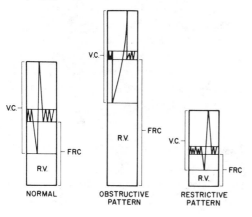

NORMAL   OBSTRUCTIVE PATTERN   RESTRICTIVE PATTERN

lung and thorax, decreasing as the total compliance falls. Thus, VC is reduced in any condition in which the compliance of the lungs is decreased, such as pulmonary fibrosis or congestion, or in which the compliance of the chest wall is decreased, such as obesity or kyphoscoliosis. However, a low VC (or TLC) is not necessarily indicative of respiratory disease. Since voluntary effort is essential to perform the VC maneuver, the VC will be lower than expected if respiratory muscle strength is reduced. Similarly, full cooperation of the patient is essential to calculate the TLC, since an inadequate inspiration will yield a falsely low TLC. Thus, it is essential that one be confident that respiratory muscle function is normal and that the test be repeated several times in order to be certain that the maximal value has been obtained.

If a reliable VC has been obtained, the value should be interpreted in the light of any associated changes in lung volume. Figure 7–8 illustrates that the VC may be low in both restrictive disease and obstructive lung disease. In restrictive disease the vital capacity is low because all components of the TLC are reduced. In obstructive disease, the vital capacity is low because the residual volume (RV) is increased (i.e., there is air trapping). In this disorder, the FRC is also increased (i.e., the individual is breathing at an increased breathing level).

FORCED VITAL CAPACITY. The forces which tend to narrow the airways (peribronchial pressure and contraction of bronchial smooth muscle) are balanced by those which tend to widen them (intraluminal pressure and the tethering action of the surrounding lung tissue). During forced expiratory efforts the pleural and the alveolar pressure become greater than atmospheric pressure, while the intrabronchial pressure decreases from values near alveolar pressure in the peripheral airways to atmospheric pressure near the airway opening. At some point in the airway, intrabronchial pressure is equal to or less than the pleural pressure, so that the airways downstream from that point (towards the mouth) tend to become narrowed, thereby limiting airflow. In patients with chronic disease of the airways, flow limitation may occur when there is a loss of tissue retraction (lung elastic recoil) or narrowing of the airway lumen by bronchospasm, mucosal thickening, or secretions, with resultant increased resistance to airflow. Obviously, both loss of tissue retraction (as in emphysema) and increased resistance within the bronchi (as in asthma or bronchitis) may be present in a single patient.

Considerable information about the flow resistance can be obtained by determination of the rate at which both an inspiratory and an expiratory vital capacity take place. As is demonstrated in Figure 7–9, the maximal inspiratory flow rate at every level of lung inflation depends primarily on the force developed. In contrast, the maximal flow rate achieved during a forced expiration depends on the degree of lung inflation. At lung volumes near the TLC, the expiratory flow rate increases as the effort (pressure) exerted increases. Clearly, then, assessment of expiratory flow resistance which is based on analysis of the flow rate at high degrees of lung inflation may be related more to patient cooperation and effort than to alterations in intrapulmonary mechanics and therefore may be fraught with inconsistency. At lesser degrees of lung inflation, expiratory flow increases with greater effort, up to a certain point, beyond which more effort fails to elicit any further increase in flow and may even result in a slight decrease in flow. At lesser degrees of lung inflation, then, the influence of the extent of patient cooperation is minimized to some degree because maximum flow ($\dot{V}_{max}$) does not require maximal effort.

The assessment of expiratory flow limitation from the forced vital capacity (FVC) can be carried out in several ways, but certain measurements are more informative

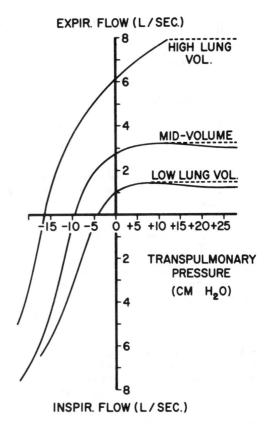

FIG. 7–9. The relationship between pressure and maximal flow at high lung volumes, mid-volume, and low lung volume. (Cherniack, R. M., Cherniack, L., and Naimark, A.: Respiration in Health and Disease, ed. 2. Philadelphia, W. B. Saunders, 1972.)

than others. During a properly performed FVC, the flow-volume relationship during the forced expiratory maneuver, and assessment of the maximal expiratory flow rate at a particular lung volume is frequently used to assess flow limitation. However, just as flow resistance varies with lung volume, $\dot{V}_{max}$ during the FVC is also related to lung volume, being greatest at high lung volumes and falling as lung volume diminishes (Fig. 7–10).

Many laboratories report the forced expired volume in the first second ($FEV_{1.0}$) while others use the forced expiratory flow during the middle half of the FVC (FEF_{25-75}) as indicators of the status of expiratory flow resistance. This is recommended over a measurement such as the peak expira-

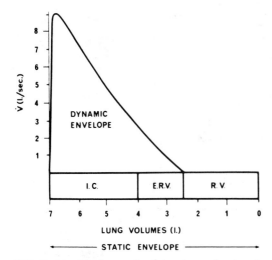

FIG. 7–10. The lung volume (*static envelope*) and the flow volume relationship during a forced expiratory vital capacity (*dynamic envelope*) in a healthy subject.

FIG. 7–11. The effect of effort on the flow-volume characteristics in an individual. (*A*) maximum effort throughout. (*B*) the initial effort was less than maximal. (*C*) the initial effort was maximal, but the maneuver was terminated abruptly.

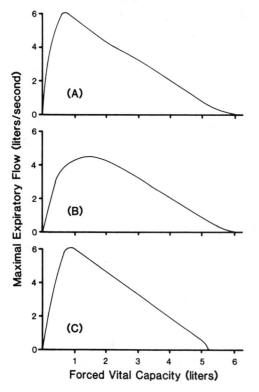

tory flow rate, but it must be pointed out that the calculated FEF_{25-75} will be in error if determined from an FVC which is ended prematurely.

Since the FVC maneuver requires total effort from the patient, the finding of low flow rates or other parameters indicative of flow resistance may reflect patient effort rather than flow limitation. For this reason, it is recommended that the FVC duration be at least 6 seconds, and that a spirogram record of volume over time should always be obtained. Observation of the flow-volume loop is particularly helpful in assessment of the extent of patient cooperation. This is seen in Figure 7–11, in which three forced vital capacity maneuvers of a patient have been illustrated. Poor effort is demonstrated by a low peak flow rate during the patient's initial effort (Fig. 7–11*B*), or by the sudden cutoff of airflow when the forced expiration is terminated before residual volume is reached (Fig. 7–11*C*).

Provided maximum effort is ensured, the finding of an FEV_1, FEF_{25-75}, $\dot{V}_{max\,50}$, or $\dot{V}_{max\,25}$ which is less than expected for a given age, height, and sex suggest the presence of expiratory airflow limitation. However, interpretation of these parameters derived must also take into account the lung volume at which the flow rates were achieved. Flow limitation can be inferred only if the airflow rate is different from that expected at an equivalent lung volume. When absolute lung volume is not known, the FEV_1/FVC ratio may be used to determine whether expiratory airflow limitation is present. This ratio is usually within normal limits (i.e., greater than 70 per cent) in patients suffering solely from restrictive disorders, while it is markedly reduced when there is airflow limitation. However, anything which affects the ability of the patient to expire fully, such as dyspnea, muscular weakness, or pain, or lack of total effort may limit the FVC maneuver, so that the FEV_1/FVC ratio may appear to be greater than it actually is and the presence of flow limitation may be missed.

In addition, it is important to remember that the very small airways (<2 to 3 mm. in diameter) contribute less than 20 per cent of the airway resistance. Even a marked increase in peripheral airway resistance will not be recognized in measurements of airway resistance or spirometric tests used as indices of airflow limitation, such as the $FEV_{1.0}$ or FEF_{25-75}, all of which are mainly a reflection of the large airways. Measurements such as dynamic lung compliance at increasing respiratory frequencies, or of gas distribution as derived from the slope of Phase III of the single-breath nitrogen curve are thought to be sensitive to alteration of mechanical properties of the peripheral lung units. The severity of impairment of distribution or mixing of gas in the lung is indicative of the degree of inequality of time constants in lung units.

As we have seen in Figure 7–9, the airflow rates achieved during a forced inspiration is related more to the degree of patient effort than to changes in intrapulmonary mechanical properties. Nevertheless, measurements of maximal inspiratory flow can be particularly useful, provided maximal effort is ensured, when upper airway obstruction is suspected. In the presence of upper airway obstruction both the maximal inspiratory flow rates and maximal expiratory flow rates are limited.

Patterns of impaired ventilatory function

Patients with restrictive pulmonary disease and those with obstructive pulmonary disease exhibit differing patterns of abnormal ventilatory function. These abnormalities are presented in Table 7–2 and Figure 7–12.

RESTRICTIVE PULMONARY DISEASE. In restrictive pulmonary disease, lung volume is reduced, and the FEV_1, FEF_{25-75}, and $\dot{V}_{max}$ at any particular volume are frequently lower than predicted values. Figure 7–12 illustrates that the flow rates are low not because of an increase in flow resistance but

TABLE 7–2

Disturbances in Ventilatory Function in Obstructive and Restrictive Disease

Test	Obstructive Disease		Restrictive Disease	
VC	↔	↓	↓	
$FEV_{1.0}$	↓		↔	↓
FEF_{25-75}	↓		↔	↓
$FEV_{1.0}$/FVC ratio	↓		↔	
RV	↑		↓	
FRC	↑		↓	
TLC	↑		↓	

↑, Increased; ↓, Decreased; ↔ Unchanged

rather because they are being achieved at a low lung volume. In fact, the low expiratory flow rates may actually be higher than would be expected at any particular lung volume in such patients, because the driving pressure (lung elastic retractive force) is increased (i.e., compliance is low). As a corollary, it is clear that if $\dot{V}_{max}$ is not higher than expected at a particular lung volume in a patient suffering from a restrictive disorder, then flow resistance is likely increased. As indicated earlier, the FEV_1/FVC ratio may help to determine whether airflow obstruction is present.

In some cases, it may be difficult to determine whether a low TLC and its compartments indicate an alteration of elastic resistance of the chest wall or of the lung. This question can be answered by evaluating the shape of the curve relating the static transpulmonary pressure to the absolute lung volume (pressure-volume relationship) over the range of the expiratory vital capacity. As seen in Figure 7–2, the pressure-volume curve is shifted downward and to the right and the slope of the curve at around FRC is decreased in the patient with pulmonary fibrosis. If the low lung volume is due solely to a disorder of the chest wall, the slope of the curve will be normal.

OBSTRUCTIVE PULMONARY DISEASE. In obstructive pulmonary disease, the TLC and

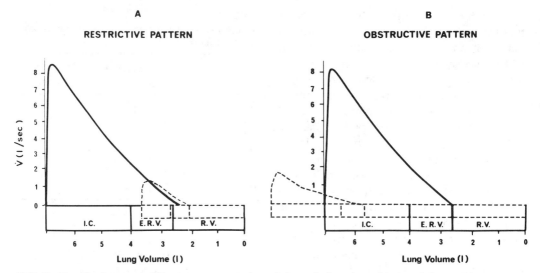

FIG. 7–12. The interrupted lines represent static and dynamic "envelopes" in restrictive and in obstructive disease. The solid line envelopes represent the expected values in a healthy person of the same age and height. In the restrictive pattern (A) the "static envelope" is reduced in size (i.e., lung volumes are diminished). The "dynamic envelope" is also reduced in size (i.e., flow rates are low), not because of increased air flow resistance, but because of the reduced lung volume. In fact, flow rates are greater than predicted at equivalent lung volumes because of the increase in elastic recoil. In the obstructive pattern (B) the static envelope is increased in size (i.e., lung volumes are high). The dynamic envelope is reduced in size, but in this case the flow rates are considerably lower than predicted at an equivalent lung volume as a result of the increase in air flow resistance.

its compartments are usually increased, and the FEV_1, FEF_{25-75}, and $\dot{V}_{max}$ at a particular lung volume (Fig. 7–12) are lower than expected, indicating the presence of flow limitation. As is the case in a restrictive disorder, assessment of the pressure-volume characteristics of the lung will help differentiate whether the elevated lung volume and low expiratory flow rates are associated with an increase in flow resistance, as in asthma or bronchitis, or a loss of lung elastic recoil as in emphysema. As seen in Fig. 7–2, the curve is shifted upward and to the left in both emphysema and asthma, but in emphysema, the slope of the curve at FRC is increased, while in asthma (or bronchitis) the slope is normal.

When airflow limitation has been demonstrated by clinical and spirometric means, the extent of improvement in expiratory flow rates following the administration of a nebulized bronchodilating agent must be determined. An increase in FEV_1,

FEF_{25-75}, or $\dot{V}_{max}$ at a particular lung volume of greater than 15 per cent following the inhalation of the bronchodilator, indicates that the expiratory airflow limitation is at least partially reversible. On the other hand, the converse may not be true, and failure of these parameters to increase does not necessarily imply irreversibility, for changes in lung volume must be taken into account before reporting a lack of improvement following bronchodilator. Figure 7–13 illustrates two examples in which a beneficial bronchodilator effect may be missed. In one patient, the FVC was unchanged after bronchodilator, but the TLC and RV both fell. The postbronchodilator FEF_{25-75} and $\dot{V}_{max\,50}$ were the same as those found before bronchodilator, but the flow rates were clearly greater at any particular lung volume (i.e., at isovolume). In the second patient, TLC was unchanged following bronchodilator, but RV fell (i.e., FVC increased). In this case, the calculated

FEF$_{25-75}$ and $\dot{V}_{max\,50}$ were also unchanged following bronchodilator, but this was because these parameters were calculated at a different lung volume after bronchodilator. Thus, in both patients, failure to take isovolume flow rates into consideration would suggest a lack of benefit from nebulized bronchodilator. When the expiratory flow rates after the aerosol are compared with those before bronchodilator at an equivalent lung volume (isovolume), there is obvious improvement following inhaled bronchodilator.

AIRWAY HYPERREACTIVITY. Improvement in expiratory flow parameters following bronchodilator indicates the presence of airway hyperreactivity. In some patients the presence (or absence) of increased bronchial reactivity may be difficult to ascertain. In such individuals, a reduction in FEV$_1$ or $\dot{V}_{max}$ following exercise or the inhalation of methacholine or histamine is indicative of airway hyperreactivity. On the other hand, it must be recognized that airway hyperreactivity develops even in healthy individuals following an acute upper respiratory infection. Nevertheless, as with bronchodilator therapy, interpretation of hyperreactivity (or its absence) should be made only when changes in lung volume are taken into consideration. The changes following bronchial challenge may be opposite to those shown in Figure 7–13, and an increase in lung volume and an alteration of isovolume flow rates may be missed.

Gas exchange

Major alterations of gas exchange in the lungs are readily assessed by analysis of the partial pressures of oxygen and carbon dioxide in the blood. More precise assessment is obtained by analysis of simultaneously collected samples of arterial blood and expired air during a *steady state* (when the metabolic respiratory quotient is be-

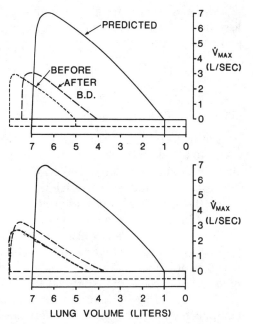

FIG. 7–13. The effect of bronchodilator on lung volume and flow-volume characteristics in two patients with chronic air flow limitation. In the first patient the forced vital capacity was unchanged, as was $\dot{V}_{max\,50}$, but total lung capacity and residual volume fell so that flow at any particular volume (isovolume) was improved. In the second patient $\dot{V}_{max\,50}$ also did not improve. However, the forced vital capacity increased and the residual volume decreased, which indicate improvement.

tween 0.7 and 1.0), at rest, during exercise, and where indicated, while breathing 100 per cent oxygen. By collecting these samples, minute ventilation, oxygen consumption, carbon dioxide production, respiratory quotient (R), physiological dead space, and the alveoloarterial oxygen tension gradient P(A-a)O$_2$ can be calculated. The P(A-a)O$_2$ is particularly important in assessing the status, or possible improvement or deterioration in a patient.

Alveoloarterial oxygen gradient

Even if expired air is not collected the P(A-a)O$_2$ can be determined by using a simplified alveolar air equation:

$$P_AO_2 = P_IO_2 - \frac{Pa_{CO_2}}{R}$$

If the patient is breathing room air, R is assumed to be 0.8, and 1.0 if breathing oxygen. Then, on room air:

$$P_AO_2 = P_IO_2 - (Pa_{CO_2} \times 1.25)$$

and

$$P(A\text{-}a)O_2 = P_IO_2 - (Pa_{CO_2} \times 1.25) - Pa_{O_2}.$$

Even in healthy persons, the $P(A\text{-}a)O_2$ is about 10 torr, as a result of mismatching of ventilation and perfusion, and some mixing of venous blood with arterialized blood in the thebesian and bronchial veins.

As we have seen earlier, the $P(A\text{-}a)O_2$ will be elevated and the PaO_2 will be lower than normal when there is increased mismatching of ventilation and perfusion, true venous admixture, a diffusion defect, or a combination of these disturbances. In uncomplicated alveolar hypoventilation, however, oxygen in the alveoli equilibrates with that in the pulmonary capillary blood so that the $P(A\text{-}a)O_2$ is within normal limits.

Clearly, then, if hypercapnia is present, which is indicative of alveolar hypoventilation, and the $P(A\text{-}a)O_2$ is greater than 10 torr, then a defect in blood-gas equilibrium must also be present. Although this may result from a diffusion defect under special circumstances, by far the most common cause of a failure of blood and gas to equilibrate is a mismatching of the distribution of ventilation and perfusion in the lungs. The extent of the ventilation/perfusion mismatching may be inferred from the size of the $P(A\text{-}a)O_2$, provided that excessive shunting (true venous admixture) has been ruled out.

The amount of true venous admixture can be assessed by having the subject breathe 100 per cent oxygen. Under normal circumstances at sea level the arterial PO_2 will rise above 500 torr when 100 per cent oxygen is being inspired, and a value of less than 500 torr at sea level suggests that there is an abnormal amount of true venous admixture. Thus, in the presence of an elevated $P(A\text{-}a)O_2$ while breathing ambient air, the finding of a PO_2 greater than 500 torr

while breathing oxygen will rule out the presence of a shunt.

The finding of a $P(A\text{-}a)O_2$ that is greater than normal when spirometric tests are relatively normal may be particularly significant, for this may indicate peripheral lung unit disease. As pointed out previously, inequalities of time constants throughout the lung will be reflected by a maldistribution of inspired air. If the poorly ventilated regions of the lung are still well perfused, the $P(A\text{-}a)O_2$ will be increased, particularly at increased respiratory frequencies.

Diffusing capacity

The diffusing capacity of the lung for carbon monoxide (DL_{CO}) is usually assessed to determine the status of diffusion of oxygen in the lung. The single breath technique is the easiest to perform, but values obtained by the steady state technique are more likely to be relevant to the actual conditions of gas exchange than are those obtained under artificial conditions of breath holding. As indicated earlier, the DL_{CO} may be reduced when the alveolar surface available for diffusion is reduced as a result of parenchymal disease (fibrosis or emphysema), or when there has been surgical removal of considerable lung tissue. However, in the vast majority of cases, a low DL_{CO} is a reflection of excessive mismatching of ventilation and perfusion rather than a true diffusion defect, particularly when the inspired air is unevenly distributed.

Acid-base balance

Since alterations in the partial pressure of carbon dioxide can affect the arterial pH, with subsequent renal compensatory measures, and conversely, metabolic disturbances lead to compensatory measures by the respiratory system, estimation of the acid-base balance is an essential component of the assessment of pulmonary function in respiratory disease. From knowledge of any two of the variables of the Henderson-Hasselbalch equation, the acid-base status can be established and proper therapy instituted. (For further details concerning acid-base relationships the reader is referred to Chap. 14.)

Cardiopulmonary response to exercise

The assessment of gas exchange during exercise, which provides information regarding exercise tolerance and the factors which may limit it, is probably the most important and informative of all assessments of respiratory function. Measurement during exercise of the arterial partial pressures of oxygen, carbon dioxide, and pH, and ventilation, oxygen consumption, carbon dioxide production, respiratory quotient, dead-space/tidal air ratio, and $P(A-a)O_2$ may demonstrate an impairment when all other pulmonary function measurements are normal, since most patients experience disability, particularly on exertion. Thus, gas exchange during exercise should be assessed in all patients complaining of dyspnea.

As indicated in Table 7–3, measurement of physiological parameters may help to differentiate the mechanism of a reduction in exercise tolerance. If the work of breathing is excessive, the reduced exercise tolerance is associated with a high O_2 consumption and CO_2 production, and as a result, a decreased alveolar and arterial PO_2 and an increased alveolar and arterial PCO_2. If there is mismatching of blood and gas distribution or if the diffusing capacity for oxygen fails to increase in proportion to the oxygen consumption, the $P(A-a)O_2$ will rise and arterial PO_2 will fall further. On the other hand, if the cardiovascular system does not satisfy the increased tissue demands during exercise, or if the individual is not physically fit, the inadequate distribution of systemic blood flow to the exercising muscles will result in tissue hypoxia, with excessive lactate production, acidemia, and increased dissociation of carbonic acid to produce CO_2 and water. Thus, the finding of an R value greater than 1.0 and a fall in pH during the exercise period suggests that there is excessive lactate production, possibly as a result of anaerobic metabolism. When patients with cardiorespiratory insufficiency undertake a program of exercise training, considerable improvement in exercise tolerance may be achieved. This training effect is often due not to improvement of ventilatory function but rather to improved circulatory function (i.e., in the cardiac response or in the distribution of peripheral blood flow to the exercising muscles).

TABLE 7–3
Gas Exchange Abnormalities During Exercise

Parameter	Ventilatory Impairment	V/Q Imbalance	Cardiovascular Impairment or Physical Unfitness
VE	↑	**♦**	**♦**
R	↓ ↔	↔	**♦**
V_D/V_T(%)	↓ ↔	↕	↑ ↔
$P(A–a)O_2$	↔	**♦**	↔ ↑
Pa_{O_2}	**♦**	**♦**	↔ ↓
Pa_{CO_2}	**♦**	↔ ↓	↔ ↓
pH	**♦**	↔ ↓	**♦**

V_E, Ventilation
R, Respiratory quotient
V_D/V_T (%) Ratio of dead space to tidal volume
↑, Increased; ↓, Decreased; ↔ Unchanged
The most characteristic abnormalities are indicated by the bold arrows.

$P(A–a)O_2$, Alveolar-arterial oxygen tension gradient
Pa_{O_2}, Arterial partial pressure of oxygen
Pa_{CO_2}, Arterial partial pressure of carbon dioxide

Airway reactivity

As indicated earlier, exercise may be used as a bronchoprovocation in patients suspected of having airway hyperreactivity. In such patients a drop in FEV_1, $\dot{V}_{max}$, or specific conductance following an exercise load is indicative of exercise-induced bronchospasm. Clearly, therapy in such cases should be directed at preventing or blocking bronchospasm induced by exercise.

References

Anthonisen, N. R., and Milic-Emili, J.: Distribution of pulmonary perfusion in erect man. J. Appl. Physiol. *21*:760, 1966.

Ball, W. C., Stewart, P. B., Newsham, L. G. S., and Bates, D. V.: Regional pulmonary function studied with xenon[133]. J. Clin. Invest. *41*:519, 1962.

Bates, D. V., Macklem, P. T., and Christie, R. V.: Respiratory Function in Disease. Philadelphia, W. B. Saunders, 1971.

Buist, A. S., and Ross, B. B.: Predicted values for closing volumes using a single breath nitrogen test. Am. Rev. Respir. Dis. *107*:744, 1973.

Cherniack, R. M.: Pulmonary Function Testing. Philadelphia, W. B. Saunders, 1977.

Cherniack, R. M.: Physiologic evaluation of patients with chronic respiratory insufficiency. Seminars in Respiratory Medicine. New York, Thieme & Stratton, July 1979.

DeRenne, J-PH., Macklem, P. T., and Roussos, CH.: The respiratory muscles: Mechanics, control and pathophysiology. Am. Rev. Respir. Dis., *118*:119SA, 373SA, 581SA, 1978.

DuBois, A. B., Botelho, S. Y., Bedell, G. N., Marshall, R., and Comroe, J. H., Jr.: A rapid plethysmographic method for measuring thoracic gas volume. J. Clin. Invest. *35*:322, 1956.

Hudson, L.: Adult respiratory distress syndrome. Seminars in Respiratory Medicine. New York, Thieme & Stratton, January 1981.

Kory, R. C., Callahan, R., Boren, H. G., and Syner, J. C.: The Veterans Administration-Army Cooperative Study of Pulmonary Function. I. Clinical spirometry in normal men. Am. J. Med. *30*:243, 1961.

Levison, M., and Cherniack, R. M.: Ventilatory cost of exercise in chronic obstructive pulmonary disease. J. Appl. Physiol. *25*:21, 1968.

Mead, J.: Mechanical properties of lungs. Physiol. Rev., *41*:281, 1961.

Mead, J., Turner, J. M., Macklem, P. T., and Little, J. B.: Significance of the relationship between lung recoil and maximum expiratory flow. J. Appl. Physiol. *22*:95, 1967.

West, J. D.: Respiratory Physiology. Baltimore, Williams & Wilkins, 1974.

8   Oxygen Transport and Cellular Respiration

Eugene D. Robin, M.D.
Lawrence M. Simon, M.D.

The lung is only one unit of a complex functional and structural arrangement that provides molecular oxygen for critical oxygen-consuming reactions occurring in cells. This chapter describes the steps involved in oxygen transport from external environment to intracellular sites of utilization; the nature of cellular metabolic processes that require oxygen; the methods available for monitoring abnormalities of oxygen transport and metabolism (Table 8–1); the consequences of oxygen depletion; and the therapy for abnormalities of oxygen transport and metabolism.

Oxygen transport and metabolism may be classified conveniently into pulmonary oxygen uptake, cellular oxygen delivery, and cellular oxygen utilization.

Abnormal cellular utilization of oxygen has been termed *dysoxia*. *Hypoxic dysoxia* refers to conditions with disordered cellular oxygen utilization due to decreased O_2 supply or transport. *Normoxic dysoxia* refers to disorders in which O_2 supply is normal but O_2 utilization is abnormal as a result of abnormalities or dysfunction of the subcellular organelles and enzymes. *Hyperoxic dysoxia* occurs when O_2 supply or transport is excessive and is also known as oxygen toxicity. This chapter deals primarily with hypoxic dysoxia, with a brief discussion of normoxic dysoxia.

Pulmonary oxygen uptake

Pulmonary oxygen uptake is described in Chapter 7. As outlined, abnormalities of pulmonary structure and function frequently result in abnormal gas exchange within the lung, leading to a reduction of arterial oxygen tensions (Pa_{O_2}). In practice, measurements of Pa_{O_2} provide a convenient and accurate estimate of the adequacy of pulmonary oxygen uptake. A normal value of Pa_{O_2} (90 $\pm$ 10 mm. Hg while breathing room air) establishes normal pulmonary oxygen uptake. In the normal subject Pa_{O_2} tends to decrease with advancing age (being perhaps 70 mm. Hg at ages greater than 60) and is several mm. Hg lower during sleep than during the waking state.

The term *hypoxemia* refers to values of Pa_{O_2} that are abnormally low and is frequently associated with *hypoxic dysoxia*. However, mild hypoxemia may be present without evidence of dysoxia, and dysoxia can be present in the absence of hypoxemia (because of abnormalities of cellular oxygen delivery or oxygen utilization). The

163

TABLE 8–1
Measurement Available for Monitoring Oxygen Transport and Metabolism

Process	Measured Variable
Pulmonary oxygen uptake	Pa_{O_2} (arterial oxygen tension)
Cellular oxygen delivery	
Oxygen supply	Arterial O_2 content
	Hb Concentration
	Sa_{O_2} (arterial oxygen saturation)
	Cardiac output
	$P\bar{v}_{O_2}$ (mixed venous oxygen tension)
	Interstitial oxygen tissue PO_2
State of binding	$P_{50}O_2$
Cellular oxygen utilization	Reflected cytoplasmic $NAD^+/NADH$ (plasma lactate/pyruvate)
	Free cytoplasmic $NAD^+/NADH$ (cytoplasmic lactate/pyruvate)
	Free mitochondrial $NAD^+/NADH$ (mitochondrial β-hydroxybutyrate/acetoacetate)
	Reflected mitochondrial $NAD^+/NADH$ (appropriate plasma redox pair)?
	Erythropoietin

precise Pa_{O_2} at which impairment of tissue oxygen utilization occurs is variable. It depends, among other factors, on the rate of development of hypoxemia. All things being equal, the more rapid the onset of hypoxemia, the more extensive the tissue abnormalities. In general, Pa_{O_2} values persistently less than 50 mm. Hg are associated with abnormal tissue oxygen metabolism. However, patients with chronic values as low as 30 mm. Hg have been reported, and such values are occasionally compatible with long-term survival.

Cellular oxygen delivery

Oxygen delivery to the tissues is determined by oxygen supply, which, in turn, depends on both arterial oxygen content and cardiac output; state of oxygen binding to hemoglobin; diffusion of oxygen from capillary plasma to intracellular sites of utilization; relationship of regional perfusion to diffusion at the tissue level; and intracellular oxygen binding.

Oxygen supply

Arterial blood is the "inspired" fluid of cells, just as ambient air is the inspired fluid of the lung. In this sense, the amount of oxygen supplied to the tissues depends upon the product of arterial oxygen content and cardiac output.*

Arterial oxygen content is determined by the value of Pa_{O_2} and the hemoglobin concentration. As previously noted, Pa_{O_2} depends on pulmonary gas exchange and, in turn, determines the amount of oxygen that will be taken up by hemoglobin. Each gram of hemoglobin can combine with 1.34 ml. of oxygen. In contrast, the absolute quantity of oxygen represented by dissolved oxygen (P_{O_2}) is small because of the relatively poor solubility in plasma (0.003 ml. of oxygen per 100 ml. of plasma per mm. Hg at 37° C). Thus, a P_{O_2} of 450 mm. Hg would be required to provide the same oxygen content in 100 ml. of arterial blood as would be achieved by a full saturation of hemoglobin at a concentration of 1 g/100 ml.! It is obvious that an adequate tissue supply of oxygen requires adequate hemoglobin concentration. Anemia (reduced red cell mass) is, therefore, an important cause of abnormal oxygen supply, and measurements of hemoglobin concentration are important in monitoring adequacy of oxygen supply.

CARDIAC OUTPUT. An adequate cardiac output is required for normal oxygen deliv-

* Pulmonary oxygen supply is equal to the volume of ventilation × concentration of oxygen in inspired air. Cellular oxygen supply is equal to volume of arterial blood flow (cardiac output) × concentration of oxygen in arterial blood.

ery. Not only must total cardiac output be quantitatively sufficient, but the regional distribution of output must match the variable oxygen requirements of the different tissues. (Specific details of normal circulatory dynamics and the alterations of tissue oxygen delivery produced by disease are described in Chap. 2.)

Interruption of circulation interferes more drastically with tissue oxygen utilization than does cessation of ventilation. In the former instance the oxygen supply of any organ is limited to that contained in its residual blood volume, whereas in the latter case the oxygen supply of the entire circulating blood volume is available (see below).

With inadequate cardiac output, there are mechanisms that provide for redistribution of blood flow so that more critical regions (i.e., brain) receive a major portion of the total output at the expense of less critical areas (i.e., skin, gut, liver).

Quantitative estimates of the adequacy of cardiac output may be obtained by direct measurement; however, this is not entirely satisfactory, since the precise level of cardiac output appropriate for a given metabolic state is difficult to estimate. Moreover, the heterogeneous distribution of cardiac output to various organs makes it difficult to infer regional deficiencies in oxygen supply from measurements of total cardiac output. The adequacy of cardiac output in terms of oxygen supply may also be monitored by measurements of mixed venous oxygen tension (Pv_{O_2}). Normal resting Pv_{O_2} is approximately 40 mm. Hg; with an inadequate cardiac output this value tends to fall as a result of increased tissue extraction. This measurement also fails to evaluate the heterogeneous distribution of blood flow and regional variations in tissue oxygen utilization. Since the mixed venous P_{O_2} is also affected by arterial oxygen content, venous pH, redistribution of blood among metabolizing tissues, and the tissue oxygen uptake, absolute values are less useful in most patients than trends of change.

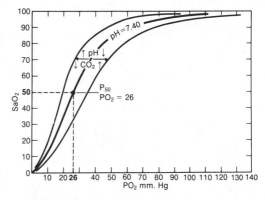

FIG. 8–1. Oxyhemoglobin dissociation curve: the effects of changes in pH and Pco_2 are indicated by the arrows. Note that the P_{50} (P_{O_2} at 50% saturation) equals 26 mm. Hg.

Oxygen binding to hemoglobin (oxyhemoglobin dissociation curve)

Tissue oxygen supply (as well as pulmonary oxygen uptake) depends critically on the affinity of hemoglobin for oxygen. Affinity is described by the relationship between P_{O_2} and hemoglobin saturation and is quantitatively defined by the oxyhemoglobin dissociation curve (Fig. 8–1). The curve is sigmoid shaped, and the physiological range of oxygen tensions encompasses values of approximately 10 to 100 mm. Hg. At the relatively high oxygen tension present in alveolar air, oxygen combines with hemoglobin, leading to a high oxygen saturation and a high oxygen content. At the relatively low oxygen tension of tissues, oxygen is released from hemoglobin; the unbound oxygen (quantitatively reflected by P_{O_2}) then diffuses from the red cell through the plasma, capillary wall, and interstitial space to enter intracellular water. Shifts of the curve to the right result in a smaller quantity of oxygen bound to hemoglobin at any oxygen tension (decreased affinity). All things being equal, a shift to the right will impair pulmonary uptake of oxygen but facilitate unloading of oxygen to the tissues. It should be noted that the upper part of the curve (P_{O_2} greater than 60 mm. Hg) is relatively flat. Thus, with a

TABLE 8–2
Conditions Associated with Alterations of
O_2-Hb Affinity

O_2-Hb Dissociation Curve	
Shift to Left	Shift to Right
Increased pH	Decreased pH
Decreased Pc_{O_2}	Increased Pc_{O_2}
Decreased temperature	Increased temperature
Decreased 2,3-DPG	Increased 2,3-DPG
1. Decreased pH	1. Increased pH
2. Stored blood	2. Hypoxemia
3. Increased ADP	3. Anemia
4. Phosphate	4. Phosphate retention
depletion	
5. Red cell pyruvate	5. Red cell pyruvate
kinase excess	kinase deficiency
6. Red cell hexo-	
kinase deficiency	
7. Chemical inhi-	
bition of glycolysis	
(e.g., monoiodo-	
acetate)	
8. Diphosphoglyc-	
erate mutase	
deficiency	
Decreased 2,3-DPG	
binding to Hb	
1. Fetal hemoglobin	
2. Diabetes mellitus	
Abnormal	Abnormal
Hemoglobins	Hemoglobins
Hereditary	Hereditary
Hb Ranier	Hb Kansas
Hb Barts	Hb Seattle
Hb H	Hb S
Hb Yakima	
Hb J Capetown	
Hb Chesapeake	
Hb Kempsey	
Hb Hiroshima	
Hb Little Rock	
Hb McKees rocks	
Acquired	
Carboxyhemoglobin	
Methemoglobin	

normal end-pulmonary capillary P_{O_2}, a shift to the right produces only a moderate decrease in arterial oxygen saturation. However, patients with abnormal gas exchange resulting in low arterial P_{O_2} have values that fall on the steep part of the curve. A shift to the right in these patients produces a substantial additional decrease of oxygen uptake by hemoglobin in the lung.

Conversely, shifts of the curve to the left (increased affinity) will increase the amount of oxygen bound to hemoglobin at any oxygen tension. A shift to the left tends to increase oxygen uptake in the lung but impairs the release of oxygen to the tissues. In patients with a low Pa_{O_2} a shift to the left could produce considerable improvement in pulmonary oxygen uptake. At a normal Pa_{O_2} the effect on pulmonary loading is of little importance; however, the effect on tissue unloading is considerable.

Affinity of hemoglobin for oxygen is regulated by a number of normal control mechanisms which may be modified by disease. In addition, there are a number of diseases unrelated to these normal regulatory processes which are characterized by alterations of oxyhemoglobin affinity. The physiological and pathophysiological factors modifying the oxyhemoglobin dissociation curve are summarized in Table 8–2.

Abnormalities in oxyhemoglobin affinity cannot be identified by measurements of Pa_{O_2}, or oxygen saturation alone. Several blood levels of P_{O_2} and saturation at known pH and temperature, or more formal measurements of the P_{O_2} at 50 per cent oxyhemoglobin saturation, are required. Fortunately, isolated abnormalities of oxyhemoglobin affinity are rarely clinically important. Several examples are discussed below. Most frequently, altered oxyhemoglobin affinity gains importance when additional abnormalities of oxygen supply are present; arterial hypoxemia, impaired cardiac output, and anemia are generally better understood clinically and can frequently be treated.

RED CELL pH, P_{CO_2}, TEMPERATURE. Decreases in pH (acidemia) decrease hemoglobin affinity for oxygen, and increases in pH (alkalemia) increase hemoglobin affinity. This effect of pH is known as the Bohr effect.* At the tissue level, at which plasma and red cell pH are relatively low, the Bohr effect serves to facilitate oxygen dissociation from hemoglobin. Conversely, in the lung, where plasma and red cell pH are relatively high, the combination of oxygen with reduced hemoglobin is facilitated. Normally, the time required for the operation of the Bohr effect is exceedingly brief, and its contribution to pulmonary and tissue oxygen exchange is of decisive importance. As a result of acid-base disturbances, the Bohr effect may operate to impair oxygen delivery. Severe alkalosis will reduce the release of oxygen from hemoglobin and may compromise oxygen delivery to the tissues, whereas severe acidosis will decrease the uptake of oxygen by hemoglobin and may impair pulmonary oxygen uptake, particularly in patients with reduced Pa_{O_2}.

Carbon dioxide influences the affinity of hemoglobin for oxygen both by affecting red cell pH and by a direct effect combining with α-amino groups of hemoglobin to form carbamate. Hemoglobin with carbamino groups binds oxygen less avidly. As a result, increased P_{CO_2} leads to decreased affinity (shift to the right), and decreased P_{CO_2} to increased affinity.

Increased temperature decreases the affinity of hemoglobin for oxygen and decreased temperature has the opposite effect. In the basal state this effect is of little consequence; however, during severe exercise, metabolically active nonpulmonary tissue may be substantially warmer than the lung and, consequently, the unloading of oxygen to the tissues may be facilitated. In febrile disorders, especially those occurring in patients with a decreased Pa_{O_2}, pulmonary uptake of oxygen may become impaired because of this mechanism. Conversely, in hypothermic states unloading of oxygen from hemoglobin at the tissue level may be impaired.

ORGANIC PHOSPHATES. A variety of anions, especially organic phosphates, have profound effects on the affinity of hemoglobin for oxygen and are involved in the physiological regulation of oxygen transport as well as its pathophysiological aberrations. Dilute hemoglobin solutions containing no phosphates show marked affinity for oxygen. Addition of organic phosphates to the solution substantially reduces this affinity.

From the quantitative standpoint the major organic phosphate in human red cells is the compound 2,3-diphosphoglycerate (2,3-DPG). Although both inorganic phosphate and other organic phosphates such as ATP have similar effects, 2,3-DPG is the major anion involved in the regulation of oxygen transport in man. Increasing concentrations of 2,3-DPG leads to progressive decreases in affinity of hemoglobin for oxygen. This effect is mediated by the ability of 2,3-DPG to combine reversibly with reduced, but not oxygenated, hemoglobin. The combination of 2,3-DPG with reduced hemoglobin results in increased stability of the deoxyhemoglobin, so that a higher P_{O_2} is required to achieve a given degree of combination between oxygen and hemoglobin. Thus, increased 2,3-DPG concentrations lead to a rightward shift of the oxyhemoglobin dissociation curve and decreased values to a leftward shift.

Red cell organic phosphates, including 2,3-DPG, are generated metabolically by anaerobic glycolysis, the major energy-providing metabolic pathway in mature mammalian erythrocytes. 2,3-Diphosphoglycerate is found in a uniquely high concentration only in the red cells of most mammals. The formation of 2,3-DPG represents a deviation from the classical glycolytic pathway (see Fig. 8–2). The glycolytic intermediate 1,3-diphos-

---

* Christian Bohr (1835–1911) was a Danish physiologist who made the study of hemoglobin, blood gases, and gas transport his life work. He was the first to describe the sigmoid shape of the O_2 dissociation curve and the shift of the curve to the right with increasing P_{CO_2}.

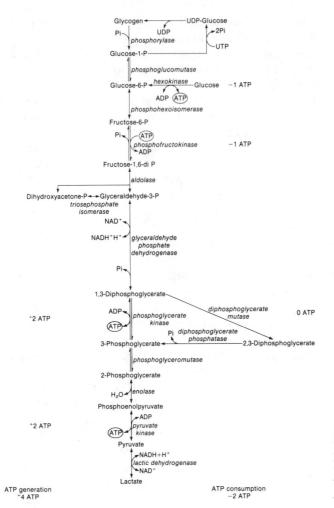

FIG. 8–2. Anaerobic glycolysis including the 2,3-DPG bypass: the usual pathway results in the net generation of 2 ATPs, whereas the 2,3-DPG pathway results in the net generation of 0 ATPs.

phoglycerate is converted to 2,3-DPG by the action of the enzyme 2,3-diphosphoglycerate mutase with 3-phosphoglyceric acid serving as a cofactor. It is degraded to 3-phosphoglyceric acid by the enzyme 2,3-diphosphoglycerate phosphatase, which splits the phosphate from the 2- position, leaving 3-phosphoglycerate, an intermediary on the main pathway of glycolysis. The concentration of 2,3-DPG in the red cell reflects the balance between its rate of formation and its rate of breakdown. The precise mechanisms that influence rate of 2,3-DPG synthesis versus rate of breakdown are not known, but factors that influence the overall rate of glycolysis in the erythrocyte affect its concentration.

Glycolytic rate in the red cell is increased by alkalosis and decreased by acidosis. These changes are at least partially related to the increased activity of phosphofructokinase (a rate-limiting step in glycolysis) that results from increased pH. Changes in glycolytic rate induced by pH changes lead to corresponding changes in 2,3-DPG concentration.

Concentrations of ADP, ATP, and inorganic phosphate likewise play important roles in the regulation of glycolysis and thus influence the metabolism of 2,3-DPG.

For example, low levels of ADP lead to the conversion of 1,3-DPG and 2,3-DPG rather than to the phosphorylation of ADP to ATP (see Fig. 8–2).

With hypoxemia there is increased binding of 2,3-DPG (and reduced free 2,3-DPG) which leads to increased diphosphoglyceromutase activity that results in augmented 2,3-DPG synthesis.

An interesting example of the effects of variable binding of hemoglobin to DPG is observed in the mother and the fetus. Maternal and fetal blood have equal levels of 2,3-DPG, but binding to fetal hemoglobin is much reduced. Consequently, the maternal unloading of oxygen in the placenta is normal and the loading of oxygen by the left-shifted, low-oxygen, fetal blood is strikingly increased.

Metabolic alkalosis increases and metabolic acidosis decreases red cell 2,3-DPG concentrations. In turn, this produces shifts in the oxyhemoglobin dissociation curve which are opposite to those produced by the direct effect of pH. Stored red cells are acidotic and have a decreased glycolytic rate. Levels of 2,3-DPG are reduced and there is increased affinity of hemoglobin for oxygen. Therefore, patients transfused with large amounts of stored blood may have initial impairment of tissue oxygen supply because of the leftward shift of the oxyhemoglobin dissociation curve; this is corrected within hours by reestablishment of normal glycolysis.

Genetic abnormalities involving red cell glycolytic enzymes may produce changes in 2,3-DPG concentration and the oxyhemoglobin dissociation curve. These are listed in Table 8–2. The most severe example is a patient with complete deficiency of diphosphoglycerate mutase with a $P_{50}O_2$ of 17 mm. Hg.

Acute and chronic hypoxemia are associated with increased 2,3-DPG and a shift of the dissociation curve to the right. This has been demonstrated in congestive heart failure, anemia, and exposure to high altitude. Patients with diabetes mellitus under poor control have increased glycosylation of hemoglobin, reduced 2,3-DPG binding, and slightly increased oxygen affinity.

ACQUIRED ABNORMAL HEMOGLOBINS. High concentrations of carboxyhemoglobin (resulting from the inhalation of carbon monoxide) and high concentrations of methemoglobin (resulting from increased oxidation of the ferrous moiety of hemoglobin to the ferric form) produce increased oxyhemoglobin affinity, with a leftward shift of the dissociation curve. Consequently, these patients have two abnormalities of oxygen delivery. The fraction of hemoglobin in the abnormal form (carboxy or methemoglobin) does not combine with oxygen, and thus the amount of hemoglobin available for oxygen transport is reduced. Like anemia, this results in a reduction in effective red cell mass. In addition, the leftward shift of the remaining normal hemoglobin results in deficient unloading of oxygen to the tissues.

HERITABLE ABNORMALITIES OF HEMOGLOBIN. A number of heritable disorders of hemoglobin structure may be associated with either increased or decreased oxygen affinity. Such changes may occur as a result of single amino acid substitutions in one or more of the hemoglobin chains, or replacement of an entire chain by an abnormal chain; also, the chains making up the hemoglobin may all be of one type or there may be structural modifications of the heme group (Chap. 27). These hemoglobinopathies not only are of great intrinsic interest but serve to demonstrate the potential pathological effects of shifts in the dissociation curve.

SHIFTS TO THE RIGHT. Heterozygotes with hemoglobin Seattle have a modest shift of the oxyhemoglobin dissociation curve to the right. At the normal arterial P_{O_2} (approximately 90 mm. Hg) hemoglobin saturation averages approximately 85 to 90 per cent. At the usual mixed venous blood P_{O_2}

(40 mm. Hg) saturation is only 45 per cent. This leads to a wide difference between arterial and venous oxygen content at normal oxygen tensions. Red cell mass is reduced in some of these patients, and it has been suggested that this results from decreased erythropoietin activity related to supranormal oxygen delivery to the tissues.

Heterozygotes with hemoglobin Kansas show a marked decrease in oxygen affinity. At normal arterial P_{O_2} values, saturation is approximately 60 per cent, and at normal mixed venous P_{O_2} hemoglobin is only 30 per cent saturated. These individuals show gross cyanosis but appear to be normal in other respects. In particular, red cell mass is not increased, suggesting no tissue hypoxia.

The hemoglobin associated with sickle cell disease (Hb S) shows a moderate decrease in oxygen affinity. However, this abnormality appears not to play an important role in the manifestations of sickle cell disease.

The practical importance of *decreased hemoglobin affinity for oxygen* can be summarized as follows: with normal arterial P_{O_2}, these patients appear to have no striking abnormality of oxygen utilization. Pulmonary and cardiac function and the general clinical status are normal. The development of lung disease with reduced Pa_{O_2} may result in a further substantial decrease in pulmonary oxygen uptake, leading to severe hypoxemia. There is a possibility that increased tissue oxygen delivery results in decreased erythropoietin activity and may account for the reduced red cell mass seen in some of these patients. Screening of patients with hypoxemia and cyanosis* should include studies of hemoglobin affinity for oxygen. The possibility of a rightward shift is particularly pertinent in

---

* Cyanosis may also be caused by an excess concentration of methemoglobin and sulfhemoglobin. In addition to the acquired forms previously described, there are several congenital types of methemoglobin (Hemoglobin M Boston, Iwate, Milwaukee, Saskatoon, and Hyde Park).

patients who are asymptomatic, with normal pulmonary and cardiac function and who, rather than being polycythemic, have a normal or reduced red cell mass. A large number of hemoglobin variants result in increased oxygen affinity. A severe left shift occurs in heterozygotes for Hb McKees rocks ($P_{50}O_2$ = 12 mm. Hg). Many of these individuals function almost normally, demonstrating polycythemia and increased erythropoietin as their major evidence of impaired oxygen supply.

Hemoglobin H and Hemoglobin Barts have profound shifts to the left; in the former the shift is so extreme that the dissociation curve resembles that of myoglobin (see below), and, therefore, this hemoglobin is useless in oxygen transport. Both of these hemoglobins are found in patients with α-thalassemia, and hemoglobin Barts is associated with death of the affected fetus.

Movement of oxygen from plasma to intracellular sites of utilization

Oxygen moves from capillary plasma to intracellular sites of oxygen utilization by the process of diffusion. The amount of oxygen that diffuses per unit time depends on: driving pressure; surface area available for diffusion; diffusion distance; character of the membranes across which diffusion occurs; and the time available for diffusion.

Driving pressure

This refers to the difference between the P_{O_2} of capillary plasma and the P_{O_2} at various intracellular sites where oxygen is utilized. This pressure differential is generated by the various oxygen-dependent metabolic processes. Thus, the greater the metabolic rate, the lower the P_{O_2} at a given site and the greater the driving pressure. This arrangement provides for the highest availability of oxygen at the intracellular sites where oxygen need is maximal. For example, during muscular exercise, mitochondrial oxygen utilization increases, the capillary-mitochondrial driving pressure is increased, and oxygen delivery is augmented. Since there are variable rates of oxygen consumption at different intracellular sites, there may be regional variations of P_{O_2} within the cell. Although exact driving pressure along the entire length of capillaries is difficult to estimate, a maximal figure can be derived. The P_{O_2} at the arterial end of the capillary is 90 mm. Hg. Estimates of mitochondrial P_{O_2} suggest values of less than 1 mm. Hg. Thus, normal maximal driving pressure has been estimated to be in excess of 89 mm. Hg. Decreases in driving pressure can occur as a result of reduction in Pa_{O_2} or intracellular oxygen utilization.

In considering cellular oxygen uptake, it is important to distinguish between oxygen content and partial pressure. The major fraction of the oxygen content of blood is combined with hemoglobin. Therefore, content may be considered as a reservoir serving the function of providing an adequate overall supply, whereas the partial pressure serves as a driving pressure to permit adequate diffusion of oxygen into the cell. Increasing of P_{O_2} values above 100 mm. Hg provides only a small increase in blood oxygen content; however, marked increases in P_{O_2} proportionately increase the driving pressure for diffusion of oxygen into the cell. Under normal circumstances, a driving pressure of 90 mm. Hg is adequate for ensuring an adequate oxygen supply, and there is little advantage to increasing P_{O_2} further. In special circumstances, such as interstitial or intracellular edema, a normal driving pressure might be inadequate, and there could be a theoretical advantage to increasing arterial P_{O_2} to supranormal values, thereby increasing driving pressure.

Surface area

A major determinant of the surface area available for oxygen transport is the number of capillaries per unit of tissue mass. Under normal circumstances, there may be an excess number of capillaries in most tissues. This excess provides an important reserve, and it is probable that not all capillaries are perfused during the resting state. With increased flow, additional (unused) capillaries may be recruited, and the surface area available for diffusion of oxygen is thereby increased. During chronic hypoxia and with exercise training, there is evidence to suggest that new capillaries may form in tissues such as skeletal muscle. This increased capillarity serves the function of maintaining oxygen diffusion in the face of a reduced driving pressure or increased oxygen requirement. In some diseases, the ratio of capillaries to unit tissue mass may be decreased. For example, hypertrophy of myocardial fibers is not generally accompanied by a corresponding increase in the number of capillaries. As a result, the hypertrophied fiber may be relatively ischemic.

Diffusion distance and pathway

The distances and pathways involved in the movement of oxygen from capillary to mitochondria have not been quantitated precisely. Estimates depend on assumptions in regard to the geometry of the transport pathway. Qualitatively, it appears reasonable that the development of interstitial edema, interstitial fibrosis, intracellular edema, or changes in the physical character of cell membranes may increase the length of the diffusion pathway or decrease the rate of diffusion and thus interfere with oxygen delivery. Theoretically, inhalation of oxygen-rich mixtures to produce supranormal Pa_{O_2} values and increase the driving pressure may be helpful in these conditions.

Physicochemical properties of the membranes

The membranes that offer resistance to oxygen flow include capillary endothelium, with its basement membrane; the plasma membranes of the various cell types throughout the body, with their basement membranes; the double mitochondrial membrane; and the limiting membranes of other subcellular structures in which oxygen-consuming reactions take place. Although no precise characterization of the diffusibility of oxygen through these membranes is available, it is reasonable to assume that structural alterations by disease could serve to impair oxygen diffusion.

Time available for diffusion

Although the rate constants of tissue oxygen diffusion have not been determined, a marked decrease in the capillary transit time might be expected to decrease capillary to tissue transport.

Mixed venous-metabolizing cell oxygen gradient (V-c oxygen difference)

In the lung, the difference between alveolar P_{O_2} and arterial P_{O_2} is an important variable. Analysis of the factors that produce this A-a oxygen difference serves to clarify the mechanism of pulmonary gas exchange (see Chap. 7). There is likewise a mixed venous-cellular P_{O_2} difference. Normal mixed venous P_{O_2} is approximately 40 mm. Hg; mean cellular oxygen tensions are difficult to measure and interpret, but the value for most the cell types is 20 mm. Hg or less. Thus, from the standpoint of whole body tissue gas exchange, a V-c P_{O_2} difference of 20 mm. Hg or greater is present. As with the A-a oxygen difference, the V-c difference arises from diffusion limitation, physiological shunting, and/or anatomical shunting. Oxygen exchange in the lung is compared to oxygen exchange in the tissues in Figure 8–3.

Diffusion limitation

This would operate to produce an oxygen gradient between systemic end-capillary blood and the mean cellular P_{O_2}. A useful integrated approach to diffusion is to consider that the time available for diffusion is equivalent to the time required for a unit volume of blood to traverse the capillary (transit time). Diffusion limitation, in this sense, could be defined as a period of time insufficient for equilibration to occur. The time spent in the capillary increases with decreasing cardiac output and is affected by both proximal (arteriolar) and distal (venular) vascular resistance. It is well known that decreases in cardiac output (increasing the time available for diffusion) result in a decreased mixed venous P_{O_2}. This presumably means that the V-c gradient is decreased and strongly suggests that diffusion limitation is an important factor in the generation of the V-c oxygen difference. Current evidence suggests that oxygen uptake in the lung at the high altitudes encountered on the top of Mount Everest is diffusion dependent. At that altitude, due to very low inspired oxygen, the driving pressure is severely reduced, and as a result of increased cardiac output the capillary transit time is also reduced. Similar conditions may well develop in various disorders to result in limited tissue diffusion of oxygen.

Physiological shunting

Oxygen exchange in the tissues involves liquid-to-liquid interfaces. In other forms of liquid-to-liquid gas exchange (gill, placenta) the dynamics of flow operate so that a portion of the flow does not participate in gas exchange. This may be considered as resulting from unevenness of flow distribution with respect to the gas exchange surface. It is likely that this is also true of capillary flow. This process, then, results in some degree of physiological shunting, with resultant inequality between end-capillary and cellular oxygen tensions.

Anatomical shunting

In some organs (e.g., skin) there are arteriovenous shunts that bypass capillaries entirely. Flow through such shunts does not participate in cellular gas exchange. In some forms of shock there also may be capillary-to-venous shunting, with failure to venulize the blood that is shunted. Such anatomical shunts would be expected to contribute to the inequality between mixed venous and cellular oxygen tensions.

The same general mechanisms that operate in the lung to produce A-a oxygen differences operate in tissues to produce V-c oxygen differences. Little quantitative information is available; but it appears that diffusion limitation is more important in the tissues than in the lung. A major difference between the two areas is that diffusion limitation, physiological shunting, and anatomical shunting produce higher P_{O_2} values and oxygen contents in the efferent blood of the tissues rather than the lower values seen in the lung (Fig. 8–3).

Tissue oxygen stores (myoglobin)

Myoglobin, an intracellular heme protein, is present in various tissues, including skeletal and cardiac muscle. Like hemoglobin, it is capable of a reversible combination with oxygen, but, unlike hemoglobin, it contains only one heme prosthetic group. As a result, the oxymyoglobin dissociation curve is hyperbolic rather than sigmoid (Fig. 8–4). Myoglobin is fully saturated with oxygen at P_{O_2} values greater than 5 mm. Hg. Although its precise physiological role is unknown, it has been suggested that it serves as an emergency storage form of oxygen. At relatively high cellular P_{O_2} (resting), oxygen is bound and represents a potential reserve. When cell P_{O_2} drops below 5 mm. Hg (during severe muscular contraction), oxygen is released from myoglobin. It is of some interest that, under conditions of chronic hypoxia in highly trained muscles, and in some animals capable of withstanding severe oxygen depletion, myoglobin concentrations may be increased.

Oxygen utilization

The utilization of molecular oxygen by cells subserves two general biochemical functions: Oxygen is the terminal electron ac-

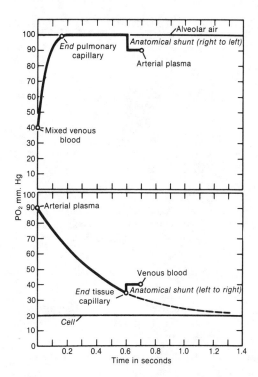

FIG. 8–3. A hypothetical comparison of pulmonary vs. tissue oxygen exchange: note diffusion equilibrium in the lung and lack of diffusion equilibrium in the tissues. Anatomical shunting produces a decrease in arterial plasma P_{O_2} but an increase in venous plasma P_{O_2}.

ceptor in substrate oxidation, a process by which energy is made available for various endergonic processes, and oxygen is an obligatory oxidant in a wide variety of biosynthetic processes. It is also involved in a number of oxidative processes that do not provide energy directly.

Oxygen and energy provision

In most forms of life the major source of energy is provided by the oxidation of substrate by molecular oxygen, with energy being made available in the form of ATP.

MITOCHONDRIAL STRUCTURE. The reactions involved in this pathway are localized in the mitochondrion, which may be regarded as a biological energy transducer. This pathway accounts for well over 75 per cent of total cellular oxygen consumption.

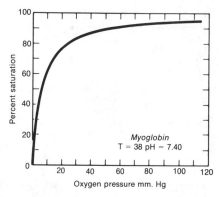

FIG. 8–4. Myoglobin dissociation curve: note the hyperbolic shape and the essentially complete saturation at oxygen tensions greater than 10 mm. Hg.

Mitochondrial structure closely subserves mitochondrial function. The chemical reactions involved in substrate oxidation are facilitated by the spatial arrangements. This cell organelle has as its limiting structure two continuous membranes; the outer one is smooth, and the inner is thrown into a series of folds called cristae. Within the inner membrane is a semiliquid substance called the matrix. The enzymes of the Krebs cycle are located in the matrix, whereas the enzymes directly involved in respiration are located on the inner membrane in the same sequence in which the chemical reactions occur. Abnormalities of mitochondrial structure have been reported in a rare variety of skeletal muscle disease (giant mitochondria) and in alveolar epithelial cells following exposure to high oxygen concentrations and various noxious gases.

The diffusion pathway of oxygen from cytoplasm to the active site of oxygen utilization (cytochrome oxidase) includes cytoplasmic water, the outer and inner membranes of the mitochondrion, and the active site on cytochrome oxidase. With intracellular edema or structural abnormalities of the mitochrondrial membranes, intracellular oxygen diffusion might be impaired.

CHEMICAL REACTIONS IN THE MITOCHONDRIA. Two general processes are involved in the reactions culminating in mitochondrial oxygen utilization. By means of the Krebs tricarboxylic acid (TCA) cycle (Fig. 8–5), the three classes of foodstuff (proteins, fats, and carbohydrates) are oxidized to carbon dioxide and water. In this cycle, one molecule of acetic acid, in a special activated form, acetyl CoA, is oxidized in eight steps to carbon dioxide, with the net provision of four pairs of electrons. The TCA cycle may be regarded as a common pathway for the ultimate use of all types of foodstuffs as a source of energy. These transformations are not involved primarily in energy provision but may be considered as providing electrons which are then used in energy transformations.

In each "revolution" of the Krebs cycle there are four steps in which electrons are provided. In three of these oxidation steps, the molecule nicotinamide adenosine dinucleotide (NAD^+) serves as the electron acceptor. Nicotinamide adenosine dinucleotide, which is found in all cells, serves as an electron carrier in a wide variety of reactions. When oxidized, NAD^+ accepts electrons and is converted to the reduced form, NADH. In the fourth oxidation step a pair of electrons is removed from succinic acid, with flavine adenine dinucleotide (FAD) serving as the electron acceptor.

The direct provision of energy in a utilizable form (ATP) occurs in the electron transport chain. Electrons enter the chain from the Krebs cycle at a fairly high energy level which decreases progressively as electrons flow down the chain. The energy released is conserved in the form of ATP. The electron transport chain consists of a series of cytochromes, heme-containing enzymes which can participate reversibly in oxidation-reduction reactions. The iron atom of a given cytochrome molecule alternates between the oxidized ferric state (Fe^{+++}) and the reduced ferrous states (Fe^{++}). In the oxidized Fe^{+++} state electrons are accepted, with reduction to the reduced Fe^{++} state. The reduced form, in turn, donates electrons to the next cytochrome, and ultimately the final cytochrome (cytochrome

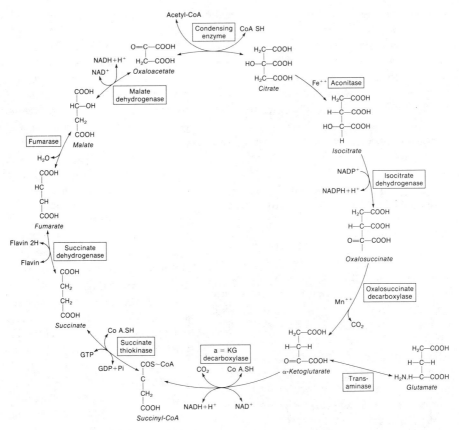

FIG. 8—5. Krebs cycle: note that no net energy conservation is involved. The Krebs cycle is linked to the electron transport chain by way of the $NAD^+/NADH$ steps and the flavin/flavin 2H steps.

oxidase) donates electrons to molecular oxygen. The sequence of reactions from substrate to molecular oxygen may be summarized as follows:

Krebs Cycle

Substrate
derived   →NADH→FAD→
electrons

Respiratory Chain

FAD→Cytochrome b→Cytochrome c→
Cytochrome oxidase (a and a_3)→
molecular O_2

The passage of electrons from one mole of glucose to oxygen, with the production of carbon dioxide and water results in the generation of 38 moles of ATP from ADP and inorganic phosphate. The theoretical energy yield from 1 mole of ATP is approximately 7,000 cals and thus the minimal energy provided by the oxidation of 1 mole of glucose to carbon dioxide and water is equal to 7,000 × 38 cal./mole or approximately 266,000 cal./mole. The overall reaction may be summarized as follows:

$$1 \text{ Glucose} + 38 \begin{array}{c} \text{Inorganic} \\ \text{phosphate} \end{array} + 38 \text{ ADP} + 6 \text{ O}_2$$
$$\rightarrow 6 \text{ CO}_2 + 44 \text{ H}_2\text{O} + 38 \text{ ATP}.$$

METABOLIC CONTROL. Total oxygen consumption may vary from approximately 250 ml./min. at rest to 3,500 ml./min. during maximal exercise. Moreover, changes in oxygen consumption vary widely from one organ to another. The mechanisms by which energy availability is geared to energy requirement are collectively termed metabolic control.

The processes involved in metabolic control are complex and, in many cases, poorly understood. It is likely that changes in metabolic rate can occur either as a result of an actual change in the concentration of a given enzyme or as a result of the alteration of activity of the enzyme independent of a change in concentration.

Regulatory mechanisms for mitochondrial oxygen utilization may involve either short-term or long-term control factors.

SHORT-TERM REGULATION. Short-term control factors are concerned with the second-to-second regulation of energy output. In general, these mechanisms involve changing concentrations of low molecular weight molecules such as ADP, ATP, and inorganic phosphate. In the case of mitochondrial oxygen utilization, emphasis has been placed on the ratio of ADP to ATP. During times of rapid energy expenditure (e.g., during muscular exercise), ATP is utilized rapidly and this ratio increases, resulting in increased mitochondrial oxygen consumption. During more basal activities, ATP utilization is low and the ADP/ATP ratio is decreased. As a result, mitochondrial oxygen uptake is decreased. Absolute levels of ADP and ATP depend not on the rate of utilization solely, but also on the rate of formation and rate of breakdown. A large number of factors other than energy utilization may influence the rate of formation or rate of breakdown.

LONG-TERM REGULATION. Although short-term control mechanisms are important, there are a number of situations in which the rate of oxygen consumption does not appear to depend on differences in the concentration of low molecular weight species. With exercise training, skeletal muscle acquires an increased mitochondrial oxygen uptake capacity. This is associated with a general increase in mitochondrial protein and specific increases in the various cytochromes, including cytochrome oxidase. Under conditions of severe hypoxemia, there may be reduction of mitochondrial mass and reduction in cytochrome oxidase activity. Therefore, the biogenesis of mitochondria and the biosynthesis of key electron transport chain enzymes may be geared to variable requirements for oxygen consumption. Recent work suggests that the biosynthesis of some of the respiratory enzymes and mitochondrial biogenesis are regulated by a distinct type of DNA present in mitochondria (mitochondrial DNA). Mitochondrial DNA is physically and chemically distinct from classical nuclear DNA and is regulated by different controls. It is possible that oxygen availability or oxygen utilization may play an important role in modifying the rate of synthesis of electron transport chain enzymes by affecting nuclear and, particularly, mitochondrial DNA metabolism.

DISORDERS AFFECTING MITOCHONDRIAL OXYGEN UTILIZATION. Abnormalities of cellular oxygen supply are the commonest type of disorder affecting mitochondrial oxygen utilization. In addition, various components of the enzymatic pathways can be inhibited by one or another chemical species, some of which are of clinical importance. In appropriate concentrations, cyanide ion inhibits cytochrome oxidase, which prevents passage of electrons to molecular oxygen, thereby stopping mitochondrial oxygen utilization. This inhibition is the basis of the extreme toxicity of cyanide (and explains its popularity as an agent for suicide, homicide, and execution). Carbon monoxide also inhibits cytochrome oxidase, but the concentrations required are so high that patients may well die of the effects of this gas on hemoglobin before the effects on cytochrome oxidase become important. A chronic form of carbon monoxide intoxication characterized by nonspecific central nervous system abnormalities may be related to low level exposure, with resultant partial inhibition of cytochrome oxidase.

Chloramphenicol is a relatively specific inhibitor of mitochondrial protein synthesis and, in pharmacological doses, inhibits the biosynthesis of mitochondrial enzymes (i.e., cytochrome oxidase). This inhibition may be the basis of the aplastic anemia that develops in some patients treated with this drug.

Thyroid hormone plays a critical role in the regulation of cellular oxygen consumption. The precise mechanism involved is not clear, but the hormone is probably involved in the coupling of oxygen consumption to the generation of ATP. With excessive amounts of hormone (hyperthyroidism), partial uncoupling may occur and excess energy may be dissipated as heat rather than being available in a useful form. With inadequate concentrations of thyroid hormone (hypothyroidism), oxygen consumption becomes inappropriately low, leading to an inadequate energy supply.

Luft's syndrome is a rare disorder in which oxygen consumption is markedly increased without evidence of thyroid abnormality. The site of the hypermetabolism is in skeletal muscle. Profuse sweating and extreme heat intolerance are important clinical features. There is excess heat production; cardiac output is high, and there is hyperventilation. The disorder is characterized by a large overgrowth of bizarre mitochondria of skeletal muscles, which micrographically show mitochondrial hyperplasia and hypertrophy with abnormal mitochondrial cristae. Biochemical examination of these mitochondria indicates some uncoupling of oxidation from phosphorylation. Chloramphenicol, which depresses mitochondrial biogenesis, produced a partial remission in one patient.

Pellagra is a disorder characterized by dementia, dermatitis and gastrointestinal symptoms. It results from inadequate availability of the vitamin nicotinamide, usually on the basis of dietary deficiency. Nicotinamide is an important constituent of NAD. It is reasonable to conclude, therefore, that pellagra results from deficient NAD supplies, leading to abnormalities in mitochondrial electron transfer.

In addition to these more or less well-defined disorders, it should be expected that other specific disorders involving the Krebs cycle and the electron transport chain will be uncovered. Given the large number of different enzymes and complex control factors, the probability of heritable disorders or environmental alterations is high.

Anaerobic glycolysis

When energy supplied by mitochondrial oxygen utilization is inadequate, supplemental energy is provided by glycolysis. During glycolysis, glucose is oxidized in a series of 11 reactions to pyruvate and lactate. (Fig. 8–2) This pathway has several general functions: it provides a supplemental source of energy under normal circumstances when the delivery of oxygen is insufficient to meet metabolic needs and during pathological hypoxic states; in some cell types (e.g., the mature mammalian erythrocyte) it represents the sole source of cellular energy; and it serves as an important source of pyruvate for the Krebs cycle and ultimate oxidation of substrate to carbon dioxide and water during aerobic glycolysis.

Structural basis

The reactions involved in glycolysis are largely localized in the cytoplasm. The glycolytic enzymes are distributed more or less randomly in the cytoplasmic brei, and, in contrast to provision of oxygen-dependent energy, no precise anatomical arrangement is required for the reactions to proceed. Exposure of substrate to the mixture of enzymes is adequate to produce the sequence of chemical reactions.

Chemical reactions

The sequential steps of glycolysis involve the conversion of glucose to pyruvate. Pyruvate may be converted to lactate by the enzyme(s) lactate dehyrogenase. The overall reaction may be summarized as follows:

glucose + 2 ATP
+ 2 phosphate + 2 ADP→
2 lactate + 2 ADP + 4 ATP.

In terms of energy supply, the oxidation of 1 mole of glucose requires 2 moles of ATP but results in the production of 4 moles of ATP. Thus, there is a net gain of 2 moles of ATP (or approximately 14,000 calories).

Metabolic control

Metabolic control of glycolysis is exceedingly complex and the potential mechanisms for regulation of mitochondrial oxygen utilization reviewed above are pertinent to anaerobic glycolysis. It is useful to consider both short-term and long-term control factors.

SHORT-TERM REGULATION. The various enzymes in the glycolytic sequence can be stimulated or inhibited by the various products of the reactions. Hexokinase activity is inhibited by its product glucose-6-phosphate. Phosphofructokinase activity is stimulated by ADP and inorganic phosphate and is inhibited by ATP. The activity of this enzyme is unusually sensitive to H^+ concentrations, increasing with alkalosis and decreasing with acidosis. The activity of glyceraldehyde phosphate dehydrogenase depends on NAD^+ and inorganic P_{O_4} concentrations. Concentrations of ADP decisively influence the activity of phosphoglycerate kinase and pyruvate kinase. Lactate dehydrogenase is strongly influenced by cytoplasmic NADH concentrations. Overall, the short-term regulation of glycolytic rate depends heavily on intracellular ADP concentrations. This, in turn, depends on the balance between ADP synthesis and ADP removal.

LONG-TERM REGULATION. A number of the glycolytic enzymes, including hexokinase, phosphofructokinase, and pyruvate kinase, are rate limiting under one condition or another. For example, there is a precise relationship between the maximal glycolytic rate in various organs and the pyruvate kinase activity of these tissues. This suggests that the biosynthesis of this enzyme must be closely geared to the requirements of the individual organ for glycolysis. Furthermore, in the chronically exercised heart, there is an increase in pyruvate kinase activity, which presumably increases myocardial glycolytic capacity. Similarly, pyruvate kinase is increased in cultured cells incubated under hypoxic conditions. These changes in enzyme activity and glycolytic capacity are independent of the short-term regulating factors described above. Cytoplasmic enzyme biosynthesis depends on the nuclear DNA–microsomal RNA system. It appears that long-term variations in the rate of glycolysis depend on modifications of this system.

Interrelationship between mitochondrial oxygen utilization and anaerobic glycolysis (the Pasteur effect)

When cells are exposed to anaerobic environment, there is a brisk increase in the rate of glycolysis. This is known as the Pasteur effect. It may be evoked not only by anaerobiosis but also by inhibition of mitochondrial oxygen utilization by agents such as cyanide. The mechanism of the Pasteur effect probably depends on the availability of ADP in the cytoplasm versus the mitochondrion. Under conditions of normal rate of oxidative phosphorylation, there is greater mitochondrial than cytoplasmic affinity for ADP. Thus, intracytoplasmic ADP concentrations are low and the glycolytic rate is minimal. With decreased mitochondrial oxygen consumption, less ADP is used in the mitochondrion, cytoplasmic concentrations rise, and anaerobic glycolysis is progressively accelerated. The precise quantitative relationship between decreases in mitochondrial oxygen utilization and increased glycolysis is not known. As a result, the precise threshold of the Pasteur effect is unclear, and it is possible that substantial abnormalities of oxygen utilization may occur before a measurable increase in glycolysis is present.

Monitoring abnormalities of mitochondrial oxygen utilization

The ability to monitor abnormalities of mitochondrial oxygen utilization in patients would be of great importance. No entirely satisfactory approach is available. Variables such as Pa_{O_2}, Pv_{O_2}, etc. (Table 8–1) reflect processes which are proximal to the energy-providing reactions. With only moderate abnormalities of these proximal processes, or in the face of adaptive changes, these measurements provide a poor quantitative estimate of the degree of abnormality of mitochondrial oxygen utilization.

Lactate

One approach to the monitoring of abnormalities of oxygen utilization has been based on measurements of lactate concentrations in blood or plasma. The chemical basis for this approach is the Pasteur effect. However, lactate concentration in plasma depends not only on its rate of intracellular generation but also on other factors: the rate at which lactate is delivered from cells to blood, which, in turn, is dependent on the permeability of cell membranes to lactate and on tissue perfusion; and the rate at which lactate is cleared from the blood, which is dependent on the rate at which lactate is metabolized (chiefly by the liver) and the rate at which it is excreted by the kidney. Moreover, lactate may be generated by processes other than the glycolytic pathway. When hepatic blood flow and liver function are normal, the capacity of the liver to metabolize lactate is extensive. As a result, in pathological conditions, increases in blood lactate may occur with normal oxygen utilization. Measurements of blood lactate concentration are most useful as an indicator of abnormal oxygen utilization in conditions associated with inadequate circulation such as shock and severe forward heart failure.

NAD$^+$/NADH (lactate/pyruvate)

The intracellular oxidation-reduction potential (redox state) is a fundamental property of biological systems, influencing the chemical behavior of all oxidizable or reducible components of the system. The importance of NAD$^+$/NADH as an electron carrier in energy conservation reactions has already been described. Because of this key position, attempts have been made to quantitate the ratio of NAD$^+$/NADH as a reflection of the redox state of individual organs. Since direct chemical measurements are not possible, attempts have been made to substitute blood levels of diffusible substrates. The theoretical basis for the use of lactate-to-pyruvate ratios is readily apparent in the following equations:

$$\text{pyruvate} + \text{NADH} \leftrightharpoons \text{lactate} + \text{NAD}^+ + \text{H}^+$$

by the mass action equation:

$$\frac{(\text{Lactate})\ (\text{NAD}^+)\ (\text{H}^+)}{(\text{Pyruvate})\ (\text{NADH})} =$$

K equilibrium constant of the reaction

or:

$$\frac{(\text{NAD}^+)}{(\text{NADH})} = \frac{(\text{Pyruvate})\ (\text{H}^+)}{(\text{Lactate})\ (\text{K}'\ \text{equil.})}$$

The absolute value of lactate/pyruvate ratios in arterial plasma does not appear to differ from that found in liver or macrophage extracts. This suggests that measurements of lactate/pyruvate ratios in plasma may be useful in detecting far advanced changes in mitochondrial oxygen utilization; unfortunately, however, this measurement does not detect early changes.

Mitochondrial NAD$^+$/NADH ratios have been calculated in alveolar macrophages by measuring the substrates involved in a reaction localized in the mitochondrial matrix:

Acetoacetate + NADH $\rightleftharpoons$
$\quad$ β-Hydroxybutyrate + NAD$^+$ + H$^+$,

so that

$$\text{mitochondrial } \frac{\text{NAD}^+}{\text{NADH}}$$
$$= \frac{(\text{Acetoacetate}) \, (\text{H}^+)}{(\beta\text{-Hydroxybutyrate}) \, \text{K}_{eq}}$$

This measurement, made on tissue samples, appears to enable detection of early mitochondrial hypoxia. Similar determinations in arterial plasma may prove useful in monitoring early abnormalities of mitochondrial oxygen utilization.

Erythropoietin
Tissue hypoxia is frequently associated with increased levels of the humoral agent erythropoietin, which stimulates hemoglobin synthesis.

Experimental prospects
Measurement of intracellular phosphates, intracellular pH, and P_{O_2}, and assessment of regulating enzymes whose concentration is controlled by oxygen supply, are all under study. New and more precise methods for assessing cellular oxygen metabolism are highly desirable.

Oxygen and nonenergy providing processes
Molecular oxygen is involved in a series of intracellular chemical reactions that are not directly involved in energy provision. Some of these reactions involve the biosynthesis of a number of biologically important molecules, including pigments, steroids, and fatty acids, and the degradation of a variety of different compounds.

Oxygen is involved in the intracellular generation of hydrogen peroxide an intracellular oxidant in certain special cells such as macrophages. The liberated perox-

ide may play an important role in the intracellular killing of various microorganisms.

The precise intracellular localization of many of these reactions is not known, nor have the exact kinetics of these types of oxygen-consuming reactions been established.

Changes in organ function with hypoxia
Abnormalities of oxygen-dependent reactions are manifest clinically as abnormal organ function. It is useful to describe the effects of hypoxia on several key organs.

Central nervous system
Brain oxygen consumption averages approximately 3 ml. of oxygen per 100 g. per minute. In an adult with a brain weighing 1,500 g., this amounts to 45 ml./min. (or approximately 20% of basal oxygen consumption). Profound oxygen depletion rapidly produces abnormal cerebral function; in fact, an abnormal electroencephalogram may be seen within seconds after oxygen deprivation. The effects of profound cerebral oxygen depletion may be illustrated by considering cardiac arrest. With cardiac arrest, the cerebral oxygen supply is limited to the oxygen contained in the residual capillary volume of blood in the brain. The total blood volume of the brain is approximately 75 ml.; assuming that capillary volume is 30 ml. and oxygen content is 20 ml./ 100 ml., this would amount to only 6 ml. of oxygen, an amount that would be exhausted in 10 to 15 seconds. It is not surprising, therefore, that consciousness is lost seconds after circulatory arrest. Cessation of ventilation, with maintenance of circulation, provides a somewhat longer period before irreversible parenchymal changes occur, because, theoretically, total body blood oxygen stores are available. Acute hypoxic dysoxia is associated with a marked decrease in cellular tyrosine hydroxylase and other neurotransmitters. This may play an important role in the

cerebral dysfunction. Additional changes in brain capillary permeability cause cerebral edema. Chronic hypoxia of moderate degree may produce impairment of judgment, psychological abnormalities, and increased neuromuscular irritability.

Heart

The basal cardiac oxygen consumption is approximately 10 ml./100 g./min. With a heart of average weight of 350 g., total myocardial oxygen consumption is about 35 ml./min. or approximately 15 per cent of total oxygen consumption. Approximately two-thirds of myocardial oxygen consumption is involved in contractility, the remaining one-third subserving noncontractile energy consuming processes. For example, conducting tissue has an unusually high oxygen requirement, and this may account for the increased irritability of the hypoxic heart. This oxygen is made available by an unusually complete extraction of oxygen from arterial blood, as indicated by extremely low coronary venous P_{O_2} and oxygen content. Even under basal circumstances, these values are lower than in any other vascular bed. Since extraction is virtually complete, increased oxygen needs of the heart can be met only by increased flow. In contrast to most systemic capillary beds, diffusion is not limiting and there is little physiological or anatomical shunting of coronary blood flow.

The most striking example of reduced myocardial oxygen supply occurs in patients with coronary artery occlusion, regional ischemia, failure of contractility, myocardial necrosis, and a strong predisposition to rhythm disturbances.

Chronic myocardial hypoxia may produce severe structural alterations, including dilatation, hypertrophy, and fibrosis, which, in turn, may further limit oxygen supply.

Pulmonary vascular bed

Hypoxia produces vasoconstriction of small pulmonary precapillary vessels, resulting in increased pulmonary vascular resistance. This effect is evoked not only by decreases in alveolar oxygen tension but by agents such as dinitrophenol, which interfere with cellular respiration. This suggests that the basis of hypoxic vasoconstriction is an interference with mitochondrial oxygen utilization in the smooth muscle cells of affected vessels. Hypoxic vasoconstriction is of particular importance in chronic hypoxia due to lung disease as well as in normal high-altitude dwellers.

Kidney

Renal oxygen consumption averages 6 ml./100 g./min. With a renal weight of approximately 300 g., renal oxygen consumption amounts to approximately 18 ml./min., or 8 per cent of basal oxygen consumption. Renal blood flow is high and oxygen extraction low, so that renal venous blood has unusually high P_{O_2} and oxygen content. A major fraction of the energy provided by renal oxygen consumption is used in the active transport of sodium, and there is a close correlation between renal oxygen consumption and net tubular reabsorption. There is little specific knowledge in regard to the effects of oxygen depletion on renal function; however, acute renal ischemia produces the structural and functional abnormalities collectively termed acute renal failure.

Liver

The liver possesses a dual blood supply which is important in determining the effects of hypoxia. Much of the hepatic parenchyma derives its blood supply from the portal vein, with its relatively low P_{O_2}. In addition, the anatomical arrangement is such that peripheral cells in the hepatic lobule receive blood before cells in the center of the lobule. As a result, centrilobular cells receive a marginal oxygen supply even under normal conditions. Therefore, under pathological conditions, particularly circulatory insufficiency, these cells are quite vulnerable to hypoxia, and necrosis and centrilobular fibrosis are common consequences of decreased hepatic oxygen supply.

Skeletal muscle

Skeletal muscle possesses a reasonably high capacity for oxygen utilization. However, vigorous muscular activity generally requires a quick source of additional energy because the rate at which oxygen can be supplied under these circumstances is limited. It is not surprising that the glycolytic capacity of skeletal muscle is higher than that of other tissues.

Certain life-essential muscles, such as the muscles of respiration, demonstrate fatigue and failure to contract when the oxygen supply is inadequate due to hypoxemia or reduced cardiac output. There is currently great interest in further clarifying critical hypoxia, particularly in patients with already diseased muscle, or those with increased work of breathing.

Therapy of hypoxia (hypoxic dysoxia)

The treatment of hypoxic dysoxia must be planned to improve tissue oxygen utilization. The following principles must be considered:

1. Reduce requirements:
   Treat infection.
   Reduce work of breathing.
   Muscle rest.

2. Improve oxygen supply:
   Increase arterial oxygen content.
   Improve cardiac output.
   Improve regional blood flow.

Many therapeutic interventions carry attendant risks. High concentrations of inhaled oxygen may result in hyperoxic dysoxia or oxygen toxicity, with severe adverse effects on the lungs. Continuous positive-pressure ventilation may help to increase arterial oxygen content, but causes a decrease in cardiac output. Therefore it requires great skill to manipulate each therapeutic modality appropriately.

References

Dysoxia and metabolism
Robin, E. D.: Of men and mitochondria: Coping with hypoxic dysoxia. Amer. Rev. Respir. Dis. *122*:517–531, 1980.
Robin, E. D.: Dysoxia and intrinsic mitochondrial diseases. *In* Robin, E. D. (ed.): Extrapulmonary Manifestations of Respiratory Disease, pp. 171–184. New York, Marcel Dekker, 1978.

Oxygen transport in blood
Bromberg, P. A., and Balcersak, S. P.: Blood oxygen transport in humans. *In* Robin E. D. (ed.): Extrapulmonary Manifestations of Respiratory Disease, pp. 13–46. New York, Marcel Dekker, 1978.

Clinical aspects of impaired oxygen supply
Guenter, C. A., and Welch, M. H. (eds.): Pulmonary Medicine, ed. 2. Philadelphia, J. B. Lippincott, 1982.

9 Metabolic and Endocrine Functions of the Lung

Sami I. Said, M.D.

The fundamental importance of the lung in providing oxygen and eliminating carbon dioxide is well known, and the effects of failure of this function are widely appreciated. The lung appears, however, to have other critical roles, which are described as *metabolic* or *nonrespiratory*. As the site of numerous important metabolic processes, the lung can regulate and modify the functions of many other organs. Some of these metabolic activities are even essential to the normal performance of other pulmonary functions. For example, the production or surfactant is essential for the maintenance of alveolar stability and normal gas exchange. Other functions, such as the removal of serotonin and prostaglandins from circulating blood or the activation of angiotensin I, may regulate or modify the functions of many other organs. Neoplasms of the lung may be associated with a wide range of endocrine activities. Abnormalities of pulmonary metabolic activity can, therefore, have far-reaching implications for the physiology and pathophysiology of many organ systems.

Cellular sites of metabolism

Along with our increasing awareness of the importance of lung metabolism, and contributing to this awareness, has been a rapid growth in our knowledge of pulmonary structure and ultrastructure. Wider use of electron microscopy and increasing application of histochemical, radioautographic, and immunocytochemical techniques have led to the elucidation of the fine structure of the lung and its correlation with function.

If a piece of lung is removed from a living animal and quick-frozen, sectioned, and incubated in appropriate medium, the presence of oxidative enzymes can be demonstrated by a special indicator (tetrazolium) which forms a colored salt (formazan) with the completion of the enzymatic reaction (Fig. 9–1). The color not only localizes the reaction but also provides an index of its intensity. By the use of different specific substrates and coenzymes, several discrete oxidative enzymes have been shown to occur in the lung (Fig. 9–2). Figure 9–3 shows the enzymes of oxidative and biosynthetic pathways to be concentrated in bronchial epithelium and in certain alveolar cells, particularly the large alveolar cell and the alveolar macrophage.

It is appropriate to review the main features of some of the various types of alveolar and bronchiolar cells and their probable or possible metabolic functions.

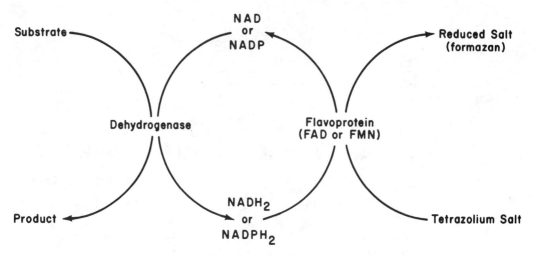

FIG. 9–1. Principle of tetrazolium reaction used for the histochemical demonstration of oxidative enzymes.

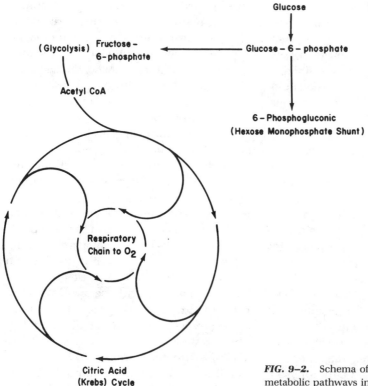

FIG. 9–2. Schema of histochemically demonstrable metabolic pathways in the lung.

LARGE (OR GREAT) ALVEOLAR EPITHELIAL CELL. This cuboidal cell (also called Granular, Type II pneumonocyte, or Corner cell) has a cytoplasm that is rich in mitochon-dria, a well-developed endoplasmic reticulum, an extensive Golgi apparatus, and multilamellar, osmiophilic inclusion bodies that are rich in phospholipids (Fig. 9–4).

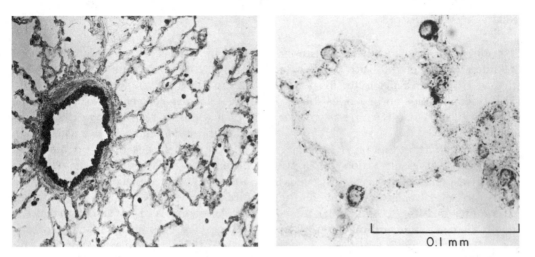

FIG. 9–3. (*Left*) Photomicrograph of quick-frozen lung section, showing reaction for NADP-diaphorase. Strongly reacting cells appear darker than others. (Dog lung, unstained, × 500) (*Right*) Higher magnification of similar section, showing reaction for NAD-diaphorase.

FIG. 9–4. Alveolar epithelial Type II cell (E II), showing numerous lamellar bodies. (*A*) Alveolar space, (*In*) interstitium (mostly collagen and fluid), (*Le*) leucocyte in capillary, (*Fb*) fibroblast. (Dog lung × 13,420). (Courtesy of Drs. Ewald R. Weibel and Joan Gil, Anatomy Institute, University of Bern.)

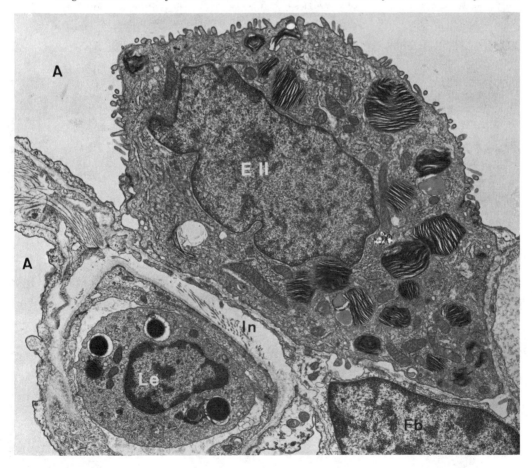

The nucleus is large and there are numerous microvilli directed toward the alveolar surface. This cell is active in the biosynthesis of phospholipids and secretes the surface-active agent that lines the alveoli (Fig. 9–5).

Another important function of the large alveolar cell is its contribution to the pulmonary response to injury. This Type II cell is the only alveolar epithelial cell that can proliferate. After acute alveolar injury, these cells regenerate and may entirely replace damaged Type I cells.

SMALL (FLAT, SQUAMOUS) ALVEOLAR EPITHELIAL CELL. This cell (Type I pneumonocyte) has a thin, elongated cytoplasm which forms the major part of the alveolar surface.

The cytoplasm is relatively deficient in organelles. With its delicate rim of cytoplasm (0.2 μm thick) and its low metabolic activity, the flat alveolar cell is ideally designed to offer the least resistance to diffusion of gases, and it consumes little oxygen itself.

Flat epithelial cells make up the major portion of the alveolar surface. They are particularly vulnerable to the toxic effects of inhaled chemical irritants (e.g., nitrogen dioxide, ozone) and oxygen, especially at high partial pressures. Incapable of mitotic division, these cells can regenerate only through the division of Type II cells.

ALVEOLAR BRUSH CELL. Also cuboidal or columnar, this cell (Type III pneumonocyte) possesses large, square microvilli and

FIG. 9–5. Lamellar body of Type II cell (E II) being secreted into a pool of extracellular lining layer. (A) Alveolar space, (TM) tubular myelin, (EI) alveolar epithelial Type I cell, (MI) mitochondrion (× 60,200). (Courtesy of Drs. Ewald R. Weibel and Joan Gil, Anatomy Institute, University of Bern.)

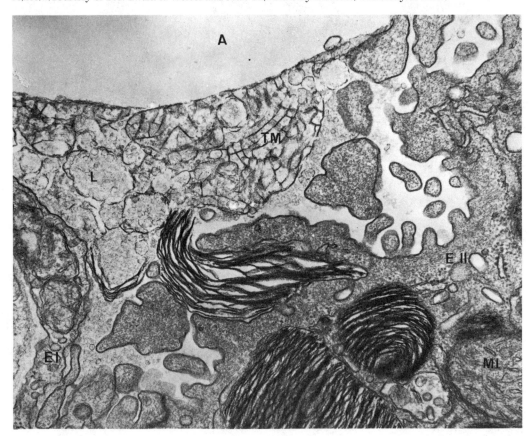

numerous vacuoles and vesicles, and is rich in glycogen. The function of this recently described cell is uncertain; a similar cell occurs in the trachea.

ALVEOLAR MACROPHAGE. This cell is morphologically similar to the great alveolar cell but is distinguishable by its location and cytochemically, by being richer in hydrolytic enzymes (e.g., acid phosphatase) and poorer in certain oxidative enzymes (e.g., glucose-6-phosphate dehydrogenase) and lipids (Fig. 9–6). The alveolar macrophage is unique among phagocytic cells in its strong dependence on oxidative phosphorylation.

A variety of proteolytic and other hydrolytic enzymes are normally contained within the membrane-bound lysosomal granules. These enzymes enable the cells to digest the foreign particles they engulf. Under abnormal conditions, however, these powerful enzymes may "leak" outside their normal confines and may then contribute to the destruction of normal lung proteins.

MAST CELL. A component of mesenchymal tissue throughout the body, mast cells are plentiful in the lung (Fig. 9–7), particularly around smaller blood vessels but also in alveolar and bronchial walls. Mast cells are packed with basophilic, electron-dense granules that are rich in histamine, heparin, and other biologically active substances. The mast cell is the chief target

FIG. 9–6. Alveolar macrophage adhering to alveolar wall. Electron micrograph of perfused rat lung, showing empty capillaries (*C*); Type I cell (*EI*) with nucleus in section; a macrophage (*MP*) overlying latter cell; fibrocyte (*Fb*) in alveolar wall; and a thin electron-dense alveolar lining (surfactant). Arrows point to cell membranes (× 24,620). (Courtesy of E. R. Weibel, Bern, Switzerland, and J. Gil, Philadelphia.)

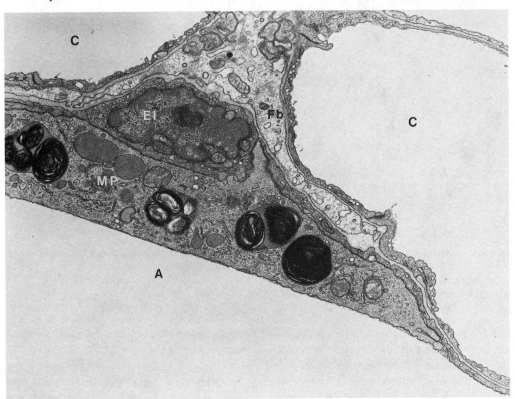

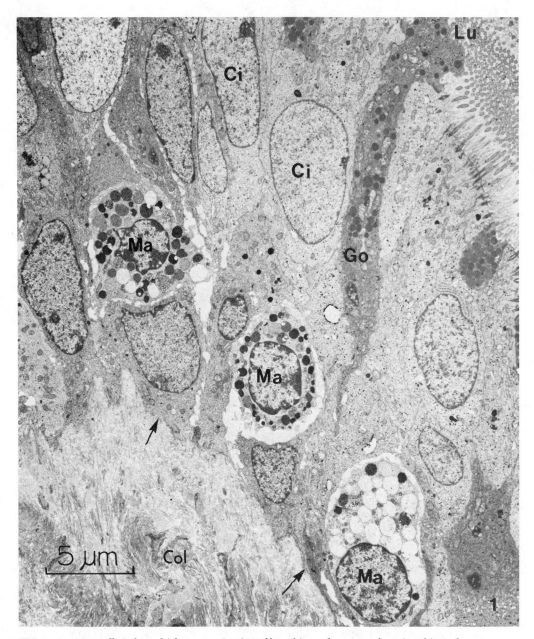

FIG. 9–7. Mast cells in bronchial mucosa. Section of lung biopsy from an asthmatic subject, showing mast cells (*Ma*) within the bronchial mucosa. The mast cells are wedged between ciliated (*Ci*) and goblet (*Go*) cells, and are distributed along the basement membrane (*arrows*). Collagen deposit beneath the basement membrane (*Col*); lumen (*Lu*) (× 6,000). (Courtesy of E. Cutz, Toronto.)

cell for the immediate hypersensitivity reactions that are typical of hay fever and extrinsic asthma. Chemicals released from mast cells may influence the airway and vascular smooth muscle, capillary permeability, inflammation, coagulation, proteolysis, and platelet aggregation.

ENDOTHELIAL CELL. On ultramicroscopic examination, the endothelial cell is relatively unimpressive in that it contains few mitochondria and other structural correlates of active metabolism (Fig. 9–8). The cytoplasm, however, is rich in pinocytotic vesicles (caveolae intracellulares). It also contains characteristic cylindrical, membrane-bound granules called Weibel-Palade bodies.

The pulmonary endothelial cell is now believed to play a key role in the metabolism of biologically active compounds, including the activation of angiotensin I to angiotensin II, the inactivation of bradykinin, the synthesis of prostacyclin (a potent inhibitor of platelet aggregation), and the elaboration of factor VIII. Numerous projections, or microvilli, and caveolae on the luminal surface of endothelial cells enlarge their area in contact with blood and thus enhance their metabolizing efficiency.

CLARA CELL. Occurring among ciliated cells in terminal bronchioles, the Clara cell is heavily endowed with large, rounded mitochondria and granular endoplasmic reticulum (Fig. 9–9) and is thus a good candidate for an active synthetic or secretory role. Its precise function, however, is not fully determined.

SMOOTH MUSCLE. The tracheobronchial tree is supplied by smooth-muscle fibers that are arranged in longitudinal and helical patterns. Spirals of smooth muscle extend to the respiratory bronchioles and the openings of alveolar ducts. The smooth-

FIG. 9–8. Electron micrograph of alveolar wall. Transmission electron micrograph of an alveolar wall from human lung showing two capillaries (C) with red cells (*black areas*) within them and an alveolar space (*Al*) between them; a Type I alveolar epithelial cell (*I*) with its nucleus in section and its thin cytoplasm covering the alveolar wall surface (*AL*); endothelial cells (*E*); and interstitial cells (*Ic*) (× 7,000). (Courtesy of R. C. Reynolds, Dallas.)

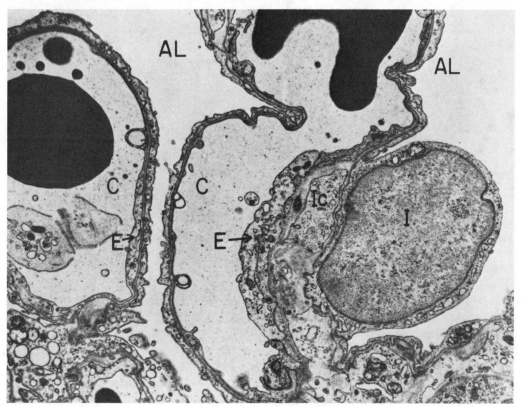

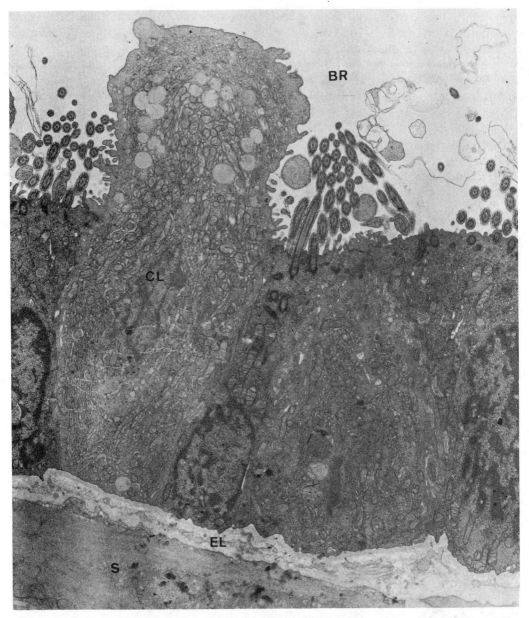

FIG. 9–9. Clara cell (*CL*) with secretory granules near the lumen of a small bronchiole (*BR*). Numerous mitochondria and prominent endoplasmic reticulum. (*EL*) Elastic fibers, (*S*) smooth muscle. (Rat lung, × 10,945) (Courtesy Drs. Ewald R. Weibel and Joan Gil, Anatomy Institute, University of Bern.)

muscle fibers contract or relax in response to neurohumoral influences. Contraction of smooth muscle constricts larger airways; it also constricts and shortens small airways. Contraction of alveolar ducts causes expulsion of alveolar air and reduces lung compliance. Airway smooth muscle is in-nervated by adrenergic and cholinergic fibers, as well as by other nerves that are neither adrenergic nor cholinergic, and appears to contain neuropeptides.

"Irritability" and hypertrophy of bronchial smooth muscle are characteristic changes in asthmatic patients.

MUCUS-SECRETING BRONCHIAL EPITHELIAL CELLS. These cells are tall, chalice-like *goblet cells* that are typically narrow at the base and distended with confluent electron-lucent secretory granules (Fig. 9–10). In man, goblet cells are present in airways before entering the lung, and they are sparse in bronchioles that are smaller than 1.0 mm. in diameter. They contribute a relatively small proportion of normal bronchial mucus, which is largely a product of submucosal glands. In conditions associated with acute or chronic bronchial irritation, goblet cells increase in number and size.

CILIATED EPITHELIUM. Ciliated epithelium covers the mucosa of the airways, extending distally to the respiratory bronchioles (Fig. 9–10). The ciliated cell has electron-lucent cytoplasm, a large Golgi apparatus, mitochondria, numerous microvilli, and about 200 cilia pulsating approximately 1,000 times per minute. The rhythmic, coordinated beating of cilia propels the overlying mucus layer toward the upper respiratory passages and the oropharynx, where the mucus is swallowed or coughed up.

BRONCHIAL GLANDS. These tubular, acinar structures, which are located in the lamina

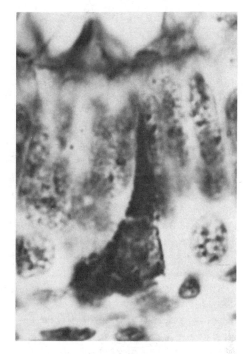

FIG. 9–11. Typical argyrophil cell with its triangular shape, cytoplasmic argyrophilia and basally located nucleus, observed in the epithelium of a bronchus of a newborn infant (× 2,350). (Courtesy of J. M. Lauweryns, Leuven, Belgium.)

FIG. 9–10. Diagrammatic representation of airway wall, showing several types of cells discussed in text. (Jeffery, P. K., and Reid, L. M.: The respiratory mucous membrane. *In* Brain, J. D., Proctor, D. F., and Reid, L. M. (eds.): Respiratory Defense Mechanisms. Vol. 5 of Lenfant, C. (ed.): Lung Biology in Health and Disease. New York, Marcel Dekker, 1977, p. 343. Reproduced with permission.)

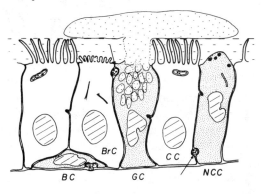

propria of the submucosa, include mucus secreting serous glands as well as other types of cells that aid in the propulsion of secretions and in fluid and ion transport. Bronchial glands are present in airways that have cartilage and are normally absent in bronchioles. As the principal source of bronchial secretions, they have a volume of secreting cytoplasm that is estimated to be 50 times that of the surface epithelium. In chronic bronchitis, these glands undergo hypertrophy.

NEUROENDOCRINE CELLS. There is considerable current interest in the neuroendocrine cells in the lung—their prevalence, distribution, physiological and pathological roles. Included in this group of cells is the Kultschitzky cell (Fig. 9–11), which is believed to be the source of bronchial carcionoid tumors and possibly also of oat-cell carcinoma. Other neuroendocrine cells, occurring in the lung as well as in other organs, are particularly active in the metabolism of biogenic amines and in the secretion of polypeptide hormones. Another name for these cells, the APUD cells, is derived from certain cytochemical characteristics they have in common: a high level of *a*mines (catecholamines, serotonin), amine *p*recursor *u*ptake, and amino-acid *d*ecarboxylases (which form the amines from their precursor amino acids). The potential ability of these cells to secrete peptide hormones may express itself in the many endocrine syndromes that sometimes complicate certain lesions of the lung, especially tumors. Clusters of APUD cells, known as "neuroepithelial bodies" have been described in mammalian bronchi; they contain argyrophilic, serotonin-rich granules and have afferent and efferent innervation (Fig. 9–12).

Of the cells named above, the great and flat alveolar cells, the macrophage, and the Clara cell are unique to the lung. The brush

FIG. 9–12. Neuroepithelial body exhibiting distinct argyrophilia. Bronchiole of neonatal rabbit lung. Van Campenhout's modification of Bodian's technique (× 1,040). (Courtesy of J. M. Lauweryns, Leuven, Belgium.)

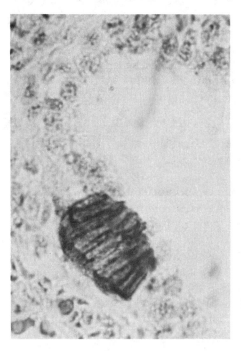

cells, mast cells, endothelial cells, and APUD cells occur in other organs. All bronchiolar and alveolar cells receive their blood supply from the pulmonary arterial system, which explains the dependence of many of the pulmonary metabolic processes on a normal pulmonary arterial blood flow, and their susceptibility to pulmonary embolism and other vascular alterations.

Some metabolic and endocrine functions of the lung

Maintenance of alveolar stability

The alveoli of all mammalian lungs are lined with a thin layer of surface-active material (alveolar surfactant), which regulates surface tension at the air-liquid interface. Research into the significance of alveolar surfactant, its composition, biosynthesis, cellular origin, secretion, and metabolism has been a major stimulus to the investigation of pulmonary metabolism as a whole.

The main component of alveolar surfactant is dipalmitoyl phosphatidylcholine, a disaturated lecithin that is probably attached to a protein in complex formation.

BIOSYNTHESIS OF SURFACTANT. The biosynthesis of dipalmitoyl phosphatidylcholine in the lung is shown schematically in Fig. 9–13. An important early reaction is the stepwise synthesis of phosphatidic acid from *sn*-glycerol-3-phosphate, which is catalyzed by acyltransferases. The conversion of phosphatidic acid to *sn*-1,2 diglycerides is a limiting step in the biosynthesis of saturated lecithin, and it is catalyzed by the enzyme phosphatidic acid phosphohydrolase. The activity of this enzyme increases during gestation, and this increase in activity precedes as increase in the content of lecithin in the lung. Lecithin is synthesized *de novo* from diglyceride and cytidine diphosphate choline, a reaction that is catalyzed by choline phosphotransferase. This newly synthesized lecithin contains one

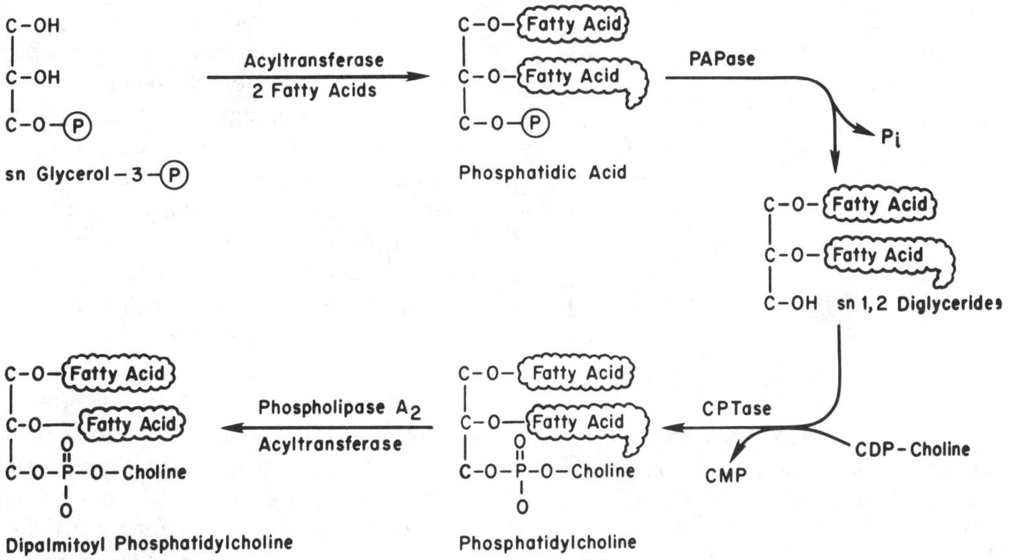

FIG. 9–13. Biosynthesis of dipalmitoyl phosphatidylcholine in the lung. (Courtesy of J. M. Johnston, Dallas.)

saturated fatty acid (position 1) and another that is unsaturated (position 2). The lecithin may be remodeled into the disaturated dipalmitoyl phosphatidylcholine through the formation of lysolecithin by the action of phospholipase A_2 and its subsequent acylation (Fig. 9–13). Disaturated dipalmitoyl phosphatidylcholine may be derived from two molecules of lysolecithin.

Another possible pathway for the biosynthesis of dipalmitoyl phosphatidylcholine is the methylation of phosphatidyl ethanolamine. However, this pathway is believed to be relatively unimportant.

Fatty acids required for these reactions may be synthesized *de novo* in the lung from acetate, pyruvate, or glucose substrates. Fatty acids may also be derived from plasma or from hydrolysis of tissue lipids.

**DEVELOPMENT AND SECRETION OF SURFAC-
TANT.** Surfactant is probably synthesized and secreted by the Type II alveolar cell (Figs. 9–4 and 9–5). Although surfactant appears in lung tissue early in pregnancy and its concentration increases steadily during the third trimester, this lipoprotein appears in the alveoli only close to term. The concentration in the newborn animal is 10 to 15 times as high as that in the term fetus.

Surfactant also appears in the amniotic fluid. This is of considerable practical importance because it makes possible a prenatal assessment of the maturity of the lung. From the measurement of the lecithin content and that of sphingomyelin in amniotic fluid, which is obtained by amniocentesis, the ratio of the lecithin content to that of sphingomyelin (L/S ratio) may be calculated. This ratio is a useful index of the state of development of alveolar surfactant. An L/S ratio of 2:1 or above signifies normal lung maturation and is rarely associated with hyaline membrane disease (Fig. 9–14). Conversely, a lower ratio indicates inadequate synthesis or secretion of surfactant, and often reflects a high likelihood of the development of respiratory distress.

**CONTROL OF PRODUCTION AND METABO-
LISM OF SURFACTANT.** The formation and maintenance of normal amounts of surfactant in the film that lines the alveoli depend on several factors. These include: the maturity of the great alveolar cells and their

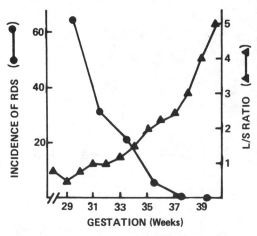

FIG. 9–14. The lecithin to sphingomyelin (L/S) ratio is a useful indicator of the state of maturity of fetal lung. Incidence of respiratory distress syndrome (RDS) as related to L/S ratio in amniotic fluid, each as a function of gestational age. (Farrell, P. M., and Avery, M. E.: Hyaline membrane disease. Am. Rev. Respir. Dis. *111*:657–688, 1975. Reproduced with permission.)

biosynthetic enzyme systems; the adequacy of the blood flow to the alveolar walls (which normally comes from the pulmonary arterial circulation); a normal rate of turnover of the surfactant; and the absence of inhibitors of its action.

The physiological factors that govern the production and secretion of surfactant remain uncertain. Among these factors, however, are certain recognized hormonal influences. For example, glucocorticoids and thyroid hormones, in pharmacological doses, can accelerate the maturation of the great alveolar cells and the secretion of surfactant in fetal lungs. Prolactin and other hormones (possibly estrogens) also may promote the biosynthesis and secretion of surfactant. The possible effects of the innervation of the alveolar cells on surfactant synthesis and release are still unknown.

FUNCTIONS OF SURFACTANT. The primary function of surfactant is to stabilize the alveoli by preventing excessive increases or unevenness in alveolar surface forces. Reduction of surface tension by surfactant reduces the pressure required to fill the alveoli during inspiration and helps to maintain alveolar patency at a given pressure during expiration. Surfactant is also a factor in guarding against transudation of fluid into the alveoli. Thus, its absence or its presence in amounts insufficient to maintain alveolar stability predictably leads to large-scale atelectasis and pulmonary edema—the pathological hallmarks of the respiratory distress syndrome of the newborn or the adult.

DEFICIENCY OF SURFACTANT. Numerous clinical and experimental situations are associated with inadequate levels of surfactant for maintenance of alveolar stability. This inadequacy could result from insufficient formation of surfactant, as in prematurity, or from its inactivation by certain constituents of serum or by certain lipids (as in pulmonary edema and alveolar proteinosis). Both insufficient formation and inactivation may occur after pulmonary arterial occlusion and in the respiratory distress syndrome of infants or adults. Excessively rapid depletion and incomplete regeneration of surfactant may complicate breathing at extremes of lung volumes. The relative importance of surfactant deficiency in these and similar states is difficult to ascertain. Even if it is not the sole or the primary lesion, however, this deficiency probably contributes to compromised lung function.

Defense against infectious agents and other foreign particles

This important function of the lung depends on the combined and integrated effects of several mechanisms: phagocytosis, through the action of the pulmonary macrophages; mucociliary transport; and immune mechanisms.

PHAGOCYTOSIS: ALVEOLAR MACROPHAGES. Because the pulmonary alveolar surface is exposed to the external environment, it is constantly threatened by the onslaught of contaminating bacteria, dusts, chemical irritants, and other particulate material. The lung, however, proves to be well prepared for defense against this challenge, and the alveolar macrophages are critically important in this defense.

As the only cells in the body that are normally present on the epithelial surface, the alveolar macrophages can be described as effective policemen of the extensive alveolar surface. Thanks to them in large measure, lung tissue normally is kept sterile. These specialized cells originate from hematopoietic elements or immature monocytes. Biochemically, the alveolar macrophages have extremely high metabolic activity, and they differ from other phagocytes (e.g., the blood monocyte) in their dependence on oxidative phosphorylation as a source of energy for phagocytosis. They are capable of protein and lipid synthesis and are especially well endowed with hydrolytic enzymes that permit them to digest material they engulf. Such material includes not only bacteria and other pathogens but also nonliving particles, such as organic and inorganic dust, and other foreign material that may reach the alveoli, such as erythrocytes, oil droplets, and desquamated epithelial cells.

Alveolar macrophages produce interferon, thus exerting antiviral activity. After completing their phagocytic function, alveolar macrophages are carried up the bronchial tree in the mucociliary stream, to be coughed up or swallowed.

The mobilization and activation of macrophages and enhancement of their bactericidal ability in immunological responses are dependent on humoral mediators (lymphokines) that are secreted by stimulated T-lymphocytes. In other words, the optimal functioning of alveolar macrophages depends on the mechanisms of cell-mediated immunity.

MUCOCILIARY TRANSPORT. This mechanism, which is responsible for clearance of the airways, depends on the presence of mucus with normal viscosity, ciliary con-

tractile proteins, and an adequate source of energy for ciliary contraction. Agents or conditions that are toxic to cilia include high concentrations of carbon dioxide, sulfur dioxide, cigarette smoke, alcohol, hypoxia, some viruses, and *Mycoplasma pneumoniae.* Despite the apparent importance of ciliary function, the immotile cilia syndrome, often associated with sterility and Kartagener syndrome, is not incompatible with life.

IMMUNE MECHANISMS. Three immune mechanisms are involved in defense of the lung against infection. The first depends on blood-borne antibodies, a major defense against highly pathogenic encapsulated bacteria such as pneumococci, *Haemophilus influenzae,* streptococci, and meningococci. These are mainly IgG antibodies, the principal circulating antibodies that are potent opsonins and that facilitate leukocyte chemotaxis and phagocytosis. Also carried in the blood are the IgM antibodies that are bacterial agglutinins and activators of the complement system.

A second immune mechanism involves the local (secretory) antibodies, principally IgA, which is the main antibacterial and antiviral antibody of mucous membranes.

The third mechanism includes the T-cell-mediated reactions constituting cell-mediated (delayed) immunity. These reactions depend on the secretion of lymphokines by activated T-lymphocytes and form the principal defense against fungal and intracellular bacterial and viral infections.

Metabolism, synthesis, and release of vasoactive hormones

The lung has the only capillary bed in the body through which the entire blood volume passes. This characteristic makes the pulmonary microcirculation uniquely suited for exercising a controlling influence on blood-borne vasoactive hormones. These hormones have many and diverse actions on all smooth muscle–containing organs, such as blood vessels, bronchi and alveolar ducts, the gastrointestinal system, and the genitourinary tract. Certain vasoactive substances (for example, the prostaglandins, catecholamines, and certain peptides) also affect various metabolic functions including lipolysis, glycogenolysis, and the cyclic nucleotide levels. Thus, the lung may influence a number of other body functions through its handling of these active agents.

Some vasoactive compounds, such as serotonin, are normal constituents of blood; others (e.g., bradykinin and prostaglandins) are generated or activated primarily as a consequence of tissue "injury" or inflammation, or other stimuli. In either case, the pulmonary handling of a given active agent can modify its systemic effects, whether they are physiological or pharmacological. The metabolic alterations of some vasoactive agents by the lung may be classified as activation, inactivation, synthesis, or release.

VASOACTIVE COMPOUNDS ACTIVATED BY THE LUNG. The conversion of angiotensin I to angiotensin II is the only known example of biological activation of a circulating hormone during its passage through the pulmonary circulation. Angiotensin II, which is the most potent pressor agent known and up to 50 times more active than its precursor, is itself normally unaffected by passage through the lung. The angiotensin-converting activity of the lung is many times greater than that of plasma. Other organs, especially the kidney and the intestine, are also rich in this enzyme.

VASOACTIVE COMPOUNDS INACTIVATED BY THE LUNG. Many vasoactive materials are partially or completely inactivated by the lung (Table 9–1). Among those that are almost completely removed (more than 80 per cent) or inactivated are serotonin (5-hydroxytryptamine); bradykinin; adenosine triphosphate (ATP); adenosine diphosphate (ADP); prostaglandins E_1, E_2, and F_{2a}; and certain steroids, including testosterone.

Vasoactive hormones that pass through the lung without significant loss in activity include the following compounds: epinephrine, histamine, prostacyclin, some bradykinin-like peptides such as eledosin and polisteskinin, angiotensin II, vasopressin (ADH), oxytocin, substance P, and vasoactive intestinal peptide (VIP). Norepinephrine and prostaglandin A compounds are taken up in small degrees.

An interesting feature of the pulmonary metabolism of vasoactive hormones is its high selectivity. As noted above, one member of a given group of substances (e.g., catecholamines, prostaglandins, kinins) may be removed in one passage, while another member of the same chemical group is permitted to pass through without change. Economy is another feature of the

---

TABLE 9–1
Degrees of Inactivation of Circulating Hormones by the Lung

Major Inactivation	Moderate Inactivation	No Significant Loss of Activity
Serotonin	Norepinephrine	Epinephrine
Bradykinin		Histamine
Prostaglandin E		Angiotensin II
Prostaglandins E and F		Vasoactive intestinal peptide
Steroids (e.g., testosterone)		Substance P
		Vasopressin
Adenosine triphosphate		Oxytocin
Adenosine diphosphate		Prostacyclin

metabolism of vasoactive substances by the lung. Thus, the same enzyme (angiotensin-converting enzyme, peptidyl dipeptidase, kininise II) that inactivates the vasodepressor bradykinin also activates angiotensin I.

In the case of compounds such as serotonin and norepinephrine, loss of activity in passage across the lung is limited by the rate of their uptake. In other instances the inactivation depends on enzymatic action, at the endothelial surface, as in the case of bradykinin. The removal of prostaglandins requires not only the action of a specific, intracellular dehydrogenase but also the existence of a transport system, to carry the substrates from the vascular lumen into the cellular site of metabolism.

Although knowledge about the cellular localization of these metabolic alterations is still incomplete, it is known that the endothelial cell, which is in intimate contact with blood, plays a dominant role. The caveolae and pinocytotic vesicles of the endothelial cell are the site of uptake of ATP and other adenine nucleotides, serotonin and other amines, as well as of the degradation of bradykinin and the activation of angiotensin. Other cells, including the neuroendocrine cells, the mast cells, and possibly also the macrophage, may participate in the metabolism of vasoactive hormones.

SYNTHESIS AND RELEASE OF VASOACTIVE HORMONES. A variety of biologically active compounds that are normally synthesized within the lung, may, under certain pathological influences, be discharged into the circulation in abnormally large quantities (see Table 9–2). Failure of normal pulmonary inactivation can also contribute to higher circulating levels of such compounds, including:

1. Histamine and serotonin (the latter mainly contained in and released from platelets within the pulmonary circulation)
2. Polypeptides, including the spasmolytic vasoactive intestinal peptide, a spasmogenic peptide not yet fully identified; substance P, a bombesin-like peptide; and other peptides
3. Proteins, such as the angiotensin-converting enzyme, thromboplastin, plasminogen activator, kallikrein, and other proteases
4. Lipids, especially prostaglandins E_2 and F_{2a}, endoperoxides, thromboxanes, prostacyclin, and leukotrienes (slow-reacting substance of anaphylaxis)

These powerful compounds may have far-reaching effects both on respiratory function and on the systemic circulation (Table 9–3).

TABLE 9–2
Biologically Active Compounds in the Lung

Biogenic Amines	*Proteins*
Histamine	Angiotensin-converting enzyme
Serotonin	Thomboplastin
Polypeptides	Plasminogen activator
Vasoactive intestinal peptide	Kallikrein
Spasmogenic lung peptide	Elastase and other proteases
Substance P	Arysulfatase B
Bombesin	*Lipids*
Adrenocorticotrophic hormone and other peptide hormones	Prostaglandins
(especially in tumors)	Endoperoxides
Eosinophil chemotactic factor of anaphylaxis	Thromboxanes
	Prostacyclin
	Leukotrienes
	Platelet-activating factor

TABLE 9–3
Nature and Actions of Biologically Active Compounds Metabolized or Released by the Lung

Compound	Action
Biogenic amines	
Histamine	Bronchoconstriction, increased fluid transudation from bronchial vessels
Serotonin	Vasoconstriction
Polypeptides	
Bradykinin	Inflammation, increased systemic capillary permeability
Angiotensin	Vasoconstriction, elevated blood pressure, aldosterone release
Vasoactive intestinal peptide	Systemic vasodilation, airway and pulmonary vascular relaxation
Spasmogenic lung peptide	Bronchoconstriction, contraction of other smooth muscle
Proteins	
Complement	Immunological injury
Lymphokines	Cell-mediated hypersensitivity
Proteolytic enzymes	
Elastase	Inflammation, hemorrhage, tissue destruction (acute); emphysema (chronic)
Kallikrein	Release of bradykinin
Other enzymes	
Thromboplastin	Intravascular coagulation
Plasminogen activator	Fibrinolysis
Arylsulfatase B	Inactivation of slow-reacting substance of anaphylaxis (leukotrienes)
Lipids	
Prostaglandins (PGE_2, PGF_{2a})	Constriction of bronchi, alveolar ducts, and pulmonary vessels (PGF_{2a}); systemic vasodilation (PGE_2)
Thromboxanes	Platelet aggregation, bronchoconstriction
Prostacyclin (PGI_2)	Inhibition of platelet aggregation, vasodilation
Leukotrienes	Bronchoconstriction, pulmonary vasoconstriction, pulmonary edema (possibly)
Platelet-activating factor	Platelet aggregation and release; other biological actions

Aside from the acute release of vasoactive hormones, alterations in the pulmonary metabolism of these hormones may have important implications in human disease. In the carionoid syndrome, for example, the usual absence of left-sided cardiac lesions may be explained by the efficient pulmonary removal of two major humoral mediators of the disease, serotonin and bradykinin. At present there is sufficient experimental evidence to suggest that either failure of inactivation or repetitive release of humoral mediators could, indeed, be an important factor in the pathogenesis of some pulmonary disorders, including shock lung and the adult respiratory distress syndrome.

Lung and hematological coagulation mechanisms

In the normal and abnormal formation of blood clots and in the dissolution of these clots, several important steps are strongly influenced through their interaction with biologically active factors in the lung.

Thus, the extrinsic path of activation of prothrombin, eventually leading to fibrin deposition, can be initiated by tissue factors (thromboplastin) that are present in higher concentration in the lung than in any other organ. Once a clot has formed, its lysis depends on the action of plasmin, which must be generated by the activation of plasminogen. Again, the lung is unusually rich in plasminogen activator and,

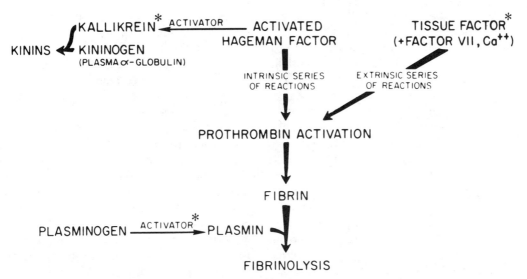

FIG. 9–15. Simplified diagram of clotting and fibrinolytic mechanisms and interactions with kallikrein-kinin system. Asterisks denote factors present in lung tissue.

hence, in fibrinolytic activity (Fig. 9–15). Several other interrelated reactions are known to occur but are not shown in the figure. For example, thrombin can accelerate some of the early reactions in the extrinsic pathway, leading to further formation of thrombin. The generation of thrombin leads not only to the conversion of fibrinogen to fibrin but also to platelet aggregation. Also the activation of the Hageman factor promotes the formation of plasminogen activator.

The megakaryocytes, parent cells of the platelets which are all-important in the intrinsic sequence of clotting, are concentrated in the pulmonary vascular bed. Trapping and aggregation of platelets in the pulmonary circulation occur in pulmonary embolism and other forms of acute lung injury. In shock there is sequestration of leukocytes as well. These changes lead to release of vasoactive materials and proteolytic enzymes, which contribute to pulmonary injury.

Heparin, a widely used anticoagulant, is a major constituent of mast cells, which are prevalent throughout the lung. The func-

tional significance of lung heparin, however, remains to be determined. The knowledge that a number of hematological mechanisms are closely related to pulmonary metabolism helps one understand the greater susceptibility of patients with a variety of pulmonary disorders to certain clotting and fibrinolytic anomalies. The higher incidence of thromboembolism in association with pulmonary malignancy, increased fibrinolytic activity following lung surgery, and disseminated intravascular coagulation (consumption coagulopathy) complicating viral pneumonitis and other pulmonary lesions associated with the adult respiratory distress syndrome exemplify the possible influence of pulmonary factors.

Protease-antiprotease balance in the lung

One of the fundamental mechanical properties of the lung—its elastic recoil—depends to a large extent on an important protein constituent, elastin. It is this protein that is characteristically lost in emphysema, with corresponding loss of lung elastic recoil. Lung proteins, including elastin, collagen, and connective tissue proteoglycans, are major determinants of pulmonary architecture. These proteins are potentially vulnerable to the action of proteolytic enzymes (proteases), but are normally protected from degradation by protease inhibitors (antiproteases).

The possible role of proteolytic enzymes in causing pulmonary disease has received considerable attention in recent years. The first indication of a possible relationship between proteolysis and lung disease came with the realization that deficiency of a serum protease inhibitor, alpha$_1$-antitrypsin, predisposes to pulmonary emphysema. Then came the discovery that intratracheal instillation of papain and other elastolytic enzymes, in experimental animals, produces lesions resembling human emphysema. The papain-treated animals also show evidence of pulmonary inflammation and hemorrhage before developing the structural changes of emphysema. The question naturally arose whether proteases, especially elastases, could be formed or released within the lung, and if they could be responsible for lung damage similar to that which follows papain instillation (or is seen in emphysema).

We now know that proteases may be derived from several sources in the lungs, the two principal sources being granular leukocytes (from the bloodstream) and alveolar macrophages. Leukocyte elastase has been demonstrated to produce emphysema in animals, and to degrade not only elastin but also other lung proteins. Tracheobronchial secretions contain an important elastase inhibitor. The main endogenous protease inhibitor at the alveolar level is alpha-$_1$-protease inhibitor (alpha$_1$-Pi or alpha$_1$-antitrypsin). This inhibitor is produced chiefly in the liver and reaches the alveolar surface through the pulmonary circulation. An imbalance in the normal protease-antiprotease relationship renders the lung vulnerable to acute injury (e.g., edema, hemorrhage) or, more commonly, to emphysema (Fig. 9–16). Such imbalance may arise from severe genetic (homozygous) deficiency of antiproteases or, alternatively, from the relative deficiency of protease inhibitors, coupled with their inactivation by such factors as cigarette smoke, oxygen excess, and free radicals.

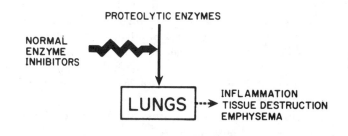

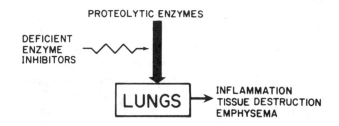

FIG. 9–16. Schema of the possible role of proteolytic enzymes in the pathogenesis of lung disease.

TABLE 9–4
Hormonal Secretion by Pulmonary Tumors: Common Clinical Features and Underlying Lesions

Hormone	Syndrome	Lesion
Adrenocorticotrophic hormone	Hypokalemic alkalosis, edema, Cushing's syndrome	Oat-cell carcinoma, adenoma
Antidiuretic hormone (arginine vasopressin)	Hyponatremia, syndrome of inappropriate secretion of antidiuretic hormone	Oat-cell carcinoma, tuberculosis, pneumonia, aspergillosis
Parathyroid hormone	Hypercalcemia	Squamous cell carcinoma
Gonadotropins	Gynecomastia (adults), precocious puberty (children)	Large-cell anaplastic carcinoma
Calcitonin	No clinical findings	Adenocarcinoma, squamous, and oat-cell carcinoma
Vasoactive intestinal peptide	Watery diarrhea or no symptoms	Squamous, oat- or large-cell carcinoma
Growth hormone	Hypertrophic osteoarthropathy	Squamous cell carcinoma
Serotonin and possibly prostaglandins	Carcinoid	Bronchial adenoma, oat-cell carcinoma
Prolactin	Galactorrhea or no symptoms	Anaplastic cell

The potential for hormone secretion: endocrine syndromes in lung disease

Malignant tumors may be hormonally active, causing ectopic or paraneoplastic endocrine syndromes, or remaining clinically silent. Tumors of the lung, notably bronchogenic carcinoma, have the capacity to produce virtually all the known polypeptide hormones. The more common examples of these hormonal syndromes are listed in Table 9–4. The most frequent offender is the oat-cell bronchogenic carcinoma, although certain syndromes (hyperparathyroidism and osteoarthropathy) are more common with squamous-cell carcinomas.

The association between endocrine syndromes and certain malignant tumors is of obvious importance in the recognition of these tumors and their management. The retention by the neoplastic cell of the ability to produce normal peptide hormones is of considerable interest to the endocrinologist, the oncologist, and the geneticist. To the student of the lung, the greater incidence of these endocrine syndromes in association with tumors of the lung than with

tumors of any other organ is of special interest.

Although hormonal secretion by the lung is largely a manifestation of malignant disease, endocrine activity may be associated with nonmalignant lesions. For example, adrenocorticotrophic hormone (ACTH) production has been reported in bronchial andenoma, and ADH secretion in pulmonary tuberculosis, lung abscess, and other conditions. Normal lung may also have endocrine functions.

References

General reviews

Bakhle, Y. S., and Vane, J. R. (eds.): Metabolic functions of the lung. *In* Lenfant, C. (exec. ed.): Lung Biology in Health and Disease, Vol. 4. New York, Marcel Dekker, 1977.

Porter, R., and Whelan, J. (eds.): Metabolic activities of the lung. Ciba Foundation Symposium 78 (new series). New York, Exerpta Medica, 1980.

Said, S. I.: The lung as a metabolic organ. N. Engl. J. Med., *279*:1330, 1968.

Said, S. I.: Endocrine role of the lung in disease. Am. J. Med., *57*:453, 1974.

Said, S. I.: Metabolic functions of the pulmonary circulation. Circ. Res., *50*:325–333, 1982.

Said, S. I.: The Endocrine Lung in Health Disease. Becker, K. L., and Gazdar, A. (eds.): Philadelphia, W. B. Saunders, (in press).

Morphologic, developmental and functional features of lung cells

Cutz, E., and Orange, R. P.: Mast cells and endocrine (APUD) cells of the lung in Asthma: Physiology, Immunopharmacology and Treatment. Vol. 22, pp. 51–76. New York, Academic Press, 1977.

Dey, R. D., and Said, S. I.: Immunocytochemical localization of VIP-immunoreactive nerves in bronchial walls and pulmonary vessels. Fed. Proc. 39:870–876, 1980.

Kuhn, C.: The cells of the lung and their organelles. In Crystal, R. G.: (ed.): The Biochemical Basis of Pulmonary Function. Lenfant, C. (exec. ed.): Lung Biology in Health and Disease, Vol. 2, pp. 3–48. New York, Marcel Dekker, 1976.

Lauweryns, J. M., and Cokelaere, M.: Hypoxia-sensitive neuro-epithelial bodies: Intrapulmonary secretory neuroreceptors modulated by the CNS. Z. Zellforsch. 145:521, 1973.

Meyrick, G., and Reid, L. M.: Ultrastructure of alveolar lining and its development. In Hodson, W. A. (ed.): Development of the Lung. Lenfant, C. (exec. ed.): Lung Biology in Health and Disease, Vol. 6, pp. 135–214. New York, Marcel Dekker, 1977.

Rhodin, J. A. G.: Ultrastructure and function of the human tracheal mucosa. Am. Rev. Respir. Dis. 93:(Suppl.):1, 1966.

Ryan, U. S., and Ryan, J. W.: Correlations between the fine structure of the alveolar-capillary unit and its metabolic activities. In Bakhle, Y. S., and Vane, J. R. (eds.): Metabolic Functions of the Lung. Lenfant, C. (exec. ed.): Lung Biology in Health and Disease, Vol. 4, pp. 197–232. New York, Marcel Dekker, 1977.

Weibel, E. R.: Morphological basis of alveolar capillary gas exchange. Physiol. Rev., 53:419, 1973.

Physiological and clinical importance of surfactant

Farrell, P. M.: Lung Development: Biological and Clinical Perspectives. Vol. 1, Biochemistry and Physiology; Vol. 2, Neonatal Respiratory Distress. New York, Academic Press, 1982.

Mechanisms of clearance of airways

Brain, J. D., Proctor, D. F., and Reid, L. M. (eds.). Respiratory defense mechanisms, Vol. 5, Parts 1 and 2. Lenfant, C. (exec. ed.): Lung Biology in Health and Disease. New York, Marcel Dekker, 1977.

Green, G. M.: The J. Burns Amberson lecture: In defense of the lung. Am. Rev. Respir. Dis., 102:691, 1970.

The lung in relation to vasoactive hormones

Vane, J. R.: The release and fate of vaso-active hormones in the circulation. Br. J. Pharmacol. 35:209, 1969.

Said, S. I.: Pulmonary metabolism of prostaglandins and vasoactive peptides. Ann. Rev. Physiol. 44:257–268, 1982.

The relationship between antiprotease deficiency and emphysema

Eriksson, S.: Studies in alpha$_1$-antitrypsin deficiency. Acta Med. Scand. 177[Suppl. 432]:1–85, 1965.

Kueppers, F.: Inherited differences in alpha$_1$-antitrypsin. Litwin, S. D., (ed.): Genetic Determinants of Pulmonary Disease. Lenfant, C. (exec. ed.): Lung Biology in Health and Diseases, Vol. 11. New York, Marcel Dekker, 1977.

Morse, J. O.: Alpha$_1$-antitrypsin deficiency. N. Engl. J. Med. 299:1045–1048, 1978.

Endocrine syndromes arising from nonendocrine tumors

Odell, W. D., Wolfsen, A. R.: Humoral syndromes associated with cancer. Ann. Rev. Med. 29:379–406, 1978.

Section III

Renal Mechanisms

Introduction

The kidney is responsible for elimination from the body of most of the nonvolatile waste products of metabolism. Equally important is its role in maintenance of a constant internal environment of electrolyte concentrations and fluid volume. By selective reabsorption and secretion of electrolytes, the normal kidney possesses an enormous capacity to maintain precisely fluid and electrolyte balance. Most fluid and electrolyte imbalances result not from loss of renal function or reserve, but rather from extrarenal pathology that initiates transmission of inappropriate or faulty information to renal regulatory mechanisms.

The kidney consists of approximately 2 million functional units or nephrons. The elaboration of urine by these nephrons can be divided, for our purposes, into two components. The first is the formation of an ultrafiltrate in the glomerular capillary bed; the second consists of the selective active and passive reabsorption and secretion of the wide spectrum of filtered solutes (and of water osmotically obligated to these solutes), as the ultrafiltrate proceeds through the tubule.

The *glomerular ultrafiltrate* is similar in composition to plasma, except that high-molecular-weight substances, primarily proteins, are largely excluded. The rate at which this filtrate is formed can be quantified by determining the clearance of any substance that is freely filterable (molecular weight less than 10,000), is not bound to protein and is neither secreted nor reabsorbed in passage through the tubule. Inulin and mannitol are the best examples of such substances.

The *filtered load* of any solute can be determined by multiplying the plasma concentration of the solute (P_{conc}) by the glomerular filtration rate (GFR). If this filtered solute passes unaltered through the tubular system, the amount excreted in the urine (measured by multiplying urine concentration [U_{conc}] by urine volume per unit of time [V]) must equal filtered load. This is expressed by the equation:

$$P_{conc} \times GFR = U_{conc} \times V.$$

Since all components other than the GFR can be measured directly, one can calculate the GFR from the following equation:

$$GFR = \frac{U_{conc}V}{P_{conc}}.$$

The rate at which glomerular filtrate is formed depends primarily upon the rate of plasma flow through the kidney. Normally about 20 per cent of the plasma circulating through the kidney is filtered, but this may vary substantially under various physiological and pathological conditions. Glomerular filtration rate depends upon glomerular capillary hydrostatic pressure minus intracapsular pressure and plasma oncotic pressure as well as on the permeability characteristics of the glomerular capillary membrane. Hydrostatic pressure depends upon the balance between afferent and efferent glomerular arteriolar constriction or resistance and renal arterial and venous blood pressures. Changes in glomerular structure, such as those seen in acute and chronic glomerulonephritis, also contribute to the rate of formation of glomerular filtrate.

The urinary clearances of some endogenous metabolic waste products, notably creatinine, approximate those of inulin and mannitol. The quantity of these metabolites formed daily remains fairly constant under normal conditions. Plasma concentrations of these metabolites remain constant only as long as daily endogenous production is matched by urinary excretion. Since the urinary excretion of these solutes depends primarily upon filtered load (determined by multiplying GFR by plasma concentration of the solute), it is apparent that a reciprocal relationship must exist between the GFR and plasma concentration. When plasma concentrations of these and other unidentified metabolites become sufficiently elevated in response to a fall in GFR, the symptom complex of *uremia* develops. Blood urea nitrogen and serum creatinine concentrations can be used to estimate the extent of renal dysfunction; however, abnormal concentrations are not responsible for the array of symptoms seen in renal failure.

Other urinary solutes have a clearance much below or above the glomerular filtration rate so that total urinary excretion is regulated not only by the amount filtered but also by the rate of tubular reabsorption or secretion. The clearance of any urinary solute can be quantified using the same technique. By comparing this clearance value with the inulin clearance, *net* reabsorption or secretion of any solute can be measured. For example, if the clearance of a specific solute was calculated to be 10 ml./min. in a person with a simultaneous inulin clearance of 100 ml./min., net reabsorption of this solute would be 90 per cent of the filtered load. Tubular reabsorptive and secretory activity are governed by local, structural, and metabolic conditions and a whole host of hormonal effects.

In 1960, Bricker and his associates presented evidence to support a rather unique pathophysiological concept. The *intact nephron* hypothesis suggested that in chronic renal disease, regardless of origin, surviving nephrons either functioned normally or did not contribute significantly to final urine formation. This report focused the attention of many investigators on the functional consequences of chronic renal disease. Their studies demonstrated that as glomerular function declined, there was a simultaneous and proportional decline in a variety of tubular functions. For example, the reduction in the tubular maximums for glucose absorption and paraaminohippurate secretion closely paralleled the decrease in glomerular filtration rate in the unilaterally diseased kidney compared to a normal control kidney.

Bricker, in updating his original hypothesis, emphasized that changes in the excretion rates for each solute by residual nephrons follow an orderly, predictable and appropriate pattern for the maintenance of homeostasis. Such balanced and regulated changes in glomerulotubular balance for sodium, potassium, hydrogen, and a whole host of other solutes do not exclude a contribution to function by diseased nephrons or nephron segments. Rather, it emphasizes that, *despite* impairment in glomerular and tubular functions in individual nephrons or nephron segments due to structural damage, the remaining intact or normal nephrons or nephron segments adapt or compensate appropriately to maintain the constant internal environment necessary for the individual to survive.

To accomplish the necessary changes in transport by the tubular epithelial cells for a host of different solutes for which biological control systems exist, mechanisms must be present (1) for detection of changes in the rate of accumulation of each solute, and (2) for transmission of this information to the tubular epithelial cells of residual nephrons so that the rate of transport of each solute can be modified. At least in some instances, these adaptations are accomplished by changes in secretion rates of peptide hormones. Bricker has raised new questions about whether or not changes in the circulating levels of these humoral agents, necessary to maintain this adaptation, might be adversely affecting extrarenal organ systems in such a way that these changes may be contributing to the symptomatology of the uremic syndrome.

An example of this *trade-off* hypothesis can be found in the phosphate control system. With a constant dietary intake of phosphorus and a decreasing number of functioning nephrons, phosphate excretion per residual nephron must increase. The trade-off for this increased excretion may be one of the major complications of uremia, namely secondary hyperparathyroidism. Similarly, could the increased levels of natriuretic substance cause any of the nervous system manifestations of the uremic syndrome?

A variety of finely tuned regulatory mechanisms maintain fluid, electrolyte, and buffer balances within a narrow, normal range. These mechanisms are discussed in detail in the five subsequent chapters. Pathological conditions are used to illustrate how normal control mechanisms become overwhelmed or ineffective and imbalances result.

We initiate presentation of major renal mechanisms by discussing the maintenance of serum protein—more specifically, albumin—concentrations. The serum proteins are the only endogenous intravascular solutes that are not readily exchangeable with the interstitial fluids and, hence, provide the colloid osmotic pressure necessary to maintain intravascular volume. One may no longer be able to maintain normal serum albumin concentrations when the liver is unable to synthesize new albumin adequately, when there are substantial losses of albumin via the gastrointestinal tract or kidney or when there is accelerated endogenous albumin catabolism. When serum albumin concentration decreases, fluid may shift from the intravascular spaces to the interstitial spaces, and edema develops. Considerable controversy has been generated over a period of years over the relative roles of glomerular versus tubular pathology in the development of proteinuria in different forms of renal dysfunction. Although our understanding of the mechanisms of proteinuria still remains limited, this chapter provides a comprehensive review of the current consensus.

Discussion of renal mechanisms is continued by a presentation of the means whereby the kidney regulates total body electrolyte stores. Maintenance of extracellular fluid volume is primarily dependent upon the osmotic effects exerted by extracellular sodium. An intricate network of renal regulatory mechanisms exists to maintain extracellular sodium within a narrow, normal range despite wide variations in oral intake and extrarenal loss. When gastrointestinal or renal salt losses exceed intake, extracellular fluid volume decreases. Volume depletion stimulates antidiuretic hormone release in an attempt to maintain volume. Continued salt loss results in development of hyponatremia.

Any decrease in intravascular volume promotes renal retention of salt and water. In a variety of edematous conditions, such as congestive heart failure and cirrhosis, local arterial and intracardiac receptors responsible for regulating intravascular volume sense a volume depletion when, in fact, total volume may be expanded but abnormally distributed. Further salt and water retention under these conditions leads to development of edema. When "apparent" intravascular volume is decreased, antidiuretic hormone release also occurs. If fluid intake continues, water in excess of sodium is retained and a dilutional hyponatremia develops.

Intracellular electrolytes play an important role in normal homeostasis and in the modifications derived from disease. Extracellular measurements of these ions are all we have available, but the information derived is limited. When practical measurement of intracellular (and bone) concentrations are available, our understanding will be advanced, and our fluid and electrolyte therapy may be quite different. Potassium is the primary intracellular solute providing the osmotic force necessary to maintain intracellular volume. Only a small portion of total body potassium is contained in the extracellular fluid compartment. Therefore, the serum concentration of potassium may fail to reflect accurately total body potassium. A potassium flux into cells occurs with cell growth, intracellular nitrogen and glycogen deposition, and increases in extracellular pH; potassium leaves the cells with cell destruction, glycogen utilization, and decreases in extracellular pH. In interpreting the significance of any given serum potassium concentration, consideration must be given to these factors, which affect the ratio of intracellular to extracellular concentrations. Other electrolytes with important intracellular (and bone) representation are calcium, magnesium, and phosphorus. Their relations to altered physiology are also considered.

The role of the renal concentrating and diluting mechanisms is discussed in terms of maintenance of body fluid volumes. Isotonic polyuria results either from structural or functional defects in the countercurrent concentrating mechanisms (e.g., hypokalemia, hypercalcemia and sickle cell disease) or from an osmotic diuresis. When the distal nephron remains impermeable to water, owing to either a deficit of or tubular unresponsiveness to antidiuretic hormone, a more striking polyuria develops, with elaboration of a hypotonic urine. Persistent polyuria leads to significant dehydration or volume depletion only when thirst mechanisms are impaired or when free access to water is denied by physical handicap or other limitation.

The section concludes with considerations of the extracellular and intracellular buffer systems and the pulmonary and renal regulatory mechanisms that are coordinated to maintain a constant serum hydrogen ion concentration or pH. Increased production of fixed acid or retention of hydrogen ions leads to development of a metabolic acidosis; retention of carbon dioxide, which is hydrated to the weak acid, carbonic acid, leads to a respiratory acidosis. Conversely, loss of hydrogen ion produces a metabolic alkalosis; loss of CO_2 (hyperventilation) produces a respiratory alkalosis. The most common of the acid-base disturbances encountered clinically is metabolic acidosis. A healthy person on a normal diet must excrete by way of the kidneys about 35 mEq. of fixed acid daily by the kidneys to eliminate the fixed acid formed by normal metabolic processes. The capacity to maintain plasma pH at 7.4 may be exceeded following administration of an exogenous acid load, increased endogenous acid production, and/or decreased renal acid excretion resulting from either a decrease in functioning nephrons or an intrinsic tubular defect (renal tubular acidosis). Although hyperventilation, with loss of the weak acid CO_2, raises the plasma pH toward normal, this partial compensation is ineffective for prolonged maintenance of pH. Permanent correction of acidemia can be achieved only by administration of a base or by renal or gastrointestinal loss of fixed acid.

Although these five chapters cover five of the most important regulatory functional areas of the kidney, numerous other examples may be cited. For example, manifestations of gout may result from an elevation of serum uric acid levels. This elevation may result from excessive endogenous production of uric acid or from an increased *net* tubular reabsorption of uric acid. Since uric acid is both secreted and reabsorbed at different levels in the tubule, increased net reabsorption could mean either increased reabsorption or decreased secretion of uric acid, or both.

Another example is stone formation in the urinary tract of the patient with cystinuria. Other proximal tubular defects in phosphate, amino acid, glucose, and bicarbonate reabsorption may be present as isolated abnormalities or in combination, resulting in a whole spectrum of disease.

The role of the kidney in maintenance of blood pressure, vitamin D metabolism, and erythropoiesis is discussed in Chapters 3, 20 and 27, respectively.

SOLOMON PAPPER, M.D.

10   Maintenance of Body Protein Homeostasis

Victor E. Pollak, M.D.
Amadeo J. Pesce, Ph.D.

It is well known that the kidney eliminates most nonvolatile waste products of metabolism. The kidney also plays a major role in body protein homeostasis. It conserves within the circulation large protein molecules such as albumin and γ-globulins. It removes from the circulation, catabolizes, and reclaims the amino acids of polypeptide molecules and of proteins of a size smaller than albumin. The anatomical and molecular architecture of the nephron seem ideally designed to perform these functions.

Direct observations relative to the movement of proteins across the glomerular filter, or their reabsorption or secretion by the tubules, require the analysis of fluid withdrawn from individual nephrons. As yet there are few such observations because nephron puncture studies and the methods for the accurate measurement of minute amounts of plasma proteins are both difficult to do. In the two following sections—on the glomerulus as a molecular barrier, and on tubular reabsorption and secretion of proteins—the views presented result largely from investigations in intact animals; thus, our concept of the role of the individual parts of the nephron in the handling of proteins is based mainly on inference rather than direct measurement.

The glomerulus as a molecular barrier

The kidney appears to act as a molecular sieve with respect to the plasma proteins. The sieving effect occurs in the glomerulus. Ions and small molecules such as water, glucose, and urea pass through the glomerulus into Bowman's space and proximal tubular lumens, virtually without hindrance, as a function chiefly of the difference between filtration pressure and osmotic pressure. By contrast, plasma protein molecules are retained by the glomerulus most efficiently and to such an extent that the normal urine contains only minimal quantities of protein.

SIEVING COEFFICIENT. Inulin, a relatively small molecule (mol. wt. 5,000), is filtered completely by the glomerulus and is not reabsorbed by the tubules. It is therefore used to measure the glomerular filtration rate, and can serve as a reference substance in considering the permeability of the glomerulus to macromolecules. It has a sieving coefficient of 100 per cent (Fig. 10–1).

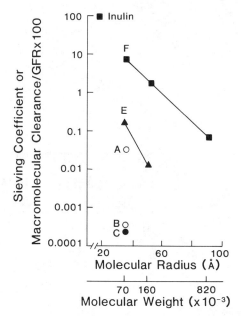

FIG. 10–1. The relationship between molecular radius (and molecular weight) and the sieving coefficient. By definition, the sieving coefficient of inulin is 100.

Three proteins are used in this figure, albumin (mol. wt. 69,000, molecular radius 35Å); IgG (mol. wt. 160,000, molecular radius 50Å); and α_2-macroglobulin (mol. wt. 820,000, molecular radius 90Å). *Point A* is the sieving coefficient for albumin in the normal rat, measured in the first part of the proximal convoluted tubule by direct puncture. *Point B* is the sieving coefficient for albumin in the normal rat, measured indirectly from the urine. *Point C* is the sieving coefficient for albumin in healthy man, measured indirectly from the urine. *Line E* is the sieving coefficient for albumin and IgG in a patient excreting a large amount of protein, but in whose glomeruli only minimal changes were found. The sieving coefficient for albumin is greatly increased as compared with the normal. *Line F* is the sieving coefficient derived from studies on a patient with severe proteinuria and a glomerular filtration rate about 5 per cent of normal. The glomerulus was highly permeable to albumin and the slope of the line indicates, as compared with *Line E*, the considerable permeability to proteins larger than albumin.

Albumin is the most abundant plasma protein (mol. wt. 69,000; molecular radius 35 Å) and has a concentration of 40 g./l. in the plasma flowing through the glomerulus. The concentration of albumin, measured *by direct puncture* of the first part of the rat proximal convoluted tubules, is less

than 10 mg./l. In this part of the proximal tubule the inulin concentration is identical to that in the plasma. The true glomerular sieving coefficient for albumin may be expressed as:

$$\left(\frac{[TF_{ALB}]V}{[P_{ALB}]} \div \frac{[TF_{IN}]V}{[P_{IN}]} \right) 100 \qquad (1)$$

where TF_{ALB} and P_{ALB} = albumin concentration in proximal tubular fluid and in plasma; TF_{IN} and P_{IN} = inulin concentration in proximal tubular fluid and in plasma; V = rate of flow through the proximal tubule. Note that V is common to numerator and denominator.

For the rat proximal convoluted tubule, Equation 1 is:

$$\left(\frac{[10]V}{[40 \times 10^3]} \div (1 \times V) \right) 100 = 0.025$$

This is indicated by Point *A* in Figure 10–1.

In practice, the effectiveness of the barrier to the passage of plasma proteins must be measured *indirectly* by a study of the urine rather than the proximal convoluted tubular fluid. In the normal 200-g. rat the albumin excretion is about 0.2 mg. per 24 hours. The glomerular filtration rate is about 1.4 l./24 hours. The sieving coefficient for albumin may be calculated as follows:

$$\left(\frac{[U_{ALB}]V}{[P_{ALB}]} \div GFR \right) 100 \qquad (2)$$

where U_{ALB} = concentration of albumin in urine, and where $U_{ALB}V$ = excretion rate of albumin; for example:

$$\left(\frac{0.2}{[40 \times 10^3]} \div 1.4 \right) 100 = 0.00035 \qquad (2)$$

In Figure 10–1 this is represented by Point *B*.

If these assumptions are correct (more data are needed), it is reasonable to assume that about 99 per cent of the albumin filtered by the glomerulus is reabsorbed by the proximal convoluted tubules. The sieving coefficient for albumin, measured indirectly in this manner, is about 1 per cent of that measured directly.

PORE THEORY. To explain the molecular sieving phenomenon, two general theories have been proposed. In the *pore theory*, developed fully and in mathematical detail by Pappenheimer, the glomerular capillary walls are envisaged as being perforated by water-permeable pores of molecular dimensions which restrict the passage of proteins and macromolecules. For a glomerular capillary wall containing a homogeneous population of such pores, the limiting size, rate, and amount passing the glomerular capillary wall could be calculated for a series of proteins. According to the pore theory, filtration is the most important factor in transfer of proteins across the glomerular capillaries, but the role of diffusion is not ignored entirely. Indeed, it is suggested that the relative concentration of a protein or macromolecule in Bowman's space is a function of both pore size and diffusion coefficient.

DIFFUSION THEORY. In the *diffusion theory*, the glomerular capillary wall is considered to be a gel that contains fibrils; the entire surface is pictured as being permeable to proteins and macromolecules as well as to water. In contrast to the pore theory, diffusion is considered to be the major mechanism of transfer; hydraulic pressure factors, although important, are assigned a lesser role. All substances are assumed to diffuse across the glomerular capillary wall at a finite rate that is a function of their intramural diffusion coefficients.

The precise role of the factors favoring restriction or passage of proteins across the glomerulus is not yet clear. The diffusion theory, proposed originally on thermodynamic grounds, predicts that protein molecules of any size can be transferred across the glomerular capillary wall; their clearance will approach asymptotically to zero with increasing molecular size or decreasing diffusion coefficient. In fact, vir-

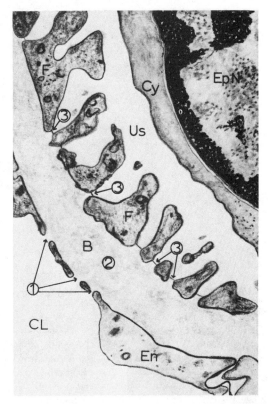

FIG. 10–2. Schematic representation of the human glomerular capillary wall. In the process of transfer from capillary lumen (*CL*) to urinary space (*US*), water, solutes and macromolecules must traverse three layers:

(*1*) The endothelial wall cytoplasm (*En*), containing numerous fenestrae (*1*) with a mean diameter of 700 Å. (*2*) The basement membrane (*B*), with a mean thickness of about 3000 Å. (*3*) The layer of foot process (*F*) of the epithelial cells (*Ep*). The foot processes are separated about 250 to 600 Å from each other by slit pores (*3*), which are lined by a distinct membrane. (After Jørgensen, F.: The Ultrastructure of the Normal Human Glomerulus. Copenhagen, Munksgaard, 1966.)

tually all plasma proteins, including minute amounts of large-molecular-weight proteins, have been found in normal urine.

Glomerular structure

In assessing the two theories it is useful to recall the unique structure of the glomerular capillary wall and the structure of the circulating plasma proteins. Electron microscopic studies have revealed three distinct layers (Fig. 10–2). The *lamina fenes-*

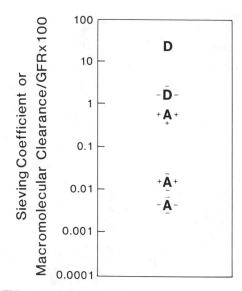

FIG. 10–3. The sieving coefficient or macromolecular clearance by the kidney of two species of polysaccharide and of three species of albumin molecules; all have a molecular radius of 35 Å, but differing ionic characteristics. Albumin molecules (A) are filtered by the glomerulus and reabsorbed by the tubules. Dextran molecules (D) are filtered but not reabsorbed, and therefore have a higher macromolecular clearance. The macromolecular clearance of the polyanionic dextran sulfate ($\bar{D}^-$) is significantly less than that of the uncharged dextran (D). Albumin is polyanionic at physiologic pH ($\bar{A}^-$). If albumin is modified to reduce the net negative charge ($+\overset{-}{A}+$), its clearance increases. When modified further to be polycationic ($+\overset{+}{A}+$), there is a considerable increase in the clearance. (Dextran data are from Chang and his colleagues, the albumin data from Purtell et al.)

trata of the endothelial cells is perforated by many pores of a size too large to restrict passage of any proteins. The *basement membrane*, approximately 3,000 Å thick in adult man, appears to be a gel-like structure containing fibrils 30 to 40 Å thick; no pores have been detected in this layer. The third layer, composed of *foot processes*, arises from the trabeculae of the epithelial cells; the foot processes are separated by slit pores, and a distinct membrane has been shown to line the slit.

It is now known that the glomerular capillary wall is highly polyanionic in charac-

ter. Let us examine this character as we move from the glomerular capillary lumen to the free urinary space. The endothelial cells are lined with polyanionic mucoproteins, which also extend over the fenestrae. The glomerular basement membrane itself has many anionic sites. These are concentrated particularly along the lamina rara interna and the lamina rara externa. Finally, the visceral epithelial cells are covered with polyanionic mucoproteins; these line the epithelial cell foot processes, and extend into the slit pores between the foot processes.

The circulating plasma proteins may be considered as globules of varying sizes. The surfaces of the plasma proteins have both positive and negative charges. Under physiological conditions (pH 7.4) the surfaces of these protein globules have a preponderance of anions. In other words, in the circulation, almost all plasma proteins, including albumin, are polyanionic.

Like electrical charges repel. Thus the glomerular capillary wall constitutes an effective barrier to the passage of circulating plasma proteins, including albumin, not only because of its anatomical structure, but also because of its highly polyanionic electrical nature. If this concept of the ionic anatomy of the glomerular wall is valid, the barrier to filtration should repel polyanionic plasma proteins, pass neutral ones, and bind polycationic proteins. Experimentally, little endogenous albumin is filtered; albumin altered to have a neutral charge is filtered more readily; when altered to be polycationic, albumin binds to the glomerular capillary wall, changing the ionic architecture and permitting passage of the endogenous (polyanionic) albumin through the barrier. Similarly, a polyanionic species of dextran molecules is retarded as compared with an uncharged species of dextran molecules (Fig. 10–3).

Sieving phenomenon

In normal man the urine albumin excretion approximates 18 mg./24 hours and the inulin clearance 180 l./per 24 hours. Thus the sieving coefficient (Equation 2) is:

$$\left\lfloor \frac{18}{40,000} \div (180) \right\rfloor 100 = 0.00025$$

This is close to the figure calculated for the rat and is represented in Figure 10–1 by Point C.

In disease, large amounts of albumin and of other plasma proteins may appear in the urine and the permeability of the glomerulus to proteins may be greatly increased (Fig. 10–1, Points E and F).

In Figure 10–4, on a different scale, is illustrated the typical curve obtained when dextran and dextran sulfate of various molecular radii are infused into rats. For both macromolecules the sieving coefficient was higher than that for proteins of identical molecular radius—perhaps because proteins are reabsorbed by the normal tubules whereas the polysaccharide macromolecules are not, or because neutral polysaccharide molecules pass more easily through the glomerulus than protein molecules do.

When dextrans were infused (Fig. 10–4), the clearance of molecules with a radius less than 23 Å approached the glomerular filtration rate (i.e., a sieving coefficient close to 100). The clearance fell off sharply as the molecular radius increased from 23 to 34 Å; for molecules above 40 Å the clearance was only a minute fraction of the glomerular filtration rate (i.e., a sieving coefficient approaching zero). With negatively charged dextran sulfate the sieving coefficient is lower than with dextran. For molecules of ~24 Å the sieving coefficient for dextran is 92, and that for dextran sulfate is 29; for molecules of ~35 Å (i.e., of a radius equivalent to albumin) the sieving coefficient for dextran is 25 and that for dextran sulfate is 2. When heterologous proteins have been injected into experimental animals, curves of a similar general form have been obtained. Clearances of small, relatively low-molecular-weight proteins such as myoglo-

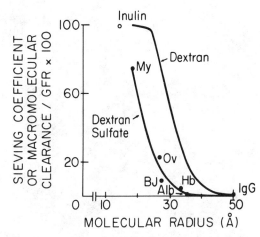

FIG. 10–4. The relationship between molecular radius and the sieving coefficient for inulin is 100. The sieving coefficients are shown for a number of proteins which have been characterized and studied. My = myoglobin; Ov = ovalbumin; BJ = light chains; Hb = hemoglobin; Alb = albumin. (The sieving coefficients for dextran and dextran sulfate are replotted from Chang, R. L. S., et al.: Kidney Int. 8:212, 1975.)

bin (mol. wt. 17,500; molecular radius 19 Å) or Bence Jones proteins (which exist as monomers or dimers; mol. wt. 22,000 or 44,000; molecular radius 22 or 28 Å) are relatively high, whereas all but a trace of the high molecular weight proteins such as albumin (molecular radius 35 Å) or IgG (molecular radius 50 Å) are retained by the kidney.

Anatomic and molecular changes in the glomerulus

In proteinuria associated with naturally occurring or experimentally induced disease of the glomeruli, many types of morphological alterations may be seen in the glomerular capillary walls. Gross rupture of glomerular capillaries may occur rarely, resulting in passage of blood into Bowman's space and, presumably, loss of the molecular sieve effect in a few capillary loops of some glomeruli. This may occur in some types of acute hemorrhagic glomerulonephritis and in necrosis of glomerular loops. Recent serial electron microscopic studies suggest that, in certain diseases, ac-

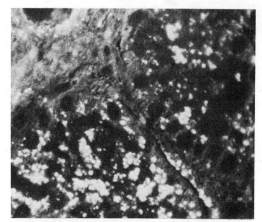

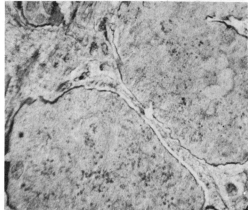

FIG. 10–5. Two proximal tubules from a renal biopsy specimen from a patient with glomerular disease and proteinuria (× 1,100): (*top*) incubated with a conjugated antiserum directed specifically against human albumin; (*bottom*) the identical section stained with periodic acid-Schiff. There are many droplets containing albumin, particularly in the tubular cytoplasm.

tual dissolution of basement membrane may occur in scattered glomerular capillary loops, presumably resulting also in a loss of the sieving effect in those loops.

The glomerular basement membrane is thickened in some disease states. It may appear normal morphologically in other conditions such as the common idiopathic nephrotic syndrome of childhood or the nephrosis induced in the rat by the aminonucleoside of puromycin. Using specific stains such as alcian blue, loss of the sialoprotein layer that lines the epithelial aspect of the basement membrane can be found. This change is associated with a decrease in the polyanionic barrier to filtration of proteins. Intravenous injection of negatively charged tracers indicates that the glomerular capillary wall is more permeable in these experimental animals.

Tubular absorption and secretion of proteins

It is clear that considerable quantities of protein must be reabsorbed by the tubules even when, as in the normal, the concentration of protein in the glomerular filtrate is extremely low. If the albumin concentration in the glomerular filtrate of normal man is 10 to 20 mg./l. (as it may well be), 1.8 to 3.6 g. of albumin must be reabsorbed by the tubules daily from the 180 liters of glomerular filtrate.

HYALINE DROPLETS. Hyaline, or colloid, droplets have long been recognized in the proximal convoluted tubules. Originally thought to be a manifestation of tubular degeneration and protein secretion, they are now recognized as protein reabsorption droplets. They occur in normal kidneys but are much more obvious in many diseases in which increased glomerular permeability occurs. With the use of specific enzymatic and immunochemical techniques it has been clearly shown that the colloid droplets contain proteins. For example, homologous proteins such as albumin have been demonstrated in colloid droplets of proximal tubules of normal human and rat kidney, and increased numbers of such albumin-containing droplets have been found in proximal tubules from kidneys with abnormal glomerular permeability (Fig. 10–5). Similar droplets may be seen in the visceral epithelial cells of the glomerulus from animals with proteinuria, an observation consistent with the view that protein which passes through the basement membrane may enter the foot processes of the epithelial cells directly.

Histochemically identifiable proteins have been injected into animals with the objective of studying the course of the protein through the nephron. Extensive studies have been made using horseradish peroxidase, a protein that readily passes into Bowman's space, as a tracer. Within a few minutes of injection, droplets of peroxidase can be detected at the luminal border of proximal tubular cells; the droplets then appear in vacuoles deeper in the cell; still later, chemical and histochemical studies indicate that vacuoles containing droplets of peroxidase become fused with lysosomes in the proximal tubular cells. It is probable that at least some of the absorbed protein is broken down by the cathepsins and other proteolytic enzymes in the lysosomes.

Tubular protein reabsorption

The amount of protein that can be reabsorbed by the normal tubules is not known. Clearance experiments suggest that the renal threshold for protein is somewhat low and the quantity of protein that the tubules can reabsorb is small—probably only slightly in excess of that filtered by the normal glomerulus. When increased amounts of protein are filtered by the glomerulus, it is likely that the capacity of the tubules to reabsorb protein increases, but in conditions with considerable proteinuria the tubular reabsorptive mechanism seems to be overloaded. This can readily be shown experimentally: when horseradish peroxidase is injected, it is reabsorbed by the proximal convoluted tubules. If the animals are given ovalbumin, the tubular reabsorption droplets become filled with ovalbumin. When horseradish peroxidase is then given to the ovalbumin-treated animals, little or no peroxidase is reabsorbed by the tubules.

It is not clear whether protein reabsorption can occur beyond the proximal convoluted tubules. In proteinuric states the development of smaller numbers of protein reabsorption droplets in the ascending limbs of the loops of Henle and distal tubules suggests that some reabsorption can occur at more distal sites.

Reabsorption of proteins seems to occur by pinocytosis at the cell membrane of the tubules. It is selective in that some proteins (e.g., albumin and β_2 microglobulin) are virtually completely reabsorbed, and others (e.g., amylase and lysozyme) are partially reabsorbed.

In the future, it should be possible to examine more effectively the quantitative and the selective aspects of tubular reabsorption, with the use of a combination of clearance, nephron puncture and stop flow techniques, and immunochemical methods which quantify individual plasma proteins. Further, it may be possible to determine whether plasma proteins are secreted by the tubules (a view for which there is little evidence at present).

The proteins of normal urine

The concentration of protein in normal urine is very low. This has made it difficult to quantify the total amount of protein excreted and to analyze the nature of the proteins in normal urine. In recent years these difficulties have been overcome, particularly as a result of the application of qualitative and quantitative immunochemical techniques to the study of normal urine proteins.

Currently, best estimates suggest that the healthy adult excretes between 40 and 150 mg. of protein in the urine every 24 hours. Some of the proteins are of plasma origin; others apparently derive from the urinary tract.

Immunochemical studies have revealed that normal urine regularly contains a large number of plasma proteins. The distribution of the various plasma proteins in urine differs radically from that in plasma. Consistent with the hypothesis that the normal glomerulus acts as a molecular sieve are the observations that a large proportion of the urine proteins are of low molecular weight, and the high-molecular-weight plasma proteins are present in normal urine only in trace amounts. Also consistent with this view are observations on the urinary excretion of the hemoglobin-binding protein haptoglobin. This protein can be detected regularly in urine from healthy subjects with haptoglobin Type 1-1, in whom the plasma haptoglobin is a monomer of 90,000 molecular weight; it cannot be found in urine from subjects with haptoglobin Type 2-2, in whom the haptoglobin exists in plasma as a series of larger, and higher molecular weight, polymers.

In healthy subjects the renal clearance of individual plasma proteins has been measured by immunochemical techniques, that of polysaccharide molecules using polydisperse dextran. The data for dextran conform to a simple relationship between molecular size (or molecular radius) and renal clearance; those for the proteins do not (Fig. 10–4). The polysaccharide molecules are not reabsorbed by the tubules. Thus if the sieving mechanism is intact for both proteins and polysaccharides, it is likely that selective or preferential reabsorption of some proteins occurs, or that some proteins penetrate into the nephron from postglomerular sites.

TAMM-HORSFALL PROTEIN. Many proteins that are excreted in normal urine are, no doubt, derived from the kidney and lower urinary tract, from turnover of cells and secretion. Little is known about most of them. An exception is *uromucoid*, a high-molecular-weight thread-like mucoprotein originally isolated from normal urine by Tamm and Horsfall, by whose names it is often known. It is predominantly the protein found in normal urine, about 40 mg. being excreted per day. It has not been identified in serum but can be found in the urine produced by the isolated perfused kidney. It is secreted by the cells of the ascending limbs of the loops of Henle, the distal tubules, and the collecting ducts. Tamm-Horsfall mucoprotein is the major component of the matrix of the casts that form in the lumens of the distal nephron and are excreted in normal and abnormal urine as hyaline casts. Uromucoid is soluble at a pH above 7.0 and precipitates at acid pH. Not surprisingly, these are the solubility characteristics known for the hyaline casts found in normal urine.

Proteinuria and exercise

Glomerular filtration rate, renal plasma flow and urine flow decrease during heavy physical exercise, but protein excretion in the urine increases manyfold. In one study, for example, the excretion of protein in healthy subjects at rest was 0.035 mg./min., whereas after exercise it was 0.21 mg./min. Increased urinary excretion of red blood cells and casts also occurs with heavy physical exercise. Urine protein excretion continues to be high after cessation of exercise and, in fact, may increase in the first 30 minutes of recuperation following exercise.

Recent observations on the nature of the urinary proteins indicate that the increased proteinuria is due almost exclusively to an increase in the amount of plasma proteins excreted. When the findings are compared with those of normal urine, a striking augmentation is noted in the excretion of albumin (mol. wt. 69,000), of proteins of similar molecular weight, and of proteins of higher molecular weight (ca. 160,000). These findings suggest that a change in glomerular permeability occurs with strenuous exercise. Normal urine does not contain haptoglobin in subjects with plasma haptoglobin Type 2-1. After exercise, this large-molecular-weight (>100,000) haptoglobin polymer is found in the urine. This observation is consistent with a change in the characteristics in the molecular sieve of the glomerulus.

The mechanism underlying the proteinuria of exercise is not well understood. The most plausible hypothesis was suggested by Poortmans. He proposed that the rise in the release of epinephrine and norepinephrine into the blood during exercise results in constriction of the afferent arterioles of the glomerulus; slowing of the glomerular filtration rate and renal plasma flow would result, thereby permitting a greater diffusion of plasma proteins through the glomerular capillary wall into Bowman's space. This cannot be the entire explanation, for, when studies of the renal clearance of plasma proteins were made, no direct relationship could be demonstrated between the molecular weights of the individual proteins and their renal clearance. An alteration in the capacity of the tubules to reabsorb proteins might also be present, such that the tubular maximum reabsorptive capacity is reached for some proteins before it is reached for others. Amylase and lysozyme clearance studies suggest that this is indeed the case. The decreased plasma flow in the peritubular capillaries might well alter the capacity of the tubules to reabsorb protein.

The large number of hyaline and other casts seen in the urine is consistent with an increased secretion of uromucoid by the cells of the distal nephron, as well as a considerable increase in the normal shedding of renal tubular cells into the tubular lumens. The increased number of red blood cells in the urine during heavy exercise seems likely to originate from glomerular or peritubular capillaries. If functional gaps or pores appear in capillaries of a size large enough to permit the passage of red blood cells, it is likely that they would also facilitate the transfer of plasma proteins directly from plasma into Bowman's space or into tubular lumens. Such a phenomenon would indicate that there is at least some heterogeneity of nephron function—that in some nephrons, at least, the molecular sieve may be temporarily bypassed and that postglomerular transfer of plasma proteins into the urine could occur in a few nephrons.

Proteinuria and posture

In healthy subjects a greater excretion of protein has been demonstrated in the upright than in the recumbent position. Using a sensitive quantitative immunochemical method to estimate the albumin excretion of apparently healthy volunteers, Robinson and Glenn found an excretion rate of 12 μg./min. while their subjects were in the upright position, as compared with 1.1 μg./min. while they were lying down. In these subjects the concentration of albumin (and protein) in the urine in both the recumbent and the upright position was too small to be detected by the qualitative tests in clinical use for the detection of protein.

When the amount of protein in the urine in the recumbent position is insufficient to be detected by the qualitative tests and protein is readily detected in the upright position while standing or during quiet ambulation, orthostatic or postural proteinuria is said to be present. This condition, which is most common during adolescence, has long been recognized. However, it remains a source of clinical controversy, for it is still unclear whether or not it is caused by or associated with significant structural abnormality of the kidney. Very limited studies, using quantitative immunochemical methods, indicate that the excretion of albumin in the recumbent position, although very small, exceeds that in healthy subjects, and that the relationship between albumin excretion in the upright position and that in the recumbent position is similar in healthy subjects and those with postural proteinuria. This suggests that in postural proteinuria there is a slight but definite exaggeration of the normal protein excretion in recumbency as well as in the upright position.

Systematic quantitative measurements of proteins other than albumin have not been made in urine and serum of subjects with postural proteinuria, so that data about the interrelationship of the clearance of individual plasma proteins and their molecular size or molecular radius are scanty. Thus, the question of whether or not there is an alteration in the molecular sieve of the glomerulus must await the results of further study.

TABLE 10–1
Proteinurias of "Prerenal" Origin

Type of Protein	Clinical Condition
Infusion of homologous plasma proteins	Experimental or therapeutic
Proteinuria of newborn ruminants: β-lactoglobulin	Physiological
Paraproteinemias—Bence-Jones proteinurias	Multiple myeloma
Hemoglobinuria	Intravascular hemolysis
Myoglobinuria	Muscle destruction
Inflammatory disease with increased low-molecular-weight proteins in plasma	Inflammation

The mechanism of postural proteinuria is unknown; however, the most plausible explanation is that there is an increased glomerular permeability to protein. This was at one time thought to be a consequence solely of the decrease in glomerular filtration rate and renal plasma flow observed in these patients when they assume the upright position, but there is little evidence that the hemodynamic alterations differ from those in healthy subjects. The demonstration of very minor structural changes in the glomeruli of half the patients with postural proteinuria suggests that there may be a minor alteration in the function of the glomerular capillary wall, facilitating protein transfer into Bowman's space in the recumbent position and reinforced by the decrease in glomerular filtration rate in the upright position.

Proteinuria in disease

In disease, proteinuria may occur when the kidney is normal if there is an unusually high level of proteins in the plasma, particularly proteins of relatively low molecular weight. It may be found when structural and functional changes in the glomeruli are associated with an increased glomerular permeability to proteins, or in diseases in which the glomeruli appear to be normal, where there are structural and functional alterations in the tubules. Also, proteinuria may develop when abnormal losses of nonplasma protein arise from cells of the kidney or ureter, lower urinary tract, and accessory glands and (at least in theory) from drainage of lymphatics into the urinary tract.

Types of Proteinuria in Disease

Prerenal, with normal glomerular permeability
Increased glomerular permeability
Renal tubular disorders
Lower urinary tract
Chyluria

Proteinuria associated with unusually high levels of plasma proteins

If the level of a plasma protein rises to such a degree that the increased amount crossing the molecular sieve of the glomerulus saturates the reabsorptive capacity of the tubules, proteinuria will be evident. Increased levels of albumin rarely occur in disease states; however, albuminuria can be demonstrated experimentally in animals and man when the plasma albumin level is raised by the infusion of albumin.

In disease, proteinuria occurs particularly when there is an increase in the plasma level of relatively low molecular weight proteins that are normally absent, or present in very low concentration (Table 10–1). For example, proteinuria is common when hemoglobin circulates in the plasma, as occurs in association with intravascular hemolysis. Initially the hemoglobin binds to the plasma haptoglobin, producing the relatively large hemoglobin-haptoglobin complex, which is retained by the glomeruli. Only when the plasma haptoglobin has been saturated does free hemoglobin circulate. This smaller molecule (mol. wt. 68,000; molecular radius 33 Å) passes through the glomerulus and saturates the reabsorptive capacity of the tubules (see Fig. 10–4). Myoglobin, the related muscle protein, is of low molecular weight (17,500) and small radius (19 Å) and appears in the circulation as a result of injury to or inflammation of muscle tissue. It binds to only a minor degree to other proteins and is readily filtered by the glomerulus; thus, myoglobinuria is the consequence of even a low level of myoglobin in plasma (see Fig. 10–4).

Another example is the light chains (Bence-Jones proteins) of the immunoglobulin molecules (see Fig. 10–4). Light chains are found in normal plasma together with the heavy chains as part of the intact immunoglobulin molecules. There are only minute traces of free light chains present in plasma, usually as monomers or dimers (mol. wt. 22,000 or 44,000). Light chains are readily filtered by the glomerulus, and minute amounts are found in normal urine. When proliferative disorders of the plasma cells such as multiple myeloma occur, the immune globulins, or their heavy or light chains, or fragments of the chains, are produced in increased amounts and circulate in the plasma. The intact molecules of IgG or IgA, for example, being of relatively high molecular weight (160,000) and large radius (50 Å), do not readily pass the glomerular molecular sieve and rarely appear in the urine, even when the plasma level is greatly increased. In contrast, when the low molecular weight free light chains are produced in large amounts, they readily pass the glomerular molecular sieve, presumably saturate the tubular reabsorptive mechanism, and are excreted in the urine.

It might at first sight be thought that these filtered low-molecular-weight proteins were passive passengers through the tubules, the reabsorptive capacity having been exceeded. In the proximal tubule the pH of the tubular fluid is that of plasma; in the distal nephron the pH of the tubular fluid is significantly lower and may be as low as 4.5. The pH of the tubular fluid has a profound effect on the ionic character of the filtered low-molecular-weight proteins. Both myoglobin and hemoglobin have isoelectric points (pI) of about 6.9; that is, at pH 6.9 there is an equal number of positive and negative charges. Under most normal conditions on a protein-rich diet, the pH of the distal nephron fluid is in the range 6.5 to 4.5, at which pH myoglobin and hemoglobin are polycationic. In this polycationic state they can interact with the polyanionic Tamm-Horsfall mucoprotein to form casts, and with the polyanionic epithelial cell surface. These interactions may in part be responsible for the occurrence of acute renal failure with myoglobinuria and hemoglobinuria. Bence Jones proteins vary between individuals; their pI varies from 4.7 to 7.4. Any individual excretes one species of Bence Jones protein with a single uniform pI. Acute and chronic renal failure occur with Bence Jones proteinuria, and are more likely to be associated with Bence Jones proteins of relatively high pI (>5.2), i.e., with those species which are more polycationic in the tubular fluid.

Proteinuria associated with increased glomerular permeability

The urinary excretion of a large quantity of protein is common in patients with diseases affecting the glomeruli, and the protein loss may exceed 20 g./day or more. Many changes may be seen in the structure of the glomeruli. Capillary loops or whole glomeruli may be obliterated, inflammation with cellular proliferation may occur, the thickness and character of the capillary basement membrane may be altered, and substances may be deposited in the mesangium and on the subendothelial and subepithelial aspects of the basement membrane, and in the basement membrane itself.

Many other structural changes appear to result in a glomerular molecular sieve that is more leaky than the normal. Since it is likely that the tubular capacity to reabsorb filtered proteins does not increase greatly, increased leakiness of the glomerulus or increased filtration of protein will be reflected in increased protein excretion in the urine. If the glomerular ultrafilter becomes blocked or obliterated, the physiological consequence is a decrease in the glomerular filtration rate; proteinuria may become quantitatively reduced or even disappear.

The leakiness of the glomerular ultrafilter can be expressed physiologically in two ways. Increased permeability results in the filtration and clearance of relatively more protein. Thus, the albumin clearance per unit glomerular filtration rate (i.e., the sieving coefficient) is greatly increased when there is functional damage to the glomerular ultrafilter (see Fig. 10–1, E and F) and may return to within the normal range if that damage can be repaired completely.

Plasma albumin is almost completely retained by the normal glomerular molecular sieve (see Fig. 10–1, A and B) and only trace amounts of proteins larger than albumin appear in the normal urine. Damage to the molecular sieve in disease might be expected to result in a diminished capacity to retain not only albumin but plasma proteins larger than albumin. In diseases associated with structural abnormalities of the glomerulus, the clearance of proteins larger than albumin is often greatly increased, and the urine may contain many serum proteins, including those of large molecular size. The relationship between the molecular size of the plasma proteins and their comparative clearances by the kidney may be used as another index of the leakiness of the diseased glomerular ultrafilter. The clearance by the kidney of significant quantities of proteins of large molecular weight is consistent with a considerable alteration in the molecular sieving action of the glomerulus.

In Figure 10-1 the line originating with Point E represents the sieving coefficients for albumin (molecular radius 35 Å) and IgG (molecular radius 50 Å) in a patient with massive proteinuria, normal glomerular filtration rate, and no evident morphological alterations in the glomerular basement membrane. The sieving coefficient for albumin is about 500 times normal (Point B), and that for IgG is one-tenth that for albumin.

In Figure 10–1 the line originating with Point F is derived from data from a patient with profound damage to the glomerular basement membrane, with resultant severe impairment of glomerular filtration rate and considerable proteinuria. The sieving coefficient for albumin was 8 (i.e., 20,000 times the normal; see Fig. 10–1). Larger proteins were also in the urine in abundance. The sieving coefficient for IgG (50 Å) was 2 and that for α_2-macroglobulin (mol. wt. 820,000; molecular radius 90 Å) was 0.08.

Changes in the glomerulus in disease are often heterogeneous, and it may well be that the glomerular filter is altered in varying degrees from glomerulus to glomerulus. The results of studies on intact kidneys may reflect only the average of a wide diversity of changes in the permeability of the glomerular filter of individual nephrons. Pores or gaps have not been found in the capillary basement membrane of normal glomeruli. Recent evidence suggests that actual dissolution of basement membrane may occur in segmental areas of a few capillary loops of diseased glomeruli at least occasionally. Such gaps in the continuity of the basement membrane could be responsible for the passage into the urine of large molecular weight proteins; their infrequent and segmental distribution emphasizes the probable heterogeneity of physiological alterations that may occur in the glomeruli in disease states.

Proteinuria associated with abnormal renal tubular function

A particular type of proteinuria occurs in association with disorders of function of the renal tubules. In addition to the proteinuria, the renal tubular functional abnormalities may include, singly or in various combinations, impairment of reabsorption of water, ions, glucose, uric acid, amino acids and organic acids, and impairment of acid excretion.

The proteinuria is of mild degree and, unlike that associated with impairment of glomerular function, rarely exceeds 1 to 2 g./24 hours. As indicated previously, the proteins excreted in states of altered glomerular permeability include albumin (mol. wt. 69,000) as the predominant protein and many plasma proteins larger than albumin. In tubular proteinuria, by contrast, proteins of relatively small size (mol. wt. 10,000 to 50,000) predominate in the urine. The origin of these proteins was, for a long time, difficult to determine with certainty. Obviously, they might be proteins originating from the kidney itself, particularly from damaged tubules. An alternative explanation is that they are proteins of low molecular weight normally present in plasma and in low concentration. The availability of monospecific antisera directed against many individual low-molecular-weight plasma proteins has clarified this issue. The proteins excreted in so-called tubular proteinuria have now been shown to originate predominantly from plasma.

Proteins smaller than albumin are filtered more readily by the normal glomerular molecular sieve. Indeed, the filtration rate of a small protein such as β-microglobulin (mol. wt. 13,000) approaches the glomerular filtration rate, at the same time that albumin and large plasma proteins are effectively retained by the glomerulus. Being more readily filtered by the glomerulus, the low-molecular-weight proteins present to the proximal tubules for reabsorption in amounts that are large compared with those of albumin. In the presence of normal glomerular function, impairment of the ability of the proximal tubules to reabsorb protein will therefore be reflected quantitatively as a relative inability to reabsorb proteins of low molecular weight.

Chyluria

Chyluria is a rare condition arising as a result of fistulous connections between the lymph vessels and the urinary tract; there is a consequent loss of lymph, which drains from the lower limbs and abdominal organs into the urine. Some proteins from the interstitial fluid are probably returned to the bloodstream via the lymphatics; so that the proteins in the urine in chyluria are probably a reflection of the permeability of the nonglomerular capillaries. As might be expected, the proteinuria is very different from that seen in association with glomerular damage. The composition of the urine proteins is very similar to that of plasma, and the clearance of all plasma proteins studied, including α_2-macroglobulin (mol. wt. 820,000) and β-lipoprotein (mol. wt. 3,600,000), is high.

The metabolism of albumin and other plasma proteins

It is now clearly recognized that individual plasma protein molecules are continuously being lost, and are continuously replaced by newly synthesized protein molecules. The balance of these two processes determines the actual body pool of a protein at any given time. In normal man it is probable that about one-tenth of the mass of circulating plasma protein is broken down each day. Protein molecules, when destroyed, are completely degraded.

Studies of plasma protein metabolism have been made, in the main, by coupling the isolated protein with a radioactive label, most commonly ^{131}I; analysis of the kinetics aims at defining the amounts of protein synthesized and broken down each day, at estimating the size of the intravascular and extravascular (or interstitial) pools of each protein, and at describing the interchanges that take place between the various pools. Precise details of these aspects of protein kinetics are the subject of continued investigation and debate.

The results obtained in any study must be interpreted in the light of the many theoretical and technical factors that may influence the data (see list below). The protein to be studied is altered chemically during the process of isolation from the plasma and may differ in varying degrees from the native protein. The body recognizes this and metabolizes the altered protein much more rapidly than it does the native protein. Thus, only proteins isolated by the mildest fractionation procedures should be used. The protein molecule may be damaged by the labeling procedure, as occurs when ^{51}Cr is used. Iodination is the most satisfactory method of labeling, provided that the degree of iodination is minimal (not more than one atom of iodine per molecule). The use of damaged protein molecules can be minimized by injecting the labeled protein into another animal that metabolizes them rapidly and, therefore, screens out the most altered of the protein molecules. Reutilization of the isotopic label after catabolism poses a technical problem which may be partially overcome by giving excess amounts of the metabolite (iodine).

Plasma proteins are distributed in two major compartments, the intravascular and the extravascular. The extravascular compartment in fact has many parts, each with its own concentration of protein and rate of transfer from the plasma. Thus, the nature of the mathematical model used to describe plasma protein kinetics in terms of the compartments affects the results, as do the efficiency of mixing and local variations in metabolic rates. In addition, most observations can be made only under steady state conditions, in which it is assumed that the rate of protein synthesis is equal to the rate of protein catabolism. An analysis of a typical plasma radioactivity decay curve is presented in Figure 10–6.

Some of the Factors Affecting the Results of Studies with Isotope-Labeled Proteins

Alteration of the protein during isolation
Alteration of the protein during the labeling
  procedure
Stability of isotope label
Lack of homogeneity of labeled protein
Reutilization of the isotope label
Mathematical model used
Prolonged or inefficient mixing
Local variations in metabolic rates
Steady state of the subject at the time of study

A recent and representative investigation using iodinated albumin to study albumin kinetics in normal subjects is summarized in Table 10–2. The total albumin mass is 3.9 g./kg. in women, and 4.7 g./kg. in men. Of the total mass, 45 per cent is intravascular in women, 42 per cent in men. Destruction and renewal of albumin appears to occur exclusively in the plasma or intravascular compartment, at a rate (the fractional catabolic rate) of about 8.5 per cent of the intravascular albumin mass per day. The rate of albumin synthesis is 0.16 g./kg./day.

Albumin biosynthesis (and that of most plasma proteins except the immune globulins) has been shown to occur mainly or exclusively in the liver. Albumin synthesis can perhaps best be studied by making use of a labeled precursor amino acid such as 6-^{14}C-arginine and observing its incorporation into albumin. This method does not necessarily require steady state conditions; it can be used in changing metabolic conditions and can also be applied to sequential studies. Using such a method, in which synthesis and catabolism could be measured independently, Tavill, Craigie, and Rosenoer demonstrated that, over a wide range of rates of synthesis (from 0.05 to 0.30

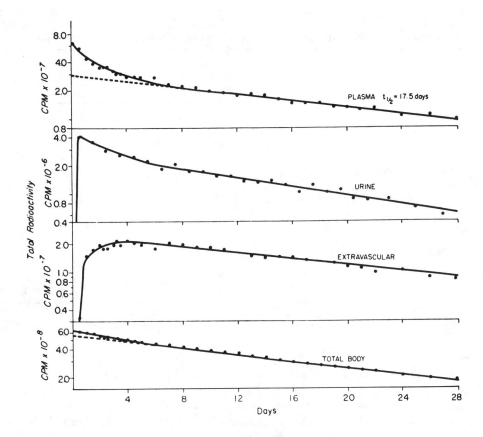

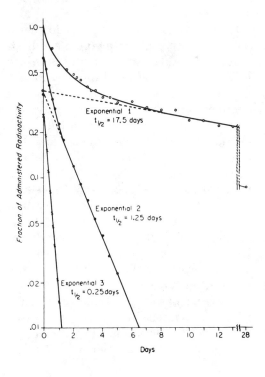

FIG. 10–6. The data obtained in a single human subject after the intravenous injection of a bolus of albumin labeled with [131]I. (*Top*) The curves of plasma and of urine radioactivity were obtained by measurement of timed plasma samples, and of 24-hour urine collections. The total body radioactivity was derived from the amount injected minus the amount excreted in the urine. The extravascular radioactivity was derived by subtracting the total plasma radioactivity from the total body radioactivity. (*Right*) A graphic analysis of the plasma radioactivity decay curve of the same subject. The open circles (*top curve*) are actual measurements plotted on semilogarithmic paper. The straight terminal portion of this curve was extrapolated back to its intersection with the vertical axis to yield exponential 1. A second curve was derived by subtracting the extrapolated curve from the original curve; exponential 2 was derived from this curve. The values in exponential 3 were obtained from residual values after the calculated data values of exponentials 1 and 2 were subtracted. This is an example of a 3-compartment model for the distribution of albumin. (Beeken, W. L., et al.: J. Clin. Invest., *41*:1312, 1962.)

TABLE 10–2
Metabolism of Albumin in Normal Subjects and in Patients with the Nephrotic Syndrome

	Normals*		Patients with Nephrotic Syndrome (15)†
	Women (19)	Men (19)	
Plasma albumin concentration (g./100 ml.)	4.3 ± 0.4	4.4 ± 0.2	1.5 (0.9–2.8)
Plasma volume (liters)	2.4 ± 0.3	2.8 ± 0.2	2.9 (1.9–3.8)
Intravascular albumin mass (g./kg.)	1.8 ± 0.3	2.0 ± 0.3	0.71 (0.36–1.28)
Total exchangeable albumin mass (g./kg.)	3.9 ± 0.3	4.7 ± 0.1	1.38 (0.72–2.42)
Ratio (intravascular/total mass) (%)	45 ± 5	42 ± 4	52 (47–60)
Fractional catabolic rate (%/day)	8.5 ± 1.2	8.4 ± 0.5	15 (5–24)
Rate of synthesis (g./kg./day)	0.15 ± 0.02	0.17 ± 0.03	0.20 (0.15–0.29)

* Mean, ±1 S.D. from Rossing, N.: Clin. Sci. *33*:593, 1967
† Mean and range from Jensen, H., Rossing, N., Andersen, S. F., and Jarnum, S.: Clin. Sci. *33*:445, 1967

g./kg./day), albumin catabolism and synthesis were equal. They also showed that in subjects with a low albumin pool the infusion of albumin to increase the plasma level resulted in no change in the rate of synthesis.

Fasting has a profound effect in that it results in a decrease in albumin synthesis of about 30 to 50 per cent. Malnutrition also leads to a decrease in albumin synthesis; this occurs rapidly as dietary intake falls. A net transfer of albumin into the intravascular pool occurs, and there is a fall in the catabolic rate. On low-protein diets, low anabolic and catabolic rates have been demonstrated, and the extravascular albumin mass is also decreased. On refeeding, anabolic and catabolic rates can return rapidly to normal.

Patients with the nephrotic syndrome lose large amounts of albumin in the urine. The metabolism of albumin has been studied in 15 such patients by Jensen, Rossing, Andersen, and Jarnum (Table 10–2). Their total albumin mass was greatly decreased to 1.38 g./kg., and a significantly higher proportion (52%) than normal was found in the intravascular compartment. In these patients both the synthetic rate and the catabolic rate were increased.

In normal man about 10 to 15 g. of albumin are catabolized daily. Destruction of plasma proteins has been shown to occur by way of the digestive tract and of the kidney, as well as by catabolism in tissue cells, particularly the liver. In the main because of difficulties in preparation and labeling discussed above, there is as yet no agreement as to the relative importance of these sites.

A constant leakage of plasma proteins occurs into the secretions of the gastrointestinal tract. There are considerable difficulties in quantifying the magnitude of this loss, but there is little doubt that it is increased in certain types of inflammatory gastrointestinal disease. The bulk of the amino acids liberated by intestinal catabolism of albumin and other plasma proteins is presumably reabsorbed and reaches the liver via the portal circulation, there to be reutilized for protein synthesis.

As discussed previously, there is much evidence that the glomerular filtrate contains small amounts of albumin and plasma proteins of molecular size smaller than albumin. If the albumin content of the glomerular filtrate is 10 mg./l., 1.8 g. of albumin would be filtered in adult man. Only one-hundredth of that amount is excreted in the urine. The remainder is presumably reabsorbed in the proximal tubules, which contain large numbers of lysosomes. The protein is absorbed into pinocytotic vacuoles, which then fuse with the lysosomes to form cytophagolysosomes in which the protein is presumably broken down by cathepsins and other proteolytic enzymes. The quantitative importance of this mechanism appears to be relatively small in the normal, but there is some evidence to suggest that in proteinuric states there is a considerable increase in albumin catabolism by the kidney. The amino acids and peptides liberated are conserved and are probably returned to the circulation via the renal lymphatics.

The other major site of albumin catabolism appears to be the liver. When carefully prepared and labeled albumin is used, after screening through another animal, the results suggest that the liver is responsible for about 10 per cent of the normal albumin catabolism.

Kinetic studies cannot be applied routinely in clinical medicine; however, the plasma albumin level is observed to be low in many disease states (see below).

Some of the Causes of a Low Plasma Albumin Level

Loss
   In the urine
   From the gastrointestinal tract
   From other sites (e.g., skin)
   From intravascular to extravascular pool
Increased fractional rate of catabolism
Decreased synthesis
Genetic analbuminemia

It may be low because a large amount of albumin is being lost—in the urine or the gastrointestinal tract or from other sites such as the skin when a large area has been burned. It may be low because the liver is so severely damaged that synthesis clearly is decreased. Many other factors such as poor nutrition, fever, tissue breakdown, etc., influence the synthetic and catabolic rates in subtle ways that the physician cannot study at the bedside.

The consequences to the body of the loss of large amounts of albumin and other proteins in the urine

When large amounts of protein are continuously lost in the urine, the plasma albumin level is decreased, the plasma cholesterol level becomes elevated, and significant edema occurs. These findings are usually observed when the urine protein loss exceeds 50 to 70 mg./kg./day, and, collectively, are known as the nephrotic syndrome. This syndrome occurs in association with a wide variety of disease processes in which the glomeruli are damaged.

In patients with the nephrotic syndrome, the plasma albumin level is correlated inversely with the degree of urine albumin (or protein) loss. As a consequence of the albumin loss, several changes in metabolism occur in an attempt to maintain homeostasis.

It will be recalled that the fractional catabolic rate of albumin is calculated as a percentage of the intravascular albumin pool, which is substantially decreased in patients with the nephrotic syndrome. An increased fractional rate of catabolism of albumin has been demonstrated clearly in the nephrotic syndrome in man (Table 10–2). The total amount of albumin catabolized (fractional catabolic rate × pool), measured in g./kg./day, is actually decreased. In the rat with experimental nephrotic syndrome, the renal tubules catabolize considerably more albumin than do the tubules of the normal animal. In the nephrotic syndrome there must therefore be a considerable decrease in albumin catabolism by degradative mechanisms in nonrenal sites. The rate of albumin synthesis is increased in many subjects with the nephrotic syndrome. In severely nephrotic patients, malnutrition occurs and may result in a relative decrease in the rate of albumin synthesis. Thus there is a decrease in the absolute rate of albumin catabolism and a very large increase in the rate of albumin loss in the urine, balanced by an increased rate of albumin synthesis. Liver slices from nephrotic rats synthesize albumin under optimal nutritional conditions *in vitro* at approximately twice the rate of those from normal rats. The rate of albumin synthesis is increased in patients with the nephrotic syndrome, but only to 1.25 times that of healthy subjects (Table 10–2). The failure to increase albumin synthesis beyond this level is probably due to the relatively poor nutritional state of the patient with the nephrotic syndrome.

Another characteristic feature of the nephrotic syndrome is the almost invariable elevation of the plasma cholesterol and phospholipid levels, which are correlated inversely with the plasma albumin level. Plasma triglycerides are elevated particularly when the plasma albumin is very low. The plasma lipids exist almost exclusively in combination with the lipoproteins. In patients with the nephrotic syndrome, significant abnormalities of the plasma lipoproteins occur. In particular, the concentration of the "low" density (D 1.019 to 1.063) and of the "very low" density (D < 1.019) lipoproteins is increased in the plasma. The concentration of high-density lipoproteins (D 1.063 to 1.125) appears to be decreased. (The density of most proteins such as albumin is about 1.3). The composition of the lipoprotein fractions is altered; they tend to contain more phospholipid and less protein than in normals. The low-density and very low density lipoproteins also contain a higher cholesterol/triglyceride ratio than do normal controls. The

cause of the alterations in plasma lipids is not yet completely understood. Lipids, particularly phospholipid and cholesterol esters, are lost in the urine in nephrotic syndrome; the apoproteins are also lost in the urine, and particularly those of high-density lipoprotein. An enhanced rate of synthesis of lipoproteins has been shown in both patients and animals with nephrotic syndrome, as well as defects in their degradation.

It is possible that the observed changes may be a reflection of alterations in the rate of hepatic production of the lipids and lipoproteins or in the rate of their disposal, elimination, or interconversion. Elevation of the plasma albumin level by the infusion of albumin leads to a decrease in the elevated plasma levels of cholesterol and phospholipids, and similar effects have been described when γ-globulin or polysaccharide molecules such as dextran and polyvinylpyrrolidone have been infused. These observations suggest the possibility that osmotic effects may control hepatic synthesis of plasma lipids. There is some evidence to suggest that the hepatic production of albumin also may be regulated by osmotic effects rather than by albumin concentration per se. Thus, the low oncotic pressure of the plasma, consequent to the decreased plasma albumin level, may be a stimulus to hepatic synthesis of both albumin and lipids.

There are changes in the levels of many plasma proteins in patients with the nephrotic syndrome. It is possible that the rates of hepatic synthesis of many plasma proteins besides albumin are increased. In general, proteins of large molecular size are retained in the plasma by the kidney; their plasma level therefore increases. An example is the large-molecular-weight (840,000) α_2-macroglobulin, which has the property of binding trypsin and is an inhibitor of α_2-trypsin and plasmin. Smaller proteins are usually excreted in large amounts in the urine, and their plasma level may be decreased. For example, the plasma transferrin (mol. wt. 90,000) level is usually decreased, leading to a reduction of the plasma iron binding capacity and, possibly, contributing to the increased susceptibility to infection which occurs in these patients. The levels of the thryoxine-binding prealbumin (mol. wt. 61,000) and of the copper-binding ceruloplasmin (mol. wt. 160,000), are also decreased. IgG (mol. wt. 160,000), which is synthesized in extrahepatic sites, is lost in the urine; the consequent reduction in plasma level may contribute to the susceptibility to infection.

The level of plasma fibrinogen is increased, as are the levels of many coagulation factors; some authors consider that there is an increase in frequency of thrombosis in patients with the nephrotic syndrome. Clotting Factor IX, on the other hand, may be lost in the urine in large amounts, and the consequent low plasma level may be associated with the bleeding tendency observed in some patients.

Edema is the characteristic clinical sign observed in the nephrotic syndrome. It occurs almost invariably when the plasma albumin level is decreased to about 16 to 18 g./l. As a consequence of the hypoalbuminemia, the plasma oncotic pressure is decreased, leading to an increased leakage of fluid from the intravascular to the extravascular compartment. A low plasma albumin level cannot be the sole cause of the edema, as edema does not occur in patients with analbuminemia in whom the plasma albumin level is usually less than 0.1 g./l. In some patients a lowered plasma volume, increased secretion of aldosterone, and increased tubular reabsorption of sodium are other important contributing factors.

Annotated references

Baxter, J. H.: Hyperlipoproteinemia in nephrosis. Arch. Intern. Med. *109:*742, 1962.

Blainey, J. D., Brewer, D. B., and Hardwicke, J.: Proteinuria and the nephrotic syndrome. *In* Black, D. A. K., and Jones, N. F.: Renal Disease, ed. 4. Brackwell, Oxford, 1979, Chap. 13, p. 383.

Caulfield, J. P., and Farquhar, M. G.: The permeability of glomerular capillaries to graded dextrans. J. Cell. Biol. *63:*833, 1974. (Morphological studies of the passage of dextran through the glomerulus).

Chang, R. L. S., Deen, W. M., Robertson, C. R., and Brenner, B. M.: Permselectivity of the glomerular capillary wall. III. Restricted transport of polyanions. Kidney Int. *8:*212, 1975. (These two papers by Chang, et al., explore in elegant detail the transport of dextran and dextran sulfate across the glomerulus.)

Chang, R. L. S., et al.: Permselectivity of the glomerular capillary wall to macromolecules. II. Experimental studies in rats using dextran. Biophys. J. *15:*887, 1975.

Chinard, F. P., Lauson, H. D., Eder, H. A., Greif, R. L., and Hiller, A.: A study of the mechanism of proteinuria in patients with the nephrotic syndrome. J. Clin. Invest. *33:*621, 1954. (The diffusion theory is considered in this paper.)

Clyne, D. H., Pesce, A. J., and Thompson, R. E.: Nephrotoxicity of Bence-Jones proteins in the rat: Importance of protein isoelectric point. Kidney Int. *16:*345–362, 1979.

Dillard, M. G., Pesce, A. J., Pollak, V. E., and Boreisha, I.: Proteinuria and renal clearances in patients with renal tubular disorders. J. Lab. Clin. Med. *78:*203, 1971. (A complete description of the proteins in urine in patients with tubular disease.)

Farquhar, M. G.: The primary glomerular filtration barrier: basement membrane or epithelial slits? Kidney Int. *8:*197, 1975. (This editorial is a critical recent review of morphological aspects of macromolecular transport across the glomerulus.)

Gherardi, E., Rota, E., Calandra, S., Genova, R., and Tamborino, A.: Relationship among the concentrations of serum lipoproteins and changes in their chemical composition in patients with untreated nephrotic syndrome. Eur. J. Clin. Invest. *7:*563–570, 1977.

Gherardi, E., Vecchia, L., and Calandra, S.: Experimental nephrotic syndrome in the rat induced by puromycin aminonucleoside; plasma and urinary lipoproteins. Exp. Mol. Pathol. *32:*128–142, 1980.

Grant, G. H.: The proteins in normal urine. II. From the urinary tract. J. Clin. Pathol. *12:*510, 1959.

Jensen, H., Rossing, N., Andersen, S. B., and Jarnum, S.: Albumin metabolism in the nephrotic syndrome in adults. Clin. Sci. *33:*445, 1967. (This paper, together with that of Rossing from the same laboratory, provides good comparative data on the normal and the nephrotic.)

Kanwar, S., and Farquhar, M. G.: Anionic sites in the glomerular basement membrane: *In vivo* and *in vitro* localization to the laminae rarae by cationic probes. J. Cell. Biol. *81:*137–153, 1979.

Katz, J., Sellers, A. L., and Bonorris, G.: Effect of nephrectomy on plasma albumin catabolism in experimental nephrosis. J. Lab. Clin. Med. *63:*680, 1964. (A paper which provides some direct experimental evidence for protein catabolism by the kidney.)

Lambert, P. P., Gassee, J. P., and Askenasi, R.: Physiological basis of protein excretion. *In:* Manuel, Y., Revillard, J. P., and Betuel, H. (eds.): Proteins in Normal and Pathological Urine, p. 67. Baltimore, University Park Press, 1970. (An excellent, brief, modern review of the physiological basis of glomerular permeability. It is an excellent starting point for the student interested in this subject.)

Lewy, J. E., and Pesce, A. J.: Micropuncture study of albumin transport in aminonucleoside nephrosis in the rat. Pediat. Res. *7:*553, 1973.

McQueen, E. G.: The nature of urinary casts. J. Clin. Pathol. *15:*367, 1962.

Oken, D. E., and Flamenbaum, W.: Micropuncture studies of proximal tubule albumin concentrations in normal and nephrotic rats. J. Clin. Invest. *50:*1498, 1971.

Pappenheimer, J. R.: Passage of molecules through capillary walls. Physiol. Rev. *33:*387, 1953. (A full review and theoretical treatment of the subject as it applies to capillaries in general and to the glomerular capillaries.)

Pesce, A. J., and First, M. R.: Proteinuria: An Integrated Review. New York, Marcel Dekker, 1979. (This is the most complete recent review of urinary protein excretion.)

Poortmans, J., and Jeanloz, R. W.: Quantitative immunochemical determination of 12 plasma proteins excreted in human urine collected before and after exercise. J. Clin. Invest., *47:*386, 1968.

Purtell, J. N., Pesce, A. J., Clyne, D. H., Miller, W. C., and Pollak, V. E.: Isoelectric point of albumin: Effect on renal handling of albumin. Kidney Int. *16:*366, 1979. (This paper describes the effect of charge on the clearance of a protein by the kidney.)

Robinson, R. R., and Glenn, W. G.: Fixed and reproducible orthostatic proteinuria. IV. Urinary albumin excretion by healthy human subjects in the recumbent and upright postures. J. Clin. Med. *64:*717, 1964.

Robinson, R. R., Lecocq, F. R., Phillippi, P. J., and Glenn, W. G.: Fixed and reproducible orthostatic proteinuria. III. Effect of induced renal hemodynamic alterations upon urinary protein excretion. J. Clin. Invest. *42:*100, 1963.

Rossing, N.: The normal metabolism of ^{131}I-labeled albumin in man. Clin. Sci. *33:*593, 1967.

Rowe, D. S.: The molecular weights of the proteins of normal and nephrotic sera and nephrotic urine, and a comparison of selective ultrafiltrates of serum proteins with urine proteins. Biochem. J. *67:*435, 1957.

Schultze, H. E., and Heremans, J. F.: Molecular Biology of Human Proteins with Special Reference to Plasma Proteins. Vol. 1, Sec. IV, Proteins of Extravascular Fluids, Chap. 2, The urinary proteins, p. 670. Section III, The Life of Plasma Proteins, Chap. 1, Synthesis of the plasma proteins, and Chap. 2, Turnover of the plasma proteins, p. 321. Amsterdam, Elsevier, 1966. (This is the most comprehensive brief review available.)

Straus, W.: Occurrence of phagosomes and phago-lysosomes in different segments of the nephron in relation to the reabsorption, transport, digestion and extrusion of intravenously injected horseradish peroxidase. J. Cell. Biol. *21:*295, 1964. (An excellent demonstration of the morphological findings in tubules presented with a protein load.)

Tavill, A. S., Craigie, A., and Rosenoer, V. M.: The measurement of the synthetic rate of albumin in man. Clin. Sci. *34:*1, 1968. (An important paper that describes the best approach to the study of protein synthesis.)

11   Maintenance of Body Fluid and Sodium Volume

Frederic C. Bartter, M.D.[*]

Abnormalities in body fluid volumes and electrolyte concentrations frequently occur, yet they often are poorly understood by physicians. The healthy kidney compensates for wide variations in intake and extrarenal loss of fluid and electrolytes; when the kidney is functionally unable to adjust for these variations, abnormalities appear in body fluid volumes and electrolyte concentrations. With a thorough understanding of fluid and electrolyte physiology, it is often possible to correct the resulting imbalances.

Serum sodium and potassium concentrations often fail to reflect total body stores. Low serum sodium concentration may occur in the face of increased extracellular fluid volume and total body sodium; conversely, hypernatremia occurs most often with dehydration and often is associated with some decrease in total body sodium. Changes in pH alter serum potassium concentrations without changing total body potassium. Therefore, therapy should be directed primarily not at correcting serum electrolyte concentrations, but rather at correction of disordered mechanisms.

* Dr. Bartter died May 5, 1983.

Normal sodium and water metabolism

Normal body fluid volumes and their measurement

The body is composed mostly of water, located predominantly in cells (Fig. 11–1). About 75 per cent of lean body mass (body mass excluding fat) is water; body fat is virtually water free. The fat content of the body varies among individuals, averaging about 15 per cent of body weight. Total body water, then, is about 60 per cent of body weight. One-third of this is extracellular and two-thirds intracellular fluid. Plasma, a portion of the extracellular fluid, represents about 4 per cent of body weight. The volume of red blood cells is normally slightly less than that of plasma. In young children, body composition is different: blood volume is greater relative to body weight, and extracellular fluid represents 30 per cent or more of body weight. Total body water, as a percentage of body weight, decreases with age and with increased weight.

Indicator-dilution techniques can measure body fluid compartments. A known

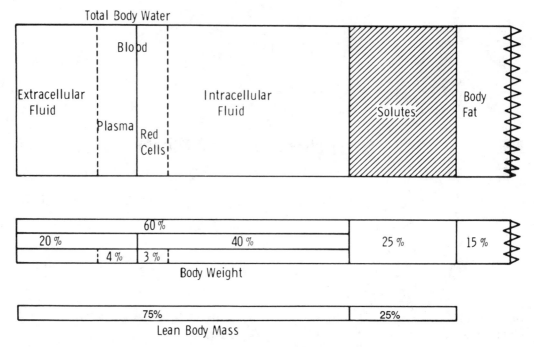

FIG. 11–1. Body composition, shown graphically. The body is made up mostly of water. Solutes in these water compartments and fat make up the remainder of the body mass. The solutes that are in solution approximate 290 mOsm./kg. of water. Body fat is obviously a variable quantity from patient to patient.

amount of a substance is injected intravenously and, after equilibration, blood is sampled and the concentration of the indicator substance is measured. Knowing the amount by which the indicator has been diluted, one can calculate the volume of distribution of the indicator by the following formula:

Volume (ml.) = Quantity of indicator injected/concentration of indicator per milliliter of sample

Albumin labeled by radioiodine or Evans blue dye (T-1824) is an indicator used for measuring plasma volume. Blood volume can be calculated if plasma volume and hematocrit are known. The albumin space exceeds plasma volume, because some albumin moves into extravascular spaces. A more nearly accurate measure of total blood volume is obtained with ^{51}Cr-labeled erythrocytes as an indicator of red cell volume, in addition to labeled albumin as an

indicator of plasma volume. Extracellular fluid volume may be estimated with radiosulfate, radiosodium, radiochloride, bromide, thiocyanate, or inulin, each having its own theoretical "space" and distribution volume. Hence, the same "label" should be used to estimate serial changes in the volume of that body fluid compartment. Total body water is measured with tritium- or deuterium-labeled water or radioantipyrine. Intracellular water is calculated as the difference between total body water and extracellular fluid volume.

The fluid compartments are dynamic, not static, with water shifting within compartments and from one compartment to another, under the influence of variables such as posture, exercise, and dietary intake. Furthermore, about 8,000 ml. enter and leave the gastrointestinal tract daily (1,500 ml. saliva, 2,500 ml. gastric juice, 500 ml. bile, 700 ml. pancreatic juice, and 3,000 ml. intestinal fluid). The glomeruli filter 180 liters of water daily, almost all of which is reabsorbed by the tubules. With minor exceptions (e.g., renal tubules, hypothalamic "osmostat") water diffuses freely across cell membranes. Thus, the osmolality of cells equals that of extracellular fluid, and large volumes of fluid enter and leave the cells daily. The osmolality of body fluids is maintained within remarkably close limits by a number of regulatory processes, particularly the thirst and renal concentrating and diluting mechanisms. The volume of extracellular fluid thus depends principally upon total body sodium, that of intracellular fluids principally upon total body potassium. However, many disorders, particularly renal and gastrointestinal diseases, may change the volume and composition of body fluids.

Composition of body fluids

Each of the various electrolytes and non-electrolytes in normal plasma (Fig. 11–2) contributes to its total osmolality (the total solute concentration per kilogram of plasma water). Plasma osmolality is normally 285 to 290 mOsm./kg. (or approximately per liter) of water. (A milliosmol [mOsm.] equals the molecular weight of a substance in milligrams and is calculated by dividing the number of milligrams of any nondissociable (nonelectrolyte) substance by its molecular weight.) In solutions of electrolytes, the ionized cations and anions each contribute to osmolality. Each nonionized cation-anion pair contributes a single milliosmol to osmotic pressure, or osmolality. For this reason, osmolality measured by the freezing-point depression is less than would be calculated by measuring the concentration of each plasma component, since not all electrolyte pairs are completely dissociated. Provided that glucose and urea concentrations are normal and no extraneous solute such as alcohol is present, plasma osmolality can be approximated by doubling the sodium concentration.

Interstitial fluid (the extravascular, extra-

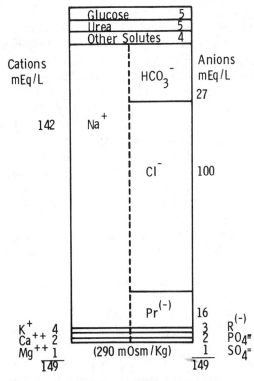

FIG. 11–2. Normal plasma composition. The sum of cations and anions ("corrected" for dissociation constants and valences) plus the nonionized compounds per liter of plasma, divided by the fraction of plasma volume composed of water, equals plasma osmolality. Glucose, urea, and "other solutes" are given in mOsm./kg. H_2O.

cellular fluid) is similar in composition to an ultrafiltrate of plasma. As it is protein free, however, it is modified by the resultant Gibbs-Donnan forces, which result from the effects of the negatively charged protein molecules upon electrolyte distribution across a semipermeable membrane. The rapid movement of water along osmotic gradients maintains an interstitial fluid osmolality almost identical to that of plasma. Plasma proteins, because of their large molecular size, contribute little to osmolality but carry many negative charges which must be balanced by positive charges of cations such as sodium. On the protein-free interstitial fluid side of the membrane, the cation (sodium) concentration is about 6 per cent lower and the non-protein anion (chloride) concentration higher than in the plasma. Osmolality of plasma is 1 to 2 mOsm./kg. of water higher than that of interstitial fluid because of the combined effects of the plasma protein concentration and the Gibbs-Donnan effect.

The composition of interstitial fluid in specialized areas such as the cerebrospinal fluid, the gastrointestinal tract, and the interstitial space of the kidney shows marked differences from that of plasma ultrafiltrate, because of the influence of active transport processes. The blood-cerebrospinal fluid and blood-brain barriers maintain gradients for some solutes such as phosphate and uric acid. Gastric juice has much higher potassium and hydrogen ion concentrations than has plasma. The interstitial fluid of the kidney is composed largely of reabsorbed glomerular filtrate. However, renal medullary interstitial fluid may have sodium, chloride, and urea concentrations, and thus total osmolality, much higher than those of plasma.

The osmolality of intracellular fluid is assumed to be identical to that of extracellular fluid. The predominant intracellular cations are potassium and, to a lesser extent, magnesium; there is little sodium in cell water. Phosphate, sulfate, and proteins are the major anions; chloride concentration is negligible. Urea diffuses freely into and out of cells. Rapid intracellular metabolism of glucose accounts for much lower glucose concentration within cells than in extracellular fluid.

Maintenance of normal fluid volume

To maintain health, there must be a balance between fluid intake and loss. Total body water depends on fluid intake minus the sum of insensible and fecal water loss and urine volume (glomerular filtrate minus renal tubular reabsorbate). Water entering the body derives from the oral intake of food and liquids or from fluids administered parenterally. Normally, the determi-

nant of fluid intake of conscious subjects with access to water is thirst. The primary stimulus to thirst is an increase in osmolality of plasma perfusing the hypothalamico-hypophyseal system. Increased osmolality of saliva or a dry mouth may also promote increased water intake. The thirst mechanism ordinarily indicates a need for "free" water (i.e., without dissolved solutes), but thirst may also be stimulated by a decreased effective blood volume, reflecting a need for isosmotic fluid. Another source of fluid, amounting to about 300 ml. daily, is the water derived from metabolism of carbohydrate and protein. Solid food contains preformed water that varies quantitatively, averaging about 300 ml./day. Alternatively, with negative caloric balance, catabolized cells release preformed water that is added to the residual cellular and extracellular fluid. Water is normally lost from the lungs, skin, bowel, and kidneys. Insensible water loss through diffusion and evaporation from the lungs and skin is about 1,000 ml./day for the average-sized adult. Water loss from the lungs can increase with hyperventilation, intubation, tracheostomy, and assisted ventilation; humidification of inspired air reduces this fluid loss. Both an increase in environmental temperature and fever increase cutaneous loss of water. Fluid loss from visible sweating may reach several l./day. Insensible fecal water loss normally is only about 100 ml./day.

The kidney is the organ primarily responsible for maintaining the balance between fluid intake and losses. Urine volume, under normal conditions may be as low as 500 ml./day, or even less if the solute load is low. Water diuresis can lead to volumes as high as 20. l./day, and even higher volumes can be achieved by solute diuresis. Oral intake ordinarily exceeds urinary output by about 500 ml./day. Since the difference between insensible gains (preformed water of solid food or tissues and water of metabolism) and losses (skin, lung, feces) of water amounts to a net loss of 500 ml./day, a balance is achieved. When water balance is maintained, body weight remains constant. Loss or gain of solid tissue is usually accompanied by water loss or gain, representing a large fraction of the change in weight.

Maintenance of sodium balance

The usual daily dietary sodium chloride intake in this country is highly variable and normally ranges from 6 to 12 g. (100 to 200 mEq. or 2,300 to 4,600 mg. of sodium). The control of sodium intake may reflect sodium appetite; hormonal control, if any, is not established. Extrarenal sodium losses are normally less than 10 mEq./day; the remaining intake must be excreted by the kidney to maintain equilibrium. Whereas wide variations in intake and extrarenal losses of sodium may occur, effective renal mechanisms exist to vary urinary excretion of sodium from several hundred to as little as 1 to 2 mEq./day. Some of these mechanisms (filtered load, aldosterone) are well understood; others (e.g., third factor) are under intensive investigation.

FILTERED LOAD OF SODIUM. Glomerular filtration rate (GFR) is determined by the area and the permeability of the filtering surface—i.e., the capillary endothelial surface—and the net hydrostatic filtering pressure, determined from the afferent arteriolar glomerular pressure, the efferent arteriolar resistance, and the opposing intratubular pressure as reflected in Bowman's space. Glomerular filtration rate correlates inversely with plasma oncotic pressure and depends in part on the action of thyroxine and carbohydrate-active steroids (e.g., cortisol). The filtered load of any substance can be determined by multiplying the GFR by the concentration of that substance in an ultrafiltrate of plasma. An approximation of ultrafiltrate concentration is given by plasma concentration. The precise ultrafiltrate concentration is estimated as the concentration of that solute in plasma water multiplied by a Gibbs-Donnan factor. As described previously, the sodium concentration in the protein-free ultrafiltrate is lower than that in plasma water and the chloride concentration is higher. Since the plasma water sodium concentration is higher than measured plasma sodium concentration, corrections for plasma water and Gibbs-Donnan distribution nearly cancel one another. The filtered load of sodium normally is about 26,100 mEq./day (145 mEq./l. × 180 l./day). Since sodium excretion varies from 2 to over 200 mEq./day, it represents from 0.01 to 1.0 per cent of the filtered load. The filtered load of sodium is the first factor to be considered in the evaluation of sodium excretion. The renal tubules must reabsorb large quantities of sodium, normally 99 to 99.99 per cent of the filtered load of about 18 mEq./min. (145 mEq./l. × 0.125 l. GFR/min.). As GFR varies, the tubular reabsorption of sodium does not remain a constant fraction of filtered sodium. A much greater natriuresis may occur with a small increase in filtered load than would be anticipated if the two parameters increased proportionately; conversely, a small decrease in filtered sodium may cause greater sodium retention than would be expected if the two variables decreased proportionately. When GFR decreases because of loss of nephrons (as occurs in most forms of chronic renal disease), other factors affect the fractional sodium reabsorption so that sodium excretion per nephron is usually relatively high.

RENAL TUBULAR REABSORPTION OF SODIUM. The tubule reabsorbs sodium against electrical and concentration (i.e., electrochemical) gradients by active transport, a process that accounts for most of the oxygen utilization by the kidney (Fig. 11–3). Between 50 and 80 per cent of glomerular filtrate is reabsorbed isosmotically in the proximal tubule. Under all conditions, proximal tubular fluid osmolality remains nearly identical to that of plasma water. Normally, some solutes, such as glucose, are completely reabsorbed in this segment. Others,

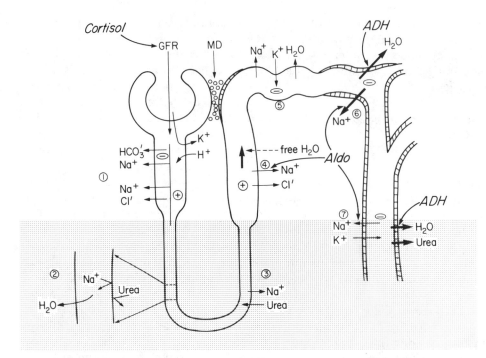

FIG. 11–3. Schematic diagram of renal tubular function. Numbers refer to (1) proximal, isosmotic reabsorption of Na HCO₃ (proximal convoluted) and Na Cl (proximal straight); (2) passive reabsorption of H₂O in urea- and Na-impermeable segment (thin descending limb); (3) passive reabsorption of Na and entry of urea in H₂O-permeable segment (thin ascending limb); (4) active reabsorption of Na and Cl (thick ascending limb); (5) active reabsorption of Na and passive reabsorption of Cl and H₂O (distal convoluted); (6 and 7) active reabsorption of Na, passive reabsorption of H₂O (cortical (6) and medullary (7) collecting duct).

⊕, ⊖ = potential of lumen, body considered "0"; ALDO = a site of action of aldosterone; ADH = a site of action of antidiuretic hormone (Vasopressin); cortisol = a site of action of cortisol; shading = area in which osmolality is above that of plasma.

such as urea, are incompletely reabsorbed and some—creatinine, for example—are not reabsorbed at all. Under normal conditions, the fractional reabsorption of all solutes except bicarbonate in the proximal tubules approximates the fractional reabsorption of sodium. (Fractional reabsorption of bicarbonate greatly exceeds that of sodium, so that 10 per cent of the filtered bicarbonate, but 20 to 60 per cent of the filtered sodium enters the loop of Henle.) Therefore, sodium concentration in proximal tubular fluid remains nearly equal to that in plasma. If the proximal tubular fluid contains a large quantity of a nonreabsorbable solute, such as mannitol, the water it obligates will dilute the sodium and lower its concentration. (This provides evidence that sodium reabsorption is *active*—i.e., against its electrochemical gradient—in the proximal tubule.) The fractional reabsorption of bicarbonate in the early proximal tubule, a carbonic anhydrase-dependent process, exceeds that of sodium. This increases the chloride concentration in the late proximal tubular fluid, increasing the gradient for chloride across the tubular epithelium, and rendering the luminal potential positive.

Proximal tubular sodium reabsorption probably depends in part on aldosterone, as demonstrated by the increased rate of disappearance, with addition of aldosterone, of a segment of tubular fluid isolated by oil drops, and by the decreased delivery of sodium to the ascending limb of Henle's loop (reflected in a decreased free water clearance in hydrated subjects) after treatment with aldosterone or other sodium-retaining corticosteroids.

Proximal tubular sodium reabsorption decreases after extracellular fluid volume expansion. Since absolute proximal sodium reabsorption normally far exceeds distal reabsorption, this virtually always results in increased urinary sodium. Such control of sodium reabsorption independently of filtered load, and unrelated to the action of sodium-retaining steroids, has been termed *third factor*. In the thin descending segment of the loop of Henle, water is removed passively by the hypertonic medullary interstitium, and luminal sodium concentration rises markedly. In the thin ascending portion of the loop, sodium diffuses out and urea diffuses in, both passively.

Sodium and chloride are reabsorbed actively without water in the thick ascending limb of the loop of Henle, a water-impermeable segment of the nephron. (The positive luminal potential in this segment was formerly thought to represent active removal of chloride. It is now thought to represent (1) active reabsorption of sodium dependent upon Na-K-ATPase, each sodium ion entering the cell by cotransport with chloride, and (2) back-diffusion of sodium through the tight junction between cells.) The reabsorption of sodium in this segment is partially dependent upon the action of aldosterone. This reabsorption of sodium and chloride generates the hypertonicity of the medullary interstitium and, conversely, hypotonicity of the luminal fluid. The water which has thus been freed of solute is termed *free water*. The distal convoluted tubule is only slightly permeable to water, the reabsorption of which is not potentiated by antidiuretic hormone in this segment. Accordingly, most of the dilute luminal contents enter the collecting

ducts. Some sodium is reabsorbed actively, and the luminal potential changes from a positive one at the macula densa to a negative one.

The collecting ducts further reabsorb sodium, actively, by a mechanism potentiated by aldosterone (which increases Na-K-ATPase activity in this segment). This creates a further negative intraluminal potential. Potassium and hydrogen ion secretion down this potential are thus stimulated by luminal sodium and its reabsorption. If antidiuretic hormone is present in the circulation, the collecting ducts become water-permeable, the free water entering them and additional water is reabsorbed, and the final urine is concentrated. The degree of concentration achieved is directly dependent upon the degree of hypertonicity of the renal medulla: if this is completely lost by washout (see below), concentrating ability is lost.

ALDOSTERONE. The adrenal cortical sodium-retaining steroids, of which aldosterone is the most potent, increase sodium transport, as first shown in the isolated toad bladder. Aldosterone secretion is stimulated by hypovolemia and hypotension, both effects largely indirect, dependent upon release of renin from the renal juxtaglomerular apparatus, and the consequent generation of angiotensin II in the plasma. It is stimulated directly by an increase of ambient and/or cellular potassium concentration.

The action of aldosterone on sodium transport in the proximal tubule and in the thick ascending limb of the loop of Henle has been noted. Quantitatively, the most important site of action of aldosterone is in the collecting ducts. Here, sodium reabsorption produces or increases negativity of the intraluminal potential. Secretion of potassium and of hydrogen ions into the lumen down this potential gradient is in part dependent upon the delivery of sodium to this site and upon its active reabsorption under the influence of aldosterone. Only in the proximal tubule does reabsorption of sodium as stimulated by aldosterone induce reabsorption of water, itself a passive process.

THIRD FACTORS. When experimental conditions maintain both a constant GFR and constant aldosterone activity, expansion of plasma or extracellular fluid volume still results in a natriuresis. Therefore, there must be a third factor (or factors) that determines, in part, sodium excretion. Possible explanations for this natriuresis attributed to a third factor are: a natriuretic hormone, effects of expanding blood volume to increase peritubular or renal interstitial pressure, effects of the fluid used for expansion to lower blood viscosity or to lower the oncotic pressure in peritubular vessels, or redistribution of renal blood from long, corticomedullary to short, cortical nephrons. Attempts have been made to demonstrate an inhibitor of sodium transport in the plasma of saline-expanded animals by a variety of experimental techniques. A natriuretic hormonal factor, possibly of cerebral origin, that acts on the proximal tubule, has not been identified definitively; however, delayed disappearance of saline from micropunctured tubules and increased free water clearance following saline loading suggest inhibition of proximal sodium transport. Results of cross-circulation experiments with extracellular volume of one partner expanded are consistent with a natriuretic hormone, perhaps of renal origin, which acts on the distal tubule. Such a factor may oppose the action of aldosterone or desoxycorticosterone and account for the phenomenon of escape from their salt-retaining effects.

Altered sodium and water metabolism

Sodium ions and their associated anions are the solutes that account for most of the osmolality of extracellular fluid. Loss of sodium from the body isosmotically causes a proportionate decrease in extracellular fluid volume; retention of sodium may result in a proportionate increase in extracellular fluid volume. Serum sodium concentration under these conditions remains normal. Later, compensatory mechanisms operating to restore extracellular fluid volume may superimpose changes in fluid balance and result in hyponatremia or hypernatremia.

Sodium depletion

Sodium may be lost by any of a variety of mechanisms (Table 11–1). Sodium depletion may result from urinary losses in disorders that impair tubular transport (e.g., obstructive uropathy, pyelonephritis, polycystic kidneys, multiple myeloma, polyarteritis nodosa, diuretic phase of acute renal failure, and occasionally, glomerulonephritis). Impaired bicarbonate reabsorption associated with impaired tubular ability to secrete hydrogen ion or to maintain a tubular fluid/plasma hydrogen ion gradient, resulting from congenital or acquired disease, also may lead to excessive loss of sodium with bicarbonate (renal tubular acidosis). In addition to extracellular fluid volume depletion, hyperchloremic acidosis results, causing mobilization of phosphate and calcium from bone, hypercalciuria and potentially nephrolithiasis, and release of intracellular potassium. Parathyroid hormone in excess also induces urinary loss of sodium bicarbonate, rendering the urine alkaline and resulting in systemic acidosis. In patients with renal failure, the osmotic diuresis resulting from retained urea and other nitrogenous end-products, and the increased solute load and GFR per residual nephron are associated with a reduced fractional reabsorption of sodium. When salt intake is restricted in such patients, salt depletion can occur. The lack of aldosterone and other sodium-retaining steroids in adrenal insufficiency can also cause renal salt loss. Because of failure of the distal exchange (sodium for potassium-or-hydro-

TABLE 11-1
Causes of Sodium Loss

Renal	Gastrointestinal	Skin	Drainage
Renal disease	Vomiting	Sweating	Peritoneal
Diuretic therapy	Diarrhea	Mucoviscidosis	Pleural
Osmotic diuresis	Drainage	Cystic fibrosis	Subcutaneous
Adrenal insufficiency		Burns	

gen ion) mechanism, potassium retention with hyperkalemia and hydrogen ion retention resulting in systemic acidosis are prominent features of salt depletion in adrenal insufficiency. Insulin directly enhances tubular reabsorption of sodium, potassium, and water. In patients with diabetes mellitus, the osmotic diuresis of glucosuria increases sodium excretion.

Most frequently, the cause of renal salt depletion is the administration of diuretics. The mercurial diuretics tend to become ineffective once hypochloremia develops, and the thiazide congeners are much less effective when extracellular fluid volume is depleted. The newer, more potent diuretics, ethacrynic acid and furosemide, may continue to induce saluresis in spite of extracellular fluid volume depletion, hyponatremia, and reduced GFR.

RENAL RESPONSE TO SODIUM DEPLETION.
Salt loss is accompanied by loss of extracellular fluid. The clinical manifestations of such volume depletion include postural hypotension, tachycardia, and tissue ischemia; severe extracellular fluid volume depletion can result in shock, coma, lactic acidosis, uremia, and hepatic failure.

Renal responses to salt depletion occur earlier than other manifestations. Through multiple mechanisms, the decreased extracellular fluid volume diminishes urinary sodium excretion. The decreased volume increases tubular sodium reabsorption by mechanisms considered under Third Factors, above. Decreased extracellular fluid volume also stimulates aldosterone secretion through the mechanism of increased

release of renin and increased angiotensin formation. As volume decreases further, renal blood flow, GFR, and filtered load of sodium decrease. Volume depletion also stimulates secretion of antidiuretic hormone, which enhances water reabsorption from the collecting ducts. The decreased filtered load of sodium may limit the capacity to generate free water in the diluting segment of the nephron and, if insufficient sodium reaches the loop of Henle, both the reabsorption of sodium to maintain the normal medullary hypertonicity and, *pari passu*, the generation of free water in the tubular lumen may be impaired, so that neither maximal dilution nor maximal concentration can be achieved. Moreover, the thirst center is stimulated in an attempt to increase fluid intake and restore volume.

If salt loss continues and the patient has access to water, this water is retained and hyponatremia may develop. The retained water is not confined to the extracellular space but is distributed in total body water. This results in dilutional expansion of cells.

When sodium is administered without water, an increase in extracellular osmolality occurs. Intracellular water shifts to extracellular sites to maintain osmotic equilibrium, thereby diluting extracellular and concentrating intracellular fluid. Thus, although sodium is confined extracellularly, it is osmotically effective throughout total body water. The amount of sodium required to restore sodium concentration to normal in the presence of hyponatremia, sodium depletion, and normal total body water is equal to the product of the total body water volume and the difference between the normal and observed sodium concentrations. For example, at a sodium concentration of 115 mEq./l. (a decrease in serum sodium concentration of 25 mEq./l.), a 70-kg. man (estimated total body water, 42 l.) would require a 1,050 mEq. of sodium (25 × 42) to raise the serum sodium concentration to normal. Of course, losses would have to be prevented or replaced simultaneously, in order to achieve the desired net sodium retention. If total body water is initially reduced, volume depletion will persist after the sodium concentration has been corrected. Repletion with isotonic saline is then necessary to replace the deficit. It is often difficult to determine the relative roles of sodium depletion and water excess in the genesis of hyponatremia. If hyponatremia results primarily from water retention, attempts to correct hyponatremia by sodium administration may cause increased extracellular fluid volume and potentially life-threatening acute pulmonary edema or cardiac failure. Sodium repletion, therefore, can be accomplished best by estimating the sodium deficit (from the product of the total body water and the difference between the normal and observed serum sodium concentrations) and replacement of a fraction of this while monitoring arterial and central venous pressures. Fluid balance may need to be adjusted independently.

Sodium retention

An excess of total body sodium results in an expanded extracellular fluid volume; this may be associated with a normal, low, or high serum sodium concentration. A clinically detectable increase in extracellular fluid volume is edema, which may be localized or generalized. Localized edema may result from diverse causes such as thermal burns, trauma, localized toxins (insect bites), and venous or lymphatic obstruction. Two pressure gradients govern the rate at which edema is formed. The first is the gradient in hydrostatic pressure between the intravascular and interstitial spaces; the second is the osmotic or oncotic pressure gradient resulting from the difference in protein concentration between plasma and interstitial fluid (i.e., the plasma oncotic pressure). At the arterial end of the capillary bed, hydrostatic pressure is sufficient to favor movement of fluid into the interstitial spaces against the oncotic pressure exerted by the plasma proteins; at the venous end of the capillary bed, the hydrostatic pressure falls below plasma oncotic pressure, so that fluid returns to the intravascular circulation. Any increase in capillary permeability or in the ratio of hydrostatic to osmotic pressure can cause movement of plasma fluid into the interstitium. When localized edema occurs, there is necessarily a transient decrease in extracellular fluid and plasma volumes throughout the remainder of the body. This initiates a renal response that causes sodium retention, restoring normal volumes outside the edematous area; thus, even localized causes of edema should be considered as having generalized effects.

CONGESTIVE HEART FAILURE. This is a term used to describe the results of a sustained lowering of cardiac output, albeit of a relatively minor degree, that is, compatible with continued survival. The lowered output is sensed by a number of mechanisms, not all of which have been elucidated. (1) Kidney perfusion decreases as the volume and pressure of the blood in the renal arteries are lowered. In the hypertensive patient with cardiac failure, the pressure in the renal artery may be actually high. However, the total peripheral resistance, which is excessively high (thus producing excessive afterload on the heart, and the cardiac failure) involves also the renal arterioles, so that renal circulation is compromised as in the normotensive patient with failure. This leads to secretion of renin and formation of angiotension I, leading in turn to increased formation of angiotension II, and thus to increased production of aldosterone. The aldosterone augments sodium reabsorption, as discussed above, tending to increase body fluid volume. (2) An increase in efferent autonomic activity from medullary centers results in vasoconstriction by means of α-receptors in arterioles, and stimulation of cardiac rate and contractility by means of β-$_1$ receptors. Important afferent stimuli to such activity result from *under*filling (with decreased cardiac output) of the great vessels and atria, through low-pressure receptors and lowering of arterial mean or pulse pressure in aorta and carotid arteries. Afferent pathways include the vagus and the glossopharyngeal nerves. If the nature of the cardiac failure precludes a full compensation for the decreased output by these sympathetic stimuli together with an increase in the volume of and pressure in central vessels (as it generally does), retention of sodium and water and the peripheral vasoconstriction continue, and venous volume continues to rise, resulting in pulmonary and peripheral edema.

Under such conditions of decreased renal blood flow, there may be a decrease in filtration rate, inducing sodium reabsorption as *fractional* tubular sodium reabsorption increases. Even without a measurable decrease in GFR, proximal tubular reabsorption of sodium and water are often markedly elevated, limiting the ability of the kidney to form and excrete free water—which, we have seen, are distal tubular functions. In addition, a decrease in arterial pressure and total blood volume may stimulate vasopressin secretion, providing another mechanism for inability to excrete free water.

In so-called forward failure—such as occurs with the heart disease associated with beriberi—cardiac output does not fall, but rather rises, in the face of a defect in or excess of tissue metabolism, such that even an increase in oxygen supply cannot match demands. Edema is prominent, with retention of salt and water, of course outside of the circulation. Renal vasoconstriction occurs, and all the mechanisms for re-

nal sodium retention are activated. Clearly these conditions require repair of the metabolic defect rather than treatment directed at cardiac function.

HEPATIC DISEASE. This may produce increased portal venous pressure, thereby causing increased formation of intraperitoneal fluid (ascites) and contraction of plasma volume. As liver failure progresses, other mechanisms for generalized edema come into play. Inactivation of antidiuretic hormone and of aldosterone by the liver is less effective. Proximal tubular sodium reabsorption may be further increased (see above), causing more sodium retention and limiting the capacity to form free water. Impaired protein synthesis and loss of protein into ascitic fluid lead to hypoproteinemia. The low protein concentration causes a shift of plasma water into interstitial spaces, leading to generalized edema.

HYPOPROTEINEMIA. This may also result from nutritional deficiency or from protein loss, as in the nephrotic syndrome or in protein-losing enteropathy. Increased degradation of protein and impaired hepatic synthesis may play secondary roles in the nephrotic syndrome. Plasma protein depletion reduces oncotic pressure and increases fluid loss through the capillaries into the interstitium. This decreases plasma volume, and the kidney responds by retaining salt and water. Since the retained sodium and water are not confined to the plasma volume, edema is worsened. Under these circumstances, even a further expansion of the plasma volume by infusion of a colloid to increase plasma oncotic pressure may promote diuresis.

RENAL FAILURE. This is characterized by an inability both to conserve sodium and to excrete maximal quantities of salt. Unless the renal circulation is impaired, renal failure must be severe before sodium retention is important clinically. Many of the causes of renal failure, however, involve the renal vasculature and the glomeruli and impair sodium excretion early in their course. Sodium retention may lead to congestive heart failure, decreased renal perfusion, and a vicious cycle that further aggravates sodium retention and edema.

Impaired tubular reabsorption of sodium

Diseases that lead to reduced glomerular filtration and enhanced tubular reabsorption of sodium and water cause inappropriate increases in body fluid volumes. Accumulation of sodium and water in the proportions found in plasma expands primarily the interstitial fluid space, causing edema. One of the major therapeutic achievements over the last several decades has been the development of pharmacological agents that block tubular sodium reabsorption and thereby reduce edema formation.

Saluresis and diuresis may also be achieved by improving cardiovascular and renal function. Thus, augmentation of cardiac output by digitalis, bed rest, salt deprivation, or corrective cardiac surgery may reverse salt and water retention secondary to congestive heart failure. A low plasma volume resulting from such causes as nephrotic syndrome or cirrhosis of the liver with rapidly forming ascites will cause sodium retention. Expansion of plasma volume with blood, albumin, dextran, or other plasma expanders or by mobilization of retained fluid by the wrapping or elevation of edematous extremities leads to saluresis. The use of vasodilators such as aminophylline may also achieve diuresis, resulting from an increased filtration rate, with increased delivery of salt and water to distal tubules.

Therapeutic diuresis is frequently achieved with drugs that inhibit tubular sodium transport. Osmotic diuresis may result from administration of a nonreabsorbable solute such as mannitol, by elevation of the filtered load of a substance such as glucose to exceed the tubular capacity for its reabsorption, or by administration of a nonreabsorbable anion such as sulfate. Because the filtrate remains isosmotic in the lumen of the proximal tubule, the addition of a nonreabsorbable solute obligates water to remain within the lumen to maintain isosmolality. The increased volume augments flow rate of tubular contents, dilates tubules, and lowers the tubular fluid concentration of sodium and other solutes, thereby increasing the gradient against which these solutes must be absorbed. These factors all contribute to the saluresis.

Acetazolamide inhibits sodium reabsorption by blocking carbonic anhydrase activity. This is one of the few oral saluretics that has a proximal tubular inhibitory effect; most inhibit predominantly distal tubular sodium transport. Chlorothiazide and its numerous congeners inhibit active sodium chloride reabsorption in the distal convoluted tubule beyond the macula densa; a concomitant loss of potassium may result from increased distal tubular flow rate and an increase of sodium-dependent electronegativity of the distal tubular lumen. In addition, a mild sodium bicarbonate diuresis results, because thiazides have weak carbonic anhydrase inhibitory effects. Ethacrynic acid and furosemide are two potent saluretics that block sodium transport primarily in the thick ascending limb of the loop of Henle. Mercurial diuretics also cause saluresis by inhibiting distal renal tubular sodium reabsorption. Spironolactone prevents aldosterone-dependent potassium secretion in the collecting ducts, and triamterene prevents non-aldosterone-dependent potassium secretion in distal convoluted tubules and collecting ducts.

Hyponatremia

Hyponatremia, that is, low sodium concentration in plasma, indicates, with rare exceptions, a low sodium concentration in plasma water and thus in extracellular water. These exceptions are found in the presence of large quantities of solutes such as glucose or mannitol or of volume-occupying colloid, such as lipids or proteins. Under these exceptional circumstances, plasma sodium may be low even when the sodium concentration in plasma water is normal or, indeed, high.

Decreased plasma sodium concentration may result from sodium loss in excess of osmotically obligated water or sodium depletion with retention of water in an attempt to maintain extracellular volume (as described above); it may also result from primary retention of water in excess of sodium (dilutional hyponatremia).

Dilutional hyponatremia

The clinical picture of hyponatremia secondary to water retention differs considerably from that of salt depletion. Coma and convulsions are common and are, at least in part, the result of cerebral edema; no evidence of ischemia of organs such as the skin and kidneys is detectable.

Dilutional hyponatremia may be caused by an extremely high fluid intake, by severe renal failure, by decreased renal perfusion, by excessive proximal tubular reabsorption of sodium and water as in cardiac failure or cirrhosis, or by inappropriate secretion or administration of antidiuretic hormone. In a normal individual, dilutional hyponatremia does not occur unless water intake exceeds 20 l./day. With only a moderate decrease in renal function, the capacity to excrete water is adequate to handle the usual fluid intake. However, severe renal failure and, obviously, anuria or pathological oliguria limit water tolerance, and dilutional hyponatremia is common under these circumstances. Decreased renal perfusion diminishes sodium and water excretion and impairs the capacity to excrete water.

A marked decrease in renal blood flow may occur from causes such as congestive heart failure, myxedema, vasoconstrictor drugs, and aortic or renal vascular disease. These disorders can be complicated by water retention and dilutional hyponatremia, which often is superimposed upon actual sodium *retention* (Fig. 11–4). Hyponatremia may also result from dilution of the extra-

FIG. 11–4. The normal body water compartments may be expanded (*a*) isotonically (extracellular fluid), causing edema, or (*b*) hypotonically, causing dilutional hyponatremia. In this example the added free water distributes itself proportionately to the abnormal volumes of each compartment and dilutes extracellular sodium and intracellular potassium to 117 mEq./1.

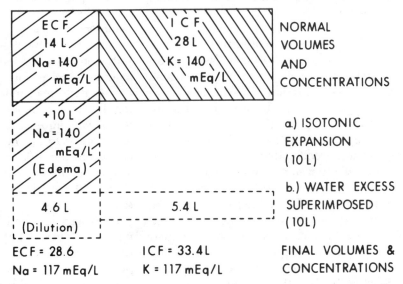

cellular fluid by cellular water. This may develop when the osmotic gradient favors the movement of water out of cells. For example, when cells become depleted of potassium, cellular osmolality must decrease. Osmotic equilibrium then demands a shift of sodium into the cell or of water out of the cell. Both mechanisms appear to be operative in the hyponatremia of potassium depletion. An increase in the concentration of extracellular solutes such as mannitol or glucose causes a similar osmotic gradient favoring the movement of water from the cells into the extracellular fluid compartment.

Often hyponatremia results from multiple causes. For example, patients with diabetes mellitus may develop hyperglycemia, causing a shift of water from the cells, with expansion of extracellular fluid volume. The hyperglycemia leads to glycosuria, which promotes an osmotic diuresis and renal salt loss. Complications, such as shock, and the therapeutic use of vasoconstrictor drugs may lead to renal ischemia and an inability to excrete free water. Potassium depletion accompanying diabetic acidosis may also enhance the development of hyponatremia. It is therefore important to identify all possible mechanisms causing the change in sodium concentration in order to correct the underlying abnormalities.

ANTIDIURETIC HORMONE (ADH). An arginine vasopressin in man, ADH increases the permeability of the collecting tubules to water, causing concentration of the urine. In central, or pituitary diabetes insipidus, the supraopticohypophyseal system is unable to secrete or release ADH and the distal tubule and collecting ducts are relatively impermeable to water. Polyuria, with urinary osmolality as low as 30 to 50 mOsm./kg. of water, results. This leads to dehydration, increased plasma osmolality, and thirst. Normally, little of the free water generated by sodium reabsoption in the ascending limb of Henle's loop is reabsorbed as a result of the osmotic gradient between the

dilute intratubular fluid in the distal convoluted tubules and the isotonic interstitial fluid of the renal cortex, because the distal convoluted tubule is poorly permeable to water. Further concentration occurs as water crosses the collecting duct epithelium in the presence of ADH in response to the osmotic gradient presented by the hypertonic medullary interstitium. Sustained increased delivery of dilute fluid to the collecting ducts may allow sufficient reabsorption of water to wash out medullary hypertonicity. The response to vasopressin, thus, may be submaximal until medullary hypertonicity is restored. In so-called nephrogenic diabetes insipidus, the tubular cells are unresponsive to ADH, and urinary osmolality remains minimal, despite plasma hyperosmolality resulting from dehydration.

The effect of aqueous vasopressin given intravenously lasts about 30 minutes. Under maximal ADH stimulation, urinary osmolality often exceeds 1,000 mOsm./kg. of water. If fluid is administered while ADH is exerting its effect, dilutional hyponatremia results. Since ADH is metabolized by the liver, impaired metabolism may potentiate water retention in hepatic failure.

Dehydration, with loss of water in excess of solute and a resultant increase in osmolality of plasma, is the normal stimulus for ADH release. Experimentally, intracerebral osmoreceptor cells also can be dehydrated and stimulated by direct intracarotid or intrahypothalamic infusion of a hypertonic solution such as sodium chloride; this indicates that the osmolality of blood perfusing the supraopticohypophyseal system is the usual stimulus for ADH release. There also appear to be volume or "stretch" receptors in the atria, and very likely in the arterial system, that are stimulated to cause ADH release when plasma volume is decreased or cardiac output is reduced. ADH is also released by stimuli such as pain and administration of barbiturates, nicotine, or morphine, which usually limit postoperative water tolerance. Any one of these stimuli is sufficient to ensure continued release of ADH in spite of a low serum osmolality—that is, SIADH. In pituitary or adrenal insufficiency, impaired diluting ability may reflect ADH release, possibly associated with low plasma cortisol concentrations and decreased extracellular fluid volume. Thus, the familiar hyponatremia found in Addison's disease is an example of SIADH. Cortisol also may have a direct effect of decreasing distal tubular and collecting duct permeability to water.

SYNDROMES OF INAPPROPRIATE ADH SECRETION. A chronic form of inappropriate ADH secretion occurs in some patients with intrathoracic disease (mediastinal tumors, tuberculosis, pulmonary carcinoma) or intracerebral disease. Sometimes the source of autonomous ADH secretion is a tumor itself, such as oat-cell or small-cell pulmonary carcinoma; often the normal physiological source is inappropriately stimulated. In such instances, where persistent ADH secretion occurs with low plasma osmolality, the term SIADH is applied. Inappropriate ("inappropriate," that is, for the accompanying low plasma osmolality) ADH activity leads to water retention, increased total body water, cellular swelling with associated cerebral edema, and hyponatremia (water intoxication). Inhibition of ADH secretion by alcohol is not useful clinically, except to identify that the source of ADH is physiological rather than an autonomous tumor. Expansion of the extracellular fluid volume resulting from ADH-mediated water retention produces hemodilution and hyponatremia, decreased aldosterone secretion, stimulation of third factor,—that is, decreased proximal tubular sodium reabsorption—and significant natriuresis. High urinary sodium concentration indicates mainly excretion of the sodium intake in a small volume; urinary sodium may exceed sodium intake, thus leading to negative sodium balance. The hyponatremia cannot be explained fully by either renal sodium loss or hemodilution resulting from retention of water. Under these circumstances, hypo-

natremia is generally not corrected by sodium administration: this merely increases sodium excretion. To correct the hyponatremia, water intake must be reduced sufficiently to achieve, with continued insensible loss, a negative fluid balance.

Patients with SIADH present with a low serum sodium concentration and serum osmolality and frequently with a urinary osmolality above the osmolality of plasma. The osmolality of the urine, however, may be below the osmolality of plasma, but it must be above a minimum of about 50 mOsm./kg. for a diagnosis of SIADH. Normal or low values for blood urea nitrogen and normal creatinine clearances help to exclude significant salt depletion. Following a water load, a substantial increase in free water clearance with a resultant dilution of the urine caused by inhibition of ADH release (resulting in turn from the decrease in plasma osmolality) excludes SIADH.

Failure of a water load to induce water diuresis, however, may indicate preexisting hypovolemia or increased proximal sodium reabsorption. Abnormally high proximal tubular sodium reabsorption is found in cirrhosis of the liver with ascites and in cardiac failure. This results in a limitation in delivery of sodium and water to the thick ascending limb of Henle's loop, and thus a limitation of free water formation and excretion. Thus, cirrhosis with ascites and cardiac failure may produce hyponatremia, inappropriate concentration of the urine, and inability to excrete a water load normally. In these conditions, only a measurement of plasma ADH will serve to establish the mechanism: if plasma ADH is normal or high (not suppressed by the hyponatremia), SIADH is present; if it is low, excessive proximal tubular reabsorption is the probable mechanism.

Hypernatremia

An increase in total body sodium usually is associated with edema, since water must be retained in order to maintain a normal sodium concentration in extracellular fluid. If excessive sodium is administered when there is limited access to water, or if water is lost without sodium, hypernatremia will occur. Sodium excesses have occurred when infants' formulas were made incorrectly. Hypernatremia resulting from loss of dilute fluids—i.e., fluids containing free water—is more frequent. Urinary water loss occurs in congenital and acquired forms of pituitary and nephrogenic diabetes insipidus. Excessive sweating or osmotic diuresis also will produce hypernatremia if water replacement is insufficient.

The most common cause of hypernatremia is the use of high-protein, high-sodium tube feedings for elderly or comatose patients. A high-protein diet causes an increase in urea production and thus, when renal function is not changed, an increase in urea excretion. This produces an osmotic diuresis, with water loss in relative excess of sodium loss, dehydration, and hypernatremia. Another common cause of hypernatremia is peritoneal dialysis with hypertonic glucose. This causes a proportionately greater removal of water than of sodium, resulting in hypernatremia. However, most causes of hypernatremia are associated with urinary salt loss that is associated with proportionately greater water loss. Hypernatremia, therefore, usually indicates dehydration and does not imply an excess of total body sodium. Primary aldosteronism may cause relatively mild hypernatremia. In this case, the sequence of events is probably as follows: (1) aldosterone-induced potassium depletion (and hypokalemia); (2) renal damage, such as to produce renal diabetes insipidus; (3) excretion of dilute urine; and thus (4) hypernatremia.

Annotated references

Bartter, F. C.: The syndrome of inappropriate secretion of antidiuretic hormone (SIADH). Disease-a-Month 2:47, 1973. (An extensive review of the literature to 1973.)

Bartter, F. C., and Fourman, P.: The different effects of aldosterone-like steroids and hydrocortisone-like steroids on urinary excretion of potassium and acid. Metabolism 11:6, 1962. (Mechanisms of kaliuresis are distinguished. An early critique of the collective term "mineralocorticoid.")

Bartter, F. C., and Schwartz, W. B.: The syndrome of inappropriate secretion of antidiuretic hormone. Am. J. Med. 42:790, 1967. (Clinical features and pathophysiology of this interesting syndrome are outlined.)

Bell, N. H., Schedl, H. P, and Bartter, F. C.: An explanation for abnormal water retention and hypo-osmolality in congestive heart failure. Am. J. Med. 36:351, 1964.

de Wardener, H. D.: Control of sodium reabsorption. Br. Med. J. 3:611, 1969. (A lucid presentation of factors that regulate sodium excretion, including the effects of volume expansion.)

de Wardener, H. D., Mills, I. H., Clapham, W. F., and Hayter, C. J.: Studies on the efferent mechanisms of the sodium diuresis which follows the administration of intravenous saline in the dog. Clin. Sci. 21:249, 1961. (The "discovery" of third factor I.)

Gross, J. B., Iwai, M., and Kokko, J. P.: A functional comparison of the cortical collecting tubule and the distal convoluted tubule. J. Clin. Invest. 55:1284, 1975. (The localization of vasopressin function in distal and collecting tubules.)

Gross, J. B., and Kokko, J.: Effects of aldosterone and potassium-sparing diuretics on electrical potential differences across the distal nephron. J. Clin. Invest. 59:82, 1977. (The localization of aldosterone function in collecting ducts.)

Katz, A. I., and Epstein, F. H.: Physiologic role of sodium-potassium activated adenosine triphosphate in the transport of cations across biologic membranes. N. Engl. J. Med. 278:253, 1968. (A basic look at the active transport system for sodium and its energy substrate.)

Leaf, A., and Cotran, R. S. Renal Pathophysiology, Ed 2. New York, Oxford University Press, 1980. 2nd edition. (A lucid summary of known and unknown factors in the control of renal salt and water excretion in health and disease.)

Mills, I. H., de Wardener, H. D., Hayter, C. J., and Clapham, W. F.: Studies on the afferent mechanism of the sodium chloride diuresis which follows intravenous saline in the dog. Clin. Sci. 21:259, 1961. (The "discovery" of third factor II.)

Oparil, S., and Haber, E.: The renin-angiotensin system. N. Engl. J. Med. 291:389–401, 446–457, 1974. (A critical, complete review with extensive bibliography.)

Orloff, J., Walser, M., Kennedy, T. J., Jr., and Bartter, F. C.: Hyponatremia. Circulation 19:284, 1959. (An analysis of the significance and pathogenesis of hyponatremia, with particular reference to the role of antidiuretic hormone).

Ross, E. J., and Christie, B. M.: Hypernatremia. Medicine, 48:441, 1969. (A thorough review, with particular details about the causes of hypernatremia, and an extensive bibliography.)

Schedl, H. P., and Bartter, F. C.: An explanation for and experimental correction of the abnormal water diuresis in cirrhosis. J. Clin. Invest. *39*:248, 1960. (The identification of the proximal tubular sodium hyperabsorption syndromes.)

Schwartz, W. B., Bennett, W., Curelop, S., and Bartter, F. C.: A syndrome of renal sodium loss and hyponatremia probably resulting from inappropriate secretion of antidiuretic hormone. Am. J. Med. *33*:529, 1957. (The original description of this pervasive syndrome.)

Seldin, D. W., Eknoyan, G., Suki, W. N., and Rector, F. C., Jr.: Localization of diuretic action from the pattern of water and electrolyte excretion. Ann. N. Y. Acad. Sci. *139*:328, 1966. (An early "dissection" of the tubule from indirect evidence.)

Verney, E. B.: Antidiuretic hormone and the factors which determine its release. Proc. Roy. Soc. Lond. B. *135*:25, 1947–1948. (The classic definition of the mechanisms for control of ADH secretion.)

12

Maintenance of Body Fluid Potassium, Calcium, Magnesium, and Phosphorus

Christian E. Kaufman, M.D.
Arnold J. Felsenfeld, M.D.
Jerry B. Vannatta, M.D.
Robert Whang, M.D.
Francisco Llach, M.D.

The maintenance of homeostasis of intracellular electrolytes such as potassium, calcium, magnesium, and phosphate is subject to a very elaborate system of controls, which are still poorly understood; such maintenance is essential for normal cell function. Similarly, concentration of these ions in the extracellular compartment plays an essential role in their intracellular concentration. In addition, the kidney plays an important role in the regulation of these ions. Transcellular shifts of these ions both at the renal and cell membrane levels are of paramount importance.

This chapter concerns the normal metabolism of these ions. In addition, the pathogenesis of various abnormalities of these ions is reviewed. Finally, the biological effects and clinical manifestations observed with alterations in potassium, calcium, magnesium, and phosphate metabolism are also examined.

Potassium

Normal potassium metabolism

Potassium is the most abundant intracellular cation. Normal concentration is 120 to 160 mEq./l. of cell water, and total body stores approximate 3400 to 4200 mEq. in healthy adults. In contrast, its usual concentration of 4 to 5 mEq./l. in extracellular fluid represents a total of less than 70 mEq. in a typical 70-kg man. This difference in potassium concentration between intracellular and extracellular fluid is essential in maintaining the intracellular electronegativity of about 70 mV that characterizes healthy cells. Potassium is also essential for the function of numerous enzymes and, together with its attendant anions, also provides the majority of intracellular osmotic solute, thus regulating cell volume.

The concentration of potassium in the extracellular fluid and the quantity of potassium within cells are closely regulated by two major homeostatic systems. First, the skeletal muscles and liver provide internal homeostasis by exchanging potassium between the intracellular and extracellular pools, stabilizing the concentration in the extracellular fluid. Second, urinary excretion is adjusted to maintain potassium balance in spite of wide variations in potassium intake.

INTERNAL POTASSIUM HOMEOSTASIS. Uptake and release of potassium by skeletal muscle and liver is essential in maintaining a relatively stable concentration of potassium in extracellular fluid. Normally, at least 50% of the potassium content of a meal is taken up by the tissues. In contrast, renal potassium excretion responds too slowly to prevent significant hyperkalemia following a potassium-rich meal. Likewise, were it not for the release of potassium from cellular stores, the extracellular potassium concentration would decline rapidly when intake is curtailed. Physiological factors that are known to influence the transcellular distribution of potassium include insulin, aldosterone and the adrenergic nervous systems. Both insulin and aldosterone are secreted in response to hyperkalemia and both enhance the cellular uptake of potassium. Likewise, the β-adrenergic system regulates potassium uptake by liver and skeletal muscle and the α-adrenergic system regulates the release of potassium into the extracellular fluid.

URINARY POTASSIUM EXCRETION. The usual dietary intake of potassium is 70 to 100 mEq./day. About 90 per cent of this is absorbed into body fluids, and therefore a similar amount must be excreted by the kidneys each day to maintain balance. Under most circumstances, potassium filtered at the glomerulus is almost totally reabsorbed by the proximal tubule and loop of Henle. The cells of the distal tubule and collecting duct secrete potassium into the tubular lumen and thereby provide the major mechanisms for urinary potassium excretion. Factors that enhance this secretory function include aldosterone, delivery of increased quantities of sodium to the distal tubule, urine flow rate, and the concentration of potassium in the cells of the distal nephron.

Hyperkalemia

Hyperkalemia is defined as a serum potassium concentration of greater than 5.0 to 5.5 mEq./l. However, early in the evaluation of a patient with hyperkalemia, one should consider pseudohyperkalemia. This term designates an elevated serum potassium concentration that results from the cellular release of potassium *in vitro*. It most commonly results from hemolysis but may also occur when extreme thrombocytosis or leukocytosis are present. Pseudohyperkalemia is easily identified by simultaneously measuring plasma and serum potassium levels. Usually there is little difference, but in pseudohyperkalemia, the plasma levels remain normal and only the serum value is elevated.

PATHOGENESIS OF HYPERKALEMIA. True hyperkalemia can result from either excessive intake, impaired excretion, or a net shift of potassium out of cells. Often more than one of these basic processes operates simultaneously (see list below).

EXCESSIVE POTASSIUM INTAKE. Large oral loads of potassium (100 to 200 mEq.) can raise transiently the serum levels to over 7 mEq./l. in apparently normal individuals. In contrast, when those factors controlling cellular uptake and urinary excretion of potassium are intact, *sustained* hyperkalemia rarely results from excessive potassium intake. However, if these homeostatic mechanisms are impaired, the rate of potassium entry into the extracellular fluid from exogenous sources becomes an important determinant of the plasma concentration. Hence, under certain circumstances each of the potential sources of

---

Mechanisms of Hyperkalemia

Exogenous load
  Oral potassium supplements
  "Salt substitutes"
  Intravenous potassium therapy
  Potassium-containing drugs
  Blood transfusions
  Geophagia
Transcellular shift
  Acidosis
  Hypoaldosteronism
  Insulin deficiency
  Alpha-adrenergic activity
  Beta-adrenergic insufficiency
  Hyperosmolality (hyperglycemia)
  Diffuse cellular injury
  Drugs (succinylcholine, arginine HCl)
  Digitalis poisoning
  Familial periodic paralysis
Impaired excretion
  Renal failure
  Hypoaldosteronism
  Potassium-sparing diuretics
  Defect in potassium secretion

---

potassium listed above may be important in the pathogenesis of hyperkalemia.

TRANSCELLULAR SHIFTS OF POTASSIUM. Movement of relatively small quantities of potassium out of cells can affect profoundly the concentration of potassium in the extracellular fluid. For example, in a 70 kg-man, a shift of 56 mEq of K^+ into the extracellular space would increase the serum concentration from 4.0 mEq./l. to 8.0 mEq./l. and reduce the transcellular concentration gradient to one-half normal. Factors that can influence net transcellular potassium flux are listed above. With the exception of changes in adrenergic activity, any of these factors alone may be sufficient to cause frank hyperkalemia. When excretion is impaired or the rate of potassium intake is large, these factors can readily contribute to the development of significant hyperkalemia.

IMPAIRED URINARY EXCRETION. Patients with acute oliguric renal failure are particularly predisposed to develop hyperkale-

mia. In this situation, the low urine flow rate and impaired distal tubule function markedly limit urinary potassium excretion. Hyperkalemia is also common in non-oliguric acute renal failure, but is usually less severe and more easily managed. With chronic renal failure, significant hyperkalemia is much less common. The maintenance of normal or increased urine flow rate in chronic renal failure is probably an important factor. Also, hyperaldosteronism and poorly understood adaptations occur and allow each remaining nephron to excrete a quantity of potassium that is greater than normal. In addition with severe renal failure, dietary potassium intake may be reduced and intestinal losses are often increased. Hence, significant hyperkalemia should not be attributed to chronic renal failure without a thorough search for each of the other factors listed at left, above. In particular, hyporeninemic hypoaldosteronism should be excluded, since patients with a variety of renal diseases (especially diabetic nephropathy) may have this syndrome.

BIOLOGICAL EFFECTS OF HYPERKALEMIA. Hyperkalemia is not associated with a concomitant increase in the intracellular concentration of potassium. In fact, to the contrary, depletion of intracellular potassium often coexists. Therefore, hyperkalemia reduces the transcellular concentration gradient for potassium, leading to depolarization of both muscle cells and neurons. Additionally, in cardiac tissue, membrane conductance (permeability) for potassium is increased as a direct consequence of hyperkalemia. These changes in cell membrane properties account for the clinical effects of hyperkalemia that involve primarily cardiac tissue and neuromuscular function.

CLINICAL MANIFESTATIONS OF HYPERKALEMIA. NEUROMUSCULAR. The neuromuscular manifestations of hyperkalemia are nonspecific. Patients may complain of weakness, paresthesias, agitation and a sense of impending doom. While these features should alert the clinician to the possibility of hyperkalemia, they are uncommon except with severe and life-threatening potassium levels.

CARDIAC. The degree of cardiac toxicity is generally related to the serum potassium level but the correlation is a crude one. Other factors that influence the effect of hyperkalemia on the heart include underlying cardiac disease, the rate of rise of serum potassium, and the concentration of sodium and calcium in extracellular fluid.

FIG. 12–1. Electrocardiographic changes of hyperkalemia. *A* is a normal record. *B–F* show serial changes typical of progressively higher plasma potassium levels. (VanderArk, C. R., et al.: Electrolytes and the electrocardiogram. *In* Fisch, C. (ed.): Complex Electrocardiography, Vol. II. Philadelphia, F. A. Davis, 1973, p. 271. Reproduced with permission.)

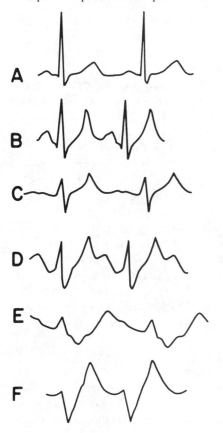

The typical electrocardiographic features of hyperkalemia are shown in Figure 12–1. Usually an increase in T-wave amplitude is the earliest electrocardiographic change. Another early change is shortening of the $Q-T_c$ interval. This may be a helpful finding since other conditions causing tall T-waves are usually associated with a prolonged $Q-T_c$ interval. T-waves and $Q-T_c$ changes may appear when the plasma potassium is only mildly elevated (i.e., 5.5 to 6.5 mEq./l. At plasma levels of 6.0 to 6.5 mEq./l., changes in atrioventricular (A-V) conduction may first appear; initially A-V conduction is accelerated, while at higher potassium levels conduction is progressively impaired. When the plasma potassium exceeds 6.5 mEq./l., progressive widening of the QRS complex occurs. The amplitude of the P-wave begins to diminish when the plasma potassium exceeds 7.0 mEq./l. As the potassium concentration increases further, the P-waves progressively flatten and broaden and the P-R interval increases. At plasma potassium levels of about 9.0 mEq./l., the P-waves may disappear entirely. With such severe hyperkalemia, S-T segment elevation occurs, leading to a sine-wave pattern on the electrocardiogram. At this point, death is imminent and may occur from either asystole or ventricular fibrillation.

Hypokalemia

PATHOGENESIS OF HYPOKALEMIA. A serum potassium concentration less than 3.5 m Eq./l. can result from either a deficit of potassium or a transient redistribution of this cation intracellularly. In chronic or moderately severe hypokalemia (serum potassium <3.0 mEq./l.) a deficit is usually present, indicating a period of either inadequate intake or excessive potassium losses (see list below).

INADEQUATE INTAKE. Since urinary excretion continues despite potassium depletion, a prolonged period with little intake can result in significant potassium deficits.

Mechanisms of Hypokalemia

Inadequate intake
Gastrointestinal loss
   Vomiting or gastric drainage
   Diarrhea
   Laxative abuse
   Ureterosigmoidostomy
Excessive urinary loss
   Diuretics
   Mineralocorticoid effect
   Tubulointerstitial renal disease
   Renal tubular acidosis
   Diabetic ketoacidosis
   Antibiotics
   Magnesium depletion
   Acute leukemia
   Bartter's syndrome
   Liddle's syndrome
Redistribution into cells
   Alkalosis
   Insulin therapy
   Treatment of megaloblastic anemia
   Periodic paralysis
   Barium poisoning

For example, 2 to 3 weeks of a potassium-free diet will reduce the plasma potassium concentration to about 3 mEq./l. (a deficit of 200 mEq. in an adult). However, such severe dietary restriction of potassium is unusual, and inadequate intake is rarely the sole cause of clinically evident potassium depletion.

EXCESSIVE GASTROINTESTINAL LOSSES. Loss of gastric juice by vomiting or nasogastric drainage is often associated with hypokalemia. Since the potassium content of gastric juice is usually less than 10 mEq./l., mechanisms other than direct loss of potassium must be involved. Most important is the metabolic alkalosis that results from the associated hydrogen ion loss. Alkalemia enhances urinary potassium excretion, which, when coupled with inadequate intake, accounts for the potassium depletion. Additionally, potassium shifts into cells in the process of buffering the alkaline extracellular fluid. In contrast, stool potassium losses from diarrhea can

account directly for potassium depletion. A large variety of bowel diseases may be responsible, but as a rule, hypokalemia does not occur unless the diarrhea is chronic or especially severe. An exception may occur with cathartic abuse: certain patients, although denying large or watery stools, nonetheless demonstrate marked potassium depletion from compulsive and often surreptitious use of laxatives. Usually, however, a careful history will identify correctly the gastrointestinal tract as the source of potassium loss. Otherwise, assistance can be gained by determining the potassium concentration in a spot urine concentration. A value of less than 10 mEq./l. implicates gastrointestinal losses. The converse is not true, since some patients with hypokalemia of brief duration or secondary to cathartic abuse may demonstrate a relatively high urine potassium (20 mEq./l.)

EXCESSIVE URINARY LOSS. If the kidneys are normal, they will respond to hypokalemia within about 1 week by decreasing urinary potassium excretion to less than 20 mEq./ 24 hours. An increase in urinary potassium in the presence of chronic hypokalemia indicates renal potassium wasting, which may be associated with a variety of conditions (see list at left, above). In addition to a careful history, the level of blood pressure usually provides the most useful clue to the mechanisms involved. The combination of untreated hypertension and hypokalemia suggests primary hyperaldosteronism, Cushing's syndrome, or aldosteronism secondary to accelerated or renovascular hypertension. And perhaps the most common cause of hypokalemia in the hypertensive patient treated with diuretics is excessive dietary sodium intake. In this situation, the secondary hyperaldosterone state facilitates potassium loss as it exchanges with the increased load of filtered sodium. The cause of renal potassium loss is usually identified by carefully considering the other conditions listed.

REDISTRIBUTION INTO CELLS. Transient hypokalemia may occur simply by movement of small quantities of potassium into cells. In acute respiratory alkalosis, for example, hypokalemia occurs quickly by intracellular movement of potassium. Likewise, if a normal person is given sufficient insulin (and glucose), a fall in serum potassium will rapidly occur. Rarely, this mechanism lowers the serum potassium to less than 3.2 mEq./l., unless potassium depletion coexists. Incorporation of potassium into erythrocytes during treatment of pernicious anemia is an analogous phenomenon. Likewise, hypokalemic periodic paralysis represents a rare situation characterized by profound hypokalemia without potassium depletion of cellular stores. A maldistribution of potassium does not explain chronic hypokalemia, since the cellular capacity for potassium is limited and factors that control the potassium concentration in the extracellular fluid establish a normal level over a period of time.

BIOLOGICAL EFFECTS OF HYPOKALEMIA. Whether the consequences of hypokalemia result from depletion of intracellular stores or from a low concentration of potassium in the extracellular fluid may not be important clinically. Both factors are probably important, and in most situations the clinical consequences of potassium depletion correlate with the serum level and the magnitude of the potassium deficit. The mechanisms by which either hypokalemia or potassium depletion impairs cell function are not entirely clear. In cardiac tissue, a decrease in membrane conductance for potassium probably underlies the repolarization abnormalities that account for the S-T segment and T-wave changes. The neuromuscular effects of hypokalemia may be related to increases in the muscle cell resting membrane potential due to an increased transmembrane potassium concentration gradient. Most of the other effects of potassium depletion are probably due to impaired enzyme functions with a subsequent decrease in synthetic rate for proteins and other intracellular macromolecules.

CLINICAL MANIFESTATIONS OF HYPOKALEMIA. Hypokalemia may be associated with a wide array of clinical manifestations (see list below). Among the most common features are electrocardiographic changes, including S-T segment depression, T-wave flattening, and the development of prominent U-waves. Hypokalemia may induce a variety of arrhythmias, including atrial and ventricular premature beats, atrial and ventricular tachycardia and ventricular fibrillation. These arrhythmias are more likely in patients with underlying cardiac disease, particularly those individuals taking digitalis; however, ectopic impulses may be induced in the absence of any apparent underlying conditions. With severe potassium depletion, myocardial dysfunction and histological changes, including fibrosis, may occur. Profound changes in skeletal muscle function and integrity may also be induced by potassium depletion. Weakness is common and on occasion may progress to frank paralysis. Rhabdomyolysis may occur, but usually it is associated with strenuous exercise as well as severe potassium depletion. The vasconstriction that also results from hypokalemia by limiting muscle blood flow probably plays an important role in the pathogenesis of such rhabdomyolysis. Impairment of intestinal motility leading to adynamic ileus is another sequela of hypokalemia. The cellular mechanism is unknown. Neuropsychiatric manifestations, including confusion and depression, may also occur but are likewise poorly understood. A defect in the urinary concentrating mechanism is a characteristic feature of potassium depletion. The mechanism involves a failure to maintain a steep solute concentration gradient in the medullary interstitium. Occasional patients demonstrate severe polyuria that suggests other mechanisms, including stimulation of thirst with resulting polydip-

Clinical Consequences of Potassium Depletion

Cardiac
    Electrocardiographic changes
    Dysrhythmias
    Myocardiopathy
Skeletal muscle
    Weakness
    Rhabdomyolysis
Smooth muscle
    Adynamic ileus
    Vasoconstriction
Neuropsychiatric disturbances
    Confusion
Renal
    Concentration defect
    Chronic renal insufficient (failure)
Metabolic
    Alkalosis
    Carbonhydrate intolerance
    Impaired protein synthesis

sia. With severe, long-standing potassium depletion, renal insufficiency and end-stage renal failure may occur. Additionally, a variety of metabolic abnormalities have been associated with hypokalemia. These include carbohydrate intolerance due to impaired insulin release and impaired protein synthesis that is related to the essential role of potassium in the function of numerous intracellular enzymes.

Calcium

Distribution and physiology
Calcium is the third most abundant ion in the human body. It has a molecular weight of 40 and a valence of $+2$. More than 99 per cent of the body's total content of calcium is located in bone. The remainder is distributed throughout the extracellular fluid and the intracellular space of soft tissues.

Normal plasma calcium is approximately 10 mg./dl. about half of which is ionized. From 40 to 45 per cent is bound to protein, chiefly albumin, and 5 to 10 per cent is complexed with other ions including bicarbonate, phosphate, and citrate. (Table 12–1). The ionized and complexed calcium will exchange freely across biological

membranes and hence is filterable at the glomerulus; the protein bound fraction is neither diffusible across biological membranes nor ultrafilterable at the glomerulus. Since many vital biochemical activities of the body are dependent upon the ionized calcium concentration, this component is regulated tightly through hormonal actions. By contrast, the protein-bound fraction may vary considerably depending on the level of plasma proteins. The plasma protein components are principally albumin and fibrinogen, and the calcium binding capability of albumin is approximately eight times that of fibrinogen. Thus, conditions affecting the albumin concentration result in alterations of total plasma calcium, but not the closely regulated ionized fraction. Albumin binding of calcium varies with ionic strength and pH but not temperature. Acidemia decreases protein binding, whereas alkalemia increases the bound fraction.

Tightly regulated ionized calcium concentration is essential for stability of several physiological processes. These include neuromuscular and cardiac function and perhaps coagulation. The action potential

TABLE 12–1
Representative Values for Plasma Calcium Fractions

	mmol./l.	Per Cent
Free ions	1.10	44
Protein bound	1.15	46
Complexed		
$CaHCO_3$	0.10	
$CaHPO_4$	0.04	
Ca Citrate	0.04	
Unidentified	0.07	
Total Complexed	0.25	10
Total diffusable*	1.35	54
Total	2.50	100

* Total Diffusible = free ions + complexed
(Parfitt, A. M., and Kleerekoper, M.: The divalent ion homeostatic system: Physiology and metabolism of calcium, phosphorus, magnesium, and bone. *In* Maxwell, M. H., and Kleeman, C. R. (eds.): Clinical Disorders of Fluid and Electrolyte Metabolism, ed. 3, p. 272. New York, McGraw-Hill, 1980. Copyright © 1980. Used with permission of McGraw-Hill Book Company.)

is elicited in a nerve fiber by a stimulus that increases sodium permeability. Restoration of the resting membrane potential is dependent upon extrusion of the sodium ion. When calcium deficiency is present, the membrane remains permeable to sodium ions, sometimes resulting in repetitive stimulation. Conversely, calcium excess decreases the permeability to sodium. Calcium is also important for muscular contraction. Calcium is released from cisternae adjacent to myofibrils producing muscular contraction, which is maintained as long as high calcium ion concentrations remain in the sarcoplasmic reticulum. With respect to cardiac tissue, calcium is also important in the actual contractile mechanism. In addition, calcium affects the cardiac membrane potential. Calcium excess produces spastic contractions while a calcium deficiency results in cardiac flaccidity. However, plasma calcium concentrations probably do not change suffi-

ciently *in vivo* to alter cardiac function. Calcium is also known to be necessary for blood coagulation. However, *in vivo*, calcium ion concentrations probably do not decrease below levels required for blood coagulation.

Calcium homeostasis depends upon the balance between dietary intake, intestinal absorption and excretion, skeletal exchange, and urinary reclamation and excretion. Dietary intake of calcium may vary considerably, although the average daily intake is approximately 1 g. of elemental calcium. Only 25 to 45 per cent of ingested calcium is generally absorbed. If 1000 mg. are ingested, approximately 400 mg. are absorbed, but simultaneously 200 mg. of calcium are excreted in the gastrointestinal secretions. Thus, 800 mg. are excreted in the feces, and the net absorption would be in the range of 33 per cent (Fig. 12–2). Intestinal calcium absorption occurs primarily in the proximal small intestine. Al-

FIG. 12–2. A schematic representation of the approximate quantities of calcium ingested, its fate in the intestine and the relative contribution of ingested and endogenously secreted calcium and total fecal calcium. (Coburn, J. W., et al.: Intestinal absorption of calcium, magnesium, and phosphorus in chronic renal insufficiency. *In* David, D. S. (ed.): Calcium Metabolism in Renal Failure and Nephrolithiasis. Somerset, N.J., John Wiley & Sons, 1977, p. 78. Reproduced with permission.)

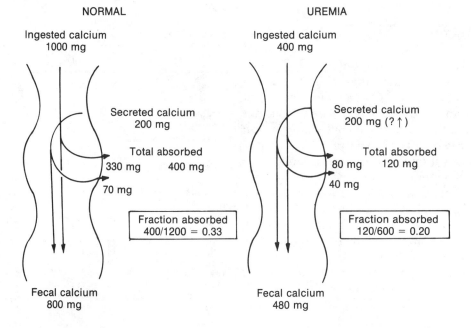

though the highest rate of calcium absorption occurs in the duodenum, the rapid transit negates the mass exchange. Thus, the principal absorption site is the jejunum.

Intestinal calcium absorption is complex and most likely reflects a combination of both passive and active transport. Movement of calcium from the intestinal lumen may occur through mucosal cells or tight junctions between cells. As discussed below, calcium transport is enhanced by vitamin D metabolites. Two proteins potentially involved in the intestinal transport of calcium have been identified. Calcium-binding protein has been identified in the surface of intestinal cells. This protein possesses a strong affinity for calcium and is concentrated in the intestinal segments with the greatest capacity for calcium absorption. Another protein, calcium ATPase, may be involved in intestinal calcium transport. It is localized in the brush border of the intestinal cells.

The skeleton is the major repository of the body's calcium. Calcium movement from the plasma to bone is bidirectional. Bone formation is a process in which minerals, largely calcium and phosphorus, are incorporated into the collagen matrix to form mineralized bone consisting of calcium hydroxyapatite. Bone resorption is a continual process during which calcium is liberated from bone and is also hormonally responsive and can be increased by parathyroid hormone and possibly certain vitamin D metabolites. The traditional view has held that increased bone resorption provides available calcium to the plasma. However, since bone formation and resorption are coupled, calcium released by bone resorption may not be available for long-term maintenance of plasma calcium. Thus, Parfitt and, more recently, Ritz, et al. have postulated that at least two independently regulated systems are present in bone, one regulating bone turnover and the other extracellular calcium homeostasis. Nonetheless, available evidence supports the concept that skeletal calcium is available to replenish plasma calcium deficits.

The exact mechanism of calcium exchange from the skeleton to the plasma has generated considerable debate. *In vitro* experiments by Neuman have suggested the bone fluid calcium concentration is less than extracellular fluid. Thus, the presence of a concentration gradient suggests that a cellular membrane must separate the compartment. Presumably the endothelial lining cells of the bone provide this property. Two possible mechanisms have been postulated and these include an active transport system in which the endothelial cells extrude calcium from the bone fluid and a passive exchange in which the membrane is affected by parathyroid hormone.

The third organ intricately involved in calcium homeostasis is the kidney. Approximately 55 to 60 per cent of total plasma calcium is filtered at the glomerulus. The filterable component of calcium consists both of the ionized fraction and the non-protein–bound complexed fraction. The magnitude of calcium handling by the kidney is illustrated by the fact that approximately 11,000 mg. of calcium are filtered daily, but, because of the tubular reabsorption of calcium, less than 300 mg. of filtered calcium is ultimately excreted in the urine. Urinary calcium excretion is affected by both hormonal and nonhormonal factors.

The intestine, skeleton, and kidney are the primary organs involved in the handling and control of body calcium. However, calcium may also be lost in sweat and hair. These losses generally are minimal, although a hot environment may increase skin losses. During pregnancy, up to 1000 mg./day of calcium are transported to the fetus and during lactation 200 to 400 mg. daily may be excreted.

Regulation of calcium

HORMONAL. PARATHYROID HORMONE. Parathyroid hormone (PTH) directly modulates calcium homeostasis through its direct effects on both the skeleton and kidney. The third organ involved in calcium regulation, the gastrointestinal tract, is indirectly affected by PTH. PTH both increases the number of osteoclasts and the activity of existing osteoclasts. Whereas the former may require 12 to 24 hours, existing osteoclasts respond to PTH by increasing metabolic activity within 15 to 30 minutes. In addition to its effect on osteoclasts, PTH has been reported by several investigators to increase the size of osteocyte lacunae. Whether the osteocytic response to PTH produces calcium resorption from bone remains a controversial issue. In a manner similar to its effects on other organs, PTH activates skeletal cyclic AMP. A PTH infusion will result in elevation of plasma calcium levels, although a small decrease may be observed shortly after the infusion of PTH.

The effect of PTH on the kidney is to increase the tubular reabsorption of calcium and hence to decrease the urinary excretion of calcium. Several studies have suggested that PTH, acting through activation of cyclic AMP, exerts its effect primarily in the distal tubule. Although PTH increases tubular reabsorption of calcium, conditions of PTH excess such as primary hyperparathyroidism may produce hypercalciuria. This apparent paradox occurs primarily because the elevated PTH levels produce hypercalcemia, which results in increased filtration of calcium. Thus, although increased tubular reabsorption occurs, hypercalciuria results because of the excessive filtered load.

PTH has no direct effect on intestinal calcium absorption. However, PTH directly influences 1,25 dihydroxycholecalciferol ($1,25(OH)_2$) production and $1,25(OH)_2D$ markedly augments intestinal calcium absorption. PTH enhances $1,25(OH)_2D$ levels through its effect on renal 1-α-hydroxylase activity and indirectly through the production of hypophosphatemia.

PTH secretion is governed by serum calcium concentration. A decrease in calcium will result in increased PTH levels and an increase in calcium will decrease PTH secretion (Fig. 12–3). A detailed discussion of PTH production and secretion is beyond the scope of this discussion and may be found elsewhere in this text (Chap. 20). Parathyroid hormone contains 84 amino acids and after secretion is metabolized to an amino terminal fragment and a carboxy terminal fragment. The amino terminal fragment is the active component of the hormone and this has been demonstrated in several *in vitro* systems. While the parathyroid gland secretion of PTH is sensitive to lowering of ionized plasma calcium, magnesium deficiency has been reported to inhibit PTH secretion. In addition, magnesium deficiency has been observed to blunt the peripheral effects of PTH. (Recent preliminary data have suggested that aluminum excess, observed in dialysis patients, may also inhibit PTH secretion.)

VITAMIN D AND ITS METABOLITES. As shown in Figure 12–4, several metabolites of vitamin D have been recognized. Most of these metabolites have been reported to possess biological activity. A complete review of vitamin D and its metabolites is beyond the scope of this section and can be found elsewhere (see Chap. 20). Distinct effects as a calcium regulating hormone have been ascribed to $1,25(OH)_2D$. Both PTH and hypophosphatemia have been shown to increase renal 1-α-hydroxylase activity and hence $1,25(OH)_2$ production. In addition, recent reports have suggested that hypocalcemia, independent of PTH and hypophosphatemia, may directly influence $1,25(OH)_2$ production.

Marked enhancement of intestinal calcium absorption is produced by $1,25(OH)_2D$. Normal subjects receiving $1,25(OH)_2D$ increased intestinal calcium absorption, decreased fecal calcium excretion, and increased urinary calcium excretion. Acting at the intestinal brush border

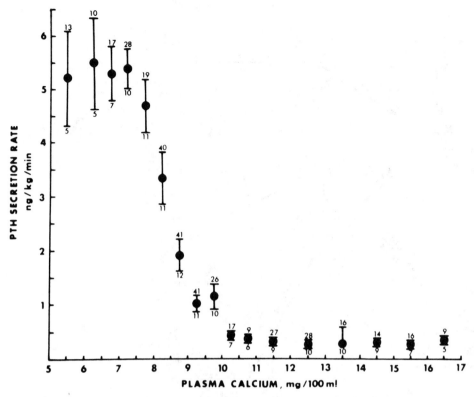

FIG. 12–3. Secretory response of bovine parathyroid glands to induced alterations of plasma calcium concentration. (Mayer, G. P., and Hurst, J. G.: Endocrinology *102*:1039, 1978. Reproduced with permission.)

membrane, $1,25(OH)_2D$ alters the properties of the microvillar surface to allow the entry of calcium into the cell. Other vitamin D metabolites enhance intestinal calcium absorption, but the magnitude of response for a comparable dose is less than $1,25(OH)_2D$.

With regard to bone, $1,25(OH)_2D_3$ independently increases calcium resorption from bone *in vitro*. In mild to severe renal failure, resistance to the calcemic action of PTH has been demonstrated, and $1,25(OH)_2D$ administration has improved the calcemic response to PTH. In all likelihood, $1,25(OH)_2D$ also participates in calcium mobilization from bone in patients with normal renal function.

Therapy with vitamin D or its metabolites with the exception of 24,25 dihydroxycholecalciferol $(24,25(OH)_2D)$, generally produces hypercalciuria, which is presumably due to increased filtered calcium secondary to increased intestinal absorption and calcium mobilization from bone. In addition, if hypercalcemia occurs, PTH levels are suppressed, resulting in decreased tubular reabsorption of calcium. Findings from early clinical studies suggested that vitamin D did not inhibit calcium transport by the nephron. Subsequent studies have provided evidence that both 25 hydroxycholecalciferol (25OHD) and $1,25(OH)_2D$ enhance tubular reabsorption of calcium.

Of considerable interest is the possible interaction between the two primary calcium regulating hormones, PTH and vitamin D metabolites. To date, conflicting data have been published. Early reports suggested that $1,25(OH)_2D$ inhibited PTH secretion, but confirmation has not been

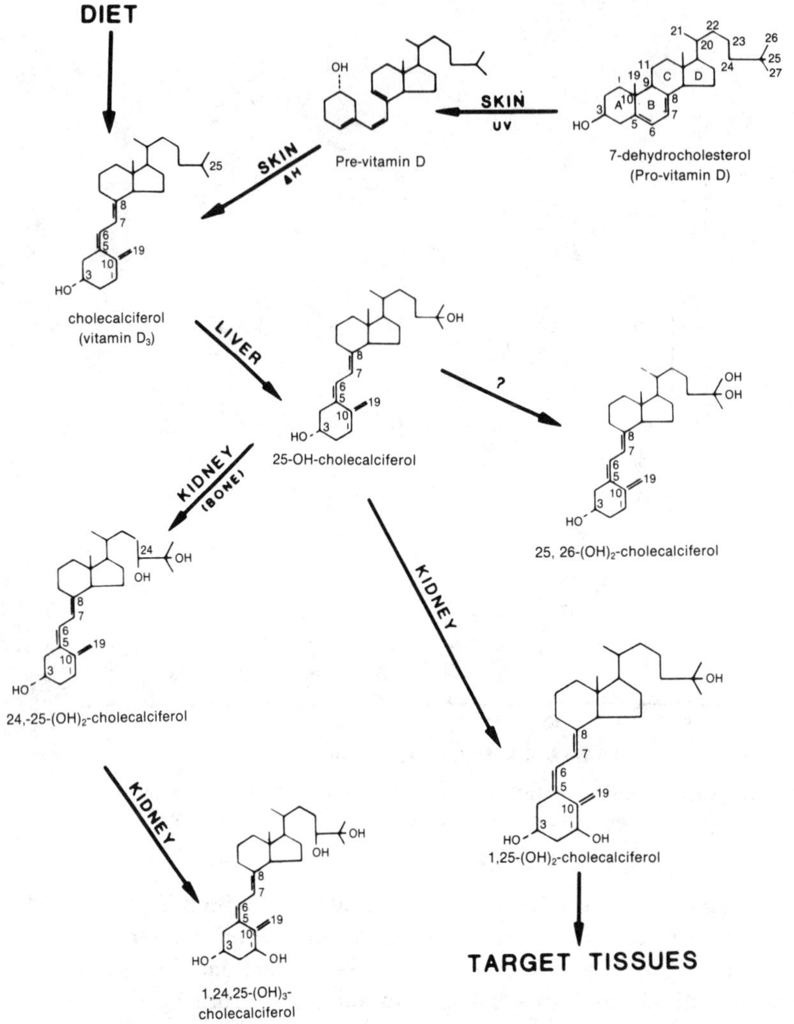

FIG. 12–4. Schematic representation of the bioactivation of vitamin D that may arise either from the skin, via 7-dehydrocholesterol, or from the diet. UV = ultraviolet light; H = increased temperature. (Coburn, J. W., and Slatopolsky, E.: Vitamin D, parathyroid hormone and renal osteodystrophy. *In* Brenner, B. M., and Rector, F. C. (eds.): The Kidney, Vol. II. Philadelphia, W. B. Saunders, 1981, p. 2215. Reproduced with permission.)

available. Findings of more recent studies have supported a role for 24,5,25(OH)$_2$D in PTH suppression.

CALCITONIN. The precise physiological role of calcitonin as a calcium regulating hormone has yet to be determined. Calcitonin is secreted by the C cells of the thyroid gland. Pharmacological doses of calcitonin are capable of inhibiting bone resorption and of decreasing renal tubular reabsorption of calcium. Calcitonin probably does not greatly influence intestinal calcium absorption.

THYROID HORMONE. Thyroid hormone also increases bone reabsorption and is capable of producing hypercalcemia. Indeed, hypercalciuria is frequently encountered in hyperthyroidism. This type of hypercalciuria may be secondary to decreased PTH levels due to hypercalcemia, an increased filtered load, or possibly a direct effect of thyroid hormone to decreased tubular reabsorption of calcium. Decreased intestinal absorption of calcium has been reported in hyperthyroidism. The possibility exists that the decreased PTH levels and the hyperphosphatemia often observed in hyperthyroidism could alter $1,25(OH)_2D$ production.

HORMONES IN ABNORMAL STATES. Hypercalcemia has been found to occur in a variety of malignancies. In many instances humoral factors have been implicated in the production of hypercalcemia. In multiple myeloma and lymphomas, a humoral agent, osteoclast activating factor, has been isolated. Osteoclast factor is a potent bone reabsorbing agent. In certain other tumors, strong evidence for a humoral etiology of hypercalcemia exists. Support for a humoral etiology includes an increased level of nephrogenous cyclic AMP, and the observation that both $1,25(OH)_2D$ and PTH levels are reduced. It is unknown whether a solitary humoral agent is responsible for the hypercalcemia in every malignancy, and the exact nature of the humoral agent or agents remain to be determined.

NONHORMONAL. PHOSPHORUS. Phosphorus and calcium metabolism are interrelated in many ways. Hyperphosphatemia decreases the plasma calcium concentration, and such a finding may account for the development of hyperparathyroidism with chronic renal failure. In addition, phosphorus therapy has been used for the treatment of hypercalcemia. Calcium-phosphorus complexes are formed with phosphorus administration, and there is increased net movement of calcium into bone. Phosphorus infusion decreases urinary calcium excretion, although the mechanism is unclear. An additional, but indirect, effect of hyperphosphatemia is decreased synthesis of $1,25(OH)_2D$.

Hypophosphatemia results in increased bone resorption. This effect may be achieved through the increased synthesis of $1,25(OH)_2D$. The increased $1,25(OH)_2D$ production will also result in increased intestinal calcium absorption. Hypophosphatemia with concomitant hypophosphaturia produces a marked reduction in the renal tubular reabsorption of calcium and hence hypercalciuria.

MAGNESIUM. Magnesium primarily exerts its effect on calcium homeostasis through parathyroid hormone. Hypomagnesemia inhibits PTH secretion and also blunts the peripheral effects of PTH. Thus, profound hypocalcemia is often observed with severe hypomagnesemia. This type of hypocalcemia does not respond to administration of intravenous calcium and resolves only after correction of the hypomagnesemia.

MISCELLANEOUS. As shown in the list below, renal calcium excretion is regulated by

---

Factors Affecting Tubular Reabsorption of Calcium

Factors that decrease tubular reabsorption
 Extracellular volume expansion
 Renal vasodilatation
 Osmotic diuresis
 Diuretic agents except thiazides
 Cardiac glycosides
 Phosphate depletion
 Acidosis
 Carbohydrate ingestion
 Alcohol ingestion
 High sodium intake
 Growth hormone
 Thyroid hormone
 Calcitonin
 Chronic mineralocorticoid effect
Factors that enhance tubular reabsorption
 Parathyroid hormone
 Phosphate loading
 Acute administration of vitamin D or its
  metabolites
 Chronic thiazide administration

(Massry, S. G., et al.: Renal handling of calcium, magnesium, and phosphate in renal failure. *In* David, D. S. (ed.): Calcium Metabolism in Renal Failure and Nephrolithiasis, p. 115. Somerset, N. J., John Wiley & Sons, 1977. Reproduced with permission.)

numerous factors. Many of these have already been discussed, particularly hormonal influences. A more detailed review of factors involved in the renal handling of calcium may be found elsewhere.

Clinical manifestations and biological effects of hyper- and hypocalcemia

HYPERCALCEMIA. Hypercalcemia produces diverse multisystemic effects. In general, the degree, duration, and rapidity of development of hypercalcemia rather than its etiology, determine the clinical manifestations.

SOFT TISSUE CALCIFICATION. Hypercalcemia, especially in the presence of an elevated serum phosphorus concentration, produces metastatic calcification of soft tissues. The greatest propensity for metastatic calcification is in the alkaline areas of the body. These include the contraluminal side of the acid-secreting epithelium, including the renal tubules and the gastric mucosa. The cornea of the eye is another preferential site for metastatic calcification. Usually the calcium crystals are deposited in the cornea in a paralimbal pattern and may be visible only by slit-lamp examination. Renal calcifications initially involve the medulla, but may enlarge to include the entire kidney. Within the renal interstitium, these deposits provoke an inflammatory response with fibrosis and scar formation.

RENAL EFFECTS. Acute hypercalcemia may induce renal failure because of vasoconstriction and tubular necrosis. Chronic hypercalcemia frequently produces renal complications. Polyuria and the inability to concentrate urine are observed early, and presumably are secondary to damage of the collecting ducts and distal tubules. The tubules become resistant to the action of vasopressin. Prolonged exposure to hypercalcemia will result in diffuse nephrocalcinosis and interstitial fibrosis. Reversibility is dependent upon the severity of existing damage.

GASTROINTESTINAL EFFECTS. Anorexia, nausea, and vomiting are frequent complications of hypercalcemia. Constipation is another common complaint. Peptic ulcer and pancreatitis have been reported to occur often with hypercalcemia. Formation of pancreatic duct calculi, activation of trypsinogen by calcium, and the occurrence of thromboarteritis secondary to intravascular coagulation have all been suggested as potential etiologies for the development of pancreatitis.

CARDIOVASCULAR EFFECTS. Cardiac effects include shortening of the systolic time intervals, sinus bradycardia, and varying degrees of A-V block. Electrocardiographic manifestations include shortening of the Q-T interval, shortened P-R interval, and QRS prolongation. Hypercalcemia also increases susceptibility of digitalis-induced arrhythmias. Hypertension is a well-recognized finding with hypercalcemia that usually reverses with correction of the calcium abnormality. Renal failure may play a contributory role, but, in all likelihood, the primary stimulus is arteriolar vasoconstriction.

NEUROMUSCULAR EFFECTS. Muscular weakness, stupor, coma, lethargy, drowsiness, confusion, memory loss, and decreased or absent reflexes have all been reported as manifestations of hypercalcemia. The neurological signs are for the most part general rather than focal, although transient aphasia and slurred speech have been observed. Electroencephalographic findings are characterized by a diffusely slow record with paroxysms of frontal dominance of 2- to 4-Hz. bursts of moderately high voltage. Resolution of both the neuromuscular and electroencephalographic findings generally occur after correction of the hypercalcemia.

HYPOCALCEMIA. Hypocalcemia primarily occurs because of the congenital deficiency of a calcium-regulating hormone such as PTH or 1,25(OH)$_2$D, or end-organ refractoriness to such a hormone. A combination of these conditions can exist in acquired conditions such as hypomagnesemia or renal failure. As previously stated, the biologically active factor is the ionized and not the total calcium. Thus, conditions such as liver disease or the nephrotic syndrome frequently have low serum albumin concentrations and hypocalcemia. However, in these states the ionized calcium concentration is usually normal and the clinical manifestations listed below are not observed.

CARDIOVASCULAR EFFECTS. Hypocalcemia produces a characteristic electrocardiographic change. The Q-T interval is prolonged as a result of delayed ventricular repolarization. Heart block as well as depressed myocardial function have been observed. Hypocalcemia may also render the myocardium resistant to the effects of digitalis. Acute hypocalcemia has been reported to produce orthostatic hypotension.

NEUROMUSCULAR EFFECTS. The most common presentation of hypocalcemia is neuromuscular. Tetany is common and may be dramatic. Children may progress to generalized seizures while adults often display only muscular cramps and carpopedal spasms. Respiratory stridor may be produced by muscular spasm of the glottis. Latent tetany may be uncovered by either the Chvostek or Trousseau sign. The Chvostek sign is elicited by tapping the facial nerve, resulting in facial contraction, and the Trousseau sign by placing a tourniquet around the arm, inducing local hypoxia and leading to contraction of the hand. Neither sign is specific for hypocalcemia and both may be found in normal people. Characteristic electroencephalographic changes include an increasing frequency of high-voltage slow waves. Hypocalcemia has also been reported to increase intracranial pressure, producing papilledema. Convulsions and unilateral neurological manifestations may also be found.

SOFT TISSUE CALCIFICATIONS. Calcification of the basal ganglia has been reported in hypoparathyroidism. The duration required to observe basal ganglia calcification, however, is many years. Lenticular cataracts are also a common complication of prolonged hypocalcemia. Hypocalcemia appears to interfere with normal lens hydration by disturbing active sodium transport.

GASTROINTESTINAL EFFECTS. Patients with hypocalcemia may present with nausea, vomiting, and abdominal pain. Intestinal malabsorption and steatorrhea have also been reported.

Magnesium

Metabolism

The importance of magnesium in electrolyte metabolism resides in the fact that it is the most significant intracellular cation, second only to potassium. The prominence of magnesium in cells is reflected by its involvement in many intracellular events, including cellular energetics (magnesium-activated ATPase) and many other enzymatic reactions, including creatine kinase, alkaline phosphate, pyruvate kinase, and enolase.

DISTRIBUTION. In a 70-kg. man there is approximately 2000 mEq. of magnesium. Its greatest concentration in the body is in bone and, secondarily, in muscle, and about 98 per cent of all body magnesium is in these two areas and in soft tissue. Only about one-third is protein bound, the remainder being diffusible or ionized. Only a small part of total body magnesium is in the extracellular compartment.

As with all nutrients, electrolytes, and minerals, magnesium is normally ingested and derived from food. It is primarily found in green, leafy vegetables, grains, and protein. Estimates of the normal daily intake of magnesium range from 24 to 36 mEq. Of the ingested magnesium, approximately one-third is absorbed by the gut in the proximal small bowel. At the cellular level, like potassium, a high intracellular gradient for magnesium is maintained and its homeostasis is maintained by the kidneys. Following filtration of free magnesium, reabsorption of this divalent cation takes place throughout the nephron. Secretion of magnesium occurs at the distal tubular site, probably governed in part by aldosterone.

Hypomagnesemia

PATHOPHYSIOLOGY. There are three conditions that account for the majority of the clinical problems associated with magnesium imbalance: (1) decreased oral intake; (2) increased gastrointestinal and renal losses; (3) renal failure with magnesium retention.

The causes of diminished magnesium intake include alcoholism, impaired bowel function, and constant nasogastric suction, protracted intravenous therapy without magnesium replacement, starvation, and impaired oral intake because of vomiting. The key clinical example in this category is the alcoholic patient. With voluntary substitution of alcohol for food, coupled with enhanced urinary magnesium losses ascribable to the tubular action of alcohol, it is evident that magnesium depletion and hypomagnesemia would supervene.

The magnesium depletion associated with increased gastrointestinal tract and kidney losses is found in patients with protracted nasogastric suction, severe diarrhea, small bowel fistulae, biliary fistula, laxative abuse, cholera, and severe malabsorption. Some examples of primary renal losses of magnesium leading to depletion include patients receiving diuretic therapy, gentamicin, patients with primary or secondary hyperaldosteronism, rarely a familial type of renal tubular magnesium loss, and the diuretic phase of acute necrosis and following cancer chemotherapy with cis-platinum. Associated biochemical alterations occurring with magnesium depletion include electrolyte perturbations such as hypokalemia, hypophosphatemia, and hypocalcemia. Hypocalcemia is explicable on the basis of the known diminished end-organ response to or secretion of parathyroid hormone in the magnesium-depleted state. Hypokalemia most likely results from increased renal losses and inability of cells to maintain the proper intracellular gradient for potassium in the magnesium deficient state. The renal effects of magnesium depletion are less clear. Experimentally, azotemia, phosphaturia, and a diminished glomerular filtration rate have been observed. The occurrence of azotemia and phosphaturia is not as clearly defined in clinical magnesium depletion.

Experimental observations indicate that magnesium is pivotal in maintaining and restoring intracellular potassium. The clinical counterpart of experimental refractory potassium repletion has been observed in burns, diet-induced magnesium depletion, and alcoholism. In these instances, restoration of potassium homeostasis appears to hinge on the concurrent treatment of the associated magnesium depletion.

CLINICAL MANIFESTATIONS. Hypomagnesemia manifests itself clinically through three organ systems: (1) central nervous system, (2) gastrointestinal, and (3) cardiovascular. The central nervous system signs and symptoms are tremor, disorientation, hyperactive deep tendon reflexes, ataxic gait, nystagmus (both vertical and horizontal) convulsions, and coma. The gastrointestinal symptoms are primarily motility problems, and are manifested by dysphagia. Cardiovascular manifestations of magnesium depletion include arrhythmias (with or without digitalis toxicity), primarily ventricular in origin, and reversible hypertension.

THERAPY. Treatment of magnesium deficiency is analogous to the correction of potassium depletion: (1) adequate renal function should be assured; (2) oral repletion is the safest route of administration; and (3) intravenous therapy is reserved for patients whose gastrointestinal tract is temporarily nonfunctional. The parenteral dosage regimen for treating the ill, convulsing, magnesium-depleted patient, as outlined by Flink, calls for administration of approximately 1 mEq./kg. body weight on the first day; 0.5 mEq./kg. is then given on the ensuing 2 to 5 days. If the magnesium-depleted patient is not convulsing or comatose, but is unable to take magnesium supplements by diet orally, we favor the intramuscular administration of magnesium, 32 to 48 mEq./day for 4 to 6 days, with daily serum magnesium determinations.

Hypermagnesemia

PATHOPHYSIOLOGY. Hypermagnesemia, in nearly every instance, is found in association with renal failure either acute or chronic. The parallel with potassium excess is quite striking since the kidneys are primarily responsible for maintaining both potassium and magnesium balance. Thus, hypermagnesemia is commonly associated with renal insufficiency, and in both acute and chronic renal failure continued ingestion of magnesium-containing food or medications is a consequence. Indeed, magnesium-containing antacids have been the chief offender among medications.

MANIFESTATIONS. The central nervous, circulatory, respiratory, and endocrine systems are the prime targets of this perturbation. Central nervous system signs and symptoms include loss of deep tendon reflexes, ataxia, dysarthria, dilated pupils, and drowsiness, progressing to coma. The circulatory system may be affected by per-

sistent and refractory hypotension that is relieved only by correction of the hypermagnesemia. Other circulatory effects include increased P-R interval, bradycardia, and heart block. The respiratory system is affected by hypoventilation which can become life threatening if the hypermagnesemia is not recognized and treated. The endocrine system is affected primarily through decreased parathyroid secretion and diminished end-organ response leading to hypocalcemia.

THERAPY. Clinical hypermagnesemia is treated by: (1) stopping all magnesium intake; (2) hemodialysis if refractory hypotension, respiratory depression, severe cardiac arrhythmias, or significant central nervous system depression is present; and (3) supportive therapy in the form of intravenous calcium chloride or gluconate.

In summary, because both hypomagnesemia and hypermagnesemia affect many organ systems, serum magnesium concentration should be determined routinely in all patients ill enough to warrant a study of serum electrolytes. More specifically, patients who are alcoholic, malnourished, and refractory to potassium repletion should be evaluated for possible magnesium deficiency; and refractory hypotension should suggest the possibility of unrecognized hypermagnesemia.

Phosphorus

Metabolism

The importance of phosphorus for normal biological processes in the human organism is best exemplified by reviewing the various forms of phosphorus in the body, i.e., phospholipids, nucleotides, carbohydrate-phosphate esters, and inorganic phosphorus. The normal 70-kg man contains approximately 712 grams (23,000 mMol.) of phosphorus; of this about 80 per cent is in bone, 9 per cent in skeletal muscle, and only 0.05 per cent in the extracellular space. Intracellular phosphorus is primarily in organic forms. Only a small fraction of the intracellular phosphorus exists inorganically, but this is very important because it is from this pool that phosphorus is derived for the synthesis of ATP and 2-3DPG.

The average daily intake of phosphorus is about 1 g; of this, 90 per cent is excreted in the urine and 10 per cent through the feces. The normal concentration of inorganic phosphorus in the plasma of man is between 2.7 and 4.5 mg./dl. The plasma phosphorus concentration is the result of net intestinal absorption, bone and soft tissue flux, and renal phosphorus excretion.

Recent evidence suggests that phosphorus is absorbed through the gastrointestinal tract by an active, vitamin D-dependent mechanism located primarily in the distal small intestine. The renal handling of phosphorus is a subject of intense current investigation. Although most data are derived from nonhuman species, it appears that 70 to 80 per cent of the filtered phosphorus is reabsorbed by the proximal tubule; a much smaller percentage is reabsorbed at the distal tubule. This reabsorption is energy requiring and under the influence of parathyroid hormone; in the proximal tubule reabsorption is also sodium dependent. Parathyroid hormone acts through cyclic AMP to cause a decreased tubular reabsorption of phosphorus, and thereby phosphaturia. Experimental animal data suggest that many other hormones (i.e., vitamin D and growth hormone) may also influence the renal handling of phosphorus.

Pathophysiological effects of phosphorus abnormalities

As in any primarily intracellular ion, abnormalities in phosphorus metabolism are usually discussed in terms of their concentrations in the extracellular space. The concentration of any ion (including phosphorus) in the extracellular space does not necessarily reflect its total body stores, nor does it necessarily parallel any pathophysiological process.

HYPERPHOSPHATEMIA. Elevated serum phosphorus concentration may result from increased absorption by the gastrointestinal tract, which, in the presence of normal renal function, is transient. This is because as the gastrointestinal absorption of phosphorus rises, the renal excretion of phosphorus increases. Hyperphosphatemia may also result from chronic renal failure when glomerular filtration rate decreases to 25 per cent of normal. In this instance, hypocalcemia develops, leading to secondary hyperparathyroidism, which in turn results in marked phosphaturia. With advanced renal failure, the kidney is no longer able to increase effectively the fractional excretion of phosphorus necessary to maintain normophosphatemia; hyperphosphatemia ensues. Hyperphosphatemia may also result from a reduction of parathyroid hormone activity that leads to increased renal phosphate reabsorption. This occurs in hypoparathyroidism, which can be primary and idiopathic or secondary to surgical removal. These conditions are usually associated with a decrease in phosphate excretion. Another condition is pseudohyperparathyroidism that is characterized by receptor resistance to PTH in those organs responsible for plasma phosphorus regulation (i.e., bone and kidney). Thus, these organs are unresponsive to PTH.

There are no symptoms of hyperphosphatemia, *per se*; but when serum phosphorus concentration is high enough for prolonged periods, the tendency for the formation of calcium phosphate compounds is increased, and these are abnormally deposited in soft tissue.

PHOSPHORUS DEPLETION. Mild hypophosphatemia is relatively common clinically, occurring in up to 2 per cent of all hospitalized patients. Severe hypophosphatemia (<1.5 mg./dl.) with or without phosphorus depletion is less common, but can represent a life-threatening abnormality.

Because phosphorus is such a ubiquitous ion in organic matter, the pathophysiological effects of its depletion affect all systems. Phosphorus is necessary for most anabolic and catabolic cellular reactions and is the key ion in such important molecules as adenosine triphosphate (ATP), 2,3-disphosphoglyceric acid (2,3-DPG), phospholipids, nucleotides, and carbohydrate-phosphate esters.

Phosphorus depletion usually results from a decrease in dietary intake or an increased renal excretion. Phosphorus depletion, however, does not necessarily lead directly to hypophosphatemia, nor does hypophosphatemia necessarily reflect body phosphorus depletion. Hypophosphatemia usually results from a shift of phosphorus from the extracellular space to the intracellular space. Prolonged, severe hypophosphatemia, in the presence of phosphorus depletion, may lead to severe pathophysiological abnormalities which may be life threatening. Renal excretion may be increased in the presence of metabolic acidosis, respiratory acidosis, magnesium depletion, excess parathyroid hormone, and extracellular volume expansion.

Shifts of phosphorus into the intracellular space from the extracellular space lead to hypophosphatemia. These shifts are stimulated by the infusion of glucose (and subsequent rise in insulin), by respiratory alkalosis, and the infusion of fructose. It is thought that when they are accompanied by phosphorus depletion, a clinical syndrome may ensue. Because phosphorus is essential for all cellular function, all organs are at risk of damage in this hypophosphatemic syndrome. Some, but not all organ systems, have been associated with this syndrome manifested by severe hypophosphatemia (<1.5 mg./dl.)

HEMATOLOGIC MANIFESTATIONS. Because blood components are so accessible, they have been well studied in patients with severe hypophosphatemia. The process most commonly studied is the glycolytic pathway, and the molecules commonly evaluated are ATP and 2,3-DPG. As regards white blood cells, Craddock found a significant reduction in phagocytic, chemotactic, and bactericidal activities in polymorphonuclear leukocytes of dogs that were experimentally depleted of phosphorus and rendered severely hypophosphatemic. He showed that these abnormalities were correlated with white cell ATP deficiencies. The effects of severe hypophosphatemia on ATP and 2,3-DPG are demonstrated best in human red blood cells. Patients have been found to become severely hypophosphatemic while undergoing intravenous hyperalimentation without adequate phosphorus added to the solution. The red blood cells of these patients developed a rise in *triose phosphates* (i.e., the sum of dihydroxyacetone phosphate and glyceraldehyde-3-phosphate). Accumulation of these triose phosphates occurs in phosphorus-depleted red blood cells because phosphorus is necessary for further metabolism of glyceraldehyde-3-phosphate to 1,3-DPG, 2,3-DPG, and eventually ATP. Decreased red blood cell 2,3-DPG produces a depression of the P_{50}, the oxygen tension at which hemoglobin is 50 per cent saturated. This, in turn, leads to the decreased ability of red blood cells to deliver oxygen to tissues. Some evidence exists in experimental animals that platelet functional abnormalities also may develop in patients with the hypophosphatemic syndrome.

CENTRAL NERVOUS SYSTEM DYSFUNCTION. The symptoms of paresthesias, dysarthria, weakness, irritability, numbness, confusion, obtundation, seizures, and coma have been described in, and ascribed to, patients with severe hypophosphatemia. Although the pathophysiological mechanisms have not been elucidated, the knowledge that the brain depends heavily on glucose metabolism, coupled with the evidence mentioned earlier in red cells of glycolytic suppression by phosphorus depletion, makes this explanation attractive.

SKELETAL MUSCLE. Skeletal muscle cells are at risk of injury and death in hypophosphatemia and phosphorus depletion. It has been observed that frank rhabdomyolysis can occur. This has been observed in patients receiving intranvous hyperalimentation, in the treatment of patients with diabetic ketoacidosis and in the treatment of alcoholics with intravenous glucose. The exact biochemical mechanism for this muscle cell injury is not known, but decreased energy supplies because of reduced intracellular ATP may occur. In addition, skeletal muscle may serve as a reservoir for phosphorus in the phosphorus-depleted dog. This phosphorus reservoir may be important in the phosphorus-depleted animal during periods of phosphorus depletion and hypophosphatemia in protecting other organs such as brain and heart muscle from the effects of phosphorus depletion. Whether this applies to man is not known at this time. However, clinical instances of skeletal muscle dysfunction leading to respiratory failure have been noted in man. These patients developed respiratory muscle weakness while severely hypophosphatemic, and their ventilatory failure has responded to phosphorus replacement. Moreover, cardiac muscle dysfunction in man has also been shown to result from hypophosphatemia, although the exact mechanism for this is not well understood.

Annotated references

Potassium

DeFronzo, R. A.: Hyperkalemia and hyporeninemic hypoaldosteronism. Kidney Int. *17*:118–134, 1980 (An up-to-date discussion of the pathophysiology of hyperkalemia with particular emphasis on hypoaldosteronism.)

Giebisch, G. H, and Thier, S. O.: Potassium: Physiological and Clinical Importance. *In* Siegel, L. (ed.): Directions in Cardiovascular Medicine. Somerville, N.J.

Hoechst-Roussel Pharmaceuticals, Inc., 1977. (An excellent review of the role of potassium in cell physiology and the factors regulating renal potassium excretion.)

Lindeman, R. D., and Pederson, J. A.: Hypokalemia. *In* Whang, R. (ed): Potassium: Its Biologic Significance. Baco Raton, C.R.C. Press (in press). (A comprehensive survey of the pathogenesis of consequences.)

Schultze, R. G., and Nissenson, A. R.: Potassium: Physiology and pathophysiology. *In* Maxwell, M. H., Kleeman, C. R. (eds.): Clinical Disorders of Fluid and Electrolyte Metabolism, ed. 3. New York, McGraw-Hill, 1980. (A scholarly and comprehensive review of the entire subject.)

Sterns, R. H., Cox, M., Feig, P. U., and Singer, I.: Internal potassium balance and the control of the plasma potassium concentration. Medicine *60*:339, 1981. (A lucid review of the factors influencing trancellular distribution of potassium including the pathophysiology.)

Calcium

Coburn, J. W., Hartenbower, D. L., Brickman, A. S., et al.: Intestinal absorption of calcium, magnesium, and phosphorus in chronic renal failure. *In* David, D. S. (ed.): Calcium Metabolism in Renal Failure and Nephrolithiasis. New York, John Wiley & Sons, 1977.

Haussler, M. R., and McCain, T. A.: Basic and clinical concepts related to vitamin D metabolism and action. N. Engl. J. Med. *297*:974–983, 1041–1050, 1977.

Massry, S. G., Friedler, R. M., and Coburn, J. W.: Renal handling of calcium, magnesium, and phosphate in renal failure. *In* David, D. S. (ed.): Calcium Metabolism in Renal Failure and Nephrolithiasis. New York, John Wiley & Sons, 1977.

Parfitt, A. M.: The actions of parathyroid hormone on bone: Relation to bone remodeling and turnover, calcium homeostasis, and metabolic bone disease. I. Mechanisms of calcium transfer between blood and bone and their cellular basis: Morphological and kinetic approaches to bone turnover. Metabolism *25*:809–844, 1976.

Parfitt, A. M., and Kleerekoper, M.: Clinical disorders of calcium, phosphorus, and magnesium metabolism. *In* Maxwell, M. H., and Kleeman, C. R. (eds.): Clinical Disorders of Fluid and Electrolyte Metabolism. New York, McGraw-Hill, 1980.

Raisz, L. G.: Bone metabolism and calcium regulation. *In* Avioli, L. V., and Krane, S. M. (eds.): Metabolic Bone Disease, Vol. 1. New York, Academic Press, 1978.

Ritz, E., Malluche, H. H., Krempien, et al.: Calcium metabolism in renal failure. Disorders of Mineral Metabolism *3*:151–250, 1981.

Rosenblatt, M.: Parathyroid hormone: Chemistry and structure-activity relations. Pathobiol Annu *11*:53–86, 1981.

Schneider, A. B., and Sherwood, L. M.: Pathogenesis and management of hypoparathyroidism and other hypocalcemic disorders. Metabolism 24:871–898, 1975.

Stewart, A. F., Horst, R., Deftos, L. J., et al.: Biochemical evaluation of patients with cancer-associated hypercalcemia: Evidence for humoral and nonhumoral groups. N. Engl. J. Med. *303*:1377–1383, 1980.

Magnesium

Estep, J., Shaw, W. A., Wattington, C., Hobe, R., Holland, W., and Tucker, St. G.: Hypocalcemia due to hypomagnesemia and reversible parathyroid hormone unresponsiveness. J. Clin. Endocrinol. Metab 29:841, 1969. (This article clearly points out the relationship between magnesium depletion and the associated hypocalcemia.)

Ferdinandus, J., Pederson, J. A., and Whang, R.: Hypermagnesemia as a cause of refractory hypotension, respiratory, depression, and coma. Arch Intern Med *141*:669, 1981. (The clinician's attention is called to the refractory hypotension and respiratory depression associated with hypermagnesemia.)

Flink, E. B.: Therapy of magnesium deficiency. Ann. N.Y. Acad. Sci. *162*:901, 1969. (A key article for the treatment of magnesium depletion based on Dr. Flink's extensive experience.)

Freitag, J. J., Martin, K. J., Conrades, M. B., Bellorin-Font, E., Teitelbaum, S., Klahr, S., and Slatopolsky, E.: Evidence for skeletal resistance to parathyroid hormone in magnesium deficiency studies in isolated perfused bone. J. Clin. Invest. *64*:1238, 1970. (An update on the magnesium deficiency-hypocalcemia interrelationship.)

Iseri, L. T., Freed, J., and Bures, A. R.: Magnesium deficiency and cardiac disorders. Am. J. Med. *58*:837, 1975. (An excellent review of the efficacy of magnesium in the treatment of refractory ventricular arrhythmias associated with magnesium depletion).

Massry, S. G.: The clinical pathophysiology of magnesium. Contr. Nephrol. *14*:64, 1978. (An excellent overview of magnesium metabolism).

Mordes, J. P.: Excess Magnesium. Pharmacol. Rev. *29*:273, 1978. (An excellent review of hypermagnesium.)

Phosphate

Agus, Z. S., Puschett, J. B., Senosky, D., and Goldberg, M.: Mode of action of parathyroid hormone and cyclic adenosine 3',5-monophosphate on renal tubular phosphate reabsorption in the dog. J. Clin. Invest. *50*:617, 1971.

Avioli, L. V.: Intestinal absorption of calcium. Arch. Intern. Med. *129*:345, 1972.

Chaimovitz, C., Spierer, A., Leibowitz, H., Tuma, S., and Better, O.: Exaggerated phosphaturic response to volume expansion in patients with essential hypertension. Clin. Sci. Mol. Med. *49*:207, 1975.

Craddock, R., Yawata, Y., Van Sorta, L., Gilberstadt, S., Silvis, S., and Jacob, H. S.: Acquired phagocyte dysfunction. N. Engl. J. Med. *290*:1403, 1974.

Knochel, J. P.: The pathophysiology and clinical characteristics of severe hypophosphatemia. Arch. Intern. Med. *137*:203, 1977.

Knochel, J. P., Barcenas, C., Cotton, J. R., Fuller, T. J., Haller, R., and Carter, N. W.: Hypophosphatemia and rhabdomyolysis. J. Clin. Invest. *62*:1240, 1978.

Knochel, J. P., Haller, R., and Ferguson, E.: Selective phosphorus deficiency in the hyperalimented hypophosphatemic dog and phosphorylation potentials in the muscle cell. Adv. Exp. Med. Biol. *128*:323, 1980.

Massry, S. G., and Fleisch, H.: Renal Handling of Phosphate, New York, Plenum, 1980.

Robertson, W. G.: Plasma phosphate homeostasis. *In* Nordin, B. E. C. (ed.): Calcium, Phosphate, and Magnesium Metabolism. London, Churchill-Livingstone, 1976.

Travis, S. F., Sugarman, H. J., Ruberg, R. L., et al.: Alterations of red cell glycolytic intermediates and oxygen transport as a consequence of hypophosphatemia in patients receiving intravenous hyperalimentation. N. Engl. J. Med. *285*:763, 1971.

13   Maintenance of Body Fluid Tonicity

Carlos A. Vaamonde, M.D.

The preservation of the volume and composition of the body fluids requires a normal thirst mechanism and precise renal regulation of water and solute excretion. Not only must the kidney respond appropriately to the daily ingested load of solutes and water, but it also has to adapt its excretory function to the extrarenal losses of water and solutes occurring in normal life (insensible water loss, sweat, fecal contents) or in disease (excessive sweating, hyperventilation, vomiting, diarrhea, fistulae, etc.). Therefore, when the kidney is diseased and renal function is impaired, body fluid homeostasis is in jeopardy.

Terminology

We are basically interested in defining two interrelated measures of the concentrating and diluting capacity of the kidney. The first is a measure of *gradient* or *concentration*—that is, the urine concentration, or total solute concentration (urine osmolality, Uosm). The others are *rate-dependent measures:* the minute volume of urine excreted (V) and its solute and water composition relative to plasma—that is, the osmolar clearance (Cosm), the solute-free water excretion or free-water clearance (CH$_2$O), and the solute-free water reabsorption (T^{c}H$_2$O).

In determining urine concentration, one is interested in the number of particles of solute dissolved in a unit volume of urine water (osmolality). This is expressed in milliosmoles per kilogram of water (mOsm./kg. H$_2$O), and the measurement is usually done by freezing point depression osmometry.* A "concentrated" urine is characterized by a concentration of total solutes higher in the urine than in plasma. Conversely, the urine is "dilute" when its total solute concentration is less than that of plasma. In the young, healthy adult, plasma osmolality is approximately 280 mOsm./kg. H$_2$O. Maximum urinary osmolality approximates 1200 mOsm./kg. H$_2$O, whereas urine may be as dilute as 30 to 35 mOsm./kg. H$_2$O.

When the urine osmolality (Uosm) is factored by plasma osmolality (Posm), urine-to-plasma osmotic ratio ($\frac{U}{P}$osm) is obtained. This is another way of expressing the capacity of the kidney to achieve concentrating or diluting gradients. Thus, the $\frac{U}{P}$osm ratios vary from 3.5 to 4.5 in maximally concentrated urine to as low as 0.12 in maximally diluted urine.

In the overall formation of urine, no net

---

* The term *osmolarity* is commonly used in reference to the osmotic concentration of a solution expressed as osmols of solute per liter of solution rather than of water. Although in clinical application there is no real difference between osmolality and osmolarity, it is preferable to express total solute concentration as osmolality.

separation of water from solutes occurs when the urine has the same osmolality as the plasma. The physiologist also expresses concentrating ability in terms of the volume of water reabsorbed from an isosmotic urine to render it hyperosmotic. This determination is made under conditions of hydropenia with or without vasopressin (Pitressin) administration. Under these conditions, the concentrated urine can be regarded as consisting of two portions: one that contains the solutes in an *isosmotic* concentration to plasma, the other the amount of water without solutes, or *free* water, reabsorbed from the isosmotic volume to make it concentrated. This may be expressed as:

$$V \text{ (Urine volume)} = \qquad (1)$$
$$\text{isosmotic urine volume}$$
$$- \text{ reabsorbed water}$$

In order to know the amount of reabsorbed water (i.e., the concentrating ability) the volume of isosmotic urine must be determined. This is accomplished by answering the question, what volume of urine would have been required to excrete the same solute load isosmotically (i.e., with a urine concentration equal to that of plasma)? Or:

(a) Solute excretion (isosmotic)
    = Solute excretion (hyperosmotic)

(b) Isosmotic volume × Posm
    = V × Uosm

(c) Isosmotic volume
    $= \dfrac{\text{Uosm}}{\text{Posm}} \times V$

This expression is the same as the "clearance" formula, and the calculated volume of isosmotic urine is referred to as the *osmolar clearance* (Cosm).

For example (Fig. 13–1), let us assume that, during hydropenia, urine volume per minute (V) is 0.5 ml. and Uosm is 1,160 mOsm./kg. H_2O and Posm is 290 mOsm./kg. H_2O (Fig. 13–1B). According to (c) above, the

Cosm is:

$$\frac{1160 \times 0.0005}{290} = \frac{580 \ \mu\text{Osm/min.}}{290 \ \mu\text{Osm/ml.}} =$$
$$2 \text{ ml./min.}$$

By substitution in (1) above:

$$V = \text{Cosm} - \text{reabsorbed water}$$

and,

$$\text{Reabsorbed water} = \text{Cosm} - V$$

Reabsorbed water is usually referred to as *solute-free water reabsorption*, or T^cH_2O, and urinary concentrating ability may be expressed in these terms.

During hydropenia and vasopressin administration, T^cH_2O will achieve its maximal value only at large rates of solute excretion (i.e., during a solute diuresis; *solute or osmotic diuresis* is due to restrained water reabsorption occurring proximal to the ascending limb of Henle's loop and to high concentration of relatively nonreabsorbable solute, such as mannitol, urea, sodium sulfate, hypertonic saline, etc.). Under these conditions, V and Cosm increase while Uosm decreases toward isotonicity (Fig. 13–1D).

In the diluting operation, the urine volume may also be considered as consisting of two portions—one that is isosmotic to plasma (Cosm), the other being free water which is *not* reabsorbed—thus rendering the urine dilute.

$$V = \text{Cosm} + \text{free water (Fig. 13–1C)} \quad (2)$$

Usually the free water is symbolized as CH_2O *(solute-free water clearance)*. T^cH_2O is the negative value of CH_2O and is sometimes referred to as *negative free water clearance* rather than reabsorbed free water. Figure 13–1 depicts graphically the concept of CH_2O and T^cH_2O, with examples of water diuresis, isosmotic urine, and antidiuresis.

The maximal CH_2O and T^cH_2O reflect, within certain restrictions, the magnitude of sodium chloride reabsorption in the as-

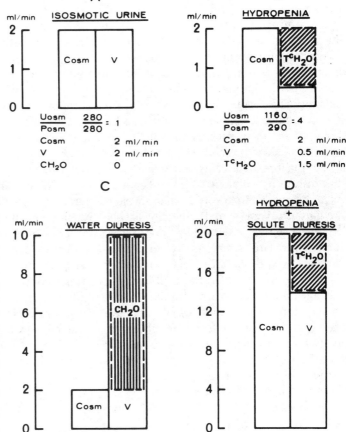

FIG. 13—1. Schematic outline of the relationship between V and Cosm during: (A) isosmotic urine excretion; (B) hydropenia; (C) water diuresis; and (D) hydropenia during solute diuresis. Note that the units in scale C are double those in A and B and the units in scale D are four times greater than those in A and B. CH_2O is the amount of water "added" to the urine to make it hypotonic (C). T^CH_2O is the amount of water "removed" from the urine to make it hypertonic (B and D). The values for Uosm, Posm, $\frac{U}{P}$osm, Cosm and CH_2O or T^CH_2O are listed below each example. Note that there is no net separation of water from solutes during excretion of an isosmotic urine (A).

cending limb of Henle's loop. Therefore, the measurement of these calculated indices of renal function has been used to examine indirectly the characteristics of sodium chloride reabsorption in this segment of the nephron under a variety of normal and abnormal conditions. These indices (Cosm, CH_2O, T^CH_2O), although of great physiological and pathophysiological significance, have no practical application in clinical medicine outside of clinical research. The simple evaluation of maximal or minimal V and Uosm and of Posm is sufficient in most clinical situations. A U/Posm greater than 1 indicates that the kidney is elaborating concentrated urine and conserving water; A U/Posm lower than 1 means that water in excess of solute is being excreted.

Regulation of thirst and water intake

Although a detailed discussion of the thirst mechanism is beyond the scope of this chapter, a brief description will follow. Thirst is largely a subjective feeling, and can be defined as a strong sensation of a desire for water. The thirst center responsible for integrating thirst sensation is located in the lateral area of the hypothalamus, in close proximity to the paraventricular and supraoptic nuclei, where antidiuretic hormone (ADH) is synthesized.

Two major stimuli are responsible for thirst: a decrease in extracellular fluid volume (ECFV) and an increase in plasma osmolality. The former is a more powerful stimulus, since healthy subjects do not become thirsty until plasma osmolality exceeds 290 mOsm./kg. H_2O, the level at which plasma osmolality increases to a concentration sufficient to produce maximal antidiuresis. In addition, high levels of circulating angiotensin II (usually associated with a decreased ECFV) may cause thirst. There are other non-volume, non-osmotic–related factors that influence thirst. Dryness of the oral and pharyngeal mucous membranes can stimulate thirst, while distension of the intestinal tract cavities has an inhibitory effect. These are temporary measures that may serve to quench thirst; for example, moistening the oral cavities when drinking or appropriate fluid replacement is not possible.

The amount of fluid that we drink depends also on many other factors: the type and quantity of food intake (solute load), the ambient temperature, the level of exercise, and habit. Table 13–1 outlines an approximation of the normal balance of water in adult man. A rather accurate (although in hospitalized patients difficult) estimation of daily fluid balance is the measurement of changes in body weight. In practical terms, the daily urine output approximates the amount of fluids drunk. Since the water losses through the skin and respiratory system maintain thermoregulation rather than fluid balance, the kidneys play a pivotal role in regulating water excretion according to the body needs for water balance. To understand clearly the importance of the thirst mechanism in maintaining water balance, it should be kept in mind that when renal or evaporative losses (skin, respiratory tract) are greater than normal, the thirst mechanisms will be activated to increase water intake and restore water balance.

Thirst can be suppressed (hypodipsia, adipsia) or exaggerated (polydipsia, resetting of thirst osmostat) in patients with organic or psychiatric disorders of the central nervous system, affecting the thirst mechanism and causing perturbations of fluid balance. The latter results in hyperosmolality and hypernatremia (adipsia), or hypoosmolality and hyponatremia, respectively.

Pathophysiology of urine concentration

The normal mechanism

It is generally accepted that urine becomes concentrated by the active tubular reabsorption of salt and the consequent passive osmotic equilibration of water.

The knowledge of the relative permeability for water and solutes (e.g., sodium, chloride, urea, bicarbonate) and of the transport characteristics of the different tubular segments is of fundamental importance for the understanding of the renal concentrating process. Table 13–2 outlines the permeability and transport characteristics of the tubular segments involved in the renal concentrating mechanism. The thin descending limb of Henle's loop is relatively impermeable to solutes but has a high osmotic permeability to water, whereas the

TABLE 13–1
Approximate Daily External Fluid Balance in Normal Man

Intake		
Drinking water, food water, endogenous production		2500 ml.
Output		
Urine		1500 ml.
Extrarenal*	Skin —600 ml.	
	Lungs—300 ml.	
	Feces —100 ml.	1000 ml.

* In practical terms, the extrarenal losses of water equal the water provided by ingestion of food and endogenous production.

TABLE 13–2
Permeability and Transport Characteristics of Tubular Segments Involved in the Renal Concentrating Mechanisms (ADH present)

	Active Salt Transport	Permeability to		
		H_2O	Urea	NaCl
Loop of Henle				
Thin Descending Limb	0	++++	±	±
Thin Ascending Limb	0	0	+++	++++
Thick Ascending Limb	++++	0	0±	0±
Distal Convoluted Tubule	+++	±	0±	0±
Collecting Duct				
Cortical Segment	++	+++*	0	0±
Outer Medullary Segment	+	+++	0	0±
Papillary Segment	+	+++	+++	0±

* When ADH is absent (urine dilution) the collecting duct permeability to water is considerably lower, (+). (The values assigned (0 to ++++) are arbitrarily chosen relative values for the different tubular segments. After Kokko, J. P., and Tischer, C. C.: Kidney Internat., *10*:64, 1976.)

thin ascending limb appears to be more permeable to salt than to urea and relatively impermeable to water. No evidence of active transport of sodium chloride has yet been found in the latter nephron segment. Thus, the function of the thin ascending limb appears to be the passive addition of salt without water to the interstitium. This permits a progressive increase in medullary osmolality from the corticomedullary junction to the papillary tip. The thick ascending limb of Henle's loop is relatively impermeable to water and solutes, but it has the greatest capacity for outward active transport of chloride, with sodium following passively down the potential gradient created by the electrogenic chloride pump. Obviously, this results in a hypotonic fluid remaining in the lumen.

The distal convoluted tubule is capable of active outward transport of salt and for the most part has a low permeability to water. The cortical and outer medullary segments of the collecting duct are capable of active outward transport of salt and are permeable to water (when ADH is present) while they are impermeable to solute (particularly urea, even in the presence of ADH). It should be pointed out that the transition from a water-impermeable and ADH-unresponsive tubular epithelium to one whose water permeability is increased by ADH occurs somewhere in the cortex (distal convoluted tubule → cortical collecting duct). Although the exact location of this transition in permeability may depend on the mammalian species studied, the cortical location assures sufficient removal of water by the high cortical blood flow, avoiding a diluting effect of the cortical and outer medullary interstitium. Finally, the papillary segment of the collecting duct is characterized by its high permeability to water and urea.

Although water reabsorption occurs along most of the length of the nephron, the final extraction of water in the elaboration of a concentrated urine takes place in the collecting duct. In this respect it was recently shown that the rise of Uosm over the last 2 mm. of the collecting duct of the rat is as great as the entire rise in Uosm up to that point.

For the concentration process to occur, the following conditions are requisite:

(1) The interstitial tissue surrounding the collecting duct must be more concen-

trated than the urine within the lumen of the collecting duct. The hypertonicity of the medullary interstitium is established primarily by the particulars of the handling of sodium chloride in the loop of Henle. This system is referred to as the countercurrent mechanism and is described below. The particular anatomical spatial distribution of the medullary structures (tubular loops, vasa recta) is known to be related to the capacity of the species (birds and mammals only) to produce a concentrated urine. It is also well recognized that the greater the number of the Henle's loops and the greater the length of the loops, the greater the renal concentrating ability of the species. About 12 to 15 per cent of the nephron population in man have long loops of Henle. These originate in the juxtamedullary region of the renal cortex (juxtamedullary nephrons) and extend deep into the inner medulla and papilla. The remaining nephrons have short loops of Henle, originating in the superficial and intermediate regions of the cortex (cortical nephrons) and extending into the outer medulla. The medullary circulation (vasa recta) and urea also participate in the maintenance of an hyperosmotic interstitium.

(2) The collecting duct must be permeable to water, so that it can be reabsorbed in accordance with the directional demands of the concentration gradient. Collecting duct permeability to water requires the action of antidiuretic hormone (ADH, vasopressin).

THE HYPERTONIC MEDULLARY INTERSTITIUM. Of the 120 ml. of plasma filtered at the glomerulus each minute, 65 per cent is reabsorbed isosmotically in the proximal tubule. Water and urea reabsorption passively follow outward active sodium chloride transport in this segment, and the tubular cell permeability to water is independent of ADH.

The reduced volume of isosmotic tubular fluid then enters the loop of Henle in the medulla. Here the concentrating operation is instituted and maintained, serving the needs of conserving water. In the loop, 25 per cent of filtered water may be reabsorbed.

The *countercurrent multiplier* (Fig. 13–2) requires the close proximity of the descending limb to the ascending limb of the loop of Henle, as well as their permeability to sodium chloride, the restricted permeability to water of the ascending limb, and the flow of fluid in opposite directions. The mechanism actually becomes operative by the active transport of chloride (sodium follows passively) *without* water from within the lumen of the thick ascending limb to the interstitial tissue surrounding the loop of Henle. This process has three important consequences: (1) The fluid leaving the ascending limb and arriving at the distal tubule is hypotonic; (2) the interstitium then becomes hyperosmotic to the fluid in the descending limb, and (3) sodium chloride diffuses passively from the hyperosmotic interstitium into the descending limb.

The mechanism whereby this process results in progressive hypertonicity as the tip of the medulla is approached is known as the *multiplier effect* and it results in the formation of a cortical-medullary osmotic gradient (that is, the cortical tissues are isosmotic while the medullary structures become progressively hyperosmotic toward the papillary tip). This mechanism requires that the ascending and the descending limbs be in close proximity and that flow be continuous. Figure 13–2 depicts the essential operation, i.e., sodium chloride being *added* continuously to the interstitium and to the fluid in the descending limb as the fluid proceeds toward the tip, producing increased concentration. Since the descending limb is highly permeable to water, the fluid within it equilibrates osmotically with the surrounding interstitial fluid (water flows passively out of the descending limb).

For the hyperosmolality to be maintained in the medulla a *countercurrent exchanger* system is necessary. This role is

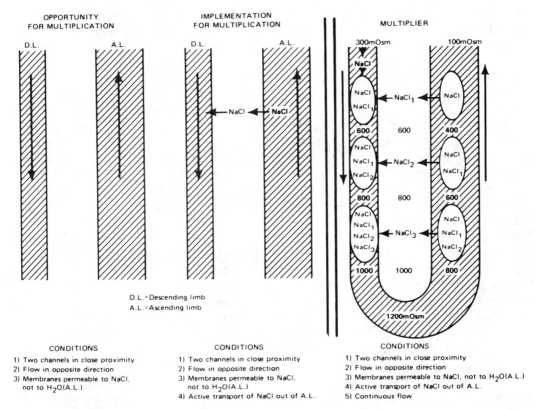

FIG. 13–2. Outline of the countercurrent multiplier concentrating mechanism. Note that in this simplified scheme the A.L. represents the thick ascending limb of Henle's loop and that this segment transports chloride actively, with the sodium ion following passively. (After Papper, S.: Clinical Nephrology, ed. 2. Boston, Little, Brown & Co., 1978.)

accomplished by the vasa recta. It is a *passive* process, depending upon diffusion of solute and water through the walls of the vasa recta capillaries. The vasa recta function as highly efficient countercurrent exchangers which serve to preserve longitudinal (corticomedullary) concentration gradients, promote horizontal mixing and uniform composition of the interstitium and remove the solute and water reabsorbed from the loops of Henle and the collecting ducts. It follows that at the papillary tip, the urine emerging from the ducts of Bellini during antidiuresis is in osmotic equilibrium with vasa recta plasma, loop of Henle fluid and interstitium. In fact, however, small differences exist (for example, the blood of descending vasa recta is slightly less concentrated—by about 75 mOsm./kg. H_2O—than the corresponding

descending limb fluid, while blood in the ascending vasa recta is about 10 to 20 mOsm./kg. H_2O higher). The blood flow in the ascending vasa recta is higher than that in the corresponding descending vessel. The effect of ADH on these vascular structures remains controversial. The net removal of fluid from the medulla which is carried out by the vasa recta equals the volume of fluid reabsorbed from the collecting duct and the descending Henle's loop. Thus, the function of the vasa recta is to *preserve the medullary hyperosmolality* by trapping of solute and removal of water.

This system can dissipate the cortical-medullary osmotic gradient if that gradient is not actively maintained by the countercurrent multiplier or if there is an accelerated removal of osmotically active solute from the medulla ("washout" effect). During water or solute diuresis medullary blood flow increases, tending to washout the corticopapillary osmotic gradient.

The countercurrent mechanism for the establishment of a hypertonic medullary interstitium is augmented by *urea entrapment.*

Urea diffuses passively from the papillary collecting duct fluid into the interstitium and fluid in Henle's loop following its concentration gradient. The progressive increase in the intraluminal concentration of urea occurring in the late distal nephron and cortical and outer medullary segments of the collecting duct is the consequence of both the low permeability of these tubular segments to urea and the outward movement of water in the presence of ADH. Vasopressin appears to increase the permeability of the papillary collecting duct to urea. This addition of urea augments the osmolality of the medullary interstitium, enhancing water reabsorption from the collecting duct. During antidiuresis urea constitutes about 40 per cent of the total solute concentration of the medullary and papillary tissues, whereas it contributes less than 10 per cent of the interstitial osmolality during water diuresis. No active tubular transport of urea appears to exist, except under extraordinary experimental conditions such as in the protein-depleted rat. There appears to be considerable recirculation of urea through the loops and the vasa recta, which also contributes to conservation of urea as a major medullary solute.

In view of the accumulation of urea in the medullary interstitium at concentrations (during hydropenia) similar to that of the collecting duct fluid, more water is not osmotically obligated in the duct lumen for the excretion of urea. It is well known that the urine volume required to excrete a given solute load is less when urea is the principal urinary solute. Thus, urea, the major solute end product of mammalian protein metabolism, has a unique role in the urine concentrating process.

PERMEABILITY OF THE COLLECTING DUCT TO WATER. Antidiuretic hormone is required for permeability to water in the distal tubule and collecting duct. ADH is produced in the supraoptic and paraventricular nuclei of the hypothalamus, then stored in the posterior pituitary gland for liberation into the bloodstream in response to stimuli for water conservation (increase in plasma osmolality, decrease in ECFV). Whereas the proximal tubule is permeable to water in the absence of ADH, ADH is required for permeability to water of the cells of the distal tubule and collecting duct.

In the cortex, during hydropenia the volume of fluid in the distal tubule is reduced by continuous extraction of water (ADH present) from the hypotonic tubular fluid into the isosmotic cortex. The solute concentration of the fluid then becomes isosmotic at the end of the distal tubule.

The achievement of the final urine concentration occurs in the collecting ducts where water leaves the lumen, following the osmotic gradient generated by the countercurrent mechanism. At this level, the final osmotic equilibration takes place. The final urine has practically the same osmolality as the structures located deep in the papilla (loops of Henle, vasa recta blood, interstitium) and is three to four times more concentrated than the plasma and equals less than 1 per cent of the volume of water filtered by the glomeruli. Figure 13–3A illustrates a simplified version of the operation of the mammalian renal countercurrent system during urinary concentration.

MODELS OF MEDULLARY CONCENTRATING MECHANISM. During the last decade theoretical models of the inner medullary concentrating mechanisms have been offered attempting to explain the hypertonicity of the inner medulla without necessitating active solute transport by the thin ascending limb of Henle's loop, a nephronal segment where active salt transport has not been experimentally demonstrated. Kokko

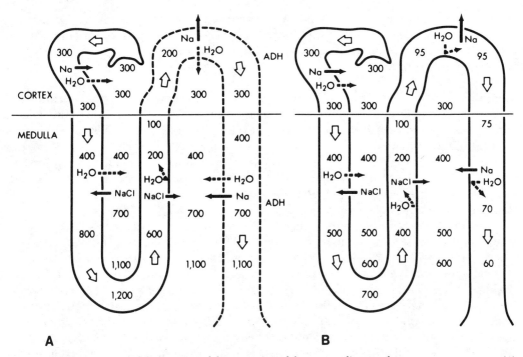

FIG. 13–3. Simplified version of the operation of the mammalian renal countercurrent system. (A) Changes in tubular fluid and interstitial osmolality when ADH is present (concentrating mechanism). (B) The same operation when ADH is absent (diluting mechanism). The numbers refer to the osmolality (mOsm./kg.H₂O) of either intratubular or interstitial fluid. The dashed line depicting the distal tubule and collecting duct in *A* represents the increased tubular permeability to water of these nephron segments in the presence of ADH. Arrows wish a dashed line represent passive reabsorption of water. The deflected dashed arrows represent a markedly decreased permeability to water, rather than an absolute warm impermeability. Note that the vasa recta vessels are not illustrated. Major unresolved questions related to the whole operation of the system are: (1) whether the osmolality of the descending limb of Henle's loop increases by primary solute entry, water removal or both; (2) whether the reabsorption of Na⁺ or Cl⁻ from the thin ascending limb of Henle's loop is active or passive, although current evidence strongly supports passive movement; (3) the exact location along the cortical distal tubule where the increase in water permeability induced by ADH takes place; (4) the role of urea (see models of concentrating mechanisms); and (5) the mechanism(s) whereby medullary blood flow is regulated. (Vaamonde, C. A.: Differential diagnosis of polyuria. *In* Strauss, J. (ed.): Pediatric Nephrology Seminar. Miami, Symposia Specialists, 1974.)

and Rector in 1972 proposed a model for countercurrent multiplication in which both the thin descending and ascending limbs operate as purely passive equilibrating segments. This model is based largely on transport and permeability characteristics (see Table 13–2) obtained by perfusing isolated segments of rabbit nephrons *in vitro*. Its salient feature is that the energy generated by active outward NaCl transport by the thick ascending Henle's limb (expressed as high urea concentration in the outer medullary collecting duct by vir-

tue of water abstraction) is transmitted to the papilla (by way of urea diffusing down its concentrating gradient). In turn, papillary interstitial urea abstracts water out of the thin descending limb, generating high intraluminal NaCl concentrations which allow the entire system to operate by passive diffusion of NaCl out of the thin ascending Henle's limb. This model major attraction stems from the fact that active NaCl transport is required in a nephron segment (thick ascending limb) where active transport has been demonstrated. Simultaneously, Stephenson outlined another model with an active and a passive mode, the latter resembling the passive model of Kokko and Rector.

Factors that influence concentrating ability

Some physiological variables influence the concentrating mechanism.

AGE. Infants generally excrete a hypotonic urine. This is related to the relatively large volume of fluid ingested and to their low urea excretion rather than to an immature concentrating mechanism; Uosm reaches adult maximal values in dehydrated infants fed large amounts of protein. Children have values for maximal Uosm similar to those accepted for adults. T^cH_2O values in young adults are between 5 to 7 ml./min./100 ml. of glomerular filtration rate (GFR) (mannitol diuresis). The aged kidney, in the absence of known renal disease, has decreased concentrating ability, along with generally reduced renal function. The mechanism for this is not known but may be related to the decrease in the number of functioning nephrons with age and the resulting increase in the relative amount of solute to be excreted by each remaining nephron (solute diuresis per nephron). Clinically it is an important point, because one cannot assume in the aged the presence of a normal ability to conserve needed water.

DIURNAL VARIATIONS also affect the concentrating mechanisms. Urine flow is lowest and osmolality highest at night. The cycle appears to be related in part to the solute (sodium) excretion cycle (lowest at night) and to the nocturnal increase in ADH activity.

ENHANCED EXCRETION OF SOLUTES in normal man results in an increase in urine flow and change in urine osmolality toward that of plasma (Fig. 13–4). If the kidney is concentrating the urine when solute diuresis occurs, the urinary concentration declines; a rise in urine osmolality occurs when a solute load is imposed on a kidney that is generating a dilute urine. It is apparent from Figure 13–4 that evaluation of maximal or minimal urine osmolality requires consideration of this important effect of solute excretion on Uosm and V. *Postural natriuresis* through a similar mechanism (increase in solute excretion during recumbency) may be accompanied by a lower maximum Uosm in the hydropenic subject. Clinical examples of this *os-*

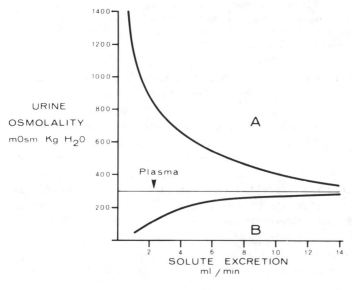

FIG. 13–4. Effect of solute diuresis on urine osmolality during renal concentration (A) or dilution (B) in normal subjects.

motic diuresis are commonly seen in diabetic patients with severe glycosuria, in patients receiving large urea loads following high-protein tube feeding or intravenous infusions of mannitol for the treatment of acute intoxications or for the prevention of acute renal failure during aortic surgery.

DIET. Another influence on the operation of the concentrating mechanism is diet. Salt deprivation in man will reduce T^cH_2O, apparently by decreasing the amount of sodium available for transport at the concentrating site. The influence of protein and nitrogen metabolism on the concentrating mechanism is apparent from the previous discussion of the role of urea. Low-protein diets reduce maximal Uosm and T^cH_2O in man and animals.

URINE pH. Osmolality is less in an alkaline than in an acid urine. The mechanism of this effect of pH on the renal concentrating capacity remains unknown, but may be related to the delivery of a bicarbonate-rich and chloride-poor tubular fluid to the distal concentrating site during alkalosis. Since bicarbonate in comparison to chloride is an anion with limited capacity to permeate the thick ascending limb of Henle's loop, this may be accompanied by a decreased transport of sodium chloride into the medullary interstitium and a less effective countercurrent mechanism.

Consequences of impaired concentrating ability

If the kidney cannot concentrate the urine at all, the urine volume will be determined largely by the filtered load of solute and its tubular reabsorption. In such a setting, water cannot be conserved selectively when there is a need to do so. The water loss may then be regarded as obligatory in the sense that it is determined by factors unrelated to any consideration of water homeostasis. Thus depending upon solute excretion, there may be inappropriate diuresis (rela-

tively large urine volume in the presence of water deficit and enhanced ADH activity) or frank *polyuria*. If the water loss is not balanced by adequate intake, water deficit or *dehydration* may result. Dehydration stimulates *thirst* as a consequence of the increase in plasma osmolality. Since the water deficit involves total body water, including the extracellular fluid, dehydration may result in *contraction of extracellular fluid volume* and *loss of body weight*. In patients with marked impairment in renal function, this results in further *reduction in filtration rate*. If the pure water loss exceeds the loss of solute, total body water hyperosmolality with increased serum sodium concentration results.

Pathophysiology of impaired concentration

Reduced concentrating ability occurs when there is an impairment in the establishment of medullary hyperosmolality, decreased permeability to water in the distal tubule or collecting duct, or a combination of both.

IMPAIRED MEDULLARY HYPEROSMOLALITY. This may occur when any of the circumstances necessary for the establishment of medullary hypertonicity are altered.

Factors that interfere with the delivery of sodium chloride for reabsorption in the loop of Henle may result in impaired concentrating ability. Conceivably, this may obtain when the filtered load of sodium is so reduced (low GFR, hyponatremia) or when the proximal reabsorption of sodium is so increased that less sodium reaches the loop of Henle. It has also been suggested that a low lumenal sodium concentration reaching the concentrating site (during mannitol diuresis) may limit sodium reabsorption by the ascending limb. A defect in the capacity of the loop for the transport of sodium chloride could result in failure to create a hypertonic medullary interstitium. For example, the loop diuretics (furosemide, ethacrynic acid) abolish

T^{c}H$_2$O in man by blocking sodium chloride reabsorption in the ascending limb of Henle's loop (in addition, recycling of urea is also greatly diminished because of the decreased urea concentration in the medullary collecting fluid). It is also possible that volume expansion (through any mechanism whatever), may decrease sodium transport at the same site and also decrease medullary hypertonicity. In addition, the presence of increased amounts of bicarbonate in the tubular fluid reaching the loop of Henle may restrict solute reabsorption by the thick ascending limb, provided that the distal delivery of chloride is concomitantly reduced, since NaCl rather than NaHCO$_3$ is the sodium salt required to generate T^{c}H$_2$O. Finally, if the flow of blood in the medulla is increased, the removal of water and solutes from the medullary interstitium may be accelerated (washout effect).

DECREASED PERMEABILITY OF THE DISTAL TUBULE OR COLLECTING DUCT TO WATER. Decreased permeability to water occurs when there is inadequate production and liberation of ADH. In addition, permeability may be reduced despite the presence of adequate ADH if the cells of the distal tubule and collecting duct are so altered that they fail to respond to the hormone or cannot actually accomplish water reabsorption. In addition, rapid flow of tubular fluid reduces the possibility for osmotic equilibration in the distal tubule and collecting duct (i.e., during osmotic diuresis due to glucose, mannitol, NaCl or urea, or augmented solute excretion induced by diuretics of any type).

Diseases associated with impaired concentrating ability

Much is known about the normal concentrating mechanism and, from that, one can reason in regard to pathophysiology. Nonetheless, in many diseases associated with impaired concentrating ability, the mechanism(s) of the alteration is not completely understood. Since in many individual disorders more than one mechanism may be responsible for the concentrating defect, no classification of the abnormalities will be attempted here; rather, they will be simply described. It should also be stated that these various mechanisms are not mutually exclusive and that it is conceivable that in a given patient or experimental model they may be present and act concomitantly or in sequence, depending on the course and severity of the disorder or experimental situation.

The term *nephrogenic diabetes insipidus* is commonly employed to identify a clinical syndrome with multiple causes characterized by the failure of the kidney to respond appropriately to ADH or exogenous vasopressin administration (*vasopressin-resistant polyuria*). In contrast, patients with pituitary diabetes insipidus (because of decreased synthesis of ADH) and patients with primary or psychogenic polydipsia (because of sustained physiologic suppression of ADH by continued excessive water ingestion) have low circulating levels of ADH and their polyuria is *vasopressin responsive*. Patients with nephrogenic diabetes insipidus who have normal circulating levels of ADH can be further separated into those of congenital origin (hereditary nephrogenic diabetes insipidus) and those associated with a variety of processes that affect the renal concentrating operation (acquired nephrogenic diabetes insipidus).

CHRONIC RENAL DISEASE. Regardless of cause, this disorder results in impaired concentrating ability. As the renal disease progresses, the decreased maximal Uosm approaches and becomes equal to the osmolality of plasma, a condition described as *isosthenuria*. For this concentrating defect to become apparent, the GFR usually should be below 50 ml./min. Initially, it may only be manifested as *nocturia* (reversal of the normal circadian rhythm of higher urine volume during daytime), subsequently by minimal polyuria, and only rarely by important polyuria.

Some patients with far advanced chronic renal failure of diverse etiology have a vasopressin-resistant *hyposthenuria* (the persistence of Uosm hypotonic to plasma in spite of the administration of supramaximal doses of vasopressin). The precise mechanism(s) whereby this occurs is not established, although at least one important factor is an enhanced solute excretion per nephron. There is a decreased number of nephron units in chronic renal disease and a compensatory increase in GFR per nephron. The latter results in increased filtered load per remaining nephron, decreased proximal reabsorption and, consequently, delivery of too much tubular fluid to the loop. An additional cause of inadequate concentrating ability relates to defective tubular responsiveness to ADH. The recently demonstrated functional unresponsiveness of the uremic cortical collecting duct to ADH gives further support to this hypothesis and may explain the presence of hyposthenuria in some patients with chronic renal failure. In other patients, structural inability to reabsorb water may be important. The latter mechanism may be particularly relevant in those renal diseases with predominant medullary involvement, i.e., chronic interstitial nephritis (including chronic [reflux] pyelonephritis, polycystic renal disease, obstructive uropathy, sickle-cell disease, analgesic nephropathy). In these disorders, structural derangement may alter the tubular, vasa recta, and interstitial spatial arrangements so vital for the normal function of the countercurrent system.

A group of uncommon clinical diseases (medullary cystic disease, cystinosis, Sjögren's syndrome) are characterized by predominant disease of the medullary structures (selective medullary destruction or fibrosis) that results in polyuria and hyposthenuria in apparent disproportion to the reduction in GFR. Also, amyloid deposits along the collecting tubule were found in a patient with nephrogenic diabetes insipidus. The concentrating defect appears to be related to a faulty water permeability of the morphologically-altered tubular medullary structures.

HYPERCALCEMIC NEPHROPATHY. The earliest renal functional abnormality in hypercalcemic nephropathy is impaired concentrating ability. Indeed, polyuria, thirst, and polydipsia may be recognized in patients with hypercalcemia before the plasma abnormality is detected. The early structural changes include damage to the loop of Henle and the collecting duct. It is, therefore, possible that the concentrating defect is due to altered ability of the ascending limb of Henle's loop to transport sodium (T^cH_2O is abolished and CH_2O is reduced) and/or decreased ability of the cells of the collecting duct to reabsorb water or to respond to ADH. There is considerable *in vitro* experimental evidence that calcium interferes with the action of vasopressin by inhibiting adenyl cyclase generation of cyclic $3'$-$5'$ adenosine monophosphate (cAMP). Finally, renal prostaglandins have also been implicated in the polyuria of hypercalcemia; their role, however, remains unproven. The reduction in GFR that frequently accompanies hypercalcemia may also contribute to the defect in renal concentration. This appears, however, after the Uosm defect is already present. Since correction of the hypercalcemia may be followed by reversal of the concentrating defect, one may postulate functional changes in the loop and the collecting duct prior to anatomical damage or healing of the structural abnormality. Hypercalcemic nephropathy may be associated with extensive renal damage, considerable calcium deposition in the parenchyma (nephrocalcinosis), and consequent renal failure.

HYPOKALEMIC NEPHROPATHY. This is a renal lesion characterized by impaired concentrating ability and only little reduction in GFR. In general, the polyuria is modest, 2 to 4 l./day, is vasopressin resistant and appears when potassium loss reaches about 200 to 300 mEq. in adult man. The diluting mechanism may be normal, but large potassium deficits reduce CH_2O both in animals and man. It is difficult to account for the defect in concentrating ability, since the major anatomical lesions in man and dog consist largely of vacuolization of the cells of the proximal tubule, while florid lesions in the papilla and collecting ducts predominate in rodents. The ultrastructural lesions both in man and rats consist of hydropic vacuolization and cytoplasmic granules containing lysosomal hydrolytic enzymes (i.e., acid phosphatase), suggesting that potassium depletion results in early increases in the formation of new cell membrane phospholipids. Although the exact mechanism remains unknown, the available evidence suggests that potassium deficiency decreases medullary hypertonicity. The early reports of increased medullary blood flow (with washout of interstitial solutes) secondary to the increased prostaglandin E_2 vasodilatory activity in experimental potassium depletion were not confirmed by subsequent work. Indeed, in the rat with prolonged hypokalemia a decreased papillary flow was found, probably the result of increased thromboxane A_2 activity (a prostaglandin with potent vasoconstrictive properties). Early in the development of the concentrating defect, when papillary plasma flow is still normal, the defect may be due to an altered generation of cAMP by the epithelium of the collecting duct resulting in vasopressin resistance. Despite recent claims that polydipsia may precede polyuria in this disorder, the available evidence suggests the opposite.

EDEMATOUS STATES. Cirrhosis of the liver, nephrotic syndrome, congestive heart failure, and other edematous states are often associated with mild reversible impairment in concentrating ability of little clinical relevance. (The situation may be more complicated in the nephrotic syndrome, in which, in some instances, there is renal functional deterioration due to renal disease.) Although the mechanism of impaired concentrating ability is not known, two possible explanations have attracted attention. First, these diseases are characterized by greatly increased proximal tubular reabsorption, resulting in decreased delivery of salt to the loop of Henle. However, increasing the distal delivery of NaCl in normokalemic decompensated cirrhotic patients did not normalize depressed T^cH_2O formation. Second, it has been suggested that edematous states are associated with a dislocation of renal blood flow from cortex to medulla. It is conceivable that this alteration in intrarenal blood flow, in some way not entirely apparent, limits the ability to establish and maintain a hypertonic medullary interstitium. One may speculate that a shift in blood flow from outer cortex to corticomedullary nephrons might increase medullary blood flow and result in a washout of medullary solutes. There is no proof, however, for this hypothesis. In addition, in cirrhosis of the liver, the reduced availability of urea (through malnutrition or decreased synthesis) further reduces maximal Uosm.

POSTOBSTRUCTIVE NEPHROPATHY. This condition, which may be accompanied by transient decreased concentrating ability, refers to the large diuresis following the relief of acute urinary tract obstruction. Associated with a solute diuresis due to the high blood levels of urea, there is a defect in sodium reabsorption, and some evidence exists for renal unresponsiveness to ADH, possibly mediated by enhanced prostaglandin E_2 activity.

ACUTE PYELONEPHRITIS. This is usually accompanied by an alteration in maximal Uosm resulting from a decreased medullary hypertonicity. This mild defect is rapidly reversed by treatment of the infection.

The mechanism of the concentrating defect remains undefined.

ACUTE TUBULAR NECROSIS. In the diuretic phase, acute tubular necrosis is associated with abnormal concentrating ability. The large urea load and its osmotic effect may be part of the explanation, but apparently it is not the entire story. A defect in sodium reabsorption, with or without impaired permeability, may also be a factor. Recent experimental work in animal models of acute renal failure supports a defect in the generation of the medullary hypertonicity, i.e., an impaired transport of NaCl by the thick ascending limb of Henle's loop, underlying the concentrating defect observed in acute tubular necrosis. This abnormality may persist for several weeks or months, with eventual recovery, or a modest reduction in Uosm may be permanent. A mechanism similar to that assumed for the post-obstructive kidney and for acute tubular necrosis is probably responsible for the transient polyuria observed after *renal transplantation*.

SICKLE-CELL DISORDERS. Patients with sickle-cell anemia are unable to concentrate the urine normally despite having elevated or normal renal plasma flow and GFR. Many explanations have been offered, mostly dealing with altered circulation in the medulla. Characteristically, this defect is more pronounced at low urine flows (decreased Uosm maximum) than during osmotic diuresis (T^cH_2O formation may be, indeed, normal or close to normal). Since in addition, CH_2O formation is normal in sickle-cell disease it has been assumed that the inherent transport capacity of the thick ascending limb is intact in this disorder.

The defect is reversible in children and young adults with the transfusion of blood containing normal hemoglobin, suggesting that the concentrating problem probably is related to the abnormal red blood cells and is not a genetic defect of the kidney. In addition, it has been demonstrated that red blood cells from patients with sickle-cell disease become sickled in hypertonic salt solutions (>1000 mOsm./kg. H_2O). Thus, it has been postulated that increased sickling in the hypertonic medulla (during hydropenia) causes impaired circulation in the vasa recta secondary to increased viscosity of the blood containing sickled red blood cells. This will decrease the supply of oxygen to the cells of the loop of Henle, resulting in decreased salt transport and diminished medullary hypertonicity and low Uosm during hydropenia. During osmotic diuresis, when medullary blood flow increases and the tonicity of the medulla is diminished below the hypertonicity of the hydropenic state, the ability of the loop of Henle to reabsorb sodium and of the kidneys to form T^cH_2O improves or approaches normal as a consequence of the diminished sickling and improved medullary circulation. This hypothesis will be operative mostly at the beginning of the disease (infants, early childhood), since Uosm rapidly deteriorates in these patients as they grow older and the papilla will seldom achieve a tonicity above 500 mOsm./Kg. H_2O because of progressive and severe structural damage.

Considerable renal papillary damage with almost complete absence of vasa recta vessels, focal scarring and distortion of tubular structures have been described in adult patients with sickle-cell anemia or sickle-cell disease, explaining why the Uosm defect is not reversible by transfusion of normal red blood cells in older patients. A milder renal concentrating defect and the papillary abnormalities have also been described in sickle-cell trait. Many patients with *hyperviscosity syndromes* (macroglobulinemia, multiple myeloma) have a defect in urine concentration which may improve concomitantly with the amelioration of the impaired microcirculation of the conjunctival and retinal vessels by plasmapheresis. It is assumed that microcirculatory changes associated with high blood viscosity of the medullary circulation are responsible for the concentrating defect.

PRIMARY OR PSYCHOGENIC POLYDIPSIA. Excessive drinking of water over a short or protracted period may result in impaired renal concentrating ability which, in some instances, is not reversible for many weeks or months. The precise mechanism in man is not known, although there is evidence in rats of a persistent decrease in medullary hyperosmolality and urea accumulation.

MALNUTRITION. Patients with severe protein malnutrition, in addition to a decreased GFR and RPF, have a decreased concentrating ability (Uosm max and T^cH_2O) with preservation of the diluting capacity, suggesting a functionally intact thick ascending limb. The concentrating defect improves with protein repletion or with exogenous administration of urea. The decreased concentration of urea in the renal medulla appears to be responsible for this defect.

ADRENAL INSUFFICIENCY. This disorder decreases maximal urinary concentrating ability. Either a defect in transport of solute by the thick ascending limb or an impaired response of cortical collecting tubules to ADH results in decreased medullary hypertonicity. It has been postulated that a high activity of phosphodiesterase in adrenal-insufficient animals decreases the intracellular generation of cAMP and impairs the ADH response. Glucocorticoid replacement improves the collecting duct responsiveness to ADH.

PITUITARY DIABETES INSIPIDUS, no matter what its cause, is characterized by partial reduction or complete absence of ADH activity. This results in impaired permeability of the distal tubule and the collecting duct to water and, therefore, limited (if any) concentrating ability with excretion of large volumes (several liters) of dilute urine. The administration of exogenous vasopressin rapidly corrects the inability to form a concentrated urine. A delay in reaching a Uosm maximal value (greater than 800 mOsm./kg. H_2O), however, may be frequently observed. Experimental work has shown that this delay is secondary to a slow reconstitution of the normal medullary hypertonicity due for the most part to delayed accumulation of urea in the medullary structures. Since, even during water diuresis, the medullary structures remain modestly hypertonic to plasma (Fig. 13–3B), it is possible that, during severe dehydration (decreased GFR, increased proximal reabsorption), urine osmolality in these patients may increase toward plasma osmolality. A decrease in tubular fluid flow rate, with increased back-diffusion of water into the interstitium, is the probable explanation. A similar mechanism explains the successful reduction of the polyuria in patients with central and nephrogenic diabetes insipidus by the use of diuretics (i.e., by inducing volume contraction, increased proximal reabsorption and decreased distal delivery of filtrate). It appears that an enzymatic defect (deficient production of cAMP after ADH stimulation) may be responsible for the vasopressin unresponsiveness of the *hereditary nephrogenic diabetes insipidus*. This is a rare X-linked recessive disorder affecting males shortly after birth, with severe polyuria, volume contraction, hypernatremia, and growth and mental retardation. Plasma arginine vasopressin (AVP) levels are normal and there is an absolute lack of response to vasopressin. Affected girls exhibit a partial response to vasopressin.

DRUG-INDUCED IMPAIRED RENAL CONCENTRATING ABILITY (DRUG-INDUCED POLYURIA). A number of drugs, some commonly used, are known to cause alterations of the renal ability to conserve water (see list below). Occasionally, this may result in clinical polyuria (drug-induced polyuria). These drugs may either interfere with the release or availability of ADH or may impair the renal concentrating process itself (the renal action of ADH may or may not be involved). Some drugs such as lithium appear to share more than one mechanism. Only a few will be described here. *Alcohol* inhibits ADH release and its administration is sometimes used to test the pituitary component of ADH release. Alcohol blocks the antidiuretic effect of hypertonic saline and nicotine, and its own diuretic effect is abolished by exogenous vasopressin. *Lithium carbonate* salts are extensively used for the treatment of manic depressive disorders. Lithium produces a reversible impairment of renal concentrating ability even with blood lithium levels maintained within the therapeutic range (0.5–1.5 mEq./l.). The mechanism of action remains undefined. A central effect (inhibition of ADH release) and a renal effect have been proposed. The major renal effect of lithium is to interfere with vasopressin-stimulated production of cAMP. However, there is evidence also supporting an effect of lithium after cAMP is generated (modest interference with protein kinase). *Demeclocycline* (Declomycin), an antibiotic commonly used for the treatment of acne, also produces a reversible, dose-dependent polyuria. All adult patients receiving 1200 mg./day of Declomycin exhibited polyuria. Declomycin antagonizes the ADH-cAMP system interaction. *Fluoride*, a metabolite of the fluorinated general anesthetic agent methoxyflurane, also reduces concentrating ability and at high blood levels is associated with polyuria. Although its mechanism of action remains unclear, interference with the vasopressin-stimulated water flow in the isolated toad bladder has been reported. Colchicine and other vinca alkaloids (vincristine) disrupt intracellular microtubules and microfilaments blocking the action of vasopressin in the isolated toad bladder. These intracellular elements form part of the cytoskeleton and play a role in cell

---

Drug-induced Impaired Renal Concentrating Ability (Drug-induced Polyuria)

Drugs that interfere with central ADH release
Alcohol*
Phenytoin
Norepinephrine
Lithium* (minor effect)

Drugs that interfere with the renal concentrating mechanism
ADH related
  Lithium* (major effect)
  Demeclocycline*
  Methoxyflurane (fluoride)*
  Colchicine, vinca alkaloids
  Amphotericin B
  Aminoglycosides
  Propoxyphene (Darvon)
ADH unrelated
  Lithium*
  Sulfonylureas (glyburide, tolazamide, acetohexamide)

---

* Of clinical relevance; i.e., capable of producing polyuria

movement; they are assumed to mediate the action of vasopressin. No polyuria has been reported in humans after the use of colchicine or vincristine; in fact, hyponatremia may develop.

Pathophysiology of urine dilution

The normal mechanism

Urine becomes diluted by the reabsorption of sodium chloride in the more distal segments of the nephron, where it is reabsorbed without accompanying water. The absence of ADH in the circulating peritubular blood causes a reduction in the permeability of the distal tubule and the collecting duct to water. Thus, tubular fluid is being "freed" of its solutes (sodium chloride), resulting in the excretion of a large volume of urine (water diuresis) of low total solute concentration (dilute urine). The concentrating and the diluting mechanism share some common steps and are closely interrelated. The essentials of the diluting process are as follows.

SODIUM TRANSPORT IN THE DILUTING SEGMENTS OF THE NEPHRON. As described in the discussion of the concentrating mechanism, above, the tubular fluid leaving the ascending limb of Henle's loop and arriving at the distal tubule in the cortex is *always* hypotonic, which results from the active transport of sodium chloride out of the lumen of the ascending limb into the interstitium (Fig. 13–3*B*). This segment of the nephron not permeable to water irrespective of the presence or absence of ADH, (i.e., whether the kidney is conserving or excreting water), has been designated the *medullary diluting segment.* Anatomically it corresponds to the thick outer medullary portion of the ascending limb of Henle's loop. Deep in the medullary structures (at the tip of the papilla), the tubular fluid flowing along the descending limb has become hypertonic to plasma (even during water diuresis, although much less so than during hydropenia), and, as it flows through the ascending limb, its osmolality decreases from hypertonic to isotonic because of the continued extrusion of NaCl (see Fig. 13–3*B*).

In the cortex, the hypotonic fluid entering the distal tubule at the corticomedullary junction becomes more and more diluted, owing to the continuing active extrusion of sodium chloride by the cells of the distal tubule. Water permeability of this segment (designated the *cortical diluting segment*) is dependent on ADH. In its absence, probably only minimal back diffusion of water occurs (from the hypotonic tubular fluid to the isotonic cortical interstitium). Thus, during water diuresis, the volume of the tubular fluid remains practically unchanged during its transit through the cortex.

Finally, the hypotonic tubular fluid enters the collecting duct, also rendered water-impermeable by the absence of ADH, and starts its journey into the medulla. Further reduction in osmolality occurs here (to the maximum diluting capacity— minimal urine osmolality), resulting from continued reabsorption of sodium chloride in this segment.

ANTIDIURETIC HORMONE. The role of ADH is central to the diluting mechanism. In the absence of circulating blood levels of ADH, (AVP is the ADH circulating in man) the permeability of the distal tubule and the collecting duct to water is reduced maximally, resulting in the excretion of a large volume of hypotonic urine (water diuresis). Some degree of back diffusion of water does occur in the absence of ADH, but this in itself does not appear to reduce tubular fluid volume to any significant extent under normal conditions. Experimentally, however, it is possible to demonstrate the production of hypertonic urine in the absence of ADH.

A detailed description of the physiology of ADH is beyond the scope of this chapter, but a brief summary follows. Under normal physiological conditions, inhibition of the release of ADH by the posterohypophysis results from a decrease in the osmolality of the blood perfusing the supraoptic nuclei or an increase in intravascular volume, or both. This decrease in blood osmolality is the result of a positive water balance, usually resulting from water drinking. Conversely, an increase in blood osmolality and/or a contraction of intravascular volume results in the release of ADH by the posterohypophysis (physiological osmotic inhibition or stimulation of ADH release). The control of plasma AVP levels by changes in plasma osmolality is very close. A 1 per cent increase in plasma osmolality caused by water deprivation increases plasma AVP concentration in normal man. Conversely, a 1 per cent decrease in plasma osmalality caused by water ingestion results in a suppression of AVP release. The nonosmotic stimuli for ADH release are less sensitive than the osmotic stimuli, necessitating approximately a 5 to 10 per cent decrement in blood volume to stimulate ADH release. Once the threshold for the volume stimulation of ADH release is reached, however, a rapid and geometric rise in AVP levels occur. The clinical implications of these relationships are important in explaining impaired diluting ability in many patients (dissociation of nonosmotic and osmotic stimuli).

Although the major factors affecting the synthesis and release of ADH are changes in blood osmolality and extracellular fluid volume, other factors also influence its release. Thus, pain, anxiety, and a variety of drugs including nicotine, central nervous system depressants such as morphine, barbiturates, and anesthetics, β-adrenergic agents (isoproterenol), vincristine, clofibrate (Atromid-S), and carbamazepine (Tegretal) are capable of stimulating the release of ADH (see list on p. 288). On the other hand, alcohol, α-adrenergic agents (norepinephrine) (see list on p. 288) and paroxysmal tachycardia (probably through stimulation of left atrial stretch receptors) may inhibit its release. There is strong experimental evidence suggesting that ADH increases tubular permeability to water by binding to receptors in the basolateral membrane of the distal tubule and collecting duct and stimulating adenyl cyclase activity. Cyclic AMP is the actual mediator of the hormonal action. Intracellular cAMP is also activated by the methylxanthine theophylline, which inhibits the enzyme, phosphodiesterase, which in turn catalyzes the conversion of cyclic AMP into inactive 5'-AMP. Like ADH, cAMP and theophylline have been shown to increase water permeability in the isolated toad bladder and in the cortical collecting duct of the rabbit kidney. Once formed, cAMP activates protein kinase, and this enzyme transfers a phosphate from ATP to the luminal membrane of the cell, somehow altering its permeability to water. The role of the cellular microtubule system in the mechanism of the cellular action of ADH remains speculative.

ADDITIONAL FACTORS. Any factor that *decreases* the osmolality of the normally hypertonic medullary interstitium will favor the diluting mechanism by minimizing or eliminating the back diffusion of water.

1. For example, during water and osmotic diuresis, the inner medullary and the papillary hypertonicity are reduced, in spite of continued transport of sodium chloride into the interstitium. An increase in the linear velocity of the tubular fluid that flows through the loop of Henle, a possible increase in the diameter of the lumen of these loops, and an increase in medullary blood flow may be some of the factors responsible for this washout of the medullary hypertonicity during the diuretic state.

2. An increase in CH_2O is found after administration of adrenal glucocorticoids. Although an increase in the glomerular filtration rate and consequent increase in the distal delivery of sodium could account for this, there also is evidence that these hormones may have a direct effect on the permeability of certain biological membranes to water in the absence of ADH. Glucocorticoids physiologically may allow the diluting segments of the nephron to become maximally impermeable to water in the absence of ADH ("permissive" role of the glucocorticoids in water excretion). The recent observations that isolated cortical collecting tubules of adrenalectomized rabbits are impermeable to water in the absence of ADH and that water-loaded glucorticoid-deficient dogs have elevated plasma ADH levels (secondary to baroreceptor estimated ADH release) mitigate against a permissive role of glucocorticoids in water permeability.

3. Since contrary to early reports it appears that angiotensin II does not stimulate the release of ADH, it may be concluded that the renin-angiotensin system does not play any important physiological role in the control of ADH release.

4. Current information suggests that renal prostaglandins (*e.g.*, PGE_2) may act as important modulators of renal water ex-

cretion. An increase in prostaglandin (PG) activity may result in diuresis by antagonizing vasopressin activity at the level of the collecting ducts and by decreasing medullary hyperosmolality. The latter results either from an increase in medullary blood flow or a decrease in NaCl reabsorption by the thick ascending limb of Henle. It has been hypothesized that a negative-feedback circuit may exist between medullary osmolality and intramedullary production of PG. A decrease in PG production would lead to an increase of NaCl and urea accumulation in the medullary tissues, and this would stimulate PG production and lower medullary solute content.

Factors that influence diluting ability

There are some physiological variables that influence the diluting process.

AGE. Infants dilute urine (Uosm) to the same extent as do older children or adults. However, the quantitative values for maximal V and CH_2O during water diuresis are limited by their low glomerular filtration rate.

CIRCADIAN VARIATIONS. The response to water loading is delayed and decreased at night.

INCREASED SOLUTE EXCRETION. When solute excretion increases during water diuresis, the result is increased V and CH_2O with Uosm rising toward isotonicity (Fig. 13–4). Conversely, decreased solute excretion (low-sodium diet) may decrease maximal V and CH_2O during water diuresis while further reducing Uosm.

POSTURAL CHANGES. In the upright posture, the response to water loading is reduced, compared to the response in recumbency. Decreased GFR, solute excretion, and release of ADH are probably determinant factors.

Consequences of impaired diluting ability

Impaired diluting capacity is the most common cause of *hypoosmolality* and *hyponatremia* observed in clinical practice. Abnormal dilution means that the kidney cannot eliminate solute-free water selectively when there is need to do so. In such a case, the defense of the body water homeostasis has to rely on extrarenal mechanisms (insensible loss, sweat) that are not physiologically equipped to meet these demands.

Impaired water diuresis results in *water retention*, with absolute or relative hypoosmolality, hyponatremia, hemodilution, and weight gain. If the positive balance of water develops rapidly or is great enough, the syndrome of *water intoxication* may develop. Thus, loss of appetite; nausea or vomiting; irritability, confusion, depression or loss of reflexes; weakness, pyramidal tract signs, stupor or convulsions may appear.

If the kidney cannot dilute the urine at all in the absence of ADH, the urine volume will be determined largely by the filtered load of solute and its reabsorption.

Pathophysiology of impaired dilution

Reduced diluting ability occurs when there is increased permeability to water of the diluting segments of the nephron in the presence or absence of antidiuretic hormone; decreased delivery of tubular fluid to the diluting segments; and decreased transport of sodium chloride by the diluting segments.

INCREASED PERMEABILITY TO WATER. This may occur when there is a sustained blood level of circulating ADH or of substances with ADH-like action increasing the permeability of the distal convoluted tubule and the collecting duct to water. The release of ADH may be appropriate or inappropriate for the stimuli present. *Appropriate release of ADH* from osmotic stimuli occurs when there is an increase in plasma osmolality caused either by a loss of water (water deprivation) or grain of salt. An appropriate release of ADH from nonosmotic stimuli occurs as a result of emotional stress and pain (usually transient), hypovolemia secondary to blood loss or extracellular volume depletion, low cardiac output, and edematous states. The mechanism(s) whereby the systemic hemodynamic changes induced by volume contraction might modulate ADH release are not completely understood. However, it has been suggested that these changes may influence ADH release through decreased stimulation via parasympathetic afferent fibers, through pressure-flow changes (baroreceptors). A change in pressure in the left atrium or at the level of the aortic arch or carotid sinus may alter parasympathetic afferent tone to the hypothalamic-neurohyposphyeal tract and thereby affect the rate of vasopressin release. It is not clear whether the receptors in the low-pressure or high-pressure system are the more sensitive. The use of certain drugs is accompanied by a nonosmotic stimulus for ADH release (morphine, nicotine, cholinergic agents, etc.—see list on p. 288). *Inappropriate release of ADH* means that elevated or nonsuppressed levels of plasma AVP (the appropriate response) persists in the presence of hypoosmolality and volume expansion. Release of ADH in the absence of appropriate stimuli is discussed under the heading Inappropriate Secretion of ADH.

Increased permeability of the diluting segments to water may occur in the absence of ADH (enhancement of back diffusion of water). This may be secondary to decreased tubular fluid flow or lack of the permissive effect of adrenal glucocorticoids.

REDUCTION IN THE DELIVERY OF FILTRATE TO THE DISTAL DILUTING SITE. This may result in impaired diluting ability. This may obtain when the volume of filtrate (low GFR) is so reduced or the proximal reab-sorption of salt is so increased (ECFV depletion) that less filtrate reaches the loop of Henle and the distal tubule.

DECREASED SODIUM TRANSPORT. Decreased sodium transport by the diluting segments of the nephron decreases their capacity to generate solute-free water (diuretics). Massive ECFV expansion, if not accompanied by hyponatremia, restricts CH_2O generation by decreasing sodium chloride reabsorption in the diluting segment. It appears that the ratio of chloride to bicarbonate delivery to the diluting segment may also affect CH_2O formation, since CH_2O decreases during administration of acetazolamide (Diamox, a carbonic anhydrase inhibitor).

Diseases associated with impaired diluting ability

Many diseases associated with poor diluting ability may involve the simultaneous operation of several abnormal mechanisms. A major obstacle to a better understanding of these mechanisms has been the difficulty of determining whether or not ADH secretion has been normally suppressed by water loading. The recent development of a sensitive and specific radioimmune assay for ADH has been of great value in resolving this problem. Nevertheless, the presence or absence of ADH activity can be clinically evaluated by using the classical criteria outlined in Figure 13–5.

EXTRACELLULAR FLUID VOLUME DEPLETION AND HYPOVOLEMIA. ECFV depletion and hypovolemia, of whatever cause, are common clinical situations which can be associated with fluid retention and hyponatremia if hypotonic fluid is ingested or administered in excess. The mechanism that underlies this usually transient decrease in diluting capacity is a reduction in renal blood flow and glomerular filtration rate with enhanced proximal reabsorption of filtrate and consequent decreased delivery of tubular fluid to the diluting sites. In addition,

	ANTIDIURESIS	WATER DIURESIS
	ADH ↑	ADH ↓
URINE FLOW	↓	↑
URINE OSMOLALITY	↑	↓
CH_2O	↓	↑
C_{OSM}	NOT ↓	NOT ↑
GFR	NOT ↓	NOT ↑
PLASMA AVP	INCREASED	DECREASED

FIG. 13–5. Analysis of urine flow and urine total solute concentration, together with glomerular filtration rate as an indirect evaluation of ADH activity.

ADH activity is usually appropriately elevated, since volume contraction is a strong stimulus for ADH release. Under these conditions the renal capacity to generate CH_2O is limited, and in response to hypotonic fluid administration water retention, hypoosmolality, and hyponatremia will ensue. If the resulting water retention is sufficiently large, even the signs or symptoms of volume deficit may be obscured.

DIURETICS. Abuse of diuretics probably represents the most common cause of impaired diluting capacity and hyponatremia. Loop diuretics (furosemide, ethacrynic acid) block NaCl reabsorption by the medullary diluting segment, limiting proximal tubule lumen hypoosmolality and thus curtailing CH_2O generation. In addition, by causing ECFV contraction (this also stimulates ADH release) they result in insufficient distal solute delivery to Henle's loop. Furthermore, concomitant potassium depletion may play a role, since potassium replacement corrects the hyponatremia. Thiazides' major effect is to block NaCl reabsorption in the cortical diluting segment. As with other potent diuretics, secondary ECFV contraction and potassium depletion will further impair diluting ability.

CHRONIC RENAL DISEASE. Chronic renal disease, regardless of cause, if sufficiently advanced, results in impaired diluting ability. Characteristically, in many of these patients, the diluting capacity persists unaltered or is moderately decreased for longer periods when concentrating ability is grossly impaired. The precise mechanism(s) of the impairment in dilution is not clear but may be related to the decrease in the number of functioning nephrons and the consequent absolute decrease in the distal delivery of filtrate.

The relative preservation of dilution over concentration may simply result because the absolute change necessary in forming, from isosmotic urine, a maximally concentrated urine is greater than the changes needed to form a maximally diluted urine (see Fig. 13–4). Many patients with chronic renal disease have a normal capacity to generate CH_2O when expressed per 100 ml. GFR, (i.e., in relation to the number of remaining functioning nephrons), suggesting that the function of the thick ascending limb of Henle's loop remains intact. This is in accord with the intact nephron hypothesis, rather than with the anatomical destruction of the diluting sites. In contrast, there is evidence that the experimentally diseased kidney always generates a greater amount of free water per unit of filtrate than does the normal organ. Perhaps those factors that decrease the back diffusion of water in the absence of ADH are relevant to the relative preservation of dilution. In terminal chronic renal disease with oliguria, the factor limiting water excretion is the extreme reduction in GFR, which severely limits the rate at which CH_2O can be generated. The ingestion of water in excess of that volume of CH_2O that can be generated daily will result in a positive water balance and hyponatremia.

ACUTE RENAL FAILURE. In oliguric acute tubular insufficiency, the limiting factor for water excretion is the marked reduction in GFR. During the oliguric phase, the urine is isosmotic.

EDEMATOUS STATES. Reversible abnormalities in water excretion are frequently encountered in patients with congestive heart failure, cirrhosis of the liver, and nephrosis and together with the abuse of diuretics represent the most common causes of hyponatremia. The mechanisms involved are complex. Impaired water diuresis has been attributed to impaired suppression of ADH during active sodium retention. A lack of suppression of AVP levels has been recently reported in decompensated cirrhotic patients with hypoosmolality. Decreased inactivation of ADH by the diseased liver also has been considered, but patients with cirrhosis apparently inactivate vasopressin as well as normals.

In the development of abnormal dilution in edematous states the most important factor appears to be a decrease in the delivery of filtrate to the distal diluting sites. Decreased GFR, with or without hyponatremia, reduces the filtered load of sodium; the decrease in distal delivery of filtrate is further enhanced by the augmented reabsorption of sodium in the proximal tubule, characteristic of these diseases. Measures that increase the distal delivery of filtrate (i.e., mannitol infusion, hypotonic saline) improve CH_2O generation in patients with both decompensated cirrhosis and congestive heart failure. In addition, increased back diffusion of water resulting from slowly flowing tubular fluid may reduce water excretion. In some cirrhotic patients with salt retention, the distal supply of filtrate is not reduced suggesting that the excessive tubular reabsorption of sodium occurs predominately beyond the proximal tubule. On the other hand, in those patients who excrete a small volume of hypotonic urine during water diuresis, excessive proximal as well as distal sodium reabsorption exists. In cirrhosis, water diuresis correlates in general with the severity of the disease. The sickest patients have the poorest responses to water administration, whereas the majority of patients without ascites or edema have normal, or near normal, diluting capacity.

ADRENAL GLUCOCORTICOID DEFICIENCY. Both primary and secondary adrenal insufficiency are characterized by an impaired ability to excrete water normally. The urine is generally hypertonic to plasma and contains a significant amount of sodium. Mineralocorticoid or alcohol administration, high sodium intake and correction of volume depletion do not entirely correct this abnormality. The excretion of water is rapidly normalized, however, by the administration of glucocorticoids. It appears that mineralocorticoids may also be necessary to assure adequate salt reabsorption and normal dilution. The exact mechanisms of the glucocorticoid action remain unknown. They may improve water diuresis by inhibiting the back diffusion of water in the diluting segments of the nephron. An abnormality in the release of ADH by the posterohypophysis (glucocorticoids normally inhibiting the release of ADH) has also been postulated. Other factors that may impair dilution in the untreated patient are reduction in GFR and solute excretion and persistence of ADH release by the stimulus of volume contraction (baroreceptor—mediated ADH release).

Irrespective of mechanisms, these patients are very sensitive to water intoxication. Thus, dilutional hyponatremia adds to the effect of salt depletion on serum sodium concentration in the patient with untreated adrenal insufficiency.

HYPOTHYROIDISM. Primary thyroid deficiency results in a reversible moderate impairment of the diluting capacity in man. Asymptomatic hyponatremia may result in some patients with severe myxedema and in about one half of the patients with myxedema coma. Possible explanations include a decreased GFR, which restricts delivery of filtrate to the distal diluting site, an altered permeability to water and sodium, and a reduced sodium transport in the diluting segments of the nephron. Furthermore, it has been shown recently that thyroid hormone influences active transport of sodium, and decreased sodium reabsorption and natriuresis have been demonstrated in hypothyroid animals. This abnormality of water diuresis in hypothyroidism is not corrected by glucocorticoids or alcohol administration; it improves with thyroid replacement. The suppressability of AVP levels after water administration in patients with myxedema remains to be defined.

INAPPROPRIATE SECRETION OF ADH. This syndrome has been reported to occur in a variety of diseases, including malignant tumors (carcinomas of the lung, duodenum and pancreas; thymomas), disorders involving the central nervous system (meningitis, head trauma, brain abscess and tumors, encephalitis, Guillain-Barré syndrome, acute intermittent porphyria), pneumonia, tuberculosis, and others. The typical features depend upon retention of water resulting from excessive intake of water coupled with the inappropriate secretion of ADH, secondary hypoosmolality and hyponatremia, and continued renal excretion of sodium. The urine osmolality is greater than that appropriate for the concomitant tonicity of the plasma. The urine may not be necessarily hypertonic to plasma (although usually it is), but it is certainly not maximally dilute. In addition, there is no clinical evidence of volume depletion or hypotension, and renal and adrenal function are normal. There is no oliguria. The patient may be asymptomatic, particularly if the serum sodium concentration is above 120 mEq./l. However, when hyponatremia is severe, symptoms of water intoxication usually appear. The syndrome of inappropriate secretion of ADH can be predictably reproduced in normal man by the daily injection of pitressin tannate in oil while maintaining free access to water. The hyponatremia is usually not corrected by large doses of mineralocorticoid steroids or hypertonic sodium chloride if fluid intake is kept high, although positive sodium balance may occur. It follows from the pathophysiological events that simple fluid restriction is the treatment of choice of this syndrome. Under certain clinical circumstances, lithium carbonate or declomycin can be used to correct the syndrome of inappropriate ADH (see above).

DRUG-INDUCED ANTIDIURESIS. It has become apparent in recent years that drugs with various pharmacological properties can exert ADH-like effects causing water retention and marked hyponatremia and oc-

---

Drug-induced Impaired Renal Diluting Ability (Drug-induced Antidiuresis)

Stimulation of thirst
  Antihistamines
  Anticholinergics
  Phenothiazines
  Tricyclic antidepressants
Release of ADH
  Nicotine
  Vincristine*
  Barbiturates
  Morphine
  Clofibrate
  Isoproterenol
  Carbamazepine
Potentiation of ADH-like activity
  Chlorpropramide*
  Tolbutamide
  Biguanides (phenformin, metformin)
  Oxytocin
  Nonsteroidal antiinflammatory agents (aspirin), indomethacin)
  Acetaminophen
Interference with renal NaCl transport
  Diuretics*
  Solute diuretics (mannitol, glucose, glycerol, urea)*
Unknown mechanisms
  Cyclophosphamide*
  Amitriptyline

---

* Of clinical importance, i.e., capable of producing water retention and hyponatremia.

casionally, the appearance of the syndrome of inappropriate ADH secretion (see list on p. 288). The mechanism whereby these drugs affect water excretion has been elucidated in part. The administration of *oxytocin* in large dosage to patients after delivery may cause an antidiuretic effect and severe hyponatremia. Of the drugs known to stimulate the release of ADH only *vincristine* has been associated with the inappropriate ADH syndrome. *Chlorpropamide* has an antidiuretic effect in partial diabetes insipidus; it apparently does so by potentiating the tubular action of vasopressin, perhaps by increasing the production of intracellular cyclic AMP. Its effect requires the presence of small amounts of vasopressin. *Cyclophosphamide* may be associated with water retention and hyponatremia in patients with malignancies undergoing chemotherapy. The risk is particularly high when these patients are given intense hydration to prevent hyperuricemic acute renal failure and hemorrhagic cystitis.

Annotated references

Anderson, R. J., Gordon, J. A., Kim, J., Peterson, L. M., and Gross, P. A.: Renal concentration defect following nonoliguric acute renal failure in the rat. Kidney Int. *21*:583, 1982. (A defect in generation of renal inner medullary interstitial solute is proposed as the mechanism of impaired urinary concentration observed in a model of nonoliguric acute renal failure in the rat: 50-minute complete occlusion of the renal artery and vein with contralateral nephrectomy.)

Bartter, F. C., and Schwartz, W. B.: The syndrome of inappropriate secretion of antidiuretic hormone. Am. J. Med. *42*:790, 1967. (Review of the subject with emphasis on pathophysiology.)

Beck, N., and Shaw, J. O.: Thromboxane B_2 and prostaglandin E_2 in the K^+-depleted rat kidney. Am. J. Physiol. *239*:F151, 1981. (An increase in renal thromboxane A_2 (thromboxane B_2 is a chemically stable metabolite of thromboxane A_2), a prostaglandin with potent vasoconstrictive activity, probably mediates the observed decrease in medullary blood flow in the potassium depleted rat [see Whinnery and Kunau].)

Berliner, R. W., and Davidson, D. G.: Production of hypertonic urine in the absence of pituitary antidiuretic hormone. J. Clin. Invest. *36*:1416, 1957. (Elegant demonstration of the production of urine hypertonic to plasma in the absence of ADH.)

Bichet, D., Szatalowicz, V., Chaimovitz, C., and Schrier, R. S.: Role of vasopressin in abnormal water excretion in cirrhotic patients. Ann. Intern. Med. *96*:413, 1982. (Seven of twelve patients with cirrhosis and ascites have non- or poorly-suppressable plasma AVP levels following an intravenous water load. The authors concluded that nonosmotic stimulation of vasopressin secondary to a decrease in effective blood volume is an important factor in the abnormal water excretion of cirrhosis.)

Boonjarern, S., Stein, J. H., Baehler, R., Osgood, R. W., Hsueh, W., Cohen, S., Yashon, D., and Ferris, T. F.: Effect of plasma sodium concentration on diluting segment sodium reabsorption. Kidney Int. *5*:1, 1974. (Normonatremic or hypernatremic extracellular volume expansion is associated with a maximal [limited] sodium reabsorption capacity in the diluting segment [and limited CH_2O generation] while in the hyponatremic dog sodium reabsorption and CH_2O generation continue to rise even at a marked distal fluid delivery.)

Chaimovitz, C., Szylman, Alroy, G., and Better, O. S.: Mechanism of increased renal tubular sodium reabsorption in cirrhosis. Am. J. Med. *52*:198, 1972. (Some decompensated patients with cirrhosis of the liver studied during combined water and hypotonic saline diuresis had CH_2O generation comparable to control subjects. These findings support the view that under these experimental conditions excessive tubular reabsorption of sodium occurs predominantly beyond the proximal tubule in some patients with cirrhosis.)

Edelman, C. M., Jr.: Maturation of the neonatal kidney. *In:* Becker, E. L.: Proceedings of the Third International Congress of Nephrology. Vol. 3. New York, Karger, 1967. (Review of the functional maturation of the kidney of young infants and comparison to adult indices of renal function.)

Eknoyan, G., Suki, W. N., Rector, F. C., Jr., and Seldin, D. W.: Functional characteristics of the diluting segment of the dog nephron and the effect of extracellular volume expansion on its reabsorptive capacity. J. Clin. Invest. *46*:1178, 1967. (Characterization of the "medullary and cortical diluting segments" of the dog nephron, with a review of T^cH_2O and CH_2O reliability as indices of sodium reabsorption in the loop.)

Fine, L. G., Schlondorff, D., Trizna, W., Gilbert, R. M., and Bricker, N. S.: Functional profile of the isolated uremic nephron: Impaired water permeability and adenylate cyclase responsiveness of the cortical collecting tubule to vasopressin. J. Clin. Invest. *61*:1519, 1978. (The isolated cortical collecting tubule of the rabbit with renal insufficiency exhibits a diminished water permeability in response to ADH.)

Goldberg, M., McCurdy, D. K., and Ramirez, M. A.: Differences between saline and mannitol diuresis in hydropenic man. J. Clin. Invest. *44*:182, 1965. (Evidence is presented in man, suggesting no maximum limit (no Tm) for T^{c}H$_2$O when hypertonic saline is used in place of hypertonic mannitol as the loading solute.)

Gottschalk, C. W., and Mylle, M.: Micropuncture study of the mammalian urinary concentrating mechanism: evidence for the countercurrent hypothesis. Am. J. Physiol. *196*:927, 1959. (Demonstration of the fundamentals of the countercurrent system.)

Imai, M.: Function of the thin ascending limb of Henle of rats and hamsters perfused *in vitro*. Am. J. Physiol. *232*:F201, 1977. (The principle of passive equilibrating models of the countercurrent multiplication system [Kokko and Rector] originally tested in the rabbit is operative in other mammalian species as well: rats, hamsters.)

Imai, M., and Kokko, J. P.: NaCl, urea and H$_2$O transport in the thin ascending limb of Henle: generation of osmotic gradients by passive diffusion of solutes. J. Clin. Invest. *53*:393, 1974. (The important demonstration of the absence of an active salt transport mechanism in the thin ascending limb of Henle. This observation gave strong support to the passive models of urinary concentration [see Kokko and Rector].)

Jamison, R. L., Bennett, C. M., and Berliner, R. W.: Countercurrent multiplication by the thin loops of Henle. Am. J. Physiol. *212*:357, 1967. (Evidence is presented that the thick and the thin segments of the ascending limb of Henle's loop are capable of active sodium transport.)

Jamison, R. L., and Oliver, R. E.: Disorders of urinary concentration and dilution. Am. J. Med. *72*:308, 1982. (Excellent review of the mechanisms responsible for the normal function of the concentrating and diluting processes. Includes a classification of disordered renal concentration by disorders altering (a) fractional reabsorption of solute delivered to the loop of Henle; (b) excretion of solute relative to the sum of solute excretion and solute delivery to Henle's loop; (c) the fraction of solute loss by vascular outflow from the medulla relative to that reabsorbed by the loop; and (d) collecting duct response to ADH.)

Kaye, D., and Rocha, H.: Urinary concentrating ability in early experimental pyelonephritis. J. Clin. Invest. *49*:1427, 1970.

Kokko, J. P., and Rector, F. C., Jr.: Countercurrent multiplication system without active transport in inner medulla. Kidney Int. *2*:214, 1972. (A model for countercurrent multiplication in which both the descending limb and thin ascending limb of Henle operate as purely passive equilibrating segments.)

Kokko, J. P., and Tischer, C. C.: Water movement across nephron segments involved with the countercurrent multiplication system. Kidney Int. *10*:64, 1976. (Review of transport and permeability characteristics of the nephron segments involved in the countercurrent system.)

Leaf, A., Bartter, F. C., Santos, R. F., and Wrong, O.: Evidence in man that urinary electrolyte loss induced by pitressin is a function of water retention. J. Clin. Invest., *32*:868, 1953. (Demonstration of effect of volume expansion [water retention] on sodium excretion—perhaps the most important mechanisms in the pathophysiology of the inappropriate ADH secretion syndrome.)

Levinsky, N. G., and Berliner, R. W.: The role of urea in the urine concentrating mechanism. J. Clin. Invest. *38*:741, 1959. (Passive accumulation of urea in the medullary interstitium increases the maximum Uosm that the dog kidney can attain. Uosm and T^{c}H$_2$O are reduced in dog and man on low-protein diets).

Oliver, R. E., Roy, D., and Jamison, R. L.: Urinary concentration in the papillary collecting duct of the rat: Role of the ureter. J. Clin. Invest. *69*:157, 1982. (When the ureter is intact, over half of the increase in urinary osmolality above isotonicity occurs in the terminal one-fourth [the last 2 mm.] of the medullary collecting duct and is due exclusively to water reabsorption [no net solute addition]. The authors concluded that it is the continuity of the ureter, rather than intermittent flow due to ureteral peristalisis, which is essential for the formation of a maximally concentrated urine. It may be that rather than supplying solute to or removing water from the papilla, the intact ureter prevents the loss of solute from the hyperosmotic papilla).

Pennell, J. P., Lacy, F. B., and Jamison, R. L.: An in vivo study of the concentrating process in the descending limb of Henle's loop. Kidney Int. *5*:337, 1974. (Demonstration in the rat that fluid concentration in the descending limb of Henle's loop is predominantly achieved by osmotic water abstraction, while solute entry contributes only one third of the increase in tubular fluid osmolality.)

Robertson, G. L., Aycinena, P., and Zerbe, R. L.: Neurogenic disorders of osmoregulation. Am. J. Med. *72*:339, 1982. (Scholarly review of water regulation with emphasis on the role of osmoregulation and ADH. Hypoosmolar and hyperosmolar syndromes are described.)

Rocha, A. S., and Kokko, J. P.: Sodium chloride and water transport in the medullary thick ascending limb of Henle: Evidence for active chloride transport. J. Clin. Invest. *52*:612, 1973. (Demonstration that the isolated perfused rabbit medullary thick ascending limb of Henle is water impermeable and has the capacity for active outward solute transport as a consequence of an electrogenic chloride pump.)

Schedl, H. P., and Bartter, F. C.: An explanation for and experimental correction of the abnormal water diuresis in cirrhosis. J. Clin. Invest. *39*:248, 1960. (Impaired response to water administration can be improved by mannitol diuresis. The same authors have demonstrated this effect in congestive heart failure. Thus, increased delivery of solute [sodium] to the distal diluting sites improves diluting ability.)

Schrier, R. W., and Berl, T.: Mechanism of effect of alpha adrenergic stimulation with norepinephrine on renal water excretion. J. Clin. Invest. *52*:502, 1973. (These studies in the dog demonstrate that the water diuresis associated with intravenous norepinephrine is mediated primarily by suppression of vasopressin release rather than by changes in renal hemodynamics, renal innervation, or an effect of norepinephrine on the water permeability of tubular epithelium. However, there may be additional effects of norepinephrine interfering with ADH action at the tubular level, since norepinephrine is known to antagonize the effect *in vitro* of ADH on water transport by the toad bladder. These and other studies explain the old observation that cold exposure in man results in transient water diuresis [cold diuresis], since catecholamine excretion is known to be increased during exposure to low environmental temperatures.)

Schrier, W. R., and Berl, T.: Nonosmolar factors affecting renal water excretion. N. Engl. J. Med. *292*:81, 1975. (Review of factors other than osmotic stimulation that can influence ADH release and water excretion.)

Schwartz, M. J., and Kokko, J. P.: Urinary concentrating mechanism defect in adrenal insufficiency. J. Clin. Invest. *66*:234, 1980.

Serros, E. R.: Prostaglandin-dependent polyuria in hypercalcemia. Am. J. Physiol. *241*:F224, 1981.

Stein, R. M., Levitt, B. H., Golstein, M. H., Porush, J. G., Eisner, G. M., and Levitt, M. F.: The effects of salt restriction on the renal concentrating operation in normal hydropenic man. J. Clin. Invest. *41*:2101, 1962.

Stephenson, J. L.: Central core model of the renal counterflow system. Kidney Int. *2*:85, 1972. (Modified hypothesis of the concentrating operation as described by Kuhn, Hargitay, and Wirz. The passive model component is essentially similar to that simultaneously proposed by Kokko and Rector.)

Stokes, J. B.: Integrated actions of renal medullary prostaglandins in the control of water excretion. Am. J. Physiol. *240*:F471, 1981. (A complete account of the role of renal prostaglandins on water homeostasis. Three physiological effects of prostaglandins have been documented: reduction of vasopressin-dependent osmotic water permeability of the collecting tubule epithelium; enhancement of medullary blood flow, and inhibition of NaCl reabsorption from the thick ascending limb of Henle's loop. The effect of increasing medullary prostaglandin production is to reduce medullary solute content and increase water excretion. Each action is additive in this regard. The medullary prostaglandins thus an-

tagonize the ultimate action of vasopressin. These actions provide an integrated mechanism for the fine tuning of water excretion.)

Strauss, M. B.: Body water in man. The acquisition and maintenance of the body fluids. Boston, Little, Brown & Co. 1957. (A scholarly review of body fluid regulation, with emphasis of volume regulation.)

Tannen, R. L., Regal, E. M., Dunn, M. J., and Schrier, R. W.: Vasopressin-resistant hyposthenuria in advanced chronic renal disease. N. Engl. J. Med. *280*:1135, 1969. (Some patients with far advanced chronic renal disease excrete urine that remained hypotonic to plasma in spite of the administration of supramaximal dose of vasopressin. A combined effect of osmotic diuresis per nephron and a defect in water permeability is the most likely explanation.)

Ufferman, R. C., and Schrier, R. W.: Importance of sodium intake and mineralocorticoid hormone on the impaired water excretion in adrenal insufficiency. J. Clin. Invest. *51*:1639, 1972. (The initial abnormality in renal diluting capacity in adrenal insufficiency may result from a deficiency of mineralocorticoid rather than glucocorticoid hormone. In adrenalectomized dogs withdrawal of glucorcorticoid hormone, for as long as 4 days, does not produce a defect in water excretion if adequate sodium intake or mineralocorticoid hormone is assured.)

Ullrich, K. J., Kramer, K., and Boylan, J. W.: Present knowledge of the countercurrent system in the mammalian kidney. Prog. Cardiovasc. Dis. *3*:395, 1961. (A detailed description of the countercurrent system with historical review of the pioneer work of Kuhn, Hargitay, and Wirz.)

Vaamonde, C. A.: Renal water handling in liver disease. *In* Epstein, M. (ed.): The kidney in liver disease, ed. 2, p. 55. New York, Elsevier, 1982. (A detailed review of the abnormalities of renal diluting and concentrating abilities in liver disease. Major focus is on pathogenetic mechanisms.)

Vaamonde, C. A., and Michael, U. F.: Renal function in thyroid dysfunction. *In* Suki, W. N., and Eknoyan, G. (eds.): The Kidney in Systemic Disease, ed. 2, p. 361. New York, John Wiley & Sons, 1981. (Review of renal abnormalities in hypo- and hyperthyroidism. Complete discussion of abnormalities of water excretion in myxedema. The published data does not support the existence of the syndrome of inappropriate secretion of ADH in severe myxedema.)

Vaamonde, C. A., Oster, J. R., and Strauss, J.: The kidney in sickle cell disease. *In* Suki, W. N., and Eknoyan, G., (eds.): The Kidney in Systemic Disease, ed. 2, p. 159. New York, John Wiley & Sons, 1981. (Detailed description of all renal abnormalities observed in sickle cell hemoglobinopathies, including impaired water conservation.)

Vaamonde, C. A., and Papper, S.: The kidney in liver disease. *In* Earley, L. E., and Gottschalk, C. W. (eds.): Strauss and Welt's Diseases of the Kidney, ed. 3,

p. 1289. Boston, Little, Brown & Co., 1979. (Review of the abnormal renal dilution and concentration capacities in patients with liver disease. Patients with cirrhosis may excrete a normally diluted urine [Uosm less than 100 mOsm./kg. H_2O] but have subnormal values for V and C_{H_2O} in response to water administration. Increased tubular sodium reabsorption and decreased urea excretion account for the low Uosm.)

Vaamonde, C. A., Presser, J. I., Vaamonde, L. S., and Papper, S.: Renal concentrating ability in cirrhosis. III. Failure of hypertonic saline to increase reduced T^cH_2O formation. Kidney Int. *1*:55, 1972. (Cirrhotic patients with low T^cH_2O formation (mannitol diuresis) did not normalize T^cH_2O generation during hypertonic saline infusion despite increased urinary sodium excretion and serum sodium concentration. The mechanism of the defect in renal concentration remains unknown, but appears not to be related to a decreased delivery of sodium to the distal nephron or to a lower tubular fluid concentration of sodium at the distal concentrating site.)

Vaamonde, C. A., Presser, J. I., and Clapp, W.: Effect of high fluid intake on the renal concentrating mechanism of normal man. J. Appl. Physiol. *36*:434, 1974.

(Review of the effect of the ingestion of large amounts of water [5–10 l./day] for 5 to 10 days on renal concentrating ability. Even 1 or 2 days of high water ingestion induced a modest concentrating defect in normal man.)

Wallin, J. D., Brennan, J. P., Long, D. L., Aronoff, S. L., Rector, F. C., Jr., and Seldin, D. W.: Effect of increased distal bicarbonate delivery on free water reabsorption in the dog. Am. J. Physiol. *224*:209, 1973. (When dogs were rendered alkalotic and hypochloremic by hemodialysis against a chloride-free bath and T^cH_2O formation was studied during a sodium bicarbonate diuresis, T^cH_2O was virtually obliterated. Thus, NaCl, not $NaHCO_3$ is the sodium salt required to generate T^cH_2O. An enhanced delivery of $NaHCO_3$ to the distal concentrating site will impair T^cH_2O formation only when there is a concomitant reduction of the distal delivery of NaCl.)

Whinnery, M. A., and Kunau, Jr., R. T.: Effect of potassium deficiency on papillary plasma flow in the rat. Am. J. Physiol. *237*:F-226, 1979. (Papillary plasma flow is decreased rather than increased in rats with prolonged potassium depletion, thus it appears that a high medullary blood flow is not responsible for the Uosm defect).

14

Maintenance of Acid-Base Homeostasis

Jack G. Kleinman, M.D.
Jacob Lemann, Jr., M.D.

The hydrogen ion concentration ($[H^+]$) of body fluids is normally maintained within a narrow range. In the blood this range in health is between 35 and 45×10^{-9} M. Even in pathological circumstances, homeostatic mechanisms usually limit deviation of $[H^+]$ between the extremes of about 20 and 150×10^{-9} M. Further deviations from normal are not compatible with life. It is certain that intracellular processes are critically sensitive to changes in $[H^+]$ since all enzymes have maximum activities within narrow ranges of $[H^+]$. Unfortunately, techniques for repetitive analysis of cell composition are not yet available in forms applicable to the evaluation and care of patients. Thus, the physician and clinical physiologist must rely on measurements in blood (including serum or plasma) and in urine for evaluation of acid-base disturbances. Based on these data and information from experimental observations, certain inferences regarding changes in intracellular composition may be drawn.

Measurement of [H⁺]

The hydrogen ion concentration of body fluids cannot, in fact, be measured directly. It is possible, however, to measure hydrogen ion activity ($\alpha[H^+]$) using glass membranes that are selectively permeable to

H^+. Such glass electrodes develop an electromotive force (EMF or voltage) which is proportional to $\alpha[H^+]$ in the solution in which they are immersed: $EMF = K \cdot \log \frac{1}{\alpha[H^+]}$, where K = a composite of physical constants. Since the value of α, the activity coefficient, has not been determined for complex biological fluids, it has become customary to assume $\alpha[H^+] \cong [H^+]$. This approximation has been felt to be appropriate because changes in the composition of extracellular fluids in clinical acid-base disorders are not radical enough to affect α significantly. Rather, it is the changes in the relative concentrations of acids and bases in solution that determine $[H^+]$ under these circumstances.

Acids, bases, and buffers

An acid is any substance containing hydrogen ions (protons) which can be liberated in aqueous solution. A base, on the other hand, is any substance that can accept or combine with hydrogen ions present in solution. It is apparent, therefore, that the dissociation of H^+ from an acid yields not only H^+ but also a substance that has some affinity for H^+. This substance is defined as a base and is known as the conjugate base of the acid. Strong acids, such as HCl, have weak conjugate bases and thus are nearly completely dissociated in dilute solutions: $HCl \rightarrow H^+ + Cl^-$. Weak acids, such as H_2CO_3, have conjugate bases with strong affinity for H^+, and thus such acids are only slightly dissociated: $H_2CO_3 \leftrightarrows H^+ + HCO_3^-$. As in the previous examples, acids may have no net charge when undissociated. However they may also either be anions, $H_2PO_4^-$ ($\leftrightarrows H^+ + HPO_4^=$) or cations, NH_4^+ ($\leftrightarrows H^+ + NH_3$).

It is evident that the total quantity of H^+ liberated by an acid will be related to its concentration ($[HA] \overset{k_1}{\rightarrow} [H^+] + [A^-]$); whereas the tendency for the conjugate base to combine with H^+ will be proportional to its concentration as well as the concentration of H^+ ($[A^-] + [H^+] \overset{k_2}{\rightarrow} [HA]$). In the steady state, the rates of dissociation of the acid to form H^+ and A^- and of association of A^- with H^+ to form HA are equal. Thus, $[HA] \overset{k_1}{\underset{k_2}{\rightleftarrows}} [H^+] + [A^-]$ and the rate coefficients for both the reactions, k_1 and k_2, can be combined. The expression can then be rewritten in the following manner: $\dfrac{[A^-][H^+]}{[HA]} = K$. This equation can also be solved for $[H^+]$: $[H^+] = \dfrac{K[HA]}{[A^-]}$. Thus $[H^+]$ is proportional to the ratio of acid to base in the solution. It is also evident from this relationship that a solution that contains both a weak acid and its conjugate base in nearly equivalent quantities will resist changes in $[H^+]$ upon addition of either strong acid or base. Such a solution provides significant reservoirs of acid to furnish H^+ and base to combine with H^+ and is said to be buffered.

The most important buffer pair present in extracellular fluid (plasma and interstitial fluid) is carbonic acid and bicarbonate (H_2CO_3 and HCO_3^-): $H_2CO_3 \leftrightarrows H^+ + HCO_3^-$ As previously described for any acid, $\dfrac{[HCO_3^-][H^+]}{[H_2CO_3]} = K$ and $[H^+] = K \cdot \dfrac{[H_2CO_3]}{[HCO_3^-]}$

The constant K is the measure of the tendency for H_2CO_3 to dissociate and the affinity of HCO_3^- for H^+. In fact, K varies with the temperature and ionic strength of the solution. However, in blood or serum at normal body temperature of 37° C, the constant for the dissociation of H_2CO_3, K', has a value of 800 when H_2CO_3 and HCO_3^- concentrations are expressed in customary units of mmol./l. (mM.) and H^+ is expressed in nmol./l. (nM; $M \times 10^{-9}$/l.). $[H^+]$ nmol./l. $= 800 \times \dfrac{[H_2CO_3] \text{ mmol./liter}}{[HCO_3^-] \text{ mmol./liter}}$ or substituting normal values: $[H^+] = 800 \times \dfrac{1.2}{24} = 40$ nmol./l.

Henderson-Hasselbalch Formulation

This dependence of $[H^+]$ on $[H_2CO_3]$ and $[HCO_3^-]$ in biological fluids was first recognized 80 years ago by L. J. Henderson so that the relationship describing this equilibrium for the dissociation of carbonic acid is termed the *Henderson equation*. He recognized that this acid-base pair is of critical importance in determining the acidity of body fluids because $[H_2CO_3]$ and $[HCO_3^-]$ can be independently regulated. It is evident from the Henderson equation that if any two of the three concentration components can be measured, the third can be calculated. The regulation of $[H_2CO_3]$ itself involves an equilibrium between dissolved CO_2 gas and H_2CO_3. This relationship can be described: $[H_2CO_3] = \alpha P_{CO_2}$, where $\alpha =$ a combined constant for the solubility of CO_2 in plasma and its reaction with H_2O to form H_2CO_3, while P_{CO_2} is the partial pressure of CO_2 in the alveolar air with which the plasma is in equilibrium. The constant, α, has a value of about 0.030 when P_{CO_2} is expressed in Torr (mm. Hg) and $[H_2CO_3]$ is expressed in mmol./l. Combining this equilibrium with the Henderson equation yields: $[H^+] = K' \dfrac{\alpha PCO_2}{[HCO_3^-]}$.

Using this relationship $[HCO_3^-]$ may be calculated from measurement of [H$^+$] using glass electrodes and of P$_{CO_2}$ using CO$_2$-sensitive electrodes as is the usual practice when arterial blood is sampled. [HCO$_3^-$] cannot be measured separately from [H$_2$CO$_3$]. However, both can be measured together as total CO$_2$ by acidification of an anaerobically collected serum or plasma sample. Bicarbonate and carbonic acid are converted to CO$_2$ gas, which is liberated and measured. When this is done and H$^+$ is also measured using the glass electrode, P$_{CO_2}$ can be calculated:

$$[H^+] = K' \frac{\alpha P_{CO_2}}{\text{Total CO}_2 - \alpha P_{CO_2}}.$$

Since the [H$^+$] of biological systems in the range of 0.00000004 mol./l. and since the glass electrode response to [H$^+$] is proportional to log $\frac{1}{[H^+]}$, the following convention has been introduced: log $\frac{1}{X} \equiv$ pX or pH $\equiv$ log $\frac{1}{[H^+]}$ (assuming $\alpha[H^+] \cong [H^+]$) and pK' $\equiv$ log $\frac{1}{K'}$. Introduction of this system allowed Hasselbalch to put the Henderson equation into inverse logarithmic form: pH = pK' + log $\frac{[HCO_3^-]}{[H_2CO_3]}$. In blood or plasma at body temperature pK = 6.1 for the dissociation of H$_2$CO$_3$. The Henderson-Hasselbalch equation can also be applied to the dissociation of any acid: pH = pK' + log $\frac{[\text{base}]}{[\text{acid}]}$. It is apparent that the pH of a solution is proportional to the ratio of base to acid. Furthermore, as this ratio approaches unity the log $\frac{[\text{base}]}{[\text{acid}]}$ approaches zero and the pH of the solution approaches the pK'. For a situation in which acid or base is being added to the solution, thus supplying H$^+$ or binding H$^+$, the ratio of any conjugate base to acid will change least when the original ratio was close to unity. In other words, the solution will be buffered against changes in pH when the pK is reasonably close to the original pH of the solution. The magnitude of buffering depends of course on the total concentration of the components of the buffer system (the base and the acid). In a complex solution containing several buffer systems, the pH of the solution will be determined by the quantity of each buffer pair present and their individual pK values.

The advantage of using the Henderson equation and [H$^+$] rather than pH is the avoidance of the potential conceptual difficulty of recalling that pH falls as a solution is acidified and has the advantage of permitting direct and simple arithmetic evaluation of the data without the need to have access to log tables or a calculator. On the other hand the use of pH and pK notation has the advantage of using directly the actual measurements of acidity as pH and of allowing an immediate appreciation of the value of buffers from their pK'. The student of acid-base physiology should appreppriate the relationships between the Henderson and the Henderson-Hasselbalch notation systems and should be able to employ either. Table 14–1 provides the interconversion of [H$^+$] and pH over the commonly observed range in blood or plasma in health and disease. It should be pointed out that for every 0.01 unit increase in pH over the range pH 7.20 to 7.50 the [H$^+$] decreases by 1 nmol./l. from about 60 to about 30 nmol./l.

TABLE 14–1
Conversion of pH to [H$^+$] NanoEq./Liter

pH	0.00	0.02	0.04	0.06	0.08
6.9	126	120	115	110	105
7.0	100	96	91	87	83
7.1	79	76	72	69	66
7.2	63	60	58	55	52
7.3	50	48	46	44	42
7.4	40	38	36	35	33
7.5	32	30	29	28	26
7.6	25	24	23	22	21

Buffers

The body actually contains a multiplicity of buffers. Within the extracellular fluid, HCO_3^-/H_2CO_3 (pK' = 6.1) is quantitatively the most important; the other buffers include the serum proteins, chiefly albumin (Alb$^-$/H alb; effective pK $\cong$ 7), and inorganic phosphate ($HPO_4^=/H_2PO_4^-$; pK$_2'$ = 6.8). Buffers of the intracellular fluid include the special case of the imidazole groups of hemoglobin within red blood cells (pK' $\cong$ 6.9), and the many proteins and organic phosphates within other cells. The mineral phase of bone, $Ca_{10}(P_{O4})_6(OH)_2$ or $Ca_{10}(P_{O4})_6 CO_3$, also provides a reservoir of base (OH^-, $CO_3^=$ and $P_{O4}^\equiv$). These buffers collectively define the pH of body fluids in health and serve to defend the organism against pH changes that result from the addition of acid or of base to the body. It must be emphasized that these buffers provide only a temporary defense mechanism. Ultimately, physiological mechanisms that result in the excretion of acid or base from the body are required to maintain normal concentrations of body buffers and thus, pH or [H$^+$].

The H_2CO_3/HCO_3^- buffer system is the fulcrum with respect to the excretion of acid or base and maintenance of normal concentrations of body buffers, since the concentrations of H_2CO_3 and HCO_3^- can be independently regulated.

Production and excretion of acid and base

In healthy adults approximately 20,000 mmol. of CO_2 are produced each day as an end-product of oxidative metabolism. Although CO_2 is itself a neutral substance, in aqueous solution it is in equilibrium with H_2CO_3. Thus the production of CO_2 represents a daily acid load that must be excreted. CO_2 diffuses into the blood (plasma and red blood cells) from cellular production sites. Red cells contain the enzyme carbonic anydrase (CA), which accelerates equilibrium of the reaction: $CO_2 + H_2O \overset{CA}{\rightleftharpoons} H_2CO_3$ and H_2CO_3 dissociates to form $HCO_3^- + H^+$. At the same time, oxygenated hemoglobin reaching the tissues gives up its oxygen to form reduced hemoglobin. The imidazole groups of reduced hemoglobin are stronger bases than those of oxygenated hemoglobin and thus serve to buffer the H$^+$ released as H_2CO_3 dissociates. The [HCO_3^-] within RBC thus increases and some diffuses into the plasma, thus accounting for the somewhat higher [HCO_3^-] in venous plasma as compared to arterial plasma. These reactions in peripheral capillaries are summarized in the upper segment of Fig. 14–1. When venous blood reaches the lung, this sequence is reversed and the CO_2 diffuses across the pulmonary alveolar membrane (lower portion of Fig. 14–1). The primary determinants of CO_2 excretion are pulmonary blood flow, the permeability of the alveolar membrane to CO_2 and alveolar ventilation. In health the equilibrium between CO_2 production and pulmonary CO_2 excretion establishes a P_{CO_2} of 40 mm. Hg and therefore a [H_2CO_3] of about 1.2 mM.

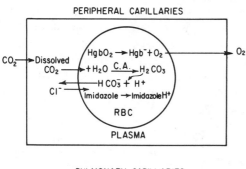

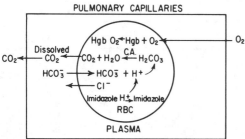

FIG. 14–1. Reactions involved in CO_2 transport and excretion.

H^+ can be added to body fluids by acids other than H_2CO_3. These acids are termed nonvolatile or fixed acids (HF) to distinguish them from H_2CO_3. The addition of such acids to body fluids results in the following reaction: $HF + HCO_3^- \rightarrow F^- + H_2CO_3$. The consequence of the consumption of HCO_3^- by this reaction is a rise in $[H^+]$ (fall in pH). Obviously, direct losses of HCO_3^- from body fluids is effectively an identical process.

Fixed acids are derived from several sources. First, neutral sulfur, contained in the amino acids methionine or cysteine, is oxidized in sequential steps to inorganic sulfate. During this process 2 mmol. of H^+ are added to body fluids for each millimole of sulfur oxidized. The magnitude of fixed acid produced by this reaction can be quantitated by the measurement of urinary inorganic sulfate and averages about 0.5 mEq./kg. body weight/day in adults eating normal diets. High-protein diets will increase fixed acid production from this source. Second, during the course of metabolism several organic acids are produced. Some are end-products of metabolism such as uric acid and hippuric acid. Others are intermediary metabolites, such as lactic acid, acetoacetic acid, and citric

acid, which ordinarily are capable of being completely oxidized to CO_2 and H_2O. These are all strong acids. The total contribution of the production of H^+ by these fixed acids can be quantitated by the measurement of daily urinary organic anion excretion. Approximately 0.5 mEq./kg. body weight/day of organic acid are produced by adults eating normal diets. Third, and depending on the type of dietary protein, hydrolysis of phosphate esters may contribute to fixed acid production.

These net gains of H^+ from fixed acids are modified by other reactions in health that consume or release H^+. Normal diets, especially in fruits and vegetables, contain sources of base as the salts of organic acids, such as K citrate and K lactate. Bicarbonate is produced when such base is absorbed and metabolized.

Base is also normally excreted in the feces as bicarbonate and in the form of salts of organic acids, principally the short-chain fatty acids acetate, propionate, and butyrate. The processes resulting in normal fixed acid production, normal dietary addition of base, and normal fecal base loss are summarized in Figure 14–2. The quantitative components of net fixed acid production in an average 70-kg. adult eating a

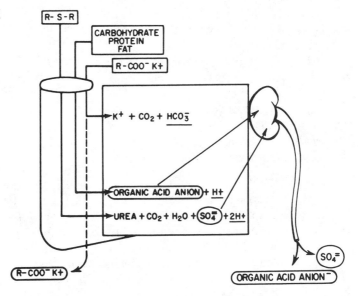

FIG. 14–2. Components of endogenous fixed acid production.

normal diet are thus:

H^+ gain from sulfur oxidation	$+35$ mEq.
H^+ gain from organic acids	$+35$ mEq.
Base loss into feces	$+35$ mEq.
minus	
Base gain from diet	-70 mEq.
	35 mEq.

or about 0.5 mEq./kg./day. This average 35-mEq. daily addition of fixed acid to body fluids results in the consumption of an equivalent quantity of HCO_3^-.

Renal acid excretion

The daily addition of 35 mEq. of fixed acid to body fluids in health would rapidly deplete body stores of HCO_3^- and lead to severe acidosis. However, the kidney is able to excrete equivalent quantities of acid into the urine each day. The kidney thus regenerates the HCO_3^- that was consumed in buffering of fixed acids as they were produced.

Obviously, glomerular ultrafiltrate contains HCO_3^- at the prevailing concentration in serum or plasma and also has a pH identical to that of the blood. Until filtered HCO_3^- is completely reabsorbed there can be no net H^+ secretion. Thus the first step in the renal excretion of acid involves the reabsorption of filtered HCO_3^-. Most of the filtered HCO_3^- is reabsorbed in proximal tubular segments by a process that involves the secretion of H^+ into the tubular lumen. The exact details of this process remain controversial but appear to involve reabsorption of Na^+ in exchange for secreted H^+. The fate of this H^+ is determined by the presence of HCO_3^- in the tubular fluid and the presence of carbonic anhydrase (CA) in the luminal membrane of the proximal tubular cells: $H^+ + HCO_3^- \rightleftarrows H_2CO_3 \overset{CA}{\rightleftarrows} CO_2 + H_2O$. Because of the presence of carbonic anhydrase, this sequence of reactions occurs almost instantaneously. The CO_2 produced by this reaction diffuses into tubule cells. Since tubules also contain carbonic anhydrase, the sequence of reactions

noted above are reversed, making H^+ available for secretion into the lumen and HCO_3^- available for diffusion into peritubular blood together with reabsorbed sodium. In a healthy adult with a serum $[HCO_3^-]$ of 24 mmol./l. and a GFR of 150 l./day, 3,600 mmol. of HCO_3^- are filtered each day. When the final urine pH is 6.0 or less the urine contains little or no HCO_3^-. All of the filtered HCO_3^- is therefore normally reabsorbed.

At the same time that H^+ secretion results in net reabsorption of filtered HCO_3^-, two other reactions occur. First, phosphate which is filtered into tubular fluid principally as $HPO_4^=$, is converted to $H_2PO_4^-$ in increasing quantity as the tubular fluid becomes acidified. The acid excreted in this manner by titration of filtered buffer is termed *titratable acidity*. Titratable acid can be measured by the amount of NaOH required to titrate the urine from its pH back to blood pH. Titratable acid can also be calculated from urine pH and blood pH, an approximate pK_2' for $H_2PO_4 \rightleftarrows HPO_4^= + H^+$ of 6.8, and the quantity of phosphate excreted in the urine. Second, the renal tubular cells supply an additional buffer, ammonia (NH_3), derived from amino acids, principally glutamine. Since NH_3 is a dissolved gas, it diffuses readily into the tubular lumen. Since NH_3 is also a strong base it can react avidly with H^+: $NH_3 + H^+ \rightarrow NH_4^+$. Since the pK for this reaction is about 9.0, formation of NH_4^+ is favored at any physiological pH. Cell membranes are relatively impermeable to NH_4^+, thus also favoring excretion of H^+ into the urine as NH_4^+. The quantity of H^+ excreted in this form is measured by the daily urinary excretion of NH_4^+.

Figure 14–3 summarizes these reactions, which result in the reclamation of filtered HCO_3^- and the regeneration of new HCO_3^-. Secretion of H^+ by tubular cells mediates all of these processes.

The daily net excretion of acid into the urine can be estimated as the sum of urinary titratable acid plus NH_4^+, minus uri-

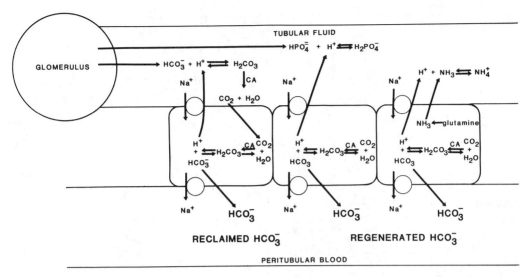

FIG. 14–3. Mechanisms for renal tubular HCO_3^- reclamation and regeneration.

nary HCO_3^-, if any. Approximately one-third of renal acid excretion is accounted for by titratable acid and the remainder by NH_4^+. In health, renal net acid excretion is quantitively equivalent to net fixed acid production and is about 35 mEq./day. The HCO_3^- consumed as fixed acids are produced and buffered is thus regenerated by the kidney, and acid balance is maintained.

Evaluation of acid-base disturbances

Acid-base disturbances can be broadly divided into processes that result in acidification of body fluids (increased $[H^+]$ or reduced pH) and processes that make body fluids more basic (reduced $[H^+]$ or increased pH). A condition falling into the first category is termed an *acidosis* and one falling into the latter is termed an *alkalosis*. As previously described, P_{CO_2} is determined by pulmonary CO_2 excretion at any given rate of CO_2 production. For this reason, P_{CO_2} is referred to as the *respiratory component* of acid-base regulation. By contrast, regulation of the serum or plasma $[HCO_3^-]$ is a result of metabolism, including the

diet, intestinal function, general metabolism, and kidney function. Thus the $[HCO_3^-]$ is termed the *metabolic component* of acid-base regulation.

When the primary event causing an acid-base disturbance resides in the respiratory component of acid-base regulation (respiratory acidosis or alkalosis), compensatory changes occur in the metabolic component which minimize the consequent changes in $[H^+]$ or pH. Similarly, when the primary event causing an acid-base disturbance resides in the metabolic component of acid-base regulation (metabolic acidosis or alkalosis), compensatory changes in the respiratory component also occur which limit the changes in $[H^+]$ or pH. The specific compensatory metabolic or respiratory events that occur in response to primary respiratory or metabolic acid-base disturbances, respectively, are discussed below. However, Figure 14–4 depicts the expected ranges of response of serum $[HCO_3^-]$ to primary increases (respiratory acidosis) or decreases (respiratory alkalosis) of P_{CO_2} in humans without metabolic or renal diseases (e.g., uncomplicated primary respiratory acid-base disorders). Figure 14–

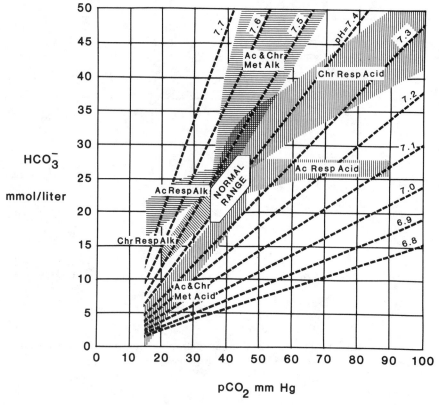

FIG. 14—4. Range of compensatory response of PCO₂ and of [HCO₃⁻] to primary acid-base disturbances. (Redrawn and expanded to include chronic respiratory alkalosis from Arbus, G. S.: Can. Med. Assoc. J. *109*:291, 1973.)

4 also shows the expected ranges of response of P_{CO_2} to primary decreases (metabolic acidosis) or increases (metabolic alkalosis) of serum [HCO₃⁻] in humans without pulmonary disease. Since the ratio of [HCO₃⁻] to P_{CO_2} determine pH (using the Henderson-Hasselbalch formulation), iso-pH lines are also included in Figure 14–4 to indicate the corresponding pH values for range of serum [HCO₃⁻] and P_{CO_2}. It is also apparent that the time course of development of primary respiratory acid-base disorders has an important effect on the magnitude of compensation for these primary disorders. This effect, as well as mixed respiratory and metabolic acid-base disorders, is considered in subsequent sections.

Respiratory acidosis

Respiratory acidosis results from CO_2 retention (hypercapnea). Any condition in which ventilatory function is inadequate to keep pace with CO_2 production will cause CO_2 to accumulate and P_{CO_2} to increase. Since a small amount of this CO_2 will combine with water to produce the weak acid H_2CO_3, [H⁺] will increase. Impaired pulmonary function may be secondary to disturbances in ventilatory drive, paralysis of the muscles of respiration, airway obstruction, or intrinsic lung disease. Unless a new steady state is reached, through both an elevation of the P_{CO_2} sufficient to increase diffusion of CO_2 and renal compensatory mechanisms, the fall in pH will be progressive and death will ensue.

Inspection of the acid-base nomogram (Fig. 14–4) reveals that blood pH falls linearly with acute elevation of the P_{CO_2}. This indicates that renal compensatory mechanisms do not begin immediately. Thus, the only acute response to the acidosis produced by the elevation in P_{CO_2} is due to the buffering action of hemoglobin in red blood cells. This will raise the plasma $[HCO_3^-]$ by only approximately 3 mEq./l. Given the information that the duration of hypoventilation is less than approximately 12 hours, the finding of a plasma $[HCO_3^-]$ significantly outside the normal range is an indication of a complicating metabolic acid-base disturbance. With hypercapnea of several days' duration, the plasma $[HCO_3^-]$ gradually rises so as to limit the acidosis. In fact, a paradoxical alkalosis has often been described, with pH values >7.4 in patients with chronic obstructive pulmonary disease (Fig. 14–4). This phenomenon, as well as the recovery of the pH in general, is due to renal compensatory mechanisms.

The renal compensation for CO_2 retention requires several days to reach a maximum response and consists of two components. First, H^+ secretion in the distal nephron is stimulated by the elevated P_{CO_2}. The mechanisms responsible for this have not been completely described, but are likely to be akin to those described for the turtle bladder, an epithelial analogue to the distal tubule. The result of this accentuation of acid secretion is generation of new HCO_3^- in excess of what is required to balance that lost through buffering of fixed acids. In order that the increment in filtered HCO_3^- not exceed the maximum reabsorptive rate for HCO_3^-, an enhanced reabsorptive capacity for HCO_3^- also develops in the proximal parts of the nephron. This appears to be a direct effect of the elevated P_{CO_2} on the proximal tubule. However, volume depletion as a result of enhanced NH_4Cl excretion may also play a part.

The recognition of acute hypercapnea in patients primarily involves the clinical detection of respiratory distress or altered mental status. Chronic elevations of P_{CO_2} are often well tolerated, but may be responsible for mild degrees of mental dysfunction such as somnolence and confusion. Since by far the largest number of patients with chronic respiratory acidosis have obstructive pulmonary disease, the clinical picture is usually characterized by the signs and symptoms of this disorder. It is obvious that the key to assessing disorders of respiratory etiology is an arterial blood gas determination. Thus, the combination of reduced P_{O_2}, elevated P_{CO_2} and low blood pH indicates respiratory acidosis. Few patients with chronic acidosis have pH values <7.25, and thus the finding of such a severe acidosis in the presence of a normal plasma $[HCO_3^-]$ suggests that the disorder is acute. Very low pH values with an elevated plasma bicarbonate signify superimposition of acute respiratory acidosis on chronic respiratory acidosis, a very common occurrence when patients with chronic lung disease develop an acute pneumonia. As indicated earlier, some patients with chronic pulmonary disease and acidosis have normal or even frankly elevated blood pH values. The cause of this is uncertain, but may involve transient diurnal variations in ventilation (and, therefore, P_{CO_2}) within time frames too small to be associated with changes in renal compensation or the frequent coexistence of metabolic factors intensifying the compensatory alkalosis—for example, the use of diuretics.

Whatever the cause, treatment of respiratory acidosis involves restoring P_{O_2} and pH to tolerable levels. With accomplishment of improvement in the former by some form of respiratory therapy, P_{CO_2} often falls to the point at which pH values increase to 7.25 or higher. If not, $NaHCO_3$ therapy may be necessary to achieve this level. Further adjustments in therapy to raise pH are probably not necessary. In fact, too aggressive an attempt to repair respiratory acidosis runs the risk of alkalosis, which can impair respiratory and central nervous system function.

Respiratory alkalosis

Respiratory alkalosis results from excretion of CO_2 at a rate greater than it can be produced. This will reduce P_{CO_2} (hypocapnea) and leads to alkalosis. Hyperventilation may be mediated through neural mechanisms, such as fear, anxiety, or central nervous system disease. Reflex mechanisms may also play a part when there are increased fluids within lung parenchyma, such as in congestive heart failure, pulmonary embolization or pneumonia. Chemicals such as endogenous pyrogens or salicylates may also stimulate respiration. Hypoxia such as encountered at high altitudes, hepatic insufficiency, sepsis, and mechanical ventilation may also produce respiratory alkalosis.

As in the case of respiratory acidosis, blood pH rises almost linearly with acute reductions of P_{CO_2} (Fig. 14–4). Thus only a modest degree of buffering by blood is involved in the immediate defense of respiratory acid-base alterations. Within hours of the onset of respiratory alkalosis there is an increase in organic acid production by cells, particularly lactic and citric acids. Over the course of several days renal acid secretion declines, producing an increased urine pH, decreases in the excretion of NH_4^+ and titratable acid and bicarbonaturia. The serum [HCO_3^-] declines, but the magnitude of the decrements are not comparable to the increments observed during compensation for respiratory acidosis. Thus, in the presence of respiratory alkalosis, a serum [HCO_3^-] below approximately 12 mEq./l. suggests the concomitant presence of metabolic acidosis.

Acute hypocapnea of significant degree generally produces lightheadedness, which may proceed to syncope owing to the subsequent respiratory depression and hypoxemia. Other signs or symptoms associated with acute hyperventilation include circumoral or distal paresthesias, positive Chvostek's or Trousseau's signs, and hypocalcemic tetany. Chronic metabolic alkalosis is generally only recognized as part of one of the conditions that give rise to the hyperventilation. In liver failure and possibly in other patients who are severely ill, alkalosis may be associated with a deterioration in central nervous system function, and thus respiratory as well as metabolic alkalosis should be suspected in these circumstances. With acute hypocapnea the pH may rise (and [H^+] fall) to barely tolerable levels, pH > 7.65 ([H^+] < 22 nEq./l.). With chronic hyperventilation pH values >7.5 are rare and when present suggest coexistence of metabolic alkalosis (Fig. 14–4). The elevated serum lactate observed in respiratory alkalosis is associated with a proportional rise in the pyruvate concentration and is thus not an indication of tissue hypoxia.

Treatment of respiratory alkalosis should be aimed at alleviating the underlying cause. This is often easier said than done. Mechanical ventilation can be decreased, fever can be lowered, and salicylate removed. Treatment of the other conditions may be more difficult and less specific in alleviating the stimulus to hyperventilation. Since chronic hyperventilation is associated with only modest alkalosis, more specific therapy is usually not required. Acute hyperventilation due to anxiety responds to the CO_2 retention induced by breathing into a paper bag or other rebreathing device.

Metabolic acidosis

Metabolic acidosis is the result of a primary decline in the serum or plasma [HCO_3^-]. This comes about through one of several mechanisms. Addition of H^+ to body fluids will titrate HCO_3^- to H_2CO_3. As pointed out above, this species is in equilibrium with P_{CO_2}. A rise in P_{CO_2} will supply an immediate stimulus to respiration, returning the pH toward normal. The result of this is a deficiency in HCO_3^-. Acidosis may also ensue from the direct loss of HCO_3^-, through loss of normally alkaline intestinal fluids or into the urine.

From the physiological point of view, as well as from the standpoint of diagnosis and treatment, H^+ addition can be in the form of two sorts of acids: those with anions (conjugate bases) that are readily reabsorbed by the renal tubule and those with poorly reabsorbable anions. The former group consists of HCl or its metabolic equivalents (NH_4Cl, arginine HCl, and lysine HCl). The latter group is much larger and consists of various sorts of organic acids of either exogenous or endogenous origin. A common cause for a primary decrease in the plasma [HCO_3^-] is dilution of the extracellular fluid with non-bicarbonate–containing solutions (usually isotonic saline). Intestinal HCO_3^- losses occur with most forms of diarrhea. Drainage of biliary, pancreatic or small bowel fistulae also leads to acidosis by loss of HCO_3^-. Renal HCO_3^- losses may be the result of one of several sorts of renal tubular acidoses, which are discussed more fully later in this section. A curious hybrid of the last two causes of HCO_3^- deficit is represented by ureterosigmoidostomy, in which HCO_3^- is added to urine by this interposed colonic segment in exchange for reabsorbed chloride.

The acid-base nomogram (Fig. 14–4) indicates that blood pH is defended to a greater degree in both acute and chronic metabolic acidosis than in the respiratory acid-base abnormalities discussed earlier. This is so because of the rapid and powerful contribution that respiratory compensation makes to the effectiveness of the HCO_3^-/H_2CO_3 buffer system, as well as to the availability of large reservoirs of buffer outside the extracellular space. Experiments involving infusion of strong acids into animals have indicated that disposition of about half of this acute H^+ load is

accounted for by buffers other than HCO_3^- in the extracellular fluid. This process is essentially complete within several hours. With more persistent acid accumulation such as in renal failure, the relative proportion of non-ECF buffering gradually increases to two-thirds or more. The sites of these buffer systems remain controversial. In chronic acidosis, bone is recognized as a reservoir of buffer. However, the importance of the skeleton in the buffering of acute acid loads is not clear. In addition, whether parathyroid hormone is necessary for the process of bone buffering remains to be elucidated. Buffering outside of the extracellular fluid may also occur within cells. The imidazole groups of hemoglobin have already been referred to in this regard. The concept of cell buffering, however, may be subject to some conceptual difficulties, since cells are, in fact, most often the source of nonvolatile acids. This is so even under conditions of toxic acidosis, in which cell metabolism usually accounts for the bulk of acid production. In situations of severe extracellular acidosis, the observation that cell pH is changed very little would appear to indicate that cells maintain efficient systems for disposition of H^+. These systems require metabolic energy either directly or indirectly and probably operate through cation exchange ($Na^+ - H^+$). Thus from the point of view of primacy, if not temporality, metabolic acidosis can be said to arise within cells, the cellular extrusion of H^+ being buffered by the extracellular or mineral compartments. Whether or not this system can operate in reverse in tissues other than those generating acid remains to be determined. Recently described systems for transporting OH^- or HCO_3^- into or out of cells could also allow for shifting of base to compartments where buffering is required.

As indicated earlier, respiratory compensation for acid loads begins immediately. This effect is mediated by chemoreceptors in the central nervous system. The increased ventilation induced by acidosis characteristically lags behind the fall in serum [HCO_3^-], presumably because of limitations to diffusion of HCO_3^- out of the brain. Because of the time required for [HCO_3^-] equilibrium across the blood-brain barrier, maximum compensation for metabolic acidosis is usually not achieved for 12 to 24 hours. At this point, in otherwise uncomplicated metabolic acidosis, the P_{CO_2} will be observed to have decreased by about 1.1 to 1.3 mm. Hg for each mEq./l. fall in [HCO_3^-] (Fig. 14–4). Finally, it should be noted that respiratory compensation is limited by the fact that P_{CO_2} values below 10 to 15 torr are not achievable.

Under circumstances in which the acidosis is not due to defective renal function, there are two components to the renal adaptation to metabolic acidosis. First, due to the fall in plasma [HCO_3^-], more complete reabsorption of HCO_3^- occurs in the proximal tubule. Decreased delivery of HCO_3^- to the distal nephron allows for an increased net acid secretion because protons are not required to reclaim residual HCO_3^-. As the urine pH falls to minimum levels of 4.5 to 5.0, the urinary excretion of titratable acid increases maximally for any given rate of phosphate excretion. Urinary NH_4^+ excretion also increases because of maximum urinary acidification. In addition, increased renal NH_3 production is a feature of metabolic acidosis. How acidosis brings this about is not clear, but the effector of this response is probably an increase in activity of the mitochondrial phosphate-dependent glutaminase pathway in renal tubule cells. The acute adaptation to acidosis may involve an increase in glutamine entry into the mitochondria. The situation in chronic acidosis is very complex; the increased ammoniagenesis may involve enzyme induction in a number of NH_3-producing reactions.

Increased production of NH_3 plays a permissive role in renal acid secretion, allowing rates of net acid excretion to reach levels that may be five- to ten-fold higher than basal levels. The factors that increase H^+ secretion itself in acidosis are imperfectly understood. They include stimulation by increased levels of aldosterone of the conductance of H^+ through the distal nephron H^+ secretory system. In addition, there is a complex relationship between H^+ secretion and the delivery of Na^+ to the distal nephron, the avidity of Na^+ absorption at distal sites and the nature of the accompanying anion. Sodium reabsorption in proximal parts of the nephron over and above that required for HCO_3^- reclamation will hinder H^+ secretion by limiting Na^+ delivery to distal nephron segments. This may lead to a situation in which renal dysfunction adds to or actually engenders the acidosis. The provision of nonabsorbable anions (such as the conjugate bases of organic acids), along with Na^+, aids distal nephron acidification, probably by an electrochemical mechanism.

Recognition of metabolic acidosis primarily involves considering its presence in situations known to be associated with it. Compensatory hyperventilation may be a clue to acidosis. However, hyperventilation may be difficult to recognize, especially in chronically acidotic patients. When recognized, it may be noted to be of the type associated with the name of Kussmaul, distinguishable by being slower or deeper than the breathing in patients with primary hyperventilation. The state of consciousness is rarely normal in severe acidosis. The blood pH, P_{CO2} and [HCO_3^-] are important parameters required in the assessment of this acid-base abnormality. It should be pointed out that properly drawn venous blood samples will provide values that do not differ importantly from arterial samples and are to be preferred unless the presence of an indwelling arterial cannula obviates the need for repeated arterial punctures or an arterial puncture is required for the simultaneous measurement of P_{O_2}. The finding of a blood pH below 7.35 ([H^+] > 45 nEq./l.) with a lower than normal serum or plasma [HCO_3^-] indicates metabolic acidosis. The P_{CO_2} may be normal, reflecting extremely acute acidosis, or lowered, indicating some degree of compensation for the acidosis. Expected values for compensatory adjustments of P_{CO_2} in otherwise uncomplicated metabolic acidosis are indicated in Figure 14–4.

Since the treatment of the various forms of acidosis may differ somewhat, an exploration of the differential diagnosis of metabolic acidosis must be undertaken. In conjunction with the clinical assessment of the patient, an evaluation of the serum unmeasured anions or anion gap is useful. The anion gap is the sum of the anions usually not measured in routine laboratory electrolyte panels. Its components are given in Table 14–2. It is usually estimated as the difference between the serum $[Na^+]$ and the sum of the $[Cl^-]$ and $[HCO_3^-]$, and its normal value is 8 to 12 mEq./l. An elevation in the anion gap indicates either an inability to excrete normal amounts of endogenously produced acids, as in renal failure, marked over production of such acids, as in diabetic ketoacidosis, or poisoning by ingested acids or their precursors.

Decreased renal function may be recognized initially by uremic symptoms or an abnormal urinalysis. An elevation in serum creatinine and blood urea nitrogen (BUN) should be sought to support the diagnosis of uremic acidosis. It should be pointed out that glomerular filtration rate (GFR) must decline below about 30 per cent of normal before retention of unmeasured anions and increases in the anion gap become significant. Lesser degrees of renal failure may be associated with hyperchloremic acidosis with a normal anion gap. The inability to excrete the anions of metabolic acids would not by itself be sufficient to generate acidosis unless there was also deficient H^+ secretion by the kidney. In renal disease this deficit is due to a reduction in ammonia production by the diseased kidney. Initially, the failing kidneys compensate by producing more NH_3 relative to residual renal function. However, at some point (usually when GFR is less than 50 per

TABLE 14–2

Calculation of the Serum or Plasma Unmeasured Anion Concentration (Anion Gap)

Cations	mEq./l.	Anions	mEq./l.
		Cl^-	103
Na^+	140	HCO_3^-	25
K^+	4	$SO_4^=$	1
Ca^{++}	5	Organic acid anion	1
Mg^{++}	2	$HPO_4^=$	2
		$H_2PO_4^-$	
		Protein (principally albumin)	19
Total	151		151

Unmeasured anion $= [Na^+] - ([Cl^-] + [HCO_3^-]) =$ 8 to 12 mEq./l. in health.

cent of normal) NH_3 production begins to decline in proportion to the fall in GFR. Therefore, despite intact distal H^+ secretory capacity as measured by a number of experimental maneuvers, net acid excretion by the failing kidney is often inadequate to maintain acid balance. Frequently this deficit is compounded by a failure of complete tubular HCO_3^- reabsorption when the serum $[HCO_3^-]$ concentration is normal. Thus, urinary pH may be inappropriately high in moderately severe renal failure but is almost always maximally acid when GFR is less than 20 to 30 ml./min.

When renal function is normal or where there are clinical grounds to suspect acid overproduction, the components of the elevated anion gap must be sought. Listed below are the unmeasured anions that may appear in the serum or plasma when fixed acid production is increased. The determinations that are usually available clinically are those for ketoacids (specifically acetoacetic acid) and lactic acid. Elevated ketoacids usually indicate diabetic ketoacidosis as the etiology of the acid-base disturbance. Ketoacids are derived from the incomplete oxidation of fatty acids. The pathophysiologically important ketoacids include the β-hydroxybutyric acid in addition to acetoacetic acid. Beta-hydroxybutyric acid, however, is not detected by the nitroprusside reaction generally used to detect ketoacids but may be the predominant metabolic acid in some patients with diabetic ketoacidosis. Thus, on occasion a weakly positive test for ketonemia may coexist with marked elevation in the serum unmeasured anions. A similar situation may exist when lactic acidosis is superimposed on diabetic acidosis. Increased production of ketoacids also occurs in starvation. However, this rarely results in significant acidosis. A more important clinical condition in which overproduction of ketoacids occurs is alcoholic ketoacidosis. In this situation the major ketoacid is β-OH butyric acid. Thus the nitroprusside test may be negative or may fail to indicate the severity of the acidosis or the magnitude of the anion gap. This diagnosis should be suspected in the alcoholic who discontinues food intake and presents with several days of vomiting. Serum glucose levels are only mildly elevated, but acidosis may be severe and lactic acidosis may coexist.

Serum Unmeasured Anions in Metabolic Acidosis

Ketoacids
  β-hydroxybutyric acid
  Acetoacetic acid
Lactic acid
Glyoxalic and oxalic acid
Formic acid
Acetic acid
Unknown

An elevation in the serum lactate levels may indicate lactic acidosis. This condition is due to an alteration in tissue metabolism such that the oxidation-reduction state of the cell is shifted in the reduced direction. A list of the causes of lactic acidosis is given below. When blood lactate concentration exceeds 4 mEq./l. and the pyruvate concentration is not proportionally elevated, the metabolic acidosis can be ascribed, at least in part, to accumulation of lactic acid.

While overproduction of lactic acid may be quite common, a more important factor in the pathophysiology of lactic acidosis is whether lactate utilization by the liver and the kidney that can metabolize lactate keeps pace with lactate production. Thus, hepatic dysfunction, alone or in combination with renal failure, is a frequent concomitant of lactic acidosis.

Causes of Lactic Acidosis

Poor perfusion of a major body segment
Shock
Hypoxemia
Carbon monoxide poisoning
Severe anemia
Lymphoma/leukemia
Toxins (ethanol, methanol, salicylate, phenformin, isonazid, ethylene glycol, fructose, sorbitol)
Hereditary disorders (glucose-6-phosphatase deficiency—Type I glycogen storage disease, fructose-1,6-diphosphatase deficiency, pyruvate carboxylase deficiency, pyruvate dehydrogenase deficiency, defective oxidative phosphorylation).

Ingested toxins may themselves be acids. However, their effects to induce metabolic acidosis and an increase in the serum unmeasured anions are more commonly due to their metabolism to strong acids or to impairment of normal metabolism. For example, in salicylate poisoning, ketoacids and lactate are usually elevated. Ingestion of ethylene glycol results in acidosis due to accumulation of oxalic and other organic acids. Methanol and paraldehyde are known to be metabolized to formic and acetic acids, respectively, but lactic acid, ketoacids, and other unidentified organic acids also contribute to the pathogenesis of these acidoses.

Acidosis without an elevation of the anion gap is, necessarily, hyperchloremic in that the serum $[Cl^-]$ must rise to offset the decrease in $[HCO_3^-]$. To all intents this represents the addition, by infusion or ingestion, of HCl to body fluids. It can also occasionally be seen under certain conditions in which anions other than Cl^- are the conjugate bases produced during increased acid production. These anions are relatively nonreabsorbable, and thus their loss in the urine may obligate sufficient Na^+ to induce volume depletion. When volume replacement is predominantly saline, as during treatment of diabetic ketoacidosis, the preexisting volume deficit may promote excess Cl^- retention. Moreover, since organic acids are metabolized to HCO_3^- during recovery from this condition, loss of organic acids may compromise the replenishment of base lost through titration of their accompanying H^+.

Most often, gain of HCl actually results from loss of HCO_3^-. At some point in the duodenum, intestinal contents become alkaline. This is primarily due to the addition of alkaline bile and pancreatic juice. These fluids are basic by virtue of their containing HCO_3^-. This would lead to a net loss of base from the body if the HCO_3^- were not largely reclaimed in more distal parts of the intestine through operation of a HCO_3^- – Cl^- exchange mechanism. Diarrhea can overwhelm this reabsorptive process. External losses of biliary tract and pancreatic secretions lead to HCO_3^- loss. Some drugs ($CaCl_2$, $MgSO_4$) can form insoluble carbonates or soaps and anion exchange resins (such as cholestyramine) can complex HCO_3^- trapping it in the intestine. These mechanisms lead to fecal HCO_3^- loss. The acidosis resulting from such intestinal HCO_3^- loss is associated with a normal anion gap. The hyperchloremia is largely due to increased intestinal and renal Cl^- reabsorption induced by extracellular fluid volume depletion. Usually the clinical circumstances make the diagnosis of acidosis due to gastrointestinal HCO_3^- loss apparent. Confusion with renal tubular acidosis may be a problem occasionally because the extracellular fluid volume depletion produces secondary hyperaldosteronism. This phenomenon will, of course, induce K^+ loss into the urine and accentuate fecal K^+ losses. The resulting hypokalemia, together with the acidoses will stimulate NH_3 production. As a consequence, the urine may not be maximally acid (pH > 5.5) in this form of metabolic acidosis, even though H^+ secretion is increased above normal. Measurement of increased renal net acid excretion should serve to differentiate this situation from the renal tubular acidoses.

The final category of metabolic acidosis with normal serum unmeasured anions is the result of renal acidification defects. These include the traditional renal tubular acidosis (RTA), proximal and distal types, and a more heterogeneous collection of disorders associated with defects in secretion of or response to aldosterone. Simply put, proximal RTA results from the kidney's inability to reabsorb normal amounts of filtered HCO_3^- when the plasma HCO_3^- concentration is normal. This disorder is considered to derive from abnormal proximal tubule function because more HCO_3^- is excreted under these conditions than can be normally reabsorbed at distal nephron sites. In addition, frequently there are associated defects in the reabsorption of glucose, phosphate, and amino acids which are known to be reabsorbed primarily in the proximal tubule (Fanconi syndrome). Moreover, distal tubule function can be demonstrated to be intact. In point of fact, these characteristics are used to arrive at a diagnosis of proximal RTA. Thus, when acidosis is observed and the serum $[HCO_3^-]$ is below 13 to 15 mEq./l., urine pH is maximally acid (pH < 5.2). Furthermore, during HCO_3^- loading the urine-minus-blood P_{CO_2} value is normal (about 40 mm. Hg.) suggesting that distal secretion of H^+ is normal. Potassium excretion is demonstrated to be intact by the absence of hyperkalemia. In fact, serum $[K^+]$ is usually low, reflecting increased delivery of Na^+ and fluid from the proximal tubule. However, when the serum $[HCO_3^-]$ is raised by HCO_3^- infusion, urine pH rises and HCO_3^- appears in the urine. When the serum $[HCO_3^-]$ is normal, excreted HCO_3^- is in the range of 10 to 30 per cent of the filtered load of this anion. Indicators of disordered proximal tubule function in addition to phosphaturia, glycosuria, and aminoacidurias include lysozymuria and increased urinary excretion of immunoglobulin light chains. Proximal

RTA may be fixed and permanent or transient. It may be a genetically transmitted disease with global tubule deficits, or it may be associated with distinct defects in reabsorption of specific amino acids. Various toxins and drugs can produce proximal RTA. It is seen in dysproteinemias and a number of systemic disorders such as amyloidosis and paroxysmal nocturnal hemoglobinuria. And, finally, this type of acidosis may be a feature of the nephrotic syndrome and some renal tubulointerstitial diseases, and is occasionally observed in the transplanted kidney.

Distal RTA results from a defect in the kidney's ability to produce a maximal acid urine pH (<5.2). Debate continues about whether the secretory capacity of the distal tubules and collecting ducts for H^+ are abnormal in this condition. Alternatively, abnormal permeability for H^+ at these sites has been proposed, leading to "back-leak" of H^+ out of the tubules and inability to sustain a high urine-to-blood pH gradient. Indeed, it may be possible that examples of both pathophysiological mechanisms exist in the spectrum of distal RTAs. Because distal acidification is abnormal, the small amount of HCO_3^- that normally escapes reabsorption in the proximal parts of the nephron is excreted and urine pH is relatively high (generally >5.8) even when the patient is markedly acidotic. In this condition, as in proximal RTA, distal K^+ secretion is normal. In fact, hypokalemia is usually present because of the increased distal $NaHCO_3$ delivery. Hypercalciuria, nephrocalcinosis, and nephrolithiasis are also usually present. Genetic, transient, or fixed idiopathic forms of this disorder occur. It may be observed in association with autoimmune diseases. Various toxins such as amphotericin B, some analgesics, and whatever agent is responsible for the Balkan nephropathy can cause distal RTA. Other interstitial disease may be associated with the disorder. An incomplete form is often seen in hepatic cirrhosis and in relatives of individuals with hereditary or congenital distal RTA.

Although RTAs have been known for a long time to be important features of adrenal insufficiency, other forms of defective acidification associated with hyperkalemia have recently been recognized. Essentially all of these conditions are caused either by deficient aldosterone secretion or an impaired renal tubular responsiveness to this hormone. In the former category are Addison's disease or other forms of adrenal insufficiency due to adrenal destruction and congenital enzyme defects at points on the steroid synthetic pathway prior to the production of glucocorticoid and mineralocorticoid. Such enzyme defects also exist for the synthesis of aldosterone alone, leading to an isolated hormonal deficiency. Hypoaldosteronism is also seen in some patients with renal disease (particularly diabetics) as a consequence of impaired renin secretion. Other patients with renal insufficiency demonstrate resistance to the effect of aldosterone despite its demonstrated presence. Resistance to the effect of the hormone is, of course, the expected and desired effect of treatment with the aldosterone antagonist, spironolactone. It has also been described in patients (particularly those with renal failure or disease) taking nonsteroidal antiinflammatory drugs. These conditions are suspected when hyperkalemia and hyperchloremic acidosis is observed in patients without renal insufficiency or in those with renal impairment too mild to account for these findings. Manifestations of adrenal insufficiency will also aid in establishing this diagnosis as well as absent or low plasma aldosterone values during hyperkalemia. Although urine pH may be quite acid and urinary HCO_3^- absent, net acid excretion is subnormal. This is primarily due to defective renal ammonia production, probably as a consequence of hyperkalemia.

The goal of therapy in metabolic acidosis is to return serum or plasma $[HCO_3^-]$ to levels that maintain $[H^+]$ or pH of body fluids at tolerable levels. If possible, the sources of acid loads must be controlled or eliminated, and normal metabolism must be reestablished in situations in which overproduction of endogenous acid occurs (such as diabetic ketoacidosis or lactic acidosis). Nonrenal HCO_3^- losses may be surgically correctible or, as in the case of diarrhea, may be amenable to medical therapy. Mineralocorticoid deficiency can be repaired through replacement therapy. Mineralocorticoid resistance may respond to removal of spironolactone or nonsteroidal inflammatory drugs. In situations of acute metabolic acidosis, it is generally appropriate to employ $NaHCO_3$ therapy in order to raise serum $[HCO_3^-]$ to 15 mEq./l. if it is below 10 mEq./l. In acute circumstances the space of distribution (in liters) of administered HCO_3^- will approximate twice the extracellular fluid volume or about 40 per cent of body weight. Thus in a 70-kg. individual the HCO_3^- space is $70 \times 0.4 = 28$ l. The amount of $NaHCO_3$ required to raise the serum bicarbonate by a given amount will be equal to the difference between the target $[HCO_3^-]$ and existing serum $[HCO_3^-]$ multiplied by the estimated space of distribution of HCO_3^-. Since restoration of central nervous system $[HCO_3^-]$ to normal lags behind the rise in plasma $[HCO_3^-]$, slowing of hyperventilation in response to acidosis will be delayed. Thus the P_{CO_2} will remain below normal for many hours. As a consequence, restoration of serum $[HCO_3^-]$ to levels of 15 to 20 mEq./l., although still below normal, will raise pH to normal. If serum $[HCO_3^-]$ were restored to normal, respiratory alkalosis would occur. This consideration has resulted in the usual clinical practice of supplying only half the calculated amount of HCO_3^- required to raise the serum $[HCO_3^-]$ to a normal level over the first 12 hours of treatment. In more severe acidosis, particularly when acid production is ongoing, the empirical method described above may grossly underestimate the amount of HCO_3^- required to correct or even keep pace with the acidosis. Thus it is imperative that frequent measurements of blood pH and $[HCO_3^-]$ be performed and the rate of infusion of $NaHCO_3$ adjusted accordingly. In chronic acidosis, the amount of $NaHCO_3$ required to maintain a given serum $[HCO_3^-]$ will be determined primarily by the rate of endogenous fixed acid production and the rate of HCO_3^- loss. Thus in renal failure or distal renal tubular acidosis, where urinary HCO_3^- losses are not significant, the quantity of HCO_3^- required to maintain a normal serum $[HCO_3^-]$ will be in the vicinity of 1 mEq./kg./day, in approximately the same range as the rate of endogenous fixed acid production. By contrast, when chronic acidosis is due to ongoing HCO_3^- losses, such as in chronic gastrointestinal HCO_3^- loss or proximal RTA, $NaHCO_3$ in amounts ranging from 3 to 10 mEq./kg./day, much in excess of the fixed acid production rate, may be necessary to maintain normal serum $[HCO_3^-]$. In acute acidosis (particularly lactic acidosis) the HCO_3^- requirement may be so large, that the associated Na^+ administration results in significant extracellular fluid volume overload. Concomitant diuretic therapy or dialysis may be required to control the volume status of such patients. In fact, dialysis may be required for the correction of acidosis in patients who cannot tolerate any volume expansion or who are already overloaded. The more recently available dialysate solutions containing HCO_3^- rather than acetate should be used in seriously ill, acidotic patients. During recovery from acute acidosis, a shift of K^+ into cells may unmask K^+ deficits produced by renal K^+ loss. Also, $NaHCO_3$ therapy may contribute to K^+ deficits by stimulating renal K^+ secretion. While in some acidoses (aldosterone deficiency or hyporesponsiveness) K^+ may have to be restricted, in others (gastrointestinal K^+ loss, proximal RTA, or during recovery from diabetic ketoacidosis) K^+ may have to be provided. When increased rates of acid production can be slowed or stopped, for example by the administration of insulin in diabetic ketoacidosis, the accumulated ketoacids can be metabolized to HCO_3^-. In this circumstance the need for supplementary $NaHCO_3$ therapy may be significantly reduced.

Metabolic alkalosis

Metabolic alkalosis results from a primary increase in the serum $[HCO_3^-]$. This may either be due to loss of H^+ from body fluids or a gain in HCO_3^-. These processes are not often clearly separable, since secretion of H^+ by stomach and kidneys is always associated with creation of HCO_3^-. Similarly, abnormal reabsorption of HCO_3^- by the kidneys must be accompanied by secretion of H^+ into tubular fluid. Production of metabolic alkalosis by these processes or as a result of exogenous HCO_3^- (or HCO_3^- precursor) administration requires additional pathophysiological processes. This is so because of the nature of the renal reabsorptive process for filtered HCO_3^-. This mechanism demonstrates a reabsorptive maximum of about 3000 μEq./min. With normal rates of glomerular filtration, the normal serum $[HCO_3^-]$ will supply a filtered load of HCO_3^- that approximates the tubular reabsorptive maximum. Thus a primary increase in the serum $[HCO_3^-]$ ought to be quantitatively excreted into the urine, preventing the development of metabolic alkalosis. It follows that in individuals with normal renal function very high rates of HCO_3^- administration (in the range of 500 to 1000 mEq./day) would be required to raise serum $[HCO_3^-]$ by more than 5 mEq./l. However, when GFR is reduced, much smaller quantities of administered HCO_3^- may significantly increase serum $[HCO_3^-]$, resulting in metabolic alkalosis. When renal function is normal, all other circumstances in which metabolic alkalosis occurs must therefore represent examples of enhanced renal tubular HCO_3^- reabsorption or regeneration (e.g., increased renal tubular H^+ secretion). Two mechanisms can result in increased renal tubular HCO_3^- reabsorption and regeneration. The first stimulus for enhanced renal tubular HCO_3^- reabsorption appears to be related to extracellular fluid volume contraction. When Cl^- losses accompany external losses of H^+, as in vomiting or drainage of acid gastric juice and in rare cases of con-

genital chloride diarrhea, or when H^+ loss into the urine accompanies Cl^- during the administration of diuretics or in Bartter's syndrome, renal HCO_3^- reabsorption capacity is enhanced. As a consequence, increments in serum $[HCO_3^-]$ are sustained. The second stimulus primarily enhances renal tubular HCO_3^- regeneration and is caused by factors that augment renal tubular H^+ secretion. These include K^+ depletion, aldosterone, glucocorticoids having mineralocorticoid activity, and other substances possessing mineralocorticoid effects (such as licorice). Metabolic alkalosis which results primarily from processes causing extracellular fluid volume depletion have been termed *chloride-sensitive* while those associated with severe K^+ depletion or mineralocorticoid excess have been termed *chloride resistent*. The causes of metabolic alkalosis, arranged according to these pathophysiological mechanisms, are listed below.

Causes of Metabolic Alkalosis

Associated with reduced GFR
    Bicarbonate loading
    Milk-alkali syndrome
Associated with ECF-volume contraction
  (chloride-sensitive)
    Vomiting
    Gastric suction
    Diuretics
    Congenital chloride diarrhea
    Bartter's syndrome
Associated with potassium depletion
  (chloride-resistant)
    Primary or secondary hyperaldosteronism
    Cushing's syndrome
    Licorice ingestion

The expected compensatory response to a primary increase in serum $[HCO_3^-]$ and a consequent fall in $[H^+]$ or rise in pH is a reduced rate of pulmonary CO_2 excretion. Since hypoventilation is limited by the hypoxic stimulus to respiration, respiratory compensation for metabolic alkalosis is limited, as illustrated in Figure 14–4. As a consequence, P_{CO_2} values above 55 to 60 mm. Hg in patients with a disorder causing metabolic alkalosis suggests the coexistence of respiratory acidosis. The detection of metabolic alkalosis that is chloride sensitive depends on detection of extracellular fluid-volume contraction. In addition, urinary Cl^- excretion is low, except when the alkalosis is the result of diuretic use or Bartter's syndrome. Since alkalosis is associated with a shift of K^+ into cells, hypokalemia does not distinguish chloride-resistant metabolic alkalosis. The symptoms of metabolic alkalosis, if any, are similar to those observed in respiratory alkalosis.

The therapy for chloride-sensitive alkalosis is provided by restoring extracellular fluid volume with NaCl and KCl and involves repair of the K^+ deficits and removal of the stimulus to excessive renal tubular HCO_3^- regeneration.

Mixed acid-base disorders

Finally, it is evident that a given patient may exhibit more than one primary acid-base disorder. Such mixed acid-base disturbances are usually recognized readily by a consideration of the measurements of P_{CO_2} and $[HCO_3^-]$ in relation to the acid-base nomogram (Fig. 14–4). Patients who exhibit P_{CO_2}–$[HCO_3^-]$ measurements that fall outside of the ranges for single primary acid-base disorders must have mixed acid-base disorders. Common examples include: (1) the coexistence of chronic respiratory acidosis and metabolic alkalosis in patients with chronic lung disease who are receiving diuretics (P_{CO_2} 60 mm. Hg and $[HCO_3^-]$ 40 mmol./l.); (2) the coexistence of chronic respiratory alkalosis and metabolic acidosis in patients with hepatic cirrhosis who develop renal failure (P_{CO_2} 25 mm. Hg, $[HCO_3^-]$ 12 mmol./l.); (3) the superimposition of acute respiratory acidosis on chronic respiratory acidosis in patients with chronic lung disease who develop acute pneumonitis (P_{CO2} 80 mm. Hg, $[HCO_3^-]$ 32 mmol./l.); and (4) combined metabolic and respiratory acidosis after cardiopulmonary arrest (P_{CO_2} 60 mm. Hg and $[HCO_3^-]$ 18 mmol./l.).

Annotated bibliography

Arbus, G. S.: An *in vitro* acid-base nomogram for clinical use. Canad. Med. Assoc. J. *109*:291, 1973. (Summarizes the responses of serum HCO_3 to primary respiratory acid-base disturbances and the Pa_{CO_2} responses to primary metabolic acid-base disturbances.)

Arruda, J. A. L., and Kurtzman, N. A.: Mechanisms and classification of deranged distal urinary acidification. Am. J. Physiol. *239*:F515, 1980. (A detailed consideration of current concepts regarding the mechanisms for H^+ secretion in the distal nephron in health and disease.)

Bates, R. G.: Determination of pH. Theory and Practice. New York. John Wiley & Sons, 1973. (A complete consideration of the measurement and meaning of pH.)

Cogan, M. G., Rector, F. C., Jr., and Seldin, D. W.: Acid-base disorders. *In* Brenner, B. M., and Rector, F. C., Jr. (eds.): The Kidney, ed. 2, Chap. 17. Philadelphia, W. B. Saunders, 1981. (A review of acid-base disorders stressing pathophysiology with comprehensive references).

Henderson, L. J.: The theory of neutrality regulation in the animal organism. Am. J. Physiol. *21*:427, 1908. (No better discussion of the $H_2CO_3/NaHCO_3$ buffer system has supplanted this original discussion.)

Lemann, J., Jr., and Lennon, E. J.: Role of diet, gastrointestinal tract and bone in acid-base homeostasis. Kidney Int. *1*:275, 1972. (A summary of the sources of fixed acid and of base from the diet and endogenous metabolism together with references to studies providing the basis for evaluating net external acid balance. The role of the skeleton in the defense against chronic metabolic acidosis is also considered.)

Relman, A. S.: Lactic acidosis, acid-base and potassium homeostasis. *In* Brenner, B. M., and Stein, J. H. (eds.): Contemporary Issues in Nephrology, Vol. 2, Chap. 3. New York, Churchill-Livingstone, 1978. (A comprehensive current review of the pathogenesis of lactic acidosis.)

Schwartz, W. B., and Cohen, J. J.: The nature of the renal response to chronic disorders of acid-base equilibrium. Am. J. Med. *64*:417, 1978. (An examination of the evidence supporting the concept that the rate of renal H^+ excretion is primarily determined by the state of sodium balance and ECF-volume.)

Sebastian, A., McSherry, E., and Morris, R. C., Jr.: Metabolic acidosis with a special reference to the renal acidoses. *In* Brenner, B. M., and Rector, F. C., Jr. (eds.): The Kidney, Chap. 16. Philadelphia, W. B. Saunders, 1976. (A thoroughly referenced review of mechanisms responsible for impaired renal tubular H^+ secretion.)

Tannen, R. L.: Ammonia metabolism. Am. J. Physiol. *235:*F265, 1978. (A critical review of the mechanisms responsible for the adaptive increases in renal ammonia production.)

Warnock, D. G., and Rector, F. C., Jr.: Renal acidification mechanisms. *In* Brenner, B. M. and Rector, F. C., Jr. (eds.): The Kidney, ed. 2, Chap. 10. Philadelphia, W. B. Saunders, 1981. (A current and exhaustively referenced review of the mechanisms for renal H^+ secretion.)

Section IV

Endocrine-metabolism Mechanisms

Introduction

The endocrinology-metabolism section of this textbook has been extensively revised for this edition to reflect the remarkable progress made over the intervening years in understanding the pathophysiology of this broad area of disease.

While the 1960s and early 1970s reflected considerable progress owing to the development of the radioimmunoassay for measurement of circulating hormone levels and its application to their physiology, the last 8 years have been dominated by the discovery, isolation, and increasing understanding of the pituitary prohormones, releasing factors, and other peptidergic messengers. These changes are extensively incorporated in the specific chapters dealing with each hormone system and include new information concerning receptor-mediated abnormalities of the target organs.

Dr. Daniels has added a great deal of new and useful information in revising the thyroid chapter. This includes information concerning the TRH-TSH relationships that have provided a model for the interaction of the central nervous system and the anterior pituitary. At the same time, the information concerning thyroid hormone conversion to its active and inactive metabolites and their relationship to disease states is of great usefulness for everyone in clinical medicine.

The chapter on diabetes and its related metabolic diseases by Dr. Gerich is remarkable for its clarity, direction, and ability to incorporate and summarize a wide body of information for the student. It will be of value whether the reader is still a student in formal training or a practicing physician who wishes to review changes in this field. Many changes in terminology as well as interaction between related hormonal systems are addressed in a manner that can be easily assimilated.

The chapter relating to bone metabolism has been completely revised. The material relating to the joint diseases has been transferred to the section on immunological mechanisms, where the greatest progress and interest in rheumatology are found. This has permitted Dr. Deftos to expand upon the information available concerning vitamin D metabolism and other calcium-active hormones. This chapter has been organized in a unique fashion, treating each disease process on the basis of the predominant underlying abnormality of bone deposition and remodeling. This provides an easily remembered approach to the multiple diseases found in the skeleton, and at the same time focuses on the predominant pathophysiological state rather than just listing certain manifestations of the disease alone. This is a signal addition to the textbook.

Dr. Lipsett has extensively revised his chapter on the reproductive system. His direct, informative style is still present and this remains a strong chapter. This chapter includes new information concerning releasing factors, receptor alterations, and the

321

extensive feedback control necessary for normal and abnormal function of the reproductive system.

The chapter on the pituitary-adrenal axis has been expanded to include regulatory mechanisms of the hypothalamic-pituitary axis, growth hormone, and prolactin. It now provides a more complete and balanced presentation of diseases resulting from derangement of this complex system that involves the interdependency of the hypothalamus, the pituitary, and the adrenal glands.

The section on genetic mechanisms of disease has been revised, and the bibliography updated by Dr. Jones. This chapter includes an important summary of the insights we need in order to understand the relationships of familial diseases. In an era of expanding expectations, this chapter provides a useful basis for reviewing the current status of these diseases.

Major revisions have been directed toward assisting the reader in relating concepts of endocrine function in total body homeostasis to diseases of hormonal and metabolic balance. Emphasis is placed on the interactions of different hormones with their metabolic effects, less attention to providing merely an abbreviated textbook of medicine.

DAVID C. KEM, M.D.

15

Regulatory Mechanisms of the Hypothalamic-Pituitary and of the Pituitary-Adrenal Axes

Tah-Hsiung Hsu, M.D.
Robert L. Ney, M.D.

THE HYPOTHALAMIC-PITUITARY AXIS

It is important to discuss the associations between the hypothalamus and pituitary at the outset of this section because of their intimate anatomical and functional relationships and their role on other hormonal functions. The hypothalamus is the common link between the central nervous system and the pituitary gland. Hypothalamic neurons have both neural and secretory activities that can be subdivided into two distinct systems: the magnocellular-neurohypophysis and the parvicellular-adenohypophysis (Fig. 15–1). The magnocellular system consists of the supraoptic and paraventricular nuclei whose axons terminate in the pituitary fossa to form the posterior pituitary. This system produces vasopressin and oxytocin. In contrast, hypothalamic hormones that control the anterior pituitary function are secreted by the parvicellular system and are delivered to the anterior pituitary by a portal venous system. The parvicellular system neurons likewise form a number of less discrete nuclei (Fig. 15–1). These neuronal secretory activities are regulated primarily through

two mechanisms: (1) synaptically transmitted information from higher central nervous system centers, and (2) hormones that are produced by the specific endocrine target organs and whose synthesis and release is regulated directly by the anterior pituitary hormone (Fig. 15–2). Neurotransmitters, such as dopamine, norepinephrine, serotonin, and acetylcholine, participate in the conveyance of information from the higher CNS centers to the hypothalamic neurons. It therefore follows that pharmacological manipulation of these neurotransmitters may influence indirectly anterior pituitary function.

The hypothalamic-pituitary interaction is analogous to a neural synapse but the effectors are small peptides, and the distance between the nerve endings and their target cells of the pituitary is relatively great. The gap between neurosecretory nerve endings and the cells of the anterior pituitary is filled by the portal venous system. In this arrangement, the median eminence of the hypothalamus, where most of the neurosecretory nerve endings termi-

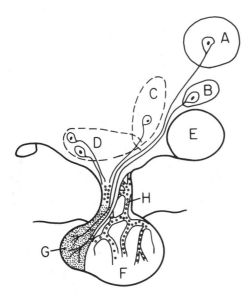

FIG. 15–1. Diagram of neurohypophyseal system. Neurons in the paraventricular nuclei (A), supraoptic nuclei, (B) and their axons constitute the magnocellular-neurohypophyseal system. These neurons are considerably larger than neurons of other hypothalamic nuclei and are stainable with Gomori's stain. The magnocellular neurons synthesize antidiuretic hormone and oxytoxin, which are stored in the posterior pituitary (G). Neurons of the parvicellular system (C, D) are small, fusiform cells scattered diffusely and form a less well-defined nuclei. Most of their axons terminate in the median eminence where hypothalamic hormones are carried to the anterior pituitary (F) by way of the portal-venous system (H). E represents optic chiasm.

nate, provides the final common pathway for the delivery of hypothalamic factors by way of the portal veins of the pituitary stalk to pituitary cells (Fig. 15–1).

Four hypothalamic hormones have been isolated and characterized: thyrotropin-releasing hormone (TRH); gonadotropin-releasing hormone (GnRH); growth hormone-release-inhibiting hormone (somatostatin); and corticotropin-releasing hormone (CRH) (Table 15–1). A fifth peptide with growth hormone releasing activity has been isolated and synthesized recently. Its significance and physiological role, however, have not been clarified. In addition, dopamine has been shown to inhibit prolactin secretion. The secretion of prolactin

(PRL), a hormone that stimulates milk production, is normally suppressed by hypothalamic factors, notably dopamine. The action of some hypothalamic hormones is not limited strictly to stimulating a single pituitary hormone. For example, TRH, at a dose needed to stimulate TSH, stimulates PRL as well. The action of some hypothalamic hormones may be mediated through cyclic AMP. In addition to their effects on the anterior pituitary, certain hypothalamic neurohormones have extrapituitary effects and are also synthesized in other sites of the body. For instance, somatostatin inhibits the secretion of insulin, glucagon, and gastrin and is also found in relatively high concentrations in the gut and pancreatic islet cells.

Most likely, additional hypothalamic factors remain to be identified as do the precise mechanism for the relese of hypothalamic hormones and the interactions among hypothalamic hormones on their target cells.

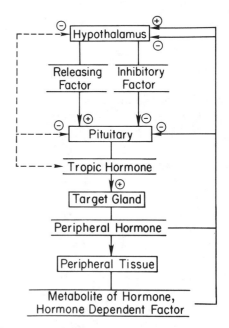

FIG. 15–2. Negative feedback control of pituitary function. The nature of the short feedback loop (*dotted line*) is still uncertain.

TABLE 15–1
Some Biological Properties of Chemically Characterized Neurohypophyseal Tropic Hormones
(Hypothalamic Hormones)

Hypothalamic Hormones	Amino Acid Content	Pituitary Effects	Extrapituitary Effects
Thyrotropin-releasing hormone (TRH)	3	Increased TSH Increased PRL Increased GH (acromegaly)	
Gonadotrin-releasing hormone (GnRH)	10	Increased LH Increased FSH Increased GH (acromegaly)	Decreased gonadal function (high dose)
Growth-hormone-release inhibiting hormone (Somatostatin)	14	Increased GH Decreased TSH response to TRH	Decreased insulin, glucagon, gastrin, secretin
Corticotropin-releasing hormone (CRH)	42	Increased ACTH	

The anterior pituitary and its hormones

The anterior pituitary gland normally weighs 0.5 to 0.7 g. and is located at the base of the skull in the sella turcia. Embryologically, the anterior pituitary arises from Rathke's pouch, originally thought to be an evagination of ectodermal tissue in the roof of the oral cavity. Recent evidence, however, suggests that this pouch is probably derived from the ventral neural ridge in the pharyngeal region, thus sharing the common embryonic origin with the hypothalamic complex. During the first three months of embryonic life, cells migrate upward toward the infundibulum to establish intimate connection with the neural projection while losing its contact with the oral-pharyngeal junction. The anterior pituitary secretes at least six physiologically important hormones (Table 15–2). They can be classified into three groups based on their chemical compositions. (1) Growth hormone (GH) and prolactin (PRL) have extensive homology with chorionic somatomammotropin (placental lactogen) in their amino acid composition. (2) ACTH and its related hormones including β-lipotropin (β-LPH) arise from a common biosynthetic precusor and serve as precusors for other related peptides. Presently, only ACTH has a clearly defined physiological role in man. Beta-melanocyte stimulating hormone (β-MSH), long believed to be a direct secretory product of the anterior pituitary, is now proved to be a fragment of γ-LPH whose parent molecule is β-LPH. (3) Pituitary hormones in the remaining group are glycoproteins and include thyroid-stimulating hormone (TSH), follicle-stimulating hormone (FSH), and luteinizing hormone (LH). Chorionic gonadotropin, although not of pituitary origin, is chemically similar to these hormones. They consist of two subunits, referred to as α and β. The α-subunits are nearly identical in each of these hormones. However, the β-subunits differ and are responsible for hormone identity and specificity of function.

Microscopic classification of pituitary cells by tinctorial staining into acidophil, basophil, and chromophobe types is clearly inadequate to ascribe the independent secretion of each pituitary hormone. Immunohistologic and electromicroscopic studies have shown that each pituitary hormone is secreted by a distinct cell type. Somatotropic or GH-secreting cells and lactotropic or PRL-secreting cells contain numerous, large, round to oval shaped acidophilic granules and are located in the posterolateral wings of the anterior pitui-

TABLE 15–2
Some Characteristics of Pituitary Trophic Hormones

Hormone	Chemical Structure	Pituitary Content (μg)	Serum Concentration
Growth hormone (GH)	191-amino acid single-chain peptide with disulfide bridges	4000–8000	0–5 ng./ml.
Prolactin (PRL)	198-amino acid single-chain peptide with disulfide bridges	100	0–25 ng./ml.
Adrenocorticotropic hormone (ACTH)	39-amino acid single chain peptide	250	50 pg./ml. or less
Thyroid stimulating hormone (TSH)	Glycoprotein (α- and β-peptide chains plus carbohydrate)	25	7 μU/ml. or less
Follicle-stimulating hormone (FSH)	Glycoprotein (α- and β-peptide chains and carbohydrate)	35	Depending on standards for radioimmunoassay and age of patient
Luteinizing hormone (LH)	Glycoprotein (α- and β-peptide chains and carbohydrate)	30	Depending on standards for radioimmunoassay and age of patient

tary. Thyrotropic cells or TSH-secreting cells, located primarily in the center wedge of the pituitary, are polyhedral in shape and have small basophilic granules. Gonadotrophs or gonadotropin-secreting cells secrete both FSH and LH and are distributed throughout the pituitary and have basophilic granules. The ACTH-secreting cells also contain basophilic granules; these cells are present in the anteromedial region of the pituitary.

Regulation of pituitary function

Anterior pituitary hormonal secretion is primarily regulated by complex negative feedback loops (Fig. 15–2). Various inputs from higher centers in the CNS are integrated in the hypothalamus to trigger neurosecretory cell secretory activity. Hypothalamic factors are transported through the portal venous system to pituitary cells to regulate the synthesis and release of tropic hormones. If the hypothalamic signal increases the output of the releasing factor (or reduces the inhibiting factor), the pituitary hormone is then released into the general circulation to stimulate the target gland to secrete its hormone. The latter hormone, when it reaches a critical concentration (a certain set point), in turn suppresses its regulators in the hypothalamus and pituitary, thereby achieving a new homeostasis. Not all tropic hormone secretion is regulated through such a simple scheme. Each has its own regulatory mechanisms but the principle of negative feedback system is operative in most cases. (The regulatory mechanisms of the secretion of TSH, FSH, LH and ACTH as well as GH and PRL will be detailed in the respective areas in this section dealing with their target organs.)

Growth hormone

Growth hormone (GH), a single peptide chain of 191 amino acids, is secreted by the somatotrophs of the pituitary. The secretory activity of these pituitary cells is mediated primarily by hypothalamic factors (i.e., GH releasing factor [GRF] and GH releasing inhibiting factor, also known as somatostatin). Somatostatin, although originally isolated from the hypothalamus, is widely distributed throughout the CNS, the gastrointestinal tract, and particularly the pancreas. In addition to inhibiting GH secretion, somatostatin has many extrapituitary actions. Neurons that mediate the production of hypothalamic factors that regulate GH secretion are believed to be distributed primarily in the ventromedial and arcuate hypothalamic nuclei and in the limbic system of the CNS. It is uncertain whether growth hormone itself has any direct feedback effect upon its own secretion. Furthermore, growth hormone stimulates the formation of a growth factor (insulin-like growth factor, also called somatomedin C) in peripheral tissues such as the liver, and it is uncertain whether this intermediary hormone has feedback to influence growth hormone secretion.

The pattern of GH secretion from the anterior pituitary is characterized by an extremely low basal activity, interrupted by discrete high secretory episodes. Between 50 to 75 per cent of the total GH secretion in children and young adults occurs within 60 to 90 minutes following the onset of sleep. GH secretion is also increased by physical or emotional stress, hypoglycemia, certain amino acids, and pharmacological agents (Table 15–3) that act primarily upon the hypothalamus.

The metabolic effects of GH on various cells and tissues are diverse but one of its major effects is to stimulate growth. It affects intermediary metabolism and synthesis of carbohydrates, lipids, and protein. Many of the acute metabolic effects of GH resemble those of insulin. GH acts like insulin to promote glucose uptake in muscle and adipose tissue. In adipose tissue, GH initially antagonizes the lipolytic effect of catecholamines. Furthermore, GH accelerates the net transport of amino acids across cell membranes and their incorpo-

TABLE 15–3

Major Physiological and Pharmacological Factors that Affect the Secretion of Human Growth Hormone

Physiological	Pharmacological
Stimulants	
Sleep	Arginine
Neonatal state	Glucagon
Stress	L-dopa
Hypoglycemia	Apomorphine
Exercise	Bromocriptine
	Pyrogens
	Estrogens
Inhibitors	
Hyperglycemia	Phentolamine
Fatty acid	Chlorpromazine
	Medroxyprogesterone
	Glucocorticoids (high dose)
	Somatostatin

ration into protein in liver and muscle. As soon as the initial effects dissipate, a series of "diabetogenic" effects on carbohydrate and lipid metabolism follow. These effects consist of inhibition of cellular glucose uptake and utilization stimulation of hepatic gluconeogenesis, enhancement of lipolytic effects of catecholamines, and stimulation of free acid release from adipose tissue.

Many of the metabolic effects of GH cannot be reproduced by exposure of tissue *in vitro* to the hormone per se; but sera from normal or GH-treated hypophysectomized animals contain growth promoting factors that are effective even *in vitro*. This has led to the hypothesis that GH acts by stimulating the production of growth factors that circulate in plasma and in turn promote cellular processes that enhance growth of both skeletal and extraskeletal tissues. These factors were first referred to as "sulfation factor" because the original assay measured the incorporation rate of radioactive sulfate into chondroitin sulfate. Subsequently, it was found that *sulfation factor(s)* stimulates a wide variety of metabolic processes in cartilage and in other tissues. The more appropriate term, *somatomedin*,

TABLE 15-4
Some Characteristics of Growth Hormone-Dependent Serum Growth Factors in Man

Factor	Molecular Weight or Amino Acids Content	Sulfation Activity	Clinical Correlation* Acromegaly	GH-Deficient State
Somatomedin A	59 amino acids, Mol. wt. 7600	+	↑	↓
Somatomedin B	Mol. wt. 5000	−	↑	↓
Somatomedin C	Mol. wt. 7600	+	↑	↓
Nonsuppressible insulin-like activity	Proinsulin-like molecules that do not react with insulin antibody	+	↑	↓
Insulin-like growth factor I	70 amino acids, similar or identical to somatomedin C	+	↑	↓
Insulin-like growth factor II	67 amino acids	+	?	?

* ↑ = increased; ↓ = decreased

was then coined to imply that this factor(s) mediated somatic growth. It is now recognized that somatomedins represent a family of polypeptides (Table 15–4) that possess insulin-like and growth-promoting effects which seem to be under the influence of growth hormone. These growth factors are carried in blood with macromolecular proteins and are synthesized chiefly in the liver. There is no somatomedin-enriched organ or tissue for extraction, but the isolation and characterization of these growth factors have been accomplished (Table 15–4). Somatomedin B no longer is considered as a growth factor but serum somatomedin A and C levels usually correlate well with growth hormone concentrations. The so-called nonsuppressible insulin-like activity (NSILA), representing insulin activity that cannot be neutralized by antibody against pancreatic insulin, also has growth promoting effects, and was considered as one of the contributors to the sulfation factor(s). Further characterization and purification of NSILA revealed two forms of NSILA: *insulin-like growth factors (IGF) I and II.* There is now strong evidence that at least IGF-I is identical to somatomedin C. (Whether the names insulin-life growth factor or somatomedin

will prevail or whether a new terminology will be developed is still uncertain.) Differences in the terminology of these growth factors primarily reflect differences in the assay system (i.e., the activity of IGF I was assayed based on its insulin-like effect on the adipose tissue, whereas the effect of somatomedin C was expressed as sulfation capacity).

The fact that somatomedins are tightly bound to proteins in circulation may account for their relatively long plasma half-life (3 to 4 hours). GH may not be the sole regulator of the blood levels of somatomedins since their blood concentrations are low in malnutrition, renal insufficiency, hypercortisolism, and hyperestrogenism.

In addition to its growth stimulating effects, GH may play an important role in short-term protection against hypoglycemia. GH increases glomerular filtration rate, renal plasma flow, and tubular reabsorption of phosphate. It also has intrinsic prolactin activity which is not surprising since these hormones have peptide sequences in common.

HYPERSECRETION OF GROWTH HORMONE. Acidophil and some chromophobe (PAS nonstaining) pituitary adenomas may sustain GH hypersecretion. Whether this disorder of growth hormone secretion is due to a hypothalamic disturbance or to an intrinsic abnormality of the pituitary gland remains to be resolved. This syndrome may be part of multiple endocrine adenomatosis, type I (adenomas of pituitary, parathyroid, pancreatic islet cells, and bronchial tree). Excess GH during the prepubertal period results in a rapid increase in skeletal growth and results in *gigantism*. With onset of puberty, a further spurt in growth occurs; then, with closure of the epiphyses of the long bones, height stabilizes. Consequently, hypersecretion of GH after epiphyseal closure in adults cannot produce increased height. Rather, it causes an overgrowth of the membranous bones and results in a condition known as *acromegaly*. For convenience, the clinical features of acromegaly and gigantism can be categorized into three major groups: (1) physical features, (2) metabolic effects, and (3) local effects of pituitary adenoma.

PHYSICAL CHARACTERISTICS. The patient with acromegaly is characterized by coarse facial features and enlarged hands, feet, and mandible. Soft tissue growth is also stimulated and the growth of hands and feet is further accentuated. In addition, most visceral organs such as kidneys, pancreas, thyroid, and adrenals are usually enlarged. The skin is thickened and there is hypertrophy of sweat and sebaceous glands resulting in hyperhidrosis and oily skin. The laryngeal cartilages are enlarged and the vocal cords thickened, commonly deepening the voice. Disproportionate enlargement of the skeletal-cartilage system may cause kyphosis, arthropathies, or carpal tunnel syndrome which is due to compression of median nerve. Hypertrophy of skeletal muscles with weakness and fatigue are also common findings in acromegaly. In patients with gigantism, the skeletal growth is fairly symmetrical and proportional. The arms and legs are abnormally long, the trunk is also long and relatively

slender. The comformation of the head and the face may be fairly normal but acromegalic features may eventually appear if hypersecretion of growth hormone persists after the complete closure of the epiphyses.

METABOLIC EFFECTS. Since GH directly inhibits peripheral glucose uptake and increases hepatic glucose production, the most notable metabolic derangements caused by excess GH are impaired glucose tolerance, compensatory hyperinsulinism, and insulin resistance. If the pancreas is able to maintain increased insulin secretion, glucose tolerance may remain normal. Ultimately, the islet cells may be unable to secrete enough insulin to offset the effect of GH, and diabetes mellitus can result. Clinically, carbohydrate intolerance is present in 50 per cent of the patients but frank diabetes mellitus occurs in only about 10 per cent of them. GH also increases renal tubular reabsorption of phosphate; hence, hyperphosphatemia is a frequent finding. Hypercalciuria and increased hydroxyproline excretion result from the multiple effects of GH on the kidneys, skeleton, and collagen. GH also increases the basal metabolic rate and contributes to the intense hyperhidrosis in acromegaly and gigantism.

LOCAL EFFECTS OF PITUITARY TUMOR. Since most patients with gigantism or acromegaly have sellar enlargement, symptoms produced by the expanding tumors are usually prominent. They include headache, bitemporal hemianopsia from the compression of otic chiasma, or hypopituitarism from hypothalamic involvement or from the compression of the remaining functioning pituitary tissue by the tumor.

LABORATORY DIAGNOSIS. Laboratory confirmation of active acromegaly or gigantism is achieved by measurements of serum GH (radioimmunoassay) at the basal condition and after glucose administration. In normal subjects, serum GH levels usually fall to less than 3 ng./ml. 2 hours after oral glucose administration. In acromegaly, the basal GH levels are elevated and cannot be suppressed to the normal range (5 ng./ml. or less) following this glucose load. Rarely, serum GH may be suppressed to a normal range but it occurs almost strictly in patients with a marginal basal GH level. Measurement of somatomedin C concentration is also useful to verify the diagnosis. Since somatomedin is relatively stable compared to serum GH, one may consider that its values reflect integrated GH secretion. This somatomedin assay is particularly useful to confirm the diagnosis in a patient with a marginal elevation of GH. A number of tests of secondary importance are available to demonstrate abnormal GH hypersecretion. For unclear reasons, the GH secretion from pituitary tumors responds inappropriately or paradoxically to some stimuli. Specifically, GH is promptly released following administration of TRH or GnRH in some patients with acromegaly but not in normal subjects, and L-dopa stimulates GH secretion in the normal subjects but suppresses it in patients with acromegaly.

Treatment of acromegaly or gigantism consists of ablation of the pituitary gland by radiation and/or surgery. Bromocriptine, a dopamine agonist, has been reported to suppress growth hormone production in some patients and may provide an alternative therapy for selected patients.

GROWTH HORMONE DEFICIENCY (INEFFECTIVE GROWTH HORMONE AND DWARFISM). Decreased GH secretion may result from panhypopituitarism or isolated failure of GH-secreting cells (somatotropic cells). In either case it may develop from a lesion of the anterior pituitary or hypothalamic disorder. As a rule, pituitary failure due to hypothalamic dysfunction is more prevalent in patients with isolated growth hormone deficiency and in children with hypopituitarism without readily demonstrable pituitary lesions. In children, GH deficiency causes short stature, but in adults it is often inconsequential or is eclipsed by more specific and striking consequences of the concurrent deficiencies of gonadotropins, TSH, and ACTH.

Many factors play an important role in promoting growth, but only growth hormone related growth retardation is discussed in this section. Since GH is one of the most important hormonal factors in stimulating linear growth in man, the consequence of GH deficiency in children is *dwarfism*. Growth in patients with isolated GH deficiency is normal *in utero* but is retarded in childhood. Puberty may be delayed but body proportions are nearly normal. Full sexual maturation is attainable and the patient becomes fertile.

Pathophysiologically, GH-related dwarfism can be classified into types (Table 15–5). Growth hormone deficiency failure may result from tumors, granulomatous lesions, trauma, irradiation, surgery, or necrosis of the hypothalamus or pituitary. In addition, psychic disturbances may produce apparent, but usually reversible, hypothalamic failure. Removal of these children from their adverse milieu restores the serum growth hormone levels and improves growth rate. Isolated GH deficiency syndrome exists in both the familial and the sporadic form.

Laron-type dwarfism is indistinguishable clinically from those due to selective GH deficiency. This syndrome is primarily heritable but sporadic occurrence has been observed. The patients are unable to release somatomedins in response to growth hormone. Lack of feedback suppression by somatomedins results in elevated serum GH. Other types of dwarfism associated with normal or elevated GH (by radioimmunoassay) include abnormal secretion of GH that is biologically inactive but measurable with a radioimmunoassay designed for normal GH, and unresponsiveness to somatomedin (characterized by normal GH and elevated somatomedin levels).

TABLE 15–5
Classification of GH-Related Dwarfism

Type of Dwarfism	Serum GH by RIA	Serum Somatomedin	Pathophysiology	Responses of Somatomedin to GH Injection
Hypothalamic Dwarf	Low	Low	Diminished GH secretion	+
Pituitary Dwarf	Low	Low	Diminished GH secretion	+
Laron Dwarf	High	Low	Failure of GH to generate somatomedin (receptor or postreceptor defect)	–
Dwarf of abnormal GH	High	Low	Biologically ineffective GH	+
Somatomedin unresponsive dwarf	Normal	High	(Receptor or postreceptor defect)	–
Pygmies	Normal	Normal	Uncertain	–

LABORATORY DIAGNOSIS. Definitive laboratory diagnosis of GH deficiency relies on a number of provocative tests. GH reserve is best evaluated by measuring GH in response to insulin-induced hypoglycemia, L-dopa, exercise, or to arginine infusion. Failure of GH response to at least two of the above stimuli is reliable evidence of GH deficiency. Measurements of somatomedin levels are helpful for further classification of dwarfism (Table 15–5). To differentiate the Laron-type dwarfism from that due to an abnormal GH, a somatomedin generation test is used. In this procedure, somatomedin concentration is measured after three consecutive daily injections of human GH. In contrast to the Laron dwarf, somatomedin levels should rise following GH injection in the dwarfism due to production of defective GH.

Prolactin

REGULATION AND PERIPHERAL EFFECTS. Prolactin (PRL) in a single chain polypeptide of 198 amino acids with two disulfide bonds. In contrast to all other pituitary hormones, secretion of PRL from lactotrophs is normally under a tonic suppression by the hypothalamus through a prolactin-inhibiting factor, probably dopamine. With hypothalamic or pituitary stalk destruction, PRL hypersecretion ensues. Agents that deplete hypothalamic dopamine or that block dopamine receptors on the lactotrophs increase PRL secretion. Conversely, drugs that increase dopamine concentration or mimic its effect decrease PRL secretion (Table 15–6). Other neurotransmitters such as serotonin, histamine, or acetylcholine also appear to be involved in regulating PRL secretion. Nocturnal and nipple suckling-induced rises of PRL secretion are likely regulated through serotoninergic fibers. The presence of a specific prolactin releasing factor(s) in the hypothalamus is suspected but its nature is still uncertain. Thyrotropin-releasing hormone (TRH) is a potent pituitary stimulus for PRL secretion from the pituitary. Its physiological importance as a regulator of PRL, however, remains to be demonstrated.

The upper limit of basal serum PRL concentration is about 25 ng./ml. During pregnancy and in neonates, PRL levels are elevated, presumably because of stimulation by estrogens. PRL is secreted episodically and in a characteristic diurnal rhythm related to the sleep-waking cycle. Thus, PRL levels begin to rise 60 to 90 minutes following the onset of sleep, reaching a peak before waking, usually 5:00 to 7:00 A.M.

Prolactin exerts most of its effect in conjunction with other hormones: PRL alone may not exert an effect, but does so indi-

TABLE 15–6
Major Factors That Affect Human Serum Prolactin Levels

	Stimulants	Inhibitors
Physiological	Sleep	
	Physical or emotional stress	
	Pregnancy	
	Postpartum	
	Neonatal state	
	Exercise	
	Hypoglycemia	
	Nipple stimulation	
Pharmacological	Tranquilizers	Dopamine
	TRH	Ergot derivatives (bromocriptine)
	Estrogens	Apomorphine

rectly by modifying the tissue response to other hormones. The major function of PRL in women is initiation and maintenance of lactation; and during pregnancy, PRL together with estrogen, progesterone, and placental lactogen, insulin and glucocorticoids promotes milk formation. Specifically, it stimulates the synthesis of milk proteins, carbohydrates, and lipids. In addition, it may (1) alter placental water and salt transport; (2) enhance renal production of 1,25-dihydroxycholecalciferol, thereby facilitating calcium absorption from the gut during pregnancy; and (3) inhibit gonadotropin secretion and interfere with reproductive function in both males and females. Since the structure of PRL resembles that of GH, prolactin also possesses certain GH-like metabolic activities, but with a much lower level of potency.

ABNORMAL SECRETION. Infiltrative diseases of the hypothalamus, tumors, and surgical or traumatic pituitary stalk damage remove the usual inhibitory hypothalamic influence on PRL secretion. Hyperprolactinemia may also occur in PRL-secreting pituitary adenomas or rarely in PRL-secreting nonpituitary tumors. The major pathophysiological consequences of hyperprolactinemia include hypogonadism in both sexes, and particularly galactorrhea and amenorrhea in women.

The causes of galactorrhea are listed below. Hyperprolactinemia is not always present in every patient with galactorrhea and, conversely, elevations in serum prolactin do not invariably cause galactorrhea. Among patients with nonpuerperal galactorrhea, disorders of the hypothalamus and/or pituitary must be considered. Hypothalamic tumors are not common but pituitary tumors are found in nearly 30 per cent of women with galactorrhea accompanied by hyperprolactinemia and menstrual abnormalities. Amenorrhea in women or hypogonadism in men, associated with hyperprolactinemia, is due primarily to suppression of gonadotropin secretion by PRL, although there may also be an important direct effect of PRL on the gonads. It is a common practice to measure serum PRL concentration in any patient with unexplained galactorrhea, amenorrhea, hypogonadism, and suspected hypothalamic-pituitary lesions. The likelihood of pituitary tumor is best correlated to the magnitude of serum PRL level. Above 300 ng./ml., the diagnosis of pituitary adenoma is virtually certain. Values less than this may be associated with other stimuli as well as with tumors and

Major Causes of Galactorrhea

Hypothalamic origin
    Destructive lesion of hypothalamus
    Postpartum
Pituitary causes
    Prolactin-secreting pituitary tumor
    Damage of hypothalamic-pituitary stalk
      (tumor, surgery, trauma)
Nonendocrine malignancy with ectopic production
  of prolactin
Neurogenic
    Nipple stimulation
    Chest wall lesions
Drug or hormone induced
Idiopathic variety

must be differentiated by other means. Bromocriptine, a dopamine agonist, effectively suppresses PRL in hyperprolactinemic syndromes with restoration of normal menses and gonadal function. The use of bromocriptine exemplifies the feasibility of using neuropharmacologic agents to control neuroendocrine disorders.

Hypoprolactinemia, or a limited PRL reserve, may occur in hypopituitarism. Inability to secrete adequate amounts of PRL has no recognizable clinical significance other than failure to lactate postpartum. In fact, inability to breast-feed is often the first manifestation of Sheehan's syndrome (postpartum necrosis of the anterior pituitary). PRL secretory capacity is best assessed by the TRH stimulation test. Failure of PRL to increase following TRH administration suggests hypofunction of the lactotrophs.

Primary and secondary hypopituitarism

Primary hypopituitarism is due to destruction or atrophy of pituitary cells, resulting in a partial or complete loss of ability to secrete hormones. Pituitary tumors, granulomatous lesions (e.g., tuberculosis, sarcoidosis), trauma, irradiation, surgery, necrosis, and infiltrative processes (e.g., hemochromatosis, amyloidosis) are major causes of pituitary failure. The hypertrophied pituitary gland during pregnancy is particularly susceptible to necrosis as a consequence of postpartum hemorrhage and obstetrical shock (Sheehan's syndrome).

The consequences of pituitary failure depend upon the age of the patient when it first occurs (prepubertal or postpubertal), the type and extent of hormone deficiency, and the rapidity of onset of the illness. Childhood hypopituitarism is often characterized by hypoglycemia, short stature, and poor sexual maturation; but in adults there are varying degrees of hypogonadism, hypothyroidism, and/or adrenal insufficiency. Most patients complain of easy fatigability and lack of appetite and sex drive. Blood pressure, heart rate, and body temperature tend to be low.

Secondary hypopituitarism results from hypofunction of the hypothalamus. Psychological disturbance also can cause usually reversible hypothalamic dysfunction. Psychogenic amenorrhea and the maternal deprivation syndrome are striking examples. Hypothalamic failure, in addition to causing pituitary insufficiency, is often manifested by disturbances of sleep, body temperature regulation, appetite, and body weight. Destructive hypothalamic lesions may produce diabetes insipidus. Laboratory confirmation of hypopituitarism is important. Strategically, one first identifies which target glands are hypofunctioning and then proceeds to prove the absence of compensatory elevations of pituitary tropic hormones, and finally, carries out studies to distinguish pituitary failure from a hypothalamic defect. GH and PRL reserve can be tested as described above. The function of the pituitary thyrotropic cells is assessed by measuring thyroid gland function (serum thyroxine) and TSH. If both are low, a disorder of the hypothalamus and/or pituitary can be suspected. In the case of a pituitary lesion, administration of TRH usually will not stimulate TSH secretion, whereas stimulation may occur after repetitive stimulation when the primary defect is in the hypothalamus because the pituitary cells may have retained their responsiveness.

Insufficient glucocorticoid secretion, accompanied by low serum ACTH levels, is indicative of adrenal failure of pituitary or hypothalamic origin. Unfortunately, the radioimmunoassay of ACTH is not as yet widely available. The integrity of the pituitary-adrenal axis can be evaluated by metyrapone administration. Metyrapone blocks normal cortisol synthesis by inhibiting 11-β-hydroxylation in the adrenal cortex. As plasma cortisol level falls, ACTH secretion normally increases and consequently results in increased steroidogenesis to that point of blockade in steroid biosynthesis. The rise of ACTH and accumulation of the cortisol precursor, 11-desoxycortisol (compound S), in plasma and urine is expected. Since metabolites of compound S are measured in the urine as 17-hydroxycorticosteroids (17-OHCS), rise in urine 17-OHCS should occur after metyrapone in normal subjects, but not in patients with hypopituitarism. Insulin-induced hypoglycemia, which normally stimulates ACTH secretion, also can be used as a test of pituitary adrenal reserve by measuring plasma cortisol levels. The usefulness of corticotropin-releasing hormone administration to distinguish hypothalamic from primary pituitary failure remains to be determined. An ACTH stimulation test is useful in diagnosing adrenal insufficiency: in primary adrenal insufficiency, adrenal function is unresponsive to ACTH; with hypopituitarism, the atrophic adrenal of hypopituitarism is usually unresponsive to a single dose of ACTH, but repeated (3 to 4 days) ACTH injections may reawaken the adrenal function. (For further discussion, the reader is referred to the part of this chapter that concerns adrenal steroidal biosynthetic defects.)

Hypogonadotropic hypogonadism (low FSH, LH, and testosterone in men, low gonadotropin and hypoestrogenism in women) is indicative of hypothalamic failure or primary hypopituitarism. Gonadotropin reserve can be assessed by administration of GnRH or clomiphene. However, it is difficult to distinguish between hypothalamic and pituitary disease using the GnRH stimulation test since an absent or subnormal response may be observed in both conditions. Clomiphene is an antiestrogen compound that blocks hypothalamic receptors for estrogen, thereby preventing normal feedback inhibition by estrogen. When a normal subject is treated with clomiphene, GnRH levels rise and stimulate secretion of gonadotropin. In contrast, in hypothalamic or hypopituitary hypogo-

nadism and prepubertal subjects, gonadotropin levels do not rise.

Treatment of hypopituitarism consists of treating the cause of pituitary failure when appropriate and hormone replacement. Hormone supplements depend on the endocrine status of the patient. For the management of panhypopituitarism, replacement of the deficient target organ hormones (e.g., glucocorticoids, thyroid hormone) is initiated. Gonadal hormones or gonadotropins may be given, depending on the patient's age and therapeutic goals. Dwarf children (lacking growth hormone) may be treated with human growth hormone. Significant growth usually occurs during the first several years of therapy.

Posterior pituitary

The posterior lobe of the pituitary contains the axons and axon terminals of the hypothalamic magnocellular system, consisting of the supraoptic and paraventricular nuclei (Fig. 15–1). Two hormones, arginine vasopressin (AVP) and oxytocin, are synthesized by separate neurons in both the supraoptic and paraventricular nuclei, but the supraoptic nucleus primarily synthesizes vasopressin, whereas the paraventricular nucleus produces predominantly oxytocin. These peptide hormones and their respective carrier proteins, neurophysins, are synthesized in the cell bodies as separate precursor proteins (propressophysin). These hormone-carrier protein complexes seem to be reduced into small biologically active molecules while transversing the nerve fibers to their site of storage in the posterior pituitary. Oxytocin causes milk ejection, and also is involved in parturition by stimulating contraction of uterine muscle. Vasopressin, the antidiuretic hormone (ADH), is an octapeptide whose primary function is to regulate water balance. In large doses, it also can cause smooth muscle contraction and can elevate arterial pressure.

Regulation of ADH secretion

The most important physiological determinant of ADH secretion is plasma osmolality. Release is controlled by osmoregulators in the anterior hypothalamus. Under normal physiological conditions, plasma osmolality is tightly regulated between 282 and 290 mOsm./kg. H_2O. A mere 1 per cent increase in osmolality above this level in plasma perfusing the osmoregulator provokes ADH release. Nonosmotic factors such as blood volume and blood pressure, when significantly altered, may also participate in the regulation of ADH secretion. There are volume receptors in the left atrium and baroreceptors in the aorta and major arteries from which impulses are related to the hypothalamus. Left atrial distention leads to a diuresis that can be abolished by high doses of ADH. Since vagotomy can abate atrial volume induced diuresis, afferent parasympathetic nerve fibers are the most likely pathway linking volume receptors to the hypothalamus.

The hypothalamic nuclei that control thirst are both anatomically and functionally related to the osmoregulatory centers. Stimulation of the thirst center by hypertonicity induces polydipsia and antidiuresis. Stimulation of the supraoptic nuclei, on the other hand, promotes ADH secretion without provoking polydipsia.

In addition to physiological factors, ADH secretion is also affected by nonphysiological factors. Barbiturates, morphine, nicotine, histamine, and acetylcholine stimulate release of ADH. Emotional disturbances, pain, and vomiting also increase ADH release. In contrast, high doses of alcohol and phenytoin inhibit ADH release.

Metabolic effect of ADH

ADH plays an important role in the regulation of osmolality and body water metabolism by stimulating cAMP generation in the distal renal tubular cells which in turn increases their permeability to water. In the absence of ADH, the distal convoluted tubules and the collecting ducts become nearly impermeable to water. Consequently, fluid leaving the distal collecting duct is extremely diluted (possibly 15 to 30 mOsm./kg.). ADH deficiency or inability to respond to ADH results in the formation of voluminous hypotonic urine. Conversely, with maximal ADH activity, active distal tubular solute reabsorption is accompanied by sufficient passive reabsorption of water to make the intraluminal fluid isotonic. As this fluid passes through the collecting ducts, it can be maximally concentrated by back diffusion of water along its osmotic gradient into the hypertonic periductal tissue to concentrations as high as 1200 to 1400 mOsm./kg. (For further discussion of urine concentration refer to Chapter 12.)

Pathophysiology of ADH secretion

When ADH is secreted excessively in the absence of physiological or compensatory stimuli, renal water excretion is impaired. Ingestion of water is followed by water retention, hypoosmolality, hyponatremia, and natriuresis. Characteristically, these changes are reversible with fluid restriction. The exact mechanism for natriuresis in the presence of hyponatremia is unclear, but it is probably a combined result of intravascular volume expansion, increased glomerular filtration rate, and diminished aldosterone secretion. ADH radioimmunoassay in the syndrome of inappropriate ADH secretion (SIADH) usually demonstrates values that are abnormally high with respect to the plasma osmolality.

SYNDROME OF INAPPROPRIATE ADH SECRETION. There are three principal origins of SIADH: hypersecretion of ADH from the supraoptic-hypophyseal system, autonomous ectopic production of ADH by malignancy, and a drug-related etiology (see list below). Virtually any continued stimulus to the supraoptic-hypophyseal system can result in ADH hypersecretion. The syndrome of inappropriate ADH was originally observed in patients with bronchogenic carcinoma; now many other malignancies associated with this syndrome have been recognized. Any drugs that stimulate ADH release, potentiate ADH effect, or possess intrinsic ADH effect may also cause SIADH.

Causes of Syndrome of Inappropriate ADH (SIADH) Secretion

Hypersecretion of ADH from
  supraoptic-hypophyseal system
    Trauma
    Infection
    CNS hemorrhage
    Psychic disorders
    CNS tumor
    Metabolic disorders (myxedema, porphyria)
Ectopic production of ADH
    Bronchogenic carcinoma (particularly oat cell)
    Carcinoma of GI tract
    Thymoma
    Tuberculosis
Drug-induced
    ADH or ADH-like substance
      Vasopressin
      Oxytocin
    Agents that stimulate ADH secretion
      Morphine
      Chlorpropamide
      Vincristine
      Clofibrate
      Carbamazepine
    Agents that potentiate ADH action (e.g., chlorpropamide)
Pulmonary diseases
    Tuberculosis
    Pneumonia
    Lung abscess

SIADH is diagnosed by demonstrating (1) hyponatremia with relatively high serum ADH levels; (2) urine that is hypertonic relative to plasma; and (3) substained urinary excretion of sodium despite persistent hyponatremia. Since adrenal insufficiency and salt-losing nephropathy can result in similar laboratory features, the following additional criteria are necessary to secure the diagnosis: (1) normal renal and adrenal function; (2) absence of dehydration or azotemia; and (3) reversal of hyponatremia with fluid restriction. Most patients with SIADH can be managed by fluid restriction alone. Rapid infusion of hypertonic saline is recommended only for a life-threatening profound hyponatremia.

DIABETES INSIPIDUS. Diabetes insipidus (DI) results from the inability of the kidney to conserve water with the resultant excretion of large quantities of dilute urine and secondary polydipsia. The disease is usually due to inadequate ADH synthesis or release. Occasionally, DI results from renal tubular unresponsiveness to ADH (nephrogenic DI). ADH hyposecretion may be caused by any lesion that damages the supraoptic neurohypophyseal system (see list below). In some patients the defect may be present early in childhood, usually due to idiopathic hypoplasia of supraoptic neurosecretory cells. Lack of sensitive and reliable techniques of evaluation of the hypothalamic area in the past may be partially responsible for the claim that more than half of all patients with ADH-deficient DI are of the idiopathic variety. Using computerized tomography, abnormalities in the hypothalamic-pituitary area are now found in many of the patients classified into the idiopathic category.

Major Causes of Diabetes Insipidus

Supraoptic-neurohypophyseal lesions
    Neoplasia (primary or metastatic)
    Head trauma
    Granulomatous infiltration (e.g., tuberculosis, sarcoidosis, syphilis, histiocytosis)
    Therapeutic hypophysectomy
    Viral or bacterial infections
Idiopathic variety
    Sporadic
    Familial

Insufficient ADH results in loss of body water with consequent hyperosmolality. The thirst center is, in turn, activated and polydipsia results. The balance between water loss and intake is restored and body fluid tonicity is stabilized. The clinical consequences of DI are polyuria, extreme thirst, often accompanied by exhaustion, weakness, dizziness, and slight weight loss. Profound dehydration, hyperosmolality, and circulatory collapse may occur when the thirst center is also destroyed or becomes insensitive. It must be recalled, however, that not all patients who excrete large quantities of urine lack ADH, nor do all patients with ADH deficiency excrete large amounts of dilute urine. Coexisting anterior pituitary deficiency (particularly adrenal insufficiency) may mask the full expression of DI, or the degree of ADH deficiency may be too mild to be symptomatic. There is evidence that glucocorticoids act directly on the renal tubules to decrease water permeability, so in their absence water resorption by the renal tubules is increased.

The hallmark of laboratory findings in patients with DI is a daily urinary output greater than 3 l. with a specific gravity less than 1.010. Urine osmolality is characteristically less than the corresponding plasma osmolality. Plasma ADH level is low relative to plasma osmolality. The most commonly employed measure to diagnose DI is based on the principle that DI patients are unable to concentrate urine maximally, despite dehydration, without receiving exogenous ADH. Thus, during water deprivation, the urine remains dilute and plasma osmolality rises. ADH injection will result in concentration of urine, but this will not occur in patients with nephrogenic diabetes insipidus.

ADH-deficient DI can be managed with (1) ADH or its derivatives such as pitressin tannate in oil or 1-desamino-8-D-arginine vasopressin (DDAVP); (2) agents that potentiate the effect of ADH such a clofibrate and chlorpropamide, or that stimulate ADH release such as chlorpropamide; or (3) diuretics that deplete body solutes. Since the modes of action of these drugs are different, their effects are additive when used concomitantly.

The clinical features of nephrogenic diabetes insipidus are similar to those of ADH-deficient DI, but its pathogenesis resides in the unresponsiveness of renal tubules to ADH. The disease is usually inherited, with the greatest frequency in males of affected kindreds. Renal concentration defects that are unresponsive to ADH are also found in patients with hypercalcemia, hypokalemia, or sickle cell traits. Nephrogenic DI may also occur due to drugs such as demeclocycline or lithium. The diagnosis is established by demonstrating the failure of exogenous ADH to cause antidiuresis in a water-deprived patient.

ADRENAL CORTEX

The adrenal gland consists of a medulla and cortex. The medulla, as part of the autonomic nervous system, produces catecholamines whose secretion is under direct control of the central nervous system. The human adrenal cortex is further divided into three zones. The outermost zone, the glomerulosa, secretes aldosterone under the predominant regulation of the renin-angiotensin system. The middle portion of the cortex is the zona fasciculata, where the most important glucocorticoid, cortisol, is produced. Various androgens and small amounts of estrogens also arise from zona fasciculata as well as from the innermost zone, the reticularis. The functions of the inner two zones (the fasciculata and reticularis) of the adrenal cortex are under the control of the anterior pituitary hormone, adrenocorticotropic hormone (ACTH).

Regulation of adrenal cortical function

There are three statements that characterize activity of the adrenal cortex: the hormones are essential for life; their secretion accelerates rapidly in response to a variety of stresses; and protracted excessive secretion rapidly produces pathological consequences. Therefore, a precise and sensitive control mechanism is necessary to provide the body with adequate basal hormone levels, to increase secretions rapidly, and subsequently to revert to maintenance levels when the need has subsided.

Glucocorticoid secretion is regulated chiefly by the hypothalamic-pituitary unit through a negative feedback mechanism. CNS-controlled corticotropin-releasing factor (CRF) of the hypothalamus regulates the synthesis and release of pituitary ACTH which is the direct stimulator of the secretory activity of the adrenal cortex. The levels of circulating corticoids, in turn, influence the secretion of its hypothalamic and pituitary regulators. This arrangement functions to decrease ACTH secretion when plasma cortisol increases beyond its *set point* (the individual's physiological needs) and, conversely, to increase ACTH when there is a decrease in plasma cortisol. There is also evidence that ACTH and/ or ACTH-like peptides may act directly upon the brain and the hypothalamus as neuroregulatory substances (the so-called short-circuit feedback loop). Thus, constant regulation of adrenal corticoid secretion depends upon adequate function of the CNS, hypothalamus, pituitary, and adrenal cortex. Extrinsic stimulation by pain, physical or emotional stress, infection, fever, and hypoglycemia may produce signals to the hypothalamus to stimulate

ACTH secretion, presumably through higher CRF secretion, until a new level of homeostasis is accomplished.

In addition to the negative feedback control mechanism and extrinsic influences, this regulatory system has an intrinsic circadian rhythmicity. In normal persons, the ACTH-cortisol secretion pattern is characterized by a sharp rise during the latter half of the usual sleep period, reaching the maximum shortly after the hour of awakening, and falling irregularly as the day progresses until the beginning of the new cycle. The circadian ACTH secretion rhythms are mediated by the hypothalamus, are dependent upon sleep-wake activity, and can be modified by altering the sleep-waking cycle.

ACTH is released episodically throughout the 24-hour day with a greater clustering of the episodes around the waking hour. The relationship between negative feedback control, intrinsic rhythmicity, and extrinsic influences is complex. These regulatory mechanisms are constantly in operation and the net effect determines the rate of ACTH, and thus cortisol secretion.

Synthesis of ACTH

ACTH, lipotropins (LPHs), melanocyte-stimulating hormones (MSHs), and endorphins are manufactured in the anterior pituitary through a common glycoprotein precursor, pro-opiomelanocortin (Fig. 15–3). The N-terminal portion of the precursor molecule is thought to serve as a hormone but a precise function for it has yet to be identified. Alpha-MSH, which is identical to ACTH 1-13, and corticotropin-like intermediate lobe peptide (CLIP, ACTH 18-39) represent two fragments of ACTH. They are found primarily in species with a prominent intermediate lobe. β-LPH contains the molecule of α-LPH and β-endorphin. β-MSH, which contains an amino acid sequence homologous to positions 4 to 10 of ACTH, is a fragment of γ-LPH. All of these LPH-related peptides except β-MSH are found in human pituitaries. In man, β-MSH

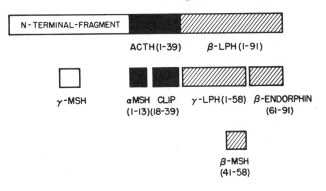

FIG. 15—3. Schematic representation of ACTH precursor molecule (also known as pro-ACTH or pro-opiomelanocortin and its relationship to ACTH, endorphin, and lipotropin. The structure and biological significance of the N-terminal portion (representing about one half of the molecule) of the prohormone is unknown. CLIP = corticotropin-like intermediate-lobe peptide.

is regarded as an *in vitro* product from proteolysis of γ-LPH that occurs during purification procedures of plasma and tissues and probably does not exist as a hormone *in vivo*.

Other than the anterior pituitary, neoplastic lesions, particularly originating from lung, pancreas, thymus, and thyroid C-cells, are capable of producing a substance very similar, if not identical, to the pituitary ACTH. The secretion of ACTH by these tumors, however, is usually independent of normal feedback control and thus causes hypercortisolism. The mechanism by which non-pituitary tissue produces a polypeptide such as ACTH is unknown. Over three-quarters of all cases of ectopic ACTH production are due to APUD (amine precursor uptake and decarboxylation) type tumors (oat cells, thymic cells, bronchial argentaffin cells, thyroid C-cells, and pancreatic islet-cells).

ACTH has a well-recognized physiological role, but the significance of the remaining peptides from pro-opiomelanocortin is still under intense investigation. MSH is probably the most potent pigmentary factor but all other related peptides possess this biological effect as well. The secretory pattern of ACTH-related peptides is generally parallel to that of ACTH.

Biosynthesis of corticosteroids

Cortisol is the most important steroid secreted by the human adrenal cortex when it is stimulated by ACTH (Fig. 15—4). ACTH acts upon the adenylate cyclase system found in the cellular membrane to convert ATP to 3′,5′-cAMP, which is then released into the cell of the zona fasciculata. How cAMP acts in the adrenal cell remains to be determined. However, the effect of ACTH and, in turn, cAMP is to accelerate the conversion of cholesterol to Δ^5-pregnenolone within adrenal mitochondria. This conversion involves hydroxylation and side-chain cleavage of the cholesterol molecule. The conversion of cholesterol to pregnenolone is the rate-limiting step in adrenal steroid biosynthesis. Pregnenolone leaves the mitochondria and enters the smooth endoplasmic reticulum where progesterone is formed. Progesterone may be hydroxylated either at the 21-position by 21-hydroxylase, or at the 17-position by the microsomal 17-hydroxylase. Corticosterone results from the 11-β-hydroxylation of 11-deoxycorticosterone. On the other hand, 11-deoxycortisol (substance S) results from 21-hydroxylation of 17-OH progesterone. 11-deoxycortisol reenters mitochondria and becomes cortisol after 11-β-hydroxylation. Under normal conditions, the chief secretory product is the substance that has been hydroxylated at the 11-, 17-, and 21-positions (i.e., cortisol). A normal adult secretes about 20 mg. of cortisol and about 2 mg. of corticosterone daily. Several androgenic steroids that have undergone side-chain cleavage, chief among them dehydroepiandrosterone, are also produced by the adrenal cortex in response to stimulation by ACTH. ACTH primarily controls glucocorticoids, whereas aldosterone pro-

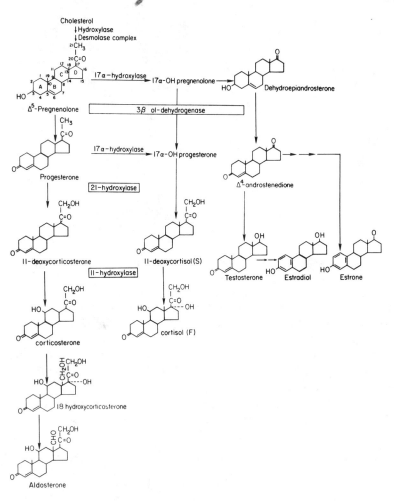

FIG. 15—4. Major pathways of steroidogenesis in human adrenal cortex.

duction is primarily dependent upon angiotensin II.

In addition to bringing about increased secretion of corticosteroids, ACTH also influences the growth of the adrenal cortex. Thus, when ACTH is absent, the gland quickly atrophies to less than a quarter of its normal size. In contrast, excessive ACTH stimulation induces adrenocortical hyperplasia producing an abnormally large gland. Increased vascularity of the gland has been observed with continued stimulation. Even acute administration of ACTH has been associated with increased adrenal blood flow.

A number of pharmacologic agents are known to inhibit steroidogenesis. Aminoglutethimide inhibits the conversion of cholesterol to Δ⁵-pregnenolone. O,P'-DDD (1,1-dichloro-2 (o-chlorophenyl)-2-(p-chlorophenyl)ethane) is adrenolytic, causing degeneration of adrenal mitochondria, thereby impairing steroidogenesis by blocking conversion of cholesterol to Δ⁵-pregnenolone. Metyrapone is an inhibitor occasionally used in treating certain forms of adrenal hyperfunction, but is most widely used for the investigation of the integrity of the hypothalamic-pituitary-adrenal axis. It inhibits 11 β-hydroxylase, and

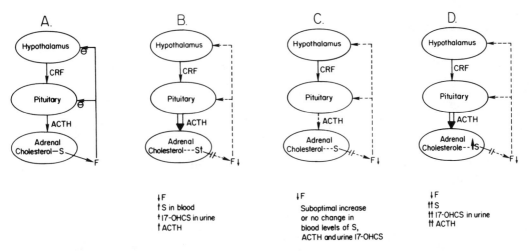

FIG. 15–5. Pituitary-adrenal response to metyrapone, an 11-hydroxylase inhibitor, in the normal subject (*B*), hypopituitarism (*C*), and pituitary Cushing's disease (*D*). (The normal steroidal biosynthetic mechanisms are shown in *A*.) Adequate inhibition of 11-hydroxylase by metyrapone results in hypocortisolism, which triggers further secretion of ACTH in the normal subject. Steroidogenesis is, therefore, increased, but only to the point of blockade. A rise in ACTH and accumulation of the cortisol (Compound F) precursor 11-deoxy-cortisol (Compound S) in plasma and urine is expected. Urinary 17-OHCS are also increased because the major metabolite of Compound S contributes to its value. No rise or a suboptimal rise in ACTH, Compound S, or urinary 17-hydroxycorticosteroids (17-OHCS) after metyrapone indicates impaired ACTH reserve (*C*). A supranormal response to metyrapone is observed in patients with pituitary Cushing's disease (*D*).

the adrenal cortex thus initially produces decreased quantities of cortisol. By reducing the plasma levels of cortisol, metyrapone triggers the hypothalamic-pituitary axis to release more ACTH which in turn stimulates steroidogenesis, leading to increased accumulation of 11-desoxycortisol (the point of blockade) (Fig. 15–5). Since the metabolites of 11-desoxycortisol are measured by the assay of 17-hydroxycorticoids (17-OHCS), an increase in either 17-OHCS in plasma and urine or ACTH in plasma is considered as evidence for a normal pituitary-adrenal axis. The methyrapone test is useful in evaluating patients with ACTH deficiency or Cushing's syndrome (see above). It is essentially a test to see if the pituitary can secrete increased amounts of ACTH, that is, a test of ACTH reserve.

Metabolism of the adrenal steroids

The adrenal steroids are secreted into the bloodstream where they are carried chiefly by transcortin (corticoid binding globulin or CBG) to their target cells. Corticoids bound to proteins are metabolically inert and it is the small unbound fraction (6 to 8 per cent of total cortisol) that is free to penetrate into the intracellular compartment to exert its biological effects. Unbound plasma cortisol is readily filtered through the glomeruli and a large portion is reabsorbed by the renal tubules. A small amount excreted into the urine constitutes the free cortisol detected in urine. The 24-hour urine free cortisol is a fairly accurate reflection of the integrated time-correlated amount of unbound cortisol in plasma.

The degradation of glucocorticoids occurs mainly in the liver, where the Δ^4, 3-ketone in the A ring is reduced and the hormone is generally conjugated with glucuronic acid, forming a highly water soluble product to be excreted in the urine. The major metabolic products of cortisol are measured in the urine as 17-hydroxycorticosteroids, whereas weak adrenal andro-

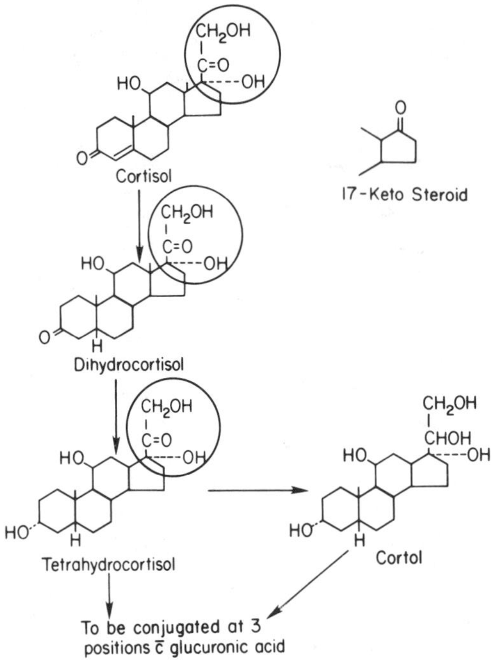

FIG. 15–6. Major catabolic pathways of cortisol by the liver and its excretion by the kidney. Compounds with dihydroxyacetone side chain in the 17 position as shown in the circles are measured as 17-hydroxycorticosteroids (17-OHCS) by the Porter-Silber method. Compounds that possess a keto group in the 17 position are measured as 17-ketosteroids (17-KS) by the Zimmermann reaction. Androsterone, dihydroepiandrosterone, and etiocholanolone are the major constituents of urine 17-KS under normal conditions.

gens, such as androsterone and etiocholanolone, are principal constituents of 17-ketosteroids (17-KS) (Fig. 15–6).

Major metabolic effects of glucocorticoids

Glucocorticoids have a variety of metabolic effects (see list below). Cortisol is the most potent of the naturally occurring glucocorticoids, although corticosterone and cortisone have "glucocorticoid" effects as well. This group of corticosteroids affect carbohydrate metabolism by promoting gluconeogenesis and deposition of liver glycogen. This involves a redistribution of metabolic fuels from lipids and proteins toward carbohydrate by increased gluconeogenesis and decreased tissue glucose up-

take. Glucocorticoids at higher doses enhance protein catabolism and inhibit protein synthesis. Glucocorticoids in excessive amounts also promote fat deposition in faciocervical and truncal areas, stimulate appetite, and stimulate hematopoiesis. Although variable in potency, all glucocorticoids suppress the synthesis and release of ACTH from the pituitary and/or CRF from the hypothalamus. In high doses, glucocorticoids possess potent antiinflammatory and immunosuppressive actions as well as some androgenic and mineralocorticoid activities. Glucocorticoids also interfere with growth in children, largely by affecting the actions of growth hormone on its target tissues, although some decrease in growth hormone secretion may occur as well. Glucocorticoids also antagonize vitamin D activity primarily by reducing intestinal calcium absorption.

Control of aldosterone secretion

Unlike glucocorticoids and adrenal sex steroids, secretion of aldosterone is primarily regulated by the renin-angiotensin system (Fig. 15–7). ACTH acts in a permissive fashion and both ACTH and serum potassium modulate aldosterone secretion. Aldosterone participates in the maintenance of body sodium balance. This hormone increases reabsorption of sodium and chloride from the distal renal tubular lumen and leads to an increased secretion of potassium and hydrogen ions into the lumen. With sodium depletion, effective blood volume contraction follows, thereby stimulating renin secretion by the juxtaglomerular apparatus. Renin, an enzyme, cleaves the protein angiotensinogen to generate angiotensin I (a decapeptide). This peptide is further hydrolyzed to the octapeptide angiotensin II by a converting enzyme (which is most abundant in the lung). Angiotensin II, in addition to being a vasoconstrictor, is a potent stimulant of aldosterone secretion. Aldosterone, in turn, increases the sodium reabsorption from the distal renal tubule, restoring the body sodium deficit and

Actions of Glucocorticoids

Intermediary metabolism
  Carbohydrate
    Decreased cellular glucose uptake
    Increased hepatic glycogen deposition
    Increased gluconeogenesis
  Lipid
    Increased lipolysis
    Increased glycerol and fatty acid production
    Decreased lipogenesis in adipose tissue
  Protein and nucleic acid
    Increased synthesis in liver
    Decreased synthesis and enhanced breakdown in muscle, skin, adipose tissue, lymphoid and
      fibroblast tissue
    Inhibition of amino acid transport
  Water and electrolyte
    Sodium retention
    Kaliuresis
    Increased free-water clearance
    Increased glomerular filtration rate
Immunologic and inflammatory responses. (Glucocorticoids in excess have immunosuppressive and
  antiinflammatory actions.)
  Immune system
    Suppression of certain antibody responses
    Suppression of predominantly cell-mediated immune response
    Lymphocyte redistribution (from circulation to other lymphoid compartments)
    Thymus regression
  Inflammatory response
    Decreased accumulation of polymorphonuclear leukocytes, macrophages, and lymphocytes at
      inflammatory sites
    Decreased liberation of effector substances (kinins, plasminogen activator, etc.) from cells
    Decreased phagocytosis mechanism
Certain organ systems
  Bone and calcium metabolism
    Decreased intestinal Ca^{++} absorption, probably by decreasing 25-(OH)-vitamin D
    Promote redistribution of extracellular Ca^{++} into intracellular compartment
    Inhibition of bone secretion
    Enhanced osteolysis
  Blood cells
    Increased neutrophils, platelets, and erythrocytes
    Decreased lymphocytes, eosinophils, and basophils
  Central nervous system
    Affect behavior and neural activity
    Stimulate appetite
    Increased wakefulness
    Altered thresholds of electrical excitability and of sensory perception
  Gastrointestinal Tract
    Enhanced gastric acid secretion in response to other stimuli
    Inhibition of DNA synthesis in gastric musosa
Other Hormones
  Increased prolactin, renin substrate, and angiotensin II generation
  Decreased ADH, growth hormone, and TSH release
  Decreased ACTH and related peptides
Growth and development
  Inhibited linear growth and skeletal maturation
  Accelerated timing and rate of tissue differentiation

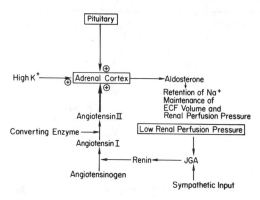

FIG. 15-7. Control of aldosterone secretion in man. JGA refers to the juxtaglomerular apparatus.

subsequently normalizing the circulating blood volume. Epithelial cells from the salivary and sweat glands and from the gastrointestinal system also respond to aldosterone by conserving body sodium. Blood loss or pooling of blood in the venous system can also provoke aldosterone secretion by stimulating the sympathetic nervous system to promote renin secretion. Increased renin-angiotensin activity, irrespective of the cause, results in increased aldosterone secretion, whereas excessive autonomous production of aldosterone suppresses the renin-angiotensin system. Recent evidence suggests that aldosterone production may also be modulated through dopaminergic mechanisms. Although the site of this effect is not established, blockade of dopamine receptors may result in rapid but not sustained elevations of aldosterone production. The clinical significance of these observations is still not clear.

Hypersecretion of glucocorticoids

Excessive production or ingestion of glucocorticoids with its metabolic consequences lead to a clinical syndrome known as Cushing's syndrome. This syndrome may occur at any age in both sexes. A typical patient with Cushing's syndrome, however, is a young to middle-aged woman with weight gain, hirsutism, plethora, menstrual irregu-

larity, hypertension, acne, weakness, easy bruisability, and emotional disturbances (Table 15-7). The metabolic consequences of hypercortisolism readily explain most of the clinical manifestations. The obesity is only moderate, but is characterized by abnormal fat redistribution to the trunk and cheeks, hence the appearance of "moon facies" and to the supraclavicular area and at the base of the neck (dorsal or buffalo hump). Because of a rather rapid weight gain and thinning of skin from the catabolic effects of steroids, violaceous striae are commonly observed, particularly over a protuberant abdomen. The skin is thin and there is a considerable increase in bruisability, causing ecchymoses. Poor wound healing is also a natural consequence of the disease. Proximal muscular weakness, most conveniently shown by the patient's inability to rise from a sitting position without assistance, results from the direct effect of glucocorticoids on protein metabolism; this is often aggravated by hypokalemia.

Acne, hirsutism, and menstrual irregularities are primarily due to excessive androgen production and androgenic effects of glucocorticoids. Moderate hypertension is common. Increased secretion of a more potent sodium-retaining hormone such as

TABLE 15-7
Clinical Manifestations of Cushing's Syndrome

Manifestation	Incidence (%)
Obesity	95
Hypertension	85
Glucosuria and decreased glucose tolerance	80
Menstrual and sexual dysfunction	76
Hirsutism and acne	72
Striae	67
Weakness	65
Osteoporosis	55
Easy bruisability	55
Psychiatric disturbances	50
Edema	46
Polyuria	16
Ocular changes	8

corticosterone, desoxycorticosterone, or more uncommonly aldosterone occasionally occurs in patients with Cushing's syndrome. The enhanced gluconeogenesis and decreased cellular glucose uptake leads to carbohydrate intolerance and secondary hyperinsulinism with eventual diabetes mellitus. Cases with excess ACTH and ACTH-related peptides may also present with slight hyperpigmentation. The combined effects of cortisol on increased catabolism of bone matrix, suppression of collagen synthesis, and antagonism of vitamin D may all contribute to development of osteoporosis. Growth hormone release and its peripheral effectiveness are inhibited by cortisol excess; therefore, growth failure is a common and early sign of childhood Cushing's syndrome. The effects of cortisol excess on the psyche may be prominent. In most cases, there is emotional lability, euphoria, irritability, and occasional frank psychosis. Glucocorticoid suppresses cellular migration, phagocytosis, and it may inhibit the immune system, as well as bactericidal actions of granulocytes, thereby rendering the patients more susceptible to chronic infection (e.g., from fungi or tuberculous).

Causes of Cushing's syndrome

Hypersecretion of glucocorticoids may result from excess ACTH production or from autonomous secretion by adrenal neoplasms (see list, p. 346). Hypersecretion of ACTH may originate either from a pituitary or a nonpituitary neoplasm such as small cell carcinoma of the lung. The term Cushing's disease (in contrast to Cushing's syndrome) is usually reserved for adrenal hyperfunction resulting from excess pituitary ACTH secretion.

Cushing's disease

Pituitary Cushing's disease is the most common cause of spontaneous Cushing's syndrome, accounting for 70% or more of all cases. Some patients have pituitary adenomas whereas others do not, although ACTH hypersecretion is present in all. One hypothesis suggests that this disorder may result from hypersecretion of hypothalamic corticotropin-releasing factor that, in turn, stimulates ACTH secretion and in some cases promotes the formation of microadenomas of the pituitary. As a result of ACTH hypersecretion the adrenal cortex appears hyperplastic in most cases, although occasionally it may appear normal despite clear hypersecretion of cortisol.

The clinical features of Cushing's disease have already been described. In patients with a large pituitary tumor, headache and visual disturbances may also supervene. Cushing's disease may be part of multiple endocrine adenomatosis (MEA), type I (adenomas of pituitary, pancreatic islet cells, parathyroid gland, and bronchial tree). However, when Cushing's disease is part of the MEA syndrome, it is more frequently caused by an ACTH-secreting bronchial adenoma (ectopic ACTH) than by a pituitary adenoma.

Biochemical changes of Cushing's disease are characterized by excessive production of cortisol and a relatively high plasma ACTH concentration (Fig. 15–8). In patients with Cushing's disease, in contrast to all other types of hypercortisolism, the pituitary-adrenal axis is essentially intact, except that it functions at a higher set point. Therefore, ACTH secretion in this circumstance is only suppressible by larger than normal amounts of endogenous or exogenous glucocorticoids. Dexamethasone, an extremely potent synthetic glucocorticoid, is often used to test pituitary-adrenal suppressibility, measuring endogenous adrenal steroids as an end point (Fig. 15–9). Since dexamethasone per se does not contribute significantly to the values of urinary free cortisol and/or 17-OHCS, the suppressive effect of dexamethasone upon the ACTH secretion is also reflected by the reduction of the values of urinary free cortisol and 17-OHCS. Conversely, when cortisol synthesis is inhibited by metyrapone, the hypothalamic-pituitary axis is activated

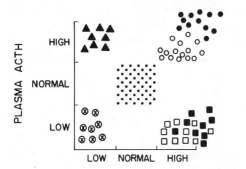

FIG. 15–8. Diagramatic representation of changes in ACTH and cortisol in Cushing's syndrome and adrenal insuffieiency.

▲ = Addison's disease; ⊗ = secondary adrenal insufficiency; ○ = pituitary Cushing's disease; ● = ectopic ACTH syndrome; □ = adrenal adenoma; ■ = adrenal carcinoma.

to release more ACTH which in turn markedly stimulates steroidogenesis up to the level of 11-deoxycortisol, the immediate precursor of cortisol (Fig. 15–5). The metabolites of 11-deoxycortisol are also measured as 17-OHCS so this value rises. Metyrapone will produce an exaggerated rise, in plasma ACTH, 11-deoxycortisol and urinary 17-OHCS in patients with pituitary Cushing's disease. Dexamethasone suppressibility and metyrapone responsiveness of the pituitary-adrenal axis are often employed as test measures to distinguish pituitary Cushing's disease from other types of Cushing's syndrome (Table 15–8).

Causes of Cushing's Syndrome

ACTH dependent
  Pituitary ACTH hypersecretion with or without
    pituitary adenoma
  Ectopic ACTH syndrome
  Adrenal nodular hyperplasia*
  Exogenous ACTH
ACTH independent
  Adrenal carcinoma
  Adrenal adenoma
  Exogenous corticoids

---

* Some cases of adrenal nodular hyperplasia are independent of ACTH.

Nodular adrenal cortical hyperplasia

This is an unusual and heterogeneous form of Cushing's syndrome. Some of these patients probably have chronic hypersecretion of ACTH that leads to diffuse hyperplasia and adenomatous lesions in the adrenals. In other patients there are primary multiple adrenal adenomas with autonomous steroid production.

Ectopic ACTH syndrome

A number of neoplasms, principally oat cell carcinoma of lung (50 per cent of cases), thymoma (15 per cent), pancreatic islet cell tumors (10%), and bronchial adenoma (5 per cent), possess the potential to make a hormone that is identical with or similar to the pituitary ACTH. Protracted hypersecretion of ACTH eventually leads to the development of Cushing's syndrome. Unlike the pituitary forms of Cushing's syndrome, secretion of ACTH from nonendocrine malignancy is highly autonomous and usually unresponsive to the suppressive effect of glucocorticoids. Moreover, secretory activity can be extremely erratic. Perhaps because of the short duration of the hypercortisolism and the concurrent catabolic effects of malignancy, classic Cushingnoid features are generally not observed. Nevertheless, these patients usually exhibit weight loss, profound hypokalemia, generalized weakness, hypertension, edema, and hyperpigmentation. When ACTH originates from relatively benign neoplasms such as bronchial adenoma or carcinoid, the clinical and biochemical presentations are frequently indistinguishable from those of pituitary Cushing's disease.

Like patients with pituitary Cushing's disease, laboratory features of patients with ectopic ACTH production include elevated plasma ACTH and plasma and urinary corticoids; and plasma ACTH levels usually exceed those of pituitary Cushing's disease (Fig. 15–8). Most importantly, ACTH secretion and thus cortisol secretion are generally not suppressible following high doses

A

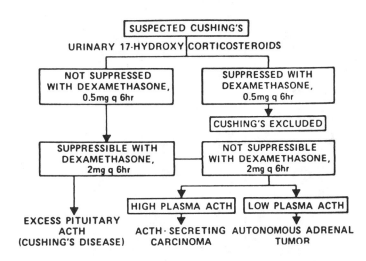

B

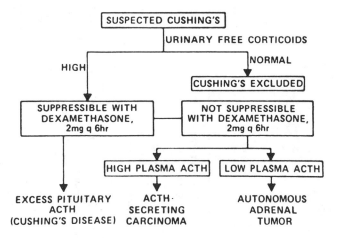

FIG. 15–9. Algorithms for use of the dexamethasone suppression test and of plasma ACTH levels in the differential diagnosis of Cushing's syndrome. Dexamethasone is given for 2 days as outlined above (*A*). The low-dose (0.5 mg every 6 hours for 8 doses) dexamethasone suppression test is usually unnecessary, since the urine free cortisol excretion rate is sensitive in separating Cushing's from the normal adrenal population (*B*).

(usually 8 mg./day) of dexamethasone that are usually sufficient to suppress ACTH secretion in pituitary Cushing's disease (Fig. 15–9). The clinical and laboratory features of patients with ectopic ACTH syndrome because of a benign neoplasm may on occasion be indistinguishable from those of pituitary Cushing's disease. Even pharmacologic stimulation or suppression maneuvers may not be sufficient to distinguish the two entities. Special measures such as

localized venous catheterization with sampling for assay of ACTH may be necessary to delineate the origin of abnormal ACTH production.

Adrenal cortical neoplasms

Functioning adrenal carcinomas and adenomas may secrete excessive amounts of adrenal hormones and cause Cushing's syndrome. The clinical features of Cushing's syndrome from adrenal neoplasia are

TABLE 15–8
Comparison of Different Types of Cushing's Syndrome

Clinical or Laboratory Changes	Pituitary Cushing's Disease	Ectopic ACTH Syndrome	Adrenal Carcinoma	Adrenal Adenoma
Main clinical features	Cushingoid appearance and symptoms	Hyperpigmentation, weight loss, hypertension, hypokalemia	Cushingoid features, hirsutism, weight gain	Cushingoid features
Urine 17-OHCS/free cortisol	Normal to high/high	High/high (may fluctuate from day to day)	High/high	High/high
Urine 17KS	Normal to high	High	Markedly elevated— 4 to 5 times the normal value	High
Serum ACTH	Normal to moderate elevation	Markedly elevated	Low	Low
High dose dexamethasone (8 mg./day)	Steroid production suppressed to 50% basal value	Not suppressed	Suppressed	Suppressed
ACTH stimulation test	Responsive	Responsive	Nonresponsive	50% of cases responsive
Metyrapone administration	Responsive with increased urine 170HCS	Variable	Unresponsive	Unresponsive
Typical radiographic appearance	Bilateral hypertrophy Bilateral nodules may appear in nodular hyperplasia	Bilateral hypertrophy	Unilateral suprarenal mass and atrophy of contralateral gland	Unilateral suprarenal mass and atrophy of contralateral gland

usually inseparable from those of other causes. However, virilization in women or feminization in men can be the dominant clinical presentation. Typically, there are elevations of plasma and urinary corticoids, low or undetectable plasma ACTH (Fig. 15–8), and radiographic evidence of unilateral adrenal mass. Marked elevation of urinary 17-KS, often to the range of 50 mg./g. creatinine (four times the upper limit of normal) and out of proportion to the elevation of 17-OHCS, is found uniquely in Cushing's syndrome due to adrenal carcinomas. Since steroidogenesis is independent of ACTH in patients with an autonomous adrenal adenoma or carcinoma, the administration of glucocorticoid (dexamethasone, for testing purpose) does not alter the secretory activity of the tumor (Fig. 15–9). Half of Cushing's patients from adrenal adenomata are capable of responding to ACTH administration by doubling glucocorticoid production; the other half are relatively unresponsive. In contrast, patients with an adrenal carcinoma are virtually always unresponsive to ACTH stimulation (Table 15–8). This suggests that cellular derangements are most severe in carcinomatous lesions and are incapable of responding to ACTH. Metyrapone administration may alter steroidogenesis qualitatively by a reduction of cortisol and an increase of 11-deoxycortisol and its precursors, but total urinary 17-OHS remain at the same level because the steroidogenesis of the adrenal tumor is independent of ACTH. Computerized axial tomography (CAT) of the adrenal glands has greatly improved the ability to determine the presence of such neoplasms but is not as effective in determining whether a gland is hyperplastic or normal in size.

Treatment of Cushing's syndrome is directed usually toward the primary lesion: surgical hypophysectomy and/or irradiation to the pituitary for Cushing's disease and removal of the tumor for adrenal neoplasia and ectopic ACTH syndrome.

For patients with adrenal tumors or the ectopic ACTH syndrome, hypercortisolism may be reduced by inhibitors of steroidogenesis such as O,P'-DDD, aminoglutethimide, or metyrapone. With the exception of OP'-DDD (which destroys adrenal corticol cells), these agents are not recommended for management of pituitary Cushing's disease since the pituitary in this situation can respond to adrenal blockade by secreting more ACTH and thereby overcoming the drug effect.

Hypersecretion of aldosterone (primary hyperaldosteronism, Conn's syndrome)

Physiology

A condition referred to as *secondary hyperaldosteronism* is characterized by high renin levels and elevated aldosterone production. It is encountered frequently in clinical practice when a patient has diminished effective blood volume (e.g., with diuretic therapy, liver disease, or congestive heart failure). Intravascular volume contraction stimulates renin secretion, thereby activating angiotensin formation, and secondarily aldosterone secretion. In contrast, adrenal lesions may autonomously secrete excessive aldosterone resulting in primary hyperaldosteronism. In this disorder, the activity of the renin-angiotensin system is markedly suppressed by the expanded extracellular volume as a result of the salt-retaining effect of aldosterone (Fig. 15–7). Pathologically, most of these patients have a solitary adrenal adenoma, between 20 to 30 per cent have bilateral adrenal nodular hyperplasia with apparent hyperactivity of the zona glomerulosa. There is little concrete evidence as to the etiology for idiopathic adrenal hyperplasia (IAH), but current speculation ranges from an unidentified hormonal stimulus to a lack of normal inhibition by dopamine-mediated events.

With respect to its clinical consequences, the cardinal manifestations of primary aldosteronism are hypertension and hypokalemia. Large quantities of aldosterone promote the reabsorption of sodium and consequent excretion of potassium primarily from the renal distal tubules but also the sweat and salivary glands and gastrointestinal tract. Chronic retention of sodium and hypervolemia lead to arterial hypertension. The sodium retention and expansion of intravascular and extracellular fluid volumes only proceed to a certain point, however, due to an escape phenomenon. Once volume expansion reaches a certain level, the patient no longer retains excessive sodium in response to aldosterone, but comes into sodium balance (intake equals output) albeit now in the presence of increased body salt and water. Because of the escape phenomenon in this syndrome, clinically detectable edema formation is characteristically absent. Hypokalemia and alkalosis give rise to muscle weakness, paraesthesia, cramps, tetany, or frank paralysis of limbs or respiratory muscles. Long-standing hypokalemia also leads to an impairment in renal concentrating ability which results in polyuria, polydipsia, and nocturia.

Biochemical characterization

The biochemical diagnosis of primary hyperaldosteronism requires proof of an increased aldosterone production that is not suppressible by intravascular volume expansion or by pharmacological interruption of angiotensin II production and a suppressed renin-angiotensin system that is not normally activated by volume contraction (usually by diuresis). From a practical point of view, the most difficult issue is to separate an aldosterone-producing adenoma (APA) from the so-called idiopathic adrenal hyperplasia (IAH) as the cause of primary hyperaldosteronism. Although a number of approaches have been designed to distinguish these two conditions, one of them deserves mention.

In normal subjects, the plasma aldosterone concentration increases upon resumption of upright posture because of activation of the renin-angiotensin system. In patients with APA, plasma aldosterone levels paradoxically decline with upright position. In contrast, plasma aldosterone levels of patients with IAH increase. The pathophysiological distinction is based on the fact that adenomas, although insensitive to angiotensin II, are partially modulated by ACTH which declines as the day progresses. In contrast, patients with IAH appear to be unusually responsive to even small increases in their suppressed renin-angiotensin system following upright posture. Although this test is useful, care must be exercised in interpreting the data since other stimuli may alter the normal circadian responses in patients with an adenoma.

The management of the patient with an APA is principally by the removal of the adenomas. This results in resolution of the hypokalemia in all of these patients and normalizes blood pressure in the majority. In patients with IAH, the hypertension usually remains even after bilateral adrenalectomy. Long-term use of diuretics as well as aldosterone antagonists such as spironolactone is the preferred treatment.

Hypofunction of the adrenal cortex

Hyposecretion of glucocorticoids

Adrenal insufficiency may result from primary adrenal failure (Addison's disease) or from ACTH deficiency. The major symptoms of adrenal insufficiency result from inadequate production of cortisol, and, in the case of Addison's disease, from aldosterone deficiency as well. Since cortisol has widespread effects on the metabolism of carbohydrates, proteins, fats, and electrolytes, severe adrenal insufficiency can be life-threatening. Patients usually complain of weakness, anorexia, nausea, vomiting, postural dizziness, and weight loss. Patients with Addison's disease usually exhibit increased skin pigmentation (notably over pressure points, palmar creases, and mucous membranes), due to high ACTH and its related peptides, some of which have melanocyte-stimulating activity (melanocyte-stimulating hormones, MSH). An occasional patient craves salt in response to salt depletion. Acute adrenal crisis often occurs following a stress, characterized by severe hypotension, dehydration, hypoglycemia, and features of shock.

The most common cause of spontaneous adrenal failure is autoimmune adrenal atrophy. Antiadrenal antibodies are often detectable in the course of the illness. Concurrence of other types of autoimmune diseases such as Hashimoto's thyroiditis, diabetes mellitus, pernicious anemia, or hypoparathyroidism may occur among patients with autoimmune Addison's disease. Other causes of adrenal insufficiency are presented in the list below.

The most commonly encountered adrenal failure is due to withdrawal of glucocorticoid therapy. The ability of corticosteroids to suppress the pituitary-adrenal axis by inhibiting ACTH secretion has been clearly established. Administration of glucocorticoids in excess of physiological amounts inhibits ACTH secretion and endogenous adrenal function, eventually resulting in adrenal atrophy. After withdrawal of the exogenous glucocorticoids, it takes a certain period of time for the pituitary-adrenal axis to recover, and during this time the patient is vulnerable to developing adrenal insufficiency and must be observed closely.

Spontaneous ACTH deficiency can occur with either hypothalamic or pituitary disease (see list at right). Clinical manifestations of secondary adrenal insufficiency differ from those of primary failure in some major aspects: hyperpigmentation is ab-

sent because the blood levels of ACTH or its related peptides are low, and aldosterone secretion is relatively undisturbed and therefore there are few electrolyte disturbances.

Differential Diagnosis of Adrenal Insufficiency

Primary adrenocortical insufficiency
  (Addison's disease)
  Autoimmune adrenal atrophy
  Infectious processes
    Granulomata (eg., tuberculosis, histoplasmosis)
    Bacterial infection (e.g., pneumococcus,
      meningococcus) or Waterhouse-Friderichsen
      syndrome)
    Viral infection (e.g., cytomegalic virus)
  Vascular
    Hemorrhage (e.g., anticoagulants, trauma)
    Infarction (e.g., thrombosis, embolism)
  Infiltrative processes
    Sarcoidosis
    Amyloidosis
    Hemochromatosis
    Lymphoma, metastatic carcinoma
  Familial or congenital
    Congenital adrenal hyperplasia
    Adrenoleukodystrophy
  Therapeutic
    Surgical adrenalectomy
    Irradiation
    Drug-induced (e.g., OP'-DDD,
      aminoglutethimide)
Secondary adrenal insufficiency (ACTH deficiency)
  Administration and withdrawal of glucocorticoids
  Hypopituitarism (e.g., pituitary tumor, infarction,
    infection, surgery, irradiation)
  Hypothalamic disorders (e.g., tumor, irradiation)

The biochemical changes of Addison's disease are characterized by reduced production of glucocorticoids with concomitantly elevated ACTH levels (Fig. 15–8). Urinary excretion rate of 17-OHCS is frequently employed as an index of glucocorticoids production rate. However, a low urinary 17-OHCS is found not only in patients with adrenal insufficiency but also in those with hypothyroidism, liver disease, or renal failure. These diseases that are characterized by low steroid excretion without adrenal disease occur primarily because of an alteration in cortisol degradation or excretion. Primary adrenal insuffi-

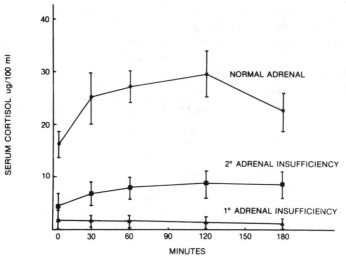

FIG. 15–10. Serum cortisol response to ACTH injection in normal subjects (●), and in patients with primary (▲) and secondary (■) adrenal insufficiency. (Speakout, P. F., et al. Arch. Intern. Med. 28:761–763, 1971.)

ciency usually is identified by unresponsive adrenal function following administration of ACTH (ACTH stimulation test) (Fig. 15–10).

The adrenal atrophy of pituitary disease is also associated with low glucocorticoid levels, but the blood ACTH concentration is also reduced (Fig. 15–8). In contrast to Addison's disease, adrenal cortical function in patients with hypopituitarism (secondary adrenal insufficiency) usually responds to ACTH administration. The magnitude of this response, as judged by steroid production, is subnormal because the adrenals are atrophic (Fig. 15–10). Repeated ACTH administration, however, will restore adrenal responsiveness. Evaluation of the reserve capacity of the pituitary-adrenal axis is important in patients with suspected hypopituitarism. Normal basal cortisol levels do not exclude partially impaired ACTH secretion. Only when such patients are stressed does such adrenal insufficiency become apparent. The pituitary-adrenal reserve capacity can be evaluated by testing responses to insulin-induced hypoglycemia or metyrapone (Fig. 15–5).

Patients with adrenal insufficiency can be treated satisfactorily with cortisol (20 to 30 mg. daily) or an equivalent amount of one of the other glucocorticoids. A salt-re-taining steroid is usually added in patients with Addison's disease who lack aldosterone. When the patient is under physical stress, supraphysiological amounts of steroids must be given to prevent hypotension and shock (adrenal crisis).

Congenital adrenal hyperplasia

Congenital adrenal hyperplasia refers to a group of inherited conditions transmitted by autosomal recessive genes resulting in enlarged adrenal glands. The basic defect of these conditions is a specific enzymatic deficiency in cortisol biosynthesis with a compensatory increase in ACTH which in turn produces the adrenal hyperplasia.

21-HYDROXYLASE DEFICIENCY. The most common defect in steroid synthesis is in 21-hydroxylation. Steroid biosynthetic intermediates prior to the site of the block are increased (Fig. 15–11). Thus, 17-hydroxyprogesterone, an important substrate for 21-hydroxylase, accumulates. Since this compound has little biological activity, the adrenal gland is overstimulated by ACTH to produce large quantities of this substance. In some patients with a mild enzyme deficiency, this ACTH stimulation may normalize cortisol production, but this occurs at the expense of increased production of

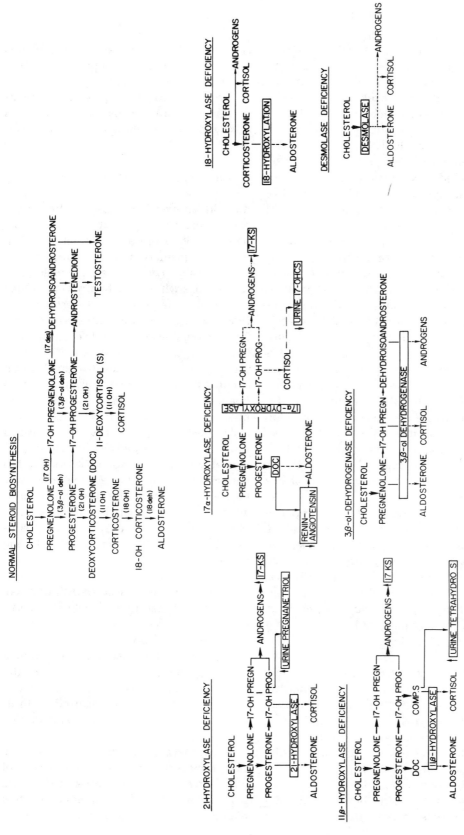

FIG. 15–11. Major types of congenital adrenal hyperplasia. The solid arrows depict the normal pathways, while the dashed lines indicate pathways that are decreased following the enzymatic blockade.

[353]

precursors. In other patients with severe enzyme deficiency, overt cortisol insufficiency may occur. As a result of the increased ACTH levels, excessive androgens having no inhibitory effect upon ACTH production are also produced. These are responsible for the clinical picture typically associated with congenital adrenal hyperplasia including virilization and, in the female infant, pseudohermaphroditism. Clitoral enlargement at birth may result in mistaken sex identification if a careful examination is not performed. Boys with the condition undergo precocious puberty, with early development of genitalia, facial and body hair growth, and increased muscle mass. Girls fail to undergo normal feminization, and the growth of both sexes is stunted owing to early epiphyseal closure secondary to high androgen levels. This adrenal hyperplasia is characterized by increased excretion of pregnanetriol, the major metabolic product of 17-hydroxyprogesterone. When the condition is recognized, it may be treated adequately with administration of physiological quantities of cortisol that suppress ACTH production and reduce adrenal size and androgen secretion. These patients may live normal lives as long as replacement therapy is carefully maintained. Some infants with the disease may tend to lose salt, since 21-hydroxylase is required for the synthesis of aldosterone. Hence, this condition often is referred to as *salt-losing congenital adrenal hyperplasia*. These patients require replacement with a salt-retaining hormone.

11-HYDROXYLASE DEFICIENCY. A much less common cause of congenital adrenal hyperplasia is a specific deficiency of 11-beta hydroxylation. This causes inadequate cortisol and corticosterone synthesis and excessive 11-deoxycorticosterone and 11-deoxycortisol production (Fig. 15–11). Since 11-deoxycorticosterone is a potent sodium-retaining steroid, these patients develop hypertension, presumably because of salt and water retention. These patients, too, may be treated by replacement therapy using cortisol, which suppresses the abnormally high ACTH production. Overproduction of androgens also accompanies this syndrome, resulting in adrenal virilization.

17-HYDROXYLASE DEFICIENCY. This condition affects 17-hydroxylation not only in the adrenal glands but also in the ovaries and testes. These patients have a female habitus, fail to undergo normal feminization, have primary amenorrhea, and have hypertension. They possess a female habitus because without 17-hydroxylation they are unable to synthesize testosterone. However, they cannot undergo normal feminization because this defect also interferes with estrogen production. Without estrogen, of course, menstruation cannot occur. With the inability to produce normal quantities of cortisol, there is an increase in production of 11-deoxycorticosterone and corticosterone (Fig. 15–11). The increased production of 11-deoxycorticosterone is considered largely responsible for the hypertension. Some of these patients may have a concomitant defect in 18-hydroxylation, which is mediated by the low levels of circulating angiotensin II, thus suppressing further the plasma aldosterone levels.

3 BETA-HYDROXYSTEROID DEHYDROGENASE DEFICIENCY. An additional defect in corticosteroidogenesis may occur at the point of conversion of pregnenolone to progesterone (Fig. 15–11). These patients produce no biologically active adrenal steroid. If the defect is complete, death is inevitable unless the condition is recognized at birth and corticosteroids are administered.

18-HYDROXYLATION DEFECTS. This rare condition results in defective conversion of corticosterone to aldosterone in the zona glomerulosa (Fig. 15–11). Infants with this disorder have no detectable abnormalities in cortisol production or genital development. The usual clinical presentation is failure to thrive, dehydration, sodium wasting, hyponatremia, and hyperkalemia.

DESMOLASE DEFICIENCY. The term congenital lipoid adrenal hyperplasia is often applied to this disorder. Pathologically, there is deposition of large amounts of cholesterol and its esters in the adrenal cortex. The prognosis is poor in patients with desmolase deficiency. Severely affected infants usually succumb to salt-wasting and adrenal corticol insufficiency. This disorder results from impaired conversion of cholesterol to pregnenolone, presumably due to a decreased activity of desmolase complex. Desmolase is responsible for hydroxylation and side-chain cleavage of cholesterol to form Δ^5-pregnenolone (Figs. 15–4, 15–11). Diminished secretion of all classes of adrenal steroids and its failure to respond to ACTH administration are the typical findings of desmolase deficiency.

Hypoaldosteronism

Hyposecretion of aldosterone usually accompanies Addison's disease. Selective aldosterone deficiency is rare but it occurs most commonly in patients with diabetes mellitus and renal failure. The major clinical manifestations reflect the lack of aldosterone stimulation of renal potassium and hydrogen ion excretion: hyperkalemia and metabolic acidosis. The pathogenesis of isolated aldosterone hyposecretion is a deficient renin secretion by the kidney, resulting in the so-called hyporeninemic hypoaldosteronism.

Diagnosis of this disorder is achieved by the establishment of subnormal aldosterone production associated with impaired renin secretion in a volume depleted state. The typical patient with this syndrome is in the fifth to seventh decade, has asymptomatic hyperkalemia with hyperchloremic metabolic acidosis, and a mild renal failure. Despite the hypoaldosteronism, symptoms of hypovalemia are not prominent, and mild hypertension is common. Hyporeninemic hypoaldosteronism usually can be managed successfully with the administration of mineralocorticoid and kaliuretic diuretics. Prostaglandin synthetase inhibitors (e.g., aspirin and nonsteroidal antiinflammatory drugs) may precipitate or aggravate this condition by further inhibiting renin production.

Adrenal medulla

Synthesis and metabolic effects of catecholamines

The adrenal medulla is derived from the neural crest as a specialized part of the sympathetic nervous system and produces catecholamines. The biologically important and naturally occurring catecholamines include dopamine, norepinephrine, and epinephrine. Dopamine, although present in only minute amounts in adrenergic nerve endings and in the adrenal medulla, is a precursor of epinephrine and norepinephrine, and is also an important neurotransmitter of the central nervous system. In man, epinephrine is located and stored almost exclusively in the chromaffin cells of the adrenal medulla. Norepinephrine, on the other hand, is widely distributed in the adrenal medulla and in the central and peripheral neurons of the adrenergic nervous system.

The major biosynthesis and degradation pathways of epinephrine and norepinephrine are illustrated in Figure 15–12. Hydroxylation of tyrosine is catalyzed by tyrosine hydroxylase and appears to be the rate-limiting step. Neural stimulation evidently enhances the activity of the enzyme tyrosine hydroxylase, and its own catalytic products dopa and norepinephrine in turn exert a negative feedback suppression of its enzymatic activity. Norepinephrine is the adrenergic neurotransmitter released from sympathetic nerve endings in the immediate vicinity of its innervated tissues such as

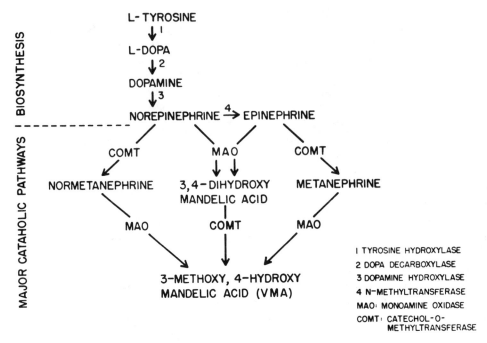

FIG. 15—12. Biosynthesis and major catabolic pathways of epinephrine (E) and norepinephrine (NE). Epinephrine is formed almost exclusively in the adrenal medulla. Hydroxylation of tyrosine (step 1) is the rate-limiting process in the biosynthesis of catecholamines. About 70 per cent of E and NE are first O-methylated prior to deamination.

blood vessels and adipose tissue. Epinephrine is released from the adrenal medulla into the circulation in bursts in response to preganglionic impulses triggered by stress, hypoglycemia, or hypotension. Of the two degradative routes, about 70 per cent of the catecholamines are first 0-methylated prior to deamination.

The metabolic effects induced by catecholamines are mediated by means of α- and β-adrenergic receptors on the surface of effector cells. Norepinephrine acts chiefly on alpha (excitatory) receptors, producing vasoconstriction, intestinal relaxation, and pupillary dilation. Epinephrine, also known as the hormone of "fight and flight," acts upon alpha- and β-receptors. Epinephrine has positive chronotropic and inotropic effects on the beta-receptors of the heart and dilates blood vessels. Medullary production of epinephrine protects against hypoglycemia by making glucose

rapidly available from the liver and free fatty acid through release from adipose tissue while inhibiting insulin secretion from the pancreas.

Hypersecretion of catecholamines

PHEOCHROMOCYTOMA AND MULTIPLE ENDOCRINE ADENOMATOSIS TYPE II (MEA II). The principal disorder of the adrenal medulla is pheochromocytoma, a neuroectodermal tumor that secretes catecholamines. Although a rare cause of hypertension, it is curable. About 90 per cent of these tumors are located in the adrenal medulla, of which 90 per cent involve one adrenal and 10 per cent are bilateral. In familial cases, bilateral and extra-adrenal tumors develop more frequently. Extra-adrenal pheochromocytomas are found in or near the sympathetic ganglia, the abdominal aorta, urinary bladder, posterior mediastinum, or at the organ of Zuckerkandl.

Hypertension is present in 98 per cent of the patients and is paroxysmal in half of them. Paroxysms are characterized by tachycardia, blood pressure elevations, severe headache, palpitation, pallor or flushing, profuse perspiration, and feeling of indigestion. Sustained hypertension could mimic essential hypertension and may be accompanied by orthostatic hypotension due to blunted sympathetic reflexes. Other features of catecholamine excess include weight loss, nervousness, carbohydrate intolerance, and even frank diabetes mellitus. In addition to catecholamines, some pheochromocytomas may produce polypeptide hormones such as ACTH causing Cushing's syndrome.

The presence of multiple endocrine adenomatosis (MEA) type II (medullary thyroid carcinoma, parathyroid adenoma, and pheochromocytoma), and type III (medullary thyroid carcinoma, parathyroid adenoma, pheochromocytoma, marfanoid habitus, mucosal neuromas, medullated corneal nerve fibers, and ganglioneuromas) should be suspected in any patient with pheochromocytoma. The basic pathogenesis of MEA remains uncertain. One hypothesis is that this syndrome is a form of neuroectodermal dysplasia. According to this theory, neural crest cells of the APUD series, characterized by their amine precursor uptake and decarboxylation, migrate to the primitive foregut derivatives that are destined to become endocrine glands. When inherited dysplasia occurs, they may form adenomas and secrete excessive quantities of hormones.

DIAGNOSIS AND MANAGEMENT. Laboratory confirmation of pheochromocytoma is achieved by measurement of catecholamines and their derivatives (metanephrines, normetanephrines, and vanillylmandelic acid) in urine (Fig. 15–12). Plasma catecholamine measurements are highly sensitive and have also been used in conjunction with radiographic techniques including CAT.

Treatment of pheochromocytoma is surgical since pre-operative use of α- and β-adrenergic receptor blockers has greatly reduced surgical risks and complications. In certain patients with metastatic disease, chronic use of the tyrosine hydroxylase inhibitor α-methyltyrosine may be effective.

Annotated references

Hypothalamic-pituitary axis

McCann, S. M.: Control of anterior pituitary hormone release by brain peptides. Neuroendocrinology *31,*355–363, 1980. (An overall review of the control of anterior pituitary hormone release by neurohypophyseal factors. Historical aspects of the discovery of the hypothalamic releasing hormones are discussed in detail.)

Krieger, D. T., and Martin, J. B.: Brain peptides. N. Engl. J. Med. *304:*876–885, 1981. (A comprehensive recent review of the hypothalamic hormones.)

Daughaday, W. H.: Growth hormone and the somatomedins. *In* Daughaday, W. H. (ed.): Endocrine Control of Growth, pp. 1–15. Amsterdam, Elsevier/North Holland, 1981.

Linfoot, J. A.: Acromegaly and gigantism. In Daughaday, W. H. (ed.): Endocrine Control of Growth, pp. 207–250. Amsterdam, Elsevier/North Holland, 1981. (An excellent review of our current understanding of growth hormone and its pathophysiology. Unfortunately, it may not be readily available to students. Alternately, Daughaday, W. D.: Adenohypophysis. In Williams (ed.): Text Book of Endocrinology, pp. 73–114, Philadelphia, W. B. Saunders, 1981, is recommended.)

Antoniades, H. N., and Owen, A. J.: Growth factors and regulation of cell growth. Ann. Rev. Med. *33:*445–63, 1982. (Authoritative review of the relationships between growth hormone, somatomedin, and insulin-like growth factor. Nongrowth hormone related growth factors are also discussed.)

Weiner, R. I., and Bethea, C. L.: Hypothalamic control of prolactin secretion. *In* Jaffe, R. B. (ed.): Prolactin, pp. 19–43. Amsterdam, Elsevier/North Holland, 1981.

Jaffe, R. B.: Physiologic and pathophysiologic aspects of prolactin production in humans. In Jaffe, R. B. (ed.): Prolactin, Amsterdam, Elsevier/North Holland, 1981. (This monograph, Prolactin, by R. B. Jaffe (ed.) provides a comprehensive overview of the current understanding of prolactin in health and in diseases.)

Frantz, A. G.: Prolactin, N. Engl. J. Med. *298:*201–207, 1978. (An important reference on hyperprolactinemia. Both physiological and pathophysiological significances of elevated serum prolactin are discussed.)

Nabarro, J. D. N.: Pituitary prolactinomas. Clin. Endocrinol. *17*:129–155, 1982. (The natural history of prolactinoma is discussed. Very importantly, the author has made an effort to distinguish between large and small prolactinomas and separately discusses their different prognoses.)

Hays, R. M.: Antidiuretic hormone. N. Engl. J. Med. *295*:659–665, 1976. (A seminar discussion on antidiuretic hormone. This article deals with the synthesis, secretion, biological actions, and metabolism of ADH.)

Brownstein, M. J., Russell, J. T., and Gainer, H.: Synthesis, transport, and release of posterior pituitary hormones. Science *207*:373–378, 1980. (One of the most recent articles on the synthesis, transport, and release of antidiuretic hormone. Physiological or pathophysiological factors that regulate ADH release, however, are not covered in this reference.)

Martinez-Maldonado, M.: Inappropriate antidiuretic hormone secretion of unknown origin. Kidney Int. *17*:554–567, 1980. (This article covers all aspects of inappropriate ADH secretion syndrome. An actual case was employed to illustrate problems associated with this syndrome.)

Moses, A. M., and Miller, M.: Drug-induced dilutional hyponatremia. N. Engl. J. Med. *291*:1234–1239, 1974. (This is a comprehensive review of drug-related hyponatremia. Mechanisms of drug-induced hyponatremia are discussed in detail.

Pituitary-adrenal axis

Eipper, B. A., and Mains, R. E.: Structure and biosynthesis of proadrenocorticotropin/endorphin and related peptides. Endocrine Rev. *1*:1–27, 1980. (Comprehensive review of current understanding of pro-ACTH [pro-opiomelanocortin] system.)

Liddle, G. W.: Pathogenesis of glucocorticoid disorders. Am. J. Med. *53*:638–648, 1972. (An important reference on the pathogenesis of glucocorticoid disorders. This article is highly recommended to students.)

Gold, E. M.: The Cushing syndrome: Changing views of diagnosis and treatment. Ann. Intern. Med. *90*:829–44, 1979. (A comprehensive review of the clinical features of different types of Cushing's syndrome. Diagnostic and therapeutic options are also included.)

Baxter, J. D., and Forsham, P. H.: Tissue effects of glucocorticoids Am. J. Med. *53*:573–589, 1972. (Detailed discussion of the metabolic effects of glucocorticoids. A good supplemental reading for Table 15–1.)

Elias, A. N., and Gwinup, G.: Effects of some clinically encountered drugs on steroid synthesis and degradation. Metabolism *29*:582–595, 1980. (Reviews the influence of drugs on steroid metabolism by altering the peripheral degradation of the steroids or inhibiting steroidogenesis. Aminoglutethimide, O,P'-DDD, metyrapone, and spironolactone are all included in its discussion.)

Crapo, L.: Cushing's syndrome: A review of diagnostic tests. Metabolism *28*:955–977, 1979. (Comparison of different laboratory tests that are commonly available for the diagnosis of Cushing's syndrome.)

Bertagna, C., and Orth, D. N.: Clinical and laboratory findings and results of therapy in 58 patients with adrenocortical tumors admitted to a single medical center (1951 to 1978). Am. J. Med. *71*:855–875, 1981. (A comprehensive report on 58 patients with Cushing's syndrome due to functioning adrenal tumors. Clinical findings, laboratory diagnosis, and pathological findings are included.)

Imura, H.: Ectopic hormone syndromes. Clin. Endocrinol. Metab. *9*:235–60, 1981. (Theories of ectopic hormone production and clinical presentations of the major ectopic-hormone syndromes are emphasized.)

Chan, J. C. M.: Control of aldosterone secretion. Nephron *23*:79–83, 1979. (An excellent review of the control of aldosterone secretion.)

Irvine, W. J.: Autoimmunity in endocrine disease. Rec. Prog. Horm. Res. *36*:509–556, 1980. (An authoritative review of autoimmune endocrine disease. A highly recommended reference.)

Irvine, W. J., and Barnes, E. W.: Adrenocortical insufficiency. Clin. Endocrinol. Metab. *1*:549–594, 1972. (A review of the etiology, clinical manifestations, and diagnosis of adrenal insufficiency.)

Kaplan, S. A.: Diseases of the adrenal cortex. II. Congenital adrenal hyperplasia. Ped. Clin. North Am. *26*:77–89, 1979. (A detailed discussion of the various forms of congenital adrenal hyperplasia.)

Schambelan, M., and Sebastian, A.: Hyporeninemic hypoaldosteronism. Adv. Intern. Med. *24*:385–401, 1979. (Pathogenesis and treatment of hyporeninemic hypoaldosteronism are included. The differential diagnosis of hyperkalemia is discussed in detail.)

Cryer, P. E.: Physiology and pathophysiology of the human sympathoadrenal neuroendocrine system. N. Engl. J. Med. *303*:436–44, 1980. (Discussion of the normal and abnormal physiology of catecholamines including their metabolic effects. This includes the use of plasma catecholamines for diagnosis of pheochromocytoma.)

16 Regulatory Mechanisms of the Thyroid

Gilbert H. Daniels, M.D.
Farahe Maloof, M.D.

The major function of the thyroid gland is to synthesize the hormones thyroxine (T_4) and triiodothyronine (T_3). Although calcitonin is produced in the thyroid gland, the significance of this hormone in normal human physiology is still unclear.

The control of thyroid gland function is tightly and directly regulated by the pituitary, by means of thyroid stimulating hormone (TSH), and indirectly by the hypothalamus. In addition, iodide is not only the substrate for thyroid hormone synthesis, but it also seems to serve as a modulator of the production and release of thyroid hormone.

Despite careful homeostatic mechanisms, overproduction (hyperthyroidism) or underproduction (hypothyroidism) of thyroid hormone is a common clinical event. Furthermore, a variety of illnesses and drug therapies can influence the metabolism of thyroid hormones, findings of uncertain clinical significance. In addition, the thyroid gland may enlarge (goiter) with, or more commonly without, overt evidence of overactivity or underactivity. Such nontoxic goiters probably represent the most common endocrine-related diagnosis in the United States.

Overview of thyroid physiology

TSH controls virtually every step in the thyroidal biosynthesis and release of the thyroid hormones, T_4 and T_3. Whereas all normal T_4 production is thyroidal in origin and hence under TSH control, the majority of circulating T_3 is produced by peripheral conversion from T_4 and is not under TSH control. In addition, TSH stimulation leads to thyroid gland enlargement through follicular cell growth and division and increased thyroidal vascularity.

Thyroid regulation can be viewed as a simple feedback loop between the thyroid gland and the pituitary gland. Decreased concentrations of thyroid hormones lead to increased pituitary production and release of TSH, apparently in order to compensate for the hormone deficiency. Decreased production of TSH leads to secondary thyroid gland failure. Exogenous administration of thyroid hormone leads to a suppression of pituitary TSH production and release, and under normal circumstances a shutdown of thyroid gland function and a decrease in thyroid gland size.

Although the hypothalamus must be in-

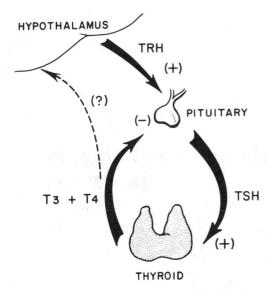

FIG. 16–1. Hypothalamic-pituitary-thyroid axis.

terposed in this simple scheme, the important negative feedback of thyroid hormone at the pituitary level is of paramount importance (Fig. 16–1). Thyrotropin releasing hormone (TRH), a tripeptide produced in the hypothalamus, can stimulate the pituitary gland to release TSH and appears to be necessary for normal TSH synthesis and release.

Hypothalamic control of thyrotropin releasing hormone

TRH isolation, chemical structure, and biosynthesis
Once thought to be the master endocrine gland, the pituitary has been shown clearly to be at least partially subservient to chemicals produced by the hypothalamus. Interruption of the hypophyseal portal blood flow from the hypothalamus or destruction of appropriate hypothalamic nuclei leads to partial or complete hypopituitarism. The first hypothalamic chemical mediator to be characterized was TRH, a tripeptide. The structure of TRH has been shown to be identical in all animal species investigated

(L-pyro-glutamyl-L-histidyl-L-proline amide—mol. wt. 362 daltons). TRH is the principal releasing factor for pituitary TSH.

TRH biosynthesis has been demonstrated in hypothalamic organ cultures. Current evidence suggests biosynthesis of a macromolecular precursor which is subsequently cleaved to yield TRH. TRH has been identified in the hypothalamus (e.g., median eminence, arcuate nuclei, and peri- and paraventricular areas) and hypophyseal portal blood. Many other areas of the brain besides the hypothalamus contain TRH, and indeed 70 per cent of total brain TRH is found outside the hypothalamus. Nonhypothalamic TRH appears to function as a neurotransmitter rather than as a releasing factor. TRH can be found in the gastrointestinal tract, pancreas, and other tissues, even in species which lack TSH; its role in these locations is being explored.

Although TRH can be detected in the peripheral circulation of man and experimental animals by exquisitely sensitive radioimmunoassays, little is known about the regulation of hypothalamic TRH. In part this is due to the rapid degradation of TRH in the serum of adult animals. In addition, the wide spread occurrence of TRH throughout the body makes assessment of site of origin impossible from peripheral measurements.

Action of TRH
Nanogram quantities of TRH stimulate the synthesis and release of TSH from the pituitary *in vitro*. High-affinity receptors for TRH are located on the pituitary gland, and pituitary TRH binding appears to lead to down-regulation of receptors. Although the mechanism for TSH release is still under investigation, TRH effects on calcium ion flux may be of major importance. The effect of TRH on TSH release appears not to involve protein or nucleic acid synthesis, since inhibitors such as puromycin, cyclohexamide, and actinomycin D have no inhibitory effect either *in vitro* or *in vivo*. In

human beings synthetic TRH elevates plasma TSH after parenteral (microgram) or oral (milligram) dosages.

The concept of a single hypothalamic releasing factor for each pituitary hormone became untenable with the discovery of prolactin release from the pituitary after TRH administration. This prolactin response can be demonstrated both *in vitro* and *in vivo*. The minimum quantity of TRH capable of releasing TSH is also capable of releasing prolactin. But TRH is of critical importance in TSH release whereas prolactin is under a dominant (tonic) negative inhibition unrelated to TRH. Clinically it is easy to dissociate the release of TSH and prolactin: pregnancy and suckling raise the serum prolactin concentration but have no effect on serum TSH levels and thyroid gland failure results in an elevated basal TSH but does not usually influence the basal serum prolactin concentration.

The effect of TRH on other pituitary hormones has been extensively studied. Under normal circumstances TRH does not cause the release of growth hormone, follicle stimulating hormone (FSH), luteinizing hormone (LH), or adrenocorticotrophic hormone (ACTH). However, in patients with acromegaly, renal failure, or anorexia nervosa, TRH may cause the release of growth hormone.

Control of TRH

In rats, cold exposure leads to increased TSH release, an effect apparently mediated by TRH and blocked by anti-TRH antibody. Anti-TRH antibody does not prevent the cold-induced release of prolactin. In human beings, cold-induced TSH release is most convincingly demonstrated in the neonate. The influence of thyroid hormone concentrations on TRH synthesis and release in humans has not been clarified.

Thyroid stimulating hormone

Thyroid stimulating hormone (thyrotropin, TSH) controls many aspects of thyroid gland structure and function: size and vascularity of the gland, height and activity of the follicular epithelium, amount of colloid, biosynthesis and release of thyroid hormone, and numerous critical biochemical processes in thyroid tissue (e.g., glucose utilization, oxygen consumption, and synthesis of phospholipid, protein, and nucleic acid). These actions vary in time of onset and may well result from the activation of an adenylate cyclase-cyclic AMP-protein kinase system, after the binding of TSH to a receptor site on the thyroid cell membrane. The resulting formation of cyclic AMP leads to the activation of a number of metabolic steps, mediating a good number, but not all, of the effects of TSH on the thyroid.

Chemistry and dynamics of TSH

TSH is a glycoprotein (mol. wt. 28,000) secreted by specific anterior pituitary cells (β_1 basophils). About 5 per cent of anterior pituitary cells are thyrotrophs. These cells contain granules which have been correlated with TSH secretion. After thyroidectomy these cells undergo marked hypertrophy with increased endoplasmic reticulum and reduced granularity, demonstrating the high rate of TSH secretion.

TSH is composed of two nonidentical, noncovalently bound polypeptide chains, α- and β-TSH. The α-chain appears to be identical to the α-chains of several other glycoprotein hormones, namely, FSH, LH, and human chorionic gonadotropin (HCG). The β-chains of these proteins convey biological and immunological specificity. The two chains are synthesized independently; the β-chain of TSH appears to be rate limiting in hormonal biosynthesis.

The pituitary of a normal human contains approximately 165 mU, with a normal TSH production rate of about 100 mU, or two-thirds of a pituitary pool per day. The circulating TSH has a half-life of 58 minutes and is cleared from the plasma at a rate of 50 ml./min. The plasma concentration is approximately 1.6 μU/ml. with a minor di-

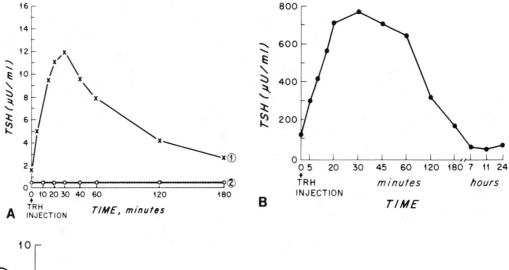

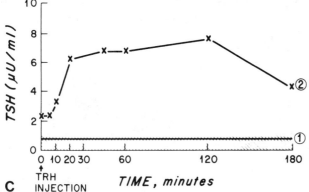

FIG. 16–2. TRH stimulation. (A) Line 1 indicates euthyroid + TRH; line 2, hyperthyroid + TRH or excess thyroid hormone administration +TRH. (B) Primary hypothyroidism + TRH. (C) Line 1 indicates pituitary hypothyroidism + TRH; line 2, hypothalamic hypothyroidism + TRH. The notched area represents the lower limit of detection of the assay.

urnal variation noted, concentrations being highest prior to the onset of sleep.

Regulation of TSH secretion

TSH secretion by the pituitary gland requires tonic stimulation by TRH but the magnitude of the TSH release is modulated by the circulating thyroid hormone concentrations. The release of TSH is exquisitely sensitive to even small changes in the concentrations of circulating thyroid hormones: a small decrease in thyroid hormone concentration leads to an increase in TSH release and a small increase in thyroid hormone concentration leads to the inhibition of TSH release.

The response of the plasma TSH concentration to TRH injections in the euthyroid, hypothyroid, and hyperthyroid states is shown in Figure 16–2. With normal circulating thyroid hormone concentrations (euthyroid state), the normal TSH concentration reaches its peak in 20 minutes. In primary thyroid gland failure (primary hypothyroidism), with decreased thyroid hormone concentrations, the basal TSH concentration is markedly elevated and rises even higher after TRH administration. This exaggerated response in hypothyroid patients suggests that the pituitary is not maximally stimulated to release TSH under basal hypothyroid conditions. With elevated serum concentrations of thyroid hormones (hyperthyroidism), the basal TSH concentration is undetectable and does not rise even after TRH injection, and thus confirms the pituitary as an important site of negative feedback.

The administration of exogenous T_3 or T_4 will result in suppression of pituitary TSH release. Exogenous T_4 appears to be more effective than T_3 in this regard, and circulating TSH concentrations in various disease states have a better inverse correlation with circulating T_4 than T_3 concentrations. The pituitary gland contains specific, saturable nuclear receptors for T_3 and T_4, but the T_3 receptors have an affinity 10 times greater.

The pituitary gland contains an active enzyme system for the conversion of T_4 to T_3. This intrapituitary $T_4 \rightarrow T_3$ conversion explains the paradox of TSH release correlating with serum T_4 (but not T_3) concentration with the pituitary receptors being relatively specific for T_3. T_3 seems to be the active hormone for pituitary TSH suppression and T_4 appears to be the most efficient way to deliver the T_3 to the pituitary.

Thyroid hormonal inhibition of TSH production and release seems to require both protein and nucleic acid synthesis. (The presumed inhibitor remains to be identified.) Like TRH itself, the thyroid hormones also decrease the number of TRH receptors.

A number of other influences on basal and stimulated TSH secretion have been described. Somatostatin (growth hormone release inhibitory factor) suppresses the nocturnal rise in TSH, decreases the elevated basal TSH levels in hypothyroid patients, and prevents the TRH-induced increase in serum TSH. These effects may explain the hypothyroidism which occasionally develops in growth hormone–deficient children treated with growth hormone. Such growth hormone therapy leads to increased somatomedin production and presumably to increased hypothalamic somatostatin release. It seems likely that the TSH deficiency and secondary hypothyroidism which develops in such patients is due to somatostatin inhibition of TSH release.

Dopamine has an important negative influence on TSH. Dopamine infusions lower basal TSH levels in normal and hypothyroid individuals and decrease the TSH response to TRH. In primary hypothyroidism dopamine antagonists lead to a further rise in the already elevated basal concentrations of TSH, suggesting an important restraining influence of dopamine under these circumstances. Whereas dopamine has only a minor influence on TSH, it appears to be the major factor controlling (inhibiting) prolactin release.

It is of interest that glucocorticoids inhibit the TRH-stimulated release of TSH, an effect presumably mediated at the pituitary level. The elevated basal TSH concentrations found in untreated patients with adrenal insufficiency may relate to the absence of glucocorticoid restraint on TSH; such elevated TSH levels fall with glucocorticoid replacement therapy.

The hypothalamic-pituitary-thyroid axis recovers from thyroid hormone suppression in a matter of 4 to 6 weeks in normal individuals, including those treated with thyroid hormone for many years. In contrast, the hypothalamic-pituitary-adrenal axis may remain suppressed for 6 to 12 months after supraphysiological doses of glucocorticoids have been administered.

Pituitary TSH reserve

Given adequate basal function for an endocrine gland, the ability to increase hormone production on demand is referred to as the "reserve function." Until recently, the only means of testing the TSH reserve function of the anterior pituitary was by decreasing the plasma concentration of thyroid hormone. This time-consuming procedure has been replaced by the TRH infusion test, currently the most sensitive test for pituitary TSH reserve. (TRH infusion is also useful as a test for prolactin reserve.)

Intravenous administration of synthetic TRH (200–400 μg.) stimulates a fivefold rise in plasma TSH in normal humans (Fig. 16–2*A*). Reaching its peak in 15 to 30 minutes, TSH returns to baseline levels after about 3 hours. The TSH response may be accompanied by a rise in the plasma levels of T_3 and less so for T_4, demonstrating that the TSH released and measured by radioimmunoassay is biologically active. Hence, TRH infusion may serve as a test of thyroid gland reserve as well. As discussed, the TSH release can be blocked by the administration of T_3 or T_4. The exact quantity required for complete inhibition is subject to individual variation; however, a blunted response may be noted with only minimal changes in the serum concentration of T_4 and T_3 in the same individual.

Since the TSH radioimmunoassay is relatively insensitive in that it cannot separate the normal range of TSH concentrations from the undetectable concentration in hyperthyroidism, the TRH test becomes an important clinical test. A comparison of the 20-minute TSH response to TRH of a euthyroid individual with the absent response of a hyperthyroid individual allows for a clear separation. The elevated basal concentration of TSH in primary hypothyroidism obviates the need for a TRH test. However, with a mild or subtle degree of primary thyroid failure, the exaggerated TSH response to TRH may be the only clue to a slightly elevated basal TSH concentration.

Patients with pituitary ablation develop secondary hypothyroidism: they cannot produce TSH and will have no TSH rise after TRH administration. Certain patients with TSH-deficient hypothyroidism demonstrate a delayed rise in TSH after TRH administration, suggesting hypothalamic TRH deficiency (tertiary hypothyroidism) as the cause of the thyroidal failure. Although classic "pituitary" and "hypothalamic" TRH tests are shown in Figure 16–2, in clinical practice the patterns of response are unreliable in secondary forms of hypothyroidism and cannot be used as a conclusive test to localize the disease at pituitary or hypothalamic sites. In actual practice, secondary hypothyroidism may be defined as thyroid hormone deficiency without an elevated TSH concentration.

The Thyroid and Thyroid Hormone Production

Embryology, anatomy and physiology

The thyroid gland is a bilobed organ which develops as an invagination of the floor of the embryonic pharynx and descends as a cellular stalk to the anterior part of the neck. By a proliferation of cells, epithelial plates and bands are formed which become the adult organ. Connections with the pharynx (the thyroglossal tract) normally disappear, but the point of origin persists in the adult as a dimple on the back of the tongue, the foramen caecum. These facts of embryonic development explain certain developmental anomalies such as development of the thyroid gland at the base of the tongue (lingual thyroid) and thyroglossal duct cysts, which form in remnants along the thyroglossal tract.

Soon after the tenth week of gestation in the human being, the thyroid gland begins to accumulate and bind iodide and forms thyroid hormones within the thyroglobulin molecule. At about this time, TSH appears in the pituitary, and the circulating blood contains TSH, T_4, and thyroxine-binding proteins.

The normal North American adult thyroid gland weighs about 20 g. and is organized into discrete follicles, the lumina of which contain colloid, the storage form of the hormone. The cuboidal lining cells of the follicle hypertrophy and multiply with stimulation and shrink with inactivity. The gland has an abundant blood supply; its blood flow is approximately 5 ml./g. of thyroid tissue per minute—greater, on a weight basis, than the kidney. The blood volume of normal man, 5 l., passes through the thyroid about once every 60 minutes, through the lungs about once every minute, and through the kidneys once in 5 minutes. In hyperthyroidism, the thyroidal blood flow may reach 2 to 10 l./min., the increased flow being audible as a bruit or

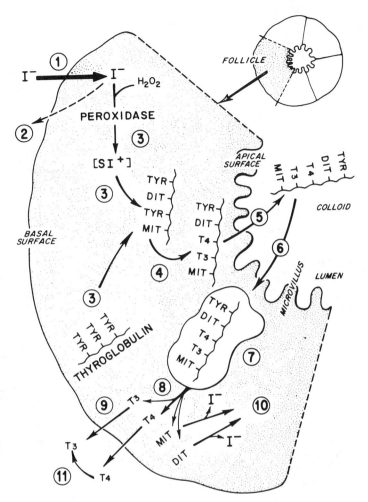

FIG. 16–3. Thyroid hormone biosynthesis. (*1*) Iodide concentration (active transport). (*2*) Passive back diffusion of inorganic iodide. (*3*) Organification of iodide (MIT + DIT production). (*4*) Iodotyrosine coupling (T$_3$ and T$_4$ production). (*5*) Release of thyroglobulin to follicular lumen. (*6*) Pinocytosis (thyroglobulin reuptake). (*7*) Phagolysosome. (*8*) Digestive release of iodotyrosines and iodothyronines (proteolysis of thyroglobulin). (*9*) Release of thyroid hormone. (*10*) Deiodination of iodotyrosines. (*11*) Peripheral deiodination of T$_4$ to T$_3$.

palpable as a "thrill" over the thyroid gland.

Thyroid hormone biosynthesis

OVERVIEW. The hallmark of thyroid hormone production and delivery is stability. The thyroid gland represents a system which insures a steady supply of hormone, rather than one designed to produce dramatic or immediate swings in hormone delivery. It is organized into discrete follicles which are capable of storing the thyroid hormone precursors in the form of thyroglobulin. Up to several months' supply of hormone is stored; this supply prevents drastic alterations in hormone release with fluctuations in iodine availability. In addition, the presence of circulating transport or binding proteins for T$_4$ and T$_3$ prevents wide swings in blood levels of active hormones and provides additional buffering capacity against adversity. Regulation of the peripheral deiodination of T$_4$ introduces some flexibility into the system.

The steps in thyroid hormone biosynthesis are: (1) iodide concentration; (2) organification of iodide (monoiodotyrosine and diiodotyrosine production); (3) iodotyrosine coupling (T$_3$ and T$_4$ production); (4) thyroglobulin reuptake and proteolysis; (5) deiodination of iodotyrosines; and (6) release of T$_3$ and T$_4$ (see Fig. 16–3).

IODIDE CONCENTRATION. Iodide ion is actively transported into the thyroid cell. The basal portion of the follicular cell actively concentrates iodide against a substantial electrochemical gradient, normal iodide concentration in the plasma being 10^{-7}M. A high-energy phosphate compound is involved, and the process is inhibited by anaerobic conditions, low temperature, oxidative poisons such as cyanide and 2,4-dinitrophenol, as well as the cardiac glycosides. The stomach, mammary, and salivary glands can also concentrate iodide from the blood. However, the thyroid uptake mechanism is the most efficient, accumulating up to 40 to 500 times the serum level of iodide. Under normal circumstances, only the thyroid uptake is responsive to TSH and only the thyroid gland incorporates iodide into organic form (thyroid hormones). Ingested iodide is cleared predominantly by the kidney (approximately two thirds), the remainder being cleared by the thyroid.

Monovalent anions such as thiocyanate, perchlorate, and pertechnetate can compete with iodide for this concentrating mechanism. The ability of the thyroid gland to trap pertechnetate is used clinically in the radioactive pertechnetate scan. Twenty minutes after an intravenous dosage, radioactive pertechnetate can be measured over the thyroid, giving a picture of iodide trapping by various regions of the thyroid.

ORGANIFICATION OF IODIDE. Under normal circumstances, very little (<10 per cent) of the roughly 8 mg. of thyroidal iodide remains as inorganic iodide. The trapped iodide is rapidly oxidized to a higher valence, which serves as the iodinating species for the thyroglobulin-bound tyrosine residues. The products of this reaction are monoiodotyrosine (MIT) and diiodotyrosine (DIT; Fig. 16–4) in peptide linkage. This then appears to be the first step that commits inorganic iodide to the synthesis of thyroid hormone. Twenty-four hours after the oral administration of radioactive isotopes of iodide, the thyroidal accumulation can be measured and the thyroid imaged. These radioiodine uptakes and scans require iodide organifiation and provide important information about normal and abnormal thyroid function.

The exact sequence of events involved in this complex biosynthetic reaction is not known, but the factors involved seem to be: (1) a thyroid peroxidase, an integral membrane-bound heme glycoprotein enzyme capable of oxidizing various substrates; (2) a source of peroxide; (3) an iodinating intermediate, which is an oxidized form of iodide, postulated to be a sulfenyl iodide species; and (4) a suitable iodide acceptor molecule, namely peptide-linked tyrosine in the thyroglobulin molecule.

Thyroglobulin is a large glycoprotein (mol. wt. 670,000) which is synthesized on follicular cell ribosomes, transported to the Golgi apparatus, and then taken by export vesicles to the apical (luminal) surface of the follicular cell. Thyroglobulin represents 50 per cent of the total protein synthesized in the thyroid and 75 per cent of the thyroid protein content. The carbohydrate moieties are added in the rough and smooth endoplasmic reticulum as well as the Golgi apparatus. The subunit structure of thyroglobulin appears to be stabilized by iodination, and the distribution of iodine between various iodotyrosines and iodothyronines depends upon substrate (iodide) availability. Other proteins such as albumin are also capable of being iodinated by the thyroid, but the conformational structure of thyroglobulin is most favorable for iodination. The thyroglobulin molecule contains approximately 125 tyrosine residues, with approximately 10 MIT and DIT residues, 2 T_4 residues and only 0.2 T_3 residues.

COUPLING REACTION. The coupling of the iodinated tyrosines to form iodothyronines

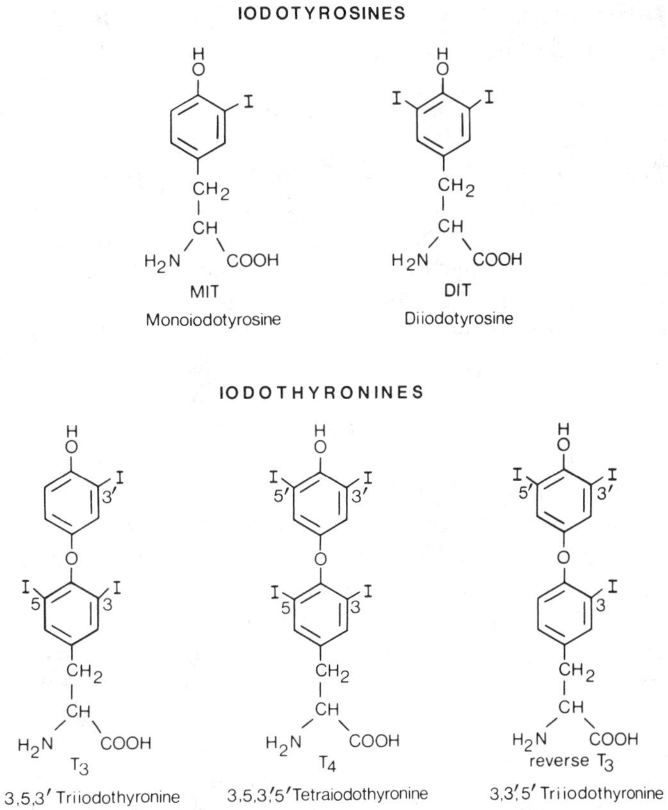

IODOTYROSINES

MIT
Monoiodotyrosine

DIT
Diiodotyrosine

IODOTHYRONINES

FIG. 16–4. Structure of iodotyrosines and iodothyronines.

T_3
3,5,3′ Triiodothyronine

T_4
3,5,3′,5′ Tetraiodothyronine

reverse T_3
3,3′,5′ Triiodothyronine

(T_3 and T_4; Fig. 16–4) probably involves peptide-linked iodotyrosines. This reaction appears to be catalyzed by the thyroid peroxidase, although a nonenzymatic conformation-induced coupling, or the involvement of other enzyme systems have not been excluded. The precise site for the iodination and the coupling reactions remains controversial; however, the authors favor an intracellular location. Iodinated thyroglobulin is secreted into the follicular lumen for storage as the thyroid colloid.

The organification and coupling reactions are particularly sensitive to inhibition by the drugs propylthiouracil and methimazole, which are used to treat hyperthyroidism. Despite blockade of these reactions, the stored thyroid hormone prevents a decline in the plasma concentration of thyroid hormones for weeks to months.

REUPTAKE. Stored thyroglobulin is available for reuptake by the cells. Thus, the microvilli at the apical cell surface imbibe colloid by pinocytosis. The combination of this colloid droplet or vesicle with cellular lysosomes leads to a phagolysosome, which contains proteolytic enzymes capable of digesting thyroglobulin.

DEIODINATION AND RELEASE. The proteolysis of thyroglobulin leads to the release of free iodotyrosines (MIT and DIT) as well as T_3 and T_4. The efficiency of the thyroid is demonstrated by the deiodinase reaction, which strips iodide from MIT and DIT (nonmetabolically active) for reutilization in the synthesis of thyroid hormones. The T_3 and T_4 (metabolically active) are not significantly deiodinated but are released into the bloodstream.

Circulating thyroid hormones and their metabolism

The major circulating thyroid hormones are T_4 and T_3. The plasma concentration of T_4 is 4 to 12 μg./dl. (about 10^{-7}M) and of T_3, 70 to 190 ng./dl. (about 3×10^{-9}M). Although 99.97 per cent of T_4 is protein-bound and 99.7 per cent of T_3 is similarly bound, it is the free, unbound hormone that appears to be metabolically active in initiating appropriate intracellular reactions. Measurements of the serum free thyroxine are in the range of 0.8 to 2.4 ng./100 ml. ($\sim10^{-11}$M); free triiodothyronine 80 to 240 pg./100 ml. (3×10^{-12}M).

Under normal circumstances, the thyroid gland is the sole source of circulating T_4, and T_4 is the major hormone secreted by the thyroid gland. Hence, the measurement of circulating T_4 or free T_4 is an important measure of thyroid gland function. Circulating T_3, on the other hand, is derived predominantly from T_4 by an irreversible deiodination reaction that takes place in many tissues of the body. Approximately 80 per cent of circulating T_3 is derived from T_4 peripheral conversion, only 20 per cent being a result of direct thyroidal secretion. Therefore, circulating T_3 concentration is subject to a number of nonthyroidal influences.

Much of the metabolic activity of thyroid hormone depends upon T_3 interactions with specific intracellular receptors. It is an open question, however, whether circulating T_3 is the major source of intranuclear T_3 or whether T_4 to T_3 conversion *in situ* is of major importance (as in the pituitary). Equally important is the question of T_4's intrinsic metabolic activity. Is T_4 merely a prohormone for T_3? The ability of thyroid hormone–sensitive tissues to deiodinate T_4 to T_3 (e.g., heart, liver, kidney) makes this question difficult to answer.

T_4 has a very long half-life (6.7 days) in part related to its tight protein binding. The thyroid gland secretes approximately 80 to 90 μg of T_4 per day. The space of distribution of T_4 approximates that of extracellular fluid (11.6 l.). T_3 has a shorter half-life than T_4 (1 day) and circulates in a lower concentration, at least in part because of less tight protein binding. The production rate of T_3 is about 30 μg./day with only 6 μg. coming from the thyroid gland under normal circumstances. The space of distribution is 38 l. Although direct comparison is difficult, administered T_3 is at least 4 times as potent as T_4 in terms of peripheral metabolic activity.

Recent studies have focused on the peripheral metabolism of T_4. T_4 is metabolized primarily (80 per cent) by mono-deiodination. Loss of an iodide atom from the outer ring of T_4 (5'-position) leads to T_3, a more potent hormone. Loss of an iodide atom from the inner ring (5-position) leads to the apparently inactive metabolite 3,3'5'-triiodothyronine, also known as reverse T_3 (rT_3) (Fig. 16–5). The choice of pathways is thus between an *activating* pathway to T_3 or an *inactivating* one to reverse T_3. The deiodination of T_4 to T_3 is not a random process. Conversion of T_4 to T_3 is inhibited by drugs (e.g., propylthiouracil, glucocorticoids, amiodarone, iodinated dyes, and others), fasting, the neonatal state, and a number of "stresses" (major illness such as renal or hepatic disease, sepsis, surgery). The liver and the kidney appear to be quantitatively the most important sites of conversion of T_4 to T_3. Although inhibition of 5'-deiodinase may account for this impaired conversion, impaired organ (liver) uptake may provide an alternative explanation. The fall in T_3 concentration may be associated with a decreased thyroid hormone metabolic activity, providing flexibility for a system whose major secretory product (T_4) has a half-life of a week. Still debatable is whether the fall in T_3 concentration in severe illness is homeostatic (i.e., diminishing catabolic rate) or harmful. Increased conversion of T_4 to T_3 occurs in chronic cold exposure and with overfeeding. In both situations increased calorigenesis may be homeostatic.

Most conditions of impaired T_4 to T_3 conversion are associated with an increase in concentration of reverse T_3 (rT_3). The rise in rT_3 concentration is probably related to decreased metabolic clearance of this compound, rather than increased production. Since rT_3 is primarily metabolized by a 5'-deiodinase (see Fig. 16–5), it has been suggested that impaired 5'-deiodinase may account for both decreased T_3 production and impaired rT_3 clearance. However, recent data suggest that T_4 and rT_3 5'-monodeiodinases are different enzymes.

rT_3 circulates at a concentration of 20 to 60 ng./dl. The rT_3 production rate of 20 to 40 μg/day is similar to that of T_3, but the rT_3 concentration is lower than T_3 because of its rapid clearance from the circulation. Almost all of the rT_3 in the circulation is derived from T_4 by peripheral conversion.

Thyroglobulin is released from the thyroid gland in small amounts but is not hormonally active (concentration 5 to 25 ng./ml.). The concentration declines after T_3 or T_4 administration and rises after TSH administration or in the usual forms of hyperthyroidism. The main use for the thyroglobulin radioimmunoassay, however, is as a marker for the growth or recurrence of well-differentiated thyroid carcinoma.

Protein binding

The major proteins that bind thyroid hormone in the blood are thyroxine-binding globulin (TBG), thyroxine-binding prealbumin (TBPA), and albumin. These account, respectively, for 75, 15, and 10 per cent of the thyroxine-binding capacity of plasma. T_3 seems to have minimal binding to TBPA. The concentration of TBG is 2 mg./dl.

TBG is rich in carbohydrate and circulates at a concentration of about 2 mg./100 ml. The binding affinity of TBG for T_4 is extremely high, $2.2 \times 10^{10} M^{-1}$. The capacity for T_4 binding, however, is rather limited,

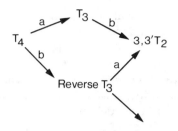

FIG. 16–5. Mechanism of hyperthyroidism or thyrotoxicosis. $a = 5'$ deiodinase; $b = 5$ deiodinase; $3,3'T_2$ = 3,3' diiodothyronine.

with one readily detectable T_4 or T_3 binding site per molecule, and a total capacity for binding T_4 of 15 to 25 μg. T_4/100 ml. ($3 \times 10^{-7} M$). Under normal circumstances, this protein is approximately one third saturated with T_4. The binding of T_3 by TBG is only one tenth as strong as that of T_4 (about $10^9 M^{-1}$), but the two hormones appear to share the same binding site. The deaminated thyronines, such as tetraiodothyroacetic acid do not bind to TBG.

It is well known that concentrations of T_4 and T_3 vary as a result of thyroid disease, but it is less well known that these concentrations may also vary in situations in which the binding proteins are changed. Alterations in the concentrations of T_4 and T_3 due to binding protein abnormalities do not alter the free hormone concentrations. TBG binding capacity can be increased by estrogens (drug therapy or pregnancy), 5-fluorouracil, acute hepatitis or as a result a genetic trait, all of which lead to an elevated total serum thyroxine. The free thyroxine does not change and the patients remain euthyroid. Decreased TBG may occur in debilitated patients, can be induced by high-dose glucocorticoids or androgens, or may be inherited. Drugs such as phenytoin compete with T_4 for binding sites, thereby lowering the binding capacity without apparently influencing the number of TBG molecules. All of these conditions cause a decrease in the total serum T_4. In general, situations with excess TBG are also associated with increased serum T_3 concentrations.

T_4-binding prealbumin is poor in carbohydrate and rich in trypotophan, and has a serum concentration of 30 mg./100 ml. It has a relatively low affinity for T_4 (1.3 × $10^8 M^{-1}$), but a much greater capacity than TBG, binding 250 to 350 μg./100 ml. T_4. It has one readily detectable binding site per molecule and binds tetraiodinated thyronines (such as T_4) and deaminated thyronines, but T_3 to a much lesser degree (1.2 × $10^7 M^{-1}$). The binding capacity of TBPA is increased in patients given androgens and corticosteroids; it is decreased immediately postoperatively, in acute and chronic illness, in patients with thyrotoxicosis, and in those treated with estrogens or drugs that compete for binding, such as salicylates or dinitrophenol.

Albumin has a low affinity for T_4 and T_3 ($10^5 M^{-1}$) but a high binding capacity. Recently families have been described with an abnormal albumin (familial dysalbuminemic hyperthyroxinemia) which avidly binds T_4 but not T_3, leading to an elevated total serum thyroxine concentration without an increase in T_3 concentrations.

Modulation of thyroid hormone biosynthesis and release

The role of iodide
Although TSH plays the major role in the regulation of thyroid hormone synthesis and release, the iodide ion, in addition to acting as a substrate for hormone synthesis, may also play a fine tuning or modulating role in this control.

If decreased iodide availability resulted in an acute decrease in thyroid hormone production, the animal or human would be at a clear disadvantage. The glandular stores of iodide and, to a certain extent, the protein-bound serum hormones and the relatively long half-life of T_4 prevent such a rapid fluctuation. In addition, the uptake of iodide by the thyroid gland becomes more efficient under the stress of iodide deficiency. Lastly, preferential thyroidal synthesis of T_3 relative to T_4 seems to occur in states of iodide deficiency. This is particularly frugal, as T_3 requires 25 per cent less iodine for its synthesis and yet is four times as potent.

Marked increases in thyroid hormone production in the presence of acute iodide excess would be equally harmful. Although this phenomenon may occur with abnormal "autonomous" thyroid tissue, normally autoregulatory mechanisms within the thyroid gland prevent this occurrence. A temporary inhibition of the release of T_4 and T_3 occurs with iodide excess, either as a direct effect on the release mechanism for these hormones, or more likely, due to increased resistance to proteolysis of the highly iodinated thyroglobulin molecule. An excess of organified iodine in the thyroid leads to a decrease in the trapping of iodide by the thyroid. Finally, the so-called Wolff-Chaikoff effect represents a decreased organification of iodide when the intraglandular inorganic iodide content is elevated. Although this effect is transient in normal individuals, iodide excess may lead to hypothyroidism in the patient with an underlying defect in the organification mechanism.

Other influences
Sympathetic nerve terminals are abundant within the thyroid gland. Recently vasoactive intestinal peptide (VIP) has been found in nerve fibers as well. The role of these neural influences in the fine regulation of thyroid gland function is currently under exploration.

Thyroid hormone action
Although nature has taken great pains to insure a constant supply of thyroid hormone, the precise role and mechanism of action of thyroid hormone is far from clear.

Thyroid hormone appears to regulate and have a profound effect on a wide variety of biological processes of virtually all tissues in higher organisms, varying from oxygen consumption to cell growth and differentiation. The hormone is certainly necessary for the growth, development and maturation of the central nervous system and the skeleton. In amphibians, the process of metamorphosis is controlled by thyroid hormone.

Thyroid hormones play a major role in the stimulation of calorigenesis, a condition that is accompanied by increased oxygen consumption in most body tissues except brain, spleen, and testis. Much of the increased energy expenditure induced by thyroid hormone is due to stimulation of membrane bound sodium-potassium-ATPase, the sodium pump. Thyroid hormone deficiency (primary hypothyroidism) leads to hypometabolism in most tissues except for the pituitary, where a striking increase in oxygen consumption, protein synthesis, and growth hormone occurs. Excess thyroid hormone (hyperthyroidism) leads to increased cellular oxidation in most tissues, a decline in oxygen consumption being found uniquely in the pituitary. Unlike cortisol, thyroid hormone concentration or requirements do not increase during stress states.

Thyroid hormones have many effects on the enzyme systems and metabolism of proteins, lipids, and carbohydrates. In children, hyperthyroidism leads to accelerated growth; in adults negative nitrogen balance is noted. Thyroid hormone excess facilitates lipolysis and increases the peripheral disposal of cholesterol. Hyperthyroidism leads to enhanced gluconeogenesis, increased carbohydrate absorption, and increased insulin degradation. It is not clear which of these actions are specific effects of thyroid hormones and which are related to nonspecific stimulation of metabolism.

Hyperthyroidism shares many features in common with sympathetic nervous system stimulation. However, catecholamine concentration is actually diminished in hyperthyroidism. The increased catecholamine effect appears to be mediated by an increase in the number of β-adrenergic receptor sites induced by the thyroid hormone. In hypothyroidism, apparent decreased sympathetic activity is mediated through a reduction in the number of β-adrenergic receptors.

The primary mechanism of action of thyroid hormone has intrigued investigators for years. Thyroid hormone–induced RNA synthesis is an early metabolic effect and this has prompted investigators to study the nucleus as a primary site of action. High-affinity, specific nuclear receptors for T_3 appear to correlate best with hormone action. Although T_4 receptors are found as well, their affinity is 10 times less than that for T_3; and 90 per cent of the hormone occupying chromatin receptors appears to be T_3. Oxygen consumption appears to correlate with T_3 receptor occupancy. The binding to nuclear receptors appears to be a direct process which does not require the intermediary step of cytosol binding and nuclear translocation, as is found with steroid hormones.

Additional evidence indicates that thyroid hormone may have primary actions at other cellular sites. It was initially believed that thyroid hormone action was mediated by uncoupling of oxidative phosphorylation at the mitochondrial level, but recent studies suggest otherwise. High-affinity T_3 receptors are located on mitochondria, and preliminary studies suggest early increases in mitochondrial ATP after T_3 administration. Furthermore, cell surface T_3 receptors, which seem to mediate translocation of amino acids into cells, have been located in rat tissues.

It should be apparent that no single mechanism of action or single reaction stimulation can account for all the metabolic actions attributed to thyroid hormone. The diversity of reactions and tissues involved leads to the striking and varied clinical manifestations in hyperthyroidism and hypothyroidism.

Hyperthyroidism

Hyperthyroidism, or thyrotoxicosis, is a syndrome caused by an excess amount of one or both of the thyroid hormones (T_4 and T_3; see Fig. 16–5). In the majority of patients, the T_4 in the blood is elevated, but recently, classical cases of thyrotoxicosis have been associated with T_3 excess alone, so-called T_3-toxicosis or triiodothyronine hyperthyroidism. T_3-toxicosis is not a discrete entity, but may be found in all of the various forms of hyperthyroidism. In T_3-toxicosis, the "extra" T_3 is secreted by the thyroid gland rather than being produced by increased T_4 to T_3 through extrathyroidal tissue conversion. The ^{131}I uptake and scan is quite useful in the differential diagnosis of thyrotoxicosis.

With a single notable exception (pituitary overproduction of TSH) the physiology of hyperthyroidism is consistent. An excess of thyroid hormone in the blood suppresses pituitary TSH release as measured in the basal metabolic state or after TRH administration.

The symptoms of thyroid hormone excess are common to all forms of hyperthyroidism but need not invariably be present. The general speeding up of metabolism results in warmth, increased perspiration, heat intolerance, weight loss with good appetite, and increased frequency of bowel movements. Palpitations, dyspnea (with or without congestive heart failure), muscle weakness, nervousness, emotional lability, and fatigue are common. The patient is typically thin, hyperkinetic, warm, with a hyperdynamic circulation. Tachycardia, increased pulse pressure, increased respiratory rate, tremor, muscle atrophy, and brisk deep tendon reflexes highlight the physical examination. Extreme cachexia or high output cardiac failure may be present. The physical examination of the thyroid depends upon the specific hyperthyroid syndrome in question.

Factitious thyrotoxicosis

Exogenous administration of excessive T_4 or T_3 by mouth will result in the typical clinical picture of thyrotoxicosis. However, the thyroid gland will be suppressed by such therapy leading to an undetectable radioactive iodine uptake and no goiter on physical examination (Fig. 16–6). Surreptitious ingestion of thyroid hormone is most common in medical and paramedical personnel. Serum thyroglobin will be undetectable as well.

Hyperfunctioning thyroid nodule

The autonomous function of a thyroid adenoma ("hot" nodule) may lead to overproduction of thyroid hormone. Suppression of pituitary TSH is clearly evident anatomically by suppression of the contralateral normal thyroid tissue and on radioisotope scanning by demonstrating all of the radioactivity on the side of the nodule. Clinically, a nodule is palpable on one side and often no thyroid tissue can be felt on the other. Pituitary TSH may also be suppressed in the patient with a hyperfunctioning thyroid nodule who is not clinically hyperthyroid and who has "normal" levels of circulating T_4 and T_3. The implication is that each individual has a set point for thyroid hormone concentration. Above that level, even within the "normal range" the pituitary will sense hormone excess. Exogenous administration of thyroid hormone does not decrease the iodide uptake by the hot nodule (i.e., it demonstrates autonomy with respect to ^{131}I uptake). Administration of exogenous TSH causes iodine uptake by the suppressed thyroid tissue.

Toxic nodular goiter

This represents a general case of the more specific hyperfunctioning nodule described above. Multiple hot nodules may be present, each of which is autonomous, but, more commonly, the autonomy is at

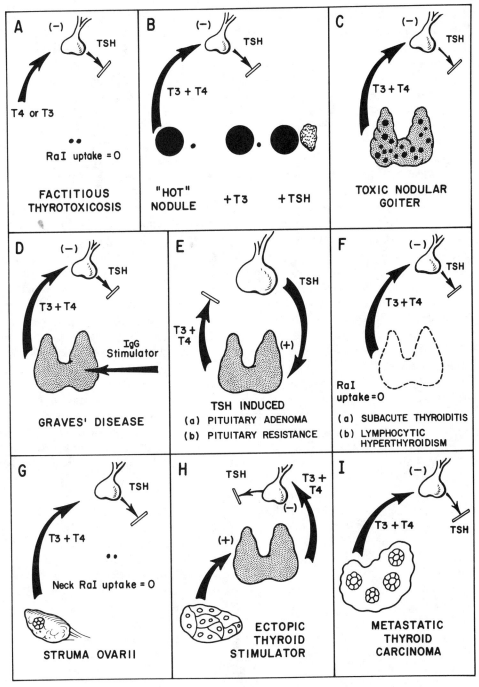

FIG. 16—6. Differential diagnosis of hyperthyroidism.

the level of the cells or follicles without evidence of discrete hot nodules on radioiodine scans. Surrounding the hot nodules are areas that are suppressed and, hence, take up less [131]I. This condition may develop slowly and insidiously in patients with long-standing multinodular (nontoxic) goiters. The development of thyrotoxicosis can be dramatic when patients with multinodular goiters are given large quantities of iodide. Iodide-induced hyperthyroidism (Jod-Basedow phenomenon) undoubtedly represents a failure of the autoregulatory mechanisms previously described and is analogous to supplying fuel to a furnace that cannot be turned off.

Graves' disease

Graves' disease (toxic diffuse goiter, hyperfunctioning hyperplastic thyroid) represents the most common cause of thyrotoxicosis, particularly in younger patients, the majority of whom are women. This is a systemic disease, representing not only thyroid hormone excess, but, in addition, unexplained effects on other organ systems. Exophthalmos or proptosis ("bug eyes") is due to the presence of retroorbital inflammatory tissue and fat, a phenomenon which does not result from thyroid hormone excess per se. In addition, lymphadenopathy and splenomegaly may be present. Pretibial myxedema, a peculiar deposition of mucopolysaccharides in the skin overlying the anterior tibia, is unique to Graves' disease. The thyroid gland is generally diffusely enlarged.

Recent evidence strongly suggests that Graves' disease is an autoimmune disease. Graves' disease demonstrates familial clustering with more frequent occurrence of certain HLA types (e.g., HLA-B8 in Caucasian, HLA-BW46 in Chinese and HLA-BW35 in Japanese). The thyroid gland behaves as if it is being vigorously stimulated, but TSH concentrations are undetectable, both in the basal state and after TRH administration. The thyroid stimulator in Graves' disease appears to be a polyclonal immunoglobin G (or several classes of immunoglobin G) which interacts with thyroid membrane receptor for TSH and "tricks" the thyroid into responding as if stimulated by TSH. It is not clear whether these autoantibodies are directed against TSH receptors or against other overlapping membrane antigens; however, the interaction of these immunoglobulins with TSH receptors in tissues as diverse as fat cells and white cells suggests that the TSH receptor is the primary target.

These thyroid stimulating immunoglobulins (TSIs) have been identified by a variety of names. The first to be described was the so-called long-acting thyroid stimulator (LATS), identified by its presence in the serum of Graves' patients and its ability to stimulate the mouse thyroid in a bioassay system. Current assays measure either thyroid stimulation (colloid droplet formation or cyclic AMP formation in slices or thyroid membranes) or membrane binding (inhibition of radio-labeled TSH binding to thyroid membranes). With the most sensitive available assays, 90 per cent of patients with untreated Graves' disease demonstrate such immunoglobulins.

The significance of these TSIs is best appreciated in neonatal Graves' disease. In this syndrome, newborn infants of mothers with Graves' disease may be hyperthyroid at birth. The hyperthyroidism is due to the transplacental passage of TSI from the mother. The hyperthyroidism disappears over a matter of months as the IgG is catabolized.

TSH-induced hyperthyroidism

The widespread application of the TSH radioimmunoassay has confirmed that TSH production is suppressed and TSH concentration is undetectable in almost all patients with hyperthyroidism. However, approximately 30 patients have been described in whom TSH excess is the cause of the hyperthyroidism. These patients fall into two groups. In about half, a pituitary adenoma has been described with a fixed output of TSH that is unresponsive to thyroid hormone suppression or TRH stimulation. The α-subunit of TSH is often released by these tumors as well, and some of the patients have an acromegaly or evidence of overproduction of other pituitary hormones. In the second group, selective pituitary resistance to the action of thyroid hormone appears to lead to the elevated TSH concentrations. The response of the peripheral tissues to the increased concentrations of T_3 and T_4 leads to clinical hyperthyroidism; however, the set point of pituitary feedback is altered. If supraphysiological concentrations of T_3 and T_4 are administered, TSH will fall; if TRH is administered, the elevated basal concentration of TSH will increase further. At present, there is no evidence that the defect resides at the level of the pituitary T_3 or T_4 receptor. The disease appears to be acquired rather than inborn, and many of the patients had been previously treated for conventional forms of hyper- or hypothyroidism. The possibility of a cryptic pituitary microadenoma in such patients has not been excluded.

Lymphocyte populations in Graves' disease have been studied and appear to be deficient in the suppressor T-cell class of lymphocytes. The disease is not fixed, however, and hyperthyroidism (and T-cell abnormalities) may wax and wane, either spontaneously or under the influence of antithyroid drugs.

The eye changes of Graves' disease do not appear to be mediated by immunoglobulins and may represent a cell mediated inflammatory process.

Occasionally, patients with Graves' disease are euthyroid. In euthyroid Graves' disease, the thyroid gland is autonomous and independent of TSH, and the administration of exogenous thyroid hormone fails to suppress thyroid function, suggesting stimulation of the thyroid gland by lower concentrations of stimulating immunoglobulins.

Subacute thyroiditis and lymphocytic hyperthyroidism

Typical subacute thyroiditis is also known as DeQuervain's or giant cell thyroiditis. It is an inflammatory condition which often follows in the wake of a viral infection. The inflammatory process results in a swollen, tender thyroid gland, an elevated erythrocyte sedimentation rate (ESR), and the leakage of T_3, T_4, and inactive iodinated compounds from the thyroid gland with resultant TSH suppression. Clinically important, but transient, hyperthyroidism may be present. The salient laboratory abnormality is the negligible or undetectable uptake of radioactive iodine at 24 hours. The hyperthyroidism is caused by an excessive release of hormones from the thyroid gland, rather than by an increase in their production. After the initial phase (lasting weeks to months), as glandular hormone stores are depleted, transient hypothyroidism often develops (lasting months), but the eventual return to the euthyroid state is almost certain.

A second syndrome has recently been described which follows a time course identical to subacute thyroiditis; hyperthyroidism with an undetectable radioactive iodine uptake, subsequent hypothyroidism, and a return to normal function in most cases. The disorder is probably autoimmune, with the thyroid gland containing lymphocytes, rather than granulomata as in subacute thyroiditis. The goiter is not painful, the ESR is only slightly elevated if at all, the disease is extremely common in the postpartum period, and it may go through multiple cycles over many years. Many patients are left with permanent thyroid abnormalities, such as goiter or mild primary hypothyroidism. Here too, the hyperthyroidism appears to result from increased hormone leakage and release, rather than from increased synthesis by the thyroid gland.

Struma ovarii

Dermoid ovarian tumors of the ovary may contain thyroid elements; and occasionally, hyperthyroidism results therefrom. The ^{131}I uptake in the neck is negligible, but scanning over the pelvis establishes the diagnosis.

Hypermetabolism associated with neoplasm

Certain patients with tumors of trophoblastic origin present with laboratory (and, occasionally, clinical) evidence of hyperthyroidism. This syndrome has been recognized with hydatidiform mole, choriocarcinoma, and in men with embryonal carcinoma. A material that has biological, but not immunological activity (assessed by radioimmunoassay of human TSH), has been described in these patients. All have had a high level of urinary chorionic gonadotropin excretion. Although human chorionic gonadotropin was thought to be the actual thyroid stimulator in these patients, recent evidence suggests that this may not be so.

Metastatic thyroid carcinoma

Rarely, metastases from follicular carcinoma of the thyroid may be sufficiently large and sufficiently active to produce enough thyroid hormone to result in hyperthyroidism.

Laboratory diagnosis of hyperthyroidism

A clinical suspicion of hyperthyroidism is necessary before it can be diagnosed. In most instances, the concentration of thyroxine (T_4) and free T_4 will be elevated. The T_3 concentration is often elevated out of proportion to the T_4 increase and in about 5 to 10 per cent of patients only the T_3 will be elevated (T_3-toxicosis). If all tests are equivocal, the TRH stimulation test can be used to exclude the diagnosis of hyperthyroidism (except, of course, TSH-induced hyperthyroidism). For practical purposes, a normal TSH response to TRH excludes hyperthyroidism. An additional test is sometimes used to prove that the thyroid gland is not hyperfunctioning: the 24-hour radioiodine uptake is measured before and after 7 to 10 days of suppressive dosages of thyroid hormone. In normal individuals, the 24-hour radioiodine uptake is reduced by at least 50 per cent during this time and often falls close to zero. This test obviously cannot be performed in patients with undetectable baseline radioiodine uptakes. Once the diagnosis of hyperthyroidism has been made, the 24-hour radioiodine uptake and scan is of major importance in the differential diagnosis of hyperthyroidism.

It is important to realize that the thyroid gland may be autonomous (i.e., independent of TSH) and yet not be overproducing thyroid hormone. This can happen in patients with multinodular goiters, hot nodules, and euthyroid Graves' disease. Such patients have abnormal TRH tests and abnormal suppression tests but are not clinically hyperthyroid and have normal circulating concentrations of T_4, free T_4, T_3, and free T_3. For this reason, the TRH test and the thyroid suppression test can be used to exclude but not to make the diagnosis of hyperthyroidism.

Therapy of hyperthyroidism

The therapy of hyperthyroidism depends in part upon the type of hyperthyroidism. The antithyroid drugs (propylthiouracil and methimazole) block thyroid hormone synthesis at the organification and coupling steps. Hence, it may require weeks to months to achieve the euthyroid state as glandular stores of thyroid hormone continue to be released. In addition, propylthiouracil can inhibit the peripheral conversion of T_4 to T_3. Both agents can be used in hyperthyroid patients with increased synthesis and release of thyroid hormone. In Graves' disease these agents appear to inhibit the production of thyroid-stimulating immunoglobulins and may help induce a remission of the disease process. After 6 months to several years of therapy with these agents in Graves' disease, approximately one-half of the patients can discontinue all medication and remain euthyroid.

The thyroid gland can be partially or

completely destroyed by radioactive iodine (^{131}I) to control the hyperthyroidism. In Graves' disease a high prevalence of long-term hypothyroidism is noted. In hyperthyroidism due to a toxic adenoma (hot nodule), only the functioning thyroid tissue takes up ^{131}I, hence hypothyroidism rarely occurs. The thyroid gland (or part of the thyroid gland, as in a hot nodule) can be removed surgically in order to control the hyperthyroidism. Beta-adrenergic (or other adrenergic) inhibiting agents provide symptomatic relief while these other more specific modes of therapy are applied.

Hypothyroidism

PRIMARY HYPOTHYROIDISM

The state of relative or absolute deficiency of circulating thyroid hormone represents the syndrome of hypothyroidism (decreased thyroxine (T_4) and triiodothyronine (T_3) synthesis). It may result from thyroid gland failure (primary hypothyroidism), or it may be secondary to pituitary failure (secondary hypothyroidism) or hypothalamic disease (tertiary hypothyroidism). Rarely, a syndrome of end-organ unresponsiveness to thyroid hormone may be associated with manifestations of thyroid hormone deficiency.

The signs and symptoms of thyroid hormone deficiency are common to all varieties of hypothyroidism and may be present to a variable degree. Therefore, it may be difficult to define the hypothyroid state on the basis of the clinical evaluation alone. When hypothyroidism is profound, myxedema is said to be present. The name myxedema is derived from the mucinous edema, due to mucopolysaccharide deposits, which presents as a nonpitting cutaneous thickening. Other effects of hypothyroidism include decreased metabolism with cold intolerance, weight gain, hypothermia, bradycardia, and decreased cardiac output. Intellectual deterioration, mental and physical lethargy, constipation, and delay in relaxation of deep tendon reflexes represent effects on the peripheral and central nervous system. Mucinous infiltration of the larynx may result in striking hoarseness, and impaired hearing is common. Menstrual abnormalities are frequently present.

Primary thyroid failure

The multiple etiologies of primary thyroid gland failure or insufficiency (hypothyroidism) are discussed below; however, the physiology is uniform. As the thyroid gland begins to fail, the pituitary attempts to compensate by increasing production of TSH. The pituitary becomes overactive with increased oxygen consumption, protein synthesis, and TSH release. The elevated serum TSH concentration is the result of accelerated pituitary synthesis and secretion of TSH (average of 4,400 mU/day) and a prolonged half-life of serum TSH (about 85 minutes). A wide range of elevated TSH values are noted, from 3.5 μU/ml. to greater than 500 μU/ml. The variability, no doubt, stems from several factors: many patients do not have complete lack of T_4; some patients with low T_4 levels may synthesize adequate amounts of T_3; or severe and prolonged hypothyroidism may lead to myxedema of the pituitary, causing less than maximal basal TSH levels and less than the expected rise after TRH administration. This latter concept can be extended to the other pituitary hormones, since some patients with hypothyroidism have impaired maximal release of growth hormone, gonadotropins and ACTH. These abnormalities are usually reversible with thyroid treatment.

Decreased thyroid reserve ("prehypothyroidism")

As the thyroid gland begins to fail, the earliest abnormality noted is a rise in the serum TSH, which may be present for months or years before clinical symptoms or low serum T_4 concentrations develop. An elevated TSH level is *de facto* evidence (and the most sensitive test) of decreased thyroid reserve. As the TSH rises, the thyroid gland produces relatively more T_3 than T_4 in an attempt to provide adequate hormonal activity for peripheral tissues.

Etiology of Primary Hypothyroidism or Decreased Thyroid Reserve

Autoimmune (Hashimoto's thyroiditis)
Substrate deficiency (iodine deficiency)
Biochemical defects
    Concentration defect
    Organification defect
    Organification defect with deafness (Pendred's)
    Coupling defect
    Iodotyrosine dehalogenase defect
    Abnormal serum iodopeptides
Goitrogens
    Dietary
    Drugs
    Iodide excess in susceptible individuals
Ablation and agenesis
    Congenital athyreosis
    Ectopic dysgenesis
    Surgery
    Radioactive iodine
    External irradiation
Subacute thyroiditis (transient) or lymphocytic
    hyperthyroidism (usually transient)

The early recognition of decreased thyroid reserve is of critical importance. Patients usually do not change dramatically from normal to abnormal, but often pass through a continuum of progressively failing thyroid function until frank hypothyroidism develops. A serum thyroxine concentration of 6 μg. per cent, for example, might be perfectly normal; however, in a patient whose usual thyroxine is 8 μg. per cent, it might represent significant thyroid hormone deficiency which would be readily detected by the pituitary. Any condition that results in frank hypothyroidism would be expected to cause, first, a state of "decreased thyroid reserve."

Elevated serum TSH concentration in the presence of apparent euthyroidism (both clinical and metabolic, as assessed by a normal serum T_4 and basal metabolic rate) suggests a "prehypothyroid" state. The elevated TSH suggests that the pituitary is already sensing a state of T_4 deficiency. Although it is not clear that all such patients will go on to develop overt hypothyroidism, they should be considered a high-risk group. This is particularly true for patients who have been treated for hyperthyroidism, or who have Hashimoto's thyroiditis.

Etiology of primary hypothyroidism

The multiple causes of primary hypothyroidism or decreased thyroid reserve are listed below. As in hyperthyroidism, the most common spontaneous cause of hypothyroidism is autoimmune disease (Hashimoto's thyroiditis, chronic lymphocytic thyroiditis, autoimmune thyroiditis). Most patients treated with radioactive iodine or surgery for Graves' disease eventually become hypothyroid; Graves' disease patients treated with antithyroid drugs alone may become hypothyroid (presumably on an autoimmune basis) years after stopping the medication. External radiation for head and neck malignancies such as Hodgkins disease often leads to hypothyroidism. Iodide deficiency is rarely if ever encountered in the United States today. Certain drugs may lead to goiter and hypothyroidism, the most common being iodine (in excess) and lithium, patients with Hashimoto's thyroiditis or post-radioactive iodine for Graves' disease being more susceptible to both drugs. Biochemical defects in the synthesis of thyroid hormone are rare, but patients with these defects have been crucial for our understanding of the thyroid hormone biosynthetic processes. The biochemical defects are inherited as autosomal recessive traits. In addition, patients may be born with primary hypothyroidism presumably related to thyroid agenesis. This occurs in one of 4000 to 5000 neonates and is easily diagnosed with neonatal screening blood tests.

HASHIMOTO'S THYROIDITIS. This autoimmune disease is predominantly found in women. The thyroid gland demonstrates marked lymphocytic infiltration and antibodies against various thyroid components (colloid, microsomes, thyroglobulin) circulate in almost all patients with this disorder. It is still undetermined how much of the thyroidal damage is mediated by antibodies rather than by lymphocytes per se. As in Graves' disease, a deficiency of suppressor T-cell populations has been noted. It is of great interest that both Graves' disease and Hashimoto's disease may occur in the same families, and both are associated with other autoimmune phenomena such as pernicious anemia and vitiligo.

Thyroid hormonal biosynthesis is impaired to a variable degree in Hashimoto's thyroiditis. Initially, the thyroid gland is enlarged and may be firm, but is rarely tender. The enlargement is due in part to TSH stimulation in an attempt to compensate for hormonal deficiency, and in part to lymphocytic and fibrous tissue infiltration of the thyroid. Profound hypothyroidism may occur in middle-aged patients in association with thyroid atrophy, a condition which is presumed to be related to Hashimoto's thyroiditis. Rarely, patients may produce TSH receptor blocking antibodies which prevent TSH from interacting with its receptor and thus manifest hypothyroidism without a goiter. The relationship of Hashimoto's thyroiditis to lymphocytic hyperthyroidism is not clear at this time.

BIOSYNTHETIC DEFECTS. Patients with hereditary biosynthetic defects do not ordinarily have a goiter at birth but rather develop it years later. Patients with an *iodide transport defect* are hypothyroid with a goiter and have very low 24-hour radioiodine uptakes. This defect is shared by salivary tissues, and a low salivary-to-plasma iodide ratio helps confirm the diagnosis. Although iodide administration may allow sufficient substrate to enter the gland for normal hormone synthesis, L-thyroxine therapy (L-T4) is usually the therapy of choice.

Iodide organification defects may be partial or complete. The thyroid gland has a high radioiodine uptake (50 to 80 per cent) which peaks in 2 to 4 hours. In the complete defect almost 99 per cent of the thyroidal iodine content remains as inorganic iodide. If a drug which blocks further iodide accumulation such as perchlorate is administered 4 hours after a tracer dose of radioiodine, almost all the thyroidal radioactivity will be discharged. The discharge of radioiodine is due to the diffusion of accumulated iodide out of the thyroid, down its electrochemical gradient. In normal individuals, over 90 per cent of accumulated iodide is organified by 4 hours and hence is not dischargable by perchlorate (so-called perchlorate discharge test). In some instances, the thyroid peroxidase is absent, in others it may be diminished or abnormally labile. If eighth-nerve deafness accompanies an organification defect, the diagnosis of *Pendred's syndrome* is made.

Patients with a *defect in the coupling reaction of iodotyrosines* present with a goiter and variable hypothyroidism. The radioiodine accumulated by the gland may be elevated but is not discharged by perchlorate. The thyroid gland contains MIT and DIT but little detectable T_3 or T_4.

Patients with a *defect in the iodotyrosine dehalogenase* are hypothyroid with elevated thyroid radioiodine uptake and a rapid turnover. After radioiodine administration a high serum concentration of labeled MIT or DIT can be measured, whereas the concentration in normal persons is negligible. After intravenous administration of radioiodine-labeled MIT or DIT, a large quantity of these iodotyrosines is excreted unchanged in the urine, suggesting a generalized tissue defect in deiodination. The importance of the dehalogenase enzyme in body iodine economy is evidenced by the iodine deficiency state which occurs in such patients. The hypothyroidism in these patients can be reversed by iodide administration.

Some patients secrete *abnormal serum iodopeptides*, leading to an elevated protein-bound iodide but decreased circulating active thyroid hormone concentration in the circulation. In some patients the acid-butanol–insoluble iodopeptides resemble albumin in their immunological properties. Thyroid tissue at surgery reveals little or no thyroglobulin but variable amounts of low-molecular-weight iodoproteins. A defect in the synthesis or structure of thyroglobulin has been postulated.

Other causes of primary hypothyroidism

Iodine deficiency has become virtually extinct in the United States. However, in other parts of the world, pockets of endemic goiter and cretinism can still be identified and prevented with iodide supplements.

A goitrogen is a substance that interferes with thyroid hormone production and leads to a goiter. Certain foodstuffs, such as cabbage, contain a natural goitrogen, an observation which is interesting but clinically unimportant. Drugs useful in the treatment of hyperthyroidism (the thioamides—propylthiouracil and methimazole) can cause hypothyroidism. Although sulfonylureas and paraaminosalicylic acid are goitrogens, they are rarely of clinical importance. Lithium, a psychoactive drug, inhibits thyroid hormone release and can lead to decreased thyroid reserve or overt hypothyroidism. Lithium-induced hypothyroidism is most common in middle-aged women with positive antithyroid antibodies, suggesting that lithium exacerbates subclinical Hashimoto's disease. Iodide excess may lead to hypothyroidism in

susceptible individuals through inhibition of hormone release and hormone synthesis. Susceptible patients include those with Hashimoto's thyroiditis, hyperthyroidism treated with radioactive iodine, organification defects, or a history of external radiation therapy in the neck for malignancies.

After radioactive iodine therapy for Graves' disease, 5 to 20 per cent of patients become hypothyroid during the first year, and an additional 3 per cent become hypothyroid each subsequent year, suggesting cell nucleus damage by ^{131}I or possibly ongoing autoimmune thyroid destruction. Similarly, long-term hypothyroidism is noted after surgery for Graves' disease. Patients with cervical Hodgkins disease treated with external radiation are susceptible to hypothyroidism, particularly if they have been studied with lymphangiograms, since these provide a long-lasting source of iodide in high concentrations.

Recent studies suggest a high incidence of hypothyroidism without goiter in workers exposed to polybrominated biphenyls, chemicals used in the fire retardant industry. Whether other industrial exposures can cause hypothyroidism is unclear.

Diagnosis of primary hypothyroidism

A clinical suspicion of hypothyroidism or the presence of a goiter should lead to the measurement of serum thyroxine, free thyroxine, and TSH concentrations. Measurement of serum TSH is of critical importance, since TSH concentration will rise long before the T_4 or free T_4 concentrations are reduced below the normal range. Measurement of serum T_3 concentrations has little utility in primary thyroid failure, since thyroidal T_3 production continues until much of the thyroid gland is destroyed. In the neonate the diagnosis of hypothyroidism may be totally unsuspected. Untreated neonatal hypothyroidism can lead to permanent mental retardation, underscoring the importance of neonatal screening programs for congenital hypothyroidism.

Therapy of primary hypothyroidism

Primary hypothyroidism is one of the easiest and most gratifying diseases to treat. Virtually all of the manifestations of the disease in the adult are reversible if recognized and treated. In neonates and children, early recognition and therapy are of critical importance to prevent irreversible brain damage.

Replacement therapy with L-T_4 mimics normal thyroid function. The long half-life of L-L_4 (7 days), the absence of increased stress requirements for T_4, and the peripheral conversion of T_4 to T_3 after secretion or exogenous administration all contribute to the ease of replacement therapy. With appropriate L-T_4 replacement therapy, the thyroidal contribution to serum T_3 is missing, and hence patients tend to have higher serum T_4 concentrations and slightly lower T_3 concentrations than normal persons. Therapy with desiccated thyroid, a preparation containing ground animal thyroids, provides variable quantities of T_4 and T_3 and has been largely replaced by L-T_4 therapy. Therapy with L-T_3 (cytomel) is nonphysiological, leading to peaks and troughs of T_3 concentration which are not normally found. Furthermore pituitary TSH secretion is relatively resistant to suppression by exogenous T_3, often requiring peripheral tissue hyperthyroidism (excess circulating T_3) for adequate TSH suppression.

SECONDARY HYPOTHYROIDISM (PITUITARY FAILURE) AND TERTIARY HYPOTHYROIDISM (HYPOTHALAMIC DISEASE)

In hypothyroidism secondary to pituitary or hypothalamic disease, multiple tropic hormone deficiencies are commonly present. For this reason, the clinical picture may differ from that of primary thyroid failure. The hypothyroidism is usually less severe in pituitary or hypothalamic disease, with less striking hair, skin, and cardiac changes. Signs of decreased ACTH and cortisol also may be evident with decreased skin pigmentation, poor ability to withstand stress, hypotension, and weight loss. The distinction between primary and secondary forms of hypothyroidism is critical because of the effect of thyroid administration on cortisol metabolism. T_4 increases the hepatic metabolism of cortisol, resulting in inactive products. In the patient with decreased ACTH reserve, administration of thyroid hormone, without cortisol, may precipitate acute adrenal insufficiency. Gonadal failure often accompanies secondary hypothyroidism as a consequence of gonadotropin deficiency. Occasionally TSH deficiency may be an isolated finding.

The presence of a goiter or high-titer antithyroid antibodies suggests primary thyroid failure. However, the ability to measure TSH levels accurately is mandatory for distinguishing primary from pituitary or hypothalamic hypothyroidism. Pituitary and hypothalamic hypothyroidism have low or normal TSH levels, associated with decreased T_4 and free T_4 levels, compared with the frankly elevated levels of TSH found in primary thyroid failure. Although an enlargement of the sella turcica suggests a pituitary adenoma and secondary hypothyroidism, sellar enlargement may occur in primary hypothyroidism as thyrotrophs enlarge and increase their TSH secretion. This emphasizes the necessity of TSH measurements in all patients with hypothyroidism.

End-organ refractoriness to thyroid hormone

Several patients with familial clustering have been reported with certain manifestations of hypothyroidism despite elevated circulating levels of T_4, free T_4, and T_3. TSH concentrations are normal or elevated in the face of high peripheral hormone concentrations and rise further after TRH administration. Although judged to be euthyroid by many measurements, deaf-mutism, goiter, and stippled epiphyses—all signs of thyroid hormone deficiency—were present. Absence of hyperthyroidism, delayed bone maturation, TSH responsiveness, or hyperresponsiveness, and failure to respond to administration of pharmacological doses of T_4 or T_3 daily suggest tissue resistance to thyroid hormone that most likely involve the pituitary as well as peripheral tissues. Abnormalities of nuclear T_3 and T_4 receptors have been suspected, but when evaluated, they appear to be normal, suggesting a postreceptor defect.

Alterations in plasma-binding proteins

Increased thyroid hormone binding

Elevated serum T_4 level, secondary to an increased TBG capacity for T_4, can lead to an erroneous diagnosis of hyperthyroidism. This abnormality can be induced by estrogens and has been described as a hereditary condition, often with an X-chromosome-linked dominant inheritance pattern. Clinical findings include elevated serum T_4 without symptoms of hyperthyroidism, normal basal metabolic rate, radioiodine uptake, TRH test, and low T_3 resin uptake. The thyroxine-binding capacity for T_4 may be increased to levels as high as 50 μg./100 ml. Although this results in an elevated level of serum T_4, there is a decreased proportion of free T_4, with a normal concentration of the free hormone. T_3 is bound to TBG also, the TBG excess resulting in increased concentrations of T_3 as well as T_4. These patients and those with TBG deficiency are clinically normal and require no therapy. Certain patients, particularly those with Hashimoto's thyroiditis, may develop antibodies which interact with T_4 and/or T_3 and which interfere with the radioimmunoassay for these hormones. Depending upon which type of radioimmunoassay is utilized, values may be spuriously high (double antibody method) or spuriously low (single antibody method). In addition, abnormal albumins may bind T_4 (but not T_3), leading to false elevations of the T_4 concentration in the familial syndrome dysalbuminemic hyperthyroxinemia.

Decreased thyroid hormone binding

Decreased serum T_4 level secondary to decreased TBG capacity for T_4 may lead to an erroneous diagnosis of hypothyroidism. TBG deficiency may occur with protein depletion. In addition, it seems to occur as a hereditary abnormality and is transmitted as an X-chromosome-linked dominant trait. Clinically, one observes a euthyroid patient with normal radioiodine uptake, normal BMR, normal TSH, and low serum T_4 with an elevated T_3 resin uptake. Such patients have pronounced abnormalities in the peripheral metabolism of T_4. The volume of distribution and the fractional rate of turnover of T_4 are increased, and hormonal clearance rates are markedly augmented. However, these data have been interpreted to mean that the daily disposal of hormone by both excretory and degradative routes is normal (i.e., the patients are euthyroid). Prolonged administration of synthetic estrogens or adrenal corticosteroids has failed to alter significantly the binding capacity of TBG in such patients.

Low T_3 syndromes and the sick-euthyroid syndromes

Many ill or starving patients demonstrate striking abnormalities on thyroid function testing: the serum T_3 by radioimmunoassay is extremely low, but concentrations of reverse T_3 tend to be reciprocally elevated. Since TSH concentrations are not elevated, these patients appear to remain euthyroid. However, a decreased basal metabolic rate may result from the decreased T_3 concentrations, a response which may be appropriate and protective for catabolic situations. The diminished T_3 concentration appears to be a consequence of impaired peripheral conversion of T_4 to T_3; the increased reverse T_3 is probably caused by impaired clearance of rT_3.

Critically ill patients with multisystemic diseases, often with accompanying sepsis and hypotension, may demonstrate abnormalities of serum T_4 concentrations which may further confuse the situation. Low concentrations of T_4 and free T_4 may be noted. The low T_4 concentrations are related in part to decreases in binding proteins, in part to inhibition of T_4 binding to binding proteins, and in part are unexplained. The low free T_4 concentrations (by many, but not all, assays) suggest that our ability to measure free T_4 *in vitro* is often useful, but is an imperfect measure of what is happening *in vivo*. These patients appear to be clinically euthyroid, do not demonstrate elevations of serum TSH, and have elevated reverse T_3 concentrations; this syndrome has been called the sick-euthyroid. Patients with primary or secondary hypothyroidism have low concentrations of reverse T_3.

Nontoxic multinodular goiter

In the United States at least 5 per cent of the living population are diagnosed as having multinodular goiters; the incidence is much higher in autopsy studies. Nontoxic nodular goiters probably represent a heterogeneous group of disorders (also called colloid or simple goiter). Iodide deficiency was once the leading cause of such goiters in the United States, but iodide deficiency has been almost entirely eliminated from this country through dietary and other iodine sources. There still remain pockets of endemic iodide deficiency goiter throughout the world. With iodide deficiency serum T_4 decreases and TSH concentration increases causing the goiter. In the clinical setting as well as under experimental conditions, iodide deficiency with TSH excess, rather than causing diffuse goiter, causes thyroid gland nodularity with adenomatous changes, cyst formation, and colloid accumulation.

The etiology of typical nontoxic nodular goiter in iodide sufficient areas is far from clear. One hypothesis begins with an inefficient thyroid gland or decreased thyroid reserve. According to this hypothesis, the thyroid gland grows and becomes nodular under the stimulus of TSH excess. Yet the concentration of TSH in patients with multinodular goiters (with the exception of iodide deficiency or congenital biosynthetic defects) is not elevated. It is common, however, to find increased concentrations of serum thyroglobulin and serum ratios of T_3 to T_4, suggesting defective thyroglobulin iodination. Before eliminating TSH as a factor in the development of these goiters, it should be noted that iodide deficiency makes the thyroid gland hypersensitive to TSH; and that similar TSH hypersensitivity might occur in nontoxic multinodular goiters. Furthermore, the goiter will frequently shrink as TSH concentrations are suppressed with exogenous thyroid hormone.

Approximately one quarter of patients with nontoxic multinodular goiter will develop thyroid gland autonomy marked by failure to suppress radioiodine uptake (or goiter size) with exogenous thyroid hormone, and blunted TSH release after TRH administration. These patients seem to be the ones who go on to develop spontaneous hyperthyroidism or who develop hyperthyroidism after exposure to excess iodides.

The majority of patients with multinodular goiters have no disability from this condition. There is occasionally a sufficient increase in size to cause local obstruction of the trachea or esophagus or to become cosmetically disfiguring. Under these circumstances surgical removal is considered. Thyroid cancer rarely develops in such gland.

Thyroid nodules and thyroid cancer

Isolated thyroid nodules are quite common and are separated into hot and cold nodules on radioiodine thyroid scanning. Hot nodules concentrate more radioiodine than surrounding normal thyroid tissue; cold nodules concentrate less isotope than surrounding thyroid tissue. Hot nodules are rarely malignant but may eventually produce hyperthyroidism. Cold nodules are benign 90 to 95 per cent of the time, but virtually all thyroid cancers fall into this scan category. The majority of cold nodules are follicular adenomas, benign tumors which may develop central degeneration and cystification. Cold nodules are cold because they have a relative decrease in ability either to trap iodide or organify iodine. Despite these abnormalities they may grow under the influence of TSH and may regress with TSH suppression.

Thyroid cancer is relatively uncommon with about 20 new cases per million persons each year. The long life expectancy of such patients suggests that 0.1 per cent of the population has or had a thyroid cancer. *Papillary Cancer* is the most common type of thyroid cancer, its name being derived from the frond-like papillae which form. Small areas of calcification known as psammoma bodies are typical of this disorder. Papillary cancer tends to spread locally to lymph glands and unlike most malignancies has an excellent prognosis despite the presence of nodal metastases. Microscopic papillary cancer is a common incidental finding at the time of thyroid surgery or at careful autopsy examination, but it is of little or no clinical consequence. It is unusual for patients to die of papillary thyroid carcinoma.

Follicular Cancer is the second most common type of thyroid cancer. Papillary thyroid carcinoma often contains small or large amounts of follicular elements; however, the presence of papillary elements defines the tumor and its biological behavior. Pure follicular thyroid cancer contains well- or poorly-differentiated thyroid follicles and spreads through vascular routes, usually to the lungs or to bones, where lytic lesions are noted. Although small follicular carcinomas with minimal vascular invasion have an excellent prognosis, larger tumors with more aggressive behavior do take lives. It is possible for a biopsy of follicular cancer metastases to resemble normal thyroid tissue, only its distant location indicating a malignancy. It is, therefore, not surprising that many of these tumors take up radioactive iodine, albeit less well than normal tissue, and thus respond to therapy with radioiodine. Especially in children, diffuse pulmonary metastases may be eradicated using repeated doses of radioiodine. How can the primary tumor be cold when the tumor is capable of taking up radioiodine? The tumor is always less efficient than normal (hence the term *cold*), and will occasionally take up radioiodine only after the normal thyroid is ablated or removed and TSH is allowed to rise. Both papillary and follicular carcinoma grow in response to TSH, and therapy with thyroid hormone to suppress TSH is an integral part of the therapeutic plan. The measurement of serum thyroglobulin may be of help in following the course of such patients, high concentrations suggesting persistence or regrowth of tumor.

RELATION TO EXTERNAL RADIOTHERAPY. Benign tumors as well as well-differentiated (papillary and follicular) thyroid cancers may occur many years after low-dose external radiation. In the past, conditions as diverse as tinea capitis, enlarged tonsils, enlarged thymus glands, and acne were treated with up to several thousand rads of external irradiation. Five to forty years after such therapy, patients develop thyroid nodules in increased frequency. Although most of these nodules are benign, up to 6 per cent of patients treated with irradiation eventually will develop thyroid cancer. It is possible that these tumors slowly develop as a result of nuclear genetic damage.

Higher doses of radiation (over 4000 rads) more commonly lead to hypothyroidism, but neoplasms are occasionally noted as well. In experimental animals, external radiation therapy to the thyroid gland is followed by increased TSH leading to thyroid cancer, the changes being prevented by TSH suppression. Whether TSH suppression will prevent the development of tumors in previously irradiated patients is unanswered. Fortunately, malignant thyroid neoplasms have not been reported in increased frequency after low (scanning) or high (therapeutic) doses of radioactive iodine.

MEDULLARY THYROID CANCER. Medullary carcinoma of the thyroid (MCT) arises from the parafollicular cells, which are derived from the ultimobranchial bodies. These cells do not concentrate iodide or synthesize T_4, but they do synthesize calcitonin. By conventional radioimmunoassays most normal human beings have undetectable blood levels of calcitonin; and no change is demonstrable with calcium infusion. Virtually all patients with medullary thyroid carcinoma have elevated basal calcitonin levels, which exhibit a striking rise after calcium or pentagastrin infusion. Although calcitonin is a hypocalcemic principle in animals, and can lower an elevated serum calcium when administered in pharmacological doses to man, patients with medullary thyroid carcinoma, with rare exceptions, have normal or elevated serum calcium concentrations. The tumor may develop sporadically. Of major import, however, is the syndrome of multiple endocrine neoplasia Type II (MEN II).

MEN II (a and b) is inherited as an autosomal dominant trait, necessitating family screening once a patient is identified with this disorder. How a single gene abnormality leads to such diverse abnormalities of the endocrine system is a fascinating but unanswered question. In MEN IIa, family members demonstrate medullary thyroid carcinoma (usually bilateral), pheochromocytoma (usually bilateral) and, less frequently, hyperparathyroidism (usually four-gland hyperplasia). Less commonly, a few families (MEN IIb) demonstrate a striking marfanoid habitus with multiple mucosal neuromas of the tongue and eyelids. Hyperparathyroidism is usually not present in the latter families despite medullary thyroid carcinoma and bilateral pheochromocytomas.

Hyperparathyroidism accounts for elevated serum calcium levels in MEN II when they occur. The medullary carcinoma itself may produce a variety of active substances: ACTH production is associated with clinical Cushing's syndrome. Prostaglandin production may explain the associated diarrhea. Histaminase is also synthesized and may serve as a tumor marker for metastatic disease.

The sensitivity of the calcitonin measurement has allowed the diagnosis of MCT, or its precursor (C-cell hyperplasia) in asymptomatic relatives of patients with documented MEN II syndrome. Although basal levels may remain undetectable in such patients, an abnormal increase after calcium or pentagastrin stimulation facilitates early diagnosis and surgical treatment at this curable stage. One hopes that future diagnosis of all malignancies by simple blood screening is presaged by these studies.

Annotated references

Bernal, J., and Refetoff, S.: The action of thyroid hormone. Clin. Endocrinol. 6:227, 1977. (A comprehensive review of metabolic processes stimulated by thyroid hormones.)

DeGroot, L. J., and Niepomniszcze, H.: Biosynthesis of thyroid hormone: Basic and clinical aspects. Metabolism 26:665, 1977. (A comprehensive review of the biosynthesis of thyroid hormone and its clinical defects.)

Edelman, I. S.: Thyroid thermogenesis. N. Engl. J. Med. 290:1303, 1974. (A discussion of one aspect of thyroid hormones mechanism of action.)

Evered, D., and Hall, R. (eds.): Hypothyroidism and goiter. Clin. Endocrinol. Metabol. 8(1):1979. (An up-to-date series of reviews on diverse aspects of hypothyroidism and goiter, including epidemiology, pathogenesis, clinical manifestations, and therapy.)

Ingbar, S. H.: Effects of iodine: Autoregulation of the thyroid. *In* Werner S. C., and Ingbar, S. H., (eds.): The Thyroid, p. 207. Hagerstown, Harper & Row, 1978. (A comprehensive review of the autoregulatory role of iodide.)

Ingbar, S. H., and Woeber, K. A.: The thyroid gland. *In* Williams, R. H. (ed): Textbook of Endocrinology. ed. 6. Philadelphia, W. B. Saunders, 1981. (An excellent general review of the thyroid including current physiology and function testing.)

Jackson, I. M. D.: Thyrotropin-releasing hormone. N. Engl. J. Med. 306:145, 1982. (A review of the clinical use of TRH and recent understanding of the extra-hypothalamic role of TRH.)

Kidd, A., Okita, N., Row, V. V., and Volpe, R.: Immunologic aspects of Graves' and Hashimoto's diseases. Metabolism 29:80, 1980. (A comprehensive but somewhat speculative review of the autoimmune nature of Graves' and Hashimoto's disease.)

Larsen, P. R.: Triiodothyronine: Review of recent studies of its physiology and pathophysiology in man. Metabolism 21:1073, 1972. (A review of the methodology and clinical importance of T3 measurements.)

Larsen, P. R., Silva, J. E., and Kaplan, M. M.: Relationships between circulating and intracellular thyroid hormones: Physiological and clinical implications. Endocr. Rev. 2:87, 1981. (A comprehensive and thoughtful review of T4 to T3 conversion and its implications for thyroid hormone action at the cellular level.)

Morley, J. E.: Neuroendocrine control of thyrotropin secretion. Endocr. Rev. 2:396, 1981. (A definitive and lucid review of the control of TSH secretion which draws from experimental animal and clinical and experimental human data.)

Oppenheimer, J. H.: Thyroid hormone action at the cellular level. Science 203:971, 1977. (A review of the mechanism of action of thyroid hormone.)

Sizemore, G. W., Health, H. III, and Carney, J. A.: Multiple endocrine neoplasia type 2. Clin. Endocrinol. Metabol. 9:299, 1980. (A recent review of MEN II and the use of calcitonin screening in affected families.)

Stanbury, J. B.: Familial goiter. *In* Stanbury, W. B., and Fredrickson, D. S. (eds): The Metabolic Basis of Inherited Disease. ed. 4, p. 206. New York, McGraw-Hill, 1978. (Review of the defects in the biosynthesis of thyroid hormones.)

Taurog, A.: Biosynthesis of iodoamino acids. *In* Greer, M. A., and Solomon, D. H. (eds.): Handbook of Physiology, Sec. 7, Endocrinology, Vol. III, Thyroid, p. 101. Baltimore, Williams & Wilkins, 1974. (A definitive review of the biochemistry of thyroid hormone biosynthesis.)

Volpe, R. (ed.): Thyrotoxicosis. Clin. Endocrinol. Metabol. 7(1):1978. (A comprehensive series of reviews on various aspects of hyperthyroidism including autoimmunity of Graves' disease, genetics of Graves' disease, laboratory and differential diagnosis of hyperthyroidism and therapy.)

Wartofsky, L., and Burman, K. D.: Alterations in thyroid function in patients with systemic illness: The "euthyroid sick syndrome." Endocr. Rev. 3:164, 1982. (An exhaustive review of alterations in thyroid hormone economy in nonthyroidal illnesses.)

Williams, E. D. (ed.): Pathology and management of thyroid disease. Clin. Endocrinol. Metabol. 10(2):1981. (A series of articles reviewing the pathology, epidemiology, and management of thyroid nodules and nodular goiters.)

17   Metabolism and Energy Mechanisms

John E. Gerich, M.D.

The average nonobese 70-kg. person consumes an 1,800- to 2,400-kcal. diet each day that is composed of approximately 50 per cent carbohydrate, 35 per cent fat, and 15 per cent protein. Following digestion in the gastrointestinal tract, these nutrients—mainly glucose, galactose, fructose, glycerol, free fatty acids, phospholipids, cholesterol, and individual amino acids—are absorbed into the circulation and routed to specific organs for storage or immediate utilization. Appetite and satiety centers within the hypothalamus regulate food intake so precisely that the weight of most individuals is relatively constant for prolonged periods. For this to happen, caloric expenditure over the long run must equal the caloric value of the food absorbed. The precision of this balance can be appreciated from the fact that a mere daily 5-per cent excess would result in a 25-kg. (55-lb.) weight gain over 5 years.

In addition to weight, the circulating and tissue concentrations of most of the major metabolic fuels are also maintained within a relatively narrow range. Considering that man eats intermittently and that energy demands can vary as much as tenfold over the course of a day, there must be regulatory mechanisms operative to ensure not only that there is efficient storage of fuels after meals but also that there is appropriate mobilization of these fuels between meals and in times of increased energy demands (e.g., exercise). Such control is of particular importance with respect to plasma glucose concentrations. The brain depends heavily on glucose for nourishment and its uptake of glucose is dependent on the plasma glucose concentration. Thus, a decrease in plasma glucose concentration, if severe enough, could result in brain damage.

Hydrolysis of adenosine triphosphate (ATP) to adenosine diphosphate (ADP) and inorganic phosphate is the immediate driving force for all the body's energy-requiring processes (e.g., muscle contraction, active transport, maintenance of membrane potentials). Most of the body's ATP is formed aerobically from ADP and inorganic phosphate within mitochondria by means of oxidation of the hydrogen generated during the catabolism of fuels in the tricarboxylic acid cycle (Fig. 17–1). An alternative mechanism is one in which monosaccharides enter the glycolytic pathway (Fig. 17–2) but do not proceed into the tricarboxylic acid. Such limited catabolism results in the formation of lactate and pyruvate rather than CO_2 and water. This process does not require oxygen (i.e., it is anaerobic) and is much less efficient in generating ATP than is oxidative phosphorylation. For example, 686 kcal. and 38 mol. of ATP can be generated for each mole of glucose that is com-

Lactate
Alanine
H2
NH2
Pyruvate
ATP
ADP
CO2
NAD+
NADH
CO2
Acetyl CoA
Oxalacetate
NADH
NAD+
Malate
Citrate
H2O
Fumarate
Isocitrate
FADH2
FAD
CO2
NADP+
NADPH
α - Ketoglutarate
Succinate
CoA
CO2
NAD+
NADH
Pi
GTP
GDP
Succinyl CoA
NADH
NAD+
FPox
FPred
CoQred
CoQox
Cytobox
Cytobred
CytcCred
Cytocox
CytoA3ox
CytoA3red
H2O
1/2O2

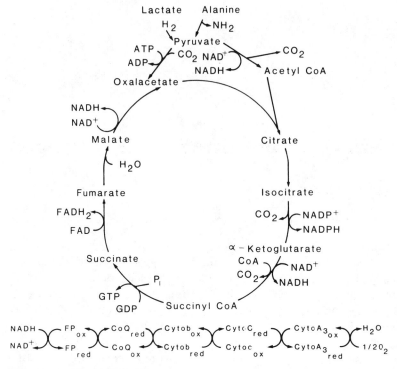

FIG. 17–1. The tricarboxylic acid cycle.

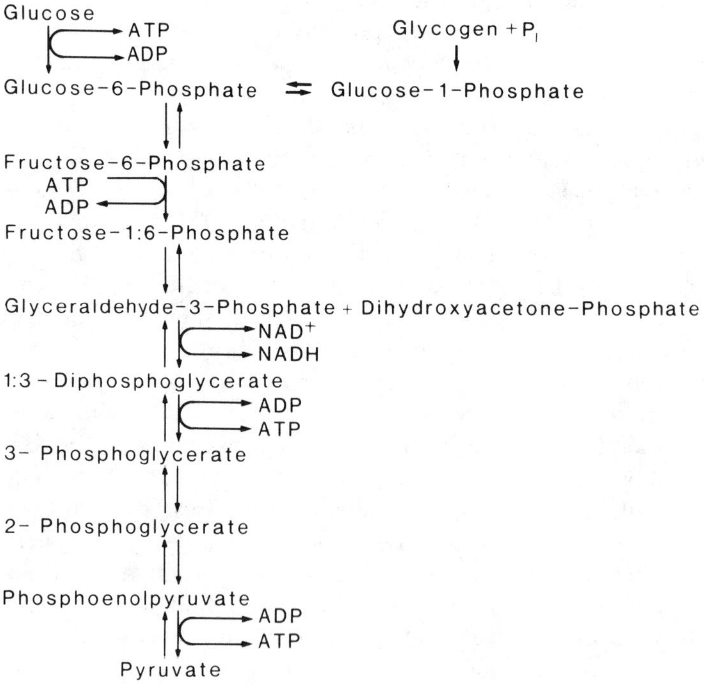

Glucose
ATP
ADP
Glycogen + Pi
Glucose-6-Phosphate
Glucose-1-Phosphate
Fructose-6-Phosphate
ATP
ADP
Fructose-1:6-Phosphate
Glyceraldehyde-3-Phosphate + Dihydroxyacetone-Phosphate
NAD+
NADH
1:3 - Diphosphoglycerate
ADP
ATP
3- Phosphoglycerate
2- Phosphoglycerate
Phosphoenolpyruvate
ADP
ATP
Pyruvate

FIG. 17–2. The glycolytic pathway.

TABLE 17–1
Stores of the Major Fuel Sources in a 70-kg Man

Fuel	Plasma Concentration (mM/dl.)	Extracellular Pool (mmol.)	Intracellular Pool (mmol.)			Total Stores (mol.)
			Muscle	Liver	Adipose Tissue	
Glucose	5.0	75	2200	360	140	2.7
Free fatty acids	0.5	7.5	720	30	34,600	35.4
Ketone bodies (acetoacetic acid + β-hydroxybutyrate)	0.2	3.0	—	—	—	0.03

pletely degraded (glycolysis plus oxidative phosphorylation), whereas only 52 kcal. and 2 mol. of ATP are generated for each mole of glucose undergoing only glycolysis.

Normally, glucose and free fatty acids are the major substrates for generation of ATP, and their metabolism is interrelated so that reciprocal changes are often observed in their rates of utilization (e.g., starvation). Utilization of these substrates and the ketone bodies (acetoacetic acid and β-hydroxybutyric acid) can account for most of the body's energy requirements. In terms of endogenous fuel stores, those potentially available from free fatty acids far exceed those from glucose and other substrates (Table 17–1). Glucose is stored as glycogen primarily in muscle and liver, whereas fatty acids are stored as triglyceride in adipose tissue. Ketone bodies can be considered partial degradation products of free fatty acids and are not stored in any tissue. Although a considerable amount of amino acids are present in protein, availability of these as potential sources of energy through proteolysis is usually limited, since it would result in the breakdown of important structural proteins and enzymes, a situation which, if excessive, would be very harmful.

Which substrates are utilized and which tissues use them depends on the circulating concentrations of the substrates (mass action effect) and the hormonal milieu. Changes in the rates of utilization of the major substrates are almost always linked to changes in their rates of production. Relative differences between these rates rather than the absolute rates themselves determine plasma and tissue concentrations. Thus, for example, if utilization of a substrate were increased twofold but production were increased threefold, the concentration of the substrate would increase. Rates of substrate production depend on the concentrations of their precursors and the hormonal milieu which determines which enzymatic pathways are activated or inhibited.

Glucose production and utilization

The liver and kidney are the only organs capable of producing glucose and releasing it into the circulation. Since muscle lacks the enzyme glucose-6-phosphatase, its glucose stored in glycogen cannot be released as free glucose; instead, the glucose in glycogen is released into the circulation as lactate and/or pyruvate generated during glycolysis. Under ordinary conditions, the contribution of the kidney is minimal (less than 10 per cent), but during prolonged fasting the kidney may contribute as much as 25 per cent. For practical purposes, however, the liver can be considered the sole source of glucose.

In the postabsorptive state, a period of approximately 6 hours after ingestion of a meal when most exogenous nutrients have been assimilated and when most meal-induced changes in plasma substrate and hormone concentrations have subsided, a relative steady state is achieved. At this point, rates of glucose production by the liver ($\approx$12 μmol./kg./min.) equal the rate of glucose utilization by all tissues. Approximately 75 per cent of hepatic glucose production is due to liberation of glucose stored in glycogen and the remaining 25 per cent is due to synthesis of glucose by a process called gluconeogenesis, which may be considered a reversal of glycolysis (Fig. 17–3), although different enzymes are involved.

The major precursors for gluconeogenesis are lactate and amino acids (Table 17–2). Muscle is the predominant source of lactate through the Cori cycle (Fig. 17–4). Alanine accounts for approximately 50 per cent of all the amino acids converted to glucose, although this amino acid represents only a small percentage of the amino acids in proteins. It is thus thought that most of the alanine converted to glucose originates from transamination of pyruvate generated during glycolysis in muscle. Glycerol is primarily derived from breakdown of triglyceride in adipose tissue (li-

polysis), which can liberate one molecule of glycerol and three molecules of free fatty acid.

The major hormonal regulators of glucose production on a moment-to-moment basis are insulin, glucagon, and catecholamines (primarily epinephrine). Within minutes, insulin can suppress glucose output by decreasing both glycogenolysis and gluconeogenesis through its inhibitory actions on phosphorylase and phosphoenolpyruvate carboxykinase. The activity of these enzymes is increased in situations in which insulin concentrations are low, such as starvation and diabetes mellitus. Insulin also stimulates glucose incorporation in glycogen by activation of glycogen synthetase. Complete suppression of glucose output by the liver can be achieved with a four- to fivefold increase in circulating insulin levels (e.g., 50 μU/ml, a value observed after meals). Glycogenolysis is more sensitive to inhibition than is gluconeogenesis.

Glucagon and epinephrine both stimulate glycogenolysis and gluconeogenesis. Their effects on glycogenolysis are probably mediated through activation of phos-

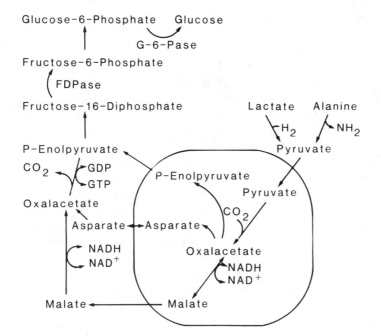

FIG. 17–3. Gluconeogenesis.

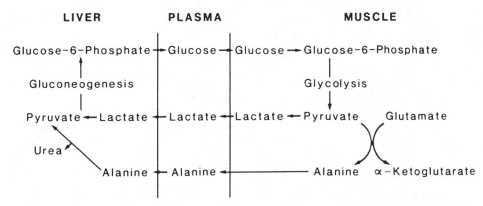

FIG. 17–4. The Cori cycle.

phorylase by a cyclic adenosine monophosphate (CAMP)-dependent mechanism. Their site of action on gluconeogenesis is still not known for certain. In man, the actions of epinephrine on glucose production are almost exclusively β-adrenergic. Ordinarily, glucagon is the more potent and more important hormone, since inhibition of its secretion results within minutes in a decrease in hepatic glucose production, whereas β-adrenergic blockade has trivial effects.

Other hormones, such as cortisol and growth hormone, do not have acute effects on glucose production. However, sustained changes in the circulating levels of these hormones affect the responses of the liver to insulin, glucagon, and epinephrine. For example, increases in plasma cortisol and growth hormone impair the ability of insulin to suppress glucose production and enhance the stimulation of glucose production by glucagon or epinephrine. Deficiency of growth hormone or cortisol have the opposite effects, and these patients with hypopituitarism or Addison's disease (adrenocortical insufficiency) are prone to develop hypoglycemia.

In addition to these hormonal and substrate influences, direct neural input to liver through both the sympathetic and parasympathetic systems can affect glucose production. Moreover, there is evidence that the liver itself can regulate its own output of glucose. However, in man, these mechanisms are substantially less important than hormonal influences.

As already indicated, in the postabsorptive state, the rate of glucose production approximates the rate of glucose utilization (12 μmol./kg./min.). Oxidation accounts for approximately 60 per cent and the remain-

TABLE 17–2
Major Gluconeogenic Precursors and Their Relative Contribution to Gluconeogenesis and Overall Glucose Production

Precursor	Rate of Conversion to Glucose* (μmol./kg./min. Glucose Equivalents)	Percent of Gluconeogenesis	Percent of Total Glucose Production
Lactate	1.8	56.0	15.0
Amino acids	0.8	25.0	6.4
Glycerol	0.2	6.4	1.6
Pyruvate	0.4	12.6	3.2
Sum	3.2	100	26.2

* Each of these precursors is a three-carbon molecule; their actual rate of conversion into glucose (in μmol.) must be divided by 2 to give the amount of glucose (in μmol) produced, since glucose is a six-carbon molecule.

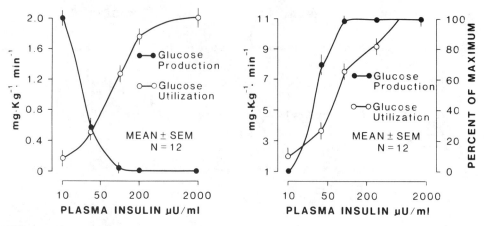

FIG. 17–5. Dose-response curves for suppression of glucose production and stimulation of glucose utilization by insulin in man.

der is due to glycolysis. Little if any glucose is converted into glycogen or lipids (triglyceride-glycerol or free fatty acids), since the postabsorptive state is a catabolic condition and storage forms of energy are generally being broken down.

The major factors determining the rate of glucose utilization are the plasma glucose and plasma insulin concentrations and the obligatory requirements of certain tissues. Over the physiological range, glucose utilization is directly proportional to the glucose concentration in plasma due to simple mass action. Insulin stimulates glucose uptake and subsequent steps in glucose metabolism in a variety of tissues, quantitatively the most important of which is muscle.

In the postabsorptive state about 50 per cent of glucose utilization is due to obligatory needs of the brain. The remainder occurs predominantly in tissues such as the formed elements of the blood, bone marrow, and renal medulla through glycolysis to satisfy their obligatory needs. Insulin contributes relatively little to the promotion of glucose utilization in contrast to its effects on restraining glucose production. This is due to two factors. First, most of the glucose utilization at this time occurs in satisfaction of obligatory needs in tissues whose glucose uptake is independent of

insulin, and, second, glucose production is more sensitive to insulin than is glucose utilization (Fig. 17–5). Thus, small increases in plasma insulin concentrations decrease plasma glucose concentrations by inhibiting glucose production without affecting glucose utilization.

Hormones such as cortisol, growth hormone, and epinephrine antagonize the ability of insulin to augment glucose utilization. In diseases characterized by excess cortisol, growth hormone, or epinephrine, such as Cushing's disease, acromegaly, or pheochromocytoma, there is impairment in insulin-stimulated glucose utilization. The availability of other substrates can also influence glucose utilization. For example, increases in plasma free fatty acids decrease glucose utilization, whereas decreases in plasma free fatty acids increase glucose utilization. Although glucose is the major substrate utilized by brain, in terms of overall fuel economy, free fatty acids are the preferred substrate for most other tissues.

We have dealt so far primarily with the postabsorptive state. After ingestion of a mixed meal containing glucose, fat, and amino acids, the increase in the plasma concentrations of these nutrients and their stimulation of insulin and glucagon secretion alter the balance between glucose production and utilization. The carbohydrate contained in the meal is hydrolyzed in the small intestine and is absorbed into the portal vein primarily as glucose. The rate at which this glucose enters the circulation plus the rate at which the liver is producing glucose exceeds the rate at which glucose is removed from the circulation for about an hour. This results in an increase in plasma glucose concentrations. Normally, for a meal containing 100 g. of carbohydrate, this increase does not exceed 3 mmol. (e.g., the plasma glucose increases from 5 mmol. to a peak value less than 8 mmol; 1 mmol. = 18 mg./dl.) and results in a six- to tenfold increase in plasma insulin concentrations whose pattern generally parallels that of the plasma glucose concentration. If the meal had contained no protein, glucagon secretion would decrease, but amino acids derived from ingested protein stimulate glucagon secretion so that normally there is a gradual increase in plasma glucagon concentrations of about 25 to 75 pg./ml. over a 1-hour interval.

Most of the glucose absorbed into the portal vein initially passes through the liver; however, about 60 per cent of it is taken up by the liver on successive passes. Moreover, the increases in plasma glucose and plasma insulin cause the liver to decrease its own production of glucose. The increases in glucagon secretion prevent the development of hypoglycemia during the period 2 to 4 hours after the meal when absorption of carbohydrate is complete and when insulin is still exerting its anabolic effects on protein and fat metabolism. The net effects of these changes is that about 60 to 80 per cent of glucose contained in a 100-g. carbohydrate meal is deposited in the liver either for glycogen formation or conversion into lipid (free fatty acids and glycerol). The remainder (20 to 40 per cent) is used by peripheral tissues to replenish glycogen stores, to provide glycerol for triglyceride formation, and to supply energy needs through glycolysis and oxidation. From the overall fate of the car-

TABLE 17–3

Disposal of a 100-g. Carbohydrate Meal in Normal Man and in Persons with Diabetes Mellitus

	Normal	Diabetic
Basal glucose production	25 g.	30 g.
Total glucose production observed after meals	45 g.	60 g.
Net increase in glucose production (B-A)	20 g.	30 g.
Hepatic glucose sequestration (100-C)	80 g.	70 g.
Peripheral glucose delivery (100-D)	20 g.	30 g.
Peripherally metabolized glucose	20 g.	10 g.
Glucose accumulated extracellular space	0 g.	20 g.*

* Assuming an extracellular volume of 15 l. and complete equilibration between the plasma and extravascular extracellular spaces, this would result in an increase in plasma glucose concentration of 7.4 mM (133 mg./dl.) at a point 3 hours after meal ingestion.

bohydrate in the meal, summarized in Table 17–3, the important role of the liver can be appreciated.

In patients with diabetes mellitus, increases in insulin secretion following meal ingestion are absent or markedly diminished, and increases in glucagon secretion are exaggerated. This results in less suppression of glucose production by the liver and less stimulation of glucose utilization by peripheral tissues. The consequent excess of glucose production over glucose utilization leads to excessive and prolonged increases in plasma glucose concentrations after meals (Table 17–3).

Free fatty acid–ketone body rates of production and utilization

In the postabsorptive state, plasma free fatty acids originate through breakdown of triglyceride from liver and adipose tissue, predominantly the latter. This lipolysis is under neural and hormonal control. Insulin decreases, while hormones such as epinephrine, growth hormone, and cortisol, as well as activation of the sympathetic ner-

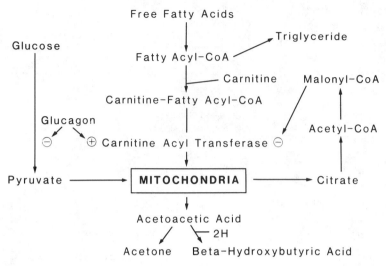

FIG. 17–6. Control of ketone body production.

vous system, increase the activity of the enzyme catalyzing this reaction. Insulin also permits reesterification of free fatty acids through its action on glucose metabolism by providing glycerophosphate. Human adipose tissue probably synthesizes little fatty acid, so that most of the fatty acids in such storage depots arise from free fatty acids made in the liver or those derived from meals. Lipid contained in the diet is completely hydrolyzed in the small intestine by lipases released by the pancreas. Short-chain fatty acids are absorbed directly in lymph. Long-chain free fatty acids (those with a carbon length greater than 10) are reesterified in intestinal cells and are carried as chylomicrons to the liver, where they are either stored as triglyceride, converted to ketone bodies, or released as lipoproteins (very low density and low-density lipoproteins) or as a complex with albumin. Free fatty acids contained in lipoproteins can be taken up by cells for storage or oxidation after their hydrolysis by an endothelial enzyme, lipoprotein lipase, the activity of which is augmented by insulin.

Ketone bodies (acetoacetic acid and β-hydroxybutyric acid) are produced only in the liver. Free fatty acids taken up by the liver are either synthesized into triglyceride or transported in mitochondria for oxidation. The acetyl CoA units derived from oxidation of free fatty acids can be condensed into acetoacetic acid. That compound can then be reduced to β-hydroxybutyric acid or undergo spontaneous decomposition to acetone. Normally, plasma concentrations of β-hydroxybutyrate are greater than those of acetoacetate. The transport of fatty acid CoA derivatives into mitochondria is under the control of carnitine acyltransferase, and this is the rate-limiting step for ketone body production when sufficient free fatty acids are provided. This enzyme is activated by glucagon and repressed by malonyl CoA (Fig. 17–6). Thus, under conditions in which glucose is being metabolized in liver, hepatic malonyl CoA is increased; ketone body production is suppressed, and free fatty acids are primarily converted into triglyceride. Conversely, under conditions in which hepatic glucose utilization is low (e.g., fasting), hepatic malonyl CoA is low, and ketone body production is derepressed. Glucagon may activate ketogenesis both by decreasing hepatic glucose utilization and by a direct effect on carnitine acyltransferase. The rate of ke-

TABLE 17–4
Potential ATP Generation from Metabolism of Carbohydrate and Fat in the Postabsorptive State

	Rate (μmol./kg./min.)	ATP Generated (μmol.)	Percent of Total ATP Generation
Glucose oxidation	6	216	} >32%
Glycolysis of glucose	10	4020	
Free fatty acid oxidation*	2.5	330	} >68%
Ketone body oxidation	5	165	

* Assumes average carbon length of 16

tone body production also depends on the rate of delivery of free fatty acids to the liver. Thus, even if the liver is not activated to switch from lipogenesis (triglyceride formation) to ketogenesis, an increase in the circulating levels of free fatty acids can result in an increase in the rate of ketone body formation.

In the postabsorptive state, free fatty acids are released into the circulation at a rate of approximately 7.5 μmol./kg./min. Approximately one-third is oxidized directly to CO_2 and water, while another one-third is converted into ketone bodies which are then oxidized. It can be seen from Table 17–4 that approximately two-thirds of the resting energy requirements are derived from metabolism of glucose.

After ingestion of a meal, there is a decrease in the rate of release of free fatty acids from storage depots due both to increased availability of glycerophosphate for immediate reesterification of free fatty acids and to inhibition of lipolysis by accompanying increases in plasma insulin concentrations. Plasma free fatty acid concentrations thus decrease. Increases in hepatic malonyl CoA decrease free fatty acid oxidation and the formation of ketone bodies, and shunt fatty acids into triglycerides in the liver. Although insulin promoted the utilization of free fatty acids and ketone bodies by tissues, the next effect of these changes is a decrease in the overall rate of utilization of these substrates as sources of energy. Fatty acids secreted by

the liver as triglyceride in lipoproteins are taken up by adipose tissue.

Altered states and diseases

Starvation (Table 17–5)
During prolonged fasting, there occurs a number of hormonally-induced changes in tissue metabolism aimed at providing mobilization of endogenous substrates to satisfy the body's energy requirements. First, insulin secretion decreases and glucagon secretion increases. This decrease in insulin secretion results in an increase in lipolysis, which provides additional glycerol as a substrate for gluconeogenesis and additional free fatty acids as a source for oxidation and conversion to ketone bodies. Hepatic malonyl CoA levels decrease, which further promotes the production of ketone bodies and decreases the synthesis of triglycerides in the liver, which coincidently enhances the efficiency of its extraction of gluconeogenic precursors from plasma. As insulin-dependent tissues decrease their utilization of glucose, the body adapts to a subsistance primarily on fat.

Despite decreases in insulin secretion and increases in glucagon secretion, the liver decreases its overall production of glucose. Thus, after about a 72-hour fast, hepatic glucose production decreases about 50 per cent (to 6 μmol./kg./min.). This is primarily due to exhaustion of glycogen stores. As shown in Table 17–1, he-

TABLE 17–5
Changes in Plasma Substrate Concentrations and Their Turnover During Fasting and Diabetic Ketoacidosis

	Postabsorptive State	3-Day Fast	3-Week Fast	Diabetic Ketoacidosis
Plasma Concentration				
Glucose (mM.)	5.0	3.5	3.0	25
Free fatty acids (mM.)	0.5	1.2	1.2	1.8
Ketone bodies (mM.)	0.2	2.0	4.0	10
Insulin (μU./ml.)	10	4.0	3.0	0–4
Glucagon (pg./ml.)	100	150	125	500
Cortisol (μg./dl.)	10	20	20	50
Growth hormone (ng./ml.)	3.0	5.0	5.0	20
Epinephrine (pg./ml.)	20	40	40	1000
*Turnover Rates**				
Glucose	12	6	5	20
Free fatty acids	7.5	15	10	25
Ketone bodies	5	10	7.5	30

* μmol./kg./min.

patic glycogen stores amount to only about 360 mmol. If glycogenolysis proceeded at its usual rate (9 μmol./kg./min.), then stores would be depleted in 10 hours. To compensate for this depletion, the liver increases its rate of glyconeogenesis. This is facilitated by greater availability of glycerol derived from lipolysis and of lactate, pyruvate, and alanine primarily derived from enhanced Cori cycle activity in muscle. Nevertheless, plasma glucose concentrations decrease from about 5 mmol. to 3.5 mmol. The brain and other such tissues compensate for this decrease by utilizing ketone bodies, which is one important consequence of the increased conversion of free fatty acids into ketone bodies. Another important consequence is the protein-sparing effect of increased circulating levels of ketone bodies. Early in the course of fasting there is a large increase in urinary nitrogen excretion, reflecting enhanced breakdown of protein and the conversion of the resulting amino acids into glucose. As the fast progresses, urinary nitrogen decreases, owing presumably to a reduction in proteolysis. It is thought that increased plasma ketone bodies may directly de-

crease muscle protein catabolism. Rates of glucose production and utilization and plasma glucose concentrations decrease somewhat further while turnover rates and plasma concentrations remain about the same or increase slightly. Finally, in addition to a decrease in glucose utilization, with prolonged fasting there is an overall decrease in the body's energy consumption.

Obesity

Obesity is not necessarily the exact opposite of starvation. Although exhaustive studies have sought a metabolic defect for the cause of this disorder (e.g., decreased energy utilization), such a defect has not been found. Indeed, glucose and free fatty acid turnovers are increased in obese individuals, since energy utilization is proportional to body surface area. Our current understanding indicates that it is simply the result of increased food intake. Why the hypothalamic and satiety centers permit this is unknown. Probably psychological and cultural factors are both involved. Since there is not an appropriate increase in energy expenditure to compensate for the number of calories taken in, energy stores (primarily triglycerides in adipose tissue) increase. Those who become obese while young have an increased number of fat cells as well as an increase in their size. This is in contrast to those who become obese as an adult, and who have primarily an increase in fat cell size. When individuals attain a weight greater than 150 per cent of their ideal body weight, their physical activity decreases due to the burden of this extra weight, and this can promote further weight gain if caloric intake continues at the same rate.

One of the early consequences of increased caloric intake is hyperinsulinemia. However, metabolic consequences which one would anticipate do not occur (e.g., decreased glucose production, decreased lipolysis), since this hyperinsulinemia rapidly leads to a state of insulin resistance. Thus, overall metabolism appears to be normal, but insulin-mediated processes require greater insulin concentrations for these to be maintained. As long as the pancreatic β-cell reserve holds out, most obese individuals have normal glucose tolerance at the expense of excessive insulin secretion. This insulin resistance is believed to be due both to a decrease in the number of insulin receptors on cells and to intracellular defects. Since hyperinsulinemia can decrease the number of insulin receptors on cells and can impair intracellular metabolism, the exact sequence of events leading to this insulin resistance is unclear.

Exercise

In the resting state, muscle primarily consumes free fatty acids, and its usage of glucose probably averages less than 0.2 μmol./kg./min. (less than 10 per cent of the total body's use). However, during exercise, carbohydrate contributes significantly to the increased energy demands of this tissue. During early exercise, muscle gylcogen is the major fuel consumed. As exercise continues, blood flow to muscle increases, and circulating substrates become increasingly

more important fuels. Thus, after about 40 minutes of exercise, glucose uptake by muscle may increase 10 to 15 times that observed in the resting state, and utilization of this glucose from the circulation contributes up to 75 per cent of the glucose consumed by muscle, which at this time provides nearly an equal amount of energy as that due to oxidation of free fatty acids.

To maintain the plasma glucose concentration constant while glucose utilization by muscle is increased during exercise, hepatic glucose production must increase two- to fivefold. Most likely this is facilitated by the observed increases in glucagon and epinephrine secretion, as well as the decrease in insulin secretion. These hormonal changes also promote lipolysis, and thus there is an increased release of free fatty acids into the circulation. Initially this increased glucose production by the liver is due mainly to enhanced breakdown of glycogen. However, with prolonged severe exercise, gluconeogenesis becomes more important, since by this time 75 per cent of hepatic glycogen stores will have been utilized.

Because of a slight excess of glucose utilization over glucose production, the plasma glucose concentration can decrease as much as 2 mmol. during moderate prolonged exercise. Frank hypoglycemia is rare but can occur in marathon runners, individuals who have been on a low-carbohydrate diet, and insulin-treated patients with diabetes mellitus. In McArdle's disease, a condition due to the lack of muscle phosphorylase, exercise provokes muscle cramps, and tolerance is decreased early due to the inability to mobilize muscle glycogen. However, if the patient persists in exercise, the increased blood flow to muscle can provide sufficient glucose and free fatty acids from the circulation to support exercise.

Pregnancy

Pregnancy, like exercise, increases energy requirements. This is because the fetus is solely dependent on its mother to provide nutrients. Amino acids and glucose are transported across the placenta, but free fatty acids are not. Since the fetal liver is not capable of gluconeogenesis, glucose is the predominant if not exclusive fuel for the fetus. The lack of gluconeogenic potential preserves amino acids of maternal origin for use in protein synthesis, which is essential for fetal growth.

The requirement for maternal glucose by the fetus increases throughout pregnancy. At term, a 3,600-g. fetus requires glucose at a rate of approximately 36 μmol./kg./min.—nearly three times the normal postabsorptive glucose production rate in a nonpregnant adult. How does the mother sustain this burden? Several mechanisms are operative. First, maternal glucose utilization is reduced and her glucose production is increased. Although part of this enhanced glucose production is due to increased gluconeogenesis, it is not advantageous for this to be the predominant process since this would tax maternal protein stores and possibly lead to a reduction in the availability of amino acids for the fetus. Probably because gluconeogenesis is restrained, glucose production by the mother does not quite keep up with her own needs and those of the fetus. The result is a decrease of about 1 mmol. in maternal plasma glucose concentrations after an overnight fast during the first two trimesters of pregnancy. Indeed, a more prolonged fast can result in plasma glucose concentrations as low as 2.5 mmol.

Concomitant with this decrease in plasma glucose, insulin secretion also decreases. The resultant reduction in plasma insulin concentrations permits an augmentation of lipolysis. The free fatty acids which are now made available are partly converted into ketone bodies and partly oxidized, substituting for glucose as an energy source in the mother and allowing her to utilize less glucose. Moreover, the glycerol released from adipose tissue during lipolysis can substitute for amino acids as gluconeogenic carbon sources and spare maternal protein. Thus, in many ways pregnancy produces adaptations similar to starvation.

Another important alteration during pregnancy occurs as a consequence of the endocrine role of the placenta. Human chorionic somatomamotropin (placental lactogen) is a polypeptide hormone produced by syncytiotrophoblasts. This hormone possesses actions similar to those of growth hormone in that it can promote lipolysis, decrease glucose utilization, and reduce protein breakdown. During the last 20 weeks of pregnancy, the circulating concentrations of this hormone increase tenfold. During this period, the placenta also secretes increasing quantities of estrogen and progesterone, which can act as insulin antagonists.

The net effect of these placental hormones is that the second half of pregnancy is characterized by insulin resistance and the need for the mother to increase her insulin secretion. Thus, during this period, plasma insulin concentrations in the postabsorptive state increase over those seen prior to pregnancy. Moreover, plasma insulin concentrations after meals are substantially elevated. In patients with insulin-dependent diabetes, insulin requirements can increase dramatically. Moreover, patients with insulin-independent diabetes may now require insulin therapy. Finally, in individuals with a decreased ability to augment their insulin secretion appropriately, hyperglycemia (gestational diabetes) may occur.

Diabetes mellitus

This disorder, which affects about 5 per cent of the American population, can be defined as a syndrome of unknown etiology characterized by hyperglycemia (postabsorptive plasma glucose concentration >7.8 mmol.) and relative or absolute insulin deficiency. There is considerable evidence that this is a heterogeneous disorder; however, by far the most common varieties are the so-called insulin-dependent Type I and the insulin-independent

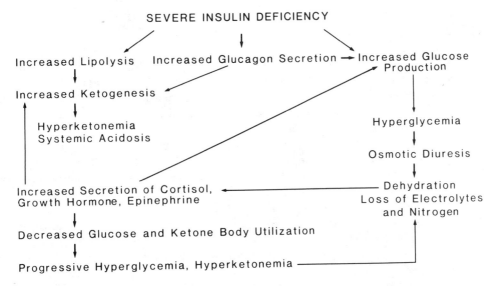

FIG. 17–7. Pathogenesis of diabetic ketoacidosis.

Type II diabetes mellitus (formally juvenile and adult-onset types, respectively). Various lines of evidence suggest that these two types of diabetes have different etiologies. Nevertheless, they can be considered to differ primarily in the severity of insulin deficiency.

In the insulin-dependent form, there is more than 90 per cent destruction of the β-cells of the pancreas, resulting in severe insulin deficiency. Without insulin treatment, patients with this disorder would develop ketoacidosis (see below) and die. In insulin-independent diabetes, the pancreatic insulin content may be only slightly reduced, but peripheral insulin resistance is frequently markedly increased. Patients with this disorder may experience severe hyperglycemia (>50 mmol.), but ketoacidosis is uncommon. It is likely that the predominant abnormality in the latter group of patients is an inherited defect in the β-cell secretory mechanism with associated insulin resistance due to obesity. In the insulin-dependent type, a genetic predisposition to viral/autoimmune destruction is probably operative.

The most extreme consequences of insulin deficiency (Table 17–4) are seen in diabetic ketoacidosis. Common initiating events are omission of insulin by the patient or failure to take additional quantities of insulin when infection or some other stress is present. Sometimes it is the presenting manifestation of diabetes. If untreated, the patient will die. The condition is characterized by severe hypoinsulinemia, although sometimes circulating insulin levels are nearly normal. However, this level is markedly inappropriate, since the attendant hyperglycemia would ordinarily result in insulin secretion, causing plasma insulin concentrations to increase 25 to 30-fold (to 250 to 300 μU./ml.). The plasma concentrations of glucagon, epinephrine, cortisol, and growth hormone are markedly increased. Thus, some of the metabolic abnormalities seen in this condition result from the actions of these hormones which are no longer opposed by insulin.

The pathogenesis of this condition can be described as follows (Fig. 17–7): As circulating insulin concentrations decrease below a critical threshold, the effects of anti-insulin hormones become prominent. Probably the first metabolic abnormality is an increase in glucose production, which at this point is mostly due to accelerated glycogenolysis. Glucose utilization also increases (because of the mass action effects of the hyperglycemia) but not as rapidly as glucose production; this results in an increase in plasma glucose concentration. About the same time, lipolysis increases and plasma free fatty acid concentrations rise. This and switching of the liver into a ketogenic mode (due to hyperglucagonemia and decreased hepatic malonyl CoA levels) result in a progressive increase in ketone body production. Ketone body utilization also increases, but not as much as ketone body production; this results in a progressive increase in plasma ketone body concentrations. Each mole of ketone body delivered into the circulation also results in delivery of 1 mol. of H^+. Once the buffering capacity of the plasma is exhausted, acidosis results. Arterial pHs below 7.0 have been observed. By now, muscle protein begins to break down with release of amino acids in the circulation. These as well as the glycerol released from adipose tissue during lipolysis are taken up at an accelerated rate by the liver and converted into glucose.

Once the plasma glucose exceeds the renal threshold ($\simeq 12$ mmol.), an osmotic diuresis starts with an obligatory loss of electrolytes (Na^+, Mg^{++}, $PO^=$, K^+). Since the thirst of the patient usually does not result in sufficient fluid intake to compensate for this diuresis, dehydration occurs. This leads to an increase in the secretion of insulin-antagonistic hormones, which reduces the utilization of glucose and ketone bodies and promotes their greater production. Now hyperketonemia, hyperglycemia, and acidosis become greater; there is further diuresis and dehydration, leading to hemoconcentration, decreased ability to excrete glucose and ketone bodies in the urine, and thus further increases in plasma glucose and ketone body concentrations. Depending on the degree of insulin deficiency and anti-insulin hormone excess, full-blown ketoacidosis may take from 12 hours to several days to develop. If not treated soon enough with insulin and fluid-electrolyte replacement, the dehydra-

tion and acidosis will lead to coma, cardiovascular collapse, and eventually death.

On the other extreme is the patient with mild insulin-independent diabetes mellitus. Many of these patients are obese, but even if they are not, this form of diabetes is usually associated with resistance to insulin. The cause of this insulin resistance is poorly understood but involves both receptor and intracellular abnormalities in insulin action. In many cases, the resistance in individuals due to obesity is not compensated for by an appropriate increase in pancreatic β-cell function, and this is the reason they become overtly diabetic. Modest weight reduction in such patients is often sufficient to restore normal glucose homeostasis. The mild insulin-independent diabetic may have only a modest increase in his fasting plasma glucose concentration (to 8 to 9 μmol.). Rates of glucose production and utilization in the postabsorptive state appear to be normal; but this "normal" rate of glucose utilization is maintained in part by the mass action effects of hyperglycemia, and the apparently normal rate of glucose production is inappropriate for the degree of hyperglycemia present.

In such patients, more marked and prolonged hyperglycemia is seen after meals. The dynamics underlying the genesis of this postprandial hyperglycemia are shown in Table 17–3. Both hepatic and extrahepatic defects in glucose metabolism are involved. Normally about 80 per cent of the carbohydrate of a meal can be considered to be sequestered in the liver. In the diabetic patient, sluggish secretion of insulin and excessive secretion of glucagon result in less of the carbohydrate in the meal being taken up by the liver (e.g., only 70 per cent). This leads to more glucose being delivered into the periphery (e.g., 30 g. instead of 20 g.). This in itself would not result in hyperglycemia if there were an offsetting augmentation of glucose utilization by peripheral tissues. But this does not occur, because of deficient insulin secretion. This unutilized glucose increases the plasma glucose concentration. It is not unusual for the postabsorptive plasma glucose concentration to increase from 8 mmol. to 15 mmol. after meals in patients with diabetes mellitus, whereas normally the plasma glucose concentration increases only from 5 mmol. to 7 mmol. transiently and returns to postabsorptive values within 2 hours.

Although patients with insulin-independent diabetes are usually not prone to develop ketoacidosis, this can occur when severe stress (e.g., infection, surgery) is superimposed on impaired insulin secretion. More commonly, these patients develop hyperosmolar nonketotic hyperglycemic coma. The same conditions prevail as in ketoacidosis except for lack of excessive release of free fatty acids or the excessive production of ketone bodies. It has been suggested that the hyperosmolarity and dehydration seen in this condition impair the mobilization of free fatty acids and that there is sufficient secretion of insulin to keep the liver from switching to a ketogenic mode. Plasma glucose concentrations are observed which are two to three times as great as those seen in ketoacidosis. This occurs primarily because the severe dehydration markedly impairs renal excretion of glucose. If not appropriately treated, this condition is fatal.

Annotated references

Dawes, G., and Shelly, H.: Physiologic aspects of carbohydrate metabolism in the fetus and newborn. *In* Dickens, F., Randle, P., and Whelan, W., (eds.): Carbohydrate Metabolism and Its Disorders, Vol. 2. New York, Academic Press, 1968. (This chapter provides information on glycogen and lipid stores and plasma concentrations of insulin during various stages of fetal development and the neonatal period based on studies in animals.)

Felig, P., and Wehren, J.: Fuel homeostasis in exercise. N. Engl. J. Med. *293*:1078–1084, 1975. (A review of the metabolism of free fatty acids and glucose in man during various degrees of severity of exercise based primarily on arteriovenous differences across the leg and the splanchnic bed.)

Freinkel, N.: Homeostasis factors in fetal carbohydrate metabolism. *In* Wynn, R., (ed.): Fetal Homeostasis,

Vol. 4. New York, Appleton-Century-Crofts, 1969. (A good review of changes in fuel metabolism in the mother and fetus during pregnancy.)

Gerich, J., Haymond, M., Rizza, R., Verdonk, C., and Miles, J.: Hormonal and substrate determinants of hepatic glucose production in man. *In* Veneziale, C. (ed.): The Regulation of Carbohydrate Formation and Utilization in Mammals. Baltimore, University Park Press, 1981. (A review of the role of glucose and on the interaction of insulin, glucagon, catecholamines, growth hormone, and cortisol in the regulation of glucose production by the liver, with primary emphasis on data derived from studies in man.)

Guyton, A.: Textbook of Medical Physiology. Philadelphia, W. B. Saunders, 1981. (The chapters on glucose and fat metabolism provide a clear description of the energy derived from the utilization of these substrates.)

Havel, R.: Caloric homeostasis and disorders of fuel transport. N. Engl. J. Med. *287*:1186–1192, 1972. (A concise review of the changes in the storage and in the pattern of utilization of glucose, free fatty acids, and ketone bodies during the transition from the postabsorptive state to the prolonged fasting state. Early and late stages of diabetic ketoacidosis are also covered.)

McGarry, J., and Foster, D.: Regulation of hepatic fatty acid oxidation and ketone body production. Ann. Rev. Biochem. *49*:395–420, 1980. (A critical review of the hormonal and substrate factors regulating ketone body formation in the liver based primarily on experiments performed *in vitro* and in animals.)

Schade, D., Eaton, B., Alberti, K., and Johnston, D.: Diabetic Coma: Ketoacidotic and Hyperosmolar. Albuquerque, University of New Mexico Press, 1981.

Sherwin, R.: Amino acid and protein metabolism in normal and diabetic man. *In* Brownlee, M. (ed.): Diabetes Mellitus, Vol. 3. New York, Garland STPM Press, 1981. (A review of amino acid metabolism in man, emphasizing the role of insulin and the fluxes of branched-chain amino acids and alanine between different organs.)

Shrago, E., and Ewart, R.: Carbohydrate metabolism and its implications. *In* Katzen, H., and Mahler, R. (eds.): Diabetes, Obesity, and Vascular Disease, Part 1. New York, Halsted Press, 1978. (This chapter provides a good description of the key role played by certain enzymes in the control of glucose metabolism and modulation of the activity of these enzymes of hormones.)

Williams, R.: Textbook of Endocrinology. Philadelphia, W. B. Saunders, 1981. (The classic textbook in endocrinology; the chapter on the endocrine pancreas provides much useful information on insulin and glucagon secretion and their actions as well as a good discussion of the metabolic abnormalities found in diabetes mellitus.)

18   Genetic Mechanisms of Metabolism

Oliver W. Jones, M.D.

In 1902, from studies on his patients, A. Garrod, a physician, provided the first application of the gene concept to a human medical abnormality, alkaptonuria. This marked the beginning of a new medical discipline, human biochemical genetics, or inborn errors of metabolism. As molecular biology of gene function unfolded between the mid-1950s through the early 1970s, a "one gene-one enzyme protein" dogma became generally accepted.

Genetic research during the past decade, especially in eukaryotic cells, including those in man, has amply demonstrated that the system is not as simple as Garrod's early suggestions indicated. The human genome is a formidable collection of genetic information amounting to some 5 billion nucleotide pairs. Assuming 1,000 nucleotides encode a single protein molecule, this would amount to some 5 million genes; however, current estimates suggest that there are only 100,000 to 300,000 genes for structural proteins. Thus, a major portion of the human genome remains genetically "silent." More than half of the deoxyribonucleic acid (DNA) in the human genome consists of nucleotide sequences repeated variable numbers of times in linear array. Information from a genetic region transcribed into the genetic message of messenger ribonucleic acid (RNA) may include intervening sequences which may be structurally silent, yet are crucial to the normal processing of the structural gene message to its final product, the complete polypeptide molecule. Thus, the phrase multiple gene segments-one protein may be more appropriate for human cells. Certain human biochemical disorders may reflect mutation in the silent intervening sequences of the gene message.

Research on biochemistry of gene function has led to discovery of precise biochemical tools for gene analysis on individual chromosomes. Genomic libraries for humans are being developed, and in the near future it may be possible to identify specific inborn errors of metabolism without the necessity of an assay for a given gene product. Precise gene analysis, and even a survey of the entire human genome, may be possible with as little as 20 ml. of peripheral blood. It is virtually a certainty that within this decade a review of genetic mechanisms of metabolism will bear little resemblance to any current review of this topic.

Initially, inborn errors of metabolism in man were thought to result from a mutationally altered enzyme affecting a single biochemical reaction in a metabolic pathway (e.g., melanin biosynthesis in individuals with albinism). The first inborn metabolic errors described were inherited as autosomal recessive disorders. Thus, the risk to children born to phenotypically normal carrier parents was 1 chance in 4, or 25 per cent with each pregnancy. It is now recognized that for certain metabolic disorders (e.g., osteogenesis imperfecta or epidermolysis bullosa) the same disorder may have different inheritance patterns in certain families: either autosomal recessive, dominant, or sex-linked. However, recessive inheritance nearly always results in the most severe form of the particular disease. For most well-characterized metabolic disorders (e.g., cystic fibrosis) there is considerable variation in the phenotypic expression of the disease. Moreover, in the example of Tay-Sachs disease, due to deficiency of β-D-N-acetylhexosaminidase, there is at least one instance of an adult with no evidence of enzyme activity who is phenotypically normal with no manifestations of clinical disease. Extensive studies on phenylketonuria have shown that a small proportion of untreated individuals with the biochemical defect still have normal intellectual development. Therefore, it would not be surprising to find that most inborn errors of metabolism, like hemoglobinopathies, may result from mutations at various loci on the genome and result in biologic diversity confronting the physician in diagnosis and treatment of these disorders. As our knowledge of genetic mechanisms has increased, it has been recognized that certain genetic disorders reflect abnormalities in transport of molecules or defects at the receptor sites on the cell surface. Other genetic disorders are associated with specific antigens on the surface of peripheral blood cells. These abnormalities are also expected to show heterogeneity in genotype and phenotype.

Some of the most important advances in our understanding and management of inborn errors of metabolism have come from detailed studies of each step in specific biochemical pathways. Invariably, as each enzyme function is characterized, an individual is discovered who is deficient in that particular function. One example is the hexose monophosphate shunt pathway in man. The initial metabolic disorder discovered in this pathway was glucose-6-phosphate dehydrogenase deficiency, a sex-linked disorder associated with drug-induced hemolytic anemia. This enzyme has been studied in great detail and we now know that certain mutants of glucose-6-phosphate dehydrogenase can be associated with another form of inheritable anemia, chronic nonspherocytic hemolytic anemia. Patients suffering from familial nonspherocytic hemolytic anemia have been described in whom an enzyme deficiency can be demonstrated at any step along the hexose monophosphate shunt (e.g., glutathione peroxidase, glutathione reductase, glutathione synthetase, 6-phosphogluconate dehydrogenase). Five additional enzymes related to red cell membrane integrity have been described in association with chronic nonspherocytic hemolytic anemia. We now can identify ten different enzyme deficiencies associated with a specific type of familial anemia and half of this number, including glucose-6-phosphate dehydrogenase, are associated with a single biochemical pathway. All except glucose-6-phosphate dehydrogenase deficiency are inherited as an autosomal recessive disorder. Another example of our increasing understanding of metabolic alterations in man is related to the biochemical steps converting the amino acid methionine to cysteine. Homocystinuria is known to result from deficiency in the enzyme cystathionine synthase and the inability to convert homocysteine to cystathionine. Three additional metabolic disorders in the same pathway have now been described, and as we learn more about this biochemical pathway, other human mutants will be discovered.

The remainder of this chapter discusses in more detail specific metabolic inheritable aberrations and their associated biological diversity in man.

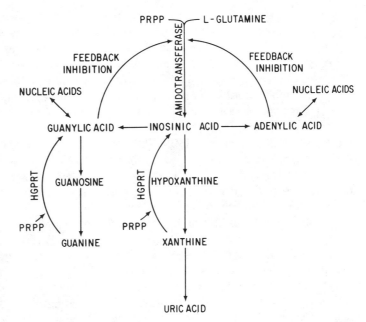

FIG. 18–1. Schematic summary of purine and urate synthesis via de novo and "salvage" pathways: PRPP = 5-phospho-ribosyl-1-pyrophosphate; HGPRT = hypoxanthine guanine phosphoribosyl transferase.

Disorders of purine metabolism

In this section the dynamics of purine metabolism are described, particularly the metabolic basis of hyperuricemia of genetic origin. The clinical and pathological observations are summarized and suggested treatment for well-established disorders of purine metabolism are given. Fig. 18–1 depicts schematically the metabolic pathway interrelating purine and urate biosynthesis. The formation of the purine nucleotides, adenylic and guanylic acid, from which nucleic acids are made, occurs in two principal ways: (1) synthesis *de novo*, in which inosinic acid is formed through a series of ten enzymatic steps using 5-phospho-ribosyl-1-pyrophosphate (PRPP), L-glutamine, glycine, formate and aspartic acid; and (2) by "salvage" pathways involving a one- or two-step mechanism for reforming purine nucleotides from adenine or guanine.

The biochemistry of purines is characterized by important feedback mechanisms which regulate synthesis (Fig. 18–1). The key reaction appears to be the formation of

β-5-phospho-D-ribosyl-1-amine (a primitive nucleotide whose amino nitrogen becomes N-9 of the purine ring) from PRPP and L-glutamine. This step is considered irreversible. The enzyme for this reaction, PRPP amidotransferase, contains two regulatory sites which bind adenosine-5'-phosphate (AMP) and guanosine-5'-phosphate (GMP). Moderate inhibition of the enzyme is obtained by the binding of one of the ribonucleotides; however, binding of AMP and GMP exerts a greater inhibitory effect than can be attributed to their additive inhibitions. Holmes has shown conclusively that PRPP amidotransferase exists in two species. The biologically active form is a small subunit of 133,000 molecular weight which is converted to the larger inactive subunit by AMP and/or GMP. An inactive subunit of 270,000 molecular weight can be converted to the active form in the presence of PRPP. Inhibition by purine ribonucleotides has not been demonstrated in any other reaction step in the de novo synthesis pathway leading to the formation of inosine-5'-phosphate (IMP).

The reutilization pathway plays a major

role in the formation of nucleotides and has two modes of action. First, the reversal of the nucleoside phosphorylase reaction causes nucleoside formation from a purine base and ribose-1-phosphate which can be converted subsequently to a nucleotide by the appropriate kinase. Second, the direct reaction of hypoxanthine, guanine, or adenine with PRPP produces IMP, GMP, or AMP, respectively. Aside from providing necessary nucleotides for cell function, the reutilization pathways contribute to the balance between substrate and inhibitor in the crucial PRPP amidotransferase reaction.

The metabolism of purine suggests three possible types of genetic abnormalities which would result in overproduction of uric acid: (1) increase in the amount of or activity of PRPP amidotransferase, (2) defects that increase the amount of PRPP or L-glutamine produced, and (3) abnormalities resulting in the decreased concentration of nucleotide inhibitors of amidotransferase.

In addition to overproduction of uric acid, hyperuricemia could also be attributed to diminished renal excretion of uric acid, to a myeloproliferative disorder involving an increase in the turnover of nucleic acids, or to combinations of overproduction and underexcretion of uric acid.

Experience has shown that some metabolic disorders which are clinically indistinguishable can be attributed to different biochemical abnormalities. These are of two types: (1) similar metabolic disorders that involve different enzymes and mutations on different genes (genocopies) and (2) different mutations involving a single gene locus (allelic series).

Mutations causing increased amounts of PRPP

HYPOXANTHINE GUANINE PHOSPHORIBOSYL-TRANSFERASE DEFICIENCIES. Hyperuricemia caused by metabolic dysfunction can be associated with abnormally high levels of PRPP. Underutilization of PRPP by the sal-

vage pathway occurs as a result of a deficiency of hypoxanthine guanine phosphoribosyltransferase (HGPRT), the enzyme which converts hypoxanthine and guanine to IMP and GMP.

Markedly elevated levels of PRPP have been demonstrated in erythrocytes and fibroblasts of patients with HGPRT deficiency, whereas concentrations of adenyl and guanyl ribonucleotides have not been low. Thus, maintenance of catalytically active amidotransferase is probably a result of increased substrate concentration rather than decrease in the feedback inhibitors, GMP and AMP.

The most severe form of the deficiency resulting in nearly complete absence of HGPRT in body tissues (as reported by Seegmiller in 1967) is associated with the *Lesch-Nyhan syndrome.* The syndrome was first described by Lesch and Nyhan in 1964 as involving choreoathetosis, spasticity, mental retardation, and compulsive self-mutilation. An X-linked recessive pattern of inheritance has been suggested by all the pedigrees so far obtained. In addition, cultured maternal fibroblasts of an affected son have been shown to be mosaic for the enzyme deficiency. This finding is consistent with the random inactivation of one X chromosome in females (Lyon hypothesis).

Since 1964 more than 150 individuals with the clinical Lesch-Nyhan features have been identified; all have hyperuricemia. These children excrete three to six times more uric acid daily in their urine than control subjects of comparable weight. This amount is actually two to three times greater per kilogram of body weight than that excreted by patients with severe clinical gout. The first signs of Lesch-Nyhan syndrome usually appear at about 6 months, when the child fails to develop at a normal rate. Orange crystals of uric acid may also be noticed in the diaper. Onset of self-mutilation tendencies may occur early or may be delayed considerably. Aggressive behavior toward oneself and others can be controlled, usually by placing the hands in mittens or splints. Once restrained, patients become relatively calm. Even though these children manifest aggressive tendencies toward others, they are generally apologetic for their actions.

Some Lesch-Nyhan patients later develop gouty arthritis and are susceptible to kidney damage. The extent of renal damage can be reduced by administration of allopurinol, which inhibits the enzyme xanthine oxidase, thus reducing the formation of uric acid from xanthine and hypoxanthine. However, even when started shortly after birth, allopurinol does not prevent the neurological degeneration associated with the disease.

Currently there is no treatment for the neurological disorders associated with the Lesch-Nyhan syndrome. Prenatal diagnosis by amniocentesis is possible, and genetic counseling is very important for support of the patient and family.

Affected children can be identified by urine screening tests and confirmed by blood sample analysis for the enzyme deficiency. Identification is desirable so that affected children can be treated for possible renal dysfunction.

Incomplete deficiencies of HGPRT as a result of different types of mutations in one genome are expressed clinically in less severe forms than the Lesch-Nyhan syndrome. All of the 28 patients reported with partial enzyme deficiency were uric acid overproducers and manifested severe primary gout.

Pedigrees for this partial HGPRT deficiency suggest an X-linked recessive mode of inheritance. The more residual enzyme activity in these patients, the less severe are the clinical Lesch-Nyhan symptoms; and as little as 1 per cent of the normal enzyme concentration in erythrocytes can prevent the severe neurologic disorders. These deficiencies are, however, sufficient to produce marked uric acid overproduction and an early onset of gouty arthritis (average age, 20 years). These patients, who comprise a small part of the total patients with primary gout, usually show more severe joint involvement, renal dysfunction, and kidney stone formation than patients with gout of later onset.

Recently, a very low, but consistently measurable, amount of HGPRT has been found in patients with the Lesch-Nyhan syndrome such that the levels sometimes even overlap those found in patients with gouty arthritis. Despite this overlap, it may still be true that the greater the enzyme activity the more likely will be the tendency to manifest only gouty arthritis and renal dysfunction. It now appears that there is a continuous spectrum of enzyme activity ranging from undetectable to 30 per cent in hemizygotes with HGPRT deficiency. Similar enzyme levels within families and correspondingly similar clinical pictures, and other abnormalities in enzyme activity have led some to suspect that at least 10 to 12 different forms of the mutant gene may exist.

Gout is a chronic disease characterized by acute inflammation, usually in the lower extremities. The inflammatory response is brought on as a result of deposition of needle-shaped crystals of monosodium urate monohydrate in the relatively avascular tissues of tendon and cartilage and in the interstitial tissues of the renal pyramid, due to supersaturation of plasma with urate ion. Diagnosis can be made on the basis of a simple test in which monosodium urate crystals from aspirated joint fluid are viewed by using cross-polarizing filters attached to a light microscope. Recurrent attacks of gouty arthritis not only involve the initially affected joint but also spread to other joints in the upper and lower extremities. The attacks may not subside completely, leaving roentgenographically detectable tophaceous deposits which cause persistent swelling, stiffness, and pain.

Treatment of the symptoms of primary gout includes administration of colchicine, a three-ringed alkaloid, which dramatically reduces joint swelling and tenderness and relieves pain. Continued small doses of colchicine seem to lead to complete, or near complete freedom from further gouty episodes, or at least to reduce markedly the frequency and severity of attacks. In patients with large tophaceous deposits, allopurinol has been shown to dissolve deposits over a period of months and minimize renal stone formation by lowering serum and urine acid levels.

SYNTHETASE MUTANTS. OVERPRODUCTION OF PRPP. Several gouty families who are overproducers of uric acid have been reported to have increased synthesis rates of PRPP. PRPP is produced from ribose-5′-phosphate and adenosine triphosphate (ATP) catalyzed by PRPP synthetase. The enzyme is allosterically activated by inorganic phosphate and inhibited by a number of nucleotides and 2,3-diphosphoglycerate (2,3-DPG). At least two PRPP synthetase mutants have been described in families with increased PRPP synthesis.

In 1971 Sperling showed normal activation of the partially purified enzyme by inorganic phosphate in two brothers with accelerated PRPP synthesis. However, the sensitivity of PRPP synthetase to inhibition by nucleotides and 2,3-DPG was markedly reduced. This indicated that the enzyme was altered structurally rendering it insensitive to inhibitors, hence causing overproduction of PRPP.

Seegmiller examined another family with an accelerated rate of PRPP synthesis and found normal sensitivity to nucleotide inhibitors and hypersensitivity to activation by inorganic phosphate. The mutant enzyme was altered structurally to respond with a two- to threefold increase in activity with normal amounts of phosphate.

Other enzyme deficiencies

Hyperuricemia and gout have also been suspected to be a result of abnormalities in L-glutamine metabolism, since L-glutamine is a second substrate required for the rate limiting reaction in uric acid production. Yu has proposed that reduced activity of glutamate dehydrogenase could cause overproduction of L-glutamine. Mutations in PRPP amidotransferase have also been proposed to explain hyperuricemia although no conclusive studies have yet been conducted. It appears that in many instances primary gout represents a close interaction between environmental and genetic factors which are polygenic in inheritance.

Hyperuricemia secondary to neoplastic disease

Purine biosynthesis may be enhanced as a result of increased utilization of ribonucleotides in proliferative tissue disorders by reducing the concentration of nucleotides inhibiting the amidotransferase reaction. Hyperuricemia can result from spontaneous excessive tissue breakdown or tissue destruction due to cytotoxic or irradiation therapy.

Disorders of amino acid metabolism

Carson and Neill, early in 1960, during a metabolic survey of mentally retarded individuals in Northern Ireland, found two mentally retarded sisters who excreted large amounts of cyst(e)ine in the urine. It eventually developed that homocyst(e)ine rather than cyst(e)ine was being excreted and that in addition, there was elevation of homocyst(e)ine and methionine in the plasma of these two retarded children. It is now established that this abnormality in amino acid metabolism is due to a defect in transsulfuration of methionine to cysteine and is inheritable as an autosomal recessive disorder.

Only approximately half of the patients with this disorder are retarded and many of these retarded individuals will reach adulthood with only mild to moderate intellectual handicap. Clinical features of this disease include, in addition to mental retardation in 50 per cent of the cases, downward dislocation of the ocular lens, malar

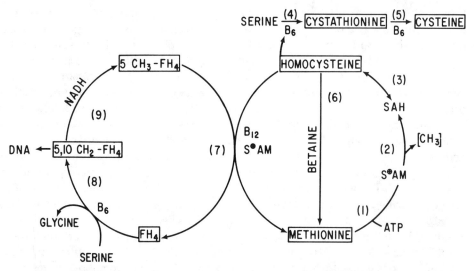

FIG. 18–2. Metabolism of sulfur-containing amino acids. Enzymatic reactions: (1) methionine-activating enzyme; (2) various specific methyltransferases; (3) s-adenosylhomocysteine hydrolase; (4) cystathionine synthase; (5) cystathionase, (6) betaine-homocysteine methyltransferase; (7) ^{5}N-methyltetrahydrofolate methyltransferase; (8) serine hydroxymethyltransferase; (9) ^{5,10}N-methylenetetrahydrofolate reductase. NADH = reduced nicotinamide adenine nucleotide; ATP = adenosine triphosphate; S $\oplus$ AM = s-adenosylmethionine; SAH = s-adenosylhomocysteine; B_6 = pyridoxal phosphate; B_{12} = methylcobalamin; DNA = deoxyribonucleic acid. (After Gaull, J. Pediat. *81:* 1014, 1972.)

flushing, marked thrombotic tendency, and skeletal features similar to Marfan's syndrome. Osteoporosis can be seen so frequently that some clinicians feel that any young individual who presents with osteoporosis should have a urine analysis for homocyst(e)ine excretion.

THE TRANSSULFURATION PATHWAY. The metabolism of sulfur-containing amino acids and the interrelationship to DNA synthesis via B_{12} and folate reactions is shown in Fig. 18–2. Classic homocystinuria results from a deficiency of the enzyme cystathionine synthase (Fig. 18–2, Reaction 4) which condenses serine with homocysteine to form cystathionine. As a result, methionine and homocyst(e)ine accumulate in the blood and are excreted excessively in the urine. Biosynthesis of cystathionine virtually ceases as noted by its absence in brain tissue where it is usually found as a major free amino acid. Mudd and associates have demonstrated that two other enzyme deficiencies may result in homocystinuria:

deficiency of ^{5}N-methyltetrahydrofolate-homocysteine methyltransferase (MTHF methyltransferase) and deficiency of ^{5,10}N–methylenetetrahydrofolate reductase (reductase; Fig. 18–2, Reactions 7 and 9).

There is no specific clinical syndrome associated with MTHF methyltransferase deficiency or with reductase deficiency although muscle weakness, seizures and abnormal electrocardiographic findings are noted. Moreover, these patients with homocystinuria do not resemble patients with cystathionine synthase-deficient homocystinuria. On the other hand, this phenomenon does represent an example of human mutations documented for several different enzymatic steps in a single biochemical pathway.

MTHF methyl transferase is a vitamin B_{12}-linked enzyme, and in this form of homocystinuria, serum methionine is decreased, presumably owing to limited capacity of the alternative pathway for remethylation of homocysteine to methionine through betaine-homocysteine

(BH) methyl transferase (Fig. 18–2, Reaction 6). Cystathionine and homocyst(e)ine accumulate in the blood, probably because the capacity for transsulfuration is not sufficient to accept the excess cystathionine synthesized. Methylmalonicaciduria also occurs because the vitamin B_{12}-linked methylmalonyl CoA carbonylmutase is decreased too. There may be a defect in the activation of vitamin B_{12} to its coenzyme forms, which would account for diminished activity of both known vitamin B_{12}-linked enzymes. In reductase deficiency (Fig. 18–2, Reaction 9) the inability to form ^{5}N-methyl tetrahydrofolate ($5CH_3$-FH_4) results in lack of $5CH_3$-FH_4 for remethylation of homocysteine to methionine. This leads to decreased plasma methionine and increased plasma homocyst(e)ine.

It is of considerable importance to note that the last two steps in transsulfuration of methionine to cysteine (Fig. 18–2, Reactions 4 and 5) are dependent upon another cofactor, vitamin B_6. Clinically, patients with homocystinuria due to cystathionine synthase deficiency may be divided into two groups: those in whom the biochemical defect is reversed by administration of large amounts of vitamin B_6 and those who do not respond to vitamin B_6. Gaull and associates have found increased hepatic synthase activity of 2 to 3 per cent in patients who respond to vitamin B_6. However, the significance of this slight increase is difficult to interpret. Thus, the role of vitamin B_6 in reversing the biochemical abnormality in homocystinuria remains unclear. Studies to date have failed to provide evidence that vitamin B_6 stimulates an alternate pathway for cysteine synthesis via cystathionine (Fig. 18–2, Reaction 4). On the other hand, Morrow and Barness have found evidence that responsiveness to vitamin B_6 may be limited by the folate-B_{12} pathway for remethylation of homocysteine to methionine. This is an interesting example, in a biochemical pathway, of the importance of interrelationships between metabolism of methionine, folate, and B_6.

In some patients with synthase deficiency, serum folate is found to be decreased, possibly because of increased utilization, and the folate deficiency appears during treatment with vitamin B_6. In a small series of patients with cystathionine synthase deficiency and low serum folate, one patient responded to vitamin B_6 but only after supplemental folic acid was given. In another patient unresponsive to B_6, there was demonstrable increased requirement for folic acid. Although these interrelationships have not been completely elucidated, it does seem clear that the clinical response to folate occurs only in combination with vitamin B_6 administration. The importance of interrelated metabolism between sulfur-containing amino acids and folic acid can be extended further. Human fetuses lack cystathionase, the last enzyme in the pathway converting methionine to cysteine (Fig. 18–2, Reaction 5). Specifically, the fetal brain is devoid of this enzyme. This would imply that cysteine is an essential amino acid in fetal tissues, particularly brain, until sometime after birth when cystathionase reaches mature activity. On the other hand, the folate-requiring methyl transferases are active during fetal development but decrease as the fetus approaches term. The lack of cystathionase coupled with increased activity of methyl transferases suggests that the transsulfuration pathway is turned off for further metabolism of homocysteine in favor of the folate-B_{12} remethylation pathway. Possibly this is in order to facilitate de novo synthesis of thymidylate required for more rapid synthesis of DNA during periods of rapid cell growth.

Although it is quite likely that we shall now have a clear understanding of the transsulfuration pathway and its relationship to remethylation via folate-B_{12}, we still have no understanding as to why a block in this pathway results in the pathophysiological changes so characteristic of patients with homocystinuria, namely osteoporosis, lens subluxation, thrombotic tendencies, skeletal abnormalities, and often mental retardation. There has been some evidence to suggest that the thrombotic tendency is related to increased platelet adhesiveness as a result of increased plasma homocyst(e)ine.

An additional abnormality relative to methionine transsulfuration in man is familial cystathioninuria due to deficiency of the enzyme cystathionase (Fig. 18–2, Reaction 5). These patients excrete large amounts of cystathionine in the urine. Excess cystathionine is also found in tissue extracts from cystathioninuric patients. Like homocystinuria, cystathioninuria is inherited as an autosomal recessive. However, in contrast to patients with homocystinuria, there is no increase in urinary homocyst(e)ine. Relatively few patients with this disorder have been described, and no specific clinical syndrome associated with cystathioninuria has been delineated. On the other hand, a variety of clinical findings have been described including mental retardation, multiple congenital defects, thrombocytopenia, and various endocrinopathies, particularly hypothyroidism. Some patients with the biochemical defect have had no demonstrable clinical abnormalities. Since acquired cystathioninuria has been observed in patients with neuroblastoma, this diagnosis should be given consideration in subjects found to have unexplained cystathioninuria. Like homocystinuria, administration of vitamin B_6 in large doses often reverts blood and urine cystathionine levels to normal. Thus the apparent deficiency of liver cystathionase may result from improper binding by the enzyme to the coenzyme, pyridoxal phosphate.

PHENYLALANINE AND TYROSINE METABOLISM. The metabolism of phenylalanine—and subsequently tyrosine—is complicated and, as yet, incompletely understood. Detailed review of the entire metabolic sequence will not be covered here; but several clinically distinct autosomal recessive abnormalities involving this path-

way are worthy of comment and will be considered briefly (Fig. 18–3).

PHENYLKETONURIA. Phenylketonuria (PKU) occurs in approximately 1 in every 43,000 live births. The abnormality results from deficiency of liver phenylalanine hydroxylase and the inability to form tyrosine. Secondarily, there is excessive transamination of phenylalanine to phenylpyruvic acid and other metabolites. There is elevation of plasma phenylalanine and urinary o-hydroxyphenylacetic acid. Clinical symptoms may appear during the first few weeks of life. Excessive irritability, vomiting, and seizures are common symptoms. Generalized eczema or dry skin may occur. Most patients with this disorder have blue eyes, blond hair and fair skin. Untreated, this disorder almost invariably results in irreversible mental retardation. This disastrous outcome may be prevented by judicious and early implementation of a phenylalanine-free diet. Rare cases have been reported of normal untreated individuals with elevated plasma phenylalanine.

GOITROUS CRETINISM. Several categories of cretinism with goiter have been described. More than one of these involves a defect (or several) in conversion of tyrosine to active components of thyroid hormone. Different biochemical variants may be characterized as follows (see also Chap. 16): Type I, failure of iodine trapping by the thyroid gland, salivary glands, and gastric mucosa; Type II, failure of organic binding of trapped iodide, with or without congenital deafness; Type III, failure of iodotyrosine coupling to form thyroxine and triiodothyroxine; Type IV, failure of deiodination of iodotyrosines; and Type V, production of an abnormal iodinated protein by the thyroid gland.

The time and onset of hypothyroidism varies with each of the categories of goitrous cretinism and will depend also upon the severity of the defect. Most commonly, the goiter becomes noticeable during early

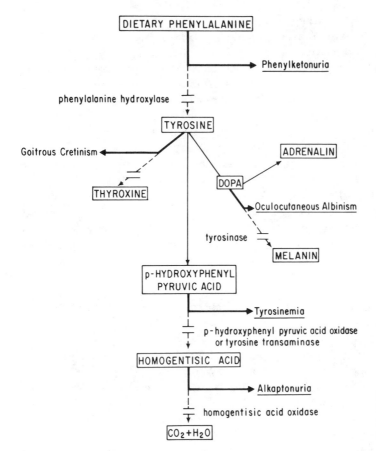

FIG. 18—3. Schematic pathway for phenylalanine and tyrosine metabolism with summary of genetic disorders in this pathway.

childhood or even in infancy. Early clinical manifestations include lethargy, feeding difficulty, constipation, respiratory distress, macroglossia, abdominal distension, hypotonia, dry skin, sparse hair, cretinoid facies, umbilical hernia, growth retardation, delayed skeletal maturation, and slow motor development. Hypothyroidism may be confirmed by a specific test for serum thyroxine. A definitive diagnosis as to specific category requires carefully chosen radioiodine tests. Early diagnosis is imperative, and replacement with adequate doses of thyroid hormone must continue throughout life.

TYROSINEMIA. The most likely pathogenesis of this disorder is a deficiency of p-hydroxyphenylpyruvic acid oxidase or tyrosine transaminase. Biochemical determinations include elevated plasma levels of tyrosine and constant hyperexcretion of p-hydroxyphenyllactic acid, p-hydroxyphenylpyruvic acid, and p-hydroxyphenylacetic acid. In addition, there may be a generalized aminoaciduria, hyperphosphaturia, and a generalized glycosuria. Roentgenograms may demonstrate bone changes consistent with rickets.

Clinically there are two variant patterns:

The acute form (Type I) represents most of the reported cases. Its onset is in early infancy and is manifested by elevated temperature, lethargy, irritability, hepatomegaly (with or without cirrhosis), and failure to thrive.

The chronic form (Type II) afflicts a smaller number of patients. In this chronic form there is more evidence of renal tubular dysfunction and less hepatic disease, although cirrhosis may occur in some cases of chronic tyrosinemia. Mental retardation and neurologic abnormalities are not constant findings for either form of tyrosinemia. Death from the chronic form usually occurs within the first decade of life. Treatment has not been successful, and no long term follow-up is available. However, a low phenylalanine, low tyrosine diet may prove to be of some benefit. At present, it appears that the response to a low phenylalanine diet in tyrosinemia cannot be compared to the response in patients with PKU.

ALKAPTONURIA. Alkaptonuria is due to a deficiency of homogentisic acid oxidase. As a result, homogentisic acid cannot be converted to CO_2 and H_2O and is excreted in the urine. The urine may appear clear when passed, but on standing or becoming alkaline, it turns dark brown or black. Children or young adults may be asymptomatic, but with aging, patients develop pigmentation of the sclerae, cartilage, or other

---

Classification of Pigment Mutations in Man

Oculocutaneous albinism
  Negative hair bulb tyrosine test
    Tyrosinase-negative oculocutaneous albinism*
    Albinism-hemorrhagic diathesis (Hermansky-Pudlak syndrome)
  Positive hair bulb tyrosine test
    Tyrosinase-positive oculocutaneous albinism*
    Chediak-Higashi syndrome
    Hypopigmentation-microphthalmiamia (Cross syndrome)
  Variable hair bulb tyrosine test (yellow type oculocutaneous albinism*)
Ocular albinism (sex-linked ocular albinism)
Cutaneous albinism
  Without deafness
    Piebaldism with white forelock
    Menkes syndrome
    Miscellaneous
  With deafness
    Waardenburg's syndrome
    Cutaneous albinism-hyperpigmentation-deafness—Ziprokowski-Margolis syndrome

---

* The term *albinism* generally refers only to these conditions. *Tyrosinase-negative* refers to lack of pigment formation when fresh, unfixed epilated hair bulbs from albinos are incubated in an L-tyrosine solution. Hair bulbs from tyrosinase-positive albinos do form pigment when this test is performed.

Modes of Inheritance of the Hypopigmentation Syndromes

Autosomal recessive
  Tyrosinase-negative oculocutaneous albinism
  Tyrosinase-positive oculocutaneous albinism
  Yellow type oculocutaneous albinism
  Chediak-Higashi syndrome
Autosomal dominant
  Piebaldism
  Waardenburg's syndrome
Sex-linked (X chromosome)
  Ocular albinism
  Menkes syndrome
  Ziprokowski-Margolis syndrome

---

fibrous tissue. During surgical procedures, tissues may darken upon exposure to air. Some patients develop a distinctive type of arthritis which resembles osteoarthritis roentgenographically but has inflammatory changes suggestive of rheumatoid arthritis. The prognosis for normal life span and intelligence is good. However, chronic arthritis may limit the patient's functional capacity.

ALBINISM. Albinism has been reserved for final discussion because, in addition to the two classic forms of oculocutaneous albinism related to tyrosinase abnormalities (Fig. 18–3), other genetic disorders of pigment are known. They are listed in tabular form, but are not described in detail.

TYROSINASE-NEGATIVE OCULOCUTANEOUS ALBINISM. The classic type of *oculocutaneous albinism* is the *tyrosinase-negative* form. In Negroids and Caucasoids the hair is snow white and the skin pink. The iris is blue or gray; the retina possesses no pigment; and the color of the hair, skin, and eyes does not increase with age. There is a pronounced red reflectance of the fundus of the eye in both children and adults. Further, nystagmus, translucence of the iris, and photophobia are pronounced; and the visual acuity is usually reduced.

TYROSINASE-POSITIVE OCULOCUTANEOUS ALBINISM. Pigmentation of the skin and hair in *tyrosinase-positive oculocutaneous albinism* varies with age and race. Negroids with tyrosinase-positive albinism may be darker than are blond Caucasoids. The hair color is yellow-white in infancy and may change to blond or red with age. The skin is white-cream color. Although a red reflectance is readily demonstrated in infants, it is not often observed in the older children and adults of the heavily pigmented races. The iris color is blue early but brown pigment increases with age. Nystagmus, photophobia, and iris translucence are present but are usually less pronounced than in persons with tyrosinase-negative oculocutaneous albinism. The severity of the eye abnormalities may decrease with age. Pigmented nevi and freckles are common. However, a number of features are common to both types of albinism and include strabismus, horizontal nystagmus, central scotomata, partial aniridia, and absent pigment in the retina.

A third type of albino has been called the *yellow mutant* type. It has been observed in 11 members of an Amish isolate in Indiana, three members of related sibships of Polish extraction in Minnesota, and in two members of isolated sibships. These albinos are born with dead-white hair, reddish-pink skin, and gray-blue eyes. In late infancy and early childhood, their hair develops a bright yellow case and pigment accumulates in the irides. By the age of three they have a distinctive bright yellow to red cast to their hair, and their skin may develop light pigmentation or tan. Photophobia and nystagmus, while present, are less severe than the tyrosinase-positive type. The intelligence of persons with oculocutaneous albinism is usually normal.

Regulation of plasma cholesterol by lipoprotein receptors

It is now recognized that, like single-gene mutations altering biochemical pathways, genetic defects in cell receptors constitute a recognizable category of genetic disorders. For example, receptor defects or receptor resistance to androgens in males may result in variable phenotype expression of male pseudohermaphroditism.

Plasma cholesterol in human beings is regulated by a mechanism for efficient transport. This transport system involves circulating lipoproteins that transport cholesterol in the peripheral circulation and receptors located on the cell surface. The transport system is under genetic control with at least six genetically related defects now known to exist. The effect of these disorders is disruption of the exogenous or endogenous transport system, leading to increased circulating levels of one or more lipoproteins. The six disorders defined thus far account for about 20 per cent of all myocardial infarctions occurring under age 60.

The lipoprotein transport system carries triglycerides and cholesteryl esters. The system is divided into two cycles, exogenous and endogenous; the liver provides overlap between the two systems. Cell utilization requires hydrolysis of triglycerides and cholesteryl esters to fatty acids and unesterified cholesterol. The fatty acids are used for energy in adipose tissue and muscle. Cholesterol is used for synthesis of steroid hormones and bile acids. Unesterified sterol is used as an integral component of plasma membranes. Both endogenous and exogenous cycles begin with secretion of lipoproteins rich in triglycerides, intestinal chylomicrons (exogenous), and hepatic very low density lipoproteins (VLDL) for the endogenous cycle.

In the exogenous cycle, excess phospholipids and free cholesterol on the chylomicrons are transferred to plasma high-density lipoprotein (HDL). *Remnants*, made up of depleted chylomicron are transported to the liver where they bind to receptors on the surface of hepatic cells. Thus, chylomicron metabolism is a two-step pathway consisting of triglyceride removal of extrahepatic tissues and cholesteryl ester uptake in the liver.

Endogenous lipid transport is characterized by conversion of carbohydrates and fatty acids into triglycerides packaged into lipoproteins and transported to adipose tissue. The liver converts lipids into VLDL particles, which transport the lipid for release in capillary beds with catalysis by lipoprotein lipase. As the lipid is lost from VLDL particles, its density decreases, producing an intermediate particle, intermediate-density lipoprotein (IDL). Intermediate-density lipoprotein lipid components, phospholipids, and cholesterol may be transferred to HDL particles. Cholesterol on HDL is esterified to cholesteryl esters, and the esters are transferred back to IDL from HDL. Intermediate-density lipoprotein is converted back to LDL; thus, in normal individuals, LDL carries approximately two-thirds of the total plasma cholesterol. LDL carries cholesterol to hepatic and extrahepatic cells, where it must bind to high-affinity specific receptors on the surface of the plasma membrane.

When dietary cholesterol is adequate and receptor function is normal, the liver uses the exogenous cholesterol as a source for sterol in lipoprotein synthesis. If dietary cholesterol is insufficient and/or an abnormality in receptors exists, then the liver synthesizes its own cholesterol by increased activity of the enzyme 3-hydroxy-3-methylglutaryl coenzyme A reductase (HMG-CoA reductase). In addition, there is increased synthesis of LDL receptors. Normal function of LDL receptors and the incorporation of LDL into the cell is the single most important control mechanism for intracellular cholesterol synthesis by the hepatic and extrahepatic cells. The role of the LDL receptor becomes especially clear in kindreds with familial hypercholesterolemia, an autosomal dominant disorder characterized by elevated plasma cholesterol and fulminant atherosclerotic disease. In this disorder there is a genetic defect in the LDL receptor. Homozygotes with this disease have virtually no LDL receptors, and heterozygotes have approximately half the normal number. LDL degradation thus is blocked, resulting in a large increase in plasma LDL cholesterol. Moreover, failure to suppress HMG-CoA reductase activity because of receptor deficiency leads to increased endogenous synthesis of cholesterol. The net result is deposition of sterol in arterial walls and progressive atherosclerosis.

Familial hypercholesterolemia due to deficiency of LDL receptors is a distinct entity. In addition, two other lipoprotein plasma elevations, chylomicron remnants and IDL, result in progressive atherosclerotic disease. VLDL and chylomicrons appear neutral in this process, neither increasing nor decreasing the atherosclerotic process, and current evidence suggests that one lipoprotein, HDL, may be protective against atherosclerosis.

This understanding of cholesterol and lipoprotein metabolism does not reduce the significance of environmental factors. Excessive weight, cigarette smoking, hypertension, and diabetes mellitus also have significant influence in the process of atherosclerotic disease. The understanding of biological derangement in certain forms of lipoprotein disorders may lead to a more rational approach to therapy while promoting awareness of the importance of environmental factors on pathophysiology of the disease process.

Disease susceptibility and the major histocompatibility system in humans

Probably as much as recombinant DNA technology, our increasing understanding of the association between certain diseases and genetically determined variation in the human histocompatibility region on chromosome 6 will expand our capability for genetic diagnosis and appropriate counseling.

The immune response in man is regulated by a group of genes clustered on the short arm of chromosome 6 (Fig. 18–4). The genes in this cluster encode information for antigens designated human leukocyte antigens (HLA) and from which is derived

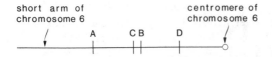

short arm of   centromere of
chromosome 6   chromosome 6

FIG. 18–4. A diagram of the linkage association of the HLA genes on the short arm of chromosome 6.

the HLA system. In addition to the four principal gene regions of the HLA system in chromosome 6 (D, B, C, and A), there are adjacent to the D region structural genes for the second and fourth components of complement and for properdin factor B. The products of the D, B, C, and A regions are antigens; these are glycoproteins which coexist on the cell surface with a β_2-microglobulin. These cell surface antigens are critical to the various functions of the body's T-cell lymphocytes.

Since gene products of the HLA system are antigens and as cell surface markers can be readily assayed on lymphocytes, the association of some HLA genes and specific diseases is identified by an increase in frequency of particular antigens in affected individuals and kindred members over antigen frequency in the general population. Two important observations justify the use of HLA-disease association in genetic evaluation. First is the remarkable polymorphism of genes in the HLA complex. Each of the four major subregions, HLA-D, HLA-B, HLA-C, and HLA-A, has 8 to 39 codominant alleles from which can be derived several million haplotype combinations. Second is the recognition of *linkage disequilibrium*. As already noted, the four major loci of the HLA complex are closely linked yet sufficiently separated on the chromosome for relatively frequent recombination, e.g., 0.8 per cent between HLA-A and HLA-B and 1 per cent between HLA-B and HLA-D/DR. To date, this phenomenon is unique to human genetics. In addition, it appears that alleles of several loci in the same region have a preferential association in the gamete. Thus, linkage disequilibrium exists between these alleles.

For some 40 diseases, the apparent susceptibility has been associated with a particular allele or alleles in the HLA region. In most instances in which the gene location for the disease in question is known, the gene is not located on chromosome 6. For example, a gene for insulin-dependent diabetes mellitus is located on the short arm of chromosome 11. Usually the disease association is with combination of particular alleles of HLA-A, B, or D/DR occurring on the same chromosome more frequently than dictated by chance (linkage disequilibrium).

Idiopathic hemochromatosis has a frequent association with HLA-A3 allele. Sacroilitis and ankylosing spondylitis have the most pronounced association with an HLA allele, HLA-B27. This marker is present in 5 per cent of the general population, but more than 90 per cent of patients with ankylosing spondylitis have this marker.

In general, population studies may be performed by typing a number of unrelated patients with a given disease and comparing the frequencies of various HLA antigens with frequencies observed in a random sample of unrelated healthy controls of the same ethnic origin as the patients. Family studies may be carried out as a form of linkage study. Only affected family members may be considered fully informative, although the rest of the family must be typed to allow HLA genotyping. At least 20 affected sib pairs must be studied before any reasonable level of significance can be reached if all the pairs share parental haplotypes in common.

The largest category of HLA-associated diseases are those in which the primary association is with an HLA-D/DR allele. One allele, HLA-Dw3, is present in approximately 20 per cent of the normal population. Nearly all patients with celiac disease have HLA-Dw3 antigen and almost 50 per cent of patients with thyrotoxicosis carry the same HLA-Dw3 allele.

The use of these genetic associations is advancing our understanding of the genetic mechanisms operative in various forms of diabetes mellitus. Type I diabetes mellitus is also referred to as *insulin-dependent diabetes mellitus (IDDM)*, whereas Type II is *non-insulin–dependent diabetes mellitus (NIDDM)*. Some medical scientists also recognize a subgroup of patients with NIDDM. These are relatively young people, usually in their midtwenties, with mild diabetes mellitus, often called *maturity-onset diabetes of young people (MODY)*. Inheritance for MODY appears to be autosomal dominant. The young affected individuals are not obese and usually do not require insulin. In NIDDM, there is an 80 to 90 per cent concordance in twin studies. NIDDM accounts for 90 to 95 per cent of all diabetes mellitus. Genetic and environmental factors influence recurrence risks, and depending upon the number of affected individuals in a family, the recurrence risk is 10 to 25 per cent. In contrast, twin concordance for IDDM rarely exceeds 50 per cent. The initial lesion in this disease may be a cytotoxic antibody for islet cells. The genetic mechanisms and recurrence risks in IDDM remain controversial, but some patterns are emerging. Thus far, two HLA alleles have been associated with IDDM—B8/B15 and Dw3/Dw4. Heterozygotes for Dw3/Dw4 have the highest association with IDDM. Heterozygosity or homozygosity for B8/B15 carries a similar risk for IDDM, but Dw3/Dw4 heterozygotes carry a greater risk than either Dw3/Dw3 or Dw4/Dw4 homozygotes. At present there may be two distinct forms of IDDM; one associated with Dw3 and the other with Dw4. There is a similar relationship with HLA-B8 (HLA-Dw3) and HLA-B15 (HLA-Dw4). The autoimmune form of IDDM is associated with HLA-Dw3 (HLA-B8). In this disorder there are increased pancreatic islet cell antibodies and a lack of antibody response to exogenous insulin. This form of IDDM accounts for a significant fraction of older patients who develop IDDM.

HLA-Dw4 (HLA-B15) is associated with a form of IDDM with no autoimmune disease

or islet cell antibodies. There is increased frequency of antibody response to exogenous insulin and onset is at an earlier age than in the HLA-Dw3 associated type of IDDM. As noted previously, the gene for insulin is on the short arm of chromosome 11 and the DNA sequence within the insulin gene region is heterogeneous.

Modern approaches to diagnosis, treatment, and prevention of genetic disorders

A brief review of genetic abnormalities involving defects in biochemical pathways, linkage with gene regions of the human immune system, and cell receptor aberrations has been used to exemplify the diversity of inheritable disorders in man. It is not possible within the scope of this text to discuss in similar detail all known inheritable metabolic disorders in man to say nothing of genetic abnormalities in which a specific enzyme defect has not as yet been established (e.g., cystic fibrosis). Even though the list of metabolic genetic diseases increases yearly, we remain woefully ignorant about the pathophysiology of many inborn metabolic errors in man.

During recent years a more systematic understanding of certain inheritable disorders has evolved. For example, delineation of all possible mucopolysaccharide defects is almost complete. Indeed, some are related to more than one enzyme alteration (e.g., Sanfilippo syndrome) and two have been described as having two separate clinical phenotypes (e.g., the sex-linked Hunter's syndrome and autosomal recessive Maroteaux-Lamy syndrome). Similarly, seven clinically distinctive variants of classic Ehlers-Danlos syndrome have been described. Thus, for the practicing physician it becomes increasingly important to be aware of the extensive phenotype and biochemical heterogeneity that is now so clearly defined in genetic disorders.

Amniocentesis

One of the major advances in prevention of specific types of genetic disorders has occurred during the past several years with the development of techniques to detect affected fetuses as early as the second trimester of pregnancy in couples at risk for specific genetic abnormalities. The procedure, amniocentesis, provides fetal cells which are utilized for analysis following a period of growth in culture. Studies thus far indicate that the technical procedure for obtaining these cells carries a low immediate risk for mother and fetus. In terms of results, all chromosome abnormalities, more than 50 biochemical abnormalities, and certain dysmorphological defects (i.e., neural tube disorders) can be diagnosed accurately within the second trimester of pregnancy.

It is beyond the scope of this chapter to deal extensively with all aspects of prenatal detection of genetic disorders (several excellent reviews on this subject are available). Certainly the diagnosis, treatment, genetic counseling, and through amniocentesis the possibility of prevention of certain serious biochemical derangements in subsequent children should be presented to any couple at risk. There are several issues involving genetic disease which virtually all physicians will face to some degree during their practice. Certain genetic disorders such as classic hemophilia, a sex-linked recessive disease, can be successfully controlled by frequent administration of a cryoprecipitate of antihemophilic globulin. Thus, these patients are restored to good health and full fertility with the net potential of doubling the disease frequency in four generations and tripling the frequency in ten. Even in view of the optimistic future of therapy in this disorder, patient counseling and education will be necessary to at least provide knowledge of options available to hemophiliacs regarding gene transmission to subsequent generations.

For a few genetic disorders, it is possible today to consider treatment by gene product (enzyme) replacement, organ and bone marrow transplants, (e.g., Fabry's disease and immune deficiency disorders). Thus far, even the best results have been only transient. Moreover, treatment for any potentially lethal genetic disorders raises serious questions. At the best level possible, such treatment is extravagantly expensive, places undue stress on the entire family, and requires a full-time team of physicians, biochemists, and technicians. This type of patient control is lifelong, and often the patient will nevertheless die at the first crisis in his condition.

Many common genetic disorders, such as cleft lip and palate, pyloric stenosis, dislocation of the hip, and several forms of congenital heart disease, are treatable by conventional forms of surgery. Untreated patients might die (except those with hip dislocation or cleft lip or palate), but now all can potentially lead a normal life and reproduce. For each of these genetic disorders the recurrence risk for children of these patients is approximately 5 per cent which, relatively speaking, is 5 to 20 times greater than that of normal unaffected parents. Although genetic counseling and patient education are important, they may have little effect on family planning, since many patients will feel that corrective surgery was made available to them and therefore would also be provided for their children.

Screening

Screening at birth for biochemical disorders and subsequent treatment also has relevant problems. PKU is considered the model for a biochemical disorder treatable through screening procedures. This disease which, if untreated, leads to severe mental retardation has a birth frequency of 1 in 43,000. A simple blood screening test followed by dietary phenylalanine restriction in positive cases has resulted in normal development in the homozygous affected children. On the other hand, it is now known that a positive screening test does not prove that a baby has PKU. Moreover, PKU variants exist which probably do not cause mental retardation. Several other biochemical disorders—maple syrup urine disease, tyrosinemia, galactosemia, and homocystinuria—can also be detected by neonatal screening tests. However, the physician must have access to skilled consultants who can advise him as to biochemical variations which could lead to inappropriate treatment of an infant who might not need treatment at all.

For the clear-cut biochemical abnormality, again using classic PKU as an example, successful treatment may lead to other kinds of problems and decisions later in life, especially if the patient is female. For example, a female infant is found to have PKU by screening test and is immediately placed on a phenylalanine-free diet. Therapy is successful. The child develops normally and later marries. She is now ready to start her family. After all these years, the physician-geneticist is still an important part of the health care team: the potential mother must be carefully instructed and supervised throughout the pregnancy because her high blood phenylalanine can cross over to her fetus and can cause mental retardation in her children, who are obligate heterozygotes. Hopefully, such education and instruction will have started years before the patient's parenthood. If the patient was no longer on a phenylalanine-free diet, consideration might be given to reinstituting the diet during pregnancy although, as of this date, no relevant studies in this area have been performed.

Genetic counseling

Although chromosomal and biochemical screening for genetic disorders plays an important role in the health care of our society, efforts to prevent most inherited disorders in man rely primarily on education and communication. In practice, this effort comes under the category of genetic counseling and is a rapidly developing discipline, especially in pediatrics, obstetrics, and internal medicine. Genetic counseling remains the most frequently used skill in efforts to prevent genetic disease. By definition, genetic counseling involves the discussion of information regarding recurrence risks for specific genetic disorders and includes the available alternatives to having subsequent affected children. Traditionally, genetic counseling is nondirective. On the other hand, it is equally important that when possible, the patient or family should fully understand the pathophysiology and natural history of a particular genetic disorder. Indeed, this type of information may be far more important and useful to the family, in consideration of future children, than a simple numerical recurrence risk. Evidence to date, suggests that it is the impact of a genetic disease rather than the recurrence risk which usually determines whether a couple will elect to have more children. Thus, it is extremely useful for the physician who provides genetic counseling to have a thorough understanding of the pathophysiology in a particular heritable disorder.

Geneticists are just beginning to study the dynamics of genetic counseling as well as to recognize how little is known about what the counselor is doing or failing to do for his patient. Eventually the disciplines of psychiatry, physiology, biochemistry, and communications must meld with pediatrics, obstetrics and medicine, in order to bring this area of medical care into an ordered discipline that can respond effectively to the health needs of society.

Annotated references

General

Bergsma, D.: Birth Defects, Atlas and Compendium. Baltimore, Williams & Wilkins, 1973. (An excellent reference source for most inheritable metabolic disorders.)

McKusick, V. A.: Mendelian Inheritance in Man, ed. 4. Baltimore, The Johns Hopkins Press, 1975. (A computer-generated data base and references for genetic disorders, including metabolic disease.)

Milunsky, A.: Genetic Disorders and the Fetus. Chap. 7. New York, Plenum, 1979. (A reference text with primary focus on prenatal diagnosis. Includes a section on metabolic disease.)

Stanbury, J. B., Wyngaarden, J. B., and Fredrickson, D. S., (eds.): The Metabolic Basis of Inherited Disease, ed. 4. New York, McGraw-Hill, 1978. (A thorough analysis of most metabolic diseases of man.)

Vogel, F., and Motulsky, A. G.: Human Genetics. Chap. 4. New York, Springer-Verlag, 1979. (An excellent text on human genetics. Chapter 4 includes metabolic diseases.)

Desnick, R. J., and Grabowski, G. A.: Advances in the treatment of inherited metabolic diseases. Adv. Hum. Genet. *11*:281, 1981. (A review of therapeutic strategy in treatment of certain inheritable metabolic diseases.)

Bondy, P. K., and Rosenberg, L. E. (eds.): Metabolic Control of Disease, ed. 8. Philadelphia, W. B. Saunders, 1980. (Another overview of metabolic disease.)

Metabolic defects of purine metabolism

Kelly, W. N., and Wyngaarden, J. B.: The Lesch-Nyhan Syndrome. *In* Stanbury, J. B., Wyngaarden, J. B., and Fredrickson, D. S. (eds.): The Metabolic Basis of Inherited Disease, ed., 4 p. 1011. New York, McGraw-Hill, 1978. (A review of clinical and biochemical abnormalities of this sex-linked disorder.)

Seegmiller, J. E.: Disorders of Purine and Pyrimidine. *In* Bondy, P. K., and Rosenberg, L. E. (eds.): Metabolic Control of Disease. ed. 8, p. 777. W. B. Saunders, 1980. (A clinical and biochemical review.)

Wyngaarden, J. B., and Kelly, W. N.: Gout. *In* Stanbury, J. B., Wyngaarden, J. B., and Fredricksen, D. S. (eds.): The Metabolic Basis of Inherited Disease, ed. 4, p. 916. New York, McGraw-Hill, 1978. (Clinical and biochemical consideration of gout.)

Homocystinuria and cystathioninuria

Carson, N. A. J., and Raine, D. N., (eds.): Inherited Disorders of Sulfur Metabolism. London, Churchill Livingstone, 1971. (An extensive review of metabolic disorders relative to sulfur metabolism.)

Gaull, G. E.: Homocystinuria, vitamin B_6 and folate: Metabolic interrelationships and clinical significance. J. Peds. *81*:1014, 1972. (Emphasizes effect of vitamin B_6 therapy for patients with homocystinuria.)

Rosenberg, L. L.: Vitamin-responsive inherited metabolic disorders. Adv. Hum. Genet. *6*:1–65, 1976. (Reviews vitamin B_6 and homocystinuria as well as several other vitamin-responsive metabolic diseases.)

Harker, L. A., Slichter, S. J., Scott, C. R., and Ross, R.: Homocystinuria: Vascular injury and arterial thrombosis. N. Engl. J. Med. *291*:537, 1974. (A study in the pathophysiology and treatment for a serious complication of homocystinuria in adults.)

Uhlemann, E. R., Tenpas, J. H., Lucky, A. W., Schulman, J. D., Mudd, S. H., and Shulman, H. R.: Platelet survival and morphology in homocystinuria due to cystathionine synthas deficiency. N. Engl. J. Med. *295*:1283, 1976. (A different viewpoint as to the etiology for thrombosis in homocystinuria.)

Metabolic defects of phenylalanine and tyrosine metabolism

Shear, C. S., Nyhan, W. L., and Tocci, P. M.: Tyrosinase and tryosinemia. *In* Nyhan, W. L. (ed.) Amino Acid Metabolism and Genetic Variation. New York, McGraw-Hill, 1967. (A fundamental review of tyrosine metabolism.)

Witkop, C. D.: Albinism. Adv. Hum. Genet. *2*:61, 1971. (An excellent summary of the various metabolic disorders associated with albinism.)

Genetic Screening: Programs, Principles and Research. Committee for the Study of Inborn Errors of Metabolism, National Research Council, B. Childs, Chairman, National Science Foundation. pp. 21–88. Washington D.C., 1975. (Practical considerations for genetic screening programs.)

Levy, H. L., Lenke, R. R., and Crocker A. C., eds.: Maternal PKU: Proceedings of a Conference. DHHS, publication No. (HSA) 81-5299, Rockville, U.S. Dept. Health and Human Services, 1982. (An excellent review of the relatively new clinical issue.)

Prenatal diagnosis

Milunsky, A., ed.: Genetic Disorders and the Fetus. Plenum, New York, 1978. (The entire text focuses on the prenatal diagnosis of genetic disease.)

Antenatal diagnosis: Report of a Consensus Developmental Conference. U.S. DHEW, Public Health Science. National Institute for Health publication No. 79-1973, April 1979. (A summary of predictors of hereditary disease and congenital defects.)

Cholesterol and lipoprotein receptors

Brown, M. S., Kovonen, P. T., and Goldstein, J. L.: Regulation of plasma cholesterol by lipoprotein receptors. Science *212*:628, 1981. (Current knowledge of metabolic regulation of plasma cholesterol.)

HLA and disease

McDevitt, H. O.: Regulation of the immune response by the major histocompatibility system. N. Engl. J. Med. *303*:1514, 1980. (Summarizes regulation of immune response in humans.)

Rotter, J. I.: The modes of inheritance of insulin dependent diabetes mellitus or the genetics of IDDM: No longer a nightmare but still a headache. Am. J. Hum. Genet. *33:835*, 1981. (A current assessment of inheritance for insulin-dependent diabetes mellitus.)

Svejgaard, A., Hauge, M., Jersill, C., Platz, P., Ryder, L. P.,

Nielsen, L. S., and Thomsen, M.: The HLA system. Monogr. Hum. Genet. (Odense), 7: 1–110 1979. (Reviews biology of the HLA system.)

McMichael, A., and McDevitt, H.: The association between the HLA system and disease. Prog. Med. Genet. 2:39, 1977. (An excellent clinical summary.)

19 Endocrine Mechanisms of Reproduction

Mortimer B. Lipsett, M.D.

Reproductive biology encompasses many diverse disciplines—from population dynamics, pregnancy, contraception, through gonadal physiology, and biochemistry of the steroid and protein hormones, to molecular biology. Since each of these topics has been the subject of extensive monographs, coverage in this chapter must necessarily be selective, brief, and somewhat didactic. In a rapidly expanding area of science such as reproductive biology, new hypotheses quickly supplant old ones, and what is fact today may be error tomorrow. With this disclaimer, I shall present fact and informed opinion concerning several aspects of *human* reproductive endocrinology with reference to other species only when evidence from the human studies is not decisive.

It may seem that to discuss reproductive endocrinology without considering sexual behavior in depth is to omit an integral, important, and certainly interesting aspect of the field. But sexual behavior is a sphere in which man clearly differs from all other species. In animals, sex drive and behavior are instinctual and depend on appropriate hormonal signals. In the human, sex behavior is conditioned by other stimuli and is largely independent of these hormones. I shall therefore ignore the behavioral aspects of sex, which have only recently been considered a legitimate topic for study, and concentrate instead on those physiological mechanisms wherein the biologist has been able to quantify response and develop sound concepts.

To the physician the word *sex* denotes differentiation at many levels. To describe the sex of an individual, one should be able to specify genetic sex, gonadal sex, hormonal sex, sex of internal and external genitalia, sex of rearing, and sex of orientation. One or several of these may be discordant, raising difficult therapeutic problems. For the individual to assume a mature and stable role in society, it is necessary for the physician to diagnose and evaluate deviations from the normal pattern and to recommend therapy. Concepts and techniques now at hand make this possible in every instance.

The hormones

Definitions

The terms *androgens, estrogens,* and *progestins* are operational definitions for classes of compounds that exert distinctive biological effects. Thus, any compound that causes growth of the prostate and seminal vesicles in the castrate male is an androgen. Similarly, estrogens cause

423

growth of the uterus and hyperplasia of the endometrium, and progestins transform the endometrium to a secretory type and maintain pregnancy after castration. The steroid chemist has synthesized hundreds of steroids with one or more of these actions and several are in therapeutic use.

The corresponding hormones circulating in the blood are testosterone, estradiol, and progesterone. In addition to these hormones, there are a number of steroids secreted by the endocrine glands that have no intrinsic biological activity but are converted by the liver and other tissues into the hormones. An example of this is the prehormone androstenedione, secreted predominantly by the adrenal cortex. This prehormone has little intrinsic activity in humans but is androgenic because of its transformation to testosterone after secretion and estrogenic because of peripheral conversion to estrone.

Gonadal steroids

As a prelude to the study of the pathophysiology of those endocrine glands important in reproductive biology, the hormones themselves should be considered briefly. The biosynthesis of steroids by the adrenal cortex, testis, ovarian follicle and corpus luteum follows the same basic pattern (Fig. 19–1). Plasma cholesterol, from low-density lipoprotein, is the preferential substrate, although the glands may also use acetate to synthesize cholesterol. This is the sterol converted by mitochondrial enzymes to the first steroid, pregnenolone. Subsequent steps leading to the synthesis of testosterone and estradiol are directed by microsomal enzymes. Steroid biosynthesis also occurs in the placenta, but the pathways slightly differ from those in the gonads and adrenal cortex.

Although the testicular Leydig cells and ovarian follicle cells have the same enzymes, the steroid secretory patterns differ. The testis poorly aromatizes androgens (inserts double bonds in Ring A; Fig. 19–1),

and only small amounts of estradiol are synthesized and secreted. In the ovary, although androstenedione and testosterone are the proximate precursors of estrone and estradiol respectively, little of each androgen is secreted normally. Aromatization of the A-ring of the steroids to produce estrogens proceeds by a series of steps in the ovary. The process of aromatization is irreversible *in vivo*. The ovary and placenta have the highest aromatizing activity, but the testis, adrenal cortex, liver, fat and brain are capable of carrying out this transformation to a limited extent.

The principal precursors of urinary 17-ketosteroids in men and women are the adrenal cortical steroids, dehydroepiandrosterone and its sulfate. Since the liver and other tissues are able to hydroxylate the steroid molecule at several sites, many metabolites of testosterone have been identified. These metabolites are present in small amounts and are of chemical interest, but have little physiological significance. Following reduction of the 3-ketone and oxidation of the 17-β-hydroxyl group of testosterone, the resulting isomeric ketosteroids are conjugated primarily with glucosiduronic acid and to a lesser extent with sulfuric acid. This conjugation reduces metabolic activity and renders the steroids water-soluble so that they are excreted by the kidney.

The principal metabolites of estradiol are estrone, estriol (16 β-hydroxyestradiol) and 2-hydroxyestrone. These steroids, too, are conjugated at the 3-position with glucosiduronic acid; conjugation with sulfuric acid and other compounds also occurs. Over 30 different metabolites of estradiol (all of minor physiological interest) have been identified in urine. Estrone, estradiol, and estriol can be measured although they are normally excreted at rates of less than 10 μg./day. However, 2-hydroxyestrone is easily destroyed; and since this may be the major pathway of metabolism in some patients, it should be recognized that the measurement of the classical urinary estro-

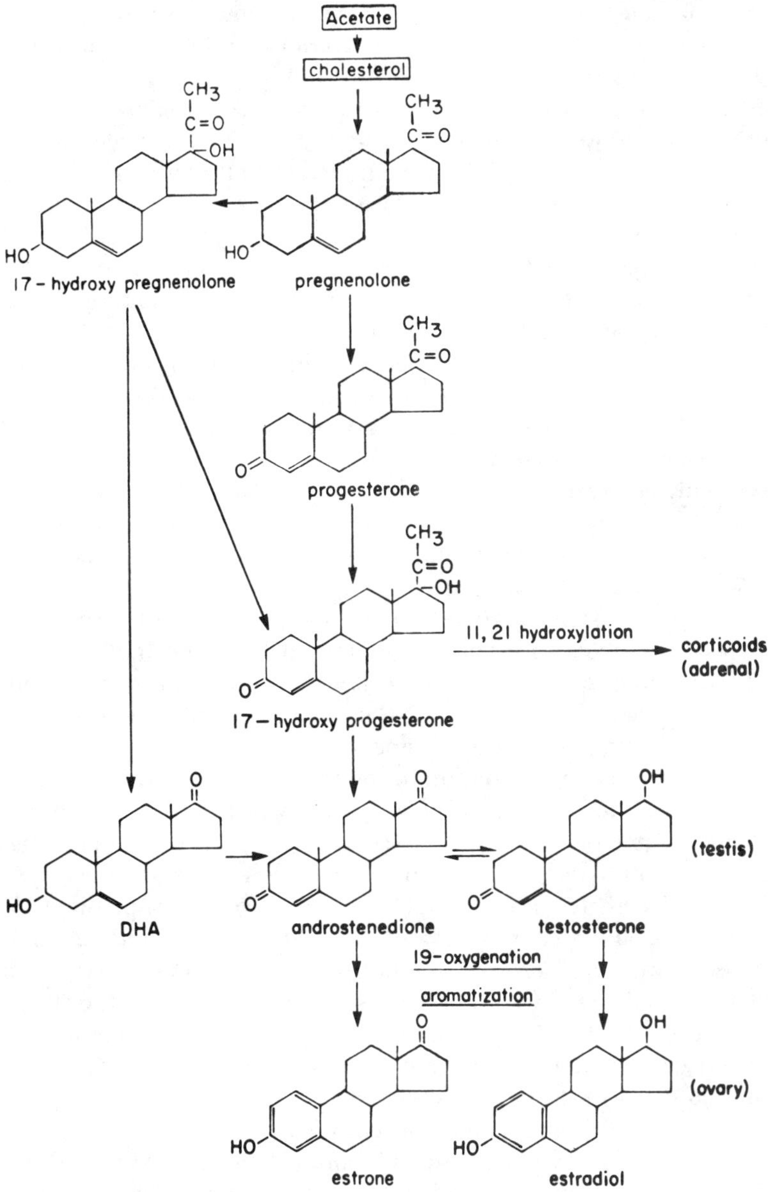

FIG. 19—1. Synthesis of testosterone.

gens may not always accurately reflect estrogen secretion rates.

Protein hormones

The five protein hormones that are of immediate relevance are follicle-stimulating hormone (FSH), luteinizing or interstitial cell-stimulating hormone (LH), human chorionic gonadotropin (hCG), human pla-

cental lactogen (hPL) and prolactin. The first two are synthesized and secreted by the anterior pituitary gland in response to hypothalamic-releasing factors. FSH is necessary for ovarian follicle growth; in the testis, it initiates spermatogenesis and, has a minor role in maintaining spermatogenesis in conjunction with androgen. LH stimulates androgen synthesis by the ovary and testis, initiates ovulation, and is responsible for maintenance of the corpus luteum during the menstrual cycle. hCG and hPL are synthesized by the placenta, by tumors originating from trophoblastic cells, and rarely by other cancers. hPL has a molecular weight of 20,000 to 30,000 and is quantitatively the most important protein synthesized by the placenta late in pregnancy. This hormone is thought to be the critical hormone concerned with fetal growth and development, exerting its influence upon the fetus through alterations in maternal metabolism, notably utilization of fat, thus sparing glucose for fetal consumption. It has activities resembling those of growth hormone and prolactin; in fact, it has considerable immunological and structural similarity to growth hormone.

hCG is the placental gonadotropin (mol. wt. 60,000) whose presence in increased amounts in the urine is responsible for positive pregnancy tests. hCG, FSH, and LH consist of two polypeptide chains, one of which the α-chain is common to the three hormones. Specificity is conferred by the β-chain, a polypeptide unique to each hormone. α- and β-chains have no intrinsic biological activity. hCG is a glycoprotein containing 30 per cent of carbohydrate. At the terminal end of each carbohydrate chain is sialic acid, which confers specific properties on hCG. A high content of sialic acid is associated with a long half-life of the molecule; thus hCG has a half-life of 10 to 20 hours in the circulation. By contrast, FSH and LH have relatively small amounts of sialic acid and are removed quickly from the blood. hCG has the important function of maintaining corpus luteum function and the high rates of progesterone secretion during the first 6 to 8 weeks of pregnancy; its role later in pregnancy is unknown.

Prolactin (mol. wt. 20,000) is the hormone associated with lactation. Although it has a luteotrophic effect in the rodent, it is apparently not involved in the normal regulation of the menstrual cycle in women. When prolactin secretion is elevated, it may exert a profound suppression in gonadotropin release and function.

Prostaglandins

The prostaglandins are 20-carbon fatty acids containing a cyclopentane ring and are synthesized *in vivo* from 20-carbon essential fatty acids by cyclization and oxidation. General descriptions of the effects of the prostaglandins are difficult, since each of the 15 to 20 naturally occurring prostaglandins may have a different effect. For example, two of these, prostaglandin F (PGF) and prostaglandin E (PGE), stimulate uterine contractions, whereas other prostaglandins have opposite effects. Furthermore, the hormonal status of the uterus influences the response to the prostaglandins. Prostaglandins may be involved in the regulation of corpus luteum function by several mechanisms: alterations of blood flow, stimulation of steroidogenesis, and competition with LH for receptors.

Since prostaglandins have been isolated from many tissues, it is difficult to characterize them as hormones. The inhibition of prostaglandin synthesis to relieve dysmenorrhea is a recent example of the practical results of prostaglandin research. Such intriguing findings as the correlation of seminal fluid prostaglandin content with male fertility and the effects of prostaglandins on fallopian tube motility and on ovarian secretory activity continue to make their study an exciting area of research.

Blood levels and transport

The development of relatively simple methods for measuring small amounts of steroid or protein hormones initiated the recent advances in reproductive endocrinology. These radioligand assays for proteins, polypeptides, steroids, and various small molecules have assumed such importance in medicine that an understanding of the principles is necessary.

A radioligand assay requires a binding substance, usually an antibody, with high binding affinity and specificity for the li-

TABLE 19–1
Plasma Steroid and Protein Hormone Concentrations (ng./100 ml.)

	Men	Women		
		Follicular Phase	Luteal Phase	Pregnancy Third Trimester
Testosterone	700	35	40	100
Estradiol	3	5	15	100
Progesterone	30	40	1,500	15×10^3
LH	15	120 ⎫ Peak level	15	—
FSH	15	30 ⎭	12	12
hCG	0	0	0	100×10^3
hPL	0	0	0	600×10^3
Estriol	<1	<1	<1	20×10^3

gand which is chemically similar or identical to the substance to be measured. Pure ligand must be labeled at high specific activity (e.g., tritium-labeled estradiol or ^{131}I-labeled LH). When labeled ligand (*L) is added to a solution containing the binding protein, B, the equilibrium, B + *L $\rightleftharpoons$ B*L is reached; the characteristic association constant is 10^8 to 10^{10}. Since labeled ligand can be displaced from the complex by unlabeled material, the radioactivity not bound to protein will vary with the amount of unlabeled material present. The assayist must then separate *L from B*L, and there are adequate and simple ways of doing this. The sorts of binding proteins that have been used in these measurements are antibodies directed against the protein hormones or against the steroids used as haptenes coupled to a protein, specific intracellular cytoplasmic steroid receptor proteins, and membrane-bound polypeptide hormone receptors. The necessity for the sensitivity achieved by these methods is emphasized by the low plasma concentrations of the protein and steroid hormones (Table 19–1). The use of monoclonal antibodies and several new enzyme-linked methods that alleviate the need for radioactive isotopes are continuing the analytic revolution.

In plasma, testosterone is bound to TeBG, a β-globulin with an association constant for this steroid of about 10^9. This binding protein has moderate specificity, complexing structurally related, but not necessarily biologically active, steroids. By contrast albumin binds steroids but the Ka is only 10^5. At physiological concentrations, most of the testosterone is bound to the specific binding protein and only a small fraction can be considered to be either free or albumin-bound in plasma. Because of the relatively tight binding, protein-bound hormone cannot be metabolized by the cell, nor can it enter cells of target tissues. Since the level of binding protein may be altered by drugs and hormones, knowledge of binding as well as of total steroid hormone concentration may be necessary to appreciate fully alterations in steroid secretion and metabolism. Estradiol has a lesser affinity for TeBG than testosterone and considerably more is bound to albumin.

Mechanism of action

This exciting frontier of endocrinology has seen the greatest progress in defining the way the sex steroids act. The steroids enter the cytoplasm of steroid-responsive tissues and are retained by a specific cytosol steroid-binding protein of high affinity and biological specificity.

Because of the presence of cytoplasmic binding receptors, estradiol and progesterone are retained within responsive cells for prolonged periods without further metabolism. The protein steroid complex then undergoes a temperature-dependent modification and is transported to the nucleus, where it is recognized by specific chromatin-binding sites. This process presumably initiates transcription of new information from the deoxyribonucleic acid (DNA), resulting in synthesis of those structural proteins and enzymes necessary for cell growth and function.

For testosterone to exert its effect on prostate and seminal vesicles, it is first reduced at the $\Delta^{4,5}$-double bond to 5α-dihydrotesterone. The dihydrotestosterone-receptor complex then initiates transcription of DNA with resulting growth and increase in function of the responsive cells. There is inferential evidence that testosterone itself is the proximate androgen in the seminiferous tubule.

These are the barest outlines of the intersection of molecular biology and endocrinology. The mechanisms of certain diseases have already been clarified by these concepts.

Testis

Fetal activity

The H-Y antigen, a protein whose synthesis is directed by a gene on the Y chromosome, organizes the undifferentiated gonad into a testis. The fetal testis then plays an important role during embryogenesis. Leydig cells are prominent by the 12th week of development and secrete testosterone that stimulates the wolffian duct to form the vas deferens, epididymis, and seminal vesicles, but dihydrotestosterone derived from the testosterone induces male external genital development—fusion of the genital folds to form the scrotum, lengthening of the genital tubercle, and fusion of the folds to form the phallus. In addition to testosterone, the fetal testis produces a peptide of unknown structure that suppresses müllerian duct structures which in the female are the anlagen of the uterus, the fallopian tubes, and the upper one third of the vagina (Fig. 19–2). In the absence of or failure to respond to testicular secretions, the genetic male develops a female phenotype.

Patient 1. An 18-year-old girl was referred because of primary amenorrhea. She was tall and well developed, but there was an absence of axillary and pubic hair. There were masses in the labia, the vagina was short and ended in a blind pouch; cervix and uterus could not be palpated. The karyotype was 46 XY. Laparotomy revealed epididymis, testes with abundant Leydig cells, tubules containing only Sertoli cells, vas deferens in the labia, and an absence of müllerian derivatives. The plasma testosterone concentration was 700 ng./100 ml.

This syndrome, termed *testicular feminization*, is characterized by a female phenotype and plasmatestosterone concentrations in the normal male range. These patients have an absence of testosterone receptor, thus preventing its action. This congenital insensitivity to androgen, which existed in fetal life, is a classic experiment of nature, revealing the role of fetal androgen in development and providing a striking demonstration of the many different aspects of sex. Although a normal fetal testis was present, the fetal androgen was ineffective and phenotypic development was therefore female. However, the inducer of müllerian duct regression was secreted and was active so that the uterus and fallopian tubes did not differentiate.

The patient, then, was a genetic, gonadal, and hormonal male; had male internal genitalia but female external genitalia; and was female in sex of rearing and orientation. Such patients are well-adjusted women who are unhappy only about their amenorrhea and their inability to have children. A clinical note—the physician should always refer to these patients as women, for they are women in the most important ways—sex of rearing and of orientation. The fact that they are genetic, gonadal, and hormonal males is of little concern.

Discussion of the nature of the problem with family or patient must be done with tact, understanding, and a careful explanation of the significance of the defect.

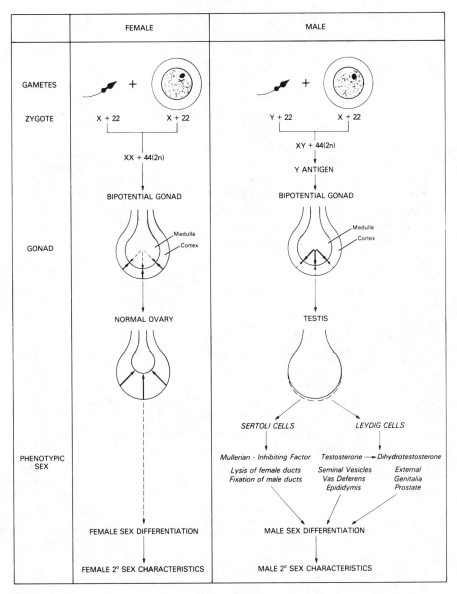

FIG. 19—2. Gonadal and genital development. Note that internal genitalia develop from separate primordia present in both sexes and external genitalia develop in a continuous transformation on anlage common to both sexes, (Lipsett, M. B.: The Testis. *In* Wyngaarden, J. B., and Smith, L. H., Jr. (eds.): Cecil's Textbook of Medicine, ed. 16. Philadelphia, W. B. Saunders, 1982.)

Pubescence

Shortly after birth, Leydig cell function becomes unmeasurable. However, at about 3 months, and for a few months thereafter, testosterone secretion increases. Leydig cell function again becomes quiescent, not to resume until pubescence. In animals and in man there is a gonadal-hypotha-lamic interplay throughout childhood. Thus, gonadectomy at any time causes an increase in the secretion of gonodotropins. Although the mechanism of pubescence is unknown, it is probable that a decreasing sensitivity of the hypothalamic-pituitary system initiates those events associated with pubescence. Through childhood FSH and LH are secreted in small amounts with a predominance of FSH. During pubescence, the levels of these hormones increase, but the rate of increase of LH is three- to fourfold whereas FSH increases about 1.5 times, so that by maturity LH is greater than FSH. At about age 8, the seminiferous tubules begin to enlarge and develop a lumen and subsequently Sertoli cells differentiate. The earliest evidence of pubescence is the increase in testicular volume, since the tubules make up 90 per cent of the volume of the testis. Six months to two years before the first clinical evidence of pubescence, the Leydig cells differentiate from mesenchymal cells under the influence of increasing LH levels, and plasma testosterone begins to increase toward the normal adult levels.

LEYDIG CELLS. The adult testis has two functions, production of the microgamete (the sperm) and secretion of testosterone. The single function of the Leydig cell is the secretion of testosterone and this is regulated by LH via the now classic cyclic AMP mechanism. Virilization requires plasma testosterone concentrations above 250 ng./100 ml.

Patient 2. A 6-year-old boy was referred because of precocious virilization. Increased height, muscular development, facial, and pubic hair, and phallic growth confirmed the impression of increased androgenic effect. The testes, however, were small. Since an increase in testicular volume is a reliable index of FSH effect, this could not be true precocious puberty due to early initiation of gonadotropin (FSH and LH) secretion. Rather, the cause of the virilization was most likely secretion of an androgen from an unusual source, either an adrenal cortical tumor, congenital adrenal hyperplasia, or a testicular tumor.

In adult men, testosterone is secreted by the Leydig cells at a rate of about 7 mg. daily, which maintains a plasma testosterone concentration of about 0.7 μg./100 ml. Testosterone is not a 17-ketosteroid although its principal metabolites are. Since only 40 per cent of testosterone is excreted as urinary 17-ketosteroid, small changes in

this moiety cannot be appreciated within the total 17-ketosteroid excretions of 15 to 20 mg./24 hr. Thus 17-ketosteroid measurement cannot be an adequate index of testosterone production. Further, since the analysis of urinary 17-ketosteroids is relatively imprecise, small changes are unreliable.

Patient 2 (cont'd). The boy's urinary 17-ketosteroid excretion was 3.5 mg./24 hr. But his plasma testosterone level was 0.35 μg./100 ml., sufficient to account for his virilization. To achieve this testosterone level in a 6-year-old boy, a testosterone secretion rate of about 1 mg. daily would be required. This would yield only 0.5 mg. of 17-ketosteroids, an amount within the error limits of the assay. Thus, the apparent paradox of high plasma testosterone levels and low urinary 17-ketosteroids is easily explained. Since patients who are virilized by adrenal tumors or congenital adrenal hyperplasia almost invariably have high urinary 17-ketosteroids, the working diagnosis became a tumor of the testis.

Normally, Leydig cell activity is strictly dependent upon pituitary LH, although endogenous or exogenous hCG can also stimulate the Leydig cells. The secretion of LH is related to Leydig cell activity by a negative feedback system. When free testosterone levels in plasma fall, hypothalamic receptors are activated and LH-releasing factor is secreted, causing the pituitary to release LH, thereby returning plasma testosterone levels toward normal. Conversely, a high dose of androgen will suppress LH secretion and plasma testosterone will decrease.

Patient 2 (cont'd). Since one possible cause of the virilization was the secretion of hCG (placental protein hormone with LH activity) by a trophoblastic tumor, plasma and urinary hCG levels were measured and none was found. The patient was given a synthetic androgen in doses high enough to inhibit LH secretion in normal men and thereby cause a 90 per cent decrease in plasma testosterone. His testosterone level did not change and his low urinary gonadotropins were unaltered. Since functional endocrine tumors are usually autonomous and maintain secretory activity irrespective of normal control mechanisms, it was probable that the virilization was the result of a tumor of the Leydig cells. The testes were explored even though no tumor was palpable. At surgery, a small benign interstitial cell tumor was found and the testis was removed. One week later, plasma testosterone concentration was 0.04 µg./100 ml. and marked regression of virilization was noted 6 months later.

Spermatogenesis is a process as complex as stem cell differentiation and proliferation in the bone marrow. Waves of spermatogenesis begin at different areas within the same tubule so that casual inspection gives an impression of haphazard maturation. The kinetics of spermatogenesis have been studied in detail, and in man development of the mature sperm from the most primitive spermatogonium takes about 70 days. The unique feature about this maturation process is the large segment of 20 to 30 days occupied by the premeiotic prophase of the spermatocyte. The duration of each step in the sequence of spermatogenesis is fixed and neither drug nor alteration of hormonal environment has altered this timing. If FSH is withdrawn, as after hypophysectomy in the rat or development of a pituitary tumor in man, the spermatogonia slowly disappear, the tubules hyalinize and the changes then become irreversible. However, if FSH secretion remains low from childhood, then the tubules remain in their immature state and can be induced to mature at any time by administration of human FSH and hCG to stimulate Leydig cell secretion.

Control of FSH secretion in men is unknown. In general, FSH and LH increase and decrease in parallel; however, the responsivity of the gonadotrophs to gonadotropin-releasing hormone can vary in response to steroid hormones so that LH and FSH secretory rates can diverge. Recent data suggest that the Sertoli cell secretes a low-molecular-weight protein, inhibin, that also modulates pituitary FSH secretion.

Normal spermatogenesis requires high concentrations of testosterone. The concentration of testosterone in the interstitial fluid surrounding the tubules is the same as that of the spermatic venous plasma or 20 to 50 times that of peripheral blood. Thus exogenous androgen at doses suffi-

cient to maintain secondary sexual characteristics will not support normal tubular function.

This physiological feature has been exploited in trials of a male contraceptive. Men were given doses of testosterone large enough to suppress LH and thereby Leydig cell function, and at the same time maintain normal peripheral blood testosterone concentrations. This caused azoospermia.

SECONDARY HYPOGONADISM. *Patient 3.* A 21-year-old man was referred because of sexual immaturity. He had not entered pubescence; there was no beard, the voice was high-pitched, the testes and penis were small, the prostate was not palpable. All tests of pituitary function were normal except that plasma and urinary FSH and LH levels were at or below the lower limits of normal for the prepubescent child. The diagnosis then was either hypogonadotropic hypogonadism or delayed puberty.

In general, pubescence begins by age 12; rarely it does not begin until age 20 or later. If spontaneous pubescence occurs, then the diagnosis must have been only delayed pubescence. At some point, however, a diagnosis of hypogonadotropic hypogonadism must be entertained. It is difficult to distinguish these patients from those with delayed puberty except by the duration of the problem. There is an interesting variant of hypogonadotropic hypogonadism in which the defect is associated with anosmia (Kallman's syndrome); this sensory modality should be tested in any patient, man or woman, with delayed pubescence. Parenthetically, the relationship between the sense of smell and reproductive function is important in many animals.

Patient 3 (cont'd). Because it was imperative for psychological reasons to induce virilization in this 21-year-old man, he was given injections of testosterone enanthate, a long-acting preparation, 200 mg. every 2 weeks. On this regimen he virilized completely, but his testes remained small. He married at age 24, and 3 years later inquired about the possibility of fertility. The measurements of gonadotropins were unchanged. Testicular biopsy showed a few primary spermatogonia and no Leydig cells. There was no tubular fibrosis. The testosterone was stopped, and the patient was given FSH, as human menopausal gonadotropin, 25 IU three times weekly, and HCG, 1,000 IU three times weekly. Ten weeks later sperm were present in the ejaculate, and after 16 weeks the sperm count had reached 3,000,000/mm.[3] The patient's wife became pregnant soon after.

In the previous unstimulated testis, FSH will produce full maturation of the seminiferous tubule in the presence of testosterone. The process of maturation involves differentiation of Sertoli cells, progression of spermatogenesis through meiotic division to mature sperm. Concurrently, elastic fibers become visible in the tunica propria of the tubule and their presence is a sign that pubescence was initiated. It should be noted that the relation between sperm count and fertility is still fuzzy. Below a total sperm count of 60 million there is relative infertility and the risk of infertility increases with decreasing counts. However, absolute infertility does not exist if sperm are present in the ejaculate. Men with hypogonadotropic hypogonadism in whom administration of FSH and HCG have brought the total sperm count to 3 to 10 million are usually fertile. Other factors such as mobility and shape may have considerable prognostic significance.

In the hypogonadotropic states, it is usual to have FSH and LH depressed simultaneously whether the cause be idiopathic, as in Case 3, genetic, as in Kallman's syndrome, or a result of pituitary destruction. Monotropic deficiency of FSH has not been recognized as yet. The reverse situation, a relative deficiency of LH, has been accorded the title of *fertile eunuch syndrome.* In patients with this disorder pubescence fails because of inadequate stimulation of Leydig cell secretions. However, the testes enlarge owing to seminiferous tubule development, and spermatogenesis is complete in a few tubules. Since these men have little libido or potentia, they are fertile in name only. When hCG is given, masculinization ensues and the sperm count improves.

One of the most common symptoms of pituitary tumors in men is loss of libido owing to low LH levels and consequent decreased testosterone secretion. Since only gonadotropin production by the pituitary may be impaired, other pituitary function tests may be normal. Following removal of the tumor and consequent hypopituitarism, androgen deficiency may be replaced by either hCG or a synthetic androgen, and FSH deficiency may be overcome by the use of human menopausal gonadotropin when fertility is desired. In contrast to patients with hypogonadotropic hypogonadism in whom testicular function may always be restored, if pituitary function in normal adult men is interrupted for a prolonged period, irreversible hyalinization and fibrosis of the seminiferous tubules occur. Prolactin-secreting microadenomas of the pituitary may cause infertility and impotence by direct suppression of gonadotropins. Surgical removal of the microadenoma or suppression of prolactin secretion by bromocriptine (Parlodel) will reverse the syndrome.

PRIMARY HYPOGONADISM. *Patient 4.* An 18-year-old man reported for a preinduction physical examination. On casual inspection he was noted to be without facial hair and to have mild gynecomastia. Pubic hair was sparse and the phallus was normal, but the testes were firm and pea-sized. The appearance of gynecomastia and small firm testes alerted the physician to the possibility of Klinefelter's syndrome, the most common cause of primary hypogonadism.

Patients with classical Klinefelter's syndrome, as described in 1942, are men with gynecomastia, aspermatogenesis, increased numbers of apparently normal Leydig cells in clumps and variable eunuchoid features. Total urinary gonadotropins are high, this being due to the invariable secretion of increased amounts of FSH and often increased amounts of LH. In the subsequent years, increasing knowledge of variants of the syndrome and its pathogenesis have changed the definition somewhat. It is accepted now that any man with more than one X chromosome in any tissue has Klinefelter's syndrome. Thus, karyotypes such as 47 (XXY), 48 (XXYY), 48 (XXXY) all define the disease. Rarely, mosaicism may produce an XXY karyotype in the testis only, so that there is disturbance of spermatogenesis alone.

The clinical features of over a thousand patients with this relatively common syndrome have been reviewed. Among the most frequent signs were small testes, azoospermia (no sperm in the ejaculate), and impaired spermatogenesis. Gynecomastia and features of eunuchoidism appeared less frequently.

The histological picture of the testis is essentially normal before pubescence, but the onset of pubescence is associated with hyalinization of the seminiferous tubules and greatly decreased numbers or absence of germinal cells. Elastic fibers do not appear in the tunica propria. The Leydig cells appear hyperplastic and are often clumped. In spite of this, plasma testosterone levels tend to be reduced and respond poorly to HCG. It has been concluded that the hyperplasia is due to continued LH stimulation, but that there is an intrinsic Leydig cell defect that prevents adequate testosterone synthesis.

Patient 4 (cont'd). Plasma testosterone was increased from 0.09 μg./100 ml. to only 0.21 μg./100 ml. after 4 days' stimulation with 4,000 IU of HCG daily. Testicular biopsy showed hyalinization and fibrosis of most of the tubules and a few intact tubules with progression to the spermatocyte stage. The karyotype was 47 (XXY). Because of inadequate virilization, the patient was placed on a depo-testosterone preparation. This caused an increase in beard growth and some deepening of the voice, but the patient never shaved more than twice a week. Curiously, and without adequate explanation, some eunuchoidal patients with Klinefelter's syndrome seem to be resistant to what are usually adequate virilizing doses of androgen.

Primary testicular insufficiency may range from anorchia to instances of spermatogenic arrest. We should note that the presence of normal male internal and external genitalia is adequate evidence that fetal testicular function was normal. Thus, the patient with anorchia suffered the loss of testes after organogenesis had been completed and before birth. The etiology of this syndrome is unknown.

The common cause of male infertility is spermatogenic arrest. This is descriptive of many probably unrelated disorders resulting in inadequate sperm maturation because of arrest of the process at any of several levels. This may not be absolute, and seminal fluid analysis may reveal either azoospermia or low numbers of normal active sperm. The cause is unknown and treatment with gonadotropins or other agents has generally proved unsuccessful. It is important to remember that men with low sperm counts may still be fertile so that a diagnosis of absolute sterility should never be made when sperm are present. Further, evaluation of such men should include at least three semen analyses at 5-day intervals, since sperm counts may fluctuate widely.

Sperm motility is essential for fertilization and several structural and enzymatic defects that affect motility and cause infertility have now been identified.

The ovary

The adult ovary is a complex organ with continually shifting relationships between specific organelles and cell types throughout the menstrual cycle. Thus, each compartment within the ovary must be considered separately. They are the follicle, containing the ovum, the granulosa cells and theca interna cells; the corpus luteum; and the interstitium, composed of stromal cells and hilus (interstitial) cells, the latter resembling the Leydig cells of the testis.

The steroid-synthetic capacities of each compartment have been studied, and the *in vitro* information can be summarized briefly as follows: the interstitium (stromal cells) is active in the synthesis of the C_{19}-steroids such as androstenedione, dehydroepiandrosterone, and testosterone; the

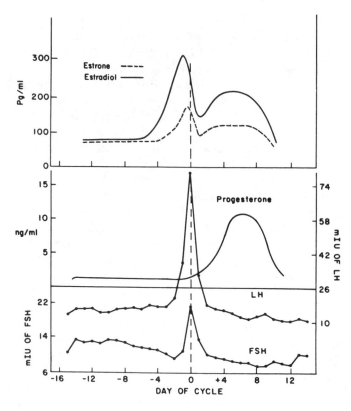

FIG. 19–3. Hormonal patterns in the normal menstrual cycle. The follicular phase lasts from day −16 to day 0; day 0 is the day of the LH peak. The luteal phase lasts from day 0 to the onset of menses, which usually begins on day +14.

corpus luteum synthesizes chiefly progesterone but is also active in estrogen synthesis; the theca interna cells, under the influence of LH, secrete androgens that are aromatized to form estrogens, predominantly in the granulosa cells. During the cycle of follicular development the capacity of the follicular cells to respond varies. Under the influence of estrogen and FSH, LH receptors appear and increase in the granulosa cell. With increasing LH effect and luteinization, FSH receptors decrease markedly. There is thus a continuing necessary sequential interaction among the hormones to achieve normal follicular growth and function.

These activities may be brought into focus by consideration of the events of the normal menstrual cycle (Fig. 19–3). The menstrual cycle has been divided into follicular and luteal phases by the LH peak, and the luteal phase is considered to end with the onset of menstrual bleeding. Toward the end of the luteal phase, the plasma concentration of FSH begins to rise and continues to the peak in the early follicular phase. It is probable that this early increased secretion of FSH is necessary for follicular development, since, if it is prevented, the menstrual cycle is postponed. It has been shown in the monkey that follicular growth begins at this time. Throughout the follicular phase the level of LH rises slowly. During this time, the dominant follicle has been increasing in size and, a few days before the LH surge, the theca interna cells begin to proliferate and take on the appearance of secretory cells. Under the influence of LH, acting in concert with FSH, the theca interna granulosa cell complex increases its secretion of estradiol. As this rises to a peak, it triggers the release of LH, probably through the LH-releasing factor, which then causes ovulation 18 to 36 hours later. The events associated with ovulation induce luteinization of the granulosa cells.

Granulosa cells, removed from the preovulatory follicle and placed in tissue culture, undergo functional and morphological changes compatible with luteinization. This suggests that, once the granulosa cells have matured sufficiently, luteinization is not hormonally directed but is a consequence of follicular rupture. At the time of rupture and follicular disorganization, the secretion rate and plasma level of the estrogens fall, to increase again as the corpus luteum begins to function.

The plasma progesterone concentration remains low until the LH surge and then rises slowly for 2 to 3 days. With full luteinization and development of the corpus luteum, the rate of increase of plasma progesterone is accelerated and it reaches a peak 5 to 8 days after the LH surge. As the progesterone again reaches normal levels the uterine endometrium sloughs and signals the onset of a new cycle.

It is now possible to construct a model of the regulation of follicular growth. The arcuate nucleus in the hypothalamus releases regular pulses of gonadotropin-releasing hormone. The pituitary gonadotrophs release LH and FSH in response to these stimuli. The magnitude of each LH pulse is influenced by estradiol, and when plasma estradiol reaches a threshold of about 150 pg./ml., LH is discharged. The high progesterone levels of the luteal phase reduce the frequency of the pulses of gonadotropin-releasing hormone.

The control of corpus luteum function varies widely among species so that the following comments apply only to women. It is now probable that maintenance of the normal corpus luteum life span of 14 ($\pm$1) days depends upon the continuous presence of LH. In a limited series of studies, it has been found that the life of the corpus luteum cannot be greatly prolonged by LH although hCG will extend it. hCG is secreted by trophoblastic cells of the blastocyst as early as 8 days after ovulation, and is the signal for continuation of corpus luteum function in pregnancy. Although the length of the luteal phase is remarkably constant, there is wide variability of follicular phase duration thus making it extremely difficult to time ovulation solely from the onset of menstrual flow.

In several animals such as the guinea pig and sheep, the presence of a normal uterus causes cessation of corpus luteum function and the start of a new estrus cycle. When the uterus is removed, or when pregnancy ensues, corpus luteum life span is extended. That this effect is local was shown by removal of one uterine horn in the sheep with consequent prolongation of function of the ipsilateral corpus luteum. It has been suggested that in these species prostaglandins may be the uterine factors that regulate corpus luteum function by their effects on local blood flow. Uterine control of luteal function is not important in humans.

However, the increase in uterine prostaglandins occurring during the waning phase of progesterone secretion is responsible for heightened uterine motility and the cramps of dysmenorrhea.

Amenorrhea

In either primary or secondary amenorrhea, the central nervous system, the anterior pituitary gland, or the ovary may be primarily involved. Although the causes may be different, investigation of the patient is similar and necessitates methodical evaluation of each gland. Thus, the integrity of the hypothalamic-pituitary system must be assessed, the adequacy of gonadotropin secretion measured, gonadal function evaluated, and the genetic milieu determined. It should be clear that these are dependent variables (e.g., gonadal dysgenesis due to genetic defects causes low estrogen secretion with consequent high gonadotropin secretion).

PRIMARY AMENORRHEA. The largest proportion of women with this disease have a

45 (XO) karyotype resulting in the typical Turner's syndrome. In the absence of either a second X or a Y chromosome, gonadal differentiation does not occur and the absence of fetal gonadal hormones results in a female phenotype. At the expected time of puberty, plasma LH and FSH levels are elevated because of the absence of normal feedback controls. The gonads in such patients demonstrate undifferentiated stromal tissue with an absence of follicles. With sequential estrogen and progestin therapy, normal menstrual cycles and development of secondary sex characteristics can be produced. The wide variety of associated anomalies in Turner's syndrome affecting skin, muscle, skeleton, kidneys, and cardiovascular system have no ready explanation but give the clue to the existence of Turner's syndrome in the prepubertal girl.

There are many other variants of intersexes that result in gonads without ova. Whether these gonads were destined to be ovaries or testes can only be surmised from the karyotype—a gonad in a woman is an ovary only when ova are present. However, in the absence of follicles, estrogen production remains low and the high gonadotropin levels are again observed.

Ovariectomized women and postmenopausal women have estrone and estradiol in urine and plasma. The facile explanation is that adrenal cortex secretes estrogens, since suppression of adrenal cortical function diminishes estrogen levels. Although it is true that the adrenal cortex does secrete a very small amount of estrogen, most of the estrone is derived from conversion of androstenedione secreted by the adrenal cortex. Thus, estrogen production will be increased when androstenedione secretion is high or the rate of conversion to estrogen is increased. The former situation occurs in feminizing adrenal cortical carcinoma in man; the latter is the cause of some instances of postmenopausal bleeding due to endometrial hyperplasia and occurs in cirrhosis.

Primary amenorrhea may also be caused by failure of the normal process of pubescence, as in the male. Differentiation of delayed pubescence from hypogonadotropic hypogonadism may also be difficult in women, but it is exceedingly rare for pubescence to occur spontaneously after age 18 in women. Anosmia or hyposmia may be a clue to hypothalamic dysfunction. Treatment with estrogen and progestin to achieve sexual maturation should be initiated by age 16. The question of subsequent fertility can be assessed by proving the presence of ovarian follicles. If they are present, then the chances are good that adequate stimulation and ovulation can be achieved later with human menopausal gonadotropin. As in the male, the follicles have an important role in the regulation of FSH secretion. When no follicle is present, as in Turner's syndrome or after the menopause, FSH is high.

SECONDARY AMENORRHEA. The preponderance of women in this category have *psychogenic* or *hypothalamic* amenorrhea, although it is clear that these two general terms do not necessarily imply the same pathophysiological mechanism. Hence, the length of the adjectives is inversely related to our knowledge of this condition. It seems probable, however, that emotional stress of many kinds can suppress the hypothalamic center responsible for cyclic variations in gonadotropins, leaving only constant low levels of gonadotropin secretion. In over 50 per cent of such patients, menses resume spontaneously within a year. However, some have prolonged amenorrhea and induction of ovulation may be necessary to restore menses and fertility.

Amenorrhea is an early sign of pituitary disease due usually to a tumor. Since only gonadotropins may be affected initially and for long periods of time, other tests of pituitary function may not be useful. Further, some small intrasellar tumors may not deform the sella turcica so that secondary amenorrhea is the only clue to their presence. These patients cannot be easily distinguished from those with psychogenic amenorrhea except by the progress of the disease. A failure to respond to administration of LH-releasing hormone is evidence of pituitary disease.

Hyperprolactinemia is associated with about 25 per cent of the cases of secondary amenorrhea, and the measurement of serum prolactin concentration is now the single most important assay to be performed in the woman with secondary amenorrhea. It is often due to a small prolactin-secreting adenoma. Surgical microdissection of the sella turcica permits removal of the tumor without destruction of the remainder of the gland and has led to resumption of menses and fertility. In other variants of the syndrome prolactin secretion is evidently a result of failure of hypothalamic secretion of prolactin-inhibiting factor. Suppression of prolactin secretion by dopamine agonists such as bromocriptine (Parlodel) will induce normal menses, irrespective of the cause of the hyperprolactinemia. Bromocriptine has also been shown to decrease tumor size and has been given safely for up to 10 years. The mechanism of the amenorrhea is direct suppression of gonadotropin secretion by prolactin.

Diseases of the ovaries cause amenorrhea. Follicular atresia may occur earlier than usual so that a postmenopausal ovary can be present in a woman under 40. The specific test for this is an increased plasma FSH, since a decrease in follicular inhibin facilitates FSH secretion.

Excess androgen production (see below) will suppress gonadotropin secretion and thereby induce amenorrhea. The other clinical manifestations will direct the diagnostic efforts.

Finally, the most common cause of secondary amenorrhea is pregnancy. Insistence on a pregnancy test has made many a diagnosis early that would have become apparent later.

INDUCTION OF OVULATION. Several methods of inducing ovulation have become available in the past several years. As in men, injections of human menopausal gonadotropin (primarily FSH) will promote development of the cells surrounding the germ cell derivatives; thus, follicular growth occurs in response to FSH. The adequacy of follicular development can be assessed by measuring urinary estrogen excretion derived presumably from the secretory activity of the increasing mass of theca interna and granulosa cells. At the appropriate time an injection of hCG is given to induce ovulation by simulating the LH surge of the menstrual cycle. When this is successful, fertilization, nidation and pregnancy can result without further medical intervention. Since the amount of FSH given is less than precise, on occasion more than one follicle will mature and superovulation and multiple pregnancies occur.

The second agent that induces ovulation is a very weak synthetic estrogen, clomiphene citrate. This compound acts as a peripheral antagonist of estrogens at any estrogen receptor. When a woman who is producing some estrogens receives clomiphene citrate, the clomiphene competes with estrogen at hypothalamic receptors, thereby inducing the release of FSH and LH, and often initiates follicular development with a subsequent ovulatory menstrual cycle. Occasional hyperstimulation with this agent has also resulted in multiple pregnancies.

OVARIAN VIRILIZING SYNDROMES. Since testosterone and testosterone prehormones are intermediates in the ovarian biosynthesis of estrogens, the occasional role of the ovary is virilizing syndromes is understandable. There are a variety of ovarian tumors, benign and malignant, that cause virilization owing to secretion of testosterone. Hyperplasia of the stromal cells of the ovary is one of the rarer causes of virilization.

Patient 5. A 32-year-old woman was referred because of increasing beard growth, minimal balding, deepening of the voice, acne, and amenorrhea of 6 months' duration. Physical examination disclosed a hirsute woman with increased facial hair, temporal hair recession, loss of female fat contours, and clitoral hypertrophy. There were no signs of Cushing's syndrome. Pelvic and abdominal examination were normal.

The severe virilization suggested that the most likely cause was a tumor of either the adrenal cortex or ovary. Urinary 17-ketosteroids were 12 mg./24 hr., making the diagnosis of adrenal tumor less likely. Plasma testosterone was 0.4 μg./100 ml., a value well within the normal male range. Since a small testosterone-secreting adrenal adenoma had not been excluded and sonography was not diagnostic, bilateral ovarian vein catheterization was performed. A high venous-arterial difference in testosterone concentration was noted on the left side. With this information, the ovaries were examined at surgery and a small yellowish tumor was found. The pathological diagnosis was benign arrhenoblastoma; the ovary was removed, and the patient resumed menses 2 months later.

In virilized women, normal urinary 17-ketosteroid excretion associated with male plasma testosterone levels almost always places the disease in the ovaries. When adrenal cortical hyperplasia or tumor causes virilization, 17-ketosteroid excretion is usually high, owing to the secretion of androstenedione and dehydroepiandrosterone and their subsequent conversion to testosterone. Ovarian virilizing tumors have the capacity to synthesize and secrete testosterone so that the urinary ketosteroids may be the same as in normal man. Tumors of the ovary can cause a high urinary ketosteroid excretion, however, so that this evidence cannot be used to localize the disease to the adrenal cortex. Additional small testosterone-secreting adenomas of the adrenal cortex have been described.

Mild virilization consisting usually of hirsutism alone frequently accompanies secondary amenorrhea, either idiopathic or due to the polycystic ovary syndrome. In each instance it has been shown that testosterone is both secreted by the ovary and produced peripherally from ovarian and adrenal androstenedione. In only a minority of these patients is adrenal cortical hyperactivity of primary importance in the genesis of the increased androgens.

Pulsatile injections of LH-releasing hormone are now being used to induce ovulation. This can provide physiological stimulation of LH release and a subsequent normal ovulation.

Polycystic ovary syndrome (Stein-Leventhal syndrome) is another cause of hirsutism and infertility. Women with this disease have the onset of amenorrhea and mild hirsutism at any time after the menarche through the third decade of life. The hirsutism is generally mild, but in a few patients appreciable beard growth and balding may occur. The ovaries are large and have multiple large cysts lying below a thickened capsule. The cysts are surrounded by luteinized theca cells. Plasma LH is chronically and inappropriately increased, although it can be easily suppressed by exogenous estrogen. The first events leading to the chronic LH secretion and hyperstimulation of ovarian androgen secretion are unknown. Patients with polycystic ovaries ovulate in response to clomiphene, but androgen production is unaltered.

Hirsutism in many women is a familial trait which is unassociated with any significant endocrine abnormality except for menstrual irregularities. The 17-ketosteroid excretion is normal or slightly elevated. Most patients with mild hirsutism have increased testosterone production rates and increased plasma testosterone levels (Table 19–2). To understand the origin of the testosterone, one must consider the origin of testosterone in normal women. About 50 per cent of testosterone is secreted, 15 per cent is derived from peripheral conversion of dehydroepiandrosterone, and the remainder results from transformation of androstenedione secreted by ovary and adrenal cortex. In some women with hirsutism or polycystic ovaries, testosterone production rates are higher and more testosterone is secreted. In addition, androstenedione secretion rates are also increased so that

TABLE 19-2
Testosterone Production Rates in Normal and Hirsute Women

	Plasma T (μg./100 ml.)	MCR$_T$* (l./24 hr.)	BPR$_T$† (μg./24 hr.)	Plasma Δ‡ μg./100 ml.	MCRΔ (l./24 hr.)	BPR† (μg./24 hr.)	$_p$ΔT§	T from Δ‖ (μg./24 hr.)
Normal women	0.04	600	240	0.15	2000	3000	0.4	120
Hirsute women	0.11	1000	1000	0.28	2000	5600	0.04	225

T = testosterone.
* MCR = metabolic clearance rate or the virtual volume of plasma cleared of steroid per day.
† BPR = blood production rate of testosterone or androstenedione, or the amount of testosterone or androstenedione, entering the bloodstream per day.
‡ Δ = androstenedione.
§ $_p$ΔT = fraction of blood androstenedione converted to blood testosterone.
‖ T from Δ = the amount of testosterone entering the circulation from peripheral conversion of plasma androstenedione.

there is an additional increment of plasma testosterone from this prehormone. In these syndromes, the ovary is usually the source of the increased androgen secretion.

In other women with hirsutism, the hair follicle has an increased ability to convert plasma steroids such as androstenedione to dihydrotestosterone. In those cases all measurements in peripheral blood are normal. This type of metabolic alteration in the hair follicle may be the cause of familial and idiopathic hirsutism. Within the next few years effective antiandrogens or inhibitors of the enzyme 5α-reductase will be available to control hirsutism.

The fetoplacental unit

Steroid metabolism

The efforts of Diczfalusy and his collaborators during the past 15 years have clarified many of our concepts of sources and routes of steroid metabolism. This knowledge is of more than theoretical interest, because measurements of steroid and protein hormones can yield information about the health of fetus and placenta.

Estriol excretion increases from 10 μg. per day in the luteal phase to 30 mg. daily during the third trimester of pregnancy (see Fig. 19–4). Other estrogens increase

proportionately. Estriol is synthesized in the placenta mainly from dehydroepiandrosterone sulfate, 70 to 80 per cent of which is secreted by the fetal zone of the adrenal cortex, the remainder coming from the maternal adrenal cortex. In the anencephalic fetus, whose adrenal cortex is atrophic, urinary estriol is low. Since estriol is easily measured in the urine at these high levels, it is feasible to follow women *sequentially* and thus look for alterations in fetal adrenal cortical activity indicating fetal disease. The large variability among women with respect to estriol excretion during the third trimester of pregnancy makes occasional determinations of little use unless the excretion is less than 3 to 4 mg./day. Below this level and depending on the laboratory, fetal death is indicated.

Patient 6. A 23-year-old juvenile diabetic, taking 40 U of NPH insulin daily, was referred for her first pregnancy. At the 30th week of gestation, estriol excretion was 11 mg./day, the mean being 20 mg. This could indicate retardation of fetal growth, but the wide scatter about the normal mean made such deductions hazardous. Weekly urinary estriol excretion continued to increase so that by the 34th week, when fetal viability seemed assured, it was 17 mg./day. At this point, urinary estriol measurements were made biweekly. In the 36th week, there was a fall from 21 to 16 mg./day. This was repeated for the next 3 days and successive values of 16, 14, and 10 mg. were obtained. Since this suggested fetal distress, a cesarean section was performed with delivery of a 5-lb. infant. The placenta showed evidence of segmental infarction.

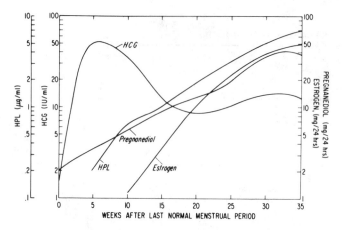

FIG. 19–4. Pattern of hormones throughout pregnancy. Human placental lactogen (hPL) and HCG are given per ml. of serum, estrogen and pregnanediol as urinary excretion per day.

In diseases such as diabetes and toxemia of pregnancy where placental insufficiency may result in decreased nutrition of the fetus and subsequent fetal distress, sequential estriol determination can signal these events before conventional signs such as alterations of fetal heart rate will be apparent. Here, then, is an outstanding example of the adaptation of fundamental knowledge to a clinical test and its translation to normal medical practice.

Progesterone is synthesized by the placenta. Curiously, the placenta cannot perform 17 α-hydroxylation, necessary for biosynthesis of androgens and estrogens. Since the corpus luteum and the placenta produce progesterone, urinary pregnanediol, the major metabolite of progesterone, might be expected to reflect contributions from both sources. The corpus luteum is the major source of progesterone during the first 6 weeks of pregnancy but its function remains low thereafter. If luteectomy is done during the first 6 weeks of pregnancy, there is an almost 100 per cent incidence of abortion; when performed after 20 weeks, abortion does not occur. Since, after 8 weeks' gestation, progesterone is synthesized almost exclusively by the placenta, a drop in progesterone excretion will signify a change in placental function. Death of the fetus does not immediately alter progesterone excretion, only estrogen excretion decreases.

Human placental lactogen (hPL) is secreted primarily into the maternal circulation, and by the 32nd week of pregnancy its blood level may be as high as 0.7 μg./100 ml. It can be detected in maternal plasma as early as the 6th week; and at term, the placenta secretes 300 to 1,000 mg. daily, suggesting that hPL is the principal placental protein synthesized during the last trimester. Of particular significance is the close correlation between placental weight and hPL blood levels. Because of this, plasma hPL levels give information concerning the state of the placenta. For example, low hPL concentrations have indicated retarded fetal growth in patients with diabetes or hypertension. Reduced hPL, associated with bleeding during pregnancy, indicates a threatened abortion. In some patients, subsequent stable or rising levels have given reassurance about the functional integrity of the placenta.

With differentiation of trophoblastic cells, the blastocyst becomes ready for implantation. At this time, hCG is synthesized by the syncytiotrophoblast and is detectable in urine 8 days after ovulation. This timing is critical, since 7 to 8 days after ovulation corpus luteum synthesis of progesterone begins to decline. It is probable that, in man, hCG is the luteotropic hormone responsible for continued function of the corpus luteum.

hCG levels in blood and urine increase

rapidly, reaching a peak of 50 IU/ml. 6 to 8 weeks after ovulation (Fig. 18–4). They slowly decline to levels of about 10 IU/ml. and remain relatively constant throughout pregnancy. Serial measurement of hCG has not been rewarding clinically. It does tend to be elevated in twin pregnancies, but the normal variations are so wide that the predictive value is low.

Contraceptives

This topic is immediately relevant to any discussion of reproductive biology. The complexity of processes of spermatogenesis, maturation of the ovum and ovulation, sperm and ovum transport, fertilization, and nidation offers countless critical points where interference would prevent pregnancy. A variety of mechanical means have been used to prevent fertilization, and the intrauterine device was foreshadowed 1,000 years ago by the insertion of an intrauterine pebble to prevent pregnancy in camels.

The use of the "pill" is, of course, the most important example of hormonal contraception. The commercially available oral contraceptives usually contain the estrogen ethinyl estradiol and a progestational agent. Although increasing doses of estradiol can induce the LH surge and thereby cause ovulation, the combination of a greater than physiological dose of estrogen and a progestational agent will effectively suppress both gonadotropins. Thus, when a normally cycling woman is given oral contraceptives, sequential measurements of LH and FSH show a steady low level of each hormone with elimination of the midcycle peaks.

These agents are remarkably effective, the failure rate being less than 0.5 per cent. This means that there is less than one pregnancy for every 200 woman-years of exposure. The effectiveness of the agents must be balanced against their side effects. Based on retrospective studies, users of the pill are at increased risk for venous throm-bosis and for death from pulmonary embolism. The absolute risk is small, and certainly no greater than that to be expected from the maternal mortality resulting from unplanned births due to the use of less effective contraceptive measures such as the diaphragm. The thromboembolic effects are due to the estrogenic component of the "pill" and it is now clear that the lower the dose of estrogen, the fewer the episodes of thromboembolism.

The oral contraceptives cause a large number of alterations of the normal physiological state; the significance of these is not clear. For example, even at the lowest effective dose of estrogen, plasma proteins such as ceruloplasmin, thyroxine-binding globulin, and cortisol-binding globulin are increased. This is good biological evidence that even 30 μg. of ethinyl estradiol is a greater than physiological amount of estrogen. The level of renin substrate, the α_2-lobulin which releases angiotensin, is uniformly increased, and reversible hypertension has been observed in about 5 per cent of women receiving oral contraceptives.

Plasma triglycerides increase and glucose tolerance is reduced—whether these changes will ultimately accelerate arteriosclerosis or diabetes remains to be seen. From a review of these additional effects, it is apparent that even small excesses of the sex steroids may exert wide-spread effects. The increasing use of combinations containing lower amounts of estrogen has decreased the incidence of serious side effects. There is also evidence that the type of progestational agent may affect plasma high-density lipoprotein (HDL) levels, those related to 19 mm. testosterone decreasing this parameter and the natural progesterone raising it.

Another example of hormonal contraception is the "minipill" containing any of several progestational agents at low doses. During therapy, the hypothalamic-pituitary-ovarian system appears normal by measurement; FSH and LH peaks appear and ovulation occurs. The reason for the effectiveness of low-dose progestins is not clear. Progesterone alters the characteristics of the cervical mucus, changing it from a profuse, watery, low-viscosity fluid favorable for sperm penetration to a scanty, thick secretion that either prevents sperm penetration or decreases sperm viability. Recent decisions about the continued clinical trial of these agents indicate the profound philosophical problems encountered by such regulatory agencies as the Food and Drug Administration. Some progestational agents that were in clinical trial were noted to cause mammary tumors in the beagle, a dog that is particularly susceptible to development of mammary carcinoma. Although the relevance to the human experience is unknown, clinical trials were stopped. However, estrogens have been known for 40 years to cause cancer in susceptible strains of mice and to produce mammary carcinoma in rats. Thus, all the current oral contraceptives contain one agent carcinogenic for at least one animal. There is no evidence that these agents are primary carcinogens, although estrogens may be viewed as promoters of carcinogenesis in their target tissues, uterus and breast.

An exciting development in contraception is the development of long-acting gonadotropin-releasing peptides. These Gn-RH agonists, either by suppressing pulsatile gonadotropin secretion or by causing down-regulation of pituitary receptors, stop ovulation. The agents are effective by nasal insufflation, and clinical trials are underway.

The prostaglandins (PGE$_1$, PGE$_2$ and PGF$_{2a}$) have been shown to be effective abortifacients in the first and second trimester of pregnancy. The mechanism is apparently the stimulation of uterine contractions. Similarly, the prostaglandins have been used to induce labor in women at term. The other effects of prostaglandin such as inhibition of progesterone secretion have not been studied adequately as yet. It should be clear, however, that any compound that prevented synthesis of progesterone for 3 days during the first 6 weeks of pregnancy would terminate that pregnancy, since maintenance of the endometrium of pregnancy requires progesterone.

Annotated references

Baird, D. T., Horton, R., Longcope, C., and Tait, J. F.: Steroid pre-hormones. Perspectives Biol. Med. *11*:384, 1967. (Discussion of endogenous steroids without specific biological activity.)

Bardin, C. W., and Lipsett, M. B.: Testosterone and androstenedione blood production rates in normal women and women with idiopathic hirsutism or polycystic ovaries. J. Clin. Invest. *46*:891, 1967.

Bardin, C. W., Ross, G. T., Rifkind, A.B., Cargille, C. M., and Lipsett, M. B.: Studies of the pituitary-Leydig cell axis in young men with hypogonadotropic hypogonadism and hyposmia: Comparison with normal men, prepuberal boys, and hypopituitary patients. J. Clin. Invest. *48*:2046, 1969. (Discussion of secondary hypogonadism with anosmia, Kallman's syndrome.)

Burger, H., and de Kretser, D.: The Testis. New York, Raven, 1980. (The best monograph on all aspects of testicular function.)

Channing, C. P.: Influence of the *in vivo* and *in vitro* hormonal environment upon luteinization of granulosa cells in tissue culture. Recent Progr. Hormone Res. *26*:589, 1970. (Fascinating tissue culture study which shows granulosa cells undergoing functional and morphological changes compatible with luteinization.)

Crosignani, P. G., and Robyn, C.: Prolactin and Human Reproduction. New York, Academic Press, 1977. (A wealth of information about disorders of prolactin secretion and their effects on the pituitary and ovary.)

Gorski, J., Taft, D., Shyamala, G., Smith, D., and Notides, A.: Hormone receptors: Studies on the interaction of estrogen with the uterus. Rec. Progr. Horm. Res. *24*:45, 1968.

Jensen, E. V., and Jacobson, H. I.: Basic guides to the mechanism of estrogen action. Rec. Progr. Horm. Res. *18*:387, 1962.

Heller, C. G., and Clermont, Y.: Kinetics of the germinal epithelium in man. Rec. Progr. Horm. Res. *20*:545, 1964. (Detailed discussion of spermatogenesis.)

Hillier, S. G.: Regulation of follicular estrogen biosynthesis: A survey of current concepts. J. Endocrinol. 3P (Suppl.) 1981. (A review of concepts and evidence for regulation of follicular estrogen secretion.)

Karolinska Symposia on Research Methods in Reproductive Endocrinology. Immunoassay of Gonadotropins. Diczfalusy, E. (ed.), Stockholm, 1969.

Karolinska Symposia on Research Methods in Reproductive Endocrinology. Steroid Assay by Protein Binding. Diczfalusy, E. (ed.), Geneva, 1970. (The above two references detail the techniques involved in immunoassay and bioassay of hormones.)

Knobil, E.: The neuroendocrine control of the menstrual cycle. Rec. Progr. Horm. Res. *36*:53, 1980. (Modern concepts of pituitary-hypothalamic-ovarian interactions.)

Lipsett, M.B.: Physiology and pathology of the Leydig Cell. N. Engl. J. Med. *303*:682, 1980. (A review of biochemistry and physiology of Leydig cell in humans in health and disease.)

McCullagh, E. P., Beck, J. C., and Schaffenburg, C. A.: A syndrome of eunuchoidism with spermatogenesis, normal urinary FSH, and low or normal ICSH: ("fertile eunuchs"). J. Clin. Endocrinol. *13*:489, 1953.

Paulsen, C. A., Gordon, D. L., Carpenter, R. W., Gandy, H. M., and Drucker, W. D.: Klinefelter's syndrome and its variants: A hormonal and chromosomal study. Rec. Progr. Horm. Res. *24*:321, 1968. (Review of 1,000 patients with this disease.)

Penny, R., Guyda, H. J., Baghdassarian, A., Johanson, J., and Blizzard, R. M.: Correlation of serum follicular-stimulating hormone (FSH) and luteinizing hormone (LH) as measured by radioimmunoassay in disorders of sexual development. J. Clin. Invest. *49*:1847, 1970. (Particular emphasis is directed to primary amenorrhea.)

Ross, G. T., Cargille, C. M., Lipsett, M. B., Rayford, P. L., Marshall, J. R., Strott, C. A., and Rodbard, D.: Pituitary and gonadal hormones in women during spontaneous and induced ovulatory cycles. Rec. Progr. Horm. Res. *26*:1, 1970. (Excellent discussion of the hormonal changes during the normal menstrual cycle.)

Ryan, K. J., and Smith, O. W.: Biogenesis of steroid hormones in the human ovary. Rec. Progr. Horm. Res. *21*:367, 1965. (Discussion of the steroid-synthetic capacities of each compartment of the ovary.)

Salhanick, H. A., Kipnis, D. M., and Vande Wiele, R. L.: Metabolic effects of gonadal hormones and contraceptive steroids. New York, Plenum Press, 1969. (An excellent review of the gamut of side effects from oral contraceptives.)

Saxena, B. N., Emerson, K., Jr., and Selenkow, H. A.: Serum placental lactogen (HPL) levels as an index of placental function. New Engl. J. Med. *281*:225, 1969. (Role of human placental lactogen in pregnancy and fetal growth.)

Southren, A. L.: The syndrome of testicular feminization. Adv. Metabol. Disord. *2*:227, 1965.

Wilson, J. D., George, F. W., and Griffin, J. E.: The hormonal control of sexual development. Science *211*:1278, 1981. (A comprehensive discussion of the role of the fetal hormones in sexual development.)

Yoshimi, T., Strott, C. A., Marshall, J. R., and Lipsett, M. B.: Corpus luteum function during early pregnancy. J. Clin. Endocrinol. *29*:225, 1969. (Study detailing the role of the corpus luteum during pregnancy.)

20   Mechanisms of Bone Disease

Leonard J. Deftos, M.D.
Julie Glowacki, Ph.D.

Normal bone physiology

Bone as a tissue

In most vertebrates the skeleton has two unique and major functions: metabolic and mechanical. The metabolic function consists of storage of minerals that can be mobilized when needed for vital body functions. The mechanism function provides the structural framework for the organism that permits support, locomotion, and protection of organs. Another important, but not unique, function of the skeleton is as a site for hematopoiesis. Bone tissue is well suited to accomplish all these skeletal functions. The mechanical properties of bone are due to the combined properties of the components of its extracellular matrix. The extracellular matrix is composed of organic osteoid (primarily collagen) and an inorganic mineral phase, largely calcium, phosphate, magnesium, and carbonate. The cells of bone are responsible for mediating the direction and rate of flow of minerals in and out. Bone tissue is also rich in nerves, blood vessels, marrow, and fibrous tissue.

The bones of the axial and appendicular skeleton, including all long bones, vertebrae, and ribs, develop as embryonic cartilage rudiments. Conversion of the cartilage to bone, a process called endochondral os-

sification, begins in the rudiment at ossification centers. In man, the first ossification centers appear in the upper limbs between the seventh and twelfth weeks of gestation. Cartilage hypertrophies, becomes mineralized, is invaded by blood vessels, and is resorbed. Proliferation, differentiation, and maturation of cartilage are promoted by growth hormone (via somatomedins), thyroid hormone, vitamins A and C, and possibly vitamin D metabolites. Osteoblasts deposit bone matrix on the cartilage remnants. The primary ossification center elongates and expands concurrent with migration of myeloid elements into vascular spaces. As the secondary ossification centers develop at the ends of bone, zones of cartilage remain at the epiphyseal growth plates, the organization of which permits the orderly lengthening of bones until late adolescence. The organization of bone tissue changes from woven to lamellar orientation as the bone matures. As the spicules of bone, called trabeculae, enlarge, some fuse to form the dense bone of the midshaft.

In contrast to the long bones, the flat bones of the skull, face, and parts of the mandible and clavicle are formed by the process of intramembranous ossification. Clusters of mesenchymal cells differentiate into osteoblasts without a cartilaginous stage. This begins in the human clavicle at 6 weeks. By a programmed process of induction, the cells enlarge, produce alkaline phosphatase, and secrete osteoid matrix. As they become mineralized and grow, these islands of bone fuse into larger trabeculae. This bone has an immature woven architecture until after birth, when it is replaced by lamellar bone by the process of remodeling.

Bone cells

The specific cells of bone are osteoblasts, osteocytes, and osteoclasts. Monocytes/macrophages and mast cells may also serve unique functions in bone.

Osteoblasts are derived from progenitor cells of mesenchymal origin. They are characterized by organelles that are typical of secretory cells, abundant rough endoplasmic reticulum, and a large Golgi zone. Their major functions are to synthesize and secrete collagen and proteoglycan complexes that constitute osteoid and to play a role in matrix mineralization. At least two products of osteoblasts, alkaline phosphatase and bone γ-carboxyglutamic acid-containing protein (BGP), may be involved in mineralization. It is also likely that, because they cover many surfaces of bony trabeculae, osteoblasts may regulate the movement of ions in and out of bone fluid.

When an osteoblast completely surrounds itself with matrix, it becomes an *osteocyte*, residing in a lacuna within the mineralized matrix but maintaining cytoplasmic connections with other osteocytes and with surface osteoblasts. This network of cells and processes in bone canaliculi provides continuity with the vascular circulation. Thus, the main function of the osteocyte is to maintain the nutrition of the bone tissue; in fact, osteocyte death is equivalent to bone death.*

Other functions of osteocytes are disputed. They may be the major cells involved in minute-to-minute mineral homeostasis and mediate this function through the acres of surface area presented by their canalicular connections. Osteocytes may also be capable of bone resorption through a process referred to as osteocytic osteolysis. The primary support for this concept comes from the histological evidence of enlarged lacunae seen in primary and secondary hyperparathyroidism, osteomalacia, thyrotoxicosis, and pregnancy, states characterized by increased bone resorption.

Osteoclasts are large, multinucleated cells found on the surfaces of bone and in Howship's lacunae. They are the major cells responsible for bone resorption and remodeling, containing a variable number of nuclei (often 20 or more) and vacuolated cytoplasm. The cells display a positive reaction for acid phosphatase. Electron microscopic evaluation reveals membrane specializations at the interface of a resorbing osteoclast and the adjacent bone surface. Hundreds of microvilli make up the "ruffled border" or active surface of the osteoclast, where dissolution of the bony matrix occurs. The adjacent bone is partially demineralized and shows frayed strands of collagen fibers.

There is little dispute that osteoclasts originate from blood-borne mononuclear cells; evidence has been provided from chick/quail chimeras, parabiosis experiments, and studies on the etiology and treatment of osteopetrosis. It appears that multinucleated osteoclasts are formed by the differentiation and fusion of monocyte-

---

* The clearest example of this is in bone transplantation. When a segment of bone is harvested for transplantation, its blood supply is interrupted and a few cells survive the procedure. When positioned in the new site, the devitalized implant is gradually reab-

sorbed as new bone grows in slowly from the edges of the defect. This process is called *creeping substitution*. Depending on its size and location, it may require years for the implant to be replaced by viable new bone. Recently, microvascular surgical techniques have been developed to reestablish circulation to the transplanted bone at the time of the operation. These vascularized bone grafts, containing viable osteocytes and osteoblasts, become united to adjacent bone by the same process of union seen in fracture repair.

derived macrophages. In addition to being the precursors of osteoclasts, monocytes themselves are capable of direct resorption of bone. Furthermore, they have been shown to produce bone-resorbing prostaglandins and to be needed for lymphocytic production of osteoclast activating factor (OAF). Although the mechanism of osteoclastic resorption is complex, certain aspects of the process are understood. Resorption of bone requires dissolution of the mineral and hydrolysis of the organic constituents of bone. Collagen fibrils are particularly resistant to lysis by most tissue proteases. This resistance is due to the triple-helical structure of the collagen molecule and intermolecular cross-linking and is enhanced further by fibril calcification. Collagenase is the specific enzyme that cleaves collagen in its helical state, but bone minerals must first be removed before collagenase can act. This local demineralization may be accomplished by accumulation of hydrogen ions or calcium chelators such as citrate. Bone collagenase has been found to act like other mammalian collagenases, cleaving the collagen triple helix at one locus to produce two fragments that are denatured (gelatinized) at body temperature. The nonhelical fragments are thus susceptible to nonspecific proteolytic digestion into small peptides or amino acids. Measurement of serum and urinary levels of hydroxyproline are useful indicators of pathological bone matrix turnover because this amino acid is almost unique to collagen. Proteolytic enzymes can also act to disperse the large proteoglycan complexes that are combined with collagen in osteoid. Acid hydrolases identified in lysosomes of osteoclasts include phosphatase, cathepsins, β-galactosidase, β-glucuronidase, deoxyribonuclease, and N-acetyl-β-glucosaminidase. Much, however, must be learned about the mechanism whereby osteoclasts digest bone.

There is much physiological and pathological evidence that bone resorption and bone synthesis are coupled in normal and pathological states. The only way that bone can grow is by having new bone form on its surface, a process called appositional growth. This is in contrast to interstitial expansion within a tissue, as in cartilage and connective tissue. In order for a long bone to expand its inner marrow space and maintain its shape during the growth of a child, old bone must be removed from the inner surface as new bone is added to its exterior surface. These processes must be coordinated for orderly growth without compromising structural and functional integrity. In some metabolic bone diseases of adulthood, the coupled processes appear to be set at abnormal rates. Paget's disease, for example, is characterized by increased bone resorption and reactive bone formation. The coupling of bone formation and resorption may complicate pharmacological attempts to alter bone metabolism. Changes induced in either bone resorption or formation by pharmacological or pathological means are almost always followed by an adjustment in relative rates to establish a new equilibrium. The mechanism of this coupling may involve chemical factors, cell-to-cell contacts, or piezoelectric phenomena.

Mast cells are associated with bone resorption. They are found in abundance in bones of patients with idiopathic osteoporosis, renal osteodystrophy, and osteoporosis accompanying mastocytosis or urticaria pigmentosa. That the mast cells may actively contribute to lytic bone disease is suggested by two lines of clinical and experimental observations: (1) osteoporosis and cystic bone lesions are seen in patients with mastocytosis or urticaria pigmentosa and (2) heparin, one of the major constituents of the secretory granules of mast cells, has dramatic effects on bone and on collagenase. Heparin has been used clinically since the 1930s to prevent vascular thrombotic episodes. It was later observed that osteoporosis developed in patients on long-term, high-dose heparin regimens. Furthermore, it has been shown that heparin enhances parathyroid hormone (PTH)-stimulated bone reabsorption *in vitro* and bone reabsorption *in vivo* even in parathyroidectomized animals. Heparin stimulates bone collagenase release and apparent activity and thus may be involved in the regulation of resorption.

Organic matrix of bone

The extracellular matrix of mature bone is 5 per cent water, one-third organic components (collagen, proteoglycan complexes, proteins), and two-thirds mineral (see list below).

Composition of Adult Cortical Bone Matrix
(Percentage Dry Weight)

Inorganic (70 per cent)
  Ca^{++}, PO_4^{---}, OH^-
  Sr^{++}, Mg^{++}, Ra^{++}, Pb^{++}, Na^{++}
  F^-, CO_3^{--}
  Citrate, phosphate esters, pyrophosphate

Organic (30 per cent)
  Collagen (85 per cent)
  Noncollagenous (15 per cent)
    Proteoglycans (protein-glycosaminoglycan
      complexes)
      Hyaluronic acid
      Core protein
      Link protein
      Chondroitin sulfates
      Keratan sulfate
    Glycoproteins
    Gamma-carboxyglutamic acid-containing
      protein
    Phosphoproteins
    Bone sialoprotein
    Plasma proteins
    Peptides
    Lipids

COLLAGEN. The organic matrix of bone is 90 per cent collagen, which is synthesized by osteoblasts as soluble polypeptide chains, called proto-α chains (Fig. 20–1). Following translation of the chains, they are modified enzymatically. Specific prolyl and lysyl residues are hydroxylated; this requires vitamin C, α-ketoglutarate, molecular oxygen, and ferrous ion as cofactors.

Certain hydroxylysyl residues are further modified in the endoplasmic reticulum by addition of carbohydrates, commonly glucosylgalactosyl. Three α chains then spontaneously assume the triple-helical configuration of the procollagen molecule. The three chains, each a left-hand helix, are coiled around one another in a right-handed superhelix to form an asymmetric, rope-like structure. The stability of the triple-helical structure at 37°C depends on the facts that: (1) every third amino acid is the small glycine residue which allows close packing of the chains; (2) the high concentration of rigid amino acids, proline and hydroxyproline, restricts rotation of the peptide backbone of the chains; and (3) the posttranslational hydroxylation of proline and lysine enhances interchain bonding. After secretion of procollagen into the extracellular space, small peptides are cleaved from the nonhelical N- and C-termini. The resultant collagen molecules (mol. wt. 300,000) spontaneously aggregate in a quarter-staggered array with the formation of "hole" zones between the head of one molecule and the tail of the next. The earliest deposition of mineral in bone occurs in these hole zones. The fibrils are further strengthened by enzymatic and nonenzymatic covalent cross-linking between collagen molecules. Errors in the complex biosynthesis of collagen can result in bone diseases. For example, in scurvy (vitamin C deficiency) collagen pro-α chains have a low hydroxyproline content and therefore cannot form the stable triple helix. Bone in these patients has poor mechanical properties. Although collagen is a remarkably conserved protein throughout a wide variety of species, molecular heterogeneity does exist in different tissues. Bone collagen is composed of two α chains of the same amino acid sequence and one different chain; the molecule can be represented by the formula $[\alpha 1(I)]_2 \alpha 2$. In contrast, the collagen in cartilage contains three identical chains and is called Type II collagen $[\alpha 1(II)]_3$.

PROTEOGLYCANS. Proteoglycan aggregates are high-molecular-weight substances composed of a long hyaluronic acid backbone and numerous proteoglycan branches. Proteoglycan subunits are composed of a core protein and sulfated and unsulfated glycosaminoglycans. In bone the major glycosaminoglycans are chondroitin 4- and 6-sulfate, keratan sulfate, and dermatan sulfate.

OTHER MATRIX COMPONENTS. Small amounts of glycoproteins, proteins, peptides, and lipids are also found in bone. An acidic sialoprotein, phosphoproteins, and phospholipids have all been implicated in bone mineralization.

The major noncollagenous protein of bone is a vitamin K–dependent, calcium-binding protein containing γ-carboxyglutamic acid (Gla), known as bone Gla protein (BGP), or osteocalcin thought to be produced by the osteoblast. Gla-containing proteins are also found associated with normal and pathological mineralized tissue. They may either be actively involved in mineralization or passively accumulate in the matrix because of their high calcium affinity. BGP can be detected in plasma by radioimmunoassay and its levels reflect the metabolic status of bone. Although not firmly established, plasma BGP levels seem to reflect bone formation directly. Thus, plasma BGP is increased when there is increased bone formation (e.g., in Paget's disease) and decreased when there is decreased bone formation (e.g., following glucocorticoid administration).

Bone mineral

Hydroxyapatite is the crystalline mineral component of bone. The unit cell of the crystal is represented by the formula $Ca_{10}(PO_4)_6(OH)_2$. The minerals precipitate as a poorly crystallized phase, with variable ratios of calcium to phosphorus. This phase "matures" to yield a more ordered structure, but the intermediates, be they amorphous calcium phosphate, dicalcium phosphate, or octacalcium phosphate, are not well defined. Other ions can substitute within the crystal lattice and on its surface. Thus, the inorganic component of bone contains significant amounts of Mg^{++}, Sr^{++}, Ra^{++}, Pb^{++}, Na^+, CO_3^{--}, F^{--}, citrate, and pyrophosphate. In order to understand the mechanisms and regulation of mineral homeostasis, it will be important to resolve questions that remain about the solid phases of calcium phosphate in bone.

The process of calcification *in vivo* is mediated by certain cell types: chondroblasts in cartilage, osteoblasts in bone, and odontoblasts in teeth. There are many factors involved in the mechanism of mineralization including $[Ca^{++}] \times [PO_4^{---}]$, alkaline phosphatase, vitamin D metabolites, and possibly BGP. The normal circulating levels of ionized calcium and phosphate are below the solution ion product necessary for spontaneous crystal nucleation. Mineralization could begin (1) if the concentrations of these ions were elevated locally; (2) if nucleators were present in the bone to overcome unfavorable crystallization conditions; or (3) if endogenous inhibitors of mineralization were removed. All three of the mechanisms are probably involved in normal and pathological mineralization. Once a critical nucleus has been formed, further crystallization can readily occur from $[Ca^{++}] \times [PO_4^{---}]$ ion products at or even below normal serum values.

DNA
  ↓ Transcription
mRNA
  ↓ Translation
Protocollagen
  ↓ Hydroxylation
  ↓ Helix formation
Procollagen
  ↓ Glycosylation
Secretion
  ↓ Procollagen peptidase
Tropocollagen
  ↓
Fibrils
  ↓ Lysyl oxidase
  ↓ Cross-linking
Insoluble fibers

FIG. 20–1. Collagen synthesis and fibril formation.

Alkaline phosphatase, produced by osteoblasts, is considered a marker of active mineralization although its precise role is unclear. Serum levels of alkaline phosphatase are elevated during growth, fracture repair, and diseases that are characterized by increased bone formation. Serum alkaline phosphatase is also increased in diseases characterized by decreased mineralization (e.g., osteomalacia); in these circumstances the increase may be due to underutilization of the enzyme. Furthermore, hypophosphatasia, a rare deficiency of alkaline phosphatase, is accompanied by osteomalacia, that is, inadequate mineralization of osteoid. Alkaline phosphatase may increase the local concentration of free phosphate ions by cleaving organic phosphate esters. The distribution of hydroxyapatite in bone has been deduced from electron microscopic observations; 80 per cent of the mineral crystals are located within the collagen fibrils. In specimens in the early stage of mineralization, electron microscopy reveals that the mineral occurs at axial repeat distances of approximately 700 Å. This corresponds to the periodic hole zones formed by the quarter-staggered model of collagen fibrillogenesis. *In vitro* studies point to collagen as the primary crystal nucleating site. Collagen's role *in vivo* is more difficult to document; at the least, collagen fibrils provide an oriented support for crystal deposition. Other substances have been proposed as endogenous nucleators in bone because of their affinity for calcium or localization in mineralized regions; these include phosphoproteins, phospholipids, bone sialoprotein, and BGP.

Another possible mechanism of mineralization concerns matrix vesicles. Hypertrophic chondrocytes have been shown to slough off membrane-bound cytoplasmic vesicles that deposit mineral within cartilage matrix. The vesicles are also rich in alkaline phosphatase. It is unclear whether matrix vesicles serve as the mediators of mineralization in bone tissue.

Mineralization of osteoid may be promoted by the removal of endogenous ionic and macromolecular inhibitors. The most potent of these inhibitors is pyrophosphate; others are citrate, Mg^{++}, nucleotides, proteoglycan aggregates, lysozyme, and osteocalcin (also referred to as BGP), several of which also enhance nucleation *in vitro*. It should be noted that pyrophosphate or other esterified phosphate inhibitors of mineralization may be the endogenous substrates of alkaline phosphatase. Thus, another role for the enzyme in calcification may be to inactivate endogenous inhibitors of crystal growth. Pyrophosphate consists of two phosphate groups joined together by a common oxygen atom. It accumulates on the surface of hydroxyapatite in vitro and blocks further crystal growth as well as retards crystal dissolution. Diphosphonates—for example, ethane-1-hydroxy-1,1-diphosphonic acid (EHDP) and dichloromethane diphosphonic acid (Cl_2MDP), are synthetic, more stable analogues of pyrophosphate that substitute a carbon atom for the oxygen. The diphosphonates have been shown to be useful therapeutic agents for clinical conditions characterized by excessive bone resorption (e.g., Paget's disease and hypercalcemia of malignancy).

The term *calcification* should not be confused with *ossification*. Ossification means bone formation and refers to the development of osteoblasts and their secretion of the unique matrix of bone. Calcification, or mineralization, refers to the deposition of crystalline calcium salts in a tissue and is only one part of the normal ossification process. Calcification can also be a pathological process if it occurs outside the skeletal system. Heterotopic calcification, or calciphylaxis, can occur in calcinosis universalis or in soft tissues secondary to trauma, tissue necrosis, vitamin D intoxication, hypoparathyroidism, and advanced renal failure. It can also occur in the vasculature (e.g., atherosclerosis). Pathological ossification can occur outside the skeleton

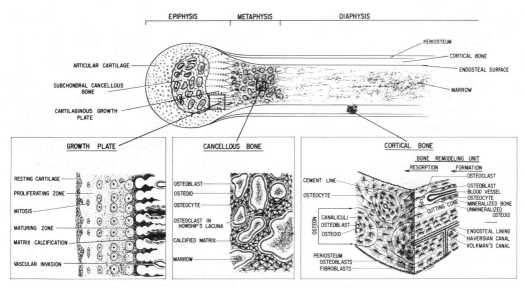

FIG. 20—2. Histological forms of skeletal tissue.

as well, although it is rarer than extraskeletal calcification; examples include myositis ossificans, paraplegia, and occasional abdominal scars. Roentgenograms reveal the presence of radiopaque minerals, be they in osseous, fibrous, or other tissue. Biopsy is necessary to determine whether heterotopic radiodensities represent calcification or ossification.

Bone as an organ

In most vertebrates, bone serves as a reservoir of exchanging and mobilizable mineral that contributes to the mineral homeostasis of the organism. Notable exceptions are the acellular bony fish, which lacks the cellular mechanisms for bone resorption. As organs, bones also function for locomotion and support; as protection for more vulnerable organs such as the brain, spinal cord, and heart; and as sites of hematopoiesis. Depending on the biomechanical forces exerted on them, the relative contribution of different functional demands, and the stage of growth of the organism, bones can have diverse shapes, composition, and organization. The shafts of long bones and the surfaces of flat bones are organized as *compact* bone, also called

cortical bone. The bone in the metaphyses and between cortical surfaces, as in the pelvic bones and the vertebrae, is called *trabecular, cancellous*, or *spongy* bone.

Fig. 20–2 represents a long bone and the different histological forms of skeletal tissue. The cartilaginous growth plate is the essential mechanism for elongation and circumferential expansion of bones by endochondral ossification. The proliferating zone is the region of cell division. This region grows by interstitial growth between cells. This is possible because the plasticity of cartilage matrix allows for expansion and reorganization from within. Cartilage matrix is composed of hydrophilic, chondroitin-sulfate rich proteoglycan complexes and only 30 per cent collagen. Maturation of the epiphysis consists of hypertrophy of the chondrocyte, intracellular accumulation of glycogen, and matrix calcification. Vascular invasion is then accompanied by the initiation of chondrolysis and replacement by bone. The chondrocytes probably degenerate, although there is some experimental evidence that chondroblasts can be modulated to become osteoblasts under the influence of increased oxygen tension. In late adoles-

TABLE 20–1
Factors That Modulate Bone Metabolism

	Resorption	Formation
Increase	Parathyroid hormone	Vitamin D metabolites
	Vitamin D metabolites	Androgens
	Osteoclast activating factor	Parathyroid hormone
	Prostaglandin E$_2$	Fluorides at low doses
	Heparin	Vitamin C
	Vitamin A	Growth hormone
	Endotoxin	Thyroid hormones
	Thyroid hormones	Insulin
	EGF	
	Cholera toxin	
Decrease	Calcitonin	Anticonvulsants
	Aspirin, indomethacin	Diphosphonates
	Mithramycin	Fluorides at high doses
	Diphosphonates	Glucocorticoids
	Glucocorticoids	
	Estrogens	
	Androgens	
	Platelet Factor IV	
	Cartilage-derived anti-invasive factor	
	Colchicine	
	Phentolamine	
	Ionophores	
	Ouabain	

cence, proliferation of the cartilage ceases, and the resultant fusion of epiphyses with the diaphysis into one osseous element brings an end to the active bone growth. Throughout adulthood, however, bone continues to undergo remodeling and metabolic turnover. By this process, calcium exchange in and out of the skeleton amounts to approximately 200 to 300 mg./ day.

The underlying metaphyseal region, composed of trabecular bone, is an area of active remodeling. It has the architecture of a spongy network of interlacing bone spicules, providing strength with lightness. The abundant marrow spaces are lined by endosteal osteoblasts or resting cells. This large surface area accounts for the great

involvement of trabecular bone in mineral homeostasis and bone turnover. Turnover in the human skeleton has been estimated up to 33 per cent per year for trabecular bone, in comparison with 10 per cent for cortical bone. In some bones, termed pneumatic bones, the trabecular cavities are devoid of marrow and filled with air. Examples are the maxillary sinus and the mastoid process of the temporal bone.

The diaphysis or midshaft of bone is surrounded by a periosteal layer that contributes to the widening of a bone by direct appositional ossification (without a cartilaginous stage). Cortical bone is characterized histologically by the organization of cylindrical, axial osteons. Concentric lamellae of bone surround haversian canals, which contain blood vessels and nerves. Haversian systems are continuous with cytoplasmic projections of osteocytes in canaliculi and permit nourishment of the tissue. The cortical shell can be thick, as in long bones, or relatively thin, as in vertebrae or iliac bones.

Regulatory factors (Table 20–1)

BONE METABOLISM AND MINERAL REGULATION. The factors controlling skeletal homeostasis and mineral homeostasis are complex and integrated. Control of blood levels of calcium and phosphate depend on normal function of three organ systems (intestine, kidney, and skeleton) and normal function of three major hormones (vitamin D, parathyroid hormone, and calcitonin). Under normal conditions, the rates and sites of bone formation and resorption are regulated in order to permit mineral homeostasis as well as orderly growth, remodeling, and repair of the skeleton. As demonstrated by the metabolic bone diseases, the structural functions of bone are often sacrificed by the organism's attempt to maintain serum mineral levels. The maintenance of a constant serum concentration of calcium ions is achieved by the integrated actions of PTH, vitamin D, and calcitonin.

PARATHYROID HORMONE. Parathyroid hormone is synthesized in the parathyroid gland in the form of a 115-amino acid precursor designated preproparathyroid hormone (preproPTH). PreproPTH is processed within the gland to the 90-amino acid proparathyroid hormone (proPTH). The proPTH is packaged in secretory granules and is subsequently processed into the biologically active 84-amino acid PTH. There may be further metabolism of PTH 1–84 by the parathyroid glands themselves and hormone fragments may thus be generated. These intraglandular species of PTH are secreted into blood and further metabolized by the liver and kidney. The peripheral metabolism of PTH results in the production of peptide fragments of the molecule. Fragments of the amino-terminal region of the molecule retain biological activity, whereas others are biologically inert. Thus, serum contains a mixture of PTH molecules, including the intact molecule and fragments from various regions of the molecule. The biologically active intact hormone and amino-terminal fragment(s) of PTH are cleared rapidly from the circulation and have a half-life of minutes. Other fragments are cleared more slowly, have half-lives of hours, and consequently are the predominant hormonal forms in blood.

The secretion of PTH is primarily regulated by the level of calcium in the blood. An increase in blood calcium leads to a decrease in PTH secretion and a decrease in blood calcium leads to an increase in PTH secretion. Phosphate and magnesium also influence the secretion of PTH, as do neuroendocrine factors.

The effect of PTH on blood minerals is to increase the concentration of calcium and decrease the concentration of phosphorus. This is a summation of its individual organ effects. Parathyroid hormone increases bone resorption, thereby increasing the release of both calcium and phosphorus from bone. Because PTH also acts to decrease clearance of calcium, the renal effect augments the hypercalcemia. However, because PTH acts to increase renal phosphate clearance, the net effect is to decrease blood phosphate. The direct effect of PTH on gastrointestinal absorption of calcium and phosphorus is not important for mineral regulation; yet, because PTH facilitates the production of $1,25(OH)_2D_3$, it has an important indirect effect of mineral absorption.

Two theories exist regarding the mechanism of action of PTH. Like many other hormones, PTH seems to exert its effect through the adenylate cyclase-cAMP mechanism. However, PTH also acts to affect the permeability of membrane cells to calcium; this translocation of calcium may also be responsible for some of the hormone's actions.

CALCITONIN. Calcitonin is a 32-amino acid peptide secreted by the C-cells of the thyroid gland. Like PTH, biosynthesis of calcitonin occurs through larger precursor forms, and other peptides are coelaborated with the hormone.

Secretion of calcitonin is also regulated by blood calcium concentration, but in a manner opposite that of PTH. An increase in blood calcium increases calcitonin secretion and a decrease in blood calcium decreases calcitonin secretion. In human beings, age and gender influence calcitonin secretion. There is a progressive decrease of calcitonin secretion in both sexes with age. The consequent loss of calcitonin's inhibition of bone resorption may thus contribute to the progressive loss of bone mass that occurs with aging. Because women are calcitonin deficient relative to men, their accelerated loss of bone mass, especially after menopause, may be related to decreased calcitonin reserves. The factors responsible for these age- and gender-related events have not been defined, but estrogens have been implicated through a stimulatory effect on calcitonin secretion. Gastrointestinal factors also play a role in calcitonin secretion, but their physiological importance has not been established.

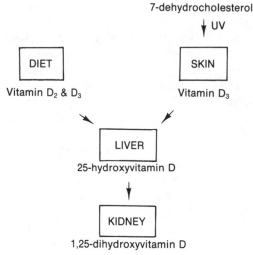

7-dehydrocholesterol

FIG. 20–3. Metabolism of vitamin D.

Calcitonin is metabolized in the kidney and liver. Unlike PTH, no active fragment of calcitonin is known to exist; the entire linear span of the molecule seems to be necessary for biological effects.

The most important action of calcitonin in human physiology is to inhibit the resorption of bone. If the egress of minerals from bone is sufficiently inhibited by calcitonin, hypocalcemia and hypophosphatemia will result. These effects may also be enhanced or produced by calcitonin's action to increase the urinary clearance of calcium and phosphorus. The inhibitory effect of calcitonin on bone resorption can be appreciated best in states of relative calcitonin deficiency and consequent increased bone resorption that occurs with aging, especially in women. Calcitonin may also promote bone formation or translocation of phosphate from blood into tissues, primarily bone. Calcitonin also inhibits a variety of other organ functions but the physiological significance of these is unknown. There are also extrathyroidal forms of calcitonin, but their significance is unknown. Unlike PTH, there is considerable amino acid structure dissimilarity among various species of calcitonin.

VITAMIN D. Vitamin D is synthesized endogenously from cholesterol to cholecalciferol (vitamin D_3) or is derived from the diet as either cholecalciferol or synthetic ergocalciferol (vitamin D_2). When the skin is exposed to sunlight, ultraviolet radiation converts 7-dehydrocholesterol to previtamin D_3, which slowly converts to vitamin D_3 over a period of 2 to 3 days (Fig. 20–3). Endogenous vitamin D, along with that ingested from the diet, is transported on serum binding proteins to the liver, where it is hydroxylated at the C^{25} position to 25-hydroxyvitamin D [25(OH)D]. The serum level of this hydroxymetabolite varies depending on the exposure to sunlight and the dietary intake of vitamin D. 25-Hydroxyvitamin D is transported to the kidney, where it is further hydroxylated, especially during challenges to calcium homeostasis, at the C^1 position to the active hormone $1,25(OH)_2D$. The enzyme that mediates this important hydroxylation step is located in the mitochondria of the renal cortex and is very tightly regulated by hormonal and other factors. The production of $1,25(OH)_2$-D_3 was thought to occur exclusively in the kidney, but extrarenal sites of production have recently been described (e.g., the placenta). $1,25(OH)_2D_3$ acts to promote the gastrointestinal absorption of calcium and phosphorus. This mineral-conserving role is also reflected in the $1,25(OH)_2D_3$ effect to decrease the renal clearance of calcium and phosphorus. In addition to providing minerals for bone formation, there is increasing evidence that $1,25(OH)_2D_3$ acts directly to promote the mineralization of osteoid. It thus seems that the physiological role of $1,25(OH)_2D_3$ is anabolic for bone, whereas its pharmacological effect is catabolic; this hypothesis needs further evaluation. Unlike the peptide hormones, PTH and calcitonin, which act through the second messengers of cAMP and calcium, $1,25(OH)_2D_3$ acts by binding with a specific cytoplasmic receptor and influencing genomic function; there may also be a cell membrane (liponomic) effect of $1,25(OH)_2D_3$.

In addition to the 1α-hydroxylase there exists another renal enzyme, 24-hydroxylase, that converts $25(OH)D_3$ to $24,25(OH)_2D_3$

or converts $1,25(OH)_2D_3$ to the trihydroxy product $1,24,25(OH)_3D_3$. The role of this hydroxylation is not clearly established. However, recent evidence points to the possibility that it represents a deactivation step for the $1,25(OH)_2D_3$ molecule. New metabolites of vitamin D have been described such as calcitroic acid produced from metabolism of $1,25(OH)_2D_3$; $25(OH)D$-26,23-lactone from $25(OH)D_3$; and cholecalcioic acid from $24,25(OH)_2D_3$. Whether these compounds represent catabolic products or have biological roles themselves is as yet unknown.

Synthesis of $1,25(OH)_2D_3$ is stimulated by PTH, low blood calcium, and low blood phosphorus. In addition, there is evidence that growth hormone, prolactin, and calcitonin may also stimulate $1,25(OH)_2D_3$ synthesis. These hormones may mediate in part the adaptation to increased calcium requirements during periods of growth, pregnancy, and lactation. The synthesis of $1,25(OH)_2D_3$, on the other hand, is inhibited by increased blood calcium or blood phosphorus, as well as increased concentrations of $1,25(OH)_2D_3$ itself. Thus, there are important interactions between the metabolism of vitamin D and the other hormones and minerals involved in skeletal homeostasis.

FACTORS THAT INCREASE BONE RESORPTION. The effects of PTH on bone are complex but center around the hormone's role in mineral homeostasis. The initial effect of PTH on bone is to promote the release of calcium from bone mineral into blood. This action is probably mediated through all the cell types in bone—osteoblasts, osteocytes, and osteoclasts. A second effect of PTH is to promote bone resorption by stimulating the development and activity of osteoclasts. Parathyroid hormone also affects the skeletal system by interacting with vitamin D metabolism. It stimulates renal 1-hydroxylase and thus the synthesis of $1,25(OH)_2D_3$. Effects of PTH on the formation of bone are not as clearly understood. Although small doses of PTH both *in vitro* and *in vivo* can increase collagen synthesis

and bone mass, respectively, under certain conditions, the importance of this effect is not known.

In vitro techniques have been developed for measuring bone resorption and its regulation. Viable animal bones prelabeled *in utero* with either ^{45}Ca, ^3H-proline or ^3H-glycine are incubated with test substances. In this manner, direct stimulation of osteoclastic bone resorption has been demonstrated for PTH as well as for OAF, prostaglandins, endotoxin, vitamin D metabolites, heparin, glucocorticoids, and vitamin A. Osteoclast activating factor is a protein produced by activated lymphocytes or malignant cells from patients with myeloma, leukemia, and Burkitt's lymphoma. Such patients often suffer from malignancy-associated hypercalcemia. Steroids can effectively inhibit OAF-mediated bone resorption both *in vitro* and clinically. Monocytes play many roles in bone resorption. They are capable of direct resorption of bone and are the precursors of osteoclasts and are also necessary for lymphocytic production of OAF. They themselves synthesize another bone-resorbing factor, PGE_2. Inhibitors of prostaglandin biosynthesis such as aspirin or indomethacin can inhibit PGE_2-mediated resorption. These inhibitors have been useful for controlling hypercalcemia in patients with PGE_2-secreting tumors.

Endotoxin acts synergistically *in vitro* with PGE_2 and PTH to augment bone resorption. This finding may be of clinical importance in periodontal disease, the rapid, progressive loss of alveolar bone around the teeth.

Vitamin D metabolites stimulate resorption when tested in *in vitro* and some *in vivo* situations. The various metabolites can be graded according to their relative efficacy on bone resorption: $1,25(OH)_2D_3$ > $25(OH)D_3$ > $24,25(OH)_2D_3$ > D_3.

Heparin is another important factor that enhances bone resorption. Heparin has been shown to stimulate osteoclast activity and collagenase activity *in vitro*. Furthermore, development of osteoporosis has been detected in patients on long-term, high-dose heparin regimens (greater than 15,000 units/day for 6 months or more). Heparin, one of the most abundant materials in mast cell secretory granules, could be the mediator of osteolytic disease in patients with mastocytosis.

The dramatic effect of glucocorticoids on bone metabolism is evident from the severe osteoporosis that develops in patients with Cushing's syndrome. Attempts to elucidate the mechanism of this effect reveal the complicated differences between *in vivo* and *in vitro* experiments and the dramatic dose dependence of the actions of adrenal steroids.

Vitamin A excess leads to rapid erosion of the growth plate and cessation of bone growth. There is controversy about the mechanism of these effects. Vitamin A may directly enhance osteoclast number and activity and may also act by stimulating PTH secretion.

FACTORS THAT INCREASE BONE FORMATION. The formation of new bone is a function of the osteoblast. Different groups of factors affect the differentiation, proliferation, and metabolic activity of osteoblasts. Recruitment of progenitor cells into the osteoblast pool depends on growth hormone and its mediators, the somatomedins. Deficiency of these factors results in impaired bone formation. Other factors are essential for anabolic functions of osteoblasts, in particular thyroid hormone and insulin. Another category of factors is needed for the synthesis of specific products of the osteoblasts. For example, vitamin C, or ascorbic acid, is a cofactor in the posttranslational hydroxylation of proline and lysine in collagen. Deficiency of vitamin C (scurvy) results in underhydroxylated collagen and weak fibrils in all connective tissues as well as bone. Thus, scorbutic bone is fragile and subject to fractures and poor healing. Administration of vitamin C readily corrects this condition. In patients with fractures and other injuries, the dietary requirement for vitamin C is raised; this is presumed to be due to increased utilization. Synthesis of BGP depends on adequate vitamin K. Vitamin K is needed for the posttranslational microsomal conversion of three glutamic acid residues to Gla. Vitamin K deficiency is a complication of warfarin therapy and, if administered during gestation, warfarin can produce skeletal abnormalites in the offspring, including stippling of bones and nasal deformities.

In osteomalacia and rickets of vitamin D deficiency, the major pathological lesion is inadequate calcification of the skeleton. It is unsettled whether this can be avoided completely or reversed solely by correction of serum calcium or whether the hydroxylated metabolites of vitamin D directly stimulate bone formation and mineralization. Although bone matrix synthesis exceeds its mineralization in vitamin D deficiency, the rate of osteoid synthesis is reduced from the normal state. Vitamin D repletion results in increased serum calcium, increased alkaline phosphatase activity, increased bone formation, and increased mineralization. The healing of osteomalacia by vitamin D or its metabolites is more uniform than the patchy mineralization that occurs with correction of the hypocalcemia without vitamin D. This issue remains controversial because of different observations made in different species and in different models *in vitro*, but many recent studies support the view that vitamin D metabolites may be essential for normal biosynthesis of bone collagen, BGP, and alkaline phosphatase.

That androgens have bone anabolic effects is suggested by the growth spurt at male puberty, the greater bone mass in adult males than females, and the relative rarity of osteoporosis in elderly men. Increase in bone formation has been documented with testosterone administration for male hypogonadism. The development of anabolic steroids without androgenic side effects may prove helpful in the management of postmenopausal osteoporosis.

Low-dose fluoride therapy has been suggested for osteoporosis because fluoride can increase bone formation. In clinical trials, fracture rate has been reduced in patients receiving fluoride in addition to more conventional therapy (calcium with or without estrogens). Unfortunately, unacceptable side effects have been experienced in nearly half the patients; these include gastrointestinal symptoms, joint pain, and skin rashes.

Parathyroid hormone, when administered at low doses, increases bone synthesis in a variety of experimental systems but has not been clinically evaluated.

FACTORS THAT DECREASE BONE RESORPTION. The major biological effect of calcitonin is to decrease bone resorption by inhibiting osteoclastic activity. Its potent hypocalcemic and hypophosphatemic effects are attributed to reduced release of minerals from bones; by simultaneously inhibiting collagenolysis, calcitonin also decreases excretion of urinary hydroxyproline. Clinical experience with calcitonin in osteoporosis and hypercalcemia of malignancy has shown its efficacy to be transient, due either to "escape" or to development of antibodies against the salmon form of calcitonin.

Aspirin and indomethacin inhibit prostaglandin-mediated bone resorption. Both have been useful clinically in certain types of hypercalcemia of malignancy.

Mithramycin, a cytotoxic antibiotic that is used for testicular tumors, inhibits osteoclastic bone resorption by an unknown mechanism. It is beneficial in Paget's disease and in hypercalcemia, but is not recommended for chronic use because of its side effects including thrombocytopenia, nephrotoxicity, and hepatotoxicity.

Diphosphonates are stable analogues of pyrophosphate which, by coating mineral surfaces of bone, inhibit osteoclastic resorption. Encouraging clinical results have been obtained using diphosphonates in Paget's disease and in hypercalcemia.

Steroids have multiple effects on bone function. Because of their anti–vitamin D actions, they are useful in hypercalcemia of sarcoidosis and hypervitaminosis D. In addition to inhibiting vitamin D-induced bone resorption, steroids also inhibit the effects of vitamin D on intestinal calcium absorption. Osteoclastic stimulation by OAF and by prostaglandins can also be inhibited by steroids. Parathyroid hormone–stimulated bone resorption is unaffected by steroids.

Estrogens inhibit bone reabsorption. It is unlikely that estrogen deficiency is the sole cause of postmenopausal osteoporosis because (1) postmenopausal osteoporotics do not have lower estrogen levels than age-matched women without bone disease and (2) estrogen therapy alone does not prevent or reverse osteoporosis in the long run.

Other factors have been shown to inhibit bone resorption by *in vitro* analysis but have not been used clinically because of side effects or unavailability. These include colchicine, phentolamine, platelet Factor IV, and cartilage-derived antiinvasive factor.

FACTORS THAT DECREASE BONE FORMATION. By interfering with hydroxyapatite crystal growth, the diphosphonates can decrease bone mineralization and are beneficial in heterotopic ossification.

Anticonvulsants, because they interfere with vitamin D metabolism, can produce osteomalacia or rickets. Increased metabolism of $25(OH)D_3$ into more polar inactive forms combined with peripheral resistance to vitamin D may explain in part the development of anticonvulsant bone disease.

The third major inhibitor of mineralization and potential cause of osteomalacia is fluoride in excess. This effect was discovered in regions of endemic fluorosis. Osteomalacia can be prevented in osteoporotic patients on sodium fluoride therapy by restricting the dose to 40 mg./day and by adding oral calcium (1,000 mg.).

Aluminum is another example of the class of agents that inhibit matrix calcification. Excess aluminum accumulates in bone and results in osteomalacia. This is a recently described complication of aluminum contamination of renal dialysate solu-

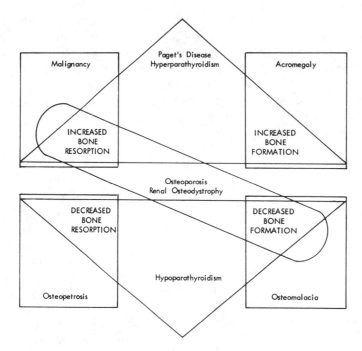

FIG. 20–4. Pathophysiology of bone diseases.

tions and total parenteral nutrition (TPN) fluids.

There is some evidence that glucocorticoids directly decrease matrix synthesis by osteoblasts as well as inhibit the differentiation and maturation of osteoblast precursors. It is important to remember that continuous cortisol treatment in children must be judicious, as it inhibits growth hormone action.

Bone disease

The pathogenesis of most bone diseases is multifactorial. This notwithstanding, bone disease can be classified according to abnormalities in bone resorption or bone formation (Fig. 20–4). Although this approach has didactic value, it should not obscure the complex pathogenetic interactions that take place in skeletal disorders.

Diseases in which the primary abnormality is increased bone resorption

PAGET'S DISEASE. Paget's disease is a disorder of unknown etiology characterized by excessive and abnormal remodeling of bone. The disease usually remains localized to one region of the skeleton such as the skull, pelvis, or ends of long bones, but occasionally becomes widespread. The primary abnormality in Paget's disease seems to be increased bone resorption by osteoclasts. These cells are found in great profusion at resorptive sites, may assume bizarre shapes, and may contain up to 100 nuclei per cell, an amount rarely found in other states with increased osteoclastic activity. The stimulus to increased numbers of osteoclasts has not been identified, although many factors have been suggested. Recently, inclusions that resemble viral nucleocapsids and antigens of the measles and respiratory syncitial virus have been identified in osteoclasts in Paget's disease. The osteoclasts present an advancing resorptive front which can take on a circular form in the skull (osteoporosis circumscripta) and an arrowhead shape in long bones. The resorptive front is followed by intense osteoblastic activity and increased bone formation. The abundant osteoblasts produce immature bone that is characterized by an irregular, nonlamellar pattern referred to as *woven bone*. Osteocytic osteolysis may also contribute to the excessive resorption of Paget's disease. The combination of excessive resorption and formation of bone produces an abnormal histological architecture. The accelerated remodeling produces incomplete osteons and irregular cement lines, which result in a "mosaic" histological pattern. Abnormal pagetic bone also has abundant vascular and fibrous tissues during this active phase of the disease. As the disease progresses, remodeling slows down and the disease process "burns out." Dense, sclerotic bone remains, with little evidence of cellular activity.

The clinical and biochemical consequences of Paget's disease can be related to these underlying pathological processes which produce the abnormal structure of pagetic bone. The architecture of pagetic bone is not well suited for mechanical support. Bone pain is common and the bone is subject to deformity and fracture. Although fracture healing appears to be normal, the mechanical and vascular properties of pagetic bone makes orthopedic procedures difficult. The mechanical stress of bony deformities contributes to the development of degenerative arthritis. The most dreaded complication of Paget's disease is the development of an osteogenic sarcoma, but fortunately this is very rare. A bony deformity at an anatomically critical site can produce serious effects. For example, bony sclerosis and deformity (platybasia) at the base of the skull can produce posterior fossa compression and impinge on the spinal tracts to produce neurological abnormalities. Compromised cranial foramina can interfere with the function of transversing nerves and produce problems such as hearing impairment. Vertebral involvement can produce nerve root or spinal cord compression. Increased bony vascularity can result in congestive heart failure.

The biochemical abnormalities of Paget's disease are direct consequences of the increased bone remodeling. Increased osteoclastic resorption of collagen releases hydroxyproline, hydroxylysine, and collagen peptides which accumulate in urine. There is concomitant release of minerals from bone. However, despite the fact that calcium turnover is greatly increased in pagetic bone, formation as well as resorption is enhanced, and the net exchange of bone calcium is usually normal. Increased osteoblastic activity is reflected by increased serum concentrations of alkaline phosphatase and BGP. If resorption exceeds formation, as may happen during immobilization, hypercalciuria and hypercalcemia may ensue.

Because the primary abnormality in Paget's disease is increased bone resorption, treatment has been aimed at inhibiting bone resorption. This has been effectively achieved with agents that inhibit osteoclasts, such as calcitonin and mithramycin, and by agents that stabilize the crystal structure of bone, such as the diphosphonates.

OSTEITIS FIBROSA CYSTICA. Osteitis fibrosa cystica is the name given to a bone disease due to parathyroid hormone excess. Such an excess can occur in either primary or secondary hyperparathyroidism. When the hyperparathyroidism is secondary, it usually has resulted from a disease state such as hypocalcemia or renal disease, which can further complicate the picture of osteitis fibrosa cystica.

The classic pattern of bone involvement in hyperparathyroidism is osteitis fibrosa cystica with increased radiolucency, bone cysts, fractures, and deformities. Osteitis fibrosa cystica is characterized by increased osteoclastic activity and increased but patchy osteoblastic activity along with increased vascularity and fibrosis. The primary abnormality appears to be increased bone resorption with a compensatory attempt to increase bone formation. However, these manifestations of the disease are now more likely to be seen in the patient with secondary hyperparathyroidism due to renal failure in the patient with primary hyperparathyroidism. The decline in the frequency of osteitis fibrosa cystica in primary hyperparathyroidism has been accompanied by the appearance of a more subtle bone disease that has the histological features of osteitis fibrosa cystica, but that is manifest by osteopenia rather than the gross deformities of classical osteitis fibrosa cystica. The most likely explanation for the lesser severity of bone disease in primary hyperparathyroidism in recent years is the early recognition and treatment of the disease. However, even this milder disease is characterized by increased resorption and formation.

The increased bone turnover and consequent hypercalcemia found in all patients with hyperparathyroidism reflect the action of parathyroid hormone on bone. These effects of parathyroid hormone may lead to mild resorptive changes in bone; examination by microradiography of bone specimens from patients with hyperparathyroidism without evidence of osteitis fibrosa may indicate signs of osteocytic osteolysis. Many enlarged osteocytic lacunae, resembling small resorption cavities, are detected; the lacunae are surrounded by areas of reduced mineral density. The changes found in classical osteitis fibrosa cystica may represent an exaggeration of the bone-remodeling effects of PTH.

Increased concentrations of parathyroid hormone are present in all patients, yet a minority of patients with equivalent chemical evidence of hyperparathyroidism have striking bone involvement. It is likely that compensatory mechanisms that operate in response to the challenge of hypercalcemia, such as production of calcitonin, or a relative failure of such compensatory mechanisms, may in some way influence the occurrence of osteitis in hyperparathyroidism. In fact, the relative calcitonin deficiency of women may account for their greater severity of bone disease when compared to men with primary hyperparathyroidism.

A number of clinical features of the bone disease are important in evaluating the symptoms of the patient, in providing diagnostic clues to the presence of the disease, or in assessing the response of the patient following surgical correction of the hyperparathyroidism. Diffuse bone pain may be encountered without the marked reoentgenographic changes of osteitis. However, when severe osteitis fibrosa cystica develops, the loss of bone may result in marked deformities and fractures; pain is often localized and quite severe. Histological examination of bone specimens from patients with severe osteitis fibrosa reveals a number of changes. These are (1) a reduction in the number of trabeculae; (2) an increase in giant multinucleated osteoclasts seen in scalloped areas on the surface of the bone (Howship's lacunae); (3) an increase in the number and activity of osteoblasts; and (4) a marked replacement of normal cellular and marrow elements by fibrous tissue. In milder forms, changes can sometimes be detected in roentgenograms of the clavicles and hands. Subperiosteal resorption in the phalanges is a very striking finding; the phalangeal tufts are resorbed and an irregular outline replaces the normally sharp cortical outline of the bone in the digits. Detection of loss of the lamina dura of the teeth is also helpful diagnostically but is found less frequently than subperiosteal resorption.

Two types of radiolucent lesions are detected in hyperparathyroidism: bone cysts and brown tumors. These lesions appear similar on x-ray, but in fact represent quite different pathological processes and evolve differently following surgical correction of the disease. A true bone cyst consists of a fluid-filled cavity lined with fibrous tissue; such cysts often occur subperiosteally. By contrast, brown tumors are highly cellular and usually occur in the mandible. Various vascular and cellular elements of bone, particularly fibroblasts, osteoblasts, and osteoclasts, are present; these latter lesions are often referred to as *osteoblastomas* or *osteoclastomas*. Following surgical correction of the hyperparathyroidism, the brown tumors resolve. True bone cysts, however, remain evident on subsequent roentgenographic examination. In addition to resolution of brown tumors, the other roentgenographic changes indicative of increased bone reabsorption resolve following adequate treatment.

HYPERCALCEMIA AND MALIGNANCY. Hypercalcemia accompanying malignant disease is the most common abnormality of calcium metabolism and one of the most common metabolic abnormalities in clinical medicine. The hypercalcemia is invariably due to increased bone resorption. The hypercalcemia per se produces a wide variety of signs and symptoms. The gastrointestinal symptoms are prominent and include anorexia, nausea, vomiting, constipation, and weight loss. Hypercalcemia has significant central nervous system and neuromuscular effects ranging from mild personality changes to central nervous system suppression and coma. Other signs include suppression of neural activity resulting in decreased reflexes or increased neuromyocardial irritability, resulting in cardiac dysrhythmias and conduction abnormalities. The latter effects of hypercalcemia are enhanced by digitalis and hypokalemia and thus are especially important in the patient with cardiac diseases. Most of the signs and symptoms of hypercalcemia can be ascribed to altered neuromuscular function.

In most patients with hypercalcemia of malignancy, it can be attributed to the increased bone resorption caused by the invasion of bone by the malignant cells. In some cases, there may be direct bone resorption by accompanying macrophages and monocytes. In 15 to 20 per cent there is no demonstrable evidence for bone

metastases to account for the increased bone resorption; in these instances the increased bone resorption is usually produced by the host osteoclasts. The numerous osteoclasts resemble those seen in primary hyperparathyroidism, but, unlike primary hyperparathyroidism, they are not accompanied by an increased number of osteoblasts. These osteoclasts are stimulated by the presence of the metastases or by substances produced by the primary or metastatic tumor. Tumors have been reported to produce a variety of factors that can stimulate osteoclastic bone resorption. Those that seem reasonably well related to the hypercalcemia of malignancy include (1) PTH or a PTH-like substance; (2) prostaglandins, especially of the E_2 series; and (3) OAF, a protein produced by lymphomas, myelomas, and perhaps other malignant cells. Other factors have also been implicated in this process, but evidence for them is less well established. These include (1) osteolytic sterols; (2) other peptides such as vasoactive intestinal polypeptide (VIP) and epidermal growth factor (EGF); and (3) a recently appreciated but unidentified substance which, like PTH, stimulates nephrogeneous cAMP but does not react in PTH immunoassays. Some of these factors, especially the prostaglandins, could also come from host cells such as macrophages, monocytes, or perhaps even bone cells themselves.

The hypercalcemia of malignancy is often complicated by other factors that accentuate the hypercalcemia, such as dehydration and immobilization, and by certain drugs used for tumor treatment, such as gonadal steroids, adrenal steroids, and tamoxifen. A small percentage of patients with hypercalcemia of malignancy have associated hyperparathyroidism.

Treatment can be directed against the mechanism of the hypercalcemia. Agents that inhibit osteoclasts, such as mithramycin and calcitonin, can be useful. Diphosphonates can be used to stabilize the crystalline structure of bone and inhibit further bone resorption. Steroids can be used to inhibit OAF-mediated resorption. Indomethacin and aspirin can be used as prostaglandin antagonists. Obviously, tumor therapy is desirable.

OSTEOPOROSIS. Osteoporosis is a disease characterized by a generalized decline in bone mass; this disorder is the most frequently occurring metabolic bone disease and is particularly common in elderly women. The progressive decrease in the amount of bone results in a marked narrowing of cortical bone and a reduction in the number and size of trabeculae in cancellous bone. However, no abnormalities in the chemical and crystal structures of osteoporotic bone have been discovered. Because loss of skeletal mass invariably occurs with advancing age, osteoporosis with pain and pathological fractures may represent an exaggeration of the normal aging process. Osteoporosis can occur in association with several disease states, notably hypercortisolism. In such instances, osteoporosis can be considered a secondary disorder. However, the most common form of osteoporosis is the form that is usually seen in the aging population, especially in postmenopausal women.

The principal clinical features of the disease are related to the loss of bone mass and the severe deformities that result from this loss, particularly in the axial skeleton. Bone pain is very common. It is often manifested as a persistent, dull low backache; the severity of the pain is usually proportional to the degree of bone loss. However, there can be a striking degree of rarefaction of the skeleton without any clinical symptoms. Vertebral compression fractures are quite common and are usually accompanied by sharp and localized pain, but in many instances compression fractures of the vertebral bodies are discovered on x-ray as an incidental finding. Severe bone loss of the axial skeleton may result in a striking loss of stature so that patients become inches shorter than their original adult height. Despite the marked involvement of the vertebral bodies, neurological complications are rare. Hip fractures are quite common in the patient with osteoporosis, and they may occur with little or no associated trauma. There is a high rate of morbidity because the structural weakness of the bone in many instances precludes internal fixation and a high mortality rate because confinement to bed leads to cardiopulmonary complications such as thromboembolic disease. Fractures of the radius, humerus, and ribs are also common.

Although the diagnosis of osteoporosis is usually made on the basis of roentgenographic findings, these procedures are not very useful for early detection of the disease. Thirty to fifty per cent of skeletal calcium must be lost before there are changes apparent on x-ray. Consequently, newer diagnostic techniques such as photon beam densitometry and computerized axial tomography (CAT) scanning are being evaluated for early diagnosis and quantification of the progression and treatment of osteoporosis.

Laboratory findings in osteoporosis are generally unremarkable. Serum and urinary calcium and phosphorus and the serum alkaline and acid phosphatase are usually normal. However, elevated levels of urinary hydroxyproline and serum BGP have been detected in some patients.

The pathogenesis of osteoporosis is heterogeneous. The decreased skeletal mass indicates a disproportionate increase in bone resorption and/or a decrease in bone formation. The imbalance need not be large; loss of bone equivalent to only 20 to 50 mg. of calcium per day for several decades can result in loss of more than 25 per cent of skeletal mass. For many years osteoporosis was considered to be primarily a failure of normal bone formation and the loss of estrogenic steroids in postmenopausal women was believed to be important in causing decreased bone matrix formation. However, newer techniques such as radioactive calcium kinetic analyses, microradiography, and improved histomorphometric procedures have demonstrated that the major abnormality in osteoporosis is increased bone resorption. Bone formation may be normal but it may be low, if not absolutely, at least relative to the increased resorption.

Many possible etiologies or metabolic abnormalities have been investigated, but none has been convincingly identified as the cause of osteoporosis. It is important to note that the calcium metabolism of most patients with osteoporosis is normal compared with that of normal subjects of a similar age. Furthermore, long-term calcium supplements, although they may inhibit the increased resorption, do not result in remineralization of the skeleton. One must therefore conclude that calcium deficiency may be a contributory factor in some cases of osteoporosis, but is certainly not the sole or even principal factor in most cases.

A number of hormonal factors have been implicated in the pathogenesis of osteoporosis. Despite the well-known importance of menopause in the pathogenesis of osteoporosis, in most reports there seem to be no differences between patients with osteoporosis and normal subjects in plasma concentrations of gonadal steroids and gonadotropins. However, in some studies, lower levels of testosterone and progesterone and even higher levels of urinary free cortisol have been reported in osteoporosis. And, more recently, decreased levels of free estradiol and testosterone levels were observed in a group of postmenopausal women presumed to have osteoporosis on the basis of hip fractures.

An etiological role in osteoporosis has been postulated for PTH, $1,25(OH)_2D_3$, and calcitonin. In most studies of PTH, plasma concentrations in patients with osteoporosis have been indistinguishable from age- and gender-matched controls; however, both high and low values have also been reported. Most studies indicate that $1,25(OH)_2D_3$ levels decrease with age and are around the lower limits of normal in patients with osteoporosis. There has been an accumulation of evidence to implicate calcitonin in the pathogenesis of the age-related loss of bone mass that occurs in human beings and perhaps even in the pathogenesis of osteoporosis. The main skeletal effect of calcitonin is to inhibit bone resorption; the main skeletal defect in osteoporosis is increased bone resorption. When assessed by radioimmunoassay procedures, sera from women are calcitonin deficient compared to men. In both sexes there is a progressive decline with age in calcitonin secretion in response to calcium stimulation tests. In each decade from ages 20 to 80, calcitonin secretion is less pronounced in women than in men. Studies of basal levels of plasma calcitonin in osteoporotic women fail to reveal any consistent differences from normal women. However, just as provocative testing is necessary to clearly demonstrate calcitonin deficiency in women compared to men, so does calcium infusion reveal a deficient calcitonin response in osteoporotic women compared to normal age-matched women. Studies of calcitonin therapy suggest that calcitonin deficiency may be one mechanism underlying the pathophysiology of osteoporosis. Administration of calcitonin in some, but not all, studies has been reported to retard and perhaps reverse the progressive and accelerated loss of bone mass that occurs in osteoporosis. Calcitonin and gonadal steroids may be linked to the pathogenesis and treatment of osteoporosis. Estrogens are used in the treatment of osteoporosis, but their mechanism of action has not been elucidated. The major therapeutic effect of estrogens is to inhibit bone resorption. Despite this effect, most investigators have failed to demonstrate the presence of estrogen receptors in bone. Thus, estrogens may exert their skeletal effects indirectly, for example, by inhibiting the sensitivity of bone to other bone-active substances like PTH and $1,25(OH)_2D_3$. It is also possible that estrogens may act by stimulating the secretion of calcitonin. It has been reported that pregnancy and oral contraceptives increase plasma calcitonin in women, that levels of calcitonin are higher during the middle of the menstrual cycle, and that administration of estrogens increases plasma calcitonin in both young and elderly women.

Thus, it is very likely that osteoporosis comprises a heterogenous group of skeletal disorders and that both sex hormones and calcemic hormones are important in pathogenesis. Further studies should elucidate the hormonal abnormalities important in the development of osteoporosis. Only then will it be possible to institute rational therapy with the many regimens that have been suggested for the treatment of osteoporosis, including those designed to inhibit bone resorption and to increase bone formation—calcium, estrogens, calcitonin, vitamin D metabolites, anabolic steroids, PTH, fluorides, and physical activity.

Diseases in which the primary abnormality is increased bone formation

Increased bone formation occurs as a secondary event in many bone diseases such as Paget's disease and primary hyperparathyroidism. Furthermore, some patients treated with fluorides at low doses also demonstrate increased bone formation. There is seldom a net increase in bone mass because there is also increased bone resorption in these disorders. In acromegaly, however, the excess of growth hormone, either directly or indirectly, produces new subperiosteal bone growth which may not be accompanied by equivalent bone resorption.

Diseases in which the primary abnormality is decreased bone resorption

OSTEOPETROSIS. Osteopetrosis is a rare, inherited disorder characterized by abnormally dense bone throughout the skeleton. The disease occurs in two distinct genetic patterns, an autosomal recessive form which occurs *in utero* and in infancy or childhood and has a rapidly fatal course, and an autosomal dominant form which becomes manifest in adolescence or adulthood and has a more prolonged and benign course. The disease may also be genetically heterogeneous.

Osteopetrosis is a disease of the osteoclast. Although these cells are present they do not reabsorb bone normally and thus do not contribute to the bone remodeling process. This defect results in the abnormally dense bone that is characteristic of the disease. Blood mineral and hormone values are normal but acid phosphatase may be elevated. Many of the clinical findings can be accounted for by the proliferation of the abnormally dense bone. The deficiency of marrow results in anemia, which is most severe in the childhood form of the disease and may account for the hepatosplenomegaly produced by a compensatory extramedullary erythropoiesis. The dense bone also obliterates the cranial nerve foramina in the skull, resulting in compression of the cranial nerves. Most commonly involved are cranial nerves 2, 3, 7, and 8; optic atrophy and blindness, oculomotor and facial palsy, and hearing loss may result. Other striking features of the disease include poor growth, mental retardation, and abnormal dentition. Despite its abnormal density, the bone is structurally weak as evidenced by the frequent occurrence of fractures in both forms of the disease. It is surprising that fracture healing occurs in an apparently satisfactory manner. Impaired bony vascularity may account for the increased incidence of osteomyelitis, especially in the mandible.

The hypothesis that osteopetrosis is a disease of the osteoclast is strongly supported by the recent reports of remarkable improvement of the disease in children treated with transplantation of hematopoietic tissue. Following such transplantation procedures there is extensive resorption and remodeling of bone and improvement of hematological status. This effect is due to the activity of donor monocytes and supports the view that the osteoclast is of hematogeneous origin and derived from the monocyte and macrophage.

HYPOPARATHYROIDISM. The bone of patients with hypoparathyroidism is characterized by decreased metabolic activity, as evidenced by calcium kinetic studies and biochemical studies. It is possible that the absence of PTH results in decreased osteoclastic activity, which in turn results in decreased osteoblastic activity. The low levels of $1,25(OH)_2D_3$ also contribute to the decreased skeletal activity. On x-ray examination, bone has normal or slightly greater than normal density, but there are no clinical abnormalities of the skeleton except in patients with pseudohypoparathyroidism who have metacarpal and metatarsal abnormalities.

Diseases in which the primary abnormality is decreased bone formation

OSTEOMALACIA. Osteomalacia is defined as deficient mineralization of the skeleton. Osteomalacia (and rickets) can result from a deficiency of factors important in the process of bone formation. Three—calcium, phosphorus, and $1,25(OH)_2D_3$—have been identified, and a fourth—alkaline phosphatase—seems likely. Because calcium and phosphorus are of obvious importance in bone mineralization, a deficiency of either can produce osteomalacia. Alkaline phosphatase is associated with the process of normal mineralization, and its deficiency in hypophosphatasia undoubtedly contributes to the osteomalacia associated with this inherited disorder.

The center stage for the pathogenesis of osteomalacia is occupied by abnormalities in the metabolism and effects of vitamin D (see list below). True vitamin D deficiency is rare in the United States because sufficient exposure to the sun protects against osteomalacia in normal persons, However,

malabsorption states causing defective absorption of vitamin D and calcium are a relatively common cause of osteomalacia. Because of the central role of the kidney in the synthesis of $1,25(OH)_2D_3$, impaired renal function also leads to osteomalacia. Only very severe decreases in the concentration of $25(OH)_2D_3$ produce a decline in levels of $1,25(OH)_2D_3$ sufficient to cause osteomalacia; this is most likely to occur when there is total vitamin D deficiency.

Abnormalities of Vitamin D That Can Result in Osteomalacia

Vitamin D deficiency
  Inadequate solar radiation
  Dietary deficiency
  Malabsorption
Decreased serum 250 HD_3
  Anticonvulsants
  Hepatic disease
  Nephrotic syndrome
Decreased serum $1,25(OH)_2D_3$
  Renal failure
  Vitamin D-dependent rickets, Type I
  Oncogenic
  Hypophosphatemic rickets
    Sex-linked
    Adult
Target tissue (bone) abnormality (unresponsiveness)
  Vitamin D-dependent rickets, Type II
  Anticonvulsants
  Hypophosphatemia
    Hypophosphatemic rickets
    Dietary—aluminum hydroxide ingestion
  Hypophosphatasia
  Prematurity

The terminology of the hereditary disorders that result in defective mineralization has been clarified by an appreciation of vitamin D metabolism. The characteristics of hereditary rickets are as follows: *vitamin D–dependent rickets Type I* is autosomal recessive and due to defective renal vitamin D_1 hydroxylase and decreased production of $1,25(OH)_2D_3$. *Vitamin D–dependent rickets Type II* is x-linked dominant and due to end-organ resistance to $1,25(OH)_2D_3$, which is increased in plasma in a presumed compensatory manner. X-linked hypophosphatemic rickets, also referred to as *vitamin D–resistant rickets (VDRR)* or hypophos-

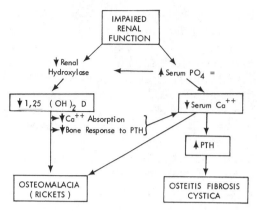

FIG. 20–5. Pathogenesis of renal osteodystrophy.

phatemic bone disease, is due to hypophosphatemia caused by defective renal tubular phosphate resorption and a defective control system for $1,25(OH)_2D_3$; in normal subjects the hypophosphatemia would produce an increase in serum $1,25(OH)_2D_3$, but in these patients it does not. Autosomal (adult) hypophosphatemic osteomalacia is also caused by diminished renal threshold for phosphate, but there does not seem to be any abnormality in vitamin D metabolism.

The clinical manifestations of the disease depend on the age of the patient. In young patients, widened, undermineralized cartilage confers structural weakness to the skeleton. The resultant inability of the skeleton to distribute the biochemical forces of motion and weight-bearing has several manifestations. In the recumbent infant these manifestations are flattening of the skull and pelvis; in the ambulatory child they are bowing of the shafts and splaying of epiphyses. The cartilaginous ends of the ribs can become protuberant in the rickety (imprecisely termed *rachitic*) rosary and there is an accompanying indentation (Harrison's groove). This constellation of clinical signs is called rickets. The hypocalcemia is occasionally severe enough to produce tetany. In the adult with a mature skeleton the bony manifestations of the disease are more subtle and often limited to radiological findings such as radiolucent

lines along the course of arteries (pseudofractures of Looser's zones). However, bone pain and muscular weakness can be a prominent feature of the disease. Any patient with rickets or osteomalacia is susceptible to fracture.

Common laboratory abnormalities in patients with rickets or osteomalacia include normal or slightly to moderately decreased serum calcium and diminished urinary calcium to less than 200 mg./24 hours. The serum level of inorganic phosphorus is considerably diminished with the exception of the osteomalacia in renal osteodystrophy. Tubular reabsorption of phosphate is decreased to values below 85 per cent of normal. In rickets, and to a lesser extent in osteomalacia, serum alkaline phosphatase is elevated but may increase further during the early phase (2 to 3 months) of therapy with vitamin D analogues. This may be due to an anabolic effect of vitamin D along with underutilization (and "spillover") of alkaline phosphatase.

The treatment of rickets or osteomalacia can be very successful if the specific deficiency is identified and corrected. This usually means administration of calcium and phosphate and, most importantly, appropriate replacement with vitamin D or its metabolites.

RENAL OSTEODYSTROPHY. Renal osteodystrophy is the term used for the complex bone disease that occurs in chronic and advanced renal failure. There are two principal components to renal osteodystrophy: osteitis fibrosa cystica due to secondary hyperparathyroidism and osteomalacia due to deficiency of $1,25(OH)_2D_3$ production. The pathogenesis of these two features is summarized in Fig. 20–5. The two key features in pathogenesis are the phosphate retention that leads to the secondary hyperparathyroidism and the decreased production of $1,25(OH)_2D_3$ that leads to the osteomalacia. Secondary hyperparathyroidism is accentuated by the presence of skeletal resistance to PTH. Both of these mechanisms are set into motion by the loss of renal function and destruction of renal mass. Treatment is directed at reversing these defects. Calcium supplementation and phosphate restriction are designed to increase the blood calcium and decrease the blood phosphate (which indirectly increases the blood calcium) and thus to suppress the hyperplastic parathyroid glands. Vitamin D supplementation, especially with $1,25(OH)_2D_3$, is designed to restore levels of this hormonal metabolite to normal.

Other factors play at least a contributory role in the development of renal osteodystrophy. Accumulation of aluminum through the dialysate or from phosphate binders may become deposited in bone or in the parathyroid gland. This may interfere with PTH secretion and further impair skeletal mineralization. Chronic metabolic acidosis also has deleterious effects on the mineralization of the skeleton. The role of other metabolites of vitamin D such as $24,25(OH)_2D_3$ is uncertain. Levels of 24,25 are undetectable. Levels of $25OHD_3$ are normal unless there is protein wasting and consequent loss of serum binding proteins as in the nephrotic syndrome. In addition to the osteitis fibrosa cystica and the osteomalacia, bone in patients with renal osteodystrophy can display osteoporosis, osteosclerosis, and a mixed histological pattern.

Annotated references

Albright, F., and Reifenstein, E.: The Parathyroid Glands and Metabolic Bone Disease. Baltimore, Williams & Wilkins, 1948. (The classic.)

Avioli, L. V., and Krane, S. M. (eds.): Metabolic Bone Disease, Vols. I and II. New York, Academic Press, 1977. (Clinical descriptions of the metabolic bone diseases.)

Boskey, A. L.: Current concepts of the physiology and biochemistry of calcification. Clin Orthop *157*:225–257, 1981. (A review article of experimental studies concerning mechanisms of mineralization.)

Broadus, A. E.: Mineral metabolism. Felig, F., Baxter, J. D., Broadus, A. E., and Frohman, L. A. (eds.): *In* Endocrinology and Metabolism, New York, McGraw Hill, 1981. (A detailed chapter on systemic mineral homeostasis and the biochemistry and physiology of the mineral-regulating hormones, including 209 references.)

Deftos, L. J.: Calcitonin secretion. *In* Bronner, F., and Coburn, J. (eds.): Disorders of Mineral Metabolism, Chap. 8, pp. 433–479. New York, Academic Press, 1982.

Deftos, L. J., and First, B. P.: Calcitonin as a drug. Ann. Intern. Med. *95*:192–197, 1981. (A review of the biology and pharmacology of calcitonin.)

Deftos, L. J., Parthemore, J. G., and Price, P. A.: Changes in plasma bone Gla protein during treatment of bone disease. Calcif. Tissue Int. *34*:121–124, 1982. (A discussion of new methods of evaluating bone disease.)

DeLuca, H. F., Frost, H. M., Jee, W. S. S., Johnston, C. C., Jr., and Parfitt, A. M.: Osteoporosis: Recent Advances in Pathogenesis and Treatment, Baltimore, University Park Press, 1981. (Current concepts of etiology and management of osteoporosis.)

First, B. P., and Deftos, L. J.: Hypercalcemia in malignancy. *In* Klastersky, J., and Staquet, M. (eds.): Medical Complications in Cancer Patients, pp. 155–169. New York, Raven Press, 1981. (A discussion of the pathogenesis of the hypercalcemia of malignancy.)

Harrison, H. E., and Harrison, H. C.: Disorders of Calcium and Phosphorus Metabolism in Childhood and Adolescence. Philadelphia, W. B. Saunders, 1982 (Clinical and biochemical aspects of pediatric calcium and skeletal metabolism.)

Judd, H. L., Meldrum, D. R., Deftos, L. J., and Henderson, B. E.: Estrogen replacement therapy: Indications and complications. Ann. Intern. Med. *98*:195–205, 1983. (A discussion of the pathogenesis and treatment of osteoporosis.)

Urist, M. R. (ed.): Fundamental and Clinical Bone Physiology. Philadelphia, J. B. Lippincott, 1980. (A detailed presentation of bone physiology by 12 authorities.)

Vaughan, J.: The Physiology of Bone, ed. 3. Oxford, Clarendon Press, 1981. (A small volume that reviews bone biochemistry and mineral homeostasis).

Section V

Gastrointestinal Mechanisms

INTRODUCTION

The past two decades have witnessed the emergence of new means to manage gastrointestinal illnesses in patients. Significant advances in diagnostic procedures and in medical and surgical therapy of these diseases have been based upon a growing understanding of their pathophysiology. The succeeding six chapters reflect the sophistication of our concepts about digestive diseases and the altered mechanisms involved in their genesis. The pathophysiology presented here assumes that the reader has been exposed to the basic sciences of medicine.

In the past two decades biomedical technology has allowed us to determine the chemical structure of gastrointestinal hormones and to develop highly sensitive radioimmunoassays to detect their presence in the plasma. Our advances have included the application of fiberoptics and computer assisted imaging to make more accurate diagnoses of gastrointestinal lesions at earlier stages. The introduction of receptor pharmacology to the control of acid secretion by the stomach has resulted in highly effective and selective drugs for peptic ulcer with even better agents which could prevent ulcers in the offing. Advances in lipid chemistry have given insight into the pathophysiology of biliary disorders and have opened new avenues for preventing the formation of gallstones. Similarly, research in viral immunology has opened up improved diagnosis and even prevention of forms of hepatitis. We have learned much about the spectrum of malabsorptive diseases in the gut and are able to manage some successfully. Motor disorders of the hollow organs are being attacked investigatively as well as clinically with better diagnostic instruments and chemicals.

Even the enigma of circulatory diseases of the gut are being explored and better understood with apparent improvement in their management. Acute splanchnic vascular disorders may be categorized as ischemic or hemorrhagic. Intestinal ischemia may impair absorptive functions of the gut or result in infarction of the bowel, while impaired blood flow to the liver may hamper a variety of metabolic functions or lead to progressive parenchymal cell necrosis (as occurs in those patients who demonstrate deteriorating hepatic function following a surgically performed portosystemic shunt). Mesenteric ischemic diseases constitute a surprisingly common and life-threatening group of disorders. In the past decade they have begun to receive the deserved attention of clinical investigators. These entities are difficult to recognize clinically and usually end in death. The patient with a mesenteric vascular disease is typically elderly and has an associated cardiovascular disorder such as congestive cardiac failure for which he is receiving digitalis. The mesenteric circulation has certain physiological characteristics that are readily disturbed by a coexisting cardiovascular disease. There is, for example, a

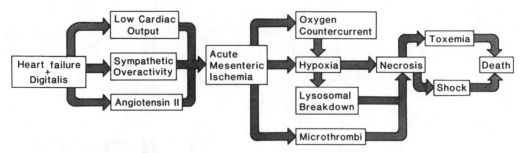

FIG. V–1. Schematic representation of pathophysiological events leading to bowel infection and death of the patient.

countercurrent exchange for oxygen in the small vessels of the villi that permits shunting of oxygen from arteriole to venule. This mechanism allows oxygen to bypass the capillary bed of the distal villi. Normally, the countercurrent does not deprive the distal villi of much oxygen, but, when mesenteric blood flow is reduced, countercurrent loss of oxygen exaggerates tissue hypoxia and leads to mucosal necrosis, starting at the villous tips. The small vessels of the mesenteric circulation can develop microthrombi during persistent low-blood-flow states, thereby further obstructing tissue perfusion. As blood flow is reduced, the viscosity of blood tends to rise. As the driving head of pressure in the arteries declines, as in shock, the critical closing pressure of small vessels in the mesenteric circulation is reached and blood flow is restricted. Norepinephrine, vasopressin, and angiotensin II are released during severe circulatory stress states: these vasoactive mediators constrict the precapillary segment of the mesenteric circulation. Furthermore, cardiac glycosides can constrict the mesenteric arterioles. The sequence of pathophysiological events leading to bowel infarction and death is shown schematically in Figure V–1. Unfortunately, in many patients the problem may remain unrecognized before death. Although various forms of treatment for intestinal ischemia have been unsuccessful in the past, there is promise in the use of vasodilator drugs infused directly into the superior mesenteric artery. Under experimental conditions, inhibition of xanthine oxidase has also been effective in preventing ischemic necrosis of villi.

Hemorrhage in the intestinal tract can be so slow as to be entirely occult or manifested clinically only by an iron deficiency anemia. Bleeding can also be so sudden and massive that death from shock and exsanguination may occur within hours. Gastrointestinal hemorrhage occurs as a major complication of many common diseases, such as portal hyptertension, peptic esophagitis, gastritis, peptic ulcer, Crohn's disease, ulcerative colitis, diverticulosis, congenital vascular malformations, vasculoenteric fistulae, pancreatitis, ischemic vascular disease that has caused infarction of intestinal mucosa, and neoplasia anywhere along the digestive tract.

Modern diagnostic radiological techniques have gone far beyond traditional use of barium; the clinician now has available angiographic methods for the diagnosis and localization of either gastrointestinal bleeding or occlusion of a major vessel. Direct intravascular infusion of vasoconstrictor drugs during angiography to slow or stop blood loss seems to be gaining wide acceptance.

Paralleling the growth of basic knowledge in gastroenterology has been the remarkable technical advance and wide use of gastrointestinal endoscopy. Within a short time, instruments and methods have been developed for the comfortable, safe, and complete visualization of the lumen and mucosal surfaces of the esophagus, stomach, duodenum, rectum, entire colon and distal ileum. Biopsies can be obtained from any visualized lesion. Methods are also available for a skilled endoscopist to cannulate the ampulla of Vater and inject radiocontrast material to obtain retrograde cholangiograms or pancreatograms. Certain lesions, particularly polyps, can now be removed from the stomach, duodenum, or colon by means of the endoscope, thereby avoiding general anesthesia and major abdominal surgery.

In the United States today, gastrointestinal diseases constitute an annual multibillion dollar economic loss, often striking people in the most productive period of life. Family physicians, internists, gastroenterologists, and surgeons care for millions who suffer from some of the most common symptoms and diseases of the modern American: ulcers, indigestion, abdominal pain, constipation, diarrhea, hepatitis, cirrhosis, gallstones, appendicitis, and colitis. However, before rational diagnosis and therapy of gastrointestinal diseases can be instituted, comprehension of their disordered mechanisms is essential.

Turner E. Bynum, M.D.
Eugene D. Jacobson, M.D.

21

Motility

James Christensen, M.D.

The flow of food and fluid through the alimentary canal is accomplished by contractions of its muscular walls. These contractions rely upon properties of the muscle, and upon responses of the muscle to autonomic nerves, substances circulating in the blood, and probably substances liberated from paracrine cells in the gut wall. The interactions of these control systems are not fully understood, but recent research has clarified much of the picture of gut movements and the place of abnormal movements in the pathogenesis of gastrointestinal disease.

The anatomy of gastrointestinal muscle

The muscle of the gut is composed of two coats, the muscularis propria or muscularis externa and the thin layer at the base of the mucosa, the mucosal muscle or muscularis mucosae (Fig. 21–1). The muscularis propria of the gut is wholly smooth muscle except at the beginning and the end of the gut; striated muscle constitutes the muscle of the pharynx and proximal esophagus and the muscle of the external anal sphincter. The muscularis propria lies in two layers: the longitudinal (external) layer contains muscle bundles which are oriented along the axis of the tubular con-

duit and the circular (internal) layer whose muscle bundles are oriented in the circumference of the tube.

Pharynx and esophagus

Striated muscle constitutes the muscle of the pharynx and of the proximal esophagus down to the level of the midesophagus. A thickened band of striated muscle, the cricopharyngeus, lies at the pharyngoesophageal junction. This muscle, which appears to be a specialized structure of the inferior constrictor muscle of the pharynx, arises from one end of the cricoid cartilage and inserts at the other end. Striated muscle constitutes both layers of the muscularis externa of the proximal third of the esophagus and interdigitates with smooth muscle in the middle third. The level of this junction varies considerably, and it is usually a littler lower in the longitudinal layer than in the circular layer. In some individuals, scattered bundles of striated muscle can be found among the smooth muscle even in the distal third of the esophagus. Both the longitudinal and circular muscle layers are continuous with the corresponding layers in the stomach without a conspicuous thickening or other anatomical demarcation at the esophagogastric junction in man, although thickening exists at this level in

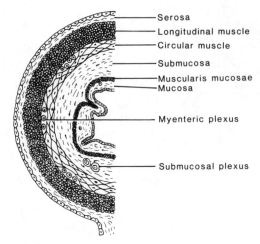

- Serosa
- Longitudinal muscle
- Circular muscle
- Submucosa
- Muscularis mucosae
- Mucosa
- Myenteric plexus
- Submucosal plexus

FIG. 21–1. A diagrammatic representation of a cross-section of the gut showing the interrelationships of muscle layers and nerve plexuses.

FIG. 21–2. A diagram showing the relative directions of the muscle fibers in the three muscle layers of the stomach: (a) esophagogastric junction (cardia); (b) fundus; (c) body; (d) antrum; (e) pylorus; (f) duodenum.

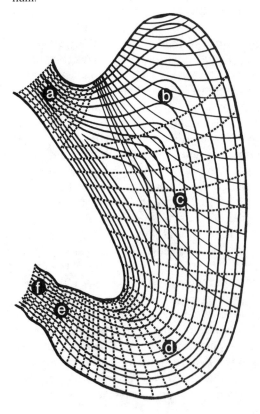

other mammals. This thickening may not be apparent in dead or formalin-fixed tissue, but it is readily seen in live muscle.

Stomach

In the stomach, the longitudinal muscle layer lies in two thick, broad bands along the greater and lesser curvatures with a thinner sheet between the bands (Fig. 21–2). The inner circular muscle layer invests the whole stomach, being thin in the fundus and thickening progressively along the stomach to the distal antrum. Inside this circular muscle coat lies a third layer, the oblique muscle layer. This layer is thickest over the fundus and extends over the body of the stomach in broad anterior and posterior sheets. The oblique layer does not cover the lesser curvature; its edges, called the gastric sling fibers, are thick bands which extend alongside the line of the lesser curvature from the cardiac orifice and fuse into the circular layer in the antrum. A concentration of fibrous tissue at the pylorus separates the circular layer of smooth muscle of the antrum from that of the duodenum. Some fibers of the longitudinal layer bridge this fibrous segment.

Small intestine

The two layers of the muscularis propria in the small intestine are uniformly thick about the circumference of the intestine. There is a progressive reduction in the thickness of these layers along the intestine.

Colon and anal canal

The arrangement of the two layers of the muscularis propria differs among species. The two layers are of uniform thickness about the circumference in some species. The longitudinal muscle layer in man lies in three bands, the taeniae coli, except in the appendix and rectum. The longitudinal muscle layer is very thin between adjacent taeniae. One taenia always lies along the line of the mesentery, while the other two are about equally distant from the mesentery. The wall of the colon bulges between the taenia. The colon is also indented by tonically-contracted bands of the circular muscle at intervals of a few segments. The net effect of these indentations is to give the colon a sacculated appearance. The sacs are called haustra. At the rectosigmoid junction the three taeniae broaden and fuse so that the rectum lacks the sacculated appearance of the rest of the colon. The circular muscle layer thickens in the anal canal to form the internal anal sphincter. This ring is surrounded by the external anal sphincter, a striated muscle that is related developmentally to the muscles of the pelvic floor.

Mucosal muscle

The mucosal muscle is composed of smooth muscle throughout the gut. In most regions, cells lie in two layers, an outer longitudinal layer and an inner circular layer, but these layers are much less well defined than are the layers of the muscularis propria. Many oblique cells are found, too. The thickness of this layer varies greatly from one organ to another. In the small intestine, the mucosal muscle gives rise to bundles that pass into the villi and presumably are responsible for villus movements.

Ultrastructure of the smooth muscle

The striated muscle of the pharynx, proximal esophagus, and external anal sphincter is like somatic striated muscle in structure. Muscle spindles are present. The smooth muscle of the gut consists of small mononuclear cells, usually described as spindle shaped, 100 to 200 μ long and 5 to 10 μ wide. The cytoplasm appears to be homogeneous by light microscopy (hence the adjective *smooth*) but electron microscopy reveals the existence of thin actin filaments and thicker myosin filaments. It is believed that a sliding filament mechanism operates in smooth muscle like that in striated muscle, differing only in the arrangement of

the filaments. The filaments pass through dark spots, amorphous bodies scattered throughout the cytoplasm, and fuse with the cell wall. The dark bodies are postulated to be analogous to the Z-lines of striated muscle. There is no t-tubule system in smooth muscle, but there is an extensive smooth endoplasmic reticulum that is thought to be analogous to the t-tubule system of striated muscle.

The smooth muscle cells abut each other very closely, although there are no cytoplasmic connections between adjacent cells. These cells, however, interact as a functional syncytium. Adjacent cells contact each other closely through processes or protrusions of the cell wall in which there is actual fusion of the cell walls or very close apposition of the walls of the two cells. These junctions constitute contact points with an extremely low electrical resistance, and are presumed to permit ions or electric currents to flow easily from one cell to the next. This would allow close correlation of activities of adjacent cells. The extracellular space between smooth muscle cells contains collagen. This probably provides mechanical linkage between muscle cells. Connective tissue septa in most gastrointestinal smooth muscle delineate bundles of muscle cells that may function as units.

The nerves of gastrointestinal muscle

Intrinsic nerves

An intrinsic nervous system exists in the gut wall throughout the alimentary tract (Fig. 21–3). The nerves of this system are arranged in laminar plexuses that are named according to their location in the wall.

The main plexus is the intermuscular or myenteric plexus that lies between the longitudinal and circular muscle layers of the muscularis propria. This plexus is composed of three networks in the same plane.

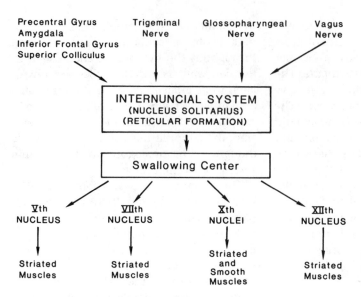

FIG. 21–3. A diagram showing the swallowing center, its inputs and its outputs.

The primary plexus consists of large bundles of nerve fibers that connect ganglia or nodes containing nerve cell bodies. A secondary plexus can be seen within the spaces in the primary plexus: it is a closer network of smaller nerve bundles that contain scattered nerve cell bodies. The tertiary plexus is a network of fine nerve fibers that enter the muscle coats.

The submucosal plexus lies in the submucosa. It is much less dense and less complex than the myenteric plexus, although it contains ganglia in most of the gut. It is scanty in the esophagus.

Most of what is known of the plexuses comes from study of the myenteric plexus. The enteric neurons in this plexus have been classified morphologically, but not functionally. In general, such cells can be classified as either argyrophilic (readily stained by silver impregnation) or argyrophobic. The proportions of argyrophilic and argyrophobic cells differ somewhat among organs.

Differential staining techniques allow identification of some cells and nerve fibers as cholinergic motor neurons, the secondary neurons of the parasympathetic motor pathways. Staining techniques also can be used to demonstrate adrenergic fibers that are very sparse in gastrointestinal muscle and exhibit a close relationship to blood vessels rather than to the layers of the muscularis propria. These catecholamine-containing fibers seen in the myenteric plexus terminate in a synaptic relationship with nerve cell bodies that are probably secondary neurons in parasympathetic motor pathways, suggesting that the sympathetic motor innervation of gastrointestinal muscle is directed mainly to the parasympathetic motor innervation rather than to the muscle itself.

Although neurons of other kinds may be present in the plexuses, including sensory and internuncial neurons, they cannot be selectively identified by morphological means. Recently, the application of staining methods that make use of antibodies have suggested that some enteric nerves may act by release of other substances. These putative neurohormonal transmitters include vasoactive inhibitory peptide, substance P, and enkephalins. Abundant evidence has indicated a neurotransmitter function for serotonin. Prostaglandins, dopamine, and histamine have also been suggested as neural messengers in the gut.

Physiological evidence accumulated in the past two decades indicates the presence of a third class of autonomic motor nerves that influence gut muscle (in addition to the parasympathetic cholinergic nerves and the sympathetic adrenergic nerves). This is a set of inhibitory nerves that can be demonstrated in the muscle of all gastrointestinal viscera. They cannot be firmly identified morphologically, and the inhibitory neurohormonal transmitter they release has not been defined. Some evidence suggests that it may be adenosine triphosphate (ATP) or another adenine nucleotide. Other evidence supports the idea that it may be vasoactive intestine polypeptide. These nerves are called the nonadrenergic inhibitory nerves or "purinergic" nerves.

Extrinsic nerves

The enteric plexuses are connected to the central nervous system through the sympathetic and parasympathetic nerve trunks. These trunks, in both systems, contain both sensory and motor fibers. Sensory fibers greatly outnumber motor fibers in the vagal trunks.

The sympathetic (or thoracolumbar) nerves are distributed to the gut by way of the perivascular plexuses that arise from the retroperitoneal ganglia. These are named according to the artery with which they are associated: the celiac, superior mesenteric, and inferior mesenteric ganglia. These ganglia, in turn, are connected to the central nervous system through the splanchnic nerves.

The parasympathetic or craniosacral nerves travel to the gut by way of the vagi and the pelvic nerves. The vagi are mixed nerves, carrying many more sensory fibers (whose cell bodies lie in the nodose ganglion) than motor fibers (whose cell bodies lie in the dorsal motor nucleus of the vagus and in the nucleus ambiguus). The motor nerves constitute at least three different systems: (1) the somatic nerves that pass without synaptic interruption to the striated muscle of the proximal esophagus; (2) the parasympathetic preganglionic nerves

that travel to the cholinergic secondary parasympathetic ganglion cells of the myenteric plexus; and (3) the preganglionic nerves that activate the nonadrenergic inhibitory neurons of the myenteric plexus. In both of the latter two cases the preganglionic nerve fibers activate the secondary neurons through the release of acetylcholine.

Autonomic receptors

Much of what is known about the actions of nerves on gut muscle function comes from observations of the actions of the established neurohormonal transmitters or their congeners on the muscle. These actions are readily explained by the presence of receptors, active sites in the membranes of nerve and muscle cells where substances act selectively.

There are two kinds of receptors which are sensitive to acetylcholine. Muscarinic receptors are selectively blocked by atropine and similar antagonists, and these receptors mediate the actions of acetylcholine and its congeners directly upon smooth muscle cells. Nicotinic receptors are selectively blocked by hexamethonium, which interferes especially with the actions of acetylcholine on nerve cells, and by d-tubocurarine which blocks those receptors in striated muscle cells that respond to cholinergic agonists.

Adrenergic (catecholamine-sensitive) receptors are also of two kinds, termed α and β receptors. Alpha receptors are selectively blocked by phentolamine and phenoxybenzamine. Beta receptors are blocked by propranolol and similar agents.

Specific sensitive receptor sites are also proposed for other agents that affect contractions in smooth muscle. These include the two kinds of histamine receptors, H-1 and H-2 receptors, and serotonin receptors. Current ideas about the possible physiological role of putative hormones in the control of gastrointestinal motility rest on observations of the actions of these agents on muscle activity *in vivo* and *in vi-*

tro. These substances include gastrin, secretin, bradykinin, angiotensin, dopamine, substance P, enkephalins, and the like. In general, the receptors activated by such agents in gastrointestinal nerve and muscle have not been so well defined as have those that respond to acetylcholine and the catecholamines.

Integration of contraction in smooth muscle

The grading and integration of contractions in striated muscle requires the organization of striated muscle and somatic nerves into a motor unit in which one motor nerve controls a fixed number of cells. This form of motor unit organization may prevail in gastrointestinal striated muscle as well. Smooth muscle, however, does not have such a fixed relationship between nerve and muscle. There do not appear to be specialized neuromuscular junctions like those in striated muscle. Nerves extend for rather long distances among muscle cells and release their transmitters from relatively long segments of terminal branches.

The grading and integration of contraction in smooth muscle depends upon (1) variation in the level of discharge of extrinsic nerves, and reflexes whose pathways pass through the extrinsic nerves; (2) variations in the level of discharge of intrinsic nerves, and reflexes using pathways confined to the intramural plexuses; (3) hormones that circulate in the blood; and (4) the metabolic and electrical properties of the smooth muscle cells themselves.

Extrinsic nerves

Variations in tonic activity of extrinsic nerves are important in control of cardiovascular muscle, but are not so important in gastrointestinal muscle. Long reflexes, making use of extramural pathways, have been demonstrated, but, except for the central reflex that is involved in defecation, they do not seem to have a major role in the regulation of gastrointestinal motility.

This conclusion is based on clinical observations. Thus, splanchnic nerve section, commonly done to relieve pain of retroperitoneal origin, does not have important effects upon gut motion. Cervical vagotomy abolishes esophageal peristalsis, but truncal vagotomy does not greatly affect gastrointestinal movements except for an alteration in gastric emptying. For this reason, the intramural nerves are considered autonomous (or autonomic).

Intrinsic nerves

Spontaneous activity of intrinsic nerves of the plexuses has been demonstrated by electrical recording from nerves *in vitro*. There are also local excitatory and inhibitory reflexes which are mediated through the plexuses. Such reflexes can be activated by movement, by stretch, or by chemical stimulation of mucosal receptors. The sensory structures have neither been identified anatomically nor characterized physiologically and their exact pathways are undefined. These reflexes are generally identified only in terms of stimulus and response. In many cases, their importance in normal physiological control remains obscure.

Circulating hormones

The effect of hormones should be broader than the effect of local reflexes. The actions of some hormones may be responsible for some specific physiological phenomena in motility. Thus, the polypeptide hormone motilin may be responsible for the governance of the remarkable pattern of contractions seen in the small intestine in fasting, but evidence to support that view is still conflicting.

Electrical properties of smooth muscle

The major mechanism integrating contractions in gastrointestinal smooth muscle is a manifestation of the electrical properties of this muscle. The electrical potential difference at rest across the cell membrane in smooth muscle in some parts of the gut

fluctuates rhythmically, reflecting alterations in the movement of ions across the
cell membrane. Rhythmic contractions in
such muscles correspond to the occurrence of these rhythmic fluctuations (Fig.
21–4). The rhythmic fluctuations are called
electrical slow waves, pacesetter potentials,
or the basic electrical rhythm. They are
found in the distal half of the stomach, in
the small intestine, and in the colon. These
electrical signals are independent of nerves
and hormones, though their frequency can
be altered somewhat by such external controls.

The tight junctions that exist between
contiguous muscle cells exhibiting slow
waves are points of low electrical resistance between cells. This is the basis for
the fact that all cells in a sheet of tissue
generate synchronous signals. When the
signals are recorded from widely separated
points in a sheet of muscle, the slow waves
appear to spread uniformly from one area
to another. Thus, slow waves are not actually simultaneous, even at two closely
spaced intervals. Instead, they spread
through the gut muscle from a source, a
pacemaker, with a fixed velocity and pattern of spread. The velocity, pattern of
spread, and frequency show little variation
and thereby establish the velocity, pattern
of spread, and frequency of the rhythmic
contractions to which they are linked.
Hence, they are called pacemaker potentials. The pacemaker function is not fixed
in one spot: if the gut is transected, a new
pacemaker site develops at the level of
transection and governs contractions beyond that point. The analogy of this system
to the cardiac pacing system is obvious.

The slow wave is not, however, an action
potential, for it recurs constantly even
though the muscle is not contracting
rhythmically. When a contraction is to occur with any specific slow wave cycle, the
muscle at the level of contraction generates
another signal, one or a few much more
rapid electrical transients that occupy a
fixed position on the slow wave. This second signal, called the spike burst, repre-

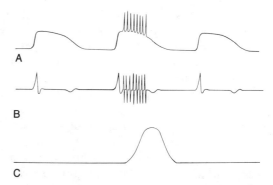

FIG. 21–4. A diagrammatic representation showing
the time relations between slow waves, spikes and
contractions. *A* represents the electromyogram of a
single cell in a tissue as recorded with an intracellular
electrode. *B* represents a corresponding record from
an extracellular electrode. *C* shows tension of the tissue under study. Note that contraction occurs only
with the slow-wave cycle that carries a spike burst.

sents a sudden movement of ions across
the muscle cell membrane that differs from
that causing the slow waves from which
spike bursts can independently be induced
or suppressed. In the distal stomach, the
spikes in this burst occur with such frequency that the phenomenon has the appearance of a second slow potential superimposed on the slow waves. Some evidence
indicates that rhythmic contractions may
sometimes be associated with slow waves
without spike bursts.

This system of slow waves and spike
bursts differs somewhat from one part of
the gut to another and from one species to
another. In the distal stomach and small
intestine slow waves are generated in the
outer longitudinal muscle layer and spread
caudad. In the colon, on the other hand,
slow waves are generated in the circular
muscle layer and spread cephalad in the
proximal colon and caudad in the distal
colon. Specific details of the slow waves in
these organs will be discussed further in
the discussion of each organ.

The pharynx and esophagus

Swallowing is initiated by a series of voluntary contractions of muscles of the mouth,
tongue, soft palate, and jaws to seal the
opening to the gut. This is followed by fur-

ther voluntary contraction of the muscles of the pharynx and the strap muscles of the neck. These contractions follow in a fixed sequence so that a bolus in the mouth is rapidly propelled to the pharyngoesophageal or upper esophageal sphincter. This sphincter, which is represented anatomically mainly by the cricopharyngeus muscle, is a segment 2 to 3 cm. long that is closed at rest to occlude the lumen with a force that can generate an intraluminal pressure of 18 to 60 cm. H_2O above atmospheric pressure. Just after contraction begins in the upper pharynx, the tonic contraction of the sphincter disappears so that little resistance is offered to the passage of the bolus into the esophageal body. It remains relaxed for 1 second or so, and then the tonic contraction that characterizes the resting state is resumed.

The esophageal body is flaccid at rest. Intraluminal pressure is slightly subatmospheric, reflecting intrathoracic pressure. In swallowing, a contraction begins just after closure of the upper esophageal sphincter. This contraction involves both muscle layers. The longitudinal layer contracts to tense and, to some degree, to shorten the esophagus. The circular muscle layer contracts to occlude the lumen, and this contraction appears first at the rostral end of the esophageal body, apparently as a direct extension of the closure of the upper esophageal sphincter, and progresses down the esophagus to the stomach. The force of this contraction is sufficient to induce a maximal intraluminal pressure of 30 cm. H_2O in the proximal esophagus and 60 cm. H_2O in the distal esophagus. The velocity of this contraction is 2 to 4 cm./sec. The length of the contracted segment is about 5 cm. in the upper esophagus and up to 12 cm. in the distal part.

The gastrointestinal junction or lower esophageal sphincter is closed at rest, to create a barrier between the stomach and the esophagus. The sphincter is 2 to 4 mm. long and represents a functionally specialized segment of the circular muscle layer. Closure of this segment is accomplished by a tonic contraction of the muscle that generates an intraluminal pressure of 10 to 40 cm. H_2O above atmospheric pressure. In the course of a swallow, this tonic contraction disappears, and the sphincter remains in a relaxed state for about 5 seconds until the peristaltic contraction that sweeps the esophageal body arrives, when the tonic contraction is resumed. This complex series of events is controlled by a combination of central and peripheral neural mechanisms (Fig. 21–5) and depends also on special properties of esophageal smooth muscle.

Contractions of the striated muscle of the mouth, tongue, jaw, palate, neck, pharynx, upper esophageal sphincter, and proximal esophagus are initiated voluntarily and also require stimulation of sensory structures in the region. Vagal afferents bear upon a swallowing center in the reticular formation of the brain stem between the caudal pole of the facial muscles and the rostral pole of the inferior olive. When activated, this center causes a stereotyped spatiotemporal pattern of contractions in the responding striated musculature through actions on motor nerves of the trigeminal, facial, hypoglossal and vagal nuclei. Sequential firing of neurons in these nuclei establishes the sequence of contractions that results in the progression of the swallowed bolus to the midesophagus. The responsible nerves are somatic nerves passing through the ninth and tenth cranial nerves directly to the muscle without synaptic interruption.

The upper esophageal sphincter is contracted because the striated muscle of this segment is tonically excited by tonic discharge of the somatic motor nerves that supply it. It relaxes because this tonic nerve activity is transiently suppressed by the action of the swallowing center on the central neurons that give rise to these motor nerves.

In the smooth muscle portions of the

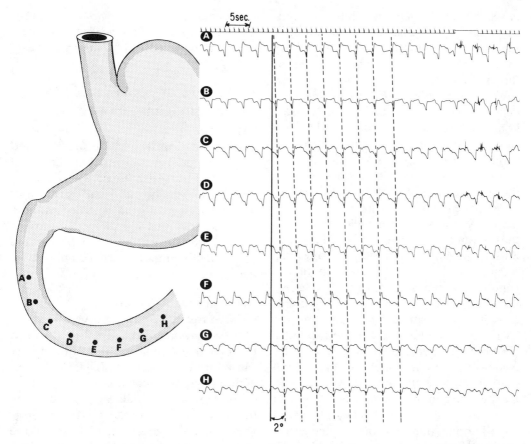

FIG. 21–5. A record of slow waves taken from the duodenum of an unanesthetized cat. Eight monopolar electrodes were sutured to the muscular wall at uniform 1-cm. intervals, indicated by letters *A–H*. The records from these experiments appear at the right. The solid vertical line indicates the time base. The dashed lines show slow waves progressing caudad. The angle, 2°, is a function of the paper speed of the recorder and of the velocity of spread of the slow waves. At the extreme right, spike bursts are shown on slow waves in channels *A–D*.

esophagus swallowing events are the consequence of the actions of autonomic nerves. Autonomic discharge is also excited by the swallowing center, since movements of esophageal smooth muscle require close coordination with the operation of the striated muscle segments. The result is that the whole apparatus, both striated and smooth muscle segments, operates as a single unit. Thus, the nerves that control the smooth muscle are autonomic in the sense that they are like the nerves of other parts of the gut, but they are not autonomous, being controlled by the central nervous system through vagal connections.

The vagal fibers are preganglionic and connect with intramural postganglionic neurons in a cholinergic synapse.

The nerves that regulate both peristalsis in the smooth muscle segment and relaxation of the lower esophageal sphincter comprise the third division of the autonomic nervous system. They are neither cholinergic nor adrenergic but inhibitory in their action on the smooth muscle, although the inhibitory transmitter has not been identified. These neural components are called nonadrenergic inhibitory nerves.

Paradoxically, the contraction of esophageal peristalsis occurs as a passive rebound of the muscle to nervous inhibition. Contraction ensues with a certain latency after the end of the period of activity of the inhibitory nerves. The swallowing center, in the course of the cycle it undergoes during swallowing, excites the inhibitory nerves to depress activity within the circular muscle layer of the distal esophagus. No relaxation can be seen in the smooth muscle because there is no tone in this muscle. The rebound contraction thus appears without the antecedent inhibition being apparent.

The progression of this contraction from the top of the smooth muscle segment to the bottom could represent a spatiotemporal sequence in the activation of these intramural nerves. The progression of the contraction, however, also probably reflects the existence of an excitatory gradient within the esophageal smooth muscle itself, that is, a gradient in its intracellular electrolyte composition which causes the resting membrane potential to decline from the top of the smooth muscle segment to the bottom. It is not clear how this gradient influences the latency of the rebound contraction, but this latency grows progressively along the esophageal body so that the rebound contraction appears to proceed caudad.

The lower esophageal sphincter constitutes a ring of smooth muscle at the esophagogastric junction that is tonically contracted at rest. This tonic contraction is caused by a special characteristic of the muscle itself rather than by special properties of the nerves: it is myogenic rather than neurogenic. This tone, however, allows the inhibitory activity of the nonadrenergic inhibitory nerves to be expressed as a relaxation.

The description of the control system given above would exclude circulating hormones from an important role in the governance of esophageal function. Gastrin and perhaps other polypeptide hormones may be of importance in regulating tone in the lower esophageal sphincter. Sphincter tone is inconstant at rest, and such hormones may induce small variations in resting tone. Another influence upon sphincter tone is tonic activity of excitatory nerves.

Disorders of pharyngeal and esophageal motor function

The motor disorders of the pharynx and esophagus are better defined than those of the rest of the gut because the processes are brief, stereotyped, and accessible. These disorders may be classified as those of the striated muscle regions and those of the smooth muscle parts. The disorders of the smooth muscle regions are more common.

Oropharyngeal dyskinesia: striated muscle dysfunction

The term oropharyngeal dyskinesia is used to encompass some of the entities that cause oropharyngeal dysphagia. This kind of dysphagia is characterized by a sense of obstruction by a swallowed bolus in the throat. When dyskinesia is at fault, there may also be frequent aspiration of the bolus or nasopharyngeal reflux, especially of liquids.

From the physiology outlined above, one can see that such dyskinesia can arise either from abnormalities of the somatic motor innervation (the more common cause) or from disease of the striated muscle itself. Years ago bulbar poliomyelitis was the most common cause; the disease attacked the vagal nuclei or the swallowing center, leaving survivors of this often fatal disease with permanent dysphagia. Today the most common cause is ischemic disease, involving the area of the fourth ventricle, where the responsible nerve centers lie. Atherosclerotic occlusion of the posterior inferior branch of the cerebellar artery causes infarction of the brain stem, thereby producing oropharyngeal dysphagia, a prominent symptom of the Wallenberg syndrome. Since only one side of the brain stem is affected, the dysphagia is rarely severe, and it is often reversible to some degree as healing occurs, presumably because some of the damage is attributable to local cerebral edema. When a generalized neuropathic process is responsible, as, for example, in diabetic cranial neuropathy, the degree of dysphagia may be much greater because the nerves from both sides may be affected.

The myasthenic syndrome is the principal disease of striated muscle that can cause oropharyngeal dyskinesia, and the symptoms in this case are often life threatening because of aspiration pneumonia. Myotonic dystrophy also may cause dysphagia, but it is rarely severe.

Hypopharyngeal diverticulum: striated muscle disorder

The hypopharyngeal diverticulum (Zenker's diverticulum) is an outpouching of the wall that arises just above the cricopharyngeus on the posterior wall of the pharynx. It is usually accompanied by mild oropharyngeal dysphagia. This is a disorder of middle age or later which is attributed to motor dysfunction of the pharyngoesophageal sphincter. Two mechanisms have been proposed: the forces of pharyngeal contraction may be increased, or there may be incoordination of sphincteric relaxation and pharyngeal contractions. In either case, an abnormally raised intraluminal pressure would occur in the hypopharynx and cause the diverticulum to form at the weakest point in the wall, posteriorly at the midline. Surgical therapy is directed toward dividing the cricopharyngeus muscle and elevating the diverticulum so that its contents will drain.

Achalasia: smooth muscle dysfunction

In achalasia there are two abnormalities of the smooth muscle segment: peristalsis in that part of the esophagus is lost and the gastroesophageal sphincter fails to relax upon swallowing. In Europe and North America, this idiopathic entity is not apparently associated with disordered motor function of other gastrointestinal organs, whereas in Chagas' disease other organs are affected as well. The etiological agent of Chagas' disease, *Trypanosoma cruzi*, attacks neurons of the myenteric plexus. This suggests that idiopathic achalasia is a disease of intramural nerves of the esophagus as well, and this idea is supported by histological evidence.

The striated muscle of the pharynx and esophagus functions normally in idiopathic achalasia, but the distal esophagus is dilated, often to a considerable degree, and often elongated. Peristalsis initiated by swallowing passes normally to the smooth muscle segment in which there is either no response or a feeble and nonprogressive contraction. Similar contractions may sometimes occur spontaneously. The gastroesophageal sphincter remains closed, and the force of closure is sometimes greater than normal.

The disease is caused by denervation of the smooth muscle. The most direct evidence for this is the histological finding of a reduced number of argyrophilic ganglion cells in the myenteric plexus. Indirect evidence is provided by the Mecholyl test, which uses an acetylcholine analogue that is resistant to destruction by cholinesterases. Injection of Mecholyl into patients with achalasia induces a brief period of powerful contractions in the smooth muscle segment of the esophagus. This is taken as evidence of postdenervation supersensitivity, since the response does not occur in normal people.

The modern concept of the genesis of esophageal peristalsis and gastroesophageal sphincteric relaxation attributes both these functions to a single set of intramural nerves, the nonadrenergic inhibitory nerves. Achalasia thus can be looked upon as a disorder confined to this one class of nerves. Whether the process attacks the secondary neurons directly, the vagal preganglionic nerves or the preganglionic neurons in the brain stem is unknown. The nature of the pathological process itself is also unknown.

Esophageal spasm: smooth muscle dysfunction

In esophageal spasm, swallowing causes simultaneous or incoordinated contractions, often of abnormally increased force, in the smooth muscle of the distal esophageal body. Spontaneous contractions are often seen. The gastroesophageal sphincter functions normally. Esophageal spasm is actually a heterogeneous group of disorders, for the patients who seem to fall in this diagnostic category show great variability in symptoms and in objective findings. Some patients are nearly asymptomatic, while others may face starvation because of dysphagia, or be incapacitated by pain. The smooth muscle of the esophagus may show any of the several motor abnormalities together or separately. Thus, spontaneous contractions, abnormally forceful contractions, or incoordinated contractions may all be present or be found in varying combinations. Furthermore, the abnormalities may be variably present or variable in quantity from one time to another. To confuse the definition of esophageal spasm further, it has been found that asymptomatic nonagenarians commonly show incoordinated contractions in the distal esophagus in response to swallowing. Some cases of esophageal spasm evolve into an achalasia-like syndrome, and some patients present features of esophageal spasm with failure of sphincteric relaxation in response to swallowing. This constellation is called *vigorous achalasia*. Also, some patients exhibit a positive Mecholyl test, suggesting a neural disease, although esophageal spasm may be a primary disorder of the smooth muscle itself, and hypertrophy of the esophageal smooth muscle is sometimes seen. In some cases, the force of resting closure of the gastroesophageal sphincter is abnormally high, and this could reflect hypertrophy as well.

Reflux esophagitis: smooth muscle dysfunction

It is generally accepted that the function of the gastroesophageal sphincter is to prevent the return of the gastric contents to the esophagus. When such reflux occurs, the corrosive action of acid gastric juice causes esophagitis. Sphincteric closure, however, is not the only motor function that opposes reflux: peristaltic contractions of the esophagus, whether produced by swallowing (primary peristalsis) or occurring as a reflex response to the reflux itself (secondary peristalsis), serve to clear the esophageal lumen of refluxed material. Thus, defective peristalsis may contribute to perpetuation of reflux esophagitis.

Weakness of the tonic closure of the gastroesophageal sphincter is common. Nicotine excites the secondary neurons of the nonadrenergic inhibitory nerves, and this could account for the weakness of sphincteric closure that is caused by smoking. The sphincteric closure is weakened by the presence of fat in the duodenum, and this could account for the relationship of heartburn to fat ingestion. There is evidence that the tone of sphincteric muscle *in vitro* depends exclusively upon oxidative phosphorylation, and so hypoxemia or vascular insufficiency might contribute to some cases.

In patients with scleroderma both defective sphincter closure and defective esophageal peristalsis are present, and these patients commonly have severe reflux esophagitis. The smooth muscle in this disorder shows degeneration and replacement with collagen, often focal but sometimes generalized. This alone would account for the motor abnormalities. The process is not confined to the esophagus, for smooth muscle in all gastrointestinal viscera may be variably affected.

Diverticula: smooth muscle dysfunction

A diverticulum of the esophagus is sometimes seen arising just above the diaphragm. It is often associated with achalasia or diffuse spasm. This suggests that, like the hypopharyngeal diverticulum, it may be a consequence of raised intraluminal pressures or a consequence of hypertrophy of the smooth muscle of this region.

The stomach

The motions of the stomach allow it to receive food, to store it, to mix it, to reduce it to a fine suspension of solids in liquid, and to regulate its delivery to the duodenum. These functions can be assigned to some degree to different parts of the stomach. The stomach is usually considered to consist of three parts, the fundus, the body, and the antrum. This classification is based mainly upon the character of the mucosa, and these three regions do not correspond fully to the differences in motor function from one part to another. Nevertheless, the three conventional terms will be retained in this discussion.

Receptive relaxation and accommodation

After a fast, the stomach contains only about 50 ml. of fluid. After a meal, it may contain a liter or more, and in some diseases it may contain several liters. This enormous increase in volume is accomplished with a very small rise in intraluminal pressure. The capacity of the stomach to expand so greatly without a rise in pressure depends upon a fall in wall tension. Relaxation of tension in the wall of the stomach can be demonstrated both when the food is swallowed (receptive relaxation) and when it is put in passively through a tube (accommodation). The fundamental phenomenon seems to be the same in both cases.

The process of relaxation is not simply an unfolding of the stomach, since the empty stomach is only moderately folded and wrinkled. The process is more accurately described as a relaxation of the muscle. This relaxation in muscle is more prominent in the proximal half of the stomach, especially in the fundus, and is neurally mediated by nonadrenergic inhibitory nerves that are connected to the central nervous system through vagal preganglionic fibers, as in the esophagus. These nerves are activated centrally during eating as well as by local reflexes, excited through mechanoreceptors in the gastric wall with pathways presumably confined to the intramural plexuses.

If relaxation occurs in the fundic muscle, then there must be some degree of tonic contraction in the muscle. This has been found to be the case, the tone being most prominent in the inner oblique muscle layer. This finding is consistent with the fact that this muscle layer is much more fully developed in the fundus than it is in the distal stomach. Neurogenic relaxation of tone in the oblique muscle layer of the proximal stomach appears to be responsible for both receptive relaxation and accommodation.

Mixing, grinding, and delivery

The reduction of large solid particles is partly, of course, a function of gastic digestion, but it is aided by wall motions that grind and mix gastric content to some degree to produce a suspension of solid particles in the swallowed and secreted fluids. Wall motions also regulate the rate of delivery of gastric contents to the duodenum. These motions reside mainly in the distal half of the stomach, the body, and antrum. Emptying of the stomach is also assisted by the fundic musculature.

The movements of the distal stomach are peristaltic contractions. In peristalsis, contraction rings first form as shallow indentations of the gastric wall at about the midpoint of the stomach that spread, slowly at first but with acceleration, toward the pylorus, deepening as they progress. The acceleration and amplification cause the lumen in the last 2 or 3 cm. of the antrum to be completely occluded essentially simultaneously. These peristaltic contractions exhibit a regular rhythm. In the state of maximal activity, they begin at intervals of 20 seconds. The period of the cycle is always some multiple of 20 seconds (40, 60, 80 seconds) in conditions of less than maximal activity.

The fixed character of these contractions (their frequency of occurrence, velocity, and direction of migration) indicate that they are tightly regulated. The control mechanism is the gastric slow wave or pacesetter potential. These slow depolarizations of the smooth muscle arise at very uniform intervals of 20 seconds near the midpoint of the stomach and spread caudad with acceleration and amplification, just as do the peristaltic contractions. The slow waves recur constantly but contractions do not. When a contraction accompanies a slow wave, the muscle generates a second electrical signal, consisting of either one rapid electrical transient (spikes) or a short burst of them or a second slow potential deflection that may represent numerous fused spikes.

The proportion of slow waves that are accompanied by peristaltic contractions varies greatly. This is presumably due to variations in the excitability of the muscle, and such variations may be imposed both by excitatory cholinergic nerves and by polypeptide hormones of gastrointestinal origin. After a meal, the level of excitability is high so that nearly all slow wave cycles generate peristaltic contractions. After a fast of 4 to 6 hours, the pattern of contraction varies systematically, short periods of maximal activity alternating with longer periods of inactivity, and the cycle is variable (90 to 120 minutes). This cycling is coordinated with similar cycling in the contractions of the small intestine and is termed the *fasted pattern of activity* or the *migrating motor complex*.

The effect of these peristaltic contractions on flow of gastric content can be observed radiographically. As each contraction ring moves toward the pylorus, it induces a flow such that material next to the wall is forced caudad and material at the core of the lumen is displaced cephalad. As the ring deepens, the magnitude and velocity of the flow increases. As the contraction ring deepens, the core of the lumen in the contraction ring becomes smaller. This pattern of flow mixes the gastric contents and helps to reduce the size of solid particles. In some way that is not wholly clear, particles above a diameter of about 0.5 cm. are always returned to the gastric body; most solid particles are smaller than about 0.1 mm. when they escape from the stomach. With each terminal antral contraction in the cycle, a small volume is delivered into the duodenum.

Emptying is also assisted by the restoration of tone in the relaxed fundic muscle, probably as a result of the graded withdrawal of tonic neurogenic inhibition by the nonadrenergic inhibitory nerves. Additional controls of fundic muscle may include the polypeptide hormones gastrin (which relaxes it) and cholecystokinin (which contracts it), released as a consequence of eating. The source of tone in fundic muscle, like that in muscle of the gastroesophageal sphincter, is not fully known.

Net gastric emptying has no fixed rate. It is regulated by several mechanisms. The rate of emptying of liquids varies with the volume of the gastric content. It is also affected by the actions of sensory chemoreceptors in the duodenal mucosa, acid, fat and osmoreceptors, each monitoring the concentration of its specific excitant in the fluid that bathes the mucosal surface. These are all inhibitory receptors, acting to reduce the rate of gastric emptying probably through reflex mechanisms. The well-known release of the intestinal mucosal hormones secretin and cholecystokinin in eating also could affect emptying of the stomach. These influences, both neural and hormonal, seem to affect its rate of emptying either by altering the rate of restoration of tone in the gastric fundus or by influencing the incidence or magnitude of antral peristalsis rather than by any important effect on the frequency of generation of the gastric slow waves.

The pylorus

The pylorus is commonly assigned a role in control of gastric emptying, but its influence is probably minor at best. The evidence for this statement is that the rate of gastric emptying in the dog is little affected when a stent is put in the pylorus to hold it open. Pyloric opening and closing do occur, however, opening taking place as an antral peristaltic contraction approaches and closure occurring as the terminal event in the antral contraction cycle. The pylorus acts to prevent duodenogastric reflux. Thus, when the pylorus is breached, either by pyloroplasty or by removal as part of a subtotal gastric resection for ulcer disease, reflux of duodenal contents into the stomach is commonly seen and often may be a cause of gastritis.

Disorders of gastric motor function

Diabetic gastroparesis

Nausea, vomiting, and a very large stomach sometimes occur in patients with diabetes. This problem is usually seen in those with long-standing, poorly controlled, insulin-dependent diabetes and is frequently seen in episodes of diabetic acidosis. Such patients often exhibit signs of peripheral neuropathy and of autonomic neuropathy, such as orthostatic hypotension. The gastric problem is probably another manifestation of autonomic neuropathy and is evident as a delay in gastric emptying. Detectable delay is often seen in the absence of symptoms, but when symptoms are present extreme slowing of emptying is found. The problem is clinically significant, since it makes control of the diabetes more difficult because the net rate of intestinal

absorption depends upon the rate at which food escapes the stomach.

Vagotomy and the dumping syndrome

Cutting the trunks of the vagus nerves is very commonly performed as a part of several different kinds of operations done for gastric and duodenal ulcer disease. Vagotomy has long been known to cause gastric dilatation and delayed emptying when performed alone. Vagotomy disrupts the pattern of generation and spread of gastric slow waves and also denervates the secondary ganglion cells of both the cholinergic excitatory pathways and the nonadrenergic inhibitory pathways. The ability of a pyloroplasty to modify the gastric retention that accompanies vagotomy alone is not wholly explained. Pyloroplasty is commonly called a drainage operation, a term that implies that the breach of the pyloric barrier has a purely mechanical effect through the elimination of a point of resistance to flow.

After vagotomy and any of the several kinds of gastric resections or pyloroplasties, patients sometimes develop a cluster of symptoms that is called the dumping syndrome. These symptoms are precipitated by eating and include weakness, flushing, sweating, abdominal cramps, and diarrhea. Various mechanisms have been proposed to account for them. The rapid delivery of a relatively hypertonic fluid to the intestine is thought, through its osmotic effect, to reduce blood volume; abnormal fluctuations in blood sugar levels are attributed to a disturbance in the normal coordination between carbohydrate delivery and insulin release; abnormalities in the timing of the stimulus to pancreatic secretion or in the mixture of pancreatic secretion and gallbladder bile with the food could produce maldigestion and consequent diarrhea. These mechanisms are all related to a disturbance in the control of the delivery of food from the stomach.

The gallbladder and bile ducts

Once the bile is secreted from the liver, its flow within the biliary tract depends upon differences in resistance to flow at various levels. In the fasting state, the resistance to flow through the ampulla of Vater is high so that bile is channeled through the cystic duct, where resistance to flow is lower, into the gallbladder. The gallbladder presumably has little tone at this time. The gallbladder reduces the volume it has received through its absorptive capacity, thereby concentrating the bile. After a meal, the gallbladder contracts to discharge its contents into the bile duct. Bile flows down the common duct into the duodenum, following the path of least resistance.

The four functional regions of this system, the gallbladder, the valve of Heister, the bile ducts, and the sphincter of Oddi, are under interrelated nervous and hormonal controls. Both the vagi and the splanchnic nerves carry excitatory and inhibitory nerves to all these structures. Intramural ganglion cells occur throughout the biliary system, lying in the submucosa, in the muscle layers, and in the adventitia. These ganglion cells are connected by nerve bundles in irregular plexuses. Afferent fibers arise from these viscera and travel in both vagal and splanchnic pathways.

Filling of the gallbladder involves a yielding of the muscle of that organ to the pressure of the secretion of bile, although vagal input to the gallbladder can modify its distensibility. There may be an inhibitory innervation that fosters this process. Contraction of the gallbladder after eating results mainly from the action of cholecystokinin, the polypeptide hormone released from the duodenal mucosa by fats and certain amino acids. The hormone acts directly upon the muscle of the gallbladder to induce contractions.

The wall of the common bile duct is mostly fibrous tissue in some species, but it contains variable amounts of smooth muscle in others. This raises the possibility that contractions can occur in this organ. There is evidence in some species that at least parts of this duct can exhibit peristaltic contractions that could influence bile flow.

The sphincter of Oddi is a distinct ring of smooth muscle buried in the duodenal wall at the ampulla of Vater. Since duodenal contractions are intermittent, whereas occlusion of the duct is not, it seems likely that the sphincter itself is mainly responsible for the maintenance of the prolonged periods of resistance to flow that characterize this structure. The relaxation of this sphincter that occurs after eating to allow bile flow to occur is also a direct effect of cholecystokinin on the smooth muscle of the sphincter.

Motor disorders of the biliary system

Cholelithiasis and cholecystitis

It is often suggested that alterations in gallbladder contraction after truncal vagotomy and in pregnancy predispose to the formation of gallstones and to the development of cholecystitis in association with gallstones. Such predisposition might be related to an alteration in motor function of the biliary system in these conditions.

Biliary dyskinesia

This poorly defined diagnostic term implies that motor dysfunction in the biliary system can lead to attacks of pain that mimic those of cholecystitis. Such attacks are sometimes seen in patients who have had cholecystectomy or who lack evidence of cholecystitis or stones. There is evidence as well that the acalculous gallbladder in some of these patients does not contract fully in response to the administration of cholecystokinin and that the attacks disappear after cholecystectomy. In some such cases an anatomical abnormality, such as cholesterolosis or adenomyomatosis of the gallbladder, is found; these lesions are presumed to limit the contractility of the gallbladder by causing abnormal rigidity of the wall. In other cases, the gallbladder is found to be histologically normal. In still other cases, motor dysfunction of the sphincter of Oddi is postulated to account for symptoms.

Ascending cholangitis

Bacterial infection of the normally sterile biliary tree is an uncommon but serious problem. It is often seen when the sphincter of Oddi has been breached, as in sphincterotomy, or bypassed, as in cholecystoduodenostomy. This suggests that an important function of the sphincter of Oddi may be to prevent reflux of duodenal content into the biliary tree.

The small intestine

The movement of chyme through the small intestine can be considered to have two components, namely net flow, which refers to the gross progression of the chyme along the conduit, and local flow, which is the movement of chyme across the mucosal surface to minimize the thickness of the unstirred layer. The unstirred layer refers to the film of fluid at the mucosal surface through which molecules must diffuse to reach the surface of absorptive cells. If this layer is very thick, the rates of diffusion of these molecules will limit the rate of absorption. Both types of flow depend upon forces generated by contractions of the muscular walls of the small intestine, including probably the muscle of the mucosa.

The contractions of the circular and longitudinal layers of the muscularis propria are rhythmical and can occur together or independently. Contractions of the circular layer occlude the intestinal lumen, while those of the longitudinal layer shorten the lumen with little effect on the diameter. Contractions of both layers occur as rather brief events, lasting a few seconds, and are repeated at intervals of several seconds or minutes. They are also quite localized, any single contraction occupying only a few centimeters of the conduit at one time.

Circular contractions are more readily recognized than longitudinal contractions. They are always progressive; the contraction can begin at any level, moves a variable distance caudad and then disappears. The distance over which it moves may be as short as 1 cm., or it may move many centimeters. For 4 to 6 hours after a meal these contractions occur with an apparently random distribution in time and space, and their distance of spread tends to be short. A different pattern develops thereafter. Contractions then begin to occur in cycles: at any single level of the intestine a long (30-minute or more) period of inactivity will be followed by a period of 20 minutes in which contractions occur in short bursts of three or four. These bursts gradually become longer and culminate in a period of 10 minutes or so in which contractions occur maximally as a continuous series that terminates abruptly to initiate another prolonged period of inactivity. There is great variability in the length of this cycle and its elements, but a full cycle usually occupies 90 to 150 minutes. This cycle is termed the *interdigestive pattern.* The cycle is also called the *migrating motor complex* because, if the cycle is recorded at separate points along the intestine, the whole cycle moves slowly caudad. The causes of this cycle are not known. It has been attributed both to cyclic variations in tonic nerve activity and to variations in levels of circulating hormones, especially motilin.

The rhythm of the circular muscle contractions, whether sporadic, as after a meal, or organized into the cyclic pattern found in fasting, reveals an underlying control system. At any single point along the duodenum, the temporal intervals between successive contractions are an integral multiple of 5 seconds. The fundamental is longer at more distal levels of the intestine, lengthening to 8 or 9 seconds in the distal ileum. This indicates the existence of a pacemaker.

The pacemaker for circular muscle contractions is the intestinal slow wave. Slow waves recur constantly in the muscle at all levels of the intestine, but their frequency varies. Throughout the duodenum, the frequency (in man) is almost exactly 5 seconds, with very little variation over time or among individuals. At more distal levels, frequency declines in steps, each frequency plateau being several centimeters long, to terminate in a frequency of 8 or 9 seconds in the last frequency plateau in the ileum. Within each plateau, slow waves spread caudad. Thus, the slow waves are driven by a series of pacemakers whose frequencies decline progressively along the intestine. The locations of these pacemakers are not fixed, however, and this indicates that all of the muscle has potential pacemaker capability. The distribution of the slow waves establishes the distribution of circular contractions. Whether or not any given ring of circular muscle contracts in response to any given slow wave depends upon factors that affect the general level of excitability of the muscle. Thus, the cycling in the incidence of contractions that characterizes the fasted state represents a cycling in the level of excitability of the muscle.

Longitudinal muscle contractions are also paced by the intestinal slow waves. They too migrate caudad.

The moving ring contractions, because they occlude the lumen, are responsible for net flow. Some degree of retrograde flow also can be produced when ring contractions, through their random distribution in time and space, happen to occur in immediate proximity to each other in such a way that the caudal contraction just precedes the cephalad one, but this is probably only an occasional event. This possibility is evident from the fact that, when it first forms, a ring contraction will displace fluid in both directions.

Circular contractions probably produce rather little mixing of chyme. Mixing of chyme is attributable to the longitudinal contractions. They induce a pattern of flow such that they tend to cause local flow, to encourage movement of chyme across the mucosal surface, and thus to reduce the thickness of the unstirred layer. Such local flow is also attributable to the contractions of mucosal muscle. Though mucosal muscle has had very little study, it seems most likely that mucosal movements occur, especially motions of the intestinal villi, that should have considerable effect on stirring of the fluid at the interface between mucosal cells and the chyme. Movements of the villi are thought to be controlled by a poorly characterized polypeptide hormone called villikinin.

The peristaltic reflex is sometimes attributed a role in the governance of contractions in the small intestine. This reflex is a response to distention of the bowel. Distention at one level leads to contraction of both muscle layers above the point of distention and inhibition of contractions below the distention. It probably has little normal physiological significance, but it may be important in overcoming the resistance to flow offered by mass lesions that partially obstruct the lumen.

Disorders of small intestinal motility

Ileus

Ileus is the prolonged absence of contractions in gastrointestinal muscle, especially in the small intestine. When ileus occurs, it often affects more of the gut than the small intestine. It is often neurogenically mediated, the consequence of reflexes excited by a variety of stimuli. In some cases, electrolyte imbalance seems to be responsible. The stimuli in neurogenic ileus may be local, related to excitation of mechanoreceptors or chemoreceptors in the bowel, stimulated, for example, by inflammation as in peritonitis. Severe pain, as in fractures or other trauma, occasionally appears to cause ileus.

Megaduodenum

Rarely one encounters patients who have a grossly dilated duodenum with other parts of the intestine variably dilated too. The abnormality is sometimes a part of the syndrome of intestinal pseudoobstruction, in which the intestine appears to be obstructed in the absence of a mass lesion that obstructs the intestinal lumen. The disorder in some cases is sporadic, but in other cases it is familial. Two general causes have been suggested: in most cases, one or both layers of smooth muscle in the muscularis propria are found to be replaced with collagen. In other cases, there is evidence of neuropathy inasmuch as ganglion cells of the myenteric plexuses are found to exhibit degenerative changes or to contain inclusion bodies that suggest a viral infection.

Scleroderma

As discussed previously, scleroderma frequently is characterized by fibrous replacement of the smooth muscle of the muscularis propria. This may occasionally lead to the clinical picture of intestinal pseudoobstruction, but it is more commonly associated with chronic diarrhea. This diarrhea is attributable to the overgrowth of colonic bacteria in the small intestine, and it disappears with antibiotic treatment. This relationship indicates that intestinal contractions are important not only in propelling chyme but also in clearing the intestinal lumen of bacteria that may reflux into the small intestine from the colon. In some cases, it has been found that the migrating motor complex is absent in fasting. This suggests that the cyclic bursts of maximal activity that sweep the bowel in the interdigestive state may be important mainly in opposing bacterial invasion. For this reason, the migrating motor complex is sometimes called the *interdigestive housekeeper.*

The colon and anal canal

From the functional point of view, the colon consists of three fairly distinct parts. In the cecum and proximal colon, the dominant pattern of fecal flow is cephalad. Stool is retained and mixed in this segment for long periods of time. In the remainder of the intraabdominal colon, stool tends to move very slowly caudad. In the descending and sigmoid colon, stool is retained for longer periods of time and moves caudad mainly only in defecation.

Most of what we know about colonic movement comes from studies in the cat and a few other experimental animals. There are significant variations in the anatomy, however, so that the extrapolation of conclusions among species may not be fully justified. Nevertheless, the limited studies that have been made in the human colon suggest that the general patterns of flow and contractions seen in the cat also occur in man.

In contractions of the proximal colon, the principal pattern of contraction seems to be analogous to the rhythmic contractions described in the small intestine. In the rest of the colon, the dominant pattern of contraction appears to be rhythmic circular contractions that move caudad, but for very short distances. The pelvic colon (the rectum) is devoid of rhythmic circular muscle contractions.

Tonic contractions are probably present in the circular muscle of the colon too. At least, that is one interpretation of the indentations that are seen in the intraabdominal colon. These indentations were formerly thought to be fixed structures, but it is evident that they can form and disappear, so they seem more likely to represent prolonged contractions of the circular muscle layer.

The rhythmic contractions of the circular muscle layer, like those of the small intestine, are paced by electrical slow waves. In the proximal colon these waves usually migrate cephalad, toward the cecum, although they may reverse to spread caudad for a short time. This could be a means for temporarily inducing contractions that will serve to empty this segment. In the more distal segments of the abdominal colon, the slow waves migrate caudad, but generally over short distances. It appears that there are a series of pacemakers, each governing a short segment of the colon. Such a pattern is consistent with the slow progression of stool through this segment. The frequencies of slow waves, and of the contractions they pace, are much slower in the colon than in the small intestine, being about 5 to 6 cycles per minute, which is consistent with the slowness of progression of the colonic content.

The movement of stool through the abdominal colon is slow except at certain times when a segment of the fecal mass is propelled quite rapidly over a considerable distance in what is called a *mass movement*. A mass movement occurs only a few times each day, usually soon after eating. In a mass movement, haustral indentations disappear in the area of the fecal mass that is to be moved. This is followed by a rapid movement of the fecal mass, after which the haustral indentations are restored. The propulsion is presumably due to a forceful moving contraction of the circular muscle layer. Mass movements seem to represent an integration of events that is neurally mediated. They may, however, be initiated by some hormonal event, since they are so clearly related to eating.

Defecation

When a portion of the fecal mass enters the rectum, it is retained for long periods of time. Defecation is a neurally mediated process involving a number of different movements. Defecation is normally initiated voluntarily, though it can be wholly involuntary.

At rest the smooth-muscled internal anal sphincter is tonically contracted, as is also, to a variable degree, the voluntarily controlled external sphincter. The nature of the genesis of this tonic contraction of the internal sphincter is not known. At rest, the rectum is relaxed. In response to stimulation of sensory mechanoreceptors in the rectum by the fecal mass, the defecation reflex is initiated. This involves first inhibition of both anal sphincters, though voluntary effort can overcome the inhibition of the striated external sphincter to achieve continence. A familiar series of efforts involving the striated musculature of the diaphragm, abdominal wall, and pelvic floor follows, and then there is a peristaltic contraction that may begin as high as the sigmoid or descending colon that ejects the fecal mass. This complex set of movements obviously requires coordination of a variety of central and peripheral nervous mechanisms. The coordinating centers include a poorly defined locus in the caudal spinal cord. Relaxation of the internal anal sphincter is a function of the nonadrenergic inhibitory nerves. The genesis of the peristaltic contraction of the bowel is not known. The somatic muscles contract through activation of their somatic nerves.

Disorders of colonic motor function

Diverticulosis of the colon

Diverticulosis is currently viewed as the immediate consequence of a motor disorder in the part of the colon where diverticula are usually concentrated, the sigmoid colon. In some cases, there is thickening of rings of the circular muscle adjacent to the diverticulum, and this thickening precedes the development of the diverticulum. The force of rhythmic contractions in the areas of muscle thickening can be greatly increased above normal. It is postulated that these increased forces produce herniations of the mucosa through the circular muscle layer. Since the hypertrophied rings are separated by short spaces, the diverticula form between adjacent rings. The diverticula are almost always paramesenteric in position, because the weakest points in the muscular wall are the points where the arteries penetrate the muscle layers on both sides lateral to the mesentery. The cause of the muscular hypertrophy is unknown.

Aganglionosis of the colon (Hirschsprung's Disease)

In this congenital disorder, a segment of the distal colon, extending a highly variable distance above the anus, is tonically contracted to obstruct the colonic lumen, producing dilatation of the colon above the contracted segment. The contracted segment is deficient in or devoid of ganglion cells in the intramural plexuses, and the plexuses contain grossly disorganized bundles of nerves that are morphologically abnormal. Evidence suggests that the tonic contraction is due to unopposed cholinergic motor nerve excitation of the circular muscle layer. The internal anal sphincter fails to relax in response to rectal distention. Thus, tonic contraction of the obstructing segment seems to be a consequence of a selective loss of inhibitory nerves, presumably nonadrenergic inhibitory nerves.

Incontinence of the elderly

It is not uncommon to encounter, in aged people, rectal incontinence that seems to represent the inability to suppress the urge to defecate. Since such suppression represents in part voluntary contraction of striated muscles of the external anal sphincter and pelvic floor, it seems likely that such incontinence represents dysfunction of that musculature. Recent evidence suggests that this is a consequence of denervation of this striated muscle from abnormal trauma to the motor nerve trunks that supply this musculature. It is proposed that, as a result of chronic constipation, frequently repeated and forceful efforts to induce defecation produce traumatic stretching of these nerves through abnormal displacements of the pelvic floor.

Functional bowel (or irritable bowel) syndrome

These ill-defined diagnostic terms are often indiscriminately applied to patients who have abdominal pain, diarrhea, and constipation in the absence of objective evidence of organic disease. The terms imply that those who use them believe that a motor abnormality of the colon is at fault. It is, indeed, quite possible that such a primary motor abnormality exists, but it has not yet been convincingly demonstrated. Recent reports suggest that the frequency of colonic slow waves in such patients is abnormally slow, but this idea remains controversial. A number of well-defined pathophysiological processes that can cause diarrhea, constipation, and abdominal cramps in various combinations have been identified in recent years. It must be obvious that patients with these disease entities, before they were defined, must have been diagnosed as having the irritable bowel syndrome. Thus, these are "wastebasket" diagnostic terms that are used to conceal ignorance rather than to clarify thinking.

The laxative colon

In this entity, constipation persists in the face of long-continued abuse of laxatives that contain anthraquinones. There is evidence that the intramural plexuses are damaged in patients who used these drugs to excess, and this is presumed to contribute to the constipation.

References

Bortoff, A.: Digestion: Motility. Ann. Rev. Physiol. 34:261–290, 1972.

Bortoff, A.: Myogenic control of intestinal motility. Physiol. Rev. 56:418–434, 1976.

Burnstock, G.: Evolution of the autonomic innervation

of visceral and cardiovascular systems in vertebrates. Pharmacol. Rev. *21*:247–324, 1969.

Burnstock, G.: Purinergic nerves. Pharmacol. Rev. *24*:509–581, 1972.

Christensen, J.: The controls of gastrointestinal movement: some old and new views. N. Engl. J. Med. *285*:85–98, 1971.

Christensen, J.: The physiology of gastrointestinal transit. Med. Clin. North Am. *58*:1165–1180, 1974.

Christensen, J.: Myoelectric control of the colon. Gastroenterology *68*:601–609, 1975.

Cooke, A. R.: Control of gastric emptying and motility. Gastroenterology *68*:804–816, 1975.

Daniel, E. E.: Digestion: Motor function. Ann. Rev. Physiol. *31*:203–226, 1969.

Daniel, E. E.: Electrophysiology of the colon. Gut *16*:298–306, 1975.

Disorders of Oesophageal Motility. Clin. Gastroenterol. *5*:(1), 1975.

Goyal, R. K., and Rattan, S. Neurohormonal, hormonal and drug receptors for the lower esophageal sphincter. Gastroenterology *74*:598–619, 1978.

Hunt, J. N.: Gastric emptying and secretion in man. Physiol. Rev. *39*:491–533, 1959.

Ingelfinger, F. J.: Esophageal motility. Physiol. Rev. *38*:533–584, 1958.

Lin, T-M.: Actions of gastrointestinal hormones and related peptides on the motor function of the biliary tract. Gastroenterology *69*:1006–1022, 1975.

Martinson, J.: Studies on the efferent vagal control of the stomach. Acta Physiol. Scand. (Suppl.) *255*:1–24, 1965.

Misiewicz, J. J.: Colonic motility. Gut *16*:311–314, 1975.

Parks, T. G.: Colonic motility in man. Postgrad. Med. J. *49*:90–99, 1973.

Prosser, C. L.: Smooth muscle. Ann. Rev. Physiol. *36*:503–535, 1974.

Smith, B.: The Neuropathology of the Alimentary Tract. London, Edward Arnold. 1972.

Truelove, S. C.: Movements of the large intestine. Physiol. Rev. *46*:457–512, 1966.

Weisbrodt, N. W.: Neuromuscular organization of esophageal and pharyngeal motility. Arch. Intern. Med. *136*:524–531, 1976.

22

Gastric Secretion

Travis E. Solomon, M.D., Ph.D.

The stomach secretes hydrochloric acid, the proteolytic enzyme pepsinogen, mucus, sodium, potassium, bicarbonate, and intrinsic factor into its lumen. The stomach is in one sense a dispensable organ, since it is possible for a person to survive after partial or total gastrectomy; in fact these operations are sometimes performed as therapy for certain diseases of the stomach. However, the secretions of the stomach serve several useful purposes: (1) gastric acid is a first line of defense against many ingested pathogens and the stomach and upper intestine are usually sterile; (2) acid and pepsin initiate digestion of protein in food, although this probably accounts for no more than 15 per cent of protein digestion; (3) recent evidence indicates that gastric mucus and bicarbonate may serve to protect the stomach from digesting itself; and (4) intrinsic factor is essential for efficient absorption of vitamin B_{12} (cobalamin). The absence of intrinsic factor leads to development of pernicious anemia.

The two most common illnesses that result primarily or in part from disorders of gastric secretory function are duodenal ulcer disease and gastric ulcer disease. An even more common illness, gastritis, may also produce changes in gastric secretion. These two forms of ulcer disease are frequently lumped together under the term *peptic ulcer disease*, since it was thought in the past that the protein digesting ability of pepsin might contribute to development of

ulcers (holes in the mucosa lining the stomach or duodenum). Although the importance of pepsin in the pathogenesis of peptic ulcer disease is uncertain, there is no argument about the importance of gastric acid as a causative factor. The occurrence of gastric ulcer is extremely rare in subjects who secrete no acid (a condition called achlorhydria), while duodenal ulcer never occurs in achlorhydric subjects. Beyond this straightforward statement, summarized in the dictum "no acid, no ulcer," there are very few generalizations to be made about peptic ulcer disease. This is probably because peptic ulcer disease is many diseases rather than a single disorder. It is now clear that there are some groups of subjects with a genetic tendency to develop peptic ulcer disease. Several genetically determined disorders of gastric function combined with possible environmental factors allow many mechanisms to underlie the formation of ulcers. This complicates the pathophysiology of ulcer disease, since we can identify only those alterations that are common to a large number of ulcer subjects or which result in very striking degrees of functional change.

Functional gastric anatomy

Parts of the stomach
The stomach has been divided into different parts based upon either the gross anat-

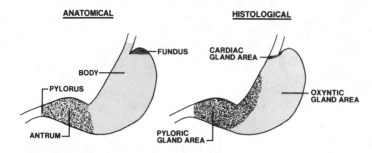

FIG. 22–1. Names of the parts of the stomach, based on anatomical (*left*) and histological (*right*) definitions. Note the lack of correlation between the two methods of naming.

omy, the histology of the gastric mucosa, and the functions of the organ. Confusion has resulted from the many names given to the parts of the stomach, several names having been assigned to the same part or cell type, or the same name having been used for different parts. Figure 22–1 illustrates the terms most frequently used today. Anatomically derived terms include the fundus, body, antrum, and pylorus. Terms derived from the histology of different parts of the stomach are the cardiac gland, oxyntic gland, and pyloric gland areas. The microscopic appearance of these areas is described below. There is only a rough correlation between the anatomically and histologically named parts of the stomach. The cardiac gland area is a narrow rim of tissue around the esophagogastric junction and thus encompasses portions of both the anatomical fundus and body. The oxyntic gland area occupies most of the fundus and body and may extend into the antrum, particularly on the greater curvature of the stomach. The pyloric gland area occupies most of the antrum but extends into the body of the stomach on the lesser curvature.

The motor and storage properties of the stomach correlate best with the anatomically derived names. The fundus and body serve partly to store swallowed food and fluid. Their muscle layers are thinner and the patterns of motility are different than in the antrum. The proximal parts of the stomach are important in the emptying of liquids, and the muscle layers produce a tonic elevation of intragastric pressure that is low but sufficient to empty large amounts of liquids from the stomach. The antrum and pylorus, on the other hand, serve primarily to grind solid food into small particles and to regulate the rate of emptying of both solid and liquid food. The muscle layers of the antrum are thickest in the most distal one or two centimeters of the stomach. There is some ambiguity about the name of this part of the stomach. The word *pylorus* is commonly used to refer to the distal one or two centimeters of the stomach or to the opening or passageway from stomach to duodenum. The term *pyloric sphincter* is also used to refer to this part of the stomach wall, although it implies a function for the most distal part of the stomach, which many physiologists doubt.

Histology of the stomach wall

The stomach wall, like the wall of almost all of the tubular digestive tract, has four distinctive layers, named (from outside to the lumen) serosa, muscularis, submucosa, and mucosa. The serosa is a thin (less than 50 μ thick) layer of loose connective tissue covered by a single layer of squamous cells (the mesothelium). Large blood vessels, nerves, and lymphatics lie between this layer and the muscularis. The smooth muscle cells of the muscularis are arranged in three approximate directions: cells of the outermost layer are oriented in the longitudinal axis of the stomach; cells of the middle layer are oriented perpendicular to the long axis (the circular layer); and cells of the inner layer lie in an oblique direction. A network of intrinsic nerve fibers is found between the circular and longitudinal layers of muscle and is called *Auerbach's plexus* (also called the *myenteric* plexus).

The submucosa is about 100 μ thick and is composed of dense connective tissue and a thin layer of smooth muscle cells, the muscularis mucosa. Blood vessels and a second network of intrinsic nerves (called either *Meissner's plexus* or the *submucosal plexus*) course throughout the submucosa. This tissue layer is used as a boundary for the definition of a gastric ulcer. An ulcer is a lesion in the mucosal lining of the stomach that extends through the muscularis mucosa into the submucosa.

The mucosa is the innermost layer of the stomach, extending from muscularis mucosa to the luminal surface. It ranges in thickness from 1 mm. in the oxyntic gland area to 0.5 mm. in the pyloric and cardiac gland areas. Gross examination of the luminal surface of the stomach does not reveal any differences in the mucosa. Thousands of tiny pits (also called foveoli) dot the surface of the mucosa. These pits average 200 μ in depth. Hence, the pits penetrate only about one-fifth of the thickness of oxyntic gland mucosa and half of the thickness of pyloric gland and cardiac gland areas. One to seven tubular glands of the stomach open into each pit, in the various areas of the mucosa. The three types of glands are described separately in this chapter.

The luminal mucosal surface and the pits are lined by a single type of epithelial cell, the surface mucous cell (Fig. 22–2). These tall, columnar cells are packed with mucus-containing granules at their apical ends. The cells are joined to each other by several junctional complexes near their apical surfaces. The gastric mucosa forms a "tight" sheet membrane, that is, the surface is almost impermeable to the movement of water soluble substances (including HCl) from the lumen. The only luminal materials that can penetrate the mucosa readily are highly lipid soluble agents, such as ethanol. The surface mucous cells and their cell-to-cell junctions provide this tight barrier to the back diffusion of secreted acid, and the surface mucous cells are also the first to be damaged when the gastric mucosal barrier is broken by injurious agents. The lamina propria is a tissue lying just beneath the surface mucous cells and surrounding the glands. The lamina propria contains blood vessels, nerves, lymphatics, and scattered plasma cells, lymphocytes, polymorphonuclear leukocytes, eosinophils, and mast cells (the latter cells are seen particularly in the oxyntic gland areas). The mast cells synthesize and store histamine, a major regulator of gastric acid secretion.

CARDIAC GLANDS. These glands are composed entirely of mucus-containing cells which resemble surface mucous cells. The glands are coiled and fairly short compared to oxyntic glands. The function of the cardiac glands is unknown.

OXYNTIC GLANDS. The oxyntic gland area occupies 85 per cent of the surface area of the stomach in man. The cell types, their relative proportions, and their products are listed in Table 22–1. The surface mucous

TABLE 22–1
Secretory Cell Types in the Oxyntic Gland

Cell	Products	Percentage of Mucosa*
Surface mucous cell	Mucus	17
Mucous neck cell	Mucus, new cells	6
Oxyntic cell	HCl, intrinsic factor	32
Peptic cell	Pepsinogens I and II, lipase	26
Endocrine cells	Glucagon, somatostatin, 5-hydroxytryptamine	1

*Lamina propria makes up the remainder of the mucosa.

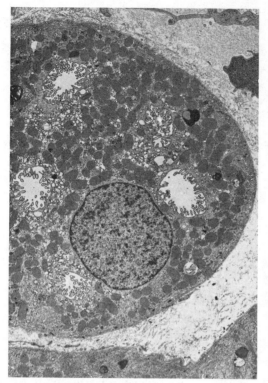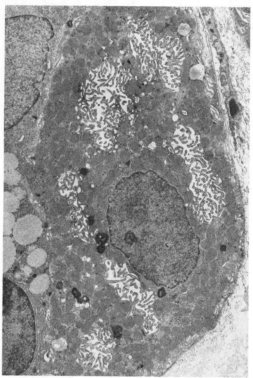

FIG. 22–2. Low-power electron photomicrograph of human oxyntic cells under basal (*left*) and stimulated (*right*) conditions. Note numerous mitochondria and the presence of tubulovesicles in basal state, as well as the pronounced increase in surface area of canalicular membrane in stimulated state.

cell has been described above. The mucous neck cell is the progenitor or stem cell which divides and proliferates to produce all of the other cell types (except endocrine cells). As its name indicates, it is located in a narrow neck-like region at the top (luminal) part of the gland, just below the junction of the gland with its pit.

The major cell type of the oxyntic gland, in terms of both number and importance of its secretory products, is the oxyntic (or parietal) cell. The oxyntic cell is the largest cell in the gastric mucosa and has the highest concentration of mitochondria of any cell in the body, reflecting the high energy consumption of the acid secreting mechanism. The oxyntic cell also has a unique system of tubulovesicles and a tortuous branching intracellular canaliculus (Fig. 22–2), both of which are important in the acid secreting function of this cell and will be described below. In human beings, the oxyntic cell has a maximal acid secretory capacity of about 50 pmol. h^{-1} cell^{-1}. In a normal stomach from a 70-kg. man there are 1 billion oxyntic cells, so the total maximal acid secretory capacity is about 50 mmol. h^{-1}. The oxyntic cell also secretes intrinsic factor, a large glycoprotein which is essential for the absorption of vitamin B_{12}. The oxyntic cells are found in greatest density in the upper part of the oxyntic glands, although a few oxyntic cells may be present throughout the length of the glands. Oxyntic cells constitute an even larger proportion of the total glandular cell

TABLE 22–2
Endocrine Cell Types in the Gastric Mucosa

	Cell Type	Product	Action
Oxyntic gland mucosa	P	Not known	?
	D_1	Not known	?
	EC(enterochromaffin)	5-hydroxytryptamine	?
	D	Somatostatin	May inhibit secretion by exocrine gland cells
	A	Glucagon	Present only in fetus
	ECL(enterochromaffin-like)	Not known	Unknown in humans
	X	Not known	?
Pyloric gland mucosa	P	Not known	?
	D_1 (few)	Not known	?
	EC	5-hydroxytryptamine	?
	D	Somatostatin	Inhibits gastrin release
	G	Gastrin	Stimulates gastric secretion

population in the more distal parts of the oxyntic gland area, until the transition zone between this area and the pyloric gland area is reached. In this transition zone, which may extend for several cm. in man, oxyntic glands gradually give way to pyloric glands until only mucus-secreting cells are present.

Peptic cells (also known as chief cells) secrete the precursor of a proteolytic enzyme. At the ultrastructural level, peptic cells exhibit the classical features of an exocrine cell which synthesizes and secretes exportable proteins. Membrane-bound storage granules fill the apical part of the cell while rough endoplasmic reticulum and Golgi complexes are prominent in the basal and middle regions of the cell.

Although endocrine cells (also called argentaffin cells because they stain with silver salts) are relatively few in number compared to the other cell types in the oxyntic gland area mucosa, they contain several important peptides that may act as local regulators of oxyntic gland function. There are at least six different classes of endocrine cells, based on their ultrastructural appearance, found in the oxyntic gland area. Some of these secrete bioactive amines (histamine, serotonin) alone or in addition to peptides. Table 22–2 lists the cell types, their products, and the known effects of these products.

PYLORIC GLAND AREA. The two major cell types in pyloric glands are mucous cells and endocrine cells. The pyloric glands are tortuous and coiled, so that a section perpendicular to the mucosal surface frequently reveals round cross-sections of glands as well as the tubular appearance seen in sections of oxyntic gland mucosa. The mucous cells of pyloric glands are similar to mucous neck cells in oxyntic gland mucosa. However, the proliferative zone in pyloric glands is limited to the neck region even though all gland cells are morphologically similar. About 1 per cent of the pyloric gland cells are endocrine cells of which there are five types (Table 22–2). G cells that synthesize and secrete gastrin constitute the majority of endocrine cells (40 to 60 per cent) in this area. G cells occur in a distinct band just below the neck of the pyloric glands. The G cell extends to the luminal surface of the gland and has microvillar extensions on its apical surface. These microvilli act as receptors, "tasting" the content of the stomach and transmitting signals which control the rate of gastrin secretion into the bloodstream. This process will be described in detail in the section on regulation of gastrin secretion.

Mucosal cell turnover

All of the various cell types in the gastric mucosa are renewed, although the rates of renewal differ between cell types. In man the entire population of surface mucous cells and mucous neck cells is replaced every 3 to 6 days, turnover being more rapid in the pyloric gland area than in the oxyntic gland area. The rapid replacement of these surface cells may be due to the buffeting which they absorb from mechanical and chemical (high acid concentration) conditions in the stomach. The glandular cells in oxyntic and pyloric gland areas have much longer life spans. Oxyntic, peptic, and G cells survive for weeks to months.

Most mucosal cells, including G cells, differentiate under normal conditions from the stem cells which are located in a narrow band occupying 10 to 20 per cent of the depth of the mucosa and extend from the base of the gastric pits down the neck of the glands. In this proliferative zone 10 to 20 per cent of the cells are always found in the S phase of the cell cycle, indicating active deoxyibonucleic acid (DNA) synthesis. Most of the newly produced cells differentiate into mucous cells and migrate upward to renew the surface cells. A small fraction differentiate into oxyntic cells, mucous neck cells, or G cells and migrate downward into the glands. The exception to this arrangement is the peptic cell, which seems to undergo autoreplication, that is, the peptic cells divide and produce more peptic cells.

The rate of stem cell division and the patterns of differentiation into various cell types are regulated by several factors. The classical trophic and metabolic hormones such as growth hormone and thyroid hormones influence cell growth in the stomach. In addition, gastrointestinal regulatory peptides also control cell proliferation in this organ. The best understood of these peptides is the hormone gastrin, which increases the rate of stem cell proliferation and also increases the proportion of oxyntic cells which differentiate from these stem cells.

Innervation

The stomach is richly innervated by the autonomic nervous system; and gastric motility, secretion, and blood flow are all influenced by postganglionic fibers which secrete acetylcholine (cholinergic nerves), norepinephrine (adrenergic nerves), and a variety of peptides (peptidergic nerves). The peptidergic nerves constitute a third division of the autonomic nervous system and have stimulated much new research in gastric physiology from which we have gained explanations about several previously unknown aspects of the neural control of gastric function.

Vagal nerves, which constitute the major extrinsic neural supply to the stomach, send bundles of preganglionic fibers directly to the stomach. These vagal branches ramify over the serosal surface before plunging into the tissue to synapse with the ganglion cells (postganglionic fibers) in Auerbach's and Meissner's plexuses. Fewer than 10 per cent of the nerve fibers in the vagus nerves are efferent (i.e., brain to gut), a total of about 2,000 fibers, while there are about 100 million ganglion cells in the stomach, small intestine, and proximal colon that are innervated by the vagus nerves. Thus, each efferent vagal fiber serves as a modulator of the activity of many thousands of postganglionic fibers. Nearly 90 per cent of the vagal trunk fibers are afferent and are involved in reflexes (vagovagal reflexes) that control many aspects of gastrointestinal function. The other major part of the extrinsic innervation of the stomach comes from adrenergic (sympathetic) nerve cells which enter the stomach accompanying the arterial supply, as discrete nerve bundles, or as a minor part of the fibers in the vagal branches. The much more numerous efferent fibers in these

TABLE 22–3
Neurotransmitters Found in the Stomach

Substance	Action
Norepinephrine	Contracts vascular smooth muscle and decreases blood flow
Acetylcholine	Stimulates secretion by oxyntic, peptic, mucus, and G cells; inhibits somatostatin secretion from D cells; increases blood flow secondary to increasing secretion; increases motility
Somatostatin	May inhibit motility and secretion
Substance P	Not known
Vasoactive intestinal peptide (VIP)	May inhibit motility and secretion
Gastrin releasing peptide	Releases gastrin
Enkephalin	May stimulate acid secretion and motility

nerves are postganglionic, norepinephrine-secreting nerves that originate in the celiac ganglion and synapse with ganglion cells in both major gastric plexuses. Adrenergic nerves act on gastric secretion and motility by modulating the activity of the ganglion cells, although there may be direct effects of norepinephrine on smooth muscle and secretory cells under certain conditions. Adrenergic nerves appear to act directly on blood vessels to regulate gastric blood flow.

This description of the course of the extrinsic innervation of the stomach is intended to emphasize the importance of the two major intrinsic nerve plexuses as the final mediators of neural regulation of gastric function. Physiologists have regained an appreciation of the importance of these plexuses (a plexus is a network of ganglion cells and interneurons) and now recognize that the "enteric nervous system" actually functions as a separate brain that can be influenced by (but is not dependent on) the central nervous system. The "business end" of this system, the ganglion cells which eventually influence gastric exocrine and endocrine secretory cells and muscle cells, consists of excitatory and inhibitory fibers. The excitatory fibers secrete acetylcholine or various peptides which act directly on the target cells (the oxyntic cell, for example). Inhibitory fibers release peptides and, possibly, ATP (the putative transmitter for a possible fourth division of the autonomic system, the purinergic nerves) which may act directly on the target cells. Acetylcholine-secreting vagal preganglionic fibers act on both excitatory and inhibitory ganglion cells. Adrenergic fibers, which are mostly inhibitory for various gastric functions, appear to cause inhibition by means of norepinephrine-induced suppression of acetylcholine release by postganglionic excitatory fibers rather than by any effect on postganglionic inhibitory fibers. Norepinephrine also increases intracellular calcium in vascular smooth muscle. Finally, there may be interneurons in the plexuses that regulate the activity of postganglionic fibers. Table 22–3 lists the postganglionic transmitters found in the stomach and their actions.

The mechanism of action of neurotransmitters found in the stomach is different than that in classical neurotransmission. In the central nervous system and at motor end-plates neurotransmission is accomplished by release of the transmitter into a narrow synaptic cleft formed by apposition of the nerve and the effector cell membrane in a highly specialized structure. Termination of stimulation by the transmitter is accomplished by reuptake into the nerve or destruction at the effector cell membrane. There are no such highly specialized nerve endings in the stomach (or in the rest of the gastrointestinal tract). Postganglionic fibers run close to their target cells but exhibit no structures that could be interpreted to exist for the purpose of transmitter release. Thus, it appears that the membrane-bound granules which contain the transmitters fuse with the neurolemma of the fiber all along its length, releasing the neurotransmitter into extracellular fluid close to target cells. Termination of transmitter action is due to diffusion of the transmitter away from the target cell, destruction by ubiquitous cholinesterases in the case of acetylcholine, and possibly by uptake and intracellular destruction of peptide transmitters by target cells or degradation of transmitters in extracellular fluid by peptidases. A consequence of this mechanism of transmission is that there is less specifity of effect, since the neurotransmitter may diffuse to many target cells or even enter the circulation and act at remote sites. This may explain the presence in the circulation of some neural peptides such as vasoactive intestinal polypeptide (VIP) and gastrin releasing peptide.

Blood supply

The stomach receives its blood supply from at least six named arteries that are all derived from the celiac axis: left and right gastric arteries, left and right gastroepiploic arteries, short gastric branches of the splenic artery, and branches from the gastroduodenal artery. There are many crossconnections among these vessels external to and on the surface of the stomach. These numerous anastomoses are the reason that all but one of the named arteries can be totally occluded without producing any lasting effect on blood flow to the stomach wall. Acute occlusion of three of the arteries does not produce even shortterm changes in total gastric blood flow. In clinical settings infarction or nonocclusive ischemia of the stomach are rare occurrences.

Branches of the arteries of the stomach pierce the serosal surface and give off vessels which supply the muscular layer of the stomach wall. The main trunk of each piercing branch continues into the submucosa where arcades of interconnected arteries are formed. The best current evidence is against the existence of anastomoses at this level. Hence, blood flow to the gastric mucosa is controlled by varying the resistance to blood flow through the terminal segment of the arterial vessels in the mucosa. These vessels are classified histologically as microscopic arteries (100 to 25 μ in diameter) and together are known as the resistance vessels, since two-thirds of the total resistance to blood flow across the wall of the stomach is developed in this segment of the circulation. By varying the internal diameter of the resistance vessels in response to sympathetic neurons or local metabolic changes, total blood flow to the tissues of the stomach is regulated. In addition, since the submucosal arterial vessels are arranged in series with the mucosal, marked alterations in the resistance to blood flow through submucosal vessels will affect blood flow to the mucosa. From the submucosal plexus of arteries single small arteries pierce the muscularis mucosa to supply the mucosa. It is believed that these vessels are end-arteries that are the only source of blood for patches of mucosa up to 50 mm.^{2} in surface area. Occlusion of these vessels would, therefore, produce an ischemic area of mucosa.

When the stomach is empty, total gastric blood flow is about 0.5 ml./min.$^{-1}$ g.$^{-1}$ of tissue, approximately two-thirds being distributed to the mucosa. When a meal is eaten total blood flow to the stomach may increase two- or threefold and almost all of the increased flow goes to the mucosa. A similar response occurs whenever gastric secretion is stimulated. Other factors regulating blood flow to the stomach are discussed elsewhere in this book.

Pathophysiology

As described, two of the most common disorders affecting the stomach are gastric ulcer disease and duodenal ulcer disease. Also to be considered because of their frequency or their illustrative value are gastritis (including pernicious anemia), gastric cancer, and gastrinoma (also called the *Zollinger-Ellison syndrome*), as well as certain surgical procedures such as vagotomy or antrectomy, and certain antisecretory drugs (anticholinergics and histamine H_2-receptor blockers).

The most common pathophysiological mechanism that alters the functional anatomy of the stomach is a change in the relative numbers of the various cell types in the mucosa. In duodenal ulcer disease the maximal capacity to secrete acid is increased substantially above normal in about one-third of the people with this disorder. In stomachs of duodenal ulcer patients removed at autopsy, the average total number of oxyntic cells per stomach is 1.8 billion, compared to 1.0 billion for normal subjects. This would account for an increased acid secretory capacity, since it is a reasonable assumption that the total number of oxyntic cells in the stomach is directly related to the maximal acid secretory capacity. Thus, the degree of reduction in acid secretory capacity is closely correlated with the number of oxyntic cells removed in duodenal ulcer subjects who undergo partial gastrectomy (removal of the distal one-half to two-thirds of the stomach). It is not known why duodenal ulcer patients have more oxyntic cells in their stomachs than normal subjects, although this increase in oxyntic cells may be due to an elevated circulating level of gastrin or to an increased sensitivity of mucosal stem cells to gastrin in duodenal ulcer patients with subsequent higher rates of oxyntic cell production. The role of increased levels of circulating gastrin in producing an increased population of parietal cells is clearly established in the Zollinger-Ellison syndrome. People with this disease have a tumor, usually in the pancreas, which is an ectopic, uncontrolled source of gastrin (a gastrinoma). Circulating gastrin levels are frequently very high and may be 100 times normal. These patients secrete enormous volumes of acidic gastric juice and have been found to have 3 to 6 billion oxyntic cells in their stomachs. Another example suggesting an important role for gastrin in regulating the size of the oxyntic cell population is the effect of antrectomy, surgical removal of the antrum in order to decrease circulating gastrin levels (basally and in response to food) and thereby decrease acid secretion in duodenal ulcer disease. Some investigators have found a substantial decrease in the number of oxyntic cells in biopsies of oxyntic gland area mucosa taken several years after this operation.

In gastric ulcer disease acid secretion is usually normal or less than normal. There is a slight decrease in the total number of parietal cells (0.8 billion per stomach). Since subjects with this disorder usually do not secrete more acid than normal, it is most likely that the gastric ulcer results from factors that lower the resistance of the surface mucosa cells to normal or even reduced amounts of acid. One such factor is the chronic backflow of duodenal contents into the stomach. This duodenogastric reflux exposes the gastric mucosa to bile salts and pancreatic digestive enzymes which kill the surface mucous cells and generate an ulcer. The reason for increased duodenogastric reflux in gastric ulcer subjects is uncertain. We are also unclear about the greater susceptibility of the gastric mucosa to the damaging effects of bile and pancreatic juice compared with the duodenal mucosa. The greater resistance of the duodenal mucosa may be partly related to the much more rapid renewal rates of duodenal mucosal cells.

Gastritis is a catchall word. Although its strict meaning is "inflammation of the stomach," the word is used to describe a spectrum of histological changes of the gastric mucosa, whether from pyloric gland or oxyntic gland areas. At one end of this spectrum are slight degrees of inflammation (infiltration of the mucosa by various inflammatory cells) but with no loss of normal architecture, through chronic atrophic gastritis in which the mucosal glands have disappeared and the mucosa consists of a thin layer of surface mucous cells and rudimentary glands. In severe and very chronic atrophic gastritis, the disease called pernicious anemia develops because of the absence of oxyntic cells and their product, intrinsic factor. Intrinsic factor is necessary for absorption of vitamin B_{12}, and in its chronic absence a deficiency of vitamin B_{12} develops. Body stores of B_{12} take several years for depletion when no B_{12} is being absorbed, explaining the long latency of onset of pernicious anemia. The most prominent clinical feature of pernicious anemia is usually a severe macrocytic anemia, which, if untreated, becomes fatal. The mechanism that leads to gastritis in pernicious anemia is not known. It may also be related to chronic duodenogastric reflux with diffuse rather than focal damage to gastric mucosa. Another possibility is an autoimmune destruction of oxyntic cells. Gastritis may be the most common chronic disorder of the gastrointestinal system, since it is present in many young people and increases in frequency to nearly 100 per cent in septagenarians. The severe form which leads to pernicious anemia occurs in less than 1 per cent of the adult population, however. It is of interest that stem cell proliferation rates are higher than normal in subjects with gastritis, even in the face of severe atrophy. When the rates of cell loss into the gastric lumen are measured by determining the amount of DNA in gastric washings, these rates are found to be very high. Thus, it appears that increased cell death, rather than decreased cell proliferation, accounts for the mucosal atrophy.

The final disease to be considered in this section is gastric cancer. We still know nothing about the cause of this devastating disease. There are several histological types of gastric carcinoma. Of those which appear to arise from mucosal precursors (adenocarcinoma) it is possible that abnormalities in cell proliferation in the gastric mucosa may predispose to their development. Some believe that chronic gastritis may increase the risk of gastric cancer, presumably because of the high rates of cell proliferation in gastritis which somehow becomes uncontrolled. Indeed, gastritis is a very common finding in subjects with gastric cancer, and cell proliferation rates are very high in the noninvolved mucosa of these patients. Whether this is a cause or an effect, however, cannot be stated with certainty. Supporting a role for gastritis as a precancerous lesion in some people is the fact that the risk of developing gastric cancer is several times higher in persons who have had partial gastrectomy as treatment for duodenal (or gastric) ulcer disease. Such patients frequently develop severe postoperative gastritis. There may be another subset of patients with gastric cancer whose tumor grows in response to gastrin; preliminary evidence indicates that giving gastrin to mice which have received a subcutaneous transplant of human gastric cancer cells causes the tumor to grow faster.

Mechanisms of gastric secretion

Patterns of gastric secretion

The human stomach secretes acid, pepsinogen, mucus, and intrinsic factor at low rates when it is empty and the person is at rest (basal secretion). The basal secretion of HCl is less than 20 per cent of maximal rates and there is much variation in these rates from person to person and from day to day in the same individual. There is also diurnal variation in basal gastric secretion with lower rates at night and higher rates during the day.

The most common method for determining maximal gastric secretory capacity in a human subject is to insert a soft rubber or plastic tube into the distal stomach through the nose or mouth 8 to 12 hours after the subject has last eaten, aspirate the residual contents of the stomach, collect

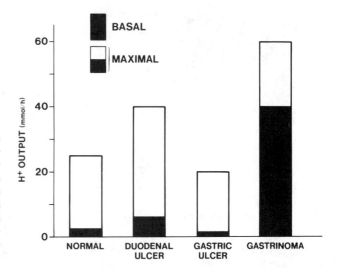

FIG. 22–3. Acid output from groups of human subjects: normal, duodenal ulcer, gastric ulcer, gastrinoma (Zollinger-Ellison syndrome). The values shown are from tests in which a single high dose of pentagastrin was administered intramuscularly or subcutaneously and gastric juice was collected for 1 hour afterwards. Continuous intravenous infusion gives higher values.

basal secretion for 30 to 60 minutes, then administer a stimulant such as histamine or pentagastrin (a synthetic fragment of gastrin) as a subcutaneous or intramuscular injection or intravenous infusion, and continue to aspirate gastric secretions for 60 to 90 minutes more. The normal adult 70-kg. man has a gastric secretory capacity of 50 mmol./hour which is positively correlated with lean body mass. Men secrete more acid than women, but this is most likely due to differences in lean body mass. Acid secretory capacity decreases in the elderly, presumably because the frequency of gastritis is higher in the elderly. There are also unexplained ethnic and racial differences in secretory capacity; for example, Chinese subjects secrete less than subjects from Scotland.

The basal and maximal acid secretory rates in normal subjects and patients with duodenal ulcer disease, gastric ulcer, and Zollinger-Ellison syndrome differ in important ways (Fig. 22–3). For example, the average basal and maximal acid secretory capacities in duodenal ulcer subjects are higher than those in normal subjects, but this is due to elevated rates in only about one-third of all duodenal ulcer patients. The same pattern of overlap, in this case with low average secretory rates, is seen with gastric ulcer patients. In Zollinger-Ellison syndrome basal acid secretion is frequently higher than maximal stimulated rates in normal subjects. Furthermore, stimulation with pentagastrin does not increase acid secretion or does so only slightly in patients with this hypersecretory state in which there is a continuously very high circulating level of tumor-derived gastrin. In patients with advanced gastric cancer, achlorhydria is frequently present, both basally and after stimulation.

Recently it has become possible to measure the gastric secretory response to food by means of a technique called *intragastric titration.* In one variant of this method a double-lumen tube is inserted into the stomach and a liquid or blenderized meal is instilled into the stomach. The meal and gastric secretions are mixed and sampled by withdrawing and injecting 20 ml. of stomach contents through one lumen of the gastric tube every 3 to 5 seconds using a "push-pull" syringe pump. An electrode senses the pH of the gastric contents; as acid is secreted, pH drops. A strong base, such as NaOH, is added to gastric contents through the second lumen of the tube to maintain intragastric pH at a predetermined set level. Thus, the amount of base added per unit time (as mol. of OH^-)

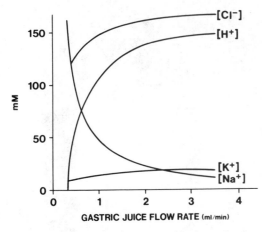

FIG. 22–4. Changes in concentration of major cations and anions of gastric juice as flow rate increases in response to stimulation.

equals the acid secretory rate. Although gastric contents are continuously emptying into the intestine, this introduces a negligible error because the mixing and sampling rate (240 to 400 ml./min.) is much greater than even the fastest gastric emptying rates (10 ml./min.). Using this technique it has been found that a protein-containing meal stimulates gastric acid secretion to 85 per cent of maximal secretory rates as determined by a standard gastric secretory test. There are marked differences in the acid secretory response to food in patients with duodenal ulcer disease.

When the composition of basal and stimulated gastric juice is analyzed, several changes are found (Fig. 22–4). In basal gastric juice the concentration of H^+ is low, ranging from 0 to 20 mM. The major cation of basal gastric juice is Na^+, with a concentration of about 140 mM. Potassium is present at about 6 mM. The major anions are Cl^- and HCO_3^-, the concentrations of which are about 130 and 25, respectively. As gastric secretion is stimulated and the flow rate (amount of gastric juice appearing per unit time) increases, the concentration of H^+ increases dramatically to 145 mM., that of Na^+ drops to 5 mM., and that of K^+ increases to 15 mM. Concentrations of Cl^- increase to 165 mM., and HCO_3^- decreases

to 0. The most likely explanation for this pattern of changes in ionic composition of gastric juice during stimulation is the two-component hypothesis. Basal gastric juice is probably secreted by surface and neck mucous cells and peptic cells and is plasma-like in its ionic composition. When oxyntic cells are stimulated to secrete, they produce a fluid containing 145 mM. HCl; the volume of this fluid is so much greater than basal juice volume that the oxyntic cell component is the major determinant of ionic composition of stimulated gastric juice.

The other components of gastric juice (pepsinogen, mucus, and intrinsic factor) have distinctly different secretory patterns. The concentration of pepsin and intrinsic factor in gastric juice collected in the basal state may be quite high. When a stimulant of acid secretion is given as a continuous intravenous infusion, the outputs (total amount secreted per unit time) of pepsin and intrinsic factor increase rapidly to a peak and then decline to a plateau value which is still higher than basal output values. This phenomenon is called *washout*. These substances are continually secreted into the lumina of gastric glands at a basal rate and are washed out by the high volume flow of acid and water secreted by the oxyntic cell. The plateau output represents the true stimulated output rate of these two substances. Mucus is a mixture of several macromolecules, the principal ones being glycoproteins (carbohydrate plus protein). Mucus is secreted at a low rate under basal conditions by the various mucus-secreting cells of gastric mucosa. Its rate of secretion is increased during stimulation of acid secretion by several agents as well as by topical applications of HCl and irritants. The mechanisms of stimulation are not well understood.

Mechanism of HCl secretion

The oxyntic cell is the source of the HCl in gastric juice. This cell is able to generate a millionfold concentration gradient for H^+

from about 10^{-7} M. in plasma and extracellular fluid to about 10^{-1} M. in gastric juice. This ionic gradient is the largest one known in biology and requires a highly specialized cell and an enormous energy consumption. The energy is directly supplied from the densely packed mitochondria in the oxyntic cell (40 per cent of cell volume) and is released from adenosine triphosphate (ATP). The high-energy phosphate bonds in ATP provide the fuel to run a special ion pump found in the membrane of the apical part of the cell. Energy consumption by this pump is so high that oxygen consumption can be used as a convenient measurement of acid secretion (the two are directly proportional) in experimentally prepared suspensions of isolated oxyntic cells.

MORPHOLOGICAL TRANSFORMATION. Another of the specializations of the oxyntic cell which is well suited for ion transport is its ability to increase the surface area of its apical membrane (Fig. 22–2). Electron micrographs of the oxyntic cell at rest show a triangular configuration (conical in three dimensions) with a narrow apex. The intracellular canaliculus and apical membrane are studded with short chubby microvilli. The cell is packed with tubulovesicles consisting of thousands of smooth membrane-bound compartments. Although it is not yet proven, it is likely that the tubulovesicles are not in continuity with the intracellular canaliculus in the resting oxyntic cell and are closed sacs of membrane. When acid secretion is stimulated these tubulovesicles fuse with the intracellular canalicular membrane and produce as much as a tenfold increase in surface area of this part of the cell. This transformation occurs in minutes after the onset of stimulation. With cessation of stimulation the reverse transformation (canalicular membrane to tubulovesicles) also occurs rapidly. Using data obtained from experimental animals it can be calculated that each stimulated oxyntic cell has about 1000 μ^2 of apical plus

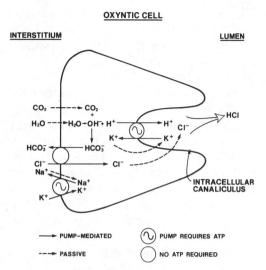

OXYNTIC CELL

FIG. 22–5. Schematic picture of mechanisms involved in secretion of HCl by the oxyntic cell. The K^+-activated ATPase is shown on the membrane of the intracellular canaliculus. It exchanges luminal K^+, derived from leakage across this membrane, for H^+, derived from H_2O. Cl^- follows H^+ passively due to the net positive charge in the lumen. The usual basolateral Na^+-K^+-ATPase pumps K^+ in and Na^+ out of the oxyntic cell.

canicular surface area. Consequently, in a human stomach stimulated to secrete, the total surface area available is about 1 m². At maximal secretory rates each cell secretes 1 billion molecules of HCl per second across its apical surface.

PROTON PUMP. The apical membrane of the oxyntic cell and the canalicular and tubulovesicular membranes contain the ATP-fueled pump which moves H^+ into the lumen of the oxyntic glands. This pump requires K^+ for activity and is called the K^+-stimulated ATPase. Present models of how this pump works assume that H^+ and K^+ are exchanged across the oxyntic cell apical and canalicular membranes with ATP supplying the energy required to maintain the H^+ gradient which is established. (Fig. 22–5). The H^+ which is secreted from the cytoplasm of the oxyntic cell comes from the reaction of CO_2 with H_2O to form H^+ and HCO_3^-; the enzyme carbonic anhydrase acts at this step to increase greatly the rate of formation of carbonic acid (H_2CO_3), the source of these ions. It is assumed that HCO_3^- cannot cross the apical cell membrane and exits the cell by being exchanged across the basolateral membrane for Cl^-. This chloride subsequently moves across the apical membrane of the cell by passive diffusion to produce HCl. The K^+ necessary for exchange with H^+ is postulated to be recycled inside the cell and to enter the pump within the cell membrane. It should be recognized that this is a highly simplified explanation of an extremely complex event. The reference list provides a starting place for those interested in more detailed explanations of the cellular H^+ secretory mechanisms.

PATHOPHYSIOLOGY. There are no known derangements of the cellular H^+ secretory mechanism; however, it has not been possible to define these events even in normal subjects until recent years. One potential disorder might be an "overachiever" oxyntic cell which has the ability to transform more than the average amount of tubulovesicular membrane into canalicular membrane or has an amount of K^+-stimulated ATPase per cell that is higher than average. If such conditions were to exist, they would explain some of the disorders of acid secretion in duodenal ulcer disease.

Mechanism of pepsin secretion

Pepsin is secreted in the form of an inactive precursor, pepsinogen. Upon exposure to an acid solution with a pH less than 5.0, several small fragments are cleared from pepsinogen, leaving the active molecule, pepsin. Pepsin is an endopeptidase which has an optimum pH ranging from 1.8 to 3.5, obviously a very unusual property but also obviously well suited for an intragastric protease. Once activated to pepsin by an acidic environment, pepsin is irreversibly inactivated if the pH increases above 5.5. There are several molecular variants of pepsin, each differing in charge and size,

and there are several corresponding variant forms of the parent molecule, pepsinogen. When antibodies are raised to crude or semipurified "pepsinogen" prepared from gastric mucosa, two major classes of pepsinogens are recognized: pepsinogen I, which can be further separated into five variants with other biochemical techniques; and pepsinogen II, which has two variants. Group I pepsinogens are found only in oxyntic gland mucosa, where they are synthesized and secreted by peptic cells and mucous neck cells. Group II pepsinogens occur in these same cells as well as in mucous cells in cardiac glands, pyloric glands, and duodenal Brunner's glands. Both group I and group II pepsinogens are found in blood plasma and their concentrations can be measured by specific radioimmunoassays. The mechanism by which the pepsinogens enter the blood is not known, but they are presumed to be derived from the stomach. There is a positive correlation between blood levels of group I pepsinogens and maximal acid secretory capacity. This is probably due to the correlation between the total number of peptic cells and oxyntic cells in the oxyntic gland area mucosa.

In peptic cells pepsinogen is synthesized, transported, stored intracellularly, and secreted by exocytosis just as in all protein-secreting cells. In humans secretion of pepsinogen is stimulated by cholinergic agents, gastrin, secretin, and histamine. The intracellular mediators, or "second messengers," for this process are unknown.

PATHOPHYSIOLOGY. Pepsin probably plays a role in the "peptic" ulcerations found in gastric and duodenal ulcer disease. There is very little reliable information about differences in pepsin secretion in patients with these diseases and normal subjects. Several attempts to treat peptic ulcers by administering inhibitors of pepsin have met with little or no success. An exciting recent discovery is that elevated blood concentrations of group I pepsinogens (hyper-pepsinogenemia I) is a genetic marker for duodenal ulcer disease in some families. The marker (elevated blood pepsinogen) is inherited as an autosomal dominant trait, and only those family members with the trait have developed duodenal ulcers. It is assumed that the mechanism for the ulcers and the elevated blood levels of pepsinogen I are the same, namely, more oxyntic and peptic cells than normal with subsequent hypersecretion of acid and pepsin. Since only a fraction of family members with the trait develop duodenal ulcers, factors other than acid and pepsin hypersecretion must be important. Conversely, the risk of developing an ulcer in persons with the trait is five times higher than that of others, supporting at least a strong predisposing role for gastric hypersecretion. Population studies suggest that about half of all patients with duodenal ulcer disease come from families with the hyperpepsinogenemic trait. Thus, it may be possible to predict an increased risk of developing duodenal ulcer disease in members of the general population by measuring blood pepsinogen levels.

Mechanisms of gastric mucus and bicarbonate secretion

The pure secretory product of the mucous cell is a polymeric glycoprotein with a molecular weight of about 2 million. There are four glycoprotein subunits of this polymer, each consisting of a protein backbone and carbohydrate sidebranches arranged like a bottlebrush: these subunits have a molecular weight of 0.5 million. A fifth subunit consists of a 70,000 molecular weight protein which links the other four subunits. The carbohydrate content of the polymer is over 70 per cent by weight. About 80 per cent of people secrete gastric glycoproteins which contain the ABH antigens of the ABO blood group system in their structure.

The glycoprotein of gastric mucus readily forms a gel with H_2O molecules, resulting in a slippery, slimy substance which adheres strongly to the gastric mucosal surface. This sticky mucous layer, 0.5 to 1.0 mm. in thickness, consists of 95 per cent H_2O and 5 per cent or less glycoprotein. It greatly impedes the diffusion of large molecules such as pepsin but has little or no effect on the diffusion of ions such as H^+. However, adherent mucus creates a thick "unstirred layer" through which H^+ must diffuse, and this diffusion requires several minutes. This effect, combined with the secretion of HCO_3^- by surface mucous cells, may offer some degree of protection against H^+ secreted by the oxyntic glands. Without the unstirred layer created by adherent mucus, the mucosal surface would be continually exposed to H^+ at the pH of bulk solution in the gastric lumen because of the constant mixing and churning created by gastric motility.

The thickness of the adherent layer of mucus is determined by the rate of secretion of mucus by cells and the rate of degradation of mucus at the luminal surface of the mucus. The most important chemical component of gastric juice which degrades mucus is pepsin. The products of pepsin digestion of mucus are the glycoprotein and protein subunits. As noted before, factors which control the rate of mucus secretion are acetylcholine, gastrin, histamine, and secretin. But because most methods which have been used to measure mucus secretion have serious flaws, it is not possible to give any details about control of mucus secretion.

Another important secretory product of mucous cells is a bicarbonate containing fluid. The concentration of bicarbonate in this fluid has been measured as 10 to 20 mM. in humans, and HCO_3^- appears to be secreted in both the oxyntic and pyloric gland areas of the stomach. The maximal rate of HCO_3^- secretion (mmol./unit time) is no more than 10 per cent of the maximal rate of H^+ secretion. However, this may still be an important protective measure for the surface mucous cells. A pH gradient across the adherent mucous layer, from pH 2.4 at the luminal surface to pH 7.0 at the mucous cell surface, has been measured in experimental animals. This is consistent with the hypothesis that adherent mucus plus bicarbonate derived from mucous cells combine to protect the mucous cells from damage by H^+. Very little is known about the factors which control the rate of bicarbonate secretion by mucosal cells; a vagal cholinergic mechanism may be important.

PATHOPHYSIOLOGY. The role of gastric mucus in protecting the stomach from its own corrosive secretions and from exogenous irritants has been postulated for a century or more. There is still no proof that mucus fulfills such a role to any important degree, however. Recent characterization of gastric HCO_3^- secretion has resurrected the mucus barrier theory, but there are still no clear examples of alterations in this mucus-plus-bicarbonate system in any human disease. It may be that gastritis and gastric ulcer are due in part to an inefficient or an overwhelmed mucus barrier. Substances known to produce gastric irritation and ulcers, such as aspirin and ethanol, decrease the secretory rates of both mucus and bicarbonate. Certain prostaglandins which protect the stomach against these damaging agents increase mucus and HCO_3^- secretion.

Mechanisms of intrinsic factor secretion

Human intrinsic factor is a glycoprotein with a molecular weight of about 45,000, 15 per cent of which consists of carbohydrate. Intrinsic factor binds very tightly to a family of compounds called cobalamins, one of which is cyanocobalamin. Other forms are hydroxycobalamin, methylcobalamin, and adenosylcobalamin. Intrinsic factor is localized to oxyntic cells in several species, including humans. However, there are no distinct secretory granules containing the intrinsic factor in oxyntic cells. Immunohistochemical techniques show that intrinsic factor is localized to membranes of the nucleus, endoplasmic reticulum, Golgi ap-

paratus, and tubulovesicular system. Further exploration will be required to understand the meaning of this unusual intracellular distribution of intrinsic factor.

As noted above, secretion of intrinsic factor is stimulated by gastrin, histamine, and cholinegric agonists. Recent studies indicate that cyclic adenosine monophosphate (cAMP) is an intracellular messenger which mediates the stimulation caused by histamine. Intracellular mediators for the other stimulants have not been elucidated. The actual mechanism of secretion of intrinsic factor by oxyntic cells is completely unknown.

PATHOPHYSIOLOGY. Any condition which abolishes intrinsic factor secretion will eventually lead to pernicious anemia. Total gastrectomy, a rarely performed operation, is one such condition. A small percentage of people with atrophic gastritis and total obliteration of oxyntic cells develop pernicious anemia.

Control of gastric secretion

General comments

It is helpful to consider three different categories of factors that regulate gastric secretion: (1) factors regulating basal secretion, (2) factors regulating the gastric secretory response to a meal, and (3) general factors which influence gastric secretion, such as emotional state, physical condition, race, nutrition, and environmental influences. We know the least about the third category of factors. For example, the long-held belief that anxiety and stress in humans may lead to ulcers by causing increased acid secretion is unsupported by any definitive data. Air traffic controllers, who would be expected to lead a stressful life, have no higher incidence of ulcers than the general public, although they may have a much greater incidence of hypertension. Chronic anxiety states may or may not cause increased acid secretion, but no careful study exists that would decide this question. There are numerous psychosocial and genetic factors that could have generalized effects on gastric secretion, but the factors themselves and their mechanisms of action are ill defined.

We know more about regulation of basal and meal-induced gastric secretion. Both stimulatory and inhibitory factors contribute to yield a net secretory response to a meal. The stimulatory factors that are best characterized are gastrin (released from antral G cells), acetylcholine (released from postganglionic cholinergic neurons in the oxyntic and pyloric gland mucosa), and histamine (released from mast-like cells in the lamina propria of oxyntic gland area mucosa). These three stimulants also exemplify the three methods of transmission by chemical messengers in the body. Gastrin is released into the bloodstream and is carried through the systemic and pulmonary circulation to return to the stomach; this is classical *endocrine* transmission. Acetylcholine is released from nerves and diffuses a very short distance to effector cells; this is *neurocrine* transmission. Histamine is released from mast cells and diffuses through extracellular fluid for variable but short distances; this is *paracrine* transmission. The roles of these three stimulants in regulating the gastric secretory response to a meal can most easily be discussed by dividing the meal response into three phases: the cephalic, gastric, and intestinal phases of gastric secretory regulation. These phases are determined by where the factors which initiate gastric secretion during that phase are acting. Gastric secretion which is initiated by factors acting in the brain is called the cephalic phase, for example. There are also important inhibitory mechanisms which regulate gastric secretion. These too will be discussed in the framework of the cephalic, gastric, and intestinal phases of gastric secretion.

Methods of study

Common physiological methods used to determine whether nerves or a hormone regulate the activity of an organ include cutting the nerves to the organ, excising the source of the hormone, or administering a specific receptor blocker to observe changes in organ function. For the stomach, several procedures can be used to estimate the relative importance of acetylcholine, gastrin, and histamine in regulating interdigestive (basal) and digestive (meal-induced or postprandial) gastric secretion.

Vagotomy involves cutting the vagus nerves, thereby removing the afferent and postganglionic efferent links of "long" (vagovagal) reflexes which affect gastric secretion. Vagotomy can be performed at several levels (in order of increasing specificity of effect on gastric secretion): (1) truncal vagotomy involves cutting the major anterior and posterior vagus nerves at their point of entry into the abdominal cavity; (2) selective gastric vagotomy involves severing only the vagal branches to the stomach; and (3) oxyntic cell vagotomy involves transection of the nerves to the body and fundus of the stomach. In experimental animals antral vagotomy has also been performed. Vagotomy at any level does not change the ability of either Auerbach's or Meissner's plexus to act as a reflex net with its own sensory receptors and intramural afferent and efferent limbs ("short" reflexes). Local anesthetics can be applied to the mucosa to block sensory receptors, although frequently such local anesthetics are highly toxic to the mucosa and thus have nonspecific effects. Muscarinic cholinergic receptor blockers, such as atropine, can block the action of acetylcholine on effector cell receptors. This is a nonspecific method if the blocker is injected into an intact animal or person, since atropine will block acetylcholine in many organs with sometimes unpredictable effects. In addition, the specific blocking effect of muscarinic anticholinergic agents occurs in a narrow and low dose range and is frequently overlooked. Less commonly used approaches include administering blockers of ganglionic transmission (the nicotinic action of acetylcholine) or axonic transmission. With care and an appreciation for the potential nonspecificity of effect of these methods, however, much useful information can be collected.

There are no well-characterized receptor blockers for the effects of gastrin. Thus, the usual methods to study its role in regulating gastric secretion are to perform an antrectomy or to use radioimmunoassay to measure the circulating levels of gastrin. This latter technique has been a great advance in the field of gastrointestinal physiology and has redefined many concepts which were based on erroneous interpretations.

Another remarkable advance in studying regulatory factors in the stomach was the development in the early 1970s of a receptor blocker for the effects of histamine on gastric acid secretion. There are two classes of cell receptors for histamine: H_1 receptors, which mediate the effects of histamine on smooth muscle of small intestine and bronchi; and H_2 receptors, which mediate the effects of histamine on gastric acid secretion and certain other actions. The classical antihistamines are H_1 receptor blockers and have little effect on histamine-induced gastric secretion. With the discovery of potent, highly specific H_2 receptor blockers, a major role for histamine in regulating gastric acid secretion was proven. The best known of these H_2 receptor blockers is cimetidine, a compound which is used clinically to decrease acid secretion. It has become the most commonly prescribed drug in the world in the few years since its introduction.

Basal gastric secretion

Basal (interdigestive) secretion of acid is reduced by 60 per cent after antrectomy, 80 per cent after truncal or highly selective vagotomy or administration of atropine, and 100 per cent by a large dose of a histamine H_2 receptor blocker. These studies were done in duodenal ulcer patients, but anticholinergic drugs and H_2 receptor blockers have about the same effects when given to normal subjects. Thus it appears that basal gastric acid secretion actually depends on stimulation of the oxyntic cell by gastrin, acetylcholine, and histamine. The observation that blocking the effect of or removing any one of the three secretory stimulants produces nearly complete inhibition of basal secretion is best explained by the property of potentiation. In studies concerning the secretory responses of an *in vitro* suspension of oxyntic cells it has been found that adding gastrin, acetylcholine, or histamine alone to the buffered solution in which the cells are kept produces only a weak acid secretory response. When two agents are added together to the solution, a potentiated response occurs and the addition of all three agents produces a response much greater than the maximal response to any single agent or combination of two agents (Fig. 22–6). These experiments provide an explanation for the enigma of apparent total dependence of basal acid secretion on each of three agents. Removal or blocking of either gastrin, histamine, or acetylcholine suppresses the potentiated interactions of all three secretory agonists under basal conditions.

PATHOPHYSIOLOGY. Although duodenal ulcer patients frequently have higher rates of interdigestive and nocturnal acid secretion than normal subjects, there is no consistent elevation of basal gastrin levels in duodenal ulcer subjects. The finding that administration of single large doses of anticholinergics or H_2 receptor blockers produces about the same degree of inhibition of basal acid secretion in normal subjects and in ulcer subjects does not rule out a possible difference in tissue levels of acetylcholine and histamine. Unfortunately, we cannot measure the concentration of these substances directly in oxyntic gland tissue.

Another possible mechanism for increased basal gastric secretory rates is the presence of increased sensitivity of oxyntic cells to stimulation. Duodenal ulcer subjects exhibit a greater proportional increase in acid secretion than normal sub-

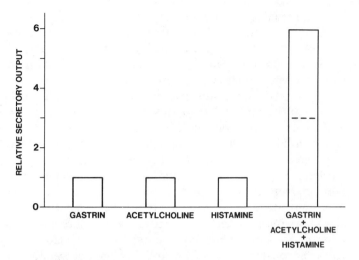

FIG. 22–6. Potentiating interactions between gastrin, histamine, and acetylcholine. The bars represent acid output resulting from administration of each stimulant singly, in maximal doses, or together. The dotted line on the far right bar indicates the amount of acid secretion that would be expected if the three-way interaction of the stimulants was additive.

jects to a given increment in circulating gastrin. This higher "bounce to the ounce" for gastrin could explain increased basal acid secretory rates in ulcer subjects who have normal basal gastrin levels. We do not know if there is also increased sensitivity to acetylcholine and histamine in duodenal ulcer subjects.

Gastric secretory response to meals

CEPHALIC PHASE. The brain influences gastric secretion through efferent fibers carried in the vagus nerves. The cephalic phase of gastric secretion consists of the secretory response to the anticipation and the act of eating appetizing food. The brain integrates a number of pleasant ocular, olfactory, and taste stimuli and converts them into neural input to the vagal nuclei; preganglionic vagal efferents are activated, then postganglionic cholinergic fibers, and gastric secretion results. Vagotomy obliterates the gastric secretory response to the cephalic phase.

The cephalic phase of gastric secretion can be studied in several ways. Subjects who have become accustomed to having a nasogastric or orogastric tube in place may be stimulated by having an appetizing meal prepared in front of them or may "chew and spit" such a meal, taking bites of food, chewing, and expectorating the food without swallowing it. This method is called sham feeding. Another method is to administer insulin or 2-deoxy-D-glucose (2-DG), a nonmetabolizable glucose analogue, both of which increase the electrical activity of the vagal nuclei either by decreasing the glucose supply to the brain (insulin) or by interfering with glucose usage by neurons (2-DG).

If sham feeding is carried out for 30 minutes in normal subjects, gastric acid secretion increases to about 50 per cent of maximal secretory capacity. Acid secretion returns to basal rates within 45 minutes after cessation of sham feeding. Administration of an optimal dose of insulin produces higher rates of acid secretion than sham feeding (up to 80 per cent of maximal). If sham feeding is combined with instillation of a blenderized meal into the stomach through the nasogastric tube, acid secretion reaches near maximal levels for a brief period and is prolonged at lower levels for several hours.

The mechanism of stimulation of acid secretion during the cephalic phase involves mainly the direct action of acetylcholine released by postganglionic efferent nerve fibers on oxyntic cells. Denervation of the oxyntic gland area nearly abolishes the acid secretory response to sham feeding or insulin injection in humans. There is a small increase in the circulating gastrin level after sham feeding. Insulin injection produces a more pronounced release of gastrin. Gastrin release contributes to the acid response, since antrectomy moderately reduces the acid response to sham feeding. H_2 receptor blockers have a greater inhibitory effect on cephalic phase acid secretion, which is probably due to blocking of a strong potentiating interaction between histamine and acetycholine. At present there is no evidence that vagal activation releases gastric mucosal histamine, so the interaction that occurs is between a constant level of histamine plus increased amounts of acetylcholine.

GASTRIC PHASE. When food enters the stomach it stimulates acid secretion by distending the stomach and by releasing gastrin. Gastrin is the major regulator of acid secretion during the gastric phase, but distention contributes to the overall response.

Distention of the stomach stretches its walls and elicits firing of stretch receptors. These activate the afferent limb of reflex arcs which prompt a release of acetylcholine from oxyntic gland area postganglionic cholinergic fibers. The resulting stimulation of acid secretion rapidly reaches a peak of 50 per cent of maximal levels. In normal subjects acid secretory rates tend to decrease over several hours even though the distention is maintained. The rate of acid secretion is related to the intragastric volume if this is held constant; as little as 50 ml. of fluid is sufficient to increase acid secretion two- to threefold over basal secretion. The reflexes that are responsible for most of the secretory response to distention consist of both long (vagovagal) and short (intramural) arcs. Vagotomy reduces, but does not abolish, distention-induced acid secretion. In humans there is little or no release of gastrin by distention of the whole stomach.

Protein-containing foods, coffee, and ionic calcium release gastrin and thereby stimulate gastric secretion. The lipid and carbohydrate components of food do not stimulate gastric secretion other than by distention. Release of gastrin by certain foodstuffs is probably the predominant mechanism explaining gastric secretion during the gastric phase. Continuous intravenous infusion of gastrin at a rate which elevates blood levels of gastrin to those seen after instilling a protein-containing meal into the stomach also reproduces the rate of gastric acid secretion seen after the meal. This provides strong evidence that increased release of gastrin is sufficient to explain the effect of a meal on gastric acid secretion. The background levels of mucosal acetylcholine and histamine also contribute an important share of the response. Removal of the participation of each of these from the three-way potentiated interaction described above is sufficient to explain the fact that vagotomy, atropine, and cimetidine strongly inhibit the gastric secretory response to food, reducing it by 50 to 80 per cent. These strong inhibitory effects are the basis for their use in treatment of duodenal and gastric ulcer disease. Antrectomy reduces the gastric secretory response by 50 per cent; the remainder is probably due to distention-induced secretion. Combination of antrectomy with vagotomy nearly abolishes the gastric secretory response to food.

Gastrin is a peptide found in antral and duodenal mucosa and in the blood in two major chemical forms. These forms are known by the abbreviations G34 and G17, G for gastrin and 34 or 17 for the number of amino acids in each peptide. G17 is identical in amino acid sequence with the carboxyl-terminal 17 amino acids of G34. It is now established that G34 is synthesized as precursor of G17, and that there are even larger precursors which are synthesized in antral and duodenal G cells (the endocrine cells which synthesize, store, and secrete gastrin). G34 is cleaved by a peptidase in the G cell to form G17. Both G34 and G17 are released into the circulation in response to stimuli. The amino-terminal 17 amino acid fragment resulting from cleavage of G34 to G17 is also found in the circulation; however, it is biologically inactive. In the interdigestive period the total serum concentration of gastrin is 10 to 20 pM. in normal subjects and is composed of about two-thirds G34 and one-third G17. After stimulation by ingestion of a protein-containing meal, the degree of gastrin release is related to the proportion of protein and peptides in the meal. A three- to fourfold increase in total gastrin concentration is the maximal response. Most of the increase in gastrin levels is due to G17. Since G17 is six to eight times more potent than G34 (expressed on a molar basis) in stimulating acid secretion, G17 can be considered the most important molecular form of gastrin. The two forms of gastrin are metabolized at different rates. The half-time for disappearance of G17 in the human circulation is 5 minutes, while that of G34 is 35 to 40 minutes. Removal and degradation of G17 from the circulation occurs throughout the body with no specific localization to kidney, liver, or other sites. Removal and degradation of G34 appears to occur in the kidney.

Intact food protein does not release gastrin. Protein must be digested to small peptides and free amino acids in order to stimulate gastrin secretion from G cells. Although several amino acids release gastrin, the most potent are tryptophan and phenylalanine. At the ultrastructural level the G cell extends to the luminal surface of the pyloric glands and has microvilli which extend into the lumen. These extensions probably allow the G cell to sense the contents of the stomach. It is not known whether there are specific receptors on the luminal membrane of the G cell for known gastrin releasers and whether the releasers diffuse or are transported into the cell to act as intracellular messengers. The release of gastrin is also strongly dependent on the pH of luminal contents; when luminal pH drops below 2.5, gastrin release is markedly inhibited. The mechanism of this H^+-de-

pendent inhibition is also unknown. There are numerous neurotransmitters and peptide hormones which can act on the basolateral surface of the G cell to regulate gastrin release. Those which increase gastrin release are acetylcholine, catecholamines (through a β-adrenergic mechanism), and a newly discovered peptide called *gastrin releasing peptide*. Inhibitors of gastrin release include somatostatin, secretin, VIP, glucagon, and calcitonin, among others. Again, it is not possible at present to provide a coherent picture which accounts for all of these substances in regulating gastrin release. The most important points to keep in mind are that gastrin is released by protein digestion products, Ca^{++}, and coffee and that low pH of gastric contents strongly inhibits gastrin release.

INTESTINAL PHASE. Food and gastric juice enter the intestine during a meal and elicit a mixture of stimulatory and inhibitory effects on gastric secretion. The inhibitory effects appear to be more important when a mixed meal is eaten. The stimulatory effect is due solely to dietary peptides and free amino acids in the lumen of the small intestine. Perfusion of the proximal small intestine with a mixture of amino acids stimulates gastric acid secretion to as much as 30 per cent of maximal levels. There is no increase in the circulating gastrin concentration, so some other mechanism must be responsible for the increase in acid secretion. When a mixture of amino acids is infused intravenously in human beings, a similar pattern is seen of increased gastric acid secretion without an increase in circulating gastrin. Thus, the intestinal phase of acid secretion is probably due to absorption of small peptides and amino acids, which then act to stimulate oxyntic cell secretion.

PATHOPHYSIOLOGY. Numerous defects have been described in the regulatory mechanisms for gastric secretion in duodenal ulcer patients. These patients frequently have increased gastric acid responses to stimulation during the cephalic, gastric, and intestinal phases. As already noted, part of this increased response may be due to a larger number of secretory cells. This increased capacity to secrete acid may obscure other defects. One method to correct for this is to normalize secretory data in duodenal ulcer patients by expressing acid output as a percentage of maximal response to an exogenous stimulant such as pentagastrin or histamine. When this is done it is found that most duodenal ulcer subjects also have a higher fractional response, i.e., they secrete more nearly maximal levels than normals during the three phases of response to a meal. One reason that may underlie the increased acid secretory response in all phases is an increased sensitivity of the oxyntic cell to many stimuli. This increased sensitivity, in turn, could be due to several factors: increased gastric mucosal levels of histamine and acetylcholine; an increased number of receptors for these stimulants; more efficient coupling of receptors to the secretory machinery; and increased release of and sensitivity to gastrin. With respect to gastrin release and sensitivity, it is frequently found that the gastrin response to a meal is increased in duodenal ulcer patients, although basal circulating levels of gastrin are not elevated. In addition, duodenal ulcer subjects have an increased serum gastrin concentration in response to cephalic stimuli, whereas normals do not. The gastric phase of the gastrin response is also elevated in duodenal ulcer disease. In one subset of such patients an enormous gastrin response to food occurs, and this condition is called antral G cell hyperfunction. This is another condition which appears to be genetic, since there is familial clustering of cases. In these and other less dramatic cases, increased gastrin release could be due to increased sensitivity of G cells to stimulation; increased amounts of gastrin per G cell; increased numbers of G cells per

antrum; or decreased inhibition of gastrin release by low intraluminal pH. At present none of these mechanisms have been established.

In several conditions marked by low rates of acid secretion, circulating gastrin levels are frequently elevated, both basally and after a meal. This observation has been made in gastric ulcer patients, and it is assumed to occur in those with acid hyposecretion, as well as in people with achlorhydria due to chronic atrophic gastritis with or without pernicious anemia. The mechanism is very likely that of decreased inhibition of gastrin release by intraluminal acid. Instilling HCl into the stomach in these patients dramatically lowers circulating gastrin concentrations.

In patients with a gastrinoma (Zollinger-Ellison syndrome), a benign or malignant growth consisting of G cells occurs in the pancreas or duodenal wall. With gastrinoma there are very high blood levels of gastrin which show little response to feeding. Such patients usually present clinically with intractable duodenal ulcer disease, multiple ulcers, even ulcers in the jejunum, or with diarrhea. Each of these is due to the massive levels of acid secretion stimulated by high gastrin levels.

Inhibition of gastric secretion

Although the net result of eating a mixed meal is stimulation of gastric secretion, the inhibitory mechanisms that are called into play are important in limiting rates of acid secretion. Thus, duodenal mucosal defense mechanisms are not overwhelmed and pancreatic and brush border digestive enzymes are not inhibited by low intestinal pH. One source of inhibition of gastric secretion is acid itself in the stomach, which bathes the antral mucosa and blocks the release of gastrin. A variety of other inhibitory factors that act in the intestine to limit gastric secretion include: (1) fatty acids and monoglycerides, the products of triglyceride digestion; (2) hypertonic solutions consisting of any solute with an effective osmotic pressure greater than 400 mOsm.; and (3) acid at pH less than 5.0.

Consider for a moment the events and time course of digestion of a meal. Ingestion of a solid, mixed meal elicits acid and pepsinogen secretion through cephalic and gastric phase mechanisms. Acid provides the proper pH for conversion of pepsinogen to pepsin and for the action of pepsin on proteins. Only the outer part of the food mass is exposed to acid and pepsin, since gastric motility is not sufficient to mix food and gastric juice completely. However, pepsin digests about 15 per cent of the protein in a meal. The resulting small peptides stimulate gastrin release and more acid and pepsinogen secretion. Peptides and proteins are strong buffers, so that they soak up much of the acid secreted in the first hour or two after eating. Triglycerides in the food are broken down by lingual and gastric lipases into fatty acids, mono- and diglycerides. Although probably less than 10 per cent of fat in the meal is digested before entering the duodenum, fat digestion products are very strong inhibitors of gastric secretion, and inhibition from this source begins almost immediately. Finally, the stomach is a very poor adjuster of osmolarity so that, depending on the amount of water taken with food, the stomach empties a hypertonic solution of digestion products into the duodenum. This also rapidly initiates inhibitory mechanisms which limit gastric secretory rates. Thus, a highly coordinated series of events acts to provide "just enough and not too much" gastric juice to produce optimal intragastric and intestinal digestion of food. Acid and pepsinogen are initially secreted at near maximal rates, but within 30 minutes inhibitory factors begin to limit secretory rates to 50 per cent or less of the maximal rate.

What are the factors that mediate the inhibitory effects on gastric secretion of various substances in the intestine? Both neural and hormonal mechanisms probably contribute, but our knowledge in this area of regulation of gastric secretion is the least complete of any. Peptides which inhibit acid secretion when given exogenously include secretin, VIP, glucagon, cholecystokinin (CCK), somatostatin, gastric inhibitory peptide (GIP), neurotensin, gastrin-releasing peptide, and urogastrone (very similar to epidermal growth factor). Of these, secretin, VIP, and somatostatin also inhibit gastrin release. Under experimental conditions in animals, products of the digestion of fat in the intestine cause measurable release into the blood of secretin, somatostatin, and GIP. In humans the amount of GIP released by a meal is insufficient to inhibit gastric secretion. Intraluminal acid releases secretin, but again there is no evidence concerning the importance of this mechanism for inhibiting gastric secretion in humans. There are no known hormonal mediators for the effects of hypertonic solutions on gastric secretion.

The neural mechanisms regulating inhibition of gastric secretion by substances in the intestine are probably reflexes whose efferent limbs release adrenergic transmitters or neuropeptides. Experiments have shown that both α- and β-adrenergic agonists inhibit gastric secretion, although α-adrenergic inhibition may be due to decreased gastric mucosal blood flow. The peptides found in nerves of the oxyntic mucosa which are known to inhibit gastric secretion are VIP and somatostatin.

PATHOPHYSIOLOGY. Duodenal ulcer disease is the major disease in which defective intestinal inhibition might interfere with the regulation of gastric secretion. Studies have shown less inhibition of gastric acid secretion by the products of fat digestion and other substances in duodenal ulcer subjects. The possible mechanisms that remain to be proven include: (1) decreased release of inhibitory peptides; (2) disordered reflex inhibition; and (3) decreased sensitivity of oxyntic cells to inhibition.

Clinical tests of gastric function

There are only a few tests commonly used in the clinical setting for assessment of gastric secretory function. By far the most common are upper gastrointestinal fiberoptic endoscopy and gastrin radioimmunoassay, neither of which directly measures gastric secretion. During endoscopy, a flexible fiberoptic endoscope is inserted through the mouth into the stomach in order to observe the gastric mucosa. This method for detecting gastric ulcers, gastritis, and other diseases has to a large extent supplanted x-ray studies.

Gastrin radioimmunoassay (RIA) provides a measurement of circulating levels of gastrin. This test is useful in detecting hypergastrinemia in the Zollinger-Ellison syndrome. It is necessary to distinguish between hypergastrinemia due to the presence of a gastrinoma and that due to achlorhydria in gastritis, pernicious anemia, and other causes of appropriate hypergastrinemia. This can usually be decided on the basis of the clinical picture, the results of acid secretory studies, or the change in blood gastrin concentration after administering secretin as an intravenous injection. Only in subjects with gastrinoma does secretin have the paradoxical effect of increasing gastrin concentration; in all other gastric diseases and in normal subjects, gastrin is decreased. Gastrin RIA may also become an important test in detecting the presence of G cell hyperfunction as a cause of duodenal ulcer.

Gastric secretory testing is much less commonly done than in past years. The methods all involve aspiration of gastric secretion through a nasogastric or orogastric tube, followed by measurement of pH or calculation of acid output in the fluid. However, it is very rare that the results of such testing have any effect on therapy, so this procedure is not usually performed.

Annotated references

Baron, J. H.: Lean body mass, gastric acid, and peptic ulcer. Gut 10:637, 1969. (A collection of data on secretory patterns in different diseases and in normal subjects.)

Donaldson, R. M., Jr.: Intrinsic factor and the transport of cobalamin. In Johnson, L. R. (ed.): Physiology of the Gastrointestinal Tract, p. 641. New York, Raven Press, 1981. (A cogent review of a complicated topic.)

Feldman, M., and Richardson, C. T.: Gastric acid secretion in humans. In Johnson, L. R. (ed.): Physiology of the Gastrointestinal Tract, p. 693. New York, Raven Press, 1981. (A useful compilation of data on human gastric function; points out differences between humans and other species.)

Forte, J. G., Black, J. A., Forte, T. M., Machen, T. E., and Wolosin, J. M.: Ultrastructural changes related to functional activity in gastric oxyntic cells. Am. J. Physiol. 241:G349, 1981. (Detailed description of evidence favoring a role for morphological transformation of oxyntic cells in production of acid secretion.)

Forte, J. G., Machen, T. E., and Obrink, K. J.: Mechanisms of gastric H^+ and Cl^- transport. Ann. Rev. Physiol. 42:111, 1980. (A general discussion of the theoretical mechanisms for production of HCl and movement of other ions in gastric juice.)

Ito, S.: Functional gastric morphology. In Johnson, L. R. (ed.): Physiology of the Gastrointestinal Tract, p. 517. New York, Raven Press, 1981. (Well-illustrated review of cell types, gland morphology, and functional correlation with ultrastructural changes.)

Kauffman, G. L.: Gastric mucus and bicarbonate secretion in relation to mucosal protection. J. Clin. Gastroenterol. 3(Suppl. 2):45, 1981. (A short, critical review of the still-theoretical mechanisms by which mucus and bicarbonate may prevent mucosal damage by acid.)

Lipkin, M.: Proliferation and differentiation of gastrointestinal cells in normal and disease states. In Johnson, L. R. (ed.): Physiology of the Gastrointestinal Tract, p. 145. New York, Raven Press, 1981. (Discusses alterations of cell proliferation in human diseases.)

Rotter, J. I., and Rimoin, D. L.: The genetic syndromology of peptic ulcer. Am. J. Med. Genet. 10:315, 1981. (A catalog of apparent genetic associations between ulcers and other diseases or markers; a good reference source for further reading.)

Solcia, E., Capella, C., Buffa, R., Usellini, L., Fiaccha, R., and Sessa, F.: Endocrine cells of the digestive system. In Johnson, L. R. (ed.): Physiology of the Gastrointestinal Tract, p. 39. New York, Raven Press, 1981.

(The most up-to-date review of endocrine cell morphology and proposed secretory products in a rapidly changing field.)

Soll, A. H., and Grossman, M. I.: The interaction of stimulants on the function of isolated canine parietal cells. Phil. Trans. R. Soc. Lond. B. *296*:5, 1981. (Discusses the interactions and intracellular mediators of gastrin, histamine, and acetylcholine on oxyntic cells.)

Weinstein, W. M.: The diagnosis and classification of gastritis and duodenitis. J. Clin. Gastroenterol. *3*(Suppl. 2):7, 1981. (A thoughtful discussion of the most common "disease" due to changes in gastric morphology.)

23

Absorption of Water and Water-Soluble Solutes

Konrad H. Soergel, M.D.

The absorptive capacity of the human intestinal tract far exceeds normal metabolic requirements of the body. This functional reserve capacity permits rapid correction of metabolic deficits by sudden increases in oral intake; however, the evidence for precise physiological autoregulation of absorptive mechanisms is scant. Excluding calcium and iron, inorganic solutes and nutrients undergo absorption with little regard to bodily needs.

Many variables affect net absorption of a particular substance, including its intraluminal concentration, digestion, contact time with the absorbing surface, integrity of the absorbing cells, blood and lymph flow, and even the rate of endogenous secretion of some substances into the intestinal tract. All factors contribute to the complete picture of intestinal absorption in health and disease.

General considerations

Digestion
Carbohydrates are absorbed only after complete enzymatic hydrolysis to monosaccharides, whereas proteins with rare exceptions require digestion to dipeptides or tripeptides or to individual amino acids. Digestive activity occurs both intra-luminally and at the cell surface. Intraluminally, digestion is carried out by salivary, gastric and pancreatic enzymes. Neurohumoral regulation of digestive secretions is complex. However, the hormone cholecystokinin (CCK), which is released from the proximal small intestine and causes pancreatic secretion of enzymes and gallbladder contraction, is the most important mediator of intraluminal digestion. At the luminal surface, digestion is carried out by enzymes located on the microvilli of the absorbing cells.

SURFACE AREA. Anatomically the small intestine is a pliable cylinder. The distance from pylorus to ileocecal valve is about 500 cm., shortening to about 280 cm. when intubated with even a very thin and flexible tube. Were its lining smooth, the surface area would be 3,100 cm.^{2}; the folds of Kerkring (valvulae conniventes) increase this value threefold and the presence of crypts and villi another tenfold. The interaction between intestinal epithelium and luminal contents is, however, believed to be limited to the villus tips, resulting in an estimated effective absorptive area of 1.12 m.^{2}. The microvilli covering the absorptive cells account for an expansion of this value to about 27 m.^{2}, comparable to the surface area of a living room. The absorptive area

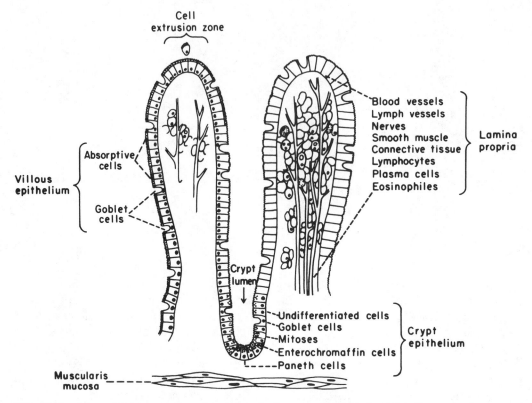

FIG. 23–1. Schematic representation of small intestinal crypt and villi. (Code, C. F. (ed.): Handbook of Physiology, Sec. 6, Vol. 3, p. 1127. Washington, D.C., Am. Physiol. Soc., 1968.)

per unit length declines from jejunum to ileum owing to the gradual decrease in number and height of valvulae conniventes and villi. The colonic surface, lacking valvulae and villi and discounting the microvilli, measures only 0.15 m.².

INTESTINAL EPITHELIUM. The absorptive cells of the small intestinal mucosa are derived from undifferentiated cells in the crypts of Lieberkuhn (Fig. 23–1). The latter cells proliferate rapidly (mean generation time: 24 hours). Absorptive cells migrate steadily toward the lumen. They appear at the villus base within 24 hours, reach the villus tips in 5 to 7 days and are then shed into the intestinal lumen. It has been estimated that 20 to 50 million epithelial cells are lost from the human intestine each minute. The absorptive cells continue to mature during their upward migration and

do not acquire their full complement of enzymes or their full absorptive capacity until they reach the upper half of the villus. A positive correlation exists between absorptive capacity and the functional maturation of these villus epithelial cells. Goblet cells, present in the epithelium of both crypts and villi, secrete mucus continuously by a merocrine process.

The crypts, which outnumber villi by a factor of 6 to 10, contain two additional cell types: the basophilic Paneth cells, which discharge prominent secretory granules into the crypt lumen, and endocrine cells, which make little or no contact with the lumen, suggesting secretion into the capillaries of the lamina propria. The latter, previously termed enterochromaffin (Kulchitsky) and argentaffin cells, synthesize serotonin and a variety of polypeptide hormones, namely gastrin, secretin, CCK, gastrointestinal inhibitory polypeptide (GIP), vasoactive intestinal polypeptide (VIP), bombesin, motilin, enteroglucagon, and substance P. They are part of the amine precursor uptake and decarboxylation (APUD) cell system believed by some to originate from the neural crest. Some of the named hormonal substances, particularly VIP, substance P, serotonin, and C-terminal fragments of CCK and gastrin, have also been identified in the neurons of the gastrointestinal intramural ganglia and in certain areas of the central nervous system.

The high turnover rate makes intestinal epithelium particularly vulnerable to alterations in cell proliferation. Ionizing radiation, radiomimetic cytotoxic drugs, and severe protein deficiency cause mitotic inhibition in the intestinal crypt, resulting in inadequate replacement of the senescent cells lost from the villus tips. A different disturbance exists in nontropical (celiac) sprue: the absorptive cells are shed abnormally fast, causing a compensatory increase in undifferentiated crypt cell proliferation and in the epithelial cell migration rate. Both processes result in flattening and eventual disappearance of intestinal villi. This is combined with shortening of the crypts in diseases with reduced cell proliferation but with crypt elongation and hyperplasia in sprue. After resection and surgically created bypass of large parts of the small intestine, the ileum adapts to the sudden drop in absorptive surface by increasing the number of cells per villus, the villus length, and its absorptive capacity. Detailed studies in the rat have shown that this compensatory hyperplasia and hypertrophy is mediated by increased exposure of the ileal mucosa to both intraluminal nutrients and pancreaticobiliary secretions.

Three additional morphologic features have important implications with regard to intestinal transport. First, adjacent lateral cell walls appear fused just below the base of the microvilli. These so-called tight junctions limit the exchange of water and solutes between the lumen and the lateral intercellular space. This space is bordered by complex folds of the lateral cell membranes; it is distended during active absorption, particularly near the villus tips. Second, the luminal microvilli are covered by a filamentous layer of glycoproteins synthesized by the underlying absorbing cell; this "fuzzy coat" is firmly attached to the cell wall and contains enzymatic proteins which hydrolyze oligosaccharides and peptides. Third, the narrow diameter of crypt orifices (100 μ) prevents contact between crypt epithelium and intestinal contents. Crypt cells probably have secretory functions and contribute to the formation of fasting intestinal contents, the succus entericus, which also contains mucus and desquamated epithelial cells. Thus, it is believed that intestinal net absorption of water and electrolytes represents a balance between secretion by the crypts and absorption by the villi.

MUCOSAL CONTACT TIME. All other factors being equal, absorption of any substance increases with the time it is in contact with the absorbing surface. The time spent by a bolus within the small intestine depends upon the mean velocity of propulsion. During fasting, as well as after a meal, intestinal contents move aborad at 1.5 to 2.0 cm./min.; hence, a bolus of food reaches the terminal ileum 2 to 3 hours after leaving the stomach. Within the physiological limits, variations in intestinal flow rate elicit no changes in aborad flow velocity. The fasting flow in the midjejunum is about 2.6 ml./min., rising to a maximum of 4.5 ml./min. after a small meal. However, the intestine can accommodate flows up to 7 to 8 ml./min. by increasing its diameter, without change in velocity. At flows greater than 8 ml./min., however, the lumen does not distend further and flow velocity rises (transit time decreases). The large reserve capacity of the small intestine for absorption tends to prevent malabsorption from occurring, even when large amounts of food and drink enter the small intestine or when the aborad flow velocity of intestinal contents is moderately increased.

INTESTINAL BLOOD FLOW. Blood supplies the absorptive cells with oxygen and energy-producing substrates (mainly ketone bodies, glutamine, and glucose), provides the water entering the intestinal lumen in response to osmotic gradients and transports absorbed water and water-soluble substances into the general circulation. Sight of a meal increases intestinal blood flow, which rises further during digestion and absorption, indicating a redistribution of flow to the splanchnic circulation. A reverse relationship (variations in blood flow affecting absorption) undoubtedly exists, but is probably of no physiological significance. In dogs, intestinal blood flow must be halved before the rate of glucose absorption decreases. To what extent, and how, intestinal mucosal blood flow and absorption are linked in normal human physiology remain unexplained. The major absorptive role of intestinal lymphatics is to transport absorbed lipids in the form of emulsion droplets (chylomicrons).

INTESTINAL BACTERIAL FLORA. More than 60 bacterial species have been isolated from the human large intestine. About 15 per cent of samples obtained from the stomach, duodenum, jejunum, and proximal ileum during fasting are sterile. The rest contain viable bacteria in counts up to 10^5/ml. This flora is predominantly gram-positive and microaerophilic (*Streptococcus viridans, Lactobacillus, Staphylococcus*). In the distal ileum, the bacterial flora changes in composition and increases greatly in number (up to 10^9/ml.). Wet stool has a colony count of 10^{11} to 10^{12}/g.; bacteria account for about one-third of dry fecal weight. The flora in the distal ileum and colon is predominantly anaerobic (bacteroides, clostridia, bifidobacteriae); in addition, enterobacteriae (*E. coli, A. aerogenes*), *Proteus, Pseudomonas, Streptococcus fae-*

calis, aerobic lactobacilli, and *Staphylococcus* are usually present. Small numbers of fungi, mainly *Candida albicans,* can be found at all levels of the intestinal tract. The mechanisms operative in preventing bacterial proliferation in the small intestine are mainly gastric acidity and mechanical cleansing by the passage of intestinal contents; bile acids and the secretory immunoglobulin A may also inhibit growth. In healthy North Americans and Europeans, viral cultures of small bowel contents and stool are usually negative.

The bacterial flora of the stomach and proximal small intestine does not alter nutrients or bile acids. The bacteria in the distal ileum and colon, however, act upon a variety of substrates present in intestinal contents; this interaction increases when abnormal quantities of nutrients enter the colon because of malabsorption. Representative examples of such bacterial alterations and the subsequent fate of the resulting metabolic products are summarized in Table 23–1.

Significantly increased numbers of aerobic and anaerobic bacteria are found in the stomach and small intestine of achlorhydric patients. Proliferation throughout the small intestine of the flora normally residing in the distal ileum occurs with intestinal stasis, as with diverticula, blind loops, gastrojejunocolonic fistula, or motility disorders, and occasionally after resection or bypass of the ileocecal sphincter. In this static situation (the blind loop syndrome) bacteria compete with the host for ingested vitamin B_{12}; bile acids are deconjugated with consequent impairment of the digestion and absorption of fat; carbohydrates are fermented intraluminally; and the bacterial enzymes destroy brushborder enzymes required for surface digestion, such as maltase and sucrase; finally, patchy areas of mucosal damage are found when multiple intestinal biopsy samples are taken.

Treatment with the antibiotics lincomycin and clindamycin results in reduction of the anaerobic flora of the colon and is associated with nonspecific diarrhea in 15 to 25 per cent of people receiving these drugs, and with life-threatening pseudomembranous colitis in about 4 per cent of patients. The latter condition is caused by a cytopathic exotoxin produced by newly appearing *Clostridium difficile* and can be treated effectively with antibiotics to which this anaerobe is sensitive. The mechanism responsible for nonspecific diarrhea without colitis has not been identified but may involve direct interference by the antibiotic with water and salt absorption. Other antibiotics, particularly tetracycline and ampicillin, occasionally cause the same complications.

Although colonic bacteria may synthesize the B vitamins, vitamin K, and folic acid, this is of negligible nutritional significance to the host.

Mechanisms of intestinal membrane transport

Absorption represents the transport of water and solutes from intestinal contents, which have a variable composition, across a membrane—the intestinal epithelium—to the body fluids, which have a relatively fixed composition. The rates of absorption and secretion of a substance for a given intraluminal concentration are determined by the properties of this biological membrane, which vary along the intestine. Conversely, membrane characteristics can be deduced from determining transport rates under defined conditions.

Membrane structure

Permeability of the epithelial barrier depends upon the lipid structure of luminal and basolateral cell walls, the action of carrier mechanisms and ion pumps, and the configuration of water-filled channels ("pores") penetrating the tight junctions. A double layer of polar lipid molecules forms the "backbone" of the cell wall which is impermeable to ions and water. Unionized solutes penetrate the membrane in direct relation to their lipid solubility. The entry of water-soluble solutes into the absorbing cell is facilitated by carrier mechanisms believed to represent special protein molecules. The solute combines with stereospecific sites on the carrier, causing the carrier-substrate complex to become mobile within the lipid cell wall. The substrate then dissociates from the carrier at the opposite face of the membrane. This process is specific for certain molecules and ions and is inhibited in the presence of a second substrate that is capable of competing for the binding sites. The direction of carrier-mediated transport is determined by concentration gradients, and it reaches a maximum velocity when all carrier sites are occupied by substrate. Ion pumps combine the specificity of a carrier with the ability to expend energy in order to transport ions against concentration gradients. A prominent example is the Na, K-activated adenosine triphosphatase (ATPase) located in the basolateral cell wall. The pores traversing the tight junctions provide an extracellular pathway for nonlipid substances. In the rabbit ileum, about 80 per cent of the transepithelial movement of Na^+, K^+ and Cl^- in response to electrochemical gradients occurs through this "shunt" pathway. These channels contain fixed negative electrical charges which facilitate permeation by cations but restrict anion movement. The effective pore radius is about 4 Å (0.4 μ) in the jejunum and decreases progressively in the ileum and colon. Thus, the pore dimensions completely exclude solutes of molecular weight greater than 150. Water, however, moves exclusively through the pores because there are no carriers to transport water across cell walls.

In summary, there exist two distinct pathways for solute transport from cell surface to blood or lymph (absorption) and in the reverse direction (secretion). The first is the transcellular path from the "fuzzy coat" (glycocalix) covering the microvilli to the

TABLE 23–1
Bacterial Biotransformation and Metabolism of Products in Man

Class of Compounds	Example	Type of Biotransformation and Product	Fate of Product	Comments*
Triglycerides	Trioleate	Lipolysis → oleic acid, monoglyceride, glycerol	Poorly absorbed	See below
Fatty acids	Oleic acid	Hydrogenation → 10-hydroxystearic acid	Poorly absorbed	If present in sufficient concentration in colon, inhibit absorption of Na⁻ and water.
		Saturation → stearic acid	Poorly absorbed	
Conjugated bile acids	Cholyltaurine	Deconjugation → cholic acid (+ taurine)	Absorbed in part; if not absorbed, 7α-dehydroxylated (see below)	After absorption, conjugated with glycine or taurine BE; EHC; rate of deconjugation may be assessed with bile acid deconjugation breath test.
	Cholylglycine	Deconjugation → cholic acid (+ glycine)		
Unconjugated bile acids	Cholic acid	7α-dehydroxylation → deoxycholic acid	Partly absorbed; conjugated; EHC	After absorption, conjugated with glycine or taurine; Be; EHC
	Chenodeoxy-cholic acid	7α-dehydroxylation → lithocholic acid	Partly absorbed; conjugated, and sulfated; BE; little EHC.	After absorption, conjugated with glycine or taurine sulfated; Be; little EHC.

Decreased 7α-dehydroxylate may be associated with high concentrations of chenodeoxycholic acid and consequent diarrhea. |

Carbohydrates	Poly-, Oligo- and Monosaccharides	Oxidation → Organic acids: acetic, propionic, butyric, lactic Fermentation gases: CO_2, CH_4, H_2	Partly absorbed; oxidized to CO_2 Partly absorbed; expired in breath	Can cause osmotic diarrhea. If absorbed, metabolized acetate, then CO_2. Increased H_2 production in the colon signalled by increased breath H_2.
Amino acids	Glycine	Oxidative deamination → CO_2, NH_3	Absorbed	CO_2 expired in breath; ammonia converted to urea, see above;
	Taurine	Oxidative deamination → CO_2, $NH_3SO_4^-$	Absorbed	sulfate excreted in urine.
	Tryptophan	Deamination → indole	Probably poorly absorbed; if absorbed, oxidized and sulfated to form indican; RE	Oxidized to indoxyl (H); sulfated to form indican. Two molecules of indican may condense to form an indigo dye.
		Decarboxylation → tryptamine	Absorbed	Oxidized (H) to indoleacetic acid (RE)
	Histidine	Deamination 5-imidazole acrylic acid (urocanic acid)	Absorbed	Converted to forminoglutamic acid.
Bile pigments	Bilirubin diglucuronide	Deconjugation and reduction urobilinogens	Poorly absorbed	BE-RE
Urea	Urea	Hydrolysis → NH_3, CO_2	Absorbed	See above
Drugs	Sulfasalazine	Reduction of azo bond → sulfapyridine / 5-NH_2-salicylic acid	Absorbed / Poorly absorbed	N-acetylation, 5-hydroxylation (H): excreted as glucuronides in urine. / Acetylation (H) → urine
	Phenolphthalein	Absorbed → 85% reexcreted as glucuronide (EHC); 15% excreted into urine as glucuronide.	Glucuronide poorly absorbed → deconjugated in colon.	Unconjugated drug reduces colonic water and Na^+ absorption (inhibition of Na, K,-ATPase).

*This listing is incomplete, but all of the transformations shown have been found to occur in man. Abbreviations: BE: biliary excretion; EHC: undergoes enterohepatic circulation; (H): hepatic, i.e., in the liver. Absorption is defined as an amount of magnitude sufficient for detection of urinary or biliary excretion of a compound or its metabolite.

cell membrane with its interspersed carrier substances, to the cytoplasm, where metabolic energy is generated for active transport steps and certain biotransformations occur, to the basolateral cell walls. The second is the paracellular or shunt path which extends from the luminal cell surface through the pores penetrating the tight junctional complex between adjoining cells. Both pathways join in the lateral intercellular space from where the solute has to traverse the basement membrane and then the wall of capillaries and lymphatics before entering blood or lymph.

The microclimate

The luminal cell wall is covered by the so-called unstirred layer (UL), a coating of water that does not mix with bulk luminal contents. Its thickness, d, has been estimated to range from 200 to 600 μm. Solutes can cross this layer only by diffusion; thus, the effective solute concentration at the absorbing surface, C_2, is less than C_1, the concentration in the bulk phase, and $C_2 = C_1 - J(d/D)$, where J is the rate of uptake into the cell, d is the thickness of the UL, and D the diffusion constant in water. Long-chain fatty acids, for example, encounter little resistance in crossing the cell wall but diffuse slowly in water. In this situation, d becomes rate limiting for absorption. The opposite is true for inorganic ions; consequently, the UL does not affect their absorption kinetics. To date, variations in d have not been shown to contribute to malabsorption and d does not change with intestinal flow rates likely to be encountered *in vivo*.

In addition, the UL traps H^+ ions secreted by the intestinal mucosa. Thus, an acid microclimate is maintained at the cell surface with pH values of 5.8 in the jejunum, 6.4 in the ileum, and 6.7 in the colon. Only the uncharged species of organic acids, including fatty acids and certain drugs, can diffuse through the cell membrane. Accordingly, their cellular uptake rate increases as the surface pH approaches their dissociation constant.

Transport processes

A major activity of the epithelial cell is the maintenance of a fixed, low intracellular sodium concentration by pumping the cation into the lateral intercellular spaces; thus, a hyperosmolar region is created. This localized area of high sodium ion concentration induces an osmotic water flow from the lumen, across the "tight" junction. As water flows, it carries with it those solute molecules that are small enough to pass through the aqueous pore channels (solvent drag). Thus, water movement is passive in response to active sodium pumping at the lateral cell wall, and this osmotic flow of water transports solutes by convection.

Both solvent drag and active transport can move solutes against a concentration gradient, that is, to a compartment with higher solute concentration. Processes such as simple or facilitated diffusion move solutes to compartments with lower solute concentration.

Active transport and facilitated diffusion

Active transport is defined as net absorption against a concentration or electrical gradient or both; it requires metabolic energy and is assumed to involve a carrier molecule in the membrane which associates reversibly with the actively transported solute. Active transport creates a serosal-to-mucosal concentration gradient which may cause part of the transported solute to diffuse back into the intestinal lumen. Net transport rate then equals active transport minus the rate of passive back diffusion. Thus, net absorption of actively transported substances, particularly of inorganic ions, decreases with increasing diffusion permeability of the membrane. When a solute is absorbed down a mucosal-serosal concentration gradient without energy expenditure but requires attachment to a carrier in order to gain entry into the cell, the process is termed *facilitated diffusion*.

The interaction between solute and car-

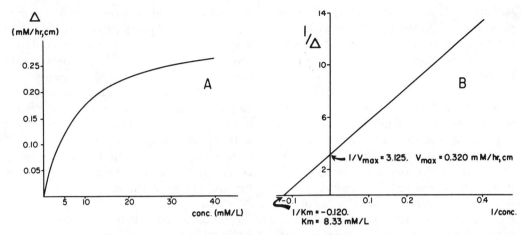

FIG. 23–2. Absorption of D-fructose from the human jejunum during continuous perfusion. The test solutions contained D-fructose in varying concentrations, a nonabsorbable marker substance, and NaCl to an osmolality of 285 mOsm./kg. Samples for analysis were obtained 15 cm. and 45 cm. distal to the site of infusion, and absorption of D-fructose was calculated from changes in D-fructose concentration relative to that of the marker. (A) Plot of mean concentration vs. absorption rate. (B) Lineweaver-Burk plot of the same data. These data represent the net result of all factors affecting D-fructose disappearance from the bulk lumen fluid, including the effects of the unstirred layer.

rier conforms to the reaction kinetics of an enzyme (the carrier) with a substrate. This relationship can be expressed in terms of first-order (Michaelis-Menten) kinetics by considering the active transport or facilitated diffusion as the velocity of an enzymatic reaction. Figure 23–2A illustrates the facilitated diffusion of D-fructose from the human jejunum. In Figure 23–2B the data have been transposed into a Lineweaver-Burk plot, which allows the calculation of the maximal absorptive rate of this sugar (V_{max}) and of the concentration at which half the maximal transport rate occurs (K_m). In general, substrates with high affinity for the carrier (low K_m) exhibit a low V_{max}.

Active transport of single ions generates electrical membrane potentials, but so many factors influence the potential difference during intestinal absorption that potential difference measurements have shed little light on the mechanism of active ion transport in man.

MOVEMENT OF SOLUTES BY OSMOTIC FLOW OF WATER. When unequal concentrations of solute exist on the two sides of a membrane, osmotic forces are generated that cause bulk flow of water through the pores. When solute particles are too large to enter the pores, they exert the full theoretical osmotic pressure (as measured in an osmometer). With small solutes, a fraction of the molecules hits the entry to the pore in dead center and thus traverses the channel of the pore. When, for example, every third solute particle enters the pore, the so-called reflecton coefficient of this solute is 2 out of 3, or 0.67. The effective osmotic pressure created by this solute then equals 0.67 times the theoretical osmotic pressure. Bulk flow through an essentially cylindrical channel such as a membrane pore is analogous to flow through a blood vessel and is governed by Poiseuille's Law, according to which small increases in pore diameter will have a marked effect on bulk flow. As water passes through the pores, it sweeps solutes along, a process termed *solvent drag*.

Osmotic permeability decreases progressively from duodenum to rectum. Thus, the jejunum is best suited for rapid equilibration of osmotic gradients generated by (for example) ingestion of a hypertonic meal. In patients with nontropical sprue, pore size is diminished, thereby interfering with bulk flow, solvent drag and, diffusion of small molecules and ions that penetrate normal pores.

SIMPLE DIFFUSION. Lipid-soluble substances, such as nonionized fatty acids and cholesterol, diffuse directly through the lipid bilayer of the membrane. The force that drives this process is the concentration gradient between the luminal cell surface and the cell interior. The absorption rate is linearly related to this gradient and does not show saturation. The permeability coefficient describes the rate of absorption per unit area and concentration gradient. Water-soluble substances diffuse through the water-filled pores if the solute particles are sufficiently small to enter these channels. Besides the radius of the pore, other factors that influence diffusion rates are the number of pores, friction determined by pore length and configuration, and, for electrolytes, the fixed negative charges which act to slow diffusion of anions and to accelerate cationic diffusion.

ION EXCHANGE. Ion exchange (i.e., a stoichiometric exchange of ions of like charge between luminal content and mucosal cell) is an important determinant of ion concentration in the chyme. In the jejunum and ileum, luminal sodium probably is exchanged for hydrogen ion formed in the absorptive cells. In the ileum and colon, luminal chloride is exchanged for cellular bicarbonate. The Na, K-ATPase continuously extrudes sodium from the cell in exchange for potassium ions. While ion exchange does not alter total ion concentration at either side of the membrane, it is capable of affecting uphill transport of a given ion species, such as sodium absorption in the ileum.

PINOCYTOSIS. Pinocytosis, the engulfing of fluid by the cell membrane which then pinches off to form a vacuole containing the imbibed fluid, is an important mechanism for absorption of protein in the newborn, particularly for the immunoglobulins contained in colostrum. Whether this transport process disappears completely ("closure") or whether it continues to function at a greatly reduced rate after the newborn period is uncertain. In any event, antigenic macromolecules can be absorbed in trace amounts by the adult, mainly by specialized epithelial microfold (M) cells overlying Peyer's patches.

INTESTINAL SECRETION. The presence of fluid within the fasted intestine, even when the entry of exocrine secretions is prevented by an occluding duodenal balloon, suggests that the intestinal mucosa has secretory as well as absorptive functions. Recent studies of the normal human jejunum revealed net secretion of sodium, chloride and water, with active chloride secretion as the driving force. It is probable that the secretory diarrhea caused by certain hormones and bacterial enterotoxins (see below) is due, in part, to excessive stimulation of this normal anion secretory mechanism. The question whether active ion secretion takes place in the same villus cells that carry out absorptive functions or whether secretion is the property of nonabsorbing crypt cells is unresolved, although available evidence favors the latter interpretation.

Transport of specific nutrients and water

Fluid and electrolyte movement during digestion

Water and salt ingested orally represent only a fraction of the amounts that the intestine must absorb to preserve homeostasis. Most of the absorptive load comes from endogenous secretions (Table 23–2). Formation of succus entericus contributed by the jejunum and, in much smaller quantities, by the ileum is estimated at 2.5 l./day; it contains 0.3 g./dl. of proteins, 0.5 mEq./l.

TABLE 23–2
Flow Rate and Composition of Intestinal Contents in Man

| Location | Fasting Flow Rate (l./day) | Fasting Composition (meq./l.) | | | | 3 Meals Daily Flow Rate (l./day) |
		Na	K	Cl	HCO₃	
Duodenum	3.5					10
Jejunum	3.7	125	5	115	10	4.6
Distal ileum	1.8	110	5	50	65	1.9
Stool	0.1	35	75	15	30	0.1

of calcium, and mucus, in addition to the major electrolytes. With food intake, the load increases by augmentation of salivary, gastric, biliary, and pancreatic secretions and by the amount ingested. By the time an ordinary meal reaches the distal jejunum, however, the total increment in luminal flow amounts to only one-third of the original meal volume. Thus, most of the meal and the digestive secretions stimulated by it are absorbed in the jejunum. The colon normally absorbs about 1.7 l./day. Hence, salt and water absorption occur at 99 per cent efficiency (9.9 of 10.0 l./day). The maximum capacity for colonic absorption is estimated at 5.0 to 6.0 l./day; consequently, diarrhea is inevitable when greater volumes are emptied into this organ from the small bowel.

ELECTROLYTE MOVEMENTS. From jejunum to colon, the electrolyte composition of intestinal contents deviates progressively from that of serum and interstitial fluid (Table 23–2). The sodium concentration in jejunal contents is 10 mEq./l. below that of serum, a difference widened to 30 in the ileum and to 105 to 140 mEq./l. in the colon. Hence, sodium is absorbed against chemical concentration gradients and, in addition, against an electrical gradient, since the luminal surface of the gut is negatively charged with respect to the blood. The aborad decrease in intestinal permeability progressively restricts passive diffusion of sodium back into the intestinal lumen and

contributes to the increasing efficiency of sodium absorption from jejunum to colon.

Potassium is passively distributed between blood and intestinal lumen; its increased concentration in the colon is a consequence of the sizeable electrical potential difference (PD) across this organ. In the jejunum, the concentration of chloride is slightly greater than that in blood, whereas the concentration of bicarbonate is less. In the ileum and colon, the concentration of chloride falls and that of bicarbonate rises reciprocally.

The mechanisms for sodium, chloride, and bicarbonate transport are as follows: in the jejunum, luminal contents are slightly acidic and sodium is absorbed in exchange for hydrogen ion produced by the absorptive cells; hydrogen ion then reacts with bicarbonate in the lumen to form CO_2, accounting for the low bicarbonate concentration of jejunal contents. Sodium absorption ceases at a luminal pH below 6.0 when no bicarbonate ions are present. In the ileum, bicarbonate is secreted in exchange for chloride, in addition to sodium—hydrogen ion exchange. Therefore, sodium and chloride absorption does not generate a potential difference. In the colon, the exchange of chloride for bicarbonate continues and the luminal bicarbonate is largely dissipated by reabsorption and by reacting with short-chain fatty acids produced by bacterial fermentation of carbohydrate. Sodium is actively absorbed, which accounts for the high lumen-negative values for PD in this organ. Colonic sodium absorption and potential difference rise in the presence of increased blood levels of aldosterone.

In the very rare syndrome of *congenital chloridorrhea* the ileal and colonic exchange of chloride for bicarbonate is defective, resulting in diarrhea, high chloride concentrations in stool water, and metabolic alkalosis owing to decreased bicarbonate loss with persistent excretion of hydrogen ion excretion (mainly in the form of ammonium ion) in stool.

Active sodium and chloride absorption is stimulated by α-adrenergic agonists and by enkephalins. Cholinergic stimulation causes secretion, as do VIP and several prostaglandins. While all these agents have been identified within the intramural ganglia, their role in the physiological regulation of the intestinal ion transport remains to be elucidated.

PASSIVE WATER MOVEMENT. Water movement across the intestinal mucosa is entirely passive, occurring only through the paracellular or shunt path in response to osmotic gradients. This mechanism efficiently maintains the isotonicity of intestinal contents with respect to plasma (280 mOsm./kg.) with one exception: the osmolality of stool water is about 400 mOsm./kg. This probably results from continuing production of solute—mainly short-chain fatty acids—in the colon, with insufficient time for osmotic equilibration across the relatively impermeable colonic and rectal mucosa. Active solute absorption is accompanied by water absorption, even in the presence of opposing osmotic gradients. In fact, when water absorption is plotted against solute absorption from isotonic solutions, a linear correlation is obtained. This coupling of water to active solute absorption represents a form of "local osmosis." The solute, after crossing the luminal cell wall with a carrier and exiting through the lateral cell wall, raises the osmotic pressure in the lateral intercellular space. A

standing osmotic gradient is thereby created which causes water to flow from the lumen, through the pore channels. The intercellular space widens owing to increasing hydrostatic pressure within it, and water with the absorbed solutes flows in the serosal direction. Basement membrane and capillary endothelium provide negligible resistance to the entry of the absorbate into the capillaries.

Carbohydrates

An American adult consumes about 350 g. of carbohydrates daily, of which about 60 per cent is starch, 30 per cent sucrose, and 10 per cent lactose. After digestion, this amount of carbohydrate yields about 2,000 mmol. of monosaccharides, by far the largest contribution to the total solute load presented to the intestine.

Amylopectin, the main component of starch, is composed of 1,4 α-linked glucose chains, connected by 1,6 α-branching points. Amylose, the minor component, is a straight chain of 1,4 α-linked glucose molecules. Salivary and pancreatic amylase hydrolyze 1,4 α-glucose-glucose links but show little or no activity against either 1,4 α-links located near the end of the glucose chain or those adjacent to the 1,6 α-branching points. Therefore, the products of starch digestion are two- and three-unit pieces of straight glucose chains (maltose and maltotriose) and α-limit dextrins (i.e., the intact 1,6 α-branch points with short 1,4 α-linked glucose chains attached). Normal amylase activity of pancreatic secretions is about 40 times greater than that required for digestion of dietary starch. Consequently, impaired carbohydrate digestion occurs rarely, if ever, with advanced pancreatic exocrine insufficiency.

Several plant polysaccharides, such as hemicellulose, cellulose, pectin, and guar, are part of the diet. Their hydrophilic properties increase fecal bulk. Guar and pectin delay the absorption of glucose, by slowing gastric emptying, as well as the diffusion of glucose to the brushborder surface. These plant polysaccharides resist intestinal digestion but undergo partial bacterial fermentation in the colon. Stachyose and raffinose, oligosaccharides present in legumes and soybeans, are not digested in the small intestine but are fermented in the colon.

OLIGOSACCHARIDASES. The digestion of dietary starch is complete in the first 10 to 20 cm. of jejunum. Since α-amylase is the only intraluminal enzyme capable of digesting carbohydrate, and since carbohydrates are only absorbed as monosaccharides, the final step in carbohydrate digestion must occur on the surface of the absorbing membrane where a number of oligosaccharidases have been identified in the glycocalix (fuzzy coat) of the small intestinal mucosa. Glucoamylase (maltase) hydrolyzes straight 1,4 α-linked glucose chains. Sucrase-dextrinase (sucrase-isomaltase), a hybrid molecule, has several enzymatic activities: one splits sucrose into glucose and fructose, the other the 1,6 α-branching points of α-limit dextrins; both enzymatic sites also possess glucoamylase activity. Lactase digests lactose to glucose and galactose. Finally, trehalase hydrolyzes a rare disaccharide found in some mushrooms and bacteriae, to glucose.

Located outside the entry barrier of the absorbing cells, these enzymes, except lactase, hydrolyze dietary oligo- and disaccharides somewhat faster than monosaccharides can be absorbed. About 75 per cent of the monosaccharides liberated are immediately transported into the cell without appearing in intestinal contents, providing a kinetic advantage to the absorption of sugars arriving at the intestinal mucosa as di- or oligosaccharides. Therefore, ingested sucrose will elevate blood glucose levels just as rapidly as ingested glucose.

Oligosaccharidase activity in biopsy samples of intestinal mucosa rises from the level found in the duodenum to a peak in the distal jejunum, falling to low levels in the distal ileum.

In normal subjects, lactase is least active, maltase most active, and sucrase and α-dextrinase activities are intermediate. The activities of jejunal maltase and sucrase rise 2 to 5 days after a high sucrose or fructose diet is begun, whereas lactase activity remains unaltered by a high lactose or galactose diet. Diets high in glucose do not affect intestinal oligosaccharidase levels.

OLIGOSACCHARIDASE DEFICIENCY. Lack of these enzymes is a frequent cause of carbohydrate malabsorption. Transient, acquired deficiencies attend acute and chronic diseases of the small intestine, such as bacterial or viral enteric infections and idopathic and tropical sprue. Lactase activity may then be nearly undetectable, while the other oligosaccharidases may be reduced to 25 per cent of their normal activity. The clinical result is lactose intolerance with diarrhea, acidic stools, and flatulence after ingestion of milk and ice cream.

The high activity of maltase, which actually represents several different enzymes, explains why isolated maltase deficiency has not been described. Deficiency of sucrase and dextrinase usually, if not always, occurs for both enzymes simultaneously. Patients with congenital sucrase-dextrinase deficiency have diminished maltase activity, as expected, but enough enzyme remains to ensure adequate digestion of maltose and maltotriose. This rare disorder can be managed adequately by a diet low only in sucrose because the lack of α-dextrin digestion generally does not cause clinical symptoms.

A marked decrease or loss of intestinal lactase activity is very common. Intestinal lactase activity decreases after weaning in all mammalian species except man. Primary lactase deficiency in adults is present in 5 to 10 per cent of non-Jewish caucasians, in about 60 per cent of Jews, Arabs, and American Indians, and in 70 to 95 per

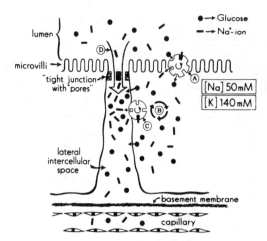

FIG. 23–3. Schematic representation of intestinal glucose and sodium absorption. (A) Glucose-sodium-carrier complex in luminal cell wall. (B) Na⁺-K⁺-dependent ATPase providing energy for active sodium extrusion from the cell. (C) Sodium-carrier complex in lateral cell wall (B and C together are often referred to as the "sodium pump"). (D) Water flow through the shunt pathway in response to the "standing osmotic gradient." Sodium and other small solute particles are carried along by the solvent drag effect (convective transport). Coupling between glucose and sodium absorption occurs at A (demonstrated mainly by studies *in vitro*) and at D. Partial uncoupling occurs by back-diffusion of sodium (opposite to the direction of D) in response to the electrical gradient generated by C.

cent of Blacks, Orientals, and Eskimos. In comparing different populations, persistent lactase activity has correlated well with dairy culture since paleolithic times. Family studies have shown lactase deficiency to be inherited as a dominant trait. The usually mild symptoms of lactose malabsorption due to lactase deficiency (excess flatus, abdominal cramping, diarrhea) can be eliminated by simple dietary adjustments.

MONOSACCHARIDE ABSORPTION. The common hexoses, D-glucose, D-galactose, and D-fructose as well as the pentose D-xylose are absorbed more rapidly from the jejunum than from the ileum (absorption rate per unit length of intestine). They exhibit a maximum absorptive velocity (V_{max}), suggesting that monosaccharide absorption is carrier mediated. Glucose and galactose compete for the same site on a carrier,

which requires the additional binding of a sodium ion before the monosaccharide and sodium are transported (bifunctional carrier). Sodium is then actively extruded into the lateral intercellular space; it is this process which provides the metabolic energy for active glucose (and galactose) absorption by maintaining a low intracellular sodium concentration. With both glucose and sodium reaching the lateral intercellular space, the resultant hypertonicity in this compartment causes inflow of water from the lumen (Fig. 23–3). In the jejunum, where permeability is high, this bulk flow carries with it small solute particles, including sodium ions, by the solvent drag effect. Hardly any solvent drag occurs in the ileum with its low effective pore radius. The potential difference becomes more lumen-negative with active sodium transport; this effect, in turn, causes back diffusion of cations to the lumen. In man, the net result is that glucose and galactose stimulate net sodium absorption in the jejunum, mainly by solvent drag. Although glucose and galactose absorption are sodium dependent, luminal sodium concentration *in vivo* never is sufficiently low enough to limit the rate of hexose transport. Postprandial intestinal glucose concentrations are higher than those in the blood. Therefore, most glucose is transported down a concentration gradient. The main role of active glucose transport is to accelerate absorption beyond the rate that would be obtained by facilitated diffusion. In enterotoxin-mediated secretory diarrhea, absorptive functions remain intact. The successful rehydration of cholera patients with oral glucose-electrolyte solutions is based on the stimulation of sodium and water absorption by glucose in the jejunal lumen.

In congenital malabsorption of glucose and galactose the carrier for these two hexoses is absent. Affected infants have profuse diarrhea and fail to thrive until all dietary carbohydrates except fructose have been eliminated.

D-Xylose is mainly absorbed from the jejunum and a constant fraction of the amount absorbed is rapidly excreted into the urine. These properties have led to its use as a test substance for assessing the absorptive capacity of the proximal small intestinal mucosa.

CONSEQUENCES OF CARBOHYDRATE MALABSORPTION. Carbohydrate entering the colon is rapidly fermented by the resident bacterial flora to short-chain fatty acids (SCFAs), and two gases: CO_2 and H_2. The concentration of SCFAs, mainly acetic, propionic, n-butyric, and lactic acid, in stool water is about 50 mEq./l. in normal subjects on a carbohydrate-free diet. Total SCFA concentration rises to 120–170 mEq./l. with a regular diet. It is estimated that about 20 g. per day of carbohydrate undergo colonic anaerobic bacterial fermentation in healthy subjects eating ordinary food. This amount is composed of dietary fiber (plant polysaccharides), incompletely digested starch granules, stachyose and raffinose in legumes, and endogenous mucins. The production of about two SCFA molecules per molecule of unabsorbed hexose equivalent has the following consequences: first, intraluminal SCFAs are neutralized by reacting with bicarbonate. With increasing rates of carbohydrate fermentation, the neutralization capacity is exceeded; eventually, colonic contents and stool water become acid (pH $<$ 5.5). At this point, the bacterial fermentation process is inhibited, and unaltered carbohydrate begins to appear in the stool. It should be noted that the discharge of acid stool water with carbohydrate malabsorption represents a net loss of bicarbonate which can result in metabolic acidosis. Second, much of the newly formed SCFAs are absorbed by the colon and then metabolized to CO_2. The process of SCFA absorption adds to intraluminal neutralization capacity by causing the appearance of bicarbonate; it also stimulates sodium and chloride absorption to a minor degree. An important consequence of the conversion of carbohydrate to SCFAs which eventually are metabolized in the body is the conservation of metabolic energy from malabsorbed carbohydrate; this process has been termed *caloric salvage.* Third, SCFAs escaping absorption increase the total solute concentration. The resulting osmotic equilibrium increases stool volume with SCFAs as the principal stool water solute. By contrast, in secretory (overflow) diarrhea, stool water contains largely inorganic ions. Hence, with equal stool volumes, osmotic diarrhea causes lower sodium, potassium, and chloride losses than secretory diarrhea. Fourth, a fraction of the hydrogen and CO_2 gas formed is absorbed and appears in the breath. After a 6-hour fast, hydrogen gas is produced at a rate of 0.6 ml./min. in the colon. A few minutes after fermentable carbohydrate has entered the colon, up to 4 ml./min. of hydrogen gas are produced. Increased hydrogen excretion in expired air is a reliable indicator of carbohydrate malabsorption. However, hydrogen gas formation increases not only during carbohydrate malabsorption, but also when nonabsorbable oligosaccharides (e.g., raffinose and lactulose) are ingested, and in the presence of bacterial overgrowth in the small intestine.

Proteins and amino acids

Of the protein load presented to the small intestine about 70 per cent is dietary protein; the rest represents endogenous proteins in digestive secretions, desquamated epithelial cells, and a small amount of plasma protein leakage. About 60 per cent of ingested protein is digested and absorbed in the duodenum and proximal jejunum. Thus, in contrast to carbohydrates, the entire length of the small intestine is required for normal protein absorption.

SURFACE DIGESTION AND ABSORPTION. The fate of the products of intraluminal protein digestion is considerably more complex than is the case for carbohydrates (Fig.

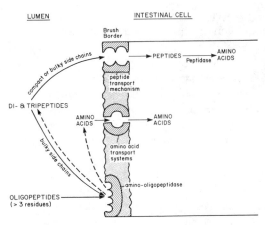

FIG. 23—4. (Gray, G. M.: Mechanisms of digestion and absorption of food. *In* Sleisinger, M. H., and Fordtran, J. S. (eds.): Gastrointestinal Disease, ed. 2. Philadelphia, W. B. Saunders, 1978, p. 247.)

23–4). There are several peptidases in the brushborder membrane which hydrolyze oligopeptides, particularly those containing amino acids with bulky side chains such as leucine and methionine. A significant fraction of amino acid uptake, however, occurs in the form of intact di- and tripeptides by a single, sodium-dependent, energy-requiring peptide carrier mechanism. Once inside the absorbing cell, these peptides are largely split by cytoplasmic peptidases and appear, therefore, as amino acids in portal venous blood.

Absorption of free amino acids shares several features with glucose and galactose absorption, including carrier-mediated, active transport and a requirement for luminal sodium. A grouping of the amino acids into different transport systems is based mainly on studies of competition for cellular uptake *in vitro*: (1) Neutral aromatic and aliphatic amino acids, including tryptophan; (2) dibasic amino acids (ornithine, lysine, arginine, and cystine); (3) dicarboxylic amino acids (aspartic and glutamic acid); and (4) imino acids (proline, hydroxyproline) and glycine. The last two transport systems are not very efficient, and small peptides containing these amino acids are preferentially absorbed intact.

The capacity for amino acid and peptide absorption is highest in the jejunum, while surface digestion by brushborder peptidases rises from the proximal to the distal small intestine. Clinically important deficiencies of brushborder peptide hydrolases have not been reported. Further, the peptide transport system is more resistant to damage by diffuse intestinal diseases than are the mechanisms for absorption of amino acids. Thus, the ability of a patient with untreated sprue to absorb free amino acids is severely reduced, while absorption of peptides is impaired to a lesser degree.

CONGENITAL DEFECTS IN AMINO ACID TRANSPORT. Studies of several congenital disorders of renal and intestinal amino acid transport support the concept of amino acid transport groups. In Hartnup's disease, the intestinal uptake and renal tubular reabsorption of Group 1 amino acids is low. Patients with this disorder present with cerebellar ataxia, a rash, and frequently mental retardation. These features, along with increased urinary excretion of indican, are presumably due to the malabsorption of tryptophan, the precursor of nicotinamide (vitamin B_6). In the blue diaper syndrome, an unexplained isolated defect in intestinal absorption of tryptophan occurs. An oxidation product of urinary indican turns the diaper blue. Patients with cystinuria have a similar but not identical transport defect of Group 2 amino acids in the intestine and renal tubules.

PROTEIN-LOSING ENTEROPATHY. In diseases associated with obstruction to mesenteric lymph flow, with ulcerative or hyperplastic mucosal disease, and in gastrointestinal hypersensitivity states, considerable transudation of plasma proteins may occur. If proteins pass into the proximal gut in this so-called protein-losing enteropathy, they will be digested and absorbed; if proteins are lost in the colon, they escape digestion, and fecal nitrogen loss will be increased. When the rate of transudation exceeds the capacity of the liver to synthesize plasma proteins, hypoproteinemia and consequent edema may occur.

Iron

Absorption of iron is regulated by the amounts of total body iron and is modified by various intraluminal factors. The average adult body contains about 4 g. of iron, of which half is in hemoglobin; the rest is stored in parenchymatous organs and the reticuloendothelial system as ferritin, in cytochrome enzymes, myoglobin, and peroxidases. The body loses iron in desquamated cells from skin, gut, and kidneys and, in women, with menstrual bleeding. Net iron absorption, occurring mainly in the duodenum, is 1 to 3 mg./day.

When the body is overloaded with iron, the amount of ferritin within the intestinal epithelial cells increases; in iron deficiency, it decreases. Part of the iron taken up by the columnar epithelial cells enters the circulation, where it is bound by transferrin; the rest remains in the epithelial cells and reenters the intestinal lumen when the senescent cells are shed from the villus tip. In iron deficiency, mucosal uptake rises and the proportion of mucosal iron entering the body becomes greater. Iron overloading evokes opposite changes. In idiopathic hemochromatosis, the mechanism normally controlling iron absorption and, hence, stabilizing metabolic iron balance goes awry. Iron continues to be absorbed despite greatly increased iron stores in the body.

Inorganic iron is absorbed in the ferrous form. Absorption of inorganic iron is decreased by intraluminal conditions that favor oxidation to the ferric form, such as absence of gastric hydrochloric acid or conditions that precipitate or sequester iron, such as in the presence of phosphate, carbonate, and plant phytates. Ascorbic acid, a reducing agent, augments iron absorption.

Most dietary iron is present in organic compounds, mainly hemoglobin, ferritin, and myoglobin, as well as phytates and siderochromes in plants. The absorption of hemoglobin iron is unaffected by the intraluminal factors influencing the absorption of inorganic iron. Probably hemoglobin iron enters the mucosal cell while still attached to the porphyrin ring.

Vitamin B_{12}

Gastric intrinsic factor, a glycoprotein, and R-protein present in gastric juice and saliva compete for binding of ingested vitamin B_{12} (cobalamin). When gastric contents are acidic, vitamin B_{12} associates preferentially with the R-protein. The latter is partially degraded by pancreatic proteases; vitamin B_{12} is set free by this process and then is bound in the duodenum and proximal jejunum to the protease-resistant intrinsic factor. The intrinsic factor–vitamin B_{12} complex attaches to specific receptor sites in the glycocalix of the ileum before vitamin B_{12} enters the mucosal cells. Transcobalamin II is required for the vitamin to leave the absorbing cells for transport in the portal venous circulation.

Malabsorption of vitamin B_{12} may result from several abnormalities: (1) gastric achlorhydria with absence of intrinsic factor, as in pernicious anemia; (2) congenital isolated absence of intrinsic factor; (3) decreased intraduodenal pancreatic protease activity in patients secreting gastric acid, as in chronic pancreatitis; (4) competition with the host for vitamin B_{12}, as in bacterial overgrowth in the small intestine or during infection with fish tapeworm; (5) accompanying extensive ileal disease or resection, as in Crohn's disease ileocolostomy, ileal resection, and idiopathic and tropical sprue; (6) selective vitamin B_{12} malabsorption, as with a defect in the transfer of vitamin B_{12} from the ileal receptor into the absorbing cell; and (7) a defect in vitamin B_{12} transport from the absorbing cells, as in congenital transcobalamin II deficiency. A radioactively labeled test dose of vitamin B_{12} given orally is poorly absorbed in all these disorders. This absorptive defect is corrected by the addition of intrinsic factor only in the first two conditions listed above.

Folic acid

This vitamin, pteroylmonoglutamic acid (PGA), exists in food predominantly as polyglutamates of reduced PGA. Folate absorption requires a brushborder peptidase, folate conjugase, to break down dietary oligopeptides into the monoglutamate form. The latter is then absorbed, mainly in the duodenum and jejunum, by a saturable, energy- and sodium-requiring transport mechanism. Oxidized and reduced forms of PGA are absorbed equally well. At high intraluminal concentrations, some PGA is also absorbed by simple diffusion. PGA uptake is maximal at pH 6.3, which is below the pH normally found in the duodenum.

Folate deficiency is often the earliest sign of diffuse mucosal disease of the proximal small intestine. Malabsorption of the vitamin is caused by decreased carrier-mediated transport rather than by impaired folate conjugase activity. Sulfasalazine, used in the treatment of Crohn's disease and chronic ulcerative colitis, and methotrexate, a folate antagonist, are competitive inhibitors of PGA absorption. Folate absorption is also diminished by duodenal alkalinization and by diphenylhydantoin, an antiepileptic drug. Increased uptake of folate occurs in patients with pancreatic exocrine insufficiency where the duodenal pH is lower than normal.

Calcium

This ion is actively absorbed by the duodenum and proximal jejunum and is passively absorbed throughout the small intestine. One of the factors influencing its uptake is the amount of a specific calcium-binding protein present in the cytoplasm of the absorbing cells. Active calcium absorption requires accumulation within the absorbing cells of 1,25-dihydrocholecalciferol. This metabolically active derivative of vitamin D is formed in the kidneys from 25-hydroxycholecalciferol, which in turn is produced in the liver by hydroxylation of the natural vitamin.

Malabsorption and diarrhea

Malabsorption refers to impaired absorption of one or several major nutritional components. The term does not differentiate between defects in digestion and absorption. Malabsorption is signalled by increased fecal excretion of the nutrient itself or of products of its intraluminal catabolism. However, increased fecal excretion of a given substance does not necessarily result from malabsorption of dietary materials, for fecal excretion represents the balance between total load (diet and endogenous secretions), digestion, and absorption. Indeed, increased fecal output of water, electrolytes, and bile acids stems exclusively from endogenous secretions. The term *malabsorption syndrome*, although widely used clinically, is not a diagnosis and should be replaced whenever possible by the name of the specific disease (e.g., tropical sprue), accompanied by identification of the particular substance malabsorbed (e.g., carbohydrate, fat). Diarrhea, defined as increased fecal loss of water and electrolytes, is characterized by a stool weight greater than 200 g./day. Since water constitutes 80 per cent of normal stools, increased stool weight represents chiefly increased fecal water. To the patient, diarrhea means either bulky or watery stools and, usually, increased fecal frequency. Diarrhea may occur alone or in association with increased fecal loss of other nutrients such as fat (steatorrhea) or nitrogen (azotorrhea).

Systemic effects

Malabsorption, as defined above, may have any of the following consequences: water and electrolyte depletion, metabolic acidosis, caloric deficiency causing inadequate weight gain in children and weight loss in adults and signs of deficiencies of specific vitamins and minerals. In addition, unabsorbed substances may interact with intestinal bacteria and products formed may influence intestinal function.

Mechanisms

Since stool water osmolality is quite constant, ranging from isotonic to moderately hypertonic (280–430 mmol./kg.), diarrhea ultimately represents an increased fecal solute load. The causes can be divided into three general categories:

INCREASED EXOGENOUS LOAD. Osmotic diarrhea results from ingestion of poorly or nonabsorbable solute, such as magnesium salts or hydroxides (antacids), lactulose (an indigestible disaccharide), sorbitol (a poorly absorbed sugar substitute), and sulfates. Another cause is abnormally rapid entry of hypertonic nutrient solutions into the intestine, as in patients given excessive amounts of tube feeding and with rapid gastric emptying of liquid after resection of a part of the stomach or after truncal vagotomy (dumping syndrome).

DECREASED ABSORPTION. Decreased absorptive capacity may result from: (1) intraluminal digestive disturbances due to reduced enzymatic activity, which may be caused by decreased exocrine secretions, lack of conversion of proenzyme to active enzyme, enzyme inhibition or denaturation, and reduced concentrations of bile acids; (2) surface digestive disturbances that may be produced by decreased enzyme activity per cell, decreased number of cells per unit area, or decreased intestinal length; and (3) mucosal disturbances that affect uptake, transport through the cell, exit from the cell, and transport from the mucosa in lymph or blood. Examples of the latter defects are a reduction in permeability, as occurs in nontropical sprue; re-

duced transport of specific substances, as in glucose-galactose malabsorption or bile acid and vitamin B_{12} malabsorption in ileal disease; loss of absorptive area because of intestinal resection or loss of intestinal villi; disturbances in biosynthetic processes, such as those required for chylomicron formation; and disturbed exit from the cell or intracellular spaces because of obstruction to lymph or blood flow.

INCREASED ENDOGENOUS LOAD. Stimulation of active secretion into the intestinal tract is a well-understood and common cause of diarrhea. This occurs in the following settings: (1) hypersecretion of digestive juices which exceeds the normal capacity for intestinal reabsorption, as in the gastrin-mediated hypersection of gastric juice found in the Zollinger-Ellison syndrome, which then induces pancreatic hypersecretion secondary to stimulation of secretin release by duodenal acidification; (2) active intestinal anion secretion, probably coupled with decreased uptake of NaCl across the brushborder membrane, which is stimulated by several bacterial enterotoxins (*Vibrio cholerae*, enterotoxigenic *E. coli*), by prostaglandins which may be synthesized locally as a reaction to injury, by some polypeptide hormones (vasoactive intestinal polypeptide, thyrocalcitonin), by a biogenical amine, serotonin, and by several cathartic drugs. Secretion caused by most enterotoxins, prostaglandins, and vasoactive intestinal peptide is mediated by activation of the enzyme, adenylate cyclase, with consequent increase in the concentration of cyclic AMP within the intestinal epithelium; (c) stimulation of colonic secretion by fatty acids and dihydroxy bile acids. Increased exudation into the intestinal tract without stimulation of specific ion transport processes occurs with mucosal damage (invasive enteric bacterial and viral infections; chronic ulcerative colitis); increased mesenteric venous pressure (strangulated loops of intestine); lymphatic obstruction; and idiopathic exudative enteropathy.

Tests of intestinal absorption

Tests of absorption can be classified as follows: balance tests, in which the fecal output of an unabsorbed substance or its metabolites is expressed as a fraction of input, and tolerance tests, in which absorption of a test substance is inferred from the appearance of the substance or its metabolic products in the blood or urine. The most reliable test of nutrient absorption is the fat balance test, which is valuable in establishing the presence of malabsorption, even if it gives no clue as to its mechanism. With the development of new techniques, as well as advances in the understanding of absorptive mechanisms, it now is possible to classify other tests of absorption according to the scheme proposed for malabsorption: (1) tests of intraluminal digestion; (2) tests of surface digestion; (3) tests of mucosal uptake; and (4) tests of bacterial overgrowth. This section does not cover the diagnostic approach to a patient with malabsorption, which, in addition to the tests discussed, generally includes appropriate endoscopy, x-ray examinations, and small intestinal biopsy.

Tests for presence of malabsorption: classical balance test

Fecal excretion of fat (and occasionally nitrogen) is measured while the patient adheres to a constant fat and nitrogen intake. Although unabsorbed food constitutes the main source of fecal fat and nitrogen, endogenous secretions, as well as bacterial production and destruction, may contribute. In addition to these factors, practical problems attend the collection of a representative fecal sample: (1) it may be difficult to obtain a complete collection during the test interval and (2) defecation may be irregular, stool excretion in a particular 24-hour period not representing the mean daily output.

Measurement of fecal water, indicated by fecal weight, is essential for defining the severity of diarrhea. The significance of increased fecal water differs from that of increased fecal fat and nitrogen, in that it is caused by increased endogenous secretion or by the osmotic action of unabsorbed intraluminal solute.

To reduce the error of fecal sampling, 48- or 72-hour stool collections are usually performed. However, with frequent bowel movements, a 24-hour sample is often adequate. In the future, measurement of fat absorption will probably feature the administration of radioactive fat together with a nonabsorbable reference marker. By comparing the ratio of marker to radioactivity in the test meal with that in the stool, the fraction of absorbed radioactive fat can be calculated. In addition, such a procedure indicates the true absorption of the administered fat, since endogenous secretions are not measured. In principle, the use of labeled test substances administered together with a marker should permit the development of a variety of useful tests of absorption.

Tests of digestion

Tests of pancreatic function

Since intraluminal digestion requires pancreatic enzymes and bile acids, tests of intraluminal digestion are concerned first with pancreatic exocrine function. To test this, one determines the volume and bicarbonate and enzyme concentrations in duodenal aspirates after intravenous administration of secretin with or without added CCK. Alternatively, one can measure enzyme concentration and extent of lipolysis in the jejunal content after a test meal (Lundh test). Pancreatic disease sufficiently severe to cause malabsorption of fat is almost invariably associated with reduced bicarbonate concentration and volume output after intravenous administration of secretin. To detect isolated lipase and co-lipase deficiency (extremely rare congenital conditions), measurement of individual enzymes in duodenal aspirates after intravenous administration of CCK is necessary. Total pancreatic enzyme output in response to endogenous stimulation may be determined by duodenal perfusion with amino acids and a nonabsorbable marker. Measurement of marker and enzyme concentration in samples aspirated from the distal duodenum permits calculation of the rate of enzyme secretion. Comparison of enzyme output after endogenous stimuli (acting through CCK-release) with that evoked by exogenous hormone permits the diagnosis of impaired CCK release, as has been observed in nontropical sprue. Fecal chymotrypsin measurements are a simple but rather unreliable test of pancreatic exocrine insufficiency.

Tests of surface digestion
The most commonly employed test of surface digestion is the lactose tolerance test. Absorption rate may be monitored qualitatively by the changes in blood glucose levels or by the appearance of diarrhea. Measurement of breath hydrogen by gas chromatography is an excellent test of lactase deficiency. Breath hydrogen also rises with intestinal bacterial overgrowth, but the rise occurs earlier after the test meal than in lactase deficiency, and it is observed with ingestion of any fermentable carbohydrate. In principle, similar tests for oligopeptide digestion could be developed. Disaccharidases can be measured quantitatively in intestinal biopsy samples obtained by peroral intubation.

Tests of mucosal uptake
Tests of mucosal uptake employ substances that require neither enzymatic di-

gestion nor micellar dispersion. Oral tolerance tests used to assess absorption are of limited value, because a flat blood tolerance curve does not distinguish delayed gastric emptying from decreased absorption or increased metabolic disposition of the test substance or its metabolic product. Tests based on urinary recovery are further subject to the vagaries of renal excretion. Investigators, recognizing the value of tests of mucosal absorptive function, have attempted to circumvent these problems. One approach is to saturate body stores before the test with a loading dose of the test substance in order to block tissue uptake. For vitamin B_{12}, for example, nonradioactive vitamin B_{12} is given parenterally before oral administration of the labeled vitamin. Another approach is to choose substances for which a constant fraction of the amount absorbed is excreted into the urine (e.g., xylose). For any tolerance test based on renal excretion, urine is collected over a period long enough that the amount recovered allows a reasonable estimate of the fraction of the test substance absorbed and is influenced little by the absorption rate.

Tests of bacterial overgrowth
Samples of the contents of the small intestine are obtained by intubation and cultured aerobically and anaerobically. Intestinal aspirates can be examined by thin-layer chromatography for the presence of free bile acids. Cholylglycine-1-^{14}C may be administered with a test meal. Bacterial, but not tissue, enzymes form $^{14}CO_2$, which may be monitored in exhaled breath. Increased $^{14}CO_2$ indicates increased bile acid deconjugation, reflecting either bacterial overgrowth or bile acid malabsorption. Alternately, ^{14}C-D-xylose may be given and the appearance of $^{14}CO_2$ measured in exhaled air. This test is based on rapid fermentation of this slowly absorbed

pentose by bacteria proliferating abnormally in the small intestine.

Most useful diagnostic tests

The tests summarized here are reliable and are based on substantial clinical experience. Several tests mentioned previously have not yet been extensively evaluated. Other tests (e.g., the vitamin A tolerance test) are popular for screening but do not distinguish digestive from absorptive disturbances.

Stool weight

Normally, less than 200 g. of feces are excreted per day. Measurement of stool weight is essential for establishing the presence and severity of diarrhea. Paint cans and plastic containers with snap-on lids have proven to be convenient for stool collections within and outside the hospital.

Stool fat

Normal stool fat is less than 7 g./day in subjects on a 100-g. fat diet, as determined by methods based on saponification and titration of the liberated fatty acids (van de Kamer method). The fraction of fecal fat in the form of triglyceride increases in patients with steatorrhea due to defective lipolysis (pancreatic insufficiency). It may be assessed by thin-layer chromatography, provided that stools are frozen immediately after collection and lipid extraction is carried out promptly. Tests based on microscopic appearance of oil droplets stained with fat-soluble dyes are insensitive and subject to sampling errors.

Oral D-xylose test

When 5-hour urinary excretion of D-xylose is measured after administration of a 25-g. oral dose, normal excretion is 5.0 g.; questionably subnormal excretion is 3.0 to 5.0 g.; with diffuse abnormality or resection of jejunum it is 3.0 g. Falsely low values are obtained with bacterial overgrowth in the proximal small intestine (bacterial utilization of D-xylose); ascites (distribution of absorbed D-xylose into an expanded extracellular space); renal disease (D-xylose clearance from the blood equals 60 to 80 per cent of the glomerular filtration rate); and urinary retention (excreted D-xylose retained with residual urine in the urinary bladder). This test, however, yields false-negative results in 10 to 15 per cent of patients with diffuse mucosal disease of the proximal small intestine (e.g., idiopathic sprue). A normal test result, therefore, should not be taken to indicate the definite absence of such a disease unless supported by a peroral small bowel biopsy.

Lactase assay

This assay is done by tolerance test (50 g. of lactose by mouth with a maximum rise in blood sugar of 20 mg.) and by direct mucosal enzyme assay (normal = 1 unit (1 mol. lactase hydrolyzed per minute per gram of mucosa, at 37° C). The tolerance test is unreliable in the presence of diabetes mellitus and may show falsely low results when gastric emptying is delayed.

Annotated references

Books

Binder, H. J. (ed.): Mechanisms of Intestinal Secretion. Kroc Foundation Series, Vol. 12. New York, Alan R. Liss, 1979. (An up-to-date review of intestinal ion transport and the mechanisms of secretory diarrhea.)

Bloom, S. R. (ed.): Gut Hormones. New York, Churchill Livingstone, 1978. (A 1977 authoritative review of the rapidly expanding knowledge on enteric polypeptide hormones.)

Code, C. F. (ed.): Handbook of Physiology, Section 6, Alimentary Canal, Vol. III (Intestinal Absorption). Washington, American Physiological Society, 1968. (Authoritative sources of information on intestinal absorption up to the time of their publication.)

Rosenberg, I. H., Selhub, J., and Dhar, G J.: Absorption and malabsorption of folates. *In* Boetz, M. I., and Reynolds, E. H. (eds.). Folic Acid in Neurology, Psychiatry, and Internal Medicine, p. 95. Raven Press, New York, 1979. (A comprehensive discussion of the physiologic and clinical aspects of folate absorption.)

Verzar, F., and McDougall, E. J.: Absorption from the Intestine. New York, Hafner, 1967. (Classical monograph on intestinal absorption, first published in 1936.)

Review articles

Diamond, J. M.: Osmotic water flow in leaky epithelia. J. Membr. Biol. *51*:195, 1979. (An analysis of the factors determining osmotic permeability and solvent drag of biologic membranes.)

Dockray, G. J.: Evolutionary relationships of the gut hormones. Fed. Proc. *38*:2295, 1979. (Fascinating information on the phylogenetic development of hormones linking brain and gut.)

Field, M.: Intestinal secretion. Gastroenterology *66*:1063, 1974. (An informative review of rapidly developing field, dealing mainly with mechanisms of ion secretion and their control by intracellular 3',5'-cyclic AMP concentration.)

Fordtran, J. S.: Speculations on the pathogenesis of diarrhea. Fed. Proc. *26*:1405, 1967. (A superb analysis of the pathophysiology of various forms of diarrhea.)

Gorbach, S. L.: The intestinal microflora. Gastroenterology *60*:1110, 1971 (Review of the literature on normal and abnormal flora written for the clinician.)

Gray, G. M.: Carbohydrate digestion and absorption: Role of the small intestine. N. Engl. J. Med. *292*:1225, 1975. (A first-rate review, mainly of carbohydrate digestion.)

Gray, G. M., and Cooper, H. L.: Protein digestion and absorption. Gastroenterology *61*:535, 1971. (A review of exemplary clarity.)

King, C. E., and Toskes, P. P.: Small intestine bacterial overgrowth. Gastroenterology *76*:1035, 1979. (A review of the clinically observable consequences of bacterial overgrowth in the small intestine.)

Larsson, L. I.: Pathology of gastrointestinal endocrine cells. Scand. J. Gastroenterol. 14, Suppl. *14*:1, 1979. (A classification of hormone-producing cells and their distribution along the gastrointestinal tract.)

Mao, C. C., and Jacobsen, E. D.: Intestinal absorption and blood flow. Am. J. Clin. Nutr. *23*:820, 1970. (A lucid, brief review of the interrelationship between blood flow and absorption.)

Newman, A.: Progress Report: Breath-analysis tests in gastroenterology. Gut *15*:308, 1974. (Review of diagnostic breath tests in gastroenterology.)

Phillips, S. F.: Diarrhea: A current view of the pathophysiology. Gastroenterology *63*:495, 1972. (Lucid summary of current views of the pathophysiology of diarrhea.)

Schanker, L. S.: On the mechanism of absorption of drugs from the gastrointestinal tract. J. Med. Pharm. Chem. *2*:343, 1960. (Relationship between state of ionization and absorption rates.)

Scheline, R. R.: Drug metabolism by intestinal microorganisms. J. Pharm. Sci. *57*:2021, 1968. (A review of bacterial transformations of drugs in vitro and in vivo.)

Schultz, S. G.: Sodium coupled transport by small intestine: A status report. Am. J. Physiol. *233*:E249, 1977. (A thorough analysis of the coupling between glucose and sodium absorption based on in vitro studies.)

Schultz, S. G., and Frizzell, R. A.: An overview of intestinal absorptive and secretory processes. Gastroenterology *63*:161, 1972. (A summary of many experimental studies about the ins and outs of ions and water across and between the absorbing cells.)

Smyth, D. H. (ed.): Intestinal absorption. Br. Med. Bull. *23*:205, 1967. (A collection of lucidly written review papers representing the British schools of thought.)

Thier, S. O., and Alpers, D. H.: Disorders of intestinal transport of amino acids. Am. J. Dis. Child. *117*:13, 1969. (Clinically oriented description of disorders of intestinal and renal tubular amino acid absorption.)

Experimental papers

Adibi, S. A., and Soleimanpour, M. R.: Functional characterization of dipeptide transport system in human jejunum. J. Clin. Invest. *53*:1368, 1974. (Neutral amino acid dipeptides are absorbed intact by specific carrier mechanisms. In intestinal mucosal disease, absorption of dipeptides is less impaired than that of free amino acids.)

Allen, R. H., Seetharam, B., Podell, E., and Alpers, D. H.: Effect of proteolytic enzymes on the binding of cobalamin to R protein and intrinsic factor: In vitro evidence that a failure to partially degrade protein is responsible for cobalamin malabsorption in pancreatic insufficiency. J. Clin. Invest. *61*:47, 1978. (A highly original description of the competition for vitamin B_{12} by two binding proteins in gastric juice.)

Ammon, H. V., and Phillips, S. F.: Inhibition of colonic water and electrolyte absorption by fatty acids in man. Gastroenterology *65*:744, 1973. (Colonic perfusion studies in man showing that long chain fatty acids induce secretion.)

Bieberdorf, F. A., Morawski, S., and Fordtran, J. S.: Effect of sodium, mannitol, and magnesium on glucose, galactose, 3-0-methylglucose, and fructose absorption in the human ileum. Gastroenterology *68*:58, 1975.

Borgstrom, B., Dahlqvist, A., Lundh, G., and Sjovall, J.: Studies of intestinal digestion and absorption in the human. J. Clin. Invest. *36*:1521, 1957. (The first demonstration that digestion and absorption of a meal is almost completed in the jejunum.)

Chung, Y. C., Kim, Y. S., Shadchehr, A., Garrido, A., MacGregor, I. L., and Sleisenger, M. H.: Protein digestion and absorption in human small intestine. Gastroenterology *76*:1415, 1979. (Observations on the fate of a protein meal in the human gut.)

Davis, G. R., Santa Ana, C. A., Morawski, S., and Fordtran, J. S.: Active chloride secretion in the normal human jejunum. J. Clin. Invest. *66:*1326, 1980. (The first critical study of the mechanism of fasting intestinal secretion in health.)

Dillard, R. L., Eastman, H., and Fordtran, J. S.: Volume-flow relationships during the transport of fluid through the human small intestine. Gastroenterology *49:*58, 1965. (Application of the dye-dilution method to the perfused small intestine which allows calculations of flow velocity and luminal radius.)

Fordtran, J. S.: Stimulation of active and passive sodium absorption by sugar in the human jejunum. J. Clin. Invest. *55:*728, 1975. (These two articles describe the regional differences of the glucose-sodium interaction in the human intestine.)

Frizzell, R. A., and Schultz, S. G.: Ionic conductances of extracellular shunt pathway in rabbit ileum. J. Gen. Physiol. *59:*318, 1972. (A conclusive experimental analysis of intestinal ion transport, contrasting movement through the epithelial cell with ionic diffusion across the "tight" junction and the lateral intercellular space.)

Hallback, D. A., Jodal, M., Sjoqvist, A., and Lundgren, O.: Villous tissue osmolality and intestinal transport of water and electrolytes. Acta Physiol. Scand. *107:*115, 1979 (Observations supporting a postulated countercurrent multiplier effect of intestinal villus blood flow. Not discussed in text.)

Levitt, M. D., and Ingelfinger, F. J.: Hydrogen and methane production in man. Ann. N.Y. Acad. Sci. *150:*75, 1968. (Methane and hydrogen gas are produced almost entirely in the colon.)

Ruppin, H., Bar-Meir, S., Soergel, K. H., Wood, C. M., and Schmitt, M. G., Jr.: Absorption of short-chain fatty acids by the colon. Gastroenterology *78:*1500, 1980. (Demonstration of the absorption of carbohydrate fermentation products by the human colon: a mechanism for "caloric salvage" from malabsorbed carbohydrates.)

Soergel, K. H.: Flow measurements of fasting contents in the human small intestine. *In* Demling, L., and Ottenjann, R. (eds.): Motility of the Gastrointestinal Tract, p. 81. Stuttgart, Georg Thieme Verlag, 1971.

(An analysis of velocity and rate of flow in the human small intestine during fasting and after a meal.)

Turnberg, L. A., Bieberdorf, F. A., Morawski, S. G., and Fordtran, J. S.: Interrelationships of chloride, bicarbonate, sodium, and hydrogen transport in the human ileum. J. Clin. Invest. *49:*557, 1970.

Turnberg, L. A., Fordtran, J. S., Carter, N. W., and Rector, F. C., Jr.: Mechanism of bicarbonate absorption and its relationship to sodium transport in the human jejunum. J. Clin. Invest. *49:*548, 1970. (Support of the hypothesis that Na^+ and Cl^- absorption in the small intestine occurs by exchange with actively secreted H^+ and HCO_{3-} is presented in these two papers.)

Wilson, F., and Dietschy, J. M.: The intestinal unstirred layer: Its surface area and effect on active transport kinetics. Biochim. Biophys. Acta *363:*112, 1974. (An exposition of the dimensions and the role of the unstirred layer in intestinal absorption.)

References for tests

Bond, J. H., Jr., and Levitt, M. L.: Use of pulmonary hydrogen (H_2) measurements to quantitate carbohydrate absorption. J. Clin. Invest. *51:*1219, 1972.

Christiansen, P. A., Kirsner, J. B., and Ablaza, J.: D-xylose and its use in the diagnosis of malabsorptive states. Am. J. Med. *73:*521, 1969.

Hofmann, A. F., and Thomas, P. J.: Bile acid breath test: Extremely simple, moderately useful. Ann. Intern. Med. *79:*743, 1973.

Jover, A., and Gordon, R. S., Jr.: Procedure for quantitative analysis of feces with special reference to fecal fatty acids. J. Lab. Clin. Med. *59:*878, 1962.

Newcomer, A. D., McGill, D. B., Thomas, P. J., and Hofmann, A. F.: Prospective comparison of indirect methods for detecting lactase deficiency. N. Engl. J. Med. *293:*1232, 1975.

Thompson, J. B., Su, C. K., and Welsh, J. D.: Fecal triglycerides. II. Digestive vs absorptive steatorrhea. J. Lab. Clin. Med. *73:*521, 1969.

Wollaeger, E. E., Comfort, M. W., and Osterberg, A. E.: Total solids, fat and nitrogen in the feces. Gastroenterology *9:*272, 1947.

24   Absorption of Lipid Solutes

Neville E. Hoffman, M.D., Ph.D.

To understand lipid absorption we must understand the biochemical and physico-chemical mechanisms which allow a group of molecules which are relatively insoluble in water to disperse in, and therefore to be transported through, an aqueous environment. The classes of molecules to be considered will be the major dietary lipids, triglyceride (triacylglycerol), cholesterol, lipid soluble vitamins (A, D, E, and K), and the group of molecules which solubilize all of these, the bile acids.

There are two major processes involved in the absorption of lipids. One is the digestion, absorption, and lymphatic transport of the water-insoluble lipids, of which triglyceride is the prime example. The other is the absorption and metabolism of the detergent or amphipathic group of molecules, the bile acids. Triglycerides will be discussed in detail because they constitute the major absorptive load, and the mechanisms involved are analogous to those for the absorption of cholesterol, phospholipids, and the vitamins A, D, E, and K. Further, impaired triglyceride absorption (steatorrhea) is the hallmark of clinical malabsorption syndromes. The steps in triglyceride absorption form the basis for constructing a differential diagnosis of malabsorption and some of the information about lipid absorption is critical to understanding clinical diagnostic tests and therapeutic modalities.

Triglyceride absorption

Triglyceride absorption can be viewed as a series of sequential steps as listed below. This provides a framework for the discussion of normal absorption as well as the differential diagnosis of malabsorption.

Steps in Fat Absorption

1. Gastric emulsification
2. Luminal digestion
3. Luminal transport
4. Membrane transport
5. Complex lipid formation
6. Chylomicron formation—extrusion
7. Lymphatic transport

Gastric emulsification

Emulsions of oil in water are common to our diet; vinegar and oil salad dressing is a prime example. If energy (shaking) is applied to the mixture, the oil can be dispersed as fine droplets within the water to produce a turbid emulsion; if allowed to stand the emulsion will separate into two layers of oil and water. In the same fashion the stomach applies energy by antral churning, producing an emulsion of oil in water. By producing fine droplets, the area of interface between oil and water has been greatly increased; this will be important in the next step of lipolysis.

Gastric emulsification would presumably be deranged following any gastric surgery which modified antral or pyloric function. This may contribute to postvagotomy and postgastrectomy diarrheas.

Luminal digestion

Triglyceride is hydrolyzed by enzymes from the pancreas and from salivary glands located on the posterior tongue. The former is of overriding importance in the adult but the latter deserves consideration in infants.

LINGUAL LIPASE. *Lingual lipase* is secreted by serous glands which surround the circumvallate papillae on the posterior aspect of the tongue. The secretion of enzyme is stimulated by suckling and declines in the postweaning period. This pregastric lipase or esterase was first described in suckling ruminants and it has properties well adapted to the needs of the immature animal. Its pH optimum is 4.0 to 5.0, which is appropriate for the neonatal stomach. It is active against triglycerides of short-, medium-, and long-chain fatty acids, although it is more active against the shorter chains which are more common in some milks. Lastly, it does not require bile acids or the pancreatic protein, co-lipase. The presence of lingual lipase has been shown to be of major importance in neonatal rats, in that the percentage absorption of a fat load falls from 89 to 58 per cent with diversion of oral secretions. Lingual lipase acts selectively on the fatty acids in the 1- and 3-positions of the triglyceride. (Pancreatic lipase has the same selectivity.) The products of lingual lipolysis are diglycerides, 2-monoglyceride (MG), and free fatty acid (FFA). One interesting facet of lingual lipase activity is its apparent ability to promote pancreatic lipase activity. The reason for this is unclear, but pancreatic lipase acts at the interface of the emulsified triglyceride droplet and water, as discussed below. It may be that the emulsified droplet of milk is stabilized by proteins. Lingual lipolysis releases FFA and MG, which may displace the protein from the surface of the emulsion, and FFA and MG, which are adequate emulsifiers, produce an oil-water interface which is better adapted for the pancreatic lipase–co-lipase–bile acid complex.

PANCREATIC LIPASE. *Pancreatic lipase* is the major lipolytic enzyme for triglycerides in adults and is also of crucial importance in infants. The enzyme is secreted by the pancreas mainly in response to the gastrointestinal hormone cholecystokinin as well as in response to vagal stimulation and to the hormones secretin and gastrin. The hydrolyzed triglyceride produced by lingual lipase would be one stimulus for cholecystokinin (CCK) release. Optimal pancreatic lipase activity occurs around pH 6.5, and the enzyme is irreversibly inactivated below pH 4, so that secretion of pancreatic bicarbonate in response to secretin is essential to maintain an appropriate luminal milieu for lipolysis. Because pancreatic lipolysis occurs at the interface of the triglyceride emulsion and water, the surface area of the emulsion determines the available substrate for lipolysis. For a fixed amount of lipid, the finer the emulsion, the greater the surface area oil-water interface, and the greater the apparent substrate concentration and rate of lipolysis. The correct orientation of pancreatic lipase at this interface requires two other molecules–bile acids and a small protein, co-lipase, which is secreted by the pancreas. It is the presence of these latter molecules which lower the pH optimum of pancreatic lipase from 8.0 to 9.0 to the pH of the duodenum, 6.5. Bile acids have an additional role in lipolysis, namely the transport of the digestive products MG and FFA away from the lipolytic surface.

The stage will be set for some derangements of fat digestion if the pH in the duodenum is altered by hypersecretion of acid from the stomach or hyposecretion of bicarbonate from the pancreas. Other causes of maldigestion of lipids include deficient secretion of enzyme or co-enzyme and deficient concentration of bile acids.

Luminal transport

The products of pancreatic lipolysis, MG and FFAs, are only sparingly soluble in water. These substances must diffuse through an aqueous environment to the microvillous membrane of the enterocyte, the site of absorption. If the concentration of these substances in water could be increased, the diffusion flux could be increased according to the Fick equation:

$$J = DA\Delta C/1$$

where J is the flux in mass units/sec., D is the diffusion coefficient, A is the absorptive area, ΔC is the concentration gradient, and 1 is the distance which must be traversed by diffusion. The process which increases the ΔC term is solubilization in bile acid micelles.

The bile acids are a group of surface-active sterols which aggregate above a limiting concentration, called the critical micellar concentration, to form thermodynamically stable particles known as micelles. The surface of the micelle is formed by the hydrophilic (water attractive) aspect of the bile acid molecule and the interior provides a hydrophobic or lipophilic environment in which are incorporated poorly soluble digestive products. In this way the aqueous concentration of fatty acids rises about a thousandfold. However, the micellar particle is much larger than the fatty acid molecule, and the diffusion coefficient is inversely related to particle size (actually to the cube root of particle weight). Thus, if we look at the Fick equation, the flux, J, will rise linearly with increase in concentration, but since D varies with the reciprocal of the cube root of particle weight, micellar solubilization may increase J. Solubilization raises ΔC by a factor of 10^3, and reduces D by a factor of only 6 to 10, so that J has increased markedly, perhaps a hundredfold.

Another term in the Fick equation requires some consideration, and that term is 1, the thickness of the diffusional layer. Intestinal motility keeps the luminal contents moderately well mixed allowing the solubilized substances to travel some distance toward the microvillous membrane by convection. However, the final distance to the absorptive surface is through a relatively unstirred layer of water. Such an unstirred layer, which has been experimentally demonstrated for lipids and other nutrients, should be considered one of the resistances to be overcome by a nutrient in its passage from the intestinal lumen to the absorptive cell. In the intestine the unstirred layer is probably on the order of 200 μM. in thickness.

Membrane transport

An examination of the dependence of the rate of fatty acid absorption on concentration, temperature, and chain length indicates that fatty acids penetrate the microvillous membrane by partition. This means that fatty acids and monoglycerides are more soluble in the lipid portion of the membrane than they are in the micellar solution. As discussed below, bile acids are minimally absorbed in the duodenum and jejunum, the major sites of fatty acid and monoglyceride absorption. Therefore, it appears that the micelle is not absorbed as an intact particle; only the solubilized load is absorbed. This inference is based on the concept that micelles are not actually diffusing as intact particles. Micelles are constantly forming and disintegrating. Although we can deal mathematically or statistically with micellar diffusion as if it were particulate, a fatty acid molecule constantly changes from one micelle to another. It is more appropriate to consider a fatty acid molecule to be diffusing through a micellar environment. Such a concept certainly simplifies an understanding of the different rates of absorption of micellar components; for example, it has been

Fatty acid + CoA $\xrightarrow[\text{ATP, Mg}^{2+}]{\overset{\text{6.2.1.3.}}{\text{Acyl CoA synthetase}}}$ Acyl CoA

2 acyl CoA + α glycerophosphate $\longrightarrow$ Phosphatidic acid $\longrightarrow$ Phospholipid synthesis

Phosphatidic acid $\xrightarrow{\overset{\text{Lα phosphatidate}}{\text{phosphohydrolase}}}$ Diglyceride + phosphate

Diglyceride + acyl CoA $\xrightarrow{\overset{\text{Diglyceride}}{\text{transacylase}}}$ Triglyceride

FIG. 24–1. The α-glycerophosphate pathway.

shown that two different lipids, a fatty acid and a monoglyceride analogue, can be absorbed at different rates from the same bile acid micellar solution.

Complex lipid formation

Having entered the microvillous membrane by partition from a micellar solution, fatty acids and monoglycerides must be extracted from the membrane on the cytoplasmic side. The mechanism for this process is not clear.

Triglyceride droplets appear deep within the enterocyte within the smooth endoplasmic reticulum and Golgi apparatus, and FFAs can be found in the cytoplasm, but not in the terminal web, the intracellular structure immediately adjacent to the microvillous membrane. An enzyme, acyl CoA synthetase is found in the enterocyte, and this enzyme, which condenses a fatty acid with CoA to form acyl CoA, renders fatty acids water soluble. This mechanism

might explain how fatty acid becomes sufficiently water soluble to leave the cell membrane on the cytoplasmic side. An alternate mechanism to render fatty acid soluble within the cytoplasm would involve binding to a protein component of the cytoplasm, e.g., the fatty acid binding protein, formerly called Z protein. However, there is no obvious acceptor for MG or cholesterol on the cytoplasmic side of the microvillous membrane, i.e., no apparent mechanism for these molecules to leave the membrane to enter intracellular water.

Triglyceride may be formed within the enterocyte by either of two pathways—the α-glycerophosphate or the MG pathway. Both pathways are shown in Figures 24–1 and 24–2, and the latter probably accounts for 60–80% of triglyceride synthesis during normal fat absorption when there is an adequate supply of absorbed monoglyceride.

Chylomicron formation and extrusion

The previous discussion considered how insoluble or poorly soluble lipids are maintained in an aqueous environment, first as the inherently unstable emulsion in the stomach, then as the thermodynamically stable mixed micelle in the intestine, and finally as acyl CoA within the enterocyte. There is another stable particle which allows transport of triglyceride and cholesterol through lymph and plasma, namely

Fatty acid + CoA $\xrightarrow{\overset{\text{6.2.1.3.}}{\text{Acyl CoA synthetase}}}$ Acyl CoA

Monoglyceride + acyl CoA $\xrightarrow{\overset{\text{monoglyceride}}{\text{transacylase}}}$ Diglyceride

Diglyceride + acyl CoA $\xrightarrow{\overset{\text{Diglyceride}}{\text{transacylase}}}$ Triglyceride

FIG. 24–2. The monoglyceride pathway.

the chylomicron. This particle varies in size from 0.1 to 0.5 μ in diameter, although at the lower end of the particle size, the chylomicron becomes indistinguishable from very low density lipoprotein. About 98 per cent of the chylomicron is lipid (90 per cent triglyceride), the remainder being phospholipid and cholesterol. The particle has triglyceride in an oily core with apolipoprotein, phospholipid, and cholesterol forming a shell which is analogous to a plasma membrane. The chylomicron is extremely stable and highly resistant to coalescence.

Chylomicra are assembled within the endoplasmic reticulum from the newly synthesized triglyceride and phospholipid and from absorbed or synthesized cholesterol along with locally synthesized apolipoprotein. The major apolipoproteins of the chylomicrons are B, CI, CII, and CIII. When protein synthesis is inhibited (or congenitally absent), triglyceride accumulates within the enterocyte. Although chylomicron formation ceases, fat absorption does continue with moderate interruption. The chylomicron is extruded probably by exocytosis through the basolateral intercellular membrane into the intestinal lymphatic channels known as lacteals.

Lymphatic transport

Over 90 per cent of the absorbed long-chain fatty acids are transported from the gut in intestinal lymphatics to the thoracic duct, which empties into the left subclavian vein. The chylomicra are transported in lymph so that, as one might predict, absorbed cholesterol and intestinally synthesized cholesterol and cholesterol esters, as well as the phospholipids (other chylomicron components) follow the same route.

The chain length of fatty acids determines whether they will be absorbed by the lymphatic or portal blood pathway. The cutoff is around C10 to C12; only a small proportion of C10 fatty acids are incorporated into triglyceride and thence

transported in lymph. More than 80 per cent of C14 fatty acids are transported as triglycerides through the thoracic duct lymph.

The enterohepatic circulation of the bile acids

The physiology of the bile acids involves rather complicated anatomical and biochemical cycles. The anatomical enterohepatic circulation starts with bile acid synthesis in the liver followed by secretion into bile, storage and concentration in the gallbladder, controlled secretion into the intestine, reabsorption from the intestine, transport in portal blood, uptake by the liver, and then secretion into bile to restart the cycle. The biochemical enterohepatic circulation starts with synthesis within the hepatocyte of a free bile acid from cholesterol, conjugation of that acid with an amino acid in the liver, deconjugation with or without dehydroxylation by bacterial enzymes in the intestine, and subsequent reconjugation in the liver. Since bile acid synthesis initiates both cycles, we will examine this process before proceeding to the anatomical cycle and then to the biochemical cycle.

Bile acid synthesis

The bile acids are 24-carbon saturated sterols which are synthesized in the liver from cholesterol. The chemical rearrangements change an almost insoluble molecule into a highly soluble substance with detergent properties. These properties allow bile acids to solubilize other lipids, including the parent molecule, cholesterol. Figure 24–3 shows the changes which cholesterol undergoes. The biochemical pathways by which this occurs remain somewhat controversial.

The three major changes shown in Figure 24–3 involve: (1) the insertion of two or three α-oriented hydroxyl groups in place

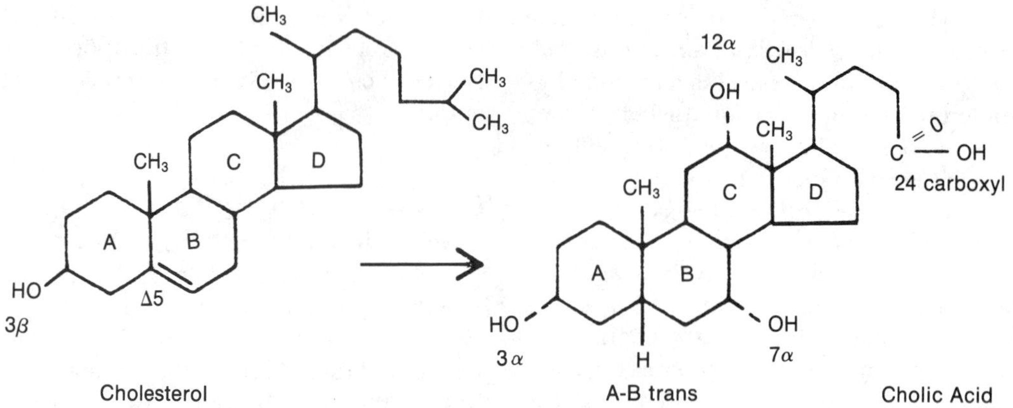

FIG. 24—3. Changes undergone by cholesterol in the synthesis of bile acid.

of one β-oriented hydroxyl group; (2) the transorientation of the A ring with respect to the B ring by stereospecific saturation of the ∂-5 double bond; and (3) the shortening of the side chain at C20 with carboxylation at C24. These changes make the molecule amphipathic by increasing its polarity (hydroxyl groups and carboxyl group) but most importantly by making a special arrangement of polar groups, in that all hydroxyls are on the same side of the molecule. The bile acid is a planar amphipath with a hydrophilic side and hydrophobic side rather than a hydrophilic head and hydrophobic tail as with detergents such as fatty acids. The transarrangement of the A ring on the B ring makes the bile acid less able to pack in monolayers and, therefore, thermodynamically more likely to form micelles.

The carboxyl group adds polarity to the molecule and also provides the site of amino acid conjugation. The pKa of a free bile acid is around pH 6.0, i.e., at this pH a bile acid such as cholic acid is 50 per cent ionized. Conjugation with glycine lowers the pKa to pH 3.0 to 4.0 and with taurine, the pKa falls to pH 2.0. Since the pH of the intestine is around 6.0 to 7.0, lowering the pKa greatly increases the proportion of the bile acid which is ionized in the intestine during digestion of food. The ionized bile acids have the necessary polarity, and hence water solubility, to form micelles.

The human liver synthesizes two primary bile acids, cholic acid (3 α-, 7 α-, 12 α-trihydroxy 5 β-cholanoic acid) and chenodeoxycholic acid (3 α-, 7 α-dihydroxy 5 β-cholanoic acid). The rate of synthesis is regulated by the return of bile acid to the liver from the intestine. If, for example, returning bile acid is diverted, bile acid synthesis rises after about 12 to 18 hours. The cellular mechanism of control is not known.

The anatomical enterohepatic circulation

Bile acid secretion
Conjugated bile acids are secreted by the liver into biliary canaliculi which drain to biliary ductules, intralobular ducts, and finally the main hepatic ducts and the common bile duct. Bile may then drain either into the intestine or be stored by passing through the cystic duct to the gallbladder.

The process of bile acid secretion by the liver is poorly understood. What is clear is that bile acid is found in portal blood at a concentration on the order of 10 to 50 μM. but is found in bile at a concentration on the order of 10 to 200 mM. Thus, there is a concentration gradient across the liver cell of about a thousandfold. Direct evidence supports the presence of a Na^+-dependent, carrier-mediated, active transport process for bile acid at the sinusoidal side of the hepatocyte. This may be the only gradient necessary to allow subsequent diffusion of bile acid across the hepatocyte into the bile. However, kinetic data suggest that there may be a secondary transport process at the canalicular side of the cell; biochemical evidence suggests that the cytoskeletal elements, myofilaments and microtubules, may play a role in transhepatic transport of bile acids. These latter elements are now under intensive study, and their importance should be clearer in the next few years.

This transhepatic flux of bile entrains the fluxes of water, phospholipid, cholesterol, steroid hormones, drug metabolites, some metallic cations, and the heme degradation products such as bilirubin. All of these may be affected by interruption of the enterohepatic circulation.

Gallbladder concentration storage and emptying

The gallbladder stores bile and concentrates it by a factor of 2 to 5; however, the gallbladder has a limited capacity, and some hepatic bile is secreted directly into the intestine. This is particularly true during the night, a period of prolonged fasting. In response to release of CCK into the circulation, the gallbladder contracts and the sphincter of Oddi relaxes, releasing bile into the intestine. It may well be that sphincter relaxation is more important than the gallbladder contraction, since the gallbladder does not contract to total emptiness. However, gallbladder discharge presents the intestine with appropriate concentrations of bile acid when a meal is being digested. Interestingly, cholecystectomy does not affect fat absorption, and some animals (rat, horse) have no gallbladder.

Intestinal absorption of bile acids

Bile acids may be passively absorbed from the intestine throughout its length or actively absorbed by a Na^+-dependent, carrier-mediated, active transport process in the terminal ileum. The major determinant of the rate of passive absorption seems to be the tendency of bile acid to enter a lipid membrane; the less polar, more lipophilic bile acids tend to be better absorbed passively. Decreasing rates of passive absorption are seen with dihydroxy acids, trihydroxy acids, glycine conjugates, and taurine conjugates, in that order. Potential passive absorption of free bile acids assumes importance with interruption of the enterohepatic circulation, as discussed below.

In the ileum there is a specific transport process for bile acids. This bile acid absorption system ensures that approximately 98 per cent of a bile acid load presented to the intestine is conserved. The relative proportions of active and passive absorption are unknown, but the active process probably ensures a high degree of conservation of bile acids, although the bulk of their absorption is passive. The bile acid carrier system in the ileum has been described in Michaelis-Menten terms used for other carrier systems and has been shown to have a Km on the order of 0.1 to 0.5 mM., Km being lower for conjugated bile acids. The V_{max} for cholic acid is greater than that of dihydroxy acids, but conjugation has no consistent effect on transport.

Hepatic uptake

Again, a carrier-mediated, Na^+-dependent transport process ensures efficient clearing of bile acids from the portal blood. Trihydroxy bile acids are cleared more avidly than dihydroxy acids, and conjugates more avidly than free bile acids. About 95 to 98 per cent of a dose of taurocholic acid is cleared on one pass through the liver. The reduced clearance of some other bile acids may relate to the characteristics of the transport process, or perhaps just to the affinity of bile acid for serum albumin. The order of increasing rates of hepatic clearance is the inverse of the order of increasing affinity of bile acid for albumin. The small component of portal blood bile acid which is not absorbed by the liver leads to the small but measurable concentration of bile acid in systemic blood. The systemic blood level reflects the rate of intestinal absorption of bile acid and the avidity of hepatic clearance. Thus, serum bile acid measurement may be an index of both intestinal and liver function.

Bile acid pool

The pool of bile acid is the total mass of bile acid present in the body at any instant. It is recirculated. In health the pool is located within the anatomical enterohepatic circulation except for a small amount in systemic blood; in disease, bile acids accumulate in systemic blood and may be deposited in skin and perhaps other tissues. Bile acid in skin is believed to cause the itch which is so prominent in cholestasis.

Since the intestine conserves bile acid so efficiently, and bile acid returning to the liver suppresses synthesis, most bile acid secreted by the liver is from the recycling pool. Each day the pool cycles 10 to 20 times to give a total flux on the order of 20 g. (pool size 2 to 3 g.). Of this, perhaps 0.5 g. is newly synthesized bile acid. With 10 cycles, each conserving about 98 per cent of the pool, the loss per day will be about 20 per cent; for example, for a pool of 2.5 g., 0.5 g./day would be lost. However, bile acid loss is matched by bile acid synthesis so that the pool is in a steady state in which pool size is constant and input (synthesis) equals output (intestinal loss).

The biochemical enterohepatic circulation

The biochemical recirculation of the bile acids involves the interaction of mammalian and bacterial enzymes. In the liver, mammalian enzymes synthesize the primary bile acids which enter the enterohepatic circulation. The mammalian enzymes also conjugate the bile acids. In the intestine, bacterial enzymes split the amide bond between the amino acid and the bile acid, thereby liberating the free bile acid. The free bile acid may be absorbed subsequently in this form or degraded by other bacterial enzymes to a secondary bile acid by removal of the hydroxyl group at position 7. This step produces either deoxycholic acid, the secondary acid, from the primary bile acid, cholic acid, or lithocholic acid from chenodeoxycholic acid. Free bile acids may be further modified, but the magnitude of such changes is very small.

When free bile acids return to the liver, approximately a third of the total returning bile acids, they are reconjugated with glycine and taurine. The liver also has the capacity to sulfate bile acids at the 3 position. This occurs mainly with lithocholic acid. Sulfated bile acids are poorly conserved by the intestine, so the pool of lithocholic acid remains very small. Since lithocholic acid has been found to be hepatotoxic in a number of animal species, sulfation may actually protect the liver from a potential toxin.

Absorption of cholesterol and lipid soluble vitamins

This group of compounds is nearly insoluble in water, and its members are even more dependent upon bile acids to attain sufficient solubility for absorption than are the fatty acids. Interestingly, however, the planar amphipaths, bile acids, alone are not effective solubilizers of cholesterol. The solubility of cholesterol and the vitamins is markedly enhanced when the hydrophobic core of the bile acid micelle is expanded by a polar amphiphile, such as the products of triglyceride digestion (fatty acids and monoglycerides) as well as phospholipids, such as lecithin, which expands the micellar core in bile.

Vitamins A, D, E, and K are absorbed intact and transported away from the intestine in the lymph within the oily core of the chylomicron. On the other hand, up to 80 per cent of cholesterol is esterified within the enterocyte, prior to incorporation into the chylomicron. About two-thirds of the daily cholesterol load to the intestine comes from the bile and about one-third from the diet. The intestine absorbs about one-third of the total cholesterol presented to the organ.

An intriguing but unexplained phenomenon is the dependence of cholesterol absorption upon the presence of cholic acid or its conjugates. Other bile acids and other detergents can solubilize cholesterol but only the cholic acids promote cholesterol esterification, normal chylomicron incorporation, and lymphatic transport.

Clinical features of malabsorption

The clinical signs and symptoms of malabsorption are listed in Table 24–1. Weight loss is a consequence of inadequate absorption of dietary intake in general, i.e., malabsorption of calories leading to metabolism of body energy stores and eventual structural protein. Malaise, the feeling of being ill, and lethargy also stem partly from the loss of calories and perhaps even more from the anemia consequent to malabsorption of specific nutrients. In intestinal mucosal disease folic acid is commonly deficient; if ileal function is lost, vitamin B_{12} will be poorly absorbed; malabsorbed FFAs precipitate in the intestine as the insoluble soaps of iron (and other divalent cations) and lead to malabsorption of this mineral. Failure to absorb folate, vitamin B_{12}, and iron, along with protein malabsorption, will lead to disordered red cell maturation and anemia.

Diarrhea is prominent in some malabsorptive states. This may be due to the osmotic effect of malabsorbed sugars and amino acids but is more commonly due to the secretory effect of fatty acids and bile acids on intestinal mucosa. In both the small and large intestine fatty acids and bile acids inhibit absorption of water and may even induce its secretion. Malabsorption of protein leads to reduced rates of protein synthesis, hypoalbuminemia, and edema. Easy bruising follows from the dis-

TABLE 24–1
Malabsorption

Clinical Feature	Nutrient Malabsorbed
Weight loss	Calories
Malaise/anemia	Fe^{2+}, folic acid, B_{12} protein
Diarrhea	Fat, bile acids, sugars
Edema	Protein
Bruising	Vitamin K
Fractures ⎱ Tetany ⎰	Ca^{2+}

organization of the blood coagulation cascade consequent to poor absorption of vitamin K. The malabsorbed FFAs may also precipitate calcium in the intestinal lumen, leading to loss of this mineral. Malabsorption of calcium would also follow from malabsorption of vitamin D if exposure to sunlight is inadequate for endogenous synthesis of the vitamin. The end result is hypocalcemia, which may induce tetany, or osteomalacia sufficient to provoke bone fractures.

Differential diagnosis of malabsorption

This outline of the process of normal fat absorption provides a framework for considering the clinical problem of intestinal malabsorption through the process of constructing a differential diagnosis which is the basis for a well-organized patient workup (see list on following page).

Step 1: Gastric emulsification may be deranged by surgical intervention such as partial gastrectomy and perhaps even vagotomy. This is speculative rather than well established, and significant malabsorption on this basis alone is an unusual consequence of gastric surgery.

Step 2: Luminal digestion, or more specifically pancreatic lipolysis, will be ineffective if either of two situations exists. There may be an insufficient enzyme content in the intestine (quantitative defect) or the conditions for pancreatic enzyme function may not be met (qualitative defect).

Physiological Differential Diagnosis of Malabsorption

Gastric emulsification
    Postantrectomy diarrhea
    Postpyloroplasty diarrhea
Luminal digestion
    Quantitative pancreatic enzyme deficiency
    Agenesis of pancreas
    Muco viscidosis
    Chronic pancreatitis
    Carcinomatous obstruction of pancreatic ducts
    Quantitative pancreatic enzyme deficiency
    Enterokinase deficiency
    Zollinger-Ellison syndrome
Luminal transport
    Quantitative bile acid deficiency
    Idiopathic bile acid malabsorption
    Ileal dysfunction
        Crohn's disease
        Resection
        Radiation enteritis
    Obstruction of bile ducts
        Carcinoma
        Gallstones
    Qualitative bile acid deficiency
    Bacterial overgrowth
        Blind loop syndrome
        Small intestinal diverticulum
    Diabetes, scleroderma
    Zollinger-Ellison syndrome
Membrane transport
    Quantitative deficiency of absorptive surface
    Postsurgical short bowel syndrome
    Qualitative/quantitative reduction of absorptive surface villous atrophy
    Celiac disease
    Tropical sprue
    Lymphoma
    Iron deficiency
    Giardiasis
    Dermatitis herpetiformis
    Ileojejunitis
Chylomicron formation
    Abetalipoproteinemia
Lymphatic transport
    Congenital lymphangiectasia
    Whipple's disease

Quantitative pancreatic enzyme deficiencies occur when pancreatic secretions do not enter the intestine; this could occur when the pancreas is absent or destroyed or when the pancreatic duct is occluded. Examples include agenesis of the pancreas, chronic pancreatitis, cystic fibrosis, or tumors of the pancreas or ampulla of Vater.

Qualitative pancreatic enzyme deficiency may occur in a variety of circumstances. The pancreas secretes proenzymes, i.e., inactive proteins. Enterokinase from the microvillous membrane activates trypsinogen to trypsin and trypsin activates the other proenzymes. Specific deficiency of enterokinase has been described as the cause of malabsorption in a few children. Malabsorption is a feature of congenital colipase deficiency. Pancreatic lipolysis is dependent on intraluminal pH, so that excess secretion of acid or deficient secretion of bicarbonate could both lead to deficient lipolysis in the face of adequate pancreatic lipase secretion. Excess acid is secreted by the stomach in the Zollinger-Ellison syndrome, which is caused by a gastrin-secreting tumor. Low production of bicarbonate is a feature of chronic pancreatitis.

Step 3: Luminal transport may be deficient with any interruption of the anatomical enterohepatic circulation of bile acids. This produces a quantitative bile acid deficiency accompanied by steatorrhea of modest degree (less than 20 g./24 hours). Such interruptions could occur as part of the ileal dysfunction of Crohn's disease, following surgical resection of the ileum, which may have been performed for Crohn's disease, or in response to intestinal ischemia or trauma. Theoretically, common bile duct obstruction or cholestatic liver disease produces malabsorption; however, patients with such disorders rarely present with a problem of malabsorption, since other features of their disease predominate.

Qualitative bile deficiency occurs when the intraluminal conditions do not allow the bile acids to form micelles in the usual way. In the Zollinger-Ellison syndrome when pH falls, the glycine-conjugated bile acids approach their pKa, a large proportion is protonated and so does not contribute to micelle formation. Further, such protonated bile acids tend to be passively absorbed. If there is bacterial contamination of the upper intestine, the conjugated bile acids are deconjugated, and at pH 6.5 the free bile acids are approximately 50 per cent protonated, passively absorbed, and, therefore, not available for micelle formation and luminal transport of lipids. Bacterial contamination may occur with the postgastrectomy, afferent loop or blind loop syndromes and even with large upper intestinal diverticula.

Step 4: Membrane transport may be deficient for several reasons. Thus, the area of intestinal mucosa available for absorption is reduced after intestinal resection. A more subtle reduction in area occurs in conditions which cause intestinal villous atrophy. The most common of many such conditions are celiac disease and tropical sprue. With villous atrophy the absorptive area is reduced both quantitatively and qualitatively. The total number of absorptive villous cells is reduced and the remaining cells have abnormal microvillous structure.

Step 5: Complex lipid formation is never deranged as a primary cause of malabsorption.

Step 6: Chylomicron formation is deficient in the condition of A β-lipoproteinemia. The β-lipoproteins which are necessary to form very low density lipoprotein and the surface coat of the chylomicron are not synthesized, and lipid droplets accumulate within the enterocyte. Malabsorption of only modest degree ensues because an alternate route of fat absorption is utilized, although the details of this route are unknown.

Step 7: Lymphatic transport is deranged with Whipple's disease (intestinal lipodystrophy) and congenital lymphangiectasia. In the former disease, there is intestinal villous atrophy and dilatation of intestinal lymphatics. In the latter the lymphatic channels are dilated and lymph tends to leak back into the intestinal lumen.

Diagnostic tests

Fecal fat

The clinical definition of malabsorption is established by finding steatorrhea or increased fat in the stool. Normal fecal fat excretion is less than 6 to 7 g./day. To allow for the variability of daily stool output, it is usual to collect feces over 3 days and calculate the average 24 hour output. It is possible to examine a single stool sample for fat by staining it with the dye Sudan red after the stool is first acidified to protonate any free acid present. The qualitative single-sample test is less sensitive than the quantitative total collection.

The 72-hour fecal fat test is used because the nonabsorbed fat is minimally modified by colonic bacteria and both triglyceride and FFA are readily extracted from stool and easily measured. Proteins and sugars are metabolized by colonic bacteria so that what is recovered from stool does not accurately reflect what was malabsorbed by the small intestine.

Breath tests

An interesting class of clinical tests depends upon labeling a molecule with a radioisotope, ^{14}C, which can be released by enzymes and recovered as $^{14}CO_2$ in the breath. This allows an assessment of the activity of that enzyme pathway.

One such test is the triglyceride breath test in which the ^{14}C-label is located on the carbonyl carbon of one of the fatty acids of triglyceride such as trioctanoin, a medium-chain triglyceride. After hydrolysis and absorption this fatty acid is in part oxidized to $^{14}CO_2$ and is measured by sampling expired air. The label could also be located on a fatty acid of longer chain length such as triolein (C^{18}). The latter would be more dependent for absorption on pancreatic lipolysis and bile acid solubilization. Thus, if the malabsorption were of pancreatic or biliary origin, the trioctanoin breath test would be more nearly normal than the triolein test. If the disease were due to a defect in the mucosal cell (e.g., celiac disease), both would be equally abnormal.

With the glycocholic acid breath test the concept is extended a little further. The primary enzyme pathway being measured is a bacterial enzyme, cholylamidase, which releases glycine from glycocholic acid. No mammalian enzyme has the capacity to hydrolyze the amide bond between a bile acid and an amino acid.

The rate of recovery of $^{14}CO_2$ is an estimate of the exposure of the bile acid pool to bacteria. This exposure may occur within the anatomic enterohepatic circulation when there is bacterial contamination, as in the blind loop syndrome. Alternatively, if the bile acid pool spills excessively into the colon, as with bile acid malabsorption, there will also be excess exposure to the pool of bacteria. The test only assesses increased exposure to bacteria whose precise cause must be established by other techniques. Nonetheless, it is a simple and useful test.

Medium-chain triglycerides

Triglycerides composed of fatty acids whose carbon chain length is 10 or fewer are more readily hydrolyzed by pancreatic enzymes than are fatty acids of longer chain length. Furthermore, medium-chain fatty acids (C_8, C_{10}) are moderately soluble in water and do not need bile acid solubilization for luminal transport. These fatty acids are transported into portal blood without incorporation into triglycerides or chylomicra. For these reasons, medium-chain triglycerides offer a useful caloric supplement to patients who cannot digest or absorb long-chain triglycerides and fatty acids.

References

Fat absorption reviews
Johnston, J. M.: Mechanism of fat absorption. In Handbook of Physiology. Sec. 6, Vol. 3, pp. 1353–1375. Washington, D.C., American Physiological Society, 1968.
Simmonds, W. J.: Fat absorption and chylomicron formation. In Nelson, G. J. (ed.): Blood Lipids and Lipoproteins. New York, John Wiley & Sons, 1972.
Hofmann, A. F., and Mekhjian, H. S.: Bile acids and the intestinal absorption of fat and electrolytes in health and disease. In Nair, P. P., and Kritchensky, D. (eds.): The Bile Acids, Vol. 2, pp. 103–152. New York, Plenum Press, 1973.

Pancreatic lipase
Borgström, B.: The action of bile salts and other detergents on pancreatic lipase and the interaction with co-lipase. Biochem. Biophys. Acta 488:381–391, 1977.

Lingual lipase
Gooden, J. M., and Lascelles, A. C.: Relative importance of pancreatic lipase and pre-gastric esterase in lipid absorption in calves 1–2 weeks of age. Aust. J. Biol. Sci. 26:625–633, 1973.
Plucinski, T. M., Hamosh, M., and Hamosh, P.: Fat Digestion in rat: Role of Lingual Lipase. Am. J. Physiol. 237:541–547, 1979.

Micelles in fat absorption
Simmonds, W. J.: The role of micellar solubilization in lipid absorption. Aust. J. Exp. Biol. Med. Sci. 50:403–421, 1972.

Unstirred layers
Dietschy, J. M., Sallee, V. L., Wilson, F. A.: Unstirred layers and absorption across the intestinal mucosa. Gastroenterology 61:932–934, 1971.

Enterohepatic circulation
Hofmann, A. F.: The enterohepatic circulation of bile acids in man. Clin. Gastroenterol. 6:3–24, 1977.
Dowling, R. H.: The enterohepatic circulation. Gastroenterology 62:122–140, 1972.

Bile acid synthesis

Mosbach, E. H., and Salen, G.: Bile acid biosynthesis: Pathways and regulations. Am. J. Dig. Dis. *19:*920–929, 1974.

Intestinal bile acid transport

Lack, L., Weiner, I. M.: Intestinal bile salt transport: Structure activity relationships and other properties. Am. J. Physiol. *210:*1142–1152, 1966.

Schiff, E. R., Small, N. C., Dietschy, J. M.: Characterization of the kinetics of passive and active transport mechanisms for bile acid absorption in the small intestine and colon of the rat. J. Clin. Invest. *51:*1351–1362, 1972.

Hepatic bile acid transport

Hoffman, N. E., Iser, J. H., Smallwood, R. A.: Hepatic bile acid transport: Effect of conjugation and position of hydroxyl groups. Am. J. Physiol. *229:*298–302, 1975.

Reichen, J., Paumgartner, G.: Uptake of bile acids by perfused rat liver. Am. J. Physiol. *231:*734–742, 1976.

Cholesterol absorption

Treadwell, C. R., Vahouny, G. V.: Cholesterol Absorption. *In* Handbook of Physiology, Sec. 6, Vol. 3, pp. 1407–1438. Washington, D. C., American Physiological Society, 1968.

Malabsorption

Sleisenger, M.: Malabsorption Syndrome. N. Engl. J. Med. *281:*1111–1117, 1969.

Wilson, F. A., Dietschy, J. M.: Differential diagnostic approach to clinical problems of malabsorption. Gastroenterology *61:*911–931, 1971.

Breath tests

Hepner, G. W.: Breath analysis: Gastroenterological application. Gastroenterology *67:*1250–1256, 1974.

25

Hepatic Mechanisms

Turner E. Bynum, M.D.

The liver performs a remarkable variety of biochemical and physiological functions. Investigations involving the hepatic parenchymal cell and other hepatic cells have long been important to our conceptions about the metabolism of living cells as well as to the emerging picture of cell function at the molecular level. An understanding of alterations in hepatic physiological and molecular mechanisms is essential to the diagnosis and management of liver disease and promises to provide insight about the liver's response to injury.

The liver is the largest organ in the body. The normal liver is situated high in the upper right abdomen, immediately beneath the diaphram, so that it is covered by the lower rib cage and its anterior-inferior edge is at the right costal margin. It has a unique double blood supply, one side of which is the high-pressure, medium flow of the hepatic artery and the other is the low-pressure, greater flow of the portal vein. The hepatic artery is usually a branch of the celiac axis but may arise from the superior mesenteric artery. The hepatic artery provides 30 per cent of the blood supply of the liver and about 50 per cent of the oxygen consumed by the organ. The portal vein drains the effluent from the stomach, intestines, spleen, and pancreas. Portal blood is approximately intermediate in oxygen con-

tent between arterial blood and systemic venous blood. The portal vein delivers 70 per cent of the blood supplied to the liver. The two inflows converge and mix completely at the hepatic sinusoid so that the oxygen content in the sinusoid reflects the averaged new value from dissimilar sources, yet the pressure in the sinusoid is slightly less than that of the portal vein (the arterial pressure having been anatomically damped and gated by progressively smaller branches, and especially the arterioles and precapillary sphincter). The sinusoids are the exchange vessels of the liver and are analogous to capillaries in other areas of the body. The sinusoids drain into the hepatic veins, which in turn empty into the inferior vena cava.

The hepatic parenchymal cell has a sinusoidal membrane across which flows an extraordinary traffic, or exchange of materials, in both directions. At the pole of the cell opposite to the sinusoidal membrane is an invagination of the cell membrane known as the canalicular membrane, which represents the beginning of the hollow conduits of the biliary system. The liver cell has a remarkable capacity for replication and regeneration, which is controlled within the cell by the nucleus. The numerous metabolic reactions that characterize liver function occur primarily in the

561

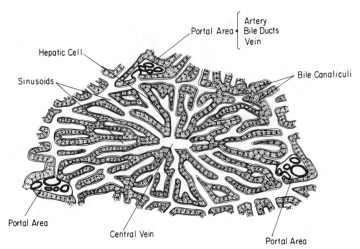

FIG. 25—1. The hepatic lobule.

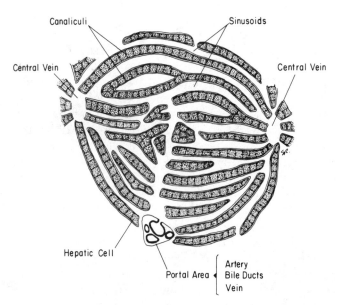

FIG. 25—2. The hepatic acinus.

mitochondria, endoplasmic reticulum, and other cytosolic organelles of the hepatocyte.

There are two ways in which the morphology of the liver has been conceptualized. The first is the hepatic lobule (Fig. 25–1), where three portal tracts make up a triangular (or, when doubled, a hexagonal) periphery with the smallest tributary of the hepatic vein at the center. In this concept the venous tributary is known as the central vein. The second conception (Fig. 25–2) involves the hepatic acinus, which has at its center the portal tract. This club- or

petal-shaped unit has at its peripheral tip the smallest tributary of the hepatic vein. In the second view the venous tributary is called the terminal hepatic vein. The lobular conception dominates pathological descriptions of the liver obtained at autopsy or by liver biopsy, while the concept of an acinar system is more applicable to hepatic developmental biology and physiology, especially as it relates to intrahepatic blood flow.

The many functions of the liver can be classified or grouped in several ways. The classification that appears to have the greatest clinical usefulness defines four major groups as follows: (1) functions that are dependent upon or relate to blood flow; (2) functions that relate to metabolism, including such functions as synthesis, oxidative detoxification, conjugation, conversion, storage, and release; (3) functions that relate to excretion or excretory capacity; and (4) reticuloendothelial system functions of the liver, which reside mostly in the Kupffer cells. (See list below.) A disturbance in any one of these functions may be expressed as a clinical disease. On the other hand a disease of the liver may variably disrupt, diminish or halt some, many, or most of these liver functions.

System of Liver Functions

Blood flow
  Portal flow
  Arterial flow
  Intrahepatic blood flow
  Sinusoidal gradient, down the acinus
  Hepatic venous flow
  Binding to blood-borne substances (e.g., albumin)
Metabolism
  Uptake from the blood
  Sinusoidal membrane activities (other
    than uptake)
  Transport within hepatocyte
  Oxidation
  Conjugation
  Conversion
  Synthesis
  Cell replication
  Storage
  Release
Excretion
  Canalicular membrane activities
  Formation of bile
  Release into blood of altered materials that can be
    rapidly excreted by the kidneys
Reticuloendothelial functions
  Phagocytosis
  Synthesis of immunoglobulin
  Extramedullary hematopoeisis

Handling of bilirubin; jaundice

The upper limit of the normal range for serum bilirubin is 1.0 or 1.2 mg./dl. Elevation of serum bilirubin beyond the normal range constitutes hyperbilirubinemia.

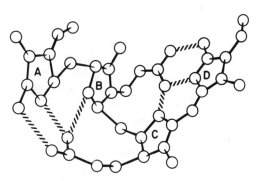

FIG. 25–3.   The bilirubin molecule.

When hyperbilirubinemia becomes sufficiently severe (usually 3 to 4 m./dl.) an experienced observer can detect clinically the state of jaundice or icterus. For the newborn, jaundice is a potential threat, since unconjugated bilirubin in particular can cross the blood-brain barrier, severely damage or even destroy immature neurons, and lead to a form of brain damage that is known as kernicterus. The older child or adult is not in such jeopardy, since there is no harm to the brain from even markedly elevated serum bilirubin levels. In mature people jaundice is merely the physical sign that indicates hyperbilirubinemia and alerts the physician to a disorder of bilirubin handling. Although nonhepatobiliary disorders may cause hyperbilirubinemia, liver and biliary tract diseases are by far the most common causes.

The major bile pigment in mammals is with a geometric isomer form (designated ZZ) such that the propionic acid side chains are internally linked to the opposite pyrrole rings by hydrogen bonds (Fig. 25–3). This hydrogen bonding shields the polar sites of propionic acid so that bilirubin IX α (ZZ) is extremely insoluble in water. Esterification of the propionic acid groups with glucuronic acid appears to prevent hydrogen bond formation, without which bilirubin becomes water soluble. Interference with bond formation, rather than glucuronide conjugation per se, represents the major mechanism for solubilizing bilirubin in water. Mammalian bile also contains trace amounts of geometric isomers of bilirubin that are more water soluble than bilirubin and that are excreted into the bile in an unconjugated form.

The heme portion of the hemoglobin molecule is the source of about 70 per cent of human bilirubin. The remainder is mostly from hepatic hemoproteins, such as cytochromes. When bilirubin metabolism is tracked by radioisotopic methods, there is an early labeled peak; this early labeled bilirubin contains material derived from nonhemoglobin source(s) as well as some from premature destruction (even within the bone marrow itself) of newly formed red blood cells. Senescent red cells are trapped in the reticuloendothelial cells of the spleen, liver, and bone marrow where microsomal heme oxygenase acts on heme, converting it to biliverdin. Biliverdin reductase completes the chemical conversion to bilirubin. In plasma, unconjugated bilirubin binds reversibly to albumin; it can be displaced from this binding by such drugs as phenobarbital, sulfonamides and other antibiotics, analgesics, diuretics, and cholangiographic contrast media.

The exact mechanism for uptake of unconjugated bilirubin is not fully understood, but there appears to be binding of the albumin-unconjugated bilirubin complex to the sinusoidal membrane. Then the unconjugated bilirubin is transported across the membrane by a carrier-mediated process. Inside the liver cell, bilirubin binds preferentially to ligandin (glutathione S–transferase B, also known as Y protein), and to Z protein when bilirubin uptake is high. Next the ligandin-bilirubin

complex passes through the liver cell (perhaps by membrane-membrane transfer) to the endoplasmic reticulum, where bilirubin is conjugated by means of esterification of the propionic side chains with glucuronic acid, mediated by UDP-glucuronyl transferase. At the endoplasmic reticulum, bilirubin is converted to bilirubin monoglucuronide. However, bilirubin diglucuronide is the major pigment fraction in human bile with monoglucuronides representing a small fraction.

There is controversy over the formation of the diglucuronide. One possibility is that glucuronyl transferase at the endoplasmic reticulum continues to act, converting bilirubin monoglucuronide to bilirubin diglucuronide. An alternate theory is that another enzyme converts 2 moles of bilirubin monoglucuronide to 1 mole of bilirubin diglucuronide and 1 mole of unconjugated bilirubin. However, there is also some evidence that bilirubin monoglucuronide is further conjugated to diglucuronide at another location in the hepatocyte, possibly by a separate enzyme at the canalicular membrane. Very little is known about transport within the hepatocyte from rough and smooth endoplasmic reticulum to the canalicular membrane, and little more is known about transport across the canalicular membrane for ultimate excretion into the bile. It is believed that transport across the canalicular membrane is the rate-limiting step in bilirubin handling, and evidence suggests that, like uptake of bilirubin at the sinusoidal membrane, it is a carrier-mediated process (Fig. 25–4).

Such a compressed review of the normal handling of bilirubin allows one to see that these reactions involve several previously mentioned hepatic functions, namely binding to a blood-borne substance, uptake within the liver, transport within the hepatocytes, conjugation, and excretion. Disorders primarily involving any one of these hepatic functions, as well as some extrahepatic events, can lead to hyperbilirubinemia and jaundice. Some of these dis-

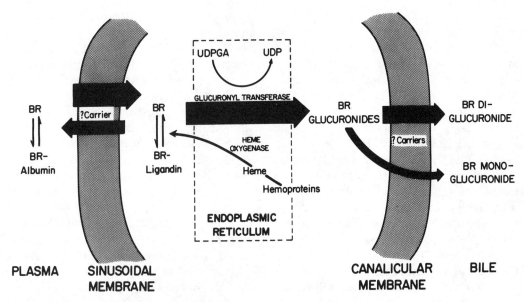

FIG. 25—4. Schematic representation depicting the handling (uptake, transport, conjugation, excretion) of bilirubin (BR) by the hepatocyte.

orders are common, some rare, some are inherited, and some acquired. Several are mentioned here to illustrate the points just made and to illustrate the pathophysiology of jaundice.

Hemolysis

Hemolysis and hemolytic anemia are discussed in detail in Chapter 27. When red blood cell breakdown occurs at a rate higher than normal, thereby presenting more heme to reticuloendothelial cells, there will be more unconjugated bilirubin delivered to the liver. The normal liver has a great reserve for handling increased amounts of unconjugated bilirubin. With the exception of relatively rare, transient hemolytic crises (as may occur in homozygous sickle cell disease or paroxysmal nocturnal hemoglobinuria), increased bilirubin production does not by itself cause a sustained increase in plasma unconjugated bilirubin to higher than 4 mg./dl. Therefore, a plasma concentration persistently greater than 4 mg./dl. implies hepatic dysfunction, whether hemolysis is present or not, and clinically detectable jaundice is

rarely due to hemolysis alone. Individuals with a plasma unconjugated bilirubin between 1.2 and 4 mg./dl. may have hemolysis or abnormal hepatic function, or both.

Drug therapy

A variety of commonly employed medications bind to albumin in the plasma and compete for bilirubin binding sites on the albumin molecule. This is not of great clinical significance to adults, but infants with hyperbilirubinemia could suffer more brain damage if therapy with albumin binding drugs were to cause more unconjugated bilirubin to become available for passage across the blood-brain barrier.

Gilbert's syndrome

Gilbert's syndrome is an inherited unconjugated hyperbilirubinemia in which there is a defect in the uptake of bilirubin at the sinusoidal membrane. In addition, persons with Gilbert's syndrome can have a mild deficiency of glucuronyl transferase. The enzyme deficiency appears to be the major factor contributing to serum unconjugated hyperbilirubinemia, so Gilbert's syndrome may be more in the spectrum of Crigler-Najjar syndromes—a Crigler-Najjar type III, if you will—rather than an example of a major defect in bilirubin uptake. In contrast to the other congenital unconjugated hyperbilirubinemias, Gilbert's syndrome is very common, occurring in 7 per cent or more of the general population. Occasionally, individuals with Gilbert's syndrome become mildly jaundiced, usually when stressed by severe fatigue, intercurrent infection, or fasting. This syndrome is entirely benign, and should not be confused with more clinically significant hematological or hepatobiliary disease.

Crigler-Najjar syndrome

Inherited unconjugated hyperbilirubinemia due to severe deficiency or total absence of glucuronyl transferase (or due to a configuration of the enzyme at the endoplasmic reticulum that prevents access of bilirubin) constitutes the Crigler-Najjar syndrome type I. Infants with this rare disorder become severely jaundiced shortly after birth and the great majority die of kernicterus in the neonatal period. A few develop brain damage but do not die until later childhood or adolescence. Children with Crigler-Najjar syndrome type II have a less severe and variable deficiency in glucuronyl transferase. After birth they can develop levels of unconjugated bilirubin that put them at risk for kernicterus, but these individuals respond to phenobarbital therapy with a dramatic fall in serum bilirubin, presumably brought about by induction of increased amounts of glucuronyl transferase. Also, such patients can be exposed to light, since phototherapy causes rotation of the outer pyrrole rings of bilirubin IX α (ZZ) at the double bonds, thereby converting the bilirubin to EZ, ZE and EE geometric isomers (photobilirubin) and causing disruption of the internal hydrogen bonds. These effects make the bilirubin more polar and water soluble, and allow increased excretion in the unconjugated form.

Dubin-Johnson syndrome

The Dubin-Johnson syndrome is a rare, inherited conjugated hyperbilirubinemia which is, like Gilbert's syndrome, a benign condition, even though clinically apparent jaundice is customary. The defect in this condition is in excretion of conjugated bilirubin at the canalicular membrane. In addition to jaundice due to conjugated hyperbilirubinemia, these individuals have an accumulation of dark pigment in their hepatocytes which can be quite dramatic on a liver biopsy specimen. This pigment has not been identified fully but seems to be a stable free radical which is not melanin nor related to melanin as was once believed. The pigment is a result of, or perhaps unrelated to, the excretion defect, rather than its cause.

Rotor's syndrome

Rotor's syndrome resembles the Dubin-Johnson syndrome in having a defect in excretion of conjugated bilirubin at the canalicular membrane. Rotor's syndrome differs in having a defect in bilirubin uptake at the sinusoidal membrane, a distinct pattern for the clearance of BSP and ICG dyes, and a general defect in hepatic storage of organic anions. Furthermore, this syndrome lacks the hepatocyte pigment.

Hepatitis

Destruction of liver parenchymal cells (hepatocellular necrosis) occurs as a result of viral infection, drug toxicity, or ischemia. Commonly, when this process is severe enough, it is associated with jaundice. Often the total serum bilirubin is quite high, with a definite increase in unconjugated bilirubin. This follows, since in hepatitis liver cells are destroyed, leaving fewer cells available for uptake and conjugation of unconjugated bilirubin. However, in hepatitis serum levels of conjugated bilirubin are as elevated as, or even more elevated than, unconjugated bilirubin. The pathophysiological explanation for this is less certain. Some of the conjugated bilirubin may have come from cells that were completely disrupted or lysed after having conjugated a considerable quantity of bilirubin, thereby releasing conjugated bilirubin into the bloodstream. Many other cells, before becoming truly necrotic, may be disabled, although their conjugating mechanism is intact. In such cells the rate-limiting step, excretion of conjugated bilirubin at the canalicular membrane, would be impaired.

Cholestasis

Cholestasis is failure of bile flow. The clinical condition of cholestasis is frequently called *obstructive jaundice*, although in about half of the cases in which jaundice is due to cholestasis, there is no mechanical obstruction to the flow of bile. For clinical purposes cholestasis may be divided into two major categories, intrahepatic and extrahepatic, although such division is somewhat artificial in the light of more modern knowledge about bile excretion. The value of this separation is that disorders causing extrahepatic cholestasis can be corrected or palliated by abdominal surgery. Extrahepatic cholestasis involves some type of mechanical blockade of bile ducts and is, therefore, obstructive. This category has also been called *surgical jaundice*, referring to the role for surgical therapy.

Intrahepatic cholestasis may also be due to mechanical obstruction of bile ducts; however, in this situation the ducts are small and buried within the liver, thereby obviating any role for major surgery other than liver transplantation. In a large number of instances, jaundice occurs because of failure of bile flow at the level of the hepatocyte, and there is no obstruction of any bile ducts, ductules, or canaliculi. In these cases cholestasis has occurred because of a cellular or metabolic problem. This form of intrahepatic cholestasis has been called hepatocellular cholestasis. All forms of intrahepatic cholestasis have also been called *medical jaundice* because the patients do not benefit from surgery.

In practice, hepatocellular cholestasis is

determined by excluding biliary tree obstruction; techniques such as endoscopic retrograde cholangiography (ERCP) or percutaneous needle transhepatic cholangiography (PTC) are used to obtain contrast radiographs of the biliary tract.

In all forms of cholestasis all the constituents of bile fail to be excreted, and many appear in abnormal quantity in the serum. As one would expect, jaundice in cholestasis represents hyperbilirubinemia that may be 80 to 90 per cent conjugated bilirubin. However, unconjugated bilirubin is also elevated, occasionally to a considerable degree. Therefore, the partition of conjugated and unconjugated bilirubin in the serum is not reliable in distinguishing cholestasis from hepatitis or from other hepatic causes of hyperbilirubinemia.

Clinical approach to jaundice (Fig. 25-5)

Consider the case of a patient with jaundice who presents to the physician. For the purpose of this discussion we will focus on the jaundice whether or not other symptoms and signs are present. Since the hyperbilirubinemia is severe enough to evoke jaundice, it is very unlikely that it is due to hemolysis alone. This allows the physician to begin considering hepatobiliary causes for jaundice. A normal hematocrit (or the presence of no more than moderate anemia) would exclude hemolysis as the primary or major cause of the jaundice. Not infrequently the patient's history discloses multiple possibilities for the cause of jaundice, and for many diseases that cause jaundice the physical findings are differentially nonspecific. Therefore, the physician seeks information from the laboratory, where only a relatively few tests need to be relied upon for the next step in identifying the type of jaundice. Quantitation of total bilirubin will confirm that hyperbilirubinemia is present. If jaundice is mild and the elevated bilirubin is all unconjugated (often reported as "indirect" fraction by the most commonly available laboratory method), if

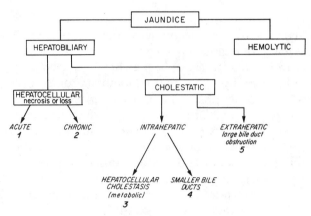

FIG. 25—5. Classification of jaundice. Examples for each pathological alteration are: (1) acute hepatocellular jaundice: acute viral hepatitis, Halothane hepatitis; (2) chronic hepatocellular jaundice: chronic active hepatitis, end-stage cirrhosis; (3) hepatocellular cholestasis: drug-induced cholestasis (Thorazine, birth control pills), cholestatic viral hepatitis; (4) smaller bile duct intrahepatic cholestasis: primary biliary cirrhosis; (5) extrahepatic cholestasis: gallstone in common bile duct, carcinoma of head of pancreas obstructing common bile duct, postsurgical stricture.

the physical examination is unremarkable, the hematocrit is normal, and all other tests relating to the liver (so-called liver function tests) are normal, the patient has Gilbert's syndrome. All other hepatic causes of unconjugated hyperbilirubinemia are very rare, especially in the adult. In adult patients the most common causes of jaundice have predominantly conjugated bilirubin, where the conjugated (or "direct") fraction ranges from 55 to 95 per cent of the total. While a very high conjugated bilirubin is compatible with cholestasis (cholestatic, or "obstructive" jaundice), this is not a reliable distinguishing feature. At this point, serum enzymes that relate to the liver become important; the two that are most helpful are the serum glutamic pyruvic transaminase and the alkaline phosphatase. If the transaminase is elevated (more than twice the normal value and usually five or more times above the normal value) and is elevated proportionally more than the alkaline phosphatase, jaundice is due to hepatocellular disease, and would be called *hepatocellular jaundice*. Viral hepatitis is a common disease that manifests as hepatocellular jaundice.

If the transaminase activity is normal or only slightly elevated, and the alkaline phosphatase is at least three times the normal value and is elevated proportionally more than the transaminase, the situation is most consistent with cholestasis. Alkaline phosphatase has many sources other than the hepatobiliary system, such as bone, placenta, neoplasms, and intestine, so there are some occasions when an elevated alkaline phosphatase needs to be defined further to be certain its origin is hepatobiliary. This can be done by performing alkaline phosphatase electrophoresis or obtaining serum 5'-nucleotidase.

There is no blood test that can further define cholestasis in the sense of distinguishing intrahepatic from extrahepatic cholestasis or identifying hepatocellular cholestasis. Therefore, if cholestasis is present, the next goal is to visualize the biliary tree. Frequently, this is initially attempted by ultrasound (sonography). If intrahepatic bile ducts are dilated, extrahepatic biliary obstruction is likely. Many times it is necessary to put radioopaque contrast material into the biliary system by ERCP or PTC and obtain definitive radiographs. When extrahepatic obstruction is present, such cholangiograms define the site and often the nature of the obstruction. The most common diseases in this category are gallstones in the common bile duct, stricture (usually due to injury sustained during a previous surgical procedure), and cancer (such as carcinoma of the head of the pancreas) which obstructs the bile duct.

If there is no obstructing lesion anywhere in the biliary tree, out to the resolution of the x-ray pictures, then the disease may be one involving only very fine intrahepatic bile ducts, as in primary biliary cirrhosis, or it may be hepatocellular cholestasis. Causes of hepatocellular cholestasis are drug toxicity (such as chlopromazine and other phenothiazines, or birth control pills), alcoholic hepatitis, the cholestatic phase in some cases of viral hepatitis, and cholestatic jaundice of pregnancy.

Mechanism of elevated serum enzymes in hepatobiliary diseases

When a patient has extrahepatic cholestasis, bile is blocked from being excreted (secreted) normally into the intestine. Conjugated bilirubin and other constituents of bile, such as bile salts, are often present in such a patient's blood in greatly elevated quantity. In such cases one may assume that the elevated quantity is due to backing up and overflowing into the circulation. Since an elevated serum alkaline phosphatase is a hallmark of cholestasis, it was easy to assume also that alkaline phosphatase was a prominent constituent of bile and its increased quantity in the blood was due to the same backing up and overflowing phenomenon. However, it has been found that in cholestasis, hepatocytes are stimulated to synthesize large amounts of new alkaline phosphatase, which then gains access to the blood by passive diffusion across cell membranes into the sinusoids. The stimulus for this excessive synthesis of alkaline phosphatase seems to be abnormally increased concentrations of bile salts within the hepatocyte.

The often striking elevation of transaminase that occurs in viral hepatitis represents release of this enzyme from hepatocytes when they become necrotic and their plasma membranes become leaky. Earlier cell injury, prior to necrosis, may stimulate synthesis of these enzymes. High serum transaminase values are observed in ischemic hepatitis also, as well as in the hepatitis that can be caused in susceptible individuals by a wide variety of medications and drugs. One notable exception is alcohol. Even when alcoholic hepatitis is accompanied by extensive hepatocellular injury and widespread necrosis, serum transaminase values are rarely more than modestly elevated, and often may be within the normal range. Recent evidence suggests that alcohol severely inhibits the synthesis of transaminase by the hepatocyte.

Pruritus of cholestasis

Patients with cholestasis, particularly of long duration, may have mild to severe itching of their skin. This can occur occasionally even when the patient has cholestasis without jaundice. In pregnancy this is called pruritus gravidarum. Elevated levels of bilirubin in the serum lead to deposition of bilirubin in the skin, which when severe enough is manifested as jaundice. Although it has long been held that bile salts might be very irritating if deposited in the skin and that bile salts were the cause of itching, recent evidence has tended to exonerate bile salts. A dialysable substance, as yet unidentified, is present in high quantity in the skin of cholestatic patients with pruritus and in the skin of renal failure patients who have pruritus but do not have cholestasis.

Portal hypertension

Increased pressure in the portal venous system results when portal blood is impeded in its flow back to the right atrium. In the United States and other highly industrialized nations, the most common impediment is intrahepatic fibrosis, sclerosis, and sinusoidal disarray (occurring in regenerating nodules) of alcoholic cirrhosis. There are other intrahepatic and extrahepatic sites of obstruction to portal flow (see list below). It is useful to consider the sinusoid as the critical center of portal blood flow. Thus, diseases which cause portal hypertension can be categorized primarily as presinusoidal (in the portal circulation before blood flow reaches the sinosoids), parasinusoidal (e.g., cirrhosis) or postsinusoidal, and secondarily as intrahepatic or extrahepatic.

Portal hypertension has several consequences. One of the most dramatic, and one that is a common clinical problem, is bleeding from esophageal varices. Obstruction to portal flow and the resulting increased pressure in the portal system leads to dilatation of portal system veins as well as flow of blood through collateral veins as portal blood circumvents the site of obstruction en route to the heart. The most important of these varicose collateral veins are in the distal esophagus, where, presumably because of their location very close to the surface epithelium, they tend to rupture and bleed torrentially into the upper gastrointestinal tract. Other evidence of the same type of vascular changes are enlarged periumbilical veins (the caput Medusae), other visibly enlarged veins on the surface of the abdomen, gastric varices, duodenal varices, vaginal varices, and hemorrhoids (rectal varices).

Another consequence of portal hypertension relates to shunting of splanchnic venous blood away from the liver, which occurs extrahepatically, as well as intrahepatically, within the fibrotic bands in cirrhosis. This, combined with hepatocyte dysfunction due to continuing, active hepatocellular disease in some patients, prevents the liver from taking up and detoxifying a variety of substances that originate in the intestines. A number of changes happen as a result, among the most prominent being central nervous system dysfunction (hepatic encephalopathy) and alterations in intrarenal hemodynamics (which can lead to the hepatorenal syndrome) both of which are discussed in greater detail below.

Increased portal pressure is transmitted back to the spleen, which progressively enlarges. The enlarged spleen often sequesters platelets and white blood cells (hypersplenism). The enlarged spleen requires more arterial blood flow, thereby increasing blood flow into the portal system through the splenic vein and aggravating the congestion and elevated pressure in the portal system.

Causes of Ascites

Cirrhosis
Budd-Chiari syndrome
Venoocclusive disease
Tuberculous peritonitis
Malignant neoplasm metastasized to peritoneum
Nephrotic syndrome
Mesothelioma of peritoneum
Meigs' syndrome
Pancreatogenous ascites
Chylous ascites
Very severe congestive heart failure
Starch peritonitis
Constrictive pericarditis
Myxedema

Ascites

Normally a small amount of fluid is present free within the abdominal cavity. This fluid is constantly, if slowly, exchanged but the volume remains small since the forces putting small amounts of fluid into the peritoneal cavity are balanced by the absorption of fluid back into the systemic circulation at the surface of the peritoneum. Ascites is the accumulation of an abnormally large volume of fluid within the peritoneal cavity. The absolute quantity of ascites may vary from a few hundred milliliters to six liters or more. There are several distinct causes of ascites (see list below), among the most common of which, after primary liver disease, are a malignant neoplasm which has metastasized to the peritoneum, the nephrotic syndrome, and tuberculosis of the peritoneum.

Portal Hypertension Classified by Site of Portal Venous Obstruction

Presinusoidal
    Extrahepatic (e.g., portal vein thrombosis)
    Intrahepatic (e.g., schistosomiasis)
Parasinusoidal (and immediate postsinusoidal [e.g., cirrhosis])
Postsinusoidal
    Intrahepatic (e.g., venoocclusive disease, Budd-Chiari disease)
    Extrahepatic (inferior vena cava thrombosis, subatrial web of inferior vena cava [Budd-Chiari syndrome])

By far the most common liver disease that is associated with clinically detectable ascites is cirrhosis. In cirrhosis the peritoneum is normal, so the structure responsible for reabsorbing fluid from the peritoneal cavity is intact. However, the reabsorbing process is quite limited and is overwhelmed by diminished oncotic pressure in the general circulation and by increased rates of peritoneal fluid formation.

Several mechanisms have been proposed to explain how ascites forms in cirrhosis. Two key elements in the development of cirrhotic ascites are (1) portal hypertension with an intrahepatic site of obstruction and intrahepatic portal venous congestion; and (2) hypoalbuminemia. Other factors are also involved, however. There are several ways of considering the roles of each factor in the pathogenesis of ascites. Ascites has been considered merely an instance of edema formation, that is, as purely an imbalance in Starling forces. Cirrhosis causes portal hypertension and the high pressure is transmitted back into the venules of the peritoneal cavity, thereby forcing more fluid out of the capillaries and exceeding the capacity for reabsorption. Furthermore, reabsorption is impaired by low oncotic pressure in the circulating blood due to the low concentration of serum albumin. As ascites develops there is a transient decrease in effective plasma volume, which (1) activates the renin-angiotensin-aldosterone system, leading to increased secretion of aldosterone; and (2) stimulates release of the antidiuretic hormone (ADH). These hormones serve to promote renal retention of sodium and water. As a result, plasma volume is expanded, further increasing portal pressure and further lowering the plasma oncotic pressure by diluting albumin. This cycle continues until a new steady state exists, usually including large amounts of ascites and moderately increased total plasma volume, at the time the patient is first seen by a physician.

An additional factor leading to ascites is the impaired liver function of cirrhosis that results from both the alcoholic insult to hepatocytes and the shunting of blood away from the liver cells. To aggravate this condition further, the circulating angiotensin II, aldosterone, and ADH (as well as a postulated peptide produced in the intestine that blocks renal sodium excretion) are not metabolized to inactive products as efficiently by the malfunctioning liver. Consequently, these agents act on the kidney to bring about increased retention of sodium and water; the resulting increased plasma volume contributes to both the portal hypertension and the ascites. Finally, we must consider the imbalance between splanchnic/hepatic lymph formation and the capacity for lymph reabsorption as an important factor in the pathogenesis of ascites. With portal hypertension due to an intrahepatic, parasinusoidal site of obstruction there is congestion of blood in the liver and gastrointestinal tract with edema of these organs. When the lymphatics no longer can return a sufficient amount of the increased tissue fluid to the general circulation, the result is decreased effective plasma volume and further stimulation of ADH and renin release. With generation of angiotensin and secretion of aldosterone, sodium and water are retained, plasma volume is expanded, and there is a further increase in lymph formation, leading to ascites. Ascites may be largely intraperitoneal lymph resulting in part from some abnormality in the structure of lymphatic channels or in the kinetics of lymph flow.

Almost all of the above mechanisms play varying roles in different patients with ascites and cirrhosis, and perhaps in the same patient at different times. Since severe portal hypertension due to a presinusoidal site of obstruction is only rarely associated with ascites, a parasinusoidal or postsinusoidal block seems quite crucial. Furthermore, some experimental information suggests that fluid does "weep" from the liver capsule in such forms of portal hypertension. A sequence of events that ex-

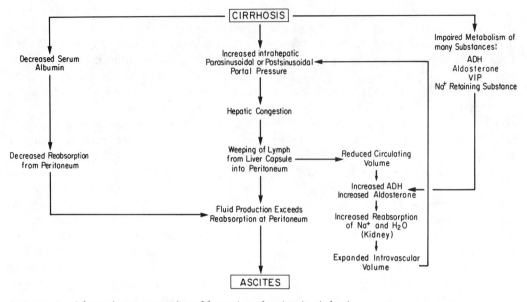

FIG. 25—6. Schematic representation of formation of ascites in cirrhosis.

plains the initiation of ascites in terms of the preceding factors is that loss of tissue fluid from the congested liver, combined with diminished oncotic pressure (hypoalbuminemia) in the blood of the peritoneum, starts the accumulation of ascites. Early in the process the accumulation is slow. Any contraction of plasma volume is compensated for by the usual mechanisms, and plasma volume remains essentially normal for a long time. However, as portal hypertension worsens and more portal blood is shunted away from the liver, substances which cause sodium retention and vasodilation are not inactivated by the liver, leading to increased water uptake by the kidney and an augmented plasma volume. There is formation of even more ascites. Eventually, so much fluid accumulates that the obviously distended abdomen leads the patient to seek a physician, who encounters the ascites only at a time when multiple mechanisms are playing a role and it is difficult to identify a single mechanism as the primary culprit (Fig. 25–6).

Hepatorenal syndrome

Patients with a severe disease of the liver may develop renal failure. This is most common in terminal cirrhosis with portal hypertension and ascites but may also occur in acute fulminant hepatitis. One form of renal failure seems to be particularly, even uniquely, associated with severe liver disease, and the term *hepatorenal syndrome* should be used only for this specific clinical situation. True hepatorenal syndrome is characterized by progressive oliguric (or, occasionally, anuric) renal failure, manifesting initially with blood test evidence for azotemia but later as fully developed uremia. Urine sodium is usually low, less than 10 mEq./l., but is rarely absent. Kidneys from such patients are grossly and histologically normal, and function normally when transplanted to recipient patients who have normal liver function. If a patient with hepatorenal syndrome receives a liver transplantation, the kidneys quickly begin to function normally and continue to do so as long as the transplanted liver functions well. Hepatorenal syndrome represents intrarenal circulatory failure where cortical glomeruli are not perfused adequately despite a normal total blood flow to the kidney. The reason for this abnormal distribution of intrarenal blood flow is uncertain. It may be that the ailing liver in these patients fails to metabolize or detoxify a substance from the intestines that is a potent vasoconstrictor of renal cortical vessels. Another possibility is that a vasodilator substance from the intestine, such as vasoactive intestinal polypeptide (which is normally metabolized by the liver) gains access to the general circulation and dilates the medullary portion of the renal circulation, thereby redistributing blood flow away from the cortex. Among the evidence supporting the concept of a release of vasodilator material in hepatorenal syndrome is the observation of other circulatory phenomena that characterize the cirrhotic state, namely increased heart rate, increased cardiac output, decreased peripheral vascular resistance, palmar erythema, telangiectasia, spider angiomas, intrapulmonary shunting, and low systemic arterial blood pressure. Furthermore, a renal medullary dilator would be expected to enhance sodium retention by the kidney and would link the pathogenesis of ascites to that of hepatorenal syndrome.

Patients who develop hepatorenal syndrome have an extremely poor prognosis, and usually die of combined liver and kidney failure. Those who improve, or ultimately survive, do so because liver function improves. There is no effective therapy for the renal failure itself. A few patients have been reported to come out of the renal failure of apparent hepatorenal syndrome after placement of a peritoneovenous shunt through which ascitic fluid is infused back into the general circulation. Unfortunately, this therapy has not been consistently effective and is commonly complicated by severe disseminated intravascular coagulation.

Cirrhotic patients develop diminished intravascular volume due to hemorrhage into the gastrointestinal tract (e.g., from esophageal varices) or due to excessively vigorous therapy with diuretic agents in an attempt to reduce the ascites. Circulating blood volume may become so impaired that oliguria results. In this situation urine sodium will be very low, perhaps undetectable; hepatorenal syndrome may then result, especially if liver failure is severe. Another common possibility, if the severely contracted volume persists, is acute renal failure (so-called acute tubular necrosis), and this will be expressed as azotemia or uremia, severe oliguria, and a high urine sodium (greater than 20 mEq./l.). It is important not to confuse this form of acute renal failure with the hepatorenal syndrome. Not all renal failure in liver disease is the hepatorenal syndrome, and thus the prognosis may be better.

Hepatic encephalopathy

Patients with severe liver failure, particularly cirrhosis with portal hypertension, or fulminant hepatic necrosis due to viral hepatitis or some profound hepatotoxin, often have impairment of brain function. In rare instances this is expressed predominantly as disordered motor function. Most commonly it is expressed as part of the continuum of declining cortical function, starting with subtle disturbances in the sensorium (altered sleep pattern, drowsiness, memory loss, personality change) and progressing in severity through intermediate states to deep coma. The most advanced stage is called *hepatic coma*, while the entire spectrum is known as *hepatic encephalopathy*.

Many metabolic abnormalities have been identified in hepatic encephalopathy. As with ascites, it has been difficult to identify a single cause of this disorder. The major factors will be considered in the following paragraphs.

Because hepatic encephalopathy resembles nonspecific toxic or metabolic encephalopathy in many of its features, particularly findings from the neurological examination and electroencephalographic changes, a neurotoxic agent may be the cause. The source of this toxin may be from the intestines, and this substance has gained access to the general circulation and the brain because it was not metabolized or detoxified by the diseased liver or was shunted away from the liver, as in portal hypertension or when portal-systemic shunts have been created surgically. One candidate for this neurotoxic cause of hepatic encephalopathy is the ammonia that is generated in the gut lumen by the action of normal enteric bacteria on a variety of nitrogen-containing substrates. The ammonia is rapidly and efficiently absorbed into the portal blood and then carried to the liver. Normally, the ammonia is taken up avidly by hepatocytes and metabolized to urea. However, in severe liver disease the ammonia "bypasses" the liver to serve as a direct suppressant of cortical neuron function.

Not all patients with hepatic encephalopathy have greatly elevated blood (or cerebrospinal fluid) ammonia levels, and even if elevated, the degree of elevation of ammonia may correlate poorly with the severity of hepatic encephalopathy, which suggests the role of at least one other neurotoxic substance. Several candidates have been proposed such as mercaptans, fatty acids, and amino acids; but each has failed to correlate with the state of coma as well as ammonia levels. The best correlation between the severity of hepatic encephalopathy and a chemical substance has been the level of spinal fluid glutamine and α-ketoglutaramate. These directly reflect the quantity of ammonia in the brain. Ammonia appears to act in concert with other toxins and with nonspecific metabolic derangements that potentiate the neurotoxicity of the ammonia. Therefore, the toxins and metabolic abnormalities are more than additive; their combined effect is greater than the sum of the individual effects: they act synergistically.

As a result of either extensive hepatic cell necrosis or the shunting of portal blood contents to the systemic circulation, patients with severe liver disease have both increased levels of amino acids in the plasma and abnormal proportions of various types of amino acids, leading to an increased uptake of aromatic amino acids by the brain. Increased tryptophan, phenylalanine, and tyrosine in the brain prompt excess production of serotonin, an inhibitory neurotransmitter in the brain, accompanied by deficient production of excitatory neurotransmitters such as norepinephrine and dopamine, contributing to encephalopathy.

False neurotransmitters are chemicals such as octopamine that bind to brain receptor sites and block access to those sites by the normal neurotransmitters. Because false neurotransmitters stimulate little or no activity from the receptor to which they bind, brain function is impaired. The source of false neurotransmitters in severe liver disease might be an intestinal peptide or a material formed in the brain as a result of increased amounts of aromatic amino acids in brain tissue. The colon has been suggested as the source for γ-aminobutyric acid, which upon gaining the systemic circulation, travels to the brain in quantities much greater than normal, thereby acting as an inhibitory neurotransmitter.

Hepatic encephalopathy may occur if the liver fails to supply some substance essential to brain function. The brain must metabolize glucose for almost all of its energy, with a very small contribution from the metabolism of fatty acids. The initiating event in hepatic encephalopathy could be insufficient energy from substrate to maintain the brain's oxidative metabolism. The severely diseased liver may not supply enough glucose because of depleted glycogen stores and impaired glycogenolysis and gluconeogenesis. The failure of energy

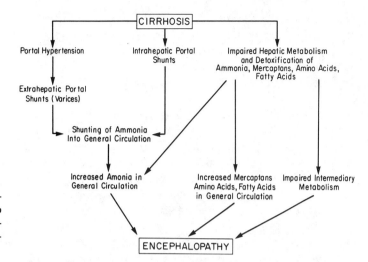

FIG. 25–7. Schematic representation of the interrelationship of the various mechanisms implicated in the development of hepatic encephalopathy.

metabolism would lead to failure of other metabolic systems, including the formation of neurotransmitters. Faulty neurotransmitter generation would result in accumulation of precursor compounds which behave as false neurotransmitters and exacerbate the situation. Severely depressed oxidative metabolism has been identified in the brains of patients who have been in coma for some time, although oxidative metabolism is normal early in comatose states due to liver disease. Impaired oxidative metabolism is probably the result of depressed brain function rather than its cause.

Clinical experience and laboratory data indicate that hepatic encephalopathy is multifactorial (Fig. 25–7). The view that ammonia plays a central role fits the observation that the most successful therapy for hepatic encephalopathy involves elimination or reduction of ammonia production. These therapeutic approaches are the use of broad-spectrum or nonabsorbable antibiotics to reduce the quantity of ammonia-producing bacteria in the colon; administration of KCl to reduce the kidney's production of ammonia that occurs in hypokalemic alkalosis; oral administration of lactulose to reduce ammonia absorption either by lowering intraluminal pH or altering the bacterial flora; and reducing pro-

tein in the diet to reduce nitrogen-containing substrate and thus lower ammonia production.

Nutrients and liver disease

Proteins
Protein deficiency was long believed to cause liver disease, particularly cirrhosis. Fatty liver is well established to be the result of severe protein malnutrition. However, even the extreme protein deprivation that is manifested as kwashiorkor, in which afflicted children can have massive hepatomegaly due to steatosis, does not progress to cirrhosis of the liver. Certainly, if extensive hepatocellular necrosis is induced by a toxin or hepatitis virus infection, severe protein deficiency may impair hepatic recovery and regeneration. There is some evidence that the result of acute liver injury is more likely to be cirrhosis if the patient is protein deficient, and that is what is now meant by a "cirrhotogenic diet."

The liver is responsible for the synthesis of many serum proteins. In severe liver disease any or many of these may be low because of failure of hepatic synthesis. Most common is low serum albumin, which may impair the oncotic pressure of plasma sufficiently to prompt formation of ascites and peripheral edema. Several clotting factors are synthesized by the liver, of which the most important is prothrombin. A prolonged prothrombin time that fails to improve after the administration of parenteral vitamin K indicates severe hepatocellular failure and signals a poor prognosis. In acute severe liver disease, such as fulminant hepatitis, unresponsive hypoprothrombinemia indicates that the patient is at very high risk of dying during that hospitalization, especially if the patient is over 40 years of age. In chronic liver disease, such as cirrhosis, unresponsive hypoprothrombinemia means that the patient has a significant chance of dying within the next 2 years.

Fats

Fat metabolism is compromised in several ways in liver disease. Cholestasis is often associated with elevated serum cholesterol, most of which is present in lipoprotein X, a lipoprotein that is characteristic for cholestasis. It is not known why this lipoprotein is present in both intrahepatic and extrahepatic cholestasis, although it is possible that cholesterol is not secreted into bile in cholestasis and the accumulated lipid is packaged and then released into the blood as lipoprotein X. Also, it may be that increased quantities of bile salts in the hepatocyte more specifically stimulate the synthesis of this lipoprotein. Patients with chronic cholestasis, such as those with primary biliary cirrhosis, can develop xanthelasmas and xanthomas of earlobes, elbows, buttocks, and hands due to the elevated serum lipid.

Hepatocytes may accumulate large amounts of fat, which is called fatty metamorphosis or steatosis, in a variety of conditions including alcohol abuse, diabetes mellitus, starvation, morbid obesity, adrenocorticosteroid therapy, and drug toxicity. The mechanism that most commonly underlies steatosis is failure of the hepatocyte to make enough of the protein moiety to get the lipid that is presented to the liver from the blood packaged and released back into the blood as lipoprotein. As a result, lipid accumulates within the liver cell.

Carbohydrates

Because the liver has a major role in carbohydrate metabolism, it was natural to expect evidence of disturbed carbohydrate handling in liver disease. Clinically, this abnormality is very rare. Some patients with cirrhosis have an abnormal glucose tolerance test; however, these patients have normal fasting blood sugar values and no glycosuria, and fail to develop any of the complications of diabetes mellitus. Hypoglycemia is rarely due to liver disease alone and its occurrence in patients with extremely severe liver disease signals imminent death. Alcoholics with underlying liver disease (alcoholic hepatitis, cirrhosis) may become severely hypoglycemic during a bout of heavy alcohol intake. In this circumstance the requisite elements that lead to hypoglycemia are that the alcoholic has not been eating, the diseased liver has diminished glycogen stores (and perhaps impaired glycogenolysis), and the liver's obligate metabolism of the heavy alcohol load blocks gluconeogenesis by tying up the electron transfer system.

Vitamins

The liver stores a large quantity of vitamin B_{12}. The size of this store accounts for the long time (approximately 3 years) that is required for an individual to manifest signs of vitamin B_{12} deficiency (as anemia or neurological abnormalities) after the onset of malabsorption of the vitamin in pernicious anemia, ileal resection, or bacterial overgrowth in the small bowel. Furthermore, if liver cells are destroyed in hepatitis or hepatic abscess, vitamin B_{12} is released into the circulation, and measured levels of serum vitamin B_{12} can be very high.

The liver has an avid capacity to store other vitamins, such as vitamin A. The unbridled storage of vitamin A can lead to severe liver damage if a large amount of the vitamin is ingested, for example during misguided "megavitamin" therapy. Bear liver contains a very large store of vitamin A; hunters who have eaten it, particularly polar bear liver, have suffered significant illness, with liver damage, due to vitamin A toxicity.

Alcoholics with cirrhosis often manifest multiple vitamin deficiencies, particularly of folate, thiamine, and other water-soluble vitamins of the B complex. Most of the deficiencies can be attributed to poor intake or an inhibitory effect of alcohol ingestion on vitamin absorption or utilization. Nevertheless, some of these patients who have severe cirrhosis appear to be resistant to therapeutic replacement of the vitamins, which may represent impaired hepatic conversion to active compounds. Vitamin K deficiency because of poor intake is rare, and hypoprothrombinemia in liver disease is usually due to impaired absorption of fat-soluble vitamins in cholestasis or to impaired synthesis of prothrombin in severe hepatocellular disease.

Trace metals

Primary hemochromatosis is an inherited disease of iron handling in which there is excessive iron absorption relative to total body iron stores, as well as faulty handling of iron by reticuloendothelial elements. The result is excessive deposition of iron in parenchymal cells of several organs, such as the pancreas, heart, skin and especially the liver, which becomes cirrhotic and may develop a hepatoma (primary hepatocellular carcinoma). Secondary hemochromatosis refers to iron loading disease without the inherited defect. Here, iron overload occurs in refractory anemia where the anemia is not due to iron deficiency or blood loss but is due to hemolysis or failure to incorporate available iron. Iron overload comes about because of multiple blood transfusions or misguided vigorous iron therapy. Secondary iron loading also occurs in alcoholics with liver disease in

whom cirrhosis and probably some pancreatic damage allows excess absorption of iron. In some of these individuals when they present with established cirrhosis, it is difficult to determine if hemochromatosis is primary or secondary. In other words, did alcohol abuse accelerate the development of cirrhosis in an individual who has inherited hemochromatosis, or is the alcohol abuse the primary factor that led to cirrhosis and thence to secondary hemochromatosis? At this point, one is usually forced to rely on family history, or the quantitation of iron in samples of liver tissue.

Wilson's disease is an inherited deficiency of ceruloplasmin, the serum protein which binds and transports copper. Copper gets deposited excessively in the brain, eyes, and liver, with severe hepatocellular degeneration if removal of copper is not initiated early.

Other trace metal aberrations that occur in liver disease are usually deficiencies that come about as a result of poor intake. Zinc deficiency is common in patients with alcoholic cirrhosis.

Hepatocyte injury and necrosis

The normal, unchallenged, liver has a very low turnover of its parenchymal cells and an enormous capacity for regeneration. If an otherwise healthy individual suffers severe blunt trauma to the liver, requiring for hemostasis resection of all of one lobe and a good part of the other lobe, the liver will regenerate back to normal size, volume, and function, and essentially to normal contour, within months.

As mentioned above, patients with Dubin-Johnson syndrome accumulate dark pigment in their hepatocytes. If one of these patients has a mild-to-moderate episode of acute viral hepatitis and subsequently has liver biopsy after beginning recovery, none of the hepatocytes seen have any pigment at all. Then, if biopsy is repeated months and years later, pigment gradually and progressively reaccumulates in the liver cells. These events imply that every hepatocyte dies during the course of even mild acute viral hepatitis. This is another testimony to the prodigious regenerative capacity of the liver.

Relatively recent investigations have been initiated to give information about the mechanisms by which liver cells die. The known hepatitis viruses are not directly cytotoxic. In viral hepatitis, and in some types of toxic hepatitis, cell death seems to be caused by "killer" lymphocytes. During cell infection or injury by a virus or drug, a cell protein on the membrane becomes altered or exposed, thereby eliciting a cellular immune response to what is now perceived by the lymphocytes as foreign antigen.

Some drugs are directly toxic to hepatocytes. Other agents are metabolized by the liver, and the metabolic product is toxic. Increasing the metabolism of these drugs increases the resultant toxicity. One mechanism by which toxic agents cause cell death is alteration of membrane permeability characteristics, particularly to calcium, sodium, and potassium. This leads to disrupted cell function or swelling and disruption of membrane-bound organelles of the cell or of the cell itself. The exact mechanism by which alcohol abuse causes cell necrosis is not well established but seems to involve mitochondrial dysfunction and disruption as well as some immune-mediated events.

Reticuloendothelial function

Generalized diseases of the reticuloendothelial (RE) system affect the RE components and functions of the liver. Examples are leukemia, Hodgkin's disease, lymphoma, myeloid metaplasia, and lipid storage diseases. On the other hand, when cirrhosis is present a patient may suffer consequences or complications which relate to defective liver RE function or to shunting of splanchnic blood away from the liver and its RE cells. Patients with cirrhosis are more susceptible to infection (especially blood-borne infections with enteric organisms) and septicemia. Hyperglobulinemia, which is predominantly a broad, polyclonal increase in γ-globulins, mostly IgG, is a regular feature of advanced cirrhosis of any cause and seems to be produced partly by hepatic RE activity and partly by excess antigens that gain access to the general circulation that occurs in the presence of portal hypertension.

References

Bilirubin and jaundice

Gollan, J. L., et al.: Cholestasis and Hyperbilirubinemia. *In* Glitnick, G. (ed.): Current Hepatology. Boston, Houghton-Mifflin, 1980.

Scharschmidt, B. F., and Gollan, J. L.: Current concepts of bilirubin metabolism and hereditary hyperbilirubinemia. *In* Popper, H., and Schaffner, F. (eds.): Progress in Liver Disease, Vol. VI. New York, Grune & Stratton, 1979.

Schmid, R.: Bilirubin Metabolism: Gastroenterology *74*:1307, 1978.

Bynum, T. E.: Jaundice. Wolf, S. (ed.): Abdominal Diagnosis. Philadelphia, Lea & Febiger, 1979.

Portal hypertension

Marks, C.: The Portal Venous System. Springfield, Ill., Charles C Thomas, 1973.

Child, C. G. (ed.): Portal Hypertension. Philadelphia, W. B. Saunders, 1974.

Sherlock, S.: Portal circulation and portal hypertension. Gut *19*:70, 1978.

Ascites

Bicket, D., et al.: Role of vasopressin in abnormal water excretion in cirrhotic patients. Ann. Intern. Med. *96*:413, 1982.

Shear, L.: Ascites. Postgrad. Med. *53*:165, 1973.

Lieberman, F. L., et al.: The relationship of plasma volume, portal hypertension, ascites, and renal sodium retention in cirrhosis: The overflow theory of ascites formation. Ann. N.Y. Acad. Sci. *170*:202, 1970.

Witte, C. L., et al.: Lymph imbalance in the genesis and perpetuation of the ascites syndrome in hepatic cirrhosis. Gastroenterology *78*:1059, 1980.

Shear, L., et al.: Compartmentalization of ascites and edema in patients with hepatic cirrhosis. N. Engl. J. Med. *282*:1391, 1970.

Hepatorenal syndrome

Conn, H. O.: A rational approach to the hepatorenal syndrome. Gastroenterology *65:*32, 1973.

Wong, P. Y., et al.: The hepatorenal syndrome. Gastroenterology *77:*1326, 1979.

Hepatic encephalopathy

Schenker, S., et al.: Hepatic encephalopathy. Gastroenterology *66:*121, 1974.

Hoyumpa, A. M., et al.: Hepatic encephalopathy. Gastroenterology *76:*184, 1979.

Zieve, L.: The mechanism of hepatic coma. Hepatology *1:*360, 1981.

Nutrients and liver disease

Mezey, E.: Liver disease and nutrition. Gastroenterology *74:*770, 1978.

Gabuzda, G. J., and Shear, L.: Metabolism of dietary protein in hepatic cirrhosis. Am. J. Clin. Nut. *23:*479, 1970.

Russell, R. M., et al.: Hepatic injury from chronic hypervitaminosis A. N. Engl. J. Med. *291:*435, 1974.

Webber, B. L., and Freiman, I.: The liver in Kwashiorkor. Arch. Pathol. *98:*400, 1974.

Edwards, C. Q., et al.: Hereditary hemochromatosis. N. Engl. J. Med. *297:*7, 1977.

Simon, M., et al.: Idiopathic Hemochromatosis. Gastroenterology *78:*703, 1980.

Strickland, G. T., and Leu, M. L.: Wilson's disease. Medicine *54:*113, 1975.

Sternlieb, J.: Diagnosis of Wilson's disease. Gastroenterology *74:*787, 1978.

Injury and necrosis

Edmondson, H. A., et al.: The early stage of liver injury in the alcoholic. Medicine *46:*119, 1967.

Isselbacher, K. J.: Metabolic and hepatic effects of alcohol. N. Engl. J. Med. *296:*612, 1977.

Khanna, J. M., et al. (eds.): Alcoholic Liver Pathology. Toronto, Addiction Research Foundation of Ontario, 1975.

Meyer zum Buschenfeld, K. H., et al.: Immunologic liver injury. *In* Popper, H., and Schaffner, F. (eds.): Progress in Liver Disease, Vol. VI. New York, Grune & Stratton, 1979.

26 Biliary Tract Mechanisms

Carlos A. Pellegrini, M.D.
Lawrence W. Way, M.D.

The gallbladder

Transport mechanisms of the gallbladder mucosa

The absorptive capacity of the gallbladder mucosa is one of the highest of all epithelia. Between 15 and 25 per cent of the gallbladder volume can be absorbed per hour, so that within 6 hours volume decreases by 90 per cent. Concentration of organic solutes (e.g., bile acid anions) increases approximately tenfold during this period; Na^+ concentration doubles and the Ca^{++} and K^+ concentrations increase three- to fivefold; chloride and bicarbonate concentrations decrease; and the bile becomes more acidic. The asymmetrical distribution of Cl^-, HCO_3-, and bile acid anions across the gallbladder epithelium generates a small electrical potential difference (up to 12 mV. in man) negative to the lumen (Table 26–1).

Hepatic bile is an isotonic solution. Because bile salts form micellar aggregates, the osmolality of gallbladder bile is about half what might be expected from its molar concentration; thus, with a Na^+ concentration twice that of plasma, gallbladder bile is still isotonic.

This work was supported by the Veterans Administration. We are indebted to James R. Gorring for his technical assistance.

ANATOMY. The gallbladder mucosa is composed of tall, large, homogeneous, columnar cells arranged on a basal membrane (Fig. 26–1). These cells are connected at their apices by bridges called *tight junctions* and are separated laterally by intercellular spaces. The tight junctions divide the cell membrane into two portions that are morphologically and functionally different: the apical membrane and the basolateral membrane. The apical membrane has sparse microvilli. The basolateral membrane has evaginations that interdigitate with adjacent cells. As a consequence, the lateral intercellular space is irregularly shaped, and its total length is three to four times that of the cell itself. It is through these spaces that much of the water and electrolyte movement occurs. The gallbladder epithelium is "leaky" like the renal proximal tubular epithelium and small bowel epithelium. These epithelia share the following characteristics: an extracellular pathway for water and electrolyte transport; a low transepithelial potential difference, owing to a low electrical resistance to current flow; and high permeability to the flow of water along an osmotic gradient.

TRANSEPITHELIAL MOVEMENT OF WATER AND ELECTROLYTES. Sodium and chloride are rapidly absorbed at equal rates by the gallbladder epithelium, which displays a

TABLE 26–1
Bile Characteristics

	Hepatic Bile	Gallbladder Bile
Na^+ (mEq./l.)	155–165	260–280
K^+ (mEq./l.)	3–6	12–14
Cl^- (mEq./l.)	90–95	13–15
HCO_3^- (mEq./l.)	42–45	5–8
Bile acid anions (mEq./l.)	28–32	280–300
Ca^{++} (mEq./l.)	4–6	12–16
pH	8.2	6.5
Osmolality (mOsm./l.)	280–300	280–300

high selectivity for Na^+, but Cl^- and HCO_3^- may substitute for one another. The presence of both Na^+ and either Cl^- or HCO_3^- on the luminal side of the gallbladder epithelium is essential, and replacement of HCO_3^- and either Na^+ or Cl^- by a nontransportable ion stops active transport (and water movement) through the epithelium.

Studies using intracellular microelectrodes have shown that NaCl enters the cell through the apical membrane as a neutral complex, an example of *coupled transport.* Since Na^+ influx is directed down its electrochemical gradient, it provides energy for the accompanying Cl^- influx, which moves against chemical and electrical forces.

The extrusion of Na^+ from the cell takes place through the basolateral membrane and is the result of an active transport process involving a Na-K ATPase exchange pump. The mechanism of Cl^- extrusion from the cell is unknown.

The first step in water absorption is NaCl transport into the intercellular space. This creates an osmotic gradient that provides the force necessary to drive water into the space; then the resulting hydrostatic force in the intercellular space produces flow towards the basal membrane. Water from the gallbladder lumen reaches the intercellular space both through the tight junctions and through the cell, but the proportion moving along each of these paths is unknown.

The original "standing gradient" theory of Diamond assumed that Na^+ transport from the cell to the intercellular space occurred preferentially through the apical part of the lateral membrane; this is no longer thought to be correct. For example, distribution of the Na-K ATPase pump is uniform throughout the basolateral membrane.

Bicarbonate is absorbed by a mechanism similar to that of Cl^- and, as a result of HCO_3^- absorption, the pH of the bile decreases. The presence of HCO_3^- in the medium bathing the mucosa enhances the ability of the epithelium to absorb NaCl.

K^+ and Ca^{++} concentrations are governed by passive redistribution following the electrochemical gradients created by the asymmetric distribution of Na^+, Cl^-, and HCO_3^- across the epithelium.

REGULATION OF WATER ABSORPTION. Classical physiology describes the function of the gallbladder mucosa as involving continuous absorption of water and electrolytes. However, several studies have shown that the absorptive function of the gallbladder mucosa is affected by hormonal, neural, and luminal stimuli. For example, stimulation of the splanchnic nerves or infusion of norepinephrine increases the rate of water absorption. Gastrin and cholecystokinin (CCK) decrease fluid absorption in the isolated canine but not the rabbit gallbladder. Secretin decreases water absorption in the gallbladder of the anesthetized cat. Taurine conjugated dihydroxy bile acids inhibit absorption in the canine gallbladder, an effect that can be reversed by adding lecithin.

Sodium chloride efflux and water absorption may be inhibited by increased intracellular levels of cyclic $3'5'$-adenosine monophosphate (cAMP). Stimulation of adenylate cyclase activity (e.g., by administration of cholera toxin or prostaglandin E_1) markedly inhibits NaCl and water transport and under certain circumstances may induce secretion of water into the lumen. Vasoactive intestinal polypeptide (VIP) re-

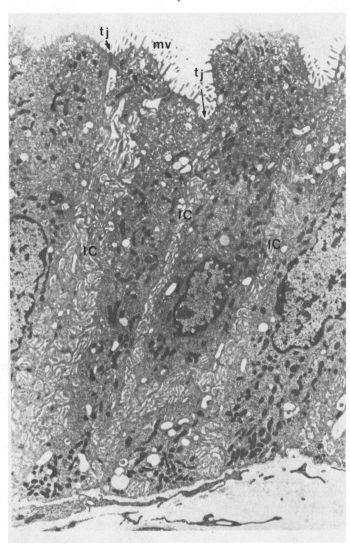

FIG. 26–1. Transmission electron micrograph of several gallbladder epithelial cells. The apical membrane has abundant microvilli (*mv*) and mucous secretory granules of varying size are prominent in the apical region of the cell. The elaborate interdigitation of lateral cell membranes forms the intercellula space (*IC*) sealed in the apical region by the tight junctions (*tj*) ($\times 8300$). (Courtesy of J. C. Mueller, A. L. Jones and J. A. Long.)

ceptors are present on gallbladder epithelial cells. VIP stimulates intracellular cAMP production and produces active secretion of fluid in the isolated guinea pig and intact cat gallbladder. In human beings with the WDHA (pancreatic cholera) syndrome, a condition characterized by high blood levels of VIP, the gallbladder is usually distended, and its contents have increased concentrations of Cl^- and HCO_3-, suggesting impaired absorption.

These observations support the notion that the degree of absorption may fluctuate and that under certain circumstances the mammalian gallbladder may even secrete fluid and electrolytes into the lumen. It is

possible that VIP-induced secretion may dilute gallbladder bile enough that it serves to facilitate gallbladder evacuation after a meal.

TRANSEPITHELIAL MOVEMENT OF NONELECTROLYTES. Substances other than electrolytes and water are minimally absorbed by the normal gallbladder mucosa. Wright and Diamond, in their extensive review of this subject, found that mucosal permeability to nonelectrolytes follows closely the lipid solubility of the molecules; lipid-soluble compounds diffuse with relative ease into the apical membrane, but only very small water-soluble molecules (e.g., urea, acetamide) are able to penetrate the cell. Conjugated bilirubin, which is water soluble, is not absorbed, but unconjugated bilirubin, which is lipid soluble, is easily absorbed by the gallbladder epithelium.

The normal gallbladder mucosa absorbs small amounts of bile acid anions by passive nonionic diffusion. Bile acids with a higher pKa, such as chenodeoxycholic acid, are more readily absorbed than are those with a lower pKa. Conjugation of bile acids with taurine or glycine lowers pKa and makes absorption negligible. Conversely, bacterial deconjugation of bile salts or a low gallbladder bile pH enhances bile salt absorption. There is evidence that deconjugated bile acids may be able to injure the mucosa, which would increase permeability of the gallbladder mucosa to organic solutes. Absorption of bile acids by the gallbladder does not seem to play a role in the pathogenesis of cholesterol gallstones, contrary to some older theories.

Cholesterol is minimally absorbed by the normal gallbladder mucosa, but cholesterol absorption becomes significant when the cystic duct is obstructed, especially when the relative concentration of bile salt and phospholipid decreases. Under these circumstances most of the absorbed cholesterol is found in the mitochondrial and microsomal fractions of the gallbladder epithelial cells. This has been thought to explain the development of cholesterolosis of the gallbladder. Inflammation of the gallbladder mucosa may result in secretion of cholesterol into the lumen.

SECRETION OF MUCUS BY THE GALLBLADDER EPITHELIUM. Mucus-secreting cells are rare in the human gallbladder mucosa, but mucus secretory granules have been identified by electron microscopy in the apical cytoplasm of the columnar cells (Fig. 26–1). These granules contain glycoproteins assembled in the Golgi apparatus, which upon reaching the apical membrane are excreted into the lumen by reversed pinocytosis.

In patients with cholelithiasis, mucus secretion by the columnar cells increases, the Golgi apparatus hypertrophies, and numerous goblet cells appear in the gallbladder mucosa (a process known as intestinal metaplasia). As a result, gallbladder bile from patients with cholelithiasis has increased amounts of mucus that may represent an early stage in cholesterol gallstone formation. Several findings support this hypothesis. Synthesis and secretion of glycoprotein by the gallbladder epithelium of mice fed a lithogenic diet increases before gallstones form. During diet-induced gallstone formation in dogs, large amounts of mucus accumulate in the epithelial crypts, and the number of secretory vesicles in the gallbladder epithelium increases. After exposure to a lithogenic diet, the prairie dog gallbladder secretes increased quantities of mucus, and cholesterol gallstone formation can be prevented in this model by administration of aspirin. The aspirin is thought to inhibit mucus secretion by blocking prostaglandin synthesis.

Therefore, experimental and clinical evidence suggests that exposure to lithogenic bile causes structural changes of the gallbladder mucosa associated with increased mucus production. Since mucoprotein mixed with pigment and cholesterol is found at the center of most gallstones, this material may serve as the nucleus for cholesterol precipitation during gallstone formation.

Motility of the biliary tract

Bile is produced continuously by the liver and then flows down the hepatic and common bile ducts. During fasting, a considerable amount of hepatic bile is diverted through the cystic duct into the gallbladder. After eating, the sphincter of Oddi relaxes and the gallbladder contracts, evacuating bile into the duodenum. Regulation of bile flow is governed by its rate of secretion and absorption, and the motor behavior of the sphincter of Oddi, gallbladder, and duodenum.

GALLBLADDER FILLING AND EMPTYING. The resting pressure in the common bile duct, about 10 mm. Hg, results from the combined effects of the rate of bile secretion in the liver and the resistance imposed by tone in the sphincter of Oddi (Table 26–2). Under basal conditions, cystic duct resistance and gallbladder pressure are low, so the gallbladder fills. In human beings studied after an overnight fast, about 70 per cent of the bile acid pool is sequestered in the gallbladder, and 55 to 70 per cent of bile being secreted by the liver is entering the gallbladder, the rest passing into the duodenum.

Normal resting pressure in the gallbladder is about 5 mm. Hg (Table 26–2). Distention of the gallbladder during fasting is influenced by the tone of its muscle (which is under active neural and hormonal control) and compliance of its wall (a passive component related to the elasticity of the tissues). Normally, muscle tone is the most important factor, but with scarring (as in chronic cholecystitis) compliance may be greatly impaired.

Pressure-volume curves during gallbladder filling and emptying illustrate the great compliance of this organ, a phenomenon that allows for large changes in volume with little change in pressure. Muscular tone in the gallbladder wall does not increase with increasing volume, for if it did, gallbladder filling would not be possible within the pressure range normally present in the common duct.

In normal subjects an intravenous injection of CCK (0.2 U./kg./min.) initiates gallbladder contraction within a few minutes. Within 15 minutes of stimulation about 50 per cent of gallbladder bile is evacuated, and the degree of emptying plateaus at 75 per cent by 30 minutes. This response is dose related.

In vivo, emptying of the gallbladder is thought to be the result of stimulation by endogenous CCK and perhaps motilin. During fasting, the gallbladder contracts periodically emptying 15 to 30 per cent of

TABLE 26–2
Normal Human Biliary Pressures

Anatomical Region	Pressure (cm./H_2O)
Gallbladder (unstimulated)	3–10
Gallbladder (stimulated)	25–30
Common bile duct	10–14
Sphincter of Oddi (passage pressure)	15–16

its contents. These periodic contractions, which are thought to represent the biliary counterpart of the interdigestive migrating motor complex, occur at the end of duodenal phase II of the complex. Bile, therefore, enters the duodenum immediately preceding the initiation of the duodenal activity front. Motilin is thought to be responsible for the motor activity of the gallbladder during fasting. Postprandially, pressure in the sphincter of Oddi decreases to about one-third of its resting value and pressure within the gallbladder doubles. Gallbladder contraction is smooth and continuous (i.e., there is no peristalsis), but phasic activity of the sphincter of Oddi produces intermittent passage of bile into the duodenum. The postprandial rate of gallbladder emptying varies in relation to the character (e.g., osmolality, caloric density, solid or liquid) of the meal. An average meal of solid food results in 30 per cent emptying of the gallbladder within 30 minutes and a maximum of 80 per cent emptying by 120 minutes. Sham feeding also induces emptying, which in the first 60 minutes closely matches the rate induced by a meal, but maximum emptying after sham feeding (67 per cent) falls short of the maximum in response to food.

Resistance imposed by the sphincter of Oddi is the key factor governing bile flow. During fasting, resting pressure within the sphincter lumen measures 10 to 15 mm. Hg, with phasic increases to 95 to 140 mm. Hg occurring at a frequency of 3 to 8/min. The phasic motor activity increases preceding duodenal phase III of the interdigestive motor complex. The phasic contractions are confined to the lower 2 to 3 cm. of the bile duct (i.e., the sphincter) and in normal subjects appear to affect first the hepatic end of the sphincter and progress toward the duodenum. Up to 25 per cent of contractions, however, occur simultaneously throughout the sphincter, and a few propagate from the distal to the proximal end. Motor activity of the sphincter is accompanied by corresponding rhythmic electrical activity. Thus, the sphincter exhibits a form of peristalsis, which explains why bile enters the duodenum in small spurts. Postprandially, the resting pressure within the sphincter and the phasic motor activity decrease, probably as a result of CCK release.

The cystic duct is the narrowest portion of the ductal system and in primates is convoluted by a system of prominent mucosal folds called the valve of Heister. These folds do not seem to constitute a valve as was once thought, but there is some debate as to whether the cystic duct possesses a sphincter-like function as a result of activity of the muscle in its wall. Resting pressure within the cystic duct has been measured at about 35 mm. Hg, substantially greater than simultaneously measured gallbladder or common duct pressure, and in one report resistance to flow through the cystic duct appeared to increase following the administration of CCK, epinephrine, or morphine. Other observers have concluded, however, that any resistance imposed by the cystic duct is more likely to result from passive mechanical factors than from muscular contraction. In animals fed a lithogenic diet, resistance to flow through the cystic duct was said to increase, which was thought possibly to contribute to cholesterol gallstone formation by impeding gallbladder emptying.

The wall of the common duct contains sparse muscle fibers in only 20 per cent of human beings, and any active role for the duct in biliary dynamics is unlikely. The duct never exhibits anything resembling peristaltic activity, and pressures within the duct remain steady following various stimuli that might be expected to affect a responsive organ. Although extrahepatic ducts behave as passive conduits, actin filaments are present within the hepatocytes and may endow the canaliculi with contractile properties and a weak ability to pump bile toward the larger system.

In the presence of sudden complete ductal obstruction, bile pressure rises to 25 to 35 mm. Hg, reflecting the activity of the bile secretory apparatus within the liver. With time, however, ductal pressure falls to normal. In the absence of a gallbladder, the pressure in the duct must reach 12 to 15 cm H_2O for bile to flow through the normal sphincter. Passage pressure may be increased in patients with choledocholithiasis as a result of mechanical obstruction to flow by the stones, but usually the pressure is normal. Common duct pressure is also normal in patients with cholelithiasis confined to the gallbladder.

NEURAL CONTROL OF BILIARY MOTILITY. Electrical stimulation of the vagus increases gallbladder tone. The effect of vagotomy is less clear; some investigators have found a decrease in resting gallbladder pressure, whereas others have not. Experiments have shown that parasympathetic stimulation contracts the gallbladder in dogs and cats, and sham feeding in man produces about three-quarters of the contractile response that follows a steak meal. Whether vagotomy decreases or increases choledochal sphincter tone is unclear.

Adrenergic control of gallbladder motility and the sphincter of Oddi is less important. Alpha-receptor stimulation produces weak contraction and β-receptor stimulation produces weak relaxation of gallbladder strips *in vitro*. However, there is evidence for inhibitory innervation of the gallbladder musculature mediated by nonadrenergic transmitters. Candidates for this inhibitory neurotransmitter include adenosine triphosphate (ATP) and VIP.

Whether vagotomy predisposes to gallstone formation by causing stasis of bile has been the subject of much debate and experimentation. Truncal vagotomy has been reported by numerous workers to result in a doubling of the resting gallbladder volume. The effect is not seen after selective or parietal cell vagotomy, procedures that spare the hepatic branches of the vagus. Theoretically, gallbladder dilatation following vagotomy could be a consequence of decreased gallbladder tone, increased resistance to flow through the sphincter of Oddi, or inadequate sphincteric relaxation in response to CCK. However, experiments involving HIDA cholescintigraphy have detected no differences in gallbladder filling and emptying between patients with truncal vagotomy and nonvagotomized controls. Since this technique appears to be more accurate than those used previously, these results create doubts that vagotomy has an appreciable effect on gallbladder dynamics. There is a variety of other, less well documented changes purported to follow vagotomy, such as thickening and chronic inflammation of the gallbladder wall, increased bile viscosity, and increased predisposition to acute cholecystitis.

HORMONAL CONTROL OF BILIARY MOTILITY. Peptones, amino acids, fat and fatty acids, meat, and other food in the duodenum produce gallbladder contraction. In 1928 CCK, the mediator of this effect, was extracted by Ivy and Oldberg from the upper gastrointestinal mucosa of the hog. CCK, a linear polypeptide normally produced within the mucosa of the proximal intestine, circulates in at least three different molecular forms: a 39-amino-acid polypeptide (CCK-39); a 33-amino-acid polypeptide (CCK-33) that consists of the carboxy-terminal 33 amino acids of CCK-39; and an octapeptide (CCK-8), also a carboxy-terminal fragment of CCK-39. On a molar basis, CCK-8 is 4 to 6 times more potent than CCK-33.

It is believed that CCK acts directly on the muscle of the gallbladder and the sphincter of Oddi, probably by inducing changes in cyclic nucleotides. Gallbladder muscle is relaxed by cAMP and agents that increase the intracellular content of cAMP. CCK activates intracellular phosphodiesterase, decreasing cAMP content in gallbladder muscle cells. CCK increases intracellular levels of cAMP in the sphincter of Oddi, and although CCK also stimulates the duodenal musculature to contract, the overall resistance to flow through the common bile duct is markedly decreased.

The effects of CCK on gallbladder smooth muscle *in vitro* are not blocked by anticholinergic drugs or α- or β-blocking agents. Behar found that parasympathetic blockade *in vivo* by hexamethonium and atropine or tetrodotoxin, however, decreases CCK-induced gallbladder contraction in the cat. He concluded that gallbladder muscle has two kinds of CCK receptors, one on muscle cells and one on intramural postganglionic cholinergic neurons. Loss of vagal innervation, thus, would partially decrease the gallbladder response to CCK. There also appears to be two kinds of CCK receptors on the musculature of the sphincter of Oddi, one inhibitory and one excitatory.

Motilin, a 22 amino acid polypeptide isolated by Brown in 1971, contracts the gallbladder when given in physiological doses, and its effectiveness is greatest during fasting, a time when motilin is normally secreted. Gallbladder contractions induced by motilin differ from those induced by CCK because they are not dose related and are transient even with continuous motilin infusion. During fasting, plasma levels of motilin follow a cyclic pattern, reaching a peak during phase II of the interdigestive motor complex. Thus, motilin may have a role in maintaining gallbladder tone, producing partial gallbladder emptying and ensuring the continued enterohepatic circulation of bile acids between meals. Since vagal stimulation releases motilin, motilin may be responsible for some of the actions currently thought to be due to direct vagal action on the biliary tract.

Gastrin, whose C-terminal tetrapeptide is identical to that of CCK, has some cholecystokinetic effect, but its potency is low, and in physiological levels gastrin is not likely to influence biliary kinetics. Secretin does not alter gallbladder pressures at doses that stimulate pancreatic secretion, but when administered with CCK, it enhances the cholecystokinetic effect.

Histamine increases pressure in the gallbladder and decreases resistance of the sphincter by stimulating H^1 and H^2 receptors on the smooth muscle cells. In contrast, VIP and glucagon relax the gallbladder muscle. The presence of VIP nerves in the gallbladder wall suggests that VIP may have a physiological role in gallbladder motility or in the secretory behavior of the epithelium. Prostaglandins contract the gallbladder muscle and can increase intraluminal pressure.

During pregnancy gallbladder emptying is sluggish. After the first trimester, the resting gallbladder volume doubles, and in response to stimulation by a liquid meal, its rate of emptying is decreased by 30 per cent and the residual volume is twice normal. Gallbladder motility is not affected by contraceptive steroids.

Biliary dyskinesia may be reflected by abnormal gallbladder emptying and may produce biliary pain. Broden, in 1958, first described the use of CCK to test extrahepatic motor function. Because it stimulates contraction of the gallbladder, CCK given during the course of an oral cholecystogram may be used to relate biliary motor activity to the patient's subjective complaints. Griffen observed that patients whose gallbladders failed to empty normally and who experienced right upper quadrant pain following CCK stimulation of the gallbladder usually had histological evidence of chronic cholecystitis and were relieved of their symptoms by cholecystectomy. Other studies have failed to substantiate these observations. The advent of cholescintigraphy has intensified interest in the use of CCK as a clinical test of motor function of the gallbladder in patients with abdominal pain but no gallstones.

Pathogenesis of acute cholecystitis

The gallbladder of patients with acute cholecystitis is distended and filled with colorless bile. Its wall is thickened by submucosal and subserosal edema and is hyperemic, and in a later stage the fundus may become necrotic and may perforate.

The pathological change most characteristic of acute cholecystitis is edema. Because edema tends to disappear with fixation, the gross specimen is usually more strikingly abnormal than are histological sections. Mucosal ulcerations are common at the site of impaction of the stone, but the epithelial lining otherwise may look remarkably intact. The submucosa is infiltrated with polymorphonuclear leukocytes, and the venules and lymphatics are dilated. Occasionally small vessels are thrombosed and fibrin deposits are found on the serosal surface. In addition to acute inflammation, the gallbladder wall in 95 per cent of patients with acute cholecystitis has evidence of chronic inflammation: flattened columnar epithelium, metaplasia of the mucosal lining, an increased number of goblet cells, and lymphocytic infiltration of the submucosal layer.

CYSTIC DUCT OBSTRUCTION. Acute cholecystitis is associated with cholelithiasis in the majority of patients, and cystic duct obstruction is almost uniformly present. In approximately two-thirds of cases the cause is a gallstone firmly impacted in Hartmann's pouch, producing external compression of the cystic duct. In a few patients, a small stone is found inside the cystic duct obturating the lumen. Because obstruction is the first step in pathogenesis, acute cholecystitis is in some respects comparable to acute appendicitis or acute diverticulitis. Stasis produces inspissation of bile, a crucial step in the pathogenesis of the less common acute acalculous cholecystitis. Obstruction of the cystic duct changes the absorptive capacity of the gallbladder mucosa. Twelve hours after ligation of the cystic duct the gallbladder ceases to absorb fluid (rabbit), or produces a net secretion (dog) into the lumen. Three days after cystic duct ligation calcium bilirubinate particles (sludge) form in the gallbladder of the dog. Since stasis may also play a role in the transformation of lecithin into lysolecithin (see below), it could precipitate a cascade of events leading to acute cholecystitis.

Obstruction of the cystic duct in dogs, however, most often produces hydrops or chronic inflammation instead of acute cholecystitis. The inability to reproduce acute cholecystitis consistently by obstruction of the cystic duct in animals is not surprising because hydrops, chronic inflammation, and acute cholecystitis are also among the spectrum of results that follows cystic duct obstruction in man. Thus the outcome of obstruction must depend on other factors.

MEDIATORS OF THE INFLAMMATORY RESPONSE. Flexner first called attention to the inflammatory potential of bile salts, and Womack showed that the severity of gallbladder inflammation that followed cystic duct ligation in dogs was related directly to the concentration of bile in the gallbladder lumen. Animals that fasted for 48 hours before the duct was ligated consistently developed acute cholecystitis. If gallbladder bile was replaced by sterile saline, however, ligation of the cystic duct produced little inflammation.

Injection of a highly concentrated solution of bile acids into dog gallbladder may cause acute cholecystitis, but the concentrations required are very high, beyond any physiologically attainable level. Studying the rabbit gallbladder *in vitro*, Heuman found that permeability to passage of dextran through the gallbladder wall was markedly increased by exposure to taurodeoxycholate and to a lesser degree to taurocholate, effects that could be inhibited by the addition of lecithin. In the dog gallbladder, taurodeoxycholate can completely abolish transepithelial water movement, an effect also eliminated by lecithin addition.

Numerous studies have confirmed the damaging potential of bile acids to other human epithelia. At low concentrations these molecules alter transport mechanisms in the cell membrane, and at higher concentrations they produce membrane lysis and release of intracellar enzymes, which may further damage adjacent cells. Lecithin may protect against these effects, probably by tying up the bile salts in mixed micelles. Since bile from patients with gallstones contains more deoxycholate and less lecithin than that from control subjects, these findings could be relevant to the pathogenesis of cholecystitis.

In prairie dogs, administration of a cholesterol-supplemented diet causes the bile to become lithogenic. Gallbladders exposed to lithogenic bile become inflamed, and if the cystic duct is ligated, acute cholecystitis develops in this model even before gallstones form. These findings may be relevant to the pathogenesis of the occasional case of acute cholecystitis in the absence of gallstones (acalculous cholecystitis) in man. This condition is usually observed in patients who have been fasting for a prolonged period (e.g., patients with pancreatitis, bowel fistula, prolonged hyperalimentation) in whom the bile is usually supersaturated with cholesterol. Theoretically, cholesterol crystals, which have been shown to produce inflammation in other tissues, could be the injurious agent.

Lysolecithin may induce structural changes in biological membranes, and after ligation of the cystic duct injection of lysolecithin into the gallbladder lumen causes acute cholecystitis. Lysolecithin originates from hydrolysis of lecithin and is present in small amounts in normal bile. In patients with acute cholecystitis the ratio of lysolecithin to lecithin in bile is increased, and it has been postulated that lysolecithin is a key factor in the pathogenesis of acute cholecystitis. After damaging the gallbladder epithelium the lysolecithin is thought to be partially absorbed by the damaged mucosa. High concentrations of lecithin protect against the effects of lysolecithin.

Conversion of lecithin to lysolecithin is catalyzed by phospholipase A, an enzyme present in lysosomes of human gallbladder mucosal cells but normally absent from gallbladder bile. However, mechanical trauma from an impacted stone could theoretically release phospholipase A from the mucosal cells. Incubation of bile in the presence of extracts from gallbladder epithelium increases the lysolecithin:lecithin ratio, suggesting that bile stasis may also release phospholipase A. Lysolecithin activity in bile is also increased by exposure of the mucosa to β-glucuronidase, and increased β-glucuronidase activity in the bile from *Escherichia coli* or other sources could augment lysolecithin release.

Prostaglandins are another candidate for a role in the pathogenesis of cholecystitis: they mediate inflammation in a number of human tissues; they are present in a high concentration in the wall of the inflamed gallbladder; and prostaglandin E_1 can convert absorption in the gallbladder to secretion, a finding that could explain gallbladder distention seen in acute cholecystitis. Indomethacin, a prostaglandin synthetase inhibitor, reverses mucosal secretion in experimental acute cholecystitis and relieves the pain of acute cholecystitis in humans, presumably by ameliorating gallbladder distention. The role of gastric and pancreatic secretion, drugs, and other factors in the pathogenesis of cholecystitis remains unclear.

Bacteria may be cultured from the gallbladder in about 20 per cent of patients with cholelithiasis (and no acute inflammation) and in 50 per cent of patients with acute cholecystitis. Injection of *E. coli*, clostridia, and other bacterial species into the gallbladder lumen, however, fails to produce acute cholecystitis when the cystic duct is patent, although when the duct is ligated, suppurative cholecystitis results. The weight of evidence therefore favors a secondary role for bacteria in human cholecystitis. Nevertheless, the suppurative complications, such as empyema, perforation, pneumocholecystitis, and septicemia cause much of the morbidity of this disease.

VASCULAR FACTORS. The early edema and late necrosis of the gallbladder wall in acute cholecystitis suggest that vascular factors may be involved in the pathogenesis. Edema is thought to stem from obstruction of venous and lymphatic outflow owing to inflammation of the cystic duct pedicle caused by the impacted stone.

As acute cholecystitis worsens, the fundus may become necrotic, a complication rarely seen before the third day of the illness. The gallbladder wall shows gross and microscopic changes of ischemic necrosis, and it has been suggested that inflammation in Hartmann's pouch may produce occlusion of the cystic artery. Attempts at producing necrotizing cholecystitis by cystic artery ligation in experimental animals, however, have not been successful, probably because there are so many arterial anastomoses between the terminal branches of the cystic artery and the liver capsule. On the other hand, intraarterial injection of thrombin induces necrosis of the gallbladder wall, presumably by thrombosing these collateral vessels. Arteriography during acute cholecystitis in humans shows the cystic artery to be patent and the gallbladder to be hypervascular. It does not appear, therefore, that cystic artery or vein occlusion contributes to the early events in acute cholecystitis in man.

There is another way in which ischemia might be involved. With increasing gallbladder distention owing to secretion of fluid into the lumen, pressure in the gallbladder rises to as high as 25 to 55 cm. H_2O, and blood flow drops progressively with increasing intraluminal pressures (e.g., at 60 mm. Hg, mucosal flow is only 25 per cent of baseline levels). The decrease in flow is disproportionately greater in the mucosa. The predilection for perforation to occur in the fundus may be due to its large diameter and the greater tension that would result in this area according to LaPlace's law. In fact, injection of arterial vessels of acutely inflamed human gallbladders have shown small vessel thrombosis adjacent to areas of necrosis in the fundus.

In patients with cholelithiasis and chronic cholecystitis the gallbladder wall contains extensive arterial changes characterized by tortuosity, occlusion of small vessels, and loss of the normal anastomotic net. Since these changes do not correlate with age of the patient or with the presence of coronary thrombosis, hypertension, or atherosclerosis, it is possible that chronic cholecystitis might predispose the gallbladder to acute ischemia. Diseases with preexisting lesions of the small blood vessels (e.g., diabetes mellitus) also predispose to perforation.

In summary, the pathophysiology of acute cholecystitis in humans has not been deciphered entirely, and the etiology is probably multifactorial. Empirical evidence suggests that sudden occlusion of the cystic duct in the presence of saturated bile in the gallbladder can alter absorptive function of the mucosa, activate inflammatory mediators, and damage the mucosa. The acutely inflamed gallbladder then becomes susceptible to secondary infection. Finally, the blood supply may be compromised by high intraluminal pressures, and necrosis or perforation may ensue.

The bile ducts

Bile flow

The liver produces 500 to 1500 ml. of bile daily. Each hepatocyte functions as an individual secretory unit. Solutes move from the sinusoids into the hepatocyte through the lateral cell membrane, and are then transported across the cell and excreted into the biliary canaliculus through the canalicular membrane, which is a specialized part of the hepatocyte membrane. The bile canaliculus has a diameter of 1 μm. at the level of the hepatocyte and is sealed from the rest of the intercellular space by a junctional complex (Fig. 26–2). Tight junctions are located in the midportion of the hepatocyte instead of being at the apical

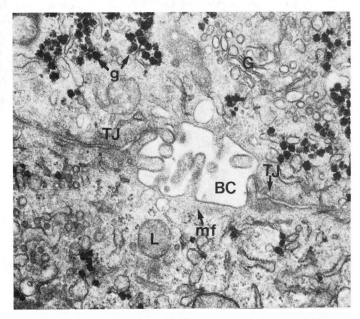

FIG. 26–2. Electron micrograph of a normal human liver. The bile canaliculus (*BC*) is sealed from the intercellular space by the tight junctions (*TJ*). Abundant microfilament (mf) surrounds the canalicular membrane. Golgi apparatus (*G*), lisosomes (*L*) and glycogen granules (*g*) are seen in the adjacent cytoplasm. (Courtesy of A. L. Jones and I. S. Goldman.)

FIG. 26–3. Schematic representation of the filamentous network of a normal hepatocyte tight junction. (After Desmet.)

pole, as in most secretory epithelia. They are formed by particles of two adjacent cell membranes aligned in parallel strands of fibrils with frequent interconnections (Fig. 26–3). The number of fibrils and their geometric pattern determine the tightness of the junction. The anatomy of the tight junction is quite variable, suggesting that it consists of points of higher and lower permeability, and it has been suggested that the conductance of the tight junctions may change under the influence of various factors (e.g., concentration of cAMP).

Hepatocytes are similar in structure and are grouped by the distribution of blood vessels into acinar lobules. The acinar lobule, the functional microcirculatory unit of the bile secretory apparatus, is supplied by an arteriolar-portal pedicle. These pedicles reach the periphery of the lobule and give origin to the sinusoidal capillary net. Blood flows through the sinusoids toward the center of the unit, where the hepatic venule originates. As a consequence of this anatomical arrangement, the more peripheral hepatocytes of the unit are exposed to the highest concentration of solutes, and the concentration decreases toward the center of the lobule as solutes are cleared by the more peripheral hepatocytes. Bile flows from the center to the periphery, opposite to the direction of blood flow. Although it has been technically impossible to obtain samples of canalicular bile for analysis, this arrangement appears to ensure that more concentrated bile does not contact hepatocytes with lesser solute concentrations, and thus it prevents back-diffusion of secreted solutes into the liver cells.

CANALICULAR BILE FLOW. Bile acids enter the hepatocyte in cotransport with Na^+ on a carrier energized by Na-K ATPase (Fig. 26–4). Histochemical studies have shown that in the hepatocyte, Na-K ATPase is largely confined to the basolateral membrane (i.e., sinusoidal and intercellular surfaces).

Within the hepatocyte cytoplasm, bile acids are probably bound to protein carriers (e.g., ligandin) and are then transported to the apical membrane in Golgi-derived vesicles. Delivery into the canaliculus may be by exocytosis from the cytoplasmic vesicles, but the evidence is inconclusive. Transport of bile acids across the canalicular membrane increases osmolality in the canaliculus, and water flows passively from the sinusoids through the tight junctions and intercellular spaces (rather than

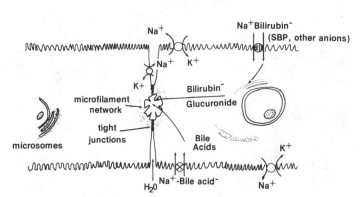

FIG. 26–4. Schematic representation of bile acids, bilirubin and water transport and the ion pumps in the hepatocytes and biliary canaliculus.

through the hepatocyte) into the canaliculus (Fig. 26–4). The notion that water flows through a paracellular pathway is supported by the observation that plasma concentrations of relatively large molecules (e.g., insulin, sucrose, polyethylene glycol 4,000) equilibrate with bile before they reach appreciable concentrations in the hepatocytes.

Bile acid secretion is the major determinant of the volume of flow, and the two are related curvilinearly. At moderate to high rates of bile acid secretion, bile flow correlates closely with bile acid output, but at lower rates the slope of the line relating bile acid and water secretion becomes steeper, probably because the osmolality of canalicular bile is greater when the amount of bile acid falls short of the critical micellar concentration. As bile acids aggregate into micelles, they lose osmotic potential. Indeed, bile acids that do not form micelles are more potent choleretics. Whether the aggregation into micelles takes place inside the cell or in the canaliculus is not known. The fraction of bile flow that is the direct result of bile acid secretion is known as the *bile acid–dependent flow.*

Other organic anions, such as bilirubin, sulfobromophthalein, indocyanine green, etc., share a common transport system different from that of bile acids. These other organic anions have a minimal influence on water transport. Although sulfobromophthalein and iodipamide may induce a choleresis, bilirubin does not. On the other hand, excretion of these organic anions increases as bile acid–dependent bile flow increases.

As bile acid output decreases (e.g., with depletion of the bile salt pool), bile flow decreases, but even at the lowest levels of bile acid output a residual bile flow remains. These findings indicate that mechanisms other than bile acid secretion are involved in bile production. This fraction of bile flow is known as *bile acid–independent flow.* Further proof of a bile acid–independent fraction is provided by the action of

drugs (e.g., phenobarbital, theophylline, glucagon) that increase canalicular bile flow without affecting bile acid secretion. Because at low bile acid output the solids in bile consist for the most part of inorganic sodium salts, sodium transport is believed to play a major role in governing bile acid–independent flow. Although evidence is sparse and indirect, it is thought that the bile acid–independent flow probably results from the action of Na-K ATPase located in the basolateral membrane. Because the two forms of canalicular bile production are partly linked, the term *bile acid–independent flow* is now recognized as being imprecise. For example, increased bile acid output results in increases in bile acid–independent flow as well as bile acid–dependent flow.

DUCTULAR BILE FLOW. About 30 per cent of bile flow is the result of ductular secretion (10 per cent from extrahepatic ducts and 20 per cent from intrahepatic ducts) and is generated by active Na^+ and HCO_3^- transport by the ductal cells. As in the gallbladder, the driving force for ductular secretion is thought to consist of a sodium pump that establishes an osmotic gradient, which in turn results in water transport. Secretin increases the ductular secretion and raises the pH of bile. It has no effect on canalicular bile flow as measured by erythritol clearance or the transport maxima for BSP.

Under certain circumstances the ductal system may also absorb fluid. In the Rhesus monkey, for example, the estimated canalicular bile flow exceeds the actual bile flow, and in the dog the common bile duct dilates following cholecystectomy and is capable of concentrating bile by absorbing water, Na^+, and Cl^-. The role of absorption in ductal physiology is limited, however, and whatever its extent, net secretion is usually the end result.

THE ENTEROHEPATIC CIRCULATION OF BILE SALTS. The enterohepatic circulation refers to the recirculation of substances, such as bile acids, between the liver, bile, and gut. In the gut they are reabsorbed and transported in portal vein blood to the liver, where they are extracted by the hepatocytes and secreted once again into bile.

In humans there are two primary and two secondary bile acids. The primary bile acids are formed in the liver from cholesterol, principally from cholesterol newly synthesized in hepatocytes. Cholesterol has a stable pool, additions to which may come from absorbed and synthesized cholesterol. Cholesterol catabolism follows two pathways: conversion to bile acids and secretion into bile.

The primary bile acids are *cholic* (3α, 7α, 12α, trihydroxy 5β cholanic) acid and *chenodeoxycholic* (3α, 7α, dihydroxy 5β cholanic) acid. They are synthesized in the liver by a process beginning with hydroxylation of cholesterol by microsomal enzymes followed by conversion of the resulting dihydroxy cholesterols to dihydroxy and trihydroxy coprostanes. Mito-

FIG. 26–5. Molecular models for the lecithin-rich type of mixed micelles (*top*) and for the bile-salt-rich type (*bottom*). The bile salts are thought to form pairs to avoid contact of the hydroxyl groups (solid circles) with the apolar environment. (Muller: Biochemistry 20:404–414, 1981.)

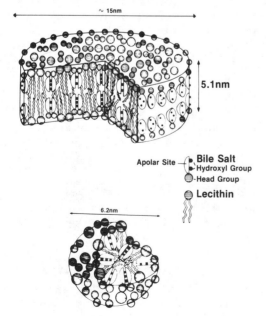

~ 15nm

5.1nm

Apolar Site — Bile Salt
Hydroxyl Group
Head Group

Lecithin

6.2nm

chondrial enzymes degrade the side chain of coprostanes to form the bile acids. The rate-limiting enzyme in this process is cholesterol 7-α-hydroxylase. Primary bile acids are converted to secondary bile acids in the intestine, mainly by bacterial dehydroxylation: cholic acid is converted to *deoxycholic* (3-α-, 12-α-, dihydroxy 5-β-cholanic) acid, and chenodeoxycholic acid is converted to *lithocholic* (3-α-hydroxy 5-β-cholanic) acid. Small amounts of the tertiary bile acid, ursodeoxycholic acid, formed from chenodeoxycholic acid by way of an intermediate, 7-ketolithocholic acid, are also present in humans. The normal bile acid pool in man consists of 40 per cent cholic, 40 per cent chenodeoxycholic, 15 per cent deoxycholic, and small amounts of lithocholic and ursodeoxycholic acids.

All bile acids become conjugated with either glycine or taurine (in a 3:1 ratio in man) before being secreted into the biliary canaliculus. Bile acids in aqueous solution spontaneously aggregate to form micelles (Fig. 26–5). The shape of each micelle and the number of molecules of biliary lipids in each micelle is determined in part by the lecithin-to-bile salt ratio. At physiological ratios the "mixed disc" micelle (Fig. 26–5) and simple micelles coexist in variable proportions. In the micelles, bile acid molecules are arranged with the hydrophylic group toward the outside (the water) and the lipophilic group toward the inside. Biliary cholesterol, a lipid, is carried inside the micelles. Lecithin enhances the capacity of bile acid micelles to solubilize cholesterol. In the intestinal lumen, bile acid micelles release the cholesterol, which precipitates or is reabsorbed; the micelle then participates in digestion by solubilizing fatty acids, monoglycerides, and other lipids.

Primary and secondary bile acids are absorbed in the intestine, extracted from the portal circulation by the hepatocyte, conjugated, and resecreted along with newly synthetized bile acids. Lithocholic acid is sulfated, becomes insoluble, and is excreted in the feces. Lithocholic acid constitutes a small fraction of the bile acids in bile because it undergoes almost no enterohepatic circulation. Owing to efficient clearing by the hepatocytes, bile acids absorbed by the intestine appear in the portal blood in concentrations 5 to 10 times greater than in the systemic circulation.

The rate at which bile acids are returned to the liver governs the rate of bile acid synthesis by the hepatocyte; total diversion of bile salts (e.g., by a biliary fistula) increases bile acid synthesis about tenfold. With a fractional diversion of 5 to 20 per cent of bile flow, synthesis of new bile acid in the liver matches the amount being lost, and the total size of the bile acid pool remains the same. If diversion exceeds 20 per cent of bile flow, synthesis can no longer compensate for the loss, and a new steady state occurs with a lower pool size (Fig. 26–6). It is thought that synthesis is regulated by the intracellular concentration of bile acids in hepatocytes, which determines the concentration of cholesterol 7-α-dehydroxylase, the rate-limiting enzyme for conversion of cholesterol to bile acids.

The total mass of bile acid within the enterohepatic circulation constitutes the bile acid pool. Pool size multiplied by the rate at which the pool circulates (cycles/day) gives the rate of hepatic bile acid secretion. The total pool, 3 to 5 g., circulates 6 to 8 times a day, so total daily secretion amounts to 18 to 40 g. The cycling frequency of chenodeoxycholate and deoxycholate is about 1.5 times greater than that of cholic acid, probably because chenodeoxycholate and deoxycholate are absorbed passively in greater amounts in the proximal gut. The fractional daily turnover rate for cholic acid (0.3) is slightly greater than that for chenodeoxycholic acid (0.2). Pool size times fractional turnover rate equals the daily synthesis rate, which is about 0.5 mmol. for chenodeoxycholic acid and 1.0 mmol. for cholic acid. The lower turnover rate for chenodeoxycholic acid is a consequence of more efficient intestinal conservation of this molecule, but the rela-

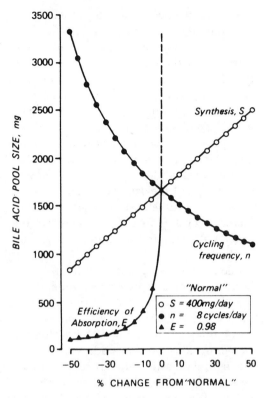

FIG. 26–6. Theoretical effects of change in bile acid synthesis, absorption efficiency and cycling frequency on bile acid pool size (P) calculated according to the equation $P = S/n(E)$. (Mok: Gastroenterology 78:1023–1033, 1980.)

tive amounts of the two primary bile acids in the bile acid pool remain because the liver synthesizes more cholic acid.

During fasting, about two-thirds of the bile acid pool is contained in the gallbladder. Intermittent partial gallbladder emptying maintains an enterohepatic circulation of bile salts during this period. Following a meal, the gallbladder delivers bile with a high concentration of bile acids to the intestine, which then contains most of the bile acid pool. In animals without a gallbladder (and in cholecystectomized humans) the bile acid pool is contained almost entirely in the intestine. Perhaps surprisingly, cholecystectomy does not decrease the rate of entry of bile acids into the intestine in response to eating, nor does it affect the efficiency of fat digestion and absorption.

Because the synthetic rate of bile acids is determined by negative feedback from the returning bile acids, impaired gallbladder emptying, which sequesters a portion of the bile acid pool, stimulates increased synthesis and expansion of the total pool. Increased amounts in the gut (and perhaps in the gallbladder) explains why the bile acid pool increases after vagotomy or the administration of atropine. Conversely, increased gallbladder contraction (e.g., with more frequent meals), cholecystectomy, or a decreased threshold to CCK stimulation increases cycling frequency, decreases bile acid synthesis, and results in a smaller pool.

Bile acid reabsorption from the intestine occurs principally by active transport in the distal 100 cm. of ileum. The location of the active transport mechanism in the lower intestine ensures that intraluminal concentrations of bile acids will be adequate until digestion and absorption of dietary lipids are complete. The carrier requires the presence of Na^+, an ion cotransported with bile acid, and thus is quite similar to the mechanism responsible for absorption in the gallbladder. It transports deconjugated as well as conjugated bile acid molecules. As noted previously, some passive absorption also occurs throughout the intestine, mainly of unconjugated dihydroxy bile acids (i.e., chenodeoxycholate and deoxycholate).

Uptake of bile acids by the ileum is highly efficient, for only 5 per cent of the load passing down the intestine reaches the colon. Following absorption, the bile acids are transported to the liver bound to albumin in portal vein blood, where another highly efficient mechanism in hepatocytes clears about 80 per cent of the entering bile acid molecules in a single passage. Hepatic uptake of cholic acid is more efficient than that of chenodeoxycholic and deoxycholic acids (90 per cent of cholic acid is cleared from the plasma in a single passage). Fasting portal vein levels of bile acids (about μmol/L) increase about

threefold 15–60 minutes after a meal. The systemic blood levels of bile acids (about 2.5 μmol/L) double after a meal, but the total fraction of the pool in the circulation never exceeds about 1 per cent.

In the intestine about one-fourth of the bile acids are deconjugated by bacteria from their amino acid moiety, which is eventually replaced as the steroid portion of the molecule traverses the liver. Intestinal bacteria also convert daily about 30 per cent of the pool of primary bile acids to secondary bile acids. Conditions that increase the concentrations of bacteria within the small intestine (e.g., surgical removal of the ileocecal valve), increase the proportion of deoxycholic acid relative to cholic and chenodeoxycholic acid in the bile acid pool.

Biliary tract obstruction
MORPHOLOGICAL CHANGES SECONDARY TO BILIARY OBSTRUCTION. Obstruction of the bile duct produces structural changes in the canalicular system and the liver parenchyma, the extent and severity of which depend on the degree, duration, site of obstruction, and the species being studied. Connective tissue proliferation with permanent hepatic functional damage is the end result of sustained and prolonged biliary tract obstruction.

Increased biliary pressure produces dilatation of the common bile duct up to two and one-half times the normal diameter. When high intrabiliary pressure is sustained for a prolonged period, the final extent of dilatation is related to the cause of the obstruction. In patients with cholangitis (e.g., patients with biliary stricture or choledocholithiasis), dilatation is rarely as marked as in those without cholangitis (e.g., patients with neoplastic obstruction). The latter patients also have more complete obstruction, however.

If the point of obstruction is below the junction of cystic duct and common duct, the gallbladder may dilate enough that it can be felt on abdominal examination.

Courvoisier's law states that a palpable, nontender gallbladder in a jaundiced patient indicates that the obstruction is from a neoplasm rather than from a gallstone or some other benign lesion. This is usually true because the obstruction in neoplastic disease is more complete and continuous, and the gallbladder is usually unscarred. With gallstone disease, common duct obstruction is less complete and intermittent, and gallbladder dilatation is inhibited by the effects of chronic inflammation in its wall.

A number of investigators have studied the relationship between the structural changes in the bile ducts and liver and the duration and degree of obstruction. In rats, total obstruction produces marked dilatation of the common bile duct, with the appearance of focal necrosis around small ducts within 48 hours. Early signs of bile duct proliferation, such as syncytial buds from the bile ducts, appear after 48 to 72 hours of total obstruction. If obstruction persists and the animal survives, the dilatation of the extrahepatic biliary tree becomes saccular, the hepatic parenchyma is compressed, and cirrhosis develops in 5 to 6 weeks. Extrapolation of these changes to man is difficult since the extent of dilatation of the common bile duct is never as great as in the rat.

In most animals, including man, long-standing obstruction leads to changes in hepatic microvascular anatomy. This may decrease blood flow through the sinusoids and explain the decreased clearing of organic anions.

Both intrahepatic and extrahepatic cholestasis produce canalicular bile plugging. In extrahepatic biliary obstruction in humans the most specific histological feature is plugging of interlobular biliary canaliculi. Centrilobular bile stasis and dilatation of biliary radicals result in periductular extravasation of bile, which causes reactive edema and infiltration of polymorphonuclear leukocytes. With progressive dilatation of the canalicular system the hepatic cells become flattened, and degenerative changes may occur. The inflammatory reaction appears to stimulate proliferation of bile ducts, and may lead to periportal and intralobular fibrosis, which give the picture of biliary cirrhosis. Infection may lead to pericholangitis and occasionally to hepatic abscess formation.

Ultrastructural studies of the liver parenchyma in patients with obstructive jaundice show subcellular changes in the canalicular membrane and the adjacent hepatocyte cytoplasm. In the early stage, the microfilaments located within the pericanalicular cytoplasm increase, while intracellular membranes (i.e., endoplasmic reticulum and Golgi complex) and the volume density of mitochondria and lysosomes decrease. After a longer period (i.e., 3 weeks) the bile canaliculi become very dilated and the microvilli of the canalicular membrane flatten and become fewer (Fig. 26–7). Complete bile duct obstruction leads to loosening of the tight junction network, increasing the depth of the fibril texture and producing marked irregularities of its pattern (Fig. 26–8). The increased number of discontinuities observed in the fibrils and the higher number of free-ending loops suggest that the barrier function of the tight junctions decreases during biliary obstruction.

FUNCTIONAL CHANGES SECONDARY TO BILIARY OBSTRUCTION. The normal pressure in the common duct, 10 to 15 cm. H_2O, results from the resistance of the sphincter of Oddi and the rate of bile flow produced by the bile secretory apparatus. Obstruction results in increased common bile duct pressure proportional to the degree and duration of obstruction. Complete obstruction gives a pressure of 25 to 35 cm. H_2O. If the gallbladder is present, the rate at which common duct pressure climbs is slowed owing to absorption of water in the gallbladder.

When canine common duct pressure exceeds 15 cm. H_2O, bile flow decreases. At

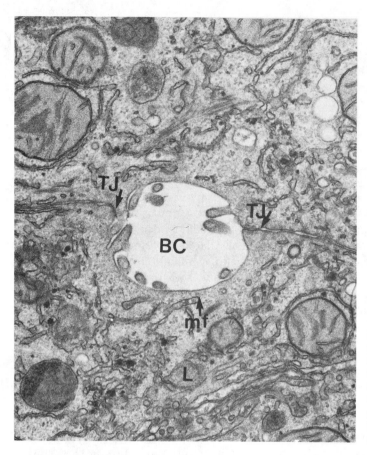

FIG. 26—7. Electron micrograph of a rat liver 3 weeks after common bile duct ligation. The bile canaliculus (*BC*) is very dilated, microvilli are flattened and there is sparse microfilament (*mf*) network. (*TJ*) Tight junction; (*L*) lisozome. (Courtesy of A. L. Jones and I. S. Goldman.)

bile pressures in excess of 30 cm. H$_2$O, net bile secretion stops. Experiments in monkeys show that at pressures up to 14 cm. of H$_2$O there is no change in bile flow. Pressures between 17 and 22 cm. of H$_2$O decrease bile flow and bile acid secretion to approximately half normal, suggesting that

FIG. 26—8. Schematic representation of the tight junction strands in cholestatic liver. (After Desmet.)

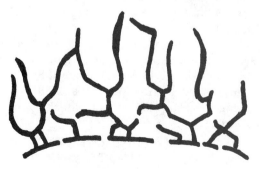

the effect is principally on the bile salt–dependent fraction of flow. Partial obstruction of the common bile duct in the dog results in inspissation of bile and occasionally in pigment stone formation.

When the ducts are obstructed, bile continues to be secreted into the canaliculus. Increased hydrostatic pressure in the biliary canaliculi distorts the junctional complexes (Fig. 26–8), increasing its permeability. Therefore, the liver has two main pathways for reflux of bile during cholestasis: transcellular reflux back into the portal sinusoids and transjunctional reflux into the paracellular space and thence to the lymphatics and portal sinusoids.

About 10 per cent of canalicular bile enters the systemic circulation by lymphatics. After complete common bile duct obstruction in dogs, bile acid and bilirubin concentrations in the hepatic lymphatics increase for 6 hours, plateau for a short period, and then drop slightly to reach a new plateau at a level higher than normal. The eventual decrease in lymphatic bilirubin concentration is due in part to a decreased capacity of the hepatocyte to secrete organic anions.

The maximal biliary excretion of bilirubin (bilirubin $T_m = 0.65 \pm 0.24$ mg./min. in normal subjects) decreases during biliary tract obstruction. Excretion of bromosulphthalein (i.e., anion transport) and the capacity of the liver to store anionic compounds also decrease. Bilirubin metabolism by the hepatocyte may be increased by the administration of phenobarbital and spironolactone, presumably owing to their ability to induce microsomal enzymes. Pretreatment with these drugs does not decrease serum bilirubin levels in rats with bile duct obstruction.

Decreased anion excretion is aggravated in late obstruction by decreased uptake of bilirubin by liver cells and by alterations in the conjugate pattern of bilirubin in bile. This suggests that, in addition to canalicular transport, uptake and conjugation of bilirubin are altered by obstruction.

Whether the reduced bile salt secretion that follows rises in intraductal pressure results from decreased secretion by the hepatocyte (due to impaired canalicular membrane transport) or from reabsorption of bile salts from canalicular or bile is not clear. Bile salt concentrations in the hepatocyte and in the systemic circulation increase, thereby inhibiting synthesis of bile salt and decreases the size of the bile salt pool. Phospholipid and cholesterol secretion also decline. When biliary tract pressure is restored to normal, the cholesterol secretion returns to normal faster than does phospholipid or bile acid secretion, and the lithogenic index transiently increases.

Complete common bile duct obstruction increases serum bilirubin, alkaline phosphatase, 5'-nucleotidase, and leucine aminopeptidase levels. Serum glutamic oxalacetic transaminase (SGOT) and serum glutamic pyruvic transaminase (SGPT) levels also rise, but these changes are less specific for obstruction. The increased serum bilirubin and alkaline phosphatase levels may be seen as early as 1 hour after total common bile duct occlusion. In complete obstruction in man, the serum bilirubin and alkaline phosphatase concentrations rise for days to weeks and then plateau. The average ultimate height of bilirubin concentration with complete obstruction ranges from 20 to 30 mg./dl. At this point, the daily bilirubin load is excreted by the kidneys. Higher bilirubin values suggest concomitant hemolysis or renal dysfunction.

In general, neoplastic obstruction produces the highest bilirubin values, and those caused by benign (e.g., posttraumatic) strictures and choledocholithiasis are substantially lower (Fig. 26–9). The cause of the jaundice is often obvious in

FIG. 26–9. Serum bilirubin values in 178 patients with biliary obstruction. CBD = common bile duct.

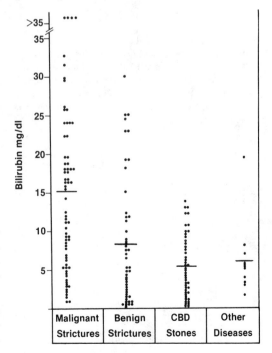

patients with a history of bile duct stricture due to a ductal injury. Most of the remaining patients with obstructive jaundice have either neoplastic obstruction or gallstone obstruction. In general, it is reasonable to assume that the diagnosis is neoplasm if the bilirubin is greater than about 13 mg./dl., and it is choledocholithiasis if the bilirubin value is below 10 mg./dl. and the syndrome also includes short-term fluctuations in the bilirubin level, cholangitis, or episodes of colic. Of course, additional tests are usually needed before the evaluation can be considered complete, but these simple guidelines permit a fairly accurate estimate of the cause of obstructive jaundice in the average patient.

The alkaline phosphatase concentration is a more sensitive indicator of the presence of biliary obstruction than is the bilirubin concentration. Its rise is rapid, precedes that of bilirubin, and occurs in almost every case of bile duct obstruction. In dogs, the serum alkaline phosphatase concentration begins to rise when common bile duct pressure reaches 15 cm. H_2O, but the serum bilirubin concentration

does not change until the pressure exceeds 15 cm. H_2O, and the bilirubin remains normal in some animals with pressures above 18 cm. H_2O. The absolute height of the alkaline phosphatase level does not correlate closely with the degree of obstruction.

The alkaline phosphatase level usually increases with segmental obstruction (e.g., solitary parenchymal metastasis, solitary hepatic abscess), a situation in which the bilirubin level is often normal. The reason is as follows. Serum bilirubin concentrations rise during obstruction because excretion is blocked, and bilirubin is regurgitated into the hepatic sinusoids and lymphatics. Hepatocyte uptake and excretion are impaired with prolonged obstruction. Bilirubin excretion is not impaired, however, until about half of the biliary tree is blocked. On the other hand, serum alkaline phosphatase values rise because of increased synthesis of alkaline phosphatase by hepatocytes proximal to the point of obstruction and regurgitation into the sinusoids.

Within a few days after the onset of bile duct obstruction, the cholesterol concentration in plasma rises. This is the result of increased cholesterol synthesis, not decreased excretion as was once thought, but the mechanism is unknown.

CHANGES FOLLOWING RELEASE OF BILIARY OBSTRUCTION. After a brief period of biliary obstruction in dogs the bilirubin returns to normal in 4 to 7 days. With a longer period of obstruction, the rate of descent does not correlate with the duration of obstruction. Nor is there any correlation between the duration of obstruction and the rate of return of alkaline phosphatase values to normal.

In human beings the serum bilirubin value decreases exponentially at an average rate of 8 per cent per day (Fig. 26–10) and is unaffected by the duration of obstruction, height of the preoperative bilirubin value, or cause of the obstruction. This rate is slower than the 14.5 per cent per day disposal of an additional bilirubin load in nor-

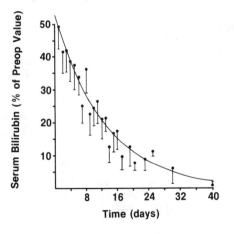

FIG. 26–10. Exponential curve showing postoperative rate of decrease in the serum bilirubin level expressed as a percentage of the preoperative serum bilirubin level in 98 patients operated on for biliary obstruction who eventually had a normal postoperative bilirubin level. The standard error of the mean is shown for each point.

mal subjects, the difference probably reflecting persistent impairment of the excretory mechanism following relief of the obstruction. In support of this view, both man and dog exhibit abnormal bromosulphthalein excretion for some time following relief of bile duct obstruction, even after the serum bilirubin level has become normal. Decreased organic anion clearance may also be the result of decreased sinusoidal blood flow induced by long-standing biliary obstruction.

After release of obstruction in rats and dogs, bilirubin concentration returns to normal earlier than does the alkaline phosphatase concentration. The decline in alkaline phosphatase values after release of obstruction in humans is sluggish compared with bilirubin, and a transient rebound elevation often occurs after the initial decline. These differences reflect the different pathways of elimination of bilirubin and alkaline phosphatase. The liver does not excrete alkaline phosphatase from serum to bile but catabolizes it like other plasma proteins, and the disappearance of alkaline phosphatase from serum is largely independent of patency of the biliary tract. Because the catabolic pathway for alkaline phosphatase is not dependent on biliary excretion and its half-life in the circulation is long (7 days), the serum concentration changes slowly in response to the reduced production following release of obstruction.

Release of biliary obstruction is accompanied by an immediate choleresis consisting of increases in both bile acid–dependent and –independent fractions of bile flow. Bile acid excretion is initially reduced, but recovers in about 4 days in man. In rats, after obstruction is relieved bile acid excretion is similar to that in control animals, and an infusion of sodium taurocholate induces a choleresis. The maximal rate of bilirubin excretion, however, is decreased and is not improved by infusions of bile salt that increase bile flow. These findings suggest that bilirubin transport is more sensitive than is bile acid transport to the effects of biliary tract obstruction.

Cholangitis

Acute cholangitis is infection of bile in the bile ducts and inflammation of the duct walls. Its clinical manifestations include right upper quadrant pain, jaundice, chills, fever, and leukocytosis. The severity of the syndrome varies from mild, in which pain and low-grade fever are the only manifestations, to severe, with overwhelming sepsis and altered mental status, hypotension, and shock. In more than 30 per cent of cases the clinical syndrome is incomplete in the sense that one or more of the major findings (e.g., pain, jaundice, chills) is absent. Acute cholangitis is responsible for many of the deaths caused by biliary tract disease.

ETIOLOGY. Two main factors contribute to the development of acute cholangitis: obstruction of the flow of bile and the presence of bacteria in bile. Obstruction of the bile duct is a requirement for the development of acute cholangitis, and the degree of biliary obstruction correlates with the severity of the infection. *Gallstones* are the most common obstructing lesion causing acute cholangitis, probably because gallstones are so common and the bile in patients with choledocholithiasis nearly always contains bacteria. Obstruction in these cases is usually partial, owing to the ball-valve action of the obstructing stone, and intermittent, as the stones move about within the duct or pass into the duodenum. The bilirubin value with gallstone obstruction is usually in the range of 5 mg./dl. and rarely exceeds 15 mg./dl.

Benign strictures of the bile duct almost invariably give rise to cholangitis at some time. Most benign strictures result from trauma to the duct during cholecystectomy. Although the stricture itself is relatively fixed, the obstruction comes and goes as a result of edema and plugging by precipitated pigment debris and mucus.

Cholangitis complicating *malignant strictures* has been reported more frequently in recent years. In 90 per cent of patients with malignant obstruction the bile is sterile, but when bacteria do gain a foothold, the resulting cholangitis is often overwhelming because the ductal obstruction caused by these lesions is usually complete.

Less common causes of cholangitis are sclerosing cholangitis; compression of the duct by periampullary metastatic tumors; scar formation in chronic pancreatitis; and transhepatic, retrograde, or T-tube cholangiography. In some areas of the world, *parasites* such as nematodes (*Ascaris lumbricoides*), trematodes (*Clonorchis sinensis*), *Fasciola hepatica*, and hydatid cysts gain access to the bile ducts and cause cholangitis. Congenital choledochal cysts, ampullary stenosis, and hemobilia are other rare causes.

BACTERIOLOGY. Bile is normally sterile, but approximately 30 per cent of patients with gallbladder stones, 75 per cent with choledocholithiasis, and 90 per cent with biliary stricture have positive bile cultures. *E. Coli* is the most common cause of cholangitis, followed by *Klebsiella*, *Enterococcus*, and *Proteus*. In about 50 per cent of patients with cholangitis, bile cultures yield two or more species. Occasionally the biliary flora may change during the course of antibiotic therapy, necessitating a change of chemotherapeutic agents. *Bacteroides fragilis* and other anaerobes (e.g., *Clostridium perfringens*) can be detected in about 25 per cent of properly cultured specimens. Anaerobes are more often present in association with severe symptoms, multiple previous biliary operations (often including a biliary enteric anastomosis), and a high incidence of suppurative postoperative complications. When anaerobes do occur in bile they usually accompany aerobic organisms, predominantly *E. coli* and *B. fragilis*.

Theoretically, bacteria may enter the biliary tree by (1) ascending directly from the duodenum through the sphincter of Oddi; (2) hematogenous transport by portal venous or hepatic arterial blood; or (3) transport in the lymphatics that drain the liver and the gallbladder (although there is no evidence to support this).

The theory of ascending contamination is supported by the similarities between the flora of bile and that of the duodenum. If the duodenum really is the most important source, this would explain why bile is usually sterile in patients with high-grade malignant obstruction.

Intestinal bacteria are commonly found in portal blood in dogs, which led to the theory of blood-borne infection. According to this explanation, bacteria from the intestine are thought to reach the liver by the portal vein, where they are cleared from the sinusoids by the hepatocytes and then excreted into bile. However, portal blood rarely contains intestinal organisms in humans, and the hepatic Kupfer cells are very effective at clearing blood-borne bacteria. Hematogenous spread may explain recurrent bacterial cholangitis in some Oriental cultures, where eating pickled vegetables highly contaminated with *E. Coli* is common, and the affected individuals are known to have bacteria in their portal blood.

Biliary obstruction and foreign bodies, such as gallstones or drainage tubes left after biliary surgery, foster the growth and multiplication of bacteria. Even though patients with biliary enteric anastomoses invariably have intestinal organisms in their ducts, they do not manifest cholangitis in the absence of obstruction. Once the bile is contaminated, the degree of biliary obstruction is the most important determinant of the course of the disease.

Cholangitis probably involves bacteremia in nearly all cases. Blood cultures are positive in about half of the patients with acute cholangitis, but this figure would undoubtedly be much greater if cultures were taken more frequently. The organisms isolated from the blood are the same as those isolated from the bile duct, and the same organisms may also appear in urine, sputum, and wound cultures. The balance between the amount and virulence of bacteria and bacterial toxins reaching the circulation, on one hand, and host defenses, on the other, determines the clinical severity of the cholangitis, which may range from mild fever to septic shock.

Experiments have shown that as intraductal pressure is raised, bacteria in bile reflux into the hepatic venous blood and thence into the systemic circulation. This reflux occurs at pressures slightly above the hepatic secretory pressure of the liver, and the absolute height of the pressure is proportional to the number of bacteria recovered from hepatic venous blood. Lymph and blood remain free of organisms as long as the pressure in the common bile duct is below 20 cm. H_2O; however, if the pressure is above 25 cm. H_2O bacteria are always recoverable from blood. The route of bacterial reflux is not known, but probably consists of abnormal communications between the bile capillaries through hepatocytes or tight junctions into the space of Disse.

The relationship of increased intrabiliary pressure to bacteremia is important when performing cholangiography since sepsis is always a risk of this procedure, particularly if the patient has recently had cholangitis. During injection of contrast media, intraductal pressures may reach 80 to 90 cm. H_2O, far above the threshold for bile-to-blood reflux.

One of the most severe complications is renal failure, so-called uremic cholangitis, which develops in about 5 per cent of cases. Hypotheses to explain the occurrence of renal failure include impaired ability of the obstructed liver to detoxify substances harmful to the kidneys, direct nephrotoxicity of high plasma levels of conjugated bilirubin, and the most plausible, decreased renal perfusion resulting from septicemia.

PATHOLOGY. The diameter of the common bile duct in humans averages 7.0 mm. The wall is composed of compact connective tissue and elastic fibers and is lined by a simple epithelium of tall columnar cells. Mucous glands are also present. In acute cholangitis, the duct becomes distended and thickened. Histological findings include edema, congestion, and leukocytic infiltration. The mucous glands become hyperplastic and hypertrophic, and mucus secretion is increased. Ulceration and epithelial necrosis may be seen in late stages. Cholangitis from any cause may be aggravated by secondary effects of the infection, such as increased mucus secretion and sloughing of ductal epithelium. Strands of debris of varying thickness may create multilocular spaces within the biliary ducts, which may make the disease more refractory to antibiotic therapy.

The contents of the common bile duct range from thick bile in early cases to pus. Severe hypotension and endotoxic shock may develop in either case, although severe sepsis (i.e., suppurative cholangitis) is more common when the duct contains pus. Other pathological changes occasionally observed are liver abscess formation and acute pancreatitis.

Cholelithiasis

Pathogenesis of gallstones

The etiology of gallstones is unquestionably multifactorial. It involves alterations in composition of bile and in motor function of the biliary tract leading to precipitation and agglomeration of crystals and other particles, and growth and maturation of stones.

CLASSIFICATION OF STONES. Gallstones may be classified as cholesterol or pigment stones. Cholesterol stones are light, tan, smooth or faceted and on cross section have a dark nucleus surrounded by a laminated or crystalline portion. On chemical analysis these stones contain more than 50 per cent cholesterol with minimal amounts of insoluble material. They represent the stones found in about 75 per cent of patients with gallstones in Western cultures. Female sex, obesity, Western culture, and intestinal resection predispose to cholesterol gallstone disease. American Indians are genetically more susceptible than Caucasians, who in turn are more susceptible than blacks.

Pigment stones are black or dark brown, irregular, and amorphous on cross section. They contain less than 50 per cent (usually less than 25 per cent) cholesterol and variable amounts of pigment solids. Pigment stones may be divided into two groups according to whether the pigment is principally calcium bilirubinate or complex pigment polymers. Pigment stones generally contain 25 to 50 per cent mucopolysaccharide by weight, less than 5 per cent cholesterol, and trace amounts of bile salts and phospholipids.

In Western cultures about 25 per cent of patients with gallstones have the pigment variety. Pigment stones are found most often in patients who are elderly or alcoholic and in those who have hemolysis, extrahepatic cholestasis, or malaria. Unlike cholesterol stones, pigment stones are equally common in both sexes, and the incidence is unrelated to obesity. Pigment stones are especially common in certain rural populations in the Orient.

Since most cholesterol stones contain a nucleus of pigment and mucoprotein resembling a small pigment stone, the initial stage of formation of both kinds of gallstones may be similar.

CHOLESTEROL STONES. Normal bile is a micellar solution in which the predominant lipids are bile acids, lecithin, and cholesterol (Table 26–3). Cholesterol is almost totally insoluble in water, and bile is an aqueous medium. Biliary cholesterol is solubilized by bile acid micelles. Lecithin itself does not form micelles and is relatively insoluble in water, but lecithin becomes incorporated into bile acid micelles where it enhances the micellar capacity for cholesterol. Bile acids aggregate into micelles because they are amphipathic; the molecules possess hydrophilic and hydrophobic surfaces, and in water they form spherical or disc-shaped clusters with their hydrophobic surfaces facing each other and the hydrophilic poles facing the water (Fig. 26–5). Whether the cholesterol is carried within the interior of the micelle or absorbed onto its surface is as yet unclear.

Cholesterol gallstone formation involves three steps: (1) the development of bile su-

---

TABLE 26–3
Composition of Human Gallbladder Bile *

	Percent of Total Solid Content	
Substance	Subjects Without Gallstone Disease	Subjects With Gallstone Disease
Cholesterol	3.08 ± 0.23	4.69 ± .52
Lecithin	19.72 ± 1.65	18.58 ± 1.06
Bile Salts	55.77 ± 5.54	55.39 ± 4.38
Total Proteins	4.57 ± 0.61	3.31 ± 0.34
Bilirubin	0.72 ± 0.13	0.79 ± 0.08
Glucose	0.55 ± 0.10	0.85 ± 0.17
Total Solids	12.02 ± 1.66 g./dl.	8.03 ± 1.52 g./dl.

*(After Carey, M. C., and Small, D. M.: The physical chemistry of cholesterol solubility in bile: Relationship to gallstone formation and dissolution in man. J. Clin. Invest. *61*:998–1026, 1978.)

persaturated with cholesterol; (2) formation of cholesterol crystals; and (3) nucleation and growth of macroscopic stones.

DEVELOPMENT OF SUPERSATURATED BILE. SECRETION OF BILE LIPIDS. The capacity of bile to solubilize cholesterol is determined by the relative proportions of bile acids, lecithin, and cholesterol, a relationship that can be depicted on a triangular phase diagram (Fig. 26–11). The curved boundary in the lower left portion of the triangle separates a zone where all the cholesterol is in solution from zones where it is partly crystalline or is in an unstable supersaturated concentration. Other factors that influence the solubility of cholesterol are bile salt species, temperature, ionic strength, and most important from the physiological standpoint, the total lipid concentration. Total lipid concentration in hepatic bile ranges from 1 to 4 g./dl. and that in gallbladder bile from 8 to 28 g./dl. The effect of changes in total lipid concentration on the limits of cholesterol solubility is illustrated in Figure 26–12 and Table 26–4.

TABLE 26–4
Mean Bile Lipid Compositions (mmol./l.) and Mean Cholesterol Saturation Indices (CSI) for North American and Swedish Patients

	Control Biles (n = 57)	
	North Americans (n = 37)	Swedes (n = 20)
Cholesterol	17.72	29.25
Lecithin	50.83	41.55
Bile Salt	165.31	150.21
CSI ± SD	1.12 ± 0.33	2.01 ± 0.82
CSI ± SD for all normals = 1.42 ± 0.68		

	Biles of Cholesterol Gallstone Patients (n = 63)	
	North Americans (n = 43)	Swedes (n = 20)
Cholesterol	16.93	19.56
Lecithin	39.95	22.52
Bile Salt	112.47	90.89
CSI ± SD	1.45 ± 0.63	2.56 ± 1.32
CSI ± SD for all gallstone patients = 1.80 ± 1.02		

(Holan, K. R., Holzbach, R. T., Hermann, R. E., et al.: Nucleation time: A key factor in the pathogenesis of cholesterol gallstone disease. Gastroenterology 77:611–617, 1979.)

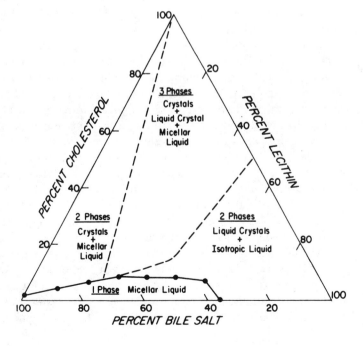

FIG. 26–11. Triangular phase diagram showing the physical state of all combinations of sodium taurocholate, lecithin and cholesterol as 20-g./dl. solutions in 0.15 M NaCl at 24°C (polarizing microscopy and x-ray analysis). The one-phase region gave a diffuse x-ray scattering profile. The two-phase zone on the left gave an x-ray diffraction profile consistent with cholesterol monohydrate crystals in addition to micellar scattering. The three-phase zone in the middle gave a mixed cholesterol monohydrate and lamellar liquid-crystalline pattern and the two-phase zone on the right gave a liquid crystalline profile with isotropic scattering. (Carey and Small: J. Clin. Invest. 61:998–1026, 1978.)

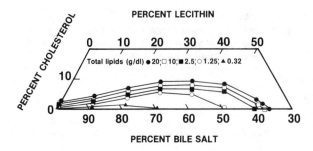

FIG. 26–12. Graphic illustration of the influence of total lipid concentration on cholesterol solubility. (Carey and Small: J. Clin. Invest. *61*:998–1026, 1978.)

The extent to which a bile specimen is saturated with cholesterol (using the boundary line in Fig. 26–11 as a reference) is termed the *cholesterol saturation index (CSI)*. A bile on the line would have a CSI of 100 per cent; a bile below the line would be undersaturated and have a CSI <100 per cent; and a bile above the line would be supersaturated and have a CSI >100 per cent. Both hepatic and gallbladder bile from persons with cholesterol gallstones is invariably supersaturated with cholesterol. Humans in general tend to have highly saturated bile, and gallbladder bile from about 50 per cent of individuals without gallstone disease also exceeds the limits of cholesterol solubility. Nevertheless, supersaturation is a prerequisite for cholesterol gallstone formation, and there is a strong correlation between the absolute magnitude of the CSI and the incidence of cholesterol gallstone disease.

The relative amount of cholesterol in bile could be increased by either an increase in cholesterol secretion or a decrease in bile acid or phospholipid secretion. In fact, both factors contribute to the pathogenesis of cholesterol gallstones in patients.

Cholesterol in bile originates from plasma-free (HDL) cholesterol (60 per cent), newly synthesized hepatic cholesterol (25 per cent), and other unknown sources. Increased cholesterol intake does not increase cholesterol secretion in bile.

Normally the rate of secretion of the three biliary lipids is fairly closely coupled: phospholipid output is proportional to bile acid output; and cholesterol output is usually tightly coupled to phospholipid output, at least over most of the physiologic range of bile acid secretory rates. At lower rates of bile acid secretion, however, cholesterol secretion no longer decreases in proportion to phospholipid and bile acid secretion, so during periods of low bile flow (e.g., fasting), bile becomes more saturated with cholesterol. The effect in one study on normal subjects was to increase the CSI from 89 ± 4 per cent during steady state stimulation (high flow) to 133 ± 10 per cent during low flow. The degree of uncoupling between cholesterol and bile acid output during low bile flow is greater for individuals with greater absolute cholesterol output.

Obesity is associated with marked increases in the amount of cholesterol secreted into bile; and this, in turn, increases the CSI and predisposes to gallstone formation. The increased output is the result of obesity per se and is not reproduced by short-term feeding of diets high enough in calories to cause obesity eventually. Furthermore, hypocaloric diets do not decrease the CSI until the patient has lost the excess weight. In fact, weight-reducing diets slightly increase the CSI as a result of decreased secretion of the solubilizing lipids. The increased cholesterol secretion in obesity is directly proportional to the percentage of overweight. Each kilogram of overweight is associated with an increment of about 1.1 mg. of cholesterol output in bile per hour. Neither the serum levels of cholesterol nor the amount of dietary cholesterol intake directly affects the output of cholesterol in bile.

The abnormal bile in markedly obese

TABLE 26–5
Biliary Lipid Secretion in Patients With Gallstones and Control Subjects

| Group | Secretion Rate | | |
	Bile Acid (μmol./hr.)	Phospholipid (mg./hr.)	Cholesterol (mg./hr.)
Normal controls	1236 ± 91	256 ± 13	27 ± 2
Nonobese gallstone patients	660 ± 91	178 ± 12	25 ± 3
Morbidly obese patients	987 ± 99	352 ± 41	66 ± 8

(Shaffer, E. A., and Small, D. M.: Biliary lipid secretion in cholesterol gallstone disease: The effect of cholecystectomy and obesity. J. Clin. Invest. *59*:828–840, 1977.)

subjects is due to increased cholesterol secretion. The size of their bile acid pool and their rate of bile acid secretion is normal (Table 26–5). Cholesterol gallstone disease in nonobese persons, on the other hand, is associated with a small pool and decreased bile acid secretion. Thus, in these patients gallstones stem from a shortage of solubilizing lipids. The total pool of bile acids in normal subjects (about 5 g.) is decreased by about half (to about 3 g.) in nonobese patients with cholesterol gallstones, and bile acid output from the liver is also about 50% of normal. In these patients the rate of cholesterol secretion is no different from controls (Table 26–5). Undoubtedly, the various defects (i.e., increased cholesterol and decreased bile acid secretion) exist to different degrees in different patients, and mixed defects explain the abnormal CSI in some.

Certain drugs have been associated with an increased incidence of cholesterol gallstones. Clofibrate, a serum cholesterol lowering drug, for example, increases the lithogenic index of hepatic bile. Although the specific mechanism is unclear, biliary cholesterol secretion is increased, probably a result of the mobilization of cholesterol tissue stores, and bile acid synthesis is decreased. Agents that bind bile acids, such as cholestyramine, do not appreciably change the cholesterol saturation of the bile, probably because the liver increases synthesis of bile acids enough to balance losses. Other potential sources of increased

hepatic cholesterol secretion include mobilization of cholesterol from tissue pools and decreased conversion of cholesterol into bile acids.

GALLBLADDER MOTILITY DISTURBANCES. Bile acid secretion may be affected by perturbations at any point in the enterohepatic circulation. Ileal disease or ileal resection may lead to bile acid wasting, diminished pool size, decreased output, and thus an increase in the CSI. The physiological storage of bile in the gallbladder partially interrupts the enterohepatic circulation (EHC) and decreases hepatic secretion of bile acids. Thus, changes in gallbladder dynamics may change bile composition, bile flow, the rate of bile acid synthesis, and the size of the bile acid pool.

If gallbladder filling is increased owing to decreased gallbladder tone or increased sphincteric resistance, or if gallbladder emptying is slowed (e.g., during pregnancy), an increasing fraction of the bile acid pool is sequestered within the gallbladder, lowering the rate of bile acid secretion. This increases the cholesterol saturation of bile because the rate of cholesterol secretion remains about the same. Conversely, increased gallbladder emptying decreases the size of the bile acid pool, probably because of increased cycling. In the prairie dog model of cholesterol gallstone formation, gallbladder stasis develops during cholesterol feeding and anticipates the development of a smaller bile salt pool and stone formation. This has been attributed to a direct effect of lithogenic bile on gallbladder emptying.

Northfield reported that in patients with gallstones the gallbladder appeared to have a lower threshold and an increased sensitivity to stimulation by CCK. However, Shaffer found that gallbladder evacuation was slower in patients with gallstones, and it slowed even further in one patient receiving chenodeoxycholic acid as treatment for cholesterol gallstones. Fischer reported that in patients with gallstones gallbladder emptying in response to a meal was decreased, but gallbladder contraction in response to intravenous CCK was similar to that in control subjects. Since gallbladder contraction depends on CCK stimulation, a diet rich in sucrose and low in fat, common in some North American Indian tribes, may decrease the cycling frequency of the bile acid pool. The idea that abnormalities of gallbladder function may contribute to the formation of lithogenic bile is supported by the finding that when the gallbladder ceases to function or following cholecystectomy for cholesterol gallstone disease, hepatic bile becomes less lithogenic.

CHOLESTEROL CRYSTAL FORMATION AND NUCLEATION. Nucleation involves a change from a single aqueous phase supersaturated with cholesterol to a system with two or more phases, one solid (cholesterol crystals) and one liquid (saturated with cholesterol). One of two types of nucleation may occur depending on the degree of bile saturation. In highly supersaturated bile (above the metastable-labile line), *homogenous nucleation* is common. In this process cholesterol molecules combine spontaneously to form crystals of cholesterol monohydrate. If the bile is only mildly supersaturated (*metastable supersaturation*), spontaneous precipitation of cholesterol occurs slowly, but precipitation is accelerated by the presence of other solids in the bile (Fig. 26–13). Mucin globules, for example, when introduced into metastable bile

cause precipitation of cholesterol crystals. Other substances that could provide a nucleus for crystallization are calcium bilirubinate, other calcium salts, cells detached from the gallbladder wall, bacteria, parasite fragments, ova, or proteinaceous material. Once nuclei are formed, additional precipitation of cholesterol may occur around them. This type of nucleation is probably important in gallstone formation in man, especially in the presence of inflammation, infection, stones, or parasitic infestation.

There are reasons to believe that mucoprotein may be an important nucleating agent in human gallstone disease. Mucoprotein provides an acidic microenvironment and a physical structure especially conducive to cholesterol (or bilirubin) precipitation. It accumulates within mucosal crevasses or free in the lumen during stasis. Microscopically, mucoprotein may be seen to form globules within which liquid crystals and solid crystals form. Small globules have been observed coalescing into larger globules and accumulating on the surface of existing stones.

Patients with gallstones have more mucin in their bile than have patients without biliary disease, and experimental evidence suggests that increased mucus production precedes stone formation. The possibility has not been excluded, however, that the increased mucin production is a coincidental phenomenon rather than an integral step in gallstone formation.

Although the incidence of cholesterol gallstone formation is closely related to the degree of cholesterol saturation, there are many persons with supersaturated bile who do not have gallstones. Another important factor that influences gallstone formation in patients with supersaturated bile is the relative tendency to crystal formation (independent of the CSI value). In the laboratory, bile from persons with cholesterol gallstone disease exhibits crystal formation after a mean of 2.9 ± 0.67 days, whereas bile from normal subjects exhibits crystal formation only after 15 days or longer.

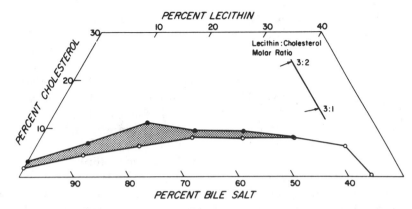

FIG. 26–13. Triangular coordinate plot of the metastable-labile limit and the maximum equilibrium cholesterol solubility in sodium taurocholate lecithin-cholesterol systems (20-g./dl., 0.15 M NaCl, 24°C). The stippled area represents the metastable region. Maximum lecithin: cholesterol molar ratios corresponding to each curve are indicated by arrows. (Carey and Small: J. Clin. Invest. *61*:998–1026, 1978.)

There is a correlation between the CSI and the time of onset of crystallization in bile from normal subjects but not in bile from patients with gallstone disease. In addition, gallbladder bile aspirated from patients with gallstones almost always is found to contain cholesterol monohydrate crystals, while crystals are invariably absent from individuals without cholesterol gallstone disease irrespective of the saturation index of their bile. These observations demonstrate that factors in addition to the CSI strongly influence the rate of crystal formation in bile. The possible explanations include something that stabilizes supersaturated bile in persons without gallstones, or conversely something that accelerates cholesterol precipitation in patients with gallstones (i.e., nucleating agents). The former is a more logical hypothesis, since it would explain why some individuals with markedly supersaturated bile do not exhibit crystal formation.

STONE GROWTH. Growth from crystals to a macroscopic stone is the last step in stone formation. If the gallbladder emptied completely with each meal, gallstones would not form because the small cholesterol particles would be flushed into the duodenum. But maximal gallbladder emptying in normal subjects is approximately 85 per cent, and it decreases in conditions causing gallbladder stasis.

As bile is concentrated in a poorly functioning gallbladder, stratification occurs. During fasting, stratification may keep more concentrated bile closer to the gallbladder mucosa and bile freshly arrived from the common duct unmixed. Abnormal gallbladder dynamics may lead to retention of supersaturated bile, which would enhance the growth of stones.

PIGMENT GALLSTONES. The relative amounts of bile salt, phospholipid, and cholesterol are normal in biles of patients with pigment stone disease. Pigment stone formation involves a different process. The relative amounts of unconjugated and monoconjugated bilirubin in biles from patients with pigment gallstones is increased. Pigment stones contain calcium bilirubinate and bilirubin polymers precipitated in a mucopolysaccharide matrix. As with cholesterol gallstones, pigment gallstone formation may be initiated by mechanisms affecting bilirubin solubility in bile.

BILIRUBIN SOLUBILITY. Unconjugated bilirubin is only slightly soluble in water. Conjugated bilirubin is water soluble, but under physiological circumstances almost all the bilirubin in bile is complexed with bile salt:phospholipid micelles. The ability of bile to solubilize bilirubin depends upon several factors, the most obvious being the total output of bilirubin and the proportion that is unconjugated. Since the normal rate-limiting step in hepatic bilirubin excretion is transport into the canaliculus (not conjugation), most bilirubin is excreted as the diglucuronide. Normally less than 1 per cent of bilirubin in bile is unconjugated. Up to 3 per cent of the bilirubin in bile is unconjugated in patients with Crigler-Najjar syndrome, sickle-cell anemia, and other hemolytic diseases. Ethanol increases the conjugation of unconjugated bilirubin in human bile. For example, 2 hours after the intravenous administration of ethanol (0.7 grams/kilo/body weight) unconjugated bilirubin in bile increased from 0.65% to 2.4% of total bilirubin. Increased total bilirubin excretion into bile plus increased proportions of unconjugated bilirubin may be sufficient to saturate the bilirubin-carrying capacity of bile and lead to bilirubin precipitation.

The conjugation state of bilirubin may change after it is secreted by the liver. For example, increased β-glucuronidase, an enzyme able to hydrolyze conjugated bilirubin, has been observed in bile from patients with pigment stones in Japan and the United States. Increased β-glucuronidase activity in the biles from patients in the Orient may originate from bacteria in bile, especially *E. coli*. In the United States, some patients with pigment stones apparently have in their biles increased amounts of an endogenous, nonbacterial β-glucuronidase with a broader pH range.

There are other factors that might affect bilirubin solubility in bile. For example, *in vitro* studies suggest that lecithin inhibits and taurocholate enhances the solubility of unconjugated bilirubin in model bile systems. Corrections were not made for total lipid ratio, which are known to affect cholesterol solubility. Nonetheless, patients with hepatic cirrhosis secrete less

than normal amounts of bile salt. Although phospholipid secretion is to some extent bile salt–dependent, the molar ratio of bile salt to phospholipid may vary among individuals. Decreased micellar capacity may contribute to bilirubin instability, and may help explain the increased incidence of pigment stones in cirrhotic patients. Changes in calcium concentration do not affect bilirubin solubility *in vitro*, although there is some evidence that Ca^{++} may be important in initiating gallstone formation by decreasing bilirubin solubility. Unconjugted bilirubin is more soluble in basic than acidic solutions, and gallbladder pH is known to decrease at night and with stasis or prolonged fasting. This suggests that pH changes coincident with decreased motility could contribute to bilirubin precipitation. Finally, since the glycoprotein content of pigment stones is considerably higher than that of cholesterol stones, mucin may contribute to growth of pigment stones as well as to nidation. The mucin may provide an acidic microenvironment and a thermodynamically favorable structure within which bilirubin may preferentially deposit, independent of the solubilizing capacity of the aqueous medium.

The ability of bile to keep bilirubin in solution appears to be quite sensitive to a variety of conditions that have biliary stasis as a common feature (e.g., prolonged fasting). Under these circumstances, ultrasound examination often demonstrates a precipitate ("sludge") in the gallbladder lumen. Analysis of this material shows it to consist of calcium bilirubinate, mucoprotein, and perhaps other lipids. Most often, sludge is of no clinical consequence, since it usually vanishes when the patient resumes eating.

Because sludge formation is so common, the development of bilirubin-mucoprotein nidi must also be common, but the nidi probably become clinically important only when other factors allow them to evolve into stones. Impaired gallbladder emptying would retain sludge in the gallbladder where it could serve as starter material for gallstones. Patients with bile supersaturated with cholesterol would form cholesterol stones, and those with increased amounts of unconjugated bilirubin would form pigment stones. In the absence of sludge, these other abnormalities would be insufficient to initiate stone formation.

CHOLEDOCHOLITHIASIS. Classical teaching states that most common duct stones originate in the gallbladder and escape through the cystic duct (i.e., they are *secondary* stones). There is little direct evidence to support this theory, however, and a few workers hold a contrary belief that the majority of ductal stones form in the duct primarily (i.e., primary stones). A study comparing the physical appearance and chemical composition of common duct stones with gallbladder stones from the same patient has shown that in cholesterol stone disease the stones from the two sites are virtually identical. In pigment stone disease the gallbladder and common duct stones are usually very similar, but there is a greater range of disparity than in cholesterol stone disease. In 95 per cent of cases common duct stones are of the same basic type (i.e., cholesterol or pigment) as the stones from the gallbladder. Finally, whenever the clinical situation dictates that the ductal stones must be primary (e.g., stones proximal to a biliary stricture), they are always composed of pigment. Although the evidence is inconclusive, these findings suggest that cholesterol common duct stones are secondary in nearly all cases, that pigment common duct stones are more often secondary than primary, but that in chronic biliary stasis pigment common duct stones may form primarily in the duct.

As mentioned earlier, normal function of the sphincter of Oddi is characterized by peristalsis; contractile activity produces a wave of pressure that progresses from the hepatic end of the sphincter toward the duodenal end, and this results in bile entering the duodenum in spurts instead of in a continuous stream. There is some evidence that the normal direction of these peristaltic contractions is reversed—a majority appear to progress from the duodenal end toward the hepatic end of the sphincter—in patients with choledocholithiasis. If this is a valid observation it is still unclear whether these retrograde contractions should be considered a cause or a result of the choledocholithiasis.

Bile acid therapy for gallstone dissolution

Approximately 75 per cent of patients with gallstones in Western cultures have cholesterol gallstones. Because cholesterol stone disease is associated with a shrunken bile acid pool, investigators have studied the effects of feeding bile acids on cholesterol solubility and the size of the pool. Oral administration of chenodeoxycholic acid (CDC) (but not cholic acid) increased the cholesterol-solubilizing capacity of bile and eventually resulted in stone dissolution.

CDC (mg./kg./day) desaturates previously saturated bile in most patients and doubles the size of the pool (to normal), but does not increase the 24-hour output of bile acid. The efficacy of bile acid therapy depends on an unanticipated result, decreased cholesterol secretion. Three mechanisms could explain this finding: decreased cholesterol synthesis, decreased cholesterol absorption, or alteration in the degree of coupling between bile acid and cholesterol secretion. The data indicate that cholesterol synthesis and secretion decrease, and the total cholesterol pool does not change. The rate of cholesterol synthesis is difficult to measure in patients, but during CDC therapy the activity in liver biopsies of hydroxymethylglutaryl CoA (HMG-CoA) reductase, the rate-limiting enzyme for cholesterol synthesis, is decreased to one-third of the pretreatment value. Although there have been reports to the contrary, the best evidence indicates that CDC has no effect on cholesterol absorption.

The presence of unsaturated bile brings about a slow dissolution of cholesterol stones. Large stones take longer to disappear—the thickness of the stone decreases by about 1 mm./month. A constant flow of bile that allows fresh unsaturated hepatic bile to bathe the stone is essential for success. Nonopacification on oral cholecystography, a reflection of low or absent flow of hepatic bile into the gallbladder, is an indication that dissolution therapy would be unsuccessful. Obese patients are relatively refractory to treatment (their bile does not readily desaturate), and thus obesity is a relative contraindication to dissolution therapy. Other contraindications are known pigment stones, calcified gallstones (which fail to dissolve whether they are pigment or cholesterol stones), active liver disease, severe symptoms, and fertile women intending to become pregnant.

Ursodeoxycholic acid (600 mg./dl.), the 7-β-hydroxy epimer of CDC, also dissolves cholesterol stones by a similar mechanism. Ursodeoxycholate has the advantage of greater biochemical efficacy, less diarrhea (a side effect of CDC treatment), and less hepatotoxicity, and is favored over chenodeoxycholate in countries where both drugs are available.

Since these bile acids are dehydroxylated and converted to lithocolic acid (which is hepatotoxic in several species). The rate of biotrans formation to lithocolic acid is the same for chenodeoxycholic and ursodeoxycholic acids, and since increased concentrations of lithocolic acid in the colon have a potential carcinogenic effect, the long-term side effects of dissolution therapy are of concern.

The results of the National Cooperative Gallstone Study (NCGS), a randomized trial of placebo, high-dose (750 mg./day), and low-dose (325 mg./day) chenodeoxycholate therapy in 916 patients with gallstone disease, were reported in 1981. Patients with a nonopacifying oral cholecystogram, calcium containing stones, or severe symptoms, or women who were at risk of becoming pregnant were excluded from the trial. Unfortunately, the highest dose used in the NCGS was less than what is now known to be an optimal dose (12 to 15 mg./kg./day) in all but 25 per cent of the patients in the high-dose group, so the full potential of dissolution therapy may not have been demonstrated.

As judged by serial oral cholecystograms, patients on the high-dose regimen in the NCGS showed a 15 per cent incidence of total gallstone dissolution after 2 years of therapy. Little effect was noted with the low dose. A similar trial in Canada demonstrated that oral cholecystograms fail to detect some gallstones demonstrable by ultrasound, however, so the rate of success reported in the NCGS is probably an overestimate. The efficacy of therapy was confined to gallstone dissolution, since no clinical efficacy was demonstrated in the form of decreased biliary colic or decreased frequency of withdrawal from the trial in favor of surgical therapy. Side effects of treatment consisted of an average 10 per cent increase in serum cholesterol concentrations and abnormal liver enzyme determinations in 8 per cent of patients. Eleven patients were dropped from the trial because of the latter problem.

Thus, the efficacy of CDC therapy was much lower than desirable, and it appears critical at this point to devise methods for selecting patients with a higher likelihood of experiencing complete dissolution and to demonstrate bona fide clinical efficacy. Otherwise treatment may not seem justifiable in relation to risks and monetary costs.

Changes in composition of duodenal bile after the initiation of therapy correlate with the chances of a successful outcome, duodenal intubation and bile analysis could be used as an early means of detecting unresponsive patients and discontinuing therapy. Other efforts are directed at finding clinical and radiological criteria that distinguish between patients with cholesterol and those with pigment gallstones.

Another weakness of dissolution therapy is the strong tendency for gallstones to recur after treatment is withdrawn. Dowling observed a 50 per cent recurrence rate over a median follow-up of 23 months after discontinuation of therapy, and it seems inevitable that the figure will climb higher with time. This is a serious drawback because it means that continued expensive medical care may be necessary to preserve the gallstone-free state. There is some hope that these patients may be placed on diet modifications that decrease the CSI, but at the moment no clinically effective regimen has been found.

Other drugs have been tried in attempts to improve the cholesterol holding capacity of bile. At present however, the only one that has proven efficacy is soybean lecithin, the administration of which produces a decrease in serum and bile cholesterol. In addition, soybean lecithin corrects type IV lipoproteinemia, but further studies are needed to explain its mechanism of action and to elucidate its clinical potential as a stone dissolving agent.

DISSOLUTION OF RETAINED COMMON DUCT STONES. Various solutions are able to dissolve retained common duct gallstones when infused into the T-tube. Chloroform, ether, heparin, bile salts, and others have been tested. A commercial emulsifying agent, glycerol-l-monooctanoate, a medium chain monoglyceride, is at present the best available agent; it dissolves cholesterol stones twice as fast as do cholic acid solutions *in vitro*, although the clinical results are about the same. Flowing solutions are probably more effective than are stagnant ones. The efficacy of these solvents is related directly to their ability to dissolve cholesterol, and at present there is no solvent that is safe to use in the treatment of retained pigment stones.

Pathophysiology of biliary pain

The gallbladder, common bile duct, liver, and peritoneum are insensitive to cutting, pinching, and crushing. Rapid distention of the gallbladder, common bile duct, or liver capsule is the principal stimulus for acute biliary pain, and the most common cause is obstruction of the gallbladder or common bile duct by a stone. Very gradual obstruction may give dilatation (e.g., neoplastic obstruction) of the bile duct without pain.

Stimulation of the gallbladder smooth muscle by endogenous CCK in patients with cystic duct obstruction produces forceful contraction and an unusually high tension of the muscle, which generates the pain of biliary colic. This is why patients so often experience biliary colic after eating.

Neoplasia may involve the biliary sensory nerves and cause pain. The common bile duct does not have a well-defined muscle coat in humans, and therefore pain originating at this level is almost always the result of passive distention produced by distal obstruction. Occasionally distention of the liver capsule secondary to bile duct obstruction may produce pain.

The intensity of the pain stimulus necessary to evoke perceived pain varies from person to person and is influenced by previous experience, psychological factors, and cultural background. There are no specific pain receptors; the free nerve endings located in the subserosal area of the gallbladder and biliary ducts act as nociceptors. These fibers have a higher threshold than other receptors, and cover smaller fields. The threshold to pain can be altered by the presence of inflammation or ischemia due to local release of bradykinin, prostaglandins, serotonin, and histamine from the damaged cells.

Sensory afferents of the gallbladder and common bile duct travel in the visceral sympathetic nerves. By way of the celiac plexus and the greater splanchnic nerve, stimuli from the gallbladder, common bile duct and inferior aspect of the liver reach the sixth to tenth (mainly the ninth) thoracic segments of the spinal cord. Some stimuli from Glisson's capsule, the hepatic ligaments, and the central portion of the diaphragm course to the fourth segment of the cervical cord.

The afferent visceral impulses from the

biliary tract are transmitted mainly through unmyelinated fibers of Gasser class C, whereas those originating in the adjacent peritoneum are transmitted mainly by thinly myelinated fibers of the Gasser class A-delta. The afferent visceral fibers that accompany sympathetic nerves reach the sympathetic chain and continue to the dorsal root ganglia (where the neuronal bodies of these fibers are located) by the white rami communicantes. From the dorsal root ganglion they enter the cord and Lissauer's tract, ascend for a variable distance, and synapse with neurons of the gray matter of the posterior horn. The axons from the posterior horn neurons cross at the anterior commissure of the spinal cord and reform in the lateral spinothalamic tract, which ascends to the posterolateral and posterior thalamic nuclei. From there the impulses travel to the region of the postcentral gyrus in the cerebral cortex.

In contrast to old ideas of the straight-through transmission of impulses, it is now clear that there is a high degree of modulation at every level of the neuraxis. The dorsal horn is not a simple relay station, but a highly complex structure containing many kinds of neurons and synaptic arrangements that allows for a high degree of sensory processing at this level. Some nociceptive impulses that reach the dorsal horn pass through internuncial neurons to the anterior and anterolateral horn cells, where they stimulate somatomotor neurons and preganglionic sympathetic neurons, producing segmental nociceptive *reflex responses* (e.g., spasm of abdominal wall muscles, decreased gastrointestinal motility, etc.). There is now much less certainty about the specificity of fibers in relation to pain transmission, and it is clear that the nerve pathways for biliary pain overlap with sensory impulses for other abdominal viscera. In addition, afferent visceral and somatic fibers that innervate the adjacent peritoneum and overlying skin often communicate by synapsing with

the same neuron in the posterior spinal horn.

Upon receiving stimuli the cortex interprets them. Impulses originating from supraspinal neurons can modulate spinal somatic sensory input through descending pathways. Neurons in the cerebral cortex and reticular formation, for example, can exert inhibitory effects on somatosensory nerve fibers at the spinal cord level. Neurons in the periventricular gray and reticular formation of the mesencephalon possess opioid receptors, and high concentrations of endorphins are present in these areas, suggesting that they constitute a major inhibitory mechanism for pain. Cerebral neurons can also directly influence impulse generation in ascending ventrolateral or dorsal column cells at spinal, brain stem, or thalamic levels, exerting a modulating influence at several points in the system. This is why the sensation of pain can be colored by subjective factors.

Pain originating from the biliary tract may be classified as visceral, somatic, or referred pain. *Visceral pain* originates in the diseased organ, is felt in the epigastrium, and because of interconnections between sympathetic afferents and efferents from other abdominal viscera and lack of cortical training, is poorly localized. *Parietal pain* originates from somatic structures or the parietal peritoneum and is caused by irritation from inflammation or mechanical trauma. It is usually perceived as being sharp and well localized and is aggravated by motion. Parietal pain is occasionally accompanied by muscular rigidity and cutaneous hyperesthesia. Rebound pain, a form of parietal pain, is a reflection of involvement of somatic afferents in the parietal peritoneum and abdominal wall by the disease process.

Referred pain is felt in some remote area innervated by fibers from the same dermatome as those supplying the diseased viscus. Referred pain occurs because of overlap in the neural pathways (mainly at level V in Lissauer's tract) between visceral and somatic afferents and because the somatic distribution has more nerve endings. In addition, the cortex is trained to interpret stimuli from that particular dermatome as somatic pain. Referred pain is dull to sharp and is well localized. A typical example is the shoulder or scapular pain occasionally observed in conjunction with biliary colic.

The sequence of pain associated with acute cholecystitis typifies the different kinds of pain. Early, distention and inflammation of the gallbladder may be felt as a dull, diffuse epigastric discomfort (visceral pain). As the adjacent parietal peritoneum becomes inflamed, the pain intensifies and becomes localized to the right subcostal area (parietal pain). Deep inspiration while the examiner presses in the right upper quadrant pushes the inflamed gallbladder against the examiner's hand and produces more severe pain and inspiratory arrest (Murphy's sign). In many cases pain is also felt in the upper back (referred pain). This area is innervated by sensory afferents from T-9, the segment that supplies the afferent fibers to the gallbladder.

Selected References

Allen, B. L., Deveney, C. W., and Way, L. W.: Chemical dissolution of bile duct stones. World J. Surg. 2:429–437, 1978. (This is a review of the agents used to dissolve common bile duct stones and their efficacy.)

Behar, J., and Biancani, P.: Effect of cholecystokinin and the octapeptide of cholecystokinin on the feline sphincter of Oddi and gallbladder: Mechanisms of action. J. Clin. Invest. 66:1231–1239, 1980. (Excellent study of the neural mechanisms that control biliary motility.)

Bennion, L. J., and Grundy, S. M.: Risk factors for the development of cholelithiasis in man. N. Engl. J. Med. 299:1161–1167; 1221–1227, 1978. (Excellent review of the risk factors involved in the development of gallstones.)

Blitzer, B. L., and Boyer, J. L.: Cellular mechanisms of bile formation. Gastroenterology 82:346–357, 1982. (This excellent update studies the sinusoidal, the intracellular, and the canalicular events that lead to the formation of bile. The article contains over 150 references.)

Boey, J. H., and Way, L. W.: Acute cholangitis. Ann. Surg. 191:270, 1980. (This article reviews the etiology, pathogenesis, diagnosis and management of acute cholangitis.)

Boonyapisit, S. T., Trotman, B. W., and Ostrow, J. D.: Unconjugated bilirubin, and the hydrolysis of conjugated bilirubin in gallbladder bile of patients with cholelithiasis. Gastroenterology 74:70–74, 1978. (This paper shows that the increase in unconjugated bilirubin observed in patients with pigment stones in the Western hemisphere is the result of increased secretion of unconjugated bilirubin by the liver as well as nonbacterial hydrolysis of conjugated bilirubin in bile.)

Bourgault, A. M., England, D. M., Rosenblatt, J. E., Forgacs, P., and Bieger, C.: Clinical characteristics of anaerobic bactibilia. Arch. Intern. Med. 139:1346–1349, 1979. (This is an excellent study of biliary bacteriology with particular emphasis in the role of anaerobic bacteria in biliary infection.)

Carey, M. C., and Small, D. M.: The physical chemistry of cholesterol solubility in bile. Relationship to gallstone formation and dissolution in man. J. Clin. Invest. 61:998–1026, 1978. (This is a very elaborate study of the factors that affect cholesterol solubility in bile and the predominant forces that play a role in the precipitation of cholesterol.)

DeVos, R., and Desmet, V. J.: Morphologic changes of the junctional complex of the hepatocytes in the rat liver after bile duct ligation. Br. J. Exp. Pathol. 59:220–227, 1978. (An electron microscopic study of the mechanisms of extravasation of bile from the biliary canaliculus and its regurgitation into the sinusoids during biliary obstruction.)

Erlinger, S.: Does Na-K ATPase have any role in bile secretion? Am. J. Physiol. 243:G243–G247, 1982. (A review of evidence supporting a key role of Na-K ATPase in bile secretion.)

Erlinger, S.: Hepatocyte bile secretion: current views and controversies. Hepatology 1:352–359, 1981. (This study revises the classic concept of bile-acid-dependent and bile-acid-independent bile flow with special emphasis on the mechanisms of bile salt secretion.)

Everson, G. T., McKinley, C., Lawson, M., Johnson, M., and Kern, F.: Gallbladder function in the human female: Effect of the ovulatory cycle, pregnancy and contraceptive steroids. Gastroenterology 82:711–719, 1982. (Excellent review of gallbladder physiology in women with special reference to motility disturbances associated with pregnancy and its possible relationship to gallstone pathogenesis.)

Frizzell, R. A., and Heintze, K.: Transport functions of the gallbladder. Int. Rev. Physiol. 21:221–247, 1980. (Excellent review of gallbladder mucosal function with emphasis in the mechanisms of water absorption.)

Jones, A. L., Schmucker, D. L., Renston, R. H., et al.: The architecture of bile secretion. A morphological perspective of physiology. Dig. Dis. Sci. 25:609–629, 1980. (This is an elaborate electron microscopy study of the morphological basis of bile formation and flow.)

Krishnamurthy, G. T., Bobba, V. R., and Kingston, E.: Radionuclide ejection fraction: A technique for quantitative analysis of motor function of the human gallbladder. Gastroenterology 80:482–490, 1981. (This is a study of the motor function of the human gallbladder using radiolabeled bile. Gallbladder filling and gallbladder contraction in response to CCK are analyzed noninvasively.)

LaMorte, W. W., Schoetz, D. J., Jr., Birkett, D. H., and Williams, L. F., Jr.: The role of the gallbladder in the pathogenesis of cholesterol gallstones. Gastroenterology 77:580–592, 1979. (This is a very detailed paper on the role of the gallbladder in gallstone formation, including a historical review and a detailed analysis of all previous papers on the subject. Numerous references.)

Lin, T.: Actions of gastrointestinal hormones and related peptides on the motor function of the biliary tract. Gastroenterology 69:1006–1022, 1975. (Overview of the influences of gastrointestinal hormones on biliary motility with over 100 references.)

Mok, H. Y. I., Von Bergmann, K., and Grundy, S. M.: Kinetics of the enterohepatic circulation during fasting: Biliary lipid secretion and gallbladder storage. Gastroenterology 78:1023–1033, 1980. (This is a study of biliary lipid secretion by the hepatocytes and the mechanisms that lead to cholesterol saturation of bile during fasting. It is of particular interest to those interested in the enterohepatic circulation of bile salts.)

Muller, K.: Structural dimorphism of bile/salt lecithin mixed micelles: A possible regulatory mechanism for cholesterol solubility in bile? X-ray structure analysis. Biochemistry 20:404–414, 1981. (Revises the classical concept of cholesterol solubility and proposes a new model for the molecular structure of the bile salt micelle.)

Pellegrini, C. A., Thomas, M. J., and Way, L. W.: Bilirubin and alkaline phosphatase values before and after surgery for biliary obstruction. Am. J. Surg. 143:67–73, 1982. (This paper studies the effects of biliary obstruction on liver function in humans and the changes observed following relief of obstruction.)

Roslyn, J. J., DenBesten, L., Thompson, J. E., Jr., and Silverman, B. F.: Roles of lithogenic bile and cystic duct occlusion in the pathogenesis acute cholecystitis. Am. J. Surg. 140:126–130, 1980. (This study suggests that lithogenic bile, in addition to obstruction of the cystic duct, is important in the development of acute cholecystitis. The etiology of acute cholecystitis in patients with gallstones is probably multifactorial.)

Ruppin, D. C., and Dowling, R. H.: Is recurrence inevitable after gallstone dissolution by bile-acid treatment? Lancet 1:181–185, 1982. (This follow-up study of patients successfully treated with bile-acid therapy shows that recurrence is high when treatment is discontinued.)

Schoenfield, L. J., Lachin, J. M., et al.: Chenodiol (chenodeoxycholic acid) for dissolution of gallstones: the National Cooperative Gallstone Study. A controlled trial of efficacy and safety. Ann. Intern. Med. 95:257–282, 1981. (This is the summary report of the results of the cooperative study in the United States.)

Schein, C.: Acute Cholecystitis. Hagerstown, Harper & Row, 1972. (This book contains considerable data on etiology and pathogenesis of acute cholecystitis.)

Sedaghat, A., and Grundy, S. M.: Cholesterol crystals and the formation of cholesterol gallstones. N. Engl. J. Med. 302:1274–1277, 1980. (This study analyzes the role of cholesterol crystals in cholesterol gallstone formation. Cholesterol crystallization appears to be a prerequisite for the development of cholesterol stones. However, in subjects without stones, marked supersaturation of bile with cholesterol may occur without the formation of crystals. Dr. Donald Small discusses the article in an accompanying editorial. [Cholesterol nucleation and growth in gallstone formation. N. Engl. J. Med. 302:1305–1307, 1980].)

Sjodahl, R., Tagesson, C., and Wetterfors, J.: On the pathogenesis of acute cholecystitis. Surg. Gynecol. Obstet. 146:199–202, 1978. (This study looks at the role of prostaglandins in the development of acute cholecystitis.)

Soloway, R. D., Trotman, B. W., and Ostrow, J. D.: Pigment gallstones. Gastroenterology 72:167–182, 1977. (One of the best papers on the subject: discusses epidemiology and pathogenesis of pigment stones with over 300 references.)

Takahashi, I., Suzuki, T., and Aizawa, I.: Comparison of gallbladder contractions induced by motilin and cholecystokinin in dogs. Gastroenterology 82:419–424, 1982. (Gallbladder contraction in response to motilin is different than that induced by CCK. Motilin probably plays a physiological role in partial gallbladder emptying during fasting, whereas CCK is responsible for gallbladder emptying following meals.)

Way, L. W.: The national cooperative gallstone study and chenodiol. Gastroenterology 84:648–651, 1983. An editorial analysis of the final results of the national cooperative gallstone study.

General

Bennion, L. J., and Grundy, S. M.: Risk factors for the development of cholelithiasis in man, Parts 1 and 2. N. Engl. J. Med. *299*:1161–1167, 1221–1227, 1978.

Javitt, N. B.: Hepatic bile formation, Parts 1 and 2. N. Engl. J. Med. *295*:1464–1469, 1511–1516, 1976.

Oshio, C., and Phillips, M. J.: Contractility of bile canaliculi: Implications for liver function. Science *212*:1041–1042, 1981.

Schoenfield, L. J., Lachin, J. M., et al: Chenodiol (chenodeoxycholic acid) for dissolution of gallstones: The national cooperative gallstone study: A controlled trial of efficacy and safety. Ann. Intern. Med. *95*:257–282, 1981.

Staehelin, L. A., and Hull, B. E.: Junctions between living cells. Sci. Am. *238*:140–152, 1978.

Section VI

Hematological Mechanisms

Introduction

(The blood) is the life of all flesh—*Leviticus 17:14*

Hematology, the systematic study of blood, has ancient origins. The Hippocratic school, for example, believed that a basic cause of disease was the separation of the blood into its four constituent "humors," normally blended together. Our modern discipline of hematology was pioneered by two British physicians, William Hewson and John Hunter, just two centuries ago, and the first textbook for the practicing physician appeared only in 1843. The growth of our knowledge about blood has been so rapid in recent years that currently as many as 30 or 40 per cent of the text of some general medical journals is devoted to this subject.

Strictly speaking, hematology should encompass all aspects of its subject, blood. By what is little more than convention, however, modern hematologists have focused principally upon the cellular elements of the blood and upon the devices through which loss of blood after vascular injury is controlled, that is, the process of hemostasis. Following no particular logic, hematologists pay relatively little attention to other important functions of blood, such as transport of anabolic and catobolic agents and the maintenance of the ionic and osmotic composition of the body. On the other hand, for a variety of reasons hematologists have extended their concerns to such diverse processes as the immunological defenses of the body and its aberrations and the nonhematologic neoplastic diseases. Indeed, the development of oncology, the study of tumors, is so firmly grounded in hematology that most present-day hematologists include the diagnosis and treatment of neoplasms within their province.

The chapters that follow provide only a small range of the subject matter of hematology, chosen to illustrate a variety of physiological processes and their alterations in disease. The reader will appreciate the hematologist's penchant for absorbing the technologies of many disciplines to solve the riddles of normal and abnormal physiology. The next few years should see a great expansion in our knowledge as the methodologies of molecular biology, protein chemistry, and immunology are increasingly applied. It is both the strength and weakness of all textbooks that they seem to pause in the midst of a rushing stream of new information. But the neophyte must begin somewhere. These chapters should provide a look at the way hematologists think, and what they are thinking about at the time this volume appears.

OSCAR D. RATNOFF, M.D.

621

27 Erythropoiesis

Sylvia S. Bottomley, M.D.
Dilip L. Solanki, M.D.

The erythron

The erythrocyte arises within the bone marrow from a stem cell which, under the influence of erythropoietin, differentiates into a precursor cell having the capacity to synthesize hemoglobin. Within 3 to 5 days, the red cell precursor (erythroblast) undergoes further differentiation and division. Its concurrent maturation is characterized by a coordinated sequence of maximal ribonucleic acid (RNA) synthesis and non-globin protein synthesis, followed by maximal synthesis of hemoglobin and a parallel progressive reduction in deoxyribonucleic acid (DNA) synthesis, finally leading to condensation and extrusion of the nucleus. The young erythrocyte (reticulocyte) then enters the vascular compartment, where, within 24 hours, it loses the residual RNA and mitochondria as well as some membrane in the spleen (splenic conditioning). Hence, the reticulocyte content in the peripheral blood reflects the "birth rate" of new erythrocytes and is a quantitative index of red cell production. The mature erythrocyte remains intravascularly for approximately 120 days, at which time, with its energy supply exhausted, it is removed by the reticuloendothelial system of the spleen, the liver, and the bone marrow.

This cell system, termed the *erythron* by Boycott, is a dynamic one. The erythroid marrow maintains, in the average adult, a population of 25×10^{12} circulating erythrocytes, which contain 750 g. of hemoglobin. With maximal stimulation, the erythroid marrow is capable of increasing the production rate six- to eightfold and can produce every day a volume of red cells equivalent to that contained in one-half pint of whole blood. Normally, aged red cells are removed from the blood at a rate of about 1 per cent per day. In marrow failure states due to disease, insufficient red cells are produced to replace this daily loss, and anemia results. Lack of cell differentiation and proliferation is reflected in a reduction or absence of erythroid cells in the marrow (hypoproliferative anemia), while impaired erythroid maturation is associated with an increase in red cell precursors which fail to develop (ineffective erythropoiesis). In either case, the peripheral blood reticulocytes are reduced or do not increase in response to anemia. If red cells are removed from the circulation at an increased rate (e.g., from bleeding or cell destruction), production increases in order to maintain the optimum number of circulating cells. For example, if the red cell survival time is reduced to one-fourth of normal (from 120 days to 30 days), production must increase to four times normal. However, if the survival time is reduced to one-tenth of normal (from 120 days to 12 days), the maximal production capacity is exceeded and a defi-

TABLE 27–1
Outline of the Pathophysiology of the Erythron

I. Insufficient production of erythrocytes
  A. Defects in general controlling mechanisms
    1. Erythropoietin deficiency: renal disease, anemia of chronic disorders, liver disease, erythropoietin antibody
    2. Endocrine deficiencies of pituitary, thyroid, adrenals, and gonads
  B. Defects in cell differentiation, proliferation, and maturation
    1. Stem cell disorders
      a. Aplastic anemia: congenital (Fanconi's anemia); acquired—due to drugs and toxins, viral infection, immune suppression, lack of hemopoietic cofactors, defective microenvironment, idiopathic
      b. Myelodysplastic syndromes
      c. Red cell aplasia: congenital (Blackfan-Diamond); acquired—due to drugs, antibodies, suppressor cells, idiopathic
    2. Impaired nuclear development
      a. Coenzyme deficit: vitamin B_{12}, folate
      b. Enzyme defects: congenital; acquired—due to drugs
      c. Metabolic antagonism secondary to drugs or neoplasms
    3. Impaired cytoplasmic development
      a. Defective heme synthesis: iron deficiency, anemia of chronic disorders, sideroblastic anemia, toxins (lead)
      b. Defective globin synthesis: thalassemias
II. Increased destruction of erythrocytes
  A. Inherited abnormalities: membrane defects (hereditary spherocytosis), hemoglobinopathies and thalassemias, enzyme deficiencies
  B. Acquired abnormalities without antibodies
    1. Membrane defects: paroxysmal nocturnal hemoglobinuria, liver disease, lead, bacterial sepsis (*C. welchii*)
    2. Enzyme defects: lead, uremia
    3. Physical injury: mechanical stress (microangiopathic states, cardiac hemolysis); thermal injury
    4. Intracellular inclusions: oxidant drugs, malaria
    5. Hypersplenism
  C. Acquired abnormalities with antibodies: isoantibodies (transfusion reaction), antibodies to drugs, autoantibodies
III. Increased production of erythrocytes
  A. Appropriate compensatory mechanism in response to:
    1. Increased destruction of erythrocytes (if pathophysiological mechanism affects only the mature cell segment of the erythron)
    2. Tissue oxygen deficit
      a. Environmental oxygen deficit (altitude)
      b. Cardiorespiratory insufficiency
      c. Abnormal hemoglobin with increased oxygen affinity
  B. Inappropriate
    1. Excessive erythropoietin activity: associated with tumors of kidney, liver, cerebellum, adrenals, and uterus; associated with renal lesions (cysts, hydronephrosis, renal artery stenosis); hereditary (mechanism unknown)
    2. Excessive proliferation of erythroid cells or all cells of marrow (myeloproliferative disease) without increased erythropoietin activity (i.e., polycythemia vera)

cit in circulating red cells ensues (anemia). The increased production rate is reflected in an increase of marrow erythroid cells and blood reticulocytes (effective erythropoiesis) to the extent that the marrow can respond, normally up to eightfold. The pathophysiology of the erythron can thus be categorized broadly as to anemic states due to insufficient production, excessive loss, or a combination of these. Conversely, a pathological state may be associated with excessive production of red cells, resulting in an increased circulating red cell mass (erythremia). A detailed outline of the pathophysiology of the erythron based upon these concepts is presented in Table 27–1.

The erythrocyte: a carrier for oxygen

The principal function of the erythrocyte is to provide a transport vehicle for hemoglobin, the respiratory pigment which accepts, transports, and releases oxygen to tissues. In order to accomplish this, the mature erythrocyte must also provide an environment in which hemoglobin can be maintained in its functional reduced state.

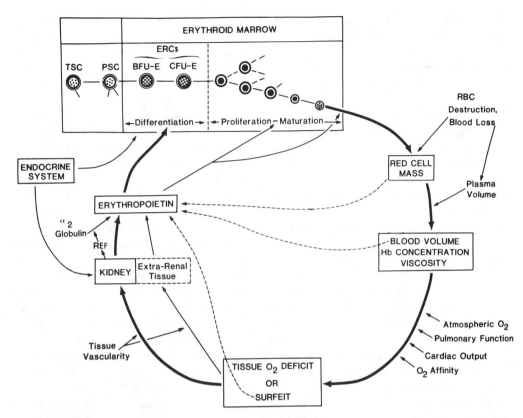

FIG. 27–1. The interrelationship of oxygen transport, the erythropoietin mechanism and the endocrine influence on erythropoiesis. TSC = totipotential stem cell; PSC = pluripotent stem cell; ERCs = erythropoietin-responsive cells; BFU-E = burst-forming unit—erythroid; CFU-E = colony-forming unit—erythroid; REF = renal erythropoietic factor.

Oxygen is accepted in pulmonary capillaries and, after being bound to hemoglobin, is transported to peripheral capillaries and released to tissues. The partial pressure of oxygen in the capillary must be sufficiently great to permit oxygen to leave the capillary and diffuse through tissues. The partial pressure of oxygen in arterial blood is about the same throughout the body, whereas the pressure of oxygen in venous blood varies, depending upon the amount of oxygen utilized by a given tissue. The quantity of oxygen delivered to a given tissue depends then upon the amount of oxygen in inspired air, the functional state of the lung, the cardiac output, the hemoglobin concentration, the affinity of the hemoglobin for oxygen, and the vascular distri-

bution of blood to a given tissue (Fig. 27–1). The mean capillary oxygen tension (and, hence, tissue oxygen) can be influenced by any one of these factors. In the event of oxygen deprivation, compensatory mechanisms function: increased cardiac output; an increased number of active tissue capillaries; a rightward shift in the oxygen-hemoglobin dissociation curve, providing for the release of oxygen at lower pressure; an enhanced production of erythropoietin to increase the number of circulating red cells and, hence, hemoglobin. A functional concept of anemia or erythremia must therefore consider the adequacy of tissue oxygen supply. For example, a healthy individual who moves to high altitude has an inadequate hemoglobin concentration for adequate tissue oxygenation and develops the symptoms of hypoxia.

Anemia and erythremia

Normally, blood contains about 15 g. of hemoglobin/100 ml., and each gram of hemoglobin carries 1.36 ml. of oxygen when completely saturated. Arterial blood carries about 21 ml. of oxygen/100 ml., while venous blood contains about 15 ml. of oxygen/100 ml., so that 6 ml. have been delivered to the tissues. In *anemia*, as the hemoglobin falls oxygen delivery is facilitated by a decrease in the affinity of hemoglobin for oxygen that is mediated by an increase in red cell 2,3-diphosphoglycerate (2,3-DPG) concentration (see Fig. 27–8 and Chap. 8). If the hemoglobin concentration falls below 5 g./100 ml., less than 6 ml. of oxygen/100 ml. of circulating blood is available to tissues, the reduced affinity of hemoglobin for oxygen is no longer sufficient, and major adjustments in other systems must occur. Cardiac output remains relatively constant until the hemoglobin level falls to about 7 or 8 g./100 ml., but then it increases in proportion to the hemoglobin deficit. Blood flow becomes redistributed to the most vital centers. Renal blood flow may be reduced by 50 per cent, and cutaneous flow is also markedly decreased. Liver function is usually fairly well maintained in severe anemia, provided the disease process causing the anemia has no direct effect on the liver per se.

The symptoms of anemia depend upon its severity, its rate of development, and its duration, and moreover, upon the ability of other physiological mechanisms to compensate for the reduction in circulating hemoglobin, principal among which is the status of the cardiovascular system. Signs and symptoms common to all anemias are pallor, shortness of breath, increased heart rate, palpitation, angina and heart failure, irritability, dizziness, fatigue, headache, and tinnitus. In general, these effects are the result of the decreased oxygen-carrying capacity of the blood. However, it is impor-

tant to realize that, if the anemia develops at a slow rate, remarkable compensation and adaptation may occur with the appearance of few, if any, symptoms in the absence of physical exertion.

In contrast to anemia, an increase in red cell production results in *erythremia*, the blood volume and hence cardiac output increase, and oxygen transport is in excess of demand. If the hemoglobin increases over 20 g./100 ml., blood viscosity rises sharply, cardiac output decreases, and oxygen transport becomes reduced. A ruddy complexion is associated with symptoms due to the expanded blood volume, slowed blood flow as a result of the increased blood viscosity, and hypoxia; that is, the cardiovascular stress may be manifest as chest pain or fullness, increased heart rate, shortness of breath, claudication; the effect on the central nervous system may result in irritability, light-headedness, fatigue, tinnitus. In addition, symptoms and signs of arterial and venous thrombotic phenomena, a consequence of the increase in blood viscosity, may occur. Thus, like anemia, the complaints of erythremia (too many erythrocytes) may be referable to several organ systems.

Regulation of erythropoiesis

Erythropoietin

Inasmuch as the erythron occupies a central position in the organism's mechanisms of cellular respiration, one might anticipate that a major factor influencing the control of the erythron would be the relation between oxygen supply and requirement (Fig. 27–1). Indeed, the major stimulus to erythroid cell production is tissue oxygen deficit; moreover, in the presence of oxygen excess, the rate of erythropoiesis is diminished or ceases entirely. Red cell production is principally regulated by erythropoietin (Ep), which is produced or activated by the kidney in response to tissue oxygen requirements. A proportional relationship has been demonstrated between the degree of induced hypoxia or anemia and the level of Ep activity in the plasma and urine. In man, plasma and urine Ep activity increase to maximum between 24 and 60 hours of induced hypoxia and then decrease to near normal levels within 4 to 5 days. The reason for this wave of plasma Ep activity rather than a sustained level is not clear, but there is evidence that a shift in the oxyhemoglobin dissociation curve results in the delivery of 20 to 30 per cent more oxygen prior to any actual increase in the number of circulating red cells. However, sustained increases in Ep are seen in severe chronic anemia, presumably because the capacity of other compensatory mechanisms has been exceeded.

The principal source of Ep in man is the kidney and by fluorescent antibodies Ep has been identified in the renal cortex. That other sites of origin exist is apparent from the observation that Ep activity can be detected in the plasma of anephric individuals. Extrarenal Ep accounts for about 20 per cent of the total Ep produced, and its source appears to be the liver macrophage (the Kupffer cell). There is evidence suggesting that Ep is formed by the enzymatic action of subcellular fractions of renal tissue (renal erythropoietic factor) on a globulin substrate in plasma. Other evidence indicates that the kidney releases a pro-erythropoietin which is activated by a plasma factor or that the kidney releases the active hormone. The stimulation of Ep production by hypoxia appears to be mediated by a medullary release of prostaglandin E_2, which elevates renal cortical cyclic adenosine monophosphate (cAMP) levels. The newly formed cAMP then may activate a protein kinase that leads to increased Ep production, possibly through phosphorylation of an Ep precursor. Erythropoietin has not been completely characterized chemically. Its molecular weight is about 46,000, of which 30 per cent is carbohydrate, and it migrates as an α-globulin.

The principal mode of action of Ep is to induce differentiation and development of committed erythroid precursor cells (Fig. 27–1). Recent studies indicate a continuum of EP-responsive cells (ERCs), the most primitive cell (assayed as the burst-forming unit erythroid, or BFU-E, in cell culture systems) being the least responsive and the most mature (in cell cultures termed the colony-forming unit erythroid, or CFU-E), probably the immediate precursor of recognizable proerythroblasts, being the most responsive. The precise mode of action of Ep is not known. There is evidence that it acts on the cell membrane to produce a cytoplasmic protein intermediate. Sequential activation of several DNA-dependent RNA polymerases is followed by new synthesis of ribosomal and transfer RNAs. Subsequent appearance of globin mRNA is associated with iron uptake and globin synthesis. In addition, Ep stimulates histone and chromosomal protein formation but stimulation of DNA synthesis and mitosis seems to follow the effects on transcription. The ERCs have been shown to have receptors for β-adrenergic agents which may modulate the responsiveness of the ERCs to Ep (e.g., cAMP potentiates the stimulation of erythropoiesis by Ep). Erythropoietin also accelerates development of differentiated erythroid cells and the release of reticulocytes into the circulation.

In summation, the production rate of erythroid cells is governed for the most part by EP, which in turn is elaborated in response to differences in tissue oxygen requirements and the delivery of oxygen to the tissues. In a larger context, the Ep system may be viewed as one facet of the total mechanism having to do with blood volume regulation. The role of the kidney in plasma volume regulation is discussed elsewhere in this volume (see Chaps. 3 and 11).

CLINICAL EXAMPLES. Thus, far, anemias resulting from abnormalities of the Ep mechanism have been related to inadequate production and inactivation of Ep (see I.A.1

in Table 27–1). In the former, renal disease provides the best example. The mechanisms of the anemia of renal disease are complex but regularly involve decreased production and, to a much smaller extent, increased destruction of red cells. Erythropoietin activity in plasma and urine is decreased in these patients, but whether this decrease is due only to insufficient production or to inactivation by inhibitors as well remains to be clarified. Erythropoietin activity is also reduced in the anemia of chronic disorders (e.g., chronic infection, inflammation) and in liver disease. That Ep antibodies can occur in certain diseases (e.g., thymoma) seems to be established. In this instance, the combination of Ep and the antibody prevents the biological activity of the hormone in an assay animal.

Erythremia resulting from abnormalities of the Ep system may be categorized as that which is associated with inappropriate Ep activity, in contrast to the erythremia resulting from hypoxia and an appropriate increase in Ep (see item III in outline). The former is associated with tumors of the kidney (hypernephroma), the liver (hepatoma), the cerebellum (cerebellar hemangioblastoma), the adrenals (adrenal carcinoma, pheochromocytoma) and the uterus (massive fibromata). Whether such tumors secrete Ep per se or stimulate the formation of Ep remains to be clarified. Certain benign lesions (renal artery stenosis, renal cysts, hydronephrosis) appear to act by inducing ischemia of the oxygen sensor in the renal parenchyma. Hereditary erythrocytosis results from inappropriately increased levels of Ep, for which the mechanism remains unknown.

The endocrine system

Most hormones may also be viewed as regulators of erythropoiesis (Fig. 27–1; see also I.A.2 in Table 27–1). Hypofunction of the pituitary, thyroid, testes, or adrenals results in diminished red cell production, and hyperfunction of the thyroid or the adrenal glands is associated with increased rates of erythropoiesis. Administration of testoste-

rone and certain analogues induce erythremia in animals and man, and various estrogenic preparations induce anemia. The mechanisms of action of these hormones have not been completely elucidated but appear to involve effects on metabolism and oxygen requirements of the organism, "growth-stimulating" effects, and specific effects on erythroid cellular production and function which are not well defined at present. Awareness of these effects has provided a stimulus for the investigation of the interrelationship of testosterone, estrogen, and thyroxine on Ep production or its mechanism of action.

It is clear that one of the mechanisms by which testosterone stimulates erythropoiesis is by inducing the elaboration of Ep, presumably by a direct renal effect, although it also increases oxygen affinity of hemoglobin and decreases red cell 2,3-DPG. Administration of fluoxymesterone to eunuchoid males or to females results in increased levels of urinary Ep and a concomitant increase in circulating erythrocytes. Evidence is accumulating that suggests an effect of androgenic compounds and their metabolites also on the stem cell compartment(s) by increasing the number of ERCs (Fig. 27–1). Estrogens may decrease the stem cell response to Ep and in large amounts may suppress Ep production. There is correlation between the hypermetabolism and increased oxygen requirements of hyperthyroidism and an increased rate of erythropoiesis. Thyroid hormone also modulates the β-adrenergic receptor activity on erythroid cells. Erythremia does not occur because the plasma volume increases and the red cell life span is shortened. On the other hand, patients with hypothyroidism have a decreased oxygen requirement and exhibit reduced levels of circulating erythrocytes.

In general, the anemia of endocrine deficiency (ovarian deficiency excluded) is characterized by normal appearing erythrocytes and hemoglobin levels of about 10 g./100 ml. Usually, the symptoms and signs of the primary disease predominate and the anemia is defined by laboratory methods. The anemia is corrected by hormone replacement therapy over a period of weeks. It is important to realize that clinically nonapparent endocrine deficiency may be the cause of anemia of obscure etiology.

Negative feedback

The absence of Ep (induced by Ep antisera) in normal animals is accompanied by cessation of erythropoiesis. Less clear, however, is the observation that erythropoiesis may be markedly increased in the absence of anemia and of demonstrable tissue oxygen deficit when the red cell life span is shortened (i.e., compensated hemolytic anemia). This suggests that control mechanisms other than Ep are operative. It has been postulated that a negative feedback mechanism (an inhibitor to erythropoiesis) related to aged red cells may exist; thus, when aged red cells are removed from the circulation, erythropoiesis may proceed. Erythropoietic inhibitory activity is demonstrable in the plasma of animals transfused to excess with red cells, in erythremic man returning from high altitude, in polycythemic newborn infants of nonanemic mothers, and in animals that have been exposed to high oxygen. Also, various protein preparations from urine, kidney, and other tissues with the capacity to inactivate Ep have been described. More work is required before the clinical significance of the control of erythropoiesis by negative feedback can be assessed. This concept is illustrated in Figure 27–1 by the broken lines, indicating that Ep activity is diminished in the presence of excess blood volume, red cells, or oxygen transport.

Differentiation, proliferation, and maturation of the erythroid cell

Current concepts hold that the erythroid cell line differentiates from a set of unipotential or committed stem cells, the latter to be distinguished from those unipotential stem cells which are destined to differentiate into granulocytes and megakaryocytes (platelet-forming cells) as well as the totipotential and pluripotent stem cells, the precursors of the committed stem cells (Fig. 27–1). The committed erythroid stem cells differentiate into the most immature, morphologically identifiable red cell precursors, the proerythroblasts. These young precursors then develop by nearly imperceptible gradations through an orderly sequence, arbitrarily divided (on the basis of morphological and staining characteristics) into basophilic, polychromatophilic, and orthochromatic erythroblasts, reticulocytes, and finally, mature erythrocytes. When circulating erythrocytes are present in excess, erythropoietic differentiation ceases. In that case those cells to the right of the committed stem cells (Fig. 27–1) mature in a few days and normally enter the circulating blood and are not replaced. Thus, examination of bone marrow 5 days after transfusion of excess red cells reveals no erythroid cells. If the excess red cells are removed, or if the animal is made hypoxic or is given Ep, erythropoiesis is resumed.

It is apparent that the erythron is not a self-maintaining population of cells; rather, the existence of erythrocyte precursors is dependent upon the differentiation of the committed stem cells in response to physiological requirements of the organism, mediated by Ep. On the other hand, the uncommitted stem cells are considered to be a population of cells which are self-maintaining. The mechanism that controls the growth of this stem cell compartment remains unknown for the most part. Evidence is accumulating that its microenvironment is of importance (e.g., in the mouse, commitment of pluripotent stem cells to ERCs occurs more readily in the spleen than in the bone marrow). Self-regulation occurs in other cell systems and has been attributed to inhibitors of cell growth which are contained within the cells themselves and reach a critical level in dense cell populations. Knowledge of regulation of the stem cell compartments is important for our further understanding of disorders of defective hematopoiesis, including anemia and polycythemia of obscure etiology.

Examples of diseases in which the pluripotent or totipotential stem cell is defective are aplastic anemia and polycythemia vera. Patients with *aplastic anemia* present with anemia and with decreased platelets (thrombocytopenia) and decreased leukocytes (leukopenia) (see I.B.1 Table 27–1), and Ep levels are very much increased. In this group of diseases a spectrum of pathophysiological mechanisms, although still poorly defined, is being recognized. Thus, the number of normal totipotential or pluripotent stem cells may be reduced, or abnormal stem cells with a decreased capacity for self-renewal may evolve. Recent observations indicate that in some instances stem cells are inhibited by humoral (antibody) or cellular (suppressor T lymphocytes) mechanisms, or stem cell function is impaired because of the lack of humoral (colony stimulating factor) or cellular (helper T-cells) hematopoietic "cofactors." Finally, damage to the marrow microenvironment can at least contribute to impaired stem cell function. Isolated erythroid aplasia occurs because of antibody directed against erythroid committed and differentiated cells (pure red cell aplasia) or, rarely, against Ep, and suppressor lymphocytes may also inhibit erythroid development. In contrast to normal marrow, the marrow of patients with *polycythemia vera* in cell cultures differentiates and proliferates in the absence of added Ep. Cell marker studies in such individuals who also are doubly heterozygous for electrophoretically distinct glucose-6-phosphate dehydrogenase variants, have demonstrated only a single isozyme in the erythrocytes, granulocytes, and platelets, in contrast to other tissues, thus supporting a clonal origin of the disorder at the pluripotent stem cell level. Patients with this disorder exhibit erythremia without increased

levels of Ep and also have variably increased numbers of platelets and leukocytes (see III.B.2 in Table 27–1). Here, the pluripotent stem cell of the new hematopoietic clone fails to recognize the normal inhibition to growth and proliferates autonomously.

Developing erythroid cells undergo well-recognized morphological changes. The proerythroblast is a cell characterized by a large nucleus containing a nucleolus and having a dark blue cytoplasm. During progressive maturation, the nucleolus disappears, and the nucleus contracts and becomes gradually pyknotic (basophilic erythroblast); the cytoplasm changes from dark blue to red (polychromatophilic and orthochromic erythroblasts) as the concentration of hemoglobin increases. Three to five mitotic divisions occur during the evolution from the proerythroblast to the orthochromic erythroblast leading to progressive decrease in cell size. The time required for maturation to occur has been estimated to be 30 hours for proerythroblasts, 12 to 64 hours for basophilic erythroblasts, 8 to 25 hours for polychromatophilic erythroblasts, and about 19 hours for orthochromatic erythroblasts, or a total of about 3 to 5 days. It has been estimated that 10 to 15 per cent of the cells normally die at various stages of development.

During this process of erythroid development, cell division (or proliferation) is closely coordinated with the cells' maturation. The proliferating capabilities depend mainly upon biosynthesis of nuclear material, while maturation results from a progressive accumulation of cytoplasmic hemoglobin. The latter may, in turn, influence the progressive decrease in nuclear biosynthetic events. A failure of nuclear or cytoplasmic growth, as discussed below, results in asynchrony of these coordinated events, and aberrant or ineffective erythropoiesis occurs. A large proportion of such cells are recognized as defective and are removed from further development by

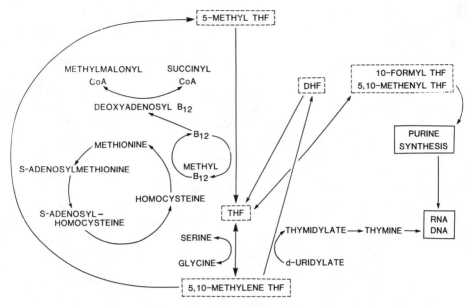

FIG. 27–2. The metabolic pathways of folate and vitamin B_{12}. THF = tetrahydrofolate.

phagocytosis within the marrow (intramedullary hemolysis).

Nuclear maturation

The biosynthesis of nuclear material (DNA and RNA) of the hemopoietic cell requires not only availability of the basic structural building blocks of nucleic acids but also intact enzyme and coenzyme systems. Certain enzymes may be deficient on a genetic basis (e.g., orotidylic decarboxylase, B_{12}-reducing enzymes) or may be inhibited by drugs (e.g., inhibitors of folate reductase such as methotrexate or pentamidine). Two critical coenzymes, folate and vitamin B_{12}, are particularly indispensable for DNA synthesis in hemopoietic cells (see I.B.2 in Table 27–1).

FOLATE. The biological function of folate is dependent upon its availability and its conversion to active (coenzyme) forms. Unable to synthesize it, man depends upon absorption of dietary folate. Most ingested natural folate compounds are in the form of polyglutamates and must be converted to monoglutamates for absorption. In the intestinal mucosa, the monoglutamates are largely reduced to tetrahydrofolate (THF) and methylated to 5-methyl THF; the latter represents the principal circulating form of folate. In tissues, folate is again polyglutamated, and current evidence indicates that its subsequent conversion products (the active coenzyme forms) are also polyglutamates.

The biochemical role of folate is that of a coenzyme in a number of reactions in which single carbon atoms are transferred to or built into nucleic acid intermediates. The important metabolic reactions dependent upon folate coenzymes are the synthesis of thymine (by 5,10-methylene THF), purines (by 10-formyl THF and 5,10-methenyl THF), and methionine (by 5-methyl THF) (Fig. 27–2). A lack of THF may result not only from folate deficiency but also from B_{12} deficiency, since B_{12} is necessary to transform 5-methyl THF to THF in the methyl transferase reaction, which converts homocysteine to methionine. Thus, lack of either vitamin would reduce the availability of THF for the subsequent generation of 5,10-methylene THF for thymine synthesis and hence DNA synthesis. This relationship provides an explanation for the correction of anemia due to B_{12} deficiency by administration of large amounts of folate.

Folate occurs naturally in yeasts, vegetables, liver, fruits, grains, nuts, and eggs. Most diets contain more than 1,000 μg./day; the minimal human adult requirement is about 50 μg./day. The monoglutamates are absorbed in the duodenum and proximal jejunum. Serum folate levels range from 6 to 21 ng./ml. and total stores from 5 to 10 mg., of which 3.5 to 7.5 mg. are stored in the liver. Because the daily requirement of folic acid is about 50 times that of B_{12}, the stores of folate can be depleted more readily (in weeks) than those of B_{12} (in years). Folate deficiency is recognized as being second only to iron deficiency as a cause of nutritional anemia. Deficiency states result from inadequate dietary intake, inadequate absorption associated with intestinal disorders, alcohol abuse inhibiting absorption and metabolism or the use of certain drugs (diphenylhydantoin) which may enchance folate catabolism, and, finally, increased requirements associated with increased red cell production rates (e.g., in hemolytic anemias and pregnancy) or with the administration of antagonists of folate metabolism (methotrexate).

VITAMIN B_{12}. Vitamin B_{12} (cyanocobalamin or hydroxocobalamin) in the diet and as the medicinal preparation contains trivalent cobalt, which must be reduced to the monovalent form to be metabolically active. The two active coenzyme forms are methyl B_{12} and deoxyadenosyl B_{12}. The participation of B_{12} in the biosynthesis of methionine from homocysteine first requires methylation of B_{12} to methyl B_{12} with the conversion of 5-methyl THF to THF (Fig. 27–2). Because most circulating and tissue folate occurs as 5-methyl THF, its conversion or recycling into THF is diminished in B_{12} deficiency ("methyl THF trap"). It is this impairment of folate metabolism which is, to date, the best explanation for the inhibited DNA synthesis in B_{12} deficiency. In addition, B_{12} is required for normal priopionate metabolism and in the form of deoxyadenosyl B_{12} serves as a coenzyme for methylmalonyl-CoA isomerase in the transformation of methylmalonyl-CoA (a product of propionate) to succinyl-CoA (Fig. 27–2). This role of B_{12} was thought to be relevant to the importance of B_{12} in neural metabolism. Recent experimental work, employing an animal model in which B_{12} deficiency is induced with nitrous oxide, which inactivates B_{12} by oxidizing the cobalt, indicates that B_{12} neuropathy is due to lack of methionine and thus S-adenosylmethionine for transmethylation of myelin in the nervous system because homocysteine cannot be remethylated to methionine (Fig. 27–2).

Vitamin B_{12} is synthesized by microorganisms and not by mammalian cells. The usual dietary intake (3 to 30 μg./day) is in the form of animal protein, and deficiency states (with rare exceptions) are the result of impaired absorption. Normally the absorption of B_{12} is dependent upon the presence of Castle's intrinsic factor, a complex glycoprotein (mol. wt. 50,000) secreted by the parietal cells in the fundus of the stomach. Recent evidence indicates that B_{12} released from food is first bound to a salivary R protein (*R*apid migrating protein on electrophoresis) at the acid pH of gastric juice. Following degradation of the R protein by pancreatic proteases in the duodenum, released B_{12} becomes bound to intrinsic factor and the complex is absorbed by specific receptor sites in the microvilli of the distal ileum. The absorbed B_{12} is transported to tissues by a β-globulin (transcobalamin II) in the plasma. The total body stores of B_{12} are estimated to be about 5 mg., of which one-third resides in the liver. Because the daily requirement is only about 1 μg., a period of 3 to 10 years is required for B_{12} deficiency to become manifest if absorption fails to occur. Vitamin B_{12} deficiency anemia and neurological deficits may occur, then, from malabsorption resulting from lack of intrinsic factor (e.g., pernicious anemia, resection of gastric fundus); because of disease of the ileum (e.g., tropical sprue, regional enteritis, ileal resection); from inadequate amounts available for absorption because of reduced liberation of food B_{12} due to gastric dysfunction (e.g., atrophic gastritis, partial gastrectomy) or utilization of B_{12} by intestinal bacteria in blind loops of bowel or diverticula; or, rarely, from lack of transport protein (transcobalamin). Patients with chronic pancreatitis have been shown to malabsorb vitamin B_{12} by absorption tests because of failure to degrade the R protein. However, clinical B_{12} deficiency and anemia do not develop, probably because treatment of the pancreatic insufficiency is instituted sufficiently early in the course of the disease, and enough dietary B_{12} is transferred to intrinsic factor for daily needs. Deficiency of B_{12} from inadequate dietary intake, even by food fadists who avoid animal products entirely and for long periods of time, is extremely unusual and probably impossible unless subtle intestinal malabsorption is present.

In B_{12} or folate deficiency as well as other states of impaired nuclear (DNA) synthesis, red cell (as well as leukocyte and platelet) production is decreased. The impaired nuclear synthesis slows the S (synthetic) phase of the dividing red cell precursors during mitosis while cytoplasmic development proceeds unimpeded or may be accelerated under the high Ep drive. Consequently, the number of cell divisions is reduced, resulting in a cell that is larger than normal and has an excessive amount of cytoplasm, with a disproportionately immature nucleus. This cell is called the megaloblast. Its ultimate progeny (if it matures) is a large erythrocyte (macrocyte).

Because the appearance of the erythroid cells (megaloblasts) and the erythrocytes (macrocytes) is the same in B_{12} and in folate deficiency, additional chemical and functional tests are required to distinguish between the two deficiencies. Administration of physiological doses of either substance will elicit a hemopoietic response (reticulocytosis) only in the specific deficiency. Measurement of serum concentrations of B_{12} and/or folate usually pinpoints the existing deficiency. Detection of increased amounts of methylmalonic acid in the urine also distinguishes untreated vitamin B_{12} deficiency. The physiological cause of vitamin B_{12} deficiency is established by assessing the absorption of radiolabeled B_{12} with and without intrinsic factor (the Schilling test). Folate absorption can also be assessed, but a standardized method has not been developed. The importance of distinguishing between deficiency of vitamin B_{12} and folate deficiency lies in the awareness of the neurological disease which may develop if B_{12} deficiency is not recognized and in providing useful clues about the underlying disease process which produced the deficiency.

Cytoplasmic maturation

Two major biosynthetic pathways, namely heme and globin synthesis, determine the unique cytoplasmic maturation of developing erythroid cells (Fig. 27–3). Alterations in the supply of building blocks, enzymes, and cofactors for, as well as a nuclear control over, these pathways may lead to impaired cytoplasmic development (see I.B.3 in Table 27–1).

HEME SYNTHESIS. The heme biosynthetic pathway is initiated with the condensation of glycine and succinyl-CoA (derived from the tricarboxylic acid cycle) in the presence of the cofactor (coenzyme) pyridoxal-phosphate and under the influence of the rate-limiting enzyme Δ-aminolevulinic acid (ALA) synthase to form ALA (Fig. 27–3). Two molecules of ALA are joined to form porphobilinogen, a reaction catalyzed by the enzyme ALA dehydrase. Four molecules of porphobilinogen are converted to uroporphyrinogen III in the presence of two enzymes, porphobilinogen deaminase and uroporphyrinogen cosynthase. Uroporphyrinogen III is then transformed to

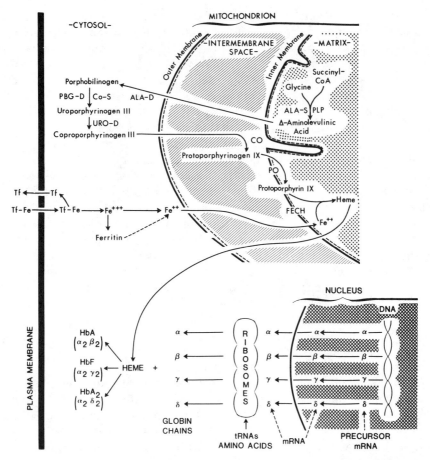

FIG. 27–3. The pathways of heme and globin synthesis. ALA-S = Δ-aminolevulinic acid synthase; PLP = pyridoxal-phosphate; ALA-D = Δ-aminolevulinic acid dehydrase; PBG-D = porphobilinogen deaminase; Co-S = uroporphyrinogen co-synthase; URO-D = uroporphyrinogen decarboxylase; CO = coproporphyrinogen oxidase; PO = protoporphyrinogen oxidase; FECH = ferrochelatase; Tf = transferrin; mRNA = messenger RNA; tRNA = transfer RNA.

coproporphyrinogen III by uroporphyrinogen decarboxylase. Protoporphyrinogen IX is formed by coproporphyrinogen oxidase and then oxidized to protoporphyrin IX. Iron is inserted into protoporphyrin IX by the enzyme ferrochelatase to form heme.

The various steps in the heme biosynthetic pathway are divided between the mitochondrial and cytosol compartments (Fig. 27–3). The first step and the last three steps proceed in the mitochondria and, hence, are present only in immature red cells, while the three steps converting ALA to coproporphyrinogen occur in the cyto-

sol and persist in the mature red cell. The principal control of the pathway is exerted at the initial and rate-limiting step catalyzed by ALA synthase. This enzyme is inhibited directly and its synthesis is repressed by heme, the end product of the pathway. Thus, an accumulation of heme results in a decrease in ALA synthase activity, and a decrease in heme stimulates the enzyme. Since pyridoxal-phosphate (the active form of vitamin B_6) is the coenzyme for ALA synthase, the availability of vitamin B_6 also controls ALA synthesis. Finally, the rate of globin synthesis may influence heme production and, conversely, heme influences globin synthesis (see below).

Defects at various steps in the heme biosynthetic pathway result in hypochromic anemia (see I.B.3.a Table 27–1). Deficiency of vitamin B_6 is not associated with anemia in man, but certain hypochromic (sideroblastic) anemias occasionally improve with the administration of large amounts of vitamin B_6. In the sideroblastic anemias accumulation of enormous amounts of iron in mitochondria, characterizing the ring sideroblasts, is associated with defective heme synthesis; the pathogenetic factors leading to these abnormalities are not defined. Certain drugs, such as isonicotinic acid hydrazine, which is used in the treatment of tuberculosis, inhibit pyridoxal-phosphate metabolism and may induce hypochromic anemia of the sideroblastic type that is reversible. Alcohol abuse can also produce reversible sideroblastic anemia, presumably through inhibition of heme synthesis. Lead has a particular affinity for several of the enzymes (ALA dehydrase, coproporphyrinogen oxidase, and ferrochelatase), ALA dehydrase being most severely affected in chronic lead poisoning. Iron depletion reduces the formation of heme from protoporphyrin, and, hence, hemoglobin; the decreased feedback inhibition of the rate-limiting ALA synthase by the lack of heme results in accumulation of free erythrocyte protoporphyrin (e.g., iron deficiency anemia, anemia of chronic disorders).

IRON BALANCE. The crucial role of iron in normal hemoglobin synthesis is readily appreciated because this element is a major structural component of the hemoglobin molecule. Inasmuch as an average of 6.3 g. of hemoglobin containing 21 mg. of iron are synthesized and degraded in each 24 hours (Fig. 27–4), inadequacy of the supply of iron to developing erythroid cells becomes apparent in a short period of time in diminished erythropoiesis and the formation of small erythrocytes which have a decreased hemoglobin concentration. Cell size seems to be determined by the number of cell divisions which continue until a critical level of hemoglobin concentration is reached in the cell. The supply of iron to the erythroid cell is dependent upon adequate iron transport and iron stores, and adequate stores depend upon the availability of iron and its absorption, as well as upon the loss of iron from the body. In the absence of excess iron loss, the iron content of the body remains relatively constant.

Normally, in the United States, the typical man, who consumes 2,500 calories per day, ingests about 15 mg. of iron, of which about 1.0 mg. (5 to 10 per cent) is absorbed. The range of the daily absorption has been estimated to be 0.5 mg. in the iron-replete individual to 3.0 mg. in the iron-deficient person. Loss of iron occurs through occult blood in the intestine (0.2 to 0.5 mg.), urinary excretion (0.1 mg.) and desquamation of skin and gut epithelium (0.2 to 0.3 mg.). The net obligatory loss of iron is therefore estimated to be about 1.0 mg./day with which normal absorption can keep pace. To compensate for menstrual blood loss (0.5 mg. of iron daily) the mature woman normally must absorb 0.7 to 2.0 mg./day; during pregnancy (with its requirement of about 900 mg. of iron), she must absorb about 3.5 mg. of iron per day in order to compensate for the loss. In the man, iron loss beyond the obligatory 1.0 mg. is almost always due to abnormal bleeding. Since 1.0 ml. of erythrocytes contains 1.0 mg. of iron, a small amount of blood (6 to 10 ml.) lost each day can result in negative iron balance and depletion of iron stores, for which the maximal absorption capacity of dietary iron of 3.0 mg. cannot compensate. Defective hemoglobin formation and anemia ensue. The sequence of events is: negative balance, depletion of iron stores, decreased saturation of the transport protein, decreased delivery to the erythroid cell, impaired hemoglobin synthesis, and anemia. Thus, the first evidence of iron deficiency is diminished storage iron, and the change in

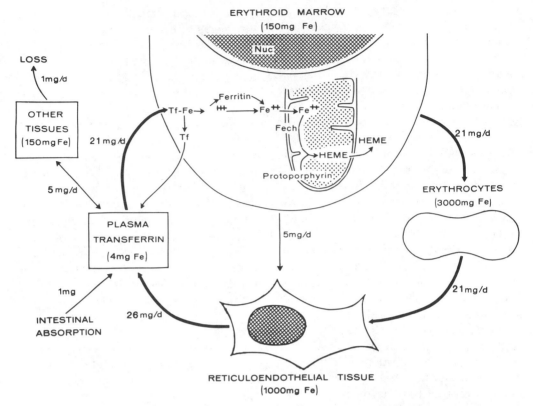

FIG. 27—4. The iron cycle. The figures show the amount (mg.) of iron in each compartment and the amount transferred along each pathway (mg./day). Tf = transferrin; Fech = ferrochelatase.

red cell size and anemia are late manifestations of iron deficiency. Depletion of essential energy-producing, iron-containing tissue enzymes may contribute to malaise, fatigue, and muscular weakness as well as surface epithelial changes (glossitis, achlorhydria, koilonychia).

IRON ABSORPTION. Absorption is greatest in the proximal portion of the small intestine, in contrast to vitamin B_{12} and folate. Control of iron absorption is relative and not absolute, since excessive soluble inorganic iron presented to the gut can be absorbed. The intestinal mucosa plays an important role in regulating iron absorption when the dietary intake of iron is within a physiological range. The mechanism is responsive to body needs and the control is apparent in relation to iron stores and the rate of erythropoiesis: absorption is increased in iron deficiency, with or without accompanying anemia, and when the rate of erythropoiesis is accelerated, as in hemolytic anemia (increased destruction of erythrocytes); excess body iron reduces absorption. The messenger to the mucosal cell remains unknown, although it is postulated that the concentration of iron within the mucosa, regulates to some extent, the absorption of iron by the cell. This messenger iron is thought to be incorporated into the cell at an early stage of development, while the cell is in the crypt of Lieberkühn, and as the cell migrates along the villus it may accept additional iron. Iron transfer across the cell is believed to be the result of a metabolic process. A portion of the absorbed iron is rapidly delivered to the plasma, and another portion is found as mucosal ferritin. Increased rates of absorption are associated with the presence of relatively small amounts and decreased rates of absorption with larger amounts of mucosal ferritin, suggesting that a diversion of iron to mucosal ferritin prevents excessive absorption. The level of plasma iron or its rate of turnover may influence iron absorption either directly or through the concentration of iron in the young crypt cells.

Luminal factors which influence but do not control iron absorption are hydrochloric acid of gastric juice, the presence of ascorbic acid, and the formation of soluble or insoluble iron complexes (phosphates, phytates); the luminal concentration of iron also affects the absorption rate. Evidence about the influence of other factors in gastric juice on iron absorption is conflicting, although a protein factor which binds iron has been described. However, absorption of dietary iron is impaired in individuals who have undergone partial gastrectomy.

IRON TRANSPORT. Iron entering the plasma is transported by an iron-binding protein, transferrin (each molecule binds two atoms of iron), which has the electrophoretic mobility of a β_1 globulin and a molecular weight near 75,000. This transport protein is normally one-third saturated with iron, the quantity of iron ranging from 60 to 180 μg./100 ml. of plasma. Plasma iron concentration represents the balance between iron delivered to plasma from the gut and hemoglobin breakdown sites and iron removed from the plasma by heme biosynthesis, cell metabolism, and deposition in storage sites (Fig. 27–4). This dynamic interchange is demonstrated by the fact that some 26 mg. of iron enter and leave the plasma pool during a 24-hour period. Of the incoming iron, about 80 per cent is transported to the developing erythron, and the remainder represents iron transport between various iron pools.

Iron (500 to 1,000 mg.) is stored intracellularly as ferritin (a large water-soluble spherical protein shell with a molecular weight of 450,000, accommodating up to 4,500 iron atoms in its central core in the form of ferric hydroxyphosphates) and hemosiderin (large aggregates of ferritin molecules, together with porphyrin, lipid, and other substances) in hepatocytes and in reticuloendothelial cells of the liver, spleen, and bone marrow. An estimate of iron stores can be obtained from observing the hemosiderin granules in smears of bone marrow stained with Prussian blue and by measurement of serum ferritin. The release of iron from its storage sites is not characterized but is believed to involve initial reduction to the ferrous form and subsequent oxidation in order to be bound to transferrin. The release of iron from the reticuloendothelial cell is impaired in the anemia of chronic disorders, constituting a major pathophysiological mechanism of the anemia.

IRON KINETICS. By using radiation detection devices, the pathway and final destination of trace amounts of radioactive iron may be determined. In the normal individual, the kinetics and pattern of distribution of iron entering the body can be summarized as follows: Ingested iron salts are absorbed quickly, and an increase in transferrin-bound plasma iron can be detected within $2\frac{1}{2}$ to 5 hours. If the iron is injected intravenously, it combines with transferrin immediately. The time required for 50 per cent of the iron to leave the plasma is about 60 to 90 minutes. This plasma iron clearance, depicted graphically, describes a curve composed of about three exponential functions, the reason being that the radioiron leaves the plasma (transferrin) transport system and later reappears from iron pools. As iron leaves the plasma it can normally be detected within the marrow in a matter of minutes. The total iron then appears in newly formed red cells in the circulating blood over a period of 1 to 10 days, depending upon the rate of red cell production. In disease, plasma iron disappearance and reappearance in red cells is rapid in states of increased erythropoiesis (e.g., hemolytic anemia) and slow in disorders of diminished erythropoiesis (e.g., aplastic anemia). The distribution of iron reflects the site(s) of active erythropoiesis; in the absence of erythropoiesis, the iron is detected in storage sites (e.g., liver and spleen). The reappearance of radioiron in circulating red cells reflects the degree of effective erythropoiesis.

IRON-DEFICIENT ERYTHROPOIESIS. Iron-deficient erythropoiesis occurs when iron is not available (iron deficiency anemia) or when iron transport to the erythroblast is impaired (anemia of chronic disorders; atransferrinemia). In either case, there is defective hemoglobin synthesis, decreased cell proliferation, impaired maturation, and the erythrocytes produced are small, pale, and misshapen, as seen in other conditions that affect hemoglobin synthesis (e.g., sideroblastic anemia and thalassemia). Iron deficiency anemia is identified by estimating iron stores and by measurement of iron and the iron transport protein in serum, and by estimation of the proportion of iron attached to the protein (serum iron, serum iron binding capacity, and percentage of saturation of transferrin). While the serum iron concentration is diminished in iron deficiency anemia as well as in anemia of chronic disorders (reduced iron recycling by the reticuloendothelial system), the transferrin level is normal or raised in the former (increased synthesis) and reduced in the latter (decreased synthesis). Having determined that the cause of the anemia is iron deficiency, the next step is to determine why it occurred. It results most often from iron loss (chronic bleeding) but may occur when iron demands exceed its supply (infancy, adolescence, and pregnancy), rarely from urinary loss (hemosiderinuria associated with intravascular hemolysis) and malabsorption, and very rarely from inadequate transport protein.

GLOBIN SYNTHESIS. Like any other developing cell, the immature red cell synthesizes a variety of proteins, among which are structural proteins and enzymes. By far the greatest quantity of protein made is globin. The genetic information required for the synthesis of a protein is coded in the DNA in the cell nucleus. Coding nucleotide sequences (codons), but not the noncoding nucleotide sequences (introns), in the gene for a specific protein dictate its amino acid sequence (structure) (Fig. 27–5). Following transcription of the full nucleotide sequence by an RNA polymerase into heterogeneous nuclear RNA, the latter is modified at the 5′ and 3′ ends and the introns are removed by splicing reactions to form mature messenger RNA (mRNA). This is called *processing*. The mRNA, now containing only codons each of which is a sequence of three nucleotides and codes for a specific amino acid, is transported to the cytoplasm. In the cytoplasm, the mRNA attaches to ribosomes to form the polysome and there serves as the template for the sequence of amino acids in the formation of the polypeptide chain (translation). Transfer RNA molecules (tRNA), specific for a given amino acid and also containing nucleotide triplets (anticodons) which are complementary to the codons of mRNA, require enzymatic activation whereupon they serve to deliver amino acids to the ribosome. Initiation of polypeptide chain formation requires that in the presence of specific protein initiation factors, an initiation tRNA is aligned with an initiation codon in mRNA along with attachment of both to ribosomes. Following this initiation step, the ribosome moves along the mRNA template, permitting exposure of successive mRNA codons to tRNA anticodons and subsequent attachment of each amino acid to the growing polypeptide chain. Synthesis of the chain stops when the ribosome reaches a termination codon and releasing factors split the completed chain from the tRNA. Before or after their release, globin

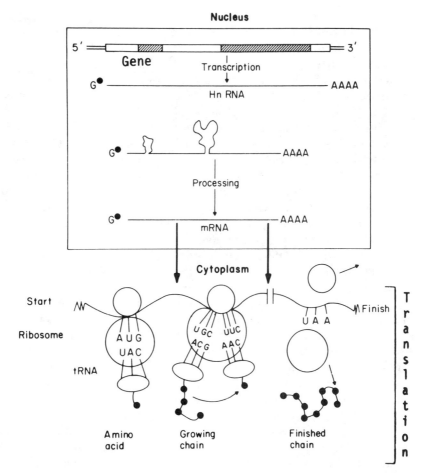

Nucleus

Cytoplasm

FIG. 27–5. Schematic representation of protein synthesis. ☐═══☐ = codon; ▨▨▨ = intron; Hn RNA = heterogeneous RNA; mRNA = messenger RNA; AUG = codon; UAC = anticodon; tRNA = transfer RNA. (After Weatherall, D. J., and Clegg, J. B.: The Thalassemia Syndromes, ed. 3, p. 42. Oxford, Blackwell Scientific Publications, 1981. Reproduced by permission of the publisher.)

chains combine with heme and with one another into stable hemoglobin tetramers by mechanisms which are not fully understood (Fig. 27–3).

Of pathophysiological importance is the genetic control of the structure as well as the quantity of globin chains formed. Separate genes control the structure of the various normal globin chains. The structure of the α-chain is controlled by two pairs of genes on chromosome 16 while the β-, γ-, and δ-chain structures are controlled by single gene pairs which are closely linked

and found on chromosome 11. The tetramers of the hemoglobin molecule are distinguished by their constituent globin chains, and two pairs of chains form the various normal hemoglobins: two α- and two β-chains form hemoglobin A; two α- and two γ-chains form hemoglobin F; two α- and two δ-chains form hemoglobin A_2. The genotype of the hemoglobins is expressed as exemplified for hemoglobin A: $\alpha_2^A \beta_2^A$, where the superscript denotes the gene controlling the structure of the respective chain or chain pair. During fetal development, α-chains (which are made from the beginning) combine early with ε-chains to form hemoglobin Gower and later with γ-chains to form hemoglobin F, while β- and δ-chain synthesis remains at a low level. At birth, the γ locus is "switched off," and the β and δ loci are activated and combine with α-chains to produce hemoglobins A and A$_2$. In the adult, hemoglobin A ($\alpha_2\beta_2$) is the major hemoglobin formed with hemoglobins A$_2$ ($\alpha_2\delta_2$) and F ($\alpha_2\gamma_2$) constituting 3 per cent and less than 2 per cent of the total, respectively.

The gene locus affecting the rate of chain synthesis is thought to be near the structural gene since rate and structural mutations appear to be inherited as alleles (e.g., sickle-β-thalassemia). (An allele is one of two or more forms of a gene which occurs at a given chromosome locus.) The quantity of globin chains formed is closely controlled and coordinated during the erythroid cell's development so that any two pairs of chains for a hemoglobin are produced in a ratio of 1:1. Excess globin chains, which are detrimental to erythrocyte function (see below), do not accumulate normally. The rate of a globin chain produced depends upon the integrity of the gene as well as the complex steps in the synthesis of polypeptide chains as outlined above, including the availability, functional integrity, and stability of mRNA. In addition, a structural alteration of a chain may reduce its stability, assembly, or rate of synthesis (e.g., in heterozygotes the abnormal hemoglobin is frequently present in lesser amounts than the normal hemoglobin). Finally, precise mechanisms exist which coordinate heme and globin production not only during the normal development of the erythroid cell but also in disorders affecting either heme or globin synthesis. Heme stimulates initiation of globin chain formation and stabilizes polysomes. Decreased globin chain synthesis leads to a decrease in heme synthesis (e.g., thalassemia), and impaired heme synthesis leads to a decrease in globin synthesis (e.g., iron deficiency, sideroblastic anemia).

Gene mutations resulting in structural alterations of the chain result in clinical disease if they alter a hemoglobin molecule so as to interfere with its molecular solubility, stability, or function. Normally, the molecule is spheroidal in shape, measuring approximately $64 \times 55 \times 50$ Å, and having a molecular weight of 64,458. Its four subunits of globin, the two α and the two β peptide chains, are organized in their secondary structure into nonhelical regions alternating with helical regions, which are stabilized by hydrogen bonds between CO and NH groups. Their tertiary structure results from folding of the chains into coils which are stabilized by hydrophobic (van der Waals) forces between the nonpolar amino acids positioned in the interior of the molecule, permitting the polar, ionizable amino acids to occupy the surface of the molecule to provide maximum solubility. The four heme groups are positioned in deep nonpolar crevices of each globin chain and between two histidines. The iron of heme is covalently bound to the proximal histidine, where the distal histidine lies opposite but not attached to the iron atom, the site which combines with oxygen. The nonpolar environment of the crevice with additional van der Waals contacts between the heme and amino acids imparts further stability to the tertiary structure of the molecule and allows the reversible association of heme iron with oxygen without significant methemoglobin formation. The quarternary structure results from contouring of the coiled chains in a complementary and compact manner but with only two main sites of contact, namely, between the α_1- and β_1-chains and the α_1- and β_2-chains, permitting spatial changes between the four globin subunits during oxygenation (subunit interaction).

The spectrum of abnormal hemoglobins is determined by combinations of structur-

TABLE 27–2
Examples of Abnormal Hemoglobins, with Pathophysiological Effects

Hemoglobin	Amino Acid Substitution	Functional Derangement	Morphology of RBC	Pathophysiological Effect
SS	$\beta^{6\,glu \to val}$	Molecular aggregation in deoxy state	Sickle cells Target cells	Rigid red cells, increased viscosity $\to$ infarctions, hemolytic anemia
CC	$\beta^{6\,glu \to lys}$	Reduced solubility with crystal formation	Target cells Spherocytes	Hemolytic anemia
Köln	$\beta^{98\,val \to met}$	Instability	Hypochromia	Hemolytic anemia Methemoglobinemia
Chesapeake	$\alpha^{92\,arg \to leu}$	Increased O_2 affinity	Normal	Erythremia
Kansas	$\beta^{102\,asn \to thr}$	Decreased O_2 affinity	Normal	Cyanosis, mild anemia
M Boston	$\alpha^{58\,his \to tyr}$	Methemoglobin	Normal	Cyanosis, erythremia in some instances

ally normal and abnormal globin chains. Individuals who inherit one abnormal structural gene will make two hemoglobins, one normal and one abnormal, e.g., hemoglobin A ($\alpha_2^A\beta_2^A$) and the abnormal hemoglobin, hemoglobin S ($\alpha_2^A\beta_2^S$). Two abnormal structural genes of the same type may be inherited (e.g., $\alpha_2^A\beta_2^S$), in which case only abnormal hemoglobin S is formed. Two abnormal genes of different type but at the same locus may occur; then only hemoglobin S ($\alpha_2^A\beta_2^S$) and hemoglobin C ($\alpha_2^A\beta_2^C$) are formed. Some mutations produce functional and clinical effects with only one chain affected, as the homozygous state in such cases would be lethal. Others are clinically silent in the heterozygous state and produce disease only in the presence of a homozygous or doubly heterozygous state.

EXAMPLES OF THE MAJOR HEMOGLOBIN-OPATHIES. Structural change may result in hemoglobins that have altered physical properties, or are unstable, or have an altered oxygen affinity, or have increased stability (Table 27–2).

Sickle hemoglobin presents the classical example of a marked change in physical properties of the hemoglobin molecule. Substitution of valine for glutamic acid in position 6 of the β-chain results in the formation of an intramolecular ring by hydro-phobic bonds between the substituted valine and the valine in position 1 of the β-chain during deoxygenation. The formation of this ring creates an abnormal site, promoting polymerization of adjacent hemoglobin molecules into rod-like structures, which in turn organize into bundles or tactoids. These deform the cells and, microscopically, they appear sickle-shaped. Such cells are rigid and fragile and have a high internal viscosity. In the homozygous state, this change in the red cell results in a disease characterized by severe hemolytic anemia and occlusion of small vessels. Patients present with a history of chronic anemia and jaundice (due to the hemolysis), beginning in childhood. They most often have episodes of pain involving joints, bones, muscles, and the abdomen, and also pneumonia-like episodes, owing to repeated vascular occlusions. Virtually every organ or tissue may be affected by this process, which may ultimately lead to organ dysfunction, best exemplified by splenic atrophy in almost all adult patients.

Hb Köln is an unstable hemoglobin resulting from the substitution of a polar (methionine) for a nonpolar (valine) amino acid in one β-chain, thereby allowing entrance of water into the normally hydrophobic heme pocket and weakening the heme-globin linkage. This results in loss of the heme group and precipitation of the globin chains that are recognized as inclusion bodies (Heinz bodies). These interfere with membrane function and deformability of the red cells, and lead to their premature removal by the reticuloendothelial system. Individuals with this hemoglobin have mild hemolytic anemia, and it is exacerbated if they are exposed to certain drugs (oxidants) which induce further precipitation of chains within the cell, leading to yet greater cell destruction.

Altered oxygen affinity can be related to the steric relationships between the α- and the β-chains, which are critical to the role of hemoglobin in oxygen transport. Single amino acid substitutions in regions of chain contacts in the hemoglobin molecule can interfere with subunit interaction, so that oxygen affinity is increased. Hb Chesapeake serves as an example (Table 27–2). The release of oxygen to tissues is impaired, and the patient develops erythremia as a compensatory mechanism to provide adequate oxygen delivery. Increased oxygen affinity also results from mutations which interfere with the Bohr effect and with 2,3-DPG binding of hemoglobin, and which increase the stability of deoxyhemoglobin. In contrast, Hb Kansas possesses a decreased oxygen affinity because the single-chain mutation reduces the hydrogen bonding between the $\alpha\beta$ dimers and is associated with a decrease in red cell production and cyanosis. In the hemoglobins M, the heme iron has been stabilized in the nonfunctional ferric state. In most instances, a heme-linked histidine is replaced by tyrosine, allowing trivalent iron to form a stabilizing intramolecular bond which prevents reduction of the iron by methemoglobin reductase. Oxygenation is impaired, and mild erythremia may result as a compensatory mechanism analogous to that occurring in patients with Hb Chesapeake.

Over 300 hemoglobin variants have so far been described, but only a minority of these produce clinical disease (syn-dromes). The biochemical defect(s) determines the altered physiology, and it is evident that patients having a hemoglobinopathy may present with anemia or erythremia. The erythrocyte may be normal in appearance or it may be markedly changed in size, shape, and hemoglobin content. Rarely, the clinical syndrome is sufficiently characteristic to suggest the hemoglobinopathy (e.g., homozygous sickle-cell disease). Red cell morphology may assist in providing clues, but the diagnosis depends upon further study of the hemoglobin by analytical methods.

THALASSEMIAS. In addition to the gene mutations that affect chain structure, mutations occur in genes which regulate the rate of synthesis of chains without altering their structure (Thalassemias, Table 27–3). Mutations in regulator genes may suppress the synthesis of one or more chains, in part or completely, because a globin gene is deleted or fails to produce the normal amount of functional mRNA. If β-chain synthesis is suppressed (β-thalassemia), production of γ-chains and/or δ-chains may be increased to varying degrees, resulting in decreased amounts of Hb A ($\alpha_2\beta_2$) and increased amounts of Hb A$_2$ ($\alpha_2\delta_2$) and/or Hb F ($\alpha_2\gamma_2$), but not sufficient to compensate for the lack of Hb A. If α-chain synthesis is suppressed (α-thalassemia), hemoglobins A$_2$ and F are not increased as both require α-chains. The clinical expression of the homozygous as well as heterozygous thalassemia states is quite heterogeneous because a variety of faults in mRNA can occur or gene deletions can be incomplete. The amount of globin chains produced which permit assembly of functional hemoglobin and, more importantly, the proportion of excess chains produced per cell by the normal gene(s) (chain imbalance) determine the pathological effects.

In homozygous β-thalassemia (β-thalassemia major), β-chain synthesis is suppressed completely or nearly so. With ap-

TABLE 27–3.
Examples of Thalassemia Syndromes, with Pathophysiological Effects

Clinical Class	Genotype	Functional Derangement	Morphology of RBC	Pathophysiological Effect
Thalassemia major	$\alpha^{thal}\beta^{thal}$	Lack of β-chain synthesis	Hypochromia, microcytes, target cells, nucleated red cells	Ineffective erythropoiesis Hypochromic and hemolytic anemia
Thalassemia minor	$\beta\ \beta^{thal}$	Decreased β-chain synthesis	Microcytes, hypochromia, target cells	None or mild hemolytic anemia
Thalassemia minima	$\alpha\alpha/\alpha-$ *	Mild decrease in α-chain synthesis	Normal	None (silent carrier)
Thalassemia minor	$\alpha-/\alpha-$ or $\alpha\alpha/--$	Decreased α-chain synthesis	Microcytes, hypochromia, target cells	None or mild hemolytic anemia
Hb H disease	$\alpha-/--$	Decreased α-chain synthesis	Hypochromia, microcytes	Hemolytic anemia
Hydrops fetalis	$--/--$	Lack of α-chain synthesis	Hypochromia, nucleated red cells	Asphyxia *in utero*
Sickle-thalassemia	$\beta^{S}\beta^{thal}$	Hemoglobin polymerization, decreased or absent normal β-chain synthesis	Hypochromia, microcytes, target cells	Infarctions, hemolytic anemia

* *Normal* α-*chain genotype,* αα/αα; − = *deleted gene*

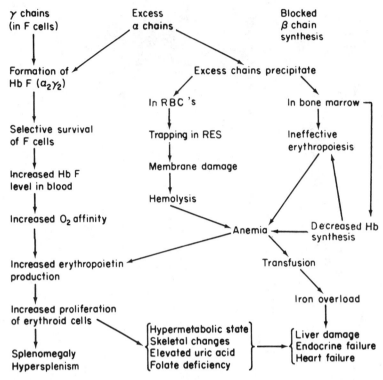

FIG. 27—6. The pathophysiology of homozygous β-thalassemia. (After Nathan, D. G.: *In* Beck, W. S. (ed.): Hematology, ed. 3, p. 159. Cambridge, MIT Press, 1981. Reproduced by permission of the publisher.)

plication of newer techniques of nucleic acid hybridization and recombinant DNA technology, the molecular pathology of many subtypes have been characterized by reduced or absent β-globin mRNA or by the fact that β-globin mRNA is not processed or translated or that it is structurally abnormal. The associated increase in γ- (and δ-) chain synthesis compensates for lack of the β-chains in part, but the amount of hemoglobin formed per cell is markedly decreased, resulting in the production of severely hypochromic-microcytic erythrocytes (Fig. 27–6). The α-chains, synthesized at a normal rate, but in relative excess, precipitate in the cell and form inclusion bodies. In the marrow, the α-chain inclusions interfere with normal precursor cell maturation, resulting in intramedullary hemolysis (ineffective erythropoiesis). In the circulating red cell, the inclusions lead to altered membrane function, premature

cell destruction, and progressive splenomegaly. In response to the severe anemia which results, an intense compensatory erythroid hyperplasia leads to impaired growth and skeletal disfigurement. An associated increased rate of iron absorption produces siderosis and consequent further increase in the size of the liver and spleen, as well as damage to other parenchymal organs. Suppression of the ineffective erythropoiesis by red cell transfusions and removal of the excess body iron by chelation have made possible partial control of these processes. In heterozygotes (β-thalassemia minor), β-chain synthesis is less severely affected, hemoglobin production is only slightly impaired, and few, if any, excess α-chains exert their adverse effects on erythroid maturation and red cell survival in the circulation. Erythroid hyperplasia in the marrow, hypochromia and microcytosis of the red cells, and a mildly

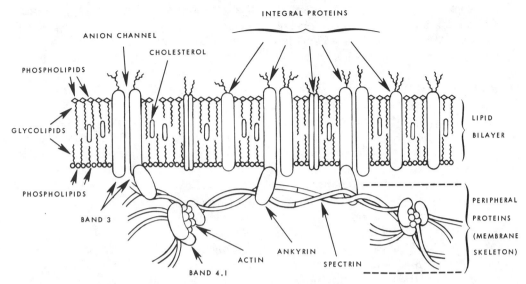

INTEGRAL PROTEINS

ANION CHANNEL

CHOLESTEROL

PHOSPHOLIPIDS

GLYCOLIPIDS

LIPID
BILAYER

PHOSPHOLIPIDS

BAND 3

ACTIN

BAND 4.1

ANKYRIN

SPECTRIN

PERIPHERAL
PROTEINS
(MEMBRANE
SKELETON)

FIG. 27–7. Diagram of the red cell membrane.

reduced cell life span can usually be detected and there is an increase in hemoglobin A_2 and, less frequently, hemoglobin F. The latter two findings provide a basis for differentiation of this disorder from other hypochromic anemias and from α-thalassemia (see below).

The α-thalassemia syndromes largely result from deletions of one, two, three, or all four genes directing the synthesis of α-chains. However, the new technology indicates a multitude of genetic determinants for these disorders, as yet unclassified. The well-defined clinical types are characterized as follows (Table 27–3): deletion of all four α-globin genes is incompatible with life because no functional hemoglobin can be made, owing to total lack of α-chains, and death occurs *in utero* (hydrops fetalis). Deletion of three α-genes leads to a moderate suppression of chain synthesis, and the excess β-chains form tetramers (Hb H) leading to cell inclusions, moderate hemolytic anemia, and splenomegaly. With two α-genes deleted the thalassemic state resembles heterozygous β-thalassemia, while deletion of only one α-gene results in a silent carrier state without recognizable red cell abnormality.

Double heterozygous forms of α- and β-thalassemia lead to a disease resembling simple heterozygous thalassemia. If one gene carries the thalassemia defect for β-chain synthesis and its allele directs synthesis of a structurally abnormal hemoglobin such as S, the majority of β-chains formed are those determined by the β^S gene: variable quantities of Hb A are formed, Hb S synthesis appears to be unimpeded, and variable increases in Hb A_2 and F may occur. The physiological effects in this instance resemble those seen in sickle cell anemia (homozygous Hb S) but tend to be less severe (Table 27–3).

The mature erythrocyte: metabolism, survival, and destruction

Membrane structure and function

The mature erythrocyte is a biconcave disc, $7\ \mu$ in diameter, with a soluble core, a membrane 75 Å thick and composed of lipids, proteins, carbohydrate, ions, and water. The structure of the membrane as currently defined is a lipid bilayer overlying a protein membrane skeleton (Fig. 27–7). The lipid bilayer consists of (1) nonidentical phospholipids intercalated with (unesterified) cholesterol and glycolipids, providing it with a viscous, two-dimensional fluid property and (2) the *integral membrane proteins* (glycoproteins) which span the lipid bilayer and have their carbohydrate portion exposed on the membrane surface. These proteins account for the acid nature (negative charge) of the membrane surface and represent red cell antigens, receptors, and ion transport proteins. The membrane skeleton is a meshwork of three interconnected *peripheral membrane proteins* (spectrin, actin, and band 4.1) that is anchored to the lipid bilayer by a fourth protein (ankyrin). This support structure stabilizes the asymmetrical phospholipids and immobilizes the integral membrane proteins and is a major determinant of the shape, flexibility, and durability of the red cell. Other functions of the membrane interrelate with the cell's interior composition and metabolism. Thus, linked cation pumps of Na^+ and K^+, which require adenosine triphosphate (ATP) and the membrane enzyme Na,K ATPase, maintain the high K^+ (100 mEq./l. red cells) and low Na^+ (10 mEq./l. red cells) concentration inside the cell, which in turn control its volume and water content. An ATP-dependent Ca^{++} ATPase serves to extrude any intracellular Ca^{++} which is deleterious to the red cell. Finally, exchange channels for anions (HCO_3, Cl^-), critical for CO_2 transport, consist of a dimer of the major intercalating membrane protein.

The red cell is unique in that by the time it has matured and left the bone marrow, it is the only cell that exists without a nucleus, ribosomes, mitochondria, and an intact Krebs cycle. Despite these losses in metabolic machinery, the cell survives approximately 120 days, during which it travels a distance of 175 miles between the heart and various tissues. Its success in withstanding the buffeting received by a free cell in constant motion and in negotiating the smallest vascular channels is determined by its unique membrane proper-ties, its fluid cytoplasm, and its unique shape (i.e., surface area to volume ratio). As a biconcave disc it is highly deformable. Decreased red cell deformability results from alterations of the membrane's protein skeleton, or loss of intracellular water (volume) leading to increased cytoplasmic viscosity, or gain of intracellular water producing a spherocyte which is less flexible, or presence of intracellular inclusions. Maintenance of the biconcave disc shape and deformability depends largely upon the integrity of the membrane and the energy metabolism of the cell.

Glycolysis

An understanding of the concepts concerning the glycolytic pathways of the mature erythrocyte provides a basis for considerations of normal red cell aging (senescence) and of genetic and acquired metabolic defects which result in shortening of the life span of the erythrocyte, manifested as hemolytic anemia (see item II in Table 27–1).

Energy requirements are met by the metabolism of glucose to lactate through two principal pathways, the Embden-Meyerhof (EM) anaerobic glycolytic pathway and the oxidative hexose monophosphate (HMP) shunt (Fig. 27–8). Normally, 10 per cent of glucose is metabolized by the HMP pathway which serves to generate NADPH. Regeneration of NADPH by glycolysis is important for the reduction of glutathione, which in turn reduces oxidized sulfhydryl groups of hemoglobin, membrane proteins, and enzymes. The maintenance of functional hemoglobin and other cellular proteins is thus dependent on this glycolytic pathway. The pathway is controlled by glucose-6-phosphate dehydrogenase (G-6-PD) activity and by the formation of oxidized glutathione (GSSG).

In the EM pathway, 1 mol. of glucose is catabolized to 2 mol. of lactate, with the net generation of 2 mol. of ATP. ATP serves as substrate for the cation-activated ATPase reactions (pumps) in the cell membrane to preserve the ionic gradients across

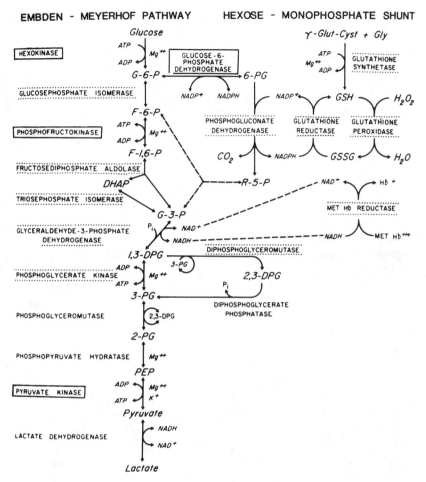

EMBDEN - MEYERHOF PATHWAY   **HEXOSE - MONOPHOSPHATE SHUNT**

FIG. 27–8. Metabolism of the erythrocyte, glycolytic pathways. Enzymes enclosed ([⎯⎯⎯⎯⎯⎯]) are those which are ratelimiting. Enzymes enclosed ([⎯⎯⎯⎯⎯⎯], ⎯⎯⎯⎯⎯⎯⎯) are those which have been found deficient due to hereditary factors. G-6-P = glucose-6-phosphate; 6-PG = 6-phosphogluconate; R-5-P = ribose-6-phosphate; F-6-P = fructose-6-phosphate; F-1,6-P = fructose-1,6-phosphate; DHAP = dihydroxyacetone phosphate; G-3-P = glyceraldehyde-3-phosphate; 1,3-DPG = 1,3-diphosphoglycerate; 2,3-DPG = 2,3-diphosphoglycerate; 3-PG = 3-phosphoglycerate; 2-PG = 2-phosphoglycerate; PEP = phosphoenolpyruvate.

the membrane. NADH, formed by reduction of nicotinamide-adenine dinucleotide (NAD^+) at the glyceraldehyde-3-phosphate dehydrogenase step, provides the reducing energy for the principal methemoglobin reductase (Fig. 27–8).

The rate-limiting steps in the EM pathway have been identified as the reactions catalyzed by hexokinase (HK), phosphofructokinase (PFK), and pyruvate kinase (PK) (Fig. 27–8). Hexokinase is inhibited by its product, glucose-6-phosphate (G-6-P),

PFK is inhibited by ATP, and PK activity is limited by the concentration of its substrate, phosphoenolpyruvate (PEP). Inorganic phosphate also serves to regulate glycolysis, inasmuch as the inhibitory effect of G-6-P and ATP can be overcome by phosphate.

The glycolytic generation of ATP may be diminished by a metabolic bypass in the conversion of 1,3-diphosphoglycerate (1,3-DPG) to 3-phosphoglycerate (3-PD) which generates 1 mole of ATP. 1,3-DPG may be mutated to 2,3-DPG, which may then be returned to the main pathway as 3-PD without the generation of the energy-rich ATP. This shunt, called the Rapaport-Luebering shunt, has been considered to be wasteful of a high percentage of glucose in the EM pathway. However, 2,3-DPG is a major determinant of hemoglobin affinity for oxygen. This important relationship and its significance in total oxygen transport in the body are discussed in detail in Chapter 8. One controlling factor of 2,3-DPG levels is the cellular concentration of hemoglobin.

Senescence and destruction

Normal aging of the red cell is associated with loss of several properties that maintain its deformability. The physical wear and tear sustained during its repeated passage through the microvasculature causes loss of membrane and hence spherocyte formation. Denaturation of enzyme proteins, which cannot be replenished, impairs glycolysis. More specifically, the activity of the rate-limiting glycolytic enzymes hexokinase and pyruvate kinase is reduced in old cells to 15 to 20 per cent of that in young cells. With the concomitant decrease of ATP, the cell loses K^+ and water and becomes internally more viscous. Progressive reduction of sialic acid from the membrane surface decreases its negative charge, promoting cell agglutination, and may uncover on the senescent red cell antigens that bind immunoglobulin (IgG) to provide a signal to the macrophage for its removal.

The removal of the senescent or injured red cells is a function of the reticuloendothelial system (RES) and its mononuclear macrophages, except for the less common occasion of intravascular hemolysis (see below). The principal organs of the RES are the spleen, liver, and bone marrow. The hepatic RES is a sinusoidal tissue in which blood passes directly into the macrophage-lined sinusoids and, thence, to the venous outflow. In contrast, the arterial

vessels of the spleen enter the pulp, where the artery is surrounded by a sleeve of loose reticular tissue packed with small lymphocytes which are immunologically competent. This sleeve also accommodates ovoid collections of lymphocytes (follicles which grossly are the Malpighian corpuscles) which have germinal centers and are the sites of antibody formation (white pulp). The sleeve, or sheath, is surrounded by a marginal zone consisting of fine meshed reticular connective tissue which receives the terminal arterioles and also contains free cells such as macrophages, lymphocytes, and plasma cells as well as the venous sinuses (red pulp). The spleen presents a greater hazard to the red cells than does the liver, although the blood circulating through the liver (35 per cent of blood volume per minute) is at least seven times the volume circulating through the spleen. The narrow passages of the spleen, namely the slits between endothelial cells lining the splenic sinuses, constitute a mechanical hazard and detain the red cells, exposing them to macrophages and to immunologically competent cells—an environment that permits the entire series of steps, from antigen encounter to antibody production. The hematocrit of the red pulp is high and the viscosity is increased; pH, glucose concentration, and partial pressure of oxygen are decreased. These metabolic conditions favor spherocyte formation. The smallest passages through this filter measure about 3 μ, and the normal healthy red cell (diameter, 7 μ) is believed to be sufficiently deformable to pass through a tube having a diameter of 2.8 to 3.6 μ without permanent injury.

In the removal of red cells from the circulation through extravascular mechanisms, the sequence of events is as follows: injury (i.e., lesion of red cell due to aging, intrinsic biochemical defect, or extrinsic cause); sequestration (i.e., the retention of injured, less deformable cells within the vascular spaces of the RES); and hemolysis by enzymatic digestion within macrophages, after phagocytosis. Whether sequestration occurs in the liver or the spleen is determined by the degree of damage sustained by the erythrocyte rather than by a selective affinity for specific types of injury. Mildly injured cells (i.e., those exposed to noncomplement-fixing antibodies, low doses of oxidant drugs, etc.) are sequestered in the spleen. Similarly, it is the degree of antibody coating rather than the type that determines the site of sequestration: a high ratio of antibody to red cell mass favors hepatic sequestration. The hepatic RES is also capable of clearance of agglutinated erythrocytes and severely injured nonagglutinated erythrocytes. These functions of the RES lead in turn to hypertrophy of the organs of erythrocyte sequestration (splenomegaly and hepatomegaly).

Following intravascular hemolysis, $\alpha\beta$ dimers of released hemoglobin complex with the α_2-globulin haptoglobin and are removed by the hepatocyte. When the haptoglobin binding capacity is exceeded, some hemoglobin may be taken up by the hepatocyte directly, but the $\alpha\beta$ dimers also readily pass through the glomerulus and are reabsorbed and catabolized by renal tubular cells, eventually resulting in hemosiderinuria as the tubular cells are sloughed. If the renal capacity to reabsorb hemoglobin dimers is exceeded, free hemoglobin appears in the urine. Free hemglobin in the plasma also becomes oxidized (methemoglobin) and releases free heme, which becomes bound to the β_1-globulin hemopexin to be removed by the hepatocyte. Heme exceeding this removal mechanism is bound to albumin (methemalbumin). Measurements of hemoglobin, methemoglobin, haptoglobin, hemopexin, and methemalbumin in the plasma, and hemoglobin and hemosiderin in the urine provide indices for intravascular red cell destruction.

Hemoglobin released within macrophages (reticuloendothelial, or RE, cells) following phagocytosis or taken up by the hepatocyte following intravascular hemoly-sis is not reutilized as such, and its component parts are metabolized separately. The protein fraction is returned to the general amino acid pool, iron is returned to the iron pool and storage sites, and the porphyrin fraction is catabolized. The methene bridge of heme is ruptured by the microsomal heme oxygenase-biliverdin reductase enzyme system, and bilirubin is formed. Bilirubin formed in macrophages is transported in plasma to the hepatocyte where it is conjugated with glucuronic acid. The bilirubin glucuronide is excreted by the biliary system into the intestine, where it is degraded by intestinal bacteria to several compounds known collectively as fecal urobilinogen. Urobilinogen is partially reabsorbed and recirculated to be reexcreted in bile, and a small fraction appears in the urine. The measurements of unconjugated bilirubin in serum and urobilinogen in the urine and feces provide indices of hemoglobin degradation and thus of red cell destruction.

Hemolytic states of long duration predispose to certain complications. The increased bilirubin production may lead to bilirubin stone formation. The precarious balance between red blood cell production and destruction may be disturbed by a transient decrease of cell production (e.g., infection, folate deficiency) or a transient acceleration of cell destruction, in either case worsening the anemia. Iron deficiency is a common consequence of chronic intravascular hemolysis due to loss of iron in the urine (hemosiderinuria). Finally, gout may rarely complicate a chronic hemolytic process because of the overproduction of uric acid, a consequence of increased nucleic acid catabolism from the rapid erythroid cell turnover in the bone marrow.

A large number of intrinsic red cell defects and abnormalities extrinsic to the red cell causing hemolysis are recognized and continue to clarify mechanisms of red cell destruction (see item II in outline). Certain examples are presented in the following sections.

INHERITED ABNORMALITIES OF THE RED CELL. Inherited abnormalities of the red cell causing hemolytic anemia all represent various defects of its membrane, its enzymes, or hemoglobin.

MEMBRANE DEFECTS. New understanding of protein interactions in the normal membrane skeleton now permit investigation of long-suspected membrane defects causing some inherited hemolytic anemias. Hereditary spherocytosis is the prototype of a hereditary red cell membrane defect and is transmitted as an autosomal dominant trait. The membrane is unstable because of a qualitative abnormality in spectrin or its association with other cytoskeletal proteins, leading to repeated loss of membrane fragments, with a gradual decrease in surface-to-volume ratio (increased spheroidicity). The resultant spherocyte has lost deformability and becomes detained in the splenic cords. In the splenic environment, the membrane loss is augmented and ATP depletion raises the cell's internal viscosity, accounting for its selective removal. Splenectomy allows red cell survival of sufficient duration to prevent anemia.

Similarly, qualitative defects in the membrane skeleton are being uncovered in hereditary elliptocytosis and pyropoikilocytosis and will likely be defined at the molecular level in the not-too-distant future.

ENZYME DEFECTS. The erythrocyte enzymes which may be deficient owing to hereditary factors are indicated in Figure 27–8, but not all such deficiencies produce clinical disturbances. The diseases which are the result of hereditary deficiencies of glycolytic and nonglycolytic enzymes have been referred to as the hereditary nonspherocytic hemolytic anemias. G-6-PD deficiency was the first to be described and serves as the most common example of an enzymopathy in the HM pathway. It is inherited as an X-linked disorder, and the fundamental defect is an instability of the variant enzyme so that its activity, although normal in reticulocytes, declines more rapidly than normal as the cell ages. This leads to its impaired ability to reduce $NADP^+$ to NADPH in the HM pathway and to maintain glutathione in the reduced form (GSH), allowing endogenous or exogenous oxidants to readily cause denaturation of membrane proteins and hemoglobin and thus premature cell destruction. The defect occurs in all races and in two principal clinical forms. In the more common form, found predominantly in blacks and peoples of Mediterranean ancestry, the enzyme deficiency is such that hemolysis occurs only upon exposure to an oxidative agent (e.g., primaquine, nitrofurantoin, sulfamethoxazole). The hemolytic event is self-limited because only the more senescent cells (most deficient in the enzyme) are unable to withstand the oxidant stress. In the much less common form, the enzyme deficiency is nearly complete, and it is characterized by a chronic hemolytic anemia, presumably due to physiological oxidant stresses alone.

The principal enzyme defect found in the EM pathway is PK deficiency, which is transmitted as an autosomal recessive abnormality. PK catalyzes a major reaction responsible for ATP production. The reticulocyte normally must endure a longer stay in the spleen (for 'conditioning') during which it has an increased requirement for ATP. This requirement is not met by the PK-deficient reticulocyte, causing the cell to gain Ca^{++} and to lose K^+ and water; it becomes rigid and is prematurely destroyed. Once the adverse metabolic environment of the spleen is removed, a striking paradoxical reticulocytosis reflects their improved survival. The anemia is well tolerated because the distal glycolytic block at PK causes a two- to three-fold increase in red cell 2,3-DPG, enhancing oxygen release from hemoglobin.

The hemolytic aspects of the hemoglobinopathies and thalassemias are discussed above.

ACQUIRED ABNORMALITIES OF THE RED CELL WITHOUT ANTIBODIES. A variety of acquired conditions produce red cell defects that in part resemble some inherited abnormalities.

MEMBRANE DEFECTS OR DAMAGE. In *paroxysmal nocturnal hemoglobinuria* a proportion of red cells produced has a distinct membrane defect causing hemolysis. Although this is a clonal disorder of the totipotential stem cell, thus affecting all hemopoietic cell lines, the abnormal red cell most prominently displays an increased susceptibility to complement-mediated lysis, causing intravascular hemolysis in the absence of a demonstrable antigen-antibody reaction. This process is frequently associated with an increased propensity to thrombosis, possibly through the activation of the complement-sensitive platelets. In *severe liver disease*, cholesterol esterification is reduced and the unesterified cholesterol in the plasma can exchange freely with red cell membrane cholesterol. Excess cholesterol accumulates in the membrane and markedly increases its surface area, producing acanthocytes (spur cells) and a clinically prominent hemolytic anemia. *Venoms* and *bacterial toxins* can cause membrane injury and spherocytosis. For example, *Clostridium welchii* sepsis is accompanied by rapidly progressive intravascular hemolysis and microspherocytosis due to attack of the lipoprotein structure of the red cell membrane by an α-lecithinase produced by the organism. *Lead* produces membrane injury by binding SH groups of membrane proteins, and inhibition of cation ATPase causes a marked and selective loss of K^+ from the cells.

ENZYME DEFECTS. In severe *lead poisoning* erythrocyte pyrimidine 5'-nucleotidase is markedly depressed, causing accumulation of pyrimidine nucleotides. These remnants of RNA degradation are normally dephosphorylated by the enzyme and then diffuse out of the reticulocyte. When retained in the cell, they are thought to interfere with glycolytic enzyme function, thereby reducing red cell survival. The lead-induced abnormality is virtually identical to the inherited deficiency of the enzyme. In a high proportion of patients with *uremia*, defective glucose metabolism through the HMP pathway has been demonstrated although the precise defect has not been identified. This abnormality explains susceptibility of the red cell to oxidant drugs and chemicals.

PHYSICAL INJURY. In the circulation, red cells may be subjected to *mechanical stress* that exceeds the tolerance of the normal membrane. They become fragmented into distorted shapes such as helmets, crescents, triangles, and microspherocytes, and are also lysed (intravascular hemolysis). Damage to the membrane and a decrease in the surface-to-volume ratio cause reduced deformability and premature destruction of the fragments. Such mechanical cell injury is associated with disorders accompanied by endothelial damage and fibrin deposition in the arteriolar microvasculature, where the red cells become entrapped and torn in the meshwork of fibrin strands (microangiopathic hemolysis). Examples of these disorders are disseminated and localized intravascular coagulation, vasculitis, and thrombotic thrombocytopenic purpura. Red cells are similarly subjected to mechanical injury at sites of excessive turbulence in association with high-pressure gradients (> 50 mm.) across the aortic valve. The principal examples of such disorders are malfunctioning Starr-Edwards aortic prosthesis and left ventricular outflow tract obstruction (e.g., aortic stenosis, idiopathic hypertrophic subaortic stenosis). Exposure of red cells to temperatures above 49° C (*thermal injury*) denatures the cytoskeletal protein spectrin, resulting in microspherocytes from marked membrane fragmentation. Spherocytic hemolytic anemia may result from third-degree burns covering more than 20 per cent of the body surface.

ERYTHROCYTE INCLUSIONS. Commonly used *drugs* with oxidative properties (e.g., phenacetin, phenazopyridine, sulfonamides), when taken in large doses or for a long time, induce cross-linking of the normal red cell membrane skeleton (resulting in membrane rigidity) but, more importantly, intracellular precipitation of hemoglobin (Heinz bodies). Removal of these precipitates by the spleen (splenic "pitting") leads to reduction of membrane surface area (spherocytes) and "bite cells." Varying degrees of intravascular hemolysis may occur with sufficient damage to the membrane. In *falciparum* malaria, hemolysis is to a large extent a direct consequence of red cell invasion by the parasite. Although the spleen removes and destroys the invaded red cells as a whole, "pitting" of the parasite from the red cell in the spleen leads to cell injury and reduced deformability.

HYPERSPLENISM. An enlarged spleen from any cause (e.g., congestive or infiltrative splenomegaly), with or without changes in its architecture, can sequester red cells excessively. In this case the red cell itself is not significantly affected.

ACQUIRED ABNORMALITIES OF THE ERYTHROCYTE WITH ANTIBODIES. Antibodies on the red cell cause its premature destruction (immune hemolysis) and may be acquired by one of three principal mechanisms: (1) *Isoantibodies* are acquired through exposure of the patient to such red cell antigens of another individual which the patient is lacking (e.g., the Rh immunization of an Rh-negative mother by her Rh-positive fetus, mismatched red cell transfusions). In this case the antibodies do not react with the patient's own red cells but only with those cells which carry the antigen. (2) Antibodies are produced against certain *drugs*. The antibody attaches to the red cell because the drug, or a metabolite of the drug, is bound to the red cell membrane, or the drug-antibody complex (immune complex) is formed in the circulating plasma and then attaches nonspecifically to the red cell. (3) Antibodies are formed against the patient's own red cell antigen(s) (*autoantibodies*) in association with diseases that predispose to autoantibody production or with administration of specific drugs that alter the red cell membrane. Failure of the mechanisms that normally permit self-recognition of the red cell antigens are believed to underlie such antibody production.

The mechanisms by which antibody-coated red cells are destroyed are determined by the immunoglobulin class of the antibody produced (e.g., IgG or IgM), and, therefore, to what extent serum complement is bound to the cell (and activated) by the antigen-antibody reaction. (Serum complement is a system of plasma proteins that react sequentially in the presence of antigen-antibody reactions). Antibodies of the IgG class may or may not bind and significantly activate complement, and react best with antigen at 37° C (and hence are called *warm antibodies*). The IgG-coated (or complement-coated) red cells are removed in part or entirely by macrophages through interaction of the Fc portion of the antibody (or complement) on the cell with the Fc (or complement) receptor on the macrophage (*extravascular hemolysis*). Partial phagocytosis removes more membrane than cytoplasm, resulting in a spherocyte. Clinical examples of this process are the Rh isoantibody-induced hemolysis causing hemolytic disease of the newborn, the immune hemolysis associated with some drugs which bind to the red cell membrane (e.g., penicillin when given in large doses), and autoimmune hemolytic anemia (e.g., in association with disseminated lupus erythematosus, lymphatic tissue malignancies, or α-methyldopa administration, or not associated with any detectable disorder).

IgM antibodies, in contrast, always bind and activate complement on the red cell surface and react best with antigen below 37° C (cold antibodies), i.e., in the cutaneous circulation. If the complement bound by IgM antibodies does not subsequently become fully activated in the central circulation, the complement-coated red cells are removed by the RES through the interaction of complement with the macrophage. If complement becomes fully activated, its components $C'8$ and $C'9$ cause the final cytolytic step of the complement sequence, producing a membrane defect 80 to 100 Å in diameter. The effect is disruption of ionic equilibrium with bypassing of the Na^+-K^+ pump and exodus of K^+ from the cell. The oncotic pressure of hemoglobin draws excess water into the cell, resulting in its lysis within the circulation (*intravascular hemolysis*). IgM antibody-mediated hemolytic anemia may occur with *Mycoplasma pneumoniae* infection, infectious mononucleosis, lymphoreticular neoplasms, or may be of unknown cause (idiopathic cold agglutinin disease). In these instances complement-coated red cells are removed by the RES. In the acute hemolytic reaction from an ABO-incompatible transfusion, the IgM isoantibody causes full activation of complement and thus intravascular hemolysis. IgM antibodies to certain drugs, such as quinine and quinidine, form a drug-antibody complex that binds complement to the red cell membrane, resulting in either partial or complete activation of complement. In either case, the hemolytic process ceases when the drug is removed.

Annotated references

Beck, W. S.: Hematology. Cambridge, M.I.T. Press, 1981. (An introduction to the pathophysiology of hematological disorders in the form of a lecture series.)

Beutler, E.: Hemolytic Anemia in Disorders of Red Cell Metabolism. New York, Plenum Medical Book Company, 1978. (An authoritative monograph providing a most accurate and comprehensive description of all the known hereditary enzyme defects causing hemolytic anemia.)

Bothwell, T. H., et al.: Iron Metabolism in Man. Oxford, Blackwell Scientific Publications, 1979. (A comprehensive text on iron metabolism and its disorders, with a section of fully detailed laboratory methods used in the study of iron.)

Bottomley, S. S.: Sideroblastic anaemia. Clin. Haematol. *11*:389, 1982. (A current assessment of the pathogenesis and clinical features of the various sideroblastic anemias.)

Bunn, H. F., et al.: Human Hemoglobins. Philadelphia, W. B. Saunders, 1977. (A concise text of the genetics, pathophysiology, and clinical aspects of normal and abnormal hemoglobins.)

Camitta, B. M., et al.: Aplastic anemia: Pathogenesis, diagnosis, treatment, and prognosis. New Engl. J. Med. *306*:645–652, 712–718, 1982. (An up-to-date discussion of all aspects of aplastic anemia.)

Eichner, E. R.: Macrocytic anemias. *In* Spivak, J. L.: Fundamentals of Clinical Hematology, Chapter 3. Hagerstown, Harper & Row, 1980. (A concise account of the biochemistry, etiologies, clinical features, diagnosis, and treatment of megaloblastic anemias presented in a physiologic perspective.)

Erslev, A. J., and Gabuzda, T. G.: Pathophysiology of Blood, 2nd ed., Chapters 1, 2, and 3. Philadelphia, W. B. Saunders, 1979. (A detailed presentation of the physiology and pathological mechanisms of the erythroid system with an extensive collection of illustrations.)

Garratty, G., and Petz, L. D.: Drug-induced immune hemolytic anemia. Am. J. Med. *58*:398, 1975. (A clear description of the various mechanisms of immune hemolytical anemia caused by drugs.)

Golde, D. W., et al.: Polycythemia: Mechanisms and management. Ann. Intern. Med. *95*:71, 1981. (A clear presentation of the current concepts of the pathophysiology and management of polycythemias.)

Graber, S. E., and Krantz, S. B.: Erythropoietin and the control of red cell production. Annu. Rev. Med. *29*:51, 1978. (A concise review providing a general background of the biology of erythropoietin as well as a current understanding of its mechanism of action and role in clinical medicine.)

Harvey, A. M., et al.: The Principles and Practice of Medicine. 20th ed. New York, Appleton-Century-Crofts, 1980. (A textbook of internal medicine emphasizing the approach to clinical problems rather than disease entities.)

Kappias, A., et al.: The porphyrias *In* Stanbury, J. B., et al.: The Metabolic Basis of Inherited Disease, 5th ed., p. 1301. New York, McGraw-Hill, 1983. (A thorough description of heme biosynthesis and its control.)

Lajtha, L. G. (ed.): Cellular dynamics of haemopoiesis. Clin. Haematol. *8*:221, 1979. (A series of authoritative articles emphasizing the fundamental aspects of hematopoietical cell differentiation, proliferation and control.)

Lindenbaum, J.: Aspects of vitamin B_{12} and folate metabolism in malabsorption syndromes. Am. J. Med. *67*:1037, 1979. (A review focusing on the clinically important disturbances of vitamin B_{12} and folate absorption.)

Miescher, P. A., and Jaffee, E. R. (eds.): Physiology and disorders of hemoglobin degradation. I. Semin. Hematol. *9*:1, 1972. (A series of articles by noted authors concerning the detailed mechanisms of hemoglobin catabolism.)

————: Polycythemia. Semin. Hematol. *12*:335, 1975. (A series of articles reviewing the various polycythemias and relevant research studies.)

————: Immune hemolytic anemias. Semin. Hematol. *13*:247, 1976. (A series of review articles on the various immune hemolytical anemias, their mechanisms, serologic features, and management.)

————: Blood cell membranes. I. and II. Semin. Hematol. *16*:1, 1979. (A comprehensive series of articles presenting the current concepts of the structure and function of the normal blood cell membrane and alterations produced in disease.)

————: Clinical aspects of iron deficiency and excess. Semin. Hematol. *19*:1, 1982. (A series of scholarly review articles presenting the current status of the clinical and diagnostic aspects of iron deficiency and iron overload.)

Mladenovic, J., and Adamson, J. W.: Erythroid colony growth in culture: Analysis of erythroid differentiation and studies in human disease states. *In* Hoffbrand, A. V.: Recent Advances in Hematology, No. 3, p. 95. London, Churchill Livingstone, 1982. (A clear description of the current assays for erythroid precursor cells and their limitations and applications to the analysis of erythroid differentiation in human disease states.)

Quesenberry, P., and Levitt, L.: Hematopoietic stem cells. N. Engl. J. Med. *301*:755, 819, 868, 1979. (A concise review of hematopoietical stem cells and the current understanding of their role in diseases of the stem cell.)

Weatherall, D. J., and Clegg, J. B.: The Thalassaemia Syndromes. 3rd ed. Oxford, Blackwell Scientific Publications, 1981. (A current reference text for the genetics, pathophysiology and clinical aspects of the thalassemias and closely related hemoglobinopathies.)

Wintrobe, M. M., et al.: Clinical Hematology. 8th ed. Philadelphia, Lea & Febiger, 1981. (A complete reference textbook of hematology.)

28   Leukopoiesis

William A. Robinson, M.D., Ph.D.

The kinetics of granulocyte production have been elucidated to a remarkable degree in the past few years, largely owing to the introduction of radioisotopes into clinical medicine. The ebb and flow of granulocytes through the bone marrow, peripheral blood, and tissues in the normal steady state are well understood. We are, however, only beginning to scratch the surface in our understanding of the factors that regulate maturation and production in the granulocyte system. A considerable body of evidence exists to indicate that this is under the control of humoral or hormonal regulatory factors acting much as erythropoietin does to regulate red cell maturation and production. It is likely that many of the diseases that afflict the granulocyte system arise as a result of alterations in these normal control mechanisms. Further elucidation in this area will be of paramount importance in understanding the pathophysiology of these disorders in man.

Theoretical considerations of granulocyte production

The majority of investigators in the past have attempted to integrate granulocyte production into a scheme of classic negative feedback inhibition using peripheral blood granulocyte counts as the signalling point for granulopoietic factor production. Although such a mechanism may be opera-

tive in the regulation of red blood cell production, careful consideration of the granulocyte system leads to the conclusion that peripheral blood granulocyte numbers alone cannot be the only regulatory arm in such a negative feedback inhibition scheme. In man, this is most apparent during times of bacterial infection. In such situations when the peripheral blood granulocyte count rises in response to infection there is a combined loss of granulocytes into the infected area and from the marrow granulocyte reserve compartment. Thus, at the point when the peripheral blood granulocyte count is highest, granulocyte production and maturation needs to be stimulated further rather than inhibited as in a negative feedback system. It is also difficult to conceive of a system in which total body granulocyte numbers regulate production rates, for once again, in situations of bacterial infection total body granulocyte numbers may be markedly increased. Various mechanisms other than peripheral blood granulocyte counts have been suggested as means of accounting for increased granulocyte production in such situations. Among these are recognition of the rate of granulocyte destruction or egress from the bone marrow, endotoxemia, and accumulation of particular cellular elements in the bone marrow, namely the most mature granulocytes. Various bits of evidence have indicated that all of these may influence

655

granulopoietic factor production and granulocyte production rates. These theoretical considerations await, however, purification of a true granulopoietic factor and clearcut determination of its cellular source.

Kinetics of granulopoiesis

There is still some disagreement about various aspects of granulocyte production rates and disappearance, but the basic system and its relationship to time have been reasonably well studied. The most widely employed techniques to evaluate the kinetics of granulopoiesis have involved cells labeled with diiopropylfluorophosphate (^{32}DFP) and tritiated thymidine. These have yielded invaluable information about the production rates and traffic of granulocytes through the bone marrow into the peripheral blood and tissues.

The developmental components of the granulocyte production system are shown in Figure 28–1. In the bone marrow the compartments that have been recognized and described are a stem cell pool; a mitotic pool, in which division and maturation both occur; a nonmitotic pool, in which maturation alone takes place; and a storage pool of mature cells, the marrow granulocyte reserve (MGR). In the peripheral blood granulocytes are equally distributed between a circulating granulocyte pool (CGP) and a marginated granulocyte pool (MGP). The latter consists of cells which are inside the vascular system but temporarily adherent to cell walls rather than circulating. Together, these two intravascular pools, which are freely interchangeable, comprise the total blood granulocyte pool (TGP). Under normal steady state conditions, the peripheral blood granulocytes enter the tissues by a process of apparent random migration, where they perform their functions and die.

The size of the stem cell pool in man is

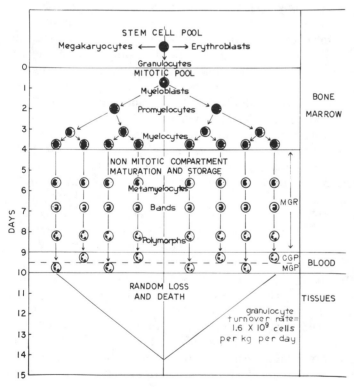

FIG. 28–1. Developmental components of the granulocyte production system.

unknown, owing to the present lack of an effective assay system. Likewise, the stem cell has not yet been clearly identified morphologically, although these cells may be similar in appearance to medium or "transitional" lymphocytes. It is clear that the stem cell is shared by other members of the hematopoietic cell lineages, namely the red blood cell, monocyte, and megakaryocytic series. The evidence for commonality of stem cells comes from those studies demonstrating that the Philadelphia chromosome, seen in chronic granulocytic leukemia, is common to all of the cell lines, and also by extensive studies using the spleen colony-forming assay systems *in vivo* in mice, in which it has been shown that single cells can give rise to colonies containing members of all of these cell lines. Under normal steady-state conditions only a small fraction, perhaps 25 per cent, of stem cells are in cycle at any one time. The factors regulating the egress of stem cells into committed granulocyte precursors have not been defined. The mitotic pool consists of myeloblasts, promyelocytes, and myelocytes. It is in this compartment that both cell division and maturation occur. The number of cell divisions between the myeloblast and nondividing forms has been variously estimated to be between 3 and 7 *in vivo*. It is uncertain whether greater numbers of cell divisions can occur during times of increased need and stress, but it has been suggested that this is one of the means whereby increased numbers of granulocytes can be produced. The minimal transit time from the myelocyte stage to mature granulocytes in the peripheral blood has been estimated at 4 days, using tritiated thymidine, with a mean transit time of 6 to 9 days. Using radioactive ^{32}DFP the mean transit time has been estimated at 11 days with a range of 8.5 to 14 days. For the present it is probably safe to assume, therefore, a mean transit time of around 10 days.

The maturation-storage pool consists of cells recognizable morphologically as metamyelocytes, bands, and segmented neutrophils. A large part of this compartment can be released on demand under situations of severe stress and has been referred to as the MGR. Although there is some discrepancy about the actual number of cells in the MGR compartment, it is clear that it is close to ten times the normal daily granulocyte production rate.

Under normal steady state conditions, granulocytes are released from the bone marrow into the peripheral blood through intracellular "pores." The most mature cells are released first, probably because they are more pliable and less sticky. Once in the bloodstream they become equally distributed between the previously mentioned circulating granulocyte pool (CGP) and the marginated granulocyte pool (MGP). These two pools are freely interchangeable. The half-life of granulocytes within the peripheral blood is extremely short, on the order of 6 to 7 hours. From the peripheral blood granulocytes enter the extravascular compartment and tissues. Once they have left the bloodstream, they do not return there, but die in the performance of their duties. Survival of granulocytes in tissues under normal steady state conditions has been estimated at 4 to 5 days. Thus, the total life span of granulocytes from the earliest recognizable form to death in the tissues is very short, on the order of 10 to 14 days. This very short life span of granulocytes within the peripheral blood and tissues necessitates the very large marrow granulocyte reserve compartment.

The production rate of granulocytes has been estimated in the range of 1.6×10^9 cells/kg./day. This estimate for granulocyte production is close to that made for daily red cell production as determined from average normal red cell mass and red cell life span. For the granulocyte system, this figure is of particular interest when considering the possibility of obtaining granulocytes for transfusion in granulocytopenic states. Assuming (for ease of calculation) a normal granulocyte count of 5,000, one would need the total WBC that could be recovered from 25 l. of blood to replace normal daily loss. Thus, the procurement of WBC for transfusion is an exceptionally difficult but not impossible task, as has been amply demonstrated.

During times of stress and increased granulocyte need, mechanisms must be called into play which can increase granulocyte production rates—and perhaps flow—through the various compartments. The possible mechanisms that have been suggested in this regard are increased stem cell input, additional mitoses within the dividing cellular compartment of the bone marrow, shortening of the generation time within the mitotic compartment, and more rapid transit time through the marrow. Various bits of evidence have suggested that all of these mechanisms may be involved.

In acute stress situations the source of cells mobilized to enter areas of inflammation are derived from the MGR. The release of cells from this compartment appears to be under the control of an as yet ill-defined *leukocytosis-inducing factor.* This factor can be found in the serum and plasma of rats and dogs with neutropenia induced by peritoneal lavage or chemotherapeutic agents. When infused into a normal animal, or one that has recovered from neutropenia, plasma obtained at the time of neutropenia will induce transient peripheral neutrophilia. The mechanism of action of such a leukocytosis-inducing factor has not been determined. There is evidence to suggest that under normal steady state conditions egress of cells from the bone marrow is dependent upon cell stickiness and deformability, with the most deformable and least sticky cells leaving the marrow first by egress through the vascular channels. It has been suggested also that the factors effecting release of cells from the marrow granulocyte reserve may act upon marrow sinusoids rather than on the cells themselves, perhaps by dilating the vascular channels to allow less deformable and more sticky cells which normally would not leave the marrow to egress into the peripheral blood.

Measurement of marrow granulocyte reserve

A number of factors, other than naturally occurring leukocytosis-inducing factors, have been shown to induce release of granulocytes from the marrow granulocyte reserve. This property has been utilized as a means of measuring bone marrow function and granulocyte reserve. The most useful of these agents clinically is hydrocortisone.

For measurement of MGR, 100 mg. of hydrocortisone is given intravenously and the total white blood cell count and neutrophil count calculated at half-hour intervals for the next 3 hours. A normal response, indicating normal MGR, is a doubling of the absolute neutrophil count within 3 hours. After 3 hours it has been shown that hydrocortisone interferes with the egress of neutrophils from the peripheral blood into the tissues and time points beyond 3 hours are thus not useful in this calculation. Bacterial pyrogens and etiocholanolone have also been used to measure MGR but have considerable side effects and the information obtained is the same as that derived using hydrocortisone. For measuring the marginated granulocyte pool, the most useful agent is epinephrine. The effect of epinephrine is to shift cells from the MGP into the circulating granulocyte pool. Epinephrine, 0.3 mg., is given subcutaneously and the neutrophil count calculated at 3, 5, 10, 15, 30, and 60 minutes. A normal response is indicated by an increase in the circulating granulocyte pool by 50 per cent at 15 to 30 minutes. A considerable proportion of the shift in pools occurs within the spleen. In patients who have had their spleen removed, a lesser rise may be found.

The possible oscillatory nature of granulopoiesis

A number of biological systems, most commonly the endocrine system, are known to function with predictable rises and falls in levels of various substances. It has been suggested that granulocyte production, and appearance of granulocytes in the peripheral blood, may also oscillate in a regular and predictable manner under normal conditions. Cycling of peripheral blood granulocytes has been noted in various abnormal hematopoietic states in both animals and man, and can be produced in normal dogs treated with low doses of cyclophosphamide. In human beings, cycling has been noted in the disease called cyclic neutropenia and also in chronic granulocytic leukemia. It is of particular interest that in the latter disease, cycling of other cellular elements, both red blood cells and platelets, also appears to occur. However, demonstration of such cycles in normal human beings has not been reported. In some strains of gray collie dogs cyclic neutropenia occurs regularly, and it appears that this results from an inherent defect within the hematopoietic system rather than in extrinsic regulatory factors. Transplantation of marrow cells from animals with cyclic neutropenia into hematopoietically normal histocompatible recipients results in the development of cyclic oscillations in the peripheral blood granulocyte count. Demonstration of cycling in normal human beings is hampered by the tremendous lability of the granulocyte system and the fact that even minor degrees of stress can markedly alter the number of cells entering and leaving the circulating blood. Computer models have been worked out to simulate possible cycling in normal human beings with two feedback loops, one controlling the production rate of granulocytes and the other controlling the rate of release of cells from the marrow. In these systems, peripheral blood granulocytes can be demonstrated to rise and fall in a sine wave fashion with a cycle length varying from 14 to 23 days. Confirmation of this finding will be of considerable interest in both normal and abnormal hematopoietic conditions.

Factors regulating granulocyte maturation and production

Over the years a great many factors have been described from animal and human sources to which have been ascribed granulopoietic activity. Until recently, however, none of these factors have been well characterized biochemically or physiologically, presumably because the majority of these studies involved *in vivo* systems in which the tremendous lability of the granulocyte system made quantitation and characterization of response extremely difficult. However, during the past 10 years a number of systems have been utilized *in vitro*. The most widely used of these have been the semisolid culture systems employing either agar-gel or methyl-cellulose as a stiffening medium into which are put single cell suspensions of human or other bone marrow. When culture plates so prepared are incubated in the presence of an appropriate granulopoietic stimulus, single bone marrow cells divide and give rise to colonies of maturing granulocytic and macrophage cells which can be counted and quantified. If no stimulus is added to the plates, little or no colony growth occurs. Using these systems as a means of detecting granulopoietic activity, a large number of tissues, organs, and body fluids have been evaluated for their ability to promote colony growth. Granulocyte colony-stimulating activity has been found in normal human serum and urine and in high content in a number of organs, most notably lung, bone shaft, and kidney. The factor responsible for colony-stimulating activity has been studied extensively for its biochemical properties and has been characterized as a glycoprotein with a molecular weight of approximately 45,000 which migrates electrophoretically as an α-globulin (see list below). These properties closely resemble those described for erythropoietin, although it has been clearly shown that the granulocyte colony-stimulating factor is distinct from erythropoietin. For want of a better name this material has been temporarily termed the colony-stimulating factor (CSF).

Properties of the Granulocyte Colony-Stimulating Factor

Stimulates granulocyte production and maturation *in vitro*
Glycoprotein
   Molecular weight about 45,000
   Electrophoretic mobility—α-globulin postalbumin range
Found in all normal human serum and urine
Produced in large part by the monocyte-macrophage system; also by T cells
   In humans in response to antigenic stimulation, granulocyte breakdown rates, and endotoxemia
Action
   1. On committed granulocyte precursors
   2. Needed for every cell division—not a "trigger substance"
   3. (?) Mechanism through cell surface phenomenon

The major cellular source of CSF in human beings appears to be the monocyte-macrophage system, accounting for its apparent tissue ubiquity. This finding resulted from the observation that peripheral white blood cells (WBCs), used as "feeder layers," were the best source of CSF for human bone marrow. Initially, it was suggested that mature granulocytes in the peripheral blood were the source of this material; but subsequently cell separation techniques (to fractionate peripheral WBC into various cellular elements) have indicated that monocytes are the major source of this activity. More recent studies demonstrated tissue-derived macrophages and T lymphocytes also produced CSF in large quantities; the factors regulating CSF production have not been elucidated precisely yet, but probably involve recognition of

granulocyte breakdown rates, perhaps through some product released or produced by granulocytes or through increased levels of endotoxin or other bacterial products. This latter mechanism is particularly applicable to neutropenic states, since when granulocyte levels are low endotoxin production increases, owing to the growth of flora normally kept in check. Bacterial products, particularly endotoxin, have been shown to stimulate CSF production *in vitro* and probably do the same *in vivo*.

Although the mechanism of action of CSF is at present unclear, considerable evidence suggests it acts primarily on committed granulocyte precursor cells rather than on uncommitted pluripotential stem cells. This material does not act simply *in vitro* as a triggering substance but is essential for every cellular division, presumably through a cell surface mechanism since its activity can be blocked by concanavalin A.

Rigorous proof that the granulocyte CSF represents a true granulopoietic substance has not, at present, been offered. Injection of partially purified material from human urine or tissue sources leads to an increase in labeled granulocytes, both in the marrow and in the peripheral blood. Rises and falls in the levels of serum and urinary CSF have been shown to occur in various human diseases, particularly the cyclic neutropenias, chronic granulocytic leukemias, and with bacterial infections. Likewise, serum levels in animals have been shown to rise abruptly following neutropenia induced by antineutrophil serum or endotoxemia. The temporal relationships between absolute neutrophil counts and the levels of colony-stimulating factor are still unclear in terms of a cause and effect relationship. Despite this present lack of concrete evidence the granulocyte CSF is the best candidate described thus far as being as true granulopoietin. Unfortunately, like erythropoietin it is a poor antigen and the usual means employed to purify hormonal substances (i.e., immune complexing for radioimmunoassay) have been unsuccessful thus far.

In addition to CSF and other factors which stimulate granulopoiesis a large number of substances which inhibit granulocyte maturation and production *in vitro* have been described. These include prostaglandin E, lactoferrin derived from neutrophils, interferon, T-cell–derived suppressive factors, acidic isoferritins derived from leukemic cells, and serum lipoproteins. The role of any or all of these factors in the regulation of granulopoiesis remains to be determined. The majority of studies describing the inhibitory effect of such factors have been *in vitro*. In the dynamic and complex process of granulopoiesis, it is likely that many such factors may be involved, but the exact role of each remains to be determined.

It has also been suggested that the major negative feedback regulator of granulopoiesis is the mature neutrophil itself. In the scheme proposed, the major driving force for granulocyte production is bacterial and other antigen stimulation through CSF. Mature neutrophils serve as the major negative regulatory arm by inactivation of these stimulatory factors. Details of this scheme will be found in several recent publications.

Abnormalities in granulopoiesis

Using the systems of granulocyte colony growth a great many studies have been carried out in the past few years, in attempts to unravel the pathophysiology of various diseases in which granulopoiesis is altered.

Aplastic anemia and granulocytopenia

This appears to be a complex group of disorders in which the pathophysiology may result from alterations in any of the various stages previously described in granulocyte production. In some patients the defect appears to result from decreased numbers of stem cells, in others altered microenvironment for stem cell growth, and in yet others from humoral and cell-mediated inhibitors of granulopoiesis. Exactly what groups of patients fit into these various pathophysiological mechanisms has not been determined clearly. The number of granulocyte colony forming cells found in the bone marrow of the majority of patients with all forms of aplastic anemia and granulocytopenia is less than the number found in normal human bone marrow. It has not been determined, however, whether this is a primary or secondary manifestation of these disorders. It has been suggested that one major defect may arise in the supporting structures (stroma) of the bone marrow. *In vitro* culture systems for human bone marrow cells with the capability of determining the interaction between stromal and hematopoietic cells have been developed in recent years. No current data are available, however, which allow conclusive statements in this area. The most interesting and potentially useful information has developed recently in the demonstration of humoral and cell-mediated suppression of granulopoiesis *in vitro*. In several patients with neutropenia, T-suppressor lymphocytes of the T-3 plus T-8 subset have been described. Further, it has been shown that at least some of these patients have responded to treatment with corticosteroids or antithymocyte globulin. Humoral inhibitors of granulopoiesis have also been described, particularly in drug-associated granulocytopenic states. In these studies it has been shown that mixing the patient's serum with the drug will inhibit not only his own bone marrow, but normal bone marrow *in vitro*. It has been suggested that this is the result of an autoimmune process. Finally, in Felty's syndrome (neutropenia associated with rheumatoid arthritis) data have suggested that there is a decreased production of granulopoietic factors which, both *in vivo* and *in vitro*, can be stimulated by the use of lithium. Lithium stimulates CSF production both *in vivo* and *in vitro*, and it has been used to treat various neutropenic disorders with limited and variable success.

Sustained leukocytosis

Granulocyte CSF levels in the serum and urine of patients with leukocytosis secondary to nonhematopoietic neoplasms have been shown to be markedly elevated with a direct relationship between the degree of the granulocytosis and the levels of CSF found in serum or urine. Attempts to extract CSFs from the neoplastic tissue in an effort to place these situations into the group of disorders known as the paraneoplastic syndrome have been unsuccessful. The reasons behind the granulocytosis and elevated CSF levels at present, therefore, remain unknown.

Cyclic neutropenia

Cyclic neutropenia occurs in both humans and gray collie dogs. The term is probably a misnomer, since in both dogs and humans, cycling of both platelets and reticulocytes may occur. The cycle periodicity is approximately 3 weeks in the majority of cases. Serum CSF and peripheral blood monocyte levels cycle with a similar periodicity but reciprocal to the peripheral blood neutrophil counts. In dogs, the pathophysiological defect appears to reside in the stem cells, since transplantation from the histocompatible normal donors eliminates the defect. However, since the major source of granulopoietic factors appears to be the monocyte-macrophage system and these cells are derived from the granulocyte lineage, this does not exclude the possibility that the actual pathophysiological mechanism of the disorder is an abnormality in granulopoietic factor production, since cells producing CSF are transplanted as well. Administration of corticosteroids to humans with cyclic neutropenia may alleviate many of the signs and symptoms but does not totally eliminate the cycles that occur. Lithium has been used with success to treat gray collie dogs but does

not appear to be of uniform benefit in human cyclic neutropenia.

Chronic granulocytic leukemia

The number of granulocyte colony-forming cells in the bone marrow and peripheral blood of patients with chronic-phase chronic granulocytic leukemia (CGL) is markedly increased. The colonies formed from these cells mature normally *in vitro*, and the Philadelphia chromosome can be recovered from the colonies grown. During the blastic phase of the disease, colony formation is markedly reduced. CSF levels in the serum and urine of patients with CGL are usually, though not consistently, elevated and may cycle in those patients with cyclic CGL. These findings indicate that the cells of patients with CGL are normally responsive to granulopoietic factors in the early phase of their disease but lose their responsiveness as blastic crisis occurs. Overproduction of granulocyte CSF has been suggested as a possible pathophysiological mechanism in this disorder, and while an attractive possibility, it remains to be proven.

Others have suggested that the major pathophysiological defect in CGL results from diseased or altered production of granulocyte inhibitors such as lactoferrin. Further evidence has been presented that CGL cells may produce factors which inhibit normal cells but not leukemic cells, thereby permitting continued leukemia cell growth while rendering normal cells dormant. This would allow for a selective proliferative advantage of leukemic cells and their continued expansion in the chronic phase of CGL, but does not explain the conversion to blast crisis. The latter seems to result from progressive unresponsiveness to normal regulators of maturation.

Acute nonlymphocytic leukemia

Three types of colony growth have been observed when culturing peripheral blood and bone marrow cells of patients with acute nonlymphocytic leukemia using normal human WBC as the source of CSF. In the majority of patients, perhaps two thirds, little or no colony growth is observed. The colonies formed are extremely small, 10 to 12 cells in number, and have been called clusters. Maturation to mature polymorphonuclear forms takes place at least in some of these clusters, and chromosomal abnormalities present in the initial bone marrow specimen can be recovered from them, indicating that they come from a leukemic cell line. In a few patients, very large colonies are formed from both the peripheral blood and bone marrow in which normal granulocytic maturation occurs. These differences in colony growth can be partially, though not fully, correlated with histological types of acute nonlymphocytic leukemia. The majority of patients whose cells form few or no colonies have diseases that appear to fall largely into the category classified histologically as acute undifferentiated or acute granulocytic leukemia, whereas those patients whose cells form very large colonies usually have diseases that fit into the histological category of acute monocytic or acute myelomonocytic leukemia. In the first two groups of patients, levels in serum and urine tend to be low, but in the latter category of disease CSF levels are usually quite high when compared to normal controls. Inhibitory factors such as neutrophil-derived lactoferrin, tend to be low during the initial or relapse phase of the disease. One might anticipate therefore that normal clones could continue to proliferate and expand. This is obviously not the case. It has been suggested that leukemia cells may provide their own selective growth advantage in this regard by the production of leukemia-associated inhibitory activity (LIA) which inhibits normal, but not leukemic, cell growth. LIA has been partially characterized as a glycoprotein acidic isoferritin with a molecular weight of approximately 550,000 daltons. It has not been determined whether LIA is a normal regulatory factor or associated only with neoplastic conditions.

Acute lymphocytic leukemia

Granulocyte colony-forming cell numbers are markedly reduced in the bone marrow of children with acute lymphocytic leukemia during the initial or relapse phases of the disease, returning to normal with induction of remission. On the other hand, peripheral blood granulocyte colony-forming numbers are markedly increased during the initial or relapse phases of the disease and return to normal during remission. Whether the latter finding represents "squeezing out" of normal granulocyte colony-forming cells from the bone marrow or production in extramedullary sites has not been determined. One of the major questions in acute lymphocytic leukemia is how leukemic cells turn off normal hematopoiesis. The possibility that these cells produce substances inhibitory for hematopoiesis such as LIA, described above, has been investigated using the colony growth technique, but no consistent inhibition has been demonstrated.

Myelofibrosis

A marked increase in the number of circulating granulocyte colony-forming cells has been found in myelofibrosis and myeloid metaplasia. As in chronic granulocytic leukemia these cells appear to respond normally to granulopoietic factors *in vitro*. No studies of CSF levels in this disorder have appeared.

Viral illnesses including infectious mononucleosis

Some of the highest levels of serum granulocyte CSF have been documented in patients having a variety of viral illnesses including infectious mononucleosis. In these situations the CSF appears to be masked by the presence of a lipid inhibitor, which can be extracted by chloroform or ether. The exact relationship and specificity of these inhibitory substances to the CSF in granulopoiesis has not been determined; but of interest in this regard is the possibility that these materials may be produced by the viruses themselves, accounting for the decrease in granulopoiesis seen in these disorders.

Also to be considered as having a possible role in virus-related neutropenia is the glycoprotein interferon. Interferon has been shown to inhibit granulocyte colony formation *in vitro* and cause neutropenia when administered to humans. No direct measurement of interferon levels correlated with granulocyte production has been reported in human viral illnesses. It seems likely, however, that interferon plays some, and perhaps the major, role in this setting.

appears to be produced in large part by the monocyte-macrophage system, but the factors controlling its production and release are poorly understood. A large number of factors which inhibit granulopoiesis have also been described. These include neutrophil-derived lactoferrin, prostaglandin E, interferon, and leukemia-associated inhibitory factor. The exact role of these in normal and abnormal granulopoiesis is as yet unclear. Abnormalities in granulocyte colony formation, CSF levels, and various inhibitors have been described in various human disease states involving the granulocyte cell line. Further understanding in this area will undoubtedly lead to recognition of the pathophysiological mechanisms involved in these disorders.

Summary

In summary, granulocytes are produced in the bone marrow from a common hematopoietic stem cell by a process of maturation and division requiring four to seven cell divisions and a 9-day transit time for entering the peripheral blood. Mature cells are released from the bone marrow under the control of an as yet illdefined leukocytosis-inducing factor. Release of cells from the bone marrow is also dependent upon cell stickiness and deformability and the leukocytosis-inducing factor may act upon marrow sinusoids. After entering the peripheral blood neutrophils have a very short life span, 6 to 10 hours, after which they enter the tissues where they die over 4 to 5 days. Theoretical considerations lead to the conclusion that the factors controlling new production and maturation of the granulocyte cell line cannot be dependent entirely upon peripheral blood neutrophil numbers in a scheme of classic negative feedback inhibition. The best category for a true granulopoietin at the present time is a so-called granulocyte CSF, a glycoprotein with a molecular weight of approximately 45,000 which can be found in all normal human and animal serum and urine. This material

Annotated references

Athens, J. W.: Neutrophilic granulocyte kinetics and granulocytopoiesis. *In* Gordon, A. S. (ed.): Regulation of Hematopoiesis. New York, Appleton-Century-Crofts, 1970. (A detailed analysis of the experimental evidence for leucocyte pools and their alterations in disease.)

Boggs, D. R.: The kinetics of neutrophilic leukocytes in health and disease. Semin. Hematol. *4*:359, 1967. (A review of leucocyte kinetics.)

Bagby, G. C., Jr.: T lymphocytes involved in inhibition of granulopoiesis in two neutropenic patients are of the cytotoxic/suppressor (T3+T8+) subset. J. Clin. Invest. *68*:1597, 1981.

Barrett, A. J., Weller, E., Rozengurt, N., et al.: Amidopyrine agranulocytosis: Drug inhibition of granulocyte colonies in the presence of patients' serum. Br. Med. J. *2*:850, 1976.

Broxmeyer, H. E., Bognacki, J., Dorner, M. H., et al.: Identification of leukemia-associated inhibitory activity as acidic isoferritins. J. Exp. Med. *153*:1426, 1981.

Dale, D. C., and Graw, R. G., Jr.: Transplantation of allogeneic bone marrow in canine cyclic neutropenia. Science *183*:83, 1974.

Deinard, A. S., et al.: Studies on the neutropenia of cancer chemotherapy. Cancer *33*:1210, 1974. (Use of hydrocortisone to measure marrow reserve.)

Greenberg, P. L.: Clinical relevance of in vitro study of granulocytopoiesis. Scand. J. Haematol. *25*:369, 1980.

Greenberg, P. L., Nichols, W. C., and Schrier, S. L.: Granulopoiesis in acute myeloid leukemia and preleukemia. N. Engl. J. Med., *284*:1225, 1971. (Experimental data on granulopoiesis in leukemia.)

Lichtman, M. A., and Weed, R. I.: Alteration of the cell periphery during granulocyte maturation: Relationship to cell function. Blood *39*:301, 1972. (Studies on factors regulating granulocyte release.)

Moberg, C., Olofsson, T., and Olsson, I.: Granulopoiesis in chronic myeloid leukaemia. I. In vitro cloning of blood and bone marrow in agar culture. Scand. J. Haematol. *12*:380, 1974.

Morley, A.: A neutrophil cycle in healthy individuals. Lancet *2*:1220, 1966.

Morley, A., King-Smith, E. A., and Stohlman, F., Jr.: The oscillatory nature of hemopoiesis. *In* Stohlman, F., Jr. (ed.): Hematopoietic Cellular Proliferation. New York, Grune & Stratton, 1970. (Theoretical and experimental considerations of granulocyte cycles.)

Olofsson, T., and Olsson, I.: Suppression of normal granulopoiesis in vitro by a leukemia-associated inhibitor (LIA) of acute and chronic leukemia. Blood *55*:975, 1980.

Quesenberry, P., Morley, A., and Stohlman, F., Jr., et al.: Effect of endotoxin on granulopoiesis and colony stimulating factor. N. Engl. J. Med. *286*:227, 1972. (Experimental data on role of endotoxin.)

Robinson, W. A., Entringer, M. A., Bolin, R. W., et al.: Bacterial stimulation and granulocyte inhibition of granulopoietic factor production. N. Engl. J. Med. *297*:1129, 1977.

Robinson, W. A., and Mangalik, Asha: The kinetics and regulation of granulopoiesis. Semin. Hematol. *12*:7, 1975. (General review of concepts and mechanisms.)

Stohlman, F., Jr., Quesenberry, P., and Tyler, W. S.: The regulation of myelopoiesis as approached with in vivo and in vitro techniques. Prog. Hematol. *8*:259, 1973. (Review of experimental granulopoiesis).

General references

Aye, M. T., Till, J. E., and McCulloch, E. A.: Growth of leukemic cells in culture. Blood *40*:806, 1972.

Boggs, D. R., Chervenick, P. A., Marsh, J. C., et al.: Neutrophil releasing activity in plasma of dogs injected with endotoxin. J. Lab. Clin. Med. *72*:177, 1968.

Bolin, R. W., and Robinson, W. A.: Bacterial, serum, and cellular modulation of granulopoietic activity. J. Cell. Physiol. *92*:145, 1977.

Bradley, T. R., Stanley, E. R., and Sumner, M. A.: Factors from mouse tissues stimulating colony growth of mouse bone marrow cells in vitro. Aust. J. Exp. Biol. Med. Sci. *49*:595, 1971.

Broxmeyer, H. E., de Sousa, M., Smithyman, A., et al.: Specificity and modulation of the action of lactoferrin, a negative feedback regulatory of myelopoiesis. Blood *55*:324, 1980.

Broxmeyer, H. E., Grossbard, E., Jacobsen, N., et al.: Persistence of leukemia-associated inhibitory activity against normal granulocyte-macrophage progenitor cells during remission of acute leukemia. N. Engl. J. Med. *301*:346, 1979.

Broxmeyer, H. E., Jacobsen, N., Kurland, J., et al.: In vitro suppression of normal granulocytic stem cells of inhibitory activity derived from human leukemic cells. J. Nat. Canc. Inst. *60*:497, 1978.

Bull, J. M., Duttera, M. J., Staskich, E. D., et al.: Serial in vitro marrow culture in acute myelocytic leukemia. Blood *42*:679, 1973.

Chan, S. H., and Metcalf, D.: Local production of colony-stimulating factor within the bone marrow: Role of nonhematopoietic cells. Blood *40*:646, 1972.

Chervenick, P. A., Ellis, L. D., Pan, S. F., et al.: Human leukemic cells: In vitro growth of colonies containing the Philadelphia (Ph) chromosome. Science *174*:1134, 1971.

Cline, M. J., Opelz, G., Saxon, A., et al.: Autoimmune panleukopenia. N. Engl. J. Med. *295*:1489, 1976.

Craddock, C. G., Perry, S., Ventyke, L., et al.: Evaluation of marrow granulocyte reserves in normal and disease states. Blood *15*:840, 1960.

Dale, D., Guerry, D., and Wewerka, J.: Chronic neutropenia. Medicine *58*:128, 1979.

Ecsenyi, M.: CSF levels and inhibitors in sera from patients with infectious diseases. *In* Van Bekkum, D. W., and Dicke, K. A. (eds.): In Vitro Culture of Hemopoietic Cells. Rijswijk, Radiobiological Institute, TNO, 1972.

Fink, M. E., and Calabresi, P.: The granulocyte response to an endotoxin (Pyrexal) as a measure of functional marrow reserve in cancer chemotherapy. Ann. Intern. Med. *57*:732, 1962.

Gatti, R. A., Robinson, W. A., Deinard, A. S., et al.: Cyclic leukocytosis in chronic myelogenous leukemia: New perspectives on pathogenesis and therapy. Blood *41*:771, 1973.

Golde, D. W., and Cline, M. J.: Identification of the colony stimulating cell in human peripheral blood. J. Clin. Invest. *51*:2981, 1972.

Golde, D. W., Finley, T. N., and Cline, M. J.: Production of colony-stimulating factor by human macrophages. Lancet *2*:1397, 1972.

Gordon, M. Y.: Circulating inhibitors of granulopoiesis in patients with aplastic anemia. Br. J. Haematol. *39*:491, 1978.

Greenberg, P. L.: Granulopoisis in vitro: Applications of clonogenic cell culture techniques and the regulation of granulocyte and monocyte proliferation. In Fairbanks, V. (ed.) Current Hematology, Vol. I. New York, John Wiley & Sons, 1981.

Greenberg, P. L., Mara, B., Steed, S. M., et al.: The chronic idiopathic neutropenia syndrome: Correlation of clinical features with in vitro parameters of granulocytopoiesis. Blood *54*:915, 1980.

Mangalik, A., and Robinson, W. A.: Cyclic neutropenia: The relationship between urine granulocyte colony stimulating activity and neutrophil count. Blood *41*:79, 1973.

Metcalf, D.: Studies on colony formation in vitro by mouse bone marrow cells. II. Action of colony stimulating factor. J. Cell. Physiol. *76*:89, 1970.

Metcalf, D.: Hemopoietic colonies, in vitro cloning of

normal and leukemic cells. New York, Springer-Verlag Berlin Heidelberg, 1977.

Metcalf, D., and Wahren, B.: Bone marrow colony stimulating activity of sera in mononucleosis. Br. Med. J. *13*:99, 1968.

Pike, B., and Robinson, W. A.: Human bone marrow colony growth in agar-gel. J. Cell. Physiol. *76*:77, 1970.

Robinson, W. A.: Granulocytosis in neoplasia. Ann. N.Y. Acad. Sci. *230*:212, 1974.

Robinson, W. A., Kurnick, J. E., and Pike, B. L.: Granulocytic colony growth in vitro from human leukemic peripheral blood. Blood *38*:500, 1971.

Robinson, W. A., and Pike, B. L.: Leukopoietic activity in human urine. The granulocytic leukemias. N. Engl. J. Med. *282*:1291, 1970.

Shadduck, R. K., and Nagabhushanam, N. G.: Granulocyte colony stimulating factor. I. Response to acute granulocytopenia. Blood *38*:559, 1971.

Stanley, E. R., and Metcalf, D.: Purification and properties of human urinary colony stimulating factor (CSF). Cell Differentiation *18*:272, 1972.

Wright, D. G., Dale, D. C., Fauci, A. S., et al.: Human cyclic neutropenia: clinical review and long-term follow-up of patients. Medicine *60*:1, 1981.

29   Thrombopoiesis

Simon Karpatkin, M.D.

Platelets

The circulating peripheral blood platelet is an anucleate cytoplasmic fragment derived from the bone marrow megakaryocyte. Its major function is to plug gaps in blood vessels following trauma. In so doing, it also acts as a catalytic surface for the efficient interaction of coagulation factor proteins, which can increase the conversion of prothrombin to thrombin 300,000-fold. For example, activated platelets release factor V, which binds to the platelet surface. Activated factor V (bound to platelets) becomes the receptor for factor Xa. Activated platelets also bind fibrinogen. The activated coagulation cascade leads to the generation of thrombin, which polymerizes fibrinogen to fibrin, leading to a more consolidated plug. Platelets also appear to be associated with three other pathophysiological conditions. Platelets contribute to the development of atherosclerosis by adhering to the basement membrane of an injured vessel and releasing a growth factor which stimulates the migration of smooth muscle cells to the luminal side of the injured vessel. A pseudoendothelial layer is developed which is indistinguishable from an early atherosclerotic lesion. Platelets contribute to the development of tumor metastases by sequestering tumor cells within the circulation (most tumor cells rapidly disappear from the circulation after intravenous injection). This permits the penetration of the endothelial surface by tumor cells (perhaps aided by platelet permeability factors) to an extravascular site where they can be stimulated to divide by platelet growth factor. Platelets are also thought to contribute to the vascular sequelae of diabetes mellitus. They have a shortened survival and are hyperaggregable owing to the presence of a diabetic plasma factor.

On EDTA-peripheral blood smear, platelets are 1 to $3\,\mu$ in diameter. In citrate-platelet–rich plasma they are discoid in shape and average 5 to $7\,\mu^3$ in volume. There are 150 to 350,000 platelets per μl. of whole blood. The number for each individual remains fairly constant over a period of time. The total platelet mass is approximately 18 ml. for a 70 kg. individual. This should be compared to 2500 ml. for the red blood cell mass. Approximately 30 to 40 per cent of the total platelet pool is sequestered in the spleen and other organs. This sequestered pool is immediately mobilizable by epinephrine secretion or exercise. Approximately half of the sequestered pool is in organs other than the spleen. The platelet which is sequestered in the spleen is, on average, larger than the nonsequestered platelet. It has been claimed that the spleen preferentially sequesters younger platelets which function better than older platelets. Platelet number is inversely related to platelet volume for the normal

667

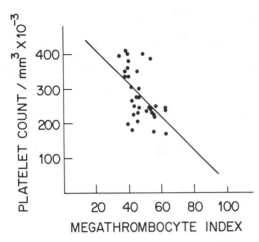

FIG. 29–1. Correlation between whole blood platelet count and megathrombocyte index. Regression line was determined by the method of least squares and is given by the formula $y = -4.80x + 508$. Correlation coefficient, $r = -0.53$ has $p < 0.001$; $N = 34$. (Karpatkin, S.: Blood 51:307, 1978.)

range of platelet counts (Fig. 29–1) as well as in pathophysiological situations of thrombocytopenia or thrombocytosis (see below). The normal platelet survival, as determined by Cr^{51}, or DFP^{32} or In (total, population labels) is 10 days. The platelet survival, as determined by Se^{75}-selenomethionine, a cohort label which tags the bone marrow megakaryocyte first (which then releases the isotopically labeled peripheral blood platelets) is 10 to 12 days (Fig. 29–2). Platelet turnover represents platelet production per day, calculated by dividing the total platelet number (including sequestered splenic and other pools) by the platelet survival. During most acquired pathophysiological conditions of increased or decreased platelet turnover, platelet volume is proportional to megakaryocyte number (Fig. 29–3) as well as immaturity of megakaryocytes. In increased peripheral destruction, characterized by autoimmune thrombocytopenic purpura, platelet volume is inversely related to platelet survival. Most noncohort platelet survival curves in normal individuals are predominantly linear (i.e., arithmetic plot of disappearance of label vs. time) suggesting

death by senescence, rather than random hits. However, these curves also have curvilinear components, suggesting a small but significant number of random hits. The platelet population is heterogeneous for size, volume, density, metabolic function and life span. It has been suggested that heavier or larger platelets are, on average, younger than lighter or smaller platelets and that this could account for some of the heterogeneity (Fig. 29–2). However, this concept has been challenged by those who believe that platelet heterogeneity is predominantly a function of heterogeneity of platelet production by megakaryocytes; i.e., different megakaryocytes give rise to platelets of different densities or size. These two theories are not mutually exclusive. It is quite possible that platelets are born heterogeneous but become lighter and smaller with age, leading to further heterogeneity, senescence, and death.

Megakaryocytes

The recognizable bone marrow megakaryocyte is a giant multinucleated polyploid cell (20 to 150 μ in diameter) which represents less than 1 per cent of all marrow cells. It undergoes repeated nuclear divisions without cytoplasmic division during maturation. The megakaryocyte contains 2 to 32 nuclei and therefore 4N to 64N chromosomes. Only immature megakaryoblasts with little or no cytoplasmic maturation are capable of undergoing DNA synthesis. The development of polyploidy by DNA replication takes place in precursor cells and in the most immature and barely recognizable megakaryocytes. Since recognizable megakaryocytes do not undergo cell division, the maintenance of this compartment is dependent upon a continuous influx of differentiating stem cells from the marrow. Unlike other hematopoietic cells, nuclear and cytoplasmic maturation do not proceed in parallel. After the cell generates its full chromosomal content, DNA synthesis terminates and cytoplasmic maturation and nuclear segmentation begins. It was once believed that cytoplasmic maturation and nuclear division occur at the same time. This is not the case. Platelets are formed from megakaryocytes of different ploidy. However, because ploidy number correlates positively with cytoplasmic size, the number of platelets produced per megakaryocyte corresponds to its degree of ploidy. A single megakaryocyte, depending upon its ploidy, can produce 4000 to 8000 platelets. Cells mature and undergo cytoplasmic maturation, fragmentation and release, in three major ploidy classes, 8N, 16N, and 32N. The residual nucleus is phagocytosed by the reticuloendothelial system. The life cycle of the megakaryocyte in man is approximately 10 days.

The sequence of bone marrow megakaryocyte maturation has been classified into three stages: Stage 1, promegakaryoblast or megakaryocytoblast; Stage II, megakaryoblast or immature megakaryocyte; Stage III, mature megakaryocyte. Stage I consists of small (10- to 20-μ diameter) cells with a thin rim of basophilic agranular cytoplasm. These cells contain negligible protein synthesis capability. The nucleus occupies most of the cell. Stage II has a lower nuclear-to-cytoplasmic ratio. Eosinophilic granules become recognizable in an area adjacent to the nucleus. Mitochondria, rough endoplasmic reticulum, a mature Golgi apparatus, and electron-dense granules are visible. In Stage III the megakaryocyte is capable of platelet production with a polyploid (8N–64N) nucleus. The cytoplasm is eosinophilic and granular. There are polyribosomes and a characteristic demarcation membrane system which is contiguous with the plasma membrane and important in platelet formation.

Cytoplasmic maturation and platelet release is associated with an extensive membrane system, the demarcation membrane. This system demarcates platelet 'zones' by enclosing and defining parts of the megakaryocyte cytoplasm which develop into circulating platelets. Under electron mi-

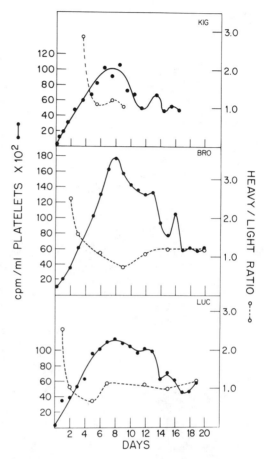

FIG. 29–2. Se[75]-selenomethionine cohort platelet survival curves on subjects KIG, BRO, and LUC. Data are expressed as c.p.m./ml. packed platelets (●). A cohort-labeled platelet should theoretically have an ascending limb, followed by a plateau, followed by a descending limb. The distance between the 50 per cent elevation of the ascending limb and 50 per cent decline of the descending limb is equivalent to the duration of the platelet survival. The squiggles in the descending limb probably represent reutilization of isotope after release from "dead" platelets or other tissues. At varying time intervals, one unit of platelet-rich plasma was removed from the subject, processed into "heavy" and "light" platelet populations and assayed for radioactivity. The heavy/light ratio is plotted as (○). (Amorosi, E., et al.: Br. J. Haematol. *21*:227, 1971.)

croscopy (EM) the demarcation system appears as round, oval, or elongated vesicles with the membrane enclosing an empty core (Fig. 29-4). In well-developed megakaryocytes transverse sections of the demarcation system appear as circular vesicles that

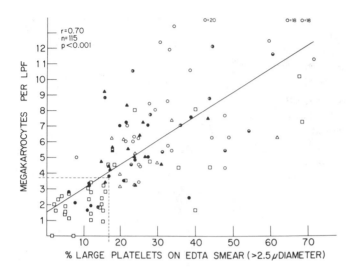

FIG. 29–3. Correlation between the percentage of large platelets on EDTA peripheral smear (diameter greater than 2.5 μ) and the number of megakaryocytes per low-power field ($\times$ 100). The open circles refer to ITP, the closed circles to SLE, the horizontal half-closed circles to DIC, the vertical half-closed circles to drug-induced thrombocytopenia, the open triangles to hypersplenism, the closed triangles to iron-deficiency anemia (normal or low platelet count), and the open boxes to other disorders. The interrupted lines include +2 standard deviations of the "control mean" (95 per cent upper confidence limit). (Garg, S. K., et al., N. Engl. J. Med., 284:11, 1971.)

FIG. 29–4. Electron micrograph of rat megakaryocyte, demonstrating demarcation membranes. Uranyl acetate and lead stain $\times$ 25,000. *Arrows* point to demarcation membranes. (G) granules; (M) mitochondria. (Behnke: Ultrastruct. Res. 24:412, 1968.)

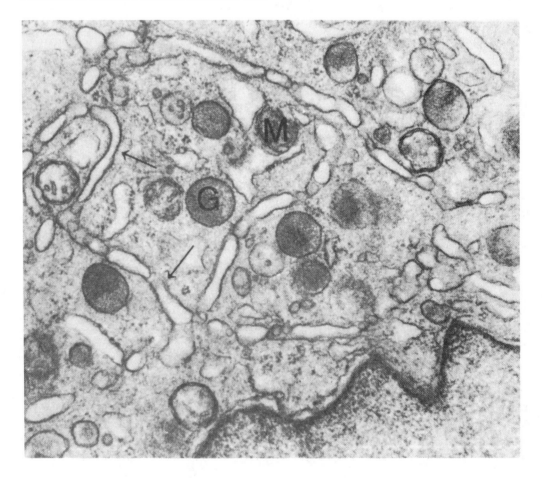

are arranged side by side in a bead-like chain forming the outline of a platelet. In other areas, stacks of longitudinally sectioned profiles are seen to be oriented in parallel arrays. The demarcation system is formed by invagination of the megakaryocyte cell membrane. Tubules are thus formed in the cytoplasm which are parallel in their long axis orientation. These fuse with each other and then separate to form two flat sheets. Upon completion of platelet formation the two sheets of membrane form cell membranes of two adjacent ready-formed platelets.

Platelet release takes place from megakaryocytes located in the subendothelial region of vascular sinuses. The cytoplasm of the megakaryocyte penetrates the endothelium to reach the luminal side of the sinus. This penetration is transendothelial (through the endothelial cell). By this method, large segments of cytoplasm, *proplatelets* (2.5 × 120 μ), enter the lumen. A mature megakaryocyte has been shown by calculations to produce 6 proplatelets, and a proplatelet may give rise to 1,200 platelets. Proplatelets are further fragmented into platelets outside the marrow, perhaps in the spleen, other reticuloendothelial organs, or the lungs. The presence of proplatelets, as well as megathrombocytes, which are further fragmented megakaryocytes, may contribute to the heterogeneity of circulating platelets as well as the concept that larger platelets (megathrombocytes) are young platelets.

Regulation of platelet production

Platelet production is regulated by feedback mechanisms which control the level of the platelet count or the total platelet mass. Several different mechanisms which may or may not be interrelated appear to control the steady state platelet level. These include platelet number, platelet mass, platelet turnover, blood loss, serum iron, and a humoral factor termed *thrombopoietin*.

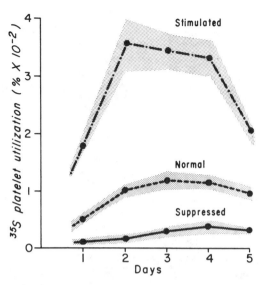

FIG. 29–5. Comparison of platelet incorporation of S^{35} in animals with normal, 4-day stimulated and 4-day suppressed thrombopoiesis. Each point represents mean of 10 animals; ± 1 SD shown by shaded area. Stimulated production involves an increase in both amount of isotope incorporation and rate of appearance. The converse occurs as platelet production is suppressed. (Harker, L.: Am. J. Physiol. *218*:1376, 1970.)

Acute thrombocytopenia, produced experimentally by exchange transfusion with platelet-poor plasma, or with antiplatelet antibody, results in the following: increased megakaryocyte number, volume, ploidy, incorporation of ^{75}Se-selenomethionine into megakaryocytes (measured by the appearance of isotope in newly formed peripheral blood platelets) (Table 29–1, Fig. 29–5), increased mean platelet volume and large platelet number (megathrombocytes); and decreased time required for megakaryocyte cytoplasmic maturation and release. The thrombocytopenia is followed by a transient thrombocytosis (compatible with a feedback mechanism). Since platelet production is a function of megakaryocyte volume and number, megakaryocyte mass (volume × number) is a more appropriate index for platelet production potential. Megakaryocyte mass correlates with platelet turnover. This can increase up to eightfold following peripheral platelet destruction or removal.

TABLE 29–1
Total Megakaryocyte Mass* and S³⁵ Platelet Incorporation

Thrombopoiesis	Mean Megakaryocyte Number	Mean Megakaryocyte Volume	Total Megakaryocyte Mass	S^{35} in Platelets
	$\times 10^6$ cells/kg.	μ^3	$\mu^3 \times 10^{10}$/kg.	$\% \times 10^{-2}$
Normal	11.0	4580	5.0	1.18
4-day suppressed	7.0	2440	1.7	0.45
10-day suppressed	4.6	2760	1.3	—
4-day stimulated	15.1	9740	14.7	3.45
10-day stimulated	36.7	9100	33.4	—

*Each value represents the mean of six animals. (Harker, L. A.: J. Clin. Invest. 47:458, 1968.)

Thus, platelet survival would have to be less than 1.25 days (assuming a normal 10 day survival) before the platelet count decreases below the normal range. Thrombocytosis, induced by hypertransfusion with platelets, leads to a decrease in megakaryocyte number, volume, ploidy; incorporation of ^{75}Se-selenomethionine into platelets; a decrease in mean platelet volume and megathrombocytes; and a delay in the time required for megakaryocyte cytoplasmic maturation and release. The thrombocytosis is followed by a transient thrombocytopenia (compatible with a feedback mechanism). Thus the peripheral blood platelet number affects platelet production.

A different mechanism must be operative for animals or patients with hypersplenism, wherein the peripheral blood platelet count is decreased two- to fivefold, yet the platelet survival is normal or near normal in the presence of a bone marrow megakaryocyte mass increased by approximately two times. Since total platelet mass (spleen and peripheral blood) is usually normal or slightly increased, despite a decrease in peripheral blood number, platelet mass must affect platelet production.

A third mechanism is illustrated by the compensated thrombocytolytic state, wherein platelet survival is one-half to one-third normal in the presence of increased megakaryocyte number (and probably volume) and normal platelet count with increased mean platelet volume and megathrombocytes (Table 29–2). In this situation, it has been postulated that platelet by-products (released during platelet destruction) must affect megakaryocyte production and volume.

A fourth mechanism must be associated with blood loss. Blood loss, in the absence of iron deficiency, can lead to a thrombocytosis associated with increased megakaryocytes and megathrombocyte number, which is not clearly understood and which appears to require an intact spleen for maximum response. Blood loss in the presence of iron replacement leads to a greater thrombocytosis than blood loss alone and is associated with a marked increase in mean platelet volume and megathrombocyte number, (Fig. 29–6).

A fifth mechanism is illustrated by the rise in platelet count noted in the presence of iron deficiency induced by diet (Fig. 29–6 and 29–7). This is also associated with increased megakaryocyte number in the absence of increased megathrombocytes. Of particular interest is the observation that iron replacement leads to a greater increase in platelet count (before return to normal) and is associated with an increase in peripheral blood megathrombocytes. Thus, iron appears to operate by a two-part system: (1) iron inhibits thrombopoiesis, perhaps by inhibition of thrombopoietin (see below), since iron loss leads to thrombocytosis; and (2) iron replacement para-

TABLE 29–2
Compensated Thrombocytolytic State (Normal Platelet Count With Increased Megathrombocytes and Megakaryocytes)

Disease	Per Cent Megathrombocytes*	No. of Megakaryocytes†	Platelet Survival, T½‡	Antiplatelet Ab Titer§
Controls	10.8 ± 3.1 (S.D.)	2.1 ± 0.8 (S.D.)	115 hr.	0–1:4
Chronic ATP‖	38.8	10.5	N.T.*	1:40
Chronic ATP	25.0	4.5	N.T.	1:4
Chronic ATP	25.0	6.2	66	1:16
Easy bruising**	25.3	4.4	N.T.	1:32
Easy bruising	54.0	11.8	42	1:48
'Normal' (R.S.)	26.7	N.T.	55	1:8
SLE	16.0	8.8	65	1:4
SLE	20.0	7.0	32	1:32
SLE	24.0	4.8	34	N.T.
SLE	16.0	N.T.	61	1:44
SLE	16.8	N.T.	71	N.T.
SLE	21.7	7.2	N.T.	1:12
SLE	27.6	5.0	N.T.	1:8
SLE	27.5	7.0	N.T.	N.T.

*Percent platelets on EDTA smear with a diameter greater than 2.5μ.
†Average number of megakaryocytes per spicule field (lower power, ×100) enumerated from bone marrow aspiration smears examining a minimum of 25 fields.
‡*In vivo* DFP[32] platelet survival studies. The per cent megathrombocytes was measured at the beginning and end of the platelet survival study and remained elevated in the 7 patients studied. The platelet counts remained constant and varied in different subjects between 210,000 and 267,000/mm.[3]
§Platelet factor 3 immunoinjury antiplatelet antibody titer expressed as the maximal dilution of globulin fraction of patient's serum giving positive test. Titers greater than 1:4 are considered significant.
‖Chronic ATP in remission.
*Not tested.
**Probable compensated ATP with history of easy bruising and negative SLE preparations.
(Karpatkin, et al.: Am. J. Med. *51*:1, 1971.)

doxically increases thrombocytosis, possibly owing to the requirement of iron as an essential component for megathrombocyte and platelet production. For example, iron is required for platelet protein synthesis. Severe iron deficiency can lead to thrombocytopenia, probably secondary to exhaustion of the essential component compartment. It is of interest that reactive thrombocytosis is also noted with infection, inflammation, or malignancy when serum iron is reduced, and postoperatively, particularly following splenectomy, when serum iron drops precipitously.

Thrombopoietin

All or some of these mechanisms may operate by their influence on a humoral thrombopoietic substance, thrombopoi-

etin. Such a material was first suggested by a female patient with severe thrombocytopenia at an early age of development, associated with immature, nonproductive megakaryocytes, who responded to fresh frozen plasma transfusion, and who required triweekly infusions of plasma to maintain a normal platelet count. Thrombopoietin has been obtained from thrombocytopenic animals (2 to 12 hours after induction of thrombocytopenia) as well as thrombocytopenic patients and patients with both reactive and autonomous thrombocytosis and has been shown to increase the incorporation of ^{75}Se-selenomethionine into peripheral blood platelets when injected into animals. Some of this isotope appears to be incorporated into mature megakaryocytes (rapid release of labeled platelets) as well as immature megakaryo-

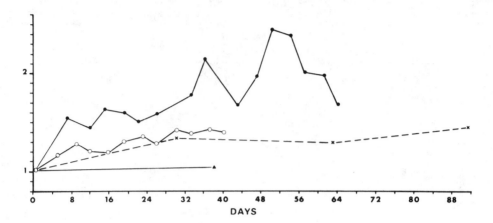

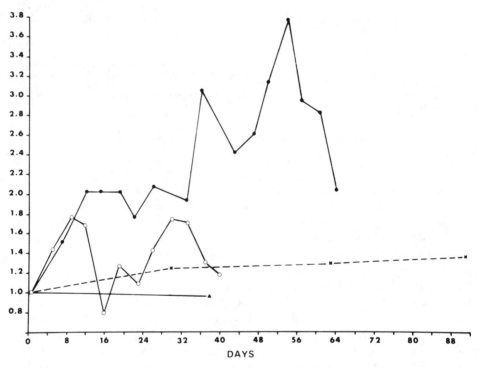

FIG. 29—6. Effect of chronic blood loss (○—○), chronic blood loss plus iron replacement (●—●), and iron-deficient diet (X---X) on thrombocyte and megathrombocyte kinetics. Guinea pigs (nine in each group) were either chronically bled by cardiac puncture, 2 ml. every 3 to 4 days, or sham manipulated as in control group (▲—▲) or iron-deficient diet group (X---X). The amount of iron replaced (imferon) was equivalent to the amount of iron lost in the blood removed. Abscissa refers to changes from values normalized to 1 on day zero. (Karpatkin, S., et al., Am. J. Med. 57:521, 1974.)

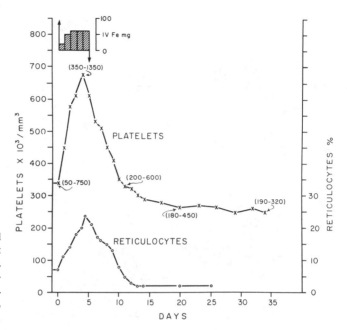

FIG. 29–7. Mean reticulocyte and platelet response in 23 iron-deficient patients receiving intravenous iron. The range in platelets is noted in parentheses at 0, 5, 10, 20, and 33 days. (Gross, S., et al.: Pediatrics *34*:315, 1964.)

cytes (delayed release of labeled platelets which peaks at 2 to 4 days in rodents). Thrombopoietin also has been shown to increase platelet count as well as mean platelet volume and megathrombocyte number. Of particular interest is an observation (as yet unconfirmed) that platelets are capable of inactivating or removing thrombopoietin from plasma. Recent studies have indicated that thrombopoietin has no effect on megakaryocyte colony formation *in vitro*, although it does result in a small increase in megakaryocyte ploidy number *in vitro* as well as *in vivo*. A megakaryocyte colony (CFU-M) stimulating factor has been recently described which has no effect on the incorporation of ^{75}Se-selenomethionine into platelets and no effect on platelet count. This second thrombopoietic factor has been obtained from the serum of three patients with aplastic anemia.

Biochemical purification of thrombopoietin, by 4,800 times, has been achieved by employing wheat germ agglutinin affinity chromatography. Controversy exists over the tissue origin of thrombopoietin. Several experiments suggest that, like erythropoietin, it is derived from the kidney. Thus, nephrectomized rats do not produce thrombopoietin following injection of antiplatelet antibody, whereas unoperated rats do produce thrombopoietin. However, ureter-ligated controls were not employed. Vinblastine-treated animals produce a serum thrombopoietic activity, whereas nephrectomized, vinblastine-treated animals do not; ureter-ligated controls also produce thrombopoietin. Human embryonic kidney cells produce thrombopoietin in the tissue media, whereas culture media of liver cells from several species do not. Another group has provided evidence that thrombopoietin is not derived from the kidney, since nephrectomy did not appear to impair the rebound response to antiplatelet serum. This group has also presented evidence that the liver may be the source of thrombopoietin, since a four-fifths hepatectomy resulted in a rapid 50 per cent decline in platelet count during the first 24 hours, a decrease in megakaryocyte diameter and ploidy, and a 50 per cent decrease in the incorporation of ^{75}Se-selenomethionine into newly formed platelets.

A thrombopoietin-like humoral factor

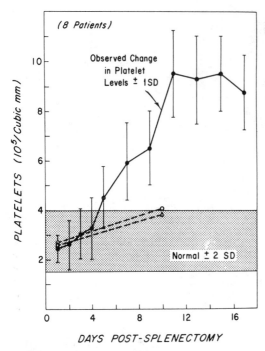

FIG. 29–8. Effect of splenectomy on platelet count in eight normal subjects requiring surgery due to trauma. Dashed lines indicate average maximum increase in platelet levels to be expected due to loss of splenic pooling (open circles) or postoperative thrombocytosis (triangles). (Aster, R. H.: *In* Williams, W. J., Beutler, E., Erslev, A. J., and Rundles, R. W., (eds.): Hematology, ed. 2. pp. 1221–1224. New York, McGraw Hill, 1977.)

has also been described in the plasma of animals subjected to acute and chronic blood loss. The acute blood loss factor is capable of increasing megakaryocyte number in recipient animals. One group studying this humoral factor noted that peak activity was reached 6 hours after blood loss, prior to the rise in platelet count of the donor animals, and disappeared from the donor 72 hours after initial blood loss. The acute blood loss factor could not be derived from nephrectomized donor animals. Another group studying the thrombopoietin-like activity of chronic blood loss plasma suggested that it may be a splenic release factor, since the recipient animal required an intact spleen for the demonstration of thrombocytosis, and the increment in platelet count and megathrombo-

cyte number was equivalent to their content in the sequestered splenic pool.

A splenic thrombopoiesis *inhibitor*, possibly a thrombopoietin antagonist, has been suggested by the work of several laboratories. In normal subjects, transient thrombocytosis occurs after splenectomy, reaching its peak by day 10, to levels 2.5 times greater than that expected from the platelet contribution of the splenic pool (Fig. 29–8). This increase is considerably greater than the 30 to 40 per cent rise that normally follows surgical procedures. A splenic humoral factor which inhibits thrombopoiesis is suggested by animal studies, wherein reimplantation of splenic tissue in semipermeable chambers prevented postsplenectomy thrombocytosis.

Pathophysiological mechanisms

Autoimmune thrombocytopenic purpura

Autoimmune thrombocytopenic purpura illustrates the pathophysiological condition of increased platelet turnover. Thrombocytopenia is secondary to increased platelet destruction of antibody-coated platelets by phagocytosis, particularly in the spleen. Platelet survival is markedly shortened to less than 10 per cent of normal. Megakaryocytes are increased in number, volume, and immaturity. Platelets are decreased in number and increased in volume.

More specifically, platelet survival is proportional to the platelet count (at low platelet survivals of 0 to 3 days). A survey of 49 patients from two studies with mean platelet counts of 30,000/μl. revealed a mean platelet survival of approximately 0.5 days; and a mean platelet turnover of approximately 3.5 times normal. In a study involving 14 patients, megakaryocyte number averaged three times normal and megakaryocyte volume 1.6 times normal. Thus the increase in megakaryocyte mass averaged 4.8 times, indicating that under suitable stress, the bone marrow can increase platelet production approximately five-fold. The ratio of platelet turnover to mega-karyocyte mass was normal, indicating effective thrombopoiesis. Thus, there is little kinetic evidence that antiplatelet antibody prevents the release of platelets by mega-karyocytes; if this were the case, thrombo-poiesis would be ineffective.

Large platelets or megathrombocytes are routinely noted. Megathrombocytes correlate with megakaryocyte number (r = 0.7, p < 0.001). Mean platelet diameter is 1.6 times greater than normal. The increase in megathrombocytes parallels the increase in megakaryocytes (effective thrombo-poiesis) and can be on the order of three- to fourfold. There is generally a shift to the right in the platelet volume distribution curve, towards larger platelets (Fig. 29–9). Approximately 50 per cent of patients in apparent clinical remission have increased megathrombocytes (as well as bound plate-let IgG), indicating increased platelet turnover despite a normal platelet count, and compensated thrombocytolytic state. There is a significant inverse correlation between platelet diameter and platelet survival (r = −0.8, p < 0.01). There is a positive correlation between platelet diameter and platelet production rate (r = 0.7, p < 0.001) and a positive correlation between platelet size and percentage of immature mega-karyocytes (r = 0.6, p < 0.005).

Patients with particularly serious disease also have evidence for intravascular throm-bocytolysis, as noted by the presence of small cytoplasmic fragments detected by volume distribution curves, as a small par-ticle spike on the left. These have been shown to contain platelet fragments, as well as unexplained red blood cell frag-ments by electromicroscopy.

Hypersplenism

Hypersplenism, discussed above (Regula-tion of Platelet Production), is commonly associated with significant splenomegaly, which may be seen in Laennec's cirrhosis (congestive splenomegaly), myelofibrosis,

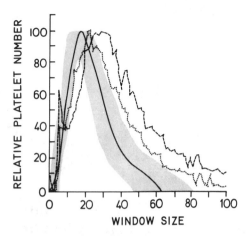

FIG. 29–9. Coulter platelet volume distribution curves in 20 normal subjects ± 2 SD (grey area). The Coulter equipment measures the resistance created by particles passing through an electric current. Resistance is proportional to particle volume. (*A*) A patient with a normal platelet count (160,000/mm.[3]) in remission from ATP (with elevated platelet IgG, 25 ng. 1 g./ 10[6] platelets). (*B*) A patient with ATP and a platelet count of 55,000/mm.[3]. Calibration: 1 window = 0.25/ cu. μ. (Karpatkin, S.: Blood *56:*329, 1980.)

chronic myelocytic leukemia, thalassemia major, Hodgkin's disease and other lym-phomas, and parasitic infestations (ma-laria, kala azar, schistosomatosis). In these conditions, 80 to 90 per cent of the total platelet pool may be sequestered in the spleen and rapidly mobilized by epineph-rine infusion.

The compensated thrombocytolytic state

The compensated thrombocytolytic state is commonly associated with patients with autoimmune thrombocytopenic purpura in remission, systemic lupus erythemato-sus, the easy bruising syndrome of women, and sometimes with vascular complica-tions of diabetes mellitus, prosthetic heart valves, rheumatic heart disease, and coro-nary atherosclerosis.

Reactive thrombocytosis

Reactive thrombocytosis (see Regulation of Platelet Production) is generally noted with blood loss or iron deficiency, malignancy,

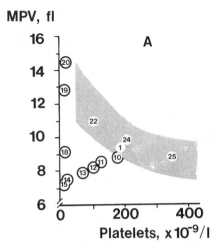

MPV, fl

FIG. 29–10. Platelet volume and count during chemotherapy and recovery in a single patient. The hatched area depicts the inverse relationship between platelet volume and count in normal subjects and patients with increased peripheral destruction of platelets. The circled numbers refer to the day on which the measurements were obtained. Before and during chemotherapy (day 1 and days 2–9 are similar) values are normal. After chemotherapy the falling mean platelet volume brings the value below the hatched area (reference nomogram), before the platelet count declines (days 10 and 11). The values are considerably below the reference nomogram during the first part of the thrombocytopenic nadir (days 12 and 15; days 16 and 17 are similar). A sudden rise in mean platelet volume is then noted, which brings the value into the reference nomogram (days 18–20), before the recovery of the platelet count. During recovery, the mean platelet volume falls as the platelet count rises (days 20–25). (Bessman, D.: Am. J. Hematol., *13*:219, 1982.)

inflammation, infection, surgery or splenectomy. The platelet count is generally fewer than 1 million/μl., although it can reach higher levels. Platelet survival is normal and platelet turnover 2 to 5 times normal. Megakaryocyte number is generally increased two- to fivefold, in the absence of an increase in megakaryocyte volume. Similarly, mean platelet volume is not increased. Thus, the normal inverse relationship between platelet count and mean platelet volume (Fig. 29–10) is maintained. These findings are significantly different from those found in autonomous thrombocytoses.

Autonomous thrombocytoses

Autonomous thrombocytoses are clinical entities in which the thrombocytosis is due to a malignancy in the megakaryocyte or stem cell line. Platelet production appears to be autonomous or unregulated. Thus, platelet counts are generally greater than 1 million/μl. and may reach as high as 5 to 10 million/μl. The common clinical disorders include such entities as thrombocythemia, polycythemia vera, and chronic granulocytic leukemia. Platelet survival is generally normal. However, platelet turnover may be 3 to 13 times normal. Megakaryocyte number is generally 2 to 8 times normal and megakaryocyte volume 2 to 3 times normal, giving a megakaryocyte mass of 4 to 24 times normal. Thus, unlike reactive thrombocytoses, megakaryocyte volume is increased. Platelet size is also increased and heterogeneous, with large megakaryocyte fragments noted on peripheral smear. In this situation the normal inverse relationship between platelet count and mean platelet volume is not maintained.

Impaired platelet production

This can be seen with either bone marrow hypoplasia or with ineffective thrombopoiesis.

Marrow hypoplasia

Marrow hypoplasia is noted with metastatic malignancies, leukemic infiltration, iatrogenic drug-induced problems, chemotherapy, viral infections (hepatitis B, congenital rubeola, measles vaccine), acute alcoholism (see below), myelofibrosis, and idiopathic causes. This can be transient or permanent. The permanent variety due to drugs is usually idiosyncratic. The platelet count is generally fewer than 50,000/μl. Platelet survival is normal to slightly decreased (5 to 10 days). Platelet turnover is generally 10 to 50 per cent of normal. Megakaryocyte number is 10 to 50 per cent of normal. Megakaryocyte volume is decreased to normal (40 to 100 per cent).

Therefore, megakaryocyte mass is 4 to 50 per cent of normal. Platelet size is generally decreased (Fig. 29–3).

The effect of chemotherapy on platelet number, mean platelet volume and megakaryocyte ploidy was carefully studied in ten adults with acute nonlymphocytic leukemia receiving 17 courses of chemotherapy. During the myelosuppression induced by chemotherapy (hydroxydaunorubicin, vincristine, cytosine arabinoside, prednisone), platelet count, mean platelet volume, and megakaryocyte ploidy all decreased in 7 to 10 days. During bone marrow recovery, megakaryocyte ploidy (not shown in Fig. 29–3) rose 1 to 2 days before the rise in platelet volume which in turn rose 1 to 2 days before the rise in platelet count (Fig. 29–10). Platelet volume became inappropriately small before the decline in platelet count and remained small during most of the thrombocytopenic period. The termination of thrombocytopenia could be predicted following the rise in platelet volume to its expected level (for the inverse relationship between platelet count and platelet volume). In three cycles of chemotherapy in which the platelet count did not recover, megakaryocyte ploidy and mean platelet volume remained depressed.

Ineffective thrombopoiesis

Ineffective thrombopoiesis is noted with acute alcoholism, megaloblastic anemias and DiGuglielmo's syndrome. The platelet count is generally fewer than 100,000/μl., and platelet survival normal to slightly decreased (7 to 10 days). Platelet turnover is generally 10 to 50 per cent of normal. However, megakaryocyte number is considerably increased (2- to 13-fold) and megakaryocyte volume slightly decreased to 30 to 85 per cent of normal, giving a megakaryocyte mass or total production capacity of 2 to 4 times normal. Since platelet production does not parallel megakaryocyte capacity, these patients have ineffective thrombopoiesis.

The effect of acute alcoholism on platelet production and bone marrow megakaryocytes has been studied in some detail and has been shown not to be due to folic acid deficiency. In two subjects, alcohol ingestion (280 to 380 g./day) prevented the normal recovery from thrombocytopenia following plateletphoresis (17 to 19 units of platelets) which generally occurs in one week. No recovery from thrombocytopenia (28 per cent of baseline) was noted in the same two subjects during 21 to 24 days of alcohol ingestion. Following cessation of alcohol the platelet count returned to baseline in 6 days. A second experiment was performed when the platelet count had returned to normal. Induction of relative thrombocytopenia required 14 days of daily ingestion of 200 to 250 g. of alcohol; the platelet count declined significantly from 350,000 to 200,000/μl. Following cessation of alcohol ingestion, the platelet count did not increase for 60 hours. It then gradually rose to 500,000/μl. 9 days after abstinence. Megakaryocyte evaluation in these subjects ingesting alcohol revealed a normal increase in nuclear lobulation after plateletphoresis. However, this was not associated with increased platelet production. Thus, ineffective thrombopoiesis had probably taken place. In another study, megakaryocyte number was shown to be depleted to 10 to 15 per cent of normal. Therefore, acute alcoholism can decrease platelet production by impairing both megakaryocyte production as well as platelet release from megakaryocytes.

Annotated references

Amorosi, E., Garg, S. K., and Karpatkin, S.: Heterogeneity of human platelets. IV. Identification of a young platelet population with Se[75]-selenomethionine. Br. J. Haematol. 21:227, 1971. (Kinetic evidence that heavy-large platelets are young platelets under basal conditions.)

Bessman, D.: Prediction of platelet production during chemotherapy of acute leukemia. Am. J. Hematol. 13:219, 1982. (Changes in megakaryocyte ploidy, platelet count and platelet volume during bone marrow suppression and recovery.)

Branehog, I., Kutti, J., Ridell, B., Swolin, B., and Weinfeld, A.: The relation of thrombokinetics to bone marrow megakaryocytes in idiopathic thrombocytopenic purpura (ITP). Blood *45*:551, 1975. (A comprehensive investigation of megakaryocyte number and volume, platelet production, platelet number, and platelet size in patients with autoimmune peripheral destruction of platelets).

Ebbe, S.: Thrombopoietin. Blood *44*:604, 1974. (A critical analysis of thrombopoiesis and thrombopoietin which raises important physiological questions.)

Garg, S. K., Amorosi, E. L., and Karpatkin, S.: Use of the megathrombocyte as an index of megakaryocyte number. N. Engl. J. Med. *284*:11, 1971. (Demonstration that platelet size correlates with megakaryocyte number in several disorders of increased peripheral destruction of platelets as well as with aplastic disorders. Demonstration of the presence of a compensated thrombocytolytic state (normal platelet count with increased platelet size and shortened platelet survival) in patients with chronic autoimmune thrombocytopenic purpura in "remission," patients with easy bruising, and patients with SLE.)

Harker, L.: Kinetics of thrombopoiesis. J. Clin. Invest. *47*:458, 1968. (Measurements of megakaryocyte number, volume, nuclear number, and cytoplasmic granulation in rats subjected to thrombocytosis and thrombocytopenia for 4 and 10 day periods.)

Harker, L., and Finch, C. A.: Thrombokinetics in man. J. Clin. Invest. *48*:963, 1969. [Measurements of platelet production and turnover are quantitated in normal man and patients with disorders of production (hypoplasia and ineffective production), distribution (hypersplenism), increased peripheral destruction, reactive thrombocytosis, and autonomous thrombocytosis.]

Karpatkin, S., Garg, S. K., and Freedman, M. L.: Role of iron as a regulator of thrombopoiesis. Am. J. Med. *57*:521, 1974. (Experimental and clinical studies demonstrating that iron regulates the reactive thrombocytosis noted with blood loss and/or iron deficiency.)

Penington, D. G., Streatfield, K., and Roxburgh, A. E.: Megakaryocytes and the heterogeneity of circulating platelets. Br. J. Haematol. *34*:639, 1976. (Evidence suggesting that the heterogeneity of platelet size and density is related to the ploidy class of the megakaryocytes from which it was derived, 8N, 16N, and 32N.)

Shreiner, D. P., and Levin, J.: Detection of thrombopoietic activity in plasma by stimulation of suppressed thrombopoiesis. J. Clin. Invest. *49*:1709, 1970. (Demonstration of thrombopoietin. Endogenous thrombopoiesis was suppressed in rabbits by transfusion-induced thrombocytosis. These animals were employed to demonstrate a humoral plasma factor, derived from donor thrombocytopenic animals. A dose-response relationship was observed between incorporation of Se^{75}-selenomethionine into newly-formed platelets and the volume of donor plasma given.)

Tavassoli, M.: Megakaryocyte-platelet axis and the process of platelet formation and release. Blood *55*:537, 1980. (A detailed discussion of platelet formation and release via the megakaryocyte demarcation membrane system into bone marrow sinusoids.)

30 Coagulation and Fibrinolysis: Mechanisms of the Fluid Interphase

PART 1 HEMOSTASIS

Oscar D. Ratnoff, M.D.

In vertebrates, blood circulates in a closed circuit of conduits, the blood vessels. Elaborate devices have evolved to minimize blood loss after disruption of these channels. Vasoconstriction, an immediate response to vascular injury, is usually transient and ineffective, but blood platelets quickly adhere to the injured endothelium and, if the vascular break is minor, these cells will stanch the flow of blood. At the same time the blood clots, making the platelet mass more compact and forming a plug that seals larger wounds. If bleeding occurs in a closed space such as the forearm, the back pressure of blood accumulating extravascularly helps to stem further loss; the rapid development of a black eye after trivial injury illustrates how bleeding progresses where back pressure is minimal. In the special case of the uterus, contraction of extravascular muscles helps to minimize hemorrhage after childbirth by narrowing vascular lumens. Failure of any of these mechanisms—and this is a relatively common occurrence—leads to undue blood loss or even death from exsanguination.

Physiology and pathology of the blood clotting mechanism

Blood clotting reflects the conversion of plasma *fibrinogen* (factor I) to an insoluble network of fibers, the *fibrin* clot. Fibrinogen is a protein (mol. wt. 340,000) synthesized in the parenchymal cells of the liver. It is probably composed of a long spindle with a nodule at each end and a third in the middle. Each molecule contains three pairs of polypeptides, the Aα, Bβ and γ chains, held together by disulfide bonds.

The conversion of fibrinogen to fibrin is brought about by *thrombin,* a protease that evolves during the clotting process and hydrolyzes arginylglycine bonds in each of the Aα and Bβ chains (Fig. 30–1). Two pairs of polypeptide fragments, fibrinopeptides A and B, are released. What remains—a monomeric unit of fibrin—polymerizes, both side-to-side and end-to-end, to form the insoluble fibrin clot. Visible clotting requires the release of fibrinopeptide A; if fibrinopeptide B is cleaved selectively, clotting does not take place. Calcium ions greatly accelerate polymerization.

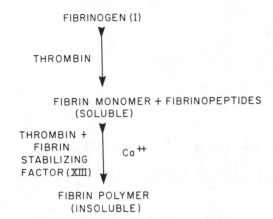

FIBRINOGEN (I)

THROMBIN

FIBRIN MONOMER + FIBRINOPEPTIDES
(SOLUBLE)

THROMBIN +
FIBRIN
STABILIZING
FACTOR (XIII)   Ca⁺⁺

FIBRIN POLYMER
(INSOLUBLE)

FIG. 30–1. The formation of fibrin in human plasma. (Beeson, P. B., and McDermott, W. (eds.): Textbook of Medicine, ed. 15. Philadelphia, W. B. Saunders, 1979.)

Fibrin derived from purified fibrinogen has poor tensile strength and dissolves in such agents as 5 M. urea or 1 per cent monochloroacetic acid, as if its molecules were loosely bound. Clots formed from normal plasma, in contrast, are stronger and do not dissolve in these agents. Such fibrin has been bonded chemically by *fibrin-stabilizing factor* (factor XIII), a plasma enzyme that forms amide links between the γ-carboxyl group of glutamine in one fibrin monomer and the ε-amino group of lysine in another. Fibrin-stabilizing factor, a protein with a molecular weight of about 300,000, functions only when activated by thrombin and in the presence of calcium ions. Teleologically, it may provide fibrin with the strength needed to hold together the edges of a wound and to resist dissolution by proteolytic enzymes.

The active site of thrombin, in common with some other proteases, contains the sequence glycyl-aspartyl-seryl-glycyl-glutamyl-alanine. Thrombin generates during clotting from its precursor, prothrombin (factor II). Two mechanisms have been distinguished through which thrombin forms, the extrinsic and intrinsic pathways of blood clotting. When shed blood comes into contact with particles of tissue, thrombin evolves rapidly. The agent or agents re-

sponsible, known generically as *tissue thromboplastin* or *tissue factor*, are lipoproteins localized to microsomes and cellular plasma membranes. Since tissue thromboplastin is not a constituent of blood, it is said to initiate clotting by way of the extrinsic pathway (Fig. 30–2). Thromboplastin contains a heat-labile protein portion and a more stable phospholipid, both needed for thrombin formation. This lipoprotein forms a complex, in the presence of calcium ions, with a plasma protein, *factor VII*, a proteolytic enzyme. The product then "activates" a second plasma protein, *Stuart factor* (factor X); the enzyme groups responsible for this activation are on the factor VII molecule. Activated Stuart factor, a proteolytic enzyme, is itself capable of converting prothrombin (mol. wt. 72,000) to thrombin (mol. wt. 32,000), but its action is greatly potentiated by another plasma component, *proaccelerin* (factor V). The effect of proaccelerin is greatly enhanced or dependent on its preliminary alteration by thrombin. Significant thrombin formation occurs only in the presence of calcium ions and phospholipids, the latter furnished by tissue thromboplastin, as if activated Stuart factor, proaccelerin, and prothrombin interacted upon the surface of phospholipid micelles.

Cell-free plasma, prepared so that it is not contaminated with tissue, nonetheless clots in glass tubes, through a process described as the intrinsic pathway of clotting (Fig. 30–2). Thrombin formation takes place as the result of a sequence of enzymatic steps, beginning with the "activation" of *Hageman factor* (factor XII), a protein with a molecular weight of about 80,000. This alteration can be brought about by contact with glass or other negatively charged surfaces, or by solutions of ellagic acid, a derivative of tannins. Glass is foreign to the body, but substances as diverse as sebum, crystalline sodium urate or calcium pyrophosphate, and possibly basement membranes or collagen may activate Hageman factor under normal or pathological condi-

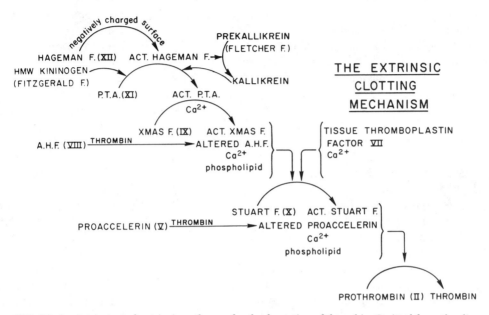

FIG. 30–2. Intrinsic and extrinsic pathways for the formation of thrombin. Omitted from the diagram are inhibitors of the various steps and the pathways of augmentation of the action of Factor VII. The phospholipid needed is furnished in the intrinsic pathway by platelets and by the plasma itself, and in the extrinsic pathways, for the most part by tissue thromboplastin. HMW = high molecular weight; PTA = plasma thromboplastin antecedent; prekallikrein = Fletcher factor; X-mas F. = Christmas factor; and AHF = antihemophilic factor. (Beeson, P. B., and McDermott, W. (eds.): Textbook of Medicine, ed. 15. Philadelphia, W. B. Saunders, 1979.)

tions. The nature of the activation process is not understood; possibly, Hageman factor changes from a water-soluble to a hydrophobic form. In the process of activation, the molecule of Hageman factor undergoes a conformational change and subsequent to this is cleaved first to a two-chain configuration and then into two fragments. The carboxy-terminal fragment possesses the enzymatic grouping needed for Hageman factor's activity.

Once activated, Hageman factor not only promotes clotting but may also initiate reactions leading to the inflammatory response, to the activation of plasminogen to form plasmin, a plasma proteolytic enzyme (see Part 3), and to a change in the first component of complement (C1) to its active form, C1 esterase (C1̄). An important property of activated Hageman factor is the acti-

vation of a plasma prekallikrein (Fletcher factor), the precursor of a plasma enzyme known as plasma kallikrein that can separate biologically active polypeptide "kinins" from their precursors in plasma, the kininogens. The kinins contract certain smooth muscles, enhance vascular permeability, dilate small blood vessels, and induce sticking of leukocytes to the walls of small blood vessels. Plasma kallikrein acts preferentially on those kininogens of higher molecular weight (HMW kininogen, Fitzgerald factor). Activation of Fletcher factor is more readily induced by fragments of Hageman factor than by the intact molecule; these fragments may be split from Hageman factor by plasma proteases, including plasma kallikrein itself or plasmin. Hageman factor can also fragment apparently "spontaneously" upon exposure of

Hageman factor to negatively charged surfaces.

The role of activated Hageman factor in clotting is to change another plasma clotting factor, *plasma thromboplastin antecedent* (PTA, factor XI), from an inert to a clot-promoting form. This transformation is *enhanced* by plasma kallikrein and in plasma *requires* the presence of HMW kininogen. PTA (mol. wt. 165, 000) is a protease; its function is the activation of a third factor, *Christmas factor* (PTC, factor IX), a step requiring calcium ions. Activated Christmas factor then interacts with *antihemophilic factor* (factor VIII), a glycoprotein with a variable molecular weight, from about 1,000,000 to 12,000,000, to form an agent that can activate Stuart factor. This step requires calcium ions and phospholipids and is accelerated by or dependent upon preliminary alteration of antihemophilic factor by thrombin. The enzyme responsible for activation of Stuart factor is activated Christmas factor.

Once Stuart factor has been activated, the subsequent steps in the intrinsic pathway are the same as those of the extrinsic pathway. In the intrinsic pathway, platelets and the plasma itself furnish the needed phospholipids.

At least in the test tube, the separation of the extrinsic and intrinsic pathways is not as sharp as has been implied. Activated Hageman factor, the first agent participating in the intrinsic pathway, enhances the clot-promoting properties of factor VII. Conversely, factor VII, complexed to tissue thromboplastin, can activate Christmas factor, circumventing the early steps of the intrinsic pathway.

The mechanisms described are relevant to clotting as studied in the test tube. In life, coagulation is probably initiated, in different situations, by exposure of blood to injured tissue or to an activator of Hageman factor such as sebum. Other stimuli may bring about clotting under special circumstances. Thus, some snake venoms are clot-promoting. The venom of the Russell's viper has components that activate Stuart factor and enhance the action of proaccelerin; the venom of the Malayan pit viper converts fibrinogen to fibrin. Certain strains of *Staphylococcus aureus* produce a "coagulase" that induces thrombin formation by acting upon prothrombin, which perhaps explains the localization of staphylococcal infections. Pancreatic trypsin also brings about clotting, through actions upon PTA, Stuart factor, and prothrombin; the physiological significance of this property is uncertain, since plasma inhibits trypsin readily.

As might be expected, plasma possesses potent inhibitors that limit the clotting process, a safeguard against thrombosis. To some extent, normal plasma itself inhibits the clot-promoting properties of such negatively charged surfaces as glass, but whether this phenomenon is of physiological importance is unknown. Plasma also contains a potent inhibitor, antithrombin III, that blocks the actions of thrombin, activated Stuart factor, and the other proteases participating in clotting, an effect greatly potentiated by heparin. $C\bar{1}$ inactivator, a plasma protein that blocks the action of $C\bar{1}$, the esterase derived from the first component of complement, also inhibits the activated forms of Hageman factor, plasma prekallikrein, and PTA. At least one additional inhibitor of PTA has been described, and α_1-antitrypsin and α_2-macroglobulin may contribute to the inhibitory properties of plasma against activated clotting factors. The coagulant properties of antihemophilic factor and proaccelerin are inhibited by a plasma agent known as protein C after this agent has been activated by thrombin. Additional, activated clotting factors may be inactivated by passage through the liver or the reticuloendothelial system.

Diagnosis of disorders of blood coagulation

The blood clotting mechanism may be inadequate because of functional deficiency of one or more clotting factors, as the result of either defective synthesis or excessive utilization. Alternatively, coagulation may be impaired by circulating inhibitors or by the proteolytic action of plasmin (Part 3). The degree of the defect in hemostasis is not related to the intensity of the abnormality measured in the test tube. Severe deficiencies of Hageman factor, HMW, kininogen or plasma prekallikrein are usually not accompanied by any discernible defect in hemostasis, whereas a deficiency of antihemophilic factor or Christmas factor, resulting in a comparable prolongation of the time elapsing until blood clots *in vitro*, is associated with a serious bleeding tendency.

In studying the patient with hemostatic disturbances, inquiry must be made as to whether bleeding has been lifelong or of recent origin, and whether it has appeared de novo or during the course of another disorder, such as hepatic or renal disease. It is important to learn whether the patient has ingested (or has been exposed to) possibly injurious agents, has recently (or concurrently) undergone infection, or has subsisted on a deficient diet. The type of bleeding sustained by the patient is diagnostically helpful. The term *purpura* encompasses both ecchymoses and petechiae. *Ecchymoses*—that is, black and blue marks—are commonplace in many forms of hemorrhagic disease, whereas *petechiae*—pinpoint dull red spots that do not blanch on pressure—are unusual in purely coagulative abnormalities but are common in disorders of platelets or blood vessels. Bleeding from the umbilicus at birth suggests a defect in the extrinsic pathway of clotting or in fibrin formation. One must determine whether there is a family history of bleeding, or whether the patient is the offspring of a consanguineous union, a hallmark of a rare autosomal recessive trait.

Differential diagnosis of disorders of blood clotting is almost entirely a laboratory exercise. A functional deficiency of any factor involved in the intrinsic pathway of clotting, or the presence of inhibitors of this pathway, may almost always be detected by measurement of the *partial thromboplastin time*. In this test, citrated plasma is mixed with calcium ions and phospholipid, and the clotting time (the interval until clotting takes place) is measured. Usually, kaolin, diatomaceous earth, or ellagic acid is added to the mixture to ensure activation of Hageman factor. The test will detect almost all cases of functional deficiency of factors participating in the intrinsic pathway, but it gives normal values in patients with deficiencies of factor VII or fibrin-stabilizing factor. Much less sensitive is measurement of the *clotting time of whole blood;* in at least one-third of patients with classic hemophilia or Christmas disease (functional deficiencies of antihemophilic factor and Christmas factor, respectively) the clotting time of whole blood is normal. Another screening test for defects in the intrinsic pathway is measurement of *prothrombin consumption* or *serum prothrombic activity.* During clotting, the conversion of prothrombin to thrombin proceeds slowly; clotting occurs when enough thrombin has formed to change fibrinogen to fibrin. Even in normal blood, appreciable prothrombin remains an hour after whole blood has been placed in glass tubes. In patients with defects of the intrinsic pathway, the formation of thrombin may be impeded, so that serum may contain excessively high amounts of prothrombin. This test is relatively insensitive, giving normal results in patients with partial deficiencies of the factors participating in the intrinsic pathway. Abnormal prothrombin consumption is also found in patients with qualitatively abnormal or decreased numbers of platelets.

Defects in the extrinsic pathway of clotting are readily detected by measuring the *prothrombin time*, that is, the clotting time of a mixture of citrated plasma, tissue thromboplastin, and calcium ions. The prothrombin time is so short that the simultaneous contributions of the intrinsic pathway to the formation of thrombin is normally negligible. The prothrombin time is abnormally long whenever a functional deficiency of any factor participating in the extrinsic pathway exists or when fibrin formation is impeded. It is also long when an inhibitor of the pathway is present, but it is normal in deficiencies of Hageman factor, plasma prekallikrein, HMW kininogen, PTA, Christmas factor, antihemophilic factor, or fibrin-stabilizing factor.

Specific localization of deficiencies of clotting factors

Specific localization of deficiencies of clotting factors other than fibrinogen or fibrin-stabilizing factor is carried out by the technique of *cross-matching*. In brief, a small amount of the plasma to be tested is mixed with a plasma known to be deficient in a single factor. The mixture is then tested by whatever method is appropriate for the factor under study—modifications of the partial thromboplastin time for assay of Hageman factor, HMW kininogen, plasma prekallikrein, PTA, Christmas factor, or antihemophilic factor, and modifications of the one-stage prothrombin time for measurement of factor VII, Stuart factor, proaccelerin, or prothrombin.

Quantitative measurement of plasma fibrinogen

Quantitative measurement of plasma fibrinogen is readily performed by converting this protein to fibrin by the addition of thrombin and assaying the fibrin chemically. Alternatively, the concentration of fibrinogen can be estimated by monitoring the evolving turbidity of a clot spectrophotometrically. Other methods, although simpler, lack accuracy. Defects in the rate of formation of fibrin are detected by measuring the *thrombin time*, the time elapsed until clotting after thrombin has been added to plasma. Hypofibrinogenemia may be suspected when the partial thromboplastin time, the prothrombin time, or the thrombin time is long; the blood is incoagulable in afibrinogenemia, the absence of detectable fibrinogen in plasma. A useful screening test for *deficiency of fibrin-stabi-*

lizing factor is determination of the solubility of a plasma clot in 5 M. urea or 1 per cent monochloroacetic acid.

This list of available procedures is, of course, incomplete. There are, among others, tests for the detection of circulating anticoagulants (agents in plasma that abnormally delay clotting) and for uncovering evidence of intravascular coagulation or fibrinolysis (Part 3).

Hereditary disorders of blood clotting

CLASSIC HEMOPHILIA. Classic hemophilia is a lifelong bleeder disorder, limited almost exclusively to males, that is characterized by a *functional* deficiency of antihemophilic factor (AHF, factor VIII) as measured in clotting assays. Antihemophilic factor is a complex of two proteins, linked by noncovalent bonds. One portion, with a molecular weight of about 200,000, possesses the coagulant properties of antihemophilic factor, and has been designated factor VIII : C. The other, whose molecular weight runs from about 800,000 to 12,000,000 or more, appears to be important for the maintenance of vascular integrity by platelets. This high-molecular-weight subcomponent of antihemophilic factor forms precipitates when it is mixed with heterologous antiserum against this factor raised in rabbits or goats, and hence is described as factor VIIIR : Ag (i.e., factor VIII-related antigen). Factor VIIIR : Ag appears to be composed of covalently linked, similar or identical polypeptide chains, each with a molecular weight of about 200,000.

Classic hemophilia is inherited as an X chromosome-linked trait. All daughters of hemophiliacs are carriers of the trait; any son of a carrier has one chance in two of being a hemophiliac, and any daughter has the same chance of being a carrier. The severity of the defect varies among different families, but the disorder is usually of about the same intensity among affected individuals within a single family. Two groups of hemophiliacs have been differentiated immunologically. As detected by heterologous antibodies against antihemophilic factor, all hemophiliacs appear to synthesize normal amounts of the higher molecular weight subcomponent (factor VIIIR:Ag). Among these patients, a small group of mildly affected hemophiliacs can be distinguished in whom the abnormal antihemophilic factor can also be detected by tests employing homologous circulating anticoagulants against the coagulant part of the molecule (factor VIII:C); this is not the case in most hemophiliacs. As would be anticipated from these considerations, the carriers of hemophilia have normal amounts of antihemophilic factor in their plasma, as determined immunologically with heterologous antiserum. Their titer of antihemophilic factor, measured in assays of clotting, varies widely, but, on the average, it is about half that of normal individuals. Measurement of the relative titers of coagulant AHF and antigenic material related to AHF thus provides a way to detect most carriers of hemophilia in the laboratory.

The deficiency of functional antihemophilic factor in the plasma of patients with classic hemophilia impairs the intrinsic pathway of clotting, so that the formation of thrombin is abnormally slow. In severe cases, the clotting time of whole blood is prolonged and prothrombin consumption is poor. In milder cases, these tests may give normal results, but the partial thromboplastin time is almost always abnormally long. Diagnosis rests upon specific assays for antihemophilic factor. Affected individuals have titers of less than 0.4 unit of activity attributable to antihemophilic factor per ml. of plasma; on the average, normal plasma contains one unit per ml. In contrast to von Willebrand's disease, the bleeding time is normal.

The symptoms of classic hemophilia are not well explained. In severe cases, the patients may have apparently spontaneous hemorrhages into the skin, muscles, soft tissues, central nervous system, and gastro-intestinal and urinary tracts and may bleed profusely after injuries, surgery, or dental extractions. Bleeding into joint spaces, particularly of the knee, hip, and elbow, are especially common, and permanent, crippling damage may ensue. In milder cases, bleeding may occur only after injury or surgery, making diagnosis perplexing. Ordinarily, trivial scrapes do not bleed profusely, perhaps because the injured tissues initiate clotting by the extrinsic mechanism. Why hemophiliacs bleed into joints is unclear, but possibly tissue thromboplastin is not readily available.

Therapy, besides local measures to improve hemostasis and minimize tissue injury, consists of temporary correction of the patient's coagulative defect by the transfusion of concentrates of normal plasma rich in antihemophilic factor. Since half of the coagulant properties of transfused antihemophilic factor disappears from the circulation within 12 hours, repeated infusions are needed. Special attention is needed for the psychological problems that beset individuals who live in daily dread of bleeding.

VON WILLEBRAND'S DISEASE. A second inherited deficiency of antihemophilic factor occurs in von Willebrand's disease, usually inherited as an autosomal dominant trait. The basic nature of the abnormality in von Willebrand's disease is only partially understood. The deficiency in antihemophilic factor in clotting assays is usually not as great as in the more severe forms of classic hemophilia. At the same time antigenic material related to AHF, detected by heterologous antiserum (factor VIIIR:Ag) is decreased in concentration, either in proportion to functional AHF or to a greater degree. Transfusion of *hemophilic* plasma into patients with von Willebrand's disease is followed by a significant increase in the recipient's coagulant titer of antihemophilic factor, maximal after 4 to 8 hours, as if hemophilic plasma contained either a precursor of antihemophilic factor, an agent stimulating its synthesis, or something inhibiting its destruction.

Besides the deficiency in antihemophilic factor, patients with von Willebrand's disease have an unexplained defect in hemostasis manifested in the laboratory by a long bleeding time and often by impairment of the normal phenomenon of retention of platelets when blood is filtered through a column of glass beads. Further, agglutination of platelets by the antibiotic ristocetin is deficient in patients with von Willebrand's disease; the property of antihemophilic factor that supports ristocetin-induced platelet agglutination has been called factor VIIIR : RC or VIIIR : RCo. These three abnormalities have been ascribed to an abnormality or deficiency of the high-molecular-weight subcomponent of antihemophilic factor. The bleeding tendency and the defect measured in the laboratory vary from patient to patient, even within a family, and not every patient has all of the abnormalities noted. In most cases, bleeding is mild, occurring only after injury or surgical procedures, or from the gastrointestinal tract. Menorrhagia is common, but childbirth is usually unaccompanied by hemorrhage, since the titer of antihemophilic factor rises during pregnancy, often to normal levels. This observation has led to the prophylactic use of oral contraceptive agents in female patients. The coagulant titers of antihemophilic factor and the concentration of antigens related to this agent (factor VIIIR :Ag) also rise to normal during certain periods of stress, but paradoxically these changes need not be accompanied by correction of the prolonged bleeding time.

CHRISTMAS DISEASE. Christmas disease, which is clinically similar to classic hemophilia and is inherited in the same way, results from a functional deficiency of Christmas factor (factor XI). A few patients with Christmas disease synthesize a nonfunctional variant of Christmas factor; most patients do not seem to synthesize this protein. Although the titer of Christmas factor in the plasma of carriers varies greatly, on the average it is about one-half that of normal individuals. Therapy for bleeding episodes is optimally performed by transfusion of whole plasma containing Christmas factor; under special conditions, fractions of plasma containing this clotting factor may be used, but paradoxically, localized or disseminated intravascular coagulation may ensue.

HEREDITARY DISORDERS OF FIBRINOGEN. Two distinct forms of hereditary disorders of fibrinogen have been described. In *congenital afibrinogenemia*, the plasma contains no detectable fibrinogen, so that the blood is incoagulable. Inherited as an autosomal recessive trait, this disorder is remarkable in that affected individuals bleed only when subjected to injury, beginning with bleeding from the umbilicus at birth. In these patients, thrombin formation is unimpaired, and platelet aggregation and adhesion to wounds are usually normal. The freedom from spontaneous bleeding is in sharp contrast to severe hemophilia or Christmas disease, in which the formation of thrombin is excessively slow.

A rare but instructive abnormality, transmitted in an autosomal dominant way, is *congenital dysfibrinogenemia*, a state in which the patient synthesizes a functionally defective fibrinogen, usually in normal amounts. The disorder ordinarily comes to notice because the prothrombin time is prolonged. Analysis of the cause for this abnormality demonstrates that the patient's fibrinogen clots excessively slowly upon the addition of thrombin either because release of fibrinopeptides by thrombin is impaired or because the polymerization of fibrin monomers is defective, or both. A growing number of abnormal fibrinogens have been described, distinguished by immunological and functional tests; each is named after the city where it was first identified. In one case, fibrinogen Detroit, an arginine residue in the N-terminal part of the Aα chain of fibrinogen was replaced by a neutral amino acid, probably serine, and the protein contained less carbohydrate than normally. Unexpectedly, patients with dysfibrinogenemia do not ordinarily have a bleeding tendency, although exceptions have been reported. In a few patients, operative wounds have dehisced, and in others, a thrombotic tendency is present. In one family, affected individuals had a fibrinogen that clotted more rapidly than normally and exhibited a high incidence of thromboembolic phenomena.

FUNCTIONAL DEFICIENCY OF FIBRIN-STABI-LIZING FACTOR (FACTOR XIII). Fibrin formation may also be impeded by this hereditary anomaly. Individuals lacking functional fibrin-stabilizing factor may have a severe bleeding tendency, beginning with umbilical bleeding after birth and often ending with central nervous system hemorrhage. Why the patients bleed is not clear, since the rate of fibrin formation, tested *in vitro*, is normal. In some instances, the patients' wounds break down repeatedly and bleed afresh, suggesting that the formed fibrin easily disrupts, since it lacks its normal chemical bonding. The plasma of patients with fibrin-stabilizing factor deficiency usually contains material antigenically related to this substance, as if a nonfunctional form were synthesized. Two groups of patients have been discerned. In some families, the defect is inherited as an autosomal recessive trait, and consanguinity is common. In others, the disorder is limited to men and consanguinity is rare, but proof of X-chromosome inheritance is lacking.

OTHER FAMILIAL DISORDERS OF BLOOD CLOTTING. A number of hereditary coagulation disorders have been described, all usually autosomal recessive. *Hageman trait*, the hereditary absence of Hageman factor (factor XII) is usually asymptomatic, the defect being detected by chance obser-

vation of a greatly prolonged clotting time or partial thromboplastin time. Suitable tests demonstrate a defect in the intrinsic pathway of coagulation. *Fitzgerald trait* (Williams trait, Flaujeac trait) is an asymptomatic disorder of the intrinsic pathway, distinguished from Hageman trait by specific tests. Two forms have been described. In both, HMW kininogen (Fitzgerald factor) is deficient in plasma, while in one form, low molecular weight (LMW) kininogen is also deficient. *Fletcher trait*, the hereditary deficiency of plasma prekallikrein (Fletcher factor), is also asymptomatic, patients being detected by the presence of a prolonged clotting time or partial thromboplastin time. The functional defect, localized to the intrinsic pathway of clotting, is corrected by prolonged incubation of plasma with kaolin (which activates Hageman factor) before measurement of the partial thromboplastin time. The defect in homozygous *PTA* (factor XI) *deficiency* is usually incomplete, so that the clotting time and partial thromboplastin time may be only marginally prolonged. Although some patients with PTA deficiency are asymptomatic, others have a mild bleeding tendency, and a few deaths have been reported. In most cases of Hageman trait, Fletcher trait, Fitzgerald trait, and PTA deficiency, the defect is apparently a failure to synthesize the missing protein, since no material antigenically related to Hageman factor, plasma prekallikrein, HMW kininogen, or PTA can be detected. In a few instances of Hageman trait and Fletcher trait, however, plasma contains immunologically recognizable but functionally incompetent forms of the factors deficient in clotting assays.

In four autosomal recessive traits, functional deficiencies of factor VII, Stuart factor (factor X), proaccelerin (factor V) or prothrombin, the prothrombin time is prolonged, and the disorders must then be differentiated by specific testing. In all, the degree of the bleeding tendency varies widely from family to family. Since autosomal recessive traits occur in both sexes, it is not surprising that menorrhagia and bleeding at childbirth are common problems. In some cases in which there are functional deficiencies of one or another of the four factors, the affected individuals have little or no detectable material related to these agents in their blood. Other patients appear to synthesize nonfunctional variants of the supposedly missing proteins.

Acquired disorders of blood clotting

The number of clinical situations in which a clotting abnormality is present is legion, but certain syndromes are of peculiar interest because they illustrate pathophysiological mechanisms.

Completion of the synthesis of four plasma clotting factors—prothrombin, Stuart factor (factor X), factor VII and Christmas factor (factor IX)—occurs only if vitamin K is available. Additionally, proteins other than these four factors are now known to require vitamin K to complete their synthesis, including protein C, a plasma protein whose activated form inhibits the coagulant properties of proaccelerin (factor V) and antihemophilic factor (factor VIII), and proteins in bone, kidney, and other organs. Although small amounts of vitamin K are furnished by the diet, the bulk of this material is synthesized by intestinal bacteria. Natural vitamin K is lipid-soluble and is absorbed only if bile salts are present in the gut. Once absorbed, it is utilized in the parenchymal cells of the liver for production of the four factors, but it does not become an integral part of their structure. Synthesis of the four factors appears to involve two stages. First, a functionally inert polypeptide is produced, after which vitamin K converts the precursor protein to a functional form before its release from the ribosomes. Vitamin K is a cofactor for an enzyme that directs the insertion of carbon dioxide into the terminal carbon of certain glutamic acid residues in the precursor proteins. In the process, vitamin K is oxidized and then conserved by enzymatic reduction. The unique tricarboxylic glutamic acid residues appear to be the sites of attachment for the calcium ions needed for activation of the four clotting factors. In the absence of vitamin K, the incomplete precursor proteins may enter the bloodstream, where they may act as competitive inhibitors of the vitamin K–dependent factors. Any interruption of this chain of events results in a deficiency of the vitamin K–dependent clotting factors.

DEFICIENCIES OF THE VITAMIN K–DEPENDENT CLOTTING FACTORS. The newborn infant's only source of vitamin K is milk; if this supply is inadequate, a severe depletion of the vitamin K–dependent factors may ensue, resulting in hemorrhagic disease of the newborn. Administration of small amounts of vitamin K to the mother just before childbirth or to the newborn infant will prevent this disorder. If the infant survives, the disorder is self-limited because the intestinal flora soon supplies adequate amounts of vitamin K. Sterilization of the gut with antibiotics, as may take place in preparation for surgery, will cut off this source. In this situation deficiency of the vitamin K–dependent factors is fostered by restriction of dietary intake of the vitamin. Impaired absorption of vitamin K because of *obstructive jaundice*, in which bile salts do not reach the duodenum, is responsible for the bleeding tendency that accompanies this disorder. The deficiency of vitamin K–dependent clotting factors is readily corrected by parenteral administration of the vitamin, greatly reducing the risk of surgery.

Impaired absorption may also reflect intrinsic disease of the bowel, such as sprue, in which lipid-soluble agents are poorly absorbed. Parenchymal hepatic disease is commonly associated with deficient synthesis of the protein precursors of the vitamin K–dependent factors; in this case, the parenteral administration of the vitamin is usually without benefit. Finally, the administration of certain anticoagulant drugs, notably those related to coumarin or phenindione, interferes with synthesis of the vitamin K–dependent factors, presumably by competitive inhibition of the vitamin. The clinical picture in all of these states resembles that of classic hemophilia, except that the onset is usually sudden. The diagnosis should be suspected whenever the prothrombin time is prolonged, since this test measures, in part, the concentrations of factor VII, Stuart factor, and prothrombin. Suitable analysis will also reveal a depression in the concentration of Christmas factor. In otherwise normal individuals in whom the prothrombin time is prolonged and multiple deficiencies of the vitamin K–dependent factors are detected, the question of surreptitious ingestion of coumarin-like compounds must come to mind; here the mechanism of disease goes beyond the alterations in hemostasis, for these patients exhibit evidence of severe emotional disturbances.

INTRAVASCULAR CLOTTING. When a massive amount of tissue thromboplastin is injected intravenously into an animal, the circulating blood clots and the animal dies because the flow of blood is obstructed. When thromboplastin is injected slowly, the animal survives and its blood is incoagulable. The mechanisms underlying these events are easily understood. Injected thromboplastin generates thrombin by way of the extrinsic pathway. A large bolus of thromboplastin induces the explosive formation of thrombin and intravascular clotting. Lesser amounts of thromboplastin bring about much slower and, perhaps, incomplete fibrin formation, so that the blood may contain soluble intermediates of fibrin formation—either monomers, soluble polymers of fibrin, or complexes of fibrin monomer and fibrinogen. These products are rapidly removed from the circulation by the reticuloendothelial system. As a result, the animal's circulating fibrinogen is depleted and its blood will not clot. At the same time, other clotting factors may be decreased in titer (as measured in tests of coagulation), particularly antihemophilic factor (factor VIII) and proaccelerin (factor V), which are inactivated by thrombin and activated protein C. The platelet count falls, perhaps because the platelets are clumped by thrombin. Sporadically, evidence of intravascular activation of plasminogen, the precursor of the plasma proteolytic enzyme plasmin, may be detected. At the same time, the serum (that is, the liquid phase of blood after defibrination) contains material antigenically related to fibrinogen and fibrin, either soluble fibrin or fibrinogen-fibrin complexes, or the products of the digestion of fibrin by plasmin (see Part 3). These fibrinogen-fibrin-related antigens may be responsible for the anticoagulant properties of blood that appear after the infusion of thromboplastin.

These experimental observations find their counterpart in many clinical situations.

In *amniotic fluid embolism*, amniotic fluid and its contaminants enter the maternal bloodstream during parturition; lanugo hairs and fetal epithelial cells are demonstrable in maternal blood. The patient may die suddenly of respiratory failure. More usually, an insidious hemorrhagic tendency appears, with bleeding from the placental site, venipuncture wounds, the gastrointestinal tract, and elsewhere. Unless vigorous treatment is instituted, death from exsanguination may result. Studies of maternal blood demonstrate depletion of fibrinogen and other clotting factors, thrombocytopenia, the presence of an anticoagulant interfering with fibrin formation, and, inconstantly, excessive fibrinolysis. These changes, similar to those seen in animals injected with thromboplastin, presumably result from initiation of the extrinsic clotting pathway by procoagulant agents in amniotic fluid.

In the same way, intravascular clotting may follow *envenoming* by poisonous snakes, the mechanism varying from species to species. For example, Russell's viper venom activates Stuart factor and alters proaccelerin so as to make it more effective; the venom of the tiger snake behaves like activated Stuart factor; and that of the Malayan pit viper clots fibrinogen directly.

Many other hypofibrinogenemic syndromes have been attributed to intravascular coagulation. In *massive transfusion of incompatible blood*, the hemolyzed erythrocytes presumably initiate the clotting process. Hypofibrinogenemia associated with *brain injury* may be related to the entrance of highly thromboplastic brain tissue into the bloodstream. The hypofibrinogenemia complicating *premature separation of the placenta* has been ascribed to the thromboplastic effect of placental tissue, which may gain entrance to the maternal circulation; in another view, hypofibrinogenemia results from depletion of plasma fibrinogen in the formation of the retroplacental clot.

In other cases, damaged vascular endothelium is thought to be the source of procoagulant material. This is most clearly seen in *purpura fulminans*, in which widespread vasculitis and thrombosis lead to patchy, usually superficial gangrene. The hypofibrinogenemia and other evidences of intravascular clotting observed in *heat stroke*, in severe *sepsis* (such as occurs after self-induced abortion), or in the *Waterhouse-Friderichsen syndrome* have similarly been attributed to vascular injury. Tumor cells or secretions may be the source of the procoagulant that induces hypofibrinogenemia in *leukemia* or *carcinoma*, particularly of the prostate. Perhaps similar mechanisms are responsible for the hypofibrinogenemia observed in association with *giant cavernomatous hemangioma*.

The view that hypofibrinogenemia is secondary to intravascular clotting may be fortified by inducing remission by the administration of heparin, a powerful anticoagulant. For example, the hypofibrinogenemia found in some patients with *intrauterine retention of a dead fetus* is temporarily corrected by administration of heparin. The detection of fibrinogen-fibrin related antigens in serum also supports the diagnosis of intravascular clotting. In experimental animals, the induction of intravascular clotting may lead to fragmentation of erythrocytes, which take on a bizarre appearance as if sheared by strands of fibrin. Demonstration of such erythrocytes in the blood of patients with hypofibrinogenemia has been taken as evidence of intravascular coagulation, but clinical evidence supporting this view is insecure.

MICROANGIOPATHIC HEMOLYTIC ANEMIA. The observations discussed above have suggested to some that other syndromes, not necessarily associated with hypofibrinogenemia, may be associated with intravascular clotting. The abnormal erythrocytes characteristic of the syndromes known collectively as *microangiopathic hemolytic anemia*, including *thrombotic thrombocytopenic purpura*, the *hemolytic-uremic syndrome* and *eclampsia*, have been ascribed to intravascular clotting secondary to vascular damage, but this is not a universal view. Similar erythrocyte changes may be seen in *acute glomerulonephritis*, *malignant hypertension*, and the hemolytic anemia complicating *cardiovascular prosthesis*. Needless to say, caution is necessary before accepting the view that these syndromes are, in fact, complicated by intravascular clotting.

Hepatic disease

Patients with chronic hepatic disease, such as cirrhosis of the liver or carcinoma, commonly have a bleeding tendency, a complication which may also accompany the more severe forms of acute hepatitis. The pathogenesis of bleeding may be complex. Commonly, a deficiency of the vitamin K–

dependent clotting factors is present. In most cases this is the result of impaired synthesis by hepatic parenchymal cells and does not respond to parenterally administered vitamin K. Less often, intrahepatic obstruction to the flow of bile may impair absorption of the vitamin, and in such cases parenterally administered vitamin K may correct the defect. Deficiencies of Stuart factor, factor VII and prothrombin can be detected by an abnormally prolonged prothrombin time. A deficiency of proaccelerin, also synthesized in the liver, is common; this abnormality also lengthens the prothrombin time. Unusually, patients with hepatic disease have hypofibrinogenemia. In some cases this has been attributed to deficient synthesis of fibrinogen by the liver; in others, it may result from disseminated intravascular coagulation. In some patients with hepatic disease, too, fibrinogen may be qualitatively abnormal. In patients with portal hypertension, the platelet count is often reduced, probably because these cells are sequestered in the enlarged spleen. Thrombocytopenia may also result from alcoholism, either as a direct toxic effect of alcohol or because of concomitant folic acid deficiency. Patients with chronic hepatic disease also often have evidence of increased plasma fibrinolytic activity, further complicating matters. This may be one of several reasons why clotting of plasma upon the addition of thrombin is delayed in patients with hepatic disease. Thus, the bleeding syndrome of patients with disorders of the liver has no simple explanation, and each case must be studied if a knowledge of the mechanisms involved is important for the patient's care.

Circulating anticoagulants

An occasional but important complication of hemophilia is the development of circulating anticoagulants directed against the coagulant properties of antihemophilic factor (factor VIII:C). These agents, detected in as many as 10 per cent or more of severely affected patients, are IgG antibodies that inactivate antihemophilic factor. In such patients, transfusions of antihemophilic factor are usually without benefit, and bleeding is therefore more difficult to control.

Rarely, similar anticoagulants may appear de novo in apparently normal adults; these have also been observed in some patients with systemic lupus erythematosus, after penicillin reactions, or after childbirth. Since the patients are deficient in coagulant antihemophilic factor, they have symptoms similar to those of severe hemophilia, and, unless the anticoagulant disappears, they may ultimately succumb to hemorrhage.

Circulating anticoagulants have also been detected in some patients with Christmas disease or hereditary deficiencies of PTA, fibrinogen, factor VII, proaccelerin or fibrin-stabilizing factor. The nature of these anticoagulants has not been well studied, but presumably they are antibodies to the specific factors the patients lack.

Another type of anticoagulant, seen most frequently in patients with systemic lupus erythematosus, interferes with the transformation of prothrombin to thrombin by activated Stuart factor (factor X). In such cases, the prothrombin time is prolonged, particularly if measured with diluted tissue thromboplastin. Hemorrhagic symptoms are minimal in this syndrome. Some evidence suggests that an antibody interferes with the action of the phospholipid required for thrombin formation.

Disorders of blood platelets

The role of the platelets in hemostasis

Platelets are small cytoplasmic fragments of megakaryocytes, large multinucleated cells found, in the adult, primarily in the red marrow. Maturation of megakaryocytes to the point at which they shed platelets into the circulation takes as long as 5 to 7 days. The maturation process is stimulated by one or more agents in plasma.

Normal peripheral blood contains 150,000 to 350,000 platelets per μl. About one-third of the total platelet mass is not in the peripheral circulation but is sequestered in the spleen. The average platelet survives about 10 days; the bulk of these is lost through senescence, although normally a small number may be destroyed at random, presumably because they are utilized in hemostasis. The aged or injured platelets are apparently removed from the circulation by the reticuloendothelial system, primarily in liver and spleen.

When vascular endothelium is damaged, platelets rapidly accumulate at the point of injury, adhering to exposed subendothelial collagen. Once this takes place, platelets discharge their cytoplasmic granules, releasing adenosine diphosphate (ADP) and other intracellular constituents. Collagen and ADP also activate one or more platelet membrane phospholipases that release arachidonic acid from membrane phospholipids. Arachidonic acid is then converted by a cyclooxygenase to cyclic derivatives, the prostaglandins, and then, by the enzymatic action of thromboxane synthetase, to a prostaglandin derivative, thromboxane A_2, that is released into the surrounding milieu. ADP, derived from platelets and probably also from injured endothelium, and thromboxane A_2, induce loose aggregation of fresh platelets to those adherent to the wound. Platelet phospholipoprotein (principally that in the platelet membrane) becomes available for local thrombin formation, which is brought about by contact of blood with thromboplastin (furnished by injured tissue) and by activation of Hageman factor, perhaps by exposed collagen. Platelets also contain proaccelerin (factor V) and antihemophilic factor (factor VIII), and these may intensify the evolution of thrombin. Proaccelerin on the platelet membrane, altered by thrombin, serves as a receptor for Stuart factor (factor X) that has been activated during the clotting process. In this situation, the thrombin-forming action of activated Stuart factor is greatly enhanced. Thrombin, formed at the site of vascular injury, stimulates local deposition of fibrin, binding the platelets into a hemostatic plug sufficient to stop bleeding from small wounds. Thrombin also induces further platelet aggregation and degranulation, with the release of intracellular components, including ADP. The action of thrombin on platelets is complex, involving participation of intracellular proteins functionally similar to plasma fibrinogen and fibrin-stabilizing factor. Further, it brings about contraction of an intracellular protein, platelet actomyosin, a substance with adenosine triphosphatase activity similar to that of muscle actomyosin. This contraction of platelets makes the platelet plug more compact and, in the test tube, is responsible for clot retraction, in which the clot shrinks, expressing the serum within its meshes. Platelet fibrinogen and calcium ions are needed both for platelet aggregation and for clot retraction.

When blood or platelet-rich plasma comes into contact with glass, some of the platelets adhere to the glass surface. This effect, perhaps related to that induced by collagen, depends upon the presence in plasma of fibrinogen, the high-molecular-weight subcomponent of AHF (factor VIIIR:Ag), and perhaps other substances. The significance of platelet retention when blood is filtered through columns of glass beads, a substance foreign to the body, is not apparent, but decreased platelet retention is a useful test for the presence of abnormal platelets or of von Willebrand's disease.

Tests of platelet function

Platelets should be counted in the blood of every patient suspected of a hemostatic defect, and a smear should be examined to see if they are morphologically abnormal. Since platelets are essential for the phenomenon of clot retraction, qualitative or semiquantitative tests of this process must be performed. Qualitative abnormalities in the clot-promoting properties of platelets may be suspected because the clotting time of whole blood, measured in plastic tubes, is prolonged, or because platelets behave defectively in the prothrombin consumption test. In these procedures, the normal conversion of prothrombin to thrombin is impeded if platelets fail to release adequate amounts of clot-promoting phospholipids. Simple tests have been devised to measure the retention of platelets by columns of glass beads through which blood is filtered (a property defective in von Willebrand's disease) and to aggregate upon the addition of collagen, ADP, epinephrine, ristocetin, or thrombin.

The bleeding time—that is, the time elapsing until bleeding stops from an incised wound—is prolonged when the platelet count is depressed below about 80,000/μl. A long bleeding time is also found in patients with thrombocythemia (with platelet counts above 800,000/μl.) or with qualitative abnormalities of platelets. Other causes of a long bleeding time are von Willebrand's disease and the dysproteinemias.

Thrombocytopenia

Individuals in whom the platelets count is less than 100,000/μl. may exhibit a bleeding tendency in rough proportion to the degree of thrombocytopenia. They bruise readily and may bleed from the nose, the gingivae or the gastrointestinal or genitourinary tracts. Petechiae are commonplace, both on the skin and on mucous membranes. Bleeding into the central nervous system, often with lethal outcome, is particularly common in patients with acute leukemia or aplastic anemia but is unusual in patients with idiopathic thrombocytopenic purpura.

Examination of aspirates of bone marrow is essential in determining the cause of thrombocytopenia in almost every case. The aspirate may demonstrate the presence of abnormal cells, such as those found in leukemia, metastatic carcinoma, or Gaucher's disease. In other cases, no abnormal cells may be seen but megakaryocytes may be absent, as in aplastic anemia. In contrast, in idiopathic thrombocytopenic purpura or the thrombocytopenia accompanying systemic lupus erythematosus, hypersensitivity to drugs, portal hypertension, lymphomas, or miliary tuberculosis, normal numbers of megakaryocytes may be present, but often the cytoplasm of these cells is scanty. In still other cases, such as agnogenic myeloid metaplasia, myelofibrosis, aplastic anemia, or the early stages of leukemia, no marrow may be obtainable by aspiration, and in these instances surgical biopsy of the marrow should be performed in an effort to reach a diagnosis.

Thrombocytopenia may result from one or another or a combination of different processes (see list on p 699). In some cases, the rate of formation of platelets is abnormally low. Sometimes, as in hypoplastic or aplastic anemia, which may occur idiopathically or after exposure to ionizing radiation, drugs, or toxins, megakaryocytes are few in number or absent, often in association with depression of other marrow elements. In other cases, the marrow may be infiltrated with abnormal cells which may crowd out the normal elements and, in this way, lead to anemia and thrombocytopenia. In still others, the normal stimuli to platelet maturation may be lacking or suppressed, as in the megaloblastic anemias of vitamin B_{12} or folate deficiencies. Suppression of normal marrow activity may account for some instances of thrombocytopenia accompanying infection.

Thrombocytopenia may also be due to excessively rapid destruction of platelets, most clearly seen in individuals sensitive to certain drugs. The list of offending agents is long. Of these, Sedormid, quinine, quinidine, and digitoxin have been studied most intensively. The most likely explanation for their effects is that the patient acquires circulating antibodies against the drugs. Upon subsequent administration of the medication, antigen-antibody complexes form within the circulation and injure the platelets through the process of immune adherence.

A somewhat similar explanation has been offered for the pathogenesis of idiopathic thrombocytopenia purpura. Two types of this disorder can be delineated: an acute, self-limited purpura, usually lasting no more than several months, and a more chronic form, which may persist for many years. In either case, the marrow contains normal numbers of megakaryocytes that nonetheless lack the budding platelets normally found at the periphery of their cytoplasm. The plasma of patients with idiopathic thrombocytopenia purpura contains an agent that will induce thrombocytopenia when introduced into normal individuals either by transfusion or across the placenta. The agent has characteristics of an antibody of the IgG type that in some cases can be detected in preparations of the patient's platelet membranes. A similar explanation has been offered for the thrombocytopenia accompanying systemic lupus erythematosus, miliary tuberculosis, lymphosarcoma, chronic lymphatic leukemia, varicella, and other disorders.

The life span of the platelets may also be decreased because of excessive peripheral utilization of platelets, the most dramatic examples being the syndromes of intravascular clotting and the thrombocytopenia that followed the use of early models of extracorporeal circulatory apparatus. A similar pathogenesis may explain the thrombocytopenia that is a feature of widespread vascular damage such as occurs in thrombotic thrombocytopenic purpura, the hemolytic-uremic syndrome, or eclampsia.

In some cases, the number of circulating platelets is reduced because these cells are sequestered within an enlarged spleen. Under these conditions, platelet production and life span are usually normal, and splenectomy often brings about remission of the thrombocytopenia. Other lesions in which a wide meshwork of small vessels is present, such as giant cavernomatous hemangioma or Kaposi's sarcoma, may be complicated by thrombocytopenia.

In still other cases, the platelet count is lowered by dilution. When patients with severe hemorrhage are transfused with many units of stored blood or packed red blood cells, the platelet count falls because those preparations are deficient in viable platelets. Other clotting factors deficient in stored blood may also decrease in concentration.

Thrombocytopenia during the course of infection may come about in different ways. For example, in rubella the platelets are destroyed prematurely, perhaps by circulating autoantibodies. In infections in which vascular damage is prominent, such as bacterial sepsis, rickettsial disease or certain other hemorrhagic fevers, thrombocytopenia may result from consumption of platelets in intravascular clotting. In still other infections—for example, rare cases of acute hepatitis—thrombocytopenia may be due to marrow aplasia. Almost certainly, in many infectious processes several different mechanisms may be operative.

This lengthy classification does not encompass all the disorders in which thrombocytopenia has been detected. Prominent among those in which the pathogenesis is unclear are various forms of hereditary thrombocytopenia. Although some of these have been ascribed to deficient formation from megakaryocytes, in others, no such explanation is forthcoming. Most interesting is the rare congenital thrombocytopenia that responds to the transfusion of normal plasma, as if patients with this disorder lacked a stimulus needed for platelet formation.

Thrombocytosis and thrombocythemia

A transient rise in the platelet count above normal is common during the period after surgical procedure, severe hemorrhage, or childbirth, or as an accompaniment to inflammatory states. A more sustained rise is more likely to be associated with iron deficiency anemia, malignancy, or one of the myeloproliferative syndromes—polycythemia vera, chronic myeloid leukemia, agnogenic myeloid metaplasia or "primary" thrombocythemia. The thrombocytosis is apparently a response to some unknown stimulus to platelet production; the platelet life span is usually normal. The administration of the oncolytic agent vincristine induces thrombocytosis under some conditions. Thrombocytosis may also follow splenectomy or atrophy of the spleen, as if platelet survival were enhanced by the absence of an organ important for the removal of these cells. Usually, the rise in the platelet count after splenectomy is transient, but it may persist for many years. When the platelet count is elevated to 800,000/μl. or more—a condition called thrombocythemia—the patient may exhibit a bleeding tendency. Although many explanations for this paradox have been offered, none is satisfactory, but many patients with thrombocythemia may have qualitatively abnormal platelets.

Qualitative abnormalities of platelets

Inherited qualitative defects in platelet function are uncommon; they serve, however, to fortify our views concerning the role of these cells in hemostasis. A bewildering array of apparently distinct abnormalities have been described. In *thrombasthenia,* or Glanzmann's disease, a bleeding tendency is correlated with impaired clot retraction, a long bleeding time, and impaired aggregation of platelets by ADP and thrombin, but the platelet count is usually normal. In *thrombopathic purpura,* the platelets lack clot-promoting activity, behaving as if they cannot release their phospholipid. The bleeding time is prolonged and aggregation of platelets by ADP and thrombin may be impaired, but clot retraction is normal. A third group of patients, with a condition called *thrombopathia,* usually has only a mild bleeding tendency. Again, the bleeding time is long and aggregation of platelets by collagen or by small amounts of ADP may be impaired, but clot retraction and clot-promoting activity are normal. In some of these patients, the storage pool of adenine nucleotides is deficient. Still other patients have been described who display a variety of functional and morphological abnormalities.

An acquired qualitative platelet abnormality may contribute to the bleeding tendency of patients with *uremia;* the defect is apparently not intrinsic to the platelet but is the result of an abnormality in the surrounding plasma. Platelet aggregation by collagen is also depressed by the ingestion of *aspirin* or *dipyridamole,* which may aggravate the bleeding tendency in patients with other forms of hemostatic abnormality. Aspirin inhibits platelet cyclooxygenase, and in this way blocks the formation of the platelet aggregating agent thromboxane A_2. Dipyridamole inhibits a platelet phosphodiesterase. The high levels of cAMP that ensure inhibit platelet aggregation. These agents have been suggested for therapy of certain forms of thrombosis (as in thrombotic thrombocytopenic purpura) to inhibit the adhesion of platelets to blood vessel walls or the local formation of platelet aggregates.

Disordered hemostasis due to vascular pathology

When the vascular wall is damaged, bleeding may occur after minor trauma or apparently spontaneously. Hemorrhage may follow damage to the endothelial surface or to the supporting structures within and around the blood vessels, and it is especially likely to occur where intravascular pressure is relatively high, as in the small blood vessels of the feet and ankles.

HEREDITARY DISORDERS OF CONNECTIVE TISSUE. Inadequate vascular support is a prominent feature of several of the hereditary disorders of connective tissue. The *Meekrin-Ehlers-Danlos syndrome* (a term probably encompassing a variety of disorders) is associated with qualitative defects of collagen, usually inherited in an autosomal dominant manner. Affected individuals may have hypermobility of joints, hyperelasticity of skin, a blue cast to the sclerae, dislocation of the lens of the eye, and diaphragmatic hernia. The patients bruise readily and may have subcutaneous hematomas, presumably because their blood vessels lack the support of normal collagen. In some families, spontaneous rupture of large arteries is a common and often lethal event. In *osteogenesis imperfecta*, usually inherited as an autosomal dominant trait, the synthesis of normal collagen is impeded. A patient with this disorder may bruise easily and have other hemorrhagic manifestations; the platelets are said to function abnormally, contributing to the bleeding diathesis. In *pseudoxanthoma elasticum*, an autosomal recessive trait, visceral, retinal, joint, and cutaneous bleeding is commonplace. In this disease, which is characterized by the presence of waxy papules in the folds of the skin, one basic defect appears to be a degeneration of the elastic fibers of connective tissue surrounding the blood vessels.

SCURVY. A diffuse lack of vascular support is also a prime feature of scurvy—an extremely common disease caused by protracted dietary deficiency of vitamin C. Hemorrhages into the skin, the gingivae and the joints are important signs. These phenomena have been attributed to a loss of vascular support as the result of defective synthesis of collagen and basement membranes.

SENILE PURPURA. A loss of connective tissue support of blood vessels, due to atrophy of collagen associated with aging, is thought to be the basis for senile purpura, the dark purple spots seen on the skin of elderly people. In this harmless but sometimes frightening problem, cutaneous hemorrhages are most likely to be present on the extensor surface and radial border of the forearms, on the backs of the hands and in the region where eyeglasses press upon the face. Characteristically, the lesions are sharply demarcated, unlike ordinary bruises, and last for days or weeks, leaving a residue of brownish pigmentation. Bleeding is thought to occur because the vessels are readily torn by shearing strains upon the skin. Similar lesions are seen in some individuals with Cushing's syndrome or who have been treated with corticosteroids, perhaps because the structure of collagen is altered. In other patients receiving steroids, typical ecchymoses may appear spontaneously or after minimal trauma. These lesions are indistinguishable from those of *simple purpura*, an unexplained tendency to bruise readily which is almost exclusively found in women. Simple purpura, or "the devil's pinches," is a cosmetic problem, but has no other significance.

ANAPHYLACTOID PURPURA. Anaphylactoid purpura (allergic or Henoch-Schönlein's purpura), a relatively common disorder in which bleeding may be an important element, is a systemic disease involving the skin, mucous membranes, joints, gastrointestinal tract, kidneys, central nervous system, and heart. Characteristically, the rash is composed of petechiae or small ecchymoses, often superimposed upon an urticarial base. The basic lesion is a diffuse angiitis, in which the arterioles, capillaries, and vessels are surrounded by neutrophils, mononuclear cells, and eosinophils; the

Mechanisms Responsible for Thrombocytopenia: Representative Disorders*

Disorders in Which Production of Platelets Is Probably Reduced

Hypoplasia or aplasia of megakaryocytes
  Depression of marrow by ionizing radiation
  Depression of marrow by drugs or toxins
    (e.g., chloramphenicol, benzene, gold salts, thiazides, cancer chemotherapeutic agents)
  Idiosyncrasy to drugs
  Congenital hypoplastic anemia
  Fanconi's familial anemia
  Congenital thrombocytopenia with absent radii
  Aplastic anemia with thymoma
  Agnogenic myeloid metaplasia or myeloid fibrosis
  Idiopathic aplastic anemia
  Congenital thrombocytopenia
  Anorexia nervosa
Infiltration of marrow by abnormal cells
  Leukemia and other lymphomas
  Metastatic carcinoma
  Multiple myeloma
  The histiocytoses
Ineffective production of platelets
  Deficiency of vitamin B_{12}
  Deficiency of folate
  Iron-deficiency anemia (inconstant)
  Azotemia
  Hyperthyroidism (?)
  Infections
  Alcoholism
  Excessive prednisone dosage
  Congenital absence of plasma factor needed for platelet production
  Familial thrombocytopenia
  Congenital thrombocytopenia with eczema and repeated infections
    (Wiskott-Aldrich syndrome) (?)
  Paroxysmal nocturnal hemoglobinuria
  Estrogens (?)

Disorders in Which the Life Span of the Platelets Is Probably Decreased

Sensitivity to drugs
  (e.g., Quinine, quinidine, digitoxin)
Isoimmunization to platelets
  Transfusion
  Pregnancy
  Neonatal
Systemic lupus erythematosus
Idiopathic thrombocytopenic purpura (ITP)
Thrombocytopenia in newborn offspring of patients with ITP

Congenital thrombocytopenia with eczema and repeated infections
    (Wiskott-Aldrich syndrome) (?)
Lymphomas
  Lymphosarcoma
  Hodgkin's disease
  Chronic lymphatic leukemia
Hemolytic anemias
Erythroblastosis
Microangiopathic anemias
  Thrombotic thrombocytopenic purpura
  Hemolytic-uremic syndrome
  Eclampsia
Intravascular coagulation
  Amniotic fluid embolism, premature separation of placenta, etc.
  Heat stroke
  Carcinoma
  Infection (bacterial, viral, Rickettsial; Candida albicans sepsis)
Extracorporeal circulation
Infections
  Miliary tuberculosis, etc.
Sarcoidosis
Chronic cor pulmonale
Onyalai

Disorders in Which Platelets Are Sequestered

Splenomegaly
  Congestive splenomegaly
  Gaucher's disease
  Miliary tuberculosis
  Agnogenic myeloid metaplasia
  Hodgkin's disease
  Sarcoidosis
Congenital hemangiomatosis
Kaposi's sarcoma

Dilution of Platelets

By transfusion of platelet-poor blood

Disorders in Which the Pathogenesis of Thrombocytopenia Is Unclear

Infections
Thermal burns
Kwashiorkor
Macroglobulinemia and other dysproteinemias
Congenital erythropoietic porphyria
Envenoming by the brown recluse spider
Hemodialysis

---

*The reader will note that a given basic disease process may, in different situations, result in thrombocytopenia through different mechanisms.

vascular walls may be plugged by aggregates of leukocytes and platelets. Particularly in children, anaphylactoid purpura may follow closely upon infection, often by hemolytic streptococci. In other cases, the syndrome appears to be initiated by the administration of one or another drug. In many ways, the lesions of anaphylactoid purpura suggest an immunological process; IgA immunoglobulins and $\beta_1 C$ (the third component of complement) have been demonstrated in renal lesions in some cases, as well as in cutaneous capillaries. In common with many vascular lesions, evidences of local intravascular clotting have been described.

AUTOERYTHROCYTE SENSITIZATION. The complexity of the mechanisms responsible for alterations in hemostasis is best exemplified by an unusual disorder of women—autoerythrocyte sensitization—which is characterized by recurrent crops of painful ecchymoses that have an inflammatory component. Because the lesions can be reproduced by the intracutaneous injection of erythrocytic stroma, the disorder was at first attributed to autosensitization to red blood cells. More recently it has become apparent that women with this disease uniformly have severe emotional disturbances, suggesting that neural influences may be important. Autoerythrocyte sensitization serves as a reminder that our focus upon biochemical alterations may teach us about the nature of disease but leaves us short of the mark when we study illness in the patient.

PART 2 COMPLEMENT

Virginia H. Donaldson, M.D.

The complement system is a group of plasma proteins intimately concerned with the defense mechanisms of the body. Complement was discovered when certain components of fresh serum that were inactivated by heating the serum and that were distinct from antibody were found to be required for the immune lysis of cells. This basic method is still used to quantify complement. Clues to the functions of complement *in vivo* in man have come to light only in recent years, in part through studies of persons with inherited deficiencies of a functional part of the complement system.

At least 20 distinct serum proteins, which can be distinguished chemically and functionally from one another, constitute the complement system (Table 30–1). During immune cytolysis, antibody becomes fixed to a cell, attaches to a portion of the first component of complement (Cl), and induces cell lysis through an orderly series of events in which later acting components of complement produce an irreversible lesion of the cell membrane.

Although complement functions in immune reactions, it can also participate in nonimmune reactions, and some pathological changes induced by complement are not directly due to its cytolytic properties. Inflammatory reactions which may be associated with immune injury, and which are induced by fragments of certain components of the complement system, are probably as important as the cytolytic consequences of complement. Fragments may be split away enzymatically from the parent protein during complement action and

TABLE 30–1
Properties of Components of Complement and Regulators of Complement Action[*]

Compound	Molecular Weight	Sedimentation Coefficient ($S_{20}W$)	Electrophoretic Mobility (Alkaline pH)	Mean Concentration Normal Serum, mg./dl.	Common Synonyms
Classical Pathway					
C1	10^6	18 S	γ	—	C1 macromolecule
C1q	400,000	11 S	γ_2	7	—
C1r	190,000	7 S	β	3	—
C1s	85,000	4 S	α	3.1	C1 esterase ($C\bar{1}s$)
C2	117,000	6 S	β_1	2.5	—
C3	185,000	9.5 S	β_1	160	β_{1c} globulin
C4	204,000	10 S	β_1	60	β_{1e} globulin
C5	180,000	9 S	β_1	8.5	β_{1f} globulin
C6	128,000	6 S	β_2	7.5	—
C7	121,000	5 S	β_2	5.5	—
C8	150,000	8 S	γ_1	5.5	—
C9	80,000	4.5 S	α	6.0	—
Alternative Pathway					
C3	185,000	9.5 S	β_1	160	β_{1c} globulin
Factor B	95,000	5.6 S	β_2	20	C3 proactivator, B, C3-PA, GBG
Factor D	25,000	—	α	—	C3-proactivator convertase, GBG-ase, D
Properdin	185,000	9.8 S	γ_2	2.5	—
Regulators					
Classical Pathway					
$C\bar{1}$-INH	105,000	4 S	α_2	20	$C\bar{1}$-Inhibitor, $C\bar{1}$-Inactivator
C4 binding protein	>500,000	—	β	—	—
Alternative Pathway					
C3b Inactivator	88,000	5 S	β	3.4	*Is also C4b Inactivator;* C3 Inactivator, conglutinogen activating factor (KAF), I
β_1H	150,000	—	β_1	50	H
S-protein	80,000	—	α	50	Membrane attack complex (MAC) Inhibitor, MAC-INH

[*]World Health Organization Terminology

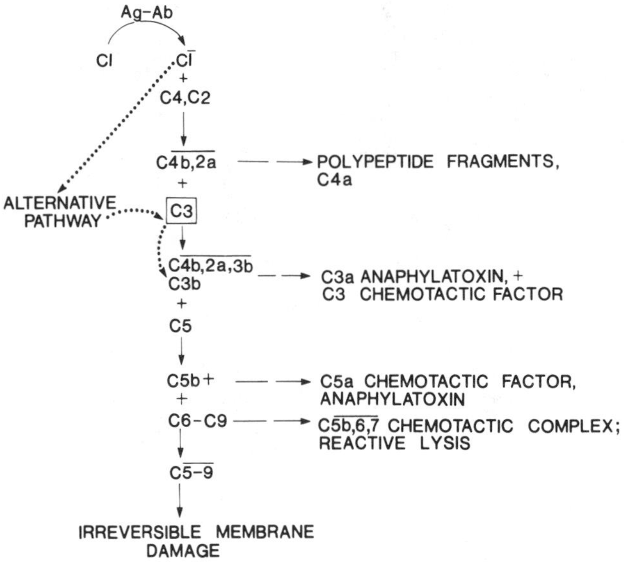

Ag–Ab

CI ...C̄I̅
+
C4,C2

C4̅b,2a ——————→ POLYPEPTIDE FRAGMENTS,
+   C4a

ALTERNATIVE......... C3
PATHWAY ···

C4̅b,2a,3b ——— → C3a ANAPHYLATOXIN, +
C3b   C3 CHEMOTACTIC FACTOR
+
C5

C5b+ ————→ C5a CHEMOTACTIC FACTOR,
+   ANAPHYLATOXIN
C6–C9 ————→ C5̅b,6,7 CHEMOTACTIC COMPLEX;
                              REACTIVE LYSIS

C5̅–9

IRREVERSIBLE MEMBRANE
DAMAGE

FIG. 30–3. Diagrammatic representation of the classical pathway of complement action in immune lysis (*left*) and some mediators of noncytolytic pathologic reactions (*right*) released by complement.

then exert potent biological activity, including the release of histamine and the attraction of phagocytes into the area of immune injury through the process of chemotaxis. Cells which are coated with the first four interacting components of complement are *opsonized* and have become "sticky," in that they adhere easily to certain other cells such as phagocytes. Through opsonization, a cell or bacterium is prepared for phagocytosis and destruction by polymorphonuclear or other phagocytic cells to which it may adhere. The third component of complement, which is present in high concentration in normal human serum, is a major opsonin because phagocytic cells may have binding sites or receptor sites for parts of this molecule. Through these reactions, pathological changes characteristic of inflammation, including vasodilatation, increased vascular permeability and the associated swelling, heat, redness, and pain, are induced.

Although complement was originally recognized because of its participation in immune lysis of bacteria or erythrocytes it can also be activated by proteolytic enzymes. Certain proteolytic enzymes may activate, or inactivate, certain components of complement by splitting the component

and releasing fragments that can then cause inflammation. For example, the fibrinolytic enzyme, plasmin, proteases in lysosomes of leukocytes or others produced by bacteria, or substances in cobra venom can attack certain components of complement. Whatever the avenue of participation of complement in response to noxious substances, its actions are ultimately directed to membranes that are altered as a consequence of cytolytic or noncytolytic effects of complement.

Physiochemical properties of components of the classical pathway of the complement system

Complement utilization may occur through either activation of its first component through the *classical pathway* (Fig. 30–3) or by an *alternative pathway* (Fig. 30–4). Thus, the complement system and blood coagulation mechanisms are similar in that both may function through at least two major pathways. The application of sophisticated biochemical procedures to studies of complement has led to the definition of separate and measurable components which are designated by number in order of their function during complement action (e.g., C1, C4, C2, C3, C5, C6, C7, C8, and C9. When a component, or a group of components, exists in an activated form, this is designated by a bar above the component in question (e.g., C$\bar{1}$). These components are all glycoproteins, of which the carbohydrate content varies from approximately 3 to over 40 per cent. Some of their chemical features are defined in Table 30–1.

The first component of complement, C1, exists in normal serum as a macromolecule, the integrity of which depends upon calcium ions which behave as ligands in maintaining a complex between subunits designated C1q, C1r, and C1s. This complex molecule sediments in solutions of graded solute concentration as if extremely heavy (18 S); the sedimentation values of the C1q, C1r, and C1s subunits separately are 11 S, 7 S, and 4.5 S, respectively. The C1s portion of the molecule bears the site of esterase activity of the C1 esterase, which evolves when C1 becomes activated. Enzymatic activity of the C1r subunit probably directs the activation of the C1s subunit during complement action; the C1q subunit is the part that can bind to the Fc portion of antibody molecules. The C1q subunit is striking in its chemical resemblance to collagen. It is a protein rich in hydroxyproline, hydroxylysine, and glycine, and consists of three disulfide-bonded pairs of similar subunits, each subunit having peripheral globular structures and a collagen-type triple helix. One IgM molecule can react with one C1q molecule and thereby activate the first component of complement. On the other hand, two or more IgG molecules are necessary for activation of C1 by reacting with C1q. Six IgG molecules are maximally bound to one C1q molecule, as if each of the subunits of C1q might, perhaps, have a binding site for the immunoglobulin.

C4 is a β-globulin with an estimated molecular weight of 204,000, and C2, also a β-globulin, has a molecular weight of about

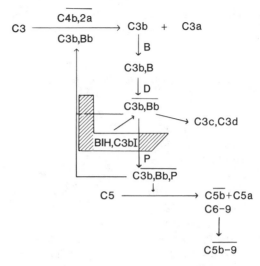

FIG. 30–4. The alternative pathway of complement action as initiated through C3b. The "positive feedback" mechanism through which C3b amplifies function of this pathway is shown to the left; the control of these actions by C3b inactivator and β1H is shown by the shaded bar.

117,000. C4 consists of three polypeptide chains (α, β, and γ) and C3, the component present in highest concentration in normal serum, consists of two polypeptide chains, an α-chain and a β-chain. It, too, is a β-globulin (mol. wt. ~ 185,000). C5, like C3, contains two polypeptide chains; the molecular weight of the intact molecule is about 200,000. C6 and C7 are also β-globulins, C8 behaves as a fast γ-globulin during electrophoresis, and C9 as an α-globulin.

Classical pathway of complement action

After C1 has been activated, it can cleave the C4 molecule, producing a large, hemolytically active portion and releasing a fragment with a molecular weight of about 10,000 from the α-chain, called C4a. The residual large C$\overline{4b}$ fragment can bind to the membrane of a cell and facilitate a reaction between C$\bar{1}$ and C2, or it can escape into solution. Only a few of the activated C4 molecules (C$\bar{4}$, or C$\overline{4b}$) bind to a cell membrane and actually participate in the reaction in which C2 is split by C$\bar{1}$, releasing a small portion of the C2 molecule (C2b). The larger portion of the C2 molecule, now activated ($\overline{C2a}$), combines with $\overline{C4b}$ to generate an active bimolecular complex designated $\overline{C4b,2a}$, which can hydrolyze the C3 molecule (Fig. 30–3). $\overline{C4b,2a}$ is, therefore, called C3 *convertase of β-1-c-convertase*, to define its action in terms of its β-globulin substrate, the C3 molecule. Once $\overline{C4b,2a}$ has formed, the $\overline{C1}$ is no longer required for action of the convertase, and can separate from the site of its interaction with these two components and can then attach to antibody on another cell membrane.

Once C1 is activated, $\overline{C1}$-inhibitor ($\overline{C1}$-INH) can bind to it and inhibit its enzymatic properties. The $\overline{C1}$-INH binds firmly to $C1\overline{r}$ and $C1\overline{s}$ subunits of $\overline{C1}$, thus blocking the enzymatic activities of both of these subunits. In addition, when $\overline{C1}$-INH combines with the $C1\overline{r}$ and $C1\overline{s}$ subunits of activated C1, it can then disassemble the molecule in a manner which allows a complex of $C1\overline{r}$, $C1\overline{s}$ and $\overline{C1}$-INH to be separated from C1q and released into solution. These released complexes consist of two $\overline{C1}$-INH molecules per $C1\overline{r}$ and $C1\overline{s}$ molecule. Laurell and her colleagues have found circulating complexes of $C1\overline{r}$, $C1\overline{s}$, and $\overline{C1}$-INH in normal serums and increased amounts in serums from patients with disorders which may be caused by immune reactions. Ziccardi and Cooper have demonstrated the same dissociation of the $\overline{C1}$ molecule by $\overline{C1}$-INH in normal serum treated with immune aggregates.

The activation of C3 and the terminal components of the complement system (C5–C9) initiates the attack on cell membranes which is characteristic of complement-mediated immune lysis. The $\overline{C2a}$ portion of the $\overline{C4b,2a}$ complex bears the catalytic site that acts upon C3, but it is unstable, and unless C3 is readily available, $\overline{C4b,2a}$ complexes deteriorate. When they do collide with C3, C3 is activated, and then binds to the surface of the target cell. In the process, C3 is split and a small fragment, C3a, escapes into solution. C3a has inflammatory properties to be noted later. The part of the C3 molecule that binds to the cell surface is called $\overline{C3b}$.

The $\overline{C3b}$ on a sensitized cell then cleaves C5 and releases a minor fragment from this component called C5a, which, like C3a, has biological activities of its own (Fig. 30–3). The major fragment, hemolytically active C5 ($\overline{C5b}$), becomes stabilized when it interacts with C6, and then combines with C7 to form a trimolecular complex ($\overline{C5b,6,7}$) upon the target cell membrane. This step creates an ultrastructural alteration in this membrane which is visible with the electron microscope. The site of attachment of the ($\overline{C5b,6,7}$) complex is probably distinct from that of $\overline{C4b,2a}$ or $\overline{C3b}$ which caused its formation. When $\overline{C5b,6,7}$ forms in solution, it can attach directly to the membrane of a cell even in the absence of antibody. $\overline{C5b,6,7}$ then interacts with C8, and a $\overline{C5b,6,7,8}$ complex forms which now has a binding site for C9. When C8 enters into this complex, the cell membrane is changed so that some intracellular contents leak out; the additional participation of C9 in the lesion accelerates this process, leading to irreversible damage and death of the cell.

The mediation of immune cytolysis through the classical complement pathway may be modified either by the intrinsic instability of activated components, or complexes thereof, or by extrinsic inhibitory substances, such as plasma inhibitors. The enzymatic property of an activated component of complement, or an activated complex, may deteriorate spontaneously, or the binding properties which allow attachment of an active component or complex to a membrane may be only transiently available. The consequences of activation of Cl are regulated by both types of mechanisms. $\overline{Cl}$-INH, the naturally occurring serum inhibitor of $\overline{Cl}$, blocks the cleavage of C4 or C2 by $\overline{Cl}$ and inhibits the hydrolysis of certain synthetic amino acid esters which are substrates of $\overline{C1s}$. This inhibitor may also regulate the activation of C1s by

C$\overline{1}$r, for it also inhibits C$\overline{1}$r. In addition, only a small number of active C4 molecules formed by C$\overline{1}$ are effectively bound to the target cell membrane, for many become inactive in solution. The C$\overline{42}$ complex is also unstable, in that the portion of the C2 molecule with the catalytic site for conversion of C3 readily deteriorates in the test tube. Membrane-bound C$\overline{3b}$ may be inactivated by another plasma enzyme called C3b inactivator (C3bI, or I) which cleaves C$\overline{3b}$ into inactive fragments, C3c, and C3d. The instability of activated C5b probably limits its function until it combines with C6 in a stabilized complex, and then participates in the generation of the lytic C$\overline{5b}$,6,7,8,9 complex. A C6 inactivator has been described in animal sera, but its significance is not yet clear. The "S protein" described by Podack and his associates can inhibit the terminal membrane attack complex (C$\overline{5b}$-9) by competing with the complex for binding sites on a target cell. The importance of C$\overline{1}$-INH and C3b inactivator in regulation of human complement in plasma has been amply demonstrated through studies of disorders resulting from their inherited deficiency.

The alternative or properdin pathway to complement utilization

The remarkable ability of certain polysaccharides to activate C3 and utilize late-acting components of complement while sparing C1, C2, and C4 was described initially by Pillemer and his colleagues in 1954 (Table 30–2). This alternative mechanism of complement action was originally noted to depend upon a serum protein called properdin and at least two other proteins called factor A and factor B. Factor A was distinguished by its inactivation during incubation with hydrazine. Factor B was inactivated during incubation at 50° C for 30 minutes. Properdin (P) has been highly purified and is a β- or γ-globulin (depending upon the gel used for immunoelectrophoretic characterization) with a molec-

TABLE 30–2
Nomenclature of the Alternative Pathway

Protein	Symbol
C3	C3
C3 Proactivator	B, C3-PA
C3 Proactivator Convertase	D
C3b Inactivator	C3b-INA, I
β1H	H
Properdin	P

ular weight estimated to be 185,000. Factor A is identical to C$\overline{3b}$, the main functional component of activated C3, and thus has a molecular weight of 185,000; it is a β-globulin. Factor B, the heat-labile protein of this mechanism, is identical with a glycine-rich β-glycoprotein, called GBG, described by Boenisch and Alper, which is the same as the C3 proactivator (C3PA) described by Müller-Eberhard and Götze. Purified factor D is a fast γ-globulin which has enzymatic activity for which a serine residue is essential; it is inhibited by diisopropylphosphofluoridate (DFP). Factor D in serum has the electrophoretic mobility of an α-globulin; it has a molecular weight of 24,000 and is found only in trace amounts (<1μg./ml.) in normal serum.

Alternative pathway action

This pathway can be activated in the absence of antibody, which is the hallmark of immunological activation of the classical pathway, and its action therefore lacks the specificity which characterizes antibody-directed classical pathway action. Because of this, the alternative pathway may be a critical defense mechanism in the absence of antibody or immune aggregates, and probably serves as an important defense against infectious agents having polysaccharide capsules, or cell walls, before an antibody response has been mounted.

In the presence of magnesium, a C$\overline{3b}$ molecule can form a complex with a factor B molecule which lacks enzymatic activity (C3b,B). Once involved in this bimolecular complex, however, the factor B is susceptible to cleavage and activation by factor D, believed to exist in an active form in plasma or serum. This results in an enzymatically active complex, designated $\overline{C3b,Bb}$. This active complex is stabilized by properdin and can in turn generate further $\overline{C3b}$ from C3 and a self-perpetuating activation reaction then proceeds. Factor D is not bound within the complex of $\overline{C3bBb}$ and is free to interact with further C3bB complexes. These events can occur on a cell membrane or in solution. This positive feedback mechanism makes the alternative pathway work. In the process, C3a, having a molecular weight of about 6,800, is cleaved from C3 and this fragment has properties of an anaphylatoxin.

Although properdin can stabilize the $\overline{C3b,Bb}$ complex, it is not required for its formation. Once properdin stabilizes the active complex of $\overline{C3b}$ and activated factor B, that is $\overline{C3b,Bb}$, upon a cell surface, the new complex $(\overline{C3b,Bb,P})$ converts further C3 to $\overline{C3b}$ and also cleaves and activates C5; it is therefore called a C3/C5 converting complex. $\overline{C3b,Bb}$ can also assemble C3 convertase in solution even in the absence of identifiable activators of the alternative pathway, and can attach to particles to which $\overline{C3b}$ is bound even in the absence of factors B and D. An activated form of properdin probably differs in its steric configuration from a relatively inactive form.

The alternative pathway is carefully regulated by at least two proteins: C3b inactivator and β1H globulin (now designated H). C3b inactivator is an enzyme, has a molecular weight of 88,000 and the electrophoretic mobility of a β-globulin, and consists of two different polypeptide chains linked together by disulfide bonds. This inactivator can cleave the α-chain of $\overline{C3b}$, thus inactivating $\overline{C3b}$ and generating C3bi, a designation for the inactive form of this molecule. C3c and C3d fragments result from this cleavage; C3d has opsonizing properties in that it can facilitate the adherence of particles upon which it resides

to certain cell surfaces. Cleavage of C3 with trypsin provides a C3e fragment which has leukocytosis-inducing activity. This C3e fragment can also form spontaneously when serum is incubated *in vitro* for a protracted period. C3b inactivator also cleaves and inactivates C4b, provided a C4-binding protein, which is a cofactor in this reaction, is available. Thus, C3b inactivator can regulate both the classical and alternative pathways of complement action.

Beta-1H globulin, an integral component of the alternative pathway, has a high binding affinity for a site on the $\overline{C3b}$ molecule and, when bound to $\overline{C3b}$, can prevent the interaction of factor B (C3 PA) and $\overline{C3b}$ as well as being able to dislodge the activated form of factor B, designated Bb, from a $\overline{C3b,Bb}$ complex. In so doing it prepares $\overline{C3b}$ for inactivation by C3b inactivator. Beta-1H enhances the inactivation of $\overline{C3b}$ by C3b inactivator where the $\overline{C3b}$ resides on cell surfaces, but in solution, separated from a cell surface, β1H globulin is an absolute requirement for the inactivation of $\overline{C3b}$ by C3b inactivator. Since β1H can compete with factor B or Bb in binding to $\overline{C3b}$, and prevent the generation of effective $\overline{C3b,Bb}$ complexes, it is an important regulator of this pathway. After $\overline{C3b}$ is complexed with β1H and then cleaved by C3b inactivator, β1H escapes into solution free to repeat this task.

Another way of regulating the action of this C3/C5 convertase complex $(\overline{C3b,Bb,P})$ is related to the environment offered by a cell membrane to which the complex becomes attached. The nature of carbohydrate molecules in a cell membrane at or near a $\overline{C3b}$ binding site may make the $\overline{C3b,Bb}$ complex more or less susceptible to inactivation by the β1H-C3b inactivator mechanism. Thus, sheep erythrocytes, which do not favor formation of $\overline{C3b,Bb}$ complexes, can be made to do so by treating them with neuraminidase, which removes sialic acid residues. Therefore, cell membrane components may alter the binding of $\overline{C3b,Bb}$ so that $\overline{C3b}$ cannot interact with β1H and then be inactivated. On the other hand, when C3̄b̄ binds to nonactivating cell surface structures, perhaps to sialic acid residues, the binding of β1H to C3̄b̄ is facilitated and C3̄b̄ is then inactivated by the C3b inactivator.

Figure 30–4 illustrates the reactions involving the alternative pathway.

The successful generation of C3̄b̄,B̄b̄ results in initiation of the action of the terminal portion of the complement system through the cleavage of C5, and its subsequent formation of an active complex with C6 and C7. Either the classical or alternative pathway can generate C5̄b̄,6,7 complexes in solution, and thereby induce reactive lysis, for these C5̄b̄,6,7 complexes can then attach to a cell which is otherwise immunologically innocent. Then, C8 and C9 participate in the formation of an ultimately lytic lesion upon this cell.

A substance in cobra venom, which can cause consumption of C3 and the late-acting components of complement (C5–C9) by a direct effect on the alternative pathway, is known as cobra venom factor. The cobra venom does not require C3̄b̄ for this action, for it behaves functionally and immunologically like C3̄b̄. In fact, Alper, et al. found that cobra venom factor appears to be a C3̄b̄ fragment in the venom which can initiate the positive feedback in the alternate pathway, as well as the consumption of C5–C9.

The role of factor D in the alternative pathway is analogous to that of Cl̄ in the classical pathway. The classical and alternative pathways can act in concert. When erythrocytes were sensitized with an excess of antibody, complement could be activated in serum from a strain of guinea pigs markedly deficient in C4, and thereby lacking one of the requirements for function of the classical pathway. Cl̄ was required for this interaction, and the components of the alternative pathway appeared to be activated. (Fig. 30–5). In addition, once C3̄b̄ is generated by classical pathway activation, this can then initiate the posi-

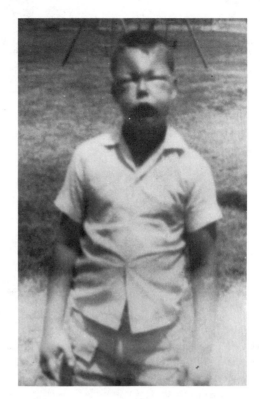

FIG. 30–5. A patient with hereditary angioneurotic edema during an attack involving his face.

tive feedback mechanism driving the alternative pathway into action.

Noncytolytic pathological functions of complement

Release of biologically active fragments

The proteolytic release of polypeptide fragments from plasma proteins, which can induce the signs of inflammation, is perhaps best exemplified by another mechanism in plasma which can act independently of complement and releases polypeptide *kinins*, such as bradykinin, from an α-globulin in plasma called kininogen. Table 30–3 notes some of the biologically active fragments released from the components of the complement system. In early experiments, kinins were released in plasma exposed to pancreatic extracts or to the

TABLE 30–3
Some Biologically Active Fragments Released from Components of Complement

Fragment	Released by	Approximate Molecular Weight	Activity
C4a	C1$\overline{\text{s}}$	8,650	Anaphylatoxin (weak)
C2b	$\overline{\text{C1,4b}}$; $\overline{\text{C1}}$, or C1$\overline{\text{s}}$	~34,000	?
?C2 Kinin	$\overline{\text{C1,4b}}$; C1$\overline{\text{s}}$, trypsin, plasmin	~1,000	Permeability-enhancing and kinin activity
C3a	$\overline{\text{C4b,2a}}$; trypsin, plasmin, $\overline{\text{C3b,Bb}}$ cobra venom factor	6,800	Anaphylatoxin: mast cell degranulation, histamine release, smooth muscle contraction, increased vascular permeability
C3b	$\overline{\text{C4b,2a}}$; $\overline{\text{C3b,Bb,P}}$	—	Opsonization
C3d	β1H + C3bI	—	Opsonization
C3e	trypsin	10,000–12,000	Induction of leukocytosis
C5a	$\overline{\text{C4b,2a,3b}}$; $\overline{\text{C3b,Bb,P}}$ alternative pathway activation	10,000–15,000	Anaphylatoxin, leukotaxin

venom of a South American crotaline snake, *Bothrops jararaca*. At first, kinins were recognized because they lowered the blood pressure of animals. Later experiments demonstrated their ability to cause pain, increase vascular permeability, contract isolated smooth muscles, dilate blood vessels, and in some situations induce leukocyte stickiness and emigration. Through these effects, the signs and symptoms of inflammation could be reproduced by bradykinin and closely related polypeptide molecules. Bradykinin can also be released through the action of enzymes normally present in an inactive form in normal human plasma. One such precursor, prekallikrein, can be converted to its active proteolytic form, kallikrein, and then can cleave kininogen to release vasoactive polypeptides such as bradykinin. Once bradykinin is released, its life in blood plasma is short, for there are two or more kinin-destroying enzymes in plasma called kininases that can inactivate this molecule by removing its carboxy-terminal portions. The kallikrein system can be activated by the early stages of blood coagulation after the activation of Hageman factor by negatively charged substances, pointing out the close relationship between these plasma enzyme systems. In fact, both plasma prekallikrein and high molecular weight kininogen are coagulation factors which function in Hageman factor–dependent reactions. These reactions lead to blood coagulation, the development of fibrinolytic activity, and, of course, the release of bradykinin.

Anaphylatoxins

Early experiments demonstrated that a snake venom destroyed serum properties needed for the phagocytosis of microorganisms by polymorphonuclear cells. More modern studies have shown that cobra venom can create this defect by initiating the destruction of C3 through the alternative pathway of complement action. When C3 is activated by $\overline{\text{C4b,2a}}$, $\overline{\text{C3b,Bb}}$, trypsin, or cobra venom, the C3a fragment which is released can increase vascular permeability, contract isolated guinea pig ileum, and degranulate rat peritoneal mast cells, properties ascribed to an *anaphylatoxin*. The ability of C3a to induce these responses in experimental animals depends largely, if not entirely, upon its capacity to release histamine, for treatment of animals with antihistaminics blocked the permeability-enhancing and muscle-contracting actions of C3a. The residual C$\overline{3b}$ is then cleaved and inactivated by C3b inactivator in a reaction which is markedly hastened by the β1H globulin if the C$\overline{3b}$ resides on a cell surface. If the C$\overline{3b}$ exists in solution, the β1H is absolutely required for the destruction of C$\overline{3b}$ by C3b inactivator.

Another anaphylatoxin is released from C5 and called C5a. This fragment has a higher molecular weight ($\sim$12,000) than C3a (6,800). C5a appears to act on an isolated smooth muscle through a different mechanism than C3a. One of the characteristics of anaphylatoxins lies in their ability to induce a state of tachyphylaxis, that is, unresponsiveness of the target tissue. When an isolated muscle has been made unresponsive to C3a, however, it can still respond to C5a as if different receptor sites in that tissue may mediate the response to these two anaphylatoxins. Both C3a and C5a are inactivated by peptidases in plasma in a manner resembling the destruction of kinin activity by kininases.

C4a, a fragment cleaved from C4 during its activation by C$\overline{1s}$, has mild anaphylatoxin activity.

Chemotaxis

Chemotaxis is a process through which cells are attracted toward certain substances. These substances, then, are chemotaxins. Although the means through which cells actually develop chemotactic activity are only partly understood, two membrane enzymes (esterases) of polymorphonuclear cells participate in chemotaxis.

At least three chemotactic substances can be generated during complement action. A chemotactic fragment that attracts polymorphonuclear leukocytes is released from C3; this fragment is apparently different from C3a anaphylatoxin but may be derived from C3a. A chemotactic fragment of C5 may be the same as C5a anaphylatoxin, and may attract mononuclear cells in preference to polymorphs. A complex of C$\overline{5b},6,7$ formed during complement action attracts polymorphs and eosinophils, and can be formed by activation of complement in serum in the absence of fixation of C5 to a cell surface. Chemotaxis of neutrophils may be impaired in states of anergy, and certain viral infections may depress chemotactic response of monocytes.

Bacteria may directly release chemotactic fragments from complement components. Thus, a proteinase from *Serratia marcescens* can release a C3 fragment which has leukotactic activity, and a streptococcal proteinase can release a fragment from C5 with leukotactic activity. Theoretically, these mechanisms might provide an effective defense against an invading organism before antibody is formed. Protein A from the cell wall of *Staphylococcus aureus* is also chemotactic for leukocytes; this action requires the cooperation of heat-labile complement components. Substances produced by *Escherichia coli*, tuberculoprotein, endotoxin, and glycogen also generate chemotactic activity in serum. Components of leukocyte lysosomes, macrophages, or platelet granules can release C5a directly from C5. (Chemotactic properties derived from complement are summarized in Table 30–4.)

Through these mechanisms, the action of certain portions of the complement system can initiate immune inflammation, which can lead to the formation of pus; lysosomal enzymes released from the polymorphs collecting in such an area may remove invading microorganisms, but they can also cause further tissue damage. The importance of C3 in generating chemotactic activity is probably best exemplified by the extraordinary susceptibility to infection of persons genetically deficient in C3 (see below).

THE NEPHRITIC FACTOR (C3-NEF). Low serum complement levels are often associated with a kind of renal disease characterized pathologically by proliferation of the mesangium and thickening of the capillary wall of glomerular blood vessels. This has

TABLE 30–4
Some Chemotactic Factors Released from Components of Complement

Factor	Released by	Approximate Molecular Weight
C3 Chemotactic fragment(s)	C4b,2a; C3b,Bb, proteases; trypsin, neutral tissue proteases, macrophage protease	?<8,000
C3 Chemotactic fragment	Leukocyte lysosomal protease	~12,400
C3 Chemotactic factor(s)	*S. marcescens* protease	?5,6000 and 34,000
C5a	C4b,2a,3b; C3b,Bb,P	15,000
C5 Chemotactic fragment(s)	Leukocyte lysosomal protease, macrophage protease, platelet protein	~15,000
C5 Chemotactic factor(s)	Streptococcal proteinase	?
C5b,6,7	C4b,2a,3b; C3b,Bb,P; trypsin	High

been called *hypocomplementemic* or *membranoproliferative glomerulonephritis*. In some cases, the decreased hemolytic complement titer is due mainly to a decrease in serum C3 concentration, despite relatively normal amounts of C1, C4, and C2. The reduction of C3 appears to be due to action of the alternative rather than the classical complement pathway. Vallota, *et al.* found that factor B, factor D, and probably properdin all participated in the destruction of C3 by a factor found in serum of patients with this variety of nephritis. In other instances of hypocomplementemic glomerulonephritis, both pathways may be involved and C1, C4, and C2 are decreased.

The factor that initiates an attack through the alternative pathway upon C3 has been isolated and appears to be distinct from other substances that can activate complement. C3-NEF is, in fact, an antibody directed against the C3b,Bb complex, and is an immunoglobulin which has an unusual structure in that its heavy chain is longer than that usually found in IgG molecules. When C3-NEF binds to the C3b,Bb bimolecular complex, it enhances the action of the active complex and prolongs the life of the complex in plasma.

The decreased C3 titer in hypocomplementemic nephritis may also reflect diminished synthesis of this protein by the liver, but the reasons for this are uncertain. It is

appealing to view this as a negative feedback effect of activation of the alternative pathway which might be protective, but there is no proof for this view.

In other forms of nephritis, complement activation leading to consumption of its components can occur. In acute glomerulonephritis and the nephritis associated with systemic lupus erythematosus, activation of the classical pathway is likely to occur, probably because of reactions following the formation of antigen-antibody complexes. The role of complement in renal disease is complex and its relative importance in actually initiating renal pathology is unclear. It is likely that the chemotactic fragments released from C3 and C5 are important in inducing tissue damage because of the release of lysosomal enzymes from leukocytes which are induced to accumulate where these fragments are being generated.

C-REACTIVE PROTEIN AND COMPLEMENT.
During the acute phase of a number of illnesses or inflammatory conditions, the concentrations of some serum proteins may be increased. A protein which may appear in significant concentrations in serum of persons during an acute illness has the unusual property of coprecipitating with the C-polysaccharide of the pneumococcus. It has therefore been called C-reactive

protein or CRP, and is an *acute-phase* protein.

When CRP forms a complex with the pneumococcal C-polysaccharide or with certain choline phosphatides, it can interact with complement. The CRP complex reacts with the C1q subcomponent of C1 and causes consumption of C4, C2, and to a smaller degree C3–C9, through activation of the classical pathway. These changes appear to occur in the absence of antibody and therefore represent a nonimmune reaction involving complement and this acute-phase protein. It is possible that CRP and phosphorylcholine may activate complement *in vivo* to release chemotactic and irritant substances which may induce or augment inflammation.

Sites of biosynthesis of complement components

Using *in vitro* tissue culture techniques, it has become apparent that cells associated with the gastrointestinal tract and cells of the reticuloendothelial system are the principal producers of complement. C1q can be synthesized by portions of the gastrointestinal tract, and also by cells in the lymph nodes, spleen, liver, and lung. Macromolecular C1 and the C1s subunit of the C1 molecule are synthesized in cells from normal human colon, adenocarcinoma of the colon, transitional epithelial cells of the bladder, urethra, and renal pelvis. $\overline{C1}$ and its subunits, C1q, C1r, and C1s, can also be synthesized by human fibroblasts and macrophages from the peritoneal cavity.

Large mononuclear cells, probably macrophages, and circulating monocytes have been shown to synthesize C2. Some of the monocyte or macrophage cells which synthesize C2 can also synthesize C4. C4 can also be synthesized by human or guinea pig fetal liver cells. The synthesis of biologically active C4 *in vitro* by some cells has been difficult to demonstrate, and present evidence indicates that a precursor form of the molecule may be synthesized within the cell. Several cultured lines of rat hepatoma cells produce large quantities of C4 in culture, and they appear to be parenchymal cells.

The synthesis of C3 by liver cells was first demonstrated by Alper and his colleagues when a patient who received a liver transplant was then found to have the genetically determined C3 allotype of the donor, rather than of his own circulating C3. Thus, the transplanted liver supplied the C3 of a new phenotype to this patient. In later studies of individuals who have received liver transplants, the recipient exhibited donor allotypes of C3, C6, C8 and factor B, and it appeared that 90 per cent of these proteins were synthesized by the liver. Even so, extrahepatic synthesis of C1, C4, C2, C3, C5, and factors B and D by monocyte/macrophage cell lines has been shown. C3 may also be synthesized by synovial tissues in patients with rheumatoid arthritis. The biosynthesis of C5 in humans has been demonstrated in lung, liver, spleen, and fetal intestinal tissue, as well as in thymus, placenta, peritoneal cells, and bone marrow cells. The peritoneal cells of mice secrete *pro-C5*, a precursor single chain molecule, as well as the two-chain C5 protein which circulates in the blood. It is not clear which cells synthesize C7, but spleen, liver, lung, intestine, and kidney tissues can produce C8, while lymph node tissues, thymus, and bone marrow do not. It is likely that C9 is synthesized by liver parenchymal cells. The hepatic parenchymal cells synthesize $\overline{C1}$-INH as shown by functional and immunochemical assays. From these observations, it is becoming clear that a number of the cells which are intimately related to the body's defense mechanisms may synthesize some of the components of the complement system.

C4 and C3 are synthesized within the cells as single chain molecules, and are later cleaved to yield the multiple-chain molecules found in plasma. Plasmin has been shown to convert single chain pro-C4 to a three-chain molecule apparently the same as that found in plasma. Liver polysomes from guinea pigs genetically deficient in circulating C4 failed to synthesize the precursor C4 molecule, but did synthesize lower molecular weight C4 polypeptides, which were not then released into the culture medium.

Genetic linkage of certain complement components and histocompatibility genes

The locations of genes which regulate the biosynthesis of certain histocompatibility (HLA) antigens are in close proximity to those regulating synthesis of certain complement components on the sixth human chromosome. The gene designating the synthesis of C2 is linked to that for factor B, and both reside in the HLA region of this chromosome. Genes for C4 antigens are also located close to the HLA locus. There are probably two structural genes involved in the synthesis of C4. The red cell antigens called Rodgers and Chido are in fact fragments of products of these two genes (C4A and C4B) bound to the red cells.

There is no clear evidence for the linkage of the genes regulating the synthesis of C3, C5, C6, C7, C8, or C9 to histocompatibility loci in humans. The clustering of genetic information regulating the production of complement proteins with those designating synthesis of histocompatibility (HLA) antigens, which are immune response genes, implies that the conservation of relationships between these genes is probably important. In studying these relationships, the different forms of the proteins which are synthesized by different individuals (allotypes) have been useful in identifying the association between histocompatibility antigens and various forms of the complement components.

Hereditary defects of the complement system

Inherited deficiencies of individual components of complement are not universally associated with a clearly defined disease picture, but deficiencies of regulators of complement action ($\overline{C1}$-INH and C3b inactivator) are accompanied by disease. Persons with hereditary angioneurotic edema have an inherited functional deficiency of serum $\overline{C1}$-INH. A patient with a hereditary deficiency of C3b inactivator is susceptible to pyogenic infections, mainly because of the resulting hypercatabolism of C3 and, therefore, low serum concentrations of C3. Inherited deficiency of C3 does predispose the affected person to pyogenic infections. Even though an inherited deficiency of other components of complement is not directly related to a specific disease picture, studies of kindred with hereditary deficiencies of complement components suggest that in many instances there may be a subtle susceptibility to ill-defined connective tissue diseases or certain infectious organisms in deficient individuals.

Hereditary angioneurotic edema: deficiency of $\overline{C1}$-INH

Hereditary angioneurotic edema is characterized by repeated attacks of subepithelial, noninflammatory edema, which may be life-threatening if the airway is involved. Episodes of gastrointestinal obstruction due to edema of the intestines occur repeatedly, but are self-limited. Swellings of the skin may be severe and cause grotesque distortions of the face, or they may be mild (Fig. 30–5). These attacks are usually limited to periods of 24 to 48 hours, after which swelling begins to disappear. They are not induced by allergic reactions but may follow emotional upsets or mild physical trauma, or they may occur for no apparent reason.

The tendency to have episodes of hereditary angioneurotic edema is inherited as an autosomal dominant trait, as is the inherited deficiency of serum $\overline{C1}$-INH, which is clearly associated with hereditary angioneurotic edema. Because of the deficiency of $\overline{C1}$-INH, complement is poorly regulated and C1 tends to become activated spontaneously in plasma obtained from these persons. This tendency to activation is significantly enhanced during attacks of edema, and it may play a role in causing the swellings. Decreased concentrations of hemolytic C4 and C2 in plasma obtained from persons having attacks of hereditary angioneurotic edema reflect the destructive effect of circulating $\overline{C1}$ upon C4 and C2 in the plasma. Remarkably, C3 and other later-acting components of the complement system are not significantly decreased. The reason for the sparing of C3–C9 and the apparent failure to generate significant C3-converting activity in the blood of persons with hereditary angioneurotic edema has not been entirely explained. It is likely that the rapid cleavage and inactivation of the hemolytic properties of C4 by the $\overline{C1}$ which forms in plasma from these individuals prevents the formation of an effective $\overline{C4b,2a}$ C3-convertase (classical pathway convertase) complex.

$\overline{C1}$-INH is an α-2 neuraminoglycoprotein which has a high carbohydrate content. Its level in normal human serum is remarkably stable during health, but can vary during illnesses. Another condition in which $\overline{C1}$-INH has been found to be markedly decreased is in serum from women late in pregnancy; this is an acquired deficiency which is corrected within a day or two of delivery. Since the concentration of inhibitor antigens in serum obtained in late pregnancy may be near normal despite loss of inhibitory function, in these cases the inhibitor has probably become complexed with other proteins and is therefore not available to block exogenous $\overline{C1}$.

The functional deficiency of $\overline{C1}$-INH in hereditary angioneurotic edema is inherited by complex mechanisms in which allotypic forms of the protein exist. The affected members of about 70 per cent of kindred with hereditary angioneurotic edema are deficient in both the functional and antigenic properties of the normal $\overline{C1}$-INH. Approximately 30 per cent have at least normal amounts of $\overline{C1}$-INH serum antigen, but the protein lacks function. Even the group of kindred with nonfunctional inhibitor proteins is heterogeneous.

The deficiency of serum $\overline{C1}$-INH in hereditary angioneurotic edema is lifelong and cannot explain the episodic swellings. $\overline{C1s}$ preparations can cause increased vascular permeability upon intradermal injection into humans; in a patient known to have hereditary angioneurotic edema, intradermally injected $\overline{C1s}$ provoked a typical attack involving the injected arm. C2 and C4 appear to be required for this reaction, for humans deficient in C2 and guinea pigs deficient in C4 developed only minimal edema around sites of injection of purified $\overline{C1s}$. The plasma of persons with hereditary angioneurotic edema can develop very high concentrations of permeability-increasing activity at the time of the attacks of swelling, or during *in vitro* incubation of plasma obtained during remissions. A unique polypeptide kinin which can increase vascular permeability has been isolated from hereditary angioneurotic edema plasma. Although it is not yet proven that this polypeptide kinin is derived from a substrate of $\overline{C1}$, it has been suggested that it may be released from C2. Other polypeptide kinins may also be released, including bradykinin.

$\overline{C1}$-INH can also inhibit other plasma enzymes which release vasoactive polypeptides including kallikrein, a permeability factor that forms when normal plasma is diluted in glass test tubes (PF/Dil), activated Hageman factor (factor XII) and plasma thromboplastin anticedent (PTA, factor XI) which are clotting factors, and the fibrinolytic enzyme plasmin. While all of these other plasma enzymes are additionally regulated by other plasma inhibitors, $\overline{C1}$-INH appears to be the sole inhibitor of $\overline{C1}$. Nonetheless, partial reduction in serum inhibition of kallikrein and Hageman factor might promote the activation of kallikrein and facilitate the release of bradykinin and similar polypeptides which can enhance vascular permeability. Thus, a number of mediators might be released and add to the symptoms of hereditary angioneurotic edema. Indeed, elevated levels of bradykinin have been found in plasma from some patients having attacks of edema.

The diagnosis of hereditary angioneurotic edema can be made tentatively on the basis of the clinical findings, past history of the patient, and family history. The swelling of the skin that these people experience does not itch, is not painful beyond the discomfort associated with distention of the tissues, and is not red, or urticarial. The edema is pale and cool and does not pit until the swelling begins to subside. Occasionally, a small knot, or serpiginous erythema, is noted in skin prior to the onset of the typical swelling of an attack involving that area. Gastrointestinal attacks may be heralded by a sense of dyspepsia a few hours before a very painful attack, but the patient is often unaware of any warning that an attack is about to occur. These episodes are not allergic in nature, and are not amenable to treatment with antihistaminics or a variety of other antiallergic medications. When the tracheobronchial tree, larynx, or lungs are involved, sudden death may occur; this is reported with alarming frequency by members of some affected kindred. Gastrointestinal episodes may not be clearly associated with skin swellings and may mislead the unwary physician to believe that he is dealing with intrinsic gastrointestinal disease necessitating surgical treatment. Vomiting and significant dehydration, sometimes associated with diarrhea as an attack subsides, are characteristic. The pain may be severe and may occasionally require opiates for relief. As suddenly as the abdominal attack begins, it subsides, after 24 to 48 hours in most instances. These abdominal episodes are quite characteristic of hereditary angioneurotic edema and in our experience offer the most significant clinical clue in making this diagnosis.

The biochemical diagnosis of hereditary angioneurotic edema can be made by directly measuring the inhibition of the enzymatic activity of preparations of C1 esterase by plasma or serum from the individual in question. In addition, the level of C4 is usually low, even during remission, because small amounts of $C\bar{1}$ may be released easily in the absence of adequate inhibition, and the $C\bar{1}$ may then inactivate the C4. If the concentration of C4 is normal in hereditary angioneurotic edema serum obtained during remission, incubation of the sample *in vitro* at 37° C for an hour will usually result in activation of C1 and destruction of hemolytic C4 activity. This incubation does not cause activation of C1 in serum which contains inhibitor, and measuring the level of C4, or the total hemolytic complement titer in incubated (as opposed to nonincubated) serum of a suspected patient may allow indirect biochemical diagnosis of this disorder.

The treatment of hereditary angioneurotic edema has been difficult, and there is no sure way of ending an attack once it has begun. The most satisfactory approach to managing patients with this disorder is to prevent the attacks of edema from occurring. In recent years, synthetic androgens which have marked anabolic effects, but very little masculinizing activity (danazol or stanozolol), have been effective in preventing the episodes of edema to which these patients are subject. In addition, this therapy corrects the biochemical defect associated with hereditary angioneurotic edema. Thus, after a week of therapy with danazol, measurable levels of serum $C\bar{1}$-INH can be identified, and the levels of C4 in the serum of these individuals will rise. This effect is secondary to the inhibition of $C\bar{1}$ and of its activation by the newly formed inhibitor. The precise mechanism of action of this therapeutic agent is not clear, but in some way it allows the synthesis of the missing normal inhibitor to occur and the normal $C\bar{1}$-INH is then released into the blood. The generation of normal $C\bar{1}$-INH during danazol therapy has also been demonstrated in patients with hereditary angioneurotic edema associated with nonfunctional $\overline{C1}$-INH proteins, which contain normal antigenic determinants, but cannot block $\overline{C1}$. During therapy these individuals develop a new component of $\overline{C1}$-INH activity in their serum which has normal electrophoretic mobility, but they continue to demonstrate the abnormal inhibitor protein in immunoelectrophoretic analysis of their serums. It is tempting to speculate that this therapy in some way induces the function of the normal gene which must exist in these heterozygous individuals but is apparently prevented from directing $\overline{C1}$-INH synthesis in the absence of such a medication. Testosterone, halotestin, and oxymethalone preparations also have this effect in hereditary angioneurotic edema.

Some patients have benefited significantly from taking antifibrinolytic agents such as ε-aminocaproic acid (Amicar) and a chemically similar compound called tranexamic acid. These agents may be effective by blocking enzymatic activity which can occur in plasma and then lead to C1 activation. Tranexamic acid, not presently available on the market in the United States, may provide significant relief to a number of these patients at a lower dose than does ε-aminocaproic acid. The side effects of chronic ingestion of ε-aminocaproic acid, which most commonly include muscle pain and weakness, can be disabling and thereby limit the usefulness of this drug. Replacement therapy with plasma, although seemingly logical when considered in the framework of replacement therapy for other deficiency states, may actually offer a disadvantage in this disease. The concentration of $\overline{C1}$-INH in normal plasma is low compared to some of the other substances in normal plasma which may behave as substrates for $\overline{C1}$, or other kinin generating enzymes in the blood. Therefore, there is theoretial risk of worsening a developing attack of hereditary angioneurotic edema by giving plasma, and we have avoided this form of therapy. Partially purified preparations of $\overline{C1}$-INH have been used, with some apparent success, for treatment of persons having attacks of edema. There has been a relatively high risk of hepatitis with some of these preparations.

The pathology of hereditary angioneurotic edema is unusual in its lack of signs of inflammation, for despite enormous amounts of edema, there are no noticeable infiltrates of acute or chronic inflammatory cells in tissues removed from patients undergoing surgery during bouts of intestinal obstruction, for example. Surgery does not benefit these patients and the amount of swelling of the intestinal tract may complicate the surgeon's task considerably. The reason for the lack of inflammatory cells may be related to the fact that in this disorder chemotactic substances are probably not generated in significant amounts during complement activation because neither C3 nor C5 is noticeably consumed, even during bouts of edema. The substances found in the blood of these patients during attacks which enhance vascular permeability and have kinin-like properties may not attract leukocytes.

Although no particular susceptibility to infection accompanies hereditary angioneurotic edema, several patients have developed syndromes resembling discoid lupus, or lupus erythematosus. Although hereditary angioneurotic edema is not a common disease, the use of biochemical methods to establish the diagnosis in persons not having attacks has led to the suggestion that it is not as rare as originally believed. The incidence of the gene for this disorder is probably greater than 1 in 160,000 persons.

C3b INACTIVATOR DEFICIENCY–HYPERCATABOLISM OF C3b. Hypercatabolism of the third component of complement was found in a patient who had had multiple overwhelming infections with pyogenic organisms including *Haemophilus influenzae*, *Meningococcus*, and *Pneumococcus*.

Because of its hypercatabolism, C3 was barely detectable in his plasma and serum. This patient has an inherited deficiency of the C3b inactivator (C3bI, KAF) and as a result of this deficiency the alternative pathway is continuously activated in his plasma, leading to continuous conversion of C3 to $\overline{\text{C3b}}$ (Fig. 30–4). In the absence of C3bI, further conversion of $\overline{\text{C3b}}$ to C3c and C3d does not occur. Therefore, when C3bI is missing, $\overline{\text{C3b}}$ remains available and can promote alternative pathway activation through a positive feedback mechanism in which factor D can activate the C3b,B complexes which form (Fig. 30–4). Because of the feedback activation effect of $\overline{\text{C3b}}$, the deficiency of C3bI indirectly facilitates the consumption of C3 so that its function is lost to the body's defenses against certain infectious organisms. Then, even in the presence of antibody to an infective organism, the full defense mechanism of the classical pathway cannot be used.

Because of the accelerated C3 destruction, significant amounts of C3a anaphylatoxin are formed, which cause intermittent episodes of urticaria, and massive amounts of histamine are excreted in this patient's urine. Infusion of a preparation of C3bI into this patient corrected the defect, halted C3 hypercatabolism, caused the disappearance of $\overline{\text{C3b}}$ from the circulation, and restored hemolytic and opsonizing functions to his serum.

HETEROZYGOUS C3 DEFICIENCY. Heterozygous C3 deficiency, in which affected persons have serum concentrations of approximately 50 per cent of normal, is not associated with susceptibility to infection or other clinical disease. This deficiency results from inheritance of a silent structural gene along with a normal structural gene; the amount of C3 synthesized in response to the normal gene appears adequate for health of the propositi.

HOMOZYGOUS C3 DEFICIENCY. Homozygous C3 deficiency, on the other hand, renders a patient susceptible to overwhelming pyogenic infections including meningitis, otitis media, and skin infections. The most extensively studied homozygous C3-deficient patient had only 1/1,000 of the normal amount of C3 in her serum by immunochemical measurement. Her parents and several siblings appeared heterozygous for the deficiency and had serum concentrations of C3 near 50 per cent of normal. Her impaired resistance to infection was accompanied by laboratory evidence that the alternative pathway response to a polysaccharide was also defective and that factor B could not be activated, as one would predict in the absence of C3. Function of the classical pathway of complement, measured with antibody-coated erythrocytes, was also impaired. Even in the presence of proven bacterial infection, she failed to mount a leukocytosis, although biopsy specimens of infected skin revealed polymorphonuclear leukocytes in the lesion, which resembled an allergic vasculitis. These findings support the view that C3, or its fragments, is required for the mobilization of leukocytes *in vivo* and that some other, still unidentified mechanism acting independently of C3 may effect chemotaxis locally. The disorder clearly demonstrates the requirement for C3 in alternate pathway function.

Other patients with severe C3 deficiency have since been identified and are also unduly susceptible to infection. At least one is a child of a consanguineous mating.

DEFICIENCIES OF C1 SUBCOMPONENTS. A deficiency of the C1q subcomponent of C1 has been found in a number of patients with immunodeficiency and agammaglobulinemia, but this has not been universally observed in immunodeficiency. This deficiency can be due to consumption or hypercatabolism of C1q. Transplantation of normal bone marrow has restored the serum levels of immunoglobulins and C1q in some of these patients.

One interesting case involves a boy presented with a lupus-like syndrome and glomerulonephritis. His serum lacked hemolytic complement activity. This defect could be remedied by adding purified C1q to his serum, which was found to contain a C1q-like protein which was antigenically deficient with respect to normal C1q. Other members of the family demonstrated abnormal C1q, which was also reflected in an electrophoretic alteration and altered amino acid composition, molecular weight, and chain structure. Therefore, this kindred has the ability to synthesize an abnormal C1q molecule. The patient died of overwhelming sepsis; he had antibodies directed against smooth muscles as well as circulating immune complexes.

Three families with inherited C1r deficiency have been found. In one case associated with chronic glomerulonephritis, the C1s levels were also depressed. Patients with inherited C1r deficiency are likely to have lupus-like diseases, focal membranous glomerulitis, arthralgias, and possibly overwhelming infections. Those homozygous for the defect have less than 15 per cent of the normal serum concentrations of C1r by immunochemical measurements; the parents and siblings are heterozygous, and the disorder appears to be inherited as an autosomal recessive trait. It is difficult to detect the heterozygous individuals with certainty because of the wide span in serum levels of C1r in the normal population. Homozygous individuals who have symptoms of lupus-like disorders do not demonstrate positive serological tests characteristic of lupus.

In the absence of C1r, the serum of a homozygous deficient patient would not sustain normal immune hemolysis; immune adherence and serum bactericidal activity were deficient. The alternative pathway function appeared normal.

C2 DEFICIENCY. A number of kinships have been reported in which deficiency of C2 is inherited as an autosomal recessive trait. Although C2 concentrations in serum of homozygous individuals are 4 per cent or less than that of normal serum, the propositi are not usually unduly susceptible to infection and may have no disease. Intermediate levels of C2 (30 to 60 per cent) can be found in serum of heterozygous relatives. Recent evidence from studies of genetic markers in affected kinships indicates linkage of the gene loci for certain histocompatibility antigens and C2.

The serum of persons profoundly deficient in C2 exhibits limited (5 to 15 per cent) immune adherence and bactericidal activity directed toward gram-negative bacilli in the presence of excess antibody *in vitro*. When antibody concentration is not excessive, immune adherence and bactericidal activity in the test tube are more effective.

Despite the fact that some individuals who are severely deficient in C2 may appear entirely well, there are a number who have connective tissue diseases resembling systemic lupus erythematosus, dermatomyositis, membranoproliferative glomerulonephritis, and anaphylactoid purpura. This group of C2-deficient persons appears to have a much higher incidence of connective tissue diseases than would occur in a population of normals.

C4 DEFICIENCY. A strain of guinea pigs deficient in C4 has been studied extensively. Despite the near absence of C4, these animals do survive in a healthy state in a laboratory environment. Except for C1 and C2, which are below normal, the other components of complement in their serum are in normal concentrations. Their capacity to produce γ-globulins is somewhat limited, but this does not appear to represent a disadvantage in resisting infection. Although the classical pathway is markedly defective, the alternative pathway of complement functions normally in serum from these animals. The defect is inherited as an autosomal recessive trait.

A patient with marked deficiency in C4 was found during evaluation of a disease resembling systematic lupus erythematosus who lacked certain serological criteria for this diagnosis. This deficiency is believed to be inherited as an autosomal recessive trait, for the patient's mother had approximately 50 per cent of the normal serum concentration of C4. The alternative pathway function was normal, but concentrations of C1 and C2 in the patient's serum were low. The level of C4 in the patient's serum was much lower than that found in most disorders in which C4 deficiency is acquired because of disease associated with immune complexes. The identification of individuals heterozygeous for C4 deficiency is difficult because of the wide range of C4 levels in the normal population.

A C4-deficient child demonstrated diminished capacity to form antibody after immunization and was unable to amplify antibody formation upon challenge or to undergo the normal transition from the production of IgM to IgG. The reasons for these abnormalities are not clear.

C5 DYSFUNCTION AND DEFICIENCY. A marked deficiency of C5 has been observed by Rosenfeld and his colleagues in a patient with severe systemic lupus erythematosus and repeated bacterial infections. The amount of C5 in her serum was less than 1/1000 of that in normal serum. A sibling, also markedly deficient in C5, had about 1 per cent of the normal serum concentration. The mother and several siblings had intermediate serum concentrations of C5. This defect appears to be inherited as an autosomal codominant trait, but the obligate heterozygotes are difficult to identify with certainty and total hemolytic complement in their serum may be normal or elevated. Serum from C5 deficient individuals does not generate optimal chemotactic activity in the presence of endotoxin or aggregated γ-globulin. One proband with C5 deficiency had a disseminated gonococcal infection. Neither she nor her affected twin sister had evidence of collagen-vascular disease.

An inherited abnormality of C5 appeared to be associated with enhanced susceptibility to infection with certain gram-negative bacteria early in life and eczema or Leiner's disease. Serum from affected persons exhibited defective phagocytosis-enhancing activity of yeast by blood leukocytes. Despite normal hemolytic properties and concentrations of C5 antigen in their serum, the defect in phagocytosis was thought to reflect abnormal C5 function.

Mice deficient in C5 were discovered a number of years ago. The C5-deficient mice have a high incidence of spontaneous leukemia, associated with a leukemogenic virus. Infusion of preparations of C5 into these mice reduced the size of the leukemic lymph nodes and spleen which suggested the destruction of leukemic cells. There is no parallel observation in the case of human leukemia.

C6 DEFICIENCY. A woman severely deficient in C6 had recurrent gonococcal infections and Raynaud's phenomenon, and another with C6 deficiency had recurrent meningococcal meningitis. Thus, these individuals seem to be subject to infections with gram-negative organisms. Bactericidal activity was virtually absent when the serum of such a person was tested against *Salmonella typhi* 0901, but complement-dependent chemotactic activity could be generated in relatively normal amounts, probably providing an important serum defense mechanism for this subject. The parents and siblings of both sexes were heterozygous for this deficiency, and it is apparently an inherited autosomal recessive trait.

In rabbits deficient in C6, a defect in whole blood coagulation involving the participation of the platelet in clotting has been described, in the absence of susceptibility to hemorrhage or infection. Although the clotting time of whole blood was delayed, that of rabbit *plasma* was normal. Humans with C6 deficiency do not demonstrate a coagulation defect. This failure to find such a defect in human C6 deficiency, in contrast to rabbit C6 deficiency, may be a reflection of one of the many differences between human and rabbit platelets. After immune adherence was originally described, it was noted that only nonprimate platelets could function in this stickiness reaction with cells coated with antibody and C$\overline{1-3}$. Therefore the human platelet probably does not react with C3 or participate in subsequent reactions which depend upon C6 to facilitate coagulation through alterations of the platelet membrane.

A combined hereditary deficiency of C6 and C7 has been found, and although the genetic implications of this combined deficiency are important, the propositus is apparently healthy.

C7 DEFICIENCY. Serum deficiency of C7 has been found in two patients. This was apparently due to an autosomal recessive trait, relatives having intermediate serum concentrations of C7. Although one patient had a disease resembling scleroderma with Raynaud's syndrome, some others in the kindred with this syndrome were not deficient in C7. Immune adherence was normal, but defects in serum bactericidal activity were detectable. The partial defect in chemotactic activity in serum from C7-deficient persons could be corrected *in vitro* with purified C7.

C8 DEFICIENCY. Persons with C8 deficiency have been reported to have disseminated gonococcal infections, and the serum of one such individual lacked bactericidal activity directed against *Neisseria gonorrhoeae*; purified C8 restored this activity. The deficiency of C8 is inherited as an autosomal trait; heterozygous individuals are difficult to identify with certainty because of the wide range of C8 concentrations in normals. Another C8 deficient proposita had xeroderma pigmentosum, and a third individual had systemic lupus erythematosus. Therefore,

this deficiency seems to be associated with a susceptibility to infections with gram-negative organisms as well as to connective tissue disorders.

C9 DEFICIENCY. A deficiency of C9, inherited autosomally, has not been associated with obvious clinical disease. Although sera deficient in C9 accomplish lysis of sensitized sheep cells more slowly than do normal serums, and bactericidal activity proceeds at a slower rate than normal, these defects do not apparently compromise the well-being of the propositus.

The clinical disorders associated with hereditary deficiencies of plasma substances can provide clues to the physiological functions of the missing proteins. In the case of deficiencies of individual complement components, there is a high incidence of connective tissue disorders of varying types in persons who are deficient in the early-acting components of the classical pathway, C1 subcomponents C4 and C2. There is probably also a tendency to overwhelming infection. Even so, deficiencies of C2 have been detected in individuals who are otherwise well. The incidence of cutaneous lupus in patients with hereditary angioneurotic edema, deficient in C$\overline{1}$-INH and secondarily deficient in C4, far exceeds that which would be predicted as a chance occurrence. One must bear in mind that in most instances complement deficiencies have been revealed because of the disease which originally led them to medical consultation. Even though this factor may influence our views of the role of complement deficiency in some diseases, it is quite clear that persons with a severe deficiency of C3—either because of failure of its synthesis or because of its hypercatabolism—are unusually susceptible to overwhelming infections. C3 is clearly an essential component in the mechanism of defense against invading microorganisms.

Table 30–5 summarizes the nature of disorders associated with hereditary de-

TABLE 30–5
Clinical Disorders Associated With Inherited Deficiencies of the Complement System

Deficient Component	Associated Clinical Disorders
C1q	Renal disease, lupus-like syndrome
C1r	Systemic lupus erythematosus (SLE), glomerulonephritis
C2	No disease in about one-third of cases, immune complex diseases, tendency to infection
C3	Syndrome of severe immune deficiency with infection
C4	SLE or lupus-like syndrome
C5	SLE, *Neisseria* infections
C6	*Neisseria* infections, glomerulonephritis, Raynaud's syndrome, rheumatoid arthritis
C6 and C7	No disease
C8	*Neisseria* infections, xeroderma pigmentosa, SLE, lupus-like syndrome
C9	No disease
C3b Inactivator (I)	Severe immunodeficiency syndrome with infection
C1-INH	Hereditary angioneurotic edema, immune complex disorders

fects of the components of the complement system.

The role of complement in *in vivo* handling of immune complexes

It is likely that inherited deficiencies in the complement system can contribute to the development of immune complex diseases. When one recognizes that phagocytic cells have complement receptors on their membranes, it is logical to reason that if the complement complexes cannot be assembled properly on immune aggregates, the phagocytes cannot then remove them from the circulation. In addition, Miller and Nussenzweig have shown that complement actually has a detergent effect upon immune precipitates. Thus, immune aggregates bearing the alternative pathway C3-convertase (C3b,Bb,P) cleaved C3 in the fluid phase, and one of the cleavage products of this reaction accumulated on the lattice structure of the immune aggregates. C2 and C4 were not absolutely required for the solubilization observed, but in the presence of the C4b,2a C3-convertase, solubilization was greatly accelerated. This reaction was probably effective because it afforded further C3 cleavage to generate more cleavage products, thus seeding a further reaction through the alternative

pathway. Upon examination of the solubilized complexes, they were found to consist of antigen, antibody, C3- and C4-derived peptides, properdin, C4-binding protein, and probably other components. They did not contain C5, and it is therefore likely that these aggregates did not bind the C5b,6,7, complex.

The most important solubilizing components are apparently peptides associated with the complex which are derived from C3 and C4. These appear to interfere with the binding of antigen to antibody, for most of the C3 peptides were associated with the H-chain of the immunoglobulin present in the complexes. Although these experiments were performed *in vitro*, similar solubilization patterns were found when antigen-antibody-complement complexes were prepared under conditions close to those which are likely to occur in the body.

It is well known from the studies of Frank and Brown and their colleagues that both C3 and immunoglobulins, through their Fc receptors, facilitate the phagocytic clearance of erythrocytes coated with antibody and complement from the circulation of animals. It is likely that complement is also important to the clearance of soluble immune complexes, for their processing by macrophages *in vitro* can be enhanced by the presence of complement. This enhancement is probably due to the interaction of C3 with its receptors upon the cell.

The concept that complement is essential for the handling of immune reactants by the body, through solubilization and removal from the circulation, is important in the context of human disease. In view of the tendencies to connective tissue disorders, immune complex diseases, and so forth in individuals who have inherited deficiencies of one or another complement component, it is logical to propose that the absence of effective solubilization and clearance mechanisms, owing to the compromised action of the complement system, may be causative factors.

Acquired defects of complement

Acquired alterations of serum hemolytic complement activity occur in numerous disorders. This summary is not intended to define each situation, but to cite some examples sufficiently well studied to provide some understanding of the mechanism of the deficiency.

In evaluating the reasons for a deficiency of a given component of complement, one must bear in mind that the serum level at any given time is a reflection of the rate of synthesis of the complement component and its rate of destruction, or catabolism, and excretion. Therefore, transient depression of synthesis of a component could be responsible for a decreased total serum hemolytic complement titer and a decrease in the titer of that specific component. The third and fifth components of complement are often depressed during the course of membranoproliferative glomerulonephritis. Since the rate of disappearance of intravenously administered purified human C3 from the blood may be normal, the low titer of C3 may be due, at least in part, to its depressed synthesis. On the other hand, consumption of C3 may occur through the alternative pathway involving the "nephritic factor" or through the classi-

cal pathway. During the course of acute poststreptococcal glomerulonephritis, the titer of complement may drop sharply early in the disease because of activation of the classical pathway and consumption of C4 and C2, presumably by immune complexes. Later in the disease, C3 may become depressed, perhaps because of impaired synthesis.

Other disorders believed to be initiated, or worsened, by the formation of immune complexes in the body can also be associated with diminished concentrations of serum complement. These disorders may reflect consumption of complement through its interaction with the antibody involved in the immune complexes characterizing the disease. Thus, in some patients with systemic lupus erythematosus, the accumulation of immune complexes in the microvasculature, particularly in the renal circulation, may initiate complement consumption and inflammation.

Certain hemolytic anemias appear to evolve through complement-mediated hemolysis. In paroxysmal cold hemoglobinuria (PCH), antibody and complement attach to the erythrocytes at low temperatures. In this disorder, the Donath-Landsteiner test provides a means of measuring the requirements for cold-induced hemolysis. The first component of complement and cold-reactive antibody attach to the erythrocyte during incubation at 0° C.; upon warming, later-acting components become active and induce lysis of the cell.

In another hemolytic disorder involving complement, paroxysmal nocturnal hemoglobinuria (PNH), the alternative pathway operates. In this disorder the destruction of erythrocytes is dependent upon the activation of C3 through the alternative pathway and consequent utilization of C5 to C9. This apparently occurs because of an abnormality of the erythrocyte, and not because of an antibody to the erythrocyte. In PNH, other blood cells, including platelets, are also unduly susceptible to the action of complement. Although the abnormality of the erythrocyte membrane is not well understood, increased amounts of C3b are bound to the surface of PNH erythrocytes, and the cells are more susceptible to the destructive action of the later-acting components of complement. Although complement may become activated and utilized in many parts of the body, alterations in serum concentrations may not occur because of enhanced synthesis. In certain persons with rheumatoid arthritis, the breakdown products of C3 can be found in synovial fluids, suggesting that complement has been utilized *in situ* despite the fact that the serum levels of complement are not lowered. In some other disorders, the formation of aggregates of γ-globulins, or of immune complexes within the vascular compartment, can lead to obvious consumption of complement. Thus, in certain patients with circulating cryoglobulins (immunoglobulins insoluble at low temperatures) the globulins may aggregate and activate complement. Urticaria may appear when the patient is exposed to cold, apparently because of complement activation in the circulation of the skin.

In overwhelming illnesses associated with shock, disseminated intravascular coagulation may occur in association with depressed serum complement levels. The activation of complement may worsen the shock syndrome because of the release of vasoactive substances such as anaphylatoxins from complement components. There are many causes of shock, and observations of changes in function of plasma mechanisms indicate involvement of multiple plasma enzyme systems active in blood coagulation, kinin release, and complement through mechanisms not necessarily involved with antigen-antibody reactions. It is difficult to determine whether altered enzymatic activities are the cause or the result of the shock syndrome.

PART 3 FIBRINOLYSIS

Charles W. Francis, M.D.
Victor J. Marder, M.D.

Fibrin formation is a central feature of inflammation, tissue repair, and hemostasis, which are basic physiological processes by which the body protects its integrity and repairs damage. These reactions are temporary, and their effects are curtailed or reversed in order to restore normal tissue function and anatomy when the inciting stimulus is removed. Thus, a fibrin clot which forms quickly in a torn blood vessel to stem the loss of blood must be remodeled and removed so as to restore blood flow. This is accomplished by a combination of cellular reorganization (revascularization) and by the fibrinolytic system, which controls the enzymatic degradation of fibrin deposits by a modulated action of plasma and tissue protein components. The coordinated action of activators, zymogens, enzymes, and inhibitors provides for local activity at sites of fibrin accumulation, while at the same time, systemic activity of the proteolytic enzyme plasmin is avoided.

Components

PLASMINOGEN. *Plasminogen* is the inactive precursor of the proteolytic enzyme plas-

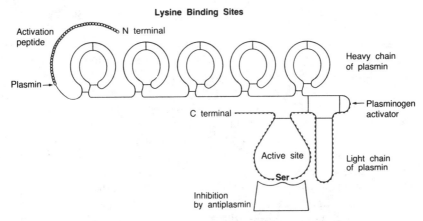

FIG. 30–6. Schematic representation of the plasminogen molecule. Plasminogen activator cleaves a single peptide bond, converting the single-chain inactive plasminogen molecule to the two-chain enzyme plasmin. The N-terminal activation peptide which is released by plasmin action converts glu to lys plasminogen, which is then more susceptible to cleavage by plasminogen activator. Lysine binding sites responsible for binding of plasminogen and plasmin to fibrin are found on the heavy chain which contains the five repetitive "kringle" structures. The light chain of the plasmin molecule contains the active enzyme site, which can be inhibited by antiplasmin. Antiplasmin also interferes with binding of plasminogen to fibrin through the lysine binding sites.

min. It is a single-chain molecule of molecular weight 93,000 and its plasma concentration is 1.5 to 2 μM.; it is also present in all body fluids and secretions in smaller amounts. The entire sequence of 790 amino acid residues has been determined, as has the location of disulfide bonds and functional regions, including those involved in binding to fibrin, activation to plasmin, and proteolysis after activation (Fig. 30–6). Plasminogen localizes in areas of fibrin deposition, primarily because of specific lysine binding sites which mediate attachment to fibrin as it polymerizes into a clot. This action has been located on the "kringle" portions of plasminogen, which are repeat loops of amino acids between residues 79 and 439. The inhibitory action of the physiological fibrinolytic inhibitor, α_2-plasmin inhibitor, and pharmacological inhibitors of fibrinolysis such as ε-aminocaproic acid (EACA) resides in their ability to bind to these lysine binding sites, thereby preventing plasminogen from attaching to its primary substrate, fibrin.

PLASMINOGEN ACTIVATORS. *Plasminogen activators* of human origin are proteolytic enzymes which are capable of activating the fibrinolytic system by converting plasminogen to plasmin. They are widely distributed in body tissues (tissue plasminogen activator) and are detectable in the blood, presumably after release from endothelial cells (vascular plasminogen activator). The tissue and vascular plasminogen activators are termed extrinsic activators because of their origin in tissues rather than the blood, have molecular weights of 60,000 to 70,000, and share the ability to bind specifically to fibrin. This is an important mechanism in directing fibrinolytic activity toward dissolution of fibrin in clots with relatively little protease effect on plasma proteins such as fibrinogen.

A second group of plasminogen activators derives from proteins of the intrinsic coagulation system present in plasma. The central component is activated Hageman factor (factor XII), which has the potential for simultaneous activation of the fibrinolytic pathway in addition to activation of the coagulation, kinin, and complement systems, thereby providing a common link between four systems involved in the inflammatory response (see Part 1 of this chapter).

Other activators of fibrinolysis can be purified and administered intravenously ("exogenous" activators), the two most commonly used agents being streptokinase (SK) and urokinase (UK). SK is a protein of 47,000 molecular weight which is purified from culture broth of the β-hemolytic streptococcus. This agent binds to plasminogen and nonenzymatically converts it to an SK-plasminogen activator complex. In contrast, UK is a trypsin-like enzyme purified either from huge volumes of human urine or from tissue culture broth of human fetal kidney cells. This protein has two major molecular weight forms of 54,000 and 32,000, both of which can directly convert plasminogen to its protease form, plasmin. The exogenous activators are less specific than the extrinsic activators, in that they bind equally well to fibrinogen and fibrin and therefore degrade both fibrin in clots and fibrinogen in blood equally well.

PLASMIN. *Plasmin* is an endopeptidase which hydrolyzes susceptible arginine and lysine bonds in proteins at neutral pH and acts upon most synthetic substrates and proteins susceptible to digestion by trypsin. Plasmin degrades fibrinogen and fibrin and can also hydrolyze other plasma proteins such as coagulation factors V (proaccelerin) and VIII (antihemophilic factor), serum complement components, adrenocorticotropic hormone (ACTH), growth hormone, and glucagon. Although the physiological sequence of *in vivo* cleavages that convert plasminogen to plasmin is still not certain, the critical bond to be broken is that between arginine 560 and valine 561. This minimal proteolytic event generates a two-chain disulfide-bonded active enzyme from the inactive zymogen. The smaller (light) chain of approximately 25,000 molecular weight encompasses the C-terminal residues 561 to 790, and contains the enzymatically active histidine/asparagine/serine site. The heavy chain possesses the lysine binding sites.

INHIBITORS OF FIBRINOLYSIS. The principal plasma inhibitor of plasmin is α_2-plasmin inhibitor, a single-chain glycoprotein of molecular weight 67,000, present at a plasma concentration of 1 μM. It inactivates plasmin very rapidly by binding irreversibly to its active site, in the process forming an equimolar inactive complex. Inhibition of fibrinolysis is perhaps facilitated by the binding of the inhibitor to the lysine binding site of plasmin as well, an action that interferes with the attachment of plasminogen or plasmin to fibrin. When the lysine binding site of plasmin is already attached to fibrin, α_2-plasmin inhibitor is an ineffective agent, either because it is unable to orient to the plasmin enzyme site properly or because the latter is "protected" by the fibrin structure.

Other plasma proteolytic inhibitors are of less importance as regulators of fibrinolysis. Although α_2-macroglobulin binds to plasmin and forms a complex with fractional residual enzymatic activity, its affinity for plasmin is much less than that of α_2-plasmin inhibitor, and it functions as an antiplasmin only if the inhibitory capacity of α_2-plasmin inhibitor is exceeded. Other plasma proteins with inhibitory activity *in vitro*, including antithrombin III, α_1-antitrypsin and Cl esterase inhibitor, have even less physiological importance as antiplasmins. The existence of inhibitors of extrinsic plasminogen activators has not been well established, and their role in the control of fibrinolysis is speculative. The plasma concentration and affinity of α_2-plasmin inhibitor for plasmin represent the principal protection of the organism against systemic fibrinolysis and the proteolytic effects of plasminemia when the fibrinolytic system is activated at a local site.

Synthetic lysine analogues such as EACA are commercially available and have been extensively applied in the treatment of many hemorrhagic disorders in which fibrinolysis may be contributing to the disordered hemostatic process. These inhibitors exert their effect solely by binding to the lysine-binding sites of plasminogen, thereby preventing attachment to fibrin. Curiously, EACA may promote the conversion of plasminogen to plasmin *in vitro*. However, even if this mechanism operates *in vivo* as well, the overall effect of the agent is to inhibit fibrinolysis, since even an increased yield of plasmin molecules would still be prevented from binding effectively to their preferred substrate, fibrin.

Activation of fibrinolysis

INTRINSIC SYSTEM. The plasminogen activators are highly specific, having restricted activity to hydrolyze the few bonds which are cleaved in the conversion of plasminogen to plasmin. As with the coagulation system, both intrinsic and extrinsic systems of plasminogen activation have been described. The intrinsic system functions with components available in plasma and involves the contact activating system, namely factor XI (plasma thromboplastin antecedent, or PTA), high-molecular-weight kininogen, prekallikrein, and Hageman factor. In a reciprocal reaction, plasmin can further activate Hageman factor and amplify the fibrinolytic process during inflammation. The physiological importance of the Hageman factor–dependent system of plasminogen activation is not certain, and no clearly identifiable pathological state of impaired fibrinolysis has been identified in patients with profound defects in their contact activating system, even though the original Hageman factor–deficient patient died of thrombotic disease.

EXTRINSIC SYSTEM. The plasminogen activators present in most human tissues are serine proteases which bind avidly to fibrin and express greater enzymatic activity in the presence of fibrin, properties which enhance the fibrinolytic potential at sites of fibrin deposition. Endothelial cells play the major role in generating plasminogen activator in response to local vascular fibrin formation, primarily because of their proximity to intravascular fibrin thrombi. Vascular plasminogen activator has been localized to the endothelial cell lining of vessels, with capillaries having the largest surface-to-volume ratio of endothelial cells, the greatest local amount of activator, and the greatest fibrinolytic potential. The capillary endothelium is, therefore, well equipped to maintain the patency of the microcirculation. The "blood activator" which can be assayed in freshly drawn plasma is thought to represent secreted vascular activator from endothelial cells. Although the basal level of activator that can be normally measured in plasma is low, its concentration can be markedly increased by physiological stimuli or pharmacological intervention.

On a molecular basis, physiological fibrinolysis begins with the attachment of both plasminogen and α_2-plasmin inhibitor to fibrin as it clots, and the relative actions of these fibrin-bound constituents may influence the rate of fibrin dissolution. The fibrinolytic effect is fostered by conformational changes in bound plasminogen that make it more susceptible to activation by vascular plasminogen activator as well as more protected against α_2-plasmin inhibitor. Plasminogen activators cleave plasminogen at the arginine 560–valine 561 bond, and with this molecular change, the protease plasmin which is capable of degrading fibrin is produced. Since such plasmin is already bound to lysine residues of fibrin, enzymatic activity resulting in clot dissolution is efficient. Plasmin also has the property of converting plasminogen from its intact glutamine form to a lysine form, by cleavage at several spots near to the N-terminus (e.g., lysine 76–lysine 77), liberating an activation peptide with an approximate molecular weight of 8,000. The

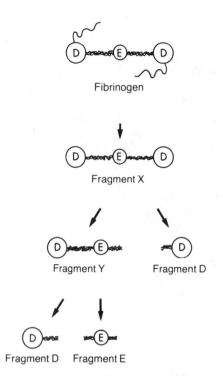

Fibrinogen

Fragment X

Fragment Y   Fragment D

Fragment D   Fragment E

FIG. 30–7. Asymmetric degradation of fibrinogen by plasmin. The fibrinogen molecule is depicted as a trinodular structure with a central E domain and two lateral D domains connected by an α-helical, coiled-coil strand. The carboxy-terminal polar appendages of the Aα chains are shown exiting from the D domains. The initial action of plasmin removes the Aα polar appendage, converting fibrinogen to Fragment X. Subsequently, cleavages occur in the coiled-coil region, initially cleaving Fragment X asymmetrically to Fragments Y and D. Cleavages in the coiled-coil region of Fragment Y further degrade it to a second Fragment D and Fragment E.

lysine-plasminogen form is much more sensitive than the glutamine form to activator conversion to plasmin, thereby providing a feedback mechanism for accelerating the formation of plasmin from plasminogen.

Fibrinogen and fibrin degradation

MOLECULAR CHANGES. Plasmic degradation of fibrinogen and fibrin is determined primarily by the number and accessibility of critical cleavage sites. Although there are potentially over 300 arginine and lysine

bonds susceptible to plasmin hydrolysis, only 50 to 60 are cleaved during degradation and an even smaller number of cleavages is responsible for breaking the intact molecule into fragments. Fibrinogen has a molecular weight of 340,000 and a plasma concentration of 200 to 400 mg./dl. It consists of three pairs of polypeptide chains, termed Aα, Bβ, and γ, with approximate molecular weights of 67,000, 56,000, and 47,000, respectively. There are three major domains, a central domain in which the two halves of the molecule are joined by disulfide bonds at their N-terminal regions (Fig. 30–7), connected on each side by α-helical "coiled coils" to the terminal domains containing the C-terminal portions of each set of three polypeptide chains. A long extension of the Aα-chain exits from each terminal domain, which consists mostly of β- and γ-chain components. The major plasmin-susceptible sites that account for the cleavage of fibrinogen into degradation products are located on the Aα-chain extension and midway along the coiled coil between central and terminal domains.

Initial plasmin cleavages occur at several points along the exposed, carboxy-terminal appendage of the Aα-chain, liberating a series of peptides and leaving behind a remnant of the chain with a molecular weight of 25,000. This Aα-chain cleavage is the principal distinction between intact fibrinogen and the fragment X group of derivatives, the smallest of which has a molecular weight of 250,000. The next cleavages involve all three of the chains in the protease-sensitive portion of the coiled coil (Fig. 30–7). Since one of the coiled coils is usually broken before the other, degradation of fragment X is asymmetrical, and it is split into unequal parts, yielding one fragment D and one fragment Y with molecular weights of 100,000 and 150,000, respectively. Fragment Y is a transient intermediate degradation product that is quickly split by plasmin into a second fragment D moiety and fragment E, which represents

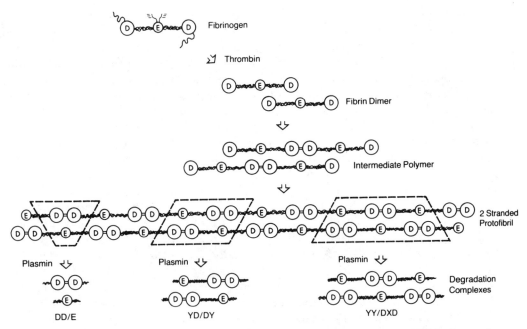

FIG. 30–8. Schematic representation of fibrin polymer formation and plasmic degradation of cross-linked fibrin. After thrombin liberation of the fibrinopeptides A and B from the N-terminal ends of Aα and Bβ chains, fibrin monomers polymerize by a half-staggered overlap sequence to form dimers, intermediate polymers, and then long two-stranded protofibrils that coalesce by lateral attractions into a fibrin fiber. Factor XIII$_a$ catalyzes the formation of crosslinks between γ chains of contiguous terminal domains, stablizing the clot. Each α chain in the polymer is also crosslinked by Factor XIII$_a$ to one or two other α chains, forming an α-polymer network that renders the clot more resistant to lysis by plasmin. For clarity, the Aα polar appendage is shown only for fibrinogen and the α polymer network is not depicted. Plasmic degradation of a long two-stranded protofibril results in a series of complexes (outlined by interrupted lines), the smallest of which is DD/E. The fragments which make up the complexes are held together by noncovalent attractions between binding sites on the D and E domains of adjacent fibrin polymer strands.

the central domain standing alone. Fragment E has a molecular weight of 50,000 and contains the disulfide-bonded N-terminal part of all six polypeptide chains of the original fibrinogen molecule.

The distinction between degradation products of fibrinogen and those of fibrin mostly reflects the formation of covalent factor XIIIa cross-links between fibrin monomers and noncovalent attractive forces between central and terminal domain polymerization binding sites. Fibrin formation is initiated by thrombin, which liberates fibrinopeptides A and B from the central domain. This changes the charge properties of the resultant fibrin monomers and at the same time exposes binding

sites on the α- and β-chains. Both of these effects result in lateral attraction and binding of central and terminal domains (Fig. 30–8), which produces a stepwise longitudinal growth of a fibrin polymer. With the end-to-end approximation of terminal domains, portions of the C-terminal regions of γ-chains interact and are covalently bonded by factor XIIIa. The growing fibrin protofibrils then coalesce by lateral attractive forces into a fibrin fiber, which is further "cross-linked" by factor XIIIa action, producing an α-chain polymer network that effectively stabilizes the clot. Plasmin degradation of a cross-linked fibrin clot differs from that of fibrinogen in two ways. First, degradation is slower because of the cross-linked bonds that join the monomers into a complex, interlacing structure and holds them together in the face of proteolytic cleavage. Second, the degradation products themselves are distinctive because of the covalent and noncovalent bonds that hold the domains together even after solubilization of the clot.

On the other hand, the unique fragments are not the result of any change in the specific cleavages caused by plasmin. As with fibrinogen, initial cleavages degrade the cross-linked α-chain polymer at the same protease-sensitive sites, following which cleavages occur in the connecting coiled coil region between central and terminal domains, again at the same location as in fibrinogen. The "backbone" degradation products consist of domains linked longitudinally by coiled coils or by cross-link bonds and laterally by noncovalent bonds into complexes that are liberated from fibrin only after the α-chain superstructure has been dismantled (Fig. 30–8). Each complex consists of two fragments, held together by the same forces between central and terminal domains that operated to form the protofibril. The constituent fragments of each complex represent portions of linked fibrin monomers, the terminal domains of which are still joined by cross-linked bonds. The smallest unique degradation product of cross-linked fibrin is fragment DD, which consists of two fragment D moieties joined by cross-linking. The smallest complex is a combination of fragment DD with fragment E, joined together by noncovalent bonds. The larger complexes are simply combinations of longer portions of each polymer chain of the protofibril. Larger complexes can be progressively degraded by plasmin *in vitro* to smaller complexes, but it is likely that the larger complexes represent the dominant forms *in vivo* following clot dissolution, since α_2-plasmin inhibitor would limit further plasmic action.

TESTS FOR FIBRINOGEN AND FIBRIN DEGRADATION PRODUCTS. Degradation products may appear in the blood as the result of two basic processes. First, lysis of a physiological or pathological blood clot releases soluble fragments of the fibrin network into the blood, which then circulate until cleared. Second, excess plasmin activity which appears in the blood in a variety of pathological conditions degrades fibrinogen to fragments. The methods which have been developed to detect these derivatives are based on their functional, immunological, or structural properties.

Functional tests for degradation products of fibrinogen or fibrin depend upon their anticoagulant properties or their propensity to form high-molecular-weight complexes with fibrinogen. Tests of clotting function such as the thrombin time, which measures the rate at which thrombin clots the plasma sample, would assess the anticoagulant effect of degradation products, since the fragments interfere with polymerization of fibrin monomers. However, such assays are nonspecific, as other abnormalities such as a decrease in plasma fibrinogen concentration similarly could produce a prolonged clotting time. Complexes of degradation products with fibrinogen or fibrin monomer are larger and less soluble than fibrinogen under certain conditions. The presence of such complexes can be inferred from the formation of precipitates in the presence of ethanol or protamine sulfate. Molecular sieve chromatography of plasma samples on calibrated columns has also been used to separate fibrinogen from its derivatives or complexes by differences in their molecular weights, but this method is too cumbersome for routine clinical application.

The most specific and sensitive methods employ immunological techniques. These assays use serum as the test material to avoid interference or cross-reaction with plasma fibrinogen. Incoagulable proteolytic derivatives of fibrinogen share antigenic determinants with the undegraded parent molecule and can be quantitatively assayed in serum after removal of fibrinogen by clotting. Blood is collected in the presence of excess thrombin to ensure clotting and removal of soluble fibrin polymers which react with the antiserum, and with EACA to prevent *in vitro* fibrinolysis. Two commonly employed tests are the tanned red cell hemagglutination inhibition immunoassay (TRCHII) and the latex particle flocculation test. The TRCHII uses fibrinogen-coated tanned erythrocytes that can be agglutinated by antifibrinogen antiserum. Interference with this reaction by prior absorption of the antiserum with the test serum is evidence that degradation products were present in the sample, and by implication, in the circulation. The test detects as little as 0.5 μg./ml. of fibrinogen or degradation product ("fibrinogen-related antigen") and may be positive in patients with a variety of disorders, in some of which the abnormality of fibrinogen or fibrin lysis is of relatively minor clinical importance. Higher levels are seen in patients with intense fibrinolysis, in which case the degradation products interfere with hemostasis. The latex particle flocculation test uses antibody-coated latex particles which are clumped by degradation products that react with the antibody. Reactions with test serum are compared with the reaction of a standard fibrinogen preparation to quantitate the concentration of degradation products. This assay is exceedingly simple and rapid, and therefore quite useful in many clinical situations.

Since fibrin formation and lysis occur with inflammation, hemostasis, and tissue repair, processes which are common to many pathological conditions, it is not surprising that low concentrations of degradation products are found in samples from many sick patients. For example, about 10 per cent of patients with systemic lupus erythematosus have elevated levels of degradation products without evidence of thrombosis or fibrinolysis, and a similar proportion of patients with malignancy have positive tests, perhaps reflecting the accumulation of fibrin on or around malignant cells. Thus, detection of degradation products does not imply any derangement of the coagulation or lytic systems, unless the concentration is markedly elevated. Clinically available tests do not distinguish between degradation products of fibrinogen and those of fibrin, since these derivatives share immunological determinants and cross-react in the assays employed. However, the presence of fibrinogen versus fibrin derivatives in the blood has important pathophysiological implication, since the former would result from plasminemia and the latter from thrombolysis. Since these conditions are managed differently, the potential to identify the source of degradation products could be of major clinical usefulness should such an assay be developed.

Control of fibrinolysis

PHYSIOLOGICAL BALANCE. The control of fibrinolysis must provide for intermittent activation at local sites of fibrin deposition in association with thrombosis or inflammation, without initiating a *systemic* state of proteolysis with damaging effects on coagulation proteins. The mechanisms which are responsible for safe but effective thrombolysis are illustrated in Figure 30–9 and include: (1) local secretion of plasminogen activator, (2) binding of both plasminogen and plasminogen activator to fibrin, and (3) highly effective circulating plasma inhibitors of plasmin. The local secretion of plasminogen activator by endothelial cells in close contact with a thrombus, or by macrophages or leukocytes in areas of inflammation, activates fibrinolysis only at the site of need within the fibrin network. Both activator and plasminogen bind specifically to fibrin and thereby increase their effective concentration locally, resulting in a greatly accelerated generation of plasmin. The plasmin produced by local activation is bound to its physiological substrate, a situation which both accelerates fibrinolytic action and prevents any systemic proteolytic effect. Both the lysine binding site and the active enzymatic center of plasmin are occupied by the fibrin substrate, making

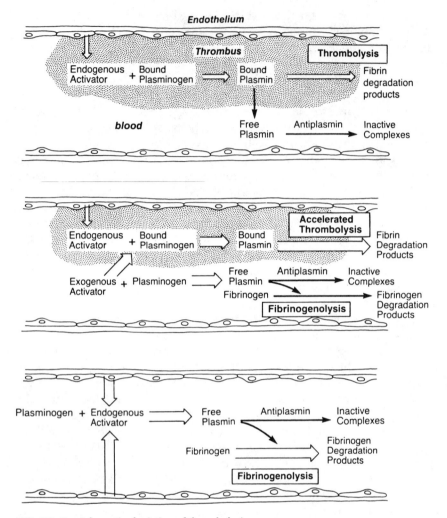

FIG. 30—9. Schematic depiction of thrombolysis.

(*Top*) Physiologic thrombolysis of a nonoccluding thrombus in a blood vessel. Endogenous activator released from the endothelial cells is bound to the fibrin matrix of the thrombus, where it converts plasminogen already attached to the clot to the active fibrinolytic enzyme plasmin. Plasmin degrades the fibrin, resulting in clot solubilization and releasing fibrin degradation products. Any plasmin that is released from the thrombus into the circulation is quickly inactivated by α-2 plasmin inhibitor to form inactive complexes.

(*Middle*) Fibrinolytic therapy. Large amounts of exogenous activator circulate in the blood, some of which diffuses into the clot, increasing the rate of conversion of plasminogen to plasmin and accelerating thrombolysis. Circulating activator converts plasminogen to free plasmin in large quantities which exceed the capacity of plasmin inhibitors to inhibit the protease action of plasmin. Plasma fibrinogen is then degraded, producing fibrinogen degradation products.

(*Bottom*) Primary hyperfibrinogenolysis. A similar pathological condition of hyperfibrinogenolysis may occur in patients with an absence of plasma α-2 plasmin inhibitor, in which case a normal amount of vascular (endogenous) plasminogen activator may be released from endothelial cells and circulate in the blood uninhibited.

them unavailable for inactivation by α_2-plasmin inhibitor in the surrounding blood. The net result of the activation of plasmin while it is attached to the fibrin surface is to increase greatly its speed of activation, augment its proteolytic activity, and prevent its inactivation. Such acceleration of activity by complex formation on surfaces has parallels in the coagulation system, in which several proteins bind to platelet surfaces to effectively increase their concentration and enhance reaction velocity (see Part 1 of this chapter).

An additional element of local control is produced by the covalent binding of α_2-plasmin inhibitor to fibrin through the action of factor XIIIa, which perhaps contributes to a controlled, slower rate of lysis than would occur with non-cross-linked fibrin clots or in patients who lack the plasma inhibitor. The prevention of systemic fibrinolytic activity is ensured by plasma inhibitors of plasmin. Although α_2-plasmin inhibitor constitutes only a small percentage of the total plasmin inhibitory capacity of plasma, it has an extremely high affinity for plasmin with the result that even small amounts of plasmin which appear in the circulation are rapidly inactivated. If larger amounts of enzyme are generated and exceed the capacity of the inhibitor, backup inhibitors are available for added protection, chief among which is α_2-macroglobulin.

SYSTEMIC HYPERFIBRINOLYSIS. Conditions associated with evidence of increased fibrinolytic activity may result from a primary increase in the fibrinolytic response, inadequate clearance or inhibition of activator, or administration of plasminogen activator. The latter case is the most straightforward example of excessive systemic fibrinolytic activity, namely, that which results from the administration of a fibrinolytic agent such as SK or UK during the therapy of thrombotic disease (Fig. 30–9, *middle*). Unlike endogenous plasminogen activators, UK and SK do not preferentially

activate plasminogen that is bound to fibrin, and circulating plasma plasminogen is converted to plasmin as well. The plasma inhibitory mechanisms are overwhelmed by the high concentrations of activator and plasmin, resulting in the degradation of fibrinogen to fibrinogen degradation products. Results of tests such as the plasma clot lysis time shorten dramatically, reflecting the high concentration of circulating plasminogen activator, and those of clotting tests such as the thrombin time, prothrombin time, or activated partial thromboplastin time (PTT) prolong because of the hypofibrinogenemia and the anticoagulant degradation products. Hemorrhagic complications may be produced by two mechanisms. First, accelerated lysis of clots may dissolve hemostatic plugs at sites of vascular injury, and second, the disordered coagulation system is unable to respond properly to a bleeding stimulus with an effective hemostatic plug.

Primary hyperfibrinogenolysis may arise by the release of large amounts of plasminogen activator into the circulation, either from neoplastic tissues that are rich in tissue plasminogen activator or as the result of potent stimuli to the endothelium, such as trauma, anoxia, or shock, that release vascular plasminogen activator (Figure 30–9, *bottom*). The former situation is usually a chronic condition that may be manifest by only laboratory abnormalities, while the latter condition often is explosive in its clinical presentation of hemorrhagic manifestations. In either case, the amount of activator entering the circulation may exceed the plasma inhibitory capacity, and a sequence of events similar to that which occurs with administration of UK or SK may result in bleeding. A related mechanism for developing primary hyperfibrinogenolysis occurs in hepatic cirrhosis. In this situation, normal amounts of vascular plasminogen activator enter the circulation, but hepatic clearance is defective to the extent that the activator accumulates and results in hyperplasminemia.

Antifibrinolytic agents such as EACA may be useful in treating primary fibrinogenolysis by supplementing the capacity of plasma inhibitory mechanisms. These agents inhibit the binding of plasminogen and plasmin to fibrinogen and fibrin by attaching to the lysine binding sites, thereby limiting the proteolytic and clinical effects of excessive circulating plasminogen activator. Enthusiasm for their application must be tempered, however, by the difficulty in distinguishing between primary fibrinogenolysis, where these agents are useful, and disseminated intravascular coagulation (DIC), for which they are potentially dangerous and contraindicated.

Rarely patients have been described with a congenital deficiency of α_2-plasmin inhibitor, a condition which resembles those produced by the presence of excess plasminogen activator (Fig. 30–9). The disorder is inherited in an autosomal recessive fashion and affected patients have a lifelong bleeding tendency that may be manifest as easy bruising and excessive bleeding after minor trauma or surgery. Routine coagulation tests are normal, as is the concentration of fibrinogen, plasminogen, and degradation products. The principal abnormality in laboratory testing is a shortening of the plasma clot lysis time. Bleeding is the result of rapid lysis of physiological hemostatic plugs which are needed at sites of vascular injury, and the lack of α_2-plasmin inhibitor cross-linked to the fibrin itself may be the mechanism for such uncontrolled or premature local fibrinolysis of hemostatic plugs. Treatment is lifelong administration of an inhibitor such as EACA which effectively prevents bleeding.

REACTIVE FIBRINOLYSIS. Activation of the fibrinolytic system occurs normally in response to the presence of fibrin. Although the fibrinolytic response is not excessive or inappropriate, the reaction may contribute to bleeding, mainly because of the large quantity of degradation product that can be released into the circulation. The pre-disposing condition can be divided into two groups, those with a single site of massive fibrin dissolution, and those with multiple or generalized sites of fibrin deposition and dissolution. Limited fibrinolysis may also contribute to bleeding in areas of injury in patients with defective hemostasis.

An example of the first situation is that of abruptio placenta, in which the local clot accumulation is massive and threatens the fetus. An example of the second situation is DIC, in which fibrin clots form in the blood and are cleared by the microcirculation, impairing multiple organ function and resulting in coma, oliguria, and skin necrosis from thrombotic occlusion of vessels in the brain, kidney, and skin. In these two processes, clotting factors such as fibrinogen, prothrombin, factors V and VIII, and platelets are consumed in the process of clot formation, resulting in a hypocoagulable state. The physiological fibrinolytic response is limited to the microcirculation, and as illustrated in Figure 30–9, *top*, leads to fibrin dissolution. Since there is an excessive quantity of fibrin dissolved, large amounts of fibrin degradation products are released into the circulation, and the hypocoagulable state and risk of hemorrhage is worsened. In the absence of an independent primary hyperfibrinolytic state (Fig. 30–9, *bottom*) there may be no increase in fibrinolytic activity in the blood.

Localized fibrinolysis of limited degree also may contribute to bleeding in patients with defective hemostasis, such as the hemophiliac. After a surgical procedure such as dental extraction, the tendency to dissolve hemostatic plugs in the injured site may exaggerate the bleeding tendency, even if the deficient clotting factor is administered. Inhibition of this local fibrinolytic response with EACA greatly improves the efficacy and decreases the required amount of clotting factor replacement treatment, greatly simplifying the treatment after surgery or injury, with a considerable saving in blood product. Upper genito-urinary bleeding occurs not infrequently in patients with congenital hemorrhagic disorders, and this condition represents a precaution for the use of EACA. In this circumstance, inhibition of UK present in the urine prevents the local fibrinolytic response necessary to clear the genitourinary tract, and may result in the formation of retained renal pelvic or ureteral clots. The use of EACA in such conditions should be undertaken only after more conservative measures have failed or in patients who may require surgical treatment of the bleeding lesion.

Consumptive thrombohemorrhagic disorders such as DIC may be manifest by both thrombotic and hemorrhagic symptoms; treatment depends primarily on removing the underlying cause. The profound coagulation abnormalities which may occur in abruptio placenta are treated by evacuation of the uterus, which stops the coagulation and thrombus formation, and allows restoration of clotting factors by transfusion or synthesis. DIC is always secondary to an underlying disorder, most often of infectious or neoplastic origin. Definitive treatment of the inciting cause such as appropriate antibiotic therapy of systemic infection removes the stimulus for further coagulation, allowing one to correct the coagulation abnormalities and sustain clinical improvement. In some cases of DIC, direct therapy of the thrombotic and hemorrhagic manifestations is necessary. This may be the case when the underlying disease is untreatable, as in advanced carcinomatosis or when the hemorrhagic or thrombotic symptoms are so prominent as to dominate the clinical picture. In such cases, heparin may be administered to curtail coagulation, prevent further thrombotic events, and halt consumption of coagulation factors. Concomitantly, replacement of platelets and coagulation factors by transfusion may be useful to treat or prevent hemorrhage due to the hypocoagulable state. In rare cases, excessive fibrinolysis is a very prominent feature in addition to DIC, contributing to the hemorrhagic symptoms. In these cases, therapy with a fibrinolytic inhibitor such as EACA may be considered, but this treatment will also block reactive fibrinolysis. Therefore, heparin should also be administered, since continued DIC will result in further fibrin formation that may be exaggerated with an inhibited fibrinolytic system and may result in even more serious, perhaps fatal, thrombosis.

DECREASED FIBRINOLYTIC ACTIVITY. The continuously active and readily stimulated normal fibrinolytic mechanism presumably limits the extent of fibrin deposition. It would therefore follow that impaired fibrinolysis would favor the growth of thrombi. There are several potential mechanisms for reduced fibrinolytic activity, such as decreased production and release of vascular plasminogen activator, accelerated clearance of activator from the circulation, deficient or defective protein components of the fibrinolytic system, and excessive amounts of inhibitor. Rare cases of a congenital abnormal plasminogen molecule have been associated in the affected patient with recurrent episodes of venous thrombosis, presumably because the abnormal plasminogen cannot be converted to plasmin, and low fibrinolytic activity results in the growth of thrombi to clinically important size. Thrombotic disease due to decreased fibrinolytic activity has also been postulated to occur in patients with inadequate release of plasminogen activator from endothelial cells, as demonstrated by an abnormal response in vascular plasminogen activator following vascular occlusion.

Unfortunately, generally available methods for the assay of the fibrinolytic system discriminate only between elevated and normal levels of activity and are not sufficiently sensitive to measure abnormally low levels. Decreased fibrinolysis may be pathogenetically important in the occurrence of venous thrombosis during immobilization, in which the perfusion of inactive muscles may not provide sufficient activator to prevent thrombosis in the deep veins. Additionally, this mechanism may be operative in congestive heart failure, malignant disease, and obesity, conditions associated with high risk of venous thrombosis which may be accompanied by an impaired fibrinolytic mechanism. The development of accurate methods to evaluate the detailed functioning of the fibrinolytic system, including the availability and functioning of activators and inhibitors, will have application to the diagnosis and treatment of patients with thrombotic disease and contribute to our understanding of fibrinolysis in the pathogenesis of thrombosis.

Annotated references

Part 1

Bloom, A. L., and Thomas, D. P. (eds.): Haemostasis and Thrombosis. New York, Churchill Livingstone, 1981. (An encyclopedic treatise.)

Colman, R. W., Hirsh, J., Marder, V. J., and Salzman, E. W. (eds.): Hemostasis and Thrombosis: Basic Principles and Clinical Practice. Philadelphia, J. B. Lippincott, 1982. (Another comprehensive text.)

Hirsh, J., Genton, E., and Hull, R.: Venous Thromboembolism. New York, Grune and Stratton, 1981. (Sensible review of pathogenesis, diagnosis, and therapy of venous thrombosis.)

Poller, L. (ed.): Recent Advances in Blood Coagulation Number Three. New York, Churchill Livingstone, 1981. (Selected topics.)

Spaet, T. H. (ed.): Progress in Hemostasis and Thrombosis. New York, Grune and Stratton. (A serial publication, now in its 6th volume, reviewing important aspects of its broad subject.)

Thompson, A. R., and Harker, L. A.: Manual of Hemostasis and Thrombosis, 3rd ed. Philadelphia, F. A. Davis, 1982. (A monograph aimed at students.)

Wintrobe, M. M. (ed.): Blood, Pure and Eloquent: A Story of Discovery, of People, and of Ideas. New York, McGraw-Hill, 1980. (Includes chapters on the history of ideas about hemostasis, thrombosis, and platelets.)

Part 2

Alper, C. A., and Rosen, F. S.: Genetic aspects of the complement system. In Dixon, F. J., and Kunkel, H. G. (eds.): Advances in Immunology, Vol. 14. New York, Academic Press, 1971.

Boenisch, T., and Alper, C. A.: Isolation and properties of glycine rich β-glycoprotein of human serum. Biochim. Biophys. Acta 221:529, 1970.

Colten, H. R., Alper, C. A., and Rosen, F. S.: Current concepts in immunology: Genetics and biosynthesis of complement proteins. N. Eng. J. Med. 304:653, 1981.

Cooper, N. R.: Isolation and analysis of the mechanism of action of an inactivator of C4b in normal human serum. J. Exp. Med. 141:890, 1975.

Donaldson, V. H.: Blood coagulation and related plasma enzymes in inflammation. Series Haematologica 3:39, 1970.

Donaldson, V. H., and Evans, R. R.: A biochemical abnormality in hereditary angioneurotic edema: Absence of serum inhibitor of C'1 esterase. Am. J. Med. 35:37, 1963.

Donaldson, V. H., and Rosen, F. S.: Action of complement in hereditary angioneurotic edema: The role of C'1 esterase. J. Clin. Invest. 43:2204, 1964.

Fearon, D. T., and Austen, K. F.: Properdin: Binding of C3b and stabilization of the C3b-dependent C3 convertase. J. Exp. Med. 142:856, 1975.

Fearon, D. T., and Austen, K. F.: Activation of the alternative complement pathway with rabbit erythrocytes by circumvention of the regulatory action of endogenous control proteins. J. Exp. Med. 146:22, 1977.

Fey, G., and Colten, H. R.: Biosynthesis of complement components. Fed. Proc. 40:2099, 1981.

Frank, M. M.: Complement. In Current Concepts. Scope publication. Kalamazoo, Upjohn, 1975.

Fu, S. M., Kunkel, H. G., Brusman, H. P., Allen, F. H., Jr., and Fotino, M.: Evidence for linkage between HL-A histocompatibility genes and those involved in the synthesis of the second component of complement. J. Exp. Med. 140:1108, 1974.

Fujita, T., Gigli, I., and Nussenzweig, V.: Human C4-binding protein. II. Roles in proteolysis of C4b by C3b inactivator. J. Exp. Med. 148:1044, 1978.

Gewurz, H., and Peters, D. K.: Alternative pathways of complement activation. In Brent, L., and Holborow, J. (eds.): Progress in Immunology II. Vol. 1, p. 296. Amsterdam, New York, Elsevier-Dutton, 1974.

Gewurz, H., Pickering, R. J., Mergenhagen, S. E., and Good, R. A.: The complement profile in acute glomerulonephritis, systemic lupus erythematosus and hypocomplementemic chronic glomerulonephritis: Contrast and experimental correlations. Int. Arch. Allergy Appl. Immunol. 34:557, 1968.

Kellermeyer, R. W., and Graham, R. C.: Kinin—Possible physiologic and pathologic roles in man. N. Eng. J. Med. 279:754, 1968.

Lachman, P. J., Lepow, I. H.: IV-Complement. Brent, L., and Holborow, J. (eds.): Progress in Immunology II. Vol. 1, p. 171. Amsterdam, New York, Elsevier-Dutton, 1974.

Laurell, A. B., Johnson, U., Martensson, U., and Sjoholm, A. G.: Formation of complexes composed of C1r̄, C1s̄ and C1̄ inactivator in human serum on activation of C1. Acta Pathol. Microbiol. Scand. [C]. *86*:299, 1978.

Lepow, I. H., Naff, G. B., Todd, E. W., Pensky, J., and Hinz, C. F., Jr.: Chromatographic resolution of the first component of human complement into three activities. J. Exp. Med. *117*:983, 1963.

Müller-Eberhard, H. J.: The serum complement system. *In* Miescher, P. and Müller-Eberhard, H. J. (eds.): Textbook of Immunopathology, Vol. 1, pp. 33–42. New York, Grune & Stratton, 1968.

Müller-Eberhard, H. J., Schreiber, R. D.: Molecular biology and chemistry of the alternative pathway of complement. Adv. Immunol. *29*:1, 1980.

Naff, G. B., and Ratnoff, O. D.: The enzymatic nature of C1r, conversion of C1s to C1 esterase and digestion of amino acid esters by C1r. J. Exp. Med. *128*:571, 1968.

Nussenzweig, V.: Interaction between complement and immune complexes: Role of complement in containing immune complex damage. Prog. Immunol. *4*:1044, 1980.

Pangburn, M. K., Schreiber, R. D., and Müller-Eberhard, H. J.: Human complement C3b inactivator: Isolation, characterization and demonstration of an absolute requirement for the serum protein β1H for cleavage of C3b and C4b in solution. J. Exp. Med. *146*:257, 1977.

Ratnoff, O. D.: Some relationships among hemostasis, fibrinolytic phenomena, immunity, and the inflammatory response. Adv. Immunol. *10*:145, 1969.

Raum, D., Donaldson, V. H., Rosen, F. S., and Alper, C. A.: Genetics of complement. *In* Yachin, S., and Piomelli, S. (eds.): Current Topics in Hematology, *3*:111. New York, Allen R. Liss, 1980.

Rosen, F. S., Alper, C. A., and Janeway, C. A.: The primary immunodeficiencies and serum complement defects. *In* Nathan, D. G., and Oski, F. A. (eds.): Hematology of Infancy and Childhood. Philadelphia, W. B. Saunders, 1974.

Rosenberg, L. E., and Kidd, K. K.: HLA and disease susceptibility: A primer. N. Eng. J. Med. *297*:1060, 1977.

Vogt, W.: Activation, activities and pharmacologically active products of complement. Pharmacol. Rev. *26*:125, 1974.

Ward, P. A., Data, R., and Till, G.: Regulatory control of complement-derived chemotactic and anaphylatoxin mediators. *In* Progress in Immunology II, Vol. 1, p. 209. Amsterdam, New York, Elsevier-Dutton, 1974.

Ziccardi, R. J., and Cooper, N. R.: Active disassembly of the first complement component, C1̄, by C1̄ inactivator. J. Immunol. *123*:788, 1979.

Part 3

Aoki, N.: Natural inhibitors of fibrinolysis. Prog. Cardiovasc. Dis. *21*:267, 1979. (A review of the naturally occurring plasma inhibitors and their relative roles in the control of fibrinolysis and contributions to disease states.)

Collen, D.: On the regulation and control of fibrinolysis. Thromb. Haemost. *43*:77, 1980. (Synopsis of molecular events involved in the interaction of all components of the fibrinolytic system, with special attention to the regulation and physiologic balance under normal conditions and its aberration in disease states.)

Marder, V. J. (ed.): Molecular aspects of fibrin formation and dissolution. Semin. Thromb. Hemostas. *88*:1–68, 1982. (An in-depth review of fibrinolysis as related to its principle substrates, fibrinogen and fibrin. The molecular events whereby fibrinogen forms a clot and is degraded by plasmin is complemented by a description of tests for detecting circulating derivatives.)

Sherry, S.: Fibrinolysis. Ann. Rev. Med. *19*:247, 1968. (A detailed summary of the ingredients and interactions of the fibrinolytic system, as organized by the scientist who conceived many of our current ideas regarding the regulation of fibrinolysis.)

Section VII

Neuromuscular Mechanisms

Introduction

The last decade has witnessed enormous progress in the neurosciences. Many intricate connections between various parts of the nervous system have been described, the physiological interactions among various initiating and controlling areas are better understood, and presently we are in the midst of learning about the variety of neurotransmitters which are the linkage between nerve cells and between nerve and muscle cells.

In this section we shall present information on normal motor activity initiated in cerebral cortex, balanced by cerebellar input, and regulated by basal ganglia function.

This computer-like function of the central nervous system ultimately sends its messages through the peripheral nerve to the neuromuscular junction, perhaps the most studied synapse in man. Activation of appropriate neuromuscular junctions results in the initiation of the excitation-contraction system of skeletal muscle. Thus electrical signals initiate the contractile system of certain muscle cells to carry out the desired motor activity. The understanding of the normal physiology of cerebral, cerebellar, basal ganglia, neuromuscular junction, and skeletal muscle function is enhanced in each chapter by examples of diseases or lesions. Hence problems like seizures, spasticity, cerebellar ataxia, myasthenia, and muscular dystrophy cannot only be understood as variations of normal physiology but also shed light on normal physiological mechanisms. With a better understanding of the pathophysiology of certain diseases, giant strides have occurred in the care of patients. Patients destined to a future of rigid shaking muscles live more productive lives through replacement of a diminished neurotransmitter dopamine.

Patients with the profound weakness of myasthenia gravis can be put into remission by diminishing the effect of acetylcholine receptor antibodies with immunosuppressive medication and they may permanently stay in remission after removing the thymus gland, which produces the harmful antibodies.

This section therefore should present details of motor activity and its regulation at all levels of the neuraxis along with the translation of the nervous system activity into mechanical muscle contraction.

FREDERICK J. SAMAHA, M.D.

31

Control of Motor Activity by the Cerebrum and Cerebellum

Eric W. Lothman, M.D., Ph.D.
Erwin B. Montgomery, Jr., M.D.

Introduction

CNS MOTOR CONTROL: AN OVERVIEW. Although our present understanding of motor physiology is far from complete, several organizing principles are known. First, if one designates as motor centers those regions of the central nervous system (CNS) that modify movements when they are activated or destroyed, it is apparent that there are many such centers spread throughout the neuraxis. Together the various centers constitute a distributed system that includes the spinal cord, certain brain stem nuclei, basal ganglia, motor cortex, and cerebellum. Second, movements can be divided into several types. Third, movements may be elicited by different mechanisms. Fourth, there is a hierarchical organization of the motor system which consists of phylogenetically older "lower" centers that generally deal with stereotyped, reflexive movements and newer "higher" centers dealing with specialized, goal-directed voluntary movements. Fifth, caution must be exercised when inferring the function of a part of the CNS from the signs and symptoms that result after its destruction. Depending on the circumstances, similar lesions of the CNS need not produce the same results. In general, lesions in younger individuals and those that evolve slowly are less deleterious, and the acute effects of CNS lesions often exceed the long-term consequences. These observations suggest the existence of mechanisms that can compensate for the loss of some functions. Sixth, the manifestations of neurological disease may be considered as either "positive" or "negative" symptoms. Negative symptoms represent the loss of a certain ability as a result of disease in a particular part of the brain or spinal cord; positive symptoms represent an excess or abnormal activity that is not normally present and is somehow released as a consequence of neurological disease. An example of the former is paralysis; of the latter, convulsive motor activity caused by a seizure.

It is the purpose of this chapter to discuss, with regard to these principles, functional aspects of the motor areas of the cerebral cortex and the cerebellum, in order to explain how their dysfunction leads to recognizable clinical syndromes. Other motor centers (the spinal cord and basal ganglia) as well as the primary movers, muscles, are considered in separate chapters.

TYPES OF MOVEMENTS. For movements to occur, there must obviously be a change in the contractile state of muscles. If enough motor units in one muscle are activated, sufficient tension will develop to produce rotation at the contiguous joint. Such *simple movements* are the basic building blocks from which more complex movements are constructed. An example is flexion at the elbow with biceps contraction. Three temporal phases of simple movements can be identified: initiation, execution, and termination. In addition, certain parameters of the movement must be specified: force, direction, and speed. Simple movements can develop from voluntary intent or as an involuntary reaction to an external perturbation.* An example of the former is contraction of the quadriceps in kicking a ball; of the latter, contraction of the quadriceps in response to tapping the patellar tendon. Electromyograph (EMG) recordings from human beings and animals have shown that there are even different sorts of simple movements. There may be a single burst of activity only in the agonist muscle, producing a rapid motion—*a ballistic movement*. In other situations there are three phases of EMG activity, first agonist, then antagonist, and finally agonist—*self-terminated movements*. Self-terminated movements may be slow with a step-like series of displacements, *discontinuous self-terminated movements*, or faster with a smooth displacement, *continuous self-terminated movements*. Either ballistic or self-terminated movements can be selected to best fit a particular situation when "higher centers" decide that a voluntary motion is desired The programming of ballistic movements is entirely determined

before they start so that the movement is initiated, executed, and terminated without modification. On the other hand, self-terminated movements are programmed to start and then modified during execution. These later control signals ensure that the overall movement is executed as intended. *Compound movements* involve motion of more than one joint, and synergistic interactions between various antagonistic and agonistic muscles. They are composites of several simple movements and can exhibit various degrees of intricacy. Bringing an index finger to the nose, standing upright, and walking are examples of compound movements distributed along a continuum of their physical complexity.

Various CNS lesions may influence one type of movement differently from others. This is especially true for the motor cortex where lesions profoundly affect fine movements of individual fingers, *fractionated movements*, while proximal movements may be spared. Thus other schemes differentiate proximal from distal movements.

THE MOTOR SYSTEM: SUBUNITS AND THEIR OPERATIONAL MODES. Although they are located at different sites in the nervous system, motor centers have multiple interconnections. One approach to this complicated distribution of neurons is to consider the functional interactions in terms of control systems theory, even though this does not afford a precise and thorough description of the behavior of individual neurons with respect to synapses and action potentials. In this context, the various linkages in the motor system allow signals then generated to complete or correct the motor activity as intended. Stated otherwise, closed loops operate as error detectors and use feedback. When information flows sequentially through nonreciprocal connections, the loop is considered open. Open loops operate in a feed-forward mode. As an illustration, consider the case when an object in the environment is recognized and an intent develops to grasp the object. The process requires complex

---

*Movements can be placed in a spectrum with "most automatic or reflexive" at one end and "least automatic or skilled" at the other. In this context it may sometimes be difficult to designate a particular motor activity as reflex or voluntary. For example, movements that were initially the result of conscious and deliberate actions may become reflexive, such as in playing the violin or riding a bicycle. Similarly, "reflexes" may be subject to conscious control. Breathing in most instances is done automatically, but can be exercised with totally volitional efforts. Recognizing this point, we will differentiate reflex from volitional movements as a didactic framework for the following discussion.

processing in the association cortex of sensory information and various motivation factors in the internal milieu. This desire is then translated into a plan for movement that must be transformed into a pattern of muscular contractions of specific force and duration. The planning occurs in the basal ganglia, lateral cerebellum, and motor cortex. Each sequential stage in the process, especially for ballistic movements, does not have direct reciprocal connections to the preceding stage. Open and closed loops have different advantages. The feedback in closed loops makes them more accurate but slower; open loops are faster but are prone to inaccuracies owing to the lack of feedback. Closed loops are subject to oscillations.

There are many loops in the motor system (Fig. 31–1). The components of these circuits are connected in series and parallel combinations. The availability of parallel routes allows for primary control signal transmission over one pathway and secondary signals over collateral pathways. By providing alternate pathways for motor programs to reach α-motoneurons these parallel routes may also account for recovery of function when one route of information flow is disrupted. The serial organization suggests a hierarchical scheme of "lower" and "higher" centers in which open loops are superimposed on closed loops.

Movements can be elicited through these loops in several ways (Fig. 31–2). They may arise in an ensemble of neurons that operates as an autonomous unit to produce patterns of discharges in motor neurons. An example of this is the central program generator of the spinal cord that produces stepping movements during locomotion. This generator can perform in total isolation although it is normally modulated by suprasegmental subcortical centers and feedback from primary spinal afferents. Alternatively, movements may result after a perturbation by an external force that activates sensory organs which, in turn, excite motor centers. Such reflexes may either be segmental (spinal) or suprasegmental. The simplest reflex is the segmental deep tendon reflex, in which action potentials in muscle spindle afferents monosynaptically lead to motoneuron discharge. An example of a more complex reflex is the suprasegmental vestibulospinal reflex, in which movement of the head causes first activation of the vestibular nerve, then vestibular nuclei neurons, and finally motoneurons. Both reflexes and central pattern generators involve lower motor centers and operate outside conscious control. In contrast, voluntary movements (Figs. 31–2 and 31–3) depend on the flow of control signals from higher centers to lower centers until the intended motion is brought about by the discharge of spinal motoneurons. This may occur by a command signal directly activating the motoneuron in an open loop mode, as for ballistic movements; by using feedback, as in the later phases of self-terminated movements; or by enlisting motor sequences generated by lower centers.

The "upper motoneuron" is a critical structure for voluntary movements. A lesion of its cell body in the motor cortex or its axon in the corticospinal tract results in an inability to perform certain goal-directed movements of the limbs. Even though the cerebellum is an integral part of the loops involved in voluntary movements it is not critical. Rather the cerebellum adds a fine texture to this type of movement. With disease of this structure voluntary movements can be done, albeit poorly. The cerebellum is also important in modulating various reflexes. The following sections will consider in further detail how the specific clinical signs that serve as markers of dysfunction of the motor cortex and cerebellum arise.

The cerebral control of movements

MOTOR CORTEX: DEFINITION AND DELINEATION. The concept of a motor cortex began with the finding of Fritsch and Hitzig in

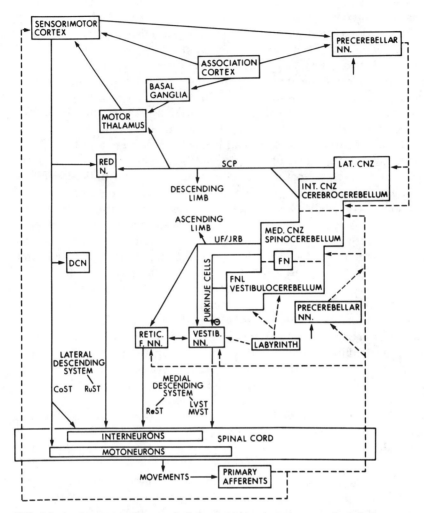

FIG. 31—1. Summary of anatomical connections of cerebrum and cerebellum in motor systems. Major efferent pathways from cortical and subcortical motor centers are denoted by solid lines and afferent pathways to these centers by broken lines. Major components of the medial and lateral descending (brain stem) systems of Lawrence and Kuypers include rubrospinal tract (*RuST*), reticulospinal tract (*ReST*) and lateral and medial vestibulospinal tracts (*LVST* and *MVST*). Note the projections of the corticospinal tract (*CoST*) directly to motor neurons and to subcortical sensory areas, shown representively for the dorsal column nuclei (*DCN*). Output from the lateral and intermediate cortical nuclear zone passes through the superior cerebellar peduncle (*SCP*), which divides into an ascending and descending system. Output from the medial corticonuclear zone passes by way of the uncinate fusiculus (*UF*) and juxtarestiform body (*JRB*) into an ascending system and descending components. The latter connects with the reticular formation (*RETIC.F.NN.*) and vestibular nuclei (*VESTIB. NN.*). Also shown are the inhibitory connections of Purkinje cells from the cerebellar cortex in the flocculonodular lobe and vermis. Some connections are deleted for clarity. Open-ended arrows denote access routes for higher-level information into cerebellar motor systems through precerebellar nuclei.

FIG. 31–2. Overview of CNS motor control loops and the means by which movements may be elicited. Components of the medial descending brain stem system arising in the vestibular nuclei *(VN)* and reticular formation *(RFN)* are on the left of the lateral descending brain stem system arising in the red nucleus *(RED N.)*, on the right. In this scheme movements may be initiated volitionally through the neocortex or involuntarily through perturbations that activate reflexes in subcortical motor centers. The open loop that effects ballistic movements is indicated by the broad stripped line and the closed loop, operating in other volitional movements by the broad line with diagonal stripes. Structures involved in stepping are connected by dotted lines, those for segmental reflexes by solid line and those for suprasegmental reflexes by solid line interrupted by dots.

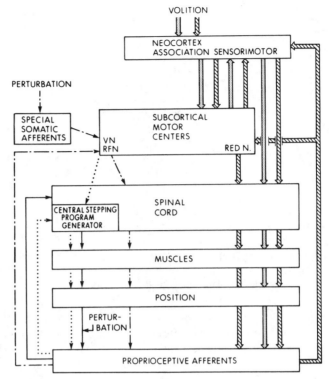

FIG. 31–3. Possible sequence of steps involved in the initiation of a volitional movement. The idea for the volitional movement is developed in the associational areas of the cerebral cortex *(ASSN CX)* after integration of various forms of information. This plan for movement initiation is then relayed to the lateral cerebellum *(LATERAL CBM)* and basal ganglia, and from these areas to the motor cortex *(MOTOR CX)*. The lateral cerebellum and basal ganglia further process information regarding the motor program. Information from the motor cortex then is relayed to the spinal cord, to produce movement. The intermediate portion of the cerebellum *(INTERMED CBM)* then functions as a follow-error correcting mechanism. (Allen, G. I., and Tsukahara, N.: Cerebro-cerebellar communication systems. Physiol. Rev. *54*:957–1006, 1974. Reproduced with permission.)

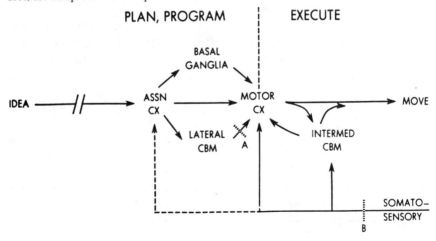

FIG. 31–4. Motor homunculus. (Kandel, E. R., and Schwartz, J. H.: Principles of Neural Science. Elsevier/North-Holland, 1981. Reproduced with permission.)

1870 and Ferrier in 1875 that application of a localized electrical stimulus to the frontal cortex of animals produced movements. At that time the concept of a localized area concerned with motor function had also emerged from the clinical observations of Jackson in humans. Since then several investigators have done careful physiological mapping of the motor cortex. Such studies delineated the well-known "motor homunculus," which is a map of where various muscle groups are represented in the cerebral cortex (Fig. 31–4). The representation in the motor homunculus is not democratic with respect to the actual body proportions. Greater representation is given to the hand (particularly the fingers) and face, suggesting that the motor cortex may be more heavily involved in movements of certain areas of the body. The lower extremities are represented in the medial aspect of the cortex. Proceeding laterally the representation shifts to involve hip, trunk, shoulder, arm, hand, fingers, and then face. Such an organization provides an anatomical foundation for the sequence of motor involvement in patients with focal motor seizures that exhibit a "Jacksonian march." For example, a seizure that begins with

jerking in the hand may then spread to involve the face and then go on to involve the lower extremity as well.

Anatomically, the motor cortex, also called the primary motor area, or MI, lies on the anterior bank of the central sulcus (the precentral gyrus). It incorporates Brodmann's architectonic area 4 (Fig. 31–5). It is a large source (though not the sole source) of fibers that comprise the corticospinal (or pyramidal) tract, which continues down the entire length of the spinal cord in man.

Mapping of motor responses also demonstrated a second homunculus, referred to as the supplementary motor cortex, MII. This occupies the medial aspect of Brodmann's area 6 (Fig. 31–5) and is discussed in more detail below. Recent studies have suggested the existence of even more representations of specific muscle groups in the cerebral cortex, but the physiological consequences of this are not yet known.

Functional considerations

SOMATIC MOVEMENTS. The exact role of the motor cortex and its efferent pathway, the pyramidal tract, in movement is not entirely clear. Some animals, like the cat, show little motor deficit with lesions of the motor cortex or pyramidal tract. Occasionally patients who have extensive lesions of the motor cortex may initially be paralyzed but recover remarkably. This illustrates that movement is not the sole franchise of the motor cortex. The recovery can be explained by other motor centers that are capable of affecting movement through descending pathways separate from the pyramidal tract. However, certain types of movements may be irretrievably lost with lesions of specific structures. The nature of that loss reflects at least part of the original function of the structure lesioned. Humans and other primates with lesions of the motor cortex do not recover fractionated finger movements. A possible explanation may be found in the pattern of motor cortex input on the α-motoneurons.

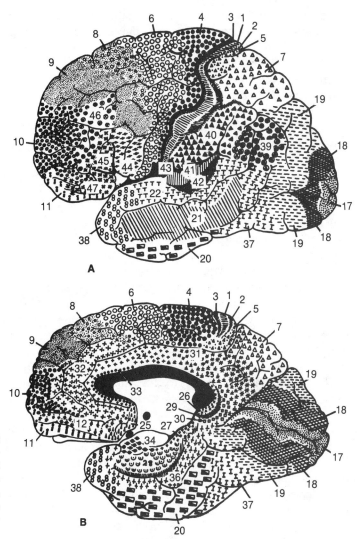

FIG. 31–5. Cytoarchitectonic map of the human cerebral cortex. (*A*) lateral surface; (*B*) medial surface. (Truex, R. C., and Carpenter, M. B.: Human Neuroanatomy, Baltimore, Williams & Wilkins, 1969. Reproduced with permission.)

Alpha-motoneurons, also referred to as *lower motoneurons*, lie in the ventral gray of the spinal cord and send their axons to the muscles. Movement ultimately depends on the activation of these motoneurons, which operate as the final common pathway. In the cat most corticospinal neurons end in the cervical enlargement, though some continue throughout the length of the spinal cord. However, none of the cat's corticospinal neurons directly contact the α-motoneurons. From these observations it is not hard to understand why a lesion of the cat's corticospinal tract

leaves movement largely unimpaired. In contrast, fibers in the corticospinal tract of man and apes have monosynaptic connections with α-motoneurons, and the corticospinal tract provides a major input to the α-motoneurons controlling the distal musculature. Only about one-third of the input to the α-motoneurons projecting to the proximal musculature comes from corticospinal neurons. Therefore, fine distal motor control is dependent on the integrity of the motor cortex and the corticospinal tract. This type of motor activity includes fine independent movements of the fingers and opposition of the thumb, fractionated movements. These movements are characteristic of man and ape, but not of lower species. The heavy projection of the corticospinal tract to the α-motoneurons of the hand correlates with the observation that the hand is represented in a relatively large area in the motor homunculus.

In addition to the corticospinal tracts, α-motoneurons (especially those projecting to the proximal musculature) have considerable input from other suprasegmental motor centers (see Fig. 31–1). These provide a basis for recovery of function following a motor cortex lesion. Lawrence and Kuypers lesioned the pyramidal tract in monkeys. Shortly after the lesions were made, the monkeys were weak but subsequently showed improvement. These monkeys were initially able to obtain food by a hooking movement which combined the arm and hand. Four to five weeks after the surgery the monkeys were able to grasp food by movements confined to the whole hand. What did not recover was fractionated movements of the fingers. When Lawrence and Kuypers placed lesions in both pyramidal tract and the lateral brain stem, the monkeys did not recover the ability to use just the hand. This implies that a descending system compensates for corticospinal tract lesions. These investigators defined two subcortical descending motor systems. The ventromedial brain stem pathway is concerned primarily with the maintenance of erect posture and control of axial and proximal musculature, especially during locomotion. Major components of this system include the vestibulospinal and reticulospinal tracts. The lateral brain stem pathway is concerned primarily with flexion movements of the extremities and provides control of independent distal musculature, particularly in the hand. The rubrospinal tract seems to be a major component of this system. The motor cortex allows for higher-level integrative control of motor responses through its projections to these two descending brain stem systems.

In addition, because of its direct contacts with α-motoneurons, the motor cortex superimposes a direct control of distal musculature, similar to that of the lateral brain stem descending system but affording even more precise control, fractionated movements. Similarly, humans occasionally recover much function following lesions of the motor cortex or pyramidal tract, but they do not recover the fine independent finger movements.

The fact that monkeys and humans can regain coarse hand movements as well as more proximal movements following motor cortex or pyramidal tract lesions does not mean that the motor cortex is not normally concerned with some aspects of these movements. Proximal muscles are represented in the motor homunculus. Lesions of the motor cortex also result in a characteristic posture (discussed below) thereby suggesting an effect on proximal musculature.

Evarts pioneered a technique in which monkeys are behaviorally conditioned to perform various motor tasks while microelectrodes record neuronal activity. In this fashion the involvement of neurons in various aspects of movements can be identified. By varying the conditions of the task, the relevance of the firing of a neuron to a specific aspect or type of movement can be tested. Recordings made in the motor cortex can be compared to similar recordings made in other areas such as the basal ganglia, cerebellum, and postcentral gyrus of the cerebral cortex (somatosensory cortex). Such studies performed on the motor cortex show that the motor cortex neurons begin to change their activity 80 to 100 msec. before the onset of muscular activity. Neurons of the postcentral gyrus begin to change their activity after the onset of movement. This type of experiment has shown that the activity of many neurons that send axons to the pyramidal tract have a firing rate proportional to the force being developed within a certain range. This latter phenomenon parallels recruitment of

α-motoneurons with increasing force of muscular contraction. Similar studies have shown that many motor cortex neurons are more related to a particular pattern of muscular contraction than to the direction of movement.

Stimulating the motor cortex with very small currents through small microelectrodes at various depths and positions in the cortex has revealed a columnar organization of motor responses. Stimulation of a given column of motor cortex neurons gives rise to isolated patterns of muscular contraction. This supports the notion that individual muscles are represented in the motor cortex and that the motor cortex is particularly concerned in the amount of force exerted by a muscle.

However, other investigators have found motor cortex neurons not clearly related to the generation of force by a specific group of muscles. Applying loads that either opposed or assisted a behavioral task performed by monkeys dissociated the direction of the movement from the patterns of muscular contraction. For example, flexion movements with an opposing load required flexor muscle activity. For flexion movements with an assisting load, only reduction of extensor muscle activity was required. In this fashion neuronal activity could be related specifically to the direction of movement, pattern of muscular activity, or the direction of the next intended movement. Neurons were found in the motor cortex which were best correlated with each one of these three features of the movement. Other behavioral tasks using an operantly conditioned isometric biting task showed a clear relationship of motor cortex neuronal activity to the force of the bite. However, these same neurons failed to change their activity during normal chewing. These studies suggest a hierarchic organization. There may be a transformation process in which information about the environment and goals are processed into a program of muscular contraction for only certain type of movements (i.e., volitional as opposed to reflexive). The different categories of units found in the motor cortex may represent samples of neurons at different levels of this transformation process.

The motor cortex receives inputs from a variety of sources. It receives information from the primary somatosensory areas (areas 1, 2, and 3) of the cerebral cortex (see Fig. 31–5). In turn, the motor cortex exerts some control over the passage of information through the sensory system, such as in dorsal column nuclei or thalamus where it exerts presynaptic inhibition.

The projection of somatosensory information into the motor cortex may act as a "closed loop" in which sensory information about a movement feeds back to the motor cortex. This is the basis for transcortical reflexes. In monkeys and humans trained to perform a movement which is then interrupted by a brief force opposing the movement, additional muscular contraction develops in the appropriate muscle to oppose this disturbing impulse. There are three bursts of EMG activity. The first, occurring shortly after the disturbing force, is mediated by the segmental reflex through spindle afferents synapsing on α-motoneurons. A second burst occurs with a latency of about 50 msec., a timing that suggests a reflex whose afferent limb passes to the somatosensory cortex and whose efferent limb arises in the motor cortex. This later phenomenon is referred to as a transcortical reflex and may act as an automatic mechanism to compensate for unexpected local disturbances during volitional movement. The origin of the third burst of EMG activity is unclear. It is abolished when the dentate nucleus of the cerebellum is cooled in monkeys, suggesting that this third response involves higher-level loops.

Higher-level sensory information also reaches the motor cortex indirectly. The primary somatosensory areas (1, 2, and 3) project to areas 4 and 5. Area 5 projects to areas 6 and 7. Area 7 does not project to either area 4 or area 6. In general, there is a direct input from primary somatosensory cortex to the motor cortex but not from the association cortex. The higher-level information processed in these regions, particularly multimodality sensory information, must gain access to the motor cortex through other routes, either through multiple corticocortical relays or through subcortical structures. Two structures that may mediate this transmission of information include the lateral portion of the cerebellum and the basal ganglia (see Figs. 31–1 and 31–3).

The concept of functionally different areas of cerebral cortex for movements has been demonstrated in humans by studying local cerebral blood flow during different types of movements. The assumption was that changes in cerebral blood flow would parallel changes in physiological activity and be confined to the areas involved in the movements. When patients performed a relatively simple movement, compressing a spring between the index finger and thumb, there was an increase in the cerebral blood flow restricted to the contralateral motor cortex. When patients performed a more complex task, touching the thumb to each finger in sequence, cerebral blood flow increased not only in the contralateral motor cortex, but also in the supplementary motor areas (area 6). When patients were asked to think about the same movement task without actually performing the task, increased cerebral blood flow appeared only in the supplementary motor cortex.

In summary, the motor cortex is preferentially involved in the control of the distal musculature, particularly fractionated movements. This region is intimately involved in those movements which are least automatic. These roles of the motor cortex correlate with its anatomical structures, physiological activity and the persistent deficits associated with lesions of the motor cortex. In addition, the motor cortex has the means to operate in a closed-loop or reflex mode to counter unexpected perturbations experienced during a self-terminated volitional movement.

ABNORMAL TONE, POSTURE, AND REFLEXES RESULTING FROM MOTOR CORTEX LESIONS. In addition to weakness or paralysis, lesions of the motor cortex cause changes in posture and tone and the appearance of pathological reflexes. Lesions of the motor cortex are followed by a characteristic *hemiplegic posture.* This posture can be recognized by the attitude in which the limbs are held at rest with the upper limb close to the body and flexed at the elbow, wrist, and fingers and the lower extremity externally rotated and extended. This posture is also referred to as an *antigravity posture* as though the limbs assumed a position to oppose the effect of gravity. The appearance of the hemiplegic or antigravity posture suggests a release phenomenon. One theory is that brain stem mechanisms facilitate the antigravity posture. Normally these mechanisms are inhibited by the motor cortex, so that its influence is augmented following lesions of the motor cortex. As such, this constitutes a positive sign. Therefore, the motor cortex may act to counter the mechanisms producing the antigravity posture (i.e., hemiplegic posture) so that volitional movements may be carried out without interference. Records of EMG activity show changes in paraspinal and proximal lower extremity muscles prior to voluntary activation of the muscles used in lifting the arms. This suggests activation of the proximal and axial muscles in anticipation of the postural disturbance caused by the distal movement. Thus, in addition to its role in fractionated movements, the motor cortex seems to participate in planning of postural responses needed to compensate in volitional movements.

Lesions of the motor cortex are also associated with an increase in muscle tone of a specific type, called *clasp-knife* or *spasticity.* This is defined as increased resistance of a limb to passive movement. As the limb of a patient with spasticity is passively moved, there is a gradual increase in tone until a point is reached where the tone is suddenly reduced. Occasionally there is a succession of such cycles in one excursion. Initially this was believed to be due to destruction of the motor cortex, but subsequent experiments showed that when the pyramidal tract was sectioned at the level of the medulla or when an area just cortical to area 4 was ablated, hypotonia developed. When the lesion was expanded to involve areas just rostral to the motor cortex, spasticity appeared. Since it is rare in clinical practice to have a lesion restricted to area 4, spasticity is a common feature of disease in cortex anterior to the central sulcus. Initially there may be hypotonia, but this is eventually replaced by spasticity. The reason for this sequence is not known.

The exact mechanisms underlying spasticity are not clear. Older theories relating clasp-knife tone to changes in muscle spindle and Golgi tendon organ sensitivity are no longer accepted. Passive flexion movement of the limb results in an increased tone of flexor muscles that is proportional to the velocity of the movement. The sudden reduction of muscle tone probably occurs at the point where the velocity of muscle stretch decreases. Neither have earlier theories that the excess γ-motoneuron drive was responsible for spasticity been supported, since muscle spindle activity does not exceed normal levels in this condition. Rather, spasticity arises from an increased excitability of the α-motoneurons to any input. This increased excitability may be due to the loss of descending inhibitory systems (disinhibition) and, therefore, spasticity can be considered a positive symptom.

The appearance of pathological reflexes are another manifestation of lesions of the motor cortex or pyramidal tract. These pathological reflexes include increased deep-tendon reflexes and the Babinski or extensor plantar responses. These reflexes reflect an increased α-motoneuron excit-ability as described above for spasticity. Increased deep-tendon reflexes arise from increased α-motoneuron sensitivity to muscle spindle input, and the Babinski reflex from on the increased response of certain α-motoneurons to nocieceptive input. Clonus, which is a rhythmic, repeated contraction in response to tonic muscle stretch, is also associated with lesions of the upper motoneurons. The following mechanism is responsible. Muscle spindles are activated by an initial muscle stretch. This results in reflex contraction of the muscle. As the muscle contracts, the spindles are unloaded and spindle activity is reduced. The reduced spindle activity cause a reduction in muscle contraction, which again allows the muscle to be stretched by the tonic load. This sets up repeat cycles which may be sustained for many seconds.

Together the signs of weakness (particularly of distal, fractionated movements), increased tone, and exaggeration of deep-tendon reflexes, and the appearance of other pathological reflexes (such as the Babinski sign) constitute the *upper motoneuron syndrome*. This term is synonymous with lesions involving the motor cortex or its descending pathways to the spinal cord.

PREMOTOR CORTEX. Motor responses can be elicited by electrical stimulation of areas of the cerebral cortex other than the motor cortex. Stimulation of area 8 results in conjugate deviation of the eyes and head toward the contralateral side. The area is referred to as the frontal eye field. Lesions of this area result in a temporary conjugate deviation of the eyes toward the side of the lesion.

Stimulation of Brodmann's area 6 (Fig. 31–5) also results in movement. However, stronger intensities of stimulation are required, and only complex responses are obtained. Stimulation results in deviation of the eyes, head, and trunk toward the opposite side. Lesions of area 6 result in the appearance of the grasp reflex, a reflexive closure of the hand to cutaneous stimuli of the palm, and spasticity. With lesions of area 6 when area 4 is not involved there is reluctance to use the affected limb not due to weakness. Based on anatomical connections, area 6 is divided into a lateral portion and medial portion. The medial portion is referred to as the supplementary motor cortex. Functional differences of these two subdivisions of area 6 have not yet been identified.

Lesions occurring in man that involve the premotor cortex while sparing the motor cortex are very rare. However, occasionally a seizure focus may arise in or near the premotor area, giving rise to a picture similar to that described for stimulation of this area. These seizures are termed *adversive* and are an example of a *positive* symptom.

MOTOR DISTURBANCE ASSOCIATED WITH OTHER AREAS OF THE CEREBRAL CORTEX. Lesions of other areas of the cerebral cortex have resulted in various motor deficits in man—apraxias and ataxias. *Ataxia* is usually linked to dysfunction of the cerebellum (see below), but in some instances patients may have similar motor deficits with cerebral lesions. Such *cerebral ataxias* are associated with lesions of various "sensory" cortexes of their associational areas. The mechanisms responsible for this are not known.

Apraxia can be defined as the inability of an alert, motivated, and comprehending subject to produce a voluntary movement sequence in the absence of paralysis, sensory loss, incoordination, or involuntary movement disorder sufficient to prevent completion of the movement. Three types of apraxia can be defined. *Limb kinetic apraxia* refers to a loss of isolated distal movements and can be considered a form of the deficit described above for the upper motoneuron syndrome. *Ideomotor apraxias* refers to an inability to produce a motor response to command with preservation of the same act carried out spontaneously.

This is associated with lesions in the supramarginal and angular gyri of the dominant parietal lobe or their subcortical pathways. Ideomotor apraxia has been explained as a disconnection syndrome in which associational pathways between the area that develops the idea of movement and the area which actually programs the execution of the movement are injured. *Ideational apraxia* refers to an inability to perform a complex sequence of gestures while the separate acts alone may be performed correctly. This is associated with lesions of areas 39, 40, 18, and 19 in the dominant hemisphere. This is believed to arise from errors in programming the entire sequence, not the individual motor task.

To identify the type of apraxia, patients can be tested for their ability to pantomine with the affected limb, first to verbal command and then by copying the examiner. In addition, they should then be asked to use common objects. Patients with ideomotor apraxia are unable to pantomime but can execute movements when handed objects. Patients with ideational apraxia are able to pantomime well but unable to use objects. The distinction has prognostic value because ideational apraxia is associated with difficulties in carrying out the activities of daily living while ideomotor apraxia is not.

The mechanisms resulting in apraxias are unclear. They surely involve the integration of higher level processed sensory and motivational information into plans for movement. Lesions of the premotor cortex in monkeys result in a temporary deficit of visually guided movement. Normal monkeys were trained to reach around and through a hole in a clear Plexiglas plate to reach food adhering to the opposite side. These monkeys subsequently had lesions placed in the premotor area. When they attempted to reach the food directly they were blocked by the plate and were unable to reach around and through the clearly identifiable hole in the Plexiglas. The monkeys did not appear weak or incoordinate. Their deficit may be a type of apraxia in which they were unable to develop an appropriate plan of movement. These deficits of visually guided movement were very similar to deficits produced by posterior parietal leucotomy and commissurotomy. This suggests that the deficit seen with lesions of the premotor area may be due to inability to integrated sensory information into a useful plan for movement.

The cerebellum

GROSS ANATOMICAL SUBDIVISIONS. The cerebellum is easily identified as a distinct part of the brain situated underneath the occipital lobes. Anterior to the cerebellum lies the brain stem, with which the cerebellum is connected by three pairs of white matter bundles (cerebellar peduncles)—the superior (brachium conjunctivum), middle (brachium pontis), and inferior (restiform body). These peduncles issue axons to and from the main cerebellar structure. The middle peduncle is entirely afferent, while the superior and inferior peduncles are mixed afferent and efferent.

The body of the cerebellum in man consists of two large hemispheres and a smaller midline structure called the vermis (Fig. 31–6). The cerebellar surface has complexly folded ridges of gray matter, folia. Below this cortex is white matter that surrounds the deep cerebellar nuclei. There are four deep nuclei on each side of the cerebellum arranged from medial to lateral as fastigial, globose, and emboliform (taken together as the interpositus) and dentate. The cortex is interrupted by multiple fissures which can be used to subdivide the cerebellum into 10 lobules. Anterior to lobule VI, the separation into a midline vermis and lateral mass is indistinct. Nonetheless, the cortex of the hemisphere is in continuity with that of the vermis in the deep fissure, and, for functional purposes, the entire cerebeller cortex may be considered to be a sheet that has been infolded.

Another way of subdividing the cerebellum is based on embryology. Part of the special somatic sensory column of the neural tube develops into the flocculonodular lobe. This is separated by the posterolateral fissure from the remainder of the cerebellum, the corpus cerebelli, which arises from the general somatic sensory column. The primary fissure develops later in embryogenesis between lobules V and VI to divide the corpus cerebelli into the anterior and posterior lobes. In the primate it is the posterior lobe that shows expansion and subdivision of its lateral portion *pari passu* to the tremendous growth of the cerebral cortex. The anatomical divisions thus obtained are the flocculonodular lobe, referred to as the archicerebellum; the anterior lobe, often referred to as the paleocerebellum; and the posterior lobe, sometimes referred to as the neocerebellum (Fig. 31–6).

CEREBELLAR AFFERENTS. Yet another scheme of subdivisions of the cerebellum is based on the afferent connections of the cerebellum. The input to the cerebellum is extensive and varied, ranging from that arising in sensory organs to that coming from higher associative or integrative areas. The former includes proprioceptive, cutaneous, vestibular, auditory, visual, and interoreceptive (visceral) information. Cerebellar inputs may come directly over first-order primary afferents or may be relayed across synapses in precerebellar nuclei in the brain stem—the trigeminal nucleus, the basal pontine and arcuate nuclei, the accessory cuneate nucleus, the inferior olivary complex, the red nucleus, the reticular formation nuclei, and the tectum.

The various afferent systems to the cerebellum can be divided into two functional sets. On the one hand, the reticulocerebellar system shows convergence of several modes of information at a precerebellar

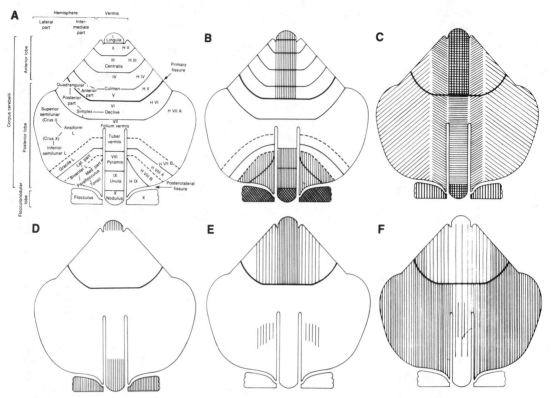

FIG. 31–6. Summary of cerebellar anatomy. (A) Schematic representation of gross anatomy of cerebellum. Descriptive terms for various areas are indicated on the left and Larsells' nomenclature on the right. (B) Subdivisions of cerebellum based on comparative anatomy with archicerebellum delineated by cross-hatched areas, paleocerebellum by vertical lines and neocerebellum by unmarked areas. (C) Corticonuclear zones. Areas of cerebellar cortex projecting to the vestibular nuclei are indicated by vertical lines, to the fastigial nuclei by horizontal lines, to the interpositus nuclei by oblique lines running upward from midline and to the dentate nuclei by oblique lines running downward toward the periphery. Terminal zones of afferent projections of the vestibulocerebellum (D), spinocerebellum (E) and cerebrocerebellum (F). Density of projection indicated by separation of lines. In C–F most fissures are deleted for clarity. (After Bell and Dow (1967).)

level, and diffuse termination within the cerebellum. These afferents primarily project to the deep cerebellar nuclei. On the other hand, specific projection systems carry one mode of information, terminate in topographically restricted loci, and project strongly to the cerebellar cortex. This discussion concentrates on three of the specific projection systems that are the best understood with respect to cerebellar function, those coming from the vestibular apparatus, spinal cord, and cerebral cortex (see Fig. 31–6).

Cutaneous and proprioceptive information from the entire body reaches the cere-

bellum. The areas in which it terminates define the spinocerebellum (Fig. 31–6E). Sensory signals from the face reach the cerebellum either directly or after relay in the trigeminal nucleus; those from the remainder of the body are carried in the spinocerebellar tracts. Spinocerebellar inputs terminate within two separate areas. One lies in the medial and intermediate aspects of lobules I to V and the anterior half of lobule VI, and the other in medial and intermediate portions of lobules VII and VIII. The densest projection is in the vermis of the anterior lobe. Within these areas there are separate somatotopic maps. In the anterior region, one map exists with the head located over lobules VI and V, the forelimbs over lobules IV and III, and the hindlimbs over lobules II and I. The representation of the trunk is medial to that of the limbs. In this map, each side of the body is represented once, on the ipsilateral side. There are two separate maps in the posterior area, one on either side of the midline in which both sides of the body are represented. Within each of the posterior maps the trunk is medial, the limbs are lateral, and the head is anterior and hindlimbs posterior.

Another major source of input to the cerebellum, that from the cerebral cortex, delineates the cerebrocerebellum (see Fig. 31–6*F*). The main projection is by way of nuclei in the basis pontis and is, in terms of numbers of fibers, the largest. Virtually the whole cerebral cortex projects to the pontine nuclei in the basis pontis. All of the cerebellar cortex, except the uvula and flocculonodular lobe, receives afferents from these nuclei through the middle cerebellar peduncle. This corticopontocerebellar projection is densest farther from the midline than the spinocerebellar projection, in the intermediate and lateral aspects of the anterior and posterior lobes. The densest projections are lateral in the posterior lobe. The inferior olivary complex provides another relay station for cerebrocerebellar pathways, projecting through the inferior cerebellar peduncle to the contralateral cerebellum. Within the cerebellum the projection zone from association areas of cerebral cortex is lateral to that from the motor cortex. The auditory and visual cortices project primarily to the medial areas of lobules VI to VIII. Visual and auditory information also reaches this area by pathways independent of cerebral cortical relays, probably the tectum.

The locus of the vestibular termination within the cerebellar cortex is medial, within the flocculonodular lobe and the vermis at the extreme anterior and posterior ends of the corpus cerebelli (Fig. 31–

6*D*). The fastigial nucleus also receives vestibular afferents.

THE CEREBELLAR CORTEX. Although there are diverse origins for cerebellar afferents, there are but three sorts of synaptic endings within the cerebellum: aminergic terminals, mossy fibers, and climbing fibers. For the purposes of this discussion, all nonolivary projections except aminergic terminals can be considered to terminate as mossy fibers, whereas the inferior olives are the source of climbing fibers. Cerebral and spinal inputs end as both mossy fiber and climbing fiber terminals. Vestibular input travels through mossy fibers. The cell bodies for fibers ending in noradrenergic terminals arise in the locus coeruleus and for serotonergic terminals in the median raphe.

The cerebellar cortex is a remarkably homogeneous laminated structure consisting of an hypocellular superficial molecular layer, a deeper cell layer with its distinctive, large Purkinje cells, and an inner, hypercellular layer packed with small granule cells. The cerebellar cortex is one of the best understood parts of the CNS from the point of view of cellular anatomy and physiology. The Purkinje cell provides the sole output of the cerebellar cortex, projecting to the deep cerebellar nuclei and to the vestibular nuclei. These cells also give off axon collaterals to other Purkinje cells, basket cells, and Golgi cells. All connections made by the Purkinje cell axons are inhibitory, probably mediated by γ-aminobutyric acid (GABA).

Excitation of Purkinje cells occurs directly by climbing fibers and indirectly by mossy fibers through granule cells and parallel fibers. While it is clear that the modes of excitation by mossy fibers and climbing fibers differ, the physiological importance of this is unknown. At present little is known of the role of aminergic terminals, but some believe they modulate the response of Purkinje cells to climbing fibers and mossy fibers.

The intricate and regular spatial organization of cerebellar cortical neurons, combined with their fixed excitatory and inhibitory couplings, results in spatial limitation of Purkinje cell activity. Thus, when granule cells are activated by mossy fiber input, they excite Purkinje cells along a strip parallel to the long axis of the folium through parallel fibers. Stellate and basket cells are also activated and distribute inhibition lateral to the excited beam. The balance of the excitation and inhibition within the Purkinje cells would result in a gradually decreasing excitation until the beam of excitation is flanked by a zone of inhibited Purkinje cells on either side. Golgi cells mediate inhibition at right angles to this beam. Powerful recurrent inhibition through the basket cells serves to temporally limit the duration of Purkinje cell discharge.

CEREBELLAR EFFERENT CONNECTIONS: COR-TICONUCLEAR ZONES AND THEIR INVOLVEMENT IN MOTOR SYSTEMS. The deep cerebellar nuclei receive Purkinje cell axons from discrete sections of the overlying cerebellar cortex. Based on these projections, the cortex of the corpus cerebelli can be divided into three parallel anteroposterior strips that subtend both anterior and posterior lobes: the most medial projects to the fastigial nucleus; the intermediate to the interpositus nucleus; and the most lateral projects to the dentate nucleus (Fig. 31–6C). Together, the strip of cortex and its nuclear projection target are called a corticonuclear zone. Purkinje cells also pass directly from the cortex of the flocculonodular lobe and parts of the anterior and posterior lobe vermis to the vestibular nuclei (Fig. 31–6C). Because of this anatomical arrangement, the flocculonodular lobe and vestibular nuclei can be considered together as a homologue of a corticonuclear zone.

As a conceptual framework, we can consider each of these functional elements (the three corticonuclear zones of the corpus cerebelli and the flocculonodular lobe–vestibular nuclei complex) to have the same basic anatomical arrangement. Climbing and mossy fibers excite the deep nuclei. The deep nuclei project excitatory signals to extracerebellar motor centers. As a superimposed loop, the mossy fiber and climbing fiber afferents activate Purkinje cells in the cerebellar cortex, which then inhibit the direct excitatory circuit through the deep nuclei.

The next step in cerebellar motor systems are projections from the deep nuclei to various nuclei in the brain stem and thalamus. These centers define two functional subsystems, the cerebellocerebral, which transmits signals rostrally to the cerebral cortex, and the cerebellospinal, which transmits signals to the spinal cord. All the deep cerebellar nuclei contribute to the cerebellocerebral system. The contribution from the fastigial nucleus is smaller and phylogenetically older than that for the interpositus and dentate nuclei, projecting through the reticular formation to intralaminary thalamic nuclei and then to widespread cortical areas. This is part of the nonspecific ascending cortical activating system. On the other hand, the interpositus nucleus contributes a large input, and the dentate even more, to a potent projection through the ventral anterior and ventral lateral nuclei of the thalamus (which together can be considered the *motor thalamus;* see Fig. 31–1) to motor cortex. This pathway has the characteristics of a specific projection system, manifest by strong excitatory coupling of the elements, and is located at higher levels in the hierarchy of motor activity.

All deep cerebellar nuclei have access to the cerebellospinal system through relays in subcortical motor centers. Both the dentate and interpositus project to the red nucleus, origin of the rubrospinal tract. The fastigial nucleus enters the cerebellospinal system through the reticular formation and vestibular nuclei, sites of origin of the reticulospinal and vestibulospinal tracts. In contrast to the ascending cerebellocerebral system, which has its major origin from lateral structures, the descending cerebellospinal system originates mainly from medial structures. These origins represent gradients rather than sharp cutoffs in terms of site of origin, and there is overlap.

Two descending pathways arising in the brain stem, the reticulospinal and vestibulospinal tracts (Figs. 31–1 and 31–2), merit special consideration and are part of the ventromedial descending brain stem system of Lawrence and Kuypers. The reticulospinal tract arises in the pontomedullary reticular formation and projects strongly to flexor and extensor motor neurons throughout the spinal cord. The vestibulospinal tract has two parts. The lateral division arises in the lateral vestibular nucleus and travels throughout the spinal cord to motoneurons of extensor muscles of the limbs. The medial division arises in the medial, descending, and lateral vestibular nuclei and projects to the cervical and thoracic cord, predominantly to motoneurons of axial musculature. The anatomical arrangement of these tracts allows for preferential control of extensor over flexor muscles and axial musculature over limb muscles. There is a potent input of vestibular information to the vestibular nuclei and proprioceptive information from the spinal afferents to the vestibular and reticular nuclei. In addition the reticular and vestibular nuclei are extensively interconnected. Altogether this anatomy provides for reflex movements in response to changes of the head, called vestibulospinal reflexes (Figs. 31–1 and 31–2). The cortex of the flocculonodular lobe and medial cortical nuclear zone of the anterior lobe project to the vestibular nuclei. Since there is vestibular input to the former and spinal afferent input to the latter, integrated modulation of this suprasegmental reflex through cerebellar loops is available. The inhibitory effect of Purkinje cells accounts for the fact that the primary effect of the cerebellum is to decrease the gain of these reflexes, the amount of response to a given input.

Functional considerations

SOMATIC MOVEMENTS. Having considered the basic anatomical features of the cerebellar systems, one may then turn to their functional roles in motor control. Consider first the most elementary motion, a ballistic displacement of a limb. This calls for pre-programming of a burst of muscle activity, which then determines the overall execution of the movement, including starting and stopping. Once the plan is developed it is carried over higher level open loops from the motor cortex to the agonist motoneurons (see above). Both clinical observations and experimental studies have shown that disease of the lateral corticonuclear zone causes errors in the execution of ballistic movements in the ipsilateral limb. In this case, the intensity and pattern of muscle activation is properly conducted, but excessively long. This results in the limb overshooting its target and accounts for one of the signs of cerebellar disease, *dysmetria*. When opposing ballistic movements are coupled together in tasks of rapid alternating movements, the prolonged agonist contractions prevent prompt reversal of direction by the subsequent antagonist burst, leading to the clinical signs of *dysrhythmia* and *dysdiadochokinesis*.

Both the intermediate and lateral corticonuclear zones, as components of cerebrocerebellar circuits, are also important for voluntary, self-terminated limb movements. It appears that the lateral is most involved in the transition from intention to move to actual initiation of movements. The medial is concerned with the later phases of phasic movements, regulating their execution and termination (braking), and in maintaining a tonic contraction. Recordings of the activity of single cells in various motor centers during goal-directed voluntary movements reveal a sequence of activation of lateral corticonuclear zone, motor cortex, and intermediate cerebellar corticonuclear zone. The discharges in the first two centers begin before the contraction of muscles, thereby excluding feedback as their cause. Rather, this reflects the general steps in the initiation of a motor program through open loops in a feed-forward manner. The same components of the cerebrocerebellar circuit react in reverse order to external forces that perturb an arm from an intended static position. This paradigm illustrates the relative role and timing of the structures in "reflex" movements that use closed loops with feedback, transcortical, and suprasegmental reflexes. Presumably the same operational modes and sequence are involved in the later phases of voluntary self-terminated movements.

With these constructs, one can account for the deficits of voluntary limb movements noted in cerebrocerebellar disease. Dysfunction of the lateral corticonuclear zone interferes with the phasic programming of the command signal in the motor cortex by way of open loops, prolonging the interval between the development of an idea to move and the discharge of precentral neurons and hence the onset of the muscle contractions. This accounts for the slow initiation of movements in patients with cerebellar hemisphere disease. In addition, the movements accelerate more slowly and reach smaller peak velocities, in part owing to delayed cessation of antagonist contraction. Moreover, voluntary movements deteriorate from continuous to discontinuous excursions with oscillations at approximately 3 Hz. because of improper timing sequences for agonist and antagonist muscle contractions. This collection of mistakes may arise from errors of feedback in closed loops. They lead to slowed voluntary movements with errors in termination and range (dysmetria) and an inability to maintain a fixed position of a limb, especially in response to an external pertubating force, signs associated with lesions of the cerebrocerebellum.

The cerebrocerebellum is also involved in the next level of voluntary motion, compound movements. The observation has been made that the intermediate and lateral corticonuclear zones are critical in smooth and efficient syntheses of complex movements from simple movements. However, little is known about the mechanisms involved. Dysfunction of these zones results in *decomposition* of movements with loss of synergy at various joints and loss of proper sequencing of component simple movements. The collective clinical picture when these corticonuclear zones malfunction is ataxia of ipsilateral limb movement.

The control of even more complicated compound movements, such as reflex postural adjustment of the trunk and head during sitting or standing, involves areas other than the cerebrocerebellum.* These activities can be considered automatic and stereotyped in that they are controlled primarily by lower centers without conscious effort. A static posture depends on mechanisms which provide antigravity support and equilibrium. The vestibulospinal and reticulospinal tracts with their linkage to the axial muscles of the trunk and neck and to extensor muscles of the limbs are important in these mechanisms. Vestibulospinal and associated tonic neck reflexes serve to position and maintain the head, trunk, and limbs in relation to each other and in relation to the force of gravity. The spinocerebellum (medial corticonuclear zone of the anterior lobe) and vestibulocerebellum (flocculonodular lobe) both regulate the intersegmental and intrasegmental synergy between these reflexes and modulate their gain. Thus, they would be expected to be important in the regulation of posture. The role of the vestibulocerebellum has been studied with lesions made in monkeys. Such ablations produce

---

*The involvement of higher centers, including the motor cortex, in a different type of postural adjustment, those that proceed voluntary movements, is mentioned above. The cerebrocerebellum may also participate in anticipating adjustments through the descending limb of the brachium conjunctivum.

deficits in posture that can be character-ized as a syndrome of disequilibrium. There is an inability to stand, and insta-bility of the head and trunk, along with dis-inclination to walk due to a staggering, wide-based gait. In contrast to the marked problems of control of axial and neck mus-culature, extremity movements are done well if proper support of the trunk is pro-vided (see Cerebrocerebellar Dysfunction, above).

Because stereotyped tonic posturing of the limbs (tonic neck and vestibular re-flexes) may appear in conjunction with such lesions, the syndrome can be ex-plained in part by a disinhibition and in-creased gain in the vestibulospinal loop. This is in accord with the observation that Purkinje cells in the flocculonodular lobe inhibit neurons in the vestibular nuclei. However, other mechanisms are probably involved as well. Patients with vestibulo-cerebellar dysfunction show greater pos-tural ataxia than those with cerebrocere-bellar or spinocerebellar disease. They are unable to sit or stand, sway at low frequen-cies (0.5 Hz.) in all directions, and have in-stability of the head and trunk. Experi-ments measuring postural reflexes in these patients have found that the abnormalities could be explained by a loss of interseg-mental synergy of reflexes and loss of the "set point" for postural fixation, resulting in multidirectional oscillations. Patients with spinocerebellar lesions have smaller degrees of postural instability, marked by anterior–posterior swaying at higher fre-quencies (3 Hz.) of the trunk and head in unison. The instability often is not evident until provoked by mechanical perturba-tions and can also be induced with stimu-lation of peripheral afferents. These find-ings indicate that alterations in reflexes are important in the pathophysiological un-derstanding of this type of postural insta-bility. Other electrophysiological studies implicate slowed transmission through long suprasegmental reflex loops along

with an increased gain in these loops as a fundamental disturbance. Patients with cerebrocerebellar disease have virtually normal postural reflexes.

Locomotion is an even more complex ac-tivity, involving many compound move-ments with phasic excursions of the limbs along with tonic postural fixations of the trunk and proximal portions of the limbs. Walking uses the antigravity and equilib-rium mechanisms cited above in combina-tion with a rhythmic pattern of compound limb movements (stepping). The motor centers responsible for stepping are lower, caudal to a plane that passes through the superior colliculus and mamillary bodies. The essential generator for stepping is in the spinal cord and produces rhythmic patterns of discharges that are integrated for individual limbs and between limbs. Sensory feedback from proprioceptors helps to shape the program. In addition, there are descending controls that pass through the reticulospinal, vestibulospinal, and rubrospinal tracts. The first two are es-pecially important. Through their interac-tions with reticular formation and vestibu-lar nuclei, both the vestibulocerebellar and spinocerebellar systems become involved in the control of gait. The postural difficul-ties already discussed account for ataxic gaits seen in vestibulocerebellar disease. Other analyses show that the spinocerebel-lum controls the stepping program through subcortical centers, producing speed, synergy, and braking for the phasic limb movements during stepping. Monkeys with anterior vermal lesions walk poorly. In addition, patients with alcoholic cerebellar degeneration are primarily recognized by their staggering gait. This disease is charac-terized pathologically by disease predomi-nantly in the rostromedial part of the ante-rior lobe, that part of the spinocerebellum where the legs are represented.

Cerebellar system disease also frequently produces tremors that are defined as in-voluntary, repetitive, rhythmic oscillatory movements of relatively constant amplitude and frequency. Tremors resulting from cerebellar pathology are of two types—postural and intention. They are distinguished from tremors at rest when muscles are relaxed. *Postural* (also called static) *tremors* associated with cerebellar disease are 3 to 5-Hz. oscillations that develop while a limb is actively kept in a fixed position, such as the arms being held at shoulder level. They usually develop a few seconds after the posture is assumed and typically are centered about proximal joints. Similar tremors can be produced in monkeys with lesions in the closed anatomical loop that connects the cerebellum to the red nucleus and the red nucleus back to cerebellum by way of the inferior olive. This circuit can also be activated pharmacologically to produce spontaneous rhythmic discharges. Cerebellar postural tremor may therefore result from disinhibition of a tremorigenic system that includes the closed cerebellorubroolivocerebello circuit, the Guillain-Mollaret triangle. Mechanisms like those discussed above for postural intabilities may also be involved.

Intention (also called *kinetic*) *tremors* are 3 to 5-Hz. oscillations that develop from improper contractions in proximal musculature during goal-directed movements of limb to a selected spatial point. They can be detected with the familiar finger-nose or heel-shin tests. Cooling the dentate nucleus in monkeys produces intention tremors. These experiments, as well as the type of movements that elicit intention tremor, implicate dysfunction loops involving the lateral corticonuclear zone. There is some evidence that intention tremors arise because of an inability to make appropriate corrections in response to a movement error. An oscillation would arise from a series of subsequent errors and inappropriate correction signals. However, abnormalities of feedback from the periphery through the spinal cord cannot be the sole reason for intention tremors, as it has been shown that intention tremor persists when dorsal roots are cut. Thus, it may be that part of the reason for the errors of intention tremor is a severe degradation of voluntary movements to discontinuous self-terminated movements (see above). Nonetheless, sensory input from the periphery can modify various aspects of the tremor. Both errors in feedback and errors in preprogramming of various parameters of the goal-directed movement seem to be involved.

REGULATION OF TONE. Hypertonia and decerebrate posturing are characteristically observed after cerebellar ablation in subprimates. These findings develop from loss of inhibition by the vermal portions of the corpus cerebelli on the reticulospinal and vestibulospinal tracts. In primates, this region seems to inhibit these reflexes less, and hypertonia and decerebrate posturing are not seen with cerebellar disease. Rather, hypotonia is observed and relates to dysfunction of the cerebrocerebellum. This is especially true for acute lesions, and, if unilateral, it produces the change ipsilaterally. Evidence has been adduced that this is a result of decreased fusimotor activity, which in turn decreases the excitability of α-motoneurons through segmental reflexes. The facilitation of α-motoneuron excitability that the cerebellum normally exerts is mediated through the corticospinal tract. Hypotonia also accounts for associated hypoactive and pendular deep-tendon reflexes.

A particularly useful test for the demonstration of cerebellar disease is the Stewart-Holmes maneuver. To perform this test, a limb is made to contract isometrically against a resistance which is suddenly removed. The standard test is to restrain the forearm during elbow flexion. When the restraint is removed, the limb normally moves rapidly a short distance along the axis of the force of contraction until it is stopped or "checked." In patients with dysfunction of the lateral cerebellum, there is a lack of the smooth and prompt stop mechanism, which results in a loss of check, and consequently in excessive movement. This can be understood as a delayed initiation of phasic motor activity. Holmes also described some patients who developed a powerful contraction in the antagonist muscle after release of the restraint, causing the limb to move back past the original set point and called this "rebound." Having so defined it, he found rebound to be a sign of spasticity and not of cerebellar disease. Subsequently others found that tapping the outstretched limbs of patients with cerebellar disease resulted in excessively wide excursions of the limbs in the direction of the tap followed by a motion of the limb back to and overshooting the original position and described this as rebound. Confusion has developed from the casual use of the single term rebound for two dissimilar phenomena. Nevertheless, the different diagnostic implications of these two types of "rebound" should be understood.

REGULATION OF EYE MOVEMENTS. The cerebellum is also important in the regulation of eye movements. Eye movements serve two purposes—to bring visual stimuli on the periphery of the retina onto the fovea and to maintain them there. With respect to the axial alignment of the two globes there are two classes of eye movements— conjugate eye movements (versions) and disconjugate movements eye movements (vergences). The motion of eyes may be horizontal, vertical, or rotatory. Versions are controlled by four functional oculomotor systems—slow, fast, vestibuloocular, and fixation (stabilization or position maintenance)—all having interactions with the cerebellum. Present understanding of these systems is most complete for horizontal versions and is emphasized in the present discussion (Fig. 31–7). Signals for horizontal versions are encoded on the paramedian pontine reticular formation (PPRF) and activate the adjacent abducens

motoneurons and contralateral medial reotus motoneurons. This produces conjugate movements of the eyes to the side of the activated PPRF.

The PPRF can, in turn, be activated by various voluntary and reflex control signals. For example, voluntary saccadic eye movements occur when a visual target falls on the periphery of the retina. A command signal originates in the frontal eye fields and travels to the contralateral PPRF; after a latency of 200 msec., eye movement is initiated. The rate of action potentials in the third and sixth cranial nerve nuclei motoneurons increases, first with a burst ("pulse") and then a sustained firing ("step"). This results in a rapid (up to 700 degrees/sec.) movement of the eyes. Like ballistic limb movements, saccades cannot be changed once started. With larger saccades, the versions are often inaccurate, and about 100 msec. later subsequent corrective saccades follow to bring the eyes to the proper point. Thus, saccadic movements are fast and preprogrammed and may be followed by subsequent saccades that deliver feedback correction based on target eye position errors. Voluntary slow eye movements (pursuits) are evoked by a moving target and begin with a control signal in the parietooccipital area which activates the ipsilateral PPRF. In contrast to saccades, slow eye movements are associated with a steady buildup in the discharge rate of oculomotor neurons ("ramps") that begins earlier (125 msec.) and produces slower eye movements (less than 50 degrees/sec.). The disparity between velocity of the eye movements and that of the target is the error signal in the slow eye movement system and generates a correction in the steady discharge rate of oculomotor neurons. Thus different forms of feedback in closed loops operates in both the fast and slow eye movement systems. For the fast eye movements it is discontinuous and corrects target eye-position errors, while for slow eye movements it is continuous and corrects target eye velocity errors.

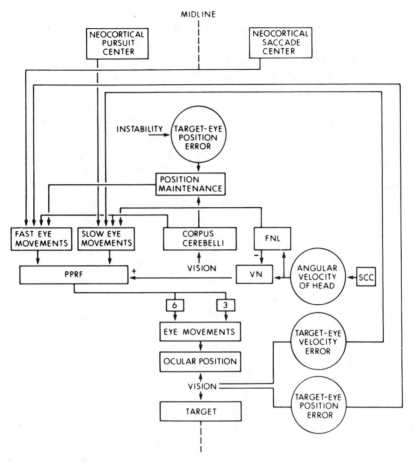

FIG. 31—7. Schematic overview of oculomotor control systems for horizontal versions and their connections with the cerebrum and cerebellum. The connections of four eye movement systems are shown along with corrective feedback controls: fast eye movements, slow eye movements, vestibuloocular reflex and position maintenance (fixation). SCC = semicircular canals; FNL = flocculonodular lobe of the cerebellum; VNC = vestibular nuclei; PPPF = paramedian pontine reticular formation; 6 = abducens nucleus and lateral rectus; 3 = oculomotor nucleus and medial rectus.

The cerebellum receives information both from spindles of oculomotor muscles and about visual stimuli. Thus, it is anatomically situated to participate in both types of eye movements. Stimulation of the corpus cerebelli produces both fast and slow eye movements. In monkeys stimulation of the vermis of lobules V to VII produce horizontal saccades, while stimuli to the intermediate and lateral corticonuclear zones produce saccades and slow eye movements. All these eye movements are to the side of stimulation. The flocculo-

nodular lobe also participates in slow eye movements.

The vestibuloocular reflex is a means to stabilize vision during head movements. For example, the right eye abducts and the left eye adducts when the head is rotated to the left. Endolymph flow in the horizontal semicircular canals is converted to a signal in the labyrinth proportional to the angular velocity of the head. Signals then spread to the vestibular nerve and nuclei and the PPRF. The organization of the labyrinths results in increased drive from one labyrinth and decreased from the other with opposing effects on opposite sides of the brain stem, in a push-pull arrangement. The overall reflex results in smooth horizontal versions with a latency of about 100 msec. The cerebellum interacts with this reflex in three ways. First, information from the labyrinth is conveyed to the flocculonodular lobe, which in turn inhibits the vestibular nuclei by means of Purkinje cell axons. Overall, this configuration provides open-loop control of the vestibuloocular reflex, the cerebellum serving to decrease its gain, the ratio of eye movement to head movement. In the most extreme case, the cerebellum integrates visual and vestibular information to decrease the gain to zero when the vestibuloocular reflex is voluntarily suppressed with fixation. Second, the cerebellum integrates visual and neck proprioceptive signals to correct errors in the vestibuloocular reflex. Third, the flocculonodular lobe is involved in long-term functional reorganization of the vestibuloocular reflex.

It is also necessary to maintain the eyes in a steady position when there is no head movement and visual attention is directed at a nonmoving target. Electrophysiological recordings show that even in normal individuals small-amplitude drifts of the eyes away from the proper position occur. The instability is maximal with the eyes at the extreme of lateral horizontal gaze (eccentric gaze position). The worsening of this phenomenon by the removal of visual input and the lessening with fixation can be attributed to a special system that compensates for this instability. This fixation system uses error signals determined by the difference between the proper and actual eye position, which then activate the fast eye movement system to produce corrective microsaccades. The cerebellum is positioned as an important element in this fixation system. In addition to these dynamic features, a stable ocular position depends on a balance of tonic excitatory drive from vestibular nuclear complexes to the contralateral PPRF.

Diseases of the cerebellum have been associated with a variety of abnormalities of eye movements (see list below). The first type of oculomotor abnormalities seen with cerebellar diseases may be considered to arise from disorders of position maintenance. Holmes reported that with unilateral cerebellar lesions the eyes no longer spontaneously rest in the midline but rather are conjugately displaced away from the lesion to a position he termed a new rest point. This may result from unilateral loss of cerebellar inhibition on the vestibular nuclei with a consequent heightening in drive of the contralateral PPRF. This sign characteristically is seen with acute lesions and is a transient phenomenon. Such unilateral cerebellar lesions may also produce a defect in horizontal versions (gaze paresis) to the side of the lesion, which is also transient (see below). The integrity of other components of the oculomotor system in these cases can be proved by the appearance of full horizontal versions with vestibular stimulation (the vestibuloocular reflex). During resolution, gaze-paretic nystagmus toward the side of the lesion may appear (see below). Conjugate vertical displacements are not features of cerebellar disease. Although Holmes described skew deviation in some of his cases, this is a nonspecific sign of posterior fossa disease.

Improper function of the position maintenance system can also result in nonvolitional spontaneous eye movements. Those produced by cerebellar disease can be grouped according to speed, symmetry, and timing of the component eye movements. The first group consists of irregular horizontal multiphasic jerks of the eyes and can be attributed to instabilities in fixation. Two small horizontal saccades, of equal amplitude and opposite direction, separated by a period of no eye movement are called square-wave jerks. They occur once or twice per second in normal individuals during visual fixation. Their amplitude is constant and so small that they usually cannot be detected clinically, being

---

Oculomotor Signs of Cerebellar System Disease

Spontaneous malposition or intabilities of
eye position
    Eccentric gaze position
    Irregular movements
        Square-wave jerks
        Macrosquare-wave jerks
        Ocular flutter
        Opsoclonus
    Regular movements
        Jerk nystagmus
           Vestibular
           Positional
           Gaze-evoked
           Rebound
           Upbeat
           Downbeat
        Pendular (acquired) nystagmus
        Ocular "myoclonus"
Disordered voluntary eye movements
    Saccades
        Ocular dysmetria; hypometric and
           hypermetric saccades
        Pendular macrooscillations
        Slowed or absent saccades
    Pursuit
        "Cogwheeling"

---

revealed only with specialized recording equipment. In cerebellar disease their frequency is increased, often appearing in bursts, and their amplitude varies greatly, producing some large-amplitude durations easily detected by observation (macrosquare-wave jerks). Ocular flutter is another sign of cerebellar disease wherein frequent ocular movements occur consisting of two horizontal saccades, first directing the eyes away from and then back to the target of fixation. In contrast to square-wave jerks, there is no latency between the two saccades. When the frequency of involuntary saccades increases so that continuous, chaotic ocular excursions in all directions develops, the condition is described as an opsoclonus. Ocular flutter and opsoclonus are considered different degrees of one pathophysiological derangement and are associated with ocular dysmetria (see below).

The third group of oculomotor signs associated with cerebellar system disease consists of spontaneous involuntary movements with a regularly recurring pattern, i.e., *nystagmus*. Daroff, Troost, and Dell Osso have outlined how different sorts of nystagmus arise in the slow eye movement system. Nystagmus can be characterized by a description of whether fast and slow movements alternate (jerk nystagmus) or there are only slow movements (pendular nystagmus). Further information is produced by the response to various maneuvers. Of particular importance for jerk nystagmus is whether it is diminished by visual fixation and whether it increases when the eyes are held away from primary position in the direction of the fast phase and decreases when they are held in the opposite direction (Alexander's law). In addition, electrophysiological techniques show that the slow eye movements of different types of jerk nystagmus have unique characteristics described below.

Vestibular nystagmus consists of horizontal jerk nystagmus (sometimes with additional rotatory vectors) appearing with the eyes straight ahead (primary position). Additional characteristics are that it obeys Alexander's law, is decreased with visual fixation, and that the slow eye movements, arising from a steady driving force, have a constant velocity. Originating from an imbalance in the vestibular inputs to the PPRF, vestibular nystagmus can be caused by disease at many points, including the vestibular end-organ, nerve, nuclei, or the cerebellum. This fact can be understood by examining Fig. 31–7. When cerebellar lesions produce vestibular nystagmus, the fast phase is toward the cerebellar lesion and, if gaze is directed to the side of the lesion, nystagmus is enhanced. This differs from nystagmus due to disease of the vestibular system proper, in which the fast phase is away from the lesion and is enhanced with eye closure. A related condition is positional nystagmus, which is a jerk nystagmus elicited by changes of head position. Although positional nystagmus has been said to occur with cerebellar lesions (especially those in the midline), it cannot be considered a specific indicator for cerebellar disease.

Gaze-evoked (gaze-paretic) nystagmus, the form of nystagmus most often clinically encountered, is elicited by maintaining an eccentric horizontal eye position. The eyes drift away from the intended position with slow movements of decreasing velocity and are returned with saccades. Fixation diminishes gaze-evoked nystagmus. Thus, nystagmus arises from an inherent instability of the position maintenance system, which causes the drift and is normally checked by a feedback curcuit involving the cerebellum (see Fig. 31–7). Persistent bilateral gaze-paretic nystagmus is characteristically seen with slowly progressive degenerative cerebellar disease. Gaze-paretic nystagmus is also associated with unilateral cerebellar lesions, either as a presenting sign or during recovery from a gaze paresis. The localizing value of gaze-paretic nystagmus is limited, since it also occurs with various extracerebellar lesions. However, a particular type of gaze-paretic nystagmus, specific for cerebellar disease, has been described. Rebound nystagmus consists of a nystagmus that is produced by lateral gaze and characterized by fast phase in the direction of gaze that fatigues or may even reverse direction over several seconds. When gaze is returned to the primary position, a nystagmus opposite in direction to the first evoked nystagmus is produced and subsequently fatigues, even though no nystagmus was initially present in primary position.

Upbeat and downbeat nystagmuses are vertical jerk nystagmuses that arise from imbalances specifically in the upward and downward pursuit systems. Other elements of the slow eye movement system are presumably intact because vertical vestibuloocular reflex movements are intact. Fixation does not reduce the nystagmus. Downbeat nystagmus is characterized by jerks of the eyes downward and does not follow Alexander's law since a combined downward and lateral deviation of the eyes maximizes the nystagmus. Cerebellar disease and craniocervical junction abnormalities are associated with downbeat nystagmus. Of note in relation to this is the fact that nystagmus can be produced by stimulation of the flocculonodular lobe. A form of upbeat nystagmus with large amplitude upward jerks, obeying Alexander's law, has been seen with lesions of anterior vermis of the cerebellum.

Pendular nystagmus is characterized by slow eye movements of alternate direction occurring in primary position and is increased with fixation. It results from excess gain and positive feedback in the slow eye movement system so that this component of the nystagmus shows increasing velocity. Acquired pendulary nystagmus is occasionally seen with lesions of the cerebellar system, particularly those caused by demyelinating disease. A related condition is *ocular myoclonus,* in which there are continuous, rhythmic pendular, conjugate oscillations of the eyes at a rate of several times per second. The direction of the movements are usually vertical, but may be in other directions. Frequently there are coexistent contractions of the palate, tongue, or pharynx. The regularity and synchrony of these movements makes them a complex tremor. Since this syndrome is associated with disease in the Guillain-Mollaret triangle, pathophysiological processes considered above for postural tremor may apply.

Cerebellar disease also causes abnormal voluntary eye movements. Saccades may be disturbed in several ways. Analogous to dysmetric limb movements, the moving eyes may fall short of (hypometric saccades) or overshoot (hypermetric saccades) the target. Corrections follow to bring the eyes to the actual target, resulting in stepwise eye movements or damped oscillations. In more severe cases several hypometric saccades are needed to complete a voluntary eye movement or there may even be total loss of saccades. With acute unilateral cerebellar lesions, one may see a disturbance triggered by voluntary saccades and characterized by large-amplitude conjugate ocular oscillations from side to side of first increasing and then decreasing amplitude at a rate of 2 to 3 times per second, called pendular macrooscillations.

The cerebellar system also participates in slow-pursuit eye movements. Defects of this system in association with cerebellar disease seem to occur less frequently but are well known. This may be demonstrated with standard tracking tasks in which the slow-pursuit eye movements become replaced by a series of saccades (cogwheeling) or by optokinetic nystagmus, in which there is a diminution of the slow phase toward the side of the lesions or in both directions with bilateral lesions. Acute and severe cerebellar lesions can also lead to total loss of slow eye movements, producing transient gaze paresis to the side of the lesion.

REGULATION OF SPEECH. *Dysarthria*, disordered speech, may be seen with severe, acute cerebellar lesions, being most marked in the early stages and gradually subsiding. It is a late sign in more slowly developing cerebellar lesions, often appearing after several of the signs discussed above are well established. Based on observations made in the early 1900s, dysarthria was believed to be a sign of midline cerebellar disease. Subsequent studies have associated dysarthria with lesions of the cerebellar hemispheres, especially those on the left, and suggested that the intermediate corticonuclear zone of lobules VI and VII was the most critical area for the development of dysarthria.

The abnormalities noted with cerebellar dysarthria have been examined in detail. Briefly, the elementary disabilities are poor articulation, resulting in improper formation and separation of individual syllables and disordered phonation, with loss of texture, tone, stress, and rhythm of individual sounds. The last is the major disturbance,

leading to slow, monotonous, and improperly measured speech, termed *scanning*, while the former adds the element of loss of clarity. Less often there may be a visible difficulty in delivering the speech with concomitant "explosiveness."

Speech in patients with cerebellar dysarthria has the proper elementary motions, but they are substantially slowed in onset and progression. In addition, there are disturbances that may be thought of as ataxia of respiratory, oral, buccal, lingual, and pharyngeal movements. Therefore, the same sort of pathophysiological mechanisms in open and closed loops described for limb movements can be involved for dysarthria. Afferent information from the cochlea and from the laryngeal nerve terminates in the medial and intermediate corticonuclear zones of lobules VI and VII. This provides a basis for the observation already mentioned that the central part of the corpus cerebelli on the anteroposterior axis is especially involved in the control of speech. In addition, the intermediate and medial corticonuclear zones at this level have a disproportionately high input from the cerebral cortex (see Fig. 31–6). Lechtenberg and Gilman have explained the greater importance of the left side of the cerebellum in producing dysarthria by its connections with the right hemisphere, which is said to be concerned with the melodic and prosodic aspects of speech, derangements of which typify cerebellar disorders of speech.

CEREBELLAR SYNDROMES. In summary, the cerebellum has a crucial role in motor activities of the trunk, head, limb, eyes, and speech. It participates in specific goal-directed tasks of individual limbs under volitional control, in reflex movements of larger parts of the body, and in phasic and tonic movements. Furthermore, the cerebellum is an important center for the maintenance of normal muscle tone. While any motor event is probably controlled to some degree by each of the four functional efferent zones (three corticonuclear zones and the flocculonodular cortex-vestibular nuclear complex), each one of these areas is intimately related to particular types and aspects of movement. From these considerations, one would expect that pathology confined to particular regions of the cerebellum would produce a particular set of signs and symptoms. Two clinically useful syndromes can be identified and are summarized below.

Cerebellar Syndromes

Lateral syndrome (lateral and intermediate corticonuclear zones)
  Disordered voluntary movements of the ipsilateral limbs
    Slowed onset and execution
    Dysmetria
    Decomposition of movement
    Dysrhythmia and dysdiadochokinesis of repetitive movements
  Hyptonia
    Abnormal attitudes of limbs
    Decreased resistance to passive movement
    Hypoactive and pendular deep tendon reflexes
  Abnormal Stewart-Holmes sign
    Impaired check and excess rebound
  Involuntary movements
    Postural tremor
    Intention tremor
  Dysarthria
  Oculomotor disturbances
Medial syndrome (medial corticonuclear zone and flocculonodular lobe)
  Abnormalities of gait and stance and head position
    (Anterior lobe: gait ≥ stance)
    (Flocculonodular lobe: stance = gait)
  Oculomotor disturbances

The *lateral syndrome* reflects disease in the intermediate and lateral corticonuclear zones of the anterior and posterior lobes. Individual limb movements on the side of the lesion are ataxic during voluntary goal-directed tasks with the signs of slow onset and progression, dysmetria, dysdiadochokinesis, and dysrhythmia. Intention and postural tremors and dysarthria may be present. The limbs are hypotonic, reflexes are pendular, and poor check and over-

shoot can be demonstrated. In addition, a number of oculomotor abnormalities may appear. The arm on the involved side may drift upward and outward to the side of the lesion if held in front of the body with closed eyes. There may be some degree of gait disturbance in proportion to the ataxia of lower limbs.

The *medial syndrome*, seen with disease of the medial corticonuclear zone and flocculonodular lobe, is manifest by abnormalities of posture and gait, titubation of the head and trunk and often a rotation or tilting of the head. However, individual limbs are not ataxic in goal-directed movements, and tone is normal. Oculomotor disturbances are common and include those seen with the lateral syndrome. The spatial separation of the spinocerebellum from the vestibulocerebellum along the anteroposterior axis of the cerebellum (see Fig. 31–6) and restriction of certain diseases to these areas account for a subdivision of this syndrome. Thus, a medulloblastoma involving the flocculonodular lobe is commonly recognized by a *vestibulocerebellar syndrome* with nystagmus and dysequilibrium of the head and trunk during static posture and locomotion. Alcoholic cerebellar degeneration, maximal in the midline of the rostral anterior lobe, leads to disordered gait without limb ataxia, except perhaps the lower extremities.

A diagnosis of cerebellar system disease is based on the presence of these findings. The specific signs and symptoms, as well as the underlying etiological process, may afford anatomical localization. However, there are several factors that interfere with the precision of localization based on this scheme. The first of these involves compensation. By mechanisms that are still not understood, the nervous system is often able to operate quite well despite destruction of parts of the cerebellum. Thus, with slowly progressive lesions or some time after acute lesions, cerebellar signs may not be apparent despite careful clinical scrutiny. The second arises because the posterior fossa, which the cerebellum shares with the brain stem, has little extra space. Thus, when brain edema occurs, cerebellar signs may be overshadowed by the effects of increased intracranial pressure in the posterior fossa. This leads to compression or ischemia of brain stem structures, resulting in other findings, as well as headache, nausea, vomiting, or depression of consciousness. Third, since the cerebellum is but a component in the loops described above, disease not within the cerebellum itself may produce cerebellar signs. For example, ataxia may be seen with spinal cord lesions and has been reported as a consequence of dysfunction of the cerebrum.

This discussion has been directed at the major deficiencies which herald cerebellar disease as correlated with physiological and anatomical facts. They are all within the motor sphere. As a final point, there is evidence that the cerebellum has a role in arousal, psychiatric, vegetative, or sensory functions, and the control of epilepsy. Although these functions are poorly understood at present, they suggest that in the future other clinical signs and symptoms may be associated with cerebellar dysfunction.

Annotated references

Allen, G. I., and Tsukahara, N.: Cerebrocerebellar communication systems. Physiol. Rev. *54*:957, 1974. (A detailed review of the connections between the cerebral cortex and cerebellum and their involvement in motor control loops.)

Brooks, V. B., and Thach, W. T.: Cerebellar control of posture and movement. *In* Brooks, V. B. (ed.): Handbook of Physiology, Sec. I: The Nervous System, Vol. II: Motor Control, Chap. 18, p. 877. Bethesda, American Physiological Society. (A comprehensive treatise on the role of the cerebellum in motor activities by two respected neurophysiologists that ranges from single cells to the organ as a whole.)

Brown, J. R., Barley, F. L., and Aronson, A. E.: Ataxic dysarthria. Int. J. Neurol. *7*:301, 1970. (These authors provide a careful description of the speech disturbance produced by cerebellar disease and how to recognize it.)

DeRenzi, E., Peiczuro, A., and Vignola, C. A.: Ideational apraxia: A quantitative study. Neuropsychologia *6*:41–52, 1968. (A summary of apraxias seen clinically and how to differentiate them.)

Dow, R. S., and Moruzzi, G.: The Physiology and Pathology of the Cerebellum. Minneapolis, University of Minnesota Press, 1958. (An encyclopedic review of the experimental literature up to the 1950s; a clinical description of cerebellar signs and symptoms is also delivered with unique clarity.)

Eccles, J. C., Ito, M., and Szentagothai, J.: The Cerebellum as a Neuronal Machine. New York, Springer-Verlag, 1967. (A Nobel laureate in neurophysiology and his colleagues collate sophisticated electrophysiological observations into a coherent account of the cellular interactions in the cerebellum and their functional meaning.)

Evarts, E. V.: Role of motor cortex in voluntary movements in primates. *In* Brooks, V. B. (ed.): Handbook of Physiology, Sec. I: The Nervous System, Vol. II: Motor Control, Chap. 23, p. 1083. Bethesda, American Physiological Society, 1981. (Extensive review of the role of motor cortex in movement including analysis of single cell recordings from monkeys during operantly conditioned motor tasks.)

Evarts, E. V., and Tanji, J.: Reflex and intended responses in motor cortex pyramidal tract neurons of monkey. J. Neurophysiol. *39*:1069, 1976. (Demonstration of the effect of intent on transcortical reflexes.)

Evarts, E. V., and Thach, W. T.: Motor mechanisms of the CNS: Cerebrocerebellar interrelations. Annu. Rev. Physiol. *31*:451, 1969. (An insightful review of the role of the motor cortex, the cerebellum and their interactions as major components of the motor system.)

Gilman, S., Bloedel, J. R., and Lechtenberg, R.: Disorders of the Cerebellum, Philadelphia, F. A. Davis, 1981. (Written by clinicians and experimental investigators, this detailed monograph first examines laboratory evidence about the function of the cerebellum and then provides a careful description of clinical syndromes and specific disease processes. This text is a useful conjunct to the work of Dow and Moruzzi, particularly in stressing current notions.)

Glaser, J. S.: Neuro-Ophthalmology. Chap. 9, Eye movement characteristics and recording techniques; Chap. 10, Supranuclear disorders of eye movements; Chap. 11; Nystagmus and related ocular oscillations. Hagerstown, Harper & Row, 1978. (A quite readable account of the basic physiology of the various types of eye movements and how lesions, such as those in the cerebrum or cerebellum, disturb these to produce distinctive clinical signs.)

Hallet, M., Shahani, B. T., and Young, R. R.: EMG analysis of stereotyped voluntary movements in man. J.

Neurol. Neurosurg. Psychiatry 38:1154, 1975a. (See next reference.)

Hallet, M., Shahani, B. T., and Young, R. R.: EMG analysis of patients with cerebellar deficits. J. Neurol. Neurosurg. Psychiatry. 38:1163, 1975b. (These two articles are careful studies done with actual patients and confirm some of the hypothesis offered from animal studies about the types of movements used by humans and how they go awry with cerebellar pathology.)

Holmes, G.: The cerebellum of man. Brain 62:1, 1939. (Written at the end of his career, this classic piece by Dr. Holmes is a masterpiece of clinical observation and lucid exposition. No better description of the basic signs and symptoms of cerebellar disease is available and it provides an index to his previous works that are equally informative.)

Jankovic, J., and Fahn, S.: Physiologic and pathologic tremors. Ann. Intern. Med. 93:460, 1980. (A tidy summary of the types of tremor encountered in neurological practice and their pathophysiologic.)

Jones, E. G., and Powell, T. P. S.: An anatomical study of converging sensory pathways within the cerebral cortex of the monkey. Brain 93:793, 1970. (Anatomical studies in monkeys describing the sequential projections from primary somatosensory cortex and other sensory cortical areas.)

Landau, W. M.: Spasticity and rigidity. In Plum, F. (ed.): Recent Advances in Neurology, Chap. 1, p. 1. Philadelphia, F. A. Davis, 1969. (A general review of these topics.)

Landau, W. M., and Clare, M. H.: The plantar reflex in man. Brain 82:321, 1959. (A review of the physiology of the plantar response in man.)

Lawrence, D. G., and Kuypers, H. G. J. M.: The functional organization of the motor system in the monkey. I. The effects of bilateral pyramidal lesions. Brain 91:1, 1968. (See next reference.)

Lawrence, D. G., and Kuypers, H. G. J. M.: The functional organization of the motor system in the monkey. II. The effects of lesions of the descending brainstem pathways. Brain 91:15, 1968. (Two part series describing two general systems descending from the brainstem that are instrumental in the control of movement.)

Lothman, E. W., and Ferrendelli, J. A.: Diseases and disorders of the cerebellum. In The Science and Practice of Clinical Medicine, Volume 5: Neurology, Rosenberg, R. N. (ed.): p. 431. New York, Grune and Stratton, 1980. (Written along the lines of this chapter with additional anatomy and physiology as well as overview of the specific etiological processes that can produce findings of cerebellar dysfunction.)

Mauritz, K. H., Dichgans, J., and Hufschmidt, A.: Quantitative analysis of stance in late cortical cerebellar atrophy of the anterior lobe and other forms of cerebellar ataxia. Brain 102:461, 1979. (A clinical neurophysiological study relating basic pathophysiology in restricted regions of the cerebellum to postural dysregulation.)

Moll, L., Kuypers, H. G. J. M.: Premotor cortical ablation in monkeys: Contralateral changes in visually guided reaching behavior. Science 198:317, 1977. (A study demonstrates motor deficits associated with lesions of the premotor cortex and how these deficits relate to apraxias.)

Phillips, C. G., and Porter, R.: The Corticospinal Neurons: Their Role In Movement. London, Academic Press, 1977. (An extensive review of the anatomy and physiology of the motor cortex and the corticospinal tract.)

Roland, P. E., Larsen, B., Larsen, N. A., and Skinhoj, E.: Supplementary motor area and other cortical areas in organization of voluntary movements in man. J. Neurophysiol. 43:118, 1980. (Studies of the changes in regional cerebral blood flow as a measure of physiological activity during different types of movement.)

Ron, S., and Robinson, D. A.: Eye movements evoked by cerebellar stimulation in the alert monkey. J. Neurophysiol. 36:1004, 1973. (Using precise electrophysiological techniques, these authors provide important information of the role of the cerebellum in producing eye movements.)

Snider, R. S., and Stowell, A.: Receiving areas of the tactile, auditory and visual systems in the cerebellum. J. Neurophysiol. 7:331, 1944. (Although this work was done almost four decades ago, it remains the standard for localization of termination for the various types of afferent input to the cerebellum and has been instrumental in subsequent interpretations of cerebellar functions.)

Thach, W. T.: The cerebellum. In Mountcastle, V. B. (ed.): Medical Physiology, 14th ed., Chap. 31, p. 837. St. Louis, C. V. Mosby, 1978. (A lucid account, amply illustrated, that provides a thorough overview of the physiology of the cerebellum.)

Thach, W. T.: Correlation of neural discharge with pattern and force of muscular activity, joint position, and direction of intended next movement in motor cortex and cerebellum. J. Neurophysiol. 41:654, 1978. (Study of the relationship of neuronal activity in the motor cortex and dentate nucleus of monkeys to various aspects of movement such as force, direction, and the next intended movement.)

Victor, M., Adams, R. D., and Mancall, E. L.: A restricted form of cerebellar cortical degeneration occurring in alcoholic patients. Arch. Neurol. 1:579–588, 1959. (A monograph on one of the "experiments of nature" that has added to our understanding of how the cerebellum works, this article correlates the pathology, physiology, and clinical features of toxic degeneration of the anterior lobe of the cerebellum.)

Wetzel, M. C., and Stuart, D. G.: Ensemble characteristics of cat locomotion and its neural control. Prog. Neurobiol. 7:1, 1976. (This review assembles many experimental observations into a concise account of what is presently known of the neural substrates for locomotion.)

Wiesendanger, M.: Organization of secondary motor areas of cerebral cortex. *In* Brooks, V. B. (ed.): Handbook of Physiology, Sec. 1: The Nervous System, Vol. II: Motor Control, Chap. 24, p. 1121. Bethesda, American Physiological Society, 1981. (Extensive review of the anatomy and physiology of the supplementary motor area.)

Wiesendanger, M., Ruegg, D. G., and Lucier, G. E.: Why transcortical reflexes? Can. J. Neurol. Sci. 2:295, 1975. (A review of the transcortical reflexes and their physiological implications.)

Wilson, V. J., and Peterson, B. W.: Peripheral and control substrates of vestibulospinal reflexes. Physiol. Rev. 58:80, 1978. (These authors neatly summarize the functions of the vestibular organs and their central connections as well as the basic role the vestibulospinal reflexes in motor programs.)

Wise, S. P., and Evarts, E. V.: The role of the cerebral cortex in movement. Trends Neurosci. 6:297, 1981. (A discussion of the relationship of the motor cortex to various types of movement.)

32   Extrapyramidal System

Stanley Fahn, M.D.

The extrapyramidal system is an anatomical and physiological unit that comprises the deep motor nuclei of the brain known as the basal ganglia and their connecting pathways. The basic science aspects of the basal ganglia are discussed in the next section of this chapter.

Lesions and diseases involving the extrapyramidal system give rise to clinical syndromes known as *movement disorders.* Movement disorders can be defined as neurological dysfunctions in which there is either an excess of movement (involuntary movements, abnormal involuntary movements, hyperkinesias, dyskinesias) or a paucity of voluntary and automatic movement, unassociated with weakness or spasticity. The latter group is referred to as bradykinesia, akinesia, or hypokinesia; when present, it indicates the disorder known as parkinsonism. Basically then, movement disorders can conveniently be divided into parkinsonism and the dyskinesias, and there are about an equal number of patients in each of the two groups. Most movement disorders are associated with pathological alterations in the basal ganglia, that group of gray matter nuclei lying deep within the cerebral hemispheres. There are some exceptions to this general rule, notably myoclonus and many forms of tremors do not appear to be related primarily to basal ganglia pathology.

In addition, it is not known which part of the brain is associated with tics, although the basal ganglia have been suggested.

We have more understanding of the physiological alterations associated with parkinsonism than those associated with the dyskinesias. In fact, discoveries in the neurochemistry and neuropharmacology of parkinsonism were extended to the dyskinesias to provide the limited knowledge that we do have in the latter group. Moreover, most of our knowledge centers on the neurotransmitter dopamine, which is one of the major neurotransmitters in the basal ganglia. The pathophysiological discussion in this chapter is concerned predominantly with the role of basal ganglia neurotransmitters, particularly dopamine, in the causation of symptoms.

Structure, function, and biochemistry of the basal ganglia

Anatomy and function

From a clinical and functional viewpoint, the basal ganglia consist of five pairs of nuclei. The three located deep within the telencephalon are the caudate nucleus, putamen, and globus pallidus. Together they are called the corpus striatum. The caudate nucleus and putamen are histologically, chemically, and physiologically similar, and they differ primarily in their somatotopically arranged afferent and efferent connections. For example, the laterally and posteriorly located putamen receives afferents from the lateral parts of the cerebral cortex, while the medially and anteriorly located head of the caudate nucleus receives afferents from the medial and anterior parts of the cortex. Essentially, the caudate nucleus and putamen can be considered to be the same structure divided by fibers of the internal capsule descending from the cerebral cortex on their way to the brain stem and spinal cord. The caudate nucleus and putamen are collectively referred to as the neostriatum, or striatum for short. Functionally, the striatum serves as the major afferent station of the basal ganglia, receiving inputs from all parts of the cerebral cortex and also major inputs from the thalamus, particularly the centrum medianum and the intralaminar nuclei.

The globus pallidus (pallidum) or paleostriatum is divided into two parts, a lateral (external) segment and a medial (internal) segment. In contrast to the predominantly small nuclei constituting the neostriatum, the pallidum consists mainly of large neurons, that can be considered motor in function. The pallidum, particularly the medial segment, is the major efferent station of the basal ganglia. Sharing this role is the pars reticulata of the substantia nigra (this nucleus is discussed below). Histologically and chemically, the pars reticulata is very similar to the medial segment of the pallidum; these two structures are separated from each other by fibers of the internal capsule. The pallidum and pars reticulata send efferents to the ventroanterior (VA) and ventrolateral (VL) nuclei of the ipsilateral thalamus, which serve as relay stations for ultimate connection to the premotor and motor cortex, respectively. Thus, the basal ganglia are linked to the motor system of the cerebral cortex and serve to regulate and modify cortical motor activity.

The pallidum and pars reticulata also send efferents to some brain stem nuclei.

The link between the afferent (neostriatum) and efferent (pallidum and pars reticulata) parts of the basal ganglia is a direct one. Fibers from the striatum enter the lateral segment of the pallidum, where some collaterals terminate, then enter the medial segment, where again some collaterals terminate, and finally go to the pars reticulata. Thus, the three terminals of the striatal efferents appear to be collaterals of the same fibers.

The remaining two nuclei of the basal ganglia, the subthalamic nucleus and the substantia nigra, are located in the diencephalon and mesencephalon (midbrain), respectively. The subthalamic nucleus has intimate connections with the pallidum. It receives fibers from the lateral segment and sends efferents to the medial segment of the pallidum. It is best to view the function of the subthalamic nucleus as modulating the activity of the pallidum. The substantia nigra can be divided into a pigmented (neuromelanin) and cellularly rich dorsal part (pars compacta) and a less dense nonpigmented ventral and lateral part (pars reticulata). The neurons in the pars compacta have dendrites that project into the pars reticulata, thereby receiving afferents from this structure, and axons that travel to and terminate in the striatum. The nigrostriatal fibers contain dopamine as the neurotransmitter, and the loss of these cell bodies and fibers represents the major pathological finding in Parkinson's disease. Since the afferents of the pars reticulata come from the striatum, as mentioned above, the substantia nigra can be viewed as a structure modulating the activity of the striatum. In this way, it is analogous to the subthalamic nucleus, which modulates the activity of the pallidum. The connections of these five nuclei are depicted in Figure 32–1.

The basal ganglia play a role in initiating voluntary movement, maintaining posture and muscle tone, regulating postural re-

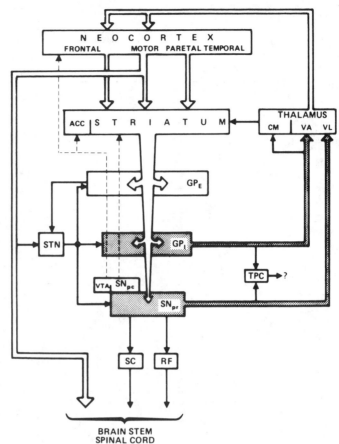

FIG. 32–1. Anatomic relations of the basal ganglia and their major connections. Gpe = external segment of the globus pallidus; GPi = internal segment of the globus pallidus; SNpr = substantia nigra pars reticulata; SNpc = substantia nigra pars compacta; STN = subthalamic nucleus; CM = centrum medianum; VA = ventralis anterior; VL = ventralis lateralis; TPC = tegmenti pedunculopontinus pars compacta; SC = superior colliculus; RF = reticular formation. (DeLong, M.R., and Georgopoulos, A.P.: Adv. Neurol. 24:131–140, 1979. Reproduced with permission from the publisher.)

flexes, and conducting automatic movements, such as swinging of the arms on walking. Much of what is known of the physiology of the human basal ganglia comes from observations on patients with diseases or lesions of the various parts of these deep nuclei. Studies in animals, including models of disease states, have greatly helped in our understanding.

Biochemistry of the dopamine system

The basal ganglia are rich in neurotransmitters and collectively have the highest concentrations of dopamine, acetylcholine, and γ-aminobutyric acid (GABA) in the brain (Table 32–1). Although the dopamine cell bodies are located in the pars compacta, the greatest concentration of dopamine is found in the nerve terminals of

the nigrostriatal fibers, and thus the striatum is the richest site of dopamine. The cellular physiological function of dopamine is complex, but the overall action appears to be one of inhibiting the activity of striatal neurons. Dopamine is synthesized in the nerve cell bodies and directly in the nerve terminals of the nigrostriatal fibers. There is a blood-brain barrier for the monoamines, and dopamine is synthesized from L-tyrosine (Fig. 32–2). The first step, the hydroxylation of tyrosine to form 3,4-dihydroxyphenylalanine (DOPA), is catalyzed by the enzyme tyrosine hydroxylase. This is the rate-limiting step in the synthesis of catecholamines. The second and final step in the synthesis of dopamine is the decarboxylation of DOPA to form dopamine, catalyzed by the enzyme DOPA decarboxylase. Both of these synthesizing

TABLE 32–1
Regional Distribution Patterns of Neurotransmitters in Human Brain

Region of Brain	Dopamine (ng./g.)	Norepinephrine (ng./g.)	Serotonin (ng./g.)	CAT* (% of Caudate)	GAD† (% of Caudate)
Caudate nucleus	3264	75	429	100	100
Putamen	4453	54	476	107	100
Globus pallidus	766	69	301	31	139
Substantia nigra	463	119	990	25	128
Pars compacta	760	60	—	—	—
Pars reticulata	340	50	—	—	—
Nucleus accumbens	3140	—	—	89	120
Red nucleus	744	267	603	14	37
Hypothalamus	540	813	560	23	80
Thalamus	155	102	50	12	41
Cerebral cortex	129	27	27	10	79
Cerebellar cortex	167	33	—	25	58
Dentate nucleus	264	35	—	10	106
Amygdala	300	115	284	25	33
White matter	30	10	250	—	0

* CAT-choline acetyltransferase, the synthesizing enzyme for acetylcholine.
† GAD-glutamic acid decarboxylase, the synthesizing enzyme for GABA.
(Data for dopamine, norepinephrine, serotonin, and CAT obtained from Fahn, S. [Adv. Neurol. 14:58–88, 1976] except for dopamine in the nucleus accumbens, which is the average of that from Price, et al. [Adv. Biochem. Psychopharmacol. 19:293–300, 1978] and Walsh et al. [Ann. Neurol. 12:52–55, 1982]. Data for GAD obtained from Maker, et al. [Ann. Neurol. 10:377–383, 1981].)

enzymes are present in nerve cell bodies and nerve terminals of catecholaminergic neurons. In the parts of the nervous system where the neurotransmitter norepinephrine is found, dopamine is an intermediary and immediate precursor of norepinephrine. Its synthesis requires the presence of the enzyme dopamine-β-hydroxylase, which is not present in neurons in which dopamine is the neurotransmitter. Hence, dopaminergic neurons are phylogenetically more ancient than are the noradrenergic neurons. This fits with the function of the basal ganglia, which represent a more archaic motor system than the more recently evolved corticospinal motor system (the pyramidal system). The basal ganglia motor system is usually referred to as the extrapyramidal system, even though this is technically a misnomer, since the basal ganglia influences motor function through the motor cortex and the pyrami-

dal tracts. The concept of the basal ganglia as phylogenetically old fits with their function as the major motor system of reptiles and birds.

Dopamine is released not only at nerve terminals in the striatum but also at nerve swellings located along the axons and referred to as *boutons en passage*. Since the nigrostriatal dopaminergic fibers are firing continuously rather than phasically, it would appear that dopamine is constantly being released all over the striatum. In this sense dopamine can be looked upon as bathing the neurons of the striatum, serving as a neurohormone to keep the neurons inhibited.

Alpha-methyltyrosine is a competitive inhibitor of tyrosine hydroxylase, and α-methyldopa inhibits DOPA decarboxylase. After its synthesis in the cytoplasm of the nerve terminal and bouton, dopamine is taken up and stored in granulated vesi-

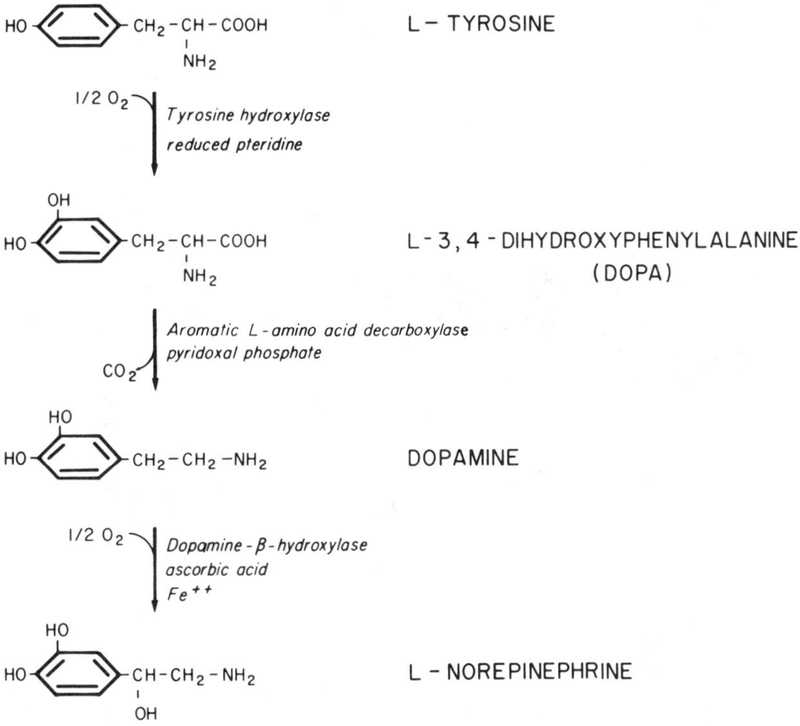

FIG. 32–2. The metabolic pathway for the biosynthesis of dopamine and norepinephrine.

cles until released following a nerve impulse. The effect of dopamine on postsynaptic receptors is terminated primarily by reuptake into the presynaptic nerve terminals and boutons, in which dopamine can again be stored within nerve granules, where it is protected from the action of monoamine oxidase (Fig. 32–3). Unstored dopamine is metabolized by monoamine oxidase and catechol-O-methyltransferase to form its major metabolite, homovanillic acid (HVA) (Fig. 32–4). The drugs reserpine and tetrabenazine block uptake of dopamine into the granules, thereby allowing the enzymatic depletion of dopamine.

The receptors of dopamine are located both postsynaptically and presynaptically. The presynaptic receptors serve as part of a feedback mechanism to regulate further synthesis and release of dopamine. There are several types of dopamine receptors, but, most conveniently, they can be divided

into those that are linked with adenyl cyclase (D1 receptors) and those that are not linked with this enzyme (D2 receptors). D2 receptors have been subdivided into D2, D3, and D4 receptors, depending on the effects of various agonists and antagonists. As a general rule, the postsynaptic receptors become supersensitive (denervation supersensitivity) if the presynaptic neurons are lost or if dopamine is blocked from reaching the receptors, such as by the action of the antipsychotic drugs (e.g., phenothiazines and butyrophenones). This concept is important because these receptors are supersensitive in both *Parkinson's disease* (in which the presynaptic dopaminergic neurons are lost) and in *tardive dyskinesia* (in which chronic treatment with antipsychotic drugs has blocked the receptors).

The highest concentration of dopamine in brain is found in the neostriatum and

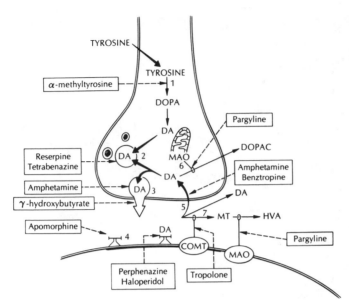

FIG. 32–3. Diagram of a dopamine synapse. DA = dopamine; DOPAC = 3,4-dihydroxyphenylacetic acid; MT = 3-methoxytyramine; HVA = homovanillic acid; MAO = monoamine oxidase; COMT = catechol-O-methyltransferase. (Cooper, J. R., Bloom, F. E., and Roth, R. H.: Biochemical Bases of Neuropharmacology, ed. 3, New York, Oxford University Press, 1978. Reproduced with permission from the publisher.)

FIG. 32–4. The metabolic pathway for the catabolism of dopamine.

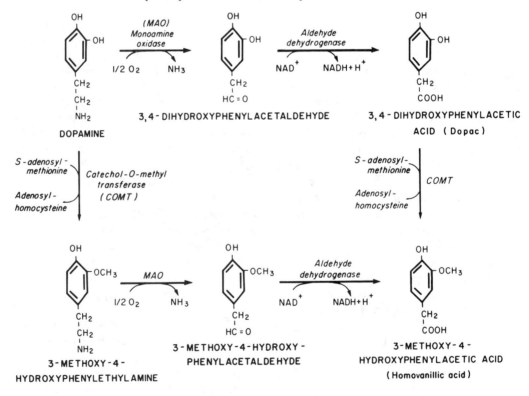

TABLE 32–2
Dopamine Pathways in Brain

System	Cells of Origin	Projections
Nigrostriatal	Substantia nigra	Neostriatum
Mesolimbic	Ventral tegmental area	Nucleus accumbens
Mesocortical	Ventral tegmental area	Suprarhinal cortex
Periventricular	Periaqueductal gray (PAG)	PAG; medial thalamus
Incertohypothalamic	Zona incerta	Hypothalamus
Tuberohypophyseal	Arcuate nucleus	Median eminence
Periglomerular	Olfactory bulb	Local dendrites
Retinal	Inner nuclear layer	Local dendrites

nucleus accumbens (part of the limbic system and geographically located in the ventromedial part of the head of the caudate nucleus). These sites contain the nerve terminals of two of the major dopaminergic pathways, the nigrostriatal and mesolimbic systems. The cells of origin of the mesolimbic (and mesocortical) dopaminergic fibers lie in the ventral tegmental area of Tsai, which is located in the midbrain medial to the substantia nigra. Dopamine is much more highly concentrated in the terminals than in nerve cell bodies because the storage sites of dopamine (i.e., granulated vesicles) are located in the terminals. Other areas in the brain contain approximately one order of magnitude less dopamine than the striatum and accumbens (Table 32–1). In contrast to dopamine, norepinephrine is most concentrated in the hypothalamus, and serotonin is more uniformly distributed in brain.

The dopamine neurons in brain have been traced by fluorescence histochemistry demonstrating dopamine, by lesioning experiments resulting in the loss of dopamine as determined by biochemical determinations, and by immunohistochemical techniques demonstrating tyrosine hydroxylase. A number of dopaminergic fiber systems have been discovered, the major ones being the nigrostriatal, the mesolimbic, the mesocortical and the tuberohypophyseal systems. Table 32–2 lists the dopaminergic pathways in brain, along with the cells of origin and the site of the terminals.

Biochemistry of other neurotransmitters

Acetylcholine also reaches its highest concentration in the striatum, where it serves as a neurotransmitter of intrinsic interneurons. In other words, cholinergic neurons and their terminals are located within the striatum, although there may be some cholinergic fibers entering the striatum from the thalamus. The exact cell-to-cell connections within the striatum are unknown. However, dopaminergic fibers terminate on cholinergic cells. They also terminate on the nerve terminals of the corticostriatal fibers that are believed to contain glutamate as their excitatory neurotransmitter. The acetylcholine in the striatum is believed to serve also as an excitatory neurotransmitter, and some of the cholinergic fibers terminate on intrinsic striatal neurons that contain GABA as their neurotransmitter. The synthesis of acetylcholine is by condensation of acetyl CoA and choline, catalyzed by the enzyme choline acetyltransferase that is located in presynaptic neurons. The synaptic activity of acetylcholine is terminated by hydrolysis to form acetate and choline; this reaction is catalyzed by acetylcholinesterase, an enzyme located in both pre- and post-synaptic neurons.

Neurons containing GABA are found

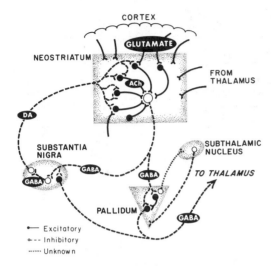

CORTEX

NEOSTRIATUM

GLUTAMATE

ACh

FROM THALAMUS

DA

SUBSTANTIA NIGRA

GABA

GABA

SUBTHALAMIC NUCLEUS

GABA

TO THALAMUS

PALLIDUM

GABA

— Excitatory
●--- Inhibitory
······ Unknown

FIG. 32–5. A simplified schematic diagram demonstrating the principal neurotransmitter pathways and their functions in the basal ganglia.

throughout the brain. They are present within the striatum, where they serve as Golgi type II cells with short axons to provide regional inhibition, and also serve as efferent neurons with long axons to innervate the pallidum and pars reticulata. This explains why GABA is found in highest concentration in these two nuclei (Table 32–1). GABA functions as an inhibitory neurotransmitter. It is synthesized through a shunt off the tricarboxylic acid cycle. Its precursor is glutamate, which is decarboxylated to form GABA, the reaction being catalyzed by the enzyme glutamate decarboxylase. The synaptic action of GABA is terminated predominantly by reuptake into the presynaptic nerve terminals. Some GABA is metabolized by glia and other structures to form succinic semialdehyde and then succinic acid, which reenters the tricarboxylic acid cycle. The two enzymes involved in the degradation steps are GABA transaminase and succinyl semialdehyde dehydrogenase.

Figure 32–5 is a simplistic diagram illustrating the major neural connections of the basal ganglia as well as pointing out their neurotransmitters and predominant physiological functions. It provides a schematic

view that permits understanding of the relationship between the neurotransmitters and of the biochemical changes that take place with the various degenerative disorders affecting the basal ganglia. To keep the diagram uncluttered and readable, other neurotransmitters have been omitted. For example, serotonergic fibers arising from the raphe nuclei in the pons and medulla innervate much of the basal ganglia, but are not depicted in Fig. 32–5. Similarly, the peptides, also present in the basal ganglia, are not shown. Since their roles are basically unknown and since they are not yet alterable pharmacologically nor have a therapeutic function in movement disorders, it seems appropriate to exclude them from the diagram at this point.

The peptides in the basal ganglia include the *enkephalins, somatostatin, cholecysto-kinin,* and *substance P.* The last compound appears to be a neurotransmitter in some of the efferent fibers from the striatum innervating the pallidum and substantia nigra. There also are substance P–containing fibers from the pallidum to the substantia nigra. Substance P is believed to act as an excitatory neurotransmitter. It is not yet known what functions the other peptides play within the basal ganglia.

Overview of movement disorders

The diagnoses of movement disorders primarily depend on their clinical features. It is important to observe (1) for the presence of involuntary movements and to describe their nature, (2) for postural changes, and (3) for the presence or the paucity of spontaneous and voluntary movements. One also needs to test for alteration of muscle tone, examine postural reflexes, and carry out a general neurological examination to determine whether other abnormalities exist. An essential feature is a knowledge of the definitions of the various types of abnormal involuntary movements so that the patient can be placed into the correct category of movement disorder; from there, the etiology can be ascertained on the basis of the history and judicial laboratory tests. As a general rule, abnormal involuntary movements are exaggerated with anxiety and disappear during sleep.

TREMOR. *Tremor* is an oscillating movement affecting one or more body parts, such as the limbs, neck, tongue, chin, or vocal cords. It is usually rhythmical and regular. The rate, location, amplitude, and constancy varies depending on the specific type of tremor and its severity. It is helpful to determine whether the tremor is present at rest (i.e., with the patient sitting or lying in repose), with posture-holding (e.g., with the arms extending in front of the body), or with action or intention maneuvers (such as bringing the finger to touch the nose). Tremors can then be classified as tremor-at-rest, postural tremor, or intention tremor, respectively. Etiologies differ according to these types. For example, the tremor of parkinsonism is characteristically a tremor-at-rest and commonly involves the distal parts of the limbs to produce a "pill-rolling" tremor. Postural tremor is seen in patients with exagerrated physiological tremor, such as in thyrotoxicosis, and in the disorders known as essential tremor and senile tremor. Intention tremor is the result of a lesion in the cerebellar outflow pathway, the superior cerebellar peduncle. This type of tremor is most common in multiple sclerosis. Wilson's disease can present with any or all types of tremor. Tremor-at-rest is associated with lesions of the dopaminergic nigrostriatal pathway, while postural tremor is, at present, not understood in terms of anatomical location.

MYOCLONUS. *Myoclonus* are shock-like contractions (or inhibitions). They are extremely brief in duration. Brief muscular inhibitions, "negative myoclonus," resembles the "positive myoclonus" of the contractions. Asterixis, which is seen in various metabolic encephalopathies, is an example of negative myoclonus because the brief flapping of the limbs is due to transient inhibition of the muscles that maintain the posture of those extremities.

Myoclonic jerks are related to seizures in that they are associated with an irritable nervous system and spontaneous electrical discharges of neurons. The anatomical location for myoclonus is widespread and ranges from the cerebral cortex, to the cerebellum and brain stem reticular formation, to the spinal cord. Myoclonic jerks usually appear at random and are frequently triggered by sudden startle of the patient, as by loud noise, visual threat, or tapping of tendon reflexes. Some forms of myoclonus are rhythmical and regular, and therein resemble tremor. Palatal myoclonus, in which the soft palate contracts at a rate of approximately 2 Hz., is the best-known example of rhythmical myoclonus. This entity is categorized as myoclonus and not tremor because it is associated with synchronous contractions elsewhere in the body when severe. Synchronization is a feature of myoclonic jerks and not other types of movement disorders. Furthermore, palatal myoclonus persists during sleep. As a general rule, all movement disorders except myoclonus disappear during sleep.

CHOREA. *Chorea*, from the Greek word meaning dance, is also a brief contraction, but not as lightning-like as myoclonus. Furthermore, choreic jerks tends to flow from one muscle to another in various parts of the body. Chorea is associated with excessive dopamine neurotransmission within the basal ganglia, either excessive concentration of dopamine at the dopamine receptors or supersensitive dopamine receptors with a normal amount of dopamine at the synapse. In the most common cause of chorea, Huntington's disease, there is neuronal loss in the caudate nucleus and putamen. (The pathophysiology of chorea is discussed in more detail below.)

BALLISM. *Ballism* is a gross form of chorea, in which the limbs are flung about in large amplitude. This movement disorder is most commonly seen on one side of the body (hemiballism) as a result of a lesion in the contralateral subthalamic nucleus. The lack of inhibitory influence on the pallidum as a result of such a lesion is believed to be responsible for releasing these involuntary movements. As with the smaller-amplitude choreic movements, decreasing dopamine neurotransmission can reduce or eliminate these involuntary movements. This implies that the disinhibited pallidum can be brought back into control by influencing (through the dopamine system) the striatum, which has the largest inhibitory input into the pallidum.

ATHETOSIS. *Athetosis* is slower than chorea, and is characterized as a continual writhing movement. It can be present in all parts of the body. When not present in certain body parts at rest, sometimes it can be brought out by having the patient carry out voluntary motor activity elsewhere on the body. This phenomenon is known as overflow. For example, speaking can induce increased athetosis in the limbs, neck, trunk, face, and tongue. Athetosis most commonly occurs as a result of injury to the basal ganglia in the neonatal period or during infancy. The speed of these involuntary movements can blend with those of chorea, and the term choreoathetosis is used. Often, athetosis is associated with sustained contractions producing abnormal posturing. In this regard, athetosis blends with dystonia. The pathophysiological basis of athetosis is not known.

DYSTONIA. *Dystonia* represents both sustained muscular contractions and twisting or torsion movements that may be sustained only for a fraction of a second. Usually the agonist and antagonist muscles contract simultaneously to cause dystonic movements. The twisting movements can be as fast as chorea and myoclonus or as slow as athetosis. When the contractions last for less than a second they are referred to as dystonic spasms. When they are sustained for several seconds, they are called dystonic movements. And when they last minutes to hours, they are known as dystonic postures. When present for weeks or longer, the postures could lead to permanent fixed contractures. Dystonia, like athetosis, can be brought out by active voluntary movements, known as action dystonia. The disease dystonia musculorum deformans (or torsion dystonia) commonly begins with action dystonia. With progression of the illness, dystonia of muscles at rest develops. Although dystonia musculorum deformans has not been linked to cellular pathology, secondary causes of dystonia (e.g., head trauma, encephalitis, and Wilson's disease) are associated with pathology of the basal ganglia. But the exact pathophysiological mechanisms are unknown.

TICS. *Tics* are rapid (and occasionally sustained) movements that appear in a sequential pattern. Usually, the same pattern is repeated at the next occurrence of the tic, which tends to appear suddenly after a brief absence of them. If the patient has several tics, i.e., several sequential patterns of complex movements, one type of tic can follow another in random occurrences. When mild, the tics are less complex and may present as very simple movements, thereby resembling single or repetitive myoclonic jerks. The presence of the more complex tics establishes the diagnosis as a tic disorder. Tics can appear as vocal expressions as well as motor movements. Vocal tics range from simple sounds and noises (coughing, throat clearing, barking sounds) to words (usually obscenities, known as coprolalia). Typically the tic is preceded by a compulsive feeling of an urge to perform it. Its performance provides relief of this inner urge. When mild, tics can be temporarily suppressed by voluntary control. But usually this is followed by an outpouring of tics. This control, however, can enable tiqueurs to be in public places without calling attention to themselves. It is not known for certain where in the brain tics originate. They have been associated with the monoamine systems because drugs that block dopamine receptors or inhibit norepinephrine-containing neurons can often suppress them.

AKATHITIC MOVEMENTS. *Akathitic movements* are quasi-voluntary because they are movements brought about from a feeling of inner motor restlessness (akathisia). With akathisia, patients feel the need to move about, such as pacing the floor, rocking the trunk, marching in place, and even making noises such as moaning. Carrying out these motor acts brings relief from akathisia. Akathitic movements can be transiently suppressed by the patient if asked to do so. The most common cause of akathisia is iatrogenic. It is a frequent complication of antipsychotic drugs, agents that block dopamine receptors. It can occur when drug therapy is initiated (acute akathisia) or after chronic treatment (tardive akathisia). Acute akathisia is eliminated upon withdrawal of the medication. Tardive akathisia is associated with the syndrome of tardive dyskinesia (discussed below) and is usually made worse on sudden discontinuation of the medication. The exact mechanism of akathisia is not known, but because the antipsychotic drugs block and supersensitize dopaminergic receptors, it seems that the dopamine system is involved, possibly in the limbic system, which also is rich in dopamine.

STEREOTYPIC MOVEMENTS. *Stereotypic movements* are rapid like chorea, but, unlike chorea, they are repetitive. They most commonly appear as oral-buccal-lingual dyskinesias, in which there are continual chewing movements with intermittent protrusions of the tongue. Repetitive movements of the fingers and toes are other manifestations of stereotypy. This disorder is the most common presentation of the syndrome of tardive dyskinesia, which is iatrogenic and is a complication of chronic treatment with antipsychotic drugs. These drugs bind with the dopamine receptor, altering it and making it supersensitive. Stereotypy is a form of excessive dopaminergic activity and can be suppressed by antidopaminergic agents. Tardive dyskinesia is discussed in more detail below.

PAROXYSMAL DYSKINESIAS. *Paroxysmal dyskinesias* are abnormal involuntary movements that occur paroxysmally, as the name indicates. The type of dyskinesia varies depending on the specific disorder. The two that have been well described are paroxysmal kinesigenic choreoathetosis and paroxysmal dystonia. The burst of choreic movements in the former is triggered by sudden body movement, hence the use of the term kinesigenic. The chorea that occurs with this disorder is brief and lasts usually less than a minute. With repeated sudden movements, habituation commonly ensues, and the chorea fails to develop. Paroxysmal dystonia is not induced by sudden movement, but rather by sudden stress, fatigue, or consumption of alcohol and caffeine. The dystonic movements that develop can last a few hours. Since seizures are also paroxysmal, these paroxysmal dyskinesias have been considered to possibly represent a "seizure disorder" of the basal ganglia.

A listing of some of the clinical features distinguishing the various types of dyskinesias is presented in Table 32–3.

Parkinsonism

Clinical features

Parkinsonism is a clinical syndrome consisting of at least two of four cardinal signs: tremor-at-rest, rigidity, bradykinesia, and loss of postural reflexes. Within the spectrum of bradykinesia are a host of clinical features denoting a paucity or slowness of

TABLE 32–3
Clinical Features of the Dyskinesias

Type	Speed of Movement	Special Features	May be Continual at Rest	Effect of Action	Agonist-Antagonist Relations	Rhythmic Contractions	Random Contractions of Specific Muscles	Sudden Burst	Transient Suppression	Synchrony	Noises
Tremor											
At-rest	3–6 Hz	Oscillating	+	↓	Alternating	+	–	–	+	–	–
Postural	7–12 Hz	Oscillating	–	↑	Alternating or simultaneous	+	–	–	–	–	–
Intention	3–6 Hz	Oscillating	–	↑	?	+	–	–	–	+	–
Myoclonus											
Random	Very rapid	↑ with startle	±	0 or ↑	Occasionally simultaneous	–	+	–	–	+	–
Rhythmical	Rapid	Oscillating	+	0	Simultaneous	+	–	–	–	+	–
Chorea	Rapid	Flowing	+	sl. ↑	–	–	+	–	+	–	±
Ballism	Rapid	Flinging, unilateral	+	0 or ↓	–	–	+	–	–	–	–
Athetosis	Slow	Writhing	+	↑	Simultaneous	–	–	–	–	–	–
Dystonia	Rapid or Slow	Sustained, twisting	+	↑	Simultaneous	±	–	–	–	–	–
Tics	Rapid	Sequence; can be complex	–	–	–	–	±	+	+	+	+
Akathitic movements	Slow	Restlessness	+	↓	Probably alternating	+	–	–	+	–	+
Stereotypy	Rapid	Repetitive	+	↓	Probably alternating	+	–	–	+	–	–
Paroxysmal	Rapid or Slow	Kinesigenic, paroxysmal	+	↑ if kinesigenic	may be simultaneous if dystonic	–	–	+	–	–	–

+ = present or yes; – = absent or no; ± = occasional; 0 = no change; ↓ = decreases; ↑ = increases

TABLE 32—4
Levels of Brain Monoamines in Parkinsonism

Region of Brain	Dopamine	Norepinephrine	Serotonin
Caudate nucleus	16	44	37
Putamen	8	77	42
Globus pallidus	20	144	57
Substantia nigra	13	38	47
Hypothalamus	19	63	41
Thalamus	55	41	50

Data are given as percent of control values. (Fahn S.: Adv. Neurol. *14*:59–88, 1976.)

movement and difficulty initiating and maintaining movement, particularly amplitude. Examples of bradykinesia include masked facies, decreased blinking, decreased swallowing with drooling of saliva, hypophonia, lack of inflection in speech (aprosody of speech), decreased rapid-succession movements, decrementing amplitude with repetitive movements, short-stepped gait, decreased arm swing when walking, difficulty rising from a chair, difficulty turning in bed, lack of spontaneous gesturing, and slowness in carrying out activities of daily living.

There are many etiologies responsible for parkinsonism. Idiopathic parkinsonism or Parkinson's disease is the most common one confronting the neurological clinician. The pathological process associated with this disorder is loss of neuromelanin-containing monoaminergic neurons (dopamine neurons in the pars compacta of the substantia nigra, noradrenergic neurons in the locus ceruleus, and serotonergic neurons in pontine and medullary raphe nuclei). Eosinophilic cytoplasmic inclusion bodies are found in many of the remaining neurons in these regions; they are called Lewy bodies. Parkinsonism can also be secondary to encephalitis, toxicity from carbon monoxide poisoning, and neuroleptic drugs (reserpine and antipsychotic agents). In all situations there is either a deficiency of dopamine or a defect of the postsynaptic dopamine receptors.

Pathophysiology

Pathologically and biochemically, Parkinson's disease results in loss of monoamine neurons. Since the biggest mass of such neurons is the dopamine-containing cells of the substantia nigra, there is a corresponding loss of dopamine in this region and especially at the site of their nerve terminals, the neostriatum. The mesocortical and mesolimbic dopaminergic neurons are also affected. Despite the fact that there is also a reduction of brain norepinephrine and serotonin in Parkinson's disease, there is considerable evidence relating the symptoms of bradykinesia, rigidity, and tremor only to the deficiency of striatal dopamine.

1. Dopamine is highly concentrated in the striatum, whereas serotonin is more widely and uniformly distributed, and norepinephrine is barely detected in the striatum and is most concentrated in the limbic system (Table 32–1).
2. The most profound reduction of any neurotransmitter in Parkinson's disease is that of dopamine in the striatum (Table 32–4).
3. In a patient with Parkinson's disease predominantly on one side of the body, the reduction of dopamine was more pronounced in the striatum contralateral to the side of symptoms. Moreover, serotonin was essentially equal on both sides.

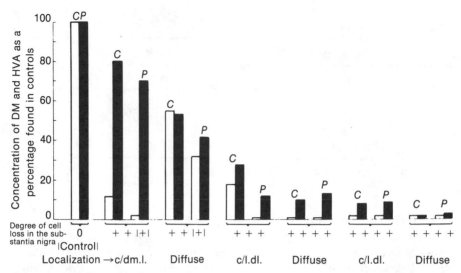

FIG. 32–6. Relationship between the concentrations of dopamine (white bars) and homovanillic acid (black bars) in the caudate nucleus (C) and putamen (P) and the degree of cell loss in the substantia nigra (along the abscissa). (Hornykiewicz, O. *In* Costa, E., Cote, L. J., and Yahr, M. D. (eds.): Biochemistry and Pharmacology of the Basal Ganglia. New York, Raven Press, 1966. Reproduced with permission from the publisher.)

4. In the striatum, the reduction of dopamine and its major metabolite, HVA, parallels the degree of cell loss in the substantia nigra (Fig. 32–6).

5. Treatment of Parkinson's disease with monoamine oxidase inhibitors results in little clinical improvement. Postmortem determinations of the monoamines in the brains of three patients treated prior to death with these agents revealed increase of norepinephrine and serotonin to above normal levels, but essentially no reversal of the reduced dopamine level.

6. Reserpine depletes the brain of its three monoamines and produces a parkinson-like effect of hypokinesia, tremor and rigidity. The clinical effect can be reversed in animals and in man with L-DOPA. The serotonin precursor, 5-hydroxytryptophan, has no effect. Although L-DOPA is the precursor of both dopamine and norepinephrine, the injection of this compound results in a marked increase of dopamine in the brain of both normal and reserpinized animals, but only a small increase in the level of norepinephrine. Furthermore, restoring the brain level of norepinephrine to normal without changing the dopamine concentration in the resepinized animal by administering dihydroxyphenylserine does not reverse the reserpine effect.

7. Administration of L-DOPA or direct-acting dopamine agonists can reverse most manifestations of Parkinson's disease in patients.

8. The bradykinesia that occurs in animals from chemical stimulation of the caudate nucleus can be alleviated by intracaudate injection of L-DOPA.

9. Phenothiazines and butyrophenones given in high dosage produce a syndrome of parkinsonism. These drugs block the postsynaptic dopamine receptor.

10. Parkinsonism is a rare complication of α-methyldopa therapy for hypertension. It is believed that its metabolite,

α-methyldopamine acts as a false transmitter and may compete with dopamine at the receptor or storage sites.

11. The effectiveness of anticholinergic agents in the treatment of parkinsonism can be explained on the relationship between the dopaminergic and cholinergic neurons within the striatum. The nigrostriatal dopaminergic neurons appear to terminate at the cholinergic intraneurons within the striatum (Fig. 32–5). Lack of dopaminergic influence on these cholinergic neurons leads to their disinhibition. Anticholinergic drugs tend to suppress the disinhibited cholinergic influence on the cells innervated by these intraneurons.

12. The effectiveness of amantadine in the treatment of parkinsonism can be explained on the basis of its dopaminergic and anticholinergic properties. Its dopaminergic function seems to be one of enhancing release of dopamine from the nerve terminals.

Postencephalitic parkinsonism

The pandemics of encephalitis lethargica that struck in Central Europe in 1918 and spread throughout the world during the 1920s left in their wake a variety of movement disorders, particularly parkinsonism. The severity of this postencephalitic disorder tended to plateau instead of being constantly progressive as in Parkinson's disease. The infectious agent responsible for the encephalitis has never been identified, but it affected the midbrain most severely and thereby destroyed the dopaminergic neurons in that region as well as the ascending noradrenergic and serotonergic pathways passing through this region from the pons and medulla. The biochemical consequences resemble the findings in Parkinson's disease, but are even more severe because of the greater destruction of nigral neurons in postencephalitic parkinsonism. Thus, the basis of the symptoms in this disorder is striatal dopamine deficiency. Treatment of postencephalitic parkinsonism is with the same medications as that for Parkinson's disease, except that patients with the postencephalitic form are more sensitive to L-DOPA and require a lower dosage.

Reserpine-induced parkinsonism

The first clue of the dopaminergic deficient nature of any type of parkinsonism came from the study of the action of reserpine. This drug produces parkinsonism in animals and human beings. Carlsson, in Sweden, showed that L-DOPA can reverse this effect and also discovered that reserpine depleted dopamine along with the other monoamines. Reserpine is now known to block uptake of all three monoamines into the granulated vesicles, leaving these compounds exposed to the action of monoamine oxidase. As a result, the brain becomes depleted of dopamine, norepinephrine, and serotonin. Parkinsonism ensues, and the term *neuroleptic* was coined for drugs that induce parkinsonism. Treatment with L-DOPA or direct-acting dopamine agonists is effective in this form of parkinsonism (as is discontinuance of reserpine).

Drug-induced parkinsonism

The term *drug-induced parkinsonism* can also be applied to that caused by reserpine, but most commonly it refers to that caused by the antipsychotic drugs, such as the phenothiazines and butyrophenones. These agents block dopamine receptors and readily induce a state of parkinsonism that resembles that of Parkinson's disease. The "zombie" state seen so often in psychiatric patients is due to the effects of these antipsychotic drugs. L-DOPA and the direct-acting dopamine agonists are not very effective in reversing this drug-induced state because there is competition for the dopamine receptor and the antipsychotic agents have a stronger affinity for the receptor than does L-DOPA or the direct-acting agonists. As a result, treatment of drug-induced parkinsonism requires the use of anticholinergic agents.

Chorea

Our understanding of the pathophysiology of chorea also comes from both neurochemical and neuropharmacological data. Huntington's disease is the prototype of the choreas because there is considerably more information available about this disorder than about other causes of chorea and because it is the most common cause of chorea. Other etiologies of chorea are streptococcal infection (Sydenham's chorea), vasculitis, oral contraceptives, and encephalitis.

HUNTINGTON'S DISEASE. *Huntington's disease* is a progressive degenerative disorder that is inherited as an autosomal dominant pattern. The major pathological change is loss of neurons with secondary gliosis in the striatum, with less severe changes in the cerebral cortex and elsewhere. Associated with this cellular alteration is a loss of the neurotransmitters contained within these neurons. The most consistent and pronounced decrease of a striatal neurotransmitter is that of GABA, accompanied by a reduction in its synthesizing enzyme, glutamic acid decarboxylase. Since some of the GABA-containing neurons make up much of the striatonigral fibers, there is also a loss of GABA in the substantia nigra. Associated with the loss of these neurons is an increase of GABA receptors in the substantia nigra due to the development of denervation supersensitivity. There is also a loss of the acetylcholine neurons within the striatum in Huntington's disease, reflected by a reduction in the activity of choline acetyltransferase, the enzyme catalyzing the synthesis of acetylcholine. Since these neurons terminate within the striatum, there is also a reduction in the acetylcholine receptors located on other lost striatal neurons. Dopamine content and the activity of its synthesizing enzyme tyrosine hydroxylase are not reduced in the striatum in Huntington's disease. This is because they are present in the nerve terminals of the nigrostriatal pathway, which does not degenerate in this disorder. Because of loss of mass in the striatum, there is actually an increase in dopamine concentration within this structure. On the other hand, there is a reduction in dopamine receptors in the striatum because these are located, in part, on dendrites of striatal neurons which have degenerated.

On the basis of the chemical changes, one could propose, as an analogy to Parkinson's disease, that the reduction in GABA and acetylcholine may be responsible for the symptom of chorea. However, studies with GABA- and acetylcholine-mimetic drugs have failed to provide benefit to patients with Huntington's disease. GABA itself does not penetrate the blood-brain barrier. Drugs that inhibit the degradation of GABA and therefore increase GABA concentration have been evaluated. These include valproate, isoniazid, and others. None are of therapeutic value in reducing choreic movements. In addition, a direct-acting GABA agonist, muscimol, has also been studied and found not to be effective in Huntington's disease. Cholinergic drugs that have been tested include choline, lecithin, deanol, physostigmine, and arecoline. None showed benefit. It is possible that the reduction of cholinergic receptors in the striatum may render the cholinergic drugs ineffective.

Of all the agents tested in patients with chorea, those affecting the dopamine system have been the most consistent and potent. Drugs that increase central dopaminergic activity (L-DOPA) increase choreic movements. Drugs that are antidopaminergic reduce chorea. The latter group are of two classes: those that block postsynaptic dopamine receptors (the antipsychotics) and those that deplete presynaptic levels of dopamine (reserpine and α-methyltyrosine). What is the explanation for the pharmacological action of the dopamine-acting drugs that suggests that there is increased activity of the dopamine system? Several hypotheses have been proposed.

1. *Increased turnover of dopamine.* This was one of the earliest suggestions. However, cerebrospinal fluid levels of homovanillic acid, the major catabolite of dopamine, has been found to be within normal limits or low in patients with Huntington's disease. These findings militate against the hypothesis of increased dopamine turnover.
2. *Supersensitive postsynaptic dopamine receptors.* Theoretically there is no explanation as to why these receptors should be supersensitive in Huntington's disease. There is no loss of presynaptic dopamine neurons, and hence no denervation supersensitivity should develop. In fact, the opposite seems to be the case. Dopamine levels in the striatum are increased approximately 50 per cent in Huntington's disease, probably because of intact nigrostriatal dopamine terminals in a region of cellular atrophy. Moreover, instead of any increase in dopamine receptors in striatum, there is, to the contrary, a 50 per cent loss of dopamine receptors. Each of these findings militate against the hypothesis of supersensitivity of the dopamine receptors.
3. *Increased dopamine activity per receptor.* All data available so far are compatible with this concept. With a 50 per cent reduction in dopamine receptors and a 50 per cent increase in dopamine concentration, there is a fourfold increase in the ratio of dopamine/receptor. Since the dopamine system in the striatum appears to be tonically, rather than phasically, active, and since the majority of dopamine terminals are boutons en passage, the striatum resembles a dopamine endocrine gland whereby there

is a general bathing of the dopamine receptors. The net effect of an increased dopamine/receptor ratio on the remaining intact postsynaptic neurons is to produce chorea through striatopallidothalamic connections. This hypothesis would predict that when the loss of postsynaptic dopamine receptors eventually falls below a critical threshold, chorea would be replaced by akinesia and rigidity, common features of far-advanced Huntington's disease.

Tardive dyskinesia

Tardive dyskinesia is a disorder induced by chronic (hence, the name tardive) administration of antipsychotic drugs. It occurs most often in elderly individuals and consists of chorea-like stereotypic movements and usually akathisia. The mouth area is the site most often involved (oral dyskinesia, oral-lingual-buccal dyskinesia), but the limbs and trunk can also be affected. In contrast to the flowing, random, involuntary movements of chorea, those in tardive dyskinesia are repetitive (stereotypy). Tardive dyskinesia can be persistent, even permanent, despite withdrawal of the offending medication. In fact, withdrawal of the antipsychotic agents may worsen the dyskinesia and akathisia. On the other hand, increasing the dosages of these drugs tends to suppress them.

Our pathophysiological understanding of tardive dyskinesia comes from pharmacological effects. High dosages of antipsychotic drugs can induce parkinsonism due to blockade of the dopamine receptors. Failure of dopamine to reach the receptors will also produce receptor supersensitivity, and this is probably an important, but not the sole, factor responsible for causing tardive dyskinesia. Reserpine also induces supersensitivity of the dopaminergic receptors, but does not induce tardive dyskinesia. However, reserpine acts on the presynaptic dopaminergic nerve terminal, and antipsychotic drugs act on the postsynaptic receptors directly. It seems likely that chronic use of antipsychotic agents, which bind to dopamine receptors, somehow chemically alter the receptors to allow them to physiologically mimic the effect of too much dopamine.

Withdrawal of the antipsychotic drugs removes the chemical blockade and leads to a worsening of symptoms. Increase in dosage would have the opposite effect. This presents a problem in treating this disorder. Ideally, one would prefer to remove the offending drug. Sometimes this can be done without much exacerbation of symptoms. In time, there can be slow disappearance of symptoms. However, many patients develop a worsening of symptoms that is intolerable to them. In this situation, treatment with reserpine can reduce the symptoms. These approaches, of course, require that the patient no longer have a persistent psychosis requiring need of therapy with antipsychotic drugs.

Conclusions

At the present time we have a working, but incomplete, understanding of the pathophysiology of some of the movement disorders. These are parkinsonism, chorea, ballism, and tardive dyskinesia. These have in common some relationship to the dopamine system in the basal ganglia, either too little dopaminergic activity (parkinsonism) or too much (chorea, ballism, and tardive dyskinesia). By knowing the mechanisms of action of drugs and how they affect the clinical conditions, we have developed some insight into the pathophysiological mechanisms underlying these disorders. Thus, our understanding is based on neuropharmacology and experimental therapeutics. Drugs affecting the dopamine system are available; by using them, we have gained information on only a few movement disorders. Although tested in other dyskinesias, they have indicated only that these other disorders are not primarily related to the dopamine system. Clearly, we need other tools to further our explorations and our understanding of the pathophysiological mechanisms of other movement disorders.

Annotated references

Chase, T. N., Wexler, N. S., and Barbeau, A. (eds.): Huntington Disease: Advances in Neurology, Vol. 23, New York, Raven Press, 1979. (This book covers the recent advances in the neurochemistry and neuropharmacology of Huntington disease.)

DeLong, M. R., and Georgopoulos, A. P.: Motor functions of the basal ganglia. In Handbook of Physiology: The Nervous System II, 2nd ed., pp. 1017–1061. Bethesda, American Physiological Society, 1982. (This is a detailed and scholarly review of the basal ganglia, correlating its anatomy with its motor physiology. There is some discussion on the pathophysiology of parkinsonian symptoms.)

Fahn, S.: Secondary parkinsonism. In Goldensohn, E. S., and Appel, S. H. (eds.): Scientific Approaches to Clinical Neurology, pp. 1159–1189. Philadelphia, Lea & Febiger, 1977. (This chapter reviews other causes of parkinsonism and relates them to the dopamine hypothesis. It provides a useful classification of the parkinsonian syndrome.)

Fahn, S., Duffy, P.: Parkinson's disease. In Goldensohn, E. S., Appel, S. H. (eds.): Scientific Approaches to Clinical Neurology, pp. 1119–1158. Philadelphia, Lea & Febiger, 1977. (This chapter provides a detailed review of the clinical, pathological, biochemical, and pharmacological data on Parkinson disease. It is a good starting point to learn about this disorder.)

Jankovic, J., and Fahn, S.: Physiologic and pathologic tremors: Diagnosis, mechanism, and management. Ann. Intern. Med. 93:460–465, 1980. (This review article classifies the various types of tremors and summarizes the known physiology and pharmacology.)

Klawans, H. L., and Weiner, W. J.: Textbook of Clinical Neuropharmacology. New York, Raven Press, 1981. (This book covers drugs used in the treatment of the various movement disorders, the mechanisms of action of these drugs, and the principles behind their use.)

33 Neuromuscular Junction

Audrey S. Penn, M.D.

Neuromuscular transmission is the best defined instance of synaptic transmission. With the recent explosion of data on the nicotinic acetylcholine receptors at neuromuscular junctions, extensive information is now available not only about axonal transmission of impulses but also about mechanisms by which transmitter is synthesized and released from nerve terminals and how it ultimately generates muscle action potentials. There are a variety of clinically significant disorders of neuromuscular transmission. "State-of-the-art" technology, using ultrastructural cytochemical and freeze-fracture techniques to visualize the neuromuscular junction and to quantitate its components, biochemical and immunochemical analysis, and electrophysiological analysis at the molecular level, has now been applied to these disorders. In some instances studies of abnormal transmission have also provided probes for further analysis of the normal state.

The motor nerve penetrates a muscle at the neurovascular hilum, then branches repeatedly. Upon entering gutters on the surface of the fiber, the nerve breaks up into a spray of myelinated preterminal axons. Just before reaching end-plates, the termi-

nal branches lose their myelin sheaths. A single terminal ends at a single end-plate on a single muscle fiber. A Schwann cell covers both motor nerve terminal and muscle end-plate. The membrane of the terminal is separated by 40 nm. to 100 nm. from the muscle end-plate, a space which constitutes the synaptic cleft. The postjunctional muscle membrane at the end-plates of mammals is thrown into multiple junctional folds so that there is not only a primary cleft ranging the entire length of the nerve terminal between terminal and muscle, but also multiple secondary clefts. This microanatomical arrangement provides a very efficient geometry for synaptic transmission, which depends on dispatch and diffusion of membrane-bound packets of transmitter across the cleft to hit and bind receptor molecules.

As in other types of cholinergic transmission which occur in the central nervous system and at sympathetic ganglia, acetylcholine (ACh) is the transmitter which interacts with a nicotinic receptor on muscle so as to generate an end-plate potential. To achieve this, ACh molecules must be synthesized, packaged, and released, in adequate amounts, within milliseconds. In addition, the reaction of receptor with transmitter must be terminated within a brief time to allow the system to respond appropriately to the next impulse.

Nerve terminal

SYNTHESIS. At this anatomical site, the arrival of a propagated nerve action potential is linked to the biochemical generation and dispatch of transmitter. ACh is synthesized by choline acetylase from choline and acetyl coenzyme A (acetyl CoA). Choline acetylase is synthesized in the perikaryon and arrives at the nerve terminal by means of axonal transport. Acetyl CoA is synthesized in the mitochondria of the terminal; the choline is derived primarily from the extracellular fluid of the synaptic cleft by a high-affinity sodium-dependent reuptake mechanism. The newly synthesized ACh is packaged into synaptic vesicles. Although all steps have not yet been elucidated, the elegant study of Heuser, Reese, Kelly, and colleagues suggests that the packaging process is achieved by recycling of nerve terminal membrane. Synaptic membrane at the edges of the nerve terminal is taken up into the terminal by endocytosis to form coated vesicles. These coalesce to form cisternae from which new synaptic vesicles bud off. It is at this point that free ACh is packaged. Evidence for these processes derives from ultrastructural analysis, use of freeze-fracture techniques, and analysis of events captured nearly instantaneously using these methods and special ultra-rapid freezing techniques.

STORAGE AND RELEASE. ACh in the terminal exists, therefore, in three major stores: newly synthesized ACh, not yet packaged into vesicles, packaged ACh, and vesicles ready for release into the synaptic cleft. Couteaux, Heuser, and Reese have demonstrated active zones on the nerve terminal membrane by ultrastructural analysis and freeze fracture. These zones appear as double rows of intramembranous particles positioned on either side of a bar of thickened

membrane so that, end on, the region resembles the Greek letter omega. In mammals, each of these *omega figures* is also directly opposite the tip of a postjunctional fold, again a very efficient geometry since the highest density of acetylcholine receptor molecules is found at these tips. According to present evidence, vesicles at the terminal membrane region of the active zones release their ACh molecules by fusion of vesicle membrane with terminal membrane, resulting in extrusion of the ACh (exocytosis). Membrane material for packaging new vesicles is then recovered by endocytosis from the lateral edges of the nerve terminal. Since synaptic vesicle membrane bears antigenic determinants distinguishable from presynaptic membrane by reaction with specific antibody, new proteins must be added during this recycling process.

The other stores of ACh in the terminal can be estimated from their readiness to be released upon stimulation, as initially studied by Birks and MacIntosh in the superior cervical ganglion of the cat. A store of ACh, presumably already packaged into vesicles, can be mobilized readily by prolonged high rates of nerve stimulation while another fraction of about 300,000 quanta is not available for release. This may represent ACh not yet packaged or in the process of being packaged. The degree of availability of this ACh fraction may also reflect the time required to recycle membrane for packaging.

Release of quanta of ACh may occur spontaneously and randomly, presumably because of a fortuitous positioning of a vesicle at the active zone membrane region, or in response to arrival of a nerve impulse. The depolarization generated by the arriving action potential opens specific voltage-dependent calcium channels which permit entry of calcium ions into the terminal. By a mechanism as yet unknown, this triggers synchronous release of about 400 quanta of ACh (which under normal circumstances results in generation of an end-plate potential large enough to trigger an action potential). The local calcium concentration is presumably restored to normal range by uptake of the excess by mitochondria or other membranes. If the calcium concentration of the extracellular fluid is reduced or the magnesium concentration increased, transmission will fail. Other divalent cations can also substitute for Ca^{++} and promote release of transmitter for the terminals. Sr^{++} achieves end-plate potentials of greater amplitude than those induced by Ca^{++}, which are maintained for a longer time period during trains of 10 impulses applied at 20 per second. Ba^{++} generates a prolonged augmentation of the end-plate potential seen after conditioning trains of 100 to 400 impulses. Each cation appears to act on the mechanisms for release at different sites from Ca^{++} and from each other. In addition, lanthanum will induce massive ACh release from frog nerve terminals.

There is also evidence to suggest that the calcium channels of the terminal are regulated by cyclic nucleotides, since agents such as sodium fluoride, theophylline and prostaglandin E_1, which inhibit adenylate cyclase, and verapamil, which inhibits calcium flux, can inhibit neuromuscular transmission.

Synaptic cleft

Each muscle fiber is sheathed by a basal lamina which passes between the nerve terminal and muscle fiber at the end-plate region, occupying a portion of the 50-nm. to 100-nm. gap between these two structures. The basal lamina also projects into the secondary postsynaptic clefts and, at the edges of the end-plate, joins the membrane of the Schwann cell which covers that terminal. The synaptic basal lamina is presumed to be composed of the collagens, glycoproteins, and glycosaminoglycans which are major constituents of other basal lamina. In addition, acetylcholine esterase (AChE) has been identified by immunocytochemistry using specific antibody on frog muscle fibers treated so that the plasma membrane was removed but the basal lamina remained. Monoclonal antibodies generated against crude embryonic muscle as well as affinity-purified mucle proteins reacted against several components of basal lamina. Identified antigens include the collagens, elastin, laminin, fibronectin, and proteoglycans. Present evidence suggests that the basal lamina at the old synaptic site of regenerating muscle serves to redirect accumulation of nerve terminal-vesicle and postsynaptic muscle membrane constituents.

AChE occurs in the synaptic cleft both in a soluble form and in a membrane-associated form. The AChE associated with the basal lamina is a collagen-tailed asymmetric form (17S species). Enzyme has also been found in the presynaptic membrane. This species is detergent-soluble and hydrophobic.

Postsynaptic membrane

Acetylcholine receptors

Acetylcholine receptor AChR is a multisubunit integral membrane glycoprotein located at a density of up to 18 to 20,000 sites/μm.^{2} in the postjunctional membrane. Deeply etched, negatively stained freeze-fracture studies of electric tissue in which the density of AChR sites approaches 10,000/μm.^{2} have revealed a rug-like mosaic of packed repeating polygonal units projecting from the membrane surface, each with a subunit structure and a clear suggestion of a central pit or core. In addition, several studies of electric tissue membrane vesicles employing anti-AChR antibodies indicate that three of the four subunits, if not all four, not only project from the membrane surface but can be detected on the cytoplasmic surface of the vesicle, indicating that the AChR is indeed a transmembrane protein which could constitute a transmembrane channel.

In neuromuscular junctions of several vertebrates, AChR are clustered at the tips of the junctional folds. Studies by Fambrough and others indicate that AChR, newly synthesized at the Golgi apparatus, are inserted into the membrane and that the presence of the nerve dictates the final assembly of AChR at neuromuscular junctions. In addition, a complex infrastructure of actin filaments, intermediate filaments, and vinculin appears to contribute to the clustering of junctional AChR. In the absence of a nerve influence or after denervation, increased numbers of a second extrajunctional type of AChR appear diffusely along the entire membrane while they are normally found only near the tendon margins. These AChR bear a set of antigenic determinants not found on junctional AChR, which are recognized in the serum of patients with myasthenia gravis. They also are more mobile within the membrane and may be "capped" by lectins and antibodies. However, most other biochemical characteristics (subunits, density-gradient sedimentation, ligand binding) and channel parameters have been found to be identical.

The quantal event: reaction of ACh with AChR

Quanta of ACh, released spontaneously or with arrival of a nerve impulse, diffuse across the synaptic cleft to junctional AChR, just opposite active zones. The reaction of these 5,000 to 10,000 molecules of ACh with AChR opens 1,000 to 2,000 ion channels at the peak of the quantal response, which can measure as a miniature end-plate current of the fiber is voltage clamped. The molecular events, the current flow through single-ion channels resulting from binding of single molecules of ACh, have also been studied.

The preparations most commonly employed are the electroplax, single cells from the tissue of electric fish, or frog and rat nerve-muscle preparations. The dose-response curve for the binding of ACh to the AChR on electroplax, as well as binding studied using solubilized AChR, is sigmoidal, indicating positive cooperativity, or an effect generated by binding of more than one molecule of AChR in which the effect produced by binding of the second is more pronounced than a simple additive process. Statistical analysis of random fluctuations of current flow (spectral analysis of noise) through the membrane, while the system is voltage clamped, has permitted estimates of the current flow through a single channel. The average channel permits 20 to 30 picosiemens (pS.) conductance for about a 1 msec. open time. Very similar estimates have also been generated using the direct patch-clamp technique on the more dispersed AChR of denervated muscle by Sakman and colleagues. The apparent conductance of about 28 pS. per channel is equivalent to a flux of about 20,000 sodium ions, yielding a potential change of about 0.3 μV. About 1,000 such channel events therefore generate the potential change of a miniature end-plate potential (MEPP), about 0.5 to 1.1 mV. In addition, the patch clamp technique has confirmed that openings of individual channels are statistically independent events whose probabilities will follow a Poisson distribution.

Binding to AChR of the 5,000 to 10,000 molecules contained in one quantum of ACh achieves a MEPP, the basic physiological unit of the end-plate potential. Randomly released MEPPs can be detected only by intracellular microelectrode recordings from end-plates. The detailed steps resulting in the generation of the miniature end-plate current (MEPC) in response to the release of a quantum of ACh have been simulated using computer models, from experimental values of the rise time, decay time of the MEPC, available measurements of the number of ACh molecules per quantum, AChR density, density of AChE sites, and kinetic of catalysis. The release of about 10,000 molecules of ACh from active zones is accepted to be instantaneous. AChR density has been measured by electron microscopic autoradiography and is close to 15,000 sites/μm.[2]. AChE sites (about 2600/μm.[2]) are scattered along the entire basal lamina between the nerve terminal and the postjunctional membrane. Next, the assumption is made that the ACh diffuses along the long axis of a cylindrical volume between terminal and end-plate whose long axis is about 0.05 μm. Thus, the binding of two molecules to one AChR so as to open the ion channel is facilitated opposite the point of release as compared to the hydrolysis of that ACh by the relatively sparse number of molecules of AChE on the basal lamina within the radius of the cylinder. As the ACh molecules come off the AChR, they are hydrolyzed as they diffuse radially out of the cleft or down into the secondary clefts where AChE sites predominate over AChR. Thus, by the time the response reaches its peak, only a few hundred of the original 10,000 ACh remain free in the synaptic cleft, several hundred are bound to AChE, and the large majority are bound to AChR, although only 10 per cent are doubly bound. All models are based on the assumptions that ACh is subject to diffusion, enzymatic hydrolysis, and binding to AChR and that the ion channel opens only when two ACh molecules bind to a single AChR.

Simulations of the consequences of varying AChR site density or inhibiting AChE have also been made, and these are instructive in considering abnormalities encountered in myasthenia gravis. Reduction in AChR density results in a decrease in the amplitude of the calculated quantal event without changing its waveform (rise time and decay time). A 10 per cent reduction in density reduces the amplitude by only 3 per cent, but further reduction in density is directly proportional to the reduction in amplitude. At 10 per cent of the normal AChR density, ACh is hydrolyzed preferentially or, if AChE is inhibited, will tend to diffuse out of the cleft rather than bind to one of the few remaining AChR. This rapidly results in inadequate numbers to

achieve opening of the ion-channel. When AChE is excluded from the system, the models predict that ACh will exit from the cleft only by diffusion and that ACh molecules will bind to other molecules of AChR as they exit so as to buffer the ACh and prolong the calculated quantal event.

AChR are characterized by their ability to bind ACh reversibly. In addition, a series of related compounds will bind so as either to open (agonist) or to close or inhibit ACh access to the ion channel (antagonists). Agonists, such as carbamylcholine, succinylcholine, and suberylcholine, although not employed clinically, bind nicotinic AChR and generate a depolarization but vary in susceptibility to AChE and in the channel open time generated. Antagonists which block the effects of ACh by binding to the AChR include tubocurarine (curare), a competitive inhibitor which does not open the channel, and hexamethonium and decamethonium, which generate depolarizing blockade. The α-neurotoxins derived from the venom of elapid snakes (cobras, kraits, and sea snakes) are the most specific ligands currently known for the AChR. These compounds block access of ACh to its binding sites on the α subunit; the action of α-bungarotoxin (α-bgt) is essentially irreversible, whereas that of α-cobrotoxin is not. The specificity and irreversibility of the binding of α-bgt to AChR permitted purification and cytochemical and ultrastructural localization and ennumeration of AChR, since both radiolabeled and fluorochrome-labeled derivatives are available. In both autoradiographic analysis of AChR density on tissue and quantitation of AChR in solution, AChR site refers to the numbers of α-bgt binding sites.

Release of multiple quanta of ACh by depolarization of the nerve terminal (arriving nerve impulse, increase in extracellular [K^+]) will produce an end-plate potential (EPP). The EPP has been shown to result from the conductance of Na^+ and K^+ and to result from the additive effects of multiple MEPPS. EPP amplitudes are progressively reduced by reducing the Ca^{++} concentration or augmenting Mg^{++} concentration but only down to a basic size equivalent to a MEPP amplitude. Statistical analysis of histograms of the amplitudes of EPPs, generated under conditions of high external Mg^{++} and low external Ca^{++}, demonstrated that the EPP amplitudes were integral multiples of the MEPP amplitude at that end-plate, measured under the same conditions. An EPP is composed of about 400 MEPPs at a normal rested mammalian end-plate. The actual quantal content of any given EPP depends upon the state of readiness of the nerve terminal and requires that adequate numbers of synaptic vesicles be ready for release from the active zones.

Clinical assessment of neuromuscular transmission

The status of neuromuscular transmission in patients is assessed by evaluation of the electrical or mechanical responses of muscle to the arrival of a nerve action potential. The procedure most commonly employed is the analysis of the amplitude of muscle action potentials (MAPs) evoked by the arrival of a train of impulses generated by supramaximal stimulation of the motor nerve at rates of 2 to 50 Hz. Electrical stimuli are applied to the nerve by means of surface, or less commonly, needle electrodes, and the compound evoked muscle action potential is recorded from a surface or needle electrode placed over the muscle belly, roughly at the motor point. The muscle should be immobilized to avoid movement artifacts and the limb maintained at $35°$ C to $37°$ C surface temperature. Nerve-muscle systems commonly studied include ulnar/hypothenar, median/thenar, axillary/deltoid, facial/nasales, or other facial muscle, or on occasion facial/orbicularis oculi. Use of three or more systems increases the diagnostic yield in myasthenia gravis to about 90 per cent. Forearm

systems, however, are usually adequate for the evaluation of disorders involving release of ACh, such as botulism. The normal human neuromuscular junction responds to stimulation rates up to 50 Hz. with minimal change in the amplitude of the evoked MAP. Because accumulation of quanta of ACh into active zones, after the first EPP is generated, requires several hundred milliseconds, the second to fifth EPPs may show a progressive decline in amplitude in a normal neuromuscular junction stimulated at 2 to 5 Hz. However this has no detrimental effect on synaptic transmission because of the large safety factor (EPP amplitudes still well above threshold for production of MAP). As repetitive stimulation continues, calcium ions will tend to accumulate in the terminal if the time required to sequester this calcium load is longer than the time interval between arrival of nerve impulses (2 Hz. or more). This may produce an increased efficiency of release of vesicles from active zones and thus a larger quantal content. This will be reflected in a larger amplitude EPP (facilitation) lasting 1 to 2 minutes at higher rates of stimulation even at the normal neuromuscular junction.

The mechanical response to electrical stimulation can also be examined. Twitch tension measurements are generally made using the adductor pollicis muscle with the forearm carefully immobilized. The mechanical responses (force, rate of force development, maximum contractile power) usually parallel the evoked electrical response at rates of stimulation up to 2 Hz. but then fuse. In normal individuals a modest enhancement (up to 142 per cent) of muscle tension may be seen at low rates of stimulation (less than 5 Hz.). This response is referred to as a *positive staircase.* Present evidence suggests that this results from a delay in the fall-off of the active state of the muscle, which could cause an increased stretch of the series elastic element and an increase in twitch force.

Single-fiber electromyography is a second method for examining neuromuscular transmission. MAPs from paired single muscle fibers are examined. Normally, there is a slight but detectable time interval between the potentials evoked in two muscle fibers from the same motor unit. This "jitter" reflects the slight variation in the transmission at the two motor end-plates involved and is measured as the mean consecutive difference of interpotential intervals. In normal individuals, the jitter ranges from 15 to 30 μsec. and is increased in only about 1 of 20 potential pairs. In about 1 per cent of potential pairs, impulse blocking, or a failure of transmission at one end-plate of a pair, occurs. Single-fiber electromyography examinations employ the extensor digitorum communis, the biceps, or less often the frontalis or quadriceps muscles. Estimates of quantal size and quantal content can be made using intracellular microelectrodes to record from intercostal muscle obtained at biopsy or operation. The mean quantum content may be calculated from amplitude distributions, assuming a Poisson distribution, of trains of EPPs studied with or without alterations of the divalent cation concentration (Ca^{++} versus Mg^{++}) of the bathing solution. MEPP amplitudes are recorded directly and a mean quantal size obtained from 50 to 100 recordings obtained in the presence of a stable resting membrane potential. Quantal content may then be calculated directly from the ratio of the mean EPP amplitude to the mean MEPP amplitude.

Presynaptic disorders of neuromuscular transmission

There are several disorders of neuromuscular transmission in which the defect has been localized to presynaptic mechanisms. Clinically important problems include botulism and the myasthenic syndrome of Lambert and Eaton (LEMS). Less common are the abnormalities generated by black widow spider bite, by high circulating lev-

els of aminoglycoside and polypeptide antibiotics, and by hypermagnesemia. Elapid snake venoms contain β-neurotoxins such as β-bungarotoxin which act on presynaptic mechanisms, as well as α-neurotoxins which act on the AChR. These also contribute to the neuromuscular paralysis that may be generated by envenomation. Most of these disorders involve the mechanisms for release of transmitter rather than synthesis or packaging.

BOTULISM. Intoxication from botulinum toxins results most commonly from the ingestion of foods contaminated by the spores of *Clostridium botulinum*, which may be found in soil-grown foods (Types A and B) or fish (Type E). If the food is not prepared at a temperature of at least 100° C, as in home-canning procedures performed at high altitudes, spores will not be destroyed. Rarely, toxin is produced *in vivo* owing to contamination of wounds by organisms and spores which then culture anaerobically because of inadequate incision or debridement. Similarly, ingestion or inhalation of spores by young infants allows the production of toxin in the gastrointestinal tract during periods of constipation. The toxin has been shown in experimental animals to destroy cholinergic nerve terminals, which results in an acute denervation of the muscle fibers supplied by that nerve.

Rodents that received toxin injections into a hind leg muscle showed complete paralysis of the leg within 18 hours which lasted up to 8 weeks. Examination of muscles obtained sequentially from 18 hours to 21 days following the administration of toxin disclosed marked reduction in MEPP amplitudes with a return toward normal by 5 to 6 days. Similarly, the amplitudes of EPPs evoked by 0.5 Hz stimulation were small, and over 95 per cent of arriving nerve impulses failed to produce release of any transmitter. At 5 Hz. however, the number of failures was markedly reduced, and initial EPPs showed facilitation occurring especially 3 days or more after toxin injection. The amplitude histograms of MEPPs and EPPs did not coincide, indicating that although many EPPs were of the same amplitude as the smallest MEPPs, the mean amplitude of MEPPs was still smaller than that of the EPPs, so that the EPPs were not made up of individual quanta. Transmitter release was much less affected by the external concentration of calcium than normal. Exposure of the muscle to a potassium-free medium produced no change in transmitter release from poisoned muscles, whereas the MEPP frequency in normal muscle rose to 8 to 10 times normal. However, when the extracellular calcium concentration was increased and a calcium ionophore was also added, nearly normal high-frequency transmitter release was induced as well as a shift of the distribution of MEPP amplitudes toward normal. These observations suggest that botulinum toxin alters the calcium channels of the terminals but not the release mechanism.

These observations help to explain observations made in patients using repetitive stimulation and single-fiber electromyography. The abnormal release of transmitter produces a low initial evoked MAP and no change or slight decrement in amplitude of succeeding potentials. At rates of 20 to 50 Hz., however, the accumulation of calcium within the terminal reaches a level which overcomes the disorder of calcium channels and permits increasing release of quanta through essentially normal release mechanisms. In addition, exercise will produce a similar marked facilitation. Single-fiber electromyography will demonstrate some degree of jitter since transmission across paired endplates is unlikely to be involved equally. Electromyography may demonstrate an increase in the number of polyphasic motor units of short duration related to the damage to motor nerve terminals, and this is particularly common in infantile botulism.

Botulinum toxin is one of the most effective poisons known. It produces nearly total paralysis of nicotinic, ganglionic, and muscarinic cholinergic transmission. Patients often succumb to respiratory paralysis. In early stages, patients complain of dry, sore mouth and throat, blurred vision, diplopia, nausea, vomiting, and generalized malaise. Signs, at maximum, include total external ophthalmoplegia and symmetrical facial, bulbar, limb, and respiratory muscle weakness. Autonomic signs include pupillary paralysis, hypohydrosis, paralysis of salivation, urination retention and constipation, and variable ileus. Not every patient shows paralysis of every system, a finding possibly related to variable intake of toxin or a variation in individual response. Therapy should be directed toward ventilatory assistance and intensive care. Antitoxins may prevent further damage if administered promptly but there is a risk of serum sickness. Drugs that promote release of ACh may be beneficial. These include guanidine hydrochloride and 4-aminopyridine, which increase the release of ACh per nerve impulse.

LAMBERT-EATON MYASTHENIC SYNDROME. The Lambert-Eaton myasthenic syndrome (LEMS) is characterized by proximal weakness, especially of leg muscles, diminished or absent deep reflexes, muscle aches, and paresthesias, as well as dry mouth, hypohidrosis, constipation, difficult micturition, and impotence. Involvement of ocular muscles is unusual, but ptosis and diplopia may occur. Strength and deep reflexes may improve after exercise. These signs are compatible with a disorder involving cholinergic neuromuscular and ganglionic transmission.

At present there appear to be at least two populations of patients. About 75 per cent of men and 25 per cent of women have an associated neoplasm, which is usually an oat cell carcinoma of the lung. Another group, mostly women, have an unexpectedly frequent occurrence of autoimmune diseases such as pernicious anemia, rheumatoid arthritis, thyroid disease, and celiac disease. Patients in this group have responded to plasmapheresis and to immunosuppressive regimens (corticosteroids with cytotoxics). The electrophysiological abnormalities in LEMS are almost identical to those found in botulism or in muscle subjected to high external magnesium and low external calcium concentrations. The first evoked MAP, in a rested muscle, is of low amplitude, and succeeding potentials may decline at low rates of stimulation. However, at 10 Hz. or above, there is a marked facilitation which may also be observed after exercise. Single-fiber electromyography demonstrates increased jitter and blocking. When intercostal muscles are examined *in vitro*, MEPP amplitudes are normal but EPP amplitudes are variable and often reduced toward the mean MEPP amplitude, so that, on average, quantal content is low. Because of this, a nerve impulse may not produce release of enough quanta to produce an EPP whose amplitude exceeds the threshold for generation of a MAP. In addition, the calcium dependence of the quantal content of the EPP is reduced, as it is in botulinum-poisoned nerve terminals. Acetylcholine content and the level of choline acetylase are within the normal range so that neither synthesis nor storage of transmitter is defective. Recent freeze-fracture studies performed on intercostal muscle from nine patients demonstrated a marked decrease in active zones and a decrease in the active zone intramembranous particles, normally arranged in parallel double rows at the active zones. Since these particles are suspected to constitute the voltage-sensitive calcium channels, these findings provide morphological support for the site of the lesion among the presynaptic release mechanisms in LEMS. Binding of either autoantibodies or toxic factors could result in damage to and loss of these elements.

Therapy of LEMS has been directed at promoting release of transmitter by administration of guanidine hydrochloride or 4-aminopyridine. In light of recent studies suggesting an autoimmune disorder in many patients, immunosuppressive therapy has been employed, with production of remission in several patients.

HYPERMAGNESEMIA. Clinically apparent derangements in neuromuscular transmission have been reported in patients receiving magnesium-containing cathartics, with and without normal renal function, or parenteral magnesium sulfate for eclampsia whose serum magnesium was 1½ to 2 times normal. In two patients, administration of magnesium salts exacerbated a pre-existing case of LEMS, resulting in life-threatening dysphagia and ventilatory insufficiency. Two pregnant women who received magnesium sulfate developed weakness resulting from neuromuscular blockade, as did their newborn infants. Repetitive stimulation of motor nerves at 50 Hz. produced up to 800 per cent facilitation in these patients. Two patients improved when the magnesium compounds were discontinued; a third received guanidine hydrochloride.

TOXINS. Although β-bungarotoxin (β-bgtx) acts to block the release of ACh, its action has not been separated from the overall toxic effect of the bite of this snake. Studies of isolated nerve-muscle preparations indicate that β-bgtx shows many similarities to the action of botulinum toxin but differs in that it is three to ten times more potent in paralyzing chick muscles, binds faster than botulinum toxin, produces an initial facilitation of release of ACh associated with an increase in quantal content in low-calcium media and spontaneous fasciculation of the muscle, and produces posttetanic potentiation. However, the paralytic actions of both toxins are preceded by latency, antagonized by high concentrations of magnesium or a reduced calcium concentration, and accelerated by stimulation of the nerve at high frequency. Pretreatment with β-bgtx completely blocked the effect of botulinum toxin; however, the presence of botulinum only partially retarded the effect of added β-bgtx.

Action of another toxin with presynaptic effects has clinical consequences. Latrotoxin is the active component toxin derived from the venom of the black widow spider. In human beings, envenomation results in abdominal rigidity, generalized painful cramps, headache, vomiting, hypertension, tremor, hyperreflexia, and fasciculations. *In vitro*, latrotoxin or whole venom produces a rapid extensive release of quanta of ACh from the nerve terminal associated with a striking increase in MEPP frequency. This results in a block of neuromuscular transmission. The effect occurs even in the absence of calcium. Indeed, the toxin markedly increases the cation conductance of artificial lipid bilayers, suggesting that it can serve as an ionophore.

Postsynaptic disorders

MYASTHENIA GRAVIS. *Myasthenia gravis* (*MG*) is a disease in which abnormal neuromuscular transmission occurs because of an autoimmune state resulting in antibody-mediated destruction of AChR. There is fluctuating weakness of ocular, facial, bulbar, neck, limb, and respiratory muscles. Not all muscle groups are equally weak in a given patient. Muscle weakness is often worsened after prolonged use and improved by rest.

Various observations since the 1870s, when MG was first described fully, have suggested that the basic defect involves neuromuscular transmission. The symptoms and signs resemble those seen in human beings after curare poisoning. Repetitive stimulation of the motor nerve and muscle, performed originally by Jolly in 1895, was reexamined by Harvey and Masland in 1940. At low frequencies (2 to 5 Hz.) a normal initial evoked MAP was seen, but a progressive drop in amplitude followed, constituting the so-called decremental response. Patients responded to the administration of anticholinesterases

(anti-AChE) with normalization of some weak muscles and considerable improvement in others as well as repair of the decremental response. Abnormalities of the thymus gland (thymoma or hyperplasia: germinal centers in the thymic medulla) had long been recognized. With the increasing appreciation of the central role of the thymus in initiating and maintaining normal cell-mediated immunity, Simpson and Smithers independently proposed an autoimmune pathogenesis for MG. In addition, an unusual number of patients with MG had either circulating autoantibodies or other autoimmune diseases. Therefore, the hypothesis of causal role for a circulating factor capable of blocking neuromuscular transmission, which had been advanced in the 1930s, was reinforced, and the factor was suspected to be an autoantibody.

Next came the important biochemical breakthroughs in the ability to isolate the nicotinic AChR, which proved feasible because of the large amounts present in electric tissue from electric fish (eels and rays) and the purification of the neurotoxins from elapid snake venoms, which proved to bind specifically to AChR. Soon, AChR was purified from electric fish tissue in milligram quantities, which permitted biochemical characterization as well as use of the protein as an antigen to immunize a variety of species.

The study of experimental autoimmune myasthenia gravis (EAMG), the disorder of neuromuscular transmission produced by immunization of frogs, chickens, rabbits, guinea pigs, rats, mice, goats, sheep, and monkeys confirmed that these animals developed clinical, pathological, and electrophysiological features of MG. The severity of weakness and rapidity of onset and the presence or absence of monocytic infiltrates at end-plates varied from species to species and with the amount of immunogen and composition of adjuvants. All showed high titers of circulating antibodies to the immunogen (electric fish or rat

AChR), and rats were shown to have detectable titers against AChR derived from non-denervated rat muscle (junctional AChR). Pathological alterations of end-plates included amputation of the tips of synaptic folds, accumulation of membranous debris in the cleft, widening and simplification of folds with IgG molecules, and Y-shaped antibody-like structures visualized by electron microscopy cytochemistry. All of these pathological features have also been found on end-plates from intercostal muscles of patients with MG.

EAMG and MG also demonstrated nearly identical electrophysiological abnormalities. The decremental response was also found in EAMG where it was shown clearly that low stimulation rates of 2 to 5 Hz. were indeed entirely adequate to produce this finding. In addition, posttetanic potentiation, a phenomenon in which EPPs, elicited 10 to 15 seconds after a conditioning tetanus, show a higher amplitude than baseline and decrement is reduced, as well as posttetanic exhaustion, a much exaggerated decrement 2 to 4 minutes after a tetanus, both characteristic of MG, were also found in EAMG. Jitter and impulse blocking, detected by single-fiber electromyography, are seen both in MG and in rats and rabbits with EAMG. Studies with intracellular microelectrodes on human intercostal muscle and diaphragm or hind-limb muscles of animals both showed either undetectable MEPPs or a marked reduction in MEPP amplitude. When MG muscle was studied under voltage clamping, the miniature end-plate current (MEPC) was also found to be reduced to about one third of normal, although decay time constants were very similar to normal. Spectral analysis of the end-plate noise obtained during steady application of ACh to voltage-clamped normal and MG end-plates, showed that the amplitude of the elementary current event (the apparent single-channel conductance) was similar, as was the average channel lifetime. However, the mean membrane currents or MEPP amplitudes were smaller than those measured in normal muscle. These measurements by Albuquerque and colleagues and Cull-Candy, Miledi, and Trautman confirm that there is a major postsynaptic disorder in MG which does not alter the ion channel but results from a loss of functional AChR.

The loss of functional AChR in MG is a consequence of the presence of circulating antibodies to AChR. These autoantibodies appear to produce their destructive effects by several mechanisms. Studies of the various EAMG models and studies of the effects of EAMG and MG sera on cultured muscle and isolated single electric tissue cells (electroplax) indicate that some antibodies can block the binding site for cholinergic ligands and for α-bgtx. These antibodies may be responsible for the acute disorder of neuromuscular transmission which causes the death of chickens within 24 hours of passive transfer and of rabbits following a second (boosting) injection of AChR. MG and EAMG serum, immunoglobulin, and divalent F $(ab^1)_2$ fragments all produce an acceleration of the normal turnover of AChR in the membrane. Normally, AChR on rat or chick myotubes is internalized and degraded in lysosomes with a half-life of 22 to 24 hours. After application of antibody, this process is accelerated to 8 hours. Since divalent but not monovalent antibody fragments can produce this effect, it is felt to be initiated by cross-linking of AChR, which are then patched and internalized. This accelerated degradation has also been detected using nerve-muscle preparations in which the AChR is in the junctional form, which is presumably attacked in MG. Immunocytochemical examination of MG muscle, using peroxidase-tagged ligands which bind to IgG (staphylococcal protein A) or to complement components 3 and 9 (specific antibody), have shown that these are bound to the tips of synaptic folds and to debris in the cleft, at more severely involved end-plates. This finding implies a contribution of complement-dependent lysis to the loss of AChR from the postsynaptic membrane which is also supported by the complement-fixing properties of both MG and EAMG antibodies. At present, there is no evidence of a direct T cell–mediated attack on AChR in MG, although in the acute phase of the EAMG induced in rats by immunization with AChR in complete Freund's adjuvant as well as extra *Bordetella pertussis*, an inflammatory infiltrate is seen. Antibody-dependent cell mediated cytotoxic attack (an attack of killer cells following opsonization of surface AChR by antibody) has, therefore, been postulated to contribute to this phase of rat EAMG.

Rash, Albuquerque, and colleagues studied sensitivity to microiontophoretic application of ACh and MEPP amplitudes in parallel with ultrastructural appearance of the end-plates of MG intercostal muscles. In some fibers, MEPPs could be recorded but EPPs were of low amplitude. This was associated with the presence of an unusual 50- to 100-Å layer, including Y-shaped structures, thought to be a coating of IgG molecules. At junctions in which no detectable MEPP or EPP could be recorded, sensitivities to applied ACh were as low as 250 mV./nC., which indicated that these end-plates did not have a high enough density of AChR to allow reaction with ACh. In addition, these were the end-plates with a marked disruption of the postjunctional membranes, widened synaptic clefts, and simplified synaptic folds, often with no secondary clefts and with coated debris.

The abnormal neuromuscular transmission at end-plates still showing postsynaptic folds in MG and EAMG appears from present evidence to result from loss of AChR. This results in small MEPP and MEPC amplitudes because a released quanta of ACh cannot bind. When ACh is applied directly by the microiontophoretic procedure, the remaining AChR in the region respond, the ion channel opens, and normal elementary current events can be detected by spectral noise analysis. However, a packet of transmitter may open only

about 600 channels instead of the 1,500 opened at normal end-plates.

At end-plates that have been damaged to the point of loss of secondary clefts and widening of primary clefts, either from modulation or complement-dependent lysis, an additional feature of the geometry that is important to generation of MEPPs is lost. AChR molecules are no longer concentrated at the tips of regularly placed synaptic folds. The ACh released in a quantum from the active zone is much more likely to be hydrolyzed by AChE in the cleft and the basal lamina before it can reach residual AChR sites. As predicted by the computer models used to analyze normal quantal events, reduction in density will be directly proportional to MEPP amplitude without changing rise time or decay time. The value of anticholinesterases in MG is therefore to prolong the time for ACh molecules to find residual AChR, but even this will not be of much value at severely damaged end-plates. Lindstrom, Engel, and Lambert, analyzing MG intercostal muscle and EAMG rat muscles, found a direct correlation between MEPP amplitude and free AChR content of the muscle. The loss of AChR and the percentage of AChR already labeled with antibody *in vivo* were also quantitatively similar.

Drugs and toxins

ORGANOPHOSPHATE COMPOUNDS. *Organophosphate compounds* are anticholinesterases whose binding to AChE is irreversible. They are used as insecticides (malathion, parathion) and petroleum additives and are the major constituents of nerve gases. Most clinically important exposure to these agents has occurred in agricultural or industrial settings or is due to accidental or deliberate exposure to insecticides and rat poisons used in the home. Inhibition of AChE results in prolonged action of ACh at all cholinergic receptors with depolarizing block. Muscanic effects include sweating, salivation, profuse bronchial secretions, and miosis. Clinical consequences from overactive nicotinic synapses include fasciculations, limb and respiratory weakness, tachycardia, central nervous system signs, headaches, and ataxia. Loss of consciousness, anxiety, emotional lability, and confusion, as well as direct central nervous system effects, may result from hypoxia and hypercapnia. The diagnosis may be suspected from the acute onset and the presence of miosis and fasciculations along with symmetrical muscle weakness. Occasionally, there may be total external ophthalmoplegia. The diagnosis may be confirmed by measurement of AChE levels in red blood cells or pseudocholinesterase in serum, both of which are reduced to less than 10 per cent of normal in severe cases. Immediate attention must be given to ventilatory assistance and clearing of secretions. Atropine (1 to 2 mg. I.V. every 20 to 30 minutes) should be administered to normalize the pupils and maintained so as to control secretions and other muscarinic overactivity. Pyridoxine aldoxime methiodide (pralidoxime 2-PAM) was specifically synthesized to regenerate AChE. Intravenous administrate of 1 g. should improve strength and reduce fasciculations but must be given promptly and in large enough doses. Its beneficial effects must be supplemented by concomitant atropinization.

Alpha-bungarotoxin (α-bgtx) is a major toxic component of *Bungarus* venom. This and other α-neurotoxins (cobrotoxin, erabutoxin) all block AChR specifically and in controlled experiments in animals, reproduce all of the major electrophysiological abnormalities of MG (decremental response, posttetanic potentiation, posttetanic exhaustion, reduced MEPP amplitudes). However, envenomation by elapid snakes, cobras, and sea snakes results in paralysis and often death because of the combined actions of the effects of phospholipases, α-toxins and β-toxins, all components of the venoms.

D-Penicillamine is a drug employed in the therapy of rheumatoid arthritis and Wilson's disease. About 2 to 3 per cent of patients with rheumatoid arthritis develop a form of MG with typical pharmacological, electrophysiological, and antibody abnormalities which, unlike typical MG, is reversible upon withdrawal of the drug. Signs appear 2 months to several years after initiation of therapy and disappear 6 months to 1 year after the drug is withdrawn. The anti-AChR titer also declines slowly. Anti-AChE therapy may be required in patients with generalized signs. D-Penicillamine has been shown to interact with sulfhydryl groups of AChR and to alter the binding characteristics of ACh which may coincide with altered antigenicity of AChR.

Disorders related to combined pre- and postsynaptic defects

Congenital myasthenic syndromes

Among patients with onset at birth of symptoms and signs suggesting MG, several different abnormalities of neuromuscular transmission have been defined recently. Involvement of eye muscles with opthalmoplegia and ptosis is common in this group and many patients have had life-threatening episodes of apnea or respiratory insufficiency. All demonstrate decremental responses at low rates of stimulation and many have received anti-AChE, corticosteroids, or thymectomy in attempts to treat their disease, assumed to be atypical myasthenia. Further definition of the exact site of the disorder of neuromuscular transmission has required intercostal biopsies to measure MEPP amplitudes, MEPC amplitude, rise time and decay time, and quantal content as well as measurements of AChR sites/μm.^{2} and any circulating or bound anti-AChR. Engel and colleagues studied one 14-year-old patient with fluctuating ptosis and strabimus and small muscle bulk in facial, bulbar, cervical, and limb muscles. Intracellular microelectrode studies showed that the duration and decay time of MEPPs and EPPs were markedly prolonged and not altered further by anti-AChE. MEPP amplitudes were in the low normal range or below normal. The quantal content was reduced, but the probability of release was normal. Normal density and distribution of AChR was detected by cytochemical, and morphometric analysis, but no AChE could be detected by cytochemical, immunochemical or biochemical means. In addition, nerve terminals were abnormally small. The absent AChE certainly accounted for the abnormally prolonged MEPPs and EPPs since individual molecules of transmitter could then bind repeatedly to AChRs before diffusing out of the cleft.

Members of two other families had ophthalmoparesis and wasting and weakness in the neck, scapula and forearm muscles from infancy. Deep reflexes were preserved. In addition to a decremental response to repetitive stimulation, a single nerve stimulus elicited repetitive MAPs. MEPP and EPP durations were markedly prolonged and anti-AChE addition produced even greater prolongation. Since AChE was present in normal amounts, a defect in the AChR ion channel resulting in prolonged open time was postulated, although this was not measured directly.

A third disorder was studied in one of the five involved siblings out of eight total. Their disease was characterized by fluctuating ptosis, weakness on exertion, and apnea which responded incompletely to anti-AChE. Along with a decremental response and weakness which developed at 10 Hz. stimulation, the MEPP amplitude decreased below normal after 10 Hz. of stimulation. Cytochemical and morphometric analysis indicated normal end-plates. The defect is suspected to involve abnormal resynthesis of ACh after exercise or several minutes of evoked neuromuscular transmission, possibly a defect in choline acetylase or packaging of ACh into vesicles. Five other patients, two in one family and one

the product of a consanguineous union, have had a similar clinical syndrome of ophthalmoplegia and bulbar, neck, and facial weakness from birth. MEPP amplitudes were reduced. AChE and ACh content of muscle were normal in all cases and there was neither circulating nor bound anti-AChR. Enumeration, by autoradiography, of radioiodinated α-bgtx binding sites per end-plate distinguished the familial cases who had reduced AChR sites but normal ACh content. Vincent, Cull-Candy, and colleagues postulate an abnormality in the AChR which does not alter the ion channel characteristics and is not immunopathological. In the two patients with normal numbers of α-bgtx binding sites but small MEPP amplitudes, one proved to respond to 4-aminopyridine, a drug which promotes release of ACh from nerve terminals and is suspected to have a presynaptic defect. Samples from the other had features also found in denervated muscle, including multiple end-plates per fiber and α-bgtx binding characteristic of extrajunctional AChR. A defect in the AChR and its α-bgtx binding site was postulated.

It is probable that other similar disorders as well as new disorders of the molecular components of neuromuscular transmission will be identified among those with congenital onset. However, exact definition of the defects requires application of the full range of cytochemical, immunochemical, and electrophysiological techniques for study of neuromuscular transmission not available in the routine laboratory.

Drugs and toxins

Aminoglycoside *antibiotics*, such as neomycin, streptomycin, kanamycin, and colistin, a polypeptide antibiotic, may produce a block in neuromuscular transmission in patients without any known neuromuscular disease or aggravate MG. The defect may occur with blood levels within the therapeutic range as well with higher than normal levels in patients with renal insufficiency. Studies of bath-applied streptomycin using *in vitro* nerve-muscle preparations disclosed inadequate release of ACh which could be antagonized by an excess of calcium ion. In addition, the sensitivity of the postjunctional membrane to ACh was reduced. Different compounds differed in their relative effects on pre- and postsynaptic events. Neomycin and colistin produced the most severe derangements. The effect of kanamycin, gentamicin, streptomycin, tobramycin, and amikacin was moderate; tetracycline, erythromycin, vancomycin, penicillin G, and clindamycin had negligible effects. Patients who fail to regain normal ventilatory effort following anesthesia or who show delayed depression of respiration following extubation and are receiving one of the more potent agents should receive ventilatory support until the agent can be discontinued or another antibiotic substituted.

Cardiovascular drugs including quinine, quinidine, procainamide, and disopyramide can unmask MG or cause clinical exacerbation. In addition, they have been reported to prolong respiratory muscle weakness generated by muscle relaxants given during anesthesia so that extubation was delayed. Intracellular microelectrode studies, by Johns and colleagues, of rat muscles perfused with procainamide revealed reduced MEPP and EPP amplitudes. The effect was dose dependent and reversed by washing with control buffer. Quantal analysis performed at 10 nM. Mg^{++} also demonstrated a reduction in the number of quanta released from the nerve terminals.

Annotated references

Albuquerque, E. X., Rash, J. E., Mayer, R. F., and Satterfield, J. R.: An electrophysiological and morphological study of the neuromuscular junction in patients with myasthenia gravis. Exp. Neurol. *51*:536–563, 1976. (A meticulous reexamination of the effects of microiontophoretic application of acetylcholine to myasthenic intercostal muscle *in vitro*, which established that reduction in MEPP amplitude resulted from a postsynaptic defect and related to loss of junctional AChR.)

Argov, A., and Mastaglia, F. L.: Disorders of neuromuscular transmission caused by drugs. N. Engl. J. Med. *301*:409–413, 1979. (A short but reasonably complete overview of important drug interactions.)

Caputy, A. J., Kim, Y. I., and Sanders, D. B.: The neuromuscular blocking effects of therapeutic concentrations of various antibiotics on normal rat skeletal muscle. J. Pharmacol. Exp. Ther. *217*:369–378, 1981. (An important study of the mechanisms of neuromuscular blockade produced by 14 different antibiotics.)

Cull-Candy, S. G., Lundh, H., and Thesleff, S.: Effects of botulinum toxin on neuromuscular transmission in the rat. J. Physiol. *260*:177–203, 1976. (Study of the low-amplitude end-plate potentials and small quantal size, which occurs after injection of toxin, strongly suggesting that the calcium sensitivity of presynaptic mechanisms for transmitter release is reduced.)

Cull-Candy, S. G., Miledi, R., and Trautmann, A.: End-plate currents and acetylcholine noise at normal and myasthenic endplates. J. Physiol. *287*:247–265, 1979. (This study of voltage-damped normal and myasthenic end-plates confirmed that the elementary current event and average channel life-time at single receptors is normal but confirmed that miniature end-plate currents were markedly reduced in myasthenia.)

de Jesus, P. V., Slater, R., Spitz, L. K., and Penn, A. S.: Neuromuscular physiology of wound botulism. Arch. Neurol. *29*:425–431, 1973. (Complete clinical electrophysiological studies of a case of wound botulism.)

Desmedt, J. E.: The neuromuscular disorder in myasthenia gravis. *In* Desmedt, J. E. (ed.): New Developments in Electromyography and clinical neurophysiology. Basel, S. Karger, A. G., 1973. (A complete and careful description of the electrical and mechanical responses to nerve stimulation in hand muscles in myasthenia with a historical overview and emphasis on posttetanic phenomena. Although not analyzed in the context of a postsynaptic loss of AChR, still very valuable.)

Engel, A. G.: Myasthenia gravis In: Vinken, P. J., and Bruyn, G. W. (eds.): Handbook of Clinical Neurology, pp. 95–145. Amsterdam, North-Holland, 1979. (A complete, meticulous, and rigorous review of all features of the disease by one of the major contributors to the field.)

Engel, A. G., Lambert, E. H., Mulder, D. M., Gomez, M. R., Whitaker, J. N., Hart, Z., and Sahashi, K.: Recently recognized congenital myasthenic syndromes (A) endplate acetylcholine (ACh) esterase deficiency (B) Putative abnormality of the ACh induced ion channel (C) Putative defect of ACh resynthesis or mobilization: Clinical features, ultrastructure and cytochemistry. Ann. N.Y. Acad. Sci. 377:614–637, 1981. (Complete report of the first three syndromes dissected from congenital myasthenia and shown not to be autoimmune.)

Fambrough, D. M., Bayne, E. K., Gardner, J. M., Anderson, M. J., Wakshull, E., and Rotundo, R. L.: Monoclonal antibodies to skeletal muscle cell surface. In Brockes, J. P. (ed.): Neuroimmunology, pp. 49–89. New York, Plenum Press, 1982. (A review of the preparation of monoclonal antibodies against a variety of muscle cell antigens and their use to probe development and organization of the membrane.)

Fukunaga, H., Engel, A. G., Osame, M., and Lambert, E. H.: Paucity and disorganization of presynaptic active zones in the Lambert-Eaton myasthenic syndrome. Muscle Nerve 5:686–697, 1982. (Provides elegant freeze-fracture ultrastructural confirmation of presynaptic abnormalities in release mechanisms in this syndrome.)

Grob, D. (ed.): Myasthenia gravis: pathophysiology and management. Ann. N.Y. Acad. Sci. 377, 1981. (The sixth international conference on myasthenia gravis recording many important developments in the study of basic mechanisms and immunotherapies.)

Heuser, J., and Reese, T. S.: Evidence for recycling of synaptic vesicle membrane during transmitter release at the frog neuromuscular junction. J. Cell Biol. 57:315–344, 1973. (Elegant ultrastructural study showing life-cycle of synaptic vesicles and evidence for recycling of vesicle membrane.)

Katz, B.: The release of neural transmitter. Springfield, Charles C Thomas, 1969. (The Tenth Sherrington Lecture by the Nobel Laureate. A very clear presentation of the mechanism of transmitter release.)

Kelly, R. B., and Hall, Z. W.: Immunology of the neuromuscular junction. In Brockes, J. P. (ed.): Neuroimmunology. New York, Plenum Press, 1982. (Complete, careful review of the use of antibodies to dissect structure and function of the neuromuscular junction, including antibodies to receptor in EAMG and myasthenia.)

Lambert, E. H., and Elmquist, D.: Quantal components of endplate potentials in the myasthenic syndrome. Ann. N.Y. Acad. Sci. 183:183–199, 1971. (Analysis using intracellular microelectrodes confirming the reduced quantal content in this syndrome and the beneficial effects of increasing external Ca^{++} concentration or addition of quanidine hydrochloride.)

Land, B. R., Salpeter, E. E., and Salpeter, M. M.: Kinetic parameters for acetylcholine interaction in intact neuromuscular junction. Proc. Nat. Acad. Sci. U.S.A. 78:7200–7204, 1981. (A study analyzing the dependency of miniature end-plate current time-courses on amplitude and acetylcholine site density, making use of computer simulation and showing that activation of the receptor occurs within a 0.3-μm radius of the point of release.)

Lang, B., Newsom-Davis, J., Wray, D., Vincent, A., and Murray, N.: Autoimmune actiology for myasthenic (Eaton-Lambert) syndrome. Lancet 1:224–226, 1981. (Describes recent evidence supporting an autoimmune pathogenesis for this syndrome.)

Lisak, R. P., and Barchi, R.: Myasthenia gravis. Philadelphia, W. B. Saunders, 1982. (An excellent monograph reviewing all phases of this disease and its immunopathophysiology.)

Neher, G., and Sakmann, B.: Nature 260:799–802, 1976. (The important early study describing the use of direct patch-clamp analysis to measure the properties of single ion-channels including the conductance of a single AChR-channel.)

Stalberg, E., Ekstedt, J., and Broman, A.: Neuromuscular transmission in myasthenia gravis studied with single-fiber electromyography. J. Neurol. Neurosurg. Psychiatry 37:540–547, 1974. (Describes the results of the application of this technique to patients with myasthenia and confirms that it is a useful and sensitive additional diagnostic test which does not require a muscle biopsy.)

Swift, T. R.: Disorders of neuromuscular transmission other than myasthenia gravis. Muscle Nerve 4:334–353, 1981. (A superb complete review of the subject with an equally complete bibliography.)

Vincent, A., Cull-Candy, S. G., Newsom-Davis, J., Trautmann, A., Molenaar, P. C., and Polak, R. L.: Congenital myasthenia: Endplate acetylcholine receptors and electrophysiology in five cases. Muscle Nerve 4:306–318, 1981. (An examination of MEPP amplitudes and currents, AChR numbers, ACh content, and antibodies to AChR in five patients with congenital, nonautoimmune myasthenia.)

Wadia, R. S., Sadagopan, C., Amin, R. B., and Sardesai, H. V.: Neurological manifestations of organophosphate insecticide poisoning. J. Neurol. Neurosurg. Psychiatry 37:841–847, 1974. (A review of the neurological findings and response to therapy in 200 cases.)

34

Physiology of Normal and Diseased Muscle

Béla Nagy, Ph.D.
Frederick J. Samaha, M.D.

Rapid advances in muscle research have contributed greatly to our understanding of its normal metabolism and physiology. As in other fields, the interactions of proteins and cellular organelles have proven to be most complex, and this is especially so with muscle contraction and relaxation. On the other hand, the structure of the contractile system is very similar in mammalian striated muscle to those in other species studied. This remarkable similarity suggests that the transducer mechanisms of the chemical energy into mechanical work is the same in different forms of muscle. Current knowledge on the structural, regulatory, and supporting proteins of vertebrate skeletal muscle and their interactions has reached a state in which a general physiological concept of muscle function has merged with biochemistry on the molecular level to explain normal muscle function.

Contractile proteins of muscle

MYOSIN. Myosin is the major protein of skeletal muscle, composed 60 per cent of total protein in myofibrils, and is located in the A-band in the sarcomere in the thick filaments. The individual myosin molecule has a molecular mass of about 480,000 daltons with a long rod-like shape and having an apparent rotational symmetry along a single axis. It consists of two apparently identical polypeptide chains, the heavy chains, with about 200,000 daltons mass each and four smaller peptides termed light chains (L), or globular chains (g), in the range of 16,000 to 30,000 daltons mass. The carboxyl end of each heavy chain is folded into an α-helical conformation which makes up half of the molecular mass. The two α-helices are twisted around each other into a coiled structure, as twines in a rope, forming a rod-like structure with dimensions about 20 Å in diameter and 1,400 Å in length. In the other half of the molecule, at the amino end, each myosin peptide is separated and forms a globular head, protruding from the rod portion, appearing like a two-headed tadpole in rotary shadowed electron micrographs. To these globular heads of myosin are attached, noncovalently, the small globular peptides, mentioned above, two to each head.

There are three different types of these small, globular peptides, also called light chains, designated L_1, L_2, and L_3 according to their migration in sodium dodecylsulfate-polyacrylamide gel electrophoresis, L_1 being the slowest component. The apparent molecular mass of each is 25,000, 20,000, and 16,000 daltons, respectively. Each myosin molecule contains two L_2s and a pair of either L_1s or L_3s. There appears to be a slight variation between the relative mobilities of the light chains in the different muscle types; however, the L_2 light chains appear to be related by their ability to undergo phosphorylation catalyzed by a light-chain kinase. The kinase is activated by the Ca^{2+}-binding protein, calmodulin. The L_2 light chain that undergoes phosphorylation has been implicated in the regulation of smooth muscle contraction. There is no apparent involvement of L_2 in skeletal muscle regulation, although in cardiac muscle a regulatory role of L_2 has been suggested. The two other light chains, L_1 and L_3, may play a role, depending on which of the two subunits is present, in the activity of actin-activated ATPase.

In a single myosin molecule the functional part is located mainly in the globular amino end of each peptides that contains the ATPase and actin-binding sites, the two main biological activities of myosin.

The myosin molecule contains regions in its polypeptide chain which, either by its structural accessibility or by its unique sequence pattern, have high susceptibility to proteolytic cleavage by proteases. Located two-thirds of the way from the carboxyl end, in the rod portion of the molecule, is one of the protease sensitive regions. Here, either tryptic or chymotryptic catalyzed proteolysis produces two fragments. One is the α-helical tail fragment from the carboxyl end of the molecule, termed light meromyosin (LMM); the other consists of two globular heads connected with about one-third of the coiled α-helical parts of the "neck" portion, called heavy meromysin (HMM). Further digestion of HMM with proteolytic enzymes splits this fragment into three pieces: the rod-like "neck" portion, consisting of the coiled α-helical fragments, termed subfragment-2 (S_2) and the two globular head portions, splitting off separately, termed subfragment-1 (S_1). Proteolysis of the whole myosin molecule with papain splits the molecule directly into the two globular heads (S_1) and the whole coiled coil, rod-like fragment, including LMM and S_2. Both HMM and S_1 fragments retain the actin-binding and ATPase activities and are useful in the understanding of the many aspects of the biological activities of contractile proteins. It is noteworthy that each protease susceptible region appears to act as a hinge region in the intact molecule. Bends were observed both in the rod portion, about two-thirds in length from the carboxyl terminus, and at the other end of the rod portion at the connection of the heads, enabling the heads to bend away from the "neck" portion of the rod. A schematic illustration of the myosin molecular structure is shown in Figure 34–1.

The thick filament in the skeletal muscle consists of an aggregate of many myosin molecules utilizing the LMM rod portions. They start to assemble with a tail-to-tail aggregate which grows to a larger structure as new molecules are attached in a helically staggered way. This aggregation of myosin molecules with the LMM portions growing away from the center, forms a bipolar thick rod with a bare middle area. On the surface of this rod, outward from the bare middle region, are the radially protruding neck and head portions of the HMM parts of individual myosins. Unknown factors terminate the growth of myosin aggregates, forming uniform lengths within the sarcomeres.

There are more than one type of myosin molecule, although the basic assemblages are similar. Differences are mainly in the types of light chains attached to the head portion, although other differences were found in the main polypeptide chain composition. Recent, but as yet unconfirmed, information indicates that embryonic skeletal muscle myosin is different from that of adult; the former is probably a multiple

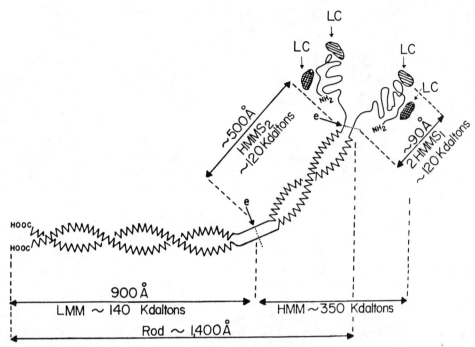

FIG. 34–1. Schematic representation of the structure and dimensions of myosin molecule based on biochemical, physical and electron microscopic data. The myosin molecule is a complex of six subunits: two large peptides, the heavy chains, of about 200K dalton mass each, and four small peptides of two identical mass of 25K daltons and another identical mass pair of either 20K or 16K daltons. The rod portion of the molecule has a coiled structure of the two helical parts of the heavy chains. The globular head portions of the heavy chains stand separate with two nonidentical small subunits attached to each head. The postulated hinge regions in the rod section are represented by parallel lines, where enzymatic cleavage of the peptides (*e*) yields the peptides denoted LMM and HMM, or LMM, $HMMS_2$ and $HMMS_1$ or the rod portion and $HMMS_1$, depending on conditions and enzymes used. The $HMMS_1$ has one chief polypeptide chain while the other fragments have two. The carboxyl and amino ends of the main polypeptide chains are marked $-COOH$ and $-NH_2$, respectively. The light chains (*LC*) are attached to the head regions, two different subunits to each $HMMS_1$.

species of slow, fast, and cardiac muscle types of myosin. Proper innervation apparently leads to the selection of one myosin type with the repression of gene expression for the other types. It is well documented that even in fully developed muscle, the types of myosin present can be changed either by shifting the frequency of stimulation of the motor nerves by external means or by changing the innervation by cross-innervating fast muscle with slow muscle nerve (or vice versa). Pattern changes in neural activity produce trophic effects influencing the pattern of myosin isoenzymes. The isoenzymes reverse to the new motor pattern type, from slow to fast

muscle or vice versa, suggesting that the myosin phenotype is determined by a neural influence.

ACTIN. Actin is the second most abundant protein in muscle, consisting of 20 per cent of the total protein in myofibrils. Structurally, actin is a single polypeptide protein folded into a globular shape. It binds strongly with one nucleotide, adenosine triphosphate (ATP) or adenosine diphosphate (ADP), and one divalent cation Ca^{2+} or Mg^{2+} per molecule. It is composed of 374 amino acids, of which the sequence is known. The molecular mass calculated from its amino acid composition is 41,785

FIG. 34–2. Schematic representation of the filametous form of actin. F-actin is a side-by-side aggregation of the doublets of the globular G-actins, forming a right-handed helical structure. One turn of the helical pitch contains 13 monomers in the single line, or the half pitch is also of 13 monomers in the double strand. This fibrous form of actin is found in thin filaments. The transition of the globular (G) single actin monomers to this fibrous (F) form is called polymerization. The G→F transformation is accompanied by a dephosphorilation of the G-actin bound ATP to ADP in F-actin. This reaction in living muscle presumably takes place during the course of development, growth or regeneration.

daltons or 42,300 daltons when bound ATP and Ca^{2+} is taken also as part of the molecule. At position 73 there is a methylation of histidine, N-methylhistidine, which occurs probably by a specific enzyme, after the whole polypeptide chain has been synthesized.

Actin has been found in every eukaryotic organism tested and is probably one of the most conservative proteins throughout the biological evolution. There are some microheterogeneities between actins of different species, but in skeletal and cardiac muscle a single type of actin, α-actin, is found. Two other species of actin, β- and γ-actin, are found in nonmuscle cells. During development in differentiating skeletal muscle of higher vertebrates in the early stages, actin species resembling β- and γ-actins are found together with α-actin. Moreover, during muscle development the β- and γ-actins decrease, and in fully differentiated muscle only the α-actin species is present. Recent studies have found that slight varieties of actin are present in different tissues of the same species. At least six different actin genes have been identified so far, differing only by a few conservative amino acid replacements in their products.

The globular form of single-actin molecules (G-actin) polymerizes at physiological salt concentrations to a filamentous form (F-actin). F-actin is the main constituent of the thin filaments in muscle sarcomers and is located in the I-band. The F-actin filament has two bead-like strands of side-by-side aggregated G-actin doublets, with a double stranded right-handed helical structure. A helical half-pitch (3,600 Å) consists of 13 monomers in the double strand. The diameter of each monomer is about 60 Å. The F-actin structures both formed from G-actins *in vivo* and *in vitro* are essentially identical. The regulation of actin polymerization to uniform lengths in the I-bands of the sarcomer in developing and regenerating muscle is influenced by factors not yet known. Fig. 34–2 shows a schematic structure of the F-actin filament.

Actin in nonmuscle cells is largely unpolymerized. This is mainly attributed to a low-molecular-mass protein, (16,000 daltons) profilin, which binds to actin in a stoichiometric ratio of 1:1 and prevents polymerization. Actin can interact with a large number of proteins. Within the myofibrils some of the interactions are essential to its biological function. It interacts with α-actinin, which in striated muscle is located in the Z-line of the sarcomere, and its role together with other Z-line proteins is probably to anchor the actin filaments. The other protein interacting with actin is tropomyosin, which is an integral part of the thin filament in myofibrils and is discussed below. The major functional interaction of actin is with myosin, specifically with the globular part of myosin. This interaction stimulates the myosin ATPase activity by accelerating the release of ADP from myosin, which appears to be the rate-limiting step in the ATPase activity. The actomyosin complex then is dissociated by ATP, and the cycle of actin and myosin interaction starts again. This process leads to contraction, which is described later.

TROPOMYOSIN AND TROPONIN. Tropomyosin and troponin are the two regulatory proteins involved with Ca^{2+}-mediated regulation of actin and myosin interactions. They are located in the thin filaments of the myofibrils with strong but noncovalent association with the double-stranded actin in the filament.

TROPOMYOSIN. Tropomyosin is a fully α-helical, two-subunit coiled-coil molecule with a molecular mass of about 65,000 daltons and is approximately 410 Å long. The molecular arrangement resembles that described for the LMM portion of myosin. There are two types of subunits designated α and β; both these chains consists of 284 amino acids. There are 39 amino acid sequence differences between the α- and β-chains. In fast skeletal muscle the ratio of α-chains to β-chains is 3:1 or 4:1. In slow red muscles and in cardiac muscle the ratio is 1:1. The two α-helical tropomyosin subunit chains interact with parallel polarity aggregations. They are also in register axially to yield a symmetrical arrangement of the side chains. At regular intervals in two series at intervals of seven residues are nonpolar residues in the sequence of each subunit. In the α-helical form these residues fall on a crestline and provide the interaction between the coiled helices. The long coiled-coil tropomyosin molecules form filamentous aggregates through overlapping amino and carboxyl ends. This long tropomyosin strand binds to F-actin in a molecular stoichiometry of tropomyosin:G-actin units of 1:7 and occupies the two flattened or grooved sides of the double stranded F-actin filament. Each tropomyosin molecule binds also the troponin-T component of troponin with a stoichiometry of 1:1. It is believed that the tropomyosin can be at either of two alternative positions on the actin filament: near or in the grooves on both sides during contraction, and on the periphery during relaxation. This allows actin and myosin interactions and sterically blocks interactions respectively. The dislocation of the tropomyosin strands on actin filament from one site to the other is thought to be regulated through troponin and the regulator being the [Ca^{2+}]. The tropomyosin-troponin-Ca^{2+}

regulation of contraction is proposed to be through tropomyosin dislocation on actin filament and is consistent with experimental observations, although it has not yet been proven firmly.

TROPONIN. Troponin is a three-subunit protein which, together with tropomyosin, confers Ca^{2+} sensitivity on pure actomyosin (i.e., activates the actomyosin ATPase in the presence of Mg^{2+}). In the absence of Ca^{2+}, troponin inhibits the "superprecipitation" of actomyosin, and in the presence of 10^{-6} M Ca^{2+} this inhibition is prevented. The "superprecipitation" phenomenon results from the formation of actomyosin from the flocculate myosin aggregates and F-actin interactions accompanied by a shrinkage of the volume of the new complex as the precipitate becomes more compact. In the thin filament troponin is located with a periodicity of 400 Å bound to tropomyosin in a molar ratio of 1:1. The molecular mass of troponin is about 70,000 daltons and consists of about 4.5 per cent of the total protein of the myofibrils, while tropomyosin is about 3 per cent. So, the thin filament consisting of actin, tropomyosin, and troponin is about 27.5 per cent of the total protein and is localized in the I-band of the sarcomere. The three subunits of troponin are denoted as troponin C (the Ca^{2+} binding component), troponin T (the tropomyosin-binding component) and troponin I (the inhibitory component, inhibiting the actin and myosin interactions in the presence of ATP).

TROPONINE C. Troponine C (TnC) is an acidic protein with a molecular mass of 17,965 daltons. The sequence of the 159 amino acids in TnC is known. There are four Ca^{2+} binding sites in the molecule, each consisting of an α-helix, a loop with chelating side chains providing oxygen ligands for Ca^{2+} and another α-helix sequentially. Two of these Ca^{2+} binding sites have an affinity constant for Ca^{2+} of 2×10^5 M^{-1}, the other two a higher one, of 2×10^7 M^{-1}.

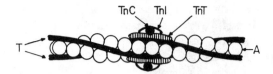

FIG. 34–3. A model of the arrangement of the proteins of the regulatory subunit assembly of the thin filaments. The double stranded F-actin filament is represented by the open circles as the 60-Å diameter single molecules of actin units (A). In the grooves of the actin double-stranded helix lies on each side the coiled double α-helix of the tropomyosin molecules. Each unit of the tropomyosin strand (T) has an attached troponin complex, consisting of a triplet of subunits. The subunit troponin T (TnT) is bound to tropomyosin and has a site of interaction with actin and with troponin C (TnC); TnC interacts with both TnT and TnI. Upon increase of [Ca²⁺] in the cytoplasm the regulatory sites on TnC bind calcium; the conformation is changed; which affects both TnI and TnT, TnI is separated from actin and actin and myosin can interact to form the actomyosin complex, resulting in contraction by sliding the thin filaments between thick filaments.

Mg^{2+} also binds to the higher affinity sites, which is why they are also known as Ca-Mg sites. Ca^{2+} binding results in a significant conformational change of the molecule. TnC interacts with both TnT and TnI. The amino acid residues 82 to 135 in the sequence appear to carry the interaction site with troponin I.

TROPONIN I. Troponin I is a basic protein (molecular mass 20,864 daltons) that has 179 amino acid residues; of these, residues 27 to 74 are needed for the interaction with TnC. Residues 70 to 151 are involved in the interaction with tropomyosin; and residues 4 to 116 interact with actin.

The third component of troponin, *troponin T*, contains 259 amino acids and has a molecular mass of 30,503 daltons. It appears that the NH_2 terminal half of the molecule has the interacting sites for tropomyosin and actin, while both halves may interact with TnC and TnI.

The troponin molecule is apparently unique to striated muscle and has not been shown to be present either in nonmuscle cells or in smooth muscle. The three subunits described above in a ratio of 1:1:1 make up troponin. Troponin, together with tropomyosin and actin in molecular ratio of 1:1:7, respectively, constitute the thin filament. The high calcium affinity of troponin as the target molecule and the Ca^{2+} as the trigger for muscle contraction have been well established. However, the molecular mechanisms by which these complexes actually operate is still a problem. The recent working hypothesis is that calcium binds to TnC. There is still a question which of the Ca^{2+} binding sites constitute the regulatory sites. The Ca^{2+} binding-controlled conformational change of TnC affects both TnI and TnT. Since TnC binds strongly to TnI, it is assumed that the conformational change in TnC separates TnI from actin, which is then able to form a complex with myosin in the presence of MgATP. This results in the sliding action between thin and thick filaments that produces contraction. A model of the arrangement of the regulatory subunit assembly of the thin filament is represented in Figure 34–3.

Besides the major proteins described, there are several other minor proteins associated with either myosin or actin. Their role is assumed to be for the regulation of the filamentous structure. One is *M-protein* associated with myosin at the M-line, the middle of the bare space of myosin (thick) filaments. It is a 165,000-dalton-mass protein, apparently together with a 43,000-dalton-mass protein identified as *creatine kinase*, in a complex with a 1:1 molar ratio. Another protein named *C protein* is found in myosin filaments, apparently bound to the shaft of the filament. α-actinin with a molecular mass of 95,000 daltons is found in the Z-line apparently as a cementing protein for the thin filaments. Another protein of the Z-line is *desmin*, with a mass of 55,000 daltons. It forms 100-Å-diameter filaments connecting Z-lines laterally. Within the Z-line a lattice structure is formed by another 55,000-dalton-mass protein named

Z-protein, which has the same mass as desmin, but immunology shows it is a different protein. There are a few other proteins found in the Z-line and also in connection with the thin and thick filaments; however, their role is not yet clear.

β-ACTININ. β-actinin is found at the free end of the actin filaments in the sarcomere. It is a heterodimer of two subunits, one 37,000-daltons, the other 34,000-daltons in mass. *Filamin*, with a molecular mass of 240,000 daltons, is located in the Z-lines and probably has an important role in the contractibility of nonmuscle cells. *Connectin*, a 42,000-dalton-mass protein, links neighboring Z-lines forming a net of very thin filaments. It is located also in I- and A-bands.

After the recent description of skeletal muscle proteins the ultrastructure of muscle can easily be described. Long known by light microscopic observations, the typical repeating structure along the length of the muscle can now be simply described as the appropriate arrangements of the proteins. The typical striation pattern of voluntary muscle is attributed to two sets of filaments with regular arrangements. The thin filaments (about 80 Å in diameter) appear to be attached to the Z-bands and constitute the I-band. A second set of filaments (with a diameter of about 150 Å), called thick filaments, occupy the A-band. The thick filaments are connected crosswise in the middle of some material that constitutes the M-zone. These arrangements repeat lengthwise and appear as the cross-striation. These repeating units of arrangements are called sarcomeres and one sarcomere measures from Z-bands to the next Z-band. Contraction is attributed to the relative sliding motion of the filaments within the sarcomere, shortening the distance between Z-bands. A schematic representation of the structure of the filament arrangements of striated muscle is shown in Figure 34–4.

On muscle contraction, the two kinds of filaments become cross-linked after excitation and the length of the whole fibril changes because of the relative motion of the two sets of filaments. The length of the muscle depends on the length of the sarcomeres. The length of the sarcomere, in turn, should be a variation in overlap between the thin and thick filaments. Since high-resolution micrographs have shown that the interaction between the filaments occurs through cross-bridges, the number of links possible would depend on the relative sarcomere length, and hence the isometrically measured length-tension diagram would reflect the changes in overlap. Figure 34–5 shows the relationship between the tension measured and the length of overlap within the sarcomere.

Interaction of actin, myosin, and ATP: the bridge cycles between filaments as bases of contraction

On contraction, the driving force for the sliding action between the actin and myosin filaments is generated by cyclic events at the individual cross-bridges. The mechanical and accompanying chemical events in a cycle are described according to our recent understanding of the contraction phenomenon and are depicted in Figure 34–6. The detailed molecular nature of the force-producing mechanism in muscle is still not fully understood. There has been convincing evidence from many studies during the last decades that muscle contraction is developed by a cross-bridge interaction between the actin and myosin filaments in muscle, thus providing a force in some way to cause the filaments to slide past each other. However, various techniques of electron microscopy and x-ray diffraction have produced thus far only static pictures of the filaments and cross-bridge interactions within the filaments. There is a need for new techniques that enable viewing of the small, cyclic, asynchronous cross-bridge movement on the millisecond time scale. An improved x-ray diffraction technique is developing with

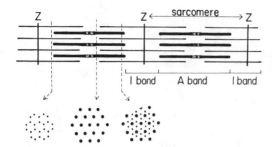

FIG. 34—4. Schematic representation of thick (myosin) and thin (actin) filaments in striated muscle sarcomere. Sarcomere is the repeating functional unit, about 23,000 Å in length in relaxed muscles as measured from Z-line to Z-line. In phase contrast light micrographs the light I bands alternate with dark A bands, resulting from diffraction pattern of the less dense material on both sides of Z-line, the actin filaments, and from the denser area in the middle, where the myosin filaments are located, respectively. In electron microscopic pictures in the central region of the A band where only myosin filaments are present there is a somewhat lighter zone, called H zone. In the middle of H zone is a darker line called M line, caused by the presence of another protein, M protein, crosslinking the myosin filaments. The Z-line material anchors the thin filaments. The thick filaments interdigitate with the thin filaments in a regular fashion. Transverse sections of the sarcomere at different levels reveal this regularity. In most instances, especially in vertebrate muscles, a regular hexagonal array is found for both thin and thick filaments as shown in the lower part of the scheme. At left is a cross-section through the thin filaments (*I band*), in the middle is a cross-section of the thick filaments (H zone), and on the right is a cross-section through the overlap array (*A band*). In the overlap array each thick filament is surrounded by six thin filaments, while each thin filament is located in the center of a triangle formed by three thick filaments. The ratio of thick to thin filaments is 1:2. In the scheme the myosin filaments in the two halves from the center zone (dark) bear the crossbridges, which can attach to the actin filaments, and as described in Figs. 34–5 and 34–6 the attachment generates force and causes shortening of the sarcomere as the filaments slide past each other. Without load, the filaments can slide past each other to about half the distance of each relaxed sarcomere length.

The nomenclature of the sarcomere bands

A band: anisotropic band, dark band in light microscope; encompasses the region filled with thick myosin filaments, and some overlaps with thin actin filaments.

I band: isotropic band, light band in light microscope; it is the region with thin actin filaments alone.

use of x-ray radiation emitted in synchrotons. The extremely high intensity of radiation with new crystal monochromator cameras expected to yield over 100-fold gains in speed which will make possible good time resolution to study the structural events in actin-myosin interactions. Our recent understanding of molecular events during muscle contraction is based on the following general knowledge.

It is convincingly established that the cross-bridges correspond to the S_1 head regions of the myosin molecule. The S_1 heads together with the S_2 stems project out laterally from the backbones of the thick filaments; the position which makes it possible to interact with the actin filaments alongside, lying parallel to the thick filaments. The ATP binding and splitting region is located in the head region of myosin S_1, and because ATPase is activated by actin there is little doubt that the energy for contraction is liberated at the cross-bridges during the interaction of myosin with actin. The sliding force shortening the muscle is generated at the same time, using the energy liberated by the ATP splitting. The force generation, according to the most accepted working hypothesis, is based on the so-called swinging cross-bridge model, which assumes that the S_1 head of myosin and the S_2 stem to which it is attached

Z-line: from the German name Zwischenscheibe, a dark line in light microscope in the middle of the I band, now called also Z band or Z disc because of the thickness of the band. The Z band consists of cross-connections linking the thin filament arrays, which have opposite polarity in each half of I Band, i.e., on each side of Z-line. The extra Z-line material is α-actinin.

H zone: a less dark region in the middle of A band resulting from the no-overlap region between actin and myosin filaments. On stretching the muscle the I band and H zone both lengthen.

M band: a dark band in the middle of H zone. It is a structure of network of high molecular mass (~150,000 dalton) nonmyosin protein connecting the myosin filaments. Several enzymes were found attached to M band, but only the M form of creatine kinase appears to be a principal structural component.

hinge out sideways from the thick filament. The S₁ head could attach to actin; and the ATP-splitting causes some structural changes in S_1, where it is attached to actin. This effectively changes the angle of attachment in the appropriate direction, producing a sliding movement between actin and myosin filaments. At the end of this work-producing stroke the cross-bridge between actin and S₁ detaches, and farther along the actin filament the work cycle with attachment starts again. When the muscle is relaxed, attachment of S_1 cross-bridges to actin is prevented and the myosin heads are located closer to the thick filaments. (See Fig. 34–6).

The basic observations that support the above hypothesis are that: (1) electron micrographs show cross-bridges attached to actin filaments in rigor but not in relaxed muscles; (2) x-ray diffraction pattern shows two basic types of equatorial diagrams in vertebrate striated muscle: one by resting muscle with a strong 1, 0 reflection and a weaker, 1, 1 reflection. Fourier reconstructions show that the reflection pattern in rigor implies a shift of density away from the myosin filaments toward the actin filaments. The equatorial x-ray reflections from isometrically contracting muscles are similar to those observed with rigor muscles. Both x-ray reflections and electron microscopic pictures support the cross-bridge formation but do not show how the attachment can generate force. Beginning new studies with high-intensity x-rays generated by synchroton radiation with fast position-sensitive detectors indicate that cross-bridges do not attach to actin with the same regular orientation during contraction as is observed in rigor, but with more than one orientation. This is considered evidence that some kind of longitudinal cross-bridge movements occur during contraction; thus the evidence so far agrees well with the proposed "swinging, tilting cross-bridge model" for force generation in connection with sliding filament contraction theory.

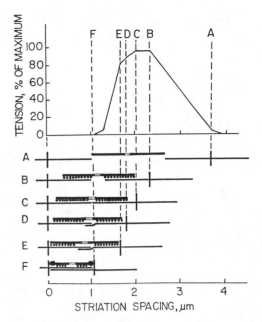

FIG. 34–5. Isometric length-tension diagram and relative sarcomere length changes of frog single-twitch fibers. In upper part of figure are the tension phases corresponding to critical sarcomere length where the overlap position between thin and thick filaments change and cause inflections in the length-tension curve. The letters indicate the corresponding changes in the curve and filament overlaps at different sarcomere lengths.

(*A*) At 3.65 μm. sarcomere length filaments are pulled apart, there is no overlap between the filaments, no crossbridge formation is possible, and there is zero tension.

(*B*) At 2.25 μm. sarcomere length maximal tension is reached as there is full overlap between the filaments where maximal number of crossbridges can form.

(*C*) At 1.90 μm. sarcomere length is the end of full overlap position, still with maximal tension. Further shortening decreases tension due to collision of ends of thin (actin) filaments.

(*D*) At 1.85 μm. sarcomere length the tension is further decreasing due to friction between opposing thin filaments and decreasing active overlap for crossbridge formation.

(*E*) At 1.65 μm. sarcomere length another tension-affecting event is the thick (myosin) filament collision at both ends with Z band material.

(*F*) At 1.05 μm. sarcomere length is the so-called supercontracture state in which opposing forces cancel out the tension generated by remaining crossbridges: actin filament ends collide with Z band material, myosin filament ends are crumbled up at the Z-line, etc.

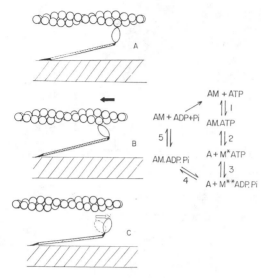

FIG. 34–6. A model of interaction of myosin and actin in the sarcomere: myosin crossbridge activity and ATP splitting cycle. On the left is the scheme of myosin bridge positions, on right the accompanying chemical events.

The description of the crossbridge cycle and chemical correlations: (*A*) Attachment of myosin to actin, accompanying step 4 in the chemical cycle (active complex). (*B*) Tilting of myosin head pulls actin filament in the direction shown by filled arrow, accompanied by release of hydrolysis products ADP and Pi, chemical step 5 (rigor complex). (*C*) Detachment of actin and myosin caused by ATP binding to myosin with two distinct conformations of myosin head, releasing the tilted angle position to assume position prior to attachment to actin (empty arrow) corresponding to steps 1, 2 and 3 of chemical cycle (relaxed state).

The whole cycle repeats as the activated myosin head attaches to actin again (*A*). The sliding movements occur when the actin-attached myosin head tilts (*B*), exerting a pulling force on actin relative to myosin filament. The two hinges of the myosin crossbridge, one at the rod attachment to the main body of thick filament, the other at the head attachment to the rod position, allow the lateral extension of the myosin bridge (see Fig. 34–1). This scheme includes the dissociating effect of ATP on actomyosin, ATP hydrolysis and the association of actin and myosin. The assumption is that the biochemical events are related to the cycle of activity of the bridges. (After Huxley, H. E.: The mechanism of muscle contraction. Science *164*:1356, 1969; Huxley, A. F., and Simmons, R. M.: Proposed mechanism of force generation is striated muscle. Nature *233*:533, 1971; Taylor, E. W.: Mechanism of actomyosin ATPase and the problem of muscle contraction. Curr. Top. Bioenergetics *5*:201, 1973.)

Calcium Transport

Calcium ion is an important cellular messenger, and a general theory of calcium signaling is slowly emerging with accumulation of Ca^{2+}-related studies. Under physiological conditions in the so-called resting or unstimulated cell, free calcium ion concentration is less than 10^{-7} M, or the pCa < 7. Following a stimulus, the calcium concentration in the cytosol rises and the calcium-dependent response is initiated. A multitude of cellular responses to calcium involve a second messenger, cyclic adenosine monophosphate (cAMP). In other cases, calcium reacts directly with target proteins to initiate the selective physiological response. The rise of free calcium concentration in the cytoplasm reaches about 10^{-5} M usually in response to stimulus, but it must be lowered again to resting levels. In most, if not all, eukaryotic organisms the lowering of the cytosolic calcium concentration to resting level of 10^{-7} M is achieved by a calcium-ATPase pump attached to the plasma membrane and to the endoplasmic reticulum. In striated muscle the calcium-ATPase pump is an integral part of the extensive sarcoplasmic reticulum (SR) system surrounding the muscle fibers. In all striated muscles this Ca^{2+}-Mg^{2+}-ATPase is responsible for recovering the calcium and relaxing the muscle.

In resting muscle cells, calcium concentration within the lumen of SR is as high as 10^{-3} to 10^{-2} M, and most of it is complexed by a calcium-binding protein or calsequestrin with low-affinity but high-capacity binding. To maintain this concentration difference between the lumen of SR and the cytosol there is a large number of calcium-ATPase protein molecules in SR. It is estimated that in striated muscle SR there are about 2×10^4 ATPase molecules per μm^2 of SR surface area (about 1 molecule per 5×10^3 A^2). Thus, about 20 per cent of the SR surface area is occupied by the ATPase protein. When muscle is homogenized, the extensive SR system is fragmented from its tubular or sleeve-shaped

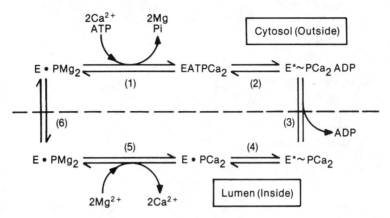

FIG. 34–7. The elementary steps of calcium transport by sarcoplasmic reticulum Ca^{2+}-ATPase. ATP = adenosine triphosphate; ADP = adenosine diphosphate; E = enzyme; Pi = inorganic phosphate; $\sim$P = phosphate covalently attached to enzyme protein.

form and reseals in response to forces of surface tension generated by its hydrophobic nature and forms globular vesicles. The SR vesicles then can be isolated by differential centrifugation and the retained calcium-pumping characteristics can be studied and inferred to the original SR properties. From such studies the mechanism of calcium transport appears to follow a cycle starting by binding two Ca^{2+} ions and a Mg^{2+} ATP to the enzyme on the cytosolic side of the membrane. The bound ATP is hydrolyzed with formation of a phosphoprotein intermediate while ADP is released. The phosphoenzyme-CA_2 is now in a different conformation, which helps the two calcium ions to be transported to the lumen where the two calcium ions are released on the interior side. The cycle is completed by the isomerization of the enzyme to its starting conformation through loss of inorganic phosphate. (A scheme of the mechanism of the calcium transport is illustrated in Fig. 34–7 as a six-step process.) There is no evidence that Mg^{2+} is a counterion for calcium transport, but after calcium release the enzyme is apparently in the $E.PMg_2$ form. Electrical neutrality is maintained by the parallel transport of anions. *In vitro*, the pump can be run backwards: SR is loaded with calcium and

phosphate and incubated in a medium containing ADP; ATP is produced as calcium leaves the vesicles. There is no evidence that this can happen *in vivo*. The large difference in dissociation constants of the Ca^{2+} enzyme complex in step (1), outside, and in step (3), inside, (see Fig. 34–8) indicates a reduced stability in the complex after the phosphorylation-induced conformational change of the enzyme protein occurred. The pK_d (Ca^{2+}) outside is 7.5, while pK_d (Ca^{2+}) inside it is 2.5.

The major proteins found in sarcoplasmic reticulum are the Mg^{2+}-Ca^{2+}-ATPase, with a 100,000-dalton mass; an 80,000-dalton-mass phosphoprotein of yet unknown significance; and two calcium-binding proteins in the 45,000- to 55,000-dalton-mass range. The weight ratio of ATPase to Ca^{2+}-binding proteins in SR is about 5:1 to 6:1.

Energetics of muscle contraction

The immediate source of energy for muscle contraction is the hydrolysis of ATP:

$$ATP \rightarrow ADP + Pi + energy$$

In resting muscle that [ATP]:[ADP] [Pi] ratio is high. This balance is quickly changed during the contraction events. Since the contractile events are very rapid,

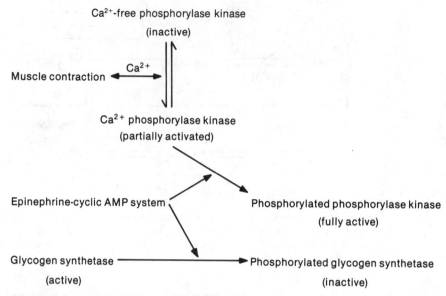

FIG. 34–8. Control of glycogen degradation and synthesis in muscle. Glycogen is a very efficient storage form of glucose. The storage consumes slightly more than one ATP per glucose-6-phosphate, the need for this amount is due to branching. The complete oxidation of glucose-6-phosphate yields 37 ATP. Comparing the ATP used up for storage with the yields from complete oxidation, the overall efficiency of storage is in the range of 97 per cent. The delicate balances of degradation and synthesis depicted in this scheme makes the two processes work according to the immediate energy needs of muscle. Obviously a defect in glycogen metabolism leads to impairment of energy utilization.

there is a time lapse before appropriate metabolic signals can reach the targets and increase metabolic rate. There is a built-in muscle buffer system which responds first and reestablishes the high [ATP]:[ADP] ratio. This high-energy phosphate buffer system utilizes the enzymes creatine kinase and myokinase. Since creatine phosphate (CrP) is the chief store of high-energy phosphates in muscle, the immediate effect of increased ADP due to ATP splitting is a disturbance of the equilibrium of creatine kinase catalized reaction:

$$CrP + ADP \rightleftharpoons ATP + creatine$$

The equilibrium is reestablished by the phosphorylation of ADP to ATP.

The other enzyme, myokinase, catalyzes the reaction:

$$2 \, ADP \rightleftharpoons ATP + AMP$$

This reaction also ensures the reestablishment of the original high [ATP]:[ADP] ratio.

A third enzyme present in the muscle, adenosine-deaminase, prevents the accumulation of AMP by deamination to form inosine monophosphate (IMP):

$$AMP \rightarrow IMP + NH_3^+$$

IMP then either returns to the nucleoside pool as inosine or is degraded further to uric acid. From the above sequence of reactions it is clear why, in a single twitch or short muscle activity, the only measurable change in the high-energy phosphate pool is a small change in [CrP].

Human skeletal muscles contain both red and white fibers, so the source of energy for rephosphorylation of ADP proceeds both through oxidative phosphorylation (red or slow fibers are rich in myoglobin and mitochondria) and by glycolysis (white or fast fibers contain little myoglobin and mitochondria). The route of glycolysis and respiration and thus ATP production are adjusted to the rate of ATP consumption by a series of feedback controls. When the [ATP]:[ADP][Pi] ratio is high (as in resting state) both the rate of glycolysis and the oxidation through tricarboxylic acid cycle is low, because they are inhibited by their allosteric negative modulator, high [ATP]. During maximal muscle activity, when the [ATP]:[ADP][Pi] ratio is greatly reduced by the abrupt breakdown of ATP to ADP and Pi, both glycolysis and tricarboxylic acid cycle are greatly increased. The increased [ADP] is a positive modulator for both glycolysis and oxidative phosphorylation through the tricarboxylic acid cycle. The AMP, formed by the adenylate kinase reaction (see above), also stimulates glycolysis, since it is a positive modulator for phosphofructokinase.

At maximal activity of skeletal muscle the oxygen uptake may increase 20-fold or more in transition from rest to full activity in response to the oxygen need for the oxidative processes. After a period of maximal activity and exertion, during which lactate appears in the blood in large amounts as the end product of glycolysis when the amount of oxygen is limiting, the breathing continues in excess of the normal resting rate and a considerable excess of oxygen is consumed. This extra oxygen during the recovery period corresponds to the oxidation of most of the lactic acid formed. Some of the excess lactic acid may be converted by the biosynthetic pathway, involving mitochondrial oxidative steps of pyruvate, through lactate-pyruvate to glucose-6-phosphate into glycogen in the liver.

The oxidative phosphorylation steps through the tricarboxylic acid cycle use also part of the pyruvate produced by glycolysis or fatty acids through oxidation, both products entering the cycle as acetyl CoA. There are three stages of fatty acid utilization before the oxidative cycle in the mitochondria begins: (1) activation of fatty acid by an enzymatic, ATP requiring esterification of free fatty acid with extra mitochondrial CoA to yield fatty acyl CoA; (2) acyl group transfer from the fatty acyl CoA

to the carrier molecule carnitine, followed by the transport of the acylcarnitine through the inner membrane of mitochondria into the matrix; (3) The reaction of the transferred fatty acyl-carnitine through a transferase with intramitochondrial CoA to form fatty acyl CoA and free carnitine. The fatty acyl CoA now becomes the substrate for the fatty acid oxidation system in the inner matrix compartment of the mitochondria. Carnitine is particularly abundant in skeletal muscle and stimulates the long-chain fatty acid oxidation by mitochondria especially in the red muscle fibers. A defect in the transferase or deficiency in carnitine impair the oxidation of long-chain fatty acids.

Glycogen is a readily mobilized stored form of glucose, a high polymer of glucose linked by α-1,4-glycosidic bonds with branches by α-1,6-glycosidic bonds. Glycogen is present in the cytosol in the form of granules with diameters of 100 to 400 Å, containing the enzymes for synthesis and degradation of glycogen. The synthesis and degradation occur by different reaction pathways. This breakdown is catalyzed by glycogen phosphorylase:

$$\text{Glycogen} + \text{Pi} \rightarrow \text{glucose-1-phosphate} + \text{glycogen}$$
$$\text{(n residues)} \qquad \text{(n-1 residues)}$$

The product glucose-1-phosphate is converted to glucose-6-phosphate and enters the glycolytic pathway. Glycogen phosphorylase exists in muscle in two forms: an active phosphorylase *a* and an inactive phosphorylase *b*. The inactive form is converted to the active form by phosphorylation by a specific enzyme. This activation is a hormone-cAMP system, but can also be activated by Ca^{2+}. Therefore, in skeletal muscle, glycogen breakdown is activated by a transient increase in the cytoplasmic Ca^{2+} level in the cytoplasm. In muscle the inactive form of phosphorylase *b* is activated also by AMP through an allosteric mechanism: AMP binds to the nucleotide binding site and alters the conformation of the inactive phosphorylase *b* to an active phosphorylase *b*. ATP acts as a negative allosteric effector by competing with AMP. Glucose-6-phosphate is also an inhibitor by binding to another site. In physiological conditions phosphorylase *b* is inactive because of inhibitions by ATP and glucose-6-phosphate. The other form, phosphorylase *a,* is fully active, irrespective of levels of AMP, ATP, and glucose-6-phosphate.

The synthetic pathway of glycogen is different from the pathway of breakdown. Glucose-1-phosphate is activated to UDP-glucose, then glycogen synthetase transfers UDP-glucose to the C-4 terminus of glycogen to form 1,4-glycosidic linkage, and UDP is formed.

UDP-glucose
+ glycogene → glycogene + UDP
(n residues)   (n + 1 residues)

Degradation and synthetic pathways each employ debranching and branching enzymes accordingly. Branching creates large numbers of nonreducing terminal residues, thus increasing the rate of glycogen synthesis and degradation. There is a coordinated control of glycogen synthesis and breakdown by the following reaction sequence:

1. Epinephrine binds to muscle cell membrane and activates adenylcyclase.
2. Adenylate cyclase catalyzes the reactions:

$$ATP \rightarrow cAMP + PPi$$

3. cAMP activates protein kinase.
4. Protein kinase phosphorylates both phosphorylase kinase and glycogen synthetase, switching on the first and inactivating the latter.

A schematic illustration of glycogen breakdown and synthesis control is in Figure 34–8.

Excitation-contraction coupling
The contractile proteins in the sarcomeres are built into units of myofibrils separated by mitochondria and sarcoplasmic reticulum. The muscle cells are enclosed by a plasma membrane, which, with various connective tissue elements and collagen filaments, in turn, forms the sarcolemma. The interior of the resting muscle cell is maintained by the active transport of ions through the plasma membrane at an electric potential of about 100 mV. more negative than the exterior. When the muscle is stimulated by its nerve an event called excitation-contraction coupling transmits the electrical nerve impulse through a transmitter molecule to the plasma membrane, where the electrical potential difference is disturbed; ionic conductance propagates along the plasma membrane into the fiber through the transverse tubular system to the triads of tubules, where the calcium stored in the terminal cysternae of the sarcoplasmic reticulum is released to the cytoplasm. A diagrammatic view of this sequence of events with a time scale is shown in Figure 34–9.

Acetylcholine as neuromuscular transmitter at motor nerve endings
Motor nerve endings do differ somewhat depending on the main fiber type. Slow-twitch fibers that respond to nerve stimulus with prolonged contracture are in general multiply innervated *en grappe*. In these fibers there is less abundant sarcoplasmic reticulum, and the T system is also very sparse. Fast-twitch fibers are innervated usually by individual end plates. In these fibers the sarcoplasmic reticulum is abundant with the regular T system in sarcomeres. At both types of motor nerve terminals acetylcholine is synthesized and stored in vesicles. These vesicles contain approximately 10^4 acetylcholine molecules and a vesicle of such content is described as being *quantum*. At the end plate there is some spontaneous release of acetylcholine, in both quantal and nonquantal forms. One or a few quanta can cause small amplitude depolarizations that remain localized to

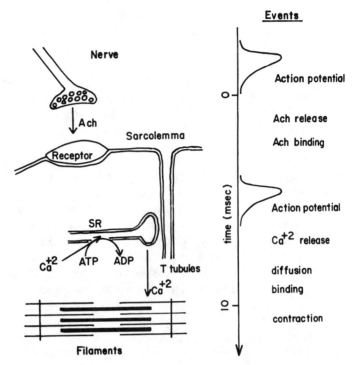

FIG. 34—9. A diagrammatic view of excitation-contraction coupling in striated muscles of vertebrates. On the left is a scheme of chronological events from excitation to contraction, while the right side indicates the time scale. Ach = acetylcholine; SR = sarcoplasmic reticulum.

In vertebrate striated muscle, including human muscle, both slow and fast types, at a single end-plate, the depolarization is initiated on the plasma membrane by acetylcholine binding to the acetylcholine binding site. The depolarization spreads as action potential to the T-tubules, the perpendicular invaginations of the plasma membrane, and down the T-tubules around the myofilaments with a speed of about 10 m. per second, the speed of slow nerve conductions. This depolarization of the T-tubules in some way affects the sarcoplasmic reticulum, close to the T-tubules, at the triad junctions. The response is a massive release of calcium, mainly at the area of terminal cysternae. The longitudinal part of the sarcoplasmic reticulum, farther away from the terminal cysterna, is apparently the area of calcium accumulation rather than release. Thus the sudden and massive coordinated release of calcium caused by a single above-threshold depolarization increases the free sarcoplasmic Ca^{2+} concentration. The Ca^{2+} rises from the resting, or unstimulated, concentrations of about 10^{-7} M to a concentration of 2×10^{-5} M. The released Ca^{2+} travels by diffusion from the T-tubule to the nearest thin filaments, where the target troponin molecules are located, in less than 1 msec. (The thin filaments are within 10,000 to 20,000 angstroms from the nearest sarcoplasmic reticulum. The hydrated calcium ion diffuses this distance to half its maximal concentration in less than 1 msec). The diffusion speed assures an almost simultaneous binding of Ca^{2+} to troponin all along the fiber, releasing inhibition to actin-myosin interaction and the entire muscle contracts synchronously.

the end-plate region (subthreshold effects). Conducted nerve impulses release large numbers of acetylcholine quanta, about 150 to 200 per impulse, giving rise to a large-amplitude end-plate potential. Such end-plate potential triggers a complex sequence of events starting with depolarization of the muscle membrane.

Fiber types in mammalian muscle

It has been recognized for over 100 years that in skeletal muscle there is a correlation between the speed of contraction and the color of muscle. It was observed, in general, that red muscles contracted more slowly than white. This simple classification as red-slow and white-fast became more complicated as more histological and chemical data accumulated, and there is more complexity now in differentiating criteria. No muscle has a single property overall because each muscle is determined by the proportion of various fiber types it contains.

The fiber classifications reported are, in general, of three types. These types are slow-twitch oxidative, fast-twitch oxidative-glycolytic, and fast-twitch glycolytic. This classification is based on physiological, histochemical, and biochemical characterizations. Some muscles are predominantly made up of one type of fiber in one species, whereas in others a different type predominates. An example is the soleus muscle in rabbit, which is a typically slow muscle; however, in rat and man, the soleus is a mixed fiber type. The different approaches employed over the years to type the fibers have revealed that a variety of factors contribute to differences in muscle types. There are differences in many enzymes that appear in different forms in different muscle types. Variations of the amino acid content and conformation of enzymes catalyzing the same reaction are called isoenzymes or isozymes. As an example, different isoenzymes were found in different muscle types in lactic dehydrogenase. The main differences in the slow and fast mus-

cles that are useful for classification were found in myosin isoenzymes. In the slow-red muscles the myosin ATPase is lower than in fast-white muscles. The slow-red muscle myosin is alkaline labile and denatures at pH 9.0 in a few minutes, while the fast-white muscle myosin does not lose activity for a much longer time period (e.g., it is not affected at all if incubated at pH 9.0). This pH stability difference is used for histochemical fiber classification. There are also differences in the light-chain components of different myosin types. Fast-twitch myosin contains all three light chains, L_1, L_2, and L_3. Slow-twitch myosin contains two types of light chains that migrate as L_1 and L_2. In cardiac myosin the light chains are also like L_1 and L_2. It is interesting to note that all types of myosin contain light chains with L_2 mobility. The light chains are also useful markers of myosin types and were used extensively to follow transformation of myosin from one type to another in cross-innervation studies. Differences also have been found in the main polypeptides of myosin, namely in the presence of methylated amino acids of lysine and histidine in different muscle types. Methyl lysine is found in all muscle types, whereas methyl histidine is present in adult fast-twitch muscle but absent from myosin in adult slow-twitch, embryonic, and cardiac muscle. Recently, differences have also been found in tropomyosin and in troponin subunits of different fiber types. The original differences based on muscle color relates to the main energy source of muscle. In fast-twitch muscle, ATP supply comes primarily from glycolytic processes, which provide a quick but rapidly exhaustible supply of energy. These muscles are white because the paucity of myoglobin and mitochondrial cytochrome gives a pale appearance to the muscle. On the other hand, muscles made up mostly from slow-twitch oxidative or fast-twitch oxidative-glycolytic fibers gain energy through oxidative processes involving the Krebs cycle. The Krebs cycle is fed by pyruvate, generated from anerobic breakdown of glucose and glycogen and from the oxidation of fatty acids. These muscles are rich in mitochondria and contain significantly higher amounts of myoglobin. The cytochromes of mytochondria and myoglobin content give these muscles a red appearance and are used for early classification.

Proteases in muscle tissue

Several kinds of proteases have been found in muscle tissue. They are generally classified by their optimal pH activity range into three categories, alkaline proteases, neutral proteases activated by Ca^{2+} ions, and acidic proteases or cathepsins.

The alkaline proteases are relatively low molecular weight proteins, their molecular mass ranging from 22,000 to 27,000 daltons. The pH optimum range for these proteases is 8.5 to 10.5. The enzymes are chymotrypsin-like, contain a serine residue at the active site, and at least one of the enzymes was localized in mast cells of muscle.

Ca^{2+} activated proteases with neutral pH optimum were localized in the cytoplasm of muscle cells and their molecular mass is in the 100,000-dalton range. One was found to be an 80,000-dalton-mass protein and is activated maximally by millimolar concentrations of calcium.

The acidic proteases or cathepsins are small or intermediate in size, with a mass of 24,000 to 45,000 daltons. The pH optimums for these enzymes, which are localized in the lysosomes, are from 4.0 to 7.0.

The apparent role of muscle proteases are in degradation of muscle proteins during normal protein turnover. In progressive muscular dystrophies, which are characterized by continuous progression of muscle wasting and weakness, the protein degradation is far greater than that of protein synthesis, although both processes are elevated above the values found in normal muscle. The role of elevated protease activities and possible use of selective protease inhibitors to slow destruction of muscle

proteins in Duchenne dystrophy and other myopathies are currently a topic of investigation in several laboratories keeping a therapeutic possibility in view.

Turnover rates and biosynthesis of muscle proteins

The biosynthesis and turnover studies of muscle proteins is of special interest for understanding of problems of developmental, regenerative and degenerative processes both in normal and pathological cases. Turnover of proteins is usually calculated from the decay rates of radioactively labeled amino acids that have been incorporated into protein. It appears from critical studies that this method overestimates the protein half-life; however, the relative rate of turnover is valid. Table 34–1 gives a comparison of the turnover rates obtained for myofibrillar proteins in rabbits. Molecular mass, amounts in myofibrils as percentages of total protein, mass of the protein relative to actin, and relative turnover rate of major myofibrillar proteins of rabbit skeletal muscle is listed in the table.

Table 34–2 lists the turnover times and measured amino acid incorporations in more detailed form. As Tables 34–1 and 34–2 indicate, there is no uniform turnover rate among the myofibrillar proteins of rabbit muscle, but a spread in relative rate over a sevenfold range. However, the difference among the turnover rates of myofibrillar proteins is not as large as that among the soluble proteins of muscle, in which the range difference is about 100-fold. Turnover rate also differs according to muscle types: slow skeletal muscle, cardiac, and smooth muscle proteins have about the same rates, but fast skeletal muscle has a rate only about half that for the others. The rate also depends on age: it is high in early stages but decreases with growth and aging. In excised muscles and in tissue cultures the turnover rate is comparable to that found in adult muscles *in vivo*.

In atrophied muscles there is an en-

TABLE 34–1

Major Myofibrillar Proteins: Molecular Mass, Amounts in Myofibrils, and Relative Turnover Rates in Rabbit Skeletal Muscle

	Actin	Myosin	Tropomyosin	Troponin	αActinin
Amounts in myofibril as percent of total protein	20	60	3	4.5	1
Molecular mass (daltons)	41,785	480,000	65,000	70,000	95,000
Mass/actin mass	1	11.5	1.5	1.7	2.3
Amounts in myofibril/actin	1.0	3.0	0.15	0.23	0.05
Molecules/actin	1.0	0.53	0.10	0.14	0.02
Relative turnover rate cpm./mg. protein*	1.0	2.5	3.3	6.5	3.0

* Turnover rate was measured by incorporation of five radioactive amino acids—glycine, leucine, valine, lysine, and alanine—assuming that the specific activity of each protein was proportioned to its glycine turnover rate.

TABLE 34–2

Turnover Times of Structural Proteins in Muscle

	Days	Relative Rates (Actin-1)	Protein Incorporation in 24 Hours (cpm./mg.)	Relative Values (cpm.)
Myosin	29	2.5	9.8	82
M protein	11	7	319	223
α-Actinin	24	3	150	98
Actin	75	1	72	32
Troponin	12	6.5	185	206
Troponin T	12	—	—	205
Troponin I	12	—	—	207
Troponin C	16	—	—	150
Tropomyosin	22	3.3	73	106

Turnover times have been measured by the rates of incorporation of five radioactive amino acids—glycine, leucine, valine, lysine, and alanine—assuming that the specific activity of each protein was proportional to its turnover rate. Relative values of incorporated amino acids are based on amino acid contents of proteins.

hanced rate of protein degradation but usually an enhanced rate of synthesis also, which is the case in dystrophic muscle. In denervated muscles the rate of synthesis is depressed first then elevated during atrophy. The slow skeletal muscle is more affected by denervation than the fast skeletal muscle. From tissue culture studies it is thought that hormones and growth factors stimulate growth, but this question is not answered sufficiently yet to explain turnover in various muscle conditions.

Duchenne muscular dystrophy

The muscular dystrophies, which are progressive and genetically determined, and which are primary degenerative myopathies, are at present classified according to clinical, genetic, and histopathological criteria. This definition, however, has changed as recent studies have conceptually broadened our understanding of normal muscle chemistry and its alteration by primary abnormalities within the muscle, by alteration in muscle function, or by changes in normal neurotrophic influences. Whether the muscular dystrophies are caused by intrinsic muscle abnormalities or by neurotrophic influences, muscle function is altered and these alterations are expressed at the molecular level. The exploration of these pathophysiological alterations has been the subject of intensive study in Duchenne muscular dystrophy.

Clinical aspects of Duchenne muscular dystrophy

This form of muscular dystrophy, which is also called progressive muscular dystrophy, is inherited as a sex-linked recessive trait. Expression of this syndrome occurs in males and, in rare instances, in patients with Turner's syndrome. The incidence of this syndrome is nearly 3 per 100,000 population and present information suggests that one-third of these cases are new mutants. The onset of this disease characteristically begins during the first few years of life with weakness of the pelvic girdle. Because of this pelvofemoral weakness, the patient has a characteristic lumbar lordosis and waddling gait. As the disease progresses, the shoulder girdle becomes involved, followed by involvement of the tibialis anterior, until, in the more advanced stages, virtually all the voluntary musculature of the body is atrophied and weak. In the earlier stages of this disease, muscle hypertrophy occurs, usually in the calves, and occasionally in other muscles such as the vastus lateralis and deltoids as well. As the disease progresses, the hypertrophic muscle degenerates and is replaced by fat and connective tissue. The muscles in this condition are described as pseudohypertrophic. The relentless progression of this disease leads to inability to walk within 10 years and death from inanition, respiratory infection, and cardiac failure, usually during the second decade. Along with the progressive atrophy of muscles, pseudocontractures with secondary skeletal deformities occur. Cardiac muscle is invariably involved, but clinical evidence of cardiac dysfunction does not usually occur until later stages of the disease. Before clinical evidence of cardiac disease is apparent, however, electrocardiographic studies show abnormalities that include tall R waves in the right precordial leads and deep Q waves in the limb and left precordial leads. Diagnostic biopsy of clinically involved muscles show random fiber degeneration, abnormal proliferation of connective tissue, and lipomatosis. Although not prominent, some evidence of regeneration of muscle fibers is usually seen.

Elevations of muscle enzymes in the serum

One of the earliest biochemical observations made in patients with Duchenne muscular dystrophy was the discovery of the release of some soluble enzymes from the muscle cell into the serum. Soon after Sibley and Lenninger noted the elevation of serum aldolase in patients with muscular dystrophy, other investigators noted serum enzyme elevations of a host of other enzymes including lactate dehydrogenase, glutamic oxaloacetic transaminase, α-glucanphosphorylase, phosphoglucomutase, glycerophosphoisomerase, creatine phosphokinase, and pyruvate kinase. Although the serum levels of muscle cell enzymes are elevated in a number of conditions in which muscle fiber degeneration or necrosis occurs, the serum of patients with Duchenne muscular dystrophy, in general, yields the highest levels of enzyme activity. High levels can also be noted in idiopathic rhabdomyolysis, but lower levels are generally observed in cases of myositis, myotonic dystrophy, facioscapulohumeral neuromuscular disease, and other conditions. In addition, more recent studies on certain isoenzymes, different forms of an enzyme catalyzing the same reaction, have proved useful diagnostically.

Creatine phosphokinase, which is confined mostly to muscle and the central nervous system, and the muscle isoenzyme of pyruvate kinase have proved to be the most sensitive enzyme assays in detecting patients with Duchenne muscular dystrophy. The creatine phosphokinase, serum enzyme levels are highest early in the disease and decline with progression of the disease. In the advanced stages, the serum creatine phosphokinase may even return to normal levels. Fewer but similar observations have been made in the case of pyruvate kinase in Duchenne muscular dystrophy; but there is evidence that the serum pyruvate kinase activity tends to remain elevated in later stages of the disease.

Since Duchenne muscular dystrophy is inherited as a sex-linked recessive trait, application of the Lyon hypothesis would indicate that a varying percentage of female muscle cells would be dystrophic, and hence carriers of Duchenne muscular dystrophy would have elevated serum enzyme levels. Serum creatine phosphokinase is elevated in two-thirds of the female carriers, so this assay is useful in genetic counseling. Unfortunately, some carriers of Duchenne dystrophy do not have an elevated creatine phosphokinase. On the basis of abnormal serum elevation alone, it is quite likely that the measurement of serum pyruvate kinase in potential carriers is a more sensitive indicator in detecting the carrier state.

Membrane abnormalities in Duchenne dystrophy

A considerable body of data has accumulated indicating the presence of membrane abnormalities in Duchenne dystrophic patients' tissue (see list below). Since morphological study of Duchenne dystrophic muscle revealed varying degrees of degenerative changes in most muscle fibers, it was reasonable to assume that these changes were responsible for the leakage of muscle enzymes into serum. Some investigators, however, have suggested that since

the serum enzyme levels are at their highest very early in the disease process in Duchenne dystrophy, there may be a primary sarcolemmal abnormality with leakage of cellular contents and subsequent degeneration. This idea, though interesting, was without support for some years until Mokri and A. B. Engel reported their findings with high-resolution phase microscopy studies of Duchenne dystrophic muscle. They demonstrated nonnecrotic fibers with one or more focal lesions. This lesion was wedge shaped with the base resting on the fiber's surface. The electron microscope revealed that the plasma membrane overlying the lesion was either absent or disrupted while the basement membrane was preserved. They suggested that this sarcolemma or plasma membrane defect resulted in an ineffective cellular barrier and possibly represented a basic abnormality in the plasma membrane of the muscle fiber in Duchenne dystrophy. In another study by Dhalla and co-workers the sarcolemmal ATP-hydrolyzing activities of human dystrophic skeletal muscle were studied and compared with animal models. As in the case of the animal models of muscular dystrophy, the sarcolemma from patients with the Duchenne dystrophy showed an abnormally elevated calcium ATPase and magnesium ATPase. In contrast, the Duchenne dystrophic sarcolemmal sodium-potassium ATPase activity was lower than control values, while in animal models the sarcolemmal sodium-potassium ATPase activity was significantly elevated.

Recently, Schotland presented some elegant freeze fracture studies on sarcolemma obtained from patients with Duchenne dystrophy. In one side of the fractured membrane (called the protoplasmic leaflet) the dystrophic sarcolemma showed a lack of uniform distribution of particles, probably representing a variety of proteins. Some areas showed extensive depletion of particles while other areas showed no particles

---

Membrane Abnormalities in
Duchenne Dystrophy

Focal gaps in sarcolemema
Leakage of soluble muscle enzymes
Increased sarcolemal calcium and magnesium
  ATPase
Decreased sarcolemal sodium-potassium ATPase
Decreased particles in freeze fractured sarcolemma
Altered adenyl cyclase activity
Increased red cell membrane protein kinase
Decreased sarcoplasmic reticulum calcium transport
Alterations in sarcoplasmic reticulum membrane
  proteins

---

at all. In the exoplasmic leaflet, a considerable decrease was noted in the number of particles in the dystrophic sarcolemma. Another abnormality was recently described in the muscle membrane of patients with several types of muscle diseases, including Duchenne muscular dystrophy by Mawatari, Miranda, and Rowland. These investigators not only confirmed previous studies, but found that the basal activity of adenyl cyclase in Duchenne myotubes was higher than in all the other studies. For some time, abnormalities in sodium-potassium-ATPase from the red blood cells of patients with Duchenne dystrophy have been the subject of controversy. Recently, however, there has been an explosion of information about abnormalities in patients with Duchenne muscular dystrophy. Using high and low substrate conditions, these same workers found that the basal adenyl cyclase activity in Duchenne erythrocyte membranes was about twice that of the controls. Epinephrine-stimulated adenyl cyclase activity of normal membranes two to three times, but did not stimulate the enzyme in Duchenne erythrocyte membranes.

In the mid-1970s, Roses and co-workers described alterations in membrane protein kinase in erythrocytes from patients with Duchenne muscular dystrophy. Studying protein II and III, defined as major peaks by electrophoresis, they found an abnormal increase in membrane protein kinase

activity. Decreases were also described in red blood cell membrane protein kinase from patients with myotonic dystrophy. The specificity or importance of this observation in Duchenne dystrophic red blood cells remains to be shown because female carriers of Duchenne dystrophy showed abnormalities quantitatively similar to those of male patients with the disease. Some confusion exists as to why females showing little evidence of clinical disease should have the same quantitative biochemical defect as patients with typical clinical signs.

Another muscle cell membrane, the sarcoplasmic reticulum, has been studied recently in Duchenne dystrophic muscle. Sugita and his colleagues revealed that the sarcoplasmic reticulum, obtained from Duchenne dystrophic skeletal muscle, had reduced calcium uptake activity. The magnesium ATPase activity appeared to be increased; but this finding must be interpreted with caution, as allowances were not made for effect of changes in Ca^{2+} concentration on ATPase activity. In a study of calcium uptake by Samaha and Gergely, the initial uptake rate over the first 2 minutes was approximately one-third of normal in eight patients with Duchenne dystrophy. Furthermore, in the presence of adequate amounts of calcium to allow removal of some calcium by the uptake process, the ATPase activity of the Duchenne dystrophic sarcoplasmic reticulum was also reduced (0.39 vs. 0.84 μmol. of phosphate per mg./min. in normal SR ATPase activity). Efficiency of the calcium uptake mechanism of the sarcoplasmic reticulum was also determined. In the normal subjects, the ratio of moles of calcium taken up to moles of ATP hydrolyzed was 0.43, and in dystrophic muscle the ratio was 0.56, clearly not less than normal. Peter and Worsfold approached more closely the physiological conditions of so-called calcium affinity in Duchenne dystrophic muscle more closely by studying the ability of sarcoplasmic reticulum to remove calcium from dilute solutions. Normal sarcoplasmic reticulum vesicles quickly reduced the calcium concentration of the medium, but none of the patients with Duchenne dystrophy showed a normal SR calcium affinity. Fifteen other cases of a variety of neuromuscular diseases showed no calcium affinity abnormality. A similar abnormality, however, was found in polymyositis and in one patient with the Becker form of Duchenne dystrophy. These studies emphasize that, while the SR showed a normal calcium accumulation, affinity for calcium in terms of the level of free calcium was reduced. It is also of interest that in three of the cases in which abnormal calcium affinity was observed, the mitochondrial fraction showed normal respiratory rates, acceptor ratios, respiratory control ratios, and ADP : oxygen ratios. It is apparent from these studies, therefore, that the sarcoplasmic reticulum is abnormal in Duchenne dystrophic muscle at a time when mitochondrial function appears to be normal.

The reduced calcium uptake by the Duchenne dystrophic sarcoplasmic reticulum is not simply related to the nonspecific degeneration of muscle tissue. The studies by Samaha and Gergely and by Peter and Worsfold have shown that the calcium uptake and calcium affinity in myotonic dystrophic muscle are normal despite the presence of degeneration changes in the muscle.

Contractile mechanism in Duchenne muscular dystrophy

The speed of contraction is slow and the contraction time is prolonged in Duchenne dystrophic muscle. Since in normal muscle there is a correlation between the speed of contraction, actomyosin ATPase activity, and superprecipitation, these studies on Duchenne dystrophic muscle have been performed with isolated actomyosin from patients with Duchenne dystrophy. In two different laboratories these studies have

shown that the initial rate of superprecipitation is slow and that the ATPase activity is low. Although the actin and myosin interaction and the ATPase activity may be the rate-limiting factor in the speed of contraction, there is no evidence that this abnormality of actomyosin is a primary factor. Other studies have shown that the pH stability of actomyosin ATPase in Duchenne dystrophy is normal and that the immunological properties of myosin from Duchenne dystrophy is also normal.

In studies on the regulatory proteins of the myofibril in Duchenne dystrophic muscle, it was found that the myofibril contained reduced amounts of tropomyosin and troponin. Study of these proteins from Duchenne dystrophic muscle on sodium dodecylsulphate polyacrylamide gels showed no abnormalities in migration and no missing or extra proteins. It is currently felt, therefore, that there are no basic abnormalities in the contractile or structural proteins of Duchenne dystrophic muscle; but there is abundant evidence of membrane abnormalities within Duchenne dystrophic muscle and in other tissue, such as the red blood cell.

Carnitine deficiency myopathies

A new syndrome, the carnitine deficiency myopathies, has recently been described. These studies have unlocked secrets of diseases involving lipid metabolism that have heretofore been unrecognized and therefore untreated. Although the title above implies a single syndrome, it is likely that it covers many diseases.

The following two cases are typical for the system and muscle forms of carnitine deficiency.

Case Report 1. This boy was born after a full-term, uneventful pregnancy and delivery. Birth weight was 7 lb. Milestones in his psychomotor development were normal. He was always a "weak and clumsy child" and his neck and head were always "floppy." At age 3½, the patient was hospitalized with acute vomiting and a major convulsive seizure. He had some abnormal hepatic function tests: serum glutamic oxaloacetic transaminase (SGOT) and serum glutamic pyruvic transaminase (SGPT) were elevated, but serum bilirubin was normal. The cause of the acute encephalopathy and the abnormal hepatic function tests was not established and the child's clinical status, electroencephalogram (EEG), and hepatic function returned to normal.

At 9 years of age he developed acute confusion and stupor and was found to have moderate hepatic enlargement. Blood sugar was normal but hepatic function was again impaired. The patient was considered to have had acute hepatic encephalopathy due to Reye's syndrome. Within 2 months his clinical status and hepatic size and function tests returned to normal; neuromuscular functioning was adequate. He could walk a mile and play baseball and hockey, albeit somewhat clumsily.

At age 11, gradual development of muscle weakness was noted. He walked slowly, often stumbled, and could no longer run. Within months he could not walk to school a few blocks away, nor could he play outdoors. He became short of breath after walking half a block. He was a thin, frail child with stunted growth. His weight of 20 kg. and his height of 130 cm. were both less than the third percentile. The liver edge was palpable 3 cm. below the right costal margin; the spleen was not enlarged. He had an elongated face with hypotonic and weak musculature and mild bilateral ptosis. The temporalis muscle showed atrophy and weakness. The neck muscles were thin and weak; he could hardly hold his head up when standing, and his head moved in a "rag doll" manner when walking. The limb muscles showed considerable atrophy and moderate weakness, more proximally than distally. Marked winging of the scapulae was noted. There was considerable lumbar lordosis, a waddling gait, and the Gower sign was positive. He could not use stairs without holding onto railings, nor could he run, stand on one leg, or hop. Tendon reflexes were present but diminished.

The following laboratory results were abnormal: SGOT was elevated; electromyography showed short duration and small amplitude and polyphasic motor units in some areas of proximal limb muscles; there was no spontaneous activity, but nerve conductions were normal.

Case Report 2. A 10-year-old girl had normal psychomotor development until age 7, when she had difficulty climbing stairs and a waddling gait. At that time, serum enzymes of muscle origin were elevated. The EEG was normal. The patient was diagnosed as having muscular dystrophy and was treated with vitamins. Physical therapy was associated with some improvement the following year; she could walk but still had marked weakness.

At age 8, she had tachycardia (125 beats per minute),

and chest roentgenogram revealed normal cardiac size. She was hospitalized at age 10 with anorexia, weight loss, and increased weakness. Although not unusually thin or frail, she had a somewhat aged appearance. There was a considerable lumbar lordosis and a positive Gower's sign. She was unable to raise her arms straight; the cervical flexor and extensor muscles were thin and weak, and there was considerable weakness of the triceps, scapular rotator, iliopsoas, and thigh abductor muscles. Deep tendon reflexes were decreased. Sensory and cranial nerve functions were intact. Routine blood and urine tests were normal for creatine phosphokinase. Electromyographic examination showed short-duration, small-amplitude, and polyphasic motor units.

Case 1 represents a systemic form while Case 2 represents a muscle form of the carnitine deficiency syndrome. In 1972, A. G. Engel and Siekert described a 19-year-old woman with severe generalized weakness whose muscle biopsy demonstrated a considerable accumulation of lipid droplets, particularly in Type I fibers. This observation alone is sufficient to characterize this disorder as a true lipid storage myopathy. The following year Engel and Angelini found that *in vitro* utilization of long-chain fatty acids by homogenate of the patient's muscle was not effective in the absence of exogenous carnitine, but became normal after addition of carnitine to incubation mixture. This led them to measure carnitine concentration, and values ranging from 8 to 31 per cent of normal were found in five different muscle biopsies. Lipid analysis showed a marked increase in tri- and diglycerides. Serum contents of triglycerides, cholesterol, and free fatty acids were normal. Ketone bodies were formed normally during fasting, suggesting that liver lipid metabolism was not changed.

More than six other cases with muscle carnitine deficiency have been reported. In all, weakness was prominent, serum enzymes were elevated, and muscle pathology was characterized by excess lipid storage, especially in Type I fibers. There was considerable variability in clinical presentation, age of onset, degree of weakness, involvement of tissue other than muscle and response to carnitine therapy. It therefore seems that this syndrome may be due to several biochemical abnormalities. In the report by Karpati, *et al.*, the patient had two episodes of acute hepatic encephalopathy at ages 3 and 9 years before progressive muscle weakness developed at age 11. Carnitine concentration was much decreased in both liver and plasma, suggesting a defect in DL-carnitine synthesis. A regimen of 2 g./day of DL-carnitine for 5 months resulted in normal plasma carnitine levels and was accompanied by dramatic clinical improvements, although hepatic carnitine concentration remained low. A slightly different report was published recently by Angelini et al. in which a 10-year-old girl with insidious muscle weakness beginning at age 7 was reported to have a lipid myopathy by histochemistry. Carnitine deficiency was found in the skeletal muscle and the patient showed rapid improvement and recovery of strength over an 8-month period with 3 g. of L-carnitine per day and a medium-chain triglyceride diet. In the case reported by Marksbury, *et al.*, a 61-year-old woman allegedly had proximal muscle weakness since she was 38 years of age. She showed widespread muscle wasting, absent tendon reflexes, and electromyographic findings of neuropathy. On biopsy, there was a lipid storage myopathy, lipid-containing vacuoles in leukocytes and vacuoles in Schwann cells. Muscle carnitine level was abnormally low but serum carnitine level was normal. In yet another report by Van Dyke, an 8-year-old boy with slowly progressive muscle weakness had a lipid myopathy predominantly involving Type I muscle fibers. Subsequent studies demonstrated markedly reduced skeletal muscle carnitine with normal serum carnitine. Although both parents were clinically normal, muscle carnitine levels were low in both the mother and the father. There was no clinical evidence of cardiac disease, but the patient showed ventricular hypertrophy on electrocardiography, vectorcardiography, and echocardiography. Treatment with prednisone resulted in clinical improvement but no change in muscle histology. The patient of Smyth, *et al.*, an 11-year-old boy, had a unique clinical picture characterized by progressive muscle weakness and central nervous system involvement with calcification of basal ganglia, increased cerebral spinal fluid protein concentration, high tone hearing loss, growth retardation, episodic vomiting, exertional dyspnea, and high serum levels of lactate, pyruvate, and alanine. Hepatic function was normal and serum carnitine was not measured.

Lipid storage myopathy has been seen in a patient with congenital muscle weakness of unknown etiology, and abnormal lipid storage was recently seen in several tissues, including muscle, in a patient with congenital ichthiosis as the sole clinical disorder. Accumulations of lipid and abnormal mitochondria in muscle have also been seen in other syndromes that are poorly described, such as oculocraniosomatic neuromuscular disease (OCS). In OCS, a combination of all or some of the following may be present: ptosis, external ophthalmoplegia, retinal degeneration, axial muscle weakness, deafness, ataxia, pyramidal tract abnormalities, small stature, mental retardation, endocrine abnormalities, and cardiac conduction defects. The abnormal mitochondria may be present not only in muscle, but also in skin and cerebellum. Unfortunately, at present there are no clues as to why abnormal lipid accumulations and mitochondria appear in this disorder.

All patients with carnitine deficiency reported so far have shown an increased number of lipid droplets in muscle. Not all cases of lipid storage myopathy, however, are due to carnitine deficiency. In the cases discussed above under the heading Carnitine Deficiency, it is likely that biochemical abnormalities of varying types will be discovered. The areas under investigation at this time involve (1) synthesis of carnitine in the liver, (2) transport of carnitine in the blood, and (3) uptake of carnitine by muscle tissue. As far as skeletal muscle is concerned, the area of lipid metabolism and its disorders is largely uncharted territory. It is likely that future studies will identify a number of abnormalities in lipid metabolism that produce disease, as has been shown with the glycogen storage diseases. In addition, it is apparent from the cases of carnitine deficiency discussed above that some patients respond to therapy with very gratifying results in an area of medicine where heretofore so little therapy has been of value.

Annotated References

Bais, R., and Edwards, J. B.: Creatine kinase. C.R.C. Crit. Rev. Clin. Lab. Sci. *16*:291–335, 1982. (A general review of properties of creatine kinase.)

Carpenter, S., and Karpati, G.: Duchenne dystrophy: Plasma membrane loss initiates muscle cell necrosis unless it is repaired. Brain *102*:147, 1979 (Detailed study between muscle cell necrosis and sarcolemal integrity in Duchenne dystrophic muscle.)

Dayton, W. R., Goll, D. E., Zeece, M. G., Robson, R. M., and Reville, W.: Some Properties of a Ca^{2+}-activated protease that may be involved in myofibrillar protein turnover. *In* Reif, E., Rifkin, D. B., and Show, E. (eds.): Protease and Biological Control, pp. 551–577. Cold Spring Harbor Laboratory, Cold Spring Harbor, 1975. (Description of a sarcoplasmic protease.)

DiMauro, S., Trevisan, C., and Hays, A.: Disorders of lipid metabolism in muscle. Muscle Nerve *3*:369, 1980. (A detailed review of disorders of lipid metabolism in muscle.)

Ebashi, S., Maruyama, K., and Endo, M. (eds.): Muscle Contraction. Its Regulatory Mechanism. New York, Springer-Verlag, 1980. (The recent state of our understanding the regulation of muscle contraction.)

Flockhart, D. A., and Corbin, J. D.: Regulatory mechanisms in the control of protein kinases. C.R.C. Crit. Revbiochem. *12*:133–186, 1982. (Regulation of protein phosphorylation.)

Foreback, C. C., and Chu J. W.: Creatine kinase isoenzymes: Electrophoretic and quantitative measurements. C.R.C. Crit. Rev. Clin. Lab. Sci. *15*:187–230, 1981. (Distribution and quantitation of creatine kinase isoenzymes in tissues.)

Goldberg, A. L., and Odessey, R.: Regulation of protein and amino acid degradation in skeletal muscle. *In* Milhorat, A. T. (ed.): Exploratory Concepts in Muscular Dystrophy. Excerpta Medica, pp. 187–199. New York, Elsevier, 1974. (Protein turnover in muscle.)

Guth, L.: Trophic influences of nerve on muscle. Physiol. Rev. *48*:645–687, 1968. (Muscle filaments develop according to the influence of nerve.)

Huxley, A. F., and Simmons, R. M.: Proposed mechanism of force generation in striated muscle. Nature *233*:533–538, 1971. (Cross bridge mediated interaction between actin and myosin filaments.)

Huxley, M. E.: Molecular basis of contraction in cross-striated muscles. *In* Bourne, G. H. (ed.): The Structure and Function of Muscle. Vol II, Pt 1. New York, Academic Press, 1972. (Detailed description of sliding filament theory as basis of muscle contraction.)

Jodice, A. A., Perker, S., and Weinstock, I. M.: Role of the Cathepsins in Muscular Dystrophy in Muscle Diseases. *In* Walton, J. N., Canal, N., Scarlato G.: Excerpta Medica, Amsterdam 1969, pp. 313–318. (Proteolytic enzymes and protein turnover in muscle.)

Korn, E. D.: Biochemistry of actomyosin-dependent cell motility: A review. Proc. Natl. Acad. Sci. U.S.A. *75*:588–599, 1978. (A recent review of cell motility.)

Korn, E. D.: Actin polymerization and its regulation by proteins from nonmuscle cells. Physiol. Rev. *62*:672–737, 1982. (Regulation of cell motility which is not troponin dependent.)

Kretsinger R. H.: Mechanism of selective signaling by calcium. Neurosci. Res. Program Bull. *19*:214–332, 1980. (A review of calcium mediated control of cellular activities.)

Mokri, B., and Engel, A. G.: Duchenne dystrophy: Electron microscopic findings pointing to a basic or early abnormality in the plasma membrane of the muscle fiber. Neurology (N.Y.) *25*:1111, 1975. (Detailed description of the sarcolemal abnormality in Duchenne dystrophic muscle.)

Perry, S. V.: The regulation of contractible activity in muscle. Biochem. Soc. Trans. *7*:593–617, 1979. (One of the recent view of regulatory proteins and their interactions.)

Rowland, L. P.: Biochemistry of muscle membranes in Duchenne muscular dystrophy. Muscle Nerve *3*:3, 1980. (A critique of membrane abnormalities described in Duchenne dystrophic muscle.)

Samaha, F. J.: Sarcotubular system in the muscular dystrophies. *In* Rowland, L. P. (ed.): Pathogenesis of Human Muscular Dystrophies, Excerpta Medica, Amsterdam-Oxford p. 630, 1977. (A summary of abnormalities in the sarcoplasmic reticulum in muscular dystrophy.)

Squire, J.: The Structural Basis of Muscular Contraction. New York, Plenum Press, 1981. (Muscle contraction: electron microscopic and X-ray diffraction studies on muscle and on muscle proteins.)

Stanbury, J. B., Wyngaarden, J. B., and Frederickson, D. S. (eds.): The Metabolic Bases of Inherited Disease. New York, McGraw-Hill, 1972. (A review of inherited diseases relevant to muscle.)

Symposium of the American Physiological Society: Protein Turnover in Heart and Skeletal Muscle. Morgan, H. E., and Wildenthal, K. (Chairman): Fed. Proc. 39:7–52, 1980. (Recent state of studies on protein turnover in muscle.)

Szent-Györgyi, A.: Chemistry of Muscular Contraction, 2nd ed. New York, Academic Press, 1951. (The first concept of muscle contraction involving actin, myosin and ATP.)

Zak, R., Martin, A. F., and Blough, R.: Assessment of protein turnover by use of radioisotopic tracers. Physiol. Rev. 59:407–447, 1979. (A critical assessment of use of radioisotopic tracers in protein turnover studies.)

Section VIII

Immunological Mechanisms

Introduction

Since the pioneering work of early microbiologists and anatomists, which resulted in the formulation of classical theories on mechanisms of infection, immunity, and resistance, the science of immunology has developed concurrently with successful application of its fundamental principles toward solution of problems in allergy, tissue transplantation, and oncology. While most of the fundamental principles of immunology were stated in early studies on resistance to infection, the complex mechanisms which underlie these principles are still not completely understood in cellular or molecular terms. Studies of clinical problems in the areas of transplantation, allergy, infection, and malignancy (to name but a few) emphasize the need for continuing basic research in immunology. In this section, we attempt to apply concepts of immunology to clinical disciplines, emphasizing the relationships between hypothetical mechanisms of immunity and mechanisms of disease. This approach will provide insight into a discipline still in a process of rapid evolution.

The scope of modern immunobiology is indeed broad (Table VIII–1). Because of this broad scope, it is not possible to cover all areas of immunology in the available space. Those topics chosen for inclusion, however, are representative of the field. An understanding of the material presented in the following four chapters will provide the basis for understanding most, if not all, immunological phenomena.

Because of the recent rapid expansion of immunology, the science has literally outgrown its vocabulary. Definitions given in this text were adopted for their working value, but students are warned that in immunology synonyms do not always have synonymous meaning, and many authors use the technical terms of immunology differently.

Traditionally, the components of immunology have been categorized into two major overlapping groups, both of which promote the integrity of "self." One of these, called *innate resistance*, is made up of all of the anatomical and physiological barriers against infection. Included are the integrity of the integument, bacteriostatic and bacteriocidal activity of secretions, ciliary action of respiratory epithelium, gastric acidity, motility of the intestine, some inflammatory responses, body temperature, sessile phagocytic elements of the reticuloendothelial system, and even the circulation of the blood. All of these factors tend, in rather obvious ways, to prevent or control infection nonspecifically (i.e., regardless of the nature of the potentially infectious agent). Similarly, the components of innate immunity tend to reject dead cells and foreign material other than infectious agents. The components of innate immunity may become secondarily involved in hypersensitivity (allergic) reactions, but they do not initiate these reactions.

TABLE VIII–1
The Scope of Contemporary Immunobiology

Subject of Study	Content
Resistance to infection	1. Innate barriers 2. Adaptive barriers to recurrent or persistent infection 3. Resistance to specific pathogens 4. Immunosuppressive action of pathogens
Fundamental relationships in cellular and homoral immunity	1. Cells and molecules basic to immunological function 2. Relation of structure to function of the immunological apparatus 3. Development of cells and molecules, tissues and organs of immunological apparatus 4. Identification and measurements of cells and molecules of immunological apparatus 5. Immunological phenomena in relationship to cellular biology 6. Relation at the cellular and molecular levels of structure to function in the immunology system 7. Defects in development of immunity, their classification, recognition and analysis as "experiments of Nature" 8. Immunological tolerance, specific negative adaptation
Tissue transplantation	1. Antigenic relationships of host and graft 2. Immune responses in graft rejection 3. Inhibition of graft rejection
Malignant adaption and immunity	1. Antigenic features of malignant cells 2. Tumor immunity 3. Chemical carcinogens and immunity 4. Malignancy in genetic and therapeutic immunodeficiency 5. Immunodeviation 6. Attempts to modify malignant adaptation by immunological means
Mechanisms of immunology injury	1. Self-recognition and autotolerance 2. Models of autoimmunity 3. Autoimmune phenomena in human disease 4. Relation of immunodeficiency, autoimmunity, virus infection, and aging 5. See also Table VIII–2 and related discussion
Allergy	1. Drug and chemical 2. Humoral and cellular immunology 3. The IgE system 4. Atopy 5. Prophylaxis and therapy

A second category of barriers against infection includes the elements of *adaptive immunity*. Adaptive immunity includes the traditional immunological phenomena associated with γ-globulin antibodies and antigen receptors on the membranes of lymphocytes. Adaptive immunity may be partially developed prior to exposure to a given foreign substance. The special elements of adaptive immunity, on the other hand, differentiate into a functional state upon initial exposure to a given foreign substance. This differentiation response (also referred to as an immune response) results in the formation of specific receptors on γ-globulin molecules or on the surface of small lymphocytes which allow the adaptive immune mechanisms to function more vigorously upon subsequent exposure to the same foreign material.

Adaptive immunity differs from innate immunity in that adaptive immunity is inducible, it manifests a high degree of specificity, and it is the system of primary concern with regard to immunological hypersensitivity. In response to initial exposure of deeper tissues to foreign substances (called antigens or allergens), immunologically "competent" tissues form antibody or become hypersensitive. These "adaptive" reactions of an individual to antigen or allergen comprise the primary immune response. Similar reactions which occur subsequent to a second exposure to the same (or chemically similar) antigen or allergen are called the secondary immunological responses. These secondary responses differ from primary immune responses in that the secondary ones develop more rapidly and, in quantitative terms, are more intense. Since primary immune responses often appear (superficially) to be transient, secondary responses are also called *anamnestic* (memory) responses.

The development of immunity and immunological hypersensitivity subsequent to exposure to antigen or allergen occurs through the same or extremely similar mechanisms. The essential difference between these two phenomena is that immunity is associated with beneficial effects (increased resistance to infection or suppression of malignancy) whereas hypersensitivity is associated with deleterious effects (allergic conditions and graft rejection).

There are at least two identifiable components of adaptive immunity and, in parallel, two kinds of immunological hypersensitivity: cell-mediated (also referred to as thymus-dependent), and humoral (referred to as thymus-independent) immunity or hypersensitivity. In general, cell-mediated immunity involves increased resistance to many infectious agents, which have antigens to which a patient has been exposed previously. Cell-mediated hypersensitivity (also called delayed hypersensitivity) is related to pathological reactivity to otherwise bland substances (in this context called allergens rather than antigens). The same is true of humoral immunity and the immediate hypersensitivities.

While these simplified distinctions are useful in defining the terms, immunity and hypersensitivity may not occur separately. The relative increase in resistance to infection with a given pathogen which results from immunity to the antigens of such an agent is usually accompanied by hypersensitivity to these same antigens (allergens). Clinically, this information is often put to practical use by assuming that hypersensitivity to antigens of a given agent (which can be detected by skin tests) is evidence of resistance to infection by agents bearing the same antigens. This is an assumption which is generally useful, but not infallible. It is also possible, for example, that a patient or experimental animal could be hypersensitive to a pathogen-associated allergen which is not crucial to that pathogen's ability to cause disease. In such cases, hypersensitivity may in fact render the patient less resistant to infection. Similarly, patients may have adaptive increases in resistance to infection without demonstrable hypersensitivity.

Table VIII–2 lists four recognized categories of hypersensitivity reactions. While this classification of hypersensitivity reactions (generally referred to as the Gell and Coombs classification) does not include all of the known immunological mechanisms resulting in hypersensitivity symptoms, it is very useful clinically, and it is important historically. According to this classification, Type I hypersensitivity, also called *anaphylactic hypersensitivity*, results when IgE antibodies on the surface of mast cells combine with allergen leading to the release of a chemotactic factor and vasoactive substances. This

TABLE VIII–2
Gell and Coombs Classification of the Mechanisms of Immunologically Mediated Hypersensitivity

Hypersensitivity Type	Major Components of the Mechanisms
I—Anaphylactic	IgE (reaginic) antibody, mast cells and basophils, release of eosinophil chemotactic factor, and vasoactive substances
II—Cytotoxic or cytolytic	Antibody (IgG and IgM)-initiated, complement-mediated cell destruction
III—Immune complex	Deposition of soluble immune complexes; local activation of complement; kinins and clotting mechanisms; inflammation
IV—Cell-mediated or delayed	Stimulation of specifically sensitized lymphocytes; recruitment and activation of lymphocytes and macrophages

mechanism seems to be important in allergic hay fever, extrinsic asthma, and other atopic diseases (Chap. 36). Type II hypersensitivity, also called *cytotoxic* or *cytolytic hypersensitivity*, occurs when antibodies combine with antigens on the surface of cell membranes. Similar reactions occur when the antigen is part of the cell membrane and when it is absorbed from other sources to a cell membrane. Symptoms of Type II reactions vary depending upon the kind of cell-bearing antigen, the kind of antibody reacting with that antigen, and the interaction of complement components with the antigen-antibody complex (Chaps. 36 and 37). Type III hypersensitivity reactions, also called *immune complex* or *innocent bystander reactions*, occur when antigen-antibody complexes are deposited in vascular components of joints, kidneys, skin, or other tissues. When relatively small amounts of antibody combine with relatively large amounts of antigen, soluble immune complexes that are not rapidly cleared by phagocytic cells of the reticuloendothelial system are formed. Such immune complexes become concentrated in certain areas where they activate the complement system and the kallikrein-kinin system. Local activation of these systems lead to tissue damage and inflammation characteristic of Type III reactions (Chap. 37). Type IV hypersensitivity, also called *delayed* or *cellular hypersensitivity*, results when specifically sensitized small lymphocytes combine with antigen in tissues. The resulting release of *lymphokines* and local activation of macrophages leads to symptoms characteristic of Type IV hypersensitivity (Chaps. 35 and 37). The final chapter (38) deals with interactions of components of immunity and hypersensitivity with oncogenic viruses and heredity related to malignancy. This topic has been included in this edition since it is becoming increasingly evident that pathophysiological mechanisms of malignancy are interwoven with the native and pathophysiological mechanisms of immunity, hypersensitivity, and infection.

Since the publication of the first edition of this book, there have been extraordinary advances in many areas of immunology, and it is not possible to list all of these achievements. Two areas will be mentioned because of their profound impact on our understanding of many of the specific matters explored in the subsequent chapters. These are immune response genes and the complex interactions of immune regulatory cells.

It has been clearly shown in animals that there exist a cluster of genes which control the quantity and to some extent the quality of immune responses. These include immune responses to single, well-defined polypeptide antigens as well as to more complex structures. These genes have also been called *disease susceptibility genes*, since in animals (and man) clear relationships are demonstrable between disease development (e.g., leukemia in certain murine strains) and certain genes on the chromosomal segments where the immune response (Ir) genes can be mapped. In all species thus far studied (including man) these genes are located among the genes which determine the fate of organ transplants. The general term for this array of genetic factors is the major histocompatibility complex (MHC) and in man this has been designated HLA since the MHC in man was defined by the typing of human leukocyte antigens (HLA). There are four loci described in man designated A, B, C, and D. Some antigens from different loci occur together in a nonrandom manner, and this has been called *linkage disequilibrium.* The antigens of the first three loci are found on the surface of all nucleated cells, while the D locus antigens are restricted to B lymphocytes and a certain macrophage subset. It is the D locus antigen associations which seem to have the greatest interest for their impact on autoimmune disease, and numerous autoimmune diseases have specific associations with various D locus antigens. Of great interest are the large number of autoimmune diseases which carry the two antigens B8, DR_w3. These are listed below. Also of some interest is the recent recognition of a defect in Fc-receptor–mediated reticuloendothelial function, which is strongly associated with the B8, DR_w3 haplotype. These D-locus associations suggest that immune response genes play a role in the expression of autoimmune disease, and the challenge of coming years is to understand how these genes affect immune responsiveness.

B8, DR_w3 *Diseases*

Myasthenia gravis
Thyrotoxicosis
Dermatitis herpetiformis
Systemic lupus erythematosus
Sjögren's syndrome
Autoimmune chronic active hepatitis
Juvenile diabetes
Addison's disease
Celiac disease

The second area to be mentioned is the growing awareness that the immune system is a network of regulating signals delivered by discrete lymphocyte subpopulations. Progress in this area has been greatly accelerated by the development of the hybridoma techniques in which antibody-producing cells are immortalized by fusion with myeloma cell lines and produce homogeneous monoclonal antibodies specific to a single antigen. Reagents have been developed, therefore, which recognize a helper or amplifier set (T4+) of T lymphocytes, while a subset with cytotoxic and suppressor activities (T5,8+) has also been identified. Monoclonal antibodies have also been made which identify macrophages, all T-cells, B-cells, etc. Clearly this technology will provide tools for a detailed classification of many lymphoid cell subsets of functional significance not previously possible. It would not be overly optimistic to state that the cellular basis for autoimmune disease should become clarified in the coming decade. With such powerful tools, it should also become apparent how important a role genetic and other factors play in the development of the immune dysregulation that seems to characterize immunologically mediated disease.

Morris Reichlin, M.D.

35

Adaptive Immunity

Morris Reichlin, M.D
John B. Harley, M.D., Ph.D.

The immune system is composed of a highly ordered set of responses, knowledge of which has developed in close association with understanding of the response of the organism to infection. Recognition of the components of the immune system in the preantibiotic era was closely linked to the practical management of human infections. For almost a century, specific antibody responses have been known to play a crucial role in determining the outcome of infections with encapsulated organisms such as pneumococci, staphylococci, streptococci and klebsiellae. Even now, certain bacterial vaccines (e.g., Pneumovax) are of great prophylactic value to certain patients. Equally important are a group of responses collectively called cell-mediated immunity, which are responsible for determining the outcome of infections with viruses, fungi, mycobacteria and a host of protozoans. Here, the monocytes and macrophage participate as accessory cells in antigen processing and presentation and as cells where certain immune-effector functions are realized, such as killing of microorganisms, tumor cells, and so forth.

This system is characterized by several properties that are similar to other well-ordered organ systems. The most dramatic of these is the specificity with which adaptive individual responses can be made to literally thousands of different microorganisms. These specificities are directed to "antigens," which are well defined chemical groupings on macromolecules that are most often proteins and carbohydrates. This flexibility and broad range of specificities make the immune system probably the most plastic and versatile of all biological response systems. A second paramount property of the system is memory. Information about specific responses is stored in cells of the immune system, and challenge of the system by an antigen previously encountered leads to a more rapid and quantitatively greater response than that experienced during the first exposure. Finally, the healthy system is exquisitely regulated. Antigen exposure usually produces activation of the system until the antigen is removed, which leads to down-regulation of the response. A network of interacting cells to be described is responsible for this regulation.

In the immune system the key cells are lymphocytes and the effector molecules are antibodies for one set of responses. Lymphokines, a nonimmunoglobulin series of protein secreted by lymphocytes, mediate the functions of cell-mediated immunity. Specificity and memory are mediated by the antigen receptors of the lymphocytes. At the cellular level both differentiation and proliferation are involved in the expression and amplification of the responses. Subsets of lymphocytes also govern the regulation of specific individual immune responses, and several soluble factors secreted by these cells are potential agents of this regulation.

In immunology as in other sections of the biology of medicine, as much or more has been learned about the structure and function of the normal system by the study of disease than by the deliberate analysis of the normal system. The study of the structure of the homogeneous products of lymphoid tumors (e.g., myeloma proteins) has been the central resource that has led to an almost complete understanding of the structure and function of normal immunoglobulins. Similarly, studies of patients with the primary immunodeficiency syndromes have provided the initial framework for our present understanding of the cellular basis of the immune response and the dual nature of lymphocyte differentiation. Here as elsewhere in medicine, physi-

ology and pathophysiology merge as their mutual elucidation enhances understanding at the boundary between the normal and abnormal.

This chapter explores the cellular aspects of humoral and cell-mediated immunity and the regulatory interactions that govern the functioning of the system as a whole.

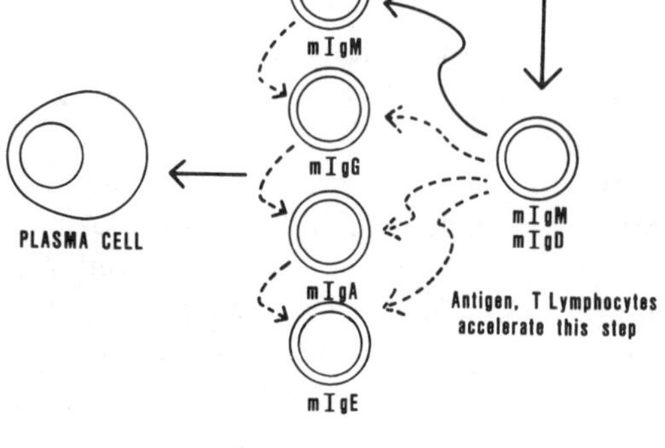

Humoral immunity

Humoral immunity is mediated by specific antibodies that are secreted by plasma cells. Plasma cells differentiate from precursor lymphocytes which are designated B because they differentiate in birds in an organ called the bursa of Fabricius. The analogous organ in mammals is probably the fetal liver. Much evidence exists that the antigen receptor on B lymphocytes has the same structure as the immunoglobulin ultimately secreted by plasma cells derived from the lymphocytes.

B-cell development

The development stages in the life of B-cells are depicted in Figure 35–1. Stem cells, which probably arise in the bone marrow, differentiate under the influence of the bursa of fabricius (hence the designation B-cells) in birds into cells that are capable of producing immunoglobulin. The bursal equivalent is not known in mammals, but evidence exists for the fetal liver and bone marrow to serve the bursal function: B-cell differentiation. There are discrete stages of B-cell development, the first of which is a cell with cytoplasmic μ-chains which differentiates to a lymphocyte with membrane-bound IgM. This lymphocyte (or a subset thereof) is then converted to a cell with membrane-bound IgM and IgD, from which point further differentiation is driven by antigen and/or T-cell-derived help. Thereafter, sequentially appearing IgM, IgG, IgA, and IgE lymphocytes can further differentiate into IgM-, IgG-, IgA-, and IgE-producing plasma cells.

FIG. 35–1. Pathway of B-cell differentiation from stem cell to plasma cell. Dotted lines suggest that cells with membrane IgM and IgD may develop into each of the mature lymphocytes with surface IgM, IgG, IgA, or IgE or alternatively that this development occurs sequentially.

These differentiated states are of great interest, not only for an understanding of B-cell development and immunodeficiency (since a block at any stage leads to a predictable arrested development of the B-cell lineage) but also for an understanding of B-cell malignancy. Lymphocytic leukemias, some lymphomas, and multiple myeloma represent cells at various stages of differentiation that have undergone malignant transformation. There are pre-B-cell leukemias and plasma cell malignancies (i.e., multiple myeloma) producing various homogeneous immunoglobulins. Chronic lymphocytic leukemia seems to be a malignancy of small lymphocytes with homogeneous surface IgM; and Waldenström's macroglobulinemia is a malignancy of a lymphocyte intermediate between a small surface Ig-bearing lymphocyte and a mature Ig-secreting plasma cell (this latter cell has been called a lymphocytoid cell and is not included in Fig. 35–1).

Immunoglobulin structure

All immunoglobulins have a basic 4-chain polypeptide structure, depicted in Figure 35–2 which shows a schematic diagram of the IgG molecule, which is the most abundant Ig class in serum and whose description can serve as prototype for an understanding of the structure of all the immunoglobulin classes. The molecule is composed of two pairs of heavy and light chains. The heavy chains are of molecular weight 50,000 daltons, whereas the light chains have a molecular weight of 23,000 daltons. The heavy chains have four intrachain disulfide loops placed at regular intervals; the light chains have two such disulfide loops, at similarly regular intervals. These divide the molecules into distinct structural regions called domains of about 105 amino acids that have functional significance. That is, each domain behaves as a functional unit. The two halves of the molecule are held together by noncovalent interactions in the carboxyl (C)-terminal halves of the heavy chains and by a variable number of interchain disulfide bonds near the midpoint of the heavy chains, which because of its molecular flexibility is called the *hinge region*. The C-terminal halves of the heavy chains (designated Fc because when separated from the whole molecule by proteolysis, they crystallize) contain the structures which determine the biological activities of the molecules such as complement-fixing capacity, catabolic rate, receptors for cells (mast cells, monocytes, etc.), and placental transfer. The amino (N)-terminal half of the heavy chains and the light chains (designated Fab, since when separated from the C-terminal half by proteoly-

FIG. 35–2. Schematic diagram of IgG molecule. Variable (V)-regions of heavy (*H*) and light (*L*) chains are shown as dashed lines. Solid lines represent constant regions (c).

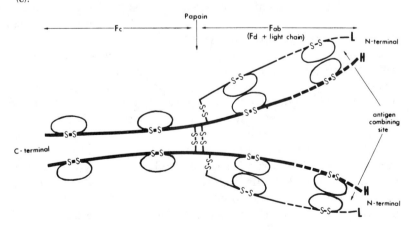

sis they contain the antigen binding or antibody activity) possess the specificity regions of the molecule. Primary sequence studies reveal that the N-terminal half of the light chain and the N-terminal quarter of the heavy chains are variable and the three-dimensional confluence of specific hypervariable regions within these variable regions constitute the antibody combining site(s). The most N-terminal domains of the heavy and light chains are designated V_H and V_L (for variable heavy and variable light), respectively, and include the structures described above that are responsible for antibody activity.

In human immunoglobulins, there are two types of light chains, designated κ and λ, and five types of heavy chains designated γ, α, μ, δ, and ε heavy chains respectively. IgA in serum is composed mainly of tetramers of two κ- or two λ-chains bound to two α-chains, but polymeric IgA may also exist in serum and contain an additional peptide, the J-chain, which is found in all polymeric immunoglobulins. In external secretions (e.g., saliva, tears, bronchial fluid, gastrointestinal fluid, and urine) a dimeric form of IgA, J-chain, and an additional polypeptide of epithelial origin (secretory component) constitute the principal immunoglobulin component. The molecular weight of this secretory form of IgA is 400,000 daltons. IgM is a pentamer of five tetrameric molecules of two light chains and two μ-chains bound together with intertetramer disulfide bonds and J-chains.

Serum immunoglobulins are composed then of the sum of these immunoglobulins, with the three principal classes, IgG, IgA, and IgM, comprising 75, 15, and 10 per cent of the total, respectively; IgD and IgE constitute less than 1 per cent. The understanding of immunoglobulin structure has provided insight into disorders of the immune system. It has long been appreciated that the immunoglobulins, which are the major constituent of the γ-globulin fraction of serum (and minor constituents of the

β-globulin fraction) are very heterogeneous. This heterogeneity is now well explained in molecular terms, since each antibody molecule has a unique amino acid sequence in its V_H and V_L regions that determine its specificity. Thus, there are literally thousands of antibody molecules with unique sequences in every human serum. In contrast, serum albumin is a homogeneous protein with a single amino acid sequence. This is in stark contrast to the situation in malignancies of the lymphocyte plasma cell series whose monoclonal expansion of a tumor cell line is almost always accompanied by secretion of a homogeneous immunoglobulin product. In multiple myeloma, this might be a tetrameric molecule such as IgG, IgA, IgD, or IgE in which only a single light chain (κ or λ) and a single heavy (μ, γ, α, δ, or ε) chain is found. In some instances, the corresponding homogeneous κ or λ light chains are found in the urine, and in 15 per cent of cases of myeloma, the light chains (κ or λ, but not both) are not detectable in the serum and are found only in the urine as the sole secreted product of the tumor. When IgM is the sole homogeneous product, Waldenström's macroglobulinemia is usually present. Diagnosis of the monoclonal protein is accomplished by serum or urine immunoelectrophoresis. This permits identification of homogeneous protein of usually γ or β mobility (rarely α_2) and of the single light and heavy chain present in the serum or the single type of light chain in the urine with specific antisera.

Properties and functions of immunoglobulins

Research performed during the past 15 years has led to the information about the structure and function of immunoglobulins listed in Table 35–1. IgG is the prototype Ig structure and has the molecular formula $\gamma_2 L_2$ (two heavy chains [γ] and two light chains [κ or λ]). This is probably the most important immunoglobulin in defense against bacterial infections; it proba-

TABLE 35–1
Properties of Human Immunoglobulins

	IgG	IgA	IgM	IgD	IgE
H-Chain class	γ	α	μ	δ	ε
H-Chain subclass	$\gamma_1, \gamma_2, \gamma_3, \gamma_4$	α_1, α_2	μ_1, μ_2		
L-Chain type	κ and λ	κ and λ	κ and λ	κ and λ	κ and λ
Molecular formula	$\gamma_2 L_2$	$\alpha_2 L^*$ or $(\alpha_2 L_2)_2$ SC $\neq$ J	$(\alpha_2 L_2)_5 J$	$\delta_2 L_2$	$\varepsilon_2 L_2$
Sedimentation coefficient	7	7	19	7–8	8
Electrophoretic mobility	γ	Fast γ to β	Fast γ to β	Fast γ	Fast γ
Complement fixation (classical)	++	0	++++	0	0
Serum concentration ($\approx$ mg./dl.)	700–1200	150–250	70–120	3	.05
Serum half-life (days)	23	6	5	2–8	1–5
Placental transfer	+	0	0	0	0
Reaginic activity	0	0	0	0	+
Lyse bacteria	+	+	+++	?	?
Antiviral	+	++	+	?	?

bly also has antiviral activity. IgG alone provides passive immunity to the newborn, since it is the only Ig class that transverses the placenta. There are four subclasses of the γ heavy chain, which means that their C-terminal halves share sequences recognizable by serological means. This has a functional corollary in that γ_1 and γ_3 are very active in complement activation; γ_2 is less active, and γ_4 is inactive in this respect. Of the IgGs, only the γ_3 subclass myeloma proteins gives rise to a clinical hyperviscosity syndrome at globulin concentrations of 5.0 g./dl. or less because of its propensity to aggregate. Finally, IgG is the most long-lived of the immunoglobulins having a half-life of 23 days. This makes it the only Ig class of practical importance in passive immunity.

IgA is the second most abundant Ig in serum, but it is the predominant Ig in secretions comprising 90 per cent of the total in normal saliva, tears, and bronchial and gastrointestinal secretions. This secretory IgA is synthesized by plasma cells at secretory sites (e.g., the gastrointestinal tract, bronchial submucosa), is probably much more resistant to proteolysis than serum Ig, and has both antibacterial and antiviral activity. It may be that the IgA system is the key defense element in local immunity against a number of viral infections, particularly poliomyelitis. Isolated IgA deficiency leads to increased sinopulmonary infections, although in some instances affected persons may be asymptomatic. In isolated IgA deficiency, IgG and IgM make up the Ig of the secretions.

IgM is the first Ig synthesized in response to antigenic challenge and in certain infections remains an important component of the native immune response. Because of its pentameric nature its binding avidity is high, and a single molecule can activate the complement system, leading to a highly efficient first-line defense mechanism against microorganisms. The differentiation of IgM-producing plasma cells is far less thymus dependent than that of any of the other Ig classes. Moreover, it is probably true that with large polymeric antigens with repeating determinants such as bacterial polysaccharides, IgM production is essentially thymus independent. The high avidity of IgM with polysaccharide antigens derives from the simple fact that since 5 of the 10 antigen binding sites are active, the maximum equilibrium constant would be $(K_A)^5$. Since

FIG. 35—3. Paper electrophoresis of normal serum (*top*), a serum with polyclonal hyperglobulinemia (*middle*), and sera with various M-proteins (*bottom*). Anode to the left and cathode to the right.

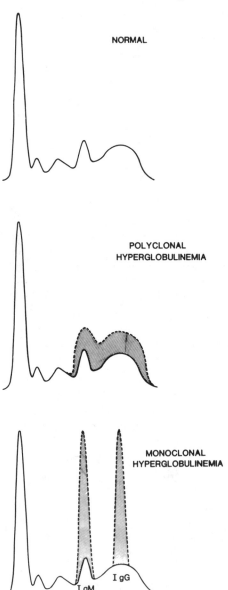

NORMAL

POLYCLONAL
HYPERGLOBULINEMIA

MONOCLONAL
HYPERGLOBULINEMIA

IgG

I gM or
I gA or
I gD or
I gE

the average K_A for numerous polysaccharide antibodies is about 10^5 M^{-1} per binding site, the maximal K_A is an astronomical 10^{25} M^{-1}. The extraordinary efficiency of IgM in complement activation is also probably a reflection of a similar argument for the interaction of IgM and C_{1q}, the first protein involved in the complement sequence.

IgD is probably more important as a cellular receptor for young lymphocytes than any function it may possess as circulating antibody. However, the special role of IgD as lymphocyte receptor is still under intensive study.

Although the serum IgE concentration is only about 0.5 μg./ml., IgE is tightly bound to mast cells and basophils throughout the body. IgE is the antibody class that mediates anaphylaxis in humans; and antigen binding of membrane-bound IgE triggers mast cell degranulation and release of all the mediators characteristic of anaphylaxis (e.g., histamine, serotonin). IgE is the antibody class responsible for reaginic activity (i.e., the activity that can be passively transferred from a sensitive donor to a normal recipient) to induce a specific immediate hypersensitivity reaction by injection of antigen at the recipient site.

Hyperglobulinemia: distinction of monoclonal and polyclonal gammopathy

A common and important clinical problem is the evaluation of excessive activity of the humoral immune system leading to marked hyperglobulinemia. This hyperglobulinemia may be either heterogeneous (i.e., polyclonal), suggesting sustained exaggerated response of many cell lines, or it may be monoclonal, reflecting the unregulated sustained secretion of Ig of a single clone of cells. In the latter situation one frequently finds a tumor of the lymphocyte-plasma cell system; it is this tumor clone that is responsible for the monoclonal globulin production. The significance of polyclonal gammopathy is that it reflects a continuing massive stimulus to many components of the B-cell repertoire. This is seen in numerous connective tissue diseases; chronic infections due to myobacteria, fungi, pyogenic organisms, and parasites; liver diseases; interstitial fibrosis of the lung; inflammatory bowel disease; and sarcoidosis.

Two laboratory techniques are essential for an analysis of serum and urine in order to distinguish polyclonal from monoclonal gammopathy: protein electrophoresis and immunoelectrophoresis. Figure 35–3 dramatizes the differences in the appearance of the electrophoretic pattern in the two situations. In polyclonal hyperglobulinemia the increased globulin has the same distribution as the normal globulin and represents an exaggerated hyperstimulated, but normal, response. The lower part of the figure shows the typical monoclonal (M)-protein and indicates that the "spike," or homogeneous band, is usually of γ mobility when it is IgG, and fast γ to β when it is IgM, IgA, IgD, or IgE. Rarely, the M-protein may have α_2 mobility. The faster mobility of the IgM, IgA, IgD, and IgE molecules is accounted for by the high carbohydrate content of those molecules (7 to 13 per cent), as compared to the 2 to 3 per cent carbohydrate content of IgG that contributes increased negative charge to the molecules of the first class. A useful simple rule for evaluating the homogeneity of these proteins is to determine the ratio of the height of the peak to the width at the peak's half-maximal value. This ratio with M-proteins is always greater than 3.0, whereas with polyclonal elevations, a ratio of 2.0 or less is usual.

The second technique for determining homogeneity and the chain composition of the M-protein is immunoelectrophoresis. Typical findings are illustrated in Figure 35–4. Normal IgG is seen as very extended flat precipitin lines reflecting the heterogeneity of the normal immunoglobulins. This is true whether the IgG molecules are precipitated by anti-γ or anti-κ or anti-λ sera.

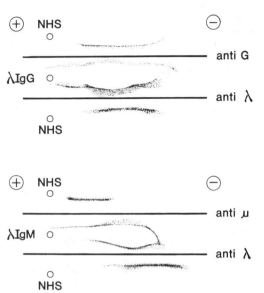

FIG. 35–4. Immunoelectrophoresis of normal serum, a λ IgG and λ IgM paraprotein. Note the flat profile of the IgM and IgG of normal serum and the characteristic deformation or bump of the precipitin arc with the paraproteins.

A λIgG M-protein is seen as a sharp bump or arc in which anti-γ and anti-λ sera yield bumps in the precipitin arcs of identical mobility. Similarly, in the lower part of the diagram is seen a homogeneous λIgM protein from a patient with Waldenström's macroglobulinemia, which gives restricted precipitin arcs with anti-μ and anti-λ sera of identical mobility revealing the chain composition of the M-protein.

Table 35–2 summarizes the data derivable from protein electrophoresis and immunoelectrophoresis that can be used to distinguish monoclonal from polyclonal gammopathy, as already discussed. Quantitative Ig class levels are of some value, since in monoclonal gammopathy there is selective increase in the involved class whereas the uninvolved classes are below normal levels in 90 per cent of cases. Thus, the patients are functionally immunodeficient, a point discussed again below. In contrast, in polyclonal gammopathy, elevation of all the Ig classes is usual and depression of even one Ig class is rare. Urine

TABLE 35–2
Differentiation Monoclonal from Polyclonal Gammopathy

Technique	Monoclonal	Polyclonal
Serum electrophoresis	Homogeneous protein of γ, β or α_2 mobility	Diffuse increase of γ and/or β fraction
Quantitative Ig levels	Selective increase of one Ig class. Remaining classes usually depressed.	Increase of all classes usual
Serum immunoelectrophoresis	Localized symmetric arc or bump which develops with one anti-heavy-class specific-chain serum and one anti-light-chain serum	Precipitin arcs increased in intensity but normal shape. Develop with both anti-κ or anti-λ serum.
Urine electrophoresis	Frequent homogeneous protein	Negative or nonspecific proteinuria
Urine immunoelectrophoresis	When homogeneous light chains, localized symmetric arc which develops with either anti-κ or anti-λ	Negative or reaction with both anti-κ or anti-λ.

electrophoresis may reveal a homogeneous protein of β or γ mobility in monoclonal gammopathy, whereas this does not occur in polyclonal gammopathy. Immunoelectrophoresis of the urine is even more useful than protein electrophoresis since it is much more sensitive in detecting M-proteins (by a factor of 10 to 20), definitively characterizes the chain composition of the M-protein when it is present, and can accomplish this characterization in the face of substantial nonspecific proteinuria as in glomerular disease.

Benign monoclonal gammopathy

In many instances in which an M-protein is demonstrated in the serum, no malignant disease of the lymphocyte plasma cell series is found. It is estimated that this occurs 3 to 4 times more frequently than M-proteins associated with malignant disease. This former condition is called *benign monoclonal gammopathy;* and in most of these instances no malignant disease occurs even after 10 to 20 years of follow-up. Some patients, however, do evolve into myeloma or Waldenström's disease that is often preceded by an increased concentration of M-protein or appearance of homo-

geneous light chains (Bence Jones protein) in urine. The biochemical and chemical features of benign monoclonal gammopathy that distinguish it from malignant gammopathy are summarized in the list below. First, homogeneous light chains in urine are rarely seen (less than 1 per cent of instances) in benign monoclonal gammopathy but occur in approximately 50 per cent of cases of myeloma. In benign monoclonal gammopathy the level of protein is low (2 g./dl. or less) and unchanging with time. An increasing concentration of serum M-protein reflects a growing tumor and strongly suggests malignancy. The plasma cell content of the marrow does not exceed 20 per cent in benign conditions, and atypical, invasive, and immature forms of lymphoid or plasma cells are not seen. Finally, the continued absence of findings

---

Findings in Benign Monoclonal Gammopathy

No homogeneous light chains in urine
Low and stable level of M-protein less than 2 g./100 ml.
Plasma cell content of bone marrow less than 20 per cent; few atypical forms and no evidence of invasive character of cells
Absence of clinical evidence of disease for several years

---

suggesting malignancy is by far the most important criterion of benign monoclonal gammopathy.

Major histocompatibility complex

Chromosome 6 in man contains the genes of the major histocompatibility complex (MHC). These genes code for proteins that determine both humoral and cellular immune responses to antigenic stimuli and also exert an important influence upon the quantitative aspects of the response.

Much of the available evidence is based upon the finding that individuals respond to particular differences in proteins (antigens) coded by the MHC genes of other individuals of the same species. Investigators have used sera from multiply transfused individuals or multiparous women and the capacity of lymphocytes from different individuals to stimulate lymphocyte division

in other individuals to identify numerous genetic loci (Fig. 35–5). HLA-A, -B, and -C are recognized by serological reactions. HLA-D is recognized by the proliferative response when cells from two individuals are mixed *in vitro*. HLA-DR antigens are based upon serological reactions against purified B-cells and monocytes. Indeed, the alleles (the alternate forms of a gene in a population) commonly found at these loci are unusually polymorphic; that is, a large number of alternate forms of these genes have been found widely distributed in the species as a whole. There are at least 19 alleles at HLA-A, 30 at HLA-B, 6 at HLA-C, 11 at HLA-D, and 7 at HLA-DR. Substantial discussion and controversy continue to surround the identification and importance of other loci (such as SB, Te, MB, and MT) as this field continues to develop. Finally, proteins that are not clearly related to the immune response have been mapped within or near to the MHC region (Fig.

FIG. 35–5. The major histocompatability complex (MHC) for the human leucocyte antigens (HLA). The position of the genes for the HLA antigens on the short arm of chromosome 6. Other markers mapping nearby include phosphoglucomutase-3 (PGM3), glyoxylase (GLO), complement components C2, C4, and factor B (FB), and urinary pepsinogen (Pg5). Distances are given in centimorgans (cM). One centimorgan is equivalent to a 1 per cent chance of a crossover event between parents and progeny.

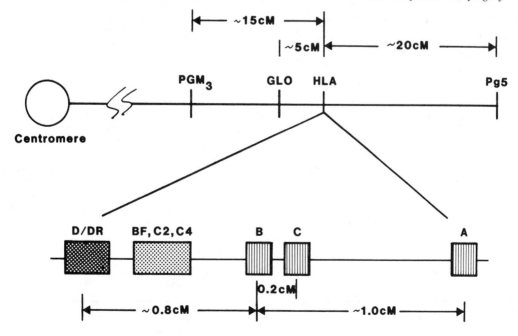

35–1). These include phosphoglucomutase-3 (PGM3), glyoxylase (GLO), complement components C2 and C4, Factor B, and urinary pepsinogen (Pg5).

The genes in the major histocompatibility locus are close to each other in the sense that the same alleles tend to travel together from generation to generation (the frequency of crossover is low). This phenomenon is called genetic disequilibrium, and it leads to certain histocompatibility alleles being found together more commonly than would be expected if they were randomly associated. For example, when an individual has the HLA-B8 marker, he is as likely as not to have the HLA-DRw3 marker as well, if he is a North American or European caucasian. These connected genetic markers that pass from one generation to the next intact are called haplotypes. A number of disorders are associated with the HLA-B8, HLA-DRw3 haplotype (Table 35–3). It is, perhaps, not surprising that the frequency of the different alleles, as well as the associations that result in genetic disequilibrium, differ from population to population.

The genetic material that controls the

TABLE 35–3
HLA-D and DR and Disease Association

| Disease | HLA Antigen | Per Cent Antigen Frequency | | Relative Risk |
		Patient	Control	
Adult Rheumatoid Arthritis				
Caucasians	Dw4	36–59	12–17	5.0
Caucasians	DRw4	70	28	5.8
Multiple Sclerosis				
Caucasians	B7	12–46	14–30	1.73
	Dw2	47–70	15–31	4.3
	DRw2	40–70	18–29	4.8
American blacks	B7	36	16	2.9
	B40	16	1	15.8
	Dw2	35	0	38.7
Optic neuritis	Dw2	40–50	18–30	2.9
Juvenile/insulin-dependent diabetes				
Caucasians	B8	19–55	2–29	2.4
	B15	4–50	2–26	1.9
	B18	5–59	5–50	1.7
	Dw3	50	21	3.8
	Dw4	48–52	19	4.01
	DRw3	36–59	11–24	5.69
Celiac disease	B8	45–89	11–29	8.6
	Dw3	63–96	22–27	10.8
	DRw3	88	25	21.1
Grass pollinosis	DRw3	43	14	4.7
Goodpasture's syndrome	DRw2	88	32	15.9
Leprosy				
Asians	DRw2	79	53	3.30
Buerger's disease				
Caucasians	A9	50	14	6.39
	B12	2	28–30	0.08
	DRw2	59	22	5.0

immune response might be expected to be important in determining the response of individuals to many diseases. Population studies have shown that various MHC loci are more (or less) commonly found than they are in normal unaffected populations. The HLA-D and -DR loci are thought to be most closely related to the immune response genes. In the mouse it has been clearly established that the immune response genes are located in a region of the chromosome called Ia. Whether HLA-D or -DR or another locus that maps close to HLA-D or -DR, such as SB, Te, MB, or MT, is the human analogue of Ia remains to be determined. As might be expected, the relationship of the MHC molecules to disease susceptibility is an area of intense interest. (Some of the known associations with HLA-D and -DR are presented in Table 35–3.)

At the HLA-A, -B and -C loci, certain alleles are also associated with the presence of disease. Why HLA-A3 is associated with hemochromatosis or why HLA-B27 is found in 50 to 90 per cent of patients with ankylosing spondylitis is not understood except to indicate that genetic factors influence the expression of these diseases.

T lymphocytes and cell-mediated immunity

T-cells in man have been traditionally defined as the lymphocytes that spontaneously rosette with sheep red blood cells. These cells were shown to mature in the thymus and, hence, have become known as T lymphocytes. T-cells do not originate in the thymus; however, in intrauterine life the first T-cell precursors migrate from the bone marrow to the thymus. This process appears to continue in man until adolescence. The T-cells migrate from the thymus to the lymph nodes, spleen, Peyer's patches, and other lymphoid areas through the blood and lymphatics. These cells do not remain in one place but recirculate among the lymphoid sites, although

particular types of T-cells may be found in higher numbers in particular areas. Indeed, surface markers probably determine lymphocyte homing, since protease treatment abolishes their accumulation in the lymphoid tissues. Also, pertussis infection may be an example of the loss of the homing in T-cells, since during this infection these lymphocytes are found in large numbers in the circulation and depleted from the lymphoid organs.

The life span of T-cells is highly variable. Long-lived T-cells, with life spans longer than 10 years, are found recirculating in the blood and lymphatics. The shorter-lived T-cells are recovered from the lymph nodes, bone marrow, and spleen.

Within the pool of postthymic lymphocytes are cells that subsume different functions. Cells that exert dampening effects on the immune response are termed suppressors; and those exerting enhancing effects are termed helper/inducers. They are functionally and phenotypically separable and play a major role in the regulation of the immune response. In addition, there are specific cytotoxic T-cells. Whether the natural killer (NK) cells are also of T-cell lineage or are more closely akin to monocytes remains controversial. The suppressor cells appear to share at least some surface antigens with the cytotoxic T lymphocytes.

In addition to surface markers related to suppressor and helper functions, T-cells also contain other unique markers. Monoclonal antibodies raised in mice have been generated and recognize almost all human T-cells. These have been used as tools for analysis and have even been used therapeutically in a few cases. The cytotoxic T-cells are very important in cell-mediated immunity. After an initial exposure (whether *in vitro* or *in vivo*) to the MHC antigens of another individual or to virally infected cells, T-cells develop that are capable of lysing the foreign cells or the infected cells. This cytotoxic T-cell immunity is largely responsible for the resistance to viral infections. What function it might serve to have T-cells proliferate or to become specific cytotoxic cells after exposure to the different MHC antigens of another individual is a subject of much conjecture.

Cellular interactions

In the adaptive immune response, antibody appears to be made against antigens in one of two basic ways. In the first and perhaps less complicated way, the antigen appears to stimulate directly B-cells. These are antigens that have a repetitive structure such as the pneumococcal polysaccharide, flagellin, and certain synthetic polymers. The IgM response predominates to these antigens and they tend to be oligoclonal (antibodies of only a few specificities are made).

Responses to ordinary protein antigens clearly involve cells other than B-cells. This interaction requires participation of monocytes (or dendritic cells) and T-cells. The macrophage "processes" the antigen and then "presents" it to the T-cell. At this level, the decision is made whether to suppress or whether to help the response. Depending on a number of factors, the responses that are helped may be influenced to make antibody or to generate cytotoxic T-cells. Many of the interactions have been shown under certain circumstances to require the same MHC molecules on both participating cell types, a phenomenon called HLA restriction.

Among the many soluble factors there are at least two which lead to full T-cell activation for helper or T-cell effector functions (e.g., cytotoxicity, delayed hypersensitivity reactions). One of these comes from the monocytes and has been known previously as leukocyte pyrogen, macrophage inhibitory factor, or leukocyte activating factor, among other activities. It has become clear that the same molecule mediates a number of effects and has been designated interleukin-1. A second factor, both

T-cell derived and T-cell stimulatory (T-cell growth factor), has been designated interleukin-2.

As shown mostly in mouse experiments, cell cooperation is also required to generate the cytotoxic T lymphocyte response. Different subpopulations of T-cells cooperate with each other to generate the cytotoxic effector cells. Cytotoxic effector cells, when exposed to viral infected cells or cells carrying the appropriate MHC locus antigens from another individual, attack and lyse those cells. It is obvious, therefore, that the adaptive immune response is capable of responding to an undetermined, relatively infinite number of antigens that results in a complicated mechanism of direction and control that is only beginning to be understood.

Immunodeficiency diseases

Recurrent infections should raise the possibility of immunodeficiency. All immunodeficiency does not lead to disease, however, as exemplified by asymptomatic selective IgA deficiency. Immunodeficiencies are termed primary when they are either genetically determined or congenital. Acquired immunodeficiencies are termed secondary when an unrelated disease or intervention compromises the capacity to resist infections. In this latter group are patients with cancer, diabetes mellitus, sickle cell anemia, cystic fibrosis, malnutrition, radiation exposure, premature birth, integumental injuries such as burns and rashes, fractures, and concurrent infection. Another large group with secondary immunodeficiency result from treatments for other diseases that are markedly immunosuppressive. These include treatment with prednisone, various cytotoxic drugs (e.g., cyclophosphamide, azothiaprine, the innumerable cytotoxic agents used in the treatment of malignancy, and radiation), and bone marrow transplantation in which dramatic immunodeficiency develops in the context of graft rejection or graft versus host disease. When a cause for suspected immunodeficiency is not known, the initial evaluation should include complete blood count, quantitative determination of immunoglobulins, skin tests for delayed hypersensitivity, Shick tests or polio virus titers (for a specific antibody response), isohemagglutinins (anti-A and anti-B are found in individuals except those with blood type AB), or antipneumococcal titers (to assess IgM function), and an adequate evaluation of any concurrent infection. Other tests are available for specific defects.

Developmental considerations

At birth, neonates have a limited capacity to synthesize specific immunoglobulin. The immunoglobulin they do have (approximately 1 g./dl.) is derived largely from transport of IgG across the placental membranes from mother to child. After birth, the neonate slowly accelerates his synthesis of immunoglobulin as the maternal IgG is metabolized. At about 3 months, a nadir is reached (average, approximately 0.5 g./dl.) in the normal child. This is the reason that defects in immunoglobulin production are not routinely a clinical problem until at least a few months after birth. Transient hypogammaglobulinemia of infancy is a disorder in which this period is prolonged. Investigators disagree on the frequency of this syndrome, and some are reluctant to make this diagnosis at all. Nevertheless, all agree that infants with this disorder spontaneously recover within, at most, a few years.

The mother also provides the breast-fed infant with immunoglobulins, cells (especially monocytes), lysozyme, and lactoferrin. All of these have antibacterial activity and presumably account for the lower incidence of diarrheal disease in the breast-fed infant than in the bottle-fed infant. Surprisingly, immunoglobulin in milk can pass through the entire gastrointestinal tract of the neonate without being degraded.

There is one disease in which it is advantageous for the infant to be bottle fed: the rare patient with Leiner's disease, which is clinically characterized by failure to thrive, recurrent sepsis, diarrhea, and seborrheic dermatitis. These infants have a defect in opsonization of yeast. Infusion of fresh plasma ameliorates the disease and, in nearly all cases carefully evaluated, it has been shown that these infants lack the functional complement component C5. Infants with this rare disorder do not appear to have as much difficulty when bottle fed because of the opsonic activity in bovine milk.

Failure of immunoglobulin synthesis

X-LINKED INFANTILE (BRUTON'S) AGAMMAGLOBULINEMIA.
In some children, serious recurrent bacterial infection is associated with low or absent immunoglobulins. Absence of detectable circulating B-cells or tissue plasma cells with normal numbers of circulating T-cells is found in X-linked infantile agammaglobulinemia, or Bruton's agammaglobulinemia. Only a few pedigrees have been analyzed to prove the X-linked nature of the disease. Multiple etiologies must exist, since 20 per cent of infants with these findings are female. In addition, cells from some patients suppress *in vitro* proliferation, differentiation, and immunoglobulin production of cells from normal individuals, implying that in these patients the defect in Ig production in some patients is not intrinsic to the B-cell lineage. These patients lack germinal centers in lymph nodes and spleen (Fig. 35–6). Cellular immunity is intact in these patients and they usually do not have difficulty with most viral, mycobacterial, or fungal infections. Difficulties with persistent viral hepatitis and enteric cytopathic human orphan (ECHO) infections, however, have been described.

An unexpectedly high proportion of individuals with infantile agammaglobuli-

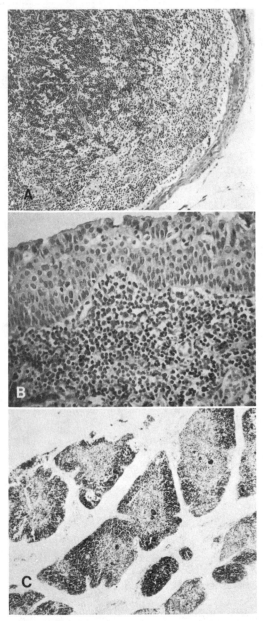

FIG. 35–6. (A) Cortical region of a lymph node from a patient with X-linked recessive immune deficiency disease (Bruton-type agammaglobulinemia). Note the sparsity of cells and absence of lymphoid germinal centers in the far cortical space (cf. Fig. 35–7). (B) Tonsil of same patient. Again note absence of germinal center and plasma cells. (C) Thymus of same patient, which is essentially normal. (Good, R. A., et al.: In Birth Defects: Original Article Series, Vol. 4, p. 17. New York, The National Foundation, 1968. Reproduced by permission of the publisher.)

nemia have autoimmune disease. A joint disease resembling juvenile rheumatoid arthritis is especially common, and dermatomyositis has developed in a few such patients.

Infantile agammaglobulinemia is treated by replacing gammaglobulin. The traditional method of multiple intramuscular injections is painful and not well accepted by young children. Newer methods using intermittent subcutaneous slow infusion by portable infusion pump should result in the maintenance of higher immunoglobulin levels in these patients. Early treatment of bacterial infection with antibiotics is also a very important aspect of management.

X-linked immunodeficiency with hyper-IgM

Other patients with recurrent bacterial infections are found to have markedly elevated levels of IgM but absent IgG, IgA, and IgE. This syndrome was first described by Dr. Fred Rosen and his colleagues and is now referred to as X-linked immunodeficiency with hyper-IgM. These patients usually have neutropenia, hypertrophic tonsils, and splenomegaly. Again, a few girls and women have been found with a similar disease indicating that the disorder is genetically heterogeneous and not always X-linked. Some of these latter cases may be related to congenital rubella infection and others may be acquired later in life.

A group of patients which is particularly difficult to identify has normal immunoglobulin levels but lack the capacity to respond to specific antigens. These patients constitute a heterogeneous group that no doubt has a variety of underlying causes for the disorder. Not surprisingly, the number of recurrent infections is decreased by gammaglobulin therapy.

Common variable immunodeficiency

Another, much larger group of patients loses the ability to make immunoglobulin later in life. Most commonly, these patients present as adolescents or young adults with recurrent infection. Their disorder is also heterogeneous and is designated *common variable immunodeficiency*. Studies of peripheral blood lymphocytes suggest a variety of underlying defects. Some patients completely lack B-cells making them similar to those discussed with the X-linked infantile agammaglobulinemia. A second subgroup has B-cells which have lost capacity to proliferate when stimulated appropriately *in vitro*, suggesting a late maturational B-cell defect. A third group of patients has B-cells that fail to secrete immunoglobulin but whose B-cells proliferate and accumulate intracytoplasmic immunoglobulin. In a fourth group there is excessive T-cell suppression, the most common finding in this disorder. T-cells from these patients suppress *in vitro* immunoglobulin production when co-cultured with B-cells from normal individuals. All of the above are *in vitro* correlates and may or may not accurately reflect the *in vivo* physiological defects, but they nonetheless identify numerous potential mechanisms for the immunodeficiency.

Common variable immunodeficiency is associated with other clinical problems, especially gastrointestinal disease. A malabsorption syndrome resembling sprue has been caused by *Gardia lamblia*. Also lymphoid nodules have been found in the small bowel of other patients, which upon pathological examination resemble giant germinal centers. One of the more common associations of this category is pernicious anemia, hypogammaglobulinemia, and lymphoid nodular hyperplasia of the small bowel. Also, disorders resembling hemolytic anemia, arthritis, dermatomyositis, systemic lupus erythematosus, among others, have been described in these patients.

Selective IgA deficiency

All of the immunodeficiency diseases thus far discussed are rare. Selective IgA deficiency is by far the most common immunodeficiency disease, affecting as many as one of every 600 adults. Affected individuals are often asymptomatic, but they may have recurrent sinopulmonary infections. Also, there is an increased incidence of autoimmune disease and allergy. Pathophysiologically, most patients lack serum and secretory IgA with apparently normal T-cell function and normal levels of secretory component. In a small minority T-cell abnormalities can be demonstrated. Of course other immunodeficiency disorders with IgA deficiency exist and should not be confused with this group of selective IgA deficiency in which no other immunoglobulin class is involved. In some individuals selective IgA deficiency has been associated with phenytoin use. IgA deficiency is usually a mild disease and no specific therapy is advisable or available. One must be cautious, however, when giving these patients blood products. They may become (or already be) sensitized to human IgA, and transfusion without the appropriate precautions can lead to a fatal allergic reaction.

Failure of cellular immunity

THYMIC HYPOPLASIA (DiGEORGE SYNDROME). DiGeorge recognized the relationship of recurrent infection, absent thymus, and hypoparathyroidism in 1965. Patients with these disorders have absent cellular immunity with relatively normal levels of immunoglobulin. Their disease is related to the abnormal development of the third and fourth pharyngeal pouch during embryogenesis. This accounts for the multisystemic nature of the disease and associated problems. They may have hypocalcemic tetany, unusual facies, and congenital heart disease. Deep cortical areas of the lymph nodes and the periarteriolar regions of the spleen are not populated with lymphocytes (Fig. 35–7). The therapy of choice is fetal thymus gland transplantation. Fetal thymus from individuals younger than 14 weeks is preferred (if obtainable), since thymus from older fetuses is capable of

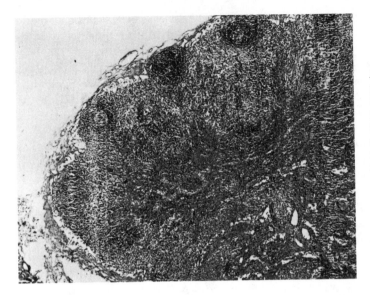

FIG. 35—7. Lymph node from a patient born without a thymus (DiGeorge syndrome). Note the presence of germinal centers in the far cortical region. Plasma cells are present near the germinal centers and in the medulla. Lymphocytes in the T-region (deep cortical area or corticomedullary junction) are sparsely distributed. (Cooper, M. D., et al.: *In* Birth Defects: Original Article Series, Vol. 4, p. 378. New York, The National Foundation, 1968. Reproduced by permission of the publisher.)

causing graft-versus-host disease. These patients are at particular risk for graft-versus-host disease because of the relative absence of cellular immunity, and even blood transfusion has been known to cause fatal graft-versus-host disease. One patient was reconstituted with thymic tissue transplanted inside a Millipore chamber, and others have received some benefit from the administration of the thymic hormone thymosin.

Severe combined immunodeficiency (Swiss agammaglobulinemia)

The severe recurrent infections found in patients with severe combined immunodeficiency are such that these children rarely survive infancy without heroic measures. Patients have neither peripheral blood B-cells nor T-cells, and they are profoundly immunodeficient. Both lymphoid aplasia and thymic dysplasia are found with loss of the normal architecture (Fig. 35–8). The diagnosis is made by the absence of IgA, IgM with very low IgG, and the absence of a proliferative *in vitro* response to specific antigens. Nezelof's syndrome is a variant, the patient demonstrating measurable immunoglobulin but without specific functional antibody. This is undoubtedly a heterogeneous disease with specific enzyme deficiencies of the purine salvage pathway leading to stem cell deficiency and thymic tissue abnormalities, thereby accounting for two of the etiologies thus far identified.

The therapy of choice is bone marrow transplantation from a histocompatible sibling. The rate of survival with this therapy continues to improve; however, most children do not have a histocompatible sibling, and for them graft-versus-host disease after transplantation is a grim alternative to the underlying disorder. Graft-versus-host disease is most severe when there is a mismatched donor at the HLA-D locus of the major histocompatibility complex. Newer strategies using lectin agglutination techniques and monoclonal antibodies directed at removing the postthymic cells (which are responsible for graft-versus-host disease) are being used with some promise of improved success.

Immunodeficiency due to specific enzyme abnormalities

Enzyme deficiency as a cause of immunodeficiency accounts for at most 15 per cent of presently recognized immunodeficiency disorders. All of the recognized enzyme deficiencies are found in the purine salvage pathway and include adenosine deaminase, purine nucleoside phosphorylase, and 5' ectonucleotidase.

Adenosine deaminase is involved in the conversion of adenosine to inosine. Several possible metabolic events such as increased cyclic adenosine monophosphate (cAMP), direct toxicity of adenosine, and the intracellular accumulation of S-adenosylhomocysteine have been mentioned as mediators of the toxicity caused by the deficiency of the enzyme. The most likely mechanism for the toxicity is due to the extraordinary accumulation of deoxyadenosine triphosphate (deoxy-ATP) which is an effective inhibitor of ribonucleotide reductase with subsequent depletion of deoxy-ribonucleoside triphosphates. This leads to cessation of deoxyribonucleic acid (DNA) synthesis and may be the toxic event for stem cells early in development. Successful treatment by bone marrow transplantation only corrects the enzyme deficiency in the transplanted tissue, but the newly infused stem cells can develop and completely correct the phenotypic immune defect.

Purine nucleoside phosphorylase catalyzes the conversion of inosine, deoxyinosine, guanosine, and deoxyguanosine to hypoxanthine and guanine. The metabolite that dramatically accumulates in this enzyme deficiency is deoxyguanosine triphosphate, deoxy-GTP. This triphosphate is also an inhibitor of ribonucleotide reductase. For unknown reasons there seems to be selective toxicity for T-cell precursors.

The clinical picture in adenosine deaminase deficiency is severe combined immunodeficiency. There may be radiological abnormalities which include concavity and flaring of the anterior ribs, abnormal contour and articulation of posterior ribs and transverse processes, platyspondylosis, thick growth arrest lines, and an abnormal bony pelvis.

Patients with purine nucleoside phosphorylase deficiency reflect an isolated T-

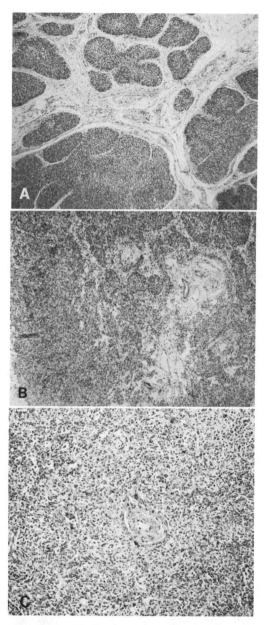

FIG. 35–8. (*A*) Thymus from patient with severe combined immune deficiency disease (Swiss-type agammaglobulinemia). Thymus is poorly developed, cortical and medullary areas are not distinguishable owing to absence of thymocytes, and Hassall's corpuscles are not present. (*B*) Lymph node from same patient. Germinal centers, lymphocytes, and plasma cells are not demonstrable. (*C*) Section of spleen from patient with the same disease. Normally present periateriolar lymphocytes are absent. (Hoyer, J. R., et al.: *In* Birth Defects: Original Article Series, Vol. 4, p. 91. New York, The National Foundation, 1968. Reproduced by permission of the publisher.)

cell deficiency with recurrent infections, normal antibody levels, and autoantibody formation. Because of their limited T-cell immunity, fatal varicella and vaccinia infections have been described. Bone marrow transplantation is effective treatment by virtue of replacing the T-cell population.

Finally, 5' ectonucleotidase deficiency has been described in both X-linked and acquired agammaglobulinemia. It may be that the 5' ectonucleotidase deficiency is not a cause of the disease but simply a B lymphocyte marker, thus reflecting the small number of B lymphocytes in the population.

Recognition of these enzyme deficiencies has introduced a new era in understanding some immunodeficiency syndromes, and better molecular definition of these disorders can be expected in the future.

Short-limbed dwarfism with immunodeficiency

The diagnosis of short-limbed dwarfism with immunodeficiency is usually made at

FIG. 35–9. Herpes simplex lesions on a patient with Wiskott-Aldrich syndrome. In immunologically competent individuals, herpes lesions are not associated with such overwhelming disease. (St. Gene, J. W., Jr., Prince, J. T., and Burke, B. A.: New Eng. J. Med. 273:229, 1965. Reproduced by permission of the publisher.)

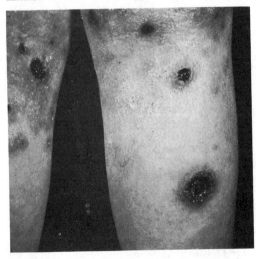

birth by the appearance of the neonate. It has an autosomal recessive inheritance, and patients may have defects involving B-cells, T-cells, or both. Viral infections are particularly severe, especially varicella, vaccinia, and poliovirus.

Wiskott-Aldrich syndrome

Wiskott-Aldrich syndrome is an X-linked recessive disorder characterized by thrombocytopenia, eczema, and recurrent infections. Patients have thrombocytopenia with small platelets at birth. Immunoglobulins are very abnormal with markedly elevated levels of IgA and IgE and low levels of IgM. These individuals seem to be unable to make isoagglutinins or antibodies to polysaccharide antigens. Patients have been described with a partial defect in which only thrombocytopenia is seen without eczema or immune attrition. The natural history is for affected boys to die in childhood of overwhelming sepsis or hemorrhage. Viral infections such as herpes simplex which are almost always contained in normal hosts frequently become disseminated in these children (Fig. 35–9). Survivors are at much increased risk for malignancy, particularly reticuloendothelioses and lymphomas. There is some evidence that thrombocytopenia and small platelets are improved by splenectomy, but this puts the patients at much increased risk for sudden, overwhelming sepsis. Some workers, however, have managed patients for long periods with prophylactic antibiotics, and there is some evidence to suggest that this strategy can prevent sudden, overwhelming infection.

Present evidence favors a T-cell defect in the immune dysfunction. Supporting this, an early effort at transplantation corrected the immune defect after engraftment of only the T-lymphocytes. The thrombocytopenia and small platelets remained. Other data indicating that the mothers show X-linked allelic exclusion separately occurring in their platelets and T-cells (with additional cell lines involved in some carriers) imply that the genetic defect independently affects platelets and T-cells. The molecular defect or defects that cause the syndrome are not known. An interesting beginning in this direction has been made by Parkman and his colleagues, who have shown that a glycoprotein is lacking in the plasma membranes of at least the lymphocytes of some patients with Wiskott-Aldrich syndrome. The therapy of choice in individuals with severe infection or recurrent bleeding is bone marrow transplantation. Patients with milder forms of the Wiskott-Aldrich syndrome may require no therapy.

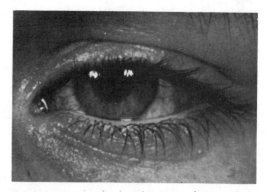

FIG. 35—10. Scleral telangiectases, characteristic of patients with ataxia-telangiectasia (similar vascular defects are widely distributed in these patients).

Ataxia-telangiectasia

Patients with an autosomal recessive disorder characterized by oculocutaneous telangiectasia and progressive ataxia were appreciated to have a complicated immunodeficiency disease which was of variable severity. They often had recurrent sinopulmonary infections, selective IgA and IgE deficiency, cutaneous anergy, and depressed *in vitro* lymphocyte proliferation. The basic defect which leads to this disease is unknown. Possible clues include the strikingly elevated levels of α-fetoprotein which led Waldman and McIntire to suggest that the fundamental defect was one of organ development and differentiation. Other investigators have shown that cells from individuals with ataxia-telangiectasia are much more susceptible to irradiation injury than are cells from normal individuals. This led to the hypothesis that ataxia-telangiectasia patients may have a defect in their capacity to repair DNA. Finally, autoantibodies to a variety of tissues are commonly found, and since the central nervous system lesions are degenerative as well as demyelinating, others have suggested that this may be basically an autoimmune disease.

The clinical course of ataxia-telangiectasia is usually relentlessly progressive. Cellular ataxia generally begins in infancy making gait, posture, and intentional movements difficult. Speech becomes slurred. Gaze becomes disconjugate. Choreoathetoid and jerky movement may develop, and mental retardation is usually present. Drooling, mask-like facies and strabismis develop later, and finally weakness and muscle atrophy ensue.

Telangiectasia is especially common in the bulbar conjuctiva (Fig. 35–10), progressing later to involve the integument. Patients are mentally immature, and rarely do the secondary sex characteristics appear. Growth failure is also a feature. Like patients with the Wiskott-Aldrich syndrome, those with ataxia-telangiectasia have a high incidence of malignancy. In this disease only supportive therapy with gammaglobulin and antibiotics has been advocated. In patients who develop malignancies care must be exercised in using radiation and radiomimetic drugs.

Chronic mucocutaneous candidiasis

Recurrent candidal infection is the hallmark of the heterogeneous group of patients diagnosed as having chronic muco-cutaneous candidiasis. Oral mucous membranes, fingernails, and toenails are most frequently chronically affected. Some patients also develop skin involvement, which may evolve into large granulomatous masses. Surprisingly, systemic candidal infection is uncommon. Subgroups of patients have an associated endocrinopathy, especially hypoparathyroidism or Addison's disease. Other patients seem to have a hereditary form, the autosomal recessive juvenile polyendocrinopathy with candidiasis.

The underlying defect in this disorder is not well defined. However, although the patients make normal amounts of anticandidal antibody, the cells from most patients fail to respond to *Candida* antigen *in vitro*. They also have cutaneous anergy when challenged with *Candida* antigen. Despite cutaneous anergy to *Candida*, they do not have increased infections, with other agents. The findings that these patients do not become sensitized to 2,4-dinitrofluorobenzene leads to the conclusion that their defect must involve more than just the specific response to *Candida*.

Therapy is directed at correcting the local *Candida* infection with antifungal agents when required. The associated endocrinopathies each requires its own separate therapies.

Hyper-IgE syndrome

A group of patients who have recurrent deep cutaneous staphylococcal abcesses (often without systemic sequellae), cavitating pulmonary disease, and recurrent otitis were shown to have very high levels of IgE (3 to 1000 times higher than the upper limit of the normal range). Buckley and colleagues described the first case of this disorder, which has become known as Job's syndrome or the hyper-IgE syndrome. Affected individuals develop characteristic coarse facial features. Laboratory studies have demonstrated a defect in polymorphonuclear leukocyte chemotaxis. All patients have eosinophilia. They tend to have normal levels of IgG, IgM, and IgA; however, IgD (as well as IgE) is elevated. They have normal T- and B-cell numbers in the peripheral blood. Finally, they tend to have positive immediate hypersensitivity reactions to bacteria, fungi, foods, and pollens.

Much of the IgE that these patients synthesize is specific for staphylococcal antigens. They must have a disorder of immune regulation, but the pathogenesis and the fundamental defect have yet to be identified. Therapy is symptomatic. Patients benefit from prophylactic chronic dicloxicillin therapy. Nevertheless, they experience much chronic morbidity.

Secondary immunodeficiencies

Any serious illness has the capacity to interfere with the function of the immune system, thereby leading to infection or autoimmune disease. Moreover, current therapies for malignancy and vasculitis employing cytotoxic drugs or steroids vastly increase the risk of opportunistic infection by interfering with the immune defenses. The use of bone marrow transplantation is profoundly immunosuppressive, and whole new areas of opportunistic infections are being defined.

Diseases such as sarcoidosis, leprosy, syphillis, and Hodgkin's disease, are frequently associated with cutaneous anergy and an apparent defect in cell mediated immunity, whereas the B-cell malignancies—myeloma, chronic lymphocytic leukemia, and Waldenström's macroglobulinemia—are accompanied regularly by a serious functional hypogammaglobulinemia due to reduced normal globulin production. A variety of dysregulatory phenomena thought to represent immunodeficiency are frequently seen in the autoimmune diseases systemic lupus erythematosus, scleroderma, rheumatoid arthritis, and Sjögren's syndrome, in which hyperglobulinemia and decreased suppressor activity are common. Advanced malignancy is often accompanied by a heterogeneous array of immunodeficiency phenomena, which can be profound. Better understanding of the interaction of the immune defects in these disorders should lead to improved management and therapy.

Acquired immunodeficiency syndrome (AIDS)

Occasionally a totally new disease arises that defies understanding. In 1979 the first case of a new disorder was recognized among the homosexual male population. These patients have a defect in cellular immunity leading to an aggressive form of Kaposi's sarcoma or opportunistic infection, especially with *Pneumocystis carinii*. A large number of other opportunistic infections have also been found, and increasing numbers of patients with autoimmune phenomena (especially autoimmune thrombocytopenia) are being reported. There have been over 700 cases reported between 1979 and 1983.

At present this is a poorly defined syndrome, recognized by the presence of an opportunistic infection or Kaposi's sarcoma in a young adult, with no other predisposition to immune deficiency. Over 70 per cent of cases have been found in homosexual or bisexual men, less than 20 per cent in heterosexual men, and less than 10 per cent in women. Other than homosexuals, intravenous drug abusers and recent Haitian immigrants are particularly at risk. Additional evidence suggests that the disease can be transmitted in blood products, and this is supported by the finding of the disorder in hemophiliacs. Kaposi's sarcoma has thus far been found only in the homosexual population. To date, therapy is supportive and the underlying immune defect is a complete mystery, although a regular finding is an inverted ratio of T-helper to T-suppressor cells with severely involved patients having few identifiable T-helper cells in their peripheral blood.

While the prevailing hypothesis is that the etiological agent is a new virus, the disease may be multifactorial with recurrent viral infections with cytomegalovirus (CMV), amyl nitrite abuse (a potential carcinogen), other drugs, and venerial infection, among other agents, playing a pathogenic role.

References

Ammann, A. J., and Fudenberg, H. H. Immunodeficiency Diseases. *In* Stites, D. P., et al. (eds.): Basic and Clinical Immunology, 4th ed., pp. 395–429. Lange Medical Publications, 1982. (Complete up-to-date analysis of enzyme deficiencies associated with immunodeficiency syndromes.)

Benacerraf, B., and Unanue, E. R. Textbook of Immunology. Baltimore, Williams & Wilkins, 1979. (Lectures in immunology for medical students, emphasizing particularly the conceptual development of the field).

Bergsma, D.: Immunodeficiency in man and animals. *In* Good, R. A., and Finstad, J. (eds.): Birth Defects. Original Article Series, Vol. XI. Sunderland, Mass., Sinauer Associates, 1975. (An extensive multiauthor work detailing much of the work on primary immunodeficiencies of man and animals appearing between 1967 and 1974.)

Capra, J. D., and Kehoe, J. M.: Hypervariable regions, idiotype, and the antibody combining site. *In* Dixon, F. J., and Kunkel, H. G. (eds.): Advances in Immunology 20:1. New York, Academic Press, 1975. (A review of progress toward a biochemical explanation of the specificity of humoral immunity.)

Friedman-Kien, A. E., Laubenstein, L. I., Rubinstein, P., Buimovici-Klein, E., Marmor, M., Stabl, R., Spigland, I., Kim, K. S., and Zolla-Pazner, S.: Disseminated Kaposi's sarcoma in homosexual men. Ann. Intern. Med. 96:693–700, 1982. (Kaposi's sarcoma as a manifestation of the newly recognized acquired immunodeficiency syndrome.)

Golub, E. S.: The cellular basis of the immune response. 2nd ed., Sunderland, Mass., Sinauer Associates, 1981. (A careful text reviewing the experimental basis for the conceptual development of cellular immunology.)

Good, R. A.: Immunological reconstitution: The achievement and its meaning. Hosp. Pract. 4:41, 1969. (Lucid description of the significance and interpretation of implications of successful therapeutic lymphoid tissue transfer.)

Goodman, J. W.: Immunoglobulins. I. Structure and function. *In* Basic and Clinical Immunology. Stites, D. P., et al., (eds.): 4th Edition, Lange Medical Publications. pp. 30–43, 1982. (Clear description of structure-function relationships of human immunoglobulins.)

Kyle, R. A.: Monoclonal gammopathy of undetermined significance. Am. J. Med. 64:814–826, 1978. (Natural history of "benign" monoclonal gammopathy. Analysis of 241 cases.)

Marx, J. L.: Spread of AIDS sparks new health concern. Science 219:42–43, 1983. (A review of in present status of AIDS research and of recent observations demonstrating that it occurs in children and in hemophiliacs.)

Mattioli, C. A., and Tomasi, T. B., Jr.: Disorders of immunoglobulin synthesis. *In* Principles of Immunology. Rose, N. R., Milgrom, F. and Van Oss, C. J. (eds.). MacMillan Publishing Co., Inc., pp. 333–357, 1973. (Excellent description of analysis of hyperglobulinemia, paraprotein analysis.)

Parker, C. W.: Clinical Immunology. Vo. 1 and 2. W. B. Saunders Co., 1980. (A wide ranging two volume treatise on virtually all aspects of clinical immunology written by multiple authors.)

Samter, M. (ed.): Immunological Diseases. Little, Brown and Co., 1979. (A major text.)

Stiehm, E. R., and Fulginiti, V. A. (eds.): Immunologic Disorders in Infants and Children. W. B. Saunders Co., 1980. (A text of immunology with special emphasis on pediatric problems.)

36

Allergic Mechanisms

James H. Wells, M.D.
William A. Cain, Ph.D.

Immunological hypersensitivity reactions include all adverse reactions produced by apparently inappropriate hyperfunction of the immune system. A variety of immunologic mechanisms may be involved in various hypersensitivity reactions. A clinically useful classification of these mechanisms has been devised by Gell and Coombs, and that classification is presented in the Introduction to this section of the textbook.

Since these hypersensitivity reactions are due to adaptive immune responses, they share the common characteristics of the immune response in general. Sensitization may follow exposure to a variety of antigens. The most antigenic sensitizing agents are in general composed completely or partially of protein. Other chemical substances including polysaccharides, lipids, and simple chemicals (acting as haptens) may also be allergens (antigens that produce a hypersensitivity reaction). Sensitization, as well as production of a subsequent hypersensitivity reaction, may follow administration of the antigen via a variety of routes including parenteral injection, inhalation, ingestion, or application to the skin or mucous membranes. The dose of antigen delivered is an important determinant of whether sensitization takes place, of the type and quantity of immune reactants that are formed, and of the type and severity of the reaction that follows challenge

with the antigen. A latent period follows the sensitizing exposure prior to the onset of the hypersensitive state, and during this period the antigen is processed and specialized types of sensitized lymphoid cells are produced. Some of these lymphocytes (B-cells) produce specific antibodies against the sensitizing allergen. Others (T-cells) may be capable of recognizing antigen as foreign, engaging in cell-mediated responses to the antigen, and regulating B-cell proliferation and differentiation. In a previously unsensitized individual, the latent period usually ranges from several days to several weeks. The latent period ends when a quantity of antibody or sensitized cells sufficient to produce a hypersensitivity reaction accumulates. The duration of the specific hypersensitivity state is variable, apparently depending on individual characteristics of the host, individual characteristics of the allergen, and the circumstances of sensitization. Sensitization may persist for only a few weeks or months or may be lifelong. When repeated exposures to the allergen cause symptoms they are usually reproducible with some types of allergic reactions, such as poison ivy or hay fever. With others, such as drug allergies, repeated reactions may vary to some extent, apparently reflecting change in the types of immune reactants involved.

Although hypersensitivity reactions are

usually initiated by reintroduction of the antigen into a previously sensitized individual, there are special circumstances in which reactions may be elicited by the original (sensitizing) dose which persists through the latent period. Examples include serum sickness and some cases of allergic contact dermatitis.

Definitions

The *immune system* has selective survival value in that it protects against infecting microorganisms and other potentially harmful foreign substances. It protects by producing intensified inflammatory responses. A *hypersensitivity state* exists when the immune system responds with an inflammatory response which is exaggerated beyond a purely protective effect or when an immune inflammatory response is apparently inappropriately directed toward a material that is not potentially harmful and is well tolerated by most of the population at large.

The term *allergy* is currently used interchangeably with hypersensitivity. Its literal translation is an *altered state of reactivity*, which refers to attainment of the hyperimmune state through the process of adaptive sensitization.

Atopy is a term coined early in this century to include a group of disorders in which a familial pattern could often be recognized. They were bronchial asthma, some forms of rhinitis, urticaria and angiodema, some anaphylactic sensitivities, and atopic eczematous dermatitis. They were thought at the time to be caused by a common type of allergic mechanism. The term *atopy* literally means strange reactivity and refers to the rather curious tendency to develop hypersensitivity to common environmental allergens such as pollens, dust, molds, animal danders, and occasionally foods.

Over the years, however, it has become apparent that immunological mechanisms are not always responsible for what were originally termed the atopic diseases.

When allergic mechanisms are involved, they are of the Gell and Coombs Type I classification, through which histamine and other pharmacological mediators are elaborated in affected tissues. The antibodies are primarily of the IgE class, and such reactions are termed immediate hypersensitivity. However, in some instances mediators can be released and reactions can occur that are apparently identical to those caused by IgE and allergen although neither may be involved. The semantic question about whether nonimmunological reactions caused by mediators should be termed *atopic* or *immediate hypersensitivity* has never been well resolved. For this discussion they will be termed *nonimmunologic immediate hypersensitivity*.

The atopic diseases frequently cluster within families, but the mode of inheritance of the atopic tendency is complex and incompletely understood. Genetic factors influencing the development of IgE-mediated allergy probably include genes which control responsiveness to specific antigens (Ir genes) as well as those that govern total production of antibodies of the IgE class. However, environmental factors are also important, for example, degree of exposure to potential allergens. Whether allergy is involved or not, immediate hypersensitivity is also an expression of the function of physiological processes such as mediator release and control mechanisms that modulate the reaction (for example, the autonomic and endocrine systems).

Immunology of atopic disorders

The presence of serum factors associated with atopic allergic reactivity was first conclusively demonstrated by Prausnitz and Küstner. When serum from Küstner (who was anaphylactically sensitive to fish) was injected into the skin of Prausnitz (who was not allergic), immediate skin tests with fish extract were positive at the site of the injected serum several days after serum injection. For the most part, this "PK" test is positive when the transferred serum is obtained from an atopic patient who is strongly skin-test positive to the allergen that is employed.

The serum factors responsible for passive transfer are antibodies which have been referred to for many years as *reagins* or skin-sensitizing antibodies. We now know that most of these reaginic antibodies are found in the IgE class, although recently skin-sensitizing IgG antibodies have been described. IgE antibodies circulate freely in the serum and are also fixed to cells in various tissues, including skin. These properties of IgE antibody form the basis of the so-called PK phenomenon, temporary specific allergic sensitivity following injection of serum from an allergic donor into the skin of a nonallergic recipient.

During the 55 years since discovery of the PK phenomenon, several functional attributes of reagins have been recognized. When the PK test can be performed with serial titrations of allergic serum, the highest serum dilution which transfers a positive skin test can be used as a measure of the reaginic potency of that serum for the test allergen. The end point of this titration varies somewhat among recipients, and this is due in part to the IgE level of the recipient. Individuals with high levels of IgE less readily accept passively transferred skin tests than do recipients with normal IgE levels, probably because the tissues of those with higher levels are more nearly saturated.

Reaginic antibodies of human beings and other species can be passively transferred only to recipients of the same or closely related species (e.g., mouse to rat or man to another primate). Thus, IgE antibodies are *homocytotropic* but not *heterocytotropic.*

Skin tests have been the major diagnostic tool available for demonstrating the presence of specific IgE antibodies fixed to the skin. These tests are carried out by injecting small quantities of dilute solutions of suspected allergens intradermally or by placing a drop on the skin and pricking or scratching the underlying site. A positive test is characterized by a local wheal and flare reaction which may be visible within a few minutes, reaches maximum intensity within 15 to 20 minutes, and is clearly waning or gone within 1 hour. The time course of this *immediate hypersensitivity* skin test involving IgE antibodies is in contrast with the skin test for Arthus reactivity, which involves IgG or IgM antibodies and complement, and with skin tests for *delayed hypersensitivity* which is initiated by sensitized lymphocytes. Positive Arthus and delayed hypersensitivity tests do not begin to react until hours after application and they subsequently reach peak intensity and fade much more slowly than do immediate hypersensitivity reactions. The presence of specific IgE in tissues other than skin may be detected *in vivo* using such procedures as inhalational bronchial challenge with aerosolized allergen extracts, or placing a drop of extract on the conjunctiva or nasal mucosa.

Quantification of total serum IgE as well as determination of the quantity of serum IgE directed toward specific allergens is now possible using sensitive techniques. Tests *in vitro* for the presence of IgE include radioimmune and enzyme marker assays and detection of mediators, such as histamine, which are released when various IgE-sensitized tissues are exposed to either specific allergen or to anti-IgE antibody.

Nonreaginic antibodies directed against allergens also occur. High levels of such antibodies are usually found in the sera of patients who have received therapeutic injections of allergens, and low levels may exist in the sera of untreated allergic individuals. If allergens are mixed *in vitro* with sera containing nonreaginic antibodies specific for them, these allergens will be bound by the antibodies and will no longer produce positive wheal and flare skin tests characteristic of Type I hypersensitivity. Antibodies responsible for this *blocking* activity are usually members of the immunoglobulin G class. Immunoglobulin A antibodies in secretions may also possess significant blocking activity.

These latter observations form part of the rationale for *hyposensitization immunotherapy*. This procedure involves a series of parenteral injections of gradually increasing doses of the allergens to which the individual is allergic. During successful immunotherapy, typical immunological changes include both production of serum IgG blocking antibodies and inhibition of production of IgE antibodies directed against the allergens employed. All of the reported beneficial results of this procedure may not be attributable to these changes, however, since there is only a general association between these laboratory parameters and clinical improvement due to this type of therapy.

The nature of IgE

IgE has a sedimentation coefficient of 7.7S and a molecular weight of approximately 200,000. The IgE molecule is composed of four polypeptide chains held together by disulfide and noncovalent bonds. As is the case with the 7S immunoglobulins, the two light chains are either κ- or λ-chains, and the heavy polypeptide chains (ε-chains) are antigenically distinct for the IgE class. IgE is a divalent antibody with two attachment sites for the antigen in the Fab portion of the molecule. Passive transfer of reaginic activity to skin and other tissues is made possible by the affinity of the Fc fragment of the IgE molecule for receptor sites on tissue mast cell membranes. Similar receptor sites are also found on blood basophils. The latent period which is required in the PK test (between injection of allergic serum into the skin and subsequent skin test with the allergen) is thought to reflect a necessary interaction time between the Fc fragment of the IgE molecule and these membrane receptor sites. Heating an atopic serum at 56° C for 1 to 4 hours renders it inactive for PK transfer. The heat lability of reaginic antibodies is due to change in the

Fc fragment of the molecule which renders IgE antibodies incapable of sensitizing cells. The ability of the Fab portion of IgE antibodies to react with antigen is not heat labile. Although IgE molecules have been shown to activate complement *in vitro*, by the C3 bypass, complement is probably not involved in the production of IgE-mediated human disease.

IgE antibodies are synthesized mainly by lymphoid tissue in close proximity to the mucosal surfaces of the respiratory and the gastrointestinal tracts, and the ratio of IgE concentration to IgG concentration is higher in respiratory secretions than in serum. It has been suggested that IgE is spilled or secreted onto mucosal surfaces more rapidly than IgG. Data of this type are difficult to interpret, since it is also possible that differential rates of immunoglobulin digestion on mucosal surfaces may account for the observations. IgE is not actively transported across the placenta (as is IgG), nor has it been proven to be actively secreted onto surfaces (as is IgA).

Serum and tissue-fixed IgE are in equilibrium. However, the attachment of the IgE molecule to membrane receptor sites is relatively stable. Serum IgE has a half-life of approximately 2.3 days, whereas the half-time of persistance of tissue-fixed IgE has been estimated to be between 8.5 and 14 days. This is consistent with the observation that skin sites sensitized with sera of high PK titer may be reactive to challenge months after initial sensitization.

Serum IgE levels may be expressed in international units (IU) or in nanograms per milliliter. One IU currently equals 2 ng. A broad range of values is present in the population, ranging from zero to many thousand. IgE synthesis begins in the perinatal period. Serum levels tend to rise gradually until about puberty and often decline later in life. Factors other than age affecting serum IgE levels include racial background, geographic source, methodology for performing the test, and presence of various disease states. In attempting to use the IgE level to discriminate between persons who have atopic allergic processes and those who do not, allergy becomes increasingly likely with increasing IgE values. However, considerable overlap is present. For example, a serum IgE level of 100 IU/ml. or greater has been reported to occur in 24 per cent of normals and 80 per cent of asthmatics. At IgE levels of around 500 IU/ml. or greater, the probability of atopic allergy or some other IgE-related process exceeds 95 per cent, but a high percentage of atopic allergic patients do not have such high levels. Thus serum IgE levels show only a general tendency to reflect the nature and severity of atopic diseases. Patients with extrinsic bronchial asthma have higher IgE levels more frequently than do those with allergic rhinitis. Many allergic asthmatic patients maintain elevated IgE levels throughout the year, while patients with seasonal allergic rhinitis often develop elevated IgE levels during their allergy season with gradual return to normal after the season. Particularly high values have been reported in patients with atopic eczema where the magnitude of IgE elevation is related both to the extent of the dermatitis as well as to the presence of concurrent atopic respiratory disease.

IgE levels may be abnormal in a number of other conditions in which the role of the IgE antibody response has not been defined clearly. Bronchopulmonary aspergillosis, a pulmonary disease due to a combination of Type I and Type III hypersensitivity to intrabronchial fungal colonization, is also characteristically associated with extremely high levels of IgE. Elevated IgE levels have been described in association with intestinal parasitism, the Wiskott-Aldrich syndrome, cirrhosis of the liver, idiopathic pulmonary hemosiderosis, celiac disease, thymic hypoplasia, and cystic fibrosis. Normal levels are usually observed in nonallergic asthma, nonallergic forms of urticaria, and nonallergic nasal disease. Low levels are usually observed in childhood, in some patients with ataxia-telangiectasia, in many patients with various other immunological deficiencies, and possibly in carcinoma and after chronic corticosteroid therapy.

The imperfect correlation between IgE levels and the presence of atopic allergy permits several speculations. First, it appears that IgE may have other functions than the production of atopic disease. Second, it is probable that factors other than the serum level of IgE are important in the production of atopic disease. The level of serum IgE, although in equilibrium with tissue-fixed IgE, may be an inadequate reflection of the amounts and activity of the IgE antibody in target tissues. Alternatively, or perhaps additionally, there may be physiological control mechanisms which are not necessarily dependent upon the amount of IgE present, but which are important in determining whether or not an atopic disease will manifest itself. The function of the autonomic system seems particularly of potential importance in this regard.

Other factors in atopic disease

Even though they are not synonymous, both IgE levels and the occurrence of atopic disease appear to depend upon interactions between familial and environmental influences. The importance of the environment upon IgE levels is illustrated by the effects of helminthic infections or pollen seasons. The ability to produce IgE antibodies is controlled by regulatory genes which have been dubbed immune responsiveness (Ir) genes. Other genes have control over serum levels of IgE and other antibodies. Population studies in human beings and breeding experiments in animals indicate that the ability to mount an immune response to a specific antigen is inherited. The importance of genetic factors in the production of atopic disease has been indicated by a number of epidemiological surveys. Most have indicated that persons with a personal history of atopy have a much higher incidence of atopic diseases among family members than do nonatopic control subjects. A number of modes of inheritance for atopic disease have been proposed, but none has been established firmly. All hypothesize a variable rate of penetrance.

Much of the difficulty in establishing genetic mechanisms is due to the important determining influence of the environment on whether or not a given individual develops atopic disease. The amount of exposure to potential allergens is often a determining factor in whether or not atopy manifests itself. For example, foreign college students coming to school in areas of the United States where they have a more intense and prolonged exposure to highly allergenic pollens and fungi than in their home countries develop allergic rhinitis and asthma much more frequently than would have been predicted from their family histories. When exposed to a relatively low allergenic load, even a strong familial atopic predisposition may never become apparent.

Many of the antigens involved in atopic allergic reactions occur naturally in the environment. Pollens, fungal spores, house dust and house dust mites, foods, animal danders, and certain insects are included in this group. Others are developed and made available through science and technology; medication, chemicals, enzymes, heterologous serum proteins, and food additives can be classified in this category.

Type I sensitization may occur after introduction of the allergen by various routes including injection, ingestion, or inhalation. One theory on allergic sensitization (the mucosal permeability theory) suggests that persons predisposed to atopy have a defect in respiratory and gastrointestinal tract mucosa which allows increased absorption of allergens and which, in turn, increases ease of sensitization for production of IgE. The evidence is conflicting, but some support exists for this theory. It is clear, however, that differential mucosal permeability does not explain all of the differences between atopic and nonatopic individuals. Another recent hypothesis suggests that a defect predisposing to atopy involves regulatory T lymphocytes, which normally suppress IgE synthesis.

Just as sensitization to potential allergens can follow a variety of initial exposure routes, a variety of routes of reexposure may elicit allergic symptoms in the sensitive individual. The manifestations and location of the reaction following challenge seem to depend on dose and route of challenge as well as on individual "shock organ" sensitivity. Thus, two atopically sensitive persons who receive identical allergen challenges may react differently. The reaction may be confined to the site of administration (local reaction), or other tissues and organs may be involved (systemic or generalized reaction). For example, although inhalation of an atopic allergen by sensitive patients almost always produces respiratory symptoms, one subject might wheeze, another develop nasal symptoms and still another might develop urticaria in addition to the respiratory reaction. Injected or ingested allergens might produce gastrointestinal symptoms, a pure respiratory allergic response without gastrointestinal symptoms, urticaria, or anaphylaxis. The basis for this individual sensitivity of shock organs is not understood. Possibilities include differences in the amount of IgE fixed to various tissues, differences in type, amount, and target organ response to the mediators acting upon these target tissues, as well as variation in absorption, distribution, and metabolic fate of the allergens.

Just as the dose of allergenic material to which an individual is exposed may determine whether or not sensitization takes place, the sum total of allergens to which he is exposed will influence the intensity of symptoms in the already sensitized individual. Thus, according to this "total allergic load" theory, allergenic exposures are much like the straws that eventually broke the camel's back: all contribute to the total burden that the patient must bear and to the symptoms he suffers as a result of it. This concept does much to explain the inconstancy in some patients with which a given allergenic exposure produces symptoms.

Another source of variability in symptoms is the so-called priming effect. With continued daily exposure to an allergen, the allergic individual frequently becomes increasingly symptomatic. Thus, exposure to a small amount of ragweed pollen late in the ragweed season is likely to produce much more severe symptoms than an equivalent exposure early in the season. This priming effect seems to depend on local mucosal factors, the physiology of which is unclear. This effect is not dependent upon an enhanced state of systemic sensitization, as under controlled conditions, it can be produced in a restricted site, for example, in only one nostril.

Chemistry of atopic syndromes

When an allergen enters the tissues of a specifically sensitive atopic individual, it combines with IgE antibody which is fixed by its Fc fragment to the surface of mast cells or basophils. If these cells are so densely sensitized with IgE specific for the allergen that the allergen molecule can bridge between two IgE molecules, a chain of events begins that is capable of producing atopic symptoms. The second step is release of naturally occurring substances referred to as the chemical mediators of atopy. Their release and physiological effects are influenced by the autonomic system, the hormonal state of the individual, and by other naturally occurring mediators capable of antagonizing their action. The physiological effects of the chemical mediators on target tissues produce the anatomical and physiological derangements which are expressed as atopic disease states. The chemical nature and tissue effects of the three most firmly established mediators of atopic allergy are listed in Table 36–1. Involvement of a number of other mediators is strongly suspected, however.

Histamine

HISTAMINE. Histamine is found (among other places) stored in preformed state in the metachromatic granules of mast cells and basophils. In these cells, it is nonco-

TABLE 36–1
Chemical Mediators of Atopy

Mediator	Chemical Nature	Major Physiological Effects in Allergic Reactions
Histamine	$HC = C - CH_2 - CH_2 - NH_2$ with N and NH joined to CH ring	Stimulation of H-receptors leading to: a. Vasodilation b. Bronchoconstriction c. Stimulation of respiratory glands d. Stimulation of gastric secretion
Eosinophile chemotactic factor of anaphylaxis (ECF-A)	Tetrapeptide (mol. wt. $\simeq$ 380)	Attracts eosinophiles which may: a. Ingest and neutralize antigen-antibody complexes b. Modulate reactions to histamine through eosinophil-derived inhibitors c. Modulates reactions to SRS-A
Slow reactive substance of anaphylaxis (SRS-A)	C_2O unsaturated carboxylic acids (leukotrienes)	Produces a characteristic gradual onset, long-lasting bronchoconstriction

valently bound to Ca^{++} and heparin. When an allergen combines with IgE on the cell surface, a series of enzymatic events results in secretion of histamine into the extracellular environment. Histamine release, once thought to involve lytic rupture of cells, is now known to be an energy requiring process which can be blocked by antagonists of glycolysis. Cell viability is not impaired and the cells that "degranulate" in the secretory process subsequently regranulate. Histamine produces a rapid, brief local physiological effect upon vascular, bronchial, and other smooth muscles as well as on glandular structures. Histamine is rapidly degraded by tissue histaminases.

EOSINOPHIL CHEMOTACTIC FACTOR OF ANAPHYLAXIS. Eosinophil chemotactic factor of anaphylaxis (ECF-A) was difficult to isolate and characterize until a sufficient quantity was obtained, from the tumor of a patient with bronchogenic carcinoma which was synthesizing ECF-A ectopically. It appears to be the major factor responsible for accumulation of eosinophils at the site of Type I reactions in target tissues. The release of ECF-A seems to be influenced by the same mechanisms that control histamine release. The role of the eosinophil in the atopic reaction is far from clear. Evidence has accumulated that eosinophils are active in phagocytosis of antigen-antibody complexes, including those containing IgE and allergens. Fractions of sonicated eosinophils have been observed to inactivate or inhibit histamine in vitro, but it has yet to be proved that these eosinophil-derived inhibitors (EDIs) have biologic activity in vivo. At one time it was thought that chemotaxis of eosinophils into the atopic reaction site was mainly stimulated by histamine itself, but this no longer seems true. It has recently been shown that eosinophils contain arylsulfatase enzyme activity which appears potentially capable of inactivating slow-reactive substance of anaphylaxis SRS-A.

SLOW-REACTING SUBSTANCE OF ANAPHYLAXIS. Slow-reacting substance of anaphylaxis (SRS-A) is an important mediator of bronchial asthma. It has recently been demonstrated to be a product of the precursor substance *arachidonic acid*, which can be derived directly from the diet or from metabolism of the dietary essential fatty acid linoleic acid. Arachidonic acid is incorporated into the phospholipid in cell membranes in all tissues. SRS-A, as well as other biologically active substances, are generated from it through the lipoxygenase pathway. SRS-A has not been identified in preformed state but is generated after interaction of an allergen and IgE in sensitized lung tissues. In human lung tissue the cellular site of production is mast cells and basophils. The major physiological effect of SRS-A is smooth muscle constriction; its glandular and vascular effects are relatively minor. Factors initiating and controlling production of SRS-A from sensitized lung tissues have been similar to those involved in the release of histamine and ECF-A.

KININS. The role of kinins in production of the atopic syndromes is not clear. These vasoactive peptides are formed by enzymic cleavage of larger precursor peptides circulating in the plasma and their activation may be accomplished by various enzymes including activated Hageman factor, Pf/dil, plasmin, and various kallikreins (kininases). The major effect of kinins apparently is on blood vessels since bronchiolar smooth muscle contraction and enhancement of glandular activity are less pronounced. Active kinins have been identified in the plasma of patients undergoing a hypotensive anaphylactoid reaction to intravenous γ-globulin, and in perfusates from allergic rhinitis and urticarial cutaneous reactions. The liberation of an arginine esterase with kallikrein activity from IgE-sensitized basophils upon in vitro exposure to specific allergen has recently been reported. The general importance of kinins relative to other mediators of Type I hypersensitivity has not been clearly established.

PROSTAGLANDINS. Prostaglandins may be regulatory mediators in some immediate hypersensitivity reactions. Like SRS-A, they are metabolic products of arachidonic acid but are derived from a different pathway, the cyclooxygenase system. They are classified in series on the basis of chemical structure. Prostaglandins of the E and F series occur in human lung parenchyma and bronchial tissue. Their chemical activities suggest the possibility that they may be determinants of lung and bronchomotor function and may play a role in bronchial asthma. Prostaglandins E1 and E2 are bronchodilators while prostaglandin F2α is a potent bronchoconstrictor.

OTHER MEDIATORS. Other mediators of immediate hypersensitivity have been proposed, but their biological relevance in humans is not firmly established. For example, during such reactions mast cells and basophils can elaborate a *platelet activating factor*, suggesting involvement of pharmacologically active platelet products. Similarly, a neutrophil chemotactic factor may contribute a neutrophilic component to the classically eosinophilic pathological infiltrate. Knowledge of these reactions promises to grow increasingly complex.

Nature of control mechanisms

The production of atopic inflammation in tissues is initiated by the interaction between allergen and specific IgE fixed to surface of certain cells which include mast cells and basophils. This initiates the energy-requiring process to produce the chemical mediators of inflammation. As described above definite roles as Type I mediators have been ascribed to histamine,

SRS-A, and ECF-A. Roles have been suggested, but not conclusively demonstrated, for kinins and prostaglandins. It is the effect of these mediators on target tissues that produces allergic inflammatory changes; and intensity of the allergic inflammation produced is under the influence of control mechanisms that can operate at several sites. Some influence the amount and rate of release of chemical mediators. Others oppose or enhance the effect of mediators on target tissues. Still others act independently on target tissues to modify the prior action of the mediators to which the target tissues have been exposed.

As an example of the interaction between mediators, control mechanisms, and target tissues, respiratory smooth muscle function in bronchial asthma has been the best studied target tissue model. However, during the discussion of the control of smooth muscle function that follows, it should be remembered that additional target tissues are involved in the production of bronchial asthma. Alteration in mucosal glandular function with hypersecretion of mucus also occurs. Blood vessels are affected with the production of edema of the bronchial walls. Inflammatory cells are also seen in the bronchial tissues on histological examination. The control of bronchial smooth muscle tone has been selected as an illustrative example, because the pharmacology of its control mechanisms is best appreciated.

BRONCHOCONSTRICTION. Bronchomotor tone of the airways depends on a balance between constrictive and relaxant forces on respiratory smooth muscle. In normal nonasthmatic individuals without bronchial disease there is slight constant bronchoconstrictive tone. This is mediated by vagal parasympathetic forces, since it can be abolished by atropine or vagotomy. Normal individuals appear to have no chronically active relaxant forces at work; but in bronchial asthma, constrictive influences are hyperactive with inadequate opposing relaxant forces. This produces narrowing of the caliber of bronchial airways which may occur variably at different levels in the bronchial tree. When large bronchial airways are constricted, marked obstruction to air flow results which is reflected in increased resistance to air flow during pulmonary function testing. When the diameter of small airways is markedly reduced the lung becomes "stiff" and loses its compliance. Obstruction at this level may also contribute to poor oxygenation of blood as it passes through the pulmonary capillary bed. Mucus plugs are also felt to be important in producing this pulmonary arteriovenous "shunting." Variation is seen from patient to patient in the caliber of airways which are most severely affected during an asthmatic attack.

Several types of constrictive forces can affect airway smooth muscle during an asthmatic attack. After an allergen is inhaled, histamine and SRS-A are released, and these mediators and perhaps others (such as prostaglandin F2-α) can act directly on airway smooth muscle to produce bronchoconstriction. There is a rich cholinergic innervation to the tracheobronchial tree, and inhaled histamine has been shown to initiate vagal bronchoconstrictive reflexes. In human beings the extent to which the histamine, liberated by inhalation of allergen, stimulates vagal reflexes is controversial. Large doses of atropine seem to be necessary to produce a protective or reversing effect against the reagin-mediated bronchoconstriction which follows allergen inhalation. This appears to be to some extent species-specific, however, since atropine is very effective in blocking or reversing reagin-mediated bronchoconstriction due to allergens in dogs.

AUTONOMIC FACTORS. In many patients asthmatic episodes can be triggered by exposure to respiratory irritants. IgE is not mechanistically involved in these asthmatic episodes. In these patients respiratory irritants such as smoke, aerosolized citric acid, air pollutants, and aerosolized histamine stimulate afferent irritant receptors in the bronchial tree to produce "irritant" or "cough" reflexes which are bronchoconstrictive. Afferent and efferent impulses pass through the vagus nerves to the medulla and back to the lung. Acetylcholine is produced at efferent nerve terminals which is capable of interacting with specific membrane receptor sites both on mediator-producing cells and on smooth muscle cells. This results in increased production of inflammatory mediators as well as a direct constrictive effect upon airway smooth muscle. Atropine is very effective in blocking or reversing this type of reflex bronchoconstriction.

Relaxant forces on airway smooth muscle include endogenous and exogenous sympathomimetic agents under most circumstances as well as some drugs which are administered therapeutically for asthma. Most sympathomimetic agents (catecholamines) are capable of attaching to two types of membrane receptors, both of which are found on many cell types throughout the body. These adrenergic receptor sites for catecholamines have been designated α and β. Most sympathomimetics can interact with either α- or β-receptor sites but vary in their relative affinities for them. For example, both norepinephrine (primarily produced by sympathetic nerves) and epinephrine (from the adrenal medulla) can stimulate both α- and β-receptors. Norepinephrine has a predominant affinity for α-receptors, while epinephrine exerts somewhat more effect on β- than α-adrenergic receptors. Varying affinities of exogenous sympathomimetic drugs for these receptor sites has been documented. Isoproterenol is a very specific and potent synthetic β-adrenergic receptor stimulant; and methoxamine is an almost pure α-adrenergic receptor stimulant.

Most sympathomimetic influence on

pulmonary function is mediated by circulating epinephrine and exogenously administered catecholamine drugs. Sympathetic innervation of the lung is sparse, although some has been described which may serve to reduce pulmonary vagal reflexes. Within the bronchial tree, α-adrenergic receptors on smooth muscle and probably mast cells are apparently vastly outnumbered by β-receptors. It is for this reason that under usual circumstances most sympathomimetic agents exert their predominant effect through stimulation of the β-adrenergic receptor sites. Stimulation of β-receptors on mediator-producing cells causes reduction in release of inflammatory mediators; their stimulation in bronchiolar smooth muscle cells causes direct bronchodilation. Under special experimental laboratory conditions effects of α-stimulation of these tissues can be revealed in the absence of opposing β-stimulation. Exposure of tracheal smooth muscle *in vitro* to the pure α-stimulant methoxamine will cause contraction of the muscle strip. Similar effects can also be seen when the β-stimulating effects of epinephrine are blocked by a β-receptor blocking agent such as propranolol; under these circumstances the unblocked α-stimulant activity of epinephrine causes constriction of the muscle strip. Using such maneuvers to produce unopposed α-adrenergic stimulation on mediator-producing tissue, the effect is to increase the production of mediators. Thus, the effect of α- and β-adrenergic receptor stimulation on both mediator production and directly on smooth muscle tone are in opposition to each other; however, under physiological conditions in the intact lung, the alpha effects of most sympathomimetic drugs are not visible.

ROLE OF ADENYLATE CYCLASE SYSTEM. The effect of cholinergic and sympathomimetic agents, as well as those from many asthma drugs, are mediated by changing the intracellular levels of cyclic nucleotide compounds in both smooth muscle and media-

tor producing tissues. The levels of the cyclic nucleotides are determined both by the rate of their formation and their degradation. When adrenergic drugs act upon membrane β-receptor sites on these cells they stimulate the receptor-associated enzyme adenylate cyclase to increase production of the cyclic nucleotide, 3'5' cyclic adenosine monophosphate (cAMP), from ATP. When levels of cAMP within smooth muscle cells rise this leads to muscle relaxation; and when (cAMP) is increased in mediator-releasing cells, the rate of release of the mediators is reduced. The effect of unopposed α-adrenergic receptor stimulation in these cells is due to reduction in (cAMP) levels. Methylxanthines, which include drugs of the theophylline category, result in elevated levels of cAMP by opposing its degradation. These drugs inhibit the phosphodiesterase enzyme from causing a breakdown of 3'5' cAMP to 5' adenosine monophosphate. Theophylline drugs are among the most clinically useful bronchodilators. They have been found to be synergistic with β-adrenergic receptor-stimulating drugs in producing bronchodilatation. This synergism has been explained by the property of these drugs to act through different pathways to increase intracellular levels of cAMP in smooth muscle cells and mediator-producing cells. Prostaglandins act through membrane receptor sites separate from adrenergic receptors and have been shown to be capable of altering intracellular cAMP levels *in vitro;* whether or not this phenomenon is physiologically significant in the production of bronchial asthma is not known at present.

Acetylcholine and other cholinergic drugs act through receptor sites on bronchial smooth muscle cells and mediator-producing cells to stimulate the receptor-associated enzyme guanylate cyclase, which leads to the production of another cyclic nucleotide, cyclic guanosine monophosphate (cGMP) from guanosine triphosphate. Elevation of cGMP in mediator-producing cells enhances mediator release. Its effects in smooth muscle cells have not been clearly identified.

Alterations in the rate of release of histamine, SRS-A, and ECF-A (with the agents so far studied) have not always been in parallel, and the rate of release of these mediators therefore seems likely to be governed by multiple control mechanisms.

ADRENAL STEROIDS. Glucocorticoids are known to be effective clinically in the treatment of bronchial asthma. While the modes of action of these drugs are not fully understood, they have an antiinflammatory effect ascribed at least partially to stabilization of lysosomal membranes. In addition, these compounds act in an as yet ill-defined manner to raise cAMP levels, possibly by "sensitizing" β-adrenergic receptors to the action of β-receptor agonists. It has also been suggested that glucocorticoids may inhibit ATPases, thereby increasing substrate availability for the adenylcyclase enzyme.

Although the action of thyroid hormone on adenylate cyclase in mediator-producing tissue has not been defined, its action in other tissues is mediated through adenylate cyclase. The clinical observation that bronchial asthma is exacerbated in the face of hyperthyroidism may be related to an effect of thyroid hormone on adenylate cyclase in mediator-producing tissues, bronchial smooth muscle, or both.

Clinical Problems

Atopic syndromes are allergic conditions that occur in certain individuals, owing at least in part to a hereditary tendency to produce IgE antibody in response to exposure to small quantities of naturally occurring antigens. The usual clinical atopic syndromes include allergic rhinitis, extrinsic bronchial asthma, and some cases of urticaria. Anaphylaxis and atopic dermatitis are also included in this group, even though the role of heredity in anaphylaxis is less clear than with the other atopic syndromes, and the direct involvement of tissue-fixed IgE antibodies in the development of the local skin lesions of atopic dermatitis is doubtful.

ALLERGIC RHINITIS. When the mediators of atopic reactions are released in or adjacent to the nasal mucosa, the symptoms of allergic rhinitis result. The differential diagnosis of nasal allergy includes other disorders that produce nasal congestion or stuffiness, fairly constant symptoms of this condition. Although not all patients with allergic rhinitis sneeze excessively, when a history of frequent attacks of repeated sneezing is obtained, the causative process is almost always (although not invariably) an atopic allergic process. Excessive rhinorrhea and itchy, watery eyes are also common. Ten to fifteen per cent of the population of the United States at some time manifest allergic nasal symptoms.

ASTHMA. Bronchial asthma is the term applied to a syndrome in which wheezing and obstruction to bronchial air flow are produced by spasm of bronchial smooth muscle, edema of the bronchial wall, and excessive secretion by mucosal mucus glands with frequent consequent formation of mucus plugs. In "pure" bronchial asthma, the disease process is potentially completely reversible. The word "potentially" is important here, since an attack of asthma is not necessarily *easily* reversible. However, complete reversibility is possible, because asthma itself does not lead to irreversible structural changes in the lung, unless some other complicating condition has been superimposed. Over 2 per cent of the population of this country have asthma.

Bronchial asthma has been divided into extrinsic, intrinsic, and mixed categories.

EXTRINSIC ASTHMA. About two-thirds of asthmatic patients wheeze largely because of immediate hypersensitivity to environmental allergens, and thus fall into the extrinsic or mixed categories. The majority of these patients began to wheeze before the age of 30, often during childhood. The major offending allergens at any age are usually inhalants, but foods are occasionally incriminated, particularly during the first few years of life. Many of these patients also have allergic rhinitis. There is a high incidence of atopy in the families of these individuals. Skin tests and careful allergy histories are frequently helpful in identifying the allergens that are producing symptoms; and IgE levels are elevated in about two-thirds of these patients. Although their asthma may be chronic and unremitting, very often it is paroxysmal. The prognosis in these patients is good, especially with well-designed hyposensitizing immunotherapy. Patients with extrinsic and mixed bronchial asthma are considered to be atopic individuals.

INTRINSIC ASTHMA. In approximately one-third of the asthmatic population, no causal role for IgE or allergens can be identified. Such patients have been commonly described as having intrinsic asthma, implying that the major etiological defect in this type of asthma is related to the constitution of the patient, rather than to interaction between patient and environment. While this premise may (or may not) be correct, and direct proof of causation is lacking, the term *idiopathic* is preferred over *intrinsic* by many allergists. Others subdivide the class of intrinsic asthma into idiopathic and infectious subcategories, since respiratory infections (particularly viral) are common initiators of asthmatic flares which appear to occur through non-reaginic mechanisms. Rather than engage further in a semantic discussion, the term *intrinsic asthma* will be employed here because of its current wide usage for those patients with asthma in whom reaginic mechanisms cannot be incriminated.

Although intrinsic asthma (like extrinsic asthma) can begin at any age, the first attack of wheezing typically begins during adult life. In some patients the attack may be precipitated by a respiratory infection. A personal and family history of other atopic problems is no more frequent in patients with intrinsic asthma than it is in the general population. They have an infrequent history of allergic rhinitis, but sinus disease, often with mucous polyp formation, is common. Skin tests and medical history generally do not point to the involvement of allergens in symptom production. The most important identifiable triggering stimuli are either respiratory infections or nonspecific irritants such as smoke and other mucosal irritants. IgE levels are generally normal in intrinsic asthmatics. Although the symptoms may be paroxysmal they show a discouraging tendency to become chronic and persistent in many of these patients. The prognosis for remission is poor, with only about 10 per cent showing significant spontaneous improvement. Hyposensitization does not seem to be of any value in intrinsic asthma.

MIXED ASTHMA. Mixed bronchial asthma is the diagnosis applied when the asthmatic process of an individual is prominently triggered both by allergens and nonallergen stimuli. A common pattern in children is that of allergic asthma which also flares with viral respiratory infections; as these children grow older, infections usually become less important as triggers. In contrast, in middle-aged adults with mixed asthma, allergens may gradually assume less importance, and the asthmatic disease becomes "more intrinsic" with advancing age.

Some asthmatic patients develop severe bronchospasm following the ingestion of acetylsalicylic acid and occasionally other analgesics (including indomethacin and opiates), tartrazine yellow dye (found in many foods and drugs), and food preservatives derived from benzoic acid. This occurs most frequently with intrinsic asthmatics with nasal polyps (aspirin triad) but is not confined to them. No immunological basis has been demonstrated and pharmacological idiosyncrasy seems the most likely cause at present. Patients with intrinsic asthma who are aspirin-sensitive often have a particularly severe asthmatic process.

Since IgE is not involved and since there is no clear-cut familial predilection in intrinsic asthma, it is not considered to be an atopic syndrome. It is produced, however, by the same or similar chemical mediators as those involved in extrinsic asthma. Since the sequence of events leading to mediator release in intrinsic asthma is not initiated by demonstrable IgE and allergen interaction, mediator release is probably the culmination of a common pathway which can be activated either by immediate hypersensitivity reactions in extrinsic asthma, by unknown events in intrinsic asthma, or by both reaginic and idiopathic mechanisms in mixed bronchial asthma.

Bronchial asthma must be differentiated from a long list of other conditions that can also cause wheezing. The list includes multiple causes of compression, constriction, inflammation, or obstruction of the airways. In making the differential diagnosis, an examination of the bronchial mucus or the nasal secretions for eosinophils is often helpful. The patient with respiratory atopic disease or an asthmatic process usually has an eosinophilia of at least 25 per cent in the respiratory secretions if: (1) the respiratory process is currently active, (2) a respiratory infection is not causing cells other than eosinophils to predominate, and (3) corticosteroid therapy has not abolished it. Eosinophils are found in the bronchial mucus in both intrinsic and extrinsic asthma. This helps to differentiate asthma in general from other conditions that cause wheezing, but it does not differentiate between reaginic and nonreaginic asthma.

URTICARIA. It has been estimated that as many as 20 per cent of the population will at some time during their lives have an attack of urticaria, with or without angioedema. Urticaria are similar in appearance to wheal and flare skin tests. The wheals are produced by localized vascular permeability and edema in the dermis; the surrounding flare is due to vasodilatation. Urticaria (also called hives or whelps) are usually pruritic. In contrast, angioedema is due to the same type of process but is less localized, involves deeper tissues, and has predilection for sites where overlying skin or membranes are thin (e.g., hands, feet, face, and oropharyngeal cavity).

Clinically, it is useful to subdivide urticarial processes on the basis of the duration of the problem into acute and chronic categories. Patients with chronic urticaria are those who have persistent or frequent problems for longer than 6 to 8 weeks. It is important to realize that not all episodes of urticaria are due to atopic allergy, although a high percentage of the acute variety seem to involve this mechanism. In retrospect, patients whose hives fall into the acute category do not become chronic, either because the event that caused their hives was self-limited (e.g., an acute viral infection) or because a cause was recognized (frequently a drug or food allergen) and subsequently avoided.

In patients with chronic urticaria, the attacks of hives are a continuing problem, either because an allergen has not been identified and eliminated from the environment or, more often, because the urticaria is not due to immediate hypersensitivity mechanisms at all. This is not to suggest that a careful history in search of an allergen should not be undertaken. However, when the history does not suggest that allergy is involved (particularly in older and middle-aged patients with no previous personal or family history of atopic allergy), the yield of useful information from extensive allergy skin tests is discouragingly low, and many allergists choose not to do them routinely in such cases.

In addition to allergy to foods and medications, urticaria can be produced by inhaled allergens. In most of these patients respiratory allergic symptoms coincide with the urticaria, although rare exceptions have been reported. Urtication may also follow exposure to a physical stimulus in susceptible individuals. Reported causes have included cold, heat, vibration, pressure, sunlight, water, and exercise; heterogeneous mechanisms are involved. Emotional stress is a factor in some patients, but probably not in the majority. Dermographism, a tendency to develop wheals at the site of very slight cutaneous trauma, must be differentiated from urticaria, since it results from a different nonurticarial process.

Occasionally chronic urticaria occurs in patients with an associated underlying disease. Examples include both infectious and noninfectious processes. Intestinal parasitism should be excluded. Other diagnoses to consider include collagen vascular diseases, occult infections, urticaria pigmentosa, systemic mastocytosis, large necrotic neoplasms, Hodgkin's Disease, and disorders associated with abnormal serum proteins such as cryoglobulins and cold agglutinins. However, in patients with chronic urticaria where a careful initial history and physical examination do not suggest any associated process and where screening laboratory tests for underlying disease are all normal, it is unlikely that any serious associated disease will later become manifest. Even after a thorough evaluation, the cause of most patients' chronic urticaria is not found.

ANAPHYLAXIS. Anaphylaxis is a systemic allergic reaction, usually severe and potentially life threatening. It most typically involves one, several, or all of the following organ systems: skin, respiratory tract, cardiovascular system, and gastrointestinal tract. In fatal reactions, death commonly occurs in one of three ways: shock due to cardiovascular collapse, asphyxiation due to angioedema of the upper airway, or severe asthmatic bronchiolar obstruction. Manifestations such as urticaria, itching, nausea, diarrhea, and miscellaneous other symptoms may coexist, but they rarely pose a threat to life. Most instances of anaphylaxis are due to immediate hypersensitivity to drugs, especially penicillin. Other occasional causes include foods (particularly eggs, nuts, berries, fruits, and seafood), stinging insects of the order Hymenoptera, and a variety of miscellaneous causes.

Although anaphylaxis is probably more common and may be more often severe in individuals with a history of atopic respiratory disease, it is by no means confined to them. Since anaphylaxis can occur by immediate hypersensitivity mechanisms in people who have no other discernible familial or personal predisposition toward atopy, the inclusion of anaphylaxis due to immediate hypersensitivity mechanisms as one of the atopic syndromes may be somewhat questionable. Furthermore, some cases of anaphylaxis are probably due to mediator release resulting through nonimmunological mechanisms (analogous to intrinsic asthma). This is thought to be so in many of the reactions to radioiodinated contrast material, for example.

ATOPIC DERMATITIS. Atopic dermatitis is an eczematous skin condition often beginning during the first year of life. A high percentage of affected children later develop respiratory atopic disease. However, the dermatitis itself is probably not allergic in nature, or at least usually involves other more important factors. Careful topical skin care, treatment of pruritus, and avoidance of scratching have long been recognized to be of more benefit than allergy-testing and hyposensitization.

Diagnosis

Diagnosis of an atopic syndrome and identification of its etiology are based largely upon history. Since symptoms may be paroxysmal, at times a presumptive diagnosis must be made from the description of typical allergic symptoms by an asymptomatic patient. If currently symptomatic, the physical examination and such procedures as pulmonary function testing and examination for eosinophils in the secretions may assist in diagnosis. If the allergen is an inhalant (pollen, house dust, mold spore, or animal dander) positive skin tests correlating with the history are generally found. In the case of food and drug allergens skin tests often do not give reliable results, and food allergy is best diagnosed by temporary elimination of the suspected food from the diet and subsequent dietary challenge with observation for correlation of symptoms. When drug allergy is strongly suspected, the drug is simply avoided if possible, even though a conclusive diagnosis of allergy to that drug may not have been made.

Presumptive confirmation of the role of a suspected allergen may be achieved by observing improvement when exposure to it is avoided or reduced. In occasional cases, bronchial challenge by inhalation of a suspected allergen is used under controlled circumstances with asthmatics. This is not done routinely by most allergists because multiple allergens are often suspected, nonspecific responses are frequent, and occasionally severe bronchospasm may be precipitated in asthmatics. Overinterpretation of allergy skin tests is a common pitfall. A positive skin test to an inhalant does not necessarily prove the presence of clinically significant hypersensitivity. Clinical correlation with the history is very important in making this judgment. Twenty to twenty-five per cent of the general population have some positive allergy skin tests, but many never develop an atopic disease process. In those who do, symptoms typically follow exposure to some but not all of the agents to which skin test positivity can be observed.

Treatment of atopic diseases

Therapy of the atopic disorders may be discussed under three categories: avoidance, drug treatment, and hyposensitization.

Avoidance

If the cause or causes of an atopic disease can be correctly identified and subsequently avoided without undue inconve-nience to the patient, this is the preferred mode of treatment. In the case of allergy to a food, animal dander, or medication, total avoidance can often be achieved and may be the only form of therapy necessary. On the other hand, in the case of allergy to most airborne inhalant allergens, complete avoidance is rarely possible. Although the use of air conditioning and filtration may be of benefit in pollen allergy, and vigorous housecleaning may help a patient who is allergic to dust, some degree of continued allergen exposure is unavoidable. Furthermore, when efforts to avoid such agents are carried to extremes, they may unnecessarily disrupt the life of the patient. Often, avoidance must be supplemented by drug therapy and/or hyposensitization.

Drug therapy

A variety of drugs are active in reducing allergic symptoms. Their intelligent use depends on an understanding of the basic mechanisms involved in the atopic disorder. The most useful drugs for the treatment of allergic rhinitis are antihistamines. Antihistamines *do not* affect the release of histamine. Rather, they compete with histamine for histamine receptor sites on cell membranes to inhibit the effects of histamine on target tissues. A side effect that all antihistamines seems to share in varying degrees is that of producing drowsiness in most patients. An additional therapeutic effect may be obtained by adding an oral "decongestant" such as phenylephrine or ephedrine to the antihistamines. Such agents have prominent α-adrenergic effects which are capable of decreasing nasal edema by producing vasoconstriction.

The antihistamines are not particularly useful in the treatment of bronchial asthma. Although occasional patients (usually children) report lessening of bronchospasm after antihistamines, this is usually a minor and infrequent effect. In some patients antihistamines may worsen asthma by causing the bronchial secretions to become dry and inspissated, thus decreasing the effectiveness of the cough.

Adrenergic drugs with prominent β activity have been used with success in bronchial asthma. Ephedrine is an orally administered drug which once was widely used and was found suitable for regular administration in the prevention and treatment of bronchial asthma. Epinephrine is frequently used by injection and exerts a rapid bronchodilatory effect of short duration. Both ephedrine and epinephrine have both α- and β-adrenergic effects. A prominent α effect is their vasoconstricting activity, which increases peripheral vascular resistance and may elevate the blood pressure. Furthermore, their β activity is not confined to the bronchial tree. In addition to reducing mediator release and producing direct bronchial smooth muscle relaxation, they affect smooth muscle in a variety of other organs and are cardiac stimulants. Isoproterenol for asthma is most frequently used by inhalation for asthma. Recently a number of new adrenergic drugs have been developed which have mainly β activity for producing bronchial smooth muscle relaxation but with less stimulant effect for the heart than isoproterenol. Some of these newer drugs are albuterol, metaproterenol, isoetharine, and terbutaline.

The methylxanthines are a class of bronchodilating drugs which act through receptors other than adrenergic receptors. Their action has been discussed above. Side effects may include GI, CNS, and cardiac disturbances, usually due to excessive dose.

Disodium cromoglycate (cromolyn) is a relatively new drug in the treatment of asthma. It is relatively insoluble and is administered as a powder by inhalation. Approximately 10 per cent of the powder reaches the bronchial tree where it acts locally. Although the mechanisms of its action are unknown, it prevents release of chemical mediators in some fashion which does not affect cAMP levels. It has no direct action of its own on bronchial smooth muscle, glandular tissue, or blood vessels, and it appears to exert its major, or perhaps sole, action through regulation of mediator release. It is used as a preventive therapeutic agent for asthma and is not active in the reversal of an acute flare.

Corticosteroids are extremely valuable in the control of asthma that has been refractory to other forms of treatment. Because of the side effects of this class of drugs their systemic administration is reserved for situations in which other therapy has not been sufficiently effective. Large amounts given in single to multiple doses daily are often required to control severe asthma flares. When corticosteroids are needed to maintain control, many patients respond to a single alternate-morning dose of a short-acting steroid, or to inhalation of one of the new poorly absorbed steroids with fewer or even no long-range side effects.

Urticaria and angioedema may often be controlled or prevented simply by oral antihistamines. Hydroxyzine seems to be particularly useful in the treatment of urticaria. When antihistamines alone do not suffice, the addition of an oral decongestant may also be of benefit. In clearing an acute episode of urticaria, injectable epinephrine is very effective. In the case of angioedema which involves the airway, measures must be taken to ensure that closure of the upper airway with asphyxiation does not occur. Injectable adrenalin should be given at once. If closure of the airway seems imminent then attempts may be made to insert an artificial airway into the trachea. However, edema of the surrounding tissues often makes this difficult or impossible, and sometimes tracheostomy must be carried out surgically as a life-saving measure. Although corticosteroid therapy for chronic urticaria is usually effective, it should be avoided in all but unusual cases because of the potential side effects.

Hyposensitization

Hyposensitization immunotherapy is sometimes useful in some of the atopic disorders where avoidance of allergens and symptomatic treatment with drugs does not give sufficient relief. It is primarily employed in the treatment of allergy to airborne inhalants such as dust, pollen, and molds and in the prophylactic treatment of patients who have had systemic reactions to stinging insects. It is not appropriate for the treatment of allergy to foods or medications. Hyposensitization involves repeated injections, given at weekly to monthly intervals, of the allergens thought to be of clinical importance. The initial dosage level is very small with gradual increments made in the dosage as the series of injections is continued. To some extent the clinical response is dependent upon dosage of allergen administered. Hyposensitization has been shown to be effective in double-blind studies with allergens such as house dust and weed and grass pollens. In such studies 65 to 90 per cent of hyposensitized patients have fewer symptoms of respiratory allergy, while control subjects show improvement in 20 to 30 per cent of patients.

During the process of hyposensitization levels of IgE antibody directed toward specific allergens are often reduced, and the seasonal increase in total IgE levels, which otherwise follows a pollen allergy season, is less pronounced or absent. In addition, treated patients produce antibodies of the IgG class which are directed toward the allergens. These IgG antibodies have been called blocking antibodies, because they can inactivate the skin test activity of allergenic extracts when they are mixed together prior to being injected into the skin. Blocking antibody levels correlate roughly with clinical improvement related to hyposensitization. In addition, blood basophils from allergic patients become less sensitive to allergen (i.e., larger amounts of allergen are required to cause these cells to release their histamine). This reduction in cell sensitivity to allergen presumably reflects reduction in the amount of allergen-specific IgE which sensitizes the cell membrane, but additional mechanisms may be involved. The cells of some patients become completely unresponsive, and such patients are usually asymptomatic upon allergen exposure.

Although the reduction in IgE antibodies, the formation of IgG antibodies and the reduction in cellular sensitivity to allergen are presumably involved in the clinical improvement which may follow hyposensitization, it is by no means clear that the clinical response can be totally ascribed to these factors. It is possible that hyposensitization has some further unrecognized effects which may contribute to the clinical result. With the present rate of discovery in this field, such mechanisms may be uncovered in the near future.

Annotated references

Austen, K. F., and Becker, E. L. (eds.): Biochemistry of the Acute Allergic Reactions, Second International Symposium. Oxford, Blackwell Scientific Publications, 1971. (A useful reference on the chemistry of allergic diseases.)

Austen, K. F., and Lichtenstein, L. M.: Asthma, Physiology, Immunopharmacology and Treatment. New York, Academic Press, 1973. (A symposium with 70 well known contributors and participants.)

Lieberman, P., and Patterson, R.: Immunotherapy for atopic disease. Adv. Intern. Med. 19:391, 1974. (A review of immunotherapy concerning historical background, efficacy, and mechanisms of action.)

McCombs, R. P.: Diseases due to immunologic reactions in the lungs. N. Eng. J. Med. 286:1186, 1245, 1972. (A short, concise review of respiratory allergic diseases.)

Middleton, E.: Autonomic imbalance in asthma with special reference to beta adrenergic blockade. Adv. Intern. Med. 18:177, 1972. (A review concerning the classification and pathophysiology of asthma.)

Middleton, E., Reed, C. E., and Ellis, E. F. (eds.): Allergy. Principles and Practice. Parts I and II. St. Louis, C. V. Mosby, 1978. (A very comprehensive text, informative even to the allergy specialist.)

Mullarkey, M. F., and Webb, D. R. (eds.): Clinical Allergy. Med. Clin. North Am. 65(5):941–1107, 1981. (A useful update on some aspects of clinical allergy.)

Nickolson, D. P.: Extrinsic allergic pneumonitis. Am. J. Med. 53:131, 1972. (An interesting article on the nature and importance of hypersensitivity pneumonitis.)

Patterson, R.: Allergic Diseases, Diagnosis and Management. Philadelphia, J. B. Lippincott, 1972. (A fairly current textbook on the scientific basis of medical diagnosis and management of allergic diseases.)

Patterson, R. (ed.): Allergic Diseases: Diagnosis and Management, 2nd ed. Philadelphia, J. B. Lippincott, 1980. (A basic textbook.)

Patterson, R., Fink, J. N., Pruzansky, J. J., Reed, C., Roberts, M., Slavin, R., and Zeiss, C. R.: Serum immune globulin levels in pulmonary allergic aspergillosis and certain other lung diseases with special reference to immunoglobulin. E. Am. J. Med. *54*:16, 1973. (A study of serum IgE levels in patients with various respiratory allergies.)

Pepys, J.: Immunopathology of allergic lung diseases. Clin. Allergy *3*:1, 1973. (A review of the pathology and pathophysiology associated with respiratory allergic diseases.)

Samter, M. (ed.): Immunological Diseases. Parts I and II, 3rd ed. Boston, Little, Brown & Co., 1978. (A text with a wide scope that does not confine itself to allergy.)

Sheldon, J. M., Lovell, R. G., and Matthews, K. P.: A Manual of Clinical Allergy, 2nd ed. Philadelphia, W. B. Saunders, 1967. (A textbook concerned primarily with the methods of practice of allergy, slightly out of date.)

Stevenson, D. D., Mathison, D. A., Tan, E. M., and Vaughan, J. H.: Provoking factors in bronchial asthma. Arch. Intern. Med. *135*:777, 1975. (A clinical study of the multiplicity of stimuli that can exacerbate asthma.)

Weiss, E. B., and Segal, M. B., (eds.): Bronchial Asthma: Mechanisms and Therapeutics. Boston, Little, Brown & Co., 1976. (A thorough discussion of many aspects of asthma.)

Wittig, H. J., Belloit, J., DeFillippi, I., et al.: Age-related serum immunoglobulin E levels in healthy subjects and in patients with allergic disease. J. Allergy Clin. Immunol. *66*:305, 1980. (A clinical study.)

37

Autoimmune Diseases and Mechanisms of Immunological Injury

Morris Reichlin, M.D.

Immunological injury can result from any of the mechanisms of immunologically mediated hypersensitivity previously presented (Gell-Coombs classification) or from a newer group of less well understood immune mechanisms. Knowledge of these phenomena can explain immunologically mediated tissue damage whether the antigen challenge is exogenous (foreign) or endogenous (originating within the self). This chapter focuses on mechanisms of immunological injury and two aspects of autoimmunity: (1) potential etiological factors and (2) description of several prototype autoimmune diseases for which pathophysiological mechanisms are becoming clearer.

Mechanisms of Immunological Injury

Immunological injury to cells and tissues results from mechanisms involving both antibodies and cellular factors. Increasing knowledge in this area emphasizes that diverse pathways exist which may act independently but as often interact in either a cooperative or antagonistic fashion. The following paragraphs detail some aspects of our knowledge of these mechanisms.

The role of antibody

Five classes of immunoglobulin are synthesized by plasma cells and lymphocytes, but only IgG and IgM appear to play important roles in immunological injury associated with the autoimmune diseases. IgM and three of the four subclasses of IgG can activate the classical complement sequence to initiate immune adherence, release of vasoactive substances and anaphylatoxins, opsonization for phagocytosis of foreign molecular complexes and cells, activation of the clotting mechanism, and enzymatic damage of cell membranes, which can result in cell death (Fig. 37–1). In this type of damage, specific antibody serves to initiate the subsequent sequence of reactions, and complement serves as a biological amplification system. The degree of complement-mediated damage is related to the density of the involved antigens on the cell membrane and to the type of antibody reacting with these antigens. Generally, IgM is a more potent activator of the classical complement system than IgG, but IgG is biologically more important in these conditions.

In recent years, the alternative pathway of the complement system has been shown

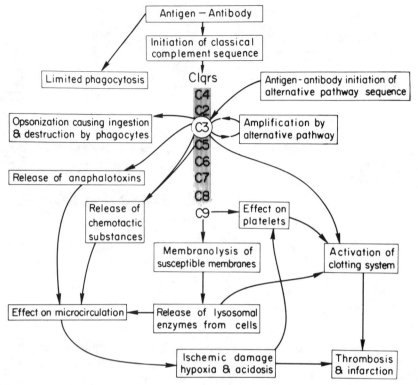

FIG. 37–1. Sequence to show that antibody-initiated injury occurs predominantly through activation of the complement system.

to participate in many immunological reactions, and function of the alternative pathway has been shown to be abnormal in at least one autoimmune disorder. However, the role of the alternative pathway in causing immunological injury in autoimmune diseases has not been established.

Specific antibody can also inactivate enzymes and may increase the removal of particulate matter by the reticuloendothelial system, even in the absence of complement.

Cell-mediated damage

Immunological damage can be caused by sensitized thymus-dependent lymphocytes (sometimes called killer cells or T_k cells) and is initiated by the direct interaction of specific receptor sites on the cell membrane of lymphocytes and an antigen on target cells, such as those of a transplanted organ. The resulting release of en-

zymes and other cytotoxic substances from the lymphocytes can cause death of target cells (Fig. 37–2). The exact biochemical sequence of this process is still poorly understood, but does not involve complement. The damage to target cells by specific immune lymphocytes can be blocked by masking antigenic sites on target cells with specific antibody or by binding receptor sites on the immune lymphocytes with free specific antigen or antigen-antibody complexes. This important process is known as efferent immunological enhancement. Enhancement is more likely to occur when: the cell membrane is resistant to complement lysis; the specific antibody binding with the antigenic sites on the cell membrane does not activate the classical complement sequence (as is the case with IgA, IgD, IgE, and IgG-4 antibodies); the density of antigens on the surface of the target cells is low; or the antigens are located improp-

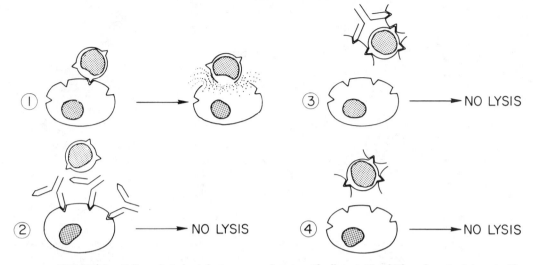

FIG. 37–2. Cell-mediated cytolysis occurs when specifically sensitized T lymphocytes interact with cell membrane antigens of the target cell (*1*). However, this reaction can be prevented by masking of antigenic sites on the cell membranes with antibody (*2*), complexes of antigen and antibody (*3*), and free antigen (*4*). Thus, inhibition of cell-mediated lysis can result from interference at either the antigenic site on the target cell or the antigen receptor site on the sensitized lymphocyte.

erly to cause activation of a sufficient amount of complement to cause lysis. Specific IgG-enhancing antibody has been shown to have a very important protective role in both tumor grafts and normal tissue grafts.

Thymus derived cells can also participate in cell-mediated damage through other mechanisms. One of these is by the synthesis and release of various mediators called lymphokines, initiated by stimulation of sensitized cells by specific antigen. The lymphokines have a variety of effects, one of which is to activate macrophages, which in turn play a role in recognition and removal of foreign antigens, whether or not they are cell associated.

Cellular-humoral cooperation

A synergistic action of damage caused by circulating antibody (IgG) and nonsensitized lymphocytes has also been observed. This process is known as antibody-dependent cell-mediated cytolysis (Fig. 37–3). It is now beginning to be recognized as an important contributing factor in histocompatibility differences, but the pres-

ence of this type of antibody is not detected in the usual direct cytotoxicity test.

Possible mechanisms of autoimmune disease

In recent years it has become clear that immune regulation is controlled by discrete subpopulations of T-lymphocytes identified by cell surface markers recognized primarily by monoclonal antibodies as described in the introduction. As is apparent from a perusal of Appendix VIIIB, autoimmune diseases are largely antigen specific. For example, various thyroid antigens (thyroglobulin, microsomes), acetylcholine receptors, red cell surface antigens, and a family of deoxyribonucleic acid (DNA) and ribonucleic acid (RNA) protein macromolecules are the target of autoimmune attack in Hashimoto's thyroiditis, myasthenia gravis, autoimmune hemolytic anemia, and systemic lupus erythematosus (SLE) respectively. Such specificity raises the question of deficiency of antigen specific regulatory T-cell populations. Gross deficiencies of functional suppressor activity and reduced numbers of T5,8+ (cyto-

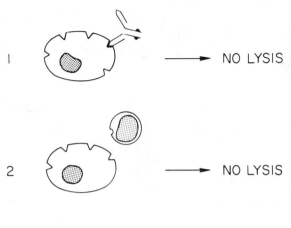

FIG. 37–3. Antibody-dependent cell-mediated cytolysis. Nonimmune lymphocytes can cause immunological damage of target cells by activation initiated by combination of an antigen-antibody activated F-c portion of an antibody molecule with an F-c receptor site on the lymphocyte.

toxic-suppressor populations) lymphocytes have been found in SLE patients. A plausible but thus far unsupported hypothesis in SLE involves a deficiency of DNA and RNA protein antigen specific suppressor clones of T lymphocytes. Such a mechanism might be operative in any of the antigen-specific autoimmune diseases. A deficiency or excess of antigen-specific T lymphocytes can contribute to human disease as evidenced by the reduced number of ragweed-specific T-suppressor cells in ragweed-sensitive hay fever sufferers and the increased number of lepromin specific T-suppressor cells in the lepromatous form of leprosy.

Another possible locus of immune dysregulation is the B-cell itself. The former view of B-cell dysfunction in autoimmunity involved the forbidden clone, a cell population with specificity to self-antigens which either escaped elimination in fetal life or arose in adult life as a result of somatic mutation in an otherwise tolerant clone. Considerable evidence has accumulated that most autoimmune clones are neither eliminated in fetal life nor transformed to an autoimmune status by a change in somatic cell lines. Instead, it appears that a large repertoire of potentially self-reactive clones exist in a down-regulated state controlled by the T-cell regulatory circuit and that what determines the emergence of autoimmune disease is the breakdown of this tightly controlled regulation.

In this view the locus of dysregulation could be deficiency of negative regulation or suppression, excessive antigen-specific positive regulation at the T-cell level, or inappropriate B-cell responses to T-cell signals due to abnormal B-cell receptors to these signals. In this latter situation, the B-cell might be insensitive to negative signals or hyperresponsive to positive signals, thus escaping its normal down-regulated state.

A description of the prototype autoimmune diseases then takes place in the context of a system in dynamic equilibrium in which autoimmune clones emerge from a network of lymphocyte interactions which have been disrupted by some external stimulus (viral infections, exposure to environmental or dietary chemicals, radiation, or drugs) acting in a genetically susceptible host. Finally the female preponderance in many of these diseases needs to be emphasized since this suggests that hormonal

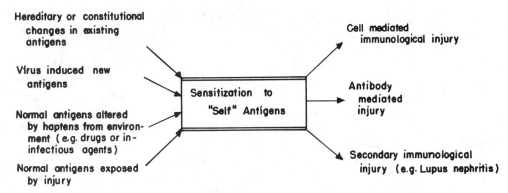

FIG. 37-4. Several possible mechanisms in the pathogenesis of autoimmune disease.

factors also modify immune responses and in some way estrogens promote autoimmunity in a number of diseases. This is strikingly true in Hashimoto's thyroiditis, SLE, and patients with myasthenia gravis in which the age-specific incidence of the disease correlates strikingly with the estrogen status of the patient. In SLE, 90 per cent of the patients between ages 15 and 45 are women while in children and the postmenopausal state the female preponderance is less striking. Similarly, there are two forms of myasthenia gravis which correlate with two peaks of age incidence. Thus, there is a young population which is predominantly female and an elderly population which is predominantly male.

Overall, self-immune reactions must be considered abnormal, whatever the basis for the reaction. In addition to the dysregulation hypothesis stated above, the more classical notion of altered self has not been excluded as a proximate cause of autoimmunity. The hypothetical causes of altered self include: unmasking of normal antigenic components not ordinarily revealed to the lymphoid system; genetic or hereditary factors resulting in the appearance of new antigens; neoantigens induced by virus infections or variant forms of bacteria; and immunity against a normal carrier protein initiated by haptene. These possibilities are represented in Figure 37–4.

A description follows of several diseases in which knowledge of immune mecha-

nisms illustrates prototype pathways for clinical expression of the autoimmune process. Antibody to cellular receptors underlies the pathophysiological basis for several diseases. Myasthenia gravis, certain cases of diabetes mellitus, and thyrotoxicosis are examples of this mechanism. Organ-specific autoimmunity is the basis of tissue destruction by a cell-mediated process in Hashimoto's thyroiditis, while tissue-specific antibody mediates shortened red cell survival in autoimmune hemolytic anemia. Finally, antibodies to a variety of soluble tissue antigens combine with these antigens to form phlogistic immune complexes which mediate vascular injury in the widespread tissue lesions of patients with SLE.

Diseases of autoimmunity

Myasthenia gravis

Myasthenia gravis is a disease in which the neurophysiological defect is a loss of acetylcholine receptors at the muscle endplate side of the neuromuscular junction with a resultant failure of the acetylcholine-induced depolarization of the endplate and consequently a failure of neuromuscular transmission. The weakness, then, is due to failure of depolarization of the muscle membrane because of the deficient neuromuscular transmission. The evidence giving rise to this concept results from direct measurement of acetylcholine receptor density in human muscle with the use of radiolabeled α-bungarotoxin, a component of snake venom which binds tightly and specifically to the acetylcholine receptor.

Availability of this specific probe for the acetylcholine receptor and the isolation from the electric eel of soluble purified acetylcholine receptor have permitted direct measurement of immune responses, both cellular and humoral, to this receptor in humans. These analytical tools have led to a flurry of investigative activity both in humans and experimental animals that have led to a completely revised concept of the pathogenesis of this disease.

It has long been recognized that myasthenia gravis had immunological features. These included the high frequency of myasthenia gravis in patients with thymomas as well as the presence of marked inflammation in the thymus in patients without thymomas. Furthermore, antibodies to the cross-striations of skeletal muscle and cells in the thymus (which in reptiles and birds contain cross-striations) were demonstrated in the sera of a high proportion of patients with the disease. These antibodies did not bind to muscle *in vivo* and could hardly be immunopathogenetic but did provide an immunological link between the thymus and skeletal muscle.

Immunization of rabbits and rats with isolated acetylcholine receptor led to the development of a disease which closely resembled human myasthenia gravis. Such studies gave immediate impetus to the search for immune responses to the acetylcholine receptor in man, and these efforts have met with great success. Antibodies to the acetylcholine receptor have been demonstrated in more than 90 per cent of patients with myasthenia gravis, and the level of such antibodies correlates roughly with disease severity. T-cell sensitization to the acetylcholine receptor has also been demonstrated in myasthenia gravis by the ability of soluble acetylcholine receptor to induce antigen-specific proliferation of lymphocytes in culture, an activity largely, if not exclusively, associated with the T-helper (T4+) class of lymphocytes.

Such findings have made more rational the various therapeutic modalities which relate to the immune status of myasthenia gravis patients. These include thymectomy, therapy with immunosuppressive agents both of the adrenal steroid and cytotoxic varieties (imuran and cytoxan), and finally the widespread use of plasmapheresis in difficult cases. The availability of a disease-related parameter (antireceptor antibody) has also permitted the development of rational approaches to treatment schedules. Preliminary studies suggest that levels of such antibodies may serve as a useful guide for the application of the above interventions.

Other diseases with antibodies to receptors

Two other diseases result from autoantibodies directed toward a physiological cell receptor. These are thyrotoxicosis and a form of insulin-resistant diabetes mellitus associated with acanthosis nigricans. Thyrotoxicosis or Graves' disease is almost always associated with a serum factor designated long-acting thyroid stimulator (LATS). This serum factor is a γ-globulin which binds to thyroid follicular cells and drives the cell to physiological hyperreactivity, presumably by activating the cell receptors normally driven by thyroid stimulating hormone (TSH). This mechanism is probably not operative in every patient with thyrotoxicosis and may not be the only mechanism operating in thyrotoxicosis even in patients who have LATS in their circulation. It almost certainly plays a role in the hyperthyroidism exhibited in the patients who have it.

A much rarer abnormality is the presence of autoantibody to the insulin receptor in patients with an insulin-resistant diabetes mellitus associated with the skin lesions of acanthosis nigricans. In this condition the γ-globulin from the serum of such patients can be shown to block the binding of insulin to the insulin receptor and provides the mechanism of the insulin resistance. Therapeutic maneuvers aimed at reducing the level of this antireceptor antibody such as immunosuppression or plasmapheresis can decrease the level of this antibody and ameliorate or even ablate the insulin-resistant diabetes. It will not be surprising if other diseases are found to be related to this mechanism of antibodies to cell receptors.

Hashimoto's thyroiditis

Focal collections of lymphocytes may be found in as many as one of four thyroid glands at autopsy, but the classical goitrous struma lymphomatosa (Hashimoto's disease) is a relatively rare disease. Hashimoto's disease is approximately 30 times more common in women than men. It is characterized by a diffuse thyroidal infiltration of lymphocytes which replace normal structures and form occasional germinal centers. In addition to the fibrosing and hypercellular variants of Hashimoto's disease, other forms of lymphocytic thyroiditis are believed to be the cause of adult primary myxedema and severe atrophic thyroiditis, as well as the much more common multifocal thyroiditis. Antithyroid antibodies may be found in a variety of other complex diseases of autoimmunity, and there is a significantly higher incidence of antithyroid antibodies in healthy relatives of patients with Hashimoto's disease.

The mechanisms responsible for the autoimmune processes in Hashimoto's disease are only partly understood, but three well defined thyroid-specific antigens are involved: a microsomal antigen, thyroglobulin, and a second colloid antigen. The autoantibodies against the microsomal antigen are organ-specific and may be of particular clinical importance, since they are complement-fixing and cytotoxic to thyroid cells in tissue culture. However, evidence for a damaging effect of the antibodies against thyroglobulin and the second colloid antigen is less impressive. Cell-mediated immunity to thyroid antigens also occurs in Hashimoto's disease. Thyroglobulin inhibits the migration of peripheral leukocytes, causes release of migration inhibition factor, and induces transformation *in vitro* of lymphocytes from patients with the disease. Lymphocytes from affected patients destroy target cells coated with human thyroglobulin and microsomal antigen. Antibody-dependent lymphocyte-mediated cytotoxicity has also been demonstrated in 29 of 39 patients in one study. For unexplained reasons, the percentage of T lymphocytes in the peripheral circulation is increased in patients with Graves' disease and Hashimoto's thyroiditis. The relative roles of antibody-mediated and cell-mediated damage have not been defined clearly, but the latter seems more heavily incriminated. Since the immunological damage in lymphocytic thyroiditis is primarily organ-specific and thyroid failure is easily corrected, immunosuppressive treatment does not seem to be warranted. Instead, therapy is directed at surgical removal of symptomatic goiters or replacement of thyroid hormones when the gland is hypoactive.

Acquired hemolytic anemias

The acquired hemolytic anemias are among the oldest recognized and best studied of the autoimmune clinical disorders. The reason for production of antibodies against erythrocyte self antigens in this group of diseases is usually unknown, but some 70 per cent are associated with other diseases which are often autoimmune in nature. The acquired hemolytic anemias have been separated into those caused by antibodies that react best at warm temperatures and those caused by antibodies that react best at cold temperatures. They are thus designated as warm or cold antibody-induced hemolytic anemia.

Warm antibodies cause the most common type, which is usually called autoimmune hemolytic anemia (AIHA). The antibodies involved are nearly always of the IgG type and can be homogenous or heterogenous, depending on the stage of the disease. The vast majority react with Rh antigens, and some of them fix complement. In 30 per cent of the cases, the patient's cells are coated only with IgG, in 50 per cent with IgG and complement, and in 20 per cent with complement only. In the 20 per cent of patients with complement only on their cell surface it has been shown that such cells contain amounts of specific anti-red cell antibody too small to be detectable by the γ Coomb's reagents but sufficiently large to activate the complement sequence leading to C_3 molecules bound to the membrane and shortened red cell survival. This has been proven by acid elution of such cells and quantitative assessment of the eluted γ-globulin. Addition of sufficient amounts of such eluates to normal cells that provide at least 500 molecules bound per cell leads to positive Coomb's tests with the γ Coomb's reagent. Quantitative determination of the eluted γ-globulin per cell from patients' cells with complement only led to the finding that 50 to 500 molecules of γ-globulin were present; an abnormal level but not sufficient for a positive Coomb's test. The lack of complement fixation by some of the antibodies may result either from the relatively large distance between Rh antigens on the cell surface or because the antigenic constituents of Rh are located on the surface in such a way that there is spatial interference with the fixation of complement. It is well recognized that two closely situated IgG antibody molecules are necessary to fix complement, in contrast to only one IgM antibody molecule. The presence of IgG and/or complement on the red cells can be detected by the direct antiglobulin (Coombs') test. Erythrocytes coated with IgG are preferentially eliminated by the spleen rather than by the liver, and early in the disease, splenectomy may be of benefit. Later in the disease, the antibody specificity becomes broader, and more complement becomes fixed to the cells, making hepatic removal of erythrocytes of greater significance. AIHA can occur either as an isolated finding or with a variety of associated diseases such as ulcerative colitis, SLE, and drug reactions. Free antibody may not be present in the serum of patients with AIHA because of complete adsorption to the erythrocytes. Indeed, the presence of free antibody may indicate a more severe form of the condition. Transfusion tends to aggravate the condition, since phenotypically identical cells become coated with antibody and are removed at the same rate as the patient's cells. In addition, the transfused cells tend to stimulate further antibody formation to autoantigens and may stimulate alloantibody production. The administration of steroids may be of benefit in treatment of the disease.

Those autoantibodies that react best at 0°C to 4°C, and become dissociated from the erythrocytes at 37°C, are referred to as the cold autoantibodies. The cold autoantibodies are predominantly IgM, but IgG forms have been reported; both fix complement. The IgM type of cold antibody is a hemagglutinin. Cold hemagglutinin is usually directed against the erythrocyte antigen I, but examples of cold hemagglutinins to antigen I and others have also been found, particularly in patients with cirrhosis, malignancies, or infectious mononucleosis (Fig. 37–5). The cold agglutinins in the latter disease are almost always directed against antigen i. Hemolytic anemia caused by cold IgM antibody is particularly common during the course of mycoplasma infections, infectious mononucleosis, sarcoidosis. Experimentally, cold agglutinin disease may be produced in rabbits by immunization with *Listeria monocytogenes*. In contrast to the hemolytic anemia due to warm autoantibodies, the affected erythrocytes are removed preferentially by the liver. Since antibody is eluted from the erythrocytes at 37°C, antiglobulin tests with anticomplement antibody are sometimes more useful in diagnosis than are tests

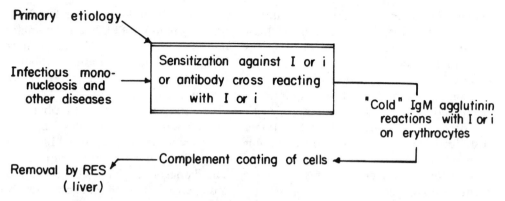

FIG. 37–5. Schematic representation of the pathophysiology for autoimmune cold agglutinin hemolytic anemia.

with anti-IgG. One puzzling feature of cold hemagglutinin disease (CHD) is that greater than 99.9 per cent of normal persons have anti-I in their serum which seems to cause no disease. However, certain differences exist to help explain this observation. For example, the antibodies in primary CHD are monoclonal for light chains (mostly κ-chains) with titers usually above 1,000 at 4°C, whereas the normal cold antibodies are polyclonal with titers usually less than 64.

Paroxysmal cold hemoglobinuria is another disease caused by cold antibodies, but, in contrast to the preceding condition, IgG is the effector molecule. Relatively large amounts of complement can be fixed by the IgG molecules, which are released from the erythrocytes at a slower rate than are the cold agglutinins. The majority of these cold hemolysins have anti-P specificity (i.e., they are directed against the very common erythrocyte antigen, designated P, which should not be confused with P_1). The disease occurs when a body part is cooled below a critical temperature. Hemolysis occurs rapidly, inducing hemoglobinemia and hemoglobinuria. Other prominent symptoms, which include fever, chills, hypotension, pain in the back and extremities, and abdominal cramps, are probably best explained as a consequence of activation of the complement system. At the beginning of the century, paroxysmal cold

hemoglobinuria was found to be associated frequently with syphilis, but recently it has more often been described following viral infections.

Systemic lupus erythematosus

Systemic lupus erythematosus (SLE) is a multisystemic disease characterized by a variety of autoimmune phenomena. It is more common in females and occurs with greatest frequency in the second and third decades of life. The etiology is unknown, but there appears to be a genetic predisposition. Identical twin studies provided the first important data that supported the idea of genetic factors operative in this disease, where the disease is expressed in concordant fashion in 50 per cent of such pairs. Recently, an increased incidence of the D-locus antigen DR_{w3} has been noted in unselected SLE patients and is especially common (as high as 90 per cent) in the subset of patients having antibodies to the soluble antigen Ro/SSA in their sera. These latter data support a role for genetic factors not only in influencing disease expression but in modulating specific autoimmune responses as well. Both viral and bacterial infections, as well as nonspecific events such as emotional stress or ultraviolet irradiation, have apparently triggered its acute onset, although it seems unlikely that they are etiologically related. Conceivably, viral infections could trigger its onset by causing somatic mutations or inducing an abnormal immune response through a haptene-carrier protein mechanism. There is substantial evidence that immune dysregulation plays a role in the production of autoantibodies. Most frequently, a decreased number of T-suppressor cells are demonstrable, and various functional tests for suppressor activity often exhibit lowered responses. As previously discussed, loss of antigen specific suppressor clones would provide a plausible mechanism for the characteristic set of autoimmune responses that are seen in this disease.

In addition, several drugs (e.g., hydralazine, procainamide) can cause a syndrome that is clinically very similar to SLE. Fortunately, most of the drug-induced lupus syndromes regress upon discontinuation of the drug. Some authors have felt that a preceding lupus diathesis is important in the pathogenesis of drug-induced lupus. At least part of this diathesis is genetic, since a slow acetylator status, the presence of a DR_{w4} antigen, or both are strongly associated with hydralazine lupus. The slow acetylator gene also predisposes to the more rapid development of antinuclear antibody and the expression of clinical disease in patients taking procainamide.

Regardless of the etiology, SLE is characterized by the development of antibodies against a wide variety of self-antigens, including DNA-histone, DNA, histone, soluble and membranous intracellular components, IgG, certain coagulation factors, platelets, erythrocytes, leukocytes, thyroglobulin, and a family of soluble RNA protein antigens, some of which are nuclear (nuclear RNP, SM, and La/SSB) and at least one of which is cytoplasmic, Ro/SSA. The clinical manifestations are protean and include fever, weight loss, various types of cutaneous eruptions, arthritis, nephropathy, hypertension, serositis, vasculitis, atypical pulmonary infiltrations, hepatomegaly, jaundice, splenomegaly, a variety of gastrointestinal symptoms, and ocular

problems. Pathological manifestations include formation of LE bodies (hematoxylin bodies), which are composed of collections of nuclei or nuclear parts, particularly in the peripheral areas of necrosis and in lymph nodes. These LE bodies are the counterpart in vivo of the LE phenomenon in vitro, which is the result of ingestion by phagocytes of the peripheral blood of nuclear material complexed with antibody. The damaged nuclei are opsonized (made more susceptible to phagocytosis) by an IgG antibody in the serum which cannot penetrate living cells. Complement is utilized in the reaction, thus increasing the opsonizing capability. The reduction in complement components may at least partially explain the susceptibility to infection found in patients with active SLE. Fibrinoid deposition is another important pathological characteristic and, most likely, it is the end result of an immunological reaction directed against damaged cells. Fibrinoid, which contains cellular debris, can be deposited in almost any portion of the body, where it stimulates the deposition of collagen fibers and fibrosis. Immune complexes are frequently deposited in the endothelium of small vessels, particularly in the renal glomeruli, resulting in the collection of irregular masses on the basement membranes that are composed of antigen, antibody, complement, and fibrinogen. Deposition of immune complexes in the kidney can result in extreme damage, often leading to death as a result of renal failure. In addition to the nephritis in which immune complexes play a major role, immune-complex-induced vascular disease may also play a role in the pleuritis, pericarditis, brain disease, serositis, and arthritis seen in these patients. A role for abnormal cellular immunity in causing tissue damage has been neither established nor excluded.

A scheme relating potential etiological factors to cell destruction, antibody formation, pathological mechanisms and disease expression is outlined in Fig. 37–6.

For unknown reasons, SLE can be associ-

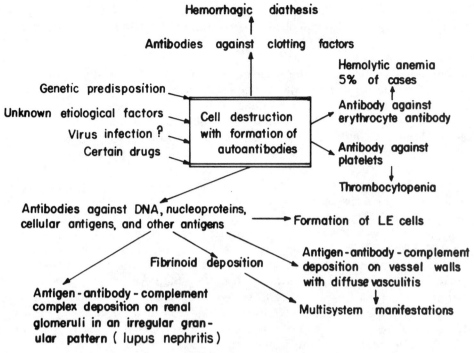

FIG. 37–6. Mechanism of disease in systemic lupus erythematosus.

ated with rheumatoid arthritis, Sjogren's disease, hemolytic anemia, pernicious anemia, lymphocytic thyroiditis, and other autoimmune diseases. Occasionally, it may present as another autoimmune disease because of a predominant manifestation such as hemolytic anemia, thrombocytopenic purpura, or arthritis. This complex disease exemplifies many of the difficulties encountered by the search for a true understanding of underlying pathophysiological mechanisms.

Immunosuppressive therapy in man

Treatment of many autoimmune diseases would logically involve attempts to decrease the patient's immunological response against his own autoantigens. Indeed, immunosuppressive therapy has proved to be quite effective in many of them. In general, cell-mediated immunity is easier to control than antibody-mediated immunity, and the IgG type of antibody is much easier to suppress than the IgM type of antibody. Autoimmune diseases involving injury caused by IgG antibody are therefore easier to control than autoimmune diseases associated with damage caused by IgM antibody.

There are numerous ways to produce nonspecific suppression of the immune response, including mechanical removal of lymphoid cells (thoracic duct fistula), destruction of lymphoid cells by ionizing radiation, blockade of antigen binding to cell receptor sites (α_2-globulin) and administration of a large variety of pharmacological agents.

A review of the pharmacological suppression of immunity is not within the scope of this presentation, but a brief discussion is pertinent. Among the drugs that have been shown to have immunosuppressive activity are the salicylates, benzene, toluene, adrenocortical steroids, purine antagonists, folic acid antagonists, certain antibiotics (such as mitomycin, puromycin, actinomycin, Azaserine, and chloramphenicol), and a variety of other drugs that inhibit protein synthesis, proteolysis, or inflammation.

The three drugs used most often are azathioprine, cyclophosphamide, and one of the adrenocorticosteroids, usually prednisone. Azathioprine primarily affects DNA synthesis and T-cell function, whereas cyclophosphamide primarily affects B-cell function and antibody formation. The adrenal steroids profoundly inhibit the development of the inflammatory lesion and can affect antigen processing and antibody formation. Unfortunately, none of the drugs are as selective as would be desirable. Bone marrow suppression is the major clinical toxic complication of azathioprine. Cyclophosphamide can also depress the bone marrow and causes loss of hair, cystitis, and other problems. Gastrointestinal bleeding, infection, and delayed wound healing are the major complications of steroid therapy. Several promising new drugs are under investigation at the present time, but their benefit is to be established.

A number of therapeutic modalities aimed at decreasing immune reactivity in patients with autoimmune diseases have come into wide use in the past few years. Foremost among these are plasmapheresis and pulse steroid therapy. Plasmapheresis has been employed most extensively in myasthenia gravis, Goodpasture's syndrome, SLE, and idiopathic thrombocytopenic purpura. It has been employed less frequently in a variety of other autoimmune diseases such as polymyositis, rheumatoid arthritis, and the various vasculitis syndromes. The rationale of the therapy is that removal of either organ-specific autoantibodies or potentially phlogistic immune complexes should lead to improvement of the patient's clinical status. Indeed, a number of glowing anecdotal reports of such therapeutic success have appeared for all the mentioned diseases, but the results of a well-controlled prospective clinical trial have yet to appear for any of the diseases in question. The procedure is quite expensive primarily because of the large volumes of replacement fluid required, either fresh frozen plasma or albumin. Widespread use will most certainly flourish in difficult cases of the above conditions if controlled data indicating a favorable effect of this therapy emerge from several controlled trials presently in progress. Because of early data, especially about SLE, suggesting a rebound of antibody titers after cessation of plasmapheresis, cyclophosphamide is usually administered during and for a short period after the plasma removal.

High-dose intravenous methylprednisolone (pulse) therapy was first used in acute rejection episodes in patients carrying renal transplants. Although many regimens have been used, doses are in the 1 g./day range with either three consecutive days or several alternating days constituting the schedule. This regimen has been employed in the treatment of rapidly progressive lupus nephritis, especially the most serious histological form, diffuse proliferative glomerulonephritis. There have been numerous anecdotal reports of the efficacy of this therapeutic regimen, but as yet no results of a controlled prospective trial comparing pulse therapy to more conventional therapies have appeared. The mechanism which underlies the effects of such extraordinary doses of steroid on the immune system is not apparent.

Finally, two other immunosuppressive regimes for severe intractable rheumatoid arthritis (RA) have appeared which have attracted considerable attention. One is lymphopheresis in which the same technology employed for plasmapheresis can be modified to remove large numbers of lymphocytes without plasma removal. This is in effect analogous to thoracic duct drainage (which has also been reported to have a favorable impact on severe RA) without the removal of lymph fluid. A second regimen recently reported to have a favorable effect in severe RA is fractionated total body x-irradiation. In both instances enthusiasm must be tempered by the lack of controlled prospective observations. It is reported that controlled trials are underway or being organized to evaluate these approaches.

Clearly, there is a great deal of effort and imagination being directed at understanding the mechanism of autoimmune diseases and finding more effective therapies. It is reasonable to expect that improved understanding of pathophysiology will lead to more rational and successful therapy.

Annotated references

Alexander, J. W., and Good, R. A.: Immunobiology for Surgeons. Philadelphia, W. B. Saunders, 1970. (A concise text containing the important principles and concepts of immunobiology, including chapters on nonspecific immunity, humoral immunity, cellular immunity, immunosuppression and specific negative adaptation, mechanisms of immunological injury, infections, cancer, and transplantation. It provides a useful overview for the student.)

Allison, A. C.: The roles of T and B lymphocytes in self-tolerance and autoimmunity. *In* Clepper, M. D., and Warner, N. L.: Contemporary Topics in Immunobiology, Vol. 3, p. 227. New York, Plenum Press, 1974. (The author presents an interesting discussion of the potential role of T cell-B cell interaction and induction of autoimmune diseases.)

Al-Sarraf, M. (ed.): Immunosuppressive Therapy Journal Articles. Flushing, Medical Examiners, 1971. (A selected collection of published articles.)

Bolte, R., Farid, N. R., VonWestarp, C., and Row, V. V.: A viewpoint. The pathogenesis of Graves' disease and Hashimoto's thyroiditis. Clin. Endocrinol. *3*:239, 1974. (The authors incriminate autoimmunity in the pathogenesis of Graves' disease, exophthalmia, and Hashimoto's thyroiditis, suggesting the diseases are caused by very closely related inherited defects in immunological surveillance. A good review of recent literature is provided.)

Burnett, F. M.: Autoimmunity and Autoimmune Disease: A Survey for the Physician or Biologist. Philadelphia, F. A. Davis, 1972. (Review of the topic by one of the founders of modern immunology.)

Cerottini, J. C., and Brunner, K. T.: Cell-mediated cytotoxicity, allograft rejection, and tumor immunity. *In* Dixon, F. J., and Kunkel, H. G. (eds.): Advances of Immunology, Vol. 18, p. 67. New York, Academic Press, 1974. (This extensive and well documented review analyzes the early events in immunological injury with 308 references.)

Dacie, J. V.: The Hemolytic Anemias, Congenital and Acquired. Part II. The Autoimmune Hemolytic Anemias. New York, Grune & Stratton, 1962. (A well referenced and complete text which is still the bible for this group of diseases.)

Dau, P. C.: Plasmapheresis and the immunobiology of myasthenia gravis. Boston, Houghton Mifflin Professional Publishers, 1979 (A collection of papers on plasmapheresis, chemistry, and immunobiology of the acetylcholine receptor.)

Dubois, E. L. (ed.): Lupus Erythematosus. New York, McGraw-Hill, 1966. (An excellent review text.)

Gabrielsen, A. E., and Good, R. A.: Chemical suppression of adaptive immunity. Adv. Immunol. *91*:1967. (A review of the suppression of immune responses by chemotherapeutic agents with an extensive bibliography.)

Good, R. A., and Yunis, E.: Association of autoimmunity, immunodeficiency and aging in man, rabbits, and mice. Fed. Proc. *33*:2040, 1974. (A strong case is made that autoimmunity occurs as a result of loss of immunoregulation, especially by T cells.)

Issitt, P. D.: Auto-immune hemolytic anemia. Am. J. Med. Technol. *40*:479, 1974. (Review article with emphasis on immunological and laboratory aspects of the disease.)

Lawley, T. J., Hall, R. P., Fauci, A. S., Katz, S. I., Hamburger, M. I., and Frank, M. M.: Defective Fc receptor functions associated with the HLA B8/DRw3 haplotype: Studies in patients with dermatitis herpetiformis and normal subjects. N. Engl. J. Med. *304*:185, 1981. (This study shows that the "autoimmune haplotype" HLA B8, DRw3 is associated with defective Fc receptor function.)

Levin, J. M., and Boshes, L. D.: Autoimmunity and the central nervous system. Dis. Nerv. Syst. *30*:273, 1969. (This brief review relates a variety of autoimmune diseases to their involvement in the central nervous system.)

Miescher, P. A., and Muller-Eberhard, H. J. (eds.): Textbook of Immunopathology, Vol. 2. New York, Grune & Stratton, 1969. (This excellent and well referenced compendium examines immunological dysfunctions and iso- and autoimmune phenomena. It is an excellent reference, still highly recommended for further study.)

Müller Eberhard, H. I., and Schreiber, R. D.: Molecular biology and chemistry of the alternative pathway of complement. Adv. Immunol. *29*:2, 1980 (Review of alternative pathway of complement.)

Paronetto, F.: Immunologic aspects of liver diseases. *In* Popper, H., and Schaffner, F. (eds.): Progress in Liver Diseases. New York, Grune & Stratton, 1970. (This review presents a cautious interpretation for the role of immunological injury in the induction and maintenance of liver diseases with 130 references.)

Reichlin, M.: Current perspectives on serological reactions in SLE patients. Clin. Exp. Immunol. *44*:1, 1981.

(A review of the current status of the serology of SLE.)

Reinherz, E., and Schlossman, S. F.: Regulation of the immune response in human beings. N. Engl. J. Med. *303*:370, 1980. (Review of the T cell circuit in man and its implications for the production of disease.)

Samter, M. (ed.): Immunological Diseases. 2nd ed. Boston, Little, Brown & Co., 1971. (A comprehensive text with numerous contributing authors.)

Schwartz, B. D., and Shreffler, D. C.: Genetic influences on the immune response in clinical immunology, Vol. 1, 49. Philadelphia, W. B. Saunders, 1980. (Update on HLA, immune response, and disease susceptibility genes; chemistry, genetics, and clinical associations.)

Schwartz, R. S.: Therapeutic strategy in clinical immunology. N. Eng. J. Med. *280*:367, 1969. (An excellent review of the problems of selective immunosuppression and specific tolerance by one of the founders of the concept of immunosuppression.)

Strauss, A. J. L.: Myasthenia gravis, autoimmunity and the thymus. *In* Snapper, I., and Stollerman, G. H. (eds.): Advances in Internal Medicine, Vol. 14, p. 24. Chicago, Year Book Medical Publishers, 1968. (This review presents the clinical and laboratory evidence for myasthenia being a disease of autoimmunity. With 260 references.)

Unanue, E. R., and Dixon, F. J.: Experimental glomerulonephritis: Mechanisms. Adv. Immunol. *6*:1, 1967. (A thorough review of the mechanisms of immunological damage in nephritis with 254 references.)

Walford, R. L.: The Immunologic Theory of Aging. Copenhagen, Munksgaard, 1969. (Interesting reading relating the problem of autoimmunity and aging.)

Watson, B. W., and Johnson, A. G.: The clinical use of immunosuppression. Med. Clin. North Am. *53*:1225, 1969. (A cautious review of the clinical value of immunosuppressive drugs in the autoimmune diseases.)

38   Mechanisms of Tumor Immunology

Samuel R. Oleinick, M.D., Ph.D.

The hypothesis of a viral etiology of malignancy has a long history. In 1908 Ellermann and Bang demonstrated that a filterable agent could cause leukemia in fowl, and in 1911 Rous transmitted sarcomas in chickens. After a hiatus of two decades, Bittner in 1936 described the mouse mammary tumor agent. Again, little generalized interest was evoked until 1951 when Gross isolated the mouse leukemia virus which bears his name. Subsequent to this observation an explosive interest in tumor virology began.

The difficulties of ascribing a viral etiology to a malignancy can be summarized in the following set of corollaries: (1) the isolation of a virus from a tumor is not sufficient evidence to incriminate that isolate in the causation of the malignancy; and (2) the failure to isolate a virus does not definitively rule out a viral etiology. In explanation of the first corollary, it is well known that viruses may exist as a latent infection in cell lines or as so-called passenger viruses with only minimal replication and little or no cytopathogenic effect on the host cell. Examples of this condition are the SV40 virus discovered in rhesus monkey kidney cultures, and the new class of xenotropic viruses which is discussed below. Thus, the isolation of a virus may be only a fortuitous event.

In explanation of the second corollary, it must be appreciated that a virus may exist in a cell in forms other than the infectious and replicative state. One form is that of a defective particle. By definition, this implies that the virus is incapable of replicating complete virus particles. The restriction of the viral replication may be a function of the species of host cell in which the virus is found, so that, for example, the Schmidt-Ruppin strain of Rous sarcoma virus can be shown to replicate in cells from chickens but not from mammals. On the other hand, it may be an innately defective virus such that it cannot direct critical events in its replicative cycle. If this defect is due to a mutation specifying a defective enzyme, such as a temperature-sensitive enzyme, manipulations of the temperature at which the host cells are grown can eventuate in the initiation of the lytic cycle. In other situations, the virus cannot direct the synthesis of enzymes and coat proteins (usually formed late in the replicative process) and requires the collaboration of a second virus in these functions. This is the case with the Bryan strain of Rous sarcoma virus, which owes its success in replication to a "helper virus" known as Rous associated virus.

In many cases, the viral genome is integrated into the host cell in a nonreplicative form. This genetic information is transmitted to daughter cells and vertically from one generation of animal to another through the germ cells. Special inducing

895

events are required to activate this "virogene" ("provirus" or "protovirus") to replicate infectious virus. These inducing agents can be ionizing radiation, chemicals, or infection with a second virus. The integration of the virus with the host cell may be at the level of the host nuclear desoxyribonucleic acid, or it may be epigenetic (extrachromosomal).

Extensive defectiveness of the virus may exist such that only an incomplete virus particle may be present in the host cell, in the form of a portion of the viral genome. In this instance, complementation of this defective viral genome with a second virus strain defective at another site has eventuated in the recovery of infectious particles. More extensive defectiveness would prevent any such recovery.

Where only partial viral genetic information resides in a host cell, or where the viral genome exists in a stable integrated form, special techniques are required to demonstrate its presence. Viral nucleic acid (DNA) can be detected by its ability to molecularly hybridize with single-stranded DNA or RNA purified from infectious or complete virus particles. A second technique is the detection of nucleic acid sequences capable of specifying virus-associated enzymes (such as the RNA-dependent DNA polymerase or "reverse transcriptase") or even identification of the enzyme itself in the host cell. Finally virus structural proteins or virus specified antigens (T antigen, early proteins, late proteins, enzymes, etc.) have been successfully identified.

If the sequence of defective viral genome contains information for cell transformation and oncogenesis, it is called an *oncogene*. The ubiquity of oncogens in animal cells has been proposed by Huebner and Todaro.

The case of human nasopharyngeal carcinoma is illustrative of the difficulty in ascribing a viral etiology to a known malignancy. This is a tumor found with greatest frequency in Chinese persons in Hong Kong and Southeast Asia. Infectious virus has not been isolated from this tumor. Studies by molecular hybridization have revealed nucleic acid sequences complementary to the nucleic acid of the *Epstein Barr virus*. However, *herpes simplex virus* antigen is also found in these tumor cells; and DNA sequences have been identified in the tumor which are homologous to oncogenic *mouse RNA viruses*. Which, if any, of these viruses is the etiological agent is still undetermined.

Classification of oncogenic viruses

There are several systems for the classification of oncogenic viruses. Animal species of origin of the virus is one criterion for classification, but it is complicated by the problem of identifying whether the agent derived from the tumor itself, or from the tissue culture species or from a laboratory animal used to isolate and identify the virus. Thus, polyoma-induced hamster tumors have been shown to replicate "C-type" murine RNA viruses, and human gastrointestinal tumors have also been shown to liberate these C-type RNA viral particles when inoculated into experimental animal hosts. Endogenous viruses which replicate poorly, if at all, in the species of origin but replicate in cells of foreign animal species have been termed *xenotropic viruses* by Levy and include certain mouse C-type viruses and the feline RD 114 virus, isolated from a human rhabdomyosarcoma inoculated into fetal kittens. As is best understood to date, these xenotropic viruses as well as other endogenous viruses are not oncogenic agents, but certainly complicate the problem of identifying a true viral etiology of malignancies.

A workable classification of tumor viruses is based upon its nucleic acid composition.

Although natural transmission of oncogenic viruses is a frequent occurrence in animals, the contagious spread of virus-induced tumors is a highly unusual phenom-

Classification of Tumor Viruses

I. RNA viruses
   A. Viper
   B. Avian leukosis-sarcoma viruses
      1. Erythroblastosis virus
      2. Myeloblastosis virus
      3. Lymphomatosis virus
      4. Rous sarcoma virus
         a. Bryan strain
         b. Schmidt-Ruppin strain
         c. Prague strain
   C. Leukemia-sarcoma viruses of lower mammalian species
      1. Gross murine leukemia virus (T lymphocyte leukemia)
      2. Friend, Moloney, Rauscher murine leukemia viruses (B lymphocyte leukemias)
      3. Graffi leukemia virus
      4. Kirsten leukemia virus
      5. Moloney sarcoma virus
      6. Harvey sarcoma virus of rats
      7. Feline leukemia viruses (Snyder-Theilen and Rickard feline leukemia viruses)
      8. Feline sarcoma viruses (ST-FSV, SM-FSV, GA-FSV)
   D. Bittner mouse mammary tumor virus
   E. Leukemia-sarcoma and other viruses of nonhuman primates
      1. Simian sarcoma virus of woolly monkeys, SSV-1, (Theilen)
      2. Simian sarcoma associated virus—SSAV-1
      3. Gibbon ape lymphoma virus, GLV or GALV (Kawakami)
      4. Mason-Pfizer virus

II. DNA viruses
   A. Papovaviruses
      1. Rabbit papilloma virus
      2. Polyoma virus of mice
      3. SV_{40} virus of monkeys
   B. Adenoviruses
      1. Avian (CELO)
      2. Nonhuman primate adenoviruses
      3. Human adenoviruses
         Highly oncogenic: Ad-12, Ad-18, and Ad-31
         Weakly oncogenic: Ad-3, Ad-7, Ad-14, Ad-16, and Ad-21
   C. Poxviruses
      1. Myxoma-fibroma viruses of rabbits
      2. Yaba virus of monkeys
   D. Herpesviruses
      1. Lucké renal adenocarcinoma of frogs
      2. Marek's disease of chickens
      3. Herpesvirus sylvilagus of rabbits
      4. Guinea pig leukemia
      5. Herpesvirus saimiri (malignant lymphomas of marmosets and owl monkeys and acute lymphoblastic leukemia of owl monkeys)
      6. Herpesvirus ateles
      7. Epstein-Barr virus (? cause of Burkitt's lymphoma, nasopharyngeal carcinoma, and Hodgkin's disease of humans)
      8. Herpesvirus hominis Type 2 (HSV-2) (? cause of human uterine cervical carcinoma)

enon. Much of the successful demonstration of a viral mechanism for malignancy relies upon the use of experimental models which, at first thought, may seem artificial. Among the conditions necessary for oncogenesis are the use of highly inbred lines of animals, parenteral inoculation of virus preparations, employment of high titers of virus, host animals that are unnatural to the virus, and the use of immunoincompetent or immunosuppressed animals. The exceptions to this generalization involve several of the herpes viruses, such as the Lucké frog tumor virus, the agent of Marek's disease of chickens, certain avian oncornaviruses, and the feline leukemia virus. However, the lack of current evidence for natural transmission of viral infection and oncogenesis is not a strong argument against the operation of this mechanism. The difficulty in identifying viral agents has already been mentioned. Newer technologies may obviate some of these problems.

The role of viruses in human malignancy

Burkitt's lymphoma

Among the earliest human malignancies attributed to a specific viral agent is the African form of Burkitt's lymphoma. The implicated viral agent is the Epstein-Barr virus (EBV) and EBV particles, virus genome, and virion antigens have been identified in cells of Burkitt's lymphoma.

Patients with Burkitt's lymphoma manifest high antibody titers to these antigens, and declines in titer have been noted with successful therapy. However, in American patients with Burkitt's lymphoma, EBV genome has not been found within the tumor cells.

The role of an infectious agent in this malignancy is supported by the observation of space-time clustering and the association with home crowding and poverty. Associated mechanisms or agents in the pathogenesis of Burkitt's lymphoma include: (1) stimulation of the reticuloendothelial system by malarial infection, (2) a second oncogenic virus, the C-type oncornavirus particle, and (3) the possible role of immunodeficiency as manifested by low levels of IgM and of antibodies to polysaccharide antigens (including the Vi antigen of *Escherichia coli*) and of impaired cutaneous delayed hypersensitivity to common "recall" antigens.

Hodgkin's disease

The evidence for the etiologic role of an infectious agent in Hodgkin's disease is based upon case clusters in schools, among spouses and other relatives, and in heroin addicts. Among first-degree relatives, the risk for Hodgkin's disease is three times the expected rate. The EBV has been implicated as a causative agent by studies that show increased EBV antibody titers in persons with Hodgkin's disease and in nasopharyngeal carcinoma. In addition, a clustering of cases of Hodgkin's disease in persons having had infectious mononucleosis, a disease caused by the EBV, has been described.

Nasopharyngeal carcinoma

As with Hodgkin's disease and Burkitt's lymphoma, the EBV has been implicated in the etiology of nasopharyngeal carcinoma (NPC). This association has been based upon the demonstration of EB viral DNA sequences in nasopharyngeal carcinoma biopsies, EBV virions in NPC cells demonstrated by electronmicroscopy, EBV antigens in cultured NPC cells, and high antibody titers against EBV capsid antigens and so-called early antigens in persons with NPC. These antibody titers decline with treatment. Oncornaviruses or nucleic acid sequences of RNA viruses have also been identified in NPC cells. Chemical environmental factors, such as nitrosamines and tea, are being investigated as contributing factors to tumor pathogenesis.

Hepatocellular carcinoma

There is strong circumstantial evidence for the association of hepatitis B virus (HBV) infection and hepatocellular carcinoma. Among such evidence is the worldwide correlation of hepatocellular carcinoma and areas of high frequency of HBV carriers, the demonstration of HBV in nonmalignant liver cells surrounding zones of malignant transformation, and the recent description of an animal model for primary liver cancer and HBV-like viral hepatitis.

Cervical carcinoma

The evidence for the etiological role of *Herpesvirus hominis* Type 2 in carcinoma of the uterine cervix is based upon: (1) the isolation of the virus from tissue culture preparations of human cervical cancer; (2) identification of HSV-2 antigens and genome in cervical cancer cells; (3) association of cervical anaplasia with genital herpes infection; (4) increased incidence of neutralizing antibodies to HSV-2 in patients with cervical cancer; and (5) identification of complement-fixing antibodies to nonvirion herpes antigens in patients with cervical cancer. Prospective studies are currently under way to confirm this etiological role of HSV-2 in cervical carcinoma.

Acute leukemia

Evidence is accumulating for the role of oncornaviruses in acute leukemia in human beings. The most likely virus candidate for such a function is a primate C-type virus related to the woolly monkey-gibbon ape group of sarcoma-leukemia viruses (SSV-GALV group). Human acute myelogenous leukemia cells have been shown to possess a reverse transcriptase and a peptide antigen which are similar, if not identical, to molecules in viruses of the SSV-GALV group. Recently, acute myelogenous leukemia cells from a patient have been shown to produce budding C-type viral particles. Techniques of cell cultivation which have been developed for replication of human myelogenous leukemia cells in tissue culture may facilitate the isolation and identification of further human oncogenic viruses.

Other human malignancies

Certain additional human malignancies have been associated with a possible viral etiology. Eilber and Morton have described a transforming agent derived from cultured human liposarcomas and osteosarcomas, and from extracts of chondrosarcomas and osteosarcomas. Giraldo, et al. obtained similar results with osteosarcomas, a chondrosarcoma, and a liposarcoma, and Gonzalez, Tzobari, Sinkovics, et al. also confirmed the presence of a transforming factor in established human sarcoma cell lines. Coalson, Nordquist, and Mohr described a transforming agent in human alveolar cell carcinomas. Finally, viral agents have been described in human breast milk and in human breast cancer.

Cell transformation

Much of the fundamental research into malignancy has involved tissue culture techniques for the explantation and growth of tumor cells and the study of the response of normal cells to mutagenic and oncogenic agents. These agents may be viruses, ionizing radiation, or chemical carcinogens. Use of these systems *in vitro* has simplified exploration of cellular structure and function in the process of malignant change since biological mechanisms of the complex host are eliminated from the interaction of oncogenic agents and target cells. Thus, the initiating event may be more completely dissected and studied. However, the delicate interaction of promoting agents of oncogenesis and of host resistance or host-directed enhancement of tumor growth cannot be studied in these systems.

Cells that have undergone morphological and functional alteration in culture and which have escaped normal regulatory growth mechanisms are called *transformed cells.* The changes that are identified in these cells run the gamut from plasma membrane structural changes to chromosomal abnormalities. Similar metamorphoses are also characteristic of tumor cells due to spontaneous events or produced by known oncogenic agents, so that characteristics *in vivo* are mimicked by the transformed cells *in vitro*. However, the absolute test for oncogenic transformation is the ability of cells transformed *in vitro* to produce tumors in susceptible host animals. This is not invariably true, and some cells which by all criteria have undergone transformation *in vitro* have failed to give tumors on animal inoculation.

Alterations of cellular genetic mechanisms are seen frequently in transformed cells. Many of these lines of cells transformed *in vitro* show discrepancies from their normal number of chromosomes. When karyotyping is done, the appearance of individual chromosomes may be distorted, suggesting fragmentation, fusion, and linkage. Giemsa and quinacrine stains ("banding") reveal changes in the expected patterns believed to correlate with specific chromosomal regions. Finally, gains or losses of individual genes may be seen in these cells, as detected by enzyme studies and growth studies utilizing media deficient in substrate components.

Another major locus of changes in transformed cells involves the plasma membrane. It has become evident that permutations of this structure have profound effects on the structure and function of the entire cell (see list below). The presence of

---

Plasma Membrane Alterations in
Transformed or Tumor Cells

Changes in oligosaccharide and sialic acid composition

Increased electronegativity

Increased electrophoretic mobility

Increased permeability to intracellular molecules and ions

Decreased intercellular "communication"

Expression of tumor-specific neoantigens and tumor associated antigens

---

tumor-specific neoantigens provides the possibility and mechanism for immunologically mediated regulation of tumor cell growth and survival. (This concept will be explored subsequently.)

Not all but many of the functional characteristics of tumor cells can be explained on the basis of plasma membrane changes. Thus, loss of normal cellular adhesiveness sets the stage for cellular sloughing from tumor nodules resulting in vascular seeding and metastasis. In addition, normal cells show intercellular communication and cell-to-cell modification of activity and regulation of cellular function. Among these functions are the migration of free cells across a glass or plastic substrate, and mitosis of cells in a sparsely colonized locale. Contact inhibition of migration (topoinhibition of migration) is a characteristic of normal cells. Following tissue trauma there is outgrowth of fibroblasts or epithelial cells to bridge and cover traumatized areas. Once the defect is repaired and the growing edges of cells make contact the outgrowth ceases. Minor abnormalities of this normal regulation occur, such as are seen in keloid formation. However, transformed cells fail to show this topoinhibition. Such cells in tissue culture do not cease their wandering upon contact with other cells, but continue to migrate over and around adjacent cells.

Another closely allied form of topoinhibi-

tion is contact inhibition of mitosis. Normal cells cease dividing when a certain saturation density is achieved. In tissue culture, this is seen as the formation of a monolayer of cells attached to the substrate. Transformed cells lack this regulation of division resulting in foci of cells many layers deep.

Morphologically, one can identify pleomorphism, cellular and nuclear immaturity, and the presence of prominent nucleoli in transformed cells. Growth rate may be rapid and mitotic index high; and aerobic glycolysis is increased. Specific enzyme levels may be increased or decreased, and new enzymes and new cellular synthetic products may appear including ectopic hormones and tumor-related embryonal molecules (α-fetoprotein, carcinoembryonic antigen, fetal sulfoglycoprotein, and Regan's isoenzyme of alkaline phosphatase). Cells transformed by chemical carcinogens show resistance to the toxic effect of the carcinogen, and cells transformed by viruses withstand the cytopathogenic effect of subsequent infection with the viral agent. Finally, normal mammalian cells grown in tissue culture show a finite life span which involves a limited number of cell divisions before senescence and death eventuates and the culture is irretrievably lost. Transformed cells have escaped this mortality and appear to possess an indefinite life span. Transformed cell lines have been in culture for 40 or more years with thousands of mitotic generations.

It has been shown recently that the adenyl cyclase system, located in the plasma membrane, and cyclic nucleotides (especially cAMP) have profound effects on cell function. Tumor cells explanted into tissue culture have been studied with agents that increase levels of intracellular cAMP. In one of these studies, a human amelanotic melanoma line was induced to synthesize melanin and acquired topoinhibition. In another, mouse neuroblastoma cells so treated demonstrated axon formation, increased acetylcholinesterase, and loss of unregulated growth. In a third study, 3T3 mouse cells transformed spontaneously or by viruses were treated with prostaglandin E_1 (PGE$_1$), a substance that activates the adenyl cyclase system. Cells treated in this manner lost their pattern of unregulated growth and showed topoinhibition. These preliminary studies suggest a mechanism for returning tumor cells to the normal regulatory mechanisms.

Tumor immunology

The interaction of the immune system with malignancies is multifaceted. In certain biological models, immunological mechanisms cause the destruction of tumor cells and enable the host animal or patient to reject these neoplasms. In other model systems, the process of the immune response favors neoplastic cell growth and the establishment of progressive tumors. The complexity of the antitumor immune response is currently being unraveled.

The basic requirement for an immune response is the existence of specific molecular configurations called antigens, which are capable of stimulating immunocompetent cells to become sensitized and to elaborate effector lymphocytes (operative in cell-mediated immunity) or antibodies. In turn, these lymphocytes or antibodies interact with the antigen molecule.

Since the 1940s it has been appreciated that tumor cells possess distinctive antigens. Prior to the work of Gross (1943), Foley (1953), Prehn and Main (1957), and Klein (1961), experimental studies on animal tumors were complicated by a lack of knowledge of transplantation immunology, and these earlier studies were therefore invalidated. However, using inbred lines of animals (syngeneic animals) or studying tumors in the autochthonous host, the above investigators demonstrated tumor immunity to chemical carcinogen-induced tumors, the specific tumor antigens in these animal malignancies. These observations were extended to virus-induced tumors by Habel (1961), Sjogren, et al. (1961), Defendi (1963), and Koch and Sabin (1963). Coincidentally and subsequently, tumor antigens and immune responses were described with human tumors, both those believed to be associated with a viral pathogenesis and those still considered to be "spontaneous" in origin.

Tumor antigens

Cellular alterations of malignant transformation involve chromosomal, nuclear, cytoplasmic, or cell-membrane structures. Mutagenic agents such as ionizing radiation and chemical carcinogens affect cell genetic information in a random fashion and theoretically may produce a nearly infinite number of genomal permutations. Accordingly, cellular alterations in such transformed cells may be nearly infinite in number, suggesting that each tumor produced by a mutagenic agent should have unique molecular structures and therefore unique ("individualistic") antigenicity. For the most part, this has been borne out in studies of tumors produced by chemical carcinogens and ionizing radiation. Tumors in different syngeneic animals, or multiple primary tumors in one animal fail to cross-immunize and therefore probably possess antigenic uniqueness. Exceptions have been noted on occasion with some chemical carcinogen-induced tumors in which cross-reactivity has been noted.

Oncogenic viruses, on the other hand, insert genetic information into the transformed cell and direct structural changes of the cell. These structural changes include new antigenic molecules in the plasma membrane, the cytoplasm, and the nucleus. Among these new antigens are tumor-specific transplantation antigens (TSTA), serologically identified cell surface antigens (S antigens), nuclear neoantigens (T antigens or CF antigens), and less commonly cytoplasmic neoantigens. Recent evidence suggests that the virus-directed

T-antigen may be present on the cell surface and in the cytoplasm as well as in the nucleus. These virus-specified or virus-specific antigens, coded by viral genetic information, are therefore common to tumors induced by that particular virus, irrespective of the animal or species in which the tumor is induced. Human cells transformed by the SV40 virus effectively immunized hamsters against SV40-induced tumors. Exceptions to the rule of virus-specific tumor antigens have been noted occasionally. Rarely, different strains of an oncogenic virus may produce tumors in a single animal species which can be shown to contain different virus-specific transplantation antigens. Conversely, on occasion, different oncogenic viruses (for example SV40 and polyoma) may produce tumors with cross-reactive antigens.

In addition to these tumor-specific antigens, other neoantigens may be demonstrated in tumor cells or in transformed cells. Certain of these antigens represent molecular configurations possibly present in the cell prior to malignant transformation but present in very low concentration or in a "cryptic" form (masked by other structures) and so not available for detection. In the tumor cell these molecules are increased in number (or are exposed) and therefore susceptible to assay. An example of these antigens is the newly appearing Forssman antigen in certain virus-transformed rodent cells. Additionally, as previously mentioned, tumor cells may regain the ability to synthesize embryonal antigens which mature, differentiated cells have repressed. In this situation, embryonal-fetal antigens are seen in the plasma membrane of tumor and transformed cells, or soluble protein antigens are synthesized by the cells and secreted as cell products. Examples of this are fetal antigens seen in SV40-transformed rodent cells and alpha fetoprotein (α-fetoglobulin) secreted by mouse hepatoma cells as described by Abelev.

The strong antigenicity of virus or chemical carcinogen-induced tumor cells contrasts with spontaneous tumors, which are generally characterized by the presence of much weaker tumor antigens or by the absence of antigens.

Human tumor cells may possess any or all of these examples of tumor-specific or tumor-associated antigens. The first recognized human tumor antigens were the carcinoembryonic antigen (CEA), and the α-fetoprotein produced by human hepatoma cells. The CEA was discovered by Gold and Freedman in colon carcinoma or in liver metastases of these tumors. Subsequently, many tumors originating from tissues of entodermal origin were noted to possess this antigen. Significantly, normal differentiated cells in the organ from which the tumor developed were shown to have very low concentrations of the antigen or none at all. However, fetal cells of entodermal origin were shown to possess the antigen, leading the authors to designate the structure as a *carcinoembryonic* antigen. The mechanism for the appearance of the antigen is believed to be the derepression of genetic information present in the adult cell but repressed until malignant transformation allowed its reexpression. The antigen is also detected in regenerating cells of entodermal origin as well as in inflammatory diseases of the gastrointestinal tract. Carcinoembryonic antigens have been described in carcinomas of the prostate, bladder, pancreas, stomach, lung, ovary, breast, uterus, and thyroid. The CEA antigen of colon carcinoma has been isolated and characterized as a relatively simple glycoprotein, located in the outermost cell membrane structure called the glycocalyx. A similar mechanism of tumor cell reexpression of fetal antigens is seen in hepatomas and germ cell testicular tumors synthesizing α-fetoprotein, in gastric carcinomas synthesizing fetal sulfoglycoprotein, in pancreatic carcinomas producing pancreatic oncofetal antigen, and in numerous human tumors producing an isoenzyme of placental alkaline phosphatase called Regan's isoenzyme. Collectively, these antigens have been designated *oncofetal antigens.*

Human tumors also elaborate new transplantation antigens (histocompatibility or HLA antigens). A human leukemia has been described in which ten HLA antigens were detected. This exceeds the maximum number of HLA antigens which individual normal cells may express. The explanation for this excess of HLA antigens is not understood, but may correspond to the expression of TL-antigens on mouse T-lymphocyte leukemia cells arising in strains of mice in which thymocytes are normally TL negative.

More importantly, human tumors demonstrate new tumor-specific antigens. Certain of these are tumor "type-specific" antigens and occur in all or the majority of tumors of the same histological type. The Hellstroms demonstrated that neuroblastoma cells possess type-specific antigens which presumptively are localized in the plasma membrane. Subsequently, they and other investigators identified similar type-specific antigens in human malignant melanomas, breast carcinomas, ovarian and testicular tumors, and sarcomas. There is also evidence that carcinomas of the lung, pancreas, skin, kidney (including the renal pelvis and Wilms' tumor), urinary bladder, prostate, uterine endometrium, and parotid possess type-specific antigens. Certain of these antigens are cytoplasmic antigens, while others are localized to the plasma membrane.

The variety of human tumors that possess type-specific antigens is very extensive. Such antigens have been additionally described in human leukemia, Burkitt's lymphoma, nasopharyngeal carcinoma, and Hodgkin's disease. The demonstration of cross-reacting antigens within malignancies is suggestive of a common etiological agent, such as a virus, responsible for the tumor. However, this viral etiology has not been proved for human malignancies, and is only suggested by current data in a mi-nority of instances. Tumor antigens *unique* to a particular malignancy also occur among melanomas from different individuals as well as in certain animal malignancies.

Immune response to tumor antigens

As previously mentioned, tumor immunity has been well demonstrated in animal model systems using assay systems *in vitro* and *in vivo.* The complexity of this immune response requires the use of multiple techniques to dissect out and identify the various elements of immunity.

The response to TSTAs has been studied extensively using transplantation assay systems. The usual procedure is to vaccinate a test animal with an inoculum of tumor cells, manipulated in such a manner that the inoculum cannot produce a malignancy (irradiation, iododeoxyuridine, hydroxylamine treatment, use of histoincompatible cells, etc.). Resistance is demonstrated by the subsequent failure of a large inoculum of untreated tumor cells ("oncogenic dose") to elicit a tumor. For the most part, immunity of this nature is a function of the cellular immune system. This can be demonstrated by adoptive transfer of the immunity by means of lymphocyte transfers between immune animals and nonimmune histocompatible recipients.

Assays *in vitro* to demonstrate cellular immunity to tumors in experimental animals have also been developed. Lymphocyte-mediated cytotoxicity is one of these approaches which involves the demonstration that lymphocytes from immune animals can kill off the appropriate tumor cells in culture. Another *in vitro* assay of tumor immunity utilizes the macrophage migration inhibitory factor (MIF). When T lymphocytes from sensitized animals are exposed to specific antigen, they elaborate and release the lymphokine MIF, which prevents macrophages from migrating across the glass or plastic culture surface.

Intact tumor cells or tumor cell homogenates have been shown to elicit the MIF response from lymphocytes of immune animals. Finally, the lymphocyte transformation assay has been applied to study tumor immunity. This test relies upon the ability of a specific antigen to activate *sensitized* lymphocytes to alter morphology (blast transformation), synthesize DNA, and undergo mitosis. Although both T and B lymphocytes can be stimulated by appropriate antigens, the test usually has been applied to tumor systems to study cell-mediated immunity.

Using techniques such as these, cellular immunity to tumor antigens has been demonstrated in numerous experimental animal models. Immunity was detected against chemical carcinogen and virus-induced tumors, and against spontaneous malignancies. In man, these techniques have also yielded evidence of cellular immunity directed against tumor antigens.

Antibody responses to tumor antigens have also been seen. These antibodies can be demonstrated by: (1) cytotoxicity studies in which antibodies plus complement cause tumor cell death; (2) cytotoxicity studies in which antibodies sensitize target tumor cells for destruction by *unsensitized* lymphocytes (antibody-dependent cell-mediated cytotoxicity: [ADCC] or LDA); (3) indirect immunofluorescence assays which detect specific antibodies binding to tumor cell plasma membranes, cytoplasmic antigens, or nuclear antigens; (4) complement fixation studies, in which specific antibodies complex with extracts of tumor cells and fix complement; and (5) "blocking" studies in which antibodies fix to tumor cells and *prevent* cytotoxic antibodies or "killer" T lymphocytes from destroying the tumor cells.

As can be seen by these test systems, tumor-specific antibodies may be demonstrated depending upon whether the antibody is destructive or protective for tumor cells. Kaliss obtained antiserum from animals bearing tumors and injected it into normal animals. When the specific tumor was inoculated as a challenge into these recipient animals, the rate of tumor growth was facilitated. This phenomenon of potentiation of tumor growth is called *enhancement.*

The majority of animals and patients having selected malignancies manifest cellular immunity to their tumors. These immune lymphocytes are capable of destroying the appropriate tumor cells *in vitro.* However, in animals or humans with progressively growing tumors, "blocking" factors are present in the serum which interpose between the attacking lymphocytes and the target tumor cells. These factors differ from the enhancing antibodies of Kaliss and appear to be soluble tumor antigens or complexes of antibodies and tumor antigens. In contrast, in animals that show regression of tumor growth, and in certain patients, there is a factor (or factors) that aborts the effect of the blocking factor. Hellstrom has called this the unblocking or deblocking factor. The mechanism of unblocking is unclear; it is not known whether the unblocking factor directly neutralizes the blocking factor; whether it is directly cytotoxic to the tumor cell; or whether it potentiates the effect of sensitized lymphocytes.

Nonspecific effector cells

Various other effector cells for tumor destruction have been identified, which either lack specificity for particular tumor cell antigens or have the specificity provided by antibodies with which these cells collaborate. The mechanism of ADCC involving the K cell has already been described. Other nonspecific cells are "natural killer" (NK) cells, "natural cytotoxic" (NC) cells, granulocytes, macrophages, and monocytes, and macrophages collaborating with natural antibodies to produce cytotoxicity against tumor cells. The progenitor cells for NK cells and K-cells are not conclusively known, since the usual cell markers are absent, but current evidence identifies these cells as large granular lymphocytes (LGLs), presumably of the T-cell lineage.

NK cells and cytotoxic macrophages can be activated by interferon and perhaps by other lymphokines or biological response modifying agents such as lipopolysaccharide. Alternatively, some of these agents, such as prostaglandins, may suppress the generation of NK cells and cytotoxic macrophages, and may indeed eventuate in suppressor macrophage or suppressor NK cells, favoring tumor growth.

Human tumor immunity

The role of immunity in human malignancy has been studied extensively. The various questions posed by clinical investigators include the following: (1) Can the level of general immune competence be correlated with tumor incidence, response to therapy, and survival? (2) Can the level of specific immunity to tumor antigens be correlated with these indices? (3) How does the malignancy escape destruction in light of the multifaceted immune response? and (4) Does manipulation of the immune system offer promise in tumor regulation?

Acute leukemia is one of the human malignancies that has been studied in this manner. Hersh, et al. found that prognosis was better for patients with acute leukemia whose general immunocompetence was satisfactory or who manifested a strong blastogenic response of their lymphocytes to autologous leukemia cells. In patients with Hodgkin's disease and other malignant lymphomas, anergy to skin testing with common antigens appeared to be correlated with the extent of disease. In lymphosarcomas, anergy was associated with an unfavorable prognosis. Failure to respond to BCG vaccination with a positive skin test was also correlated with a worsened prognosis for survival of Hodgkin's and lymphosarcoma patients. Other investigators have challenged these conclusions. Thus, while the association of anergy with advanced stages of the disease was confirmed, there was no correlation between skin test reactivity to BCG and the course of the disease or survival. Lymphocyte transformation with phytohemagglutinin also failed to correlate with survival, frequency of relapse, or duration of remission; and skin test reactivity may be conferred upon Hodgkin's patients by treatment with transfer factor, an extract of human leukocytes.

Stewart reported lymphocyte infiltration of breast carcinoma, suggesting a cellular immune response was associated with a favorable prognosis. However, he and Orizaga later found that a positive delayed hypersensitivity skin test to extracts of autologous breast carcinoma tissue was correlated with *decreased* survival rate. Thus, 2½ years after mastectomy, almost 80 per cent of patients with a negative skin test to autologous tumor were alive, whereas about half of those patients with positive skin tests had survived. In contrast to these studies with tumor cell extracts, they noted a more favorable prognosis in those patients who showed intact cellular immunity to bacterial or fungal antigens, or who could be sensitized to dinitrochlorobenzene (DNCB). Immune responses in patients with malignant melanoma have also been studied, and there was no correlation between delayed hypersensitivity skin test reactivity to autologous melanoma extract and extent of disease, presence of distant metastases, or duration of remission. In a summary of one group's experiences with skin test reactivity in various human malignancies (using DNCB as a sensitizer for contact dermatitis), it was concluded that failure to respond was a harbinger of a poor prognosis. In the case of epidermoid carcinomas, a *positive* skin test response to DNCB predicted an almost uniformly favorable prognosis following "curative" surgery. In contrast, most (24 of 25) patients with soft-tissue and skeletal sarcoma demonstrated a positive response to DNCB, regardless of extent of disease, remission, or survival. Malignant melanoma patients also demonstrated a lack of correlation between positive DNCB skin test and favorable prognosis.

Finally, when peripheral blood lymphocytes from tumor-bearing patients are compared with lymphocytes within the tumor itself or with lymphocytes in draining lymph nodes, another discrepancy becomes apparent. Thus, in a series reported by Nind, Nairn, et al., peripheral blood lymphocytes from patients with colon carcinoma and skin melanomas were cytotoxic to autologous tumor cells, whereas lymphocytes obtained from the tumors or from lymph nodes failed to demonstrate cytotoxicity. The mechanism for local impairment of lymphocyte activity remains unexplained.

Tumor immunity and tumor control

Clinicians have been aware for some time of the possibility of immune regulation of tumor growth. Histologically confirmed cases of spontaneous regression of human tumors are available; the majority of these tumors were neuroblastomas, hypernephromas, gestational choriocarcinomas, and malignant melanomas. In other instances, tumors have been noted to have periods of prolonged indolent growth before acceleration and fatal outcome. Finally, malignancies have not infrequently been shown to demonstrate the histological hallmark of a cellular immune response (lymphocyte and macrophage infiltration of the tumor). In breast and gastric carcinomas such a reaction has been correlated with a more favorable prognosis.

With the current positive laboratory and clinical evidence for tumor immunity, the critical question that requires elucidation is why this system fails to prevent tumor establishment and growth. A prominent hypothesis of the immune regulation of tumor growth is that of *immunological surveillance*. This concept was propounded by Thomas in 1959, and given the current

name by Burnet in 1961. According to the hypothesis, tumor cells may frequently arise in the host organism, spontaneously or due to oncogenic agents. The intact immune system eliminates these aberrant cells; however in the event of a breakdown in immune surveillance, an inappropriate immune response, or a change in the antigenicity of the tumor cells, the regulation of tumor growth fails and the tumor becomes established.

At the extremes of life (early infancy and old age), the immune mechanism shows less than optimal function and during these periods of life the frequency of malignancy is especially pronounced. While this association may be fortuitous, it suggests a cause-and-effect relationship.

Certain malignancies are hormone-dependent, and growth or regression may be affected by endocrine manipulation or ablation. Estrogen hormones are partially immunosuppressive, whereas progesterone compounds stimulate immunologic function. The complex relationship between tumor growth and hormone levels may involve such endocrine effects on immune function.

Immunodeficiency in humans is associated with a heightened incidence of malignancy. This is true whether the immunodeficiency is congenital, or of late onset ("common variable immunodeficiency", also called "acquired agammaglobulinemia"). For the most part, malignancy is associated with defects of cellular immunity like those seen in patients with severe combined immunodeficiency (Swiss agammaglobulinemia), ataxia telangiectasia, Wiskott-Aldrich syndrome, and common variable immunodeficiency. Children with ataxia telangiectasia may survive from the primary disease process into puberty or later, but 10 per cent of these children develop malignancies. In all of the recorded associations of primary immunodeficiency and malignancy, the majority of the tumors involve the lymphoreticular and myeloid cell systems, with lymphomas, reticulum cell sarcomas, Hodgkin's disease, and leukemias comprising better than 80 per cent of the documented malignancies. Malignancies have also been reported in patients with such defects of humoral immunity as Bruton's agammaglobulinemia and isolated IgA deficiency. Here the cause-and-effect relationship seems less clear, and the association may be fortuitous.

Physicians create immunodeficiency by the use of immunosuppressive drugs to facilitate survival of allografts (e.g., kidney, bone marrow, liver, and cardiac transplants), to treat autoimmune diseases (e.g., rheumatoid arthritis, systemic lupus erythematosus) and for therapy of malignancies and various other conditions (e.g., psoriasis). The commonly employed immunosuppressive drugs including cyclophosphamide, azathioprine, 6-mercaptopurine, methotrexate, and corticosteroids inhibit antibody production and cellular immunity, and certain of these drugs modify the effector systems of hypersensitivity reactions including monocyte-macrophage activity and inflammatory response. Antithymocyte globulin, used in conjunction with organ transplants, suppresses cellular immunity. In these cases of induced immunodeficiency, the risk of malignancy is greatly increased. Penn and Starzl reviewed the worldwide data on transplant patients receiving immunosuppression and reported an incidence of malignancy 80 times the expected natural incidence. The risk of a patient developing malignancy is 5 per cent in the first few years following successful organ transplant. Unlike the primary immunodeficiency patients, transplant patients have a majority of epithelial carcinomas (61 per cent), whereas mesenchymal tumors occur in only 39 per cent. The great majority of the mesodermal tumors are lymphomas, with an unusual number of lymphomas of the central nervous system, ordinarily a decidedly rare malignancy. Five per cent (11 of 199) patients immunosuppressed for heart transplants at Stanford University

have developed lymphomas. Of 50 kidney transplant patients immunosuppressed with the new agent cyclosporin A, 4 have developed lymphomas. Twenty patients treated with methotrexate have developed malignancies; many of these patients had been treated for psoriasis. In a study at Stanford University of treated Hodgkin's disease patients, 3 per cent were noted to have developed a second malignancy. The patients who received chemotherapy, either alone or combined with radiation therapy, had a 5-per cent incidence of leukemia.

Escape from immunological surveillance

The failure of immunological surveillance of tumors in patients who are not otherwise immunologically deficient must be explained by mechanisms other than those just described. One mechanism might be enhancement of tumor growth by noncytotoxic antibodies or by blocking factors. As described above, these humoral substances interfere with the killing role of sensitized lymphocytes or cytotoxic antibodies. Another means by which the immune system may actually potentiate tumor growth is by the progressive elimination of the more antigenic tumor cells. Immune mechanisms, by eliminating the highly antigenic tumor cells, favor the survival and outgrowth of the weakly antigenic tumor cells which are much more resistant to immunological destruction.

Criticism of the immunological surveillance theory

The strongest support for the immunological surveillance theory derives from the prevalence of tumors in immunodeficient and immunosuppressed individuals. Numerous recent experimental studies provide some criticism of this hypothesis. Thus in mice which were thymectomized at birth, *fewer* mammary tumors were produced by mammary tumor virus (MMTV) than in controls. Irradiation, splenectomy, antilymphocyte serum, and methylcholanthrene immunosuppression achieve a similar *reduction* in tumor incidence, as well as a reduction in tumors due to urethane-croton oil combinations. Moreover, chronic administration of antilymphocyte serum in mice does not produce spontaneous leukemia unless infective leukemia virus is present in the animals' environment. Congenitally thymus-deficient mice which lack cellular immunity do not develop spontaneous tumors and also resist tumor induction by certain chemical carcinogens, resist leukemogenesis by Friend and Rauscher mouse leukemia viruses, and do not develop virus-induced tumors when raised in a germ-free environment. Finally, low levels of a cellular immune response may actually stimulate tumor development. Thus if methylcholanthrene-induced tumor cells are inoculated into mice along with a small population of specifically sensitized spleen cells, tumor growth is more rapid than that achieved by equivalent numbers of tumor cells alone. Greater numbers of immune spleen cells inhibit tumor growth. In a somewhat different fashion, cellular immunity of the graft-versus-host type or the mixed lymphocyte culture reaction can activate oncornaviruses and may thereby induce malignant transformation.

Immunological effects of malignancy on oncogenic agents

The role of immunodeficiency in *facilitating* malignancy has already been described; but in other instances, malignancy may coexist with impairment of immunological responsiveness or may be the cause of the immune suppression. In many far advanced malignancies with marked debilitation and cachexia, immune responses are impaired. More subtle defects in cellular immunity have been described in patients with earlier stages of carcinoma. Multiple myeloma, with its extensive involvement of bone marrow and with the demonstrated presence of suppressor cells, is associated with impairment of functional humoral antibody production; and Hodgkin's disease is often accompanied by anergy to common delayed hypersensitivity skin test antigens. The severity of immunological impairment is correlated with the stage of the disease involvement. The anergy may be attributed to defective T-lymphocyte function, to an impairment in the nonspecific inflammatory response necessary for the skin test, to a circulating humoral inhibitor of lymphocyte function, or to suppressor T lymphocytes or suppressor macrophages. In leukemias, there is also a decrease in both cellular and humoral immunity. This is usually seen in the chronic lymphocytic leukemias; when present in other leukemias, it is ordinarily a function of the cytotoxic drugs used to treat the malignancy.

Oncogenic agents such as chemical carcinogens and viruses may also be immunosuppressive. This activity, rather than a direct carcinogenic effect, has been proposed to explain the induction of tumors in certain instances. Lymphocytotropic viruses are especially likely to impair lymphocyte function.

Immunotherapy

The goal of immunotherapy is to augment the immune response to specifically eliminate tumor cells without damaging normal tissue. Because of the complexity of the immune response it is essential to intensify immune responses that are cytotoxic to the tumor cells (cellular or humoral cytotoxicity) and to minimize enhancing antibody or blocking factor production. Current clinical immunotherapeutic procedures are empiric in nature, as the problem of achieving only the appropriate beneficial immune responses has not been solved.

Basic concepts in successful immunotherapy

There are three general concepts in tumor immunology which relate to the construction of potentially successful regimens for immunotherapy. Whenever possible, the dicta listed below should be considered in structuring experimental and clinical treatment protocols: (1) reduce the tumor burden; (2) use the rebound phenomenon of immune responsiveness; and (3) amplify cellular immunity.

The immune response is relatively ineffective against a large tumor mass. In experiments, it is most successful against tumor inocula of limited cell numbers. By extrapolation it should be most efficient in the preventive mechanism of surveillance, as well as the elimination of residual tumor cells following ablative therapy (surgery, irradiation, endocrine therapy and chemotherapy). In leukemic mice and in human leukemia, the residual tumor burden must be reduced to less than one million cells before immunotherapy is successful. Many oncological chemotherapists achieve this goal using combination-drug chemotherapy. In solid tumors, the goal should be to surgically excise primary tumors and available metastases. Radiation therapy, while capable of destroying tumor cells, has suppressive effects on the immune mechanism which might prejudice the full use of immunotherapy.

Cytotoxic chemotherapy and corticosteroid treatment achieve immunosuppression as well as tumor destruction and myelosuppression. With intermittent drug dosage, there is recovery of normal cell function and a rebound hyperreactivity of the immune system. Advantage should be taken of this rebound period to immunize specifically to tumor antigens or to augment nonspecifically the immune response. Since it is the goal of immunotherapy to maximize cellular immune responses as well as cytotoxic antibodies and unblocking factors, it is incumbent on the immunotherapist to identify appropriate agents to achieve this end.

Passive immunization
Immunotherapy by *passive transfer* of antibodies or lymphocytes from various cate-

gories of donors has been attempted on numerous occasions with negative or equivocal results. Normal donors, especially family members, have been used as sources of these materials. Patients who have demonstrated a successful clinical course following therapy for a malignancy of similar histological type have also been used as donors. Cross-immunization of pairs of patients with extensive malignancies has been performed by Nadler and Moore. This approach involved exchange of tumor cells between patients bearing histologically similar tumors. Following this immunization procedure, serum and leukocytes were collected and inoculated into the original donor of the tumor tissue. It was hoped that antibodies or lymphocytes from the immunized patient would benefit the recipient; however, the results were equivocal. In another report when viable malignant melanoma cells were used to immunize the elderly mother of a patient in expectation of using the mother as a source of immune serum or leukocytes, the mother developed fatal malignant melanoma. The many problems of human tumor research are illustrated by this latter example. Human tumor cell inocula may retain some viable cells despite extensive procedures for inactivation, including high-dosage irradiation. Some tumor preparations may contain oncogenic viruses. Recipients of tumor cells may have marginal or unrecognized immunodeficiency. Finally, closely related family members may fail to reject tumor cells used for immunization because they share a closeness of histocompatibility factors.

Active specific immunization
To heighten the immune response to tumor cells, patients have been actively immunized with preparations of their autologous cells or with pooled, inactivated cells from patients bearing histologically identical tumors. Ordinarily, inactivation is performed with irradiation or formalin fixation. More aggressive manipulations

appear to reduce or destroy the tumor-specific antigens.

The immunogenicity of native tumor cells appears to be limited. Accordingly, various investigators have employed chemical and enzymatic means to alter tumor cell membranes and amplify their immunogenic potential. In animal systems, iodoacetate, concanavalin A, dinitrophenyl-aminocaproic acid, and neuraminidase have been used for this purpose. Protective immunization has been facilitated by this approach; but immunotherapy by these methods has achieved equivocal results in cancer patients.

Another method of augmenting the immune potential of tumor cells is the use of cells disrupted by lytic viruses. Lindenmann first employed viral lysates of tumor cells for immunization, using the influenza virus strain WSA. Others have used Newcastle disease virus, vesicular stomatitis virus, and reovirus lysates of tumor cells for their immunogenic potential in animal systems. Sinkovics (1977) has investigated PR 8 influenza virus lysates of human sarcoma and melanoma cells and Wallack (1981) is currently studying vaccinia virus lysates of autologous and allogeneic human tumor cells, primarily colon and melanoma, in clinical therapeutic trials. Membrane fragments produced by viral disruption of tumor cells incorporate virus and tumor antigens; this combination of antigens offers the possibility of stimulating a more effective immune response directed against tumor cells.

Because of the possibility of adverse immune responses of an enhancing antibody or blocking factor nature when specific immunization *in vivo* is attempted, as well as the inefficiency of this form of sensitization, certain investigators have used the approach of sensitization *in vitro*, by which the patient's white cells or lymphocytes are exposed to autologous or allogeneic tumor cells in a cell culture system. Following this sensitization, the white cells are returned as an "autotransfusion." This use of autosensitized or activated lymphocytes has shown some early favorable results.

Augmentation of nonspecific immunity

Much of the recent thrust of immunotherapy has been directed at "adjuvant" therapy, or the nonspecific augmentation of immunity. For the major part, this involves methods to recruit lymphocytes and to cause lymphocyte proliferation, to activate macrophages and monocytes, and by the generation of an inflammatory response at the site of a tumor, to achieve a heightened response directed at tumor-specific antigens.

Living attenuated *Mycobacterium bovis* (bacille Calmette-Guerin, or BCG) has been used for many years as a vaccine against tuberculosis. In animal experimentation it has been employed to increase immune responsiveness and antibody synthesis. This latter role derived from observations that antigens injected into tuberculosis granulomas achieved greater antibody titers than similar antigen inoculations given at non-involved sites. Subsequently, mycobacteria have been incorporated into oil-detergent mixtures and are commonly utilized as adjuvants ("Freund's adjuvant"). Davignon and co-workers, in a study reported in 1970, analyzed leukemia deaths in Quebec in the years 1960 to 1963 and showed that children who had been vaccinated at birth or later with BCG had half as many deaths from leukemia as did unvaccinated children. Similarly, Rosenthal and co-workers analyzed leukemia deaths in Chicago for the years 1964 through 1969 and observed a lower incidence of leukemia deaths in children who had been vaccinated at birth with BCG than in an age and race-matched unvaccinated group (0.31 vs. 2.02/100,000/ year).

The investigation of BCG adjuvant therapy in human malignancy was begun by Mathé and co-workers. In 1969 these authors reported their results with combination chemotherapy and BCG in acute lymphoblastic leukemia. After remission, induced by chemotherapy, a group of patients was treated with BCG, with or without irradiated or formalin-inactivated leukemia cells. Among patients treated with BCG, five of 20 maintained remission longer than 3 years after cessation of chemotherapy. In the group treated by chemotherapy alone, none of the 10 patients maintained remission beyond 130 days. Powles in 1973 (and subsequently) has reported on patients with acute myelogenous leukemia treated with chemotherapy alone, or BCG and leukemia cells with or without maintenance chemotherapy. In the group receiving BCG and leukemia cells there was a prolongation of length of remission and a doubling of survival. BCG has also been investigated extensively in the therapy of malignant melanoma; while intralesional injection of BCG has caused direct tumor regression and some response of tumors at distant sites, response to systemic therapy has been unrewarding. Instillation of BCG into the urinary bladder has improved patients with carcinoma of the bladder. Since BCG is a living infectious organism, complications of therapy are possible and have been reported. This is especially true since the patients in whom it is used are in a state of relative immunosuppression, either because of their malignant disease or because of extensive chemotherapy. Among the reported complications are persistent BCG infections at sites of vaccination, in regional lymph nodes, or at distant sites. Other sequelae are osteomyelitis due to BCG, miliary granulomas of the liver or the lungs, and anaphylaxis or endotoxin-like shock. Disseminated BCG infections in these instances respond well to antituberculous chemotherapy.

Other augmenting agents for actively but nonspecifically heightening immunity in patients with malignancies have likewise proved disappointing after early enthusiasm. Thus whole microorganisms such as *Corynebacterium parvum*, inactivated streptococci (OK 432) and viruses, and crude microbial fractions such as the methanol extraction residue of BCG (MER-BCG), staphylococcal phage lysates, and *Brucella abortus* extraction residue have received extensive testing and have generally been discarded. Undaunted, current investigators are evaluating highly purified microbial fractions whose composition can be standardized more effectively (peptidoglycans, glycolipids, glucans and other polysaccharides, and lipopolysaccharide-endotoxin) as well as synthetic agents which induce interferon and activate macrophages (pyran-copolymer MVE2 and lysolethicin analogues).

Other approaches to immunotherapy

Several other potentially useful approaches to immunotherapy have been reported and deserve description. Klein has pursued the value of inducing local inflammatory responses at the site of superficial epithelial and mucosal tumors by sensitizing the patient to the contact allergen, dinitrochlorobenzene, and then challenging with a dilution of DNCB which is below the threshold for eliciting a contact dermatitis reaction. Sites of tumor involvement show a hypersensitivity reaction with elimination of tumor cells. Asada has used living mumps virus and vaccinia virus for therapy with suggestive positive responses. Numerous routes of inoculation were utilized, including intralesional, topical, subcutaneous, and oral.

Finally, transfer factor therapy has been under investigation by several centers in dealing with a variety of malignant diseases. Transfer factor, a dialysable extract of human lymphocytes, with a molecular weight less than 10,000 and a chemical structure of either a polypeptide or an oligoribonucleotide, has been shown by Lawrence to be capable of transferring cell-mediated immunity and to reconstitute defective cellular immunity.

The future

The future promises potentially significant developments in the basic understanding of malignant processes and effective therapy. The role of viruses in human malignancy remains to be defined, with the goal of preventive immunization to be achieved if possible. Already, treatment of persons at high risk for nasopharyngeal carcinoma with interferon or the antiviral drug acycloguanosine is being considered, as is the use of a killed EBV vaccine in infants in the geographic region of endemic Burkitt's lymphoma. The judicious use of the new HBV vaccine may have a profound effect on reducing the future incidence of hepatocellular carcinoma.

Interferon therapy for malignancies is currently entering a period of extensive investigation. Following the description by Isaacs and Lindenmann in 1957 of this polypeptide with antiviral activity, information has developed elucidating the chemical structure of this family of molecules, the cell sources and production methods for obtaining practical quantities for research and clinical trials, the cell membrane effects, and the molecular level of action of the interferons on cellular biochemistry. In addition to preventing viral replication, the interferons show antiproliferation and antidifferentiation effects on normal and malignant mammalian cells. They also have profound effects on the immune system, decreasing lymphocyte transformation and antibody synthesis while increasing cytotoxic T lymphocytes and activating macrophages and NK cells. Earlier trials in cancer patients used the synthetic double-stranded polynucleotide poly I–poly C, or poly I–poly C lysine, to induce interferon. Recent trials with various crude human interferon preparations have shown encouraging response in patients with osteogenic sarcoma, multiple myeloma, nodular poorly differentiated lymphoma, and breast cancer. Currently, more highly purified interferon preparations are undergoing clinical study in various other malignancies.

Immunodiagnosis

Screening tests for preclinical malignancies are unavailable as yet because of a lack of sensitivity or specificity. However, several tumor-associated markers have been identified in patients' plasma, cerebrospinal fluid, and effusions. These markers may be oncofetal protein antigens or enzymes, as previously described, hormones or polyamines, or surface molecules such as $\beta2$ microglobulin which are shed from the cells. Already, the carcinoembryonic antigen (CEA) is being used to follow patients for adequacy of surgical resection of carcinoma of the colon, and rising levels of CEA have prompted "second-look" operations. Pancreatic carcinoma–associated antigen, pancreatic oncofetal antigen, and galactosyl transferase isoenzyme II are being evaluated as diagnostic tests in pancreatic carcinoma. Alpha-fetoprotein is being studied in hepatoma and in gonadal germ cell neoplasms. Plasma levels of β and α subunits of human chorionic gonadotropin, calcitonin, prostatic acid phosphatase, and creatine kinase BB isoenzyme all have significant roles as markers in clinical evaluation of patients with malignancies. Immunoassay methods have simplified the quantification of some of these marker molecules. Beta 2 microglobulin and β glucuronidase in cerebrospinal fluid are potentially of value in identifying meningeal leukemia/lymphoma and carcinomatosis, respectively.

The newer technology of monoclonal antibody production, by generating monospecific antibodies in unrestricted amounts, promises to facilitate the identification of new human tumor-associated antigens and broaden the scope of clinical immunodiagnosis of malignancy.

Another application of antibodies to tumor-associated antigens is the noninvasive localization of primary malignancies or metastases. By coupling a γ-emitting radionuclide to either a heteroantibody or a monoclonal antibody, isotope localization can be achieved which can be detected by external imaging techniques. Circulating immune complexes of tumor-antigen and antibody may interfere with this localization, and nonspecific binding of labeled antibody may create technical problems. Nevertheless, preliminary human studies have been encouraging in identifying sites of tumor cells producing CEA antigen, and also in imaging metastases of testicular carcinoma which are producing human chorionic gonadotropin.

Immunotherapy

Tumor cell cloning, beginning to be used in studying *in vitro* susceptibility of human malignancies to chemotherapeutic agents, may also be applied to individual tumors to select effective immunotherapeutic agents such as monoclonal antibodies and cytotoxic T-cell lines, or cytokines such as subtypes of the interferons.

Passive immunization with monoclonal cytotoxic antibodies offers promise of therapeutic effectiveness. Moreover, highly avid and specific monoclonal antibodies can be coupled to plant or bacterial toxins, to chemotherapeutic agents, or to radioisotopes of high specific activity to chauffeur these therapeutic molecules to the target site. Passive immunity may also be conferred with clones of cytotoxic T-cells specific for the particular tumor or tumor type if this technology is successful.

Modulation or reconstitution of an effective immune system may be achieved by elimination of suppressor cells by the use of monoclonal antibodies (especially anti-idiotype antibodies), or by thymic hormones. Thymosin fraction V, especially the thymosin α-1 molecule, which appears to stimulate helper T cells preferentially, appears promising. Thymic humoral factor (THF) is undergoing limited clinical trials.

Lymphocyte and macrophage growth and maturation factors are also under study, as is the synthetic immunomodulator, levamisole.

Newer biological mediators of tumor destruction have been described. These include tumor necrotizing factor (TNF) from macrophages stimulated with endotoxin, and fibronectin from cryoprecipitated normal human plasma.

Finally, therapeutic procedures to reverse blocking activity are being investigated. Plasmapheresis has demonstrated a modest response rate in a few human tumor patients, especially in breast carcinoma. Absorption of plasma on staphylococcal protein A (SpA) columns, which removes free immunoglobulins of the IgG1, IgG2, or IgG4 subclasses or immune complexes containing these immunoglobulins, has recently been shown to produce objective response. Terman and co-workers reported that four of five patients with breast cancer who were reinfused with autologous absorbed plasma were benefitted. Indeed, SpA-absorbed plasma also produced a response when infused into a breast cancer patient different from the plasma donor.

Many of these approaches to immunotherapy of malignancy are included in the Biological Response Modifier Program supported by the Division of Cancer Treatment of the National Cancer Institute (U.S.A.). The application of newer methods of immunotherapy, in conjunction with surgery, radiation therapy, and chemotherapy, offers the expectation of significant improvement in patient survival and cure rate.

Annotated references

Books

Burnett, F. M.: Immunological Surveillance. Oxford, Pergamon Press, 1970.
Chirigos, M. A., Mitchell, M., Mastrangelo, M. J., and Krim, M. (eds.): Mediation of Cellular Immunity in Cancer by Immune Modifiers. New York, Raven Press, 1981.

Essex, M., Todaro, G., and zur Hausen, H. (eds.): Viruses in Naturally Occurring Cancers. Cold Spring Harbor Laboratory, 1980.

Klein, G. (ed.): Viral Oncology. New York, Raven Press, 1980.

Articles

Abelev, G. I., Assecritova, I. V., and Kraevsky, N. A.: Embryonal serum α-globulin in cancer patients; Diagnostic value, Int. J. Cancer 2:551, 1967. (An extensive clinical study of 308 patients analyzed for alphafetoprotein. The assay was positive in 17/28 patients with primary carcinoma of the liver and in 10/29 patients with teratoblastomas. All other patients with malignancies or with nonmalignant diseases were negative.)

Boman, B. M., and Fathman, C. G.: Monoclonal antibodies: The next attempt at tumor immunotherapy. Mayo Clin. Proc. 56:641, 1981. (A brief review and speculation regarding future developments.)

Borella, L., Green, A. A., and Webster, R. G.: Immunologic rebound after cessation of long-term chemotherapy in acute leukemia. Blood 40:42, 1972. (The authors describe the rebound of immunoglobulin and antibody synthesis in young children with acute leukemia in whom cytotoxic chemotherapy had been discontinued after chronic administration. A valuable paper stimulating research in the possible utilization of these rebound kinetics for the immunotherapy of cancer.)

Everson, T. C.: Spontaneous regression of cancer. Ann. N.Y. Acad. Sci. 114:721, 1964. (The author evaluates the world literature on spontaneous regression of human malignancy, accepts 130 cases as having adequate documentation to be considered valid, and speculates on the mechanism of tumor regression.)

Fialkow, P. J.: The origin and development of human tumors studied with cell markers. N. Engl. J. Med. 291:26, 1974. (Analysis of human tumors by enzyme and immunoglobulin markers to confirm the concept of malignancy as a clonal transformation in most instances. Burkitt lymphoma recurrences are found to be both recurrences of original tumor clonal type and establishment of new, independent clones.)

Friedman, H., and Ceglowski, W. S.: Immunosuppression by tumor viruses: Effects of leukemia virus infection on the immune response. In Amos, B. (ed.): Progress in Immunol. New York, Academic Press, 1971. (Studies of immune responsiveness in mice infected with murine leukemia virus. The site of virus effect appears to be at the level of antigen reactive cells, and perhaps antigen processing cells and T lymphocytes, although these latter two effects were not well established.)

Gresser, I.: Antitumor effects of interferon. Adv. Cancer Res. 16:97, 1972. (An extensive review of interferon, and its effects in vitro and in vivo on tumor cells.)

Gross, L.: Facts and theories on viruses causing cancer and leukemia. Proc. Nat. Acad. Sci. U.S.A. 71: 2013, 1974. (A discursive handling of the concepts of virus-induced malignancy, valuable for the author's handling of the concepts of virus latency and the oncogene hypothesis.)

Gutterman, J. U.: Clinical investigation of the interferons in human cancer. Cancer Bull. 33:271, 1981. (Current status of human interferon therapeutic trials with an introductory section describing the current knowledge regarding interferon structure, production, and biological action.)

Hellstrom, K. E., and Hellstrom, I.: Lymphocyte-mediated cytotoxicity and blocking serum activity to tumor antigens. Adv. Immunol. 18:209, 1974. (These authors, who have contributed extensively to the concept of a multiphasic immune response to tumor antigens, broadly review their work and the entire field of tumor immunology and the immunology of pregnancy, allograft transplantation, and bone marrow-induced chimerism.)

Hersh, E. M.: Augmenting agents in cancer therapy. Cancer Bull. 33:261, 1981. A comprehensive overview of approaches to immunotherapy with special emphasis on augmenting agents.)

Kaliss, N.: The elements of immunologic enhancement: A consideration of mechanisms, Ann. N.Y. Acad. Sci. 101:64, 1962. (This pioneer paper evaluates the author's experimental data on tumor enhancement by antibodies, and investigates the possible mechanisms by which tumor enhancement occurs.)

Levin, A. S., Spitler, L. E., and Fudenberg, H. H.: Transfer factor therapy in immune deficiency states. Annu. Rev. Med. 24:175, 1973. (A comprehensive discussion of transfer factor, including a brief but well-written section on the rationale for transfer factor therapy in human malignancy. Results have been inconclusive or too early to evaluate.)

Maugh, T. H., II: RNA viruses: The age of innocence ends. Science 183:1181, 1974. (This editorial summary reviews the concepts of RNA virus oncogenesis, including the oncogene, virogene, and protovirus hypotheses. The complexity of defining a specific viral etiology for human malignancy is discussed.)

Mavligit, G. M.: Recent developments in the immunodiagnosis of cancer. Cancer Bull. 33:250, 1981. (A description of human tumor-associated antigens.)

Merigan, T. C.: Virology and immune mechanisms. Cancer 47 (Suppl.):1091, 1981. (Putative human tumor viruses, EB virus and Hepatitis B virus are discussed.)

Morton, D. L., Eilber, F. R., Holmes, E. C., Hunt, J. S., Ketcham, A. S., Silverstein, M. J., and Sparks, F. C.: BCG immunotherapy of malignant melanoma: Summary of a seven-year experience. Ann. Surg. 180:635, 1974. (Extensive summary of a large, well-studied group of patients with malignant melanoma treated with BCG immunotherapy. A second group received BCG as an adjunct to surgery for metastatic disease.)

Notkins, A. L., Merganhagen, S. E., and Howard, R. J.: Effect of virus infections on the function of the immune system. Annu. Rev. Microbiol. *24*:525, 1970. (A very comprehensive summary of the various known systems of virus influence on the immune response, including depression and augmentation in both animals and human beings.)

Oettgen, H. F.: Immunological aspects of cancer. Hosp. Pract. *16*:85, 1981. (A general discussion of tumor-associated antigens, followed by discussion of strategies for tumor immunotherapy and induction of anti-tumor mediators.)

Old, L. J.: Cancer immunology: The search for specificity. Cancer Res. *41*:361, 1981. (A comprehensive survey of cell surface antigens of mouse leukemias, chemically-induced mouse sarcomas, and human cancers, especially malignant melanomas. Discusses anti-tumor agents such as immunopotentiators, tumor necrosis factor, and fibronectin from normal plasma.)

Patt, Y. Z.: Immunomodulation with thymic hormones in neoplastic disease. Cancer Bull. *33*:265, 1981. (A recent review of thymic hormones, and early clinical trials in human malignancy and histiocytosis-X.)

Penn, I., and Starzl, T. E.: Immunosuppression and cancer. Transplant. Proc. *5*:943, 1974. (Brief summary of the frequency of de novo malignancy, transplantation of malignancy, or recurrence of malignancy in organ transplant recipients and in other immunosuppressed patients. For an earlier but more extensive evaluation, the reader is referred to the authors' article in Transplantation *14*:407, 1972.)

Rapp, R.: Question: Do herpesviruses cause cancer? Answer: Of course they do! J. Nat. Cancer Inst. *50*:825, 1973. [Excellent summary of the data and arguments for the role of herpes viruses (Epstein-Barr and herpes simplex) in the etiology of human malignancies.]

Smith, R. T.: Possibilities and problems of immunologic intervention in cancer. N. Engl. J. Med. *287*:439, 1972. (A detailed monograph on the overall subject of tumor antigens, immunologic responses to tumors, and current and future procedures for utilizing the immune response to treat or prevent malignancies.)

Temin, H. M.: The protovirus hypothesis: Speculations of the significance of RNA-directed DNA synthesis for normal development and for carcinogenesis. J. Nat. Cancer Inst. *46*:3, 1971. (An editorial explicating the protovirus theory of origin of new viruses and new genetic information from mammalian cell genomes, and correlating this theory with the oncogene-virogene hypothesis of Huebner and Todaro.)

Terman, D. S., et al.: Preliminary observations of the effects on breast adenocarcinoma of plasma perfused over immobilized protein A. N. Engl. J. Med. *305*:1195, 1981. (Preliminary positive results; the mechanism of action may be the removal of tumor-directed blocking factors.)

Waldmann, T. A., Strober, W., and Blaese, R. M.: Immunodeficiency disease and malignancy: Various immunologic deficiencies of man and the role of immune processes in the control of malignant disease, Ann. Intern. Med. *77*:605, 1972. (This excellent symposium discusses immunobiology, human immunodeficiency syndromes, and the frequency and character of malignancy in these population groups. The authors additionally address the question of the reason for the association of malignancy with immunodeficiency. This article is highly recommended.)

Wolf, P., and Reid, D.: Use of radiolabeled antibodies for localization of neoplasms: State of the art. Arch. Intern. Med. *141*:1067, 1981. (Advances in monoclonal antibody production and radionuclide scintigraphy offer the promise of non-invasive tumor localization and therapy by antibody-coupled isotopes and drugs.)

Index

Page numbers followed by f indicate a figure; page numbers followed by t indicate tabular material.

acetylcholine (ACh), 502–504, 777
  motor nerve and, 789
  neuromuscular transmission by, 818
  reaction of AChR with, 792
  storage and release of, 790
acetylcholine esterase (AChE), 791
acetylcholine receptor (AChR), 791
  reaction of ACh with, 792
achalasia, 485
acid
  defined, 303
  production and excretion of, 306
  renal, excretion of, 308
acid–base balance, respiratory disease and,
      160
acid–base disturbances, 309, 310f
  mixed, 321
acid–base homeostasis, 303–321
acidosis
  defined, 309
  lactic, 316t
  metabolic, 312
  renal tubular, 317
  respiratory, 310
acquired hemolytic anemias, 887
acquired immunodeficiency syndrome (AIDS),
      859
acromegaly, 329
ACTH
  Cushing's disease and, 345
  hypopituitarism and, 333
  synthesis of, 338
ACTH syndrome, ectopic, 346
actin, muscle physiology and, 807, 808f, 811
action potentials, cardiac, 84f
active specific immunization, 909
Addison's disease, 77, 350
adenosine triphosphate (ATP)
  actin and myosin interaction with, 811
  cardiac contraction and, 6

generation of, 395t
  metabolism and, 387
adenylate cyclase system, 871
ADH. See antidiuretic hormone
adrenal cortex, 337–357
  hypofunction of, 350
  regulation of, 338–344
adrenal glucocorticoid deficiency, 293
adrenal hyperplasia, congenital, 352, 353f
adrenal insufficiency, 286
  diagnosis of, 351
adrenal medulla, 355–357
adrenal steroids
  allergy and, 872
  metabolism of, 341
adrenergic drugs, atopic diseases and,
      877
adult respiratory distress syndrome (ARDS),
      145
afibrinogenemia, congenital, 688
afterload
  cardiac performance and, 14, 60
  congestive heart failure and, 18
agammaglobulinemia
  infantile, 851
  Swiss, 854
aganglionosis of colon, 495
AIDS (acquired immunodeficiency syndrome),
      859
airway
  hyperreactivity of, 159
  obstruction of, 150
  reactivity of, 162
  ventilatory function and, 154–157
akathitic movements, 781
albinism, 413
albumin, 212
  biosynthesis of, 224
  loss of, 227
  metabolism of, 223, 226t

albumin (*continued*)
 plasma, 222
 thyroid hormone binding and, 370
alcoholism
 cirrhosis and, 576
 ineffective thrombopoiesis and, 679
aldosterone, 57
 hypersecretion of, 349
 in hypertension, 66, 70–71
 secretion of, 342, 344f
 sodium balance and, 237
aldosteronism, primary, 63, 349
alkaline phosphatase
 bone disease and, 450
 concentration, biliary obstruction and, 600
alkalosis
 defined, 309
 metabolic, 320
 respiratory, 312
alkaptonuria, 413
allergen, atopic disorders and, 865
allergic mechanisms, 861–878
allergic rhinitis, 872
alveolar epithelial cell
 large, 184, 185f, 186f
 small, 186
alveolar hypoventilation, 151, 153t
alveoli, perfusion and, 151
alveoloarterial oxygen gradient, 159
amenorrhea, genetics and, 435–439
amino acid
 absorption of, 538
 hepatic encephalopathy and, 574
 intestinal transport of, 538
amino acid metabolism, disorders of, 408–414
amniocentesis, 418
amniotic fluid embolism, 691
anaerobic glycolysis, 168f, 177–179
anal canal, 476
 function of, 493
anaphylactoid purpura, 698
anaphylatoxin(s), 708
anaphylaxis, 868, 875
anemia
 acquired hemolytic, 887
 aplastic, 630
 defined, 626
 microangiopathic hemolytic, 692
 tissue perfusion and, 37
angioneurotic edema, hereditary, 707f, 712
angiotensin, in pulmonary circulation, 196
anion gap, calculation of, 315t
antiarrhythmic drugs, 102–105

antibiotics, neuromuscular transmission and, 802
antibody(ies)
 erythrocytes with, 652
 immunological injury and, 81
anticoagulants, circulating, 693
antidiuresis, drug-induced, 294
antidiuretic hormone (ADH), 334
 inappropriate secretion of, 294
 tissue perfusion and, 30
 urine concentration and, 244–245, 275
 urine dilution and, 288
antigen
 allergic mechanisms and, 861
 tumor, 901
antihemophilic factor, 686
antihistamines, atopic diseases and, 876
anxiety states, muscle contraction and, 28
aorta
 atherosclerosis and, 39
 circulation dynamics and, 47
 coarctation of, 63
aortic stenosis, muscle contraction and, 34
aplastic anemia, 630, 661
apraxia, 752
arachidonic acid metabolism, 29
argyrophil cell, 191f
arrhythmia(s), cardiac, 101–105
arterial oxygen content, 164
arterial pressure
 aortic coarctation and, 63
 catecholamines and, 50, 51t, 64
 circulatory dynamics and, 45
 extracellular fluid volume and, 52
  control of, 53
 hemodynamics and, 47
 hormonal influences on, 57
 hypertension and, 59–71
 hypotension and, 71–79
 intravascular volume and, 53
 mechanisms controlling, 45–79
 plasma volume and, 66, 66t
 primary aldosteronism and, 63
 red cell mass and, 66t
 renal parenchymal disease, 62
 renal pressor mechanisms and, 69–71
 renal pressor system and, 54
 shock and, 72–79
arteriolar hypertrophy, 38
arteriosclerosis, 107
 tissue perfusion and, 39
artery(ies), histology of, 108
ascending cholangitis, 491

ascites, 570
asthma, 870, 872
ataxia, 752
ataxia–telangiectasia, 857
atherogenic mechanisms, 107–117
atherosclerosis, 107
  etiology of, 109
  plasma lipids and, 112
  platelets and, 110
  process of, 108
  risk factors and, 116
  tissue perfusion and, 39
athetosis, 780
atopic dermatitis, 875
atopic disorders
  diagnosis of, 875
  immunology of, 862
  treatment of, 876
atopy
  chemical mediators of, 867t
  defined, 862
ATP. *See* adenosine triphosphate
atrial tachycardia, 92
atrophic gastritis, 505
autoerythrocyte sensitization, 700
autoimmune disease
  diagnosis of, 912
  hypothyroidism and, 378
  mechanisms of, 883
  treatment of, 891
autoimmune thrombocytopenic purpura, 676
autonomic insufficiency, 79t
autonomic nervous system. *See* nervous system, autonomic
autonomic receptors, gastrointestinal, 479
autonomous thrombocytoses, 678

B-cell, immunity and, 840
bacteremeia, cholangitis and, 602
bacteria
  complement and, 709
  intestinal, 526
    overgrowth of, 543
bacterial biotransformation in man, 528t–529t
ballism, 780
basal ganglia, 771–778
  anatomy of, 772, 773f
basal gastric secretion, 514
base
  define, 303
  production and, excretion of, 306
benign monoclonal gammopathy, 846
bicarbonate secretion in stomach, 511

bile
  bacteriology of, 602
  characteristics of, 582t
  flow of, 591
    canalicular, 592
      biliary tract obstruction and, 598
    ductular, 593
    supersaturated, 605
bile acid pool, 554, 595
bile acids, 594
  enterohepatic circulation of, 551
  gallstone dissolution with, 611
  intestinal absorption of, 553
  reabsorption of, 596
  secretion of, 552
  synthesis of, 551
    cholesterol and 552f
  triglyceride absorption and, 549
bile ducts, 490, 591–603
bile lipids, secretion of, 605, 605t
bile salts
  bile duct obstruction and, 599
  enterohepatic circulation of, 593
biliary dyskinesia, 490, 588
biliary pain, 613
biliary system disorders, 490
biliary tract
  motility of, 584
    hormonal control of, 587
    neural control of, 586
  obstruction of, 596–600
bilirubin, 563
  bile duct obstruction and, 598
  solubility of, 609
  red cell destruction and, 649
  uptake of, 564
block, pacemaker impulses and, 88
blood. *See also* circulatory system
  physical characteristics of, 41
blood flow. *See also* circulatory system
  arteriosclerosis and, 39
  autonomic control of, 49
  collagen diseases and, 41
  intestinal, 526
  pregnancy and, 41
  hypertension and, 38
  renal, 24
  resistance to
    mechanical factors and, 41
    structural factors and, 38
blood loss, thrombocytosis and, 672, 674f
blood pressure, tissue perfusion and, 23. *See also* arterial pressure

blood volume, cardiac output and, 47f, 48f
body fluid
    composition, 233
    normal volumes, 231
    migration, 232
    tonicity, 271–295
body fluid volume
    extracellular, 52
        control of, 53
    intravascular, 67, 69f
    normal, 231
        maintenance of, 234
    sodium and, 231–246
bone
    cells, 446
    formation, 456, 457
    metabolism, 452t
    as organ, 451
    organic matrix, 448
    physiology, 445
    resorption, 455, 457
bone disease, 445–467
    pathophysiology of, 458f
bone marrow. *See* marrow
bone mineral, 449
borderline hypertension, 60
botulism, neuromuscular transmission and,
        795
bowel infection, 472f
bowel syndrome, functional, 495
bradykinesia, parkinsonism and, 783
bradykinin, 31–32, 708
breathing. *See also* respiration, control of *and*
        ventilation
    normal, 136, 136f
    oxygen cost of, 147
bronchial epithelial cells, lung metabolism
        and, 191
bronchial glands, 191
bronchial obstruction, 146f
bronchoconstriction, allergy and, 869
bronchodilator, lung volume and, 159
bronchopulmonary disease, chronic, 153
brush cell, alveolar, lung metabolism and, 186
Bruton's agammaglobulinemia, 851
buffer, defined, 306
Burkitt's lymphoma, 897

C-reactive protein, complement and, 710
calcification, 449
calcitonin
    bone disease and, 453

calcium regulation and, 260
    osteoporosis and, 463
calcium, 255–263
    absorption of, 540
    distribution of, 255
    homeostasis of, 256
    intestinal absorption of, 257
    physiology of, 255
    plasma fractions of, 255t
    regulation of, 257–262
        hormonal, 257
        nonhormonal, 261
    tubular reabsorption of, 261
calcium channel blockers, 8
calcium transport, muscle physiology and,
        814
calorigenesis, 371
cancer. *See also* carcinoma
    gastric, 506
    thyroid, 384
candidiasis, mucocutaneous, 857
capillary(ies)
    of lung, 195
    oxygen transport and, 171
carbohydrates
    absorption, 523
    digestion, 534
    liver disease, 576
    malabsorption, 537
carcinoma
    adrenal cortical, 347
    cervical, 898
    hepatocellular, 898
    nasopharyngeal, 898
cardiac dysrhythmia. *See* dysrhythmia
cardiac failure, muscle tissue perfusion and,
        28
cardiac muscle
    contractile state, 14
    force–velocity relationship and, 11, 11f
    muscle length and, 10f, 11
    myocardial failure and, 20
    wall stress and, 13
cardiac output, 47
    hypertension and, 68
    hypotension and, 77
    oxygen delivery and, 164
cardiac performance, 5–22
    drugs and, 7
    muscle capabilities and, 11
    muscle length and, 10f, 11
cardiac rhythm impulses. *See also* dys-
        rhythmia

generation of, 83
    abnormal sites of, 85
    propagation of, 85–88
    reflection of, 99, 100f
cardiomyopathy(ies), 21
cardiorespiratory disease, perfusion and, 150
cardiovascular system
    hypercalcemia and, 262
    hypocalcemia and, 263
    pressure in. *See* arterial pressure; hyperten-
        sion; hypotension
    tissue perfusion and, 23
carnitine deficiency myopathies, 826
carotid bodies, respiratory control and, 128
cartilage, 451
catecholamine(s) 64
    circulation and, 50
    hypersecretion of, 356
    metabolism of, 51t
    metastatic thyroid and, 376
    muscle contraction and, 26
    synthesis of, 355
cation fluxes, cardiac muscle and, 8
CCK, biliary motility and, 587
cell membranes, oxygen transport and, 171
cell transformation, tumor immunology and,
    899
cellular interactions, immunity and, 850
cellular lipid metabolism, 115
cellular lipid uptake, 114
cellular respiration, 163–182
    oxygen delivery and, 164
central nervous system (CNS)
    hypophosphatemia and, 267
    hypoxia and, 180
    motor activity controlled by, 741–768, 745f
    respiratory control and, 132–135
cerebellar cortex, 755
cerebellar syndromes, 766, 767
cerebellum, 753–768
    afferent connections of, 753
    anatomy of, 753, 754f
    disease of
        diagnosis of, 767
        oculomotor signs of, 764
    eye movement and, 761–766
    somatic movements and, 757–760
    speech regulated by, 766
cerebral cortex
    map of, 747f
    motor disturbance associated with, 752
cerebrocerebellum, somatic motion and,
    757

cerebrospinal fluid (CSF), respiratory control
    and, 128
cerebrum, movements controlled by, 743–753
cervical carcinoma, 898
chemoreceptors, respiratory control and, 128
chemotaxis, 709
chemotherapy, marrow hypoplasia and, 679
chenodeoxycholic acid, gallstone dissolution
    and, 611
Cheyne-Stokes breathing, 138
chloramphenicol, oxygen utilization and, 177
cholangitis, 601
    ascending, 491
cholecystitis, 490
    pathogenesis of acute, 588
    vascular factors in, 590
choledocholithiasis, 611
cholelithiasis, 490, 603–615. *See also* gallstones
cholestasis, 567
    pruritus of, 569
cholesterol
    absorption of, 554
    cellular transport of, 114
    gallbladder mucosa and, 584
    regulation of, 414
cholesterol crystal, formation of, 608
cholesterol saturation index, 606
cholinergic neurogenic pathway, 35
chorea, 779, 786
Christmas disease, 688
Christmas factor, 684
chylomicron, formation of, 550
chyluria, 223
ciliated epithelium, lung metabolism and, 191
circulatory failure, 16–22, 17f
    myocardial failure and, 18
    ventricular hypertrophy and, 20
    ventricular loading conditions and, 21
circulatory system
    dynamics of, 45
    neural control of, 49
    tissue perfusion and, 23
        intrinsic regulation of, 36
circus movement, cardiac, 92, 93f
    reentry and, 100
cirrhosis, 571
    formation of ascites in, 572
clara cell, 189, 190f
clonus, motor cortex lesions and, 751
clotting, 42
    intravascular, 691
clotting mechanism, 681–684. *See also* coagu-
        lation

clotting time of whole blood, 685
CNS. *See* central nervous system
coagulation mechanisms, in lung, 198, 199f
coagulation. *See also* clotting
  disorders
    acquired, 690–693
    diagnosis, 684–693
    hereditary, 686–690
  disseminated intravascular, 732
collagen, bone disease and, 448
collagen diseases, blood flow and, 41
collagenase, 447
colon
  aganglionosis of, 495
  anatomy of, 476
  function of, 493
colonic motor function, 494
common variable immunodeficiency, 852
compensated thrombocytolytic state, 672, 673t, 677
complement, 700–722
  acquired defects of, 721
  components of, 701t
    biologically active fragments from, 708t
    biosynthesis of, 711
    chemotactic factors released from, 710t
    histocompatibility genes and, 712
    immune complexes and, 720
    physiochemical properties of, 702
  deficiencies of, 712–720
  hereditary defects of, 712
    clinical disorders and, 720t
  noncytolytic pathological functions of, 707
  pathways of action of, 703–707
    alternative, 703f, 705, 705t
    classical, 703
    properdin, 705
concentrating mechanisms
  countercurrent multiplier, 277f
  impaired, 281
  medullary, 278
  physiological influences on, 280
  renal, 274–287
conduction, cardiac. *See* dysrhythmia
congenital adrenal hyperplasia, 352, 353f
congenital afibrinogenemia, 688
congenital anomalies, vascular, 41
congenital dysfibrinogenemia, 688
congenital myasthenic syndromes, 801
congestive heart failure, 17–20, 28, 240
Conn's syndrome, 63, 349
connective tissue, hereditary disorders of, 698
contraceptives, oral, 441

contractile mechanism, in Duchenne muscular dystrophy, 826
contractile state of cardiac muscle, 14–16
Cori cycle, 390, 391f
coronary artery disease. *See also* atherosclerosis
  risk factors for, 116
corpus luteum, function of, 435
corpus striatum, 771
cortex, motor. *See* motor cortex
corticosteroid(s)
  atopic diseases and, 877
  biosynthesis of, 339
cortisol, 339, 342
cough receptors, respiratory control and, 131
craniosacral nerves, 479
cretinism, goitrous, 411
Crigler-Najjar syndrome, 566
cromolyn, 877
cross-matching, 686
crypts of Lieberkuhn, 524
CSF. *See* cerebrospinal fluid
Cushing's syndrome, 344
  causes of, 346
  clinical manifestations of, 344
  diagnosis of, 347f
  types of, 348t
cyanocobalamin. *See* vitamin B12
cyclic neutropenia, 662
cystic duct obstruction, 588
cytolysis, cell-mediated, 883, 884f

defecation, muscle contraction and, 494
dermatitis, atopic, 875
desmolase deficiency, 355
DI. *See* diabetes insipidus
diabetes, genetics and, 417
diabetes insipidus
  ADH and, 336
  nephrogenic, 282
  pituitary, 286
diabetes mellitus, 40
  glucose production and, 393, 393t, 398
diabetic gastroparesis, 489
diabetic ketoacidosis, 396t
  pathogenesis of, 399f
diaphragm, 134
diarrhea, 540
  malabsorption and, 555
DIC (disseminated intravascular coagulation), 732
diet, urine concentration and, 281

diffusion
  lung capacity of, 160
  ventilation and, 151
diffusion theory of glomerular sieving, 213
DiGeorge syndrome, 853
digestion, 523–527
  electrolyte movement during, 532
  passive water movement during, 534
  tests of, 542
digitalis, 9
  fibrillation and, 101
dilutional hyponatremia, 243
disseminated intravascular coagulation (DIC),
    732
diuretics, urine dilution and, 292
diverticula, 486
diverticulosis of colon, 494
DNA, globin synthesis and, 638
dopamine
  biosynthesis of, 775f
  Huntington's disease and, 787
  parkinsonism and, 783
  pathways of, 777
  TSH and, 363
dopamine synapse, 776
dopamine system, biochemistry of, 773
downbeat nystagmus, 765
driving pressure, oxygen delivery and, 170
drugs
  cardiotonic, 7
  neuromuscular transmission and, 802
Dubin-Johnson syndrome, 566
Duchenne muscular dystrophy, 822
  contractile mechanism in, 826
  membrane abnormalities in, 824–826
  muscle enzymes in, 823
dumping syndrome, 489
duodenal ulcers, gastric secretion and, 517,
    518
dwarfism, 330, 331t
  immunodeficiency and, 856
dysarthria, 766
dysfibrinogenemia, congenital, 688
dyskinesias
  clinical features of, 782t
  oropharyngeal, 484
  paroxysmal, 781
  tardive, 787
dysoxia, 163
dyspnea, 138–140
dysrhythmia. *See also* cardiac rhythm im-
    pulses
  blocked impulses and, 88

drug therapy for, 102
  effect of, 103t
mechanisms of, 83–105
  ectopic impulses and, 85
  impulse generation and, 83–85
  impulse propagation and, 85–88
  reciprocal beats and, 88, 89f
  reentry mechanism and, 88
dystonia, 780

ectopic ACTH syndrome, 346
edema, hereditary angioneurotic, 707f, 712
edematous states
  urine dilution and, 293
  urine concentration and, 284
efferent immunological enhancement, 882
electrolyte. *See also* body fluid; body fluid
    volume; *specific electrolytes*
  hypertension and, 70–71
  movements of, 533
  water absorption and, 532
electromyography, single-fiber, 794
electrophoresis, 844f, 845
Embden-Meyerhof (EM) pathway, 646
embolism, amniotic fluid, 691
encephalopathy, hepatic, 573
endocrine syndromes in lung disease, 201
endocrine system
  erythropoiesis and, 628
  reproduction and, 423–442
endothelial cell, lung metabolism and, 189
enterohepatic circulation, 552
  biochemical, 554
enzyme(s)
  abnormality of, immunodeficiency due to,
    854
  defects of, red cell abnormalities and, 650
  in hepatobiliary diseases, 569
  muscle, Duchenne muscular dystrophy and,
    823
  proteolytic, lung disease and, 200f
eosinophil chemotactic factor of anaphylaxis
    (ECF-A), 868
epinephrine
  biosynthesis of, 356f
  glucose production and, 390
epithelium, intestinal. *See* intestinal epi-
    thelium
erythremia, defined, 626
erythrocyte, 623, 624. *See also* red cell
  mature, 645–653
  membrane structure and function of, 645

erythrocyte (*continued*)
  metabolism of, 647f
erythroid cell, 629–645
  biosynthesis of, 631
  cytoplasmic maturation of, 633
  iron in development of, 635–638
erythron, 623
  pathophysiology of, 624t
erythropoiesis, 623–653
  iron-deficient, 638
  regulation of, 626
erythropoietin, 180, 626
esophagitis, reflux, 486
esophagus
  anatomy of, 475
  diseases of, 484
  spasm of, 486
  in swallowing, 481
essential hypertension, 59–61
  in atherosclerosis, 110
  cardiac output in, 68
  diastolic arterial pressure in, 67–68
  low renin, 71
  plasma volume in, 66–67
estradiol, 424
estrogen
  arterial pressure and, 58
  reproduction and, 439
euthyroid syndromes, 382
excitation–contraction coupling, 819f
exercise
  cardiopulmonary response to, 161
  energy requirements for, 397
  gas exchange abnormalities during, 161t
  hemodynamic response to, 63
  muscle contraction and, 28
experiment autoimmune myasthenia gravis,
    798
expiration, forced, 154
extracellular fluid volume depletion, 291–292
extrapyramidal system, 771–788
extrinsic asthma, 872
eye movements, cerebellar regulation of, 761–
    766

Fanconi syndrome, 317
fasting, 396t
  albumin synthesis and, 226
fat metabolism, liver disease and, 576
fatty streak, atherosclerosis and, 108
fecal fat test, malabsorption and, 557
fecal sampling, 542

fetus, endocrine system and, 439
fibrillation
  atrial, 94f
  clinical, 97
  production, 96
fibrin
  degradation of, 726, 727f
    tests for, 728
  formation of, 682f
fibrin-stabilizing factor, deficiency, 689
fibrinogen
  asymmetric degradation of, 726
  fibrin degradation and, 726
  hereditary disorders of, 688
  plasma, measurement of, 686
fibrinolysis, 722–734
  activation of, 725
  components of, 722
  control of, 729
  inhibitors of, 724
  reactive, 732
fibroelastic lesion, atherosclerosis and, 109
fibrous plaques, atherosclerosis and, 109
fictitious thyrotoxicosis, 372
Fitzgerald trait, 689
Fletcher trait, 689
flutter, cardiac. *See also* tachycardia
  experimental, 92
folate, biological function of, 631
folic acid absorption, 539
follicle-stimulating hormone (FSH), 425, 430
  ovary and, 434
  testis and, 428
follicular cancer, 384
forced vital capacity, ventilatory function and,
    154
free fatty acid, production of, 393
fuel source, stores of in man, 389t

GABA (gamma-aminobutyric acid)
  dopamine system and, 773–778
  Huntington's disease and, 786
galactorrhea, 332
gallbladder, 490, 581–591
  bile acid pool, 595
  bile acid storage in, 553
  filling and emptying of, 585
  motility disturbances of, 607
gallbladder mucosa, 581–584
  electrolytes and, 582
  transport mechanisms of, 581
  water absorption in, 582

gallstones. *See also* cholelithiasis
    cholangitis and, 601
    cholesterol, 604
        drugs and, 607
    classification of, 603
    common duct, 613
    dissolution of, 611
    growth of, 609
    pathogenesis of, 603
    pigment, 604, 609
gammopathy, 844, 846t
ganglion, stomach innervation and, 503
gas exchange, 159
    exercise and, 161
    in lungs, 149–152
gastric cancer, 506
gastric emptying, 488
gastric emulsification, triglyceride absorption
        and, 547
gastric function, tests of, 520
gastric juice, components of, 508
gastric motor function, disorders of, 489
gastric mucus secretion, 511
gastric secretion, 497–520
    basal, 514
    control of, 512–520
    inhibition of, 518
    meal response and, 513
    mechanisms of, 506–512
    patterns of, 506
gastric ulcer disease, 505
    duodenal, 516, 518
    pepsin in, 510
gastrin, 514, 516
    gallbladder motility and, 587
gastritis, 505
gastrointestinal muscle
    anatomy of, 475
    nerves of, 477
gastrointestinal system, hypercalcemia and,
        262
gastroparesis, diabetic, 489
gaze-evoked nystagmus, 765
Gell and Coombs classification of immunologi-
        cally mediated hypersensitivity, 836t
genetic counseling, 419
genetic disorders, 417–419
genital development, 429f
gigantism, 329
Gilbert's syndrome, 565
gland(s), bronchial, 191
Glanzmann's disease, 697
globin synthesis, 634f, 638

glomerulonephritis, hypocomplementemic, 710
glomerulus, 211–216
    anatomic changes in, 215
    disease induced changes in, 222
    filtration by, 211
    permeability of, 221
    sieving by, 214
    structure of, 213
glucagon, 390
glucocorticoids, 342
    actions of, 343
    bone disease and, 456
    hypersecretion of, 344
    hyposecretion of, 350
    TSH and, 363
gluconeogenesis, 390, 391f
glucose production, 389–393
    diabetes mellitus and, 398
    exercise and, 397
    obesity and, 396
    pregnancy and, 397
    starvation and, 395
glycocholic acid breath test, 558
glycogen
    degradation of, 816f
    metabolism and, 389
    starvation and, 395
glycolysis, 646
    anaerobic, 168f, 177–179
    metabolic control of, 178
glycolytic pathway, 388
goiter. *See also* thyroid gland
    cretinism and, 411
    nontoxic multinodular, 383
    toxic nodular, 374
goitrous cretinism, 411
gonadal development, 429f
gonadal steroid(s), 424
gout, 407
granulocyte colony-stimulating factor, 660
granulocyte(s) production, 655
    factors regulating, 659
    rate of, 657
granulocytic leukemia, chronic, 662
granulocytopenia, 661
granulopoiesis
    abnormalities in, 661–664
    kinetics of, 656
    oscillatory nature of, 659
Graves' disease, 374, 378, 886
growth factors, GH-dependent, 328t
growth hormone (GH), 325, 326–331
    deficiency of, 330

growth hormone (*continued*)
 hypersecretion of, 328
 metabolic effects of, 329
 secretion of, 327t
gut, 476f. *See also* gastrointestinal muscle

Hageman factor, 682
Hashimoto's thyroiditis, 379, 887
hCG (human chorionic gonadotropin), 425
 pregnancy and, 440
HCl secretion in stomach, 508
heart. *See also* arterial pressure; cardiac *entries*; tissue perfusion
 arrythmias of, 101–105
 contraction mechanism of, 6
 failure of, 28
 hypoxia and, 181
 intact, mechanical properties of, 12
 mechanical performance of, 9–16
 mechanisms of performance of. *See* cardiac performance
 neonatal development of, 5
heme biosynthetic pathway, 633
hemiplegic posture, 750
hemochromatosis, idiopathic, 416
hemodynamics, hypertension and, 59
hemoglobin
 abnormal, 641f
 acquired abnormal, 169
 bilirubin in, 564
 heritable disorders of, 169
 oxygen affinity with, 166t
 oxygen binding to, 165
 oxygen transport and, 624
 red cell destruction and, 649
hemoglobinopathy(ies), 641
hemolysis, 565
hemophilia, 686
hemorrhage, tissue perfusion and, 34
hemostasis, 681–700. *See also* coagulation *and* clotting
 vascular disorders of, 697–700
Henderson-Hasselbalch equation, 304
heparin, lung, 199
hepatic acinus, 562f
hepatic disease, 241, 692
hepatic encephalopathy, 573
hepatic lobule, 562f
hepatic uptake, bile acids and, 553
hepatitis, bilirubin in, 566
hepatobiliary diseases, serum enzymes in, 569
hepatocellular carcinoma, 898

hepatocytes, 592
 injury and necrosis of, 577
hepatorenal syndrome, 572
hereditary angioneurotic edema, 707f, 712
herpes simplex, Wiskott-Aldrich syndrome and, 856
high altitude, exposure to, 37
Hirschsprung's disease, 485
hirsutism, female, 438
 testosterone production in, 439t
histamine, allergic mechanisms and, 867
histocompatibility, disease susceptibility and, 415
histocompatibility antigens, complement and, 710
histocompatibility complex, major, 847
Hodgkin's disease, 898
homeostasis
 acid-base, 303–321
 body protein, maintenance of, 211–229
hormone(s), 423. *See also name of hormone*
 hypertension and, 57
 secretion of by lung tumors, 201
 vasoactive, 195
human chorionic gonadotropin (hCG), 425
 pregnancy and, 440
human intrinsic factor, secretion of, 512
human leukocyte antigens (HLA), 415
 disease association and, 848t
Huntington's disease, 786
hyaline droplets, tubular absorption and, 216
hydrogen ion concentration, measurement of, 303
hydroxyapatite, bone disease and, 449
hydroxylase deficiency, 352
hyper-Ige syndrome, 858
hyperaldosteronism, primary, 349
hyperbilirubinemia, 563
 drug therapy for, 565
hypercalcemia, 262
 malignancy and, 461
hypercalcemic nephropathy, 283
hypercapnea, 160
 acute, 311, 312
hypercapnic respiratory failure, 152
hypercholesterolemia, 415
 atherosclerosis and, 113
hyperfibrinogenolysis, 730f, 731
hyperfibrinolysis, 731
hyperglobulinemia, 844
hyperglycemia, 400
hyperkalemia, 250–255
 biological effects of, 251

clinical manifestations of, 251
hypertension and, 71
mechanisms of, 251f
pathogenesis of, 252
hyperlipidemia, 40
hyperlipoproteinemia, 40
hypermagnesemia, 265
neuromuscular transmission and, 797
hypernatremia, 245
hyperosmolality, medullary, 276
impaired, 281
hyperparathyroidism, bone involvement in, 459
hyperphosphatemia, 266
hyperprolactinemia, 332
genetics and, 437
hyperproteinemia, 241
hypersensitivity state, defined, 862
hypersplenism, 652
thrombopoiesis and, 673, 677
hypertension, 59–71
atherosclerosis and, 117
blood flow resistance and, 38
borderline, 60
cardiac output and, 68
essential, 60
established essential, 59
hemodynamic characteristics of, 59t
labile, 60
muscle contraction and, 28
plasma volume in, 66t
portal, 569
red cell mass in, 66t
renal, 61
renoprival, 61
renovascular, 61
hyperthyroidism, 38, 369f, 372–377
diagnosis of, 373f, 376
lymphocytic, 375
therapy of, 376
TSH-induced, 374
hypertonic medullary interstitium, 276
hyperuricemia, 71, 406, 408
hyperventilation, metabolic acidosis and, 314
hypoaldosteronism, 355
in hypertension, 70–71
hypocalcemia, 263
hypogonadism, 431–433
hypokalemia, 252
biological effects of, 254
clinical manifestations of, 254
hypertension and, 70–71
mechanisms of, 253t

hypokalemic nephropathy, 283
hypomagnesemia, 264
hyponatremia, 242
hypoparathyroidism, 465
hypoperfusion, 16
hypopharyngeal diverticulum, 485
hypopigmentation syndrome, 413
hypopituitarism, 333
hypoplasia
marrow, 678
thymic, 853
hypoprolactinemia, 333
hyposensitization, atopic diseases and, 877
hyposthenuria, 283
hypotension, 71–79
acute, 72
chronic, 77
classification of, 72t
hemodynamics of, 74
idiopathic orthostatic, 78
orthostatic, cerebral perfusion and, 33
hypotensive shock, 27
hypothalamic disease, 381
hypothalamic hormones, 324, 325t
hypothalamic–pituitary axis, 323–337
hypothalamic–pituitary–thyroid axis, 360f
hypothalamus, 323
hypothyroidism, 377–381
diagnosis of, 380
etiology of, 378
therapy of, 381
urine dilution and, 293
hypoventilation, alveolar, 151, 153t
hypovolemia
orthostatic hypotension, 33
urine dilution and, 71, 291
hypoxanthine guanine phosphoribosyltransferase deficiency, 406
hypoxemia, 163, 169
dyspnea and, 138
hypoxemic respiratory failure, 152
hypoxia, 37
organ function changes and, 180
respiratory control and, 128f
hypoxic dysoxia, 163

idiopathic hemochromatosis, 416
idiopathic hyperdynamic state, 49
idiopathic hyperkinetic heart syndrome, 38
idiopathic orthostatic hypotension, 78
idiopathic thrombocytopenia purpura, 696
IgE, allergic mechanism and, 864

ileum
  bile acid transport in, 553
  bile acid uptake by, 596
ileus, 492
immune complexes, complement and, 720
immune cytolysis, complement pathway and, 704
immune mechanisms, lung functions and, 195
immune response, genetic regulation of, 415
immune system, 839
immunity
  adaptive, 839–859
  cellular, failure of, 853
  diseases of, 850–859
  human tumor, 905
  humoral, 840–847
  immunoglobulins and, 842
  T-cells and, 849
immunobiology, scope of, 834t
immunodeficiency diseases, 850–859
  malignancy and, 906
immunoglobulin
  failure of synthesis of, 851
  functions of, 842
  properties of, 843t
  structure of, 841
  thyroid stimulating, 374
immunological injury
  cell-mediated, 882
  mechanisms of, 881–885
immunological surveillance, 907
immunosuppressive therapy, 891
immunotherapy, 908, 913
ineffective thrombopoiesis, 679
infantile agammaglobulinemia, 851
infertility, male, 433
inotropic agents, cardiac performance and, 16
insulin
  diabetes mellitus and, 399
  glucose production and, 390, 392f
  obesity and, 397
intention tremors, 760
interferon, 912
interstitual fluid, 233
intestinal absorption tests, 542
intestinal epithelium, 524
  active transport in, 530
  facilitated diffusion in, 530
  ion exchange in, 532
  microclimate of, 530
  permeability of, 527
  pinocytosis in, 532

simple diffusion in, 532
  structure of, 527
intestinal membrane support, 527–532
intestine, small. *See* small intestine
intragastric titration, 507
intrinsic asthma, 873
inulin, 211, 215f
iodide
  concentration of, 366
  organification of, 366, 379
  thyroid hormone biosynthesis and, 370
iodine deficiency, 380
iron
  absorption of, 539
  bodily distribution of, 637
  hemoglobin synthesis and, 635
iron cycle, 636f
iron deficiency, thrombopoiesis and, 673, 675f
iron-deficient erythropoiesis, 638
irritant receptors, respiratory control and, 131
ischemic vasospastic disease, 34
isosthenuria, 282

J receptors, respiratory control and, 131
jaundice, 567
  classification of, 568f
  obstructive, 690
jejunum, osmotic permeability and, 531
juxtaglomerular apparatus (JGA), 55, 56f

kallikrein, hereditary angioneurotic edema and, 713
ketoacidosis, diabetic, 396t
ketone body, production of, 393, 394f
kidney
  albumin metabolism and, 223, 225f, 226t
  blood flow in, 24
    autoregulation of, 36
  body protein homeostasis maintenance by, 211–229
  calcium homeostasis and, 257
  erythropoietin from, 627
  hypercalcemia and, 262
  hypertension and, 61
  hypomagnesemia and, 264
  hypoxia and, 181
  parathyroid hormone and, 258
  plasma protein metabolism and, 223
kinetic tremors, 760
kinin(s)
  atopic syndromes and, 868
  complement and, 707

Klinefelter's syndrome, 432
Krebs cycle, 174, 175f

L-DOPA, 784
labile hypertension, 60
lactase assay, 544
lactate, oxygen utilization and, 179
lactic acidosis, 316
lactose tolerance test, 543
Laennec's cirrhosis, 37
Lambert-Eaton myasthenic syndrome (LEMS),
       796
Laplace law, 13
laxative colon, 495
lead poisoning, 651
Lesch-Nyhan syndrome, 406
leukemia
   acute, 898
   acute lymphocytic, 663
   acute nonlymphocytic, 662
   chronic granulocytic, 662
leukocyte antigens, human, major histocom-
       patibility complex for, 847
leukocytosis, 662
leukopoiesis, 655–664
Leydig cells, endocrine system and, 428, 430
LH (luteinizing hormone), 425
   ovary and, 434
   testis and, 428
limb movements, cerebellum and, 757
lingual lipase, triglyceride absorption and, 548
lipase
   lingual, triglyceride absorption and, 548
   pancreatic, 548
lipid-soluble vitamins, absorption of, 554
lipid solutes, absorption of, 547–558
lipid storage myopathy, 828
lipids
   cellular, 114
   complex, formation of, 550
   plasma, atherosclerosis and, 112–116
lipoprotein receptors, plasma cholesterol and,
       414
lipoproteins, atherosclerosis and, 113
liver, 561–563
   bilirubin and, 563–566
   bile acid secretion by, 552
   cirrhosis of, 571
   disease of, 575
   failure of, 573
   functions of, 563
   hypoxia and, 181

mechanisms of, 561–578
   portal hypertension and, 569
   protein deficiency and, 575
liver disease, 651
   clotting disorders and, 692
locomotion, cerebellum and, 759
low-density lipoproteins, 113
   metabolism of, 115f
Luft's syndrome, 177
lung(s)
   alveolar stability in, 192
   biologically active compounds of, 197t
       nature and actions of, 198t
   cellular metabolism sites in, 183–192
   coagulation mechanisms of, 198, 199f
   disease of, 200, 201
   endocrine functions of, 183–201
   gas exchange in, 149–152
   immune mechanisms of, 195
   metabolic functions of, 183–201
   metabolic pathways in, 184f
   mucociliary transport in, 195
   phagocytosis in, 194
   protease–antiprotease balance in, 199
   proteolytic enzymes in disease of, 200
   surfactant in, 192
   tumors of, hormonal secretion by, 201, 201t
   vasoactive hormones in, 195–198
lung volumes, ventilation and, 143, 154
lupus erythematosus, systemic, 889
luteinizing hormone (LH), 425
lymphatic transport, 551
lymphocyte(s), T, 849
lymphocytic hyperthyroidism, 375
lymphocytic leukemia, acute, 663

macrophage, alveolar, 194
   lung metabolism and, 187, 187f
magnesium, 45, 263–265
   calcium regulation and, 261
magnocellular system, 323
major histocompatibility complex, 847
malabsorption
   clinical signs of, 555, 555t
   diagnosis of, 555
   intestinal, 540
   tests for, 557
malabsorption syndrome, 540
malignancy. *See also* carcinoma
   hypercalcemia and, 461
   viral role in, 897
malnutrition, urine concentration and, 286

maltase, 535
marrow
    damage to, 630
    failure of, 623
    granulocyte production and, 656
    granulocyte reserve in, 658
    thrombocytopenia and, 695
marrow granulocyte pool (MGP), 656
marrow hypoplasia, 678
mast cell, 447
    lung metabolism and, 187, 188f
meal response, gastric secretion and, 513
    cephalic phase of, 515
    gastric phase of, 516
    intestinal phase of, 517
medullary hyperosmolality, 276
    impaired, 281
medullary thyroid cancer, 385
medullary urine concentrating mechanism,
    278
Meekrin-Ehlers-Danlos syndrome, 698
megaduodenum, 493
megakaryocyte(s), 668, 670f, 672t
    platelet release from, 671
membrane responsiveness, cardiac, 103
membranoproliferative glomerulonephritis,
    710
menstruation, hormonal patterns in, 434
metabolic acidosis, 312–319
    serum anions in, 316
metabolic alkalosis, 320
metabolism, 387–401. *See also* glucose
    genetic mechanisms of, 403–419
    glucose production and, 389
    purine, 405
    tissue perfusion and, 36
methionine transsulfuration, 410–411
metyrapone, 340, 341f
microangiopathic hemolytic anemia, 692
mineralization, bone disease and, 450
miniature end-plate potential (MEPP), 792
mitochondria
    chemical reactions in, 174
    oxygen utilization in, 173
        disorders of, 176
            monitoring of, 179
        structure of, 173
mixed asthma, 873
monoclonal gammopathy, 844
monoglyceride pathway, 550f
mononucleosis, infectious, 663
monosaccharide absorption, intestinal, 536
motilin, gallbladder motility and, 587

motility, 475–495
    gallbladder disturbances of, 607
motor activity, central nervous system control
    of, 741–768
motor cortex, 743–746
    experiments with, 748
    lesions in, 750
    somatic movements and, 746
motor homunculus, 746
motor nerve, 789
motor system
    anatomical connections in, 744f
    cerebellar efferent connections and, 756
    operation of, 742
motorneurons, somatic movements and,
    747
movement disorders, 778–781
MTHF methyl transferase, amino acid metabo-
    lism and, 409
mucin, in bile, 608
mucocutaneous candidiasis, chronic, 857
mucoprotein, gallstones and, 608
mucosal muscle, anatomy of, 477
mucosal uptake, tests of, 543
mucus
    gallbladder epithelium and, 584
    gastric, secretion of, 511
multiple endocrine adenomatosis type II (MEA
    II), 356
multiplier effect, urine concentrating mecha-
    nisms and, 276
muscle action potential (MAP), 794
muscle proteins
    biosynthesis of, 821
    turnover times of, 822t
muscle tissue, proteases in, 821
muscle tone, motor cortex lesions and, 750
muscle(s). *See also name of muscle*
    contractile proteins of, 805
    contractile state of, CNS control and, 741
    movements of, 741
    physiology of, 805–829
muscular dystrophy, Duchenne. *See*
        Duchenne muscular dystrophy
myasthenia gravis, 798, 885
myasthenic syndromes, congenital, 801
myelofibrosis, 663
myocardial contractility, 48
myocardial failure, 18
myocardial infarction, 21
    shock following, 34
myocardium, 6
    cellular structure of, 7f

failure of, 18
diseases of, 21
myoclonus, 779
myocyte(s), 7f
myofibrillar proteins, 822t
myoglobin, 173
myosin
interaction of actin with, 9f
muscle physiology and, 805–807, 811

nasopharyngeal carcinoma, 898
National Cooperative Gallstone Study, 612
NE. *See* norepinephrine
neoplasm
adrenal cortical, 347
bile duct, 599
hypermetabolism and, 376
nephritic factor, complement and, 709
nephrogenic diabetes insipidus, 282
nephron, 211
nephropathy
hypercalcemic, 283
hypokalemic, 283
postobstructive, 284
nephrotic syndrome, 226, 228
nerve. *See name of nerve*
nervous system
autonomic, control of circulation by, 49
central. *See* central nervous system
sympathetic
blood pressure and, 50
hypertension and, 64
tissue perfusion and, 32
neural reflexes, circulation and, 49
neuroendocrine cells, lung metabolism and, 191
neurohypophyseal system, 324f
neurohypophyseal tropic hormones, 325t
neuromuscular junction, 789–803
neuromuscular system
hypercalcemia and, 262
hypocalcemia and, 263
neuromuscular transmission
clinical assessment of, 794
presynaptic disorders of, 795
neurons, motor cortex, 749
neurotransmitters
biochemistry of, 777
distribution of, 774t
neutropenia, cyclic, 662
nifedipine, cardiac performance and, 8
nocturia, 282

nonlymphocytic leukemia, acute, 662
nonspecific immunity, 910
nontoxic nodular goiter, 383
norepinephrine (NE)
arterial pressure and, 52
biosynthesis of, 356f, 775f
normotension, arterial, 45–59
nystagmus
cerebellar disease and, 764
types of, 765

obesity
cholesterol in bile and, 606
glucose production and, 396
obstructive jaundice, 567, 690
obstructive pulmonary disease, 157, 158f
oculocutaneous albinism, 413
oculomotor control systems, 762f
oligosaccharidases, 535
oral D-xylose test, 544
organ(s), vascular resistance of, 24
organic phosphates, oxygen–hemoglobin affinity and, 167
organophosphate compounds, neuromuscular transmission and, 800
oropharyngeal dyskinesia, 484
orthostatic hypotension, cerebral perfusion and, 33
ossification, bone disease and, 450
osteitis fibrosa cystica, 459
osteoblast, 446
bone formation and, 456
osteoclast, 446
disease of, 464
osteocyte, 446
osteodystrophy, renal, 466
osteogenesis imperfecta, 698
osteomalacia, 456, 465
osteopetrosis, 464
osteoporosis, 462–464
hormonal factors in, 463
ovarian virilizing syndromes, 437
ovary(ies), 433–435
tumors of, hyperthyroidism and, 376
ovulation, induction of, 437
oxygen
cellular delivery of, 164–170
consumption of, 37, 148
cost of in breathing, 147
hemoglobin binding of, 165–170
measurement of metabolism of, 164
molecular, 180

oxygen (*continued*)
  movement to use site of, 170
  supply of, 164
  tissue stores of, 173
  transport of, 163–182
    measurement of, 164t
    plasma to use site, 170
  utilization of, 173–177
    mitochondrial, 174
      disorders of, 176
    monitoring of, 179
oxygen exchange
  metabolic control of, 175
  pulmonary, 173f
  in tissues, 172, 173f
oxygen gradient, alveoloarterial, 159
oxygen transport, erythropoiesis and, 625f
oxyhemoglobin dissociation curve, 165,
    165f
oxyntic gland, 499t

pacemaker activity. *See also* cardiac rhythm
    impulses
  atrial flutter and, 92
  frequency of firing of, 84
  characteristics of, 83
Paget's disease, 41, 458
pancreas, test of function of, 542
pancreatic enzyme deficiencies, 556
pancreatic lipase, triglyceride absorption and,
    548
papillary cancer, 384
paramedian pontine reticular formation
    (PPRF), 761
parasystole, 98
parathyroid hormone (PTH)
  bone disease and, 453
  calcium regulation and, 257, 259f
parietal pain, defined, 614
parkinsonism, 781–786
  brain monoamines in, 783t
  clinical features of, 781
  drug-induced, 785
  pathophysiology of, 783
  postencephalitic, 785
  reserpine-induced, 785
paroxysmal dyskinesias, 781
paroxysmal nocturnal hemoglobinuria, 651
paryoxysmal tachycardia, 90, 91f
passive immunization, 909
Pasteur effect, 178
pathological reflexes, motor cortex lesions
    and, 751

pathway(s), cardiac conduction, 94
pellagra, 177
Pendred's syndrome, 379
pendular nystagmus, 765
pepsin secretion in stomach, 510
peptic cells, stomach anatomy and, 501
perfusion
  distribution of, 150
  tissue. *See* tissue perfusion
  ventilation and, 150
peristalsis, 483
  stomach and, 487
pH, 166–167
  Henderson equation and, 304
  respiratory acidosis and, 311
  respiratory alkalosis and, 312
phagocytosis, 194
pharynx
  anatomy of, 475
  diseases of, 484
  in swallowing, 481
phenylalanine, metabolism of, 411
phenylketonuria, 411
pheochromocytoma, 27, 356
phosphates, organic, oxygen–hemoglobin
    affinity and, 167
phosphorus, 265–268
  calcium regulation and, 261
  depletion of, 266
pigment mutations in man, 413
pituitary diabetes insipidus, 286
pituitary gland
  anterior, 325–334
    regulation of, 326
  failure of, 381
  posterior, 334
  TSH production and, 363
  tumor, 329
pituitary trophic hormones, 326t
PK deficiency, enzymatic, 650
PK test, 862
PKU, screening and, 418
placenta, endocrine role of, 398
plasma fibrinogen, measurement of, 686
plasma glucose concentration, 393
  pregnancy and, 397
  starvation and, 396
plasma lipids, atherosclerosis and,
    112–116
plasma proteins
  loss of, 227
  metabolism of, 223
  proteinuria and, 220

plasma renin activity (PRA), hypertension and, 68

plasma steroid concentrations, 427t

plasma volume
  hypertension and, 66t
  physiological relationships of, 67

plasmin, 724

plasminogen, 722, 723f
  activators of, 723
  fibrinolysis control and, 729

platelet(s), 667
  atherosclerosis and, 110
  disorders of, 693–697
  hemostasis and, 693
  megathrombocyte index and, 668f
  pathophysiological mechanisms and, 677–679
  production of, 671
  tests of function of, 694

pneumotaxic center, 132

polyarteritis nodosa, 41

polyclonal gammopathy, 845

polycystic ovary syndrome, 438

polycythemia vera, 630

polydipsia, 286

polyuria, drug-induced, 286, 287t

pore theory of glomerular sieving, 213

portal hypertension, 569

postsynaptic disorders, 798

postsynaptic membrane, 791

postural tremors, 760

posture
  motor cortex lesions and, 750
  proteinuria and, 219

potassium. *See also* hyperkalemia
  in body fluid, 249–255
  depletion of, 255t
  homeostasis of, 250
  intake of, 250
  metabolism of, 249
  transcellular shifts of, 251
  urinary excretion of, 250

pregnancy
  blood flow and, 41
  energy requirements for, 397
  hormone patterns in, 440

prehypothyroidism, 378

preload, cardiac performance and, 11

premotor cortex, 751

pressure–volume relationship of lungs, 144–147
  ventilatory function and, 155

primary hyperaldosteronism, 349. *See also* aldosteronism

PRL. *See* prolactin

progesterone, in pregnancy, 440

prolactin (PRL), 324, 331–333
  serum levels of, 332t

properdin pathway, 705

prostaglandin(s), 426
  atheroscleosis and, 111f
  atopic syndromes and, 869
  tissue perfusion and, 29

protease–antiprotease balance, in lung, 199

protein
  deficiency of, liver disease and, 575
  digestion and absorption of, 537
  of normal urine, 217
  plasma, proteinuria and, 220
  synthesis of, 639f
  tubular reabsorption of, 217

protein binding, thyroid hormone and, 369

protein hormones, 425
  concentrations of, 427t

protein reabsorption, tubular, 217

protein-losing enteropathy, 538

proteinuria
  in disease, 220–223
    types of, 220
  exercise and, 218
  glomerular permeability and, 221
  posture and, 219
  prerenal, 220t
  tubules and, 223

proteoglycans, 449

proteolytic enzymes, lung disease and, 200f

prothrombin time, 685

proton pump, of oxyntic cell, 509

PRPP
  overproduction of, 408
  purine metabolism and, 405

pruritus of cholestasis, 569

pseudoxanthoma elasticum, 698

pubescence, endocrine system and, 428

pulmonary disease
  obstructive, 157
  restrictive, 157

pulmonary function, assessment of, 153–159

pulmonary oxygen exchange, 173f

purine metabolism, disorders of, 405–408

Purkinje cells, 755

Purkinje fibers. *See also* dysrhythmia
  antiarrhythmic drugs and, 103t
  premature contractions and, 100

purpura
  anaphylactoid, 698
  autoimmune thrombocytopenic, 676
  senile, 698
pyelonephritis, acute, 63, 284
pyloric gland, 501
pylorus, 489

radioligand assay, 426
radiotherapy, thyroid cancer and, 384
Raynaud's phenomenon, 34
red cell. *See also* erythrocyte
  abnormalities of
    acquired, 651
    inherited, 650
  aging of, 648
  destruction of, 649
  physical injury to, 651
  wall rigidity of, 42
red cell mass, hypertension and, 66t
red cell temperature, oxygen–hemoglobin
    affinity, 166
referred pain, defined, 614
reflex mechanisms, respiratory control and,
    131
reflexes, motor cortex lesions and, 750
reflux esophagitis, 486
refractory period
  antiarrhythmic drugs and, 104
  cardiac conduction and, 95
  fibrillation and, 97
renal acid excretion, 308, 309f
renal concentrating mechanisms, 275t. *See
    also* urine, concentration of
  impairment of, 281
    drug-induced, 286, 287t
  mammalian, 279f
renal diluting mechanisms
  impaired, 290
    drug-induced, 294, 294t
renal disease
  chronic, 282
    urine dilution and, 292
  erythropoietin and, 628
renal failure, 241
  hypercalcemia and, 262
renal function, metabolic acidosis and, 315
renal hypertension, 61, 61f
renal osteodystrophy, 466f
renal parenchymal disease, 62
renal pressor mechanisms, hypertension and,
    69

renal pressor system, 54, 55f
renal tubular acidosis, 317
renin, 54
  source of, 55
renin–angiotensin system, 29
  hyperaldosteronism and, 349
  renal pressor system and, 55
renovascular hypertension, 61
reproduction, endocrine mechanisms of, 423–
    442
respiration
  cellular, 163–182
  control of, 125–140
    feedback and, 126f, 127f
    nonchemical stimuli and, 130
respiratory acidosis, 310
respiratory alkalosis, 312
respiratory centers, 132, 133f
  output of, 137f
respiratory chemostat, 125, 126f
respiratory disease, pressure–volume charac-
    teristics and, 145
respiratory failure
  hypercapnic, 152
  hypoxemic, 152
respiratory insufficiency, 152
respiratory muscles, 134
  elastic properties of, 144, 144f
  fatigue in, 135
restrictive pulmonary disease, 157, 158f
reticuloendothelial function, 578
reticulospinal tract, 757
rickets, 456
RNA, globin synthesis and, 638
Rotor's syndrome, 566
RTA (renal tubular acidosis), 317

sarcomere, 7f
sarcoplasmic reticulum
  calcium transport by, 815f
  Duchenne muscular dystrophy and, 825
scleroderma, 41, 493
screening, genetic, 418
scurvy, 698
selective IgA deficiency, 853
senile purpura, 698
severe combined immunodeficiency, 854
shock, 72–79
  cellular alterations in, 76
  mechanisms of, 73
  tissue factors in, 75t
short-chain fatty acids (SCFA), 537

short-limbed dwarfism, immunodeficiency
    and, 856
SIADH, 335
sickle hemoglobin, 641
sickle-cell disorders, 285
sieving, glomerular, 214
sieving coefficient, glomerular, 211, 212f, 214f
skeletal muscle
    fiber types in, 820
    hypophosphatemia and, 268
    hypoxia and, 182
    physiology of, 816
skeletal tissue, 451f
skin tests, for allergy, 863
SLE (systemic lupus erythematosus), 889
slow-reacting substance of anaphylaxis (SRS-
    A), 868
small intestine, 524f
    anatomy of, 476, 523
    contents of, 533
    contractions of, 491
    function of, 491
smooth muscle
    gastrointestinal
        contraction in, 480
        electrical properties of, 480, 481f
        swallowing and, 481–484
        ultrastructure of, 477
    lung metabolism and, 189
smooth muscle contraction
    intrinsic tissue factors regulating, 36
    neurogenic factors regulating, 32–35
sodium, 235–246
    aldosterone secretion and, 342, 350
    depletion of, 238, 239
        causes of, 239t
        renal response to, 239
    proximal tubular reabsorption of, 237, 242
    renal tubular reabsorption of, 235
    renin release and, 57
    retention of, 240
    tubular reabsorption of, 242
    urine dilution and, 291
sodium balance, 235
    in hypertension, 70–71
solutes, movement of by osmosis, 531
somatomedins, 328
spasticity, motor cortex lesions and, 750
speech, cerebellar regulation of, 766
spermatogenesis, 431
splenectomy, platelet count and, 676
starvation, 395
steatosis, 576

Stein-Leventhal syndrome, 438
stereotypic movements, 781
steroid(s). *See also name of hormone*
    adrenal, allergy and, 872
    gonadal, 424
    hypertension and, 58
    metabolism of, 341, 439
steroidogenesis, 340f
Stewart-Holmes maneuver, 760
stomach
    anatomy of, 476, 497, 498f
    blood supply to, 504
    cardiac glands of, 499
    function of, 487
    gastric secretion and, 497
    innervation of, 502
    meal response and, 515
    mucosa of, 499
        cell turnover in, 502
        endocrine cell types in, 501t
    oxyntic glands of, 499, 499t, 500f
    pathophysiology of, 504
    neurotransmitters in, 503
    pyloric gland of, 501
    submucosa of, 499
    ulcer and, 504
    wall of, 498
stool fat test, 544
stool weight test, 544
stress
    granulocyte production and, 658
    hemodynamic response to, 63
stretch receptors, respiratory control and,
    131
striated muscle dysfunction, 484
struma ovarii, 376
supernormal A-V conduction, 87f
surfactant, alveolar
    biosynthesis of, 192
    development of, 193
    functions of, 194
swallowing center, 478f
Swiss agammaglobulinemia, 854
sympathetic nervous system. *See* nervous
    system, sympathetic
synapse, dopamine, 776
synaptic cleft, 791
syndrome of inappropriate ADH secretion, 335
systemic lupus erythematosus (SLE), 41, 889

T lymphocytes, cell-mediated immunity and,
    849

tachycardia
  paroxysmal, 92, 93f
  reciprocal, 90
Tamm-Horsfall protein, 218
tardive dyskinesia, 787
telangiectases, scleral, 857f
testis, endocrine system and, 428
testosterone, 425f, 427
  erythropoiesis and, 628
  hirsutism and, 439
thalassemia(s), 642, 643t
  pathophysiology of, 644f
thiamine deficiency, 37
thirst regulation, 273
thoracolumbar nerves, 479
thrombasthenia, 697
thrombin, 681–682
  formation of, 683f
thrombocythemia, 697
thrombocytolytic state, compensated, 672,
      673t, 677
thrombocytopenia, 695
  mechanisms responsible for, 699
thrombocytopenic purpura, autoimmune, 676
thrombocytosis, 672, 697
  autonomous, 678
  reactive, 677
thrombolysis, 730
thrombopathia, 697
thrombopoiesis, 667–679
  ineffective, 679
thrombopoietin, 671, 673–676
thrombosis, 42
  atherosclerosis and, 111f
thromboxane(s), atherosclerosis and, 111
thryoxine-binding prealbumin (TBPA), 370
thymic hypoplasia, 853
thyroid, 359–385. *See also names of hormones*
  cancer of, 384
  hormone biosynthesis in, 365
  iodide concentration and, 366
  physiology of, 359, 364
thyroid adenoma, hyperfunctioning of, 372
thyroid failure, primary, 377
thyroid hormone. *See also* hyperthroidism *and*
      hypothyroidism
  action of, 371
  binding of, 382
  biosynthesis of, 365
    modulation of, 370
  calcium regulation and, 260
  circulating, 368

  Hashimoto's thyroiditis and, 379
  hypertension and, 58
  iodide concentration and, 366
  oxygen utilization and, 177
  protein binding of, 369
thyroid nodules, 384
thyroid stimulating hormone (TSH), 359, 361–
      364
  hyperthyroidism and, 372
  hypothyroidism and, 377
  pituitary reserve of, 363
thyroiditis, subacute, 375
thyrotoxicosis, 886
  fictitious, 372
thyrotropin. *See* thyroid stimulating hormone
thyrotropin-releasing hormone (TRH), 324
  hypothalamic control of, 360
  stimulation of, 362
thyroxine-binding globulin (TBG), 369
tics, 780
tissue
  calcification of, 262
  oxygen exchange in, 173f
  oxygen supply to, 165
tissue perfusion, 23–42
  anxiety states and, 28
  aortic stenosis and, 34
  cardiac failure and, 28
  catecholamines and, 26
  estimation of, 24
  hemorrhage and, 34
  hypertension and, 28
  hypotensive shock and, 27
  kinins and, 31
  orthostatic hypotension and, 33
  prostaglandins and, 29
  regulation of, 23
  renin–angiotensin system and, 29
  smooth muscle contraction and, 36
  vasopressin and, 30
titratable acidity, 308
tone, cerebellar regulation of, 760
total lung capacity (TLC), 143, 154f
  ventilatory function and, 154
totl peripheral resistance (TPR), defined, 46
toxic nodular goiter, 374
trace metals, liver disease and, 577
transformed cells, malignancy and, 899
transsulfuration pathway, amino acid metabo-
      lism and, 409
tremors, 779
  cerebellum and, 759

TRH. *See* thyrotropin-releasing hormone
tricarboxylic acid cycle, 388f
triglyceride
   absorption of, 547–551
   medium-chain, 558
triglyceride breath test, 557
tropomyosin, 809
troponin, 809
troponine C, 809
TSH. *See* thyroid stimulating hormone
tubular protein reabsorption, 217
tubule(s), 216
   necrosis of, 285
   protein reabsorption by, 217
   proteinuria and function of, 223
   renal function of, 236f
   urine concentration and, 275
     impairment of, 282
tumor(s)
   adrenal cortical, 347
   pulmonary, hormonal secretion by, 201t
tumor antigens, 901
tumor immunology, 901–908
   mechanisms of, 895–913
tyrosine, metabolism, 411
tyrosinemia, 412

ulcer
   duodenal, 516, 518
   gastric, 505
upbeat nystagmus, 765
urea entrapment, renal concentration and, 278
urine
   concentration of, 274–287
     impairment of, 281
     mechanism of, 274, 275t
     physiological influences on, 280
   dilution of, 287–295
     impaired, 290–295
      drug-induced, 294
     mechanism of, 287
   normal, protein of, 217
urtigaria, 874

vagal nerves, 502
vagotomy, 489
   gastric secretion and, 513
vascular muscle, contractile state of, 26–38
vascular resistance
   control mechanisms for, 48

   measurement of, 24
   tissue perfusion and, 23
vascular system, components of, 53
vasoactive hormones, pulmonary, 195–198
   inactivation of, 196t
   synthesis of, 197
vasoconstriction
   neurogenic, 33
   tissue perfusion and, 26–38
     anxiety states and, 28
     cardiac failure and, 28
     hypertension and, 28
     hypotensive shock and, 27
     kinins and, 31
     prostaglandins and, 29
     vasopressin and, 30
vasodilator systems, 35
vasopressin, 334
   tissue perfusion and, 30
venom, clotting and, 691
ventilation, 143–149
   distribution of, 149
   elastic resistance and, 144, 144f
   flow resistance and, 146
ventilatory function
   assessment of, 154
   impaired, 157
ventilatory pump, 137
ventricle
   contractility of, 15f
   failure of, 19f
   hypertension and, 60
   hypertrophy of, 20, 60
   left, performance of, 13f
   premature beats of, 98
ventricular function curves, 14
very low density lipoproteins (VLDL), 414
vestibular nystagmus, cerebellar system disease and, 764
vestibuloocular reflex, 762
vestibulospinal tract, 757
virus(es)
   human malignancy and, 897
   oncogenic, classification, 896, 897t
   tumor immunology and, 895, 901
visceral pain, defined, 614
vital capacity, ventilatory function and, 154
vitamins
   lipid soluble, absorption of, 554
   liver disease and, 576
vitamin B_{12}, 632
   absorption of, 539

vitamin B$_{12}$ (*continued*)
    metabolic pathway of, 631f
vitamin C, bone formation and, 456
vitamin D
    calcium regulation and, 258
    metabolism of, 454
    osteomalacia and, 465
vitamin K, coagulation disorders and, 690
von Willebrand's disease, 687

water
    absorption of, 523–544
        gallbladder mucosa and, 582
    balance in adult man, 274t
water intake, regulation, 273
Wiskott-Aldrich syndrome, 856

X-linked infantile agammaglobulinemia, 851